D0882463

The Behavior Guide to African Mammals

The University of California Press
Berkeley • Los Angeles • Oxford

The Behavior Guide to African Mammals

Including Hoofed Mammals, Carnivores, Primates

Richard Despard Estes

Drawings by Daniel Otte

Foreword by E. O. Wilson

A WAKE FOREST STUDIUM BOOK

Pro Humanitate

University of California Press
Berkeley and Los Angeles, California

University of California Press, Ltd.
Oxford, England

First Paperback Printing 1992

LIBRARY OF CONGRESS CATALOGING-IN-PUBLICATION DATA
Estes, Richard
The behavior guide to African mammals : including hoofed mammals, carnivores, primates / Richard Despard Estes ; drawings by Daniel Otte.
p. cm.
Bibliography: p.
Includes index.
ISBN 0-520-05831-3 (alk. paper)
ISBN 0-520-08085-8 (ppb.)
1. Mammals—Africa—Behavior. I. Title.
QL731.A1E84 1990
599'.051'096—dc20 89-4877

Printed in the United States of America

1 2 3 4 5 6 7 8 9

The paper used in this publication meets the minimum requirements of American National Standard for Information Sciences—Permanence of Paper for Printed Library Materials, ANSI Z39.48-1984 ∞

To the memory of my uncle and aunt
Clement Lyndon and Jessie Bond Despard
who were also my foster parents
and to
Runi, Lindy, and Anna
for everything that was denied them
during the years this book was in preparation

Contents

Foreword

The Behavior Guide to African Mammals is a pioneering example of the most advanced stage in the evolution of field guides. When we first approach Nature in a serious mood we classify its occupants. It is necessary to tell the species apart, to put scientific and common names on them in order to keep a running account of local faunas and floras. Many field guides stop at this point, and so do the amateur naturalists who use them. In fact, it is a sufficient reward—a thrill—just to see a great eagle soar across the sky, or to come upon a pride of lions resting in thorn brush. The next step beyond identification is ecology, a perception of the preferred habitat of the species, its food, its enemies, and its migratory pathways. The last stage, focus of the connoisseur and the professional biologist, is behavior, the most intricate and revealing part of an animal's natural history. Behavioral profiles, including social organization, take us closest to the true nature of the species encountered in the wild. Most of animal adaptation is based on the stereotypical behavioral patterns by which animals select their habitats, identify other members of their own species, search for prey, and avoid enemies. If you know an animal's behavior well, you know its essence.

The African mammals are perfectly suited for this more sophisticated treatment. Identification poses few problems; species are few in number (compared with say, birds or butterflies) and mostly not difficult to tell apart. The animals are also often easy to observe over long periods of time, especially in the grasslands and open forests in which so much of the magnificent African mammal fauna still lives. To monitor the daily routines and life cycles of selected mammals is to engage the African wilderness intimately, to savor millions of years of evolution and the last stronghold of the ecosystem in which humanity itself originated—as a mammal species.

The greater the number of resident people and visitors who understand the African environment at the level of animal behavior, the more likely is the environment to be saved. To know a subject well, esthetically as well as factually, is to assign it greater value. In this sense *The Behavior Guide to African Mammals* is the cutting edge behind which a broadening wedge of public interest and commitment in conservation may follow. Like art and historical associations devoted to the preservation of culture, groups of naturalists act to protect this equally precious and far more ancient part of humanity's heritage. They and the African environment will be served very well by this guide, which summarizes the results of scientific field studies of some 91 different ungulates, carnivores, and primates undertaken up to 1990.

EDWARD O. WILSON
Harvard University

Preface

Conventional field guides to wildlife serve primarily for identification of species, through descriptions and illustrations. Behavioral information is usually limited to a paragraph or two under the heading of "habits." The primary purpose of this behavior guide is to describe, explain, and illustrate the behavior of the mammals commonly seen in Africa's wildlife parks and reserves.

What makes Africa's mammalian fauna so unlike that of any other continent is the wealth of "big game": its megafauna. The teeming herds of ungulates, the predators and scavengers that depend on them, the colorful array of monkeys, and the great apes that still live in largely natural conditions in Africa show us what the world was like before our species gained dominion over the rest of the animal kingdom. Moreover, to know these animals is to gain insight into what life was like while our ancestors were evolving on the African savannas.

The abundance and diversity of African big game, so realistically portrayed in Karl Akeley's dioramas at New York's American Museum of Natural History, awakened in me as a child an interest in African wildlife that eventually came to rule my life and prescribe my career. The views of savannas teeming with wildlife made me yearn to be there, to climb a kopje where I could sit and simply watch the animals day in and day out.

When many years later the dream came true, the reality exceeded my expectations. In the early 1960s, I was privileged to live on the floor of Ngorongoro Crater while doing my doctoral research on the wildebeest, one of the antelopes in Akeley's dioramas which had made an indelible impression on me.

The study of African antelopes and other large mammals has occupied me ever since, whether in the field or while living and working in the United States. Needless to say, my preference is to be among the animals, and I have managed to spend another six years observing them during fifteen trips to Africa, among which a two-and-a-quarter-year stay in Serengeti National Park stands out as the most memorable and rewarding time.

The idea of a behavior guide first occurred to me in 1967 as a way to enrich the experience of visiting Africa's parks, and thereby increase international concern for preserving African wildlife. Through daily contact with Ngorongoro visitors, I had become aware that the behavior of the animals, which I found so engrossing, was largely unnoticed by most visitors and their guides. Their main effort went into seeking out rare or glamorous creatures like rhinos, cheetahs, lions, wild dogs, and leopards, and they spent hardly any time just sitting and watching the more common species that surrounded them. How often I saw vehicles drive right past (or through) herds of plains game that were involved in interesting and even dramatic activities. Whereas the sought-after species were found most of the time doing nothing more exciting than sleeping.

Yet in talking with visitors, I found most of them interested in everything I could tell them about the behavior of the animals. Subsequent experience, as a lecturer and as leader of camping safaris to Tanzania, convinced me that most people with enough interest to attend a slide show or take an African holiday were fascinated by the same kinds of information that led people like me to study the animals. What is it, indeed, that makes animals interesting if not their behavior? The interest was there all the time, I realized; the problem was how to make the information readily available.

Although a great deal of knowledge about African wildlife has been gained through field studies over the last three decades, most of it has been published in scientific journals unseen by the general public. Nevertheless, the public appetite for this kind of information has been whetted by television programs and popular books based on some of these studies. To make this kind of information accessible and intelligible to the general public, I finally decided to produce an illustrated behavioral field guide to the mammals that are common and viewable in Africa's national parks.

In the process of researching and writing the book, I soon realized that the same information needed to give readers an adequate understanding of each species would make the guide valuable to students and behavioral scientists as a reference work. By this circular route, I thus came back to my original point of departure: the results of research on the behavior of African animals are of interest to the same audience that watches nature programs and reads books about animals, and especially those who visit African wildlife parks. And, of course, this information is even more interesting to researchers and students of animal behavior. Some colleagues have warned that the same book cannot satisfy both laymen and specialists, that it will contain too much information for the one and too little for the other. But I don't believe it. It all depends on how it's done. If the information is divided up by topic and presented in such a way that the reader can easily get to whatever topic is of immediate interest, and if the accounts are clearly written with minimal use of technical terms, both specialists and laymen can find what they want. The layman can skip over subjects and details that go beyond his interest, while the specialist who wants more detail will be guided to the key publications on which the accounts are based.

In my attempt to make this guide as useful as possible to both constituencies, I developed the standardized format outlined below in Guide to the Guide, and made a conscious effort to present all information in language readily comprehensible to literate readers. (Nevertheless, a glossary of terms has been included for readers with a limited background in biology and natural history—particularly those who may be in a park far from any dictionary.) Mainly for the benefit of specialists, but without creating any inconvenience for lay readers, all the references I have used are cited in the text and included in an extensive bibliography.

The final result is a book that contains most of the information I felt was necessary to provide an adequate understanding of the featured species. If there is more detail than some readers desire, I hope it will not deter them from finding out what they do want to know. In any case, I would prefer to produce a book that gives too much than one that gives too little information, in order to satisfy the more curious readers. The operative constraints on the contents of the *Guide* were the size and cost of the book. I was determined to keep it small enough to be conveniently carried into the field by visitors to African parks, for whose benefit the book was conceived, and hoped it would be inexpensive enough to be considered a worthwhile investment by people with only a moderate interest in African animals.

I am grateful to the following individuals for taking the trouble to read and comment on portions of the manuscript: George and Lory Frame, Laurence Frank, John Fryxell, Robert Harding, Betsey Harris, Alison Jolly, James Malcolm, Sybil Montgomery, James Moore, Jon Rood, George Schaller, Thomas Spang, Elizabeth Thomas, Fritz Walther, Jessica Warren, Peter Waser, and William Wiljanen. In addition to illustrating the book, Daniel Otte helped design it. Jeremy Anderson, Bristol Foster, George Frame, I. Golani, Brian Huntley, Salomon Joubert, Walter Leuthold, Norman Owen-Smith, Craig Packer, Jon Rood, George Schaller, Richard Schuster, and Ken Tinley kindly provided photographs of their animals, some of which were made into drawings. I am still more grateful to those who permitted their drawings to be reproduced in the *Guide,* especially to Fritz Walther and Jonathan Kingdon for contributing so many of their drawings. (The sources of the illustrations are acknowledged in the figure captions.)

I also wish to pay tribute to the editors who have guided, encouraged, and corrected me over the years since the inception of this book back in the 1960s to its final parturition in the 1990s: in chronological order, Charles Everitt, William Bennett, Eleanor-Mary Cadell, Bettyann Kevles, Diana Feinberg, Shirley Warren, David Dexter, Elizabeth Knoll, Mary Repaske, and Jess Bell. I have particularly appreciated the common sense, cooperation, enthusiasm, and bookcraft of Jess Bell and of Joyce Kachergis, the book's designer, during the final production stages. Several anonymous peer reviewers of the completed manuscript contributed thorough and very helpful critiques. Lastly, I thank Edward Wilson for writing the Foreword.

I am indebted to Lisa Gemmill for invaluable assistance in searching the literature, compiling information on various species and cataloging illustrations, while working as a volunteer in the Mammal Department of the Academy of Natural Sciences. Marina Botje and Henry Warren assisted in the collection of information used in several of the accounts. Finally, Runhild Estes spent most of her time for nearly two years working on the first version of the *Guide,* contributing translations of German-language papers and typing and proofreading chapters. She has continued to offer help since then, in the time she could spare from looking after our children, managing the household, and holding a full-time job. Without her support, this work would probably never have been finished.

Guide to the Guide

COVERAGE. The 6 orders and 16 families of hoofed mammals, carnivores, and primates covered in the *Guide* include all but two of Africa's large and medium-sized land mammals,[1] as well as some very small ones (bush babies, dwarf mongoose). A total of 91 species are featured in separate, detailed accounts.

HOW FEATURED SPECIES WERE CHOSEN. In order to hold the book within bounds and still do justice to each featured species, some of the mammals that have been studied had to be left out. After the decision was made to limit coverage to hoofed mammals, carnivores, and primates, the species to be featured in separate accounts were selected according to the following guidelines.

1. It occurs naturally in Africa south of the Sahara, the part known to zoogeographers as the Ethiopian Faunal Region.
2. It is found in well-known national parks, where it is regularly encountered, preferably by daylight in open habitats.
3. It has been studied in the wild.

The species that passed this screening process include virtually all the large to medium-sized land mammals that occur in eastern, central, and southern Africa, where nearly all the parks and game reserves that are accessible and popular are located (see map 2). Antelopes and carnivores, each with 34 species accounts, make up ¾ of the featured species.

EXCEPTIONS TO THE GUIDELINES. A small number of species have been included which do not meet the above criteria, for reasons illustrated by the following examples.

Blue duiker. One of the smallest of the forest antelopes, it is nevertheless common in many parks. Since it is also the one forest duiker that has been intensively studied, it was included to represent the largest and least-known antelope genus, *Cephalophus.*

Tree hyrax. Although this species is strictly nocturnal in most places and seldom seen, its incredible call makes it impossible to ignore.

Palm civet. This species represents a way of life that is typical of many carnivores and other solitary, polygynous mammals. Because it is one of the few such species that have been studied, an account has been included, although the only reliable way to see this arboreal, nocturnal carnivore is to wander the forest at night with a spotlight.

Dwarf mongoose. By any standards, this is a small mammal, even in comparison to rodents like the springhare or cane rat. Some of the other featured carnivores, notably other mongooses and the striped weasel, also weigh under a kilogram. The logic for including them is that they are very common and widespread, are seen quite often, and have been studied. The dwarf mongoose happens to be one of the best-studied and most interesting of all African mammals.

Bush babies. Two members of this family of small, nocturnal primates live outside the rain forest. The greater bush baby's raucous cry is one of the more common and memorable sounds of the African bush, and night drivers often see a lesser bush baby's eyes reflected in their headlights.

Great apes. Until recently, gorillas and chimpanzees, if not difficult to observe in their forest homes, lived in parks well off the beaten track. But highly publicized studies of these apes have created so much popular interest that the trickle of primate safaris has grown into a flood. Moreover, many people who never expect to see them in the wild are interested in reading about our two nearest living relatives.

PRIMARY SOURCES OF INFORMATION. Most of the information con-

[1] The little-known and seldom-seen aardvark (*Orycteropus afer*, order Tubulidentata) and the giant pangolin (*Manis gigantea*, order Pholidota).

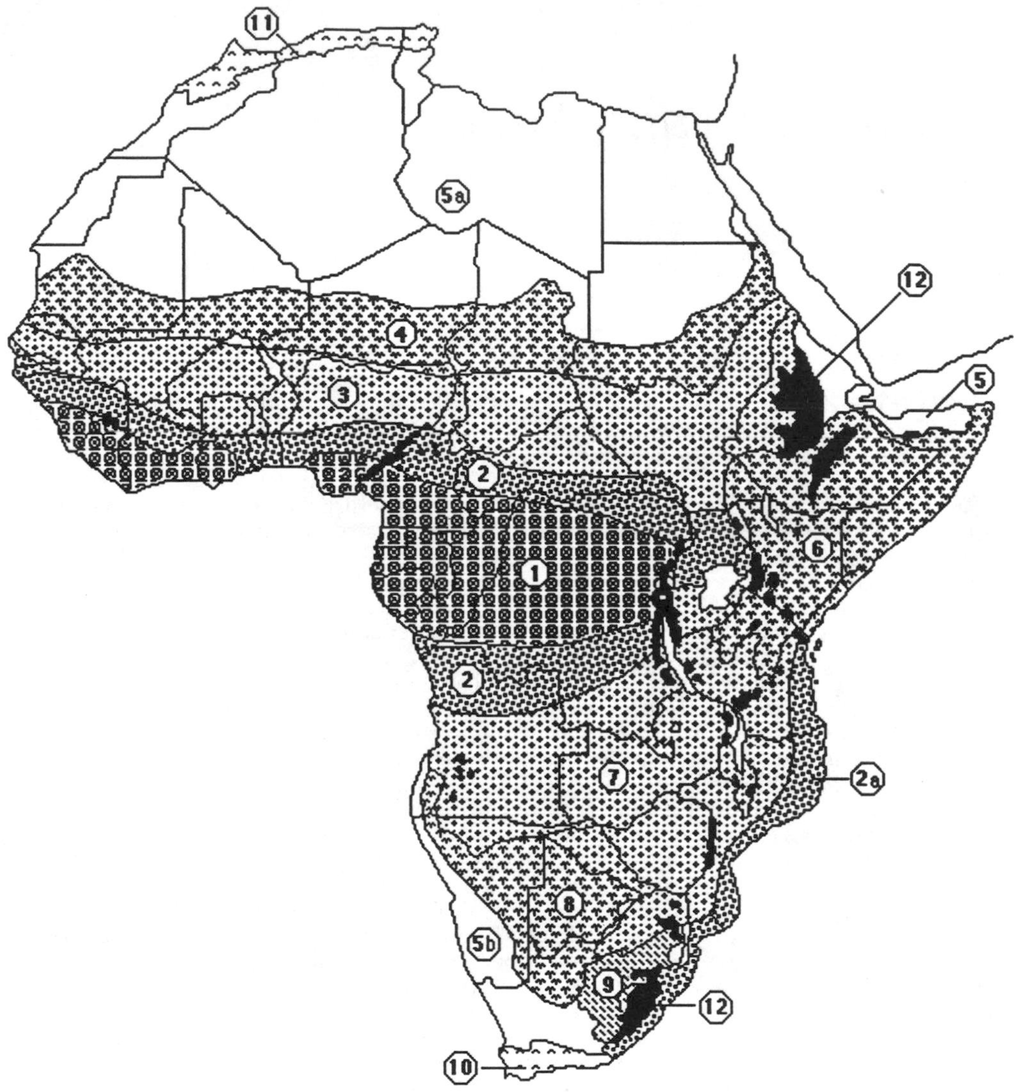

Map 1. Vegetation map of Africa. Simplified from F. White, *The Vegetation of Africa.* (Paris: UNESCO, 1983), and C. J. Tucker, J. R. G. Townsend, and T. E. Goff, "African land-cover classification using satellite data," *Science* 227 (1985): 369–75.

MAJOR VEGETATION TYPES

1. Lowland Rain Forest
2. Forest–Savanna transition and mosaic
   a. Coastal forest-savanna mosaic
3. Sudanian Savanna: includes (from south to north) Guinea or Northern Savanna and the Sudanese Arid Zone (dry savanna)
4. The Sahel (arid savanna and bushland)
5. Desert and semidesert shrubland
   a. The Sahara
   b. Namib Desert–Karoo semidesert
6. Somali–Masai Arid Zone (dry savanna)
7. *Miombo* Woodland Zone (broad-leafed deciduous woodland dominated by *Brachystegia* spp. and *Julbernardia*)
8. Kalahari savanna and bushland (Sandveld)
9. Highveld temperate grassland and Karroo grassy shrubland
10. Cape and Karroo shrubland (Cape Macchia)
11. Mediterranean vegetation
12. Afromontane vegetation (altitudinal zonation: montane forest, montane grassland, bamboo, heath, alpine, etc.)

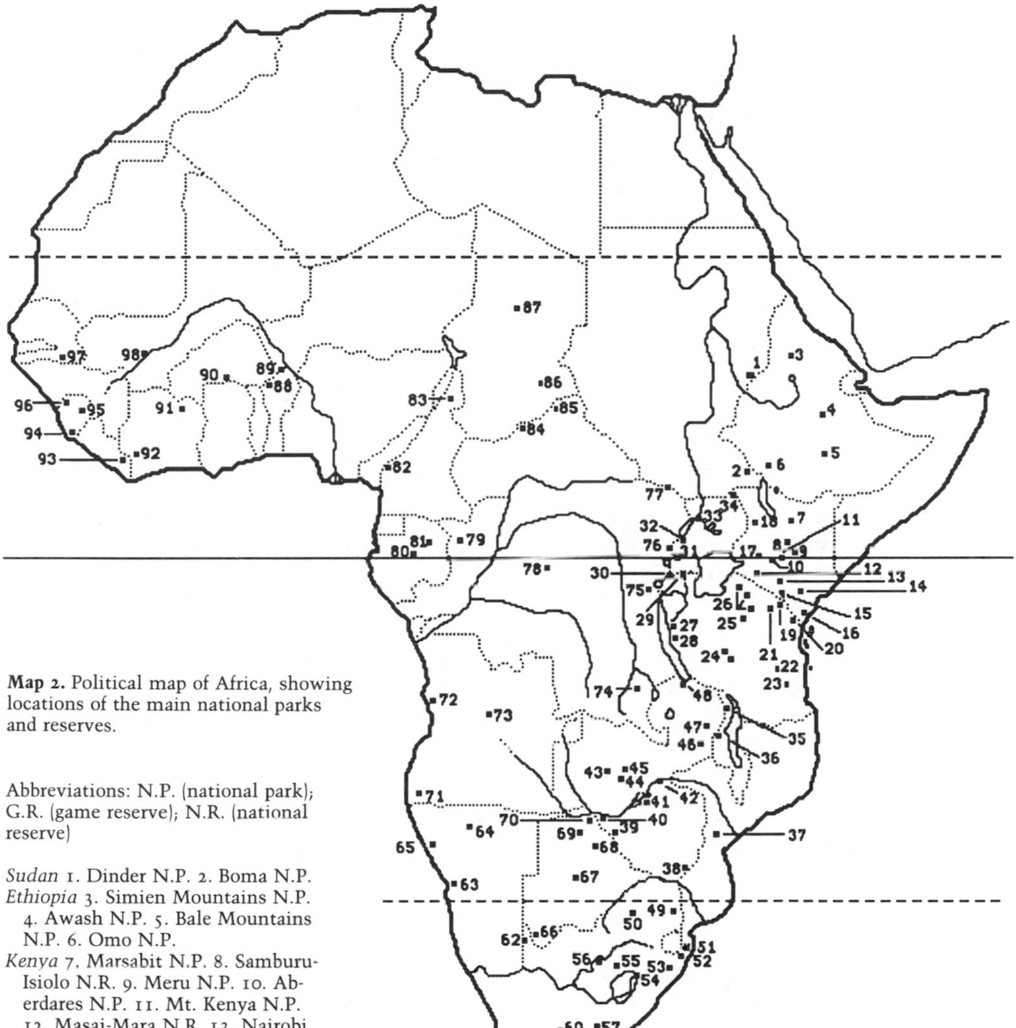

Map 2. Political map of Africa, showing locations of the main national parks and reserves.

Abbreviations: N.P. (national park); G.R. (game reserve); N.R. (national reserve)

Sudan 1. Dinder N.P. 2. Boma N.P.
Ethiopia 3. Simien Mountains N.P. 4. Awash N.P. 5. Bale Mountains N.P. 6. Omo N.P.
Kenya 7. Marsabit N.P. 8. Samburu-Isiolo N.R. 9. Meru N.P. 10. Aberdares N.P. 11. Mt. Kenya N.P. 12. Masai-Mara N.R. 13. Nairobi N.P. 14. Tsavo N.P. 15. Amboseli N.P. 16. Shimba Hills N.R. 17. Nakuru N.P. 18. Mt. Elgon N.P.
Tanzania 19. Kilimanjaro N.P. 20. Mkomazi G.R. 21. Arusha N.P. 22. Mikumi N.P. 23. Selous G.R. 24. Ruaha G.R. & Rungwa N.P. 25. Tarangire N.P. 26. Serengeti N.P. (top), Ngorongoro Conservation Area, Manyara N.P. 27. Gombe N.P. 28. Mahale Mts. N.P.
Rwanda 29. Kagera N.P. 30. Parc de Volcans.
Uganda 31. Queen Elizabeth N.P. 32. Toro (Semliki) G.R. 33. Murchison Falls N.P. 34. Kidepo N.P.
Malawi 35. Nyika N.P. 36. Kazungu N.P.
Mozambique 37. Gorongosa N.P.
Zimbabwe 38. Gonarezhou N.P. 39. Hwange N.P. 40. Zambezi & Victoria Falls N.P. 41. Matusadona N.P. 42. Mana Pools N.P.
Zambia 43. Kafue N.P. 44. Lochinvar N.P. 45. Blue Lagoon N.P. 46. Luangwa South N.P. 47. Luangwa North N.P. 48. Nsumbu N.P.
South Africa 49. Kruger N.P. 50. Pilansberg N.P. 51. Ndumu G.R. 52. Mkuzi G.R. 53. Hluhluwe & Umfolozi G.R. 54. Giant's Castle N.P. 55. Willem Pretorius G.R. 56. Sandveld Nature Reserve. 57. Mountain Zebra N.P. 58. Addo Elephant N.P. 59. Tsitsikamma Coastal N.P. 60. Karroo N.P. 61. Bontebok N.P. 62. Kalahari Gemsbok N.P.
Namibia 63. Namib-Naukluft N.P. 64. Etosha N.P. 65. Skeleton Coast N.P.
Botswana 66. Gemsbok N.P. 67. Central Kalahari G.R. 68. Nxai Pan N.P. 69. Moremi G.R. 70. Chobe N.P.
Angola 71. Iona N.P. 72. Kisama N.P. 73. Kangandala N.P. & Luando Giant Sable Reserve.
Zaire 74. Upemba N.P. 75. Kahuzi-Biega N.P. 76. Virunga N.P. 77. Garamba N.P. 78. Salonga N.P.
Congo Republic 79. Odzala N.P.
Gabon 80. Lope-Okana Reserve. 81. Ipassa-Makokou Biosphere Reserve.
Cameroon 82. Korup N.P. 83. Waza N.P.
Central African Republic 84. Bamangui-Bangoran N.P. 85. Manovo-Gounda-Saint Floris N.P.
Chad 86. Zakouma N.P. 87. Ouadi Rime-Ouadi Achim Faunal Reserve.
Benin 88. Boucle de la Pendjari.
Niger 89. "W" National Park.
Burkina Faso 90. Kabore Tambi N.P.
Ivory Coast 91. Camoe N.P. 92. Tai N.P.
Liberia 93. Sapo N.P.
Sierra Leone 94. Gola Forest Reserves. 95. Loma Mts. Forest Reserve. 96. Tiwai Island Wildlife Sanctuary.
Senegal 97. Niokolo-Koba N.P.
Mali 98. Boucle du Baoule N.P.

tained in this *Guide* is based on the results of behavioral and ecological field studies, published in scientific journals and in books over the last three decades (very few before the 1960s). A small proportion of the information comes from doctoral dissertations and masters' theses, and from popular articles. My own observations, published and unpublished (the latter identified as "author's observations"), are scattered through the book, but only the accounts of the blue wildebeest and sable are drawn mainly from my own research. Although I have done some research and published papers on a few other species (e.g., wild dog, gazelles, topi), and know something about most of the animals where I have done research, the primary purpose of this book is to present a comprehensive summary of current knowledge about each included species. Therefore priority has been given to the results of researchers who have spent years observing a species.

ORGANIZATION OF THE GUIDE

SYSTEMATIC ARRANGEMENT. The conventional systematic arrangement followed in the *Guide* groups related species in the same chapters. Apart from being the only sensible arrangement biologically, placing related animals next to one another promotes comparisons of their similarities and differences, as do the introductions to each chapter. The taxonomic arrangement of the *Guide* is actually quite simple, compared with standard mammal encyclopedias or even conventional field guides, as the following semicomplete list of the taxonomic hierarchy demonstrates. The divisions regularly used in this book are indicated in boldface, and the classification of the African buffalo is given to exemplify the hierarchical system and the rules of nomenclature that apply to the Latin names of the different taxa.

- Class Mammalia
  - Subclass (Eutheria)
    - Cohort (Unguiculata)
      - Superorder (Mesaxonia)
        - **Order** (Artiodactyla)
          - **Suborder** (Ruminantia)
            - Superfamily (Bovoidea)
              - **Family** (Bovidae)
                - **Subfamily** (Bovinae)
                  - **Tribe** (Bovini)
                    - **Genus** (*Syncerus*)
                      - Subgenus (none in this example)
                        - **Species** (*caffer*)
                          - Subspecies (*caffer*, the common buffalo)

Missing from the above list is the category of superspecies, which is used occasionally in the *Guide* to denote two or more species so closely related that some systematists consider them only different geographical forms of the same species. Most noteworthy, though, is the fact that subspecies (also called races) are only singled out in instances when there are substantial and consistent differences among populations. Even then, since the book is primarily concerned with species-typical behavior, only those subspecies in which socioecological differences have been documented receive more than passing mention (e.g., differences in the mating systems of Uganda and white-eared kob).

The *Guide* is divided into four parts, dealing in turn with ungulates, carnivores, and primates. Carnivores and primates fit neatly into single orders, but hoofed mammals are a taxonomically mixed bag. Even-toed ungulates (order Artiodactyla) contain all the ruminants plus pigs and hippos. Artiodactyls vastly outnumber all the rest, with 4/5 of all living species. Rhinos and zebras are odd-toed ungulates (order Perissodactyla). Then there are four orders of "near-ungulates" (Paenungulata), namely elephants, hyraxes, dugongs, and the aardvark (see introduction to the ungulates, Part I).

Because ruminants all have much in common, they are treated separately in Part I, and since nearly all the African species are members of the family Bovidae, this section is mainly about antelopes. The nonruminating ungulates are all lumped together in Part II.

INTRODUCTIONS TO ORDERS, FAMILIES, SUBFAMILIES, AND TRIBES. Introductions to each order and family are intended to summarize and compare the traits of related species. In the case of the bovids, each of the eight main tribes that contain the antelopes and the African buffalo has its own introduction. In the primate section (Part IV), the confusing array of monkeys and baboons is dealt with in introductions to the order (chap. 24), to the family (chap. 26), and to the two subfamilies (chaps. 27 and 28), wherein their ecology, social organization, and mating systems are compared and their elaborate communication system is outlined.

The chapter introductions include information about many species that are not the subjects of separate accounts. By reading both the introductions and the species accounts, *Guide* users can predict with some confidence the behavior of closely related, unstudied species.

The information contained in the introductory chapters is presented in a standard format that conforms closely with the arrangement of the species accounts, as shown below. In addition, the evolutionary history of the major groups is summarized, under the heading of Ancestry, in the introductions to orders and families.

ORGANIZATION OF SPECIES ACCOUNTS. The information presented in the species accounts is arranged under the following headings, with minor variations in different sections of the *Guide.*

Traits. A brief description of the species, including average shoulder height and weight (and in parentheses the ranges thereof) of adult males and females; appearance and conformation; type of teeth; coloration and markings; scent glands; and number of mammae.

Distribution. Map showing geographic range and a brief discussion of the life zone (biome) for which the species is adapted. Closely related species and well-defined subspecies, if any, are noted here.

Ecology. Degree of specialization, habitat preferences, diet, dependence on drinking water.

Social Organization. Grouping pattern, land tenure and mating system (gregarious or solitary, territorial or nonterritorial, sedentary or migratory, monogamous or polygnous reproduction); information on group size and composition; size of home range/territory; social relations within and between groups, and among different sex/age classes; social roles of males and females.

Activity. Daily pattern, periods of maximum and minimum activity, and seasonal variations.

Postures and Locomotion. Attitudes assumed while seated, sleeping, feeding, and so on; preferred gaits, climbing, digging, swimming, or other noteworthy locomotory behavior.

Foraging or Predatory Behavior. How the species finds and eats its food; daily ranging.

Social Behavior. How members of the species communicate and interact socially. Listing, description, and interpretation of signaling behavior, with illustrations of key visual displays.

COMMUNICATION. Visual, olfactory, tactile, and vocal signals.

TERRITORIAL BEHAVIOR. Behavior associated with advertising and maintaining a territory.

AGONISTIC BEHAVIOR. Behavior associated with aggression, fear, and submission.

Dominance/Threat Displays.

Defensive/Submissive Displays.

Displacement Activities. See glossary definition. Seemingly irrelevant behavior such as grooming and feeding which a species regularly performs in stressful situations (see bovid Agonistic Behavior, chap. 2).

Fighting. Techniques, frequency, and severity.

Reproduction. Seasonality, minimum breeding age, birth frequency, litter size, gestation.

SEXUAL BEHAVIOR. Male and female courtship displays and mating behavior.

Parent/Offspring Behavior. Birth, neonatal weight, behavior of mother and newborn; male and female parental roles; behavioral development of the young, age at weaning, puberty, and dispersal.

Play. Forms of play, who plays and how often.

Antipredator Behavior. Alerting/alarm signals, concealment and escape tactics; active offspring, group, or self-defense against enemies.

References. Sources of information about the species are indexed by numbers in the text, listed by author and year at the end of each account, and fully cited in the bibliography.

The bibliography is extensive but selective rather than exhaustive. A major effort was made to include the important sources of information about each featured species; to make the *Guide* as up-to-date as possible as it goes to press, sources that were unpublished or overlooked when an account was written are appended under "Added References." Minor sources of information and papers dealing with topics not covered in the *Guide* (e.g., diseases, nutrition, anatomy) have been left out on purpose, in the belief that a bibliography consisting of the most important and relevant sources will prove most valuable.

PROBLEMS OF OVERLAPPING CATEGORIES. The trouble with the above arrangement of topics—and with any attempt to categorize a species' behavior—is that most behaviors do not fit neatly into separate compartments. Some overlap is almost inevitable. Social Organization, for instance, must include information about the species' land-tenure and mating system; the latter subject comes up again under Reproduction and Parent/Offspring Behavior, whereas behavior associated with advertising and maintaining a territory belongs under the heading of Social Behavior. Subdividing Social Behavior is considerably more difficult, particularly the subject of Communication, which encompasses all sorts of signaling behavior. The problem is how to subdivide Communication so that what is said under that heading does not have to be repeated under other headings such as Agonistic or Sexual Behavior. The difficulty is greatest in the case of visual signals.

In practice, olfactory, tactile, and vocal signals have generally been described under Communication, whereas most visual signals are subdivided into Agonistic, Sexual, and Antipredator behaviors. As a general but flexibly applied practice, behavior patterns that are characteristic, distinctive, and predictable under certain conditions, such as common displays, are highlighted by the use of italics.

A MIXTURE OF LONG AND SHORT ACCOUNTS. The amount of information included under the different headings varies from species to species, depending on the complexity of an animal's social organization and behavior, on how much study it has received, and on the main focus of the research. In some species accounts, when there is nothing to add to the information given in the chapter introduction, certain subject headings (e.g. Postures and Locomotion) are omitted, in which case the reader should look under this heading in the chapter introduction.

The main focus of the species accounts is on social organization and on territorial, agonistic, sexual, parental, and antipredator behavior. Play, eating, comfort behavior (e.g., grooming), and development receive less attention.

The longest accounts in the book are devoted to the most popular, interesting, and/or well-studied African mammals: notably the impala among the antelopes, the buffalo, elephant, lion, cheetah, wild dog, spotted hyena, social mongooses, vervet monkey, baboon, gorilla, and chimpanzee.

ILLUSTRATIONS. Because this is not a conventional field guide but a behavioral guide, the illustrations were selected to show species-typical behavior. There are hardly any pictures of animals doing nothing in particular. To keep the book's size and cost within bounds, the number of illustrations had to be held in check. Nevertheless, there is at least one drawing of every featured species and an average of four pictures per species account. Fortunately, the behavioral repertoires of related species are enough alike that a pattern illustrated in one will often suffice for its near relations. When available, drawings of unfeatured species have been used in the introductory chapters to illustrate displays common to the group.

For the sake of clarity and consistency, all illustrations have been rendered as drawings. Except as noted in captions or credits, they are all by Dan Otte. To ensure accuracy, most illustrations have been redrawn from photos and drawings that appeared in publications by researchers who have studied the species in question. A considerable number of my own photographs were turned into drawings, and others were prepared from unpublished photos provided by colleagues who are thanked in the Preface.

Part I

Hoofed Mammals: Antelopes and Other Ruminants

Order Artiodactyla

True ungulates are hoofed mammals that walk on tiptoe. They are classified in 2 different orders, Perissodactyla and Artiodactyla, depending on whether they have an odd (Gr. *perisso-*) or even (Gr. *artio-*) number of toes (*dactyla*). In artiodactyls the axis of the foot passes between digits 3 and 4; in perissodactyls (zebras, rhinoceroses, tapirs) it passes through the third, middle digit. The early artiodactyls had 4 hooves (digits 2 to 5), a condition that persists in hippos and pigs. The number has been reduced to 1 pair in the ruminants, but the vestiges of digits 2 and 5 are still present as the false hooves (2, 3).

Of the approximately 200 different kinds of recent hoofed mammals, all but 17 are artiodactyls, and of these all but 20 are ruminants (suborder Ruminantia). The 20 nonruminants include the pigs and hippos, and the camelids (suborder Tylopoda). The latter are only distantly related to other artiodactyls, having separated in the Eocene and later developed a ruminating digestive system that differs in important respects from that of the true ruminants (1, 2). The camels that live in Africa descend from domesticated livestock introduced from Asia and therefore are not considered further in this *Guide.*

References

1. Janis and Jarman 1984. 2. Romer 1959. 3. Young 1962.

Chapter 1

Introduction to the Ruminants

Suborder Ruminantia

Family	Distribution center	African species
Tragulidae (chevrotains)	Southeast Asia	1
Giraffidae (Giraffe, okapi)	Africa	2
Cervidae (deer)	Old and New World temperate zone	1 (North Africa)
Bovidae (antelopes, sheep, goats, cattle)	Africa and Asia	75

TRAITS. Even-toed ungulates that chew the cud, ranging in size from the 1.5 kg royal antelope to a 1900 kg giraffe. *Teeth:* upper incisors absent, lower incisors augmented by incisiform canines to make a row of 8 chisel-shaped teeth; upper canines absent or vestigial except in chevrotains and primitive deer; cheek teeth complexly folded and sharply ridged, adapted for grinding, 2 or 3 premolars and 3 molars in each quadrant.

ANCESTRY. During the Eocene, while perissodactyls were diversifying and getting bigger, artiodactyls remained small (< 5 kg) forest frugivores and omnivores (table 1.1). But some 37 million years ago, the formerly subtropical global climate turned cooler, leading to seasonal rather than perennial production of fruits and nonfibrous vegetation. In the process of adapting to the new regime, artiodactyls became bigger and some specialized as herbivores. Pigs, camels, and ruminants were all established by the Oligocene. Ruminants reached Africa from Eurasia c. 24 million years ago in the early Miocene (7). During the same epoch, grasslands began to spread, replacing subtropical woodlands with more open savanna, and ruminants, with their superior ability to digest a fibrous diet (see chap. 2), likewise spread and diversified, becoming the dominant large herbivores.

The latest and most successful ungulate radiation was by the hollow-horned ruminants, the Bovidae. In Africa all but 3 of the

Table 1.1 Geologic Time Scale of the Cenozoic Era, the Age of Mammals

Period	Epoch	Ma[a]	Major events
QUARTERNARY	Pleistocene	1.6	Ice Ages; early man; Golden Age of Mammals: maximum no. species, giant forms, followed by extinctions = 40% of species in last 20–10,000 yrs.
TERTIARY	Pliocene	5	Giraffe, warthog, zebras, camel, elephant. *Homo erectus* and existing genera of bovids and carnivores appear in late Pliocene-early Pleistocene
	Miocene	24	Early rhinos, aardvark. Ruminants become dominant as grassland habitats spread. *Australopithecus* in Pliocene boundary Miocene
	Oligocene	37	Hyrax radiation, giraffe and elephant progenitors, early carnivores, first apes, cercopithecid monkeys
	Eocene	58	Dominance of large archaic forms, replaced in late Eocene and early Oligocene by varied, mostly small ancestors of modern mammals
	Paleocene	65	Various existing and extinct orders (e.g., creodont carnivores) arise from primitive insectivore stock and diversify
CRETACEOUS			Age of Reptiles

[a]Millions of years ago.

living ruminants belong to this family. Accordingly Part I is mainly about bovids and, in particular, about antelopes.

ADVANTAGES AND DISADVANTAGES OF RUMINANT DIGESTION.

A key advantage of the ruminant's digestive system is its superior ability to convert cellulose, the main constituent of all plant tissues and fibers, into digestible carbohydrates. Actually, all herbivores and even omnivores share this ability in some degree, including monkeys and plant-eating carnivores. The breakdown is accomplished not by the animal itself but by symbiotic microorganisms that digest cellulose by fermentation (4). In nonruminants fermentation occurs in the large intestine and an adjacent pouch, the cecum (of which our appendix is a vestige), *after* the food has passed through the stomach. In ruminants, fermentation takes place *before* gastric digestion, mainly in the rumen (fig. 1.1). Much more fiber is left undigested in the former system, as can readily be seen by comparing the coarse dung of a zebra, rhino, or elephant with the fine-grained dung of any ruminant.

The rumination process is both mechanically and biochemically complex and still not fully understood. First the animal feeds until the rumen is comfortably full, gripping foliage or grass between its lower incisors and upper dentary pad, plucking, and then swallowing after briefly chewing each mouthful. Then it settles down to ruminate, either lying or standing, chewing the cud with rhythmic side-to-side jaw movements.

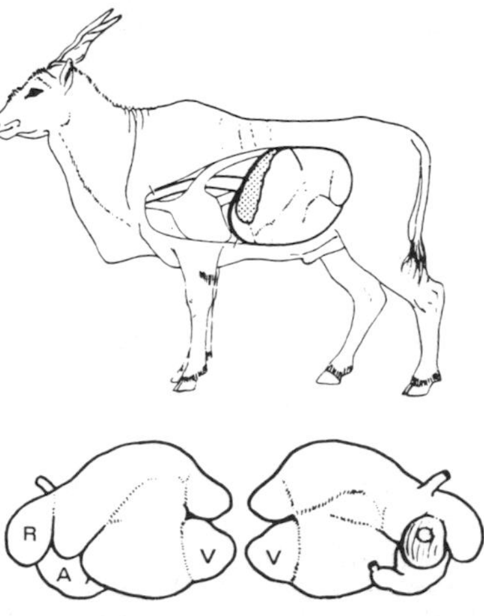

Fig. 1.1. The ruminant stomach. *Top:* an eland's rumen shown in situ, overlain by the spleen (dotted organ). *Bottom:* an eland's stomach in left and right aspect: *R*, reticulum; *A*, abomasum; *O*, omasum; *V*, ventral blind sac. The rest is the rumen. (From Hofmann 1973.)

The cud consists of the coarsest plant particles, which float on top of the semiliquid rumen contents and are regurgitated a mouthful at a time through contractions of the rumen and its annex, the reticulum (the "honeycomb tripe" relished by some gourmets). As the ruminant grinds each mouthful at a steady rate, on the same or alternate sides of the mouth, enlarged salivary glands secrete a buffered solution that helps to maintain the rumen pH preferred by the resident microorganisms. Grinding the food promotes the full extraction of nutrients by increasing the surface area that is exposed to bacterial action. Some nutrients are absorbed through the rumen wall, which is lined with tongue- or finger-shaped papillae; these both vastly increase the absorptive area and provide crannies in which bacteria and protozoans multiply.

Rhythmic contractions of the rumen and reticulum keep stirring the "vat," sorting food particles according to size and specific gravity. The smallest particles sink to the bottom, and from there are pumped through the reticulum into the omasum (fig. 1.1), also known as the "book organ" or "psalter" because of the leaflike plates that line it. Here the semiliquid ingesta are filtered once more before being pumped into the abomasum, the true stomach. Afterward, during passage through the intestines, the residue undergoes some final cellulose digestion in the cecum (4).

In addition to the more complete utilization of plant fiber in ruminant digestion, the constantly reproducing and dying rumen microorganisms that do the work provide the host with energy in the form of volatile fatty acids they excrete as metabolic wastes, and the organisms themselves become a major source of protein as they pass through the digestive tract mixed together with the rumen contents (2, 4). Ruminants possess the further important advantage of being able to recycle urea, thereby retaining and recycling inorganic nitrogen that the ruminant bacteria use to reproduce and to synthesize more protein. From this bacterial protein ruminants acquire the essential amino acids that nonruminants have to gain from their plant food. As an added bonus, recycling urea cuts down on urine excretion, helping to conserve water and contributing to the water-independence of desert-adapted ruminants (chap. 2) (7).

There is 1 major drawback to rumination: the thorough digestion of cellulose takes time, and the more fibrous the food the longer the process. It can take up to 4 days from ingestion to excretion. When protein content falls below 6 percent, ruminants cannot process their food fast enough to maintain their weight and condition (4, 7, 8). Hindgut fermenters consume and partially digest large quantities of low quality forage in half the time; they can thereby manage to obtain adequate sustenance from vegetation too tough and fibrous for ruminants to process. Thus, a horse can extract only ⅔ as much protein from a given quantity of herbage as a cow, but by processing twice as much in a given time, its assimilation of protein will exceed the cow's by 4/3 times.

Supposedly, ruminants are also less efficient than nonruminants at digesting fruits, because predigestive fermentation removes fruit sugars and other nutrients that would otherwise be digested and absorbed in the true stomach (7). Nevertheless, most small forest ruminants are largely frugivorous, and even some of the larger duikers have diets consisting of over 70 percent fruit (see below).

Behavior Associated with Rumination. All ruminants except chevrotains lie down by first kneeling on their forelegs (carpal joints), then lowering the rear end. They rise in reverse order, raising the rear end first. To enable the stomach compartments to move freely, ruminants have to lie on the brisket or stand while ruminating. They rarely lie on their sides for more than a few minutes at a time, and rolling on the back is unheard of (but see wildebeest accounts). If anesthetized ruminants are left lying on their sides, they are likely to ingest rumen contents and suffocate. Ruminants doze but do not sleep soundly like nonruminants. Possibly the need to maintain a certain position and to keep chewing the cud precludes normal sleep. Yet rumination goes together with a relaxed state (the "contented cow") and, significantly enough, the brain waves of ruminating animals resemble those associated with sleep in nonruminants (1).

Fig. 1.2. An eland bull chewing its cud, lying on the sternum in the usual ruminating position.

RESOURCE PARTITIONING AMONG AFRICAN RUMINANTS. Ruminants that live on fruit and soft, succulent vegetation high in protein and low in fiber have a small, comparatively simple rumen that recalls its origin as an S-shaped tubular blind sac serving initially merely as a food storage chamber (5).

Chevrotains, the most primitive living ruminants, have only a 3-chambered stomach (no omasum). The digestive system is most completely differentiated and complex in the grazing ruminants such as cattle, sheep, wildebeest, and oryx.

Ruminants of the second type are usually classified as *grazers;* those that eat fruit and foliage are lumped together as *browsers.* But these designations oversimplify and obscure the range of differences within each group and exclude those intermediate feeders that both browse and graze. After studying and comparing the stomach morphology of 29 wild ruminants of East Africa, Hofmann (5) proposed the following classification:

1. Concentrate selectors

a. Fruit and dicotyledon foliage selectors: duikers, dikdiks, suni, klipspringer, bushbuck, and other (mostly small) ruminants of bush and forest

b. Tree and shrub foliage eaters: giraffe, gerenuk, lesser and greater kudu, and other medium to large browsers of savanna and bush habitats

2. Bulk and roughage eaters

a. Fresh-grass, water-dependent grazers: reedbuck, kob, waterbuck, wildebeest, buffalo

b. Roughage grazers: hartebeest, topi, mountain reedbuck

c. Arid-region grazers: oryx

3. Intermediate (mixed) feeders

a. Predominantly grass (grazer/browsers): impala, Thomson's gazelle

b. Predominantly forbs, shrub or tree foliage (browser/grazers): eland, Grant's gazelle, steenbok

The small, simple stomach of concentrate selectors is designed for the quick fermentation and absorption of food rich in protein, fat, and nonfibrous carbohydrate. Ruminants in class 1a are extremely selective feeders that choose food items of the highest protein content. They feed in many short bouts, rarely filling the rumen more than half full before pausing to ruminate. The rumen contents are not stratified as in roughage feeders, and concentrate feeders generally obtain enough moisture from their food without needing to go to water. But larger (1b) species need to put away more food to sustain their greater bulk and therefore cannot afford to be as choosy. To process bulkier but still easily fermented foliage, they have larger, more subdivided stomachs designed to slow food movement and allow more time for fermentation.

Grazers have the greatest ability to subsist on vegetation that is relatively low in protein and high in fiber. A large capacity combined with food-delaying mechanisms is characteristic. The different stomach chambers are more muscular and subdivided, functioning to churn, pump, sieve, and filter the ingesta. The omasum, in particular, which is merely a strainer in concentrate selectors, is highly developed and serves a new and essential function as a site where water and soluble nutrients are absorbed. Grazers keep eating until the rumen is full, regardless of grass quality, then settle down for a long rumination bout (6–8 hours).

Roughage grazers (2b) can subsist on pastures with low nutrient availability, partly because they are equipped to feed more selectively than most bulk grazers. A relatively smaller capacity and reduced absorptive area are adaptations for a low-energy diet (6). The 3 antelopes listed in this category can go for long periods without drinking. However, the oryx is the only predominantly grass-eating ruminant in Hofmann's sample that is adapted to extremely arid conditions without surface water.

Mixed feeders have a greater ability to adapt to different habitats and seasonal vegetation changes than either concentrate selectors or grazers. When grasses are green and tender they graze, and when pastures become mature and tough or dry into hay they switch to foliage. These changes are accompanied by changes in stomach structure that reduce or enlarge the mucosal surface that is available for absorption, and also by changes in the microorganisms that are adapted to ferment the different types of plants. The basic stomach plan is similar to concentrate selectors. An animal that has adapted for a concentrate diet cannot begin digesting grass without making a gradual transition, and vice versa.

The extraordinary versatility of the ruminant stomach helps to explain how the ruminants could become so diverse, filling a greater variety of ecological niches than any other group of herbivores. The ability to structure the digestive system precisely for a given diet has enabled ruminants to partition African ecosystems into much narrower feeding niches than can nonruminants, which require a greater variety and amount of vegetation to meet their nutritional needs (7).

ECOLOGICAL SEPARATION BY SIZE. Obviously size and shape are closely connected with where and how an animal lives and what it eats (2, 8, 9). As discussed in chapter 2, the transition from concentrate selectors living in closed habitats to roughage grazers of open savanna is correlated with increasing size. But how intimately size can be connected with habitat has been revealed only recently by studies of rain-forest ruminants in Gabon. Dubost (3) found that the structure of the undergrowth predetermines shoulder heights within very precise limits. Instead of the wide spread in sizes that would be expected, the shoulder heights of 9 sympatric species clustered at a few different points: 35–40 cm (water chevrotain and blue duiker); 50–55 cm (Peter's, bay, and black-fronted duikers); 75–80 cm (yellow-backed duiker); 90–95 cm (bushbuck); and 115–125 cm (buffalo and bongo). Four of these 5 height classes turned out to be levels at which the vegetation offered least obstruction to passage in the form of horizontal and diagonal branches at chest height. Only the 2 smallest species proved to be in a disadvantageous height class. The blue duiker avoids the problem by living in places with very little undergrowth (see species account), whereas the chevrotain has a piglike shape and thickened dorsal skin that enable it to force its way through the densest thickets. Two other associated artiodactyls, the bushpig and giant forest hog, have a similar build and are also not size-limited by the vegetation structure. But most ruminants, with their light build and other adaptations for rapid flight, are very sensitive to their physical environment (3).

References

1. Bell 1971. 2. Demment and Van Soest 1985. 3. Dubost 1979. 4. Dukes 1955. 5. Hofmann 1968. 6. ——— 1973. 7. Janis and Jarman 1984. 8. McNaughton and Georgiadis 1986. 9. Underwood 1983.

Chapter 2

Antelopes and Buffalo

Family Bovidae

Tribe	African genera	African species
Cephalophini: duikers	2	16
Neotragini: pygmy antelopes (dik-dik, suni, royal antelope, klipspringer, oribi)	6	13
Antilopini: gazelles, springbok, gerenuk	4	12
Reduncini: reedbuck, kob, waterbuck, lechwe	2	8
Peleini: Vaal rhebok	1	1
Hippotragini: horse antelopes (roan, sable, oryx, addax)	3	5
Alcelaphini: hartebeest, hirola, topi, blesbok, wildebeest	3	7
Aepycerotini: impala	1	1
Tragelaphini: spiral-horned antelopes (bushbuck, sitatunga, nyalu, kudu, bongo, eland)	1	9
Bovini: buffalo, cattle	1	1
Caprini: ibex, Barbary sheep	2	2
Total	26	75

FAMILY TRAITS. Horns borne by males of all species and by females in 43 of the 75 African species. *Size range:* from 1.5 kg and 20 cm high (royal antelope) to 950 kg and 178 cm (eland); maximum weight in family, 1200 kg (Asian water buffalo, *Bubalus bubalis);* maximum height: 200 cm (gaur, *Bos gaurus). Teeth:* 30 or 32 total (see chap. 1, Ruminant Traits). *Coloration:* from off-white (Arabian oryx) to black (buffalo, black wildebeest) but mainly shades of brown; cryptic and disruptive in solitary species to revealing with bold, distinctive markings in herding plains species. *Eyes:* laterally placed with horizontally elongated pupils (providing good rear view). *Scent glands* (fig. 2.1): developed (at least in males) in most species, diffuse or absent in a few (kob, waterbuck, bovines). *Mammae:* 1 or 2 pairs.

Horns. True horns consist of an outer sheath composed mainly of keratin over a bony core of the same shape which grows from the frontal bones. Keratin, a tough, horny substance, is also the main constituent of hooves, nails, claws, hair, scales, and feathers. Horns grow slowly from an

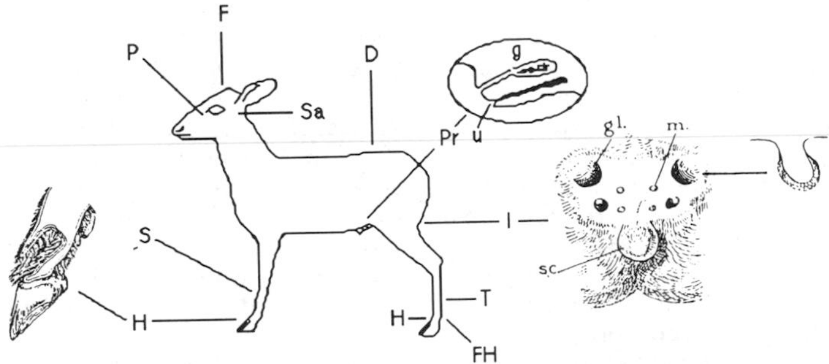

Fig. 2.1. Outline drawing of an antelope, showing locations of scent glands. *H,* hoof (interdigital) gland of a wildebeest; insert shows forefoot with one side cut away, exposing the flask-shaped gland between the digits. *S,* shin (carpal) gland (gazelles). *P,* preorbital gland (see fig. 4.1). *F,* glandular skin (impala). *Sa,* subauricular gland (reedbucks, oribi). *D,* dorsal gland (springbok). *I,* inguinal glands. Inset of reedbuck inguinal area: *gl,* gland orifice (cross-section shown in inset); *m,* mammary; *sc,* scrotum. *T,* metatarsal gland (impala only). *FH,* false-hoof gland (see fig. 9.1). *Pr,* preputial gland (inset: *g,* glandular sac; *u,* opening of penile sheath). (Inset drawings from Pocock 1910, 1918.)

epidermal layer surrounding the bony core and if broken or cut off do not regenerate. They are never branched or shed like deer antlers, which are made of solid bone. Horns evolved before hair, first in the dinosaurs, later in rhinoceroses, and at least 3 times in different families of ruminant artiodactyls. But the structural plan of bovid horns is unique to this family (18).

Horns are the weapons with which males of all species compete for dominance and reproductive success (22). Horns are present in the females of 43 African species, but are generally smaller and invariably thinner and weaker (27), reflecting the absence of sexual competition that subjects males' horns to rigorous testing in combat. There is growing fossil evidence that horns evolved first in males, and only later in females of different lineages and at different times (17, 18, E. Verba, pers. comm.). The conventional view of female horns is that they evolved as weapons of self-defense (27). But according to my own theory (10), females of polygynous species evolved horns to mimic the horns of their male offspring as a way of reducing harassment of adolescents by mature males, which leads to their separation from female herds with consequent higher mortality. When horns are present in both sexes, they cease to be badges of maleness and so young males are less likely to be singled out until they cease to resemble females. In species such as oryxes and wildebeests, the resemblance continues into adulthood and facilitates the formation of mixed herds in these migratory/nomadic antelopes. Thus, although females of all horned species surely do use their horns as weapons, in my view this is a secondary or derived function.

Like other male secondary sex characters, horns are the products of sexual selection (2) and the degree of development reflects the degree of male competition. Thus, horns are no more than simple spikes in the monogamous duikers and dwarf antelopes, but huge and elaborate in many of the highly polygynous species like the impala, sable, greater kudu, goats, and sheep.

DISTRIBUTION. Africa is the land of antelopes. Wherever vegetation grows, one species or another has adapted to eat it, from the depths of the Sahara, where the endangered addax formerly held sway, to the depths of the Equatorial Rain Forest, where duikers live; from swamps (sitatunga and lechwe) to the Afro-Alpine Biome (bushbuck, eland, duikers). Seventy-two of the 75 bovids are antelopes. The African buffalo is the only bovine that occurs naturally in the Ethiopian Faunal Region. Sheep and goats, which belong to the Eurasian fauna, have 2 African representatives, the aoudad or Barbary sheep (*Ammotragus lervia*) and the ibex (*Capra ibex*); they range as far south as the Sahel and the Abyssinian Highlands.

Only two tribes of African antelopes occur outside the continent: the Antilopini, with 9 out of a total of only 12 Asian antelopes, and the Arabian oryx (*Oryx leucoryx*) which gives the Hippotragini a foothold on the Arabian Peninsula.

What exactly is an antelope? Technically, it is the name of the Indian blackbuck, *Antilope cervicapra*, and applies to the members of its tribe, the Antilopini. In practice, bovids of all tribes apart from cattle (Bovini), sheep and goats (Caprini), and goat-antelopes (Rupicaprini) are called antelopes. But in fact, antelopes from different tribes are as unalike as are sheep and cattle. The 9 tribes of African antelopes thus represent a diversity of bovids far greater than on any other continent, either now or in the past.

THE AFRICAN BOVID RADIATION. The extraordinary diversity of African antelopes can be attributed to a number of interconnected factors (3, 18, 20, 25).

1. The African continent is unusually large and physiographically diverse. It has by far the largest tropical land mass, being the only continent that spans both tropics. The great variety of life zones (biomes) that presently exist began to differentiate during the middle Miocene, providing new and shifting ecological opportunities for the bovids to exploit. (See table 1.1.)

2. The bovid array represents the most recent ungulate radiation, which reached its peak within the last several million years in the Pliocene and Pleistocene. Although Africa, like the other continents, had its share of Pleistocene extinctions, most of the genera and species of antelopes that ever existed are still with us (14).

3. The African fauna has been repeatedly enriched by immigrations from Eurasia during extended periods when land bridges connected the 2 continents (24). The earliest known bovid, the gazelle-sized *Eotragus*, occurred both in Europe and North Africa 20 million years ago. By the late Miocene, African bovids had diversified into 9 distinct tribes, but most had Asian relatives. A massive invasion of Asian genera in the early Pliocene, followed by differentiation of African types, resulted in a major faunal rev-

olution: over 85% of the Pliocene genera are new, including 19 genera of bovids, of which all but 5 are unknown outside Africa. As late as the early Pleistocene, new genera continue to appear in the fossil record, due both to Asian immigrants (8 more bovid genera) and *in situ* evolution. The duikers, neotragines, and reedbuck/kobs are the only bovid tribes that evolved in Africa and never reached Asia (22, 24).

4. The Sahara Desert has imposed a formidable barrier to intercontinental movement since the second half of the Pleistocene, as reflected by a much higher proportion of endemic African mammals. Eurasian species could still disperse to North Africa but only desert-adapted forms could penetrate the Sahara. Most of the Eurasian tropical-savanna fauna proceeded to become extinct during the Ice Age, leaving Africa as the final refuge of Plio-Pleistocene mammals (24).

5. Speciation within Africa was promoted by expansion and contraction of the Equatorial Rain Forest during wet (pluvial) and dry (interpluvial) periods of the Ice Age. During pluvial periods the rain forest stretched from coast to coast, barring interchange between northern and southern savanna and arid biomes but facilitating the dispersal of forest forms. In succeeding interpluvials the rain forest was reduced to a number of isolated islands, and a drought corridor extending through eastern Africa connected the savanna and arid biomes; this explains the presence of some of the same mammals in the Somali and South West Arid Zones, separated by the whole *Miombo* Woodland Zone (see vegetation map), for instance the oryx, dik-dik, steenbok, bat-eared fox, and springhare (*Pedetes capensis*) (22).

6. A fundamental reason for the great diversity and success of the bovids is their ability to specialize more narrowly and efficiently than other ungulates. By tailoring size, feeding apparatus, digestive system, and dispersion pattern for a particular set of ecological conditions, antelopes have effectively partitioned African ecosystems into many small segments (see chap. 1 and refs. 3, 22, 25).

Tribal Niches. A commitment to one type of biome or another is seen in most of the tribes and genera and presumably dates back to the original differentiation of tribes in the Miocene and Pliocene.

• Gazelles and other Antilopini are specialized for arid biomes; they are medium-sized, wide-ranging gleaners which can subsist in areas too dry and poor to support larger roughage eaters. A *Gazella* species was roaming the Kenya plains 14 million years ago (14).

• The neotragines are small antelopes, allied to the gazelles, that lead sedentary, solitary lives within various closed habitats, from rain forest to subdesert.

• The duikers are specialized as forest fruit- and foliage-eaters.

• Members of the bushbuck tribe are medium and large browsers that inhabit mostly closed habitats, from forest to subdesert thornbush.

The remaining 4 tribes are grazers that have specialized for different grassland habitats.

• Cattle and the reduncines live in valley and floodplain grasslands within a short distance of water.

• Alcelaphines disperse into dry savanna during the rainy part of the year and concentrate on greenbelts near water points in the dry season (9).

• Hippotragines include the desert-adapted addax and oryx as well as the water-dependent sable and roan, which inhabit the well-watered savanna woodlands (1, 5). These ecological specializations are considered in more detail in the tribal introductions and species accounts.

PHYSIOLOGICAL ADAPTATIONS FOR ARID CONDITIONS. Vast regions of Africa and Asia are uninhabitable or only seasonally habitable by animals dependent on drinking water. Bovids able to obtain all the water they need from their food can exploit the vegetation of arid lands and thereby avoid the more rigorous competition for plants within commuting distance of water. Browsers in general are less water-dependent than grazers, mainly because bushes and trees can reach down to moist soil and produce green growth at times when grasses are parched. Moreover, many plants of arid and semiarid environments have water-storage organs such as thickened roots and tubers or succulent stems and leaves. Parts of the Kalahari sandveld produce such an abundance of melons and tubers that even water-dependent grazers like the wildebeest and hartebeest are able to live there without surface water. However, antelopes such as most gazelles, oryxes, and dik-diks require less water to survive, especially when it is extremely dry and hot. The key physiological adaptations include:

1. Allowing body temperature to rise
2. Lowered metabolic rate, decreasing with dehydration

3. Reflecting coat: flat, dense, short, smooth, light color (11)

4. Concentrating urine; extracting moisture from feces

5. Nasal panting and cooling of the blood to avoid overheating

Allowing body temperature to rise with environmental temperature—by as much as 10°C—is the most important water-conservation measure, one that is also employed by camels (29, 30). Desert-adapted species can let their body temperature go higher for longer than can other antelopes. To maintain a constant body temperature on a hot day requires water for evaporative cooling, even though the common bovid technique of closed-mouth panting is much less wasteful than sweating. Air passed across the nasal mucosa while breathing rapidly cools the blood by evaporating moisture from the nasal mucosa. The cooled venous blood then flows through a capillary network (the *rete mirabile*) surrounding the carotid arteries, where it cools the blood going to the brain (29).

CORRELATIONS BETWEEN HABITAT PREFERENCE, MORPHOLOGY, ANTIPREDATOR STRATEGY, AND SOCIAL ORGANIZATION. An analysis of the 75 species of African bovids indicates that habitat preferences, diet, size, conformation, gaits, coloration, antipredator strategy, degree of sexual dimorphism, mating system, and social organization are all coadapted (9). In table 2.1 a basic dichotomy in the traits of solitary and gregarious species is demonstrated by the results of a series of correlation tests. The significance of these differences is illustrated by comparing the traits of forest duikers, representing solitary antelopes adapted for closed habitats, with the adaptations of an oryx, an advanced open-country antelope (table 2.2).

The duikers' conformation, short, slanted horns, and gaits are designed for movement through and under dense vegetation, often on soft ground (hence the splayed hooves). The presence of undergrowth, intimate knowledge of a small home range, small to medium size, cryptic coloration, and solitary habits all support a concealment strategy. The oryx's build and gaits are adapted to traveling long distances across arid, often hard ground (hence the compact hooves). Fleet and enduring, big, well-armed and aggressive, the oryx is well-equipped to escape, and even defend itself against, predators. Although the newborn are concealed, adults avoid heavy cover. Water-independence enables the oryx to reside in arid regions that are permanently or seasonally beyond reach of water-dependent species.

Table 2.1 Correlations Between Social Organization and Selected Traits of the 75 Species of African Bovidae

	Solitary	Gregarious	χ^2
Total number of species	32	43	
Habitat preference			
Closed	30	9	
Open	2	34	38.98
Diet			
Browse or mixed (browse and grass)	31[a]	21	
Primarily grass	3[a]	20	10.98
Size			
Small to medium (1.5–45 kg)	28	10	
Medium to large (45–950 kg)	4	33	30.29
Conformation			
Rounded back, massive hindquarters, short legs	31	12	
Cursorial (long legs, back level or high shoulders)	1	31	32.91
Coloration			
Concealing (cryptic or disruptive)	31	12	
Revealing (contrasting color and/or markings)	1	31	32.91
Male horns			
Primitive: short, straight spikes or forward hooks	30	3	
Advanced: medium to large size or complex shape	2	40	52.60
Sexual dimorphism (in body size, horns, coloration, or other male secondary characters)			
Minimal	29	13	
Pronounced	3	30	27.16
Antipredator strategy of adults			
Concealment or flight to cover/sanctuary	32	13	
Avoid cover and flight in open	0	30	32.16

NOTE: All the chi-square values are highly significant ($P < .010$, 1 df).

[a]Includes 4 of the least sociable gregarious species: gerenuk, dibatag, mountain reedbuck, and sitatunga.

Table 2.2 Comparison of a Solitary Forest Antelope with a Gregarious Plains Antelope

	Forest species	Plains species
Species	Forest duiker	Oryx
Biome	Lowland and montane forest	Arid zones
Size	Small to medium (4.5–64 kg), female slightly larger than male	135–205 kg, male larger than female
Horns	Both sexes, short (5–25 cm) spikes	Both sexes, long (up to 120 cm), straight or curved
Conformation	Hindquarters more developed than forequarters, back rounded, legs short, hooves with wide splay	Limbs equally developed, back level, legs long, hooves with little splay
Gaits	Walk: cross-walk Trot: rarely observed Run: dodging, interspersed with flat leaps, head and neck low	Walk: amble Trot: long, ground-gaining Run: a fast, horselike gallop, head at shoulder level or above
Preorbital glands	Well-developed in both sexes	Vestigial or absent
Feeding habits	Concentrate selectors	Roughage feeders
Water metabolism	Water-dependent (?)	Water-independent
Coloration	Concealing	Revealing (except calves)
Social system	Solitary or monogamous, sedentary, small home range/territory	Gregarious, polygynous, nomadic, ± territorial, mixed herds and solitary males, huge home range
Breeding	Perennial; females and males mature at 1–2 years	Seasonal; females mature at 3, males at 4–5 years
Offspring	Concealed	Concealed, or calves join together in crèches
Reaction to danger	Take cover and hide	Flee in open

Group formation is a basic adaptation to life in the open; the oryx goes further and lives in mixed herds, apparently as a specialization for a subdesert biome where overall population density is very low. The long, straight horns and conspicuous markings serve as species-recognition characters that make an oryx unmistakable for any other animal.

Mating System, Age at Maturation, and Sexual Dimorphism. Sexual dimorphism in bovids, as in most mammals, is the result of male reproductive competition, which causes males to acquire physical and behavioral traits that enhance their ability to compete successfully. The greater the potential reproductive success, that is, the more females with which a male may mate—the more polygynous—the greater the sexual competition and the greater the degree of sexual dimorphism that is likely to develop (6, 10, 15, 20). The different types of social organization and mating systems are considered below.

In a monogamous system, with only one female for each male, male sexual competition is minimal and consequently there is little dimorphism—the sexes look much alike (unimorphic or unisex). In fact, females tend to be slightly larger than males in the duikers and dwarf antelopes. In all the other tribes, males mature later and end up larger than the females. The degree of size dimorphism within a species increases with the difference in age at maturation; it is much greater in species where males mature 3 or 4 years later than females than when the difference is only 1 or 2 years. This helps to explain why the most pronounced dimorphism occurs in bovids of medium and large size, for development takes longer than for smaller species (10, 15, 20, 25). Dimorphism is particularly developed in the Reduncini (kob, lechwe) and Tragelaphini (nyala, sitatunga, kudu, eland).

But the sexes look much alike in the Alcelaphini and Hippotragini, which include 8 of the 18 most gregarious antelopes. These species, though polygynous, share the tendency to form mixed herds containing adults of both sexes. Sexual segregation, conversely, increases with increasing sexual dimorphism. The correlation between these tendencies in the 43 gregarious African bovids is highly significant (table 2.3). Apparently selection for uniformity exists in species like oryx and wildebeest that are under ecological pressure to integrate, and counteracts selection for sexual dimorphism (10, 13).

Concerning the forms of sexual dimorphism, contrasts in horn development are of-

Table 2.3 Correlation Between Sexual Dimorphism and Tendency to Form Mixed Herds in Gregarious African Bovids

Sexual dimorphism	Mixed herds	Unmixed herds	χ^2
Minimal	10	3	
Pronounced	3	27	16.22[a]

[a]Highly significant ($P < 0.005$).

ten more apparent and only slightly less common than size dimorphism (9). Next commonest is color dimorphism. In the sable, kob, lechwe, nyala, eland, and other tragelaphines, for instance, adult males become much darker than adult females. Thick necks are an obvious attribute of adult male Grant's gazelles, gerenuks, impalas, kobs, reedbucks, sables, elands, and buffalos, whereas male Tragelaphini grow conspicuous beards or manes. In the eland and buffalo bulk, especially of the neck and shoulders, distinguishes mature bulls (9).

SOCIAL ORGANIZATION. Despite tremendous variation in the grouping and dispersion patterns of African bovids, all of them can be placed in 4 different categories: solitary or gregarious, territorial or nonterritorial (9). The characteristics of each type, and some basic variations, are outlined here and detailed in the tribal and species accounts.

1. **Solitary/Territorial.** The dik-diks, steenbok, klipspringer, oribi, and other dwarf antelopes, and at least some duikers, live in monogamous pairs and jointly maintain a small territory they seldom or never leave. Each member of the pair excludes outsiders of its own sex; they mark the territory with preorbital-gland secretions and/or dung middens, and maintain the pair bond by frequent tactile and olfactory interactions. Young of both sexes disperse during adolescence, the sexes develop at nearly the same rate, and as usual in monogamous mating systems, the adult sex ratio is equal. Males have to win a territory and females have to find an unmated male before they can reproduce. Two of the 3 reedbucks also live in monogamous, territorial pairs, but one of them has a tendency to polygyny and often forms loose herds away from cover, thus showing a certain gregarious tendency (see reedbuck account).

2. **Gregarious/Territorial.** Females and young associate in herds that usually contain 1 adult male, namely the owner of the territory on which the females are visiting or residing. Males without territories associate in bachelor herds, which are generally kept segregated from females and young by the intolerance of the territory owners. Female offspring are free to remain in their natal herd and traditional home range indefinitely, whereas males are forced to disperse when they acquire obvious male secondary sex characters. Higher mortality rates associated with dispersal and competition between adults for territories lead to adult sex ratios skewed in favor of females.

The gregarious/territorial form of organization is prevalent among African antelopes except for the 2 groups of solitary species (Cephalophini, Neotragini) and the spiral-horned antelopes (Tragelaphini). Probably most antelopes are distributed as resident populations, wherein males hold the same territories year in and year out. However other species, and also populations of the same species, may be migratory for some or all of the time, dispersed at one season and aggregated at another, etc. The territorial system is extremely adaptable and variable, as the following examples illustrate.

- *Conditions that intensify competition.* Generally speaking, male competition tends to be more rigorous in species and populations that have a limited rutting season than in those that breed perennially (see impala and kob accounts) (6); within populations living at high density in large herds; between territorial males who live within plain view of each other on small territories; and earlier rather than later during each individual's territorial tenure (37).
- *Lek breeding.* Males crowd together on a traditional mating ground, where females come individually to breed with the males holding the most central places. Known only in the kob, lechwe, and topi, this is perhaps the most polygynous and competitive form of territoriality. Leks only occur at high density.
- *Temporary territories.* In migratory populations, males only become territorial when the aggregation they are accompanying stops traveling. Their behavior is then perfectly comparable to that seen in resident populations, although territorial density and competition are both increased. When general movement resumes, the bulls abandon their places one by one and rejoin the migration. (See wildebeest, topi, and Thomson's gazelle accounts.)
- *Transition to rank-dominance system.* In the oryx, as already noted, herds commonly contain adult males as well as females. However, 1 male is dominant and monopo-

lizes any estrous females. The only obvious difference from a typical rank-dominance system is the presence of isolated and apparently territorial bulls which may be dominant to all males in herds.

• *Satellite males.* In the waterbuck, desirable territories may be so much in demand that the best way for a maturing male to gain one is to live as a resident satellite and wait for the owner to die or grow old.

3. **Solitary/Nonterritorial.** The bushbuck is the only African bovid in this category. Each adult has a small, exclusive core area where it rests and hides, but home ranges overlap extensively. Individuals from the same locality are loosely associated but do not herd together. The sitatunga, in which a couple of females and their young may form loose herds, represents a transition between solitary and gregarious versions of this system.

4. **Gregarious/Nonterritorial.** In addition to the spiral-horned antelopes, cattle, sheep, and goats also belong in this category. In all these tribes, male reproductive success is based on absolute dominance over other males rather than on dominance at a particular site. This more direct form of competition places a premium on size and power; to achieve them, males spend more years growing than in most territorial species, maturing typically several years later and growing much larger than conspecific females (10, 15, 20, 26). In buffalo and other Bovini, males may continue growing for most of their lives, making dominance largely a function of seniority. In kudu and nyala, as in sheep and goats, horns continue growing after maturation, so that in these species length and shape are reliable indicators of age and fighting potential. In territorial species, horns reach their maximum development at maturity; therefore their size and shape serve only to distinguish adult from younger age classes (10, 20, 22).

Segregation of the sexes except for breeding is the rule in gregarious/nonterritorial systems, but there are always exceptions. The African buffalo, for instance, lives in mixed herds.

ACTIVITY. Most bovids and other ruminants alternately feed and ruminate both day and night (62 species plus 22 probable out of 93 ruminants that were assessed) (4). Concentrate selectors tend to feed and digest in relatively short bouts, whereas roughage feeders, with a larger capacity and slower digestion, eat and ruminate at longer intervals (see chap. 1). The only strictly nocturnal or diurnal ruminants are forest frugivores, namely duikers and the water chevrotain, which are small enough (30–40 kg) to gain practically all the nutriments they need by eating fallen fruits for some hours of the day or night (4). Although some other bovids are apparently more active by night than by day (e.g., buffalo, bushbuck, reedbucks, dik-diks, grysboks), generally bovids move more in daylight, especially early and late in the day, which are also times of peak feeding and social activities. Most animals remain relatively inactive during the hottest hours. There is usually one major feeding peak at night, framed by long rest periods after dark and before dawn.

POSTURES AND LOCOMOTION. Differences in the conformation and locomotion of bovids that live in closed and open habitats have been noted (tables 2.1 and 2.2) and discussed. It should be added that bovids with a duiker-like conformation (short and/or sturdy legs, overdeveloped hindquarters), which live in cover, or in very uneven or steep terrain, characteristically walk with a diagonal stride, the *cross-walk*, whereas long-legged species and those with overdeveloped forequarters (e.g. wildebeest/hartebeest tribe) that live in open, comparatively level country typically *amble*, moving the legs on the same side of the body together, as in pacing. The latter gait is virtually unknown in bovids, which all trot in a diagonal stride. Duikers, dwarf antelopes, the bushbuck and other tragelaphines, goats and sheep, and cattle are all *cross-walkers*, whereas the tribes represented by gazelles, wildebeest/hartebeest, waterbuck, and oryx are all *amblers* (35).

Although all bovids probably can and sometimes do trot, and certain species (e.g., oryx, gerenuk, eland) even favor this gait, for most species it is the least-used gait (35). A relatively few species (e.g. wildebeest, topi, oryx, sable, dorcas gazelle) perform an exaggerated *style-trot*, in which the forelegs are lifted high, when mildly excited or alarmed, as an alternative or in addition to trotting. *Stotting*, a high, straight-legged bounding gait (fig. 2.2), is widespread among plains antelopes, including gazelles, topi, hartebeest, dik-dik, oribi, etc., most of which have a white rump patch that adds to the conspicuousness of this display. Apart from expressing excitement, stotting clearly functions as an alerting or warning signal when performed by, for instance, a gazelle fawn that has been flushed from hiding.

Fig. 2.2. The stotting gait of a Thomson's gazelle (from Walther 1984).

Differences in the way antelopes run are too complex to discuss in detail here, but follow from differences in size, general conformation, and relative limb development and length. The high, bounding gallop of a reedbuck, the labored run of a bull eland, and the ground-gaining lope of migrating wildebeests illustrate the range of differences that exist. Postures associated with displays are considered separately (see below). Resting postures are discussed in chapter 1.

Grooming and Comfort Movements. Probably all bovids scrape their coats with their lower incisors and nibble-groom with their lips; a smaller number lick themselves (duikers, cattle, and bushbuck tribes), although perhaps all clean their nostrils by inserting the tongue. The shoulders, tail, and back, followed by other areas in reach of the mouth, are the places most frequently groomed. The cheek and the side of a horn are often used to brush the withers with a sweeping movement over the shoulder. In long-horned species, the horn tips are used to scratch the back or rump. The hindfeet are employed to scratch the head, ears, eyes, between the horns, neck, chest, and upper forelegs. Insect pests are warded off by wagging or clamping the tail, raising and vigorously shaking the head and flapping the ears, hitting shoulders and sides with the head, twitching the flank skin, shuddering the legs, and stamping.

Wallowing or rolling in mud or dust is rare in antelopes. Wildebeests are the only ones that roll; these two species, hartebeests, and topi also rub their heads and horns in mud, then smear themselves. The buffalo is the one confirmed wallower among African bovids.

SOCIAL BEHAVIOR

COMMUNICATION. Bovid visual displays are considered in a separate section following Antipredator Behavior. Vocal, olfactory, and tactile communication are treated in the tribal and species accounts.

AGONISTIC BEHAVIOR. The various displays associated with aggression, fear, and uncertainty are catalogued in table 2.4.

Fighting Behavior. The different forms of fighting practiced by African bovids reflect the great morphological variety that is represented in 11 different tribes. Similarities in fighting techniques are discernible within tribes, the more closely related the more alike. Thus, all members of the hartebeest tribe always fight on their knees, and the horse antelopes usually do so. The sitatunga, bushbuck, and bongo do so occasionally (35). All other African bovids normally remain on all fours during combat. But form also affects function, and fighting behavior is particularly complex and variable. Thus, *Gazella* species of different size with horns of different proportions have different fighting techniques (37). Species with similar horns, even though only distantly related (e.g. kob and impala, sable and ibex, kudu and addax), might be expected to fight in similar fashion, and indeed may, but still within the constraints of phylogenetic differences.

The difficulty of describing and comparing fighting techniques is compounded by the fact that the repertoire of each species includes several different techniques (35), and these varied and complex behavior patterns have so far been studied in only a few species (34). Consequently only a superficial treatment can be given here. The information and descriptions are based on the work of Walther (33, 35, 37).

Horns normally have sharp points, capable of penetrating even the thick skin which forms a shield over the neck and shoulders of males in some, perhaps most bovids (e.g. ref. 19). Once I found a dead male Thomson's gazelle that had been stabbed in the heart, presumably by a rival's horn. Yet stab wounds and other serious injuries are very rare. Why?

- When bovids fight, they normally direct their attacks to the opponent's head and not the body, and alignment during the fights of all African bovids is head to head. Although several species also fight while standing parallel, these are special and not their main fighting techniques.

• Except in their most primitive form—short spikes and hooks—horns are designed as much for defense as for offense, serving to catch and hold the opponent's horns, thus preventing the tips from pressing into the skin. S-shaped horns (e.g. impala, kob, lechwe) have up to four functionally different sections: the stem or base for protecting the skull; a thickened convex section for ramming; a concave section for catching and holding the opponent's horns; and the smooth, sharp tips for stabbing (22). The ridges or annulations that occur in the majority of species not only add structural strength but also help bind opponents' horns together (12, 33).

• The purpose of fighting is to determine which of two contestants is superior. Therefore contests are normally between individuals of approximately equal size and fighting prowess. Animals that are clearly outmatched, for instance immature males threatened by individuals of older and bigger classes, will run away rather than fight. At the same time, the risk of predation resulting from even minor injuries selects in favor of individuals that know when to quit—that goes for winners as well as losers. The very act of fighting is also a time of particular danger, and individuals which engage in protracted, all-out combat may well be caught in the act by some wide-awake lion.

• Consequently, fights are normally a simple trial of strength, in which the combatants are equally adept in offense and defense and rarely gain the opportunity to break through the opponent's defenses and gore him. Fights usually end either in a draw or in the flight of the loser.

• Yet, though damaging fights are rare, probably every male is battle-tested at some point in the course of maturing and winning the right to reproduce. The incidence of broken horns, though never more than a small percentage of the adult males, is evidence of severe fights in which the quite incredible stress necessary to break a male's horns was exerted.

• The fact that attacks are rarely directed to the body may reflect some inhibition and the existence of fighting conventions. However, such inhibitions tend to disappear in serious fights and/or when one is the clear winner. The loser is at serious risk when he gives up and flees, for often the victor tries to gore a fleeing rival in the backside. But by exerting himself to the utmost, the loser nearly always manages to escape unscathed.

Fighting Techniques. The most basic forms of combat are outlined here, with examples of the species that employ them, beginning with the most common (35, 37). The position of the head, and the movements of the head and neck which are associated with these different forms, are listed and illustrated (fig. 2.3), and major variations in the form are noted. See the tribal and species accounts for more detail.

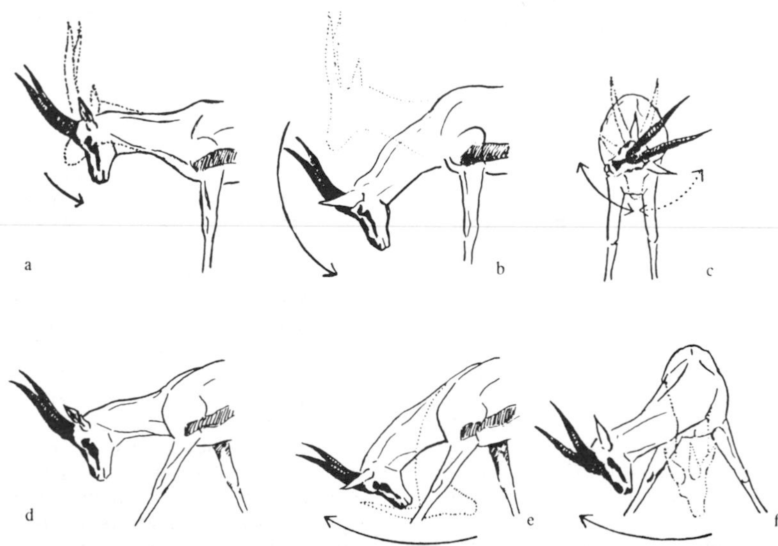

Fig. 2.3. Common fighting movements (Thomson's gazelle): a) nod-butt; b) forward-downward blow; c) horn levering (twisting); d) posture of readiness; e) forward-swing; f) sideward-swing. (From Walther 1979.)

Boxing. Low-intensity fighting in which the chin is drawn in until the horns of two opponents come into contact, a movement called *nod-butting.* Examples: sparring gazelles, topis, hartebeests.

Clash-fighting and fencing. Contestants exchange hard blows at short range, jumping backward between butts. Associated head and neck movements: *forward-downward blows, push-butting* and *forward-pushing* (less commonly forward-swinging, sideward-pushing and -swinging). This style becomes *front-pressing* when combatants lock horns. Examples: the prevalent style of small gazelles (subgenus *Gazella*), also topi/blesbok and waterbuck. Variations: fencing oryxes exchange short but strong forward and sideward blows of their long, straight horns while confronting with lowered heads, usually while remaining in one spot. The ibex and other goats rise on their hindlegs as a prelude to clashing horns, thereby intensifying their *forward-downward blows.*

Push-fighting. A transitional form between *clash-fighting* and *front-pressing* (the next style). Combatants press forward instead of disengaging, but without locking horns. Associated fighting movements are chiefly *push-butting* and *forward-pushing.*

Front-pressing. With horns crossed and firmly anchored near their bases and heads lowered parallel to the ground so that the horns point forward, combatants strain to push one another backward, employing the full array of fighting movements (forward, sideward, upward, twisting, and even backward). Examples: best developed in species with long horns, especially common and pronounced in combats of Grant's gazelle, impala, kob, lechwe, waterbuck, oryx, kudu, and eland.

Thrust-fighting. All-out fighting, in which combatants jump into contact and keep thrusting forward with all their might. *Thrust-fighting* develops from clash-fighting or *front-pressing.* Examples: unusually severe fights between male gazelles, alcelaphines, and hippotragines.

Ramming. A form of clash-fighting practiced by species whose horns and skulls are specially adapted to absorb shock, notably bighorn sheep and bovids with massive horn bosses such as the African buffalo. Combatants take a running jump and bash heads with tremendous force (cf. combat of giant forest hog).

Hooking. Side-to-side (pendulum) head movements. This is not a separate technique but performed in combination with other fighting styles, as during *thrust-fighting* and *front-pressing.* Example: wildebeests during hard fights.

Lever-fighting. The same movements as in hooking, when combatants' horns are crossed and anchored. Seen mostly in long-horned species such as Grant's gazelle, waterbuck, kudu, and other tragelaphines during *front-pressing* fights.

Fight-circling. Pivoting, meanwhile violently pulling and jerking, that occurs when combatants fighting with horns firmly locked have difficulty disengaging. Observed instances: Grant's gazelle, greater kudu, springbok—and once only, two wildebeests (*C. taurinus*) (author's observ.).

Horn-pressing. Two combatants standing close together in *medial-horn presentation* engage the front surfaces of their horns. Each then tries to force the other's horns back onto its neck. This technique carries over into *front-pressing* when the horns become interlocked. Quite a widespread technique but most developed in longer horned species such as Grant's gazelle and especially antelopes with convoluted horns such as the kudus (see fig. 9.9).

Parallel fighting. A few species stand side by side or at an angle, and with horns interlocked, neck-wrestle or push sideways with horns, and/or attempt to hook sideways and backward, over or under the opponent's guard. Examples: oryx, roan, and sable all fight in this way, but mainly as a form of sparring between immature males (author's observ.). The aoudad and some other bovids with hooked horns that extend sideways also engage in parallel fighting, with their inside horns interlocked.

Air-cushion fighting (or *shadow-boxing*). Two combatants may perform all the different attacking and parrying maneuvers that are used in real fights, but without actually making contact—as though an invisible cushion had been placed between their horns (35).

REPRODUCTION. All African bovids bear one young and breed at least once a year (small species often twice), giving birth to well-developed young after gestation periods of about 6 months (dwarf antelopes and gazelles) to 8 or 9 months in medium and large species. A female's maximum lifetime production of offspring has been found to be about 14 for both small and large species (15). Most species have extended but definite breeding seasons during the rainy part of the year and, in equatorial regions with two rainy seasons, many have a secondary calving peak. Females of perennial and semiannual breeders reenter estrus within a month or two after calving (postpartum estrus).

Sexual displays are described under Communication (see table 2.4).

SEXUAL BEHAVIOR. Estrus generally lasts for no more than a day in territorial species. But in nonterritorial species (bovines, tragelaphines), in which males seek out and attempt to maintain exclusive *tending bonds* with estrus females, estrus lasts for several days, at least. Female receptiveness is usually signaled simply by readiness to permit mounting and by moving the tail aside to permit intromission. Copulation takes a few seconds at most, consisting of a single ejaculatory thrust following intromission. Male copulatory postures differ between tribes but tend to be the same within tribes (see tribal accounts).

PARENT/OFFSPRING BEHAVIOR. In the monogamously paired duikers and dwarf antelopes, the male may come to the defense of his own offspring. In polygynous species, males usually provide no parental care. The offspring of all but two tribes of African bovids are *hiders*: calves remain concealed anywhere from a week or less up to 2 months, depending on species, emerging only when summoned by their mothers to nurse; afterward the calf seeks a new hiding place on its own initiative (23).

An elaborate concealment strategy reduces the chances of predators finding hidden calves (e.g. see Grant's gazelle account). Newborn young lie still unless in imminent danger of discovery. Their scent glands remain inactive during the concealment stage, and body wastes are retained until the calf is stimulated to void them by the mother's licking. The mother typically remains on guard but stays away from the hiding place and retrieves her offspring no more than 2–4 times in a 24-hour period. When moving with the mother, a *hider* calf does not travel beside her but alternately runs ahead and lags behind her. Even after the concealment period when they join the herd, *hider* calves spend relatively little time in their mothers' company. The more sociable species stay together in peer subgroups (*crèches*) and seek out their mothers only to nurse and when the herd undertakes general movements.

Follower calves accompany their mothers from the time they first gain their feet and do not go through a lying-out stage. The only examples among antelopes occur in the Alcelaphini, where the wildebeests have evolved a unique system as an adaptation to migratory habits and dense feeding aggregations. Groups of calves serve as "cover" for newborn individuals during an abbreviated postnatal feeble period, and the year's calf crop is produced within a few weeks (see tribal and wildebeest accounts). Only the mother actively defends her calf against predators, whereas the buffalo and other bovines have evolved a group defense. Yet another system is deployed by sheep and goats, which rely on cliffs as refuges for both adults and young.

ANTIPREDATOR BEHAVIOR. See Alerting and Alarm Signals, table 2.4, and tribal and species accounts.

BOVID COMMUNICATION

VISUAL DISPLAYS AND OTHER VISUAL SIGNALS. Strictly speaking, a display is a particular type of signal that is designed to transmit a message and elicit a social response (28). Usually it involves the performance of some distinctive behavior pattern—distinctive in the sense of being unlike the displays of associated species and unlike other signals in the species' own repertoire. However, not all of the distinctive behaviors performed by bovids elicit any discernible response, making it difficult to say what message is intended or even, in some cases (e.g. *urine-testing, displacement grooming*), whether a behavior is actually a display at all. In the catalogue of common bovid displays presented in table 2.4, the list is not limited to behaviors that meet the strict definition of displays, but also includes various postures and actions that are distinctive, common, and consistently associated with a particular situation.

Compared to higher carnivores and primates, or even to zebras, antelopes have a very limited ability to show facial expression beyond opening the mouth and (in a few species, notably cattle and impala) extending the tongue. But postures, positions and movements of the neck, head, horns, ears, legs, and tail, erection of hair, and so on are all employed to send a variety of signals concerning species, sex, social and reproductive status, emotional state, and intentions.

SPECIES RECOGNITION SIGNALS. Though not classed as displays, traits that serve to identify a species serve an important signal function by helping animals to recognize their own kind and avoid confusion with other associated species. Such traits include overall body and horn configuration, which tend to be unique, color-

Table 2.4 The Common Visual Displays of African Bovids

Category/Display	Description	Examples	Figures
ADVERTISING SOCIAL STATUS	Includes postures and movements that advertise presence, status, and motivational state. Performed primarily by territorial males and alpha males in rank-hierarchy systems.		
Static-optic advertising	Visual advertising of position/status by standing or lying apart from conspecifics, often in conspicuous place and pose (e.g. *erect posture*).	Topi on mound.	2.4, 4.8, 6.1, 8.21
Scent-marking	1) Urination and/or defecation in visually distinctive way; 2) marking objects or conspecifics with facial glands.	Most territorial species.	(1) 2.5, 4.3, 4.6, 5.2. (2) 3.2, 4.6, 4.10, 5.7, 8.16
Object-horning	Fighting movements directed to the substrate (ground, vegetation, or other objects), when not addressed to any specific individual (cf. horning as Offensive Threat).	Universal.	9.2, 9.15
Herding and chasing	Rounding up females and driving out males.	Most gregarious bovids.	5.6, 5.16, 8.19, 8.22
Rocking canter	A special gait of active territorial males.	Wildebeest, topi, hartebeest.	8.17
High-stepping, nodding, horn-sweeping, swinging tail, etc.	Special movements that call attention to the performer.	Alcelaphini, oryx, various other species.	7.3, 8.10, 8.19
Urine-testing	Checking female reproductive status (see under Courtship Displays).	Nearly universal, mainly by dominant males.	2.12, 7.8
AGONISTIC BEHAVIOR			
Dominance Displays	Intimidating as opposed to threatening acts.		
Erect posture (also called *head-up* or *proud posture*)	Neck raised above horizontal. Numerous species-specific variations in position and movements of head and neck:	A universal bovid display.	2.4, 2.6, 6.1, 6.4, 7.2, 7.10, 8.11, 8.21
Nose level	Muzzle horizontal	Gerenuk, *Tragelaphus, Damaliscus, Kobus, Gazella.*	6.4
Nose lifted (chin-up)	Muzzle above horizontal	Gerenuk, impala, oryx.	5.10
Nose lowered (chin-in)	Neck moderately erect, muzzle pointing forward-downward. (Cf. *high-horn presentation.*)	Impala, kob, wildebeest, hartebeest, sable, etc.	6.14, 8.21
Head-turned-away	The static pose, not the movement of *looking away,* which may signify submission.	Impala, kob, gerenuk, oryx, wildebeest, topi.	
Head movements	*Head-shaking* and *nodding* when performed in lateral orientation (see under Offensive Threat).	Common in Alcelaphini, widespread but less common in other bovids.	2.6
Humped-back posture	Head lowered, hunched posture, in *lateral presentation.*	Bushbuck, nyala, greater kudu.	2.7, 4.7, 9.7

Fig. 2.4. Static-optic advertising: a territorial topi standing on a termite mound in the *erect posture*.

Fig. 2.5. Dunging ritual: a territorial sable bull scraping preparatory to defecating.

Fig. 2.6. Aggressive displays of tsessebe: *head-casting* performed by adult males standing in the *erect posture* in reverse-parallel orientation.

Fig. 2.7. Nyala *lateral display*.

Fig. 2.8. *Medial-horn presentation* (Thomson's gazelles).

Table 2.4 *(continued)*

Category/Display	Description	Examples	Figures
ORIENTATION TO OPPONENT			
Lateral presentation	Standing sideways to opponent, either parallel (head-to-head) or reverse-parallel (head-to-tail) while performing dominance or threat displays.	Almost as widespread as, and typically combined with, the *erect posture.*	2.6, 2.7, 7.2, 7.10, 7.13, 9.5, 9.7
Facing	Facing opponent head-on (more threatening than presenting side).		
Supplanting	Displacing a subordinate simply by walking toward it.	All gregarious species.	
Slow, stiff-legged approach, including *high-stepping* and *prancing*	Given in combination with dominance or threat displays.	Bushbuck, nyala, kob, topi.	8.10, 9.7
Offensive Threat	Actions that signal readiness to attack, usually performed in facing orientation.		
Medial-horn presentation	Neck at about shoulder level, horns pointing upward.	A display posture common to possibly all bovids.	2.8, 6.2, 7.4, 7.9, 7.12, 10.4
Low-horn presentation (standing on all fours)	Head lowered, chin in, horns pointing at opponent.	Comparatively rare: eland, ibex.	2.9, 8.20
Kneeling	Species that fight on knees.	Alcelaphini, sable, roan.	
High-horn presentation	Neck raised, nose level or angled downward.	Thomson's and other small gazelles, *Kobus* spp., oryx, sable, roan.	
Angle-horn	Head and horns tilted toward opponent, often from *lateral presentation.*	Wildebeest, roan, sable, oryx, waterbuck, Grant's gazelle.	6.14, 7.13, 8.2
Staring	Looking directly at opponent.	Widespread.	
Symbolic butting	Pronounced, rhythmic nodding and head-throwing (exaggerated nodding).	Medial-to-low of gazelles, kob; medial-to-high of topi.	2.3, 2.6
Head-shaking or *-twisting*	Side-to-side head movements used in fighting by some species (often seen during cavorting).	Wildebeest, topi, waterbuck, impala.	
Horn-sweeping	Abrupt sideward gesture to shoulder or flank (cf. *fly-shooing* under Displacement).	Alcelaphini, Hippotragini.	7.3
Threat jump	Jumping forward with forelegs, in combat attitude, feinting an attack.	Gazelles, Alcelaphini.	
Rushing/charging	Despotic aggression of superior against inferior opponent.	Universal.	
Chasing	Typical response of dominant to low-ranking opponent which takes flight.	Territorial male chasing an adolescent male; winner of a fight.	5.16, 8.22
Cavorting	Jumping, bucking, spinning, and similar antics signifying defiance or threat (often with head-shaking). Seen in high-intensity interactions.	Alcelaphini, especially wildebeests and topi.	8.3

Table 2.4 *(continued)*

Category/Display	Description	Examples	Figures
Pursuit march	Driving inferior opponent at a walk, while displaying dominance.	Occurs in perhaps all gregarious species.	
Object-horning	A threat when addressed to a specific individual.	Sable, Grant's gazelle, springbok, buffalo, etc.	9.2, 9.15
Defensive Threat	Displays associated with self-defense (can include responding in kind to offensive threats).		
Head-low posture	Chin drawn in more or less, depending on the direction of anticipated attack, with horns usually pointing upward.	Nearly universal.	2.10, 7.7, 7.15
Mainly females of hornless-female species:		Tragelaphini (inc. eland), Reduncini.	
Symbolic biting	Snapping, usually while facing away from opponent.	Waterbuck, Tragelaphini, duikers, klipspringer.	6.15
Pushing with snout	Usually without contact.	Tragelaphini.	
Neck-winding	Writhing movements with neck raised.	Tragelaphini.	9.6
Submissive Behavior	Behavior given in response to dominance and threat displays, or sexual harassment, that functions to appease aggressor.		
Head-low/chin-out posture	Horns point backward as chin is lifted, to varying degree by different species.	Practically universal.	2.10, 4.7, 6.2, 6.15
Lying-out[a]	Lying prone like a hiding calf, in response to aggression or sexual harassment.	Wildebeest, sable.	
Kneeling	As a preliminary to *lying out.*	Wildebeest, sable.	8.20
Head-in posture	Posture of immature and other low-ranking individuals intimidated by dominants.	Hartebeest and topi.	8.4
ACTIVE AVOIDANCE	In response to aggressive displays by dominants.	Universal.	
Passing behind	When a dominant animal stands in *lateral presentation* blocking the way.	Alcelaphini, Hippotragini.	7.7
Turning tail	Withdrawal (walk or run away).	Universal?	
Looking away	The act of turning the head aside.	Widespread.	2.10, 7.12
Turning aside	Moving out of the way—being supplanted.	Universal.	2.9
DISPLACEMENT ACTIVITIES[b]	Maintenance activities performed in stressful situations. Walther (36) subdivides these into *excitement activities* and *space-claim displays.*		

Fig. 2.9. *Low-horn presentation* (eland).

Fig. 2.10. The difference between *head-low posture* (*left*) and turning away in *head-low/chin-out posture* (*right*). (From Walther 1978*a*.)

Fig. 2.11. *Displacement* or *agonistic grazing* during a wildebeest challenge ritual.

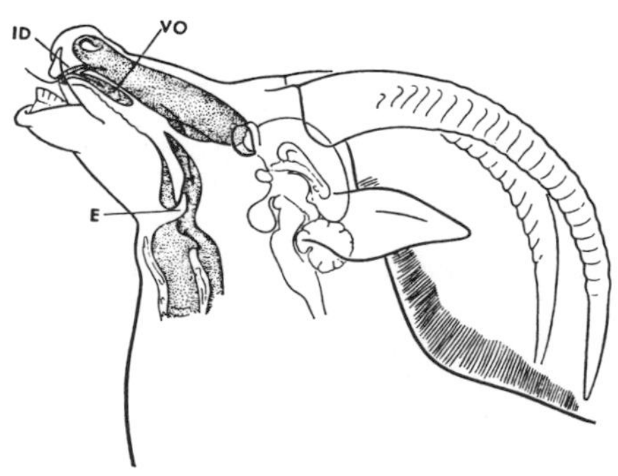

Fig. 2.12. Diagram to illustrate the pathway of urine molecules drawn into the vomeronasal organ (*VO*) through the incisive ducts (*ID*) while a sable bull performs the *flehmen* grimace, with the external nares and epiglottis (*E*) closed. (See fig. 7.8.)

Fig. 2.13. *Foreleg-lifting* during sable mating march.

Table 2.4 *(continued)*

Category/Display	Description	Examples	Figures
Grazing	Especially during encounters of territorial males. *Space claim* of (36), *agonistic grazing* (7).	Alcelaphini, *grazing duels* of Thomson's gazelle.	2.11, 5.8
Grooming	Species-typical movements.	Universal.	5.11, 8.1
Fly-shooing	Abrupt, sweeping head-to-shoulder or -flank movements, perhaps identical to *horn-sweeping.*		7.3
Alerting/alarm signals	See under Antipredator Behavior.	Alcelaphini in particular.	8.5
COURTSHIP BEHAVIOR			
Lowstretch (*neck-stretch* of (36))	Male approach/herding posture to females, with neck and chin outstretched in line.	Very widespread.	5.6, 8.12, 8.14
Responsive urination	Common female response to *lowstretch.*	Widespread.	4.4, 7.8, 9.16
Urine-testing	The posture and grimace adopted after sampling urine (termed *flehmen*).	All bovids (?) but *Alcelaphus* and *Damaliscus*.	2.12, 4.5, 7.8, 9.16
Pursuit	Early stages of courtship, when female attempts to escape.	Widespread; most pronounced in solitary species.	
Mating march	Male following female closely in pre-mating and mating stage, usually at a walk, driving her like an inferior male while performing other displays.	Universal courtship behavior?	3.7, 4.4, 5.4, 6.3, 9.6, 9.8
Foreleg-lifting (Ger. *Laufschlag;* foreleg-kick of (36))	Performed by males during the *mating march.*	Reduncini, Hippotragini, some Neotragini and Antilopini.	2.13, 6.7, 7.6
Genital-licking	By male while following female closely.	Especially duikers and Neotragines; impala, *Kobus.*	3.7, 6.8
Empty or *Symbolic licking*	Licking performed without contacting the female.	Impala, goats and sheep.	
Prancing	Quick-stepping, high-stepping, or goose-stepping during *mating march.*	Kob, Grant's gazelle, topi.	6.6, 8.10
Dominance Displays	*Erect posture,* etc. (listed under Agonistic Displays).		
Nose-lifting	Abrupt raising of the muzzle by male following female in *erect posture.*	Thomson's gazelle, kob.	5.9, 6.6
ESTROUS BEHAVIOR	Signs female is receptive to courtship.	All species.	
Coquettish behavior	Playful or skittish response.		
Mounting	Mounting or being mounted by other females.		
Token avoidance	Slowly moving away, especially with tail extended or deflected.	Widespread.	2.13, 5.9, 6.3, 7.6, 9.6, 9.8
Full estrus	Barely moving, standing or backing when male attempts to mount, holding tail out, up, or to one side.	Universal.	6.7, 6.8

Table 2.4 *(continued)*

Category/Display	Description	Examples	Figures
ALERTING AND ALARM SIGNALS			
Alert posture	Standing with head high and staring intently (sometimes combined with approach or circling, at sight of predator).	Universal.	4.8, 8.5
Stamping	With forefoot.	Most bovids.	
Stotting or *pronking*	Stiff-legged bounding, typically with white rump patch or scut displayed.	Antilopini, Reduncini, hartebeest, topi.	2.2, 5.12
Style-trotting	High-stepping, springy trot.	Alcelaphini, Hippotragini.	
High-jumping	Leaping and scattering flight.	Impala, lechwe, eland, kudu.	9.14
Freezing	Arrested movement, often with one leg raised.	Forest bovids: duikers, tragelaphines, gerenuk.	

NOTE. Terminology follows Walther (35, 36) for agonistic and courtship displays, with some exceptions.

[a]*Lying-out* as used by Walther (pers. comm.) refers to the active separation of the young from its mother. In the *Guide*, the term refers to the posture of hiding, with head on the ground.

[b]Walther (36) presents cogent arguments against the conventional view that an animal performing irrelevant-looking, "displaced" activities must be caught between two relevant but opposing drives, e.g. simultaneous "fight or flight" tendencies. This admittedly unsatisfactory term and concept is retained in the *Guide* simply because it is well-known and serves as a useful category.

ation/marking patterns, and in some cases stereotyped tail movements (continual tail-swinging of Thomson's and some other gazelles, oryx, and eland, and the up-and-down tail movement of the blue duiker).

AGONISTIC DISPLAYS. Aggression is involved in practically every aspect of bovid social life and even in sexual relations. For the most part it is symbolic, commonly expressed through dominance and threat displays. (Fighting is discussed above under Agonistic Behavior.)

Dominance/Threat Displays. The two principal types of aggressive displays are dominance displays and threat displays. Although both occur in the same situations and have a similar intimidating or challenging effect upon the addressee, depending on whether it is inferior or equal to the performer, there is an important difference between them (36). Dominance displays are designed to make the performer look as impressive as possible, especially taller and often more massive and powerful. Associated movements tend to be very deliberate, slow or stiff, and poses are "frozen" in place—in short, they are highly ritualized. The message is, "I am the greatest," or "I am superior to you," but there is no threat or immediate probability of an attack. Threat displays, in contrast, feature the same postures and movements, ritualized (exaggerated) to a varying degree, that are associated with fighting technique. The performer is saying, in effect, "I am ready to fight you," and indeed a threatening animal is much likelier to attack than one displaying dominance (36). Displays which represent intention movements that are "frozen" in their initial stages are more ritualized than performing the same movements and gestures (nodding, head-shaking, etc.) that are used in fighting; these are symbolic but scarcely ritualized.

Unfortunately, it is not always obvious into which category an aggressive display falls. The dominance display of one species may be the threat display of another. It depends on whether or not the display is closely related to fighting technique. *High-horn presentation,* for instance, is a severe threat to strike a downward blow in Thomson's gazelle, but in an oryx (fig. 7.2) it can be hard to tell whether a bull in this pose is threatening a downward blow or simply displaying dominance in the *erect posture* (36). To compound the confusion, there are displays that combine elements of both dominance and threat (dominance-threat displays of 36).

The distinction between offensive and defensive threats. An animal giving a defensive threat, typically standing in the *head-low posture,* is saying in effect, "I will defend myself if you attack," which is close to an admission of inferiority. Indeed the *head-low posture* is sometimes hard to distinguish from, and may lead into or alternate with, the submissive *head-low/chin-out* or *-raised posture* (fig. 2.10).

Defensive/Submissive Displays. There are two types of submissive displays: a) postures that are the opposite of threat or intimidation, especially lowering the head with chin raised (*head-low/chin-out posture*) so that the horns blend into the body silhouette (fig. 2.10, oryx); b) infantile behavior, including bleating or crying like a calf in distress, often heard when a young male is chased by an adult, and kneeling and *lying-out* with head outstretched like a calf in concealment.

DISPLACEMENT ACTIVITIES. Grooming behavior and other maintenance activities when performed in stressful circumstances appear out of place and have been termed *displacement activities.* Those that consistently appear during aggressive interactions between bovids clearly play a role in communication. For instance, the wildebeest, Thomson's gazelle, and many other species graze during confrontations (figs. 2.11, 5.8). (Walther [36], who questions the reality of displacement behavior, terms this *agonistic grazing.*) Alcelaphine and hippotragine antelopes make an aggressive-looking sweeping movement to the shoulder or flank (*horn-sweeping*) that resembles the movements of fly-shooing (fig. 7.3). Confronting Grant's gazelles scratch their necks and between the horns with a hindfoot as though tormented by insects, and territorial wildebeests shake their heads violently during challenge rituals. They may also give alarm signals (*alert posture, stamping, snorting*) for no obvious reason (7).

COURTSHIP DISPLAYS. Courtship displays are given almost exclusively by males (see female estrous behavior in table 2.4). How elaborate and the frequency and vigor of the performance vary greatly between species; for instance, the energetic pursuit of a courting Thomson's gazelle contrasts markedly with the stately march of a Grant's gazelle. Then too, the variety of displays performed by many gregarious antelopes is very different from the courtship behavior of duikers, dwarf antelopes, and other ruminants of closed habitats, which consists mainly of hard chasing and avid licking of the female's genitalia (4).

The prevalence of dominance displays during courtship (table 2.4) is noteworthy, and raises the question of why males should try to intimidate the females with which they are seeking to mate. The significance of performing such displays during courtship is that females will usually only mate with males which demonstrate the ability to dominate them (author's observations).

Urine-testing. Males of nearly all ungulates sample the urine of conspecific females, typically as part of a ceremony in which the male's approach stimulates a female to squat and urinate. After sniffing and/or licking the urine, the male retracts his upper lip in a grimace, known as *flehmen* (German), while the urine sample is drawn into the vomeronasal or Jacobson's organ for urinalysis (fig. 2.12). If the level of sex hormones in the urine signals approaching estrus, the male begins courtship displays, usually before the female becomes sexually receptive (8). The hartebeest and topi represent the only known bovid genera that do not perform the urine test.

References

1. Bell 1982. 2. Darwin 1871. 3. Demment and Van Soest 1985. 4. Dubost 1983. 5. East 1984. 6. Emlen and Oring 1977. 7. Estes 1969. 8. ——— 1972. 9. ——— 1974*a*. 10. ——— in press. 11. Finch 1972. 12. Geist 1966. 13. ——— 1974. 14. Gentry 1978. 15. Georgiadis 1985. 16. Gosling 1981. 17. Heintz 1969. 18. Janis 1982. 19. Jarman 1972*b*. 20. ——— 1983. 21. Kiley 1972. 22. Kingdon 1982. 23. Lent 1974. 24. Maglio 1978. 25. McNaughton and Georgiadis 1986. 26. Owen-Smith 1977. 27. Packer 1982. 28. Smith 1978. 29. Taylor 1969. 30. ——— 1970*b*. 31. Taylor and Lyman 1972. 32. Tinbergen 1964. 33. Walther 1958. 34. ——— 1978*a*. 35. ——— 1979. 36. ——— 1984. 37. Walther, Mungall, Grau 1983.

Added Reference

Spinage 1986.

Chapter 3

Duikers

Tribe Cephalophini

Genus *Cephalophus*, forest duikers

Small Duikers:

C. monticola, blue duiker*
C. maxwellii, Maxwell's duiker*

Medium Duikers:

C. adersi, Ader's duiker†
C. callipygus, Peter's duiker* (The red duiker complex)
C. natalensis, red duiker* (The red duiker complex)
(*C. n. harveyi*, Harvey's red duiker) (12) (The red duiker complex)
(*C. n. weynsi*)† (9, 12, 16) (The red duiker complex)
C. nigrifrons, black-fronted duiker
(*C. n. rubidus*, Ruwenzori duiker)† (12)
C. rufilatus, red-flanked duiker
C. dorsalis, bay duiker*
C. leucogaster, white-bellied duiker*
C. ogilbyi, Ogilby's duiker*
C. zebra, zebra duiker
C. niger, black duiker

Big Duikers:

C. silvicultor, yellow-backed duiker*
C. spadix, Abbott's duiker*†
C. jentinki, Jentink's duiker†

Genus *Sylvicapra*:

S. grimmia, gray, common, or bush duiker

*Probable superspecies or very closely related species
†Local and/or very rare species
( )Subspecies treated as separate species in some classifications

TRAITS. Cover-dependent antelopes of small to medium size (4 kg, blue duiker, to 80 kg, yellow-backed duiker) and minimal sexual dimorphism, females slightly bigger than males. Stocky build with short legs, pointed hooves with wide splay; rounded back, hindquarters more developed and higher than forequarters; short tail with terminal tuft; head proportionally large with bare, moist muffle, wide gape, small ears, and a tuft of hair (*Cephalophus* means "head crest") partially or wholly concealing short, spikelike, back-slanted horns, present in both sexes but often lacking in female *monticola, maxwellii,* and *rufilatus* (5, 14). *Coloration:* concealing, from light gray to black, commonly reddish-brown, markings generally subdued; sexes alike, young similar (Maxwell's and zebra duikers) to darker (Jentink's and bay duikers) than adults. *Scent glands:* between hooves of all feet; inguinal glands present in some species (1), preorbital glands developed in both sexes, opening into a slit. *Mammae:* 4.

Differences between *Sylvicapra* and *Cephalophus* are noted in the gray duiker account.

Preorbital Glands. The form of the preorbital glands sets duikers apart from all other antelopes. They are situated further ahead of the eyes and secrete through a row of pores which open into a narrow, hairless slit on either side of the nose. The glands make prominent swellings on the cheeks, especially in adult males. They are compound, consisting of 2 or 3 different secretory layers surrounded by a capsule of connective tissue (17): an outer, white zone composed of coiled, unbranched tubular apocrine glands and a middle layer composed of branching holocrine sebaceous glands. The presence of melanin, secreted in the middle layer, makes the substance black and sticky in some duikers, but in the blue and Ogilby's duikers the secretion is clear and oily (17).

DISTRIBUTION. Virtually every African forest and woodland is occupied by at least 1 species of duiker. However, the great majority of species live in the Equatorial Rain Forest and West Africa is clearly the distribution center, the majority of species being found in countries fronting the Gulf of Guinea. There are actually only 4 species that are widespread outside the forest block: the red-flanked duiker, which lives along the northern edge of the forest; the red duiker of eastern Africa; the blue duiker, which occurs in virtually all forested habitats and is the most widespread *Cephalophus* species; and the common duiker, which occurs in virtually all wooded habitats outside the rain forest.

Species with very limited and peripheral distributions are not among the ecologically dominant or important rain-forest ruminants. The following duikers are all rare and listed as threatened species (11): Jentink's duiker, the second largest member of the tribe (70 kg), known only from the lowland rain forests of Liberia and western Ivory Coast and Sierra Leone; the rare zebra duiker, confined to mountain forests from western Sierra Leone to central Ivory Coast; in East Africa, Ader's duiker, confined to the island of Zanzibar and the adjacent mainland; and Abbott's duiker, which may represent an isolated form of the yellow-backed duiker and occurs only in islands of montane forest in Tanzania, including Kilimanjaro (12).

There are really only half-a-dozen duikers that are widely dominant ruminants within the rain-forest block: *monticola, callipygus, nigrifrons, leucogaster, dorsalis,* and *silvicultor.* The rest are either isolated, peripheral, or questionably distinct species. The six dominant duikers may all be found within the same area, for instance in the Gabon rain forest (table 3.1) (3). No other antelope genus has that many overlapping species.

ANCESTRY. Duiker ancestry is unknown. No fossils have been found that predate the Pleistocene, by which time the living genera were already well established. The only clues to their relationships are their morphology and behavior, which may be variously interpreted. Forest duikers have usually been considered closest to the early bovids in form, diet, and social organization (6, 16, 21). But some anatomical and behavioral traits raise doubts that duikers really are more primitive than other bovids, and suggest that the Cephalophini may represent a branch of neotragine antelopes which adapted secondarily for a forest-frugivore niche (5, 12).

The duiker radiation is unequaled among forest ruminants on any other continent (5). Specializing as consumers of fallen fruits, duikers have partitioned rain-forest resources to the fullest possible extent by producing species to fill every available size class and type of habitat. Their only competitors were nonruminants (bushpig and forest hog) and the water chevrotain (*Hyemoschus aquaticus,* the only African representative of the Tragulidae, the most primitive ruminants. All the other rain-forest ruminants are folivores, including not only the okapi, bongo, and buffalo, which are too big anyway to subsist on fruit, but also the smallest of all ruminants, the pygmy and royal antelopes (see chap. 4).

The duikers' monopoly of the ruminant-frugivore niche, the array of *Cephalophus* species and their close similarity in form and basic diet, all speak of the success and high degree of specialization of this tribe (5, 12).

Duiker speciation was no doubt profoundly influenced by the waxing and waning of the rain-forest belt during pluvial and interpluvial periods of the Ice Age (12). During dry periods while the forest was fragmented, isolated populations of the same widespread species diverged as they adapted to different ecological conditions. During pluvial periods, the best-adapted among them spread most widely and went on to displace older, ecologically equivalent species that had been dominant before them. In general, it seems that the succession has been from smaller to larger forms and that the radiation has spread from the most complex and competitive forest communities of West Africa and the Zaire basin into eastern Africa. The remnant forests of eastern Africa harbor the smallest, oldest, and most generalized duikers; these may represent isolated relict populations of species that were replaced by larger, more progressive types. Thus the very small Ader's duiker of Zanzibar and the Kenya coast is seen as a relict of the earliest radiation of red duikers; the larger Natal red duiker represents a later line of coarse-necked red duikers (with longer hair on neck and shoulders) which was replaced in turn by the mainly West African group of larger, more specialized smooth-necked duikers (*C. callipygus,* etc.) now dominant in the main forest block (12).

How is it, then, that the smallest and presumably one of the oldest of all the duikers is also the most widespread and numerous *Cephalophus* species? (See species account for the answer.) Unlike Ader's duiker, which is related to the group that includes all the other dominant species, the blue duiker is phylogenetically isolated and relatively specialized: it has a number of anatomical peculiarities, such as a pointed muzzle and reduced premolars (12), and prefers more open forest than other duikers.

ECOLOGY. The fact that duiker morphology is so standardized, and that a similar conformation occurs in forest herbivores of other families and continents, would indicate that this conformation is a prototype for life in the forest understory (3, 12). Evidence that the sizes of duikers and other African rain-forest ruminants have been largely predetermined by the undergrowth structure, which permits animals of certain specific shoulder heights to move about with minimum interference from horizontal and diagonal branches, is discussed in chapter 1 (under Ecological Separation).

Both this conformation and the relatively simple stomach typical of concentrate selectors (see chap. 1) should be considered as adaptations for the niche of forest frugivore rather than as primitive traits. Compared to a chevrotain, the duiker's digestive system is much more advanced (5, 10). If the low-crowned (brachydont) cheek teeth are not very different from those of Miocene bovids, the duikers' large mouth and wide gape, which enable them to pick up and crush fruits too large or hard for primates and other competing frugivores to consume, clearly are specializations (8, 12). Their sawlike cheek teeth are used to chew bark and roots, their mobile lips and long, pointed tongues to pluck and strip foliage, and they dig with forefeet and snouts to unearth food from the forest floor (4, 12).

ECOLOGICAL SEPARATION. The sizes, diets, habitat preferences, and activity periods of 6 duikers found in the Gabon lowland rain forest are compared in table 3.1. The range of possible sizes is limited, as noted above, by undergrowth structure; in addition, species that occur in the same area are size-graded (8). Duikers of the same size class are separated either by preference for different habitats or else they use the same habitat on a different schedule. Thus, the black-fronted duiker is associated with swampy habitat wherever it lives (in lowland, montane, or alpine habitat) and has elongated hooves that help prevent it from sinking into muck. Peter's duiker prefers primary, high-canopy forest with thin undergrowth; the white-bellied duiker frequents broken-canopy and secondary forest with denser undergrowth. The bay duiker is found in both types of forest but is strictly nocturnal and conceals itself in the densest thickets by day (4). *C. silvicultor,* despite its wide distribution in the rain forest, is localized in patches of dense undergrowth adequate to conceal this largest of all duikers, whereas the smallest, the blue duiker, frequents the most open forest (see species account).

The preponderance of fruit in the diets of the 6 species in table 3.1 probably holds true of all forest duikers. The types of fruit eaten are partly a function of size: the larger the duiker, the larger the fruits it can eat. However, some 70% of all the fruits found on the Gabon forest floor are small (0.5–2 cm diameter); fruits larger than 5 cm, although amounting to nearly ⅓ by weight, are only 1/10 as abundant as smaller fruits (2). Unable to find enough large fruits or to glean enough small fruits to meet their maintenance requirements, the larger species have to eat more foliage and spend more time foraging. Accordingly, the yellow-backed duiker eats the largest proportion of foliage (29%) and has to forage both night and day, whereas smaller duikers can afford to be inactive half the time (2).

In addition to fruits and foliage, duikers eat a variety of other foods in small quantity, including flowers, roots, rotting wood, fungi, and animal food (predominantly insects, especially ants, rarely vertebrates, though captives have been known to catch and eat birds). The degree of water-dependence of the various species is largely unknown, but at least some can go without drinking for long periods, for example, blue, gray, and red duikers (3, 12).

SOCIAL ORGANIZATION. All duikers are thought to be sedentary and territorial (12). Blue and Maxwell's duikers are known to live in monogamous pairs within small territories that are jointly marked and defended against outsiders of the same sex (see blue duiker account). There is evi-

Table 3.1 Size and Ecological Separation of Sympatric Forest Duikers and the Water Chevrotain
(WS = waterside; DL = dry land; D = diurnal; N = nocturnal; F = frugivore)

Species	Habitat		Activity		Diet	Weight, kg
Cephalophus monticola		DL	D		F	4.9
Hyemoschus aquaticus	WS			N	F	10.8
Cephalophus leucogaster		DL	D		F	12.7
Cephalophus nigrifrons	WS		D		F	13.9
Cephalophus callipygus		DL	D		F	20.1
Cephalophus dorsalis		DL		N	F	21.7
Cephalophus silvicultor		DL	D	N	F	68.0

SOURCE. Gautier-Hion et al. 1980, table 1.

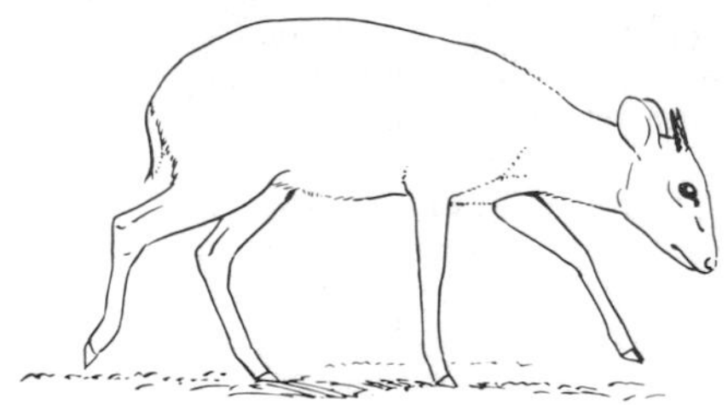

Fig. 3.1. Bay duiker moving in typical *cross-walk* and *skulking attitude* (redrawn from Dubost 1983*a*).

dence that the red, yellow-backed, and gray duikers have a similar social organization (10, 12, 13). However, too little is known about most species to say whether pair territories are the prevailing mode in the whole tribe. At any rate, pairs of blue and Maxwell's duikers apparently associate more closely than most other duikers, being more vulnerable to predators than larger species that live in denser cover in more spacious home ranges. The 12-ha home range/territory of a pair of red duikers, for instance, is 3–4 times the size of a blue duiker's. The nocturnal bay duiker, which has been kept in large enclosures under seminatural conditions and studied concurrently with the blue duiker, may be the least sociably inclined duiker: family members associate with, play with, and groom one another, communicate, and coordinate their activities much less than blue duiker families (5). The social organization and behavior of other duikers probably lie somewhere between these 2 extremes; the more nocturnal species and those that live in the densest cover may be expected to be least active and sociable.

ACTIVITY. Some duikers are entirely diurnal, for example, *monticola, callipygus, leucogaster,* and *nigrifrons.* They are completely inactive at night, during which they characteristically lie up in dense cover, often in recesses beneath logs or between buttress roots. The wholly nocturnal bay duiker spends only 33–54% of the night actively foraging, making it much less active than diurnal duikers. Thus, *callipygus,* though equal in size, moves 1.6–4 times farther than *dorsalis* during its activity period (4, 5).

POSTURES AND LOCOMOTION. Like other forest ruminants, duikers carry their heads low as they move through the understory in a *cross-walk* (fig. 3.1). Flushed from hiding, they rush for cover in a dodging run, interspersed with long, low bounds, body and limbs at full stretch. They appear to dive into the undergrowth, whence the Afrikaans name *duikerbok* ("diving buck"). The movements of foraging diurnal species are quick and "nervous", in contrast to the nocturnal bay duiker, which moves very slowly and cautiously (5).

SOCIAL BEHAVIOR

COMMUNICATION. Scent and vocal signals are most developed in this tribe, as would be expected of animals living in dense cover. Visual signals are more important in blue, Maxwell's, and gray duikers, which prefer relatively open habitats. Rapid movements of the tail which cause the white undersurface to blink on and off help keep family members in visual contact. The yellow-backed duiker's dorsal crest, erected in response to any disturbing event, is another very conspicuous visual signal, the significance of which remains unclear, however (13). The bay duiker appears to depend very largely on olfactory communication, and has minimal vocal and visual contact (5).

Tactile Communication: *social licking/nibbling, preorbital-gland-rubbing/pressing,* contact calls, social play.

Social licking and *nibbling* of the head, neck, and shoulder is frequent between mates, and mothers and calves, of diurnal species (*monticola, maxwellii, nigrifrons, callipygus,* and *silvicultor*). Adults also rub their preorbital glands on one another's body and legs, and mates press their glands together; the latter behavior is apparently of prime importance in maintaining the pair bond and is unique to this tribe. More intense mutual gland-pressing can also be the prelude to combat (see fig. 3.4) (15). Bay duikers lick but do not mark one another. At least some species exchange faint contact calls (constantly in the case of blue duikers, only as a prelude to meetings between bay duikers). Play (running, jumping, rushing, circling, chasing, and fighting) by both adults and young appears to be most frequent in blue duiker families and rare in the bay duiker.

TERRITORIAL BEHAVIOR: *preorbital-gland-marking, horning vegetation.*

Probably all duikers rub their preorbital glands against tree branches, and other objects in their territories. Males and females mark in the same manner, but males do so far more frequently. Duikers do not maintain dung middens, but defecate at random throughout the territory. Blue duikers make visible marks by horning trees

Fig. 3.2. Bay duiker marking with preorbital gland (redrawn from Dubost 1983*a*).

along territorial boundaries. Male bay duikers were found to scent-mark 19–20 times per hour (fig. 3.2), versus 5–7 times by male blue duikers (4, 5).

AGONISTIC BEHAVIOR

Dominance/Threat Displays: *low-horn* or *medial-horn presentation* with tail raised, *snorting* and *whistling, symbolic biting.*

Defensive/Submissive Displays: *head-low posture, standing or kneeling with neck extended, distress bleat.*

Fighting: *butting, chasing.*

Snorting and *whistling,* normally employed as alerting and alarm signals, may be used to discourage the approach of another duiker (see blue duiker account). A bleat of distress may be given in response to aggressive or sexual harassment. Black-fronted duikers have been seen to snap at one another as a threat (21). Vigorous *mutual gland-pressing* was invariably the prelude to combat between captive male Maxwell's duikers (fig. 3.4), followed by butting so violent that the back feet often left the ground; sometimes a combatant turned a complete flip (fig. 3.3) (15). Fights ended in wild pursuits, during which the victor endeavored to stab the loser (5).

REPRODUCTION. Whether most duikers breed perennially or seasonally remains unknown, but blue and gray duikers are thought to breed year-round in southern Africa, where any tendency to seasonal breeding should appear (18). Apparently the gray duiker matures and breeds earlier and more frequently than forest duikers (see species account) (12, 18). Yearly reproduction may be the rule in the Equatorial Rain Forest, as is the case in Gabon even for the blue duiker.

SEXUAL BEHAVIOR: *pursuit, persistent following, urine-testing, lowstretch, foreleg-lifting, genital licking/biting,* aggression, *calling, courtship circling* (female evasion).

The common denominator of duiker courtship is *persistent following* (see fig. 3.7) and licking of the female's vulva (see species accounts). Urinating females squat deeply, and males make a weak grimace while *urine-testing* (5). Although estrus lasts no more than a day, males may show sexual interest several days earlier. It is common (at least in captivity) for a male to drive his mate relentlessly, nudging, licking, and biting her posterior until the vulva becomes red and swollen; to chase her hard when she flees, threatening and sometimes butting her; and to prod her with feet and horns when she lies down to gain a respite. Unreceptive females try to ward off a driving male by presenting their horns or by *symbolic biting,* and may bleat in distress. Both sexes may give soft calls during courtship (e.g., *dorsalis* males hum and *silvicultor* males utter piglike grunts; also see

Fig. 3.3. Serious fight between captive Maxwell's duikers; male on right turned a forward flip (redrawn from Ralls 1975).

species accounts) (5, 12). *Foreleg-lifting* has been recorded in *monticola, maxwellii, nigrifrons, silvicultor,* and *Sylvicapra* (5, 12, 13), but not in *dorsalis,* which is the only duiker known to rest its chin on the female's rump as a prelude to mounting. Males hold their heads up during the very brief (<1–3 sec.) copulation (5).

PARENT/OFFSPRING BEHAVIOR. Estimates of gestation vary widely and are often conflicting, for example, 4–7 months for the yellow-backed duiker (12), 3–5½ months for the gray duiker. Gestation in captive zebra duikers (weight 9–16 kg) has been established at 221–229 days, suggesting that forest duikers may actually have a rather long gestation (18). Weight at birth is approximately 10% of the mother's weight. Duikers have the typical hider-calf system (see species accounts), but males may tend to be more protective of their offspring than in polygynous species: both sexes respond immediately to a calf's distress bleat and the hornless female gray duiker is known to attack small predators.

ANTIPREDATOR BEHAVIOR: *alert posture, sneaking away,* concealment, *flight, snorting, whistling, stamping* with hindfeet, *distress bleating.*

Like other ungulates whose safety depends on concealment, duikers can *freeze* in midstride, with one leg uplifted, when startled. They sink down in place or sneak into cover to avoid an approaching enemy that has not yet detected them. When flushed from hiding, a duiker dashes away a short distance in a dodging run, then stops in cover and tries to see or smell what scared it. Alarm signals are used most frequently by *monticola* and *maxwellii,* the most closely bonded species, and least by *dorsalis. Alarm stamping* is performed with the hindfeet and not, as usual in savanna antelopes, with the forefeet (5). When captured, duikers bleat stridently, a sound that attracts other duikers—and also predators. Imitations of the cry are used by hunters to lure duikers.

References

1. Ansell 1971. 2. Dubost 1979. 3. ——— 1980. 4. ——— 1983*a*. 5. ——— 1983*b*. 6. Estes 1974*a*. 7. Feer 1979. 8. Gautier-Hion, Emmons, Dubost 1980. 9. Groves and Grubb 1974. 10. Hofmann 1973. 11. IUCN 1986. 12. Kingdon 1982. 13. Kranz and Lumpkin 1982. 14. Packer 1982. 15. Ralls 1975. 16. Ralls and Kranz 1984. 17. Richter 1973. 18. Schweers 1984. 19. Smithers 1983. 20. Walther 1979. 21. ——— 1984.

Added Reference

Dubost 1984.

Fig. 3.4. Male Maxwell's duikers pressing preorbital glands together in fight prelude (redrawn from Ralls 1975).

Blue and Maxwell's Duikers

Cephalophus monticola and *C. maxwellii*

TRAITS. Two of the smallest antelopes. *Height and weight:* males 35.5 cm and 4.6 kg, females 36.2 cm and 5.4 kg (3). *Horns:* c. 5 cm (shortest of all duikers), sometimes absent in females, bases strongly ridged. Hindquarters more developed than forequarters, legs slender, false hooves reduced. *Coloration:* geographically variable, slate gray to maroon, with bluish gloss on back; face darker brown; chest, belly, and backs of haunches white; tail relatively full, black above and white below with white tip and edges; coat may lighten and darken seasonally (1). *Scent glands:* large preorbital glands, pores in a curved slit (straight in other duikers); no inguinal glands.

DISTRIBUTION. The most widespread of the forest duikers, *C. monticola* occurs throughout the Lowland Rain Forest east of the Cross River in Nigeria; in relict rain forests of East Africa; montane forest up to 3000 meters; gallery forest; forest-savanna mosaic, and dense thickets through most of the Southern Savanna in the coastal plain and hinterland; and also on offshore islands (Pemba, Zanzibar, Fernando Po).

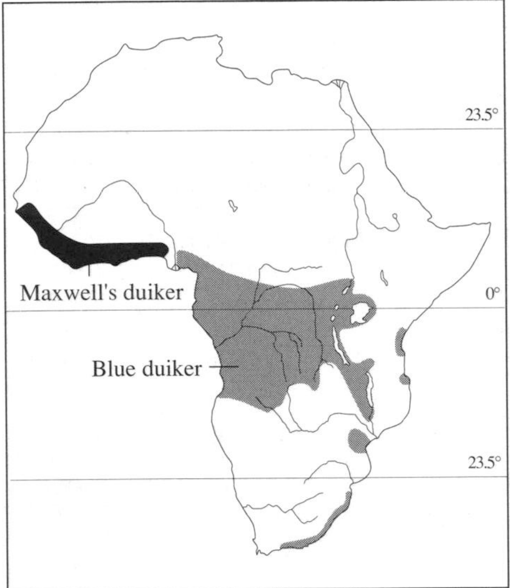

RELATIVES. Maxwell's is enough like *monticola* to be considered a superspecies if not a subspecies. Although slightly larger (8–9 kg), it fills the same niche in the rain forests to the west of the Cross River (2, 6).

ECOLOGY. The blue duiker is a diurnal forest frugivore that occupies areas with comparatively little undergrowth to obstruct its movements or vision (significance of size in forest ruminants is discussed in chap. 1). In Gabon, over ¾ of the blue duiker's diet consists of small fruits (0.5–2 cm diameter). Foliage (leaves, shoots, herbs) also has been found in all stomachs examined and accounts for 20% of the diet; flowers (0.6%), animal matter (0.5%, mainly ants and other insects), and occasional fungi round out the menu (5). Probably duikers gain enough water from their food and rarely have to drink, but are occasionally seen licking rainwater off foliage (1).

SOCIAL ORGANIZATION: mated pairs on small, permanent territories.

Patches of open forest floor suitable for blue duikers are partitioned into exclusive territories averaging only 2.5–4 ha (3). Ranges tend to be compact, and natural or accidental features such as pathways, watercourses, and logs often serve as distinct territorial boundaries. Adjacent ranges only overlap where terrain or cover discourages frequent visits, making vague buffer zones into which brief excursions by members of neighboring families may appear as sharp projections of the otherwise compact domain (see fig. 3.5). The occasion for such an excursion may be the presence of fallen fruits, but duikers that venture outside their normal beat always appear ill at ease and never stay for long (3).

Blue and Maxwell's duikers live in couples with 1 and sometimes 2 offspring of different ages. By repeatedly capturing (with nets), marking, and recapturing the population on a 74-ha study area within the Gabon rain forest, Dubost (3) found:

- A density of 62–78 blue duikers/km^2, of which 62% generally were reproducing adults and subadults, the balance being young of the year.
- An adult sex ratio close to unity, with no surplus of one sex in the percentages of solitary duikers (25% male, 23% female).
- That 75% of the adult males (N=44) and 77% of the females (N=43) lived in couples.
- A 27% annual mortality rate (7.3% reproducing males and 10.4% females), and a 66% survival rate in the first year, giving

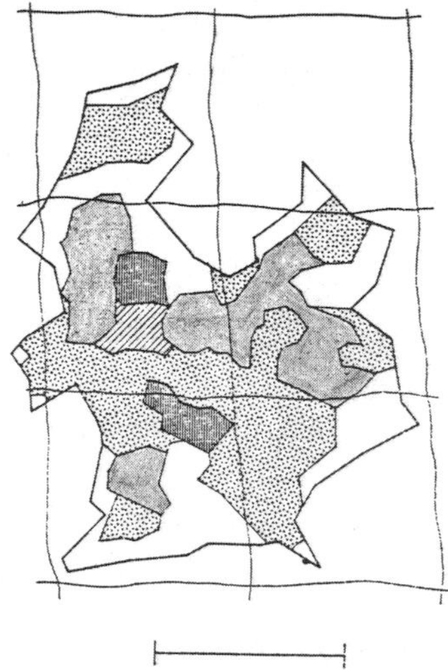

Fig. 3.5. Home range use by young-adult male blue duiker during 1 month (from Dubost 1980). Unshaded areas denote less than 2 sightings per 20 × 20 m quadrat; dotted areas, 2–4 sightings; gray areas, 4–6 sightings; diagonal hatching, 6–8 sightings; vertical hatching, more than 8 sightings. Squares (100 m/side) show the layout of transects through the rain forest.

an annual surplus of 7–10% males and 12% females.

The average group numbered 1.8 duikers and only twice were groups of 4 captured. There was never more than 1 young of the same age. Sometimes 2 adults of the same sex were caught in the same range, but invariably one was younger and almost certainly an offspring that had failed to emigrate or, in the case of males, occasionally a subadult sojourning in a place temporarily vacated by the territorial male (see below).

Pair Bonds. Observations of captive blue and Maxwell's duikers maintained under seminatural conditions confirm that pair bonds are exclusive and probably lifelong. Although an adult male will tolerate and mate with extra females, he maintains close social bonds with only 1 (1, 4, 7). Males are more pugnacious and generally dominant over females as adults, although there is rarely any evidence of competition. The male initiates most social interactions: 58% of the approaches, 64% of mutual-grooming sessions. He plays and marks 3–12 times more often than the female. However, most joint movements are initiated by the female and the male follows 92% of the time (4).

The members of a pair keep closer together (average 47.5 m, range 37–55 m) than to other family members, even a mother and fawn (66 m, 55–74 m), and the association between a pair becomes closer with time (3). Pair bonds are reinforced mainly by pressing the preorbital glands together (as in fig. 3.4). A couple also rubs preorbital secretion on each other and engages in *social licking,* but offspring also participate in these forms of social grooming, which are not specifically associated with pair-bonding (see Communication).

As an indication of the permanence of pair bonds, only 1 female (a subadult) in a sample of 11 free-living males and 12 females that were followed in different years abandoned her mate and settled elsewhere (3).

Offspring Dispersal. Young blue duikers emigrate as yearlings and try to establish themselves in vacant territories, apparently on their own initiative. Aggression within the family is very rare and it seems that offspring may stay indefinitely without being subjected to any severe harassment by either parent.

In fact, males typically stay home until nearly 2 years old and full-grown (adult dentition complete), whereas females usually leave at 1–1.5 years. Four males and 6 females, nearly ⅓ of the adolescent duikers in the study area, actually settled in their natal home ranges. But this only happened in the absence of an adult of the same sex, and 11 of 13 females that emigrated did so despite the absence of a resident adult female. It may be that blue duikers have a natural tendency to leave their birthplace once they become adolescent, and that the tendency is reinforced when an adult of the same sex is present (3).

Sex Ratio. The proportion of males in the population declines from gestation to weaning, and again during the subadult stage (46:54), partly reflecting the more aggressive competition between males for territories and mates. Among adults, however, the balance finally shifts to males (51:49), apparently reflecting the greater risks of bearing and suckling young compared to the risks faced by established territorial males (3).

The unusual stability of the blue duiker territorial system could account for these otherwise surprising findings: (*a*) A male vacates the territory for a month after his mate calvcs, coming back only on visits and spending no morc than 25% of the time with her. Meanwhile, one or two subadult males may move in and keep his mate company; neighboring territorial males also encroach on the property, though without directly contacting the female. (*b*) Resident males at least sometimes tolerate unrelated subadult males on their property. However, interlopers usually withdraw when an absent territorial male returns home, and resident males are completely intolerant of other adult males—in fact, two adults of either sex cannot be placed in the same enclosure without provoking a brutal, decisive battle (3). But penned female Maxwell's duikers were content to dominate, without actually attacking, introduced strange females (8).

ACTIVITY. This strictly diurnal duiker has the usual bimodal activity peaks: 0600–0800 and 1600–1800 h (3). In Gabon it rests between 0900 and 1400 h except for brief, irregular bouts of activity. Foraging and other activities occupy 67–76% of a duiker's day (4). Although they forage 3 hours in the morning and up to 4 hours in the afternoon during the dry season, blue duikers are able to find adequate amounts of fallen fruits and foliage during daylight to satisfy their needs (3). They spend the whole night lying, apart from brief intervals when one raises, stretches, licks itself, and resettles. Each animal has numerous resting places dispersed through its territory. Blue duikers lie in the open 40% of the time, or at the base of a tree (44%), and spend only 16% of the time in coverts, which are mainly used as refuges from danger.

Daily Ranging. Adults range over about 40% of their territories every day, traveling an average minimum of 979 m (650–1770 m). Movements while foraging may be in any direction, except when large quantities of fruit are dropping from a particular tree, which may then be visited several times a day. The central part of the range is used most, the more peripheral parts progressively less (fig. 3.5). Yet duikers of either sex spend ¼–⅓ of their time near the border. No significant seasonal difference in range or habitat preference has been found.

POSTURES AND LOCOMOTION. A duiker moving in its normal, quite rapid *cross-walk* keeps its head and neck at shoulder level. It passes beneath the lowest branches by lowering its hindquarters nearly to the ground. A foraging individual holds its head low and may dig in the soil with its forefeet and even its incisors. Sometimes it stands bipedally while browsing, with forefeet propped against a support. Blue duikers gallop with quick bounds and buck-jump in play.

SOCIAL BEHAVIOR

COMMUNICATION

Tactile and Olfactory Communication: *mutual gland-pressing; marking partner; social licking.*

Mutual gland-pressing by members of a mated pair is the second-most frequent social behavior after territorial scent-marking (see Territorial Behavior). Different captive blue duiker pairs gland-pressed at the rate of 0.05–5.6 times an hour (4). Couples scent-marked each other or their offspring (in the same manner as trees or other objects) much less frequently (0.2–1.3 times an hour), preferably on projections such as the forehead, tail base, and hocks, but not the horns or ears.

Reciprocal licking is the most important and frequent form of social contact within the family, occurring 0.9–2 times an hour in bouts averaging nearly 4 minutes. A bout begins when one duiker approaches and assumes the submissive *head-low/chin-out posture.* When the other responds, the petitioner offers the part it wants groomed, that is, areas that cannot be licked by the animal itself (ears, neck, head). After some licks, the licker stops and solicits the same service in return. Social grooming not only helps maintain family bonds; but licking is also used to appease aggression, as when a juvenile is threatened by an adult or an unreceptive female attempts to divert a pursuing male's sexual interest. But within the mated pairs observed, there was no obvious connection with dominance or gender; certain individuals simply sought contact more than others (4).

The scent trails left by the hoof glands and the preorbital secretions left on trees, also urine and possibly feces, apart from protecting the territory against intruders, presumably also help family members keep track of one another, telling not only where and when but also which animal left the marks and what its reproductive state was. With all the olfactory information available within a small space permanently occupied by animals with regular habits, family members need not remain in visual contact to know exactly where to find one another (3).

Vocal Communication: contact and alarm calls, *stamping, object-horning.*

Snorting, whistling, and *stamping* are mainly alarm signals that can also express social and sexual excitation (see Antipredator, Threat, Submissive, and Sexual Behaviors). A duiker seeking physical contact emits short, soft groaning cries that are audible no further than 10 m; a longer, slightly louder cry is made by an animal seeking to increase its distance from another. The sound of a duiker horning a tree and even the faint swishing of the flicking tail are also audible at short range. All these sounds play a role in duiker communication, supplementing and extending visual signaling as channels for instantaneous transmission of information. Conversely, olfactory signals are long-lasting but often have delayed effects (4).

Visual Communication: *tail-flicking.*

The white underside and edge of a blue duiker's tail contrast with the dark color of the hindquarters when the tail is raised. This species flicks its tail as regularly as a Thomson's gazelle, but up and down instead of side to side. The flickering white scut is a highly visible signal to nearby family members.

TERRITORIAL BEHAVIOR: *preorbital-gland marking, tree-horning,* alarm signals.

Direct interactions between neighbors are rare, but territorial occupancy is continually advertised by visual, olfactory, and probably also by auditory means. Urine and feces are deposited at random, as are scent traces of the hoof glands. Rubbing the preorbital glands on tree trunks and other objects, wherever the duiker goes, is the main form of demarcation. It is strictly a male prerogative in the blue duiker and different males may mark anywhere from 0 to 274 times a day, but the average rate is 5 to 7 times an hour. Since the secretion is colorless and odorless to the human nose, it is hard to plot the distribution of marking sites, although the white crust that forms after drying is discernible. Females only mark objects when excited (e.g., by the presence of a potential rival). Female Maxwell's duikers may play a more active role in preorbital marking (7).

The second most important form of territorial demarcation is *tree-horning,* in which both sexes engage from the time their horns emerge at 3 months (4). Permanent, visible marks are made on target trees, chiefly young saplings with stems small enough to fit into the fork between the horns. The duikers always horn at shoulder level, after first sniffing and licking and sometimes scraping the spot with the incisors. Performers end about ⅓ of the bouts by scraping the ground with one or both forefeet. Horning is concentrated toward the edge of the property, and adults engage in this activity 2–16 times a day, especially during visits to the border area before or after rest periods. Horning, unlike other marking, may be infectious; the sight or scraping sound can stimulate other family members to follow suit. Audible alarm signals may also serve as an acoustic reminder that a territory is occupied.

Blue duiker territorial defense is thus accomplished indirectly through different forms of advertising. There is no joint or family defense of the property, but each resident adult and subadult selectively excludes interlopers of its own sex (4).

AGONISTIC BEHAVIOR

Dominance/Threat Displays: *snorting, low-horn presentation, rushing, object-horning, pawing, preorbital-gland marking.*

Defensive/Submissive Displays: *head-low/chin-out posture, kneeling, lying, crying, whistling.*

Fighting: *mutual gland-pressing* (fig. 3.4), *butting, stabbing.*

The duiker's aggressive displays derive from territorial marking and alarm signals, whereas *low-horn presentation* with raised tail is the same as the combat attitude (fig. 3.6). If a duiker takes offense, say, at the approach of another animal while it is feeding, the first warning consists of a series of snorts. Should the other fail to withdraw, the offended duiker then presents its horns with head lowered and may back up this threat with a short rush. Normally an animal that is defending a resource prevails over other family members. But if a territorial male chooses to assert its dominance, females and young adopt thc *head-low/chin-out posture.* The same attitude is adopted by a duiker soliciting grooming, and often elicits licking even by an aggressive male, proving its appeasement value (4). *Kneeling* and *lying* in response to threats express increasing fear and submission. Unreceptive females sometimes respond to persistent courtship pursuit by actually taking cover; they also cry and whistle in alarm (more under Sexual Behavior). Territorial males normally never display submission; when they lower their heads it is a threat or the prelude to attack.

Fights within a family are rare, brief, undamaging, and preceded by the threat display.

Fig. 3.6. Maxwell's duikers facing off in combat attitude (redrawn from Ralls 1975).

Serious combat has only been seen when staged by putting two males in the same enclosure (4, 8), although occasional violent chases have been observed in the wild on territorial borders. Fights between blue duikers took place suddenly, without preliminaries, whereas bouts between Maxwell's duikers were preceded by intense *gland-pressing.* In both cases the combatants rammed one another with great force, as if attempting to knock each other off balance, and indeed one or both flipped in 8 of 9 fights between Maxwell's males (fig. 3.3), sometimes landing hard on their sides (8). The victor always chased the loser wildly. Fights between the male blue duikers usually ended abruptly and indecisively after a few collisions; but one could also become infuriated enough to stab an opponent repeatedly if it fled (4).

REPRODUCTION

SEXUAL BEHAVIOR: *following* in *low-stretch, pursuit, anogenital licking, preorbital-gland marking, foreleg-lifting, groaning, snorting.* Female: *groaning, snorting, whistling,* avoidance (*circling,* turning, *buck-jumps, flight, hiding*).

In a study of penned blue duikers, males engaged in sexual pursuit regardless of their mate's reproductive state, 3 out of every 4 days at the rate of 1–3 chases per hour (3). Pursuits averaged 1½ minutes and lasted up to ¼ hour (4). Approaching in *low-stretch,* the male would try to lick the female's croup, tail, and genitalia, following her closely and occasionally *moaning* or *snorting.* When the female stopped, he would make a mounting-intention movement, or begin *foreleg-lifting* often while licking her rump. An unreceptive female would keep her tail clamped and try to evade the male, meanwhile *moaning* or *snorting* without pause, and *alarm-whistle* if the pursuit was prolonged.

When his mate is nearing estrus, the male becomes far more aggressive than usual, snorts continuously, and in captivity may viciously attack other animals, large or small, in his efforts to keep them at a distance from the female. Sexual pursuits increase in frequency to 7–8/hr. Instead of attempted mounts, the male advances with aggressive gestures: blows with the forehead or horns, biting at the tail base. During the *mating march,* he licks the female's croup and marks her often with his preorbital glands. The female moans slightly and advances a little in circles or figure eights. If the male tries to mount her before she is ready, she crouches and slinks away from him.

The male clasps the female's loins and keeps his head up during the very brief copulation, jumping off the ground during ejaculation. After dismounting he licks his penis, then begins another generally calmer pursuit. Multiple copulations are common, as many as 5 in 3 minutes, interrupted by *marking* the female, but without further *foreleg-lifting.* Mating activity may be prolonged until after dark.

PARENT/OFFSPRING BEHAVIOR. Despite their small size and relatively short gestation period (c. 4 months), blue and Maxwell's duikers of the rain forest reproduce only once a year (1). Females calve in their second year and males may be mature enough at 2 years to win territories and begin reproducing (3).

In 2 observed captive births of Maxwell's duiker, delivery took only 20–30 minutes. After eating the membranes, the mother licked the fawn thoroughly and within 20 minutes of birth it was able to run. The typical tail movement began as soon as it could stand. Birth weights of 4 calves averaged 867 g (710–954 g) (1).

For the first couple of weeks a mother blue duiker maintains a mean distance of 74 m from her carefully concealed offspring. Between 3 and 4 weeks the calf becomes more mobile and tries to accompany her after nursing, leading to some reduction of the average distance, which reaches a minimum of 55 m at 2 months. It then increases progressively until weaning at 2½–3 months, whereafter a mean distance of 73 m is maintained up to the time of its voluntary departure.

The mother retrieves and suckles her calf three times a day in the first month (usually between 0800–0900, 1100–1200, 1400–1600 h), skips the midday session in the second month, and thereafter suckles only in the evening up to weaning. Except for a month or two following the initial lying-out period, mother and calf sleep in different parts of the territory.

ANTIPREDATOR BEHAVIOR. Why the father takes a monthlong furlough each time his mate calves (discussed under Social Organization) is an intriguing mystery. Dubost (3) argues that there is less risk of predators discovering helpless young if the mother is the only one that approaches the calf.

If living in pairs is actually atypical of forest ruminants, then the male's withdrawal while the calf's survival depends totally on effective concealment may indeed have been selected for, through the improved survival of offspring whose sires departed.

Dubost (3) argues convincingly that the blue duiker's family organization is correlated with its small size, diurnal habits, and comparatively open habitat, a combination that makes it vulnerable to practically the whole range of forest predators, but above all to the crowned hawk eagle (*Stephanoaetus coronatus*), which takes prey weighing up to 6 kg. Blue duikers rely primarily on their eyes and ears to detect danger (4). In theory and practice, 2 active duikers are much likelier to spot an eagle or other predator than 1: observations showed that one or the other of 2 duikers had its head up monitoring its surroundings 70% of the time, compared to 48.5% surveillance by 1 duiker alone. With 3, surveillance rose to 91.5% (3).

Having spotted danger, blue duikers have several ways of communicating different degrees of alarm: *snorting* (short-range), *whistling*, and *stamping* with the hindfeet, the latter 2 representing high-intensity alarm. *Snorting* and *whistling* are graded signals that increase in strength and rate with the degree of excitation. Other family members respond to high-intensity alarm signals by running into cover and hiding. Running itself can trigger flight; so can any sudden loud noise and the alarm signals of other species (4).

References

1. Aeschlimann 1963. 2. Ansell 1971. 3. Dubost 1980. 4. ——— 1983*a*. 5. Gautier-Hion, Emmons, Dubost 1980. 6. Kingdon 1982. 7. Ralls 1974. 8. ——— 1975.

Fig. 3.7. Gray duiker courtship: male *following* female closely while trying to sniff and lick her genitalia.

Gray, Bush, or Common Duiker
Sylvicapra grimmia

TRAITS. Compared to forest duikers: longer legs and flatter back, *horns* more upstanding and present in males only, 11 cm (7–18) (5), round and strongly ridged at base, and longer ears (largest in dry, open habitats). *Height and weight:* males 50 cm and 12.9 kg, females 52 cm and 13.7 kg in eastern Zambia (15); in a Botswana sample, males 18.7 kg (15–21), females 20.7 kg. (17–25) (11). *Coloration:* individually and geographically variable, from pale buff in arid biome, to grizzled gray in Northern Savanna, to rich chestnut in Angola; underparts paler; facial blaze, head crest, and feet dark brown or black; sexes alike, young calves darker. *Scent glands:* hoof, inguinal (7), and preorbital.

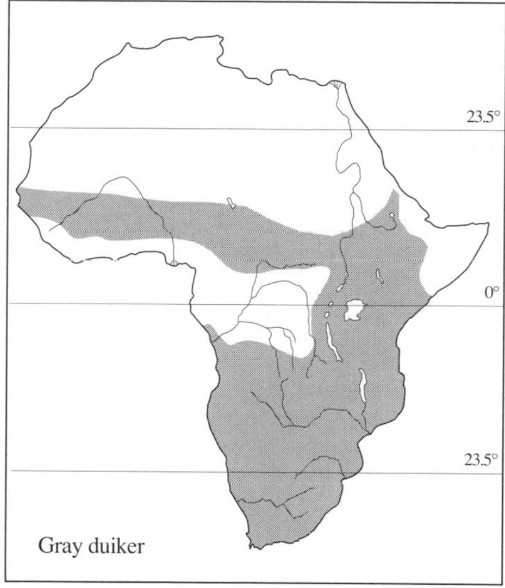

DISTRIBUTION. Virtually ubiquitous below the Sahara apart from the rain forest and desert, from sea level up to the snow line, wherever there is cover. There are at least 8 different races (5).

ECOLOGY. *Sylvicapra's* success stems from its adaptability to a broad spectrum of habitats and food plants. Though it does not venture onto open plains, it can live in virtually any habitat that affords enough cover for concealment, reaching greatest abundance in savanna and woodland. Its variable coloration may enable not only populations but also individuals to match a given habitat closely, in response to prevailing temperature and humidity conditions, being darker in moister and lighter in drier areas (5). Gray duikers manage to live on the outskirts of villages and even in suburbs, substituting garden produce and ornamentals for their natural diet. Provided they have green fodder, duikers do not have to drink: penned individuals went for up to 3 months without drinking, and 1 male ignored water offered for half an hour every 4 days through the dry season when there was no appreciable dew (14).

DIET. In a sample of 191 stomachs collected monthly for 2 years in a Zambian tsetse-control area, leaves of 45 different species were present, fruits and seeds of 33 species, and flowers of 15 species (14). Duikers of the Ethiopian highlands live mainly on herbs (87% of diet) they select in montane grassland, the foliage of shrubs and bushes making up but 11% of the diet and sprouting grass 2% (2). Crop-raiding duikers dig up sweet potatoes and peanuts, and munch vegetables. Most of the fruit they eat is fallen, often scavenged beneath trees where baboons and monkeys are feeding. When feeding on fruiting bushes they extend their reach by standing on their hindlegs. Some of the more unusual items found in duiker stomachs include tree resin, bark, insects such as ants and caterpillars, a small lizard, rodents, and birds. Duikers have been known to stalk and catch birds, and one was seen catching and eating a striped mouse (4, 14).

SOCIAL ORGANIZATION: territorial, monogamous.

The social ecology of this duiker remains largely unstudied. Of 195 sightings in Zambia, 126 were singles, 54 were pairs (29%), and 5 were trios (15). In a sample of shot animals the sex ratio was close to unity: 99:94, males: females. It was also equal in the Ethiopian study, but there 86% of 232 sightings were singles and only 8% were pairs, compared to 85% of klipspringer sightings (3). Observations of known individuals over a period of 4½ months suggested that the home ranges of males and females were equal in size, that ranges of duikers of the same sex overlapped only a little, whereas male ranges partially overlapped the ranges of at least 2 females. A male appeared to associate with whichever female was present in the part of his range where he happened to be, and no evidence of pair bonds was noted. Territorial behavior by males but not by females was observed, including *preorbital-gland marking, vegetation-horning,* fights, and high-speed chases.

Under captive conditions, male and female gray duikers behave in ways recalling mated pairs of blue and Maxwell's duikers, *grooming* one another and *pressing preorbital glands* together with some regularity (V. Wilson, personal communication and author's observations). Both rub their preorbital glands on tree trunks and other objects between 13 and 35 cm high (5), the male as usual more frequently. Each is aggressive to intruders of the same sex, especially the male: 2 adult females but not 2 males can be penned together, as in blue and Maxwell's duikers (1, 8).

Observations of gray duikers in Zaire's Upemba N.P. indicated that males and females had particular, separate resting places where they retired to lie up and ruminate. Females preferred lower, more closed hiding places, often beside a log or tree trunk, whereas males selected more open, elevated vantage points such as mounds and slopes (12).

ACTIVITY. The common duiker seems to be active both night and day, but may become almost entirely nocturnal where heavily hunted. In northern Botswana, this species has early evening (up to 2000 h) and late night (beginning c. 0400 h) activity periods (11). Zambian tsetse-control hunters who went out at midnight with spotlights rarely saw any duikers, and these were usually lying down (14). As usual, daytime activity peaks early and late in the day, and may be extended during overcast and cool conditions. In hot weather duikers seek shade and avoid exertion (11, 14).

POSTURES AND LOCOMOTION. The common duiker's conformation indicates it has considerably greater speed and endurance than forest duikers (5). When flushed from cover by dogs, it usually outruns them and loses them in the nearest covert (11).

SOCIAL BEHAVIOR

AGONISTIC BEHAVIOR

Dominance/Threat Displays: *stalking, low-horn presentation, alarm-snorting.*

Defensive/Submissive Displays: *head-low/chin-out posture, kneeling with neck on ground, lying prone.*

Fighting: *chasing, butting, stabbing.*

Aggression is no more ritualized in *Sylvicapra* than in other duikers. The response of a territorial male to another male's intrusion is to stalk and charge it, leading to a furious chase and stab wounds if the pursuer manages to overtake the fleeing animal (3). When pursued or beaten in a fight and unable to escape, even adult males have been known to *lie-out* in the most extreme submission (5). One observed fight involved an older and a younger male that had shared the same enclosure peaceably until the adult's mate came into heat (10). It began with apparently playful cavorting, butting, and fencing, followed by intense *alarm-snorting* ("wheezing"), and a strong smell of preorbital-gland secretion. Combat was then renewed with greater ferocity: instead of butting, they charged one another, dropped to their knees, and with heads tilted tried to stab to the body while parrying or avoiding the opponent's thrusts. The dueling increased in intensity and speed until the combatants were panting and frothing at the mouth. It ended when the bigger male landed a hard jab, whereupon the yearling dropped cowering to the ground.

REPRODUCTION. Lactating and pregnant females have been collected in every month in the southern part of the common duiker's range, indicating perennial breeding with no clear evidence of seasonal peaks (11).

SEXUAL BEHAVIOR: persistent *following* and *licking, urine-testing, chasing, lowstretch, foreleg-lifting, prancing, social grooming* and *gland-pressing, calling.*

Courtship tends to last some days and to consist mainly of persistent following and genital licking, interspersed with energetic chases and other aggressive behavior by the male. During courtship of a captive female, the male nudged her hard every time she stopped, and performed a quick *foreleg-lift* that looked more like a nervous twitch than a kick (fig. 3.7). Occasionally he gave a strangled sort of cry, and when nudged the female uttered a mewing call that reminded me of a gray squirrel's bark. She urinated frequently and the male put his nose in the stream prior to testing it. When not actively courting her, he fed and rested near her.

When the female is nearly ready to mate, the male may *prance* along beside her tilting his head from side to side, whether in threat or in invitation to press preorbital glands is unclear (10). He nibbles her neck, and when the female stands facing him, he presses his preorbital glands to hers, alternating sides, accompanied by a release of scent. Copulation usually follows within ½ hour (10).

PARENT/OFFSPRING BEHAVIOR. Conflicting estimates of gestation range from 3 months (11), to 5 months and 20 days (10), to 7 months (210 days) (17). The first and last estimates seem short and long, respectively. As females typically seclude themselves in dense cover prior to calving, observations of birth are almost impossible to make in the wild. A calf born in captivity was not dropped until 3 hours after the enveloping chorion became visible (10), but within 15 minutes of the first visible contractions and 10 minutes after the head was passed. Although the mother cleaned the calf thoroughly and nibbled its umbilical cord, the calf only raised its head 15 minutes after birth, gained its feet in 35 minutes, and took its first steps at 50 minutes. It found the udder but failed to nurse after 80 minutes. The afterbirth was dropped and eaten 65 minutes *post partum.* Calves can run within 24 hours and are hard to catch after 3 days, but continue to lie out for several weeks (duration unspecified). The mother retrieves and feeds the calf 2 or 3 times in a 24-hour day (11).

Common duikers may mature sooner and reproduce considerably more quickly than forest duikers. The young reach nearly adult size at ½ year and at least some females calve at 1 year (9, 11). The mean calving interval of a South African captive pair that produced 11 young between 1966 and 1973 was 259 days (232–298 days) (6). Populations may actually increase while being subjected to tsetse-control shooting designed to eradicate all game (16).

ANTIPREDATOR BEHAVIOR: *alarm snorts, stamping, skulking, lying out,* running, *distress bleats, attack.*

As in the blue duiker, adults of both sexes approach in response to the distress bleat, suggesting that fathers may defend their offspring. Though hornless, females not only attack small predators but have been known also to butt a large male baboon and an 8-foot python that had seized their fawns (13, 14). Predators, too, respond to duiker *distress bleats:* an artificial call used by tsetse-control hunters to "call up" duikers to the gun sometimes attracted spotted hyenas and leopards instead (14). Several instances have been recorded of pythons dying after swallowing male duikers whose horns pierced their stomachs (14).

References

1. Dubost 1983*b*. 2. Dunbar 1978*a*. 3. ——— 1979. 4. Hofmann 1973. 5. Kingdon 1982. 6. Ketelhodt 1977. 7. Pocock 1910. 8. Ralls 1975. 9. Riney and Child 1960. 10. Sikes 1958. 11. Smithers 1983. 12. Verheyen 1951. 13. Willis 1946. 14. Wilson 1966*a*. 15. Wilson and Clarke 1967. 16. Wilson and Roth 1967. 17. Zaloumis and Cross 1974.

Chapter 4

Dwarf Antelopes

Tribe Neotragini

Neotragus, dwarf antelopes
N. pygmaeus, royal antelope
N. batesi, dwarf antelope
N. moschatus, suni

Raphicerus, steenbok, grysboks
R. melanotis, Cape grysbok
R. sharpei, Sharp's grysbok
R. campestris, steenbok

Madoqua, dik-diks*
M. saltiana, Salt's dik-dik
(includes *phillipsi* and *swaynei*)
M. piacentinii, Piacentini's dik-dik
M. guentheri, Guenther's dik-dik
M. kirkii, Kirk's dik-dik

Oreotragus oreotragus, klipspringer
Ourebia ourebi, oribi
Dorcatragus megalotis, beira

*Dik-dik classification according to (13).

TRIBAL TRAITS. The smallest antelopes, ranging from 20 cm tall, 1.5 kg royal antelope to the 67 cm oribi and 26 kg beira. Harelike, with hindlegs much longer than forelegs (*Neotragus, Madoqua*), to gazellelike conformation (oribi, steenbok). Narrow muzzle and incisor row, muffle bare and moist (most species) or hairy, with slitlike nostrils (dik-diks, beira); ears medium or large (steenbok, dik-diks, klipspringer, oribi); tail rudimentary to short (3.5–13 cm). *Horns:* in males only (except one klipspringer race), short (2–19 cm), sharp spikes, ringed at least basally (except *Raphicerus*), slanted backward (*Neotragus*) as in duikers, or more upstanding (steenbok, oribi). *Coloration:* pale gray to dark brown, cryptic with generally inconspicuous markings apart from white rump patch or undertail in oribi and *Neotragus* species. *Mammae:* 4 (2 in beira).

Scent Glands. The distribution of scent glands in the different genera, as presently known, is shown below. Out of 7 different sets of glands, the oribi possesses 6. However, some species and even subspecies may not have all the glands their congeners have: thus, preorbital glands may be undeveloped in *N. batesi* (1), and the southern race of Kirk's dik-dik lacks hoof glands (11).

Preorbital glands are exceptionally developed in the Neotragini, so big in the male oribi, suni, klipspringer, and dik-diks that the preorbital depressions in the skulls are about the size of the eye sockets. Dik-dik and klipspringer females are known to deposit preorbital secretions; the glands may be inactive in the female steenbok, oribi, and possibly other species, although the skulls of female suni, royal antelope, and steenbok I examined had distinct preorbital fossae (the depressions were absent or poorly developed in immature individuals of either sex—author's unpub. observ.). The glands are round or oval-shaped bodies that discharge a black, sticky secretion through a central duct, situated typically in a patch of bare dark skin. At rest the duct is concealed beneath a purselike fold of skin which is opened wide preparatory to depositing secretion. *Neotragus* species have simpler glands that lack the skin fold; the preorbital glands are otherwise quite similar in both the Neotragini and Antilopini (8, 9).

When a dik-dik preorbital gland is longitudinally sectioned, it is seen to consist of a completely black middle layer surrounded by a white layer (fig. 4.1). The black layer

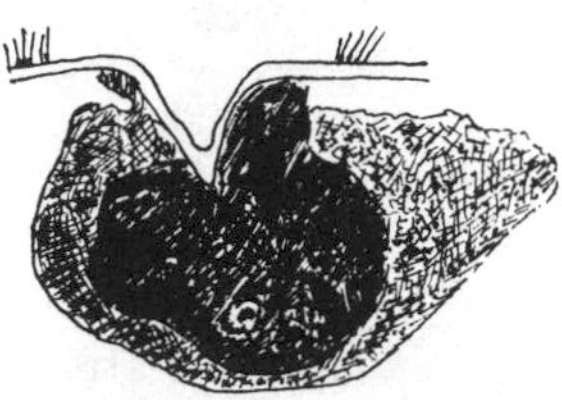

Fig. 4.1. Dik-dik (*Madoqua guentheri*) preorbital gland in cross-section (based on fig. 3 in Richter 1971).

Table 4.1 Distribution of Scent Glands in the Neotragini

Genus	Preorbital M	Preorbital F	Subauricular	Inguinal	Preputial	Hoof	Carpal and metatarsal
Neotragus	+	+	–	+	?	+	–
Raphicerus	+	?	–	–	+	+	–
Madoqua	+	+	–	–	–	+	–
Oreotragus	+	+	–	–	+	–	–
Dorcatragus	–	–	–	–	–	+	–
Ourebia	+	?	+	+	–	+	+

SOURCES: Based on Ansell 1971 and references cited in text.

consists of polygonal cells whose cytoplasm is packed with granules of melanin (black pigment). The glandular tissue, composed of strongly branched holocrine glands, produces 2 different kinds of secretion: larger drops containing melanin and whitish, transparent, fatty droplets rich in lipoids. The white layer contains strongly coiled, unbranched apocrine glands which produce a clear secretion that exits via wide, winding ducts into the middle, pigmented layer. All the secretions eventually empty into and exude from the central duct as a black, sticky compound (10).

DISTRIBUTION. In contrast to the duikers, which are much alike, this tribe includes such a variety of different types that taxonomists have divided the 13 or 14 different species into at least 6 different genera. This diversity reflects adaptations for very different niches: lowland rain forest (royal and pygmy antelopes); forest/savanna mosaic and thornbush of eastern Africa from Somalia to Natal (suni); savannas (oribi, steenbok); arid bushland of the Somali-Masai and South West Arid Zones (dik-diks); cliffs (klipspringer); rocky hills and mountains of North East Africa (beira); and rolling, rocky terrain with dense cover (grysboks). Apart from the 2 rain forest species, the others are all specialized for arid environments (more under Ecology).

RELATIVES. The neotragines are generally classified with the gazelles as a separate subfamily, the Antilopinae, on the basis of various anatomical and behavioral similarities among some of the species. In the following list of shared traits, those marked with an asterisk are largely limited to these two tribes and tend to bear out their relatedness.

Anatomical and physiological similarities: *morphology of the preorbital glands; presence of *carpal and inguinal glands in oribi (not other Neotragini), which also looks somewhat gazelle-like; structural plan of the feet and pedal glands; skull structure, so similar in dik-diks and small gazelles that young gazelle and adult dik-dik skulls may be confused; similar adaptations for arid environments (5, 8, 9).

Behavioral similarities: **linked urination/defecation,* often preceded by pawing (territorial advertising); *stotting* gait; *foreleg-lifting* in courtship; and *mounting without grasping or resting weight on female.

The unstarred shared traits are more widespread, such as inguinal glands, *foreleg-lifting,* and *stotting.* The Neotragini also have in common with other tribes and not with Antilopini certain morphological traits, such as the dik-dik's forehead crest, which otherwise occurs only in the duikers, and the preputial gland, which also occurs in the rhebok (6).

ANCESTRY. Only forms very closely related or identical to existing species, dating from the Pliocene, have been found as fossils (3). Nevertheless, according to one recent theory (5), the dwarf antelopes represent the most conservative and ancient line of bovids, from which all the other tribes except the subfamily Bovinae subsequently evolved. Sharp's grysbok and the beira, both seemingly relics of an earlier neotragine radiation, may be closest to the ancestral form, while the pygmy *Neotragus* species are thought to have invaded the forest after the duiker radiation, to fill a still-vacant niche for an understory folivore (5).

ECOLOGY. Because of its relatively large surface area, a small antelope is more vulnerable to desiccation and temperature extremes than a big one. To live in hot, waterless lands, dwarf antelopes need to avoid overheating with the least possible loss of precious water—that is, they cannot afford to sweat. Evaporative cooling of the blood by nasal panting (described in chap. 2), an efficient mechanism for lowering body temperature employed by most plains antelopes, has been developed to an extraordinary degree by the dik-diks (see species account) (5).

DIET. Dwarf antelopes are all concentrate selectors, except for the grazing oribi; they eat mainly foliage, shoots, herbs, seeds, pods, and some fruits. With their very narrow muzzles and incisor rows, coupled with small size, they can select the most nutritious available growth and glean a living within a minimal space on resources too sparse to support larger browsers.

SOCIAL ORGANIZATION. The dik-diks, klipspringer, and oribi are known to live, and the suni almost certainly lives, in monogamous couples, together with 1 or 2 immature offspring, within territories varying in size (both within and between species) from under 10 ha up to 1 km^2. From 10% to 20% of klipspringer and oribi families include 2 (very rarely 3 or more) females. The identity and relationship of the extras remain unstudied; they are most likely daughters that have failed to emigrate during adolescence. Although males are invariably more actively territorial, females participate in territorial advertising and behave aggressively toward female intruders. A dunging ceremony (see figs. 4.3, 4.5), performed by the couple and their offspring, usually on established dung middens, serves to maintain pair and family bonds (see species accounts). *Alarm-whistling* in duet apparently serves the same function in the klipspringer and perhaps dik-diks. It is comparable to the duetting of many pair-bonded birds. Social grooming, on the other hand, so important in many other monogamous mammals and birds, is comparatively infrequent (except perhaps in the suni).

The steenbok and grysboks are more often seen alone than in pairs, and it is possible that the sexes remain separate most of the time. It seems more likely, however, that these species also live in monogamous but less-cohesive pairs than some other neotragines. It is already clear that such gradations exist, for dik-dik and klipspringer pairs keep much closer company than oribi pairs. The pygmy and presumably the royal antelope, for their part, are essentially solitary, like most folivores (2). The male *N. batesi* defends a territory of 2–4 ha which usually includes the ranges of at least 2 females. The females maintain spacing of 150–190 m; sometimes two, presumed to be mother and daughter, share the same range and may stay within 50–100 m of one another.

ACTIVITY. Concentrate selectors generally feed and ruminate in alternating, fairly short bouts compared to roughage feeders, and forage actively day and night (4). Activity reaches a peak in early morning and late afternoon, as usual, continues at night up to c. 2200, and is followed by another bout beginning at 0400 h. The steenbok, grysboks, and suni appear to be more active by night, or it may be simply that they only venture into the open during darkness, as proved to be the case in the dwarf antelope (2, 5).

POSTURES AND LOCOMOTION. *Neotragus* species and dik-diks have remarkably long hindlegs, which are kept flexed with the femur and tibia folded up to the body, and the metatarsals, as long as the forelimbs and pencil-thin, extend to the ground, bearing the weight of the hindquarters upon the braced hocks. When the hindquarters are fully extended, as when stretching, the rump rises far above the shoulders (fig. 4.2). Presumably this arrangement is adapted to jackrabbit starts, long leaps, and quick turns. A pet royal antelope standing 26 cm at the shoulder could jump the 55 cm sides of its box and make 3 m horizontal leaps (7). The klipspringer has specialized hooves and peculiar conformation (see species account). The steenbok and oribi, which live in the most open habitats, have more evenly proportioned limbs, which may endow them with more speed and endurance than the rest.

Fig. 4.2. Royal antelope standing normally and stretching, showing its rabbit-like limb proportions (redrawn from Owen 1973).

SOCIAL BEHAVIOR

COMMUNICATION. The variety and size of scent glands in this tribe (table 4.1), and the time put into distributing preorbital-gland secretions and droppings, suggest that olfactory signaling is of primary importance in neotragine communication. This is undoubtedly true for antelopes living in dense cover. However, for those that live in more open habitats, notably the dik-diks, klipspringer, and oribi, visual signals are important, especially the *alert stance* and *alarm flight,* also postures associated with territorial marking. White rumps accentuate alarm flight in the steenbok and oribi, and the oribi also stots. Even the suni, which lives in dense cover, has a white undertail, the wagging of which often draws attention to animals that would otherwise remain unseen (5). Nevertheless, acoustic signals may be more important in antipredator behavior than visual displays, certainly at night or in dense cover. A reedy whistle is characteristic of the oribi, klipspringer, and dik-diks; the dik-dik actually has a structure resembling a whistle in its nose (see species account).

TERRITORIAL BEHAVIOR: *preorbital-gland marking, linked urination/defecation* and *dunging ceremony,* prolonged, loud *alarm calls* (including klipspringer duets); *static-optic display* (e.g., klipspringer standing in silhouette).

In all species the male plays the main role in marking and defending the territory. Only dik-dik and klipspringer females are known to make preorbital deposits. However, *linked urination/defecation* (fig. 4.3), a territorial display limited to males (as is territoriality) in the gazelle tribe, is also performed by female and even by juvenile neotragines in a regular *dunging ceremony,* although differences in urination posture (males straddle, females squat) distinguish the sexes.

The male's performance seems clearly to function in territorial advertising, and does not depend on the female's presence. Whether the female's performance has any separate territorial significance is unclear, but it does trigger the pair or family *dunging ceremony.* The male's tendency to approach and test the output of a urinating female is nearly universal in ungulates (see chap. 2). But in the Neotragini the male adds his own excrement to that of his mate and offspring, often in conjunction with vigorous pawing and *preorbital-gland marking.* The whole ceremony follows a set pattern, which varies slightly between species, and typically takes place at territorial boundaries on established dung middens. Whatever reproductive and social-bonding functions it may serve, the neotragine *dunging ceremony* therefore clearly includes territorial advertising.

Fig. 4.3. *Linked urination/defecation* by territorial steenbok.

AGONISTIC BEHAVIOR

Dominance/Threat Displays: *erect posture; object-horning; medial-horn presentation;* jabbing motions, grinding teeth (reported in suni) (5).

Defensive/Submissive Displays: *head-low/chin-out posture; udder-seeking* (mainly by adolescent offspring, which sometimes approach and poke groin area in response to paternal aggression); *lying-out* (as when hiding).

Fighting: *chasing, air-cushion fighting;* horn-contact (on knees: suni) (5); *stabbing* to body (mainly during chases).

Serious fighting is unusual in the whole tribe, perhaps because the danger of being wounded by their stiletto-like horns selects strongly for males that go no further than displaying. This is illustrated by the oribi, in which the hornless females butt heads but males, even when fighting mad, alternately attack and flee from one another and only rarely make contact. Dik-diks go through the motions of butting while maintaining an air cushion (12). Aggression is most common between father and adolescent son, and toward intruders into a territory.

REPRODUCTION: extended or perennial breeding with or without definable peaks, gestation 5–7 months, postpartum estrus, up to 2 offspring a year (dik-diks, steenbok, perhaps suni and others).

SEXUAL BEHAVIOR: approach in *low-stretch, close following* and persistent *genital licking, foreleg-lifting, nose-to-nose greeting* with or without grooming (nibbling, licking); *mounting* with forelegs folded, without resting on or clasping female, neck forward (dik-diks) or partially raised (oribi). Receptive females respond to genital licking by raising the tail and squatting slightly (like a calf licked anogenitally by its mother). (See species accounts for details and variations.)

PARENT/OFFSPRING BEHAVIOR. Birth, care, and development of neotragine offspring are apparently typical of *hider* species (chap. 2). The gestation period of 5–7 months seems long for such small antelopes and could turn out to be shorter in at least some species. The few descriptions of birth and early development (based mainly on observations in captivity) suggest that calves are quite precocious (see oribi account); yet they do not regularly accompany the mother for 2 or 3 months. Fathers show parental care by exercising vigilance against predators, while their mates spend extra time feeding in order to sustain lactation (see klipspringer account). It is also possible, though unproven, that fathers actively intervene in response to a calf's distress bleating. Mothers give faint calls to summon calves from hiding and suckle them. Weaning takes place by about 3 months, or around the time calves begin foraging with their mothers. Offspring become adolescent by 6 months in small species, by 10 months in the larger ones, and are adult-sized at 1 year. However, the horns may develop more slowly: klipspringer and dik-dik horns only reach full size at 1½ and 2 years, respectively. Harassment of male offspring may coincide with the appearance of horns—at 5–6 months in the klipspringer. The horns emerge at 2 months in dik-diks, but are initially concealed by the crest. Male offspring are forced to emigrate earlier than females, some of which may stay at home and even breed with their fathers (see klipspringer and oribi accounts).

ANTIPREDATOR BEHAVIOR: *lying prone, skulking* into or beneath cover, standing watch in *alert posture,* whistling *snorts, stotting, rocking-horse run* (klipspringer, oribi), sudden *flight* and *dodging run, distress bleat.* (Details in species accounts.)

References

1. Dragesco, Feer, Genermont 1979. 2. Feer 1979, 1982. 3. Gentry 1978. 4. Hofmann 1973. 5. Kingdon 1982. 6. O'Regan 1984. 7. Owen 1973. 8. Pocock 1910. 9. ——— 1918. 10. Richter 1971. 11. Tinley 1969. 12. Walther 1984. 13. Yalden 1978.

Fig. 4.4. Steenbok courtship: male sniffing urinating female.

Steenbok

Raphicerus campestris

TRAITS. A small reddish antelope with big round ears, rudimentary tail, slender build with long legs and well developed hindquarters. *Height and weight:* 50 cm (45–60) (8); males 10.9 kg (9–13.2), females 11.3 kg (9.7–13.2) (Botswana population) (13). *Horns:* sharp upstanding spikes 9–19 cm, ringed only at base. No false hooves; bare muffle, smooth, glossy coat.

Coloration: geographically variable; bright rufous (East Africa) to rufous-fawn and rufous-brown (13); underparts, undertail, and inside ears white; triangular black marking on nose, dark crescent on crown, and dark-fringed ears; sexes alike, newborn same color but woolier coat. *Scent glands*: smallish preorbital glands, larger in male; pedal glands in all feet, no inguinal glands, and possibly a throat gland (13). *Mammae:* 4.

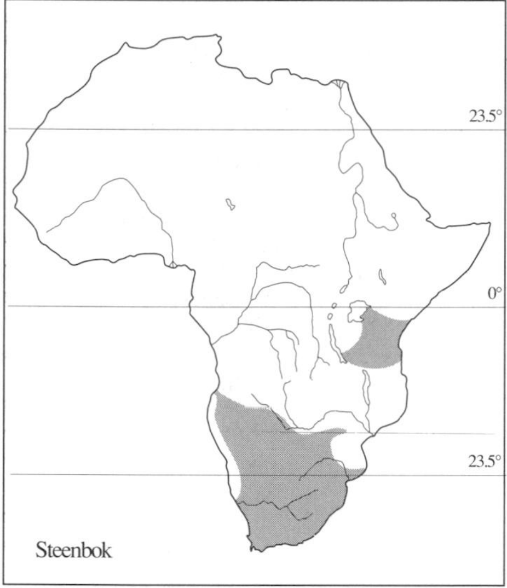

DISTRIBUTION. Like the dik-diks and various other arid zone mammals, the steenbok has a disjunct distribution on either side of the *Miombo* Woodland Zone. It was formerly found to the Uganda and Somali borders, but Mount Kenya is now the northern limit of its range (15). In southern Africa it is common in suitable habitat practically everywhere south of the Zambezi and southern Angola, including the Kalahari but only along major rivers in the Namib Desert (13).

RELATIVES. The two grysboks, *Raphicerus sharpei* and *R. melanotis*, are enough alike to be considered a superspecies and are geographically isolated. The Cape grysbok has a very restricted distribution at the tip of South Africa, and Sharp's grysbok replaces the steenbok in the *Miombo* Zone. The steenbok's range overlaps that of both grysboks, notably in most of Zimbabwe, but the grysboks frequent much denser bush and appear to be mainly nocturnal. Accordingly a steenbok is more likely to be seen on the same ground, and to be confused with, an oribi than with a grysbok. A grysbok is somewhat smaller and quite different in appearance and movements from the steenbok: it has more rounded, higher hindquarters, shorter legs, a darker coat sprinkled with white hair, and moves in a crouch. Male grysboks also have a preputial gland, which the steenbok lacks (1, 10).

ECOLOGY. For a small antelope that depends on cover for safety, the steenbok inhabits surprisingly open country, including grasslands dotted with bush and light woodland, from sea level up to 4750 m (8). However, it is more frequently associated with open plains in the Southern Savanna, where it overlaps with Sharp's grysbok, than north of the *Miombo* Zone where it does not. In Kenya it is also common in denser cover, e.g. on stony, well-wooded hills and in acacia groves. The steenbok benefits from destruction of woodland, whether by man or beast, as colonizing and regenerating vegetation creates ideal cover and browse for it (8). This antelope is thus associated with transitional and unstable conditions, especially in areas of low rainfall.

DIET. The steenbok is a concentrate selector that feeds on the leaves and shoots of a wide variety of low shrubs and trees, on forbs, seeds and seed pods, berries and fruits, and grasses in a young and tender stage. It also digs for roots and tubers in the Kalahari.

In a Kenya sample of 21 stomachs, the proportion of grass rose to as much as ⅔ after rain or fire stimulated new growth (5), and in Botswana the contents of 25 stomachs were divided equally between browse and grass. But in Zimbabwe grass amounted to less than 30% (N = 91) (12). The simple anatomy of the digestive tract makes it extremely unlikely that the steenbok could exist for long on roughage alone (ruminant digestion is discussed in chap. 1) (5).

The steenbok is water-independent but has been known to drink when the opportunity arises (13).

SOCIAL ORGANIZATION: territorial, monogamous pairs.

No comprehensive study of the steenbok has been published. Although seen more often singly than in pairs, it seems probable that this species is distributed in couples which share the same territory but hide, rest, and forage separately most of the time, unlike the dik-diks or klipspringer but much like the gray duiker, for instance. Perhaps the openness and size of steenbok territories, in which cover is typically patchy with intervening open terrain, make it safer for family members to remain dispersed rather than clustered. Even when encountered in pairs, the male and female typically flee in opposite directions (11). Estimates of home range/territory size vary from as little as 4–5 ha (8) up to 1 km² (5). The latter estimate was based on observations of steenboks living in the Kenya Highlands in comparatively open grassland. There, in 8 out of 10 sightings, a male and female were seen within 200–300 m of each other. There are no records of a male associating with more than one adult female. Indeed, a captive male placed with two females mated with only one and fought the other. The mated pair sometimes rubbed or nibbled one another's faces (4). Reports that both sexes participate in maintaining dung middens, if substantiated, would be further evidence that steenboks live in couples (8) (but see under Territorial Behavior).

Some published sex-ratio data for steenboks suggest a proponderance of males, notably in Kruger N.P., where 40 males and only 9 females were seen along the main tourist routes over a three-year period (18); and fetuses collected in a Rhodesian tsetse-control area suggested that even the natal sex ratio might be skewed toward males (21). However, sizeable samples in Natal and Namibia yielded the even adult sex ratio expected in monogamous species, and tests of most other samples indicate no significant difference from a 1:1 ratio (9, 16). Males, being more territorial and less timid than females, may well be seen more often along roadsides.

ACTIVITY. The steenbok is known to be active at night. In Matapos N.P. pairs often came out then to feed on lawns of short grass (20). Steenboks living in settled areas may become very largely nocturnal, but otherwise they are also day-active, with the usual early-morning and late-afternoon peaks, and may move about and feed at all hours on cool, overcast days (13). When lying up they apparently use regular resting places, preferably in cover.

POSTURES AND LOCOMOTION. A steenbuck flushed from hiding dashes off suddenly at great speed on the usual zigzag course, making great bounds every few strides, with its outstretched head at or slightly below shoulder level. After reaching another patch of cover, it drops from sight. Grysboks carry the head lower and dash straight into cover without bounding (14).

SOCIAL BEHAVIOR

COMMUNICATION. Largely unstudied; see tribal account.

The relative importance of olfactory, vocal, and visual signals has yet to be determined. The fact that the steenbok lives in relatively open habitats and has contrasting white markings (especially eye ring, ear lining, and rump patch) indicates that visual communication may be more important in this antelope than in, for instance, the grysboks.

TERRITORIAL BEHAVIOR. Both sexes deposit dung on middens that may serve as territorial boundary markers (8). The performance resembles that of dik-diks but differs in detail: reportedly both sexes scrape the ground, not only before but also in between and after urinating and defecating. Moreover, steenboks sometimes scrape with the hindfeet after defecating (19). But a male I watched first scraped with his forefeet, straddled and urinated, scraped again, squatted and defecated on the same spot, then scraped again, never using his hindfeet. Numerous sandy little piles showed the locations of other recent deposits. Though he was accompanying and courting a female at the time, she did not participate in the ceremony. Because couples often live separated, Walther (19) says the female establishes her own dung middens. The suggestion that covering the excrement would help retard drying and thereby prolong the scent in arid climates raises the question of why dik-diks leave theirs uncovered. Neither sex has been seen marking with the preorbital glands (8, 16, 19).

AGONISTIC BEHAVIOR. Threat and dominance displays remain undescribed, except for the use of an *erect posture* in courtship (see below).

Fighting. A penned male became so aggressive when his mate came into estrus that he attacked all intruders, human or animal, running alongside and stabbing at them with sideward head jerks (4, 8). Since the steenbok's upright, sharp horns are as dangerous as the oribi's, it would be interesting to know whether it also engages in air-cushion fights and counter-chasing, thereby avoiding horn contact (see oribi account).

REPRODUCTION. Reproductive tracts of 188 female steenboks shot on tsetse control in the Zimbabwe Lowveld indicated that ovulation occurred year round, with a possible birth peak early in the rains

(November–December) (21). Similar non-seasonal breeding was shown by 109 grysboks shot in the same area (7). Testes weights of adult males showed no seasonal variation and epididymal smears indicated that spermatogenesis continues through the year. Captive males can mate at 8½ months, when they are adolescent but still immature. Captive females have conceived at 6–7 (up to 9½) months, may calve at 1 year of age, and reproduce at intervals of about eight months (4). Gestation is estimated at 166–177 days (2, 6).

SEXUAL BEHAVIOR: *close following, genital-licking, lowstretch, foreleg-lifting, erect posture* (19).

Courtship features the usual *close following* in *lowstretch, genital licking* and nudging, with frequent *foreleg-lifting* during which the female's hindlegs are often contacted (3; author's observations). Rubbing faces together and/or mutual facial nibbling may coincide with estrus (4, 8).

PARENT/OFFSPRING BEHAVIOR. A captive female dropped a calf ½ hour after the amniotic sac became visible, while lying down (2). The fawn (weighing c. 1 kg) was up and nursing within 5 minutes (surprisingly quickly). At 2 weeks it began nibbling plants; the mother stopped lactating after 3 months. Calves remain concealed until well grown before beginning to accompany the mother, sheltering in tall grass or shrubs and *lying prone* with ears flattened to the neck—which is also the practice of hiding adults. Fawns have been reported to hide in abandoned holes (14), but this is unlikely to be a habit.

ANTIPREDATOR BEHAVIOR: close concealment; *dodging run; alarm snorts.*

The young are vulnerable to all predators down to the size of eagles and adults have to contend with carnivores as small as caracals, and possibly even jackals. They place primary reliance on concealment, *lying prone* in cover until discovery is imminent, then dash away at high speed to disappear in another patch of cover. Sudden flight may startle and thereby delay a predator. The steenbok can also change direction abruptly, a good tactic against the cheetah, for instance, and may have more endurance than most small antelopes (8).

References

1. Ansell 1964. 2. Bigalke 1963*a*. 3. Cade 1966. 4. Chalmers 1963. 5. Hofmann 1973. 6. Hofmeyer and Skinner 1969. 7. Kerr and Wilson 1967. 8. Kingdon 1982. 9. Mentis 1970. 10. Pocock 1918. 11. Shortridge 1934. 12. Smithers 1971. 13. —— 1983. 14. Stevenson-Hamilton 1947. 15. Stewart and Stewart 1963. 16. Stuart 1975. 17. Tinley 1969. 18. Van Bruggen 1964. 19. Walther 1984. 20. Wilson 1969. 21. Wilson and Kerr 1969.

Fig. 4.5. Part of dik-dik *dunging ceremony:* male testing mate's urine as she defecates on border dung midden.

Kirk's Dik-dik

Madoqua kirkii

TRAITS. A dwarf antelope of the arid zone, with pointed, mobile snout and hairy muffle with slitlike nostrils, large eyes and ears, and prominent erectile crest; very long hindlegs, and rudimentary tail. *Height and weight:* 60–72 cm; males 59 cm, 5.1 kg (3.8–6), females 62.5 cm, 5.5 kg (4.5–7.2) (8). *Horns:* 7.5 cm (6–11.4), slanted backward, strongly ridged. *Coloration:* paler in arid regions, grizzled gray to gray-brown above, flanks and legs tan, whitish eye-ring and underparts. *Scent glands:* large preorbital glands, more developed in males, hoof glands (absent in southern race [14]). *Mammae:* 4.

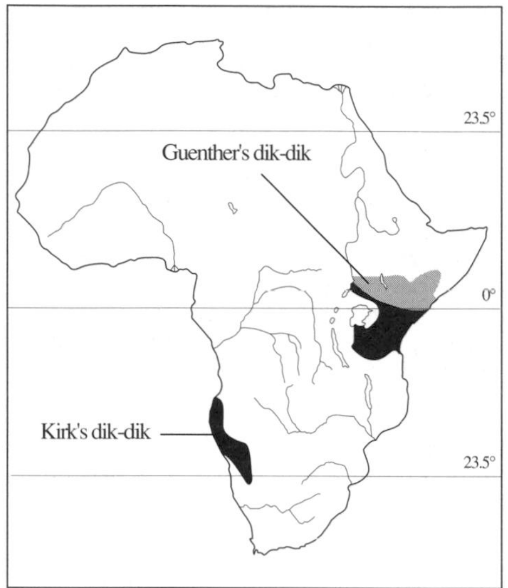

DISTRIBUTION. The Horn of Africa, with all 4 species, 3 of which are endemic, is the center of dik-dik evolution.

RELATIVES. Dik-diks may have started out in the dense evergreen thickets that encircle the base of the Ethiopian Highlands, a habitat with which *M. saltiana* is closely identified (8, 16). Although this and the other 2 endemic species inhabit the driest part of the Somali Arid Zone, they are probably exposed to less heat and desiccation than the 2 dik-diks whose ranges extend south into Kenya and beyond in more open bush. *M. kirkii* and *guentheri* have more elongated snouts, which are thought to represent a later and more specialized refinement of the dik-dik's evaporative-cooling system (discussed under Ecology).

Kirk's dik-dik, the less specialized and apparently older of these 2 sibling species (which are often treated as a separate subgenus, *Rhynchotragus*), also occurs in the South West Arid Zone, having managed to extend its range along the drought corridor that existed during Pleistocene dry periods (8). The dividing line between the 2 species is drawn along the 500 mm rainfall isohyet, but in fact they coexist along the boundary between savanna and dry bush habitat in Kenya's Samburu region, although intermediate forms (e.g. in Tsavo N.P.) indicate that hybridization can occur (8).

In western Kenya and Tanzania Kirk's dik-dik occurs in savanna with rainfall of up to 1000 mm, whereas the southern race, *M. k. damarensis,* is not found in country with more than 500 mm of rain, though absent from the coastal Namib Desert. This race has a rubbery digital pad behind the hooves which is apparently adapted for rocky ground, and lacks hoof glands (14).

In addition to having the largest range, Kirk's is also the largest of the dik-diks. The smallest is Swayne's (*M. saltiana swaynei*), weighing but 2.7 kg.

ECOLOGY. Kirk's dik-dik lives in various habitats with good cover and browse but unencumbered with tall herbage. Families have been known to shift their range after rains caused annual grasses and herbs to grow high enough to obstruct their view. *M. kirkii* is particularly associated with stony ground, where grass is sparse, but does not ascend kopjes and mountain ridges into klipspringer habitat (14). Dense aloe and sansevieria patches separated by open glades are ideal habitat, providing a maximum variety of food and cover. Conditions favorable for the dik-dik often result from overgrazing and trampling by wild and domestic herbivores, and from clearing and cultivation, which cause grassland and wooded savanna to be replaced by bushes and shrubs. Thus, in Etosha N.P. dik-diks tend to be most abundant in areas that have been overgrazed by wildebeests and zebras (14). In eastern Ethiopia, dik-diks are cultivation followers, finding shelter in the dense hedges of prickly pear (*Opuntia*) planted around villages, emerging at night to browse on crops and orchards (7).

To survive in very hot, arid country the year round, dik-diks use all the conventional methods of avoiding heat stress and conserving water (discussed in chap. 2), and have evolved an especially efficient mechanism for cooling the blood by nasal panting (11, 13). The extended snout enlarges the nasal vestibule and the area of moist black mucosa available for evaporative cooling; the flexible trunk with tiny nostrils and hairy muffle is shaped and functions like a bellows to increase the rate of air flow, while at the same time minimizing water loss to the outside (6). An experimental animal, subjected to 40°C temperature, continued breathing at a normal 50–60 times per minute for 8 hours before beginning to pant (11). Then, much to the researchers' surprise, its energy consumption while panting at 400 cycles per minute was 42% *less* than normal, partly because of a 50% reduction in blood flow to all but the muscles involved in respiration, and partly because panting is sustained almost effortlessly by matching the natural oscillation rate of the diaphragm (6).

DIET. Typical concentrate selectors, dik-diks eat the leaves, shoots, and fruits of most of the edible species within reach. They may nibble the tips of growing grass but otherwise do not graze. The extremely narrow muzzle and incisor row, combined with small body size, enables them to select the smallest food items. The flexible upper lip and tongue are also used to pluck foliage (8) and they will stand on their hindlegs to extend their reach. The larger herbivores also help dik-diks to obtain food: elephants push over trees and tear off branches, giraffes drop branchlets from the tops of trees, and browsing by larger antelopes helps keep food plants from outgrowing a dik-dik's reach.

When all fresh food was kept from captive Phillip's dik-diks as an experiment, they were seen to take small amounts of proffered water; otherwise they never drank (12).

SOCIAL ORGANIZATION: territorial, closely associated monogamous pairs.

Studies of free-ranging Kirk's dik-diks, and of captive Phillip's dik-diks kept under seminatural conditions in Somalia, indicate that dik-diks live in closely associated pairs on territories defended by males against intruders of either sex (3, 4, 12). With rare exceptions, males are dominant over their mates, although females normally initiate and lead family movements, while males bring up the rear (3). The territory has to be large enough to provide adequate food through the dry season and a choice of secure refuges. The average area occupied by 6 pairs of dik-diks that lived in a group of isolated kopjes in Serengeti N.P. was c. 5 ha (2.5–12) and their density was 20 dik-diks/km^2 (3). A few kilometers away in *Acacia-Commiphora* woodland, where dik-dik density was only 5/km^2, territories ranged from 10–35 ha (15).

Three years later the same kopjes contained 5 pairs and a different arrangement of territories (3). Four of the same females and 2 of the same males, now 6–9 years old, were still there, and 4 males and 2 females had disappeared, succeeded by 3 new males and 1 female. Two pairs remained intact, 1 of which still occupied the same territory; 1 of the 2 widowed females, which had paired with new males, had also stayed in place. The boundaries of the other territories had all changed, due to expansion by neighbors into territories whose occupants had disappeared. From these changes, and from observations that couples which venture onto unoccupied land stay especially close together, Hendrichs concluded that pair bonds exist independently of attachment to a territory and last for life, whereas attachment to a particular site does not (3).

Pair and family bonds are maintained and evinced by proximity, coordinated activities, a *dunging ceremony* (described below), *vigilance behavior* including possible *alarm-call duetting* (described under Antipredator Behavior), play (limited mainly to youthful individuals), and *social grooming.* Contact between couples and between parents and offspring takes the form of sniffing and touching with the nose, especially the preorbital area, the neck near the shoulder, and the rump. The female of a hand-reared pair sometimes licked the male's preorbital glands and face (17). A 5-month male Phillip's dik-dik was seen to treat his father in the same manner, which was interpreted as submissive/appeasement behavior (12). Dik-diks groom themselves but not one another with their lower incisors.

Young dik-diks disperse during adolescence (6–9 months). Male but not female offspring are forced to leave by their father's aggression; whether mothers are instrumental in the departure of daughters is unknown. Soon thereafter the young are mature enough to establish pair bonds and males may become territorial by one year, even though their horns are only fully developed at 2 years (3).

ACTIVITY. Preliminary evidence suggests species-differences in dik-dik activity cycles: *saltiani* may be largely nocturnal and *guentheri* predominantly diurnal, whereas *kirkii* is intermediate but more active by night than by day (8, 12). Four Serengeti *kirkii* that were observed around the clock spent 12½ hours on their feet and 11½ hours resting; 7½ hours feeding and 6 hours ruminating (4). Times of peak activity were from before dawn to c. 0900 h and from late afternoon until nearly midnight. Main rest periods were between 0900 h and noon and from midnight to pre-dawn. The dik-diks spent half their time resting between 0700 and 1900 h, compared to only 5 hours between 1900 and 0700 h. They also moved around more at night, especially by moonlight.

Hot weather, too, tends to make dik-diks more nocturnal. In the midday heat they may be seen standing in the shade and panting with mouth partly open and tongue slightly protruding (12). Often bothered by flies during the day, standing dik-diks vigorously twitch their ears and tail, stamp their feet, and ripple their skin (14).

POSTURES AND LOCOMOTION. A dik-dik runs with lowered head and abrupt changes of direction in the manner typical of small, cover-dependent antelopes. One was clocked at 42 kph (12). Resting dik-diks lie up next to a tree trunk or a rock, or inside a thicket, where they are very hard to see (8). They stretch either with legs straddled and tails sticking straight up, or with backs arched like a cat (15).

SOCIAL BEHAVIOR

COMMUNICATION. Olfactory and vocal communication are necessary for animals that are night-active and spend part of the time in concealment. Visual displays are also quite prominent in dik-dik communication, in keeping with their relatively open habitat (see under Territorial and Agonistic headings). Tactile signals take the form of *social grooming* (described under Social Organization) and *anogenital licking* (see under Sexual Behavior).

Vocal Communication. Family members use a soft whistle as a contact call and a loud, breathy *zik-zik* alarm call from which the name dik-dik derives. A turbinal in the dik-dik's nose has been modified into a reasonable facsimile of a whistle. The mobile proboscis, which protrudes downward past the mouth when a dik-dik whistles, amplifies the call (8). In addition, dik-diks emit bleating and squealing distress calls, and sometimes a horselike snort while feeding (4).

TERRITORIAL BEHAVIOR: *dunging ceremony, preorbital-gland marking, alarm signaling,* aggressive displays.

The perimeter of a dik-dik territory is demarcated by 6 to 13 regularly used dunging areas, each with an average of 4 middens. A midden may be up to 2 m across and 10 cm deep, the ground beneath soaked with urine to a depth of 15 cm (4). Neighbors use the same areas (though rarely at the same time) and their middens may lie close together on opposite sides of the invisible line.

In a typical *dunging ceremony,* first the female urinates and defecates in a sequence in a deep crouch, without pawing (4). Often the male, standing right behind her, tests the urine (fig. 4.5). When she moves off, he sniffs her droppings, paws vigorously, and proceeds to urinate, after which he crouches to defecate (fig. 4.6). As the male covers the dung of his mate and also any offspring (see under Parent/Offspring Behavior), his own excrement always ends up on top (8, 14). Immediately after the dunging ceremony both sexes mark nearby twigs with their preorbital glands.

For a young male attempting to stake a territorial claim, the process begins with establishing a dung midden (4). Until he does, he is subject to chasing by established males who, not content merely to expel trespassers, try to forestall potential rivals by coming off their property to attack them. A persistent immigrant keeps coming back

Fig. 4.6. Dik-dik scent-marking: territorial male scrapes to cover mate's excrement (*left*), defecates (after urinating) (*middle*), then deposits preorbital secretion (*right*).

and, having learned from experience that he is chased only a certain distance by the resident males, proceeds to establish a dung midden at that point. The presence of his own excretions gives him confidence and thereafter he cannot be chased beyond his midden. Day by day his territory grows, until eventually he establishes his middens on the borders with his neighbors. He has then joined the territorial establishment.

AGONISTIC BEHAVIOR

Dominance/Threat Displays: *erect posture, hunched-back posture, stotting, erected crest, medial-horn presentation* and *head-ducking, object-horning.*

Defensive/Submissive Displays: *head-low/chin-out, sucking-intention, lying out.*

Fighting: *air-cushion fights, chasing, stabbing.*

Virtually the same head-high posture assumed at the sight of a predator is also adopted at the sight of an intruding dik-dik, possibly in an exaggerated form (4). Often it is the female, who is at least as alert as the male against trespassers, that adopts the *erect posture* first and thereby alerts her mate, who then also stands rigidly with neck arched as he watches the intruder approach. In addition to the *alert posture,* another distinctive posture has been described in Phillip's dik-dik: standing with back hunched and neck drawn in (fig. 4.7). Both postures have been interpreted as dominance displays in this context (4, 12). If an intruder reaches the border the resident male suddenly bounds forward in high *stotting* jumps, his crest erected; this is a strong threat display. Strange females usually respond by running away and bleating with fear as the male pursues and tries to horn them. Strange males also flee but the resident male apparently avoids overtaking them (4).

Encounters between territorial neighbors may lead to border skirmishes featuring *air-cushion fights,* in which males rush at each other with crests erected but brake to a halt about a meter apart, at the same time stabbing upward (cf. oribi). The game of bluff may continue for some time, without the contestants ever coming into contact. A skirmish ends either with one male fleeing or with both moving farther apart after each rush, followed by feeding during which they gradually withdraw. Encounters typically terminate with intense marking bouts (4).

Males become particularly pugnacious when their mates are in estrus. Along with intensified territorial marking, they frequently slash at the vegetation with their horns (4).

Fig. 4.7. Submissive posture of an adolescent male dik-dik in response to father's *hunched posture* (threat display).

Submissive behavior is displayed most often by adolescent males in response to paternal aggression. Adopting a *head-low/chin-out posture* is the standard response (fig. 4.7), or the juvenile may jump toward his father wriggling like a puppy, sometimes poking his snout in the adult's flank or under the neck as though to nurse. When chased, young males and sometimes even intruding adults may *lie out* (12, 15). At lower intensity the male may follow his son in *lowstretch,* punctuated with short rushes (4).

REPRODUCTION. Females calve first at 15–18 months, come into heat again within 12 days for at most 48 hours, and if unbred cycle every 3–4 weeks (12, 17). Gestation is 25 weeks (3). Serengeti dik-diks reproduce twice a year, most at the end of the rains and toward the end of the dry season (4).

SEXUAL BEHAVIOR: following in *lowstretch, urine-testing, genital-licking.*

The complete mating sequence of Salt's (Phillip's) dik-dik has been observed in confined individuals (12). First the male sniffed and lightly touched the female's preorbital glands, head, neck, and flanks, then began following her and licking her vulva. She responded by arching her back, raising her tail, and urinating, very like a fawn when licked anogenitally. The male tested the urine and soon afterward tried to mount; as the female walked away, tail erect, he followed bipedally, forelegs bent and neck stretched forward, penis and head crest both erect. But the female moved too quickly for him, meanwhile whistling softly. He dropped back to all fours and followed in *lowstretch,* nosing her vulva. When she stopped for a moment, he mounted and succeeded in copulating. Gazelle-like, he did not clasp nor even rest his forelegs on the female, although his neck was held horizontal rather than erect. Immediately after dismounting, the male licked his still-erect penis, then resumed his normal activities. The full sequence was repeated at intervals of about ¾ hour.

PARENT/OFFSPRING BEHAVIOR. Although the fetus has been seen moving inside pregnant females 2 months before birth, the udder starts to develop only about 8 days before calving (2). No description of a complete birth sequence is available, but 6 zoo births all took under 1 hour (2). The afterbirth was expelled within 1–2 hours and eaten in each case. Four female calves averaged 624 g and a male weighed 795 g. All gained their feet in under 15 minutes, immediately sought the udder, and managed to nurse within 2 hours. Mothers suckled their young usually 30–40 seconds, meanwhile licking them anogenitally and consuming their wastes in the usual way (see chap. 2).

Wild *M. kirkii* mothers were observed to visit their young around sunrise, noon, sundown, and midnight, spending less than 1 hour in 24 with recently born calves (4). After feeding slowly toward the hiding place, which was often near a customary family resting spot, the mother called (inaudible to observers) and the calf immediately emerged and began nursing, poking strongly to bring down the milk. The mother remained with it for 10–15 minutes after suckling, while the calf moved and frisked about; then she led it a short distance, whereupon it went off and found a new hiding place.

ANTIPREDATOR BEHAVIOR: *alert posture, stotting, alarm-whistling, freezing.*

For dik-diks, vulnerable to some 20 different predators, the price of survival is eternal vigilance, intimate knowledge of the home range, secure hiding places, and the ability to hit top speed within the first few jumps. A comparatively low (10–20%) adult mortality rate is a sign of efficient antipredator tactics (3).

Dik-diks *freeze* at the slightest disturbance, often with 1 leg raised. The male moves his head cautiously, trying to identify the danger, while the female remains immobile except for the questing tip of her nose (12). Once a predator is detected and identified, dik-diks behave in 1 of 2 ways, depending on whether it is a stalker or a courser. If it is a cat, they flee just far enough to be safe, emitting loud, explosive whistles at the first bounds. They then proceed to keep watch and sound the alarm—possibly in a duet comparable to the klipspringer's (q.v.) (author's observations). The male is the principal caller. Cats that are hunting usually give up at once and try their luck elsewhere. When the predator is a hyena, wild dog, or other courser, dik-diks immediately sink to the ground and lie still until the enemy passes. If discovered, they flee without whistling (4).

The most dangerous time for Serengeti dik-diks comes during the plains-game migration, as it is attended by a host of predators, which often lie up in the kopjes above where the dik-diks live. Lions, leopards, and hyenas drag their prey into the shade and this brings still other predators into the dik-diks' refuges. Even vultures and small eagles are potential predators on concealed calves. Baboon troops can be particularly dangerous, as they tend to comb an area very thoroughly while foraging. Paradoxically, a leopard's presence can actually be beneficial by keeping smaller predators away.

References

1. Ansell 1971. 2. Dittrich 1967. 3. Hendrichs 1975*a*. 4. Hendrichs and Hendrichs 1971. 5. Hoppe 1977*a*. 6. Hoppe 1977*b*. 7. Kellas 1955. 8. Kingdon 1982. 9. Kurt 1964. 10. Pocock 1910. 11. Shoen 1972. 12. Simonetta 1966. 13. Taylor 1970*a*. 14. Tinley 1969. 15. Walther 1972*a*. 16. Yalden 1978. 17. Ziegler-Simon 1957.

Added Reference

Tilson and Tilson 1986.

Fig. 4.8. Klipspringer male standing sentinel while advertising his own presence. Note truncated hooves adapted for rock-climbing.

Klipspringer

Oreotragus oreotragus

TRAITS. A small antelope that lives on cliffs and rock outcrops. *Height and* *weight:* up to 60 cm and 18 kg; males 49–52 cm, 10.6 kg (9.11–11.6), females 50–53.5 cm, 13.2 kg (5–15.9) (6, 14). Stocky build, with massive hindquarters; stands on tips of truncated hooves; tail rudimentary; short neck, wedge-shaped head and large, rounded ears. *Horns:* wide-set, ringed, upstanding, spikelike (average 10 cm, record 15.9 cm), present in males only and occasionally in females of East African race, *O. o. schillingsi* (8, 13, 21). *Coat:* thick and dense, hairs hollow, brittle, and loose, unique among African bovids (similar to American pronghorn and white-tail deer). *Coloration:* cryptic, geographically variable, yellowish or grayish to brown, more or less grizzled through banding of hairs; no conspicuous markings except ears, white inside with dark lines and a black outside border. *Scent glands:* very large preorbital glands, more developed in males, orifice surrounded by bare, black skin; no hoof glands; preputial gland in male (1).

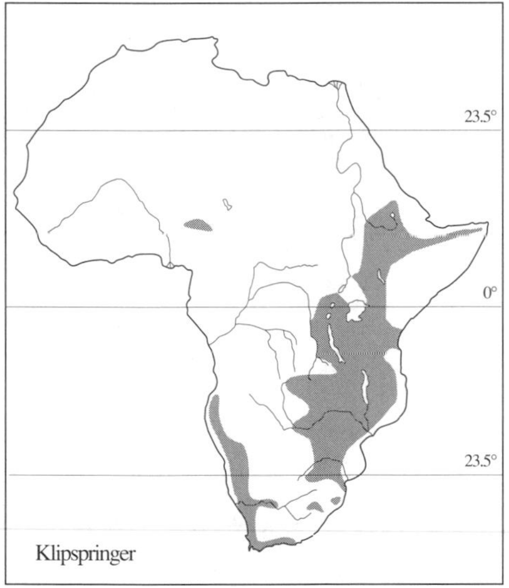

DISTRIBUTION. Eastern Africa from the Red Sea hills to the tip of South Africa and north to southern Angola, wherever its specialized habitat requirements are met, except in the rain forest. Its vertical range extends from coastal hills up to 4500 meters (peak of Mt. Meru, Tanzania). An ancient genus that ranged into Eurasia during the Pliocene, relict populations on isolated massifs in Nigeria and the Central African Republic (now possibly extinct) indicate that the klipspringer ranged across sub-Saharan Africa in the recent past (8). The most extensive regions of good klipspringer habitat are the highlands of Ethiopia, where the species may well have evolved, and of southern Africa (8).

ECOLOGY. The klipspringer is closely associated with steep, rocky terrain where it can find refuge from predators and an adequate food supply. Home may be a small inselberg in the heart of a vast woodland or a kopje in the middle of a plain, reachable only after a dangerous traverse. It may also be open screes of loose rock, a cinder cone, or recent lava flow, for here too a klipspringer's climbing and jumping abilities enable it to escape terrestrial predators. But the most extensive areas of suitable habitat, where klipspringers are most abundant, are mountain ranges and deeply dissected plateaus, rocky escarpments and cliffs, and the gorges of major rivers. In Ethiopia's Simien Mountains a density of 46.7 klipspringers/km^2 was recorded on the edge of the escarpment, an area of relatively high rainfall (>1000 mm) and lush vegetation (3). Here, at over 3300 meters elevation, temperatures regularly go below freezing. The klipspringer's specially insulated coat may be adapted to withstand extreme cold and heat, for daytime temperatures in gorges and other heat traps often reach 40°C in the shade (3).

DIET. To be able to live at high and low elevations, in areas of high and low rainfall, the klipspringer must be very adaptable in its diet. The vegetation that grows on kopjes between the rocks and at the bases of cliffs is often lush, rich in evergreen shrubs and bushes, succulents such as aloes, sansevieria, euphorbias, and forbs that provide feed of high nutritional value and easy digestibility. The klipspringer's digestive system is typical for a small concentrate selector (see chap. 1) (6). According to most accounts it eats leaves, new shoots, berries, fruits, seed pods, and flowers of woody plants, forbs and herbs, and a little new green grass (12, 14, 22).

Klipspringers may have to leave their sanctuaries to forage and will venture up to 0.5 km away in the dry season to feed on a recent burn, drawn by new grass and the sprouting of other monocotyledons and dicotyledons. Although individuals have been seen drinking when water was readily available, klipspringers normally obtain all the moisture they need from their food, especially succulents such as those mentioned above (13).

SOCIAL ORGANIZATION: territorial, closely associated monogamous pairs.

The basic klipspringer social unit is a mated pair that shares a permanent home range/territory. The mean group size of 265 klipspringers classified in Namibia was 2.6 (18). In the Ethiopian Highlands, 74%–81% of 67 classified groups consisted of a male and female pair, and less than 10% were singles (12.5% in 1 area) (4). About half the females (45.5%) were accompanied by young of the year (3).

Between 10% and 20% of the klipspringer sightings, both in Ethiopia and in the Zambezi Valley, were of 1 male and 2 adult females (7). An extra female is most likely to be a grown daughter, for female offspring commonly remain with their parents until they are yearlings (4). But at least one extra female is known to have bred, for 2 members of a trio both had calves. Apparently male offspring disperse earlier, although no parental aggression toward young of either sex has been noted (4).

Territory size varies according to rainfall and the distribution/density of the vegetation, as would be expected of a system based on resource ownership. In the lush Ethiopian Highlands (1300 mm rainfall), territories averaged 8 ha (7.5–9.4 ha) (4), compared to 15 ha in the South African Karroo (400 mm rainfall), and 49 ha in the northwestern Cape Province (160 mm) (10).

It is probably normal, though not invariable, for mated individuals to spend their adult lifetimes within the same territory and to venture outside its boundaries only when attracted to a specific feeding site or salt lick. (Up to 9 adults from neighboring hills have been seen on the same greenflush [8]). A klipspringer with a distinctive white mark, already full-grown when first sighted, was known to live on the same kopje for nine years (9). However, if fires burn off their cover and food, klipspringers may evacuate their territories and take up temporary residence in the nearest available refuge, even away from cliffs in stream-side vegetation (8). A possible seasonal pattern of occupancy was noted in Serengeti N.P., where klipspringers that lived in a group of kopjes bordering the plains were absent in the dry season (21). After the rains began, a party of 6 klipspringers arrived at the kopjes and within several days had reestablished pair territories.

Klipspringer pairs and offspring past the concealment stage stay close together. The median distance between pair members in the Ethiopian study was only 2 meters (5). The female usually initiated progressions and the male followed her 78% of the times, whereas the female only followed 52% of the male's initiatives. When not followed the male would often come back, whereas the female would keep going. However, when through feeding, she would come back to the male, who usually finished sooner, before beginning to rest. In general, then, the male plays the more active part in maintaining proximity with his mate and in standing guard against predators and rival klipspringers. This arrangement may be partly explained simply by the difference in activity budgets (see table 4.2 and Activity): a lactating female has to spend more hours feeding, and the male, with time to spare, tags along. Nevertheless, some positive advantage is needed to account for the male's investment of attention and energy in staying near his mate (5).

During rest periods males typically choose good vantage points and spend significantly more time on their feet than do females (table 4.2), who often rest in the shade and tend to make themselves inconspicuous. Males regularly stand like statues for up to ½ hour except for occasional head turns; one was recorded to stand, intermittently chewing the cud, from 1000 to 1550 h (8)!

The cohesiveness of klipspringer families and the male's vigilant behavior are both adaptive in their open habitat, for they must be prepared to flee from predators that penetrate their sanctuaries; the young are

Table 4.2 Percentage of Daylight Hours Klipspringers Spent Feeding and Resting in Two Ethiopian Highlands Sites

	Sankabar Escarpment				Sankabar Gorge			
	Males		Females		Males		Females	
Season	Wet	Dry	Wet	Dry	Wet	Dry	Wet	Dry
Feeding	18.2	9.5	31.0	24.1	16.0	5.8	26.2	11.9
Resting								
Lying	13.6	26.9	13.8	24.5	22.2	21.6	27.4	28.9
Standing	54.5	48.3	31.0	36.3	39.5	58.3	23.8	45.9

SOURCE. Based on Dunbar 1979, tables 2 and 3.

particularly vulnerable to large eagles. By standing guard in his free time, the male enables his mate to devote extra time to feeding and caring for their offspring with less risk than if he were more distant and less vigilant. At the same time he is also keeping an eye out for potential rivals and advertising his territorial status. So the simple explanation for the conspicuous vigilance of the male klipspringer is the safeguarding of his own genetic investment (4).

Although family members seem to be constantly aware of one another's whereabouts and behavior, actual contact appears to be minimal: only 8 such interactions were observed between Ethiopian klipspringers over a period of 15 months (4). The contacts involved approach and sniffing of face or rump by males and females, and 1 pair rubbed their muzzles together. Instead of mutual grooming, territorial marking with the preorbital glands and alarm-calling in duet may be most important in maintaining pair bonds (described under Territorial and Antipredator Behavior) (10, 17).

ACTIVITY. Klipspringers are known to move and feed at night, but only daytime activity has been studied. The influence of microclimate on activity pattern is well-illustrated by the consistent differences that were found between klipspringers living on the escarpment and gorge sides of the same ridge in the Simien Mts. (table 4.2) (3). Both subpopulations had 4 feeding peaks in early and mid-morning, early and late afternoon, but the gorge klipspringers began the morning feeding periods an hour earlier. This side was warmed by the morning sun while the escarpment side remained in deep shadow; the klipspringers there remained lying in cover, sheltered from the cold wind and conserving body heat, until the temperature rose. After mid-morning, gorge klipspringers spent increasing amounts of time in the shade of rocks and bushes to avoid the hot sun, whereas animals on the cooler escarpment side remained in the sunlight until late afternoon, when rising winds and falling temperatures drove the klipspringers of both locations to seek shelter. Although the 2 subpopulations spent equal time moving, the gorge animals spent significantly more time resting and less time feeding during the dry season than did the escarpment animals. That they mostly stood rather than lay in the shade during these hottest months (January—April) was probably to dissipate body heat efficiently (see table).

The same study showed that klipspringers in 3 different areas spent a greater part of the day feeding during the wet than during the dry season, and that females spent more time feeding than males, due to the greater demands of bearing and suckling young (table 4.2).

The greater availability of high-quality forage, especially forbs, on the escarpment accounts for the longer feeding time of escarpment klipspringers. Instead of putting more effort into finding food, the gorge klipspringers apparently chose to conserve energy by resting more, perhaps drawing on fat reserves accumulated during the rains. The higher nutritional plane enjoyed by escarpment animals was reflected in a higher reproductive rate (3).

A study of klipspringer feeding ecology in South Africa's arid Cape Province again gives a contradictory result: here the animals spent more time feeding during the dry season (10).

POSTURES AND LOCOMOTION. The klipspringer's muscular, rounded hindquarters and short body and neck are adapted to bounding up and down steep slopes. Even on the flat, it runs in jerky bounds reminiscent of a mule deer (15). It leaps surefootedly from rock to rock, landing *en pointe.* The hooftips are like paired hobnails, the outer edges kept sharp by more rapid wear of the softer inner hoof surface, while the 2 digits are prevented from splaying by a thick connecting integument (11). Because it stands on its hoof ends, the klipspringer's leg and hoof form a nearly straight line, enabling it to group all 4 feet within a very small compass. The profile of *Oreotragus* is accordingly unlike any other antelope's (fig. 4.8). The forelegs of a lying klipspringer can be extended straight forward like a dog's, a position otherwise known only in dik-diks and goats among ungulates (20). This unusual flexibility enables a klipspringer to lower its forequarters without bending its legs during combat and play (20).

SOCIAL BEHAVIOR

COMMUNICATION. Klipspringers can see and be seen better than other neotragine species. Therefore a major role for visual signaling would be expected, as demonstrated by the male's sentinel and position-advertising behavior. Communication by scent and sound are likewise important, whereas physical contact, in the form of social grooming, is not. (See under Social Organization, Territorial Behavior, and Antipredator Behavior.)

TERRITORIAL BEHAVIOR: *static-optic advertising* (male sentry duty), *object-horning, scent-marking,* including preorbital-gland and dung deposits by both sexes, *alarm-call duetting* (?).

Although visual displays of the male's presence may be of primary importance, regularly used dung middens up to 3 m in diameter and 10 cm deep are a prominent feature of klipspringer territories. Middens are located preferably on flat ground and spaced from 10 to 100 meters apart, with smaller piles in between (4, 19). Although not limited to the outer perimeter of the property, middens often seem to run in lines along the boundaries (4). However, no ritualized family *urination/dunging ceremony* has been observed and klipspringers do not paw before or after excreting (cf. dik-dik and steenbok). They do assume a deep, visually distinctive *defecation crouch* similar to the dik-dik's (illustrated in ref. 8). Tarlike deposits of preorbital-gland secretion are at least twice as common as middens, and klipspringers often mark nearby twigs after defecating (19). A tame male used to bite off the ends of grass stems preparatory to inserting the tip into the gland's duct (cf. oribi) (8). Males do most of the marking, and when the female marks, the male usually overmarks the same spot (10).

AGONISTIC BEHAVIOR

Dominance/Threat Displays: *medial-horn presentation* with or without lowered forequarters, *object-horning, chasing.*

Defensive/Submissive Displays: *head-low/chin-out posture, biting* (21).

Fighting: *butting* and *stabbing; biting* by females.

Intruders of either sex are liable to be chased by the resident male, sometimes assisted by his mate (10). In 1 of only 2 border violations witnessed during the Ethiopian study, a fight developed when the male of an intruding family was chased to within 15 m of his mate and offspring, which had stopped to look back. While the female and calf resumed their flight, the 2 males faced one another for about 15 seconds, then slowly approached with horns presented and proceeded to butt heads about 10 times, following which the intruding male ran after his family (4). Another fight ended with the loser so exhausted and badly wounded that instead of fleeing he took refuge under a bush (10).

In captivity, fighting females bite each other, often ripping out tufts of brittle hair (20).

REPRODUCTION. Breeding may be seasonal in Ethiopia (4), vary according to local conditions in Cape Province (10), and be perennial in Zambia, where pregnant females were collected in every month (22).

SEXUAL BEHAVIOR: *close following, courtship-circling, foreleg-lifting, mutual grooming, humming.*

Detailed descriptions of courtship are unpublished. Judging from the behavior of a male toward 2 females in Matopos N.P., courtship may be more elaborate than in most neotragine antelopes (author's observ.). When first seen, the male was following an adult female closely, foreleg-lifting (from behind and also from the side). As she circled and moved down a wide rock ledge, he passed and stood facing her with neck arched, a possible dominance display, whereupon she apparently bit him. Then they circled tightly. When this female lay down, the male then approached the second, slightly smaller one and stood facing her; she nibbled or licked his face. Mutual nibbling of preorbital-gland secretion has been observed in zoos (20). Close-following males may make a soft humming sound, inaudible beyond a few meters. A captive pair was seen copulating 5 or 6 times over a week (2), an unusually long estrus.

PARENT/OFFSPRING BEHAVIOR. An unusually long gestation of 7½ months (214 days) has been reported by the same author (2). A likelier estimate, also based on a captive animal, is 5 months (10). Females retire into a rocky recess or dense vegetation to calve. Birth weight is about 1130 g (22). The escape response is reportedly suppressed in infants, which continue to *lie out* after being discovered (10). It seems the feeble stage is prolonged in *Oreotragus,* as a calf born in a zoo remained unsteady on its feet for almost a month (2), and calves may remain hidden for 2–3 months (10), after which they stay constantly with the mother. Weaning is completed at 4–5 months (2, 10). The horns become visible by 6 months and are fully developed at 17–18 months, half a year after reaching adult body size (10).

ANTIPREDATOR BEHAVIOR: male sentry duty, *alert stance* and *fixed stare, stamping, alarm-whistling,* bounding flight, *jumping in place.*

Klipspringers always seem to be on the alert for danger, an indication that steep terrain alone does not guarantee safety. Although guarding the territory against trespassers may account for part of the male's vigilance, klipspringers react immediately to the sight of predators and to the alarm calls of other animals. In 74 encounters between klipspringers and potential predators (jackal, baboon, spotted hyena, human observer), klipspringers gave alarm whistles 47 times and fled without calling 27 times (18). When a predator is first sighted, the initial response is a single alarm call, given by any member of the family, followed by flight for 30 to 50 m, usually to a higher place, the male bringing up the rear. Having reached a relatively safe vantage point, the male and then the female stops and turns to stare at the source of the disturbance. They may proceed to give a whole series of alarm calls, especially if the predator remains in view, making forceful exhalations through the nose which are audible up to 700 m and sound like a toy trumpet (15). Either sex may begin, but after the first 2 to 4 blasts the female synchronizes her calls so that they come just after her mate's.

The importance of duetting in various pair-bonded birds is well-established (16). Duetted alarm calls may help to maintain the pair bond in klipspringers, too (17). But the primary function of *alarm-duetting* may be to discourage pursuit by predators (18).

References

1. Ansell 1969. 2. Cuneo 1965. 3. Dunbar 1979. 4. Dunbar and Dunbar 1974. 5. ——— 1980. 6. Hofmann 1973. 7. Jarman 1974. 8. Kingdon 1982. 9. Lownds 1956. 10. Norton 1980. 11. ——— 1984. 12. Smithers 1971. 13. ——— 1983. 14. Smithers and Wilson 1979. 15. Stevenson-Hamilton 1947. 16. Thorpe 1972. 17. Tilson 1977. 18. Tilson and Norton 1981. 19. Tinley 1969. 20. Walther 1972*a*. 21. ——— 1984. 22. Wilson and Child 1965.

Fig. 4.9. Male oribi approaching dung midden, stimulated by the female's elimination posture; he then adds his deposit to hers.

Oribi

Ourebia ourebi

TRAITS. The most gazelle-like of the Neotragini. *Height and weight:* up to 67 cm and 21 kg; males 58 cm (51–63.5), 14 kg (10.5–17.4), females 59 cm (51–63.5), 14.2 kg (7.5–17) (Transvaal series, ref. 14). Hindquarters rounded and slightly higher than withers, long legs and neck, tail short (5–15 cm) but conspicuous; bare muffle; sizeable ears (smaller and less rounded than steenbok's). *Horns:* thin, erect spikes 8–19 cm long, ringed at base. *Coloration:* geographically variable, generally bright rufous with contrasting white underparts, rump patch, chin, and line over eye; black tail and a black spot below each ear as in reedbucks; sexes alike except females and young have dark forehead crescent; young fawns dark brown. *Scent glands:* very large preorbital glands in males (unused and possibly undeveloped in females); black spot below ear underlain by apocrine glands which diffuse scent into air (8); deep inguinal pouches; carpal glands beneath brushes of long hair on front legs (as in gazelles), shorter brushes below hocks; well-developed glands between all hooves. *Mammae:* 4.

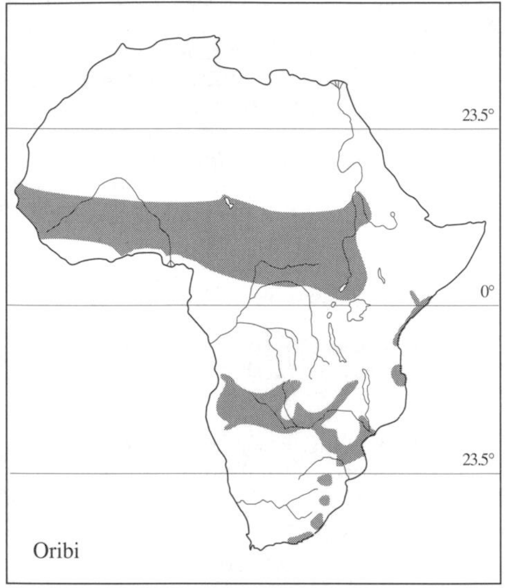

DISTRIBUTION. Discontinuous in better-watered parts of Northern and Southern Savanna: across Guinea Savanna to Ethiopia and south through western East Africa into Tanzania, with enclaves further east; savannas of Zambia and Zimbabwe; eastern Africa from Malawi to southeastern Cape Province (8).

ECOLOGY. The only neotragine that is primarily a grazer, the oribi avoids habitats dominated by woodland or bush. It likes open grassland that is not too tall or dense, affording good visibility, but needs some hiding places; therefore it often favors mosaic grassland and ecotones where bush grades into open plains (16). It reaches highest density on floodplains and in montane grassland, especially in association with large herbivores such as hippo, buffalo, cattle, zebra, topi, and kob, which help to keep pastures short during the rains. It avoids steep slopes. In Rwanda's Akagera N.P. up to 45 oribi/km^2 occur on hilltop *Loudetia simplex* medium-length grassland, and tree savanna dominated by redoat grass (*Themeda triandra*) is also preferred (10). Burns that result in greenflush during the dry season are important if not necessary for the oribi's welfare, although territories that lose all cover may expose the occupants to greater predation pressure. In the absence of fire, oribis may move locally during the dry season to feed in low-lying places where green pasture persists longer (8, 14, 16).

DIET: mixed feeder (grazer/browser). Although classified as a fresh-grass grazer dependent on water by Hofmann (6), who considered it the most primitive of the grazing ruminants, there is evidence that it also browses, especially in the dry season, and is water-independent. Thus, in Mozambique's Gorongoza N.P., oribis were recorded to eat 11 different herbs and the foliage of 7 different trees (16). Both here, in Akagera N.P., and in southeastern Transvaal, oribis have been reported to be water-independent (10, 11, 14). Nevertheless, this antelope subsists primarily on grass as long as pastures are green. It is among the first herbivores to utilize a postburn flush, when the growth is still too short and sparse to satisfy larger grazers. Akagera oribis also visit mineral licks every 1–3 days (10).

SOCIAL ORGANIZATION: territorial, monogamous pairs with a tendency to polygyny.

The oribi has been considered a possible connecting link between monogamous and polygynous mating systems, solitary and gregarious forms of social organization (3, 7, 8, 9). Males maintain territories which they share with 1 and sometimes 2 or more resident females. They use cover for concealment like other neotragines as long as cover is available. However, wildfires sweep through most medium and long grasslands every dry season, leaving little and sometimes no cover in oribi territories. Faced with the same conditions as "plains" antelopes that rely on flight alone to escape predators, oribis too depend on early detection and increased flight distance to avoid capture and show some tendency to band together. Groups of up to a dozen have been recorded (13), but such parties scatter in all directions when disturbed and even when undisturbed seemingly lack the cohesiveness of true herding species (17).

To determine whether *Ourebia* actually has the makings of a gregarious species requires closer study of groups. The evidence suggests that groups of up to 5 or 6 may be resident families, whereas larger parties represent temporary gatherings of families or individuals on neutral ground. Such gatherings are very common in Akagera N.P., where families regularly leave their territories to visit mineral licks. But families show no tendency to integrate when they meet on neutral ground; on the contrary, they maintain a minimum spacing of 5–7 m, violation of which is met with aggression by adults of either sex (10). No tendency to form even temporary herds has been seen in this population.

In Uganda's Kidepo N.P., though, Kingdon (8) observed daily formation of temporary herds on a 1.5 ha lawn surrounded by tall grassland. Half the lawn was defended by 1 male, while the other half was shared by up to 4 males and 3 females at the same time. On the defended half females came out of the high grass only after the male emerged into the open at dawn. Four females and a juvenile spent 1 whole day with him, grazing and resting as a compact group until evening, when the females went their separate ways. The male behaved throughout like a typical territory-owner; he even blocked the way when his "harem" seemed inclined to stray to the other side of the lawn.

The proportion of territorial males with more than one female also varies in different populations. In Akagera only 11% of 604 oribi groups included a male with 2 or more females (10), compared to half of the 32 groups which included females in samples of oribi in Kidepo and Murchison Parks (8), and an average of 2.1 females in such groups in Serengeti N.P. (7). In Gorongoza N.P. an adult sex ratio of 1:1.4 also suggests that about half the males had 2 females (16), whereas the adult sex ratio was equal in a Natal sample (11).

Oribi pairs associate less closely and constantly than dik-dik or klipspringer pairs but more closely than steenbok pairs. (However, data on oribi social distance are too few to make meaningful comparisons.) The large size of oribi territories and the time males devote to patrolling and marking may partially account for the frequent separation of couples. Mean territory size has been variously estimated at c. 1 km^2 in Akagera, 30 ha on a Kenya cattle ranch (6), 34 ha in southeastern Transvaal (14), and 49 ha (42–62) in the Natal Drakensberg Range (11).

Even though family members are often separated, they may seldom be out of communication thanks to the oribi's battery of scent glands, which are most useful when visibility is restricted by high grass. In particular, the male's ubiquitous preorbital-gland deposits serve to define and delimit the territory. A soft whistle (*phi-phi-phi*) helps keep family members in vocal contact, not to mention the more obvious alarm whistle. But the main form of social bonding in this species depends on visual signaling: a distinctive crouch assumed by the female during and after urinating and defecating causes the male to hasten to her and perform a *dunging ceremony* (fig. 4.9) (10, 16):

(a) The male sniffs the female's anogenital area, whereupon she moves off a few meters. (b) He marks a grass stem with both preorbital glands, (c) smells the female's deposit, (d) paws it vigorously, and (e) finally urinates and defecates in sequence on the same spot (16). The female's *excretory crouch* also evokes a similar reaction in other family members, which add their own urine and feces before or after the male, but without pawing or preorbital-marking (10). The female may thus initiate most social contacts within the family, and the frequent exchanges of olfactory information effected through and during the ceremony help maintain their social bonds (8).

Females actively defend the territory against intruders of their own sex. Although they are less active in advertising and defending property, once females have settled on a territory they probably remain for life. Individuals whose mates have died have been known to stay put and quickly establish a pair bond with an immigrant male (10) (cf. dik-dik).

One of many unanswered questions about oribi social organization concerns the identities of extra females and the resident male's relations with them. Does he maintain close bonds with all or only 1? The aggressiveness of resident females to female strangers suggests that the extras are daughters, which are known to remain until nearly full-grown. Sons, in contrast, are chased out when they reach adolescence. The built-in intolerance of potential rivals is illustrated by the account of a male, whose territory was an island in a river, killing 2 adolescent sons in succession who failed to vacate the premises when chased (8).

ACTIVITY. Information on nocturnal activity is sketchy. Typical activity peaks early and late in the day have been reported, and Transvaal oribis were found to spend 27% (20%–38%) of each day grazing (11).

SOCIAL BEHAVIOR

COMMUNICATION. See also under Social Organization. The presence of 6 different kinds of scent glands in *Ourebia* indicates the importance of olfactory signaling, although the function and significance of the different glands remain obscure. Visual displays, too, are developed more than in most neotragines, whereas vocal communication, though still important, ranks third. Tactile communication, in the form of *social grooming*, is rare except perhaps in courtship, although family members often touch noses when they meet (10). The prominence of visual signals correlates with good visibility during

Fig. 4.10. Oribi male biting off a grass stem (*top*) preparatory to depositing preorbital-gland secretion (*bottom*). (From Gosling 1972, fig. 1.)

at least part of the year, and olfactory and vocal signals are necessary under conditions of poor visibility.

TERRITORIAL BEHAVIOR: marking with preorbital-gland secretions (adult males), *linked urination/defecation, erect-posture, rocking canter.*

Territorial males spend much of their free time scent-marking and advertising their property. They deposit the copious black secretion of their huge preorbital glands on grass stems up to 16 times an hour while making regular border patrols, and also in the territorial interior with or without any apparent external stimulus (10). They do not overmark the same spot and, when the grassland is tall, prepare stems by biting off the stalks at a convenient height for insertion in the gland orifice (fig. 4.10) (4, 16). A patrolling male walks briskly with head high and neck slightly arched (*erect-posture*), often sniffs the ground, and at intervals urinates and defecates in sequence, followed by a brief bounding run, recalling an exaggerated stotting, in which the hindquarters are thrown high displaying the rump patch and erected tail (fig. 4.11) (cf. impala). Repeated visits to particular places (e.g., bare ground near borders) result in an accumulation of separate dung piles, but oribis do not make dung middens.

AGONISTIC BEHAVIOR

Dominance/Threat Displays: *erect posture* with arched neck; *head-low with chin out and ears cocked,* confronting or with head turned away, *raising neck abruptly* from *head-low posture*; *object-horning.*

Defensive/Submissive Displays: *U-neck* (head-low) *with ears back, displacement grazing* and *grooming.*

Fighting: attack, alternating flight and pursuit, rarely combat.

Dominance and threat displays are frequent, especially in the following situations:

Territorial defense: border confrontations between territorial males, male-male, male-female, and female-female aggression toward intruders.

Confrontations between families on neutral ground: male-male, especially when an estrous female approaches or is approached by another male, and female-female.

Intrafamily: father-adolescent son, male-female, to stop aggression between 2 resident females (rare) (10).

The reaction of a female to a male's threat displays depends on her status: a resident assumes the *urination/defecation posture,* thereby triggering the male's *marking ceremony*; a nonresident lowers her neck, with chin turned up (*U-neck*) and ears back, or runs away. Figure 4.12 illustrates the response of an adolescent male confronted by his father in the *erect posture* with arched neck. When territorial males confront one another across their boundary line, they both assume the *erect posture,* engage in marking bouts, and sometimes horn the ground or vegetation before withdrawing toward the interior of their property, scent-marking along the way. When an oribi approaches another with neck lowered, chin out, horns vertical, ears cocked, and tail erect, it is a strong threat;

Fig. 4.11. Oribi *high jump:* takeoff and landing (from a sketch by L. Tinley).

Fig. 4.12. Adolescent male (*left*) submitting to adult male's dominance display (from a sketch by L. Tinley).

low-horn presentation, with horns aimed at the adversary, is not in the oribi repertoire (10). If a threatening oribi suddenly raises its neck, it usually attacks immediately afterward.

The usual response to a male's attack is flight, even when 2 equally matched and highly aggressive males are involved. Combat involving butting is common enough between females but is rare and can be fatal between males (19). Tinley (16) saw a single brief clash after one male, in hot pursuit of an adolescent male, intruded into another's territory. After clashing, they pulled back and stood with heads lowered but turned aside. Apparently oribi horns are so dangerous that chasing normally substitutes for fighting (10).

REPRODUCTION. The breeding season is extended but usually with distinct birth peaks in the rainy months: August–November in Zambia and Zimbabwe (1, 15), November—December in South Africa (11), May–July in Rwanda (10). Females may conceive at 10 months and males become sexually active by 14 months (8). Gestation is 6.4 or 7 months (12).

SEXUAL BEHAVIOR: approach in *lowstretch, genital licking* and sniffing, urination/defecation sequence without *testing grimace* (?), *foreleg-lifting.*

An absence of male aggression (dominance displays and chasing) and of typical *urine-testing behavior* (see chap. 2) is an unusual feature of oribi courtship. The male routinely smells the female's genitals, urine, and feces whenever she assumes the distinctive *excretory crouch,* but only very rarely grimaces in the process (never seen by [10] but reported by [16]). Instead he goes through the *marking/scraping/dunging ritual.* Nevertheless, the male presumably monitors the female's reproductive condition in the process, for he sometimes shows sexual excitement after sniffing the female's deposit and begins following her closely in *lowstretch,* tail out, at the conclusion of his obligatory marking ritual, and proceeds to lick her genitals, often while *foreleg-lifting.* His persistent licking may induce her to crouch with tail erected, like a calf when licked anogenitally by its mother. The male sometimes prods or even lifts the female's hindquarters with his muzzle, which may also derive from infantile (nursing) behavior. If the female is unreceptive she walks or runs away, and can stop courtship by running and then lying (10). During mating the female holds the *excretory crouch*: legs slightly spread, head up, ears back, and tail raised. As in other neotragines, the male mounts with forelegs folded back, with neck raised to c. 45°, and copulates without resting on the female.

PARENT/OFFSPRING BEHAVIOR. Except to suckle, calves remain concealed for about a month and begin spending full time with their mothers only after 3 months (11, 14).

The birth and development of oribi housepets is the only available account (12). The female's udder began to enlarge and the baby's movements became visible a month before birth. Following hours of restless wandering, the mother delivered after laboring 30 minutes, while lying on her side. She licked the fawn, cleaned up all the fluids and membranes, and ate the afterbirth when it dropped 2½ hours *post partum.* The father also licked the fawn and ate part of the afterbirth. The fawn (weight 2.3 kg, height 37.5 cm, color dark gray) stood after ½ hour and suckled after 3¼ hours (between the hindlegs). Within 5 hours it was able to jump a 26-cm barrier. It nursed for ½ minute 3 times the first day, increasing to 5-minute bouts at 1 week. The mother always licked it anogenitally while it suckled. The fawn nibbled grass on day 5, butted and cavorted on day 6, ruminated and straddled to urinate (male attitude) the 4th week, marked with its preorbital glands and had assumed adult coloration by 5 weeks, but was still unweaned after 6 weeks. Wild oribis may nurse until 4–5 months old (2).

ANTIPREDATOR BEHAVIOR: concealment, *stotting,* zig-zag run, short and long *alarm whistles.*

When oribis sight a predator undetected, they often sink to the ground and hide, *lying prone* with ears folded (as do steenboks). They flush from cover at the last moment and run at 40–50 kph, sometimes whistling in several short, shrill blasts. They often stop and look back after running 200 m or so. When a pair flees the female is usually in the lead. Oribis also stott when alarmed, with white rump and black tail displayed, perhaps especially in tall grass, the high jumps with raised head affording unimpeded glimpses of their surroundings. A sustained, high sharp whistle represents a higher state of alarm than a series of short (*is-is-is*) whistles. When caught or wounded, and sometimes when pursued by another oribi, an oribi bleats (8, 10, 16).

References

1. Ansell 1963. 2. Bigourdan and Prunier 1937. 3. Estes 1974*a*. 4. Gosling 1972. 5. Hendrichs 1972. 6. Hofmann 1973. 7. Jarman 1974. 8. Kingdon 1982. 9. Leuthold 1977*a*. 10. Monfort and Monfort 1974. 11. Oliver, Short, Hanks 1978. 12. Riche 1970. 13. Roosevelt and Heller 1914. 14. Smithers 1983. 15. Thomson, 1973. 16. Tinley, K. 1972 (unpub. MS) 17. Verheyen 1951. 18. Viljoen 1977.

Added Reference

19. Arcese and Jongejan 1990.

Chapter 5

Gazelles and Their Allies

Tribe Antilopini

Genus *Gazella*, African Species

(Subgenus *Gazella*)

- *G. dorcas*, dorcas gazelle
- *G. rufifrons*, red-fronted gazelle
- *G. thomsonii*, Thomson's gazelle
- *G. spekei*, Speke's gazelle
- *G. leptoceros*, slender-horned gazelle or rhim
- *G. cuvieri*, Atlas or Edmi gazelle

(Subgenus *Nanger*)

- *G. granti*, Grant's gazelle
- *G. soemmeringii*, Soemmering's gazelle
- *G. dama*, dama or addra gazelle

Antidorcas marsupialis, springbok
Litocranius walleri, gerenuk
Ammodorcas clarkei, dibatag

With some 8 genera and 19 species, the Antilopini are the largest bovid tribe. The genus *Gazella*, with 11–13 species (7), is outnumbered only by the forest duikers (15 species). Gazelles are also by far the most widely distributed genus and tribe of antelopes, ranging from South Africa across Asia to Siberia and China.

TRIBAL TRAITS. African representatives: slender antelopes of medium size (60–90 cm high, 18–73 kg) built for speed, with long, evenly developed limbs, level back, and long neck. Tail short (gazelles) to medium length (gerenuk and dibatag). Moderately long head and narrow muzzle, muffle hairy, ears narrow and medium length (long in gerenuk). *Horns:* present in both sexes except dibatag and gerenuk, strongly ridged and typically *S*-shaped (lyrate) in males; thinner, more weakly ringed, and 10%–60% shorter in females, varying in shape from short spikes (Thomson's gazelle) to approximately the same shape as the male's horns (dama, Soemmering's, slender-horned, and Speke's gazelle, and some springbok populations). Coat short and sleek. *Coloration:* from pale fawn to dark plum, individually and geographically variable (desert forms lightest); white underparts and rump patch, a dark flank stripe present or absent, head more or less boldly marked with alternating white and dark lines, spots, or blotches. *Scent glands:* preorbitals similar to dwarf antelopes' (fig. 4.1), well-developed in males of some species, inactive in others and non-functional in females of most (perhaps all) species; hoof glands in all feet (absent in dibatag and possibly springbok); a single pair of inguinal pouches in most species and probably in females of all species; carpal or shin glands on forelegs, beneath tufts of longer, darker hair (absent in springbok and *G. rufifrons*, possibly other species) (1, 10, 14). *Mammae:*2 in gazelles, 4 in springbok, gerenuk, and dibatag.

The preorbital glands are usually employed to daub secretion on twigs, grass stems, or other objects, but are also opened wide during agonistic and sexual interactions (24). Melanin pigment makes the secretion black and pitchy in most or all of the species that employ preorbital secretions in territorial demarcation (most small gazelles and gerenuk). Pigment is lacking in the springbok, whose secretion is yellow and oily, and perhaps in dorcas and slender-horned gazelles. Large gazelles (subgenus *Nanger*) do not use their preorbital glands, and the glands of adult male Grant's gazelles that I have handled appeared undeveloped.

DISTRIBUTION. Most of the African gazelles live on the perimeter of, and a few live inside, the Sahara Desert. The Somali Arid Zone is the center of distribution, as it is for the related dwarf antelopes: 7 of the 12 African Antilopini occur there, of which 6 are endemic. Only the gerenuk and Grant's and Thomson's gazelles range south of the equator, penetrating to central Tanzania via the southern salient of the Somali-Masai Arid Zone. Gazelles lived in southern Africa during the Pleistocene, but the only surviving representative of the tribe is the springbok, which ranges throughout the South West Arid Zone.

ANCESTRY. The Antilopini are an ancient group that branched off from neotragine ancestors in the Miocene: the remains of a *Gazella* found at Ft. Ternan, Kenya, have been dated at 14 million years (7). The tribe either originated in Africa, or still-unknown Eurasian ancestors immigrated at an early date. Existing Asian and African lineages are distinctively different; for instance, females are hornless in 5 of the 6 Asian species (10). The only species that occurs both in Africa and Asia (the Arabian Peninsula) is the dorcas gazelle (7).

ECOLOGY. Adaptations for arid conditions enable the Antilopini to exploit food resources beyond the range of water-dependent herbivores and too meager to support larger competitors such as addax and oryx. Among these adaptations are the following.

• *Mechanisms for conserving body water* (discussed in chap. 2). Labile body temperature, evaporative cooling by nasal panting in extreme heat, concentrated urine and dry feces, reflective coat, and hairy muffle (17).

• *Feeding adaptations.* Narrow muzzle and incisor row, mobile lips, and digestive system all designed for highly selective feeding on protein-rich foliage and herbage; standing bipedally to increase vertical feeding range (especially gerenuk); feeding at night or early morning when vegetation has highest water content, and remaining inactive in the shade during the hottest times of day (9, 17).

• *Structural adaptations for long-distance travel and rapid flight.* Long limbs (especially the lower, medapodial, limbs), level back, and various less obvious specializations.

• *Nomadism and seasonal migrations.* Nomadism enables gazelles to exploit ephemeral, localized plant growth that flourishes for up to a year after a rare heavy thunderstorm in otherwise barren wastelands. Where bushes and trees provide a reliable food supply, gazelles may be resident in small, more or less localized herds. Gazelles also move seasonally into subdesert and even desert to eat the highly nutritious herbage that grows during the brief rainy season, and back to better-watered savanna in the prolonged dry season. Thomson's, Grant's, Soemmering's, and dama gazelles are all known to be or to have been migratory in the past, aggregating and moving in herds of hundreds or thousands of animals. Nomadic populations are more likely to be dispersed, but may also concentrate on greenflush produced by localized rainfall. The same species and even the same individuals may, of course, be nomadic, migratory, and resident in turn, depending on the distribution and condition of their food supply.

SEPARATION AND ASSOCIATION OF GAZELLES. Table 5.1 compares gazelle ecology and dispersion patterns. As none of the sub-Saharan species has been closely studied in the wild, the available information is very limited; accordingly, observations of the dorcas gazelle and mountain gazelle (*G. gazella*) in Israel have been included. The most and least desert-adapted *Gazella* species have been placed at the beginning and end of the list, but the ranking of those in between is only provisional. Different tolerances effect a measure of spatial and temporal separation among geographically associated species, as clearly seen in the gazelles that live around the Sahara, where *leptoceros, dorcas, dama,* and *rufifrons* make a series to the south, and *leptoceros, dorcas,* and *cuvieri* to the north, of most to least desert-adapted species (8). *G. leptoceros* penetrates furthest into the Sahara, the dorcas gazelle less deeply (9). The dama gazelle may only enter the Sahara during migrations in the rainy part of the year, whereas *G. rufifrons,* which shares with *thomsonii* the distinction of being most water-dependent, ranges between the Northern Savanna and the Sudanese Arid Zone.

Among sympatric species, several big and small gazelles are associated in the same manner as Grant's and Thomson's gazelles in East Africa: *dama* with *rufifrons* or *dorcas, soemmeringii* with *dorcas* or *spekei.* Regarding the last pair, Soemmering's tolerates taller, denser vegetation than Speke's gazelle. Accordingly, the 2 associate mainly on open plains—which is also the case with Grant's and Thomson's gazelles. The springbok, in contrast, has all to itself the kind of country that is partitioned among the Antilopini of eastern and northern Africa. As might be expected, it occupies a broader ecological niche and may reach comparatively high population density. It also comes in 2 different sizes (though not in the same place), as though attempting to fill the niches of small and large gazelles (see species account).

Table 5.1 Socioecological Comparison of Some African Gazelles

Species	Environmental tolerance: Desert Subdesert Arid Semiarid	Food and habitat preferences	Social organization: Dispersion pattern	Social organization: Group size	Social organization: Bisexual aggregations	Source
G. *leptoceros*	+++++++?	Stony plateaus and sand dunes	Strongly nomadic	Small	?	4, 7
dorcas	++++++++++++	Flat, stony open plains, sometimes sand dunes, associated with acacias. A browser, especially on *Acacia* and other leaves. Reported to eat locusts.	Nomadic, may migrate	Small: 3–40 (up to 60); bachelor herds up to 50	?	3, 4, 9, 11, 13, 25
dama	++++++++++++?	Browser/grazer, acacias, desert shrubs and grasses; rocky plains and dunes (open country)	Nomadic/migratory Sahara–Sahelian	Small herds up to 15	up to 600	4
soemmeringii	++++++++++?	Browser/grazer(?), thornbush, also open, grassy plains in hilly country	Migrates N–S	Seasonally variable: 27 (6–40) wet season; 3.9 (1–9) dry season	100–200 in rainy season	4, 21
granti	+++++++++++++	Browser/grazer, thornbush, open grass plains, montane grasslands; 2500 m–sea level	Migratory, seasonally resident–nomadic	$\overline{X} = \pm 6$ (2–46); bachelor herds up to 37	up to 400	21
spekei	?+++++++++	Browser(?), barren open plains on plateau 1000–2000 m, also coastal zone of NE Somalia	?	5–12; up to 20	?	4
cuvieri	?++++++++++?	Barren mountains up to 2000 m	Mostly resident(?)	4–5	?	4, 7, 10
thomsonii	++++++++++	Short-grass plains and savanna; mixed feeder, browses more in dry season, grazes when grass green and short. Rarely occurs below 1000 m.	Migratory, from short-grass plains in rains to wooded savanna in dry season; also resident	Small–large	thousands	5, 25
rufifrons	+++++++++++	Savanna grasslands to arid zone in rains, prefers open country, browser(?), feeding mainly on thorny shrubs and trees (esp. *Acacia* spp.)	Migratory	5–6, not over 15	?	4
Antidorcas marsupialis	++++++++++++++++	Karroo, Highveld and Sandveld open country, firm footing (avoids sand dunes), high to low elevation. Mixed feeder, mainly grass during rains and dicotyledons in dry season	Migratory/nomadic	Highly variable, small–large; bachelor herds of 2–50 or more	thousands; also mixed small herds	2, 16

SOCIAL ORGANIZATION: gregarious, polygynous, males territorial, sexes typically segregated in resident phase, but less so in open habitat and more or less integrated in migratory phase.

All species are gregarious and territorial at least while breeding. Females are not bound to any particular territory or male. The following social groupings are characteristic: territorial males; bachelor (all-male) herds; females with or without young, may be accompanied by a territorial male; solitary females (mothers guarding concealed young); maternity groups (associations of mothers guarding offspring); mixed herds (25).

Least gregarious and most sedentary are the gerenuk and dibatag, which live in thornbush habitats singly or in groups of 2 or 3. Young males associate in groups, but adult males do not.

Small herds are typical of the tribe as a whole, reflecting the dispersed resources in arid biomes (see table 5.1). Group size varies within species according to habitat and population density: invariably, the largest groups occur on the open plains. Thus, hundreds of Grant's gazelles join mixed herds on the Serengeti Plains, whereas herds average less than 10 in the woodland zone (22).

Since some gazelles are not known to aggregate, it is uncertain to what extent their social organization can be adjusted to meet variable environmental conditions. However, it is significant that those recorded only in small groups, notably *cuvieri, spekei,* and *leptoceros,* are also among the rarest and most endangered species. Long-continued persecution has drastically reduced their population density and probably transformed their distribution patterns. Evidence of this is found in Israel, where mountain gazelles were distributed mainly in ones and twos after years of relentless hunting. With protection they began to increase, expand their range, and form larger herds (13).

Aggregations of gazelles in the hundreds must have been commonplace in the past at normal population densities. But concentrations of many thousands have only been reported for the Thomson's gazelle and springbok among African species. On the basis of population density, ecological dominance, and tendency to aggregate, these 2 species may be the most gregarious and successful of the African Antilopini.

Very little is known about antilopine social relations. Even in the 2 African gazelles that have been studied, extended observations of groups of known individuals have yet to be made (25). However, it appears that gazelle society is generally open, with herds freely exchanging members and crossing territorial borders. Thus, there is remarkably little antagonism—or attachment—between female Thomson's gazelles, suggesting there may be no regular social hierarchy. A rank hierarchy based on age exists in bachelor herds, which, however, are also open and of varying composition. Beginning at puberty, males but not females are expelled from female herds by territorial males.

ACTIVITY. No species is primarily nocturnal, but probably all move and forage to some extent at night, and are most active early and late in the day.

POSTURES AND LOCOMOTION. Gazelles walk in an ambling gait, and trot in a diagonal stride (19). Trotting is a preferred gait in the gerenuk and dibatag, but comparatively rare in most other species. Gazelles are among the fleetest of all mammals, reaching maximum speeds of 82 kph (50 mph) (6, 15, 20). The big gazelles have proportionally longer legs but smaller gazelles compensate by making quicker strides; they look like hares, taking long bounds in the air with back arched and all 4 legs fully extended, landing with back flexed and hindlegs ahead of the forelegs. The Antilopini can make horizontal leaps of up to 10 m but are not high-jumpers (19). Thus, the springbok and mountain gazelle are adroit at slipping under and between strands of barbed-wire fences but do not jump over them. *Stotting* is particularly common in the Antilopini (fig. 2.2).

SOCIAL BEHAVIOR

COMMUNICATION. Like other open-country antelopes, gazelles appear to be visually oriented, relying less on smell and least on hearing—at least for detection of danger (13, 20). However, the variety of scent glands and the prominence of scent-marking in territorial behavior indicate that olfactory communication is important, whereas sound plays a role in alarm signaling, male herding, and courtship. The bulbous enlargement of the nasal chamber in the Speke's and to a lesser extent the dorcas gazelle amplifies their alarm snorts (fig. 5.1).

Social Grooming. Mutual grooming is comparatively unimportant in the Antilopini and even rare in some species (19). Licking and nibbling are the commonest

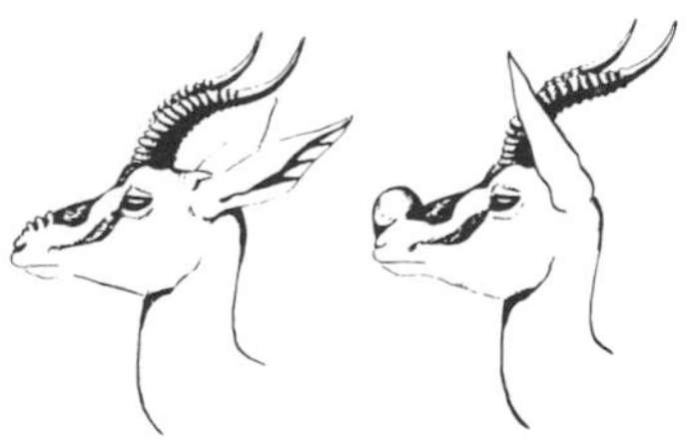

Fig. 5.1. The inflatable nose of Speke's gazelle (from Walther 1958).

forms of social grooming, while incisor-scraping and rubbing with the muzzle are rare. Attention is concentrated on the head. Licking of the hindquarters appears primarily in a sexual or maternal context. Reciprocal grooming is most likely to follow *nose-to-nose greeting* between 2 individuals. Grooming is solicited by proffering the desired part, and intention to groom by stretching the head forward and making nibbling movements (which may also serve an appeasing function). *Reciprocal grooming* occurs most often between mothers and their older fawns.

TERRITORIAL BEHAVIOR. Territorial size, demarcation, pugnacity, and occupancy vary widely within the tribe, and also intraspecifically according to environmental conditions and physiological state. Spacing between territorial Thomson's gazelles, for example, may be as close as 200–300 m, whereas Grant's gazelles defend areas of up to 3 km^2 and gerenuk males up to 4 km^2 (5, 12, 21). The territories of these larger species, furthermore, are often isolated, whereas "tommy" territories are usually part of a territorial network in which each buck has several neighbors. The proximity of several neighbors and a higher population density could account for the fact that Thomson's gazelle is more actively territorial than the other 2 species.

Individual differences in territorial activity may often reflect different phases of a regular territorial cycle, as demonstrated in studies of Grant's and Thomson's gazelles (21, 22, 23). Males of this tribe probably do not hold the same territory through their adult lifetime, as antelopes resident in more equable climates may often do. Rather, their territorial and nonterritorial periods seem to alternate with the seasons. In any event, males are most actively and aggressively territorial in the early stages of territorial establishment, and lose vigor toward the end (23).

Advertising: *preorbital-gland marking, linked urination/defecation, static-optic advertising,* calling, object-*horning.*

Urination followed by defecation on the same spot (preferably bare ground or a regular dung midden), in distinctive postures that vary little between species, advertises territorial status in all the Antilopini (fig. 5.2). In addition, species that produce a black, sticky preorbital secretion deposit it on grass stems and twigs in the same manner as the dwarf antelopes. Females do not mark in either way or display other signs of territoriality. Territorial males also advertise their presence by keeping at a distance from females and standing or lying alone for long periods. Grant's gazelle and springbok males habitually advertise their presence and dominance by *vegetation-horning. Herding* and *chasing* behavior also serve to identify territorial males. Bucks engaged in these activities emit sputtering, quacking, or snoring noises through their noses; rutting springboks are the loudest of all Antilopini (25).

Fig. 5.2. The antilopine *linked urination/defecation display* performed by a springbok.

AGONISTIC BEHAVIOR

Dominance/Threat Displays: *erect posture* ± raised nose, *head turned away,* and *lateral presentation, medial-horn* and *high-horn presentation* ± nodding motions, *head-shaking, vegetation-horning,* direct approach, *pursuit march.*

Defensive/Submissive Displays: *head-low/chin-out posture, turning away, lying-out.*

The repertoire of dominance and threat displays differs from species to species. The springbok has the least elaborate threat and dominance displays in the whole tribe, while the most elaborate dominance displays are seen in Grant's gazelle (see species accounts) (25). The *erect posture,* with or without the listed modifications, is the commonest dominance display in the whole tribe. *Lateral presentation* is next most frequent, typically combined with the *erect posture.* The commonest threat display is *medial-horn presentation.* The *pursuit march,* the mildest form of chasing, typically involves a dominant male walking and displaying behind a retreating subordinate male.

In mixed and bachelor herds *contagious aggression* may often be seen as a herd changes from resting to moving and feeding activity: adult males who are already up display to those who are still lying, and once afoot the latter do likewise. Aggression in this situation helps to coordinate activity (23, 25).

Antilopini seldom show greater submissiveness than lowering the neck with chin out, an intention movement to lie down (extreme submission) which is sometimes carried a step further by going to the knees (25).

Displacement Activities ("excitement activities" of [24]): *incisor-grooming* of trunk and limbs, *scratching neck and head with a hindfoot, scratching shoulders or back with horns, shaking all over, tail-wagging, stamping, grazing.*

Different species emphasize different movements, reflecting their preferences for the same patterns performed in normal maintenance activities. Grant's gazelles, for instance, scratch between their horns or front legs much more often than Thomson's gazelles during aggressive encounters. In the case of *displacement* or *excitation grazing,* it seems to signify uncertainty or even inferiority in *G. granti,* whereas it is an assertive, even aggressive action in *thomsonii* that features in territorial border encounters (fig. 5.8). For most of the tribe, grazing during an aggressive interaction may indicate waning aggressiveness and therefore be interpreted as a peace offer (19).

Fighting. Fighting behavior has been exhaustively studied in Thomson's and Grant's gazelles, and compared with that of other Antilopini (23, 25). The following list is arranged in order of increasing intensity (gazelle fighting styles are described in chap. 2; see fig. 2.3).

Boxing. Mild *nod-butting* while confronting in *medial-horn presentation,* common in females and young.

Horning. Low-intensity sparring with no intensive neck movements. Quite frequent in immature males, occasional in adult bachelors, unknown in territorial males.

Clash-fighting. Predominant in fights between territorial male Antilopini (see Thomson's gazelle account).

Front-pressing (or forehead-pressing) (fig. 5.3). Present in all Antilopini, is very common in Grant's gazelle but uncommon in *thomsonii* and probably other small gazelles.

Fig.5.3. Springbok fighting technique: *front-pressing* with levering and twisting.

Fight-circling. Violent pulling and jerking that occurs when combatants have trouble unlocking their horns, leading to broken horns and even necks (see fig. 5.14).

Air-cushion fighting.

Fights last anywhere from a few seconds up to 5 minutes. Few last up to 10 minutes, but contests have been known to continue intermittently for half a day (24).

REPRODUCTION. Strict breeding seasons are the exception (but see springbok account) in tropical Antilopini. However, there are usually 1 or 2 birth peaks in the year, the main one in the rains and a secondary one about half a year later. Gestation is 5½–6½ months, depending on body size. Twins are common in a couple of Asian species but unknown in African Antilopini. The minimum age for conception is about 1 year in small gazelles; males mature at 2 years, a year later in the largest species. Postpartum estrus within a week or 2 of calving is usual, making it possible for well-fed females to reproduce twice a year or 3 times in 2 years. Estrus lasts ½–1 day and recurs at 2-to-3-week intervals until conception.

SEXUAL BEHAVIOR: *lowstretch* ("neck-stretch" [25]), *urine-testing,* dominance displays, *foreleg-lifting, nose-lifting, prancing, goose-stepping, sputtering.*

Comparison of male courtship displays in different Antilopini shows that the main differences between species lie in the presence or absence of certain components, and in the frequency, emphasis, and elaboration of the shared basic repertory (cf. species accounts). For instance, male Thomson's gazelles and other small gazelles chase and herd individual females (never groups) far more actively than most of the larger Antilopini (e.g., Grant's gazelle and gerenuk). The much greater importance of *lowstretch* in the courtship behavior of small gazelles reflects this difference. There are also minor differences in the way different species perform the same displays. For instance, male Grant's gazelles hold the tail out during the mating march, whereas male Thomson's gazelles do not (5). Springbok males curl their tails over their backs; dorcas gazelle and springbok males hold their ears out at the beginning of courtship; more often the ears are held back by both sexes (25).

The progress of courtship from initial approach to copulation proceeds in stages that are more or less clearly discernible (25).

1. First contact and testing phase. The male approaches and displays (*lowstretch, erect posture,* or *nose-forward-upward*) until the female urinates, then tests the urine (mouth opened but no lip curl)(26). If the test is positive, he proceeds to the next phase.

2. Demonstrative-driving phase. The male follows the female in an *erect posture* closely resembling the dominance display. In some species (dama and Soemmering's gazelle, possibly dibatag), the male simply holds his neck erect and head level (fig. 5.4); in others, the male elevates his chin, often while *goose-stepping* or *prancing.* The goose step may represent a modified partial *foreleg-lifting.* The movements are much more pronounced in Thomson's gazelle than in the springbok, and *granti* males rely almost solely on this one display. Sputtering noises uttered through the nose are often made during this phase. Dorcas gazelle males produce a growling/snoring noise by inflating their nose pouch.

3. Mating march and premounting phase. As the female becomes more receptive to the male's advances, her efforts to escape

Fig. 5.4. Dama gazelle courtship: male following female in *erect posture.* (From a photo in Mungall 1980.)

Fig. 5.5. Antilopine copulation (Thomson's gazelle): the male has to walk bipedally behind the moving female, without clasping or leaning against her.

(including circling, sharp turns, running, sometimes defensive horn threat or poking, but mainly just walking away) slacken, and so do the male's courtship displays. The couple now walks in tandem, the male often in a normal posture. But if the female stands still, the male resumes displaying until he gets the female moving again. Thomson's gazelle performs only high *foreleg-lifting* at this stage. Antilopini rarely lick or nose the female's genitalia, in contrast to the related dwarf antelopes.

4. Mounting phase and copulation. Copulation follows a series of preliminary mounts, typically 10–20, not infrequently 30–40 (range 4–164). The female carries head and neck raised and unlike all other antelopes (except the oribi) either keeps moving or walks forward the moment the male starts to mount. Standing bolt upright, with bent forelegs dangling, the male follows bipedally and eventually achieves intromission, without either clasping or resting upon the female (fig. 5.5). Courtship ends immediately after the ejaculatory thrust, without any particular postcopulatory behavior, and does not resume for at least a half hour, when the whole process may begin over again. A mating sequence lasts 5–45 minutes, provided there are no major interruptions; a maximum of 6 copulations by a male have been recorded in a day (25).

PARENT/OFFSPRING BEHAVIOR.

Antilopini young are typical *hiders* which rely on concealment to escape predators for up to 6 weeks (*hider* strategy is described in chap. 2). Descriptions of birth and behavior of mother and young are given in the species accounts.

ANTIPREDATOR BEHAVIOR: *alert posture, snorting* and *stamping, flank twitching, stotting* (fig. 2.2), *alarm trot* (gerenuk and dibatag only), *cooperative maternal defense.*

At the moment of taking flight, gazelles shake or twitch the skin of their flanks. In species with a side stripe, the up-and-down movement produces a semaphore effect. The only other antelope outside the Antilopini in which this behavior (*flight and flicker*) has been observed is the dik-dik (24). *Cooperative defense* of the young, in which 2 and sometimes 3 or 4 females team up to drive away small predators (typically jackals) that have caught or are seeking concealed fawns, is equally unusual among antelopes. The identity of the females who assist the mother remains an open question (see gazelle accounts).

Although gazelles are faster than any predator but the cheetah, they lack the endurance of wild dogs or hyenas and can be run to exhaustion within 2–3 miles (17). Against the cheetah, their strategy is to turn sharply when overtaken; quick turns take so much out of a sprinting cheetah that it can no longer overtake the quarry.

References

1. Ansell 1971. 2. Bigalke 1972. 3. Carlisle and Ghobrial 1968. 4. Dorst and Dandelot 1970. 5. Estes 1967*a*. 6. Gambaryan 1974. 7. Gentry 1970. 8. ——— 1971. 9. Ghobrial 1974. 10. Groves 1969. 11. Harrison 1968. 12. Leuthold 1971*a*. 13. Mendelssohn 1974. 14. Pocock 1910. 15. Schaller 1972*b*. 16. Smithers 1983. 17. Taylor 1970*a,b*. 18. Taylor and Lyman 1972. 19. Walther 1968. 20. ——— 1969. 21. ——— 1972*b*. 22. ——— 1972*c*. 23. ——— 1978*a*. 24. ——— 1984. 25. Walther, Mungall, Grau 1983.

Added Reference

26. Hart and Hart 1987.

Fig. 5.6. *Lowstretch* posture of a herding Thomson's gazelle.

Thomson's Gazelle

Gazella thomsonii

TRAITS. The smaller and more colorful of the 2 East African gazelles, also the largest and most southerly of the small gazelles. *Height and weight:* 62 cm (58–70), males 17–29 kg, females 13–23.5 kg (7). *Horns:* S-shaped in male, 25–43 cm, strongly ridged and nearly parallel; 8–15 cm in female, pencil-thin, commonly malformed, broken, or missing. *Coloration:* cinnamon brown with prominent dark side stripe and facial markings, a short, dark, bushy tail and white rump patch; sexes colored alike, newborn darker with indistinct markings. *Scent glands:* preorbitals prominent in adult males; shin, hoof, and inguinal glands. *Mammae:* 2.

DISTRIBUTION. Somali-Masai Arid Zone and adjacent dry savanna. Largely restricted to the high plains and acacia savanna of East Africa, though occasionally found down to 600 m.

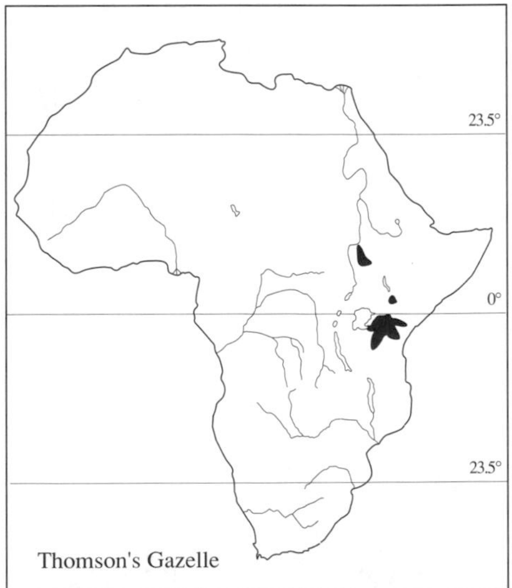

RELATIVES. A separate population, *G. t. albonotata,* occurs in the Sudan and was formerly treated as a separate species. Females of this subspecies have normal horns, whereas females of the East African race appear to be losing theirs (2). The red-fronted gazelle inhabits somewhat drier but otherwise similar habitats across the Northern Savanna and Sudanese Arid Zone.

ECOLOGY. The "tommy," as it is familiarly known in East Africa, has more narrow habitat preferences than most of the abundant species of plains game. It is largely confined to short grassland with dry, firm footing, though it enters tall grassland and fairly densely wooded habitats during migration, and to drink, feed on new grass, or avoid large wildebeest concentrations. Along with the wildebeest and zebra, the tommy has been highly successful at exploiting the seasonal productivity of semiarid savanna, dispersing over naturally short grasslands like the Eastern Serengeti Plain during the rains and migrating to areas of higher rainfall during the dry season. Here it finds water, the tall grassland is reduced by larger grazers (8), and fires and irregular thunderstorms create areas of greenflush (1).

In terms of population density and biomass, the Thomson's gazelle is clearly one of the most successful members of its tribe, though within a comparatively limited geographical range. It has accomplished this by becoming an intermediate feeder that grazes as long as grasses are green and tender, supplementing its diet with green browse (foliage and pods of leguminous subshrubs and bushes, forbs, clovers) to an increasing degree when grazing deteriorates (6). Its teeth (short premolar, long molar row) and convertible digestive system indicate the tommy is the most specialized for grazing of the African Antilopini (5). It is also the most water-dependent with the possible exception of *G. rufifrons*. In the Serengeti ecosystem, the great bulk of the tommy population spends the dry season on the edge of, or on islands of grassland within, the woodland zone and travels up to 16 km to water every other day or so (2, 9). Yet some (mainly males) remain on the parched, waterless eastern plains, so evidently this small gazelle also has the capability of water-independence. In fact *thomsonii* experimentally subjected to dehydration lost less water than *granti* did (10). However, when subjected to prolonged desertlike high temperatures, the tommies began panting—and thereby losing water through evaporative cooling—at a considerably lower body temperature than the *granti* (43° vs. 46°C).

Thus it appears that Thomson's gazelle has secondarily adapted to the savanna biome, where it has filled an available niche for a small, selective, migratory mixed feeder. It remains more drought-resistant than the other dominant savanna ungulates, most of which are pure grazers and drink regularly even when pastures are green. As a member of the Serengeti grazing succession, along with the zebra and wildebeest, the tommy comes last, being able to find enough high-quality forage on pastures the others have abandoned. It can also utilize a greenflush before larger grazers. Thus the tommy is the first to arrive and the last to leave the short-grass plains (1, 9).

SOCIAL ORGANIZATION: gregarious, territorial, migratory.

Aggregations of many thousands of Thomson's gazelles are still to be seen on the Serengeti Plain, a reminder of how plains game once thronged East Africa's grasslands. All the different types of social groups may be represented after a big concentration has moved onto green pastures: females and young on the best range, contained within a network of territorial males; groupings of bachelor males relegated to more marginal areas; large mixed herds that include numbers of adult, presently nonterritorial males; and single females and small maternity bands keeping watch over hidden fawns.

Social groupings of Thomson's gazelles are very loose and open, especially during migrations and at high densities. Female herds are usually unstable, with continual exchange between those in adjacent territories, despite often determined efforts of the proprietors to prevent their departure. Indeed sexual segregation and discrete female herds are largely the creation of territorial males competing to round up females while weeding out bachelor males. This imposed social order breaks down when the aggregation decamps. Territorial males revert to bachelor status and the sexes mingle in a moving horde, although a tendency for animals of the same class to associate may be discerned even during migration.

Social Distance. Aggressive and affiliative behaviors are both rare between females, and mutual grooming is largely limited to mothers and offspring (14). Females typically maintain individual distances of 1–3 m while filing and resting; up to ⅓ may be spaced less than a meter apart while lying, but rarely make contact. Grazing females mostly stay more than 3 meters apart. Bachelor males invariably maintain a wider individual distance: nearly 60% remain 7–12 m apart while grazing, and whether moving or resting, rarely come within one meter without provoking an aggressive response (16). Spacing between territorial males, however, is close compared to most comparably sized and larger species; from 300 down to 100 m at high density; typical territories include 10–30 ha (3, 19). With rivals so near, a territorial male must remain ever alert to keep females in and encroaching males out of his bailiwick (see under Territorial Behavior). But territories may be as small as 2.5 or as large as 200 ha, and are of irregular shape (19).

Small populations or subpopulations may be resident and dispersed, for at least part of the year. Several hundred Thomson's gazelles were studied on a small plain in the Serengeti woodland zone while they remained resident for 3 months (15). The females circulated within definite home ranges of 1–3 km diameter in separate herds. However, their home ranges overlapped and when herds met they combined, exchanged members, and often fragmented into subgroups. One large herd varied in numbers from 50 to 200 head and ranged over an area of about 10 km^2, encompassing the territories of some 24 males, sharing this range with 3 other female herds and 1 bachelor herd. The large herd met and mingled with the bachelor herd and with another female herd almost daily in late afternoon. In Ngorongoro Crater female herds averaged 23 animals, compared to an average of 13 males in bachelor herds (3).

ACTIVITY. Round-the-clock observations of Thomson's gazelle activity in May, at the end of the rainy season, revealed that adults spent a little less than half the time lying down, in bouts lasting from ½–5 h maximum (15). A healthy animal slept ½–1 h per day, dozing with eyes closed for ½–5 minutes at a time.

Resting. There were 4 long, well-defined lying periods interspersed between bouts of feeding: from 0900–1030, 1345–1700, 1930–2300, and 0330–0600 h. During the morning rest period tommies often sunbathe; all the gazelles in sight may be lying facing in the same direction, with hindquarters or flanks oriented toward the sun (15). Activity resumes in a gradual transition from resting to moving and grazing. The afternoon rest period coincides with the hottest part of the day, when gazelles seek any available shade, until cooling temperatures signal the late-afternoon feeding peak. The beginning of the evening rest period coincides closely with nightfall. The percentage of lying animals goes from 0% to 80% within ¼–½ hour, whereas the period ends raggedly and variably, depending on how many animals get up and feed before midnight. The final rest period is equally ragged at the start but ends abruptly shortly after dawn. Rest periods at night tend to be more unanimous and protracted than by day, with all the gazelles in sight lying down, especially in the predawn period. Up to 30% of the animals continued resting even during foraging hours, compared to less than 10% during diurnal activity peaks.

General movements tend to occur before and after feeding, and during midday, late-afternoon, and midnight activity peaks. The incidence of aggression, running games, and sexual behavior is highest during peaks of moving and grazing and lowest during rest periods. Play and aggression peak early and late in the day, sex in early morning, midday, and midnight (15).

POSTURES AND LOCOMOTION. See tribal account. The constant pendulum movement of this gazelle's tail is noticed by many people—and undoubtedly by other Thomson's gazelles. The sweep of the black tail across the white rump patch is probably a species-recognition signal that operates most effectively at some distance, reiterating and amplifying the message: "Here is a Thomson's gazelle." The fact that Grant's gazelle has a white tail may not be a coincidence (3).

SOCIAL BEHAVIOR

COMMUNICATION. Discussed in tribal account.

TERRITORIAL BEHAVIOR. Territorial males make themselves highly conspicuous through their energetic *herding* and *chasing*. A herding buck moves like a sheepdog as he tries to maneuver a female away from the border and toward the center of his territory (fig. 5.6). If a female tries to escape, he tries to head her off, leading to high-speed chases, during which he often makes sputtering calls; he utters the same calls while chasing male rivals. A male often succeeds in stopping or forestalling a female's escape, but cannot control groups of females once they decide to go.

Scent-Marking: *preorbital-gland marking, linked urination/defecation.*

Males regularly patrol and mark their territories in a purposeful way, pausing every few meters or so to renew or leave new preorbital secretions on the tips of grass stems or twigs (fig. 5.7). Every so often a buck stops to urinate and defecate in se-

Fig. 5.7. Thomson's gazelle depositing preorbital secretion.

quence on a dung pile, sometimes after preliminary pawing. Walther (19) mapped 110 secretion sites and 18 dung middens in an 8-ha territory. Most of the scent marks and all but 1 midden, located near the territorial center, were placed along and helped define the 310-m territorial perimeter. The owner had several regular resting places, one centrally located and the others near the border. Scent marks were concentrated in a broad belt along all border areas that were shared with neighbors, but formed only a thin line along an open section. The border was precisely defined to within a meter in the immediate vicinity of marking sites and landmarks, but only as a vaguer 2–10 m zone between marks.

Investigation of other territories confirmed that the above arrangements are typical of actively maintained tommy territories. At the beginning of a territorial period, it takes a male 1–2 weeks to establish a full marking system; after 3–5 months, the territorial drive often slackens, with a corresponding decline in demarcation and other territorial activities. Territorial occupancy for as long as a year appears to be unusual in this gazelle, at least in the Serengeti population (19).

AGONISTIC BEHAVIOR

Dominance/Threat Displays: *high-horn* and *medial-horn presentation,* head-ducking, *pursuit-march.*

Defensive/Submissive Displays: *head-low/chin-out posture.*

Displacement Activities: *grazing duels;* grooming (discussed in tribal introduction).

Fighting: *boxing, clash-fighting, air-cushion fighting.*

The aggressive behavior of Thomson's gazelle has been studied in great detail (18, 20). Females are rarely aggressive. The frequency of fighting increases in developing males, reaches a peak in subadults (1–2 years), and declines drastically among nonterritorial adults. Subadults resort to fighting to determine dominance, whereas encounters between adults are more ritualized. *Vegetation-horning* first appears in adolescence. *High-horn presentation,* a strong threat (to butt downward) in this species, appears and becomes highly ritualized only in later adolescence. *Grazing duels* and *chasing* without preliminary fighting are adult patterns, seen mainly in territorial males.

About half of the aggressive encounters of territorial males involve trespassing bachelor males, which typically respond to aggression by fleeing. The rest involve territorial neighbors, usually take place on the border, and are mostly limited to displays, with occasional brief skirmishes. Their main function is to "ratify" the territorial status quo. Apart from the listed aggressive displays, territorial encounters commonly include *grazing duels* (ritualized *displacement grazing*), *displacement grooming, marking* (especially after an encounter), and *air-cushion fighting,* and occasional fights of the types listed, above all *clash-fighting.* Grazing during territorial encounters is remarkably prominent in Thomson's gazelle, amounting to regular "duels" that may continue for up to ½ hour (fig. 5.8). It appears to be aggressive, as

Fig. 5.8. *Grazing duel* of Thomson's gazelles during territorial challenge ritual (from Walther 1978*d*).

though the animal was thereby asserting proprietary rights (space-claiming behavior), and during high-intensity encounters opponents may alternately fight and graze. The intensity of a grazing duel and the relative standing of the contestants can be judged by proximity and orientation to one another, by whether one retreats as the other advances, etc. A duel ends as both contestants gradually graze away from the border (16, 17, 18).

REPRODUCTION. Peak calving and mating occur during the rains, and a secondary peak follows about six months later. Gestation is 5½–6 months. (See tribal introduction.) Females typically come into estrus within 2 weeks of calving and may accordingly produce two young in just over a year.

SEXUAL BEHAVIOR: *lowstretch, urine-testing, erect posture* with muzzle raised (= "nose-forward-upward posture" of [21]); *nose-lifting* and *prancing* ("drum-roll"), *foreleg-lifting, sputtering.*

As mothers with hidden fawns often remain isolated, courting males run less risk of driving such females from their territory than would otherwise be the case (21). During the demonstrative-driving phase (explained in tribal introduction), the male follows the female at a distance of up to several meters in the *erect posture,* often punctuated by *nose-lifting* and *prancing* (fig. 5.9). Meanwhile he may utter the same sputtering noise given during chasing. He reverts to herding in *lowstretch* whenever the female attempts to escape. Her behavior throughout is characterized by efforts to avoid the male, starting at a run and ending in a walk (3). During this precopulatory phase, the male follows closely in lockstep, and when she stops, he raises a foreleg, causing her to move forward. When ready to copulate, the female holds her tail out; the male's tail hangs (cf. Grant's gazelle). Copulation occurs while the female is walking, as in other Antilopini (see tribal introduction).

Fig. 5.9. Thomson's gazelle courtship: male *nose-lifting* during the *mating march.*

PARENT/OFFSPRING BEHAVIOR. Preparatory to calving, a female seeks isolation and an area with ground cover, often ending up in a no-man's-land between occupied territories where the taller vegetation is avoided by other tommies. When cover is unavailable, females drop their offspring on bare ground. Fawns are surprisingly hard to spot while *lying-out,* nonetheless, and nearly odorless (see *hider* strategy in chap. 2) (13). When approaching to suckle her fawn, the mother stops about 20 m from the hiding place. Except for an hour or 2 a day spent in company with its mother, the calf spends its first 2 weeks lying still (12). During the next 8 weeks it spends progressively more time in feeding and social contacts. All this time the mother may remain apart from other females, but will associate with other mothers of concealed offspring or join a nearby herd as long as it remains within 100–300 m of her calf's hiding place. Fawns are weaned as early as 3 to as late as 6 months (2), but remain with their mothers, or at least in the same herd, until 8 months old or so, when young males are apt to be driven away by territorial bucks and join bachelor herds. Female offspring may continue to associate with their mothers as yearlings.

ANTIPREDATOR BEHAVIOR: *alert posture, approach* and *escort responses, alarm snorting, alarm circling* ("volte" of [13]), *flank twitching* (semaphore signal), *stotting.*

Because of its small size, Thomson's gazelle is vulnerable to a wide range of predators, from lions down to large vultures and eagles (newborn young), and its abundance makes it a primary resource for the larger carnivores. Wild dogs and cheetahs of the East African plains depend mainly on this antelope, while jackals are major predators of concealed fawns. Post concealment, a gazelle's survival depends on its ability to avoid capture. Although a tommy can run faster (up to 80 kph) than anything but a cheetah, it lacks the endurance of cursorial predators, begins to tire after 1–2 km, and becomes exhausted within 4–6 km, due at least partly to a 5–6°C rise in body temperature (11, 13). Wild dogs and hyenas, which keep their temperature down while running by panting, can usually run down a quarry they can keep in sight until its sprint is finished.

Thomson's gazelles maintain flight distances corresponding to the degree of danger that a predator is perceived to represent (13). This distance varies not only according to species but also to the number of predators and what they are doing; the time of day, weather, and terrain; the physical condition, sex, age, social status, and activity of the gazelle; the presence of other gazelles; and so on. The avoidance distance for predators such as jackals and baboons, which rarely bother adult gazelles, ranges from 5 to 50 m, whereas 50–300 m is the normal avoidance distance for lions or leopards, which tommies can readily outrun in a fair chase. It is over 300–1500 m for cheetahs, wild dogs, and hyenas hunting in packs (13). Gazelles in large herds tend to have shorter flight distances than those in small herds, as usual. Lone animals, however, which are usually females guarding fawns and territorial males, tend to have shorter flight distances and to circle back when pursued—habits that should make them more vulnerable to wild dogs and hyenas (4, 13).

The alerting and alarm signals given by Thomson's gazelle are standard for the tribe (see Antilopini introduction). In the above inventory, *approach* is the response of tommies (and other plains game) to a strange or suspicious object that cannot be identified from a distance. The animals come closer in the *alert posture*, ready for instant flight. Still stronger curiosity or "fascination" leads them to approach a known source of danger, such as a cheetah or lion. On the open plain, tommies often gather in a large mob to stare at and even "escort" a passing cat, while maintaining a distance of 20–50 m (author's observations). The intent in both cases may be simply to keep the predator in view; and since this behavior attracts the attention of every other gazelle in sight, it undoubtedly spoils the predator's chances of sneaking up on or ambushing prey. *Volte* is Walther's term to describe the way a tommy often walks back and forth in a small circle while intently watching a predator (13); it probably expresses conflicting tendencies to stay and to flee.

Stotting, by far the most conspicuous alerting signal, expresses excitation but may not convey any more specific message, being given in play as well as in alarm and during flight (fig. 2.2) (13). Nevertheless, by attracting the attention of other gazelles, it serves an important antipredator function. *Stotting* by young fawns signals their mothers that they have been flushed from hiding. *Stotting* is also a quite specific response to cursorial predators, especially wild dogs, and the contagion of this display as it spreads ahead of a hunting pack like a bow wave must help to alert more distant animals to the approaching danger (4).

References

1. Bell 1971. 2. Brooks 1961. 3. Estes 1967*a*.
4. Estes and Goddard 1967. 5. Gentry 1964.
6. Hofmann 1973. 7. Kingdon 1982.
8. McNaughton 1976. 9. Maddock 1979.
10. Taylor 1970*a*. 11. Taylor and Lyman 1972.
12. Walther 1968. 13.——— 1969.
14.——— 1972*c*. 15.——— 1973.
16.——— 1974. 17.——— 1977.
18.——— 1978*a*. 19.——— 1978*b*.
20.——— 1978*d*. 21. Walther, Mungall, Grau 1983.

Added Reference

Fitzgibbon 1989.

Fig. 5.10. Grant's gazelle *head-flagging* dominance display.

Grant's Gazelle

Gazella granti

TRAITS. A large, handsome gazelle, paler and with bigger horns than *G. thomsonii. Height and weight:* males 84–91 cm and 65 kg (58–81.5), females 75–83 cm and 45 kg (38–67) (6). *Horns:* long, thick, and heavily ringed in male, 50–80 cm, widest with most curvature in the western race (*G. g. robertsi*); females with thin but symmetrical horns, 30 cm up to 45 cm long in older animals (4). Erect bearing and long neck, very muscular in mature males. *Coloration:* pale to darker tan with extensive white areas, especially the lower trunk and inner legs, throat patch, a rump patch that extends over the tail onto the croup (the most reliable way to distinguish *granti* from *thomsonii*), the upper ⅔ of the tail, and facial markings; dark markings include thigh (pygal) stripes, the tassel and undersurface of the tail, the anogenital area and udder, facial blazes and patches over the eyes, and a side stripe, usually developed in the young, variably developed or lacking in adult females, and absent in adult males (except for an isolated Kenya race, *G. g. notata*) (4); young fawns darker than adults. *Scent glands:* preorbitals small and seemingly inactive; hoof and shin glands.

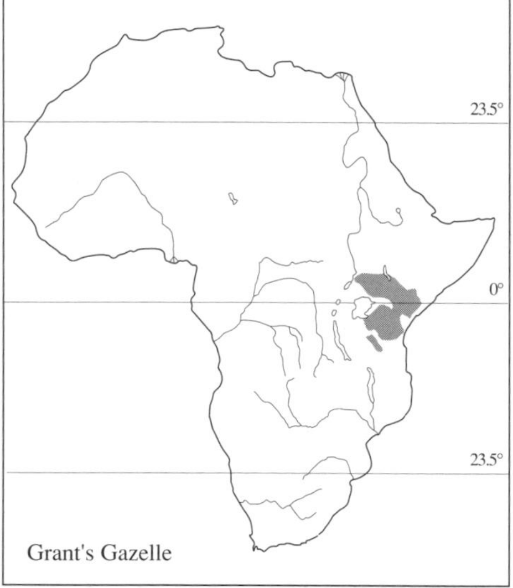

DISTRIBUTION. The Somali-Masai Arid Zone, from northern Tanzania to southern Sudan and Ethiopia, from the Kenya coast to Lake Victoria, in lowland thornbush, savanna woodland, open plains, and montane grassland up to 2500 m. In northern Kenya and Ethiopia it inhabits subdesert.

ECOLOGY. Grant's gazelle is a migratory animal but often moves in the reverse direction of the Thomson's gazelle, wildebeest, zebra, topi, and other water-dependent species. By occupying plains like the Eastern Serengeti and Masai Steppe during the dry season, this gazelle avoids competition and capitalizes on its ability to subsist on the vegetation that is available under waterless, semidesert conditions. Conversely, many Grant's gazelles reside on small plains in the woodland zone during the wet season while the main migratory populations are on the short-grass plains. However, probably the majority participate in the main Serengeti migration, moving out of the woodland zone onto the medium- and short-grass plains at the onset of the rainy season, and withdrawing as far as the edge of the woodlands when the dry season begins, like most Thomson's gazelles.

Thomson's and Grant's gazelles not only range the same plains but, unlike most other sympatric species, quite often mingle in the same herds. Mingling is most frequent when one or both species are present in small numbers; it is particularly common to find a male in a female herd of the other species. Apart from the difference in size, however, the 2 gazelles clearly differ in habitat preferences and diet. *Granti* tolerates both drier and more closed habitats and, although also a mixed feeder, it is more of a browser than a grazer (5), and rarely if ever drinks. Within Ngorongoro Crater differences in habitat preference are clearly seen during the rainy season, when most *granti* frequent the taller, bush-infested hill and floor grassland, while *thomsonii* is largely confined to short, alkaline grassland. Some Grant's gazelles move out of the Crater onto the surrounding montane grassland where tommies rarely venture (2).

SOCIAL ORGANIZATION: gregarious, territorial, migratory.

A study of the Serengeti *granti* population demonstrates that size and composition of social groups are correlated with habitat and also vary seasonally according to food availability and quality (11). The differences in dispersion pattern are illustrated in table 5.2. Most striking is the fact that nearly half of all the herds observed on the Serengeti Plain were mixed, compared to only 12% of the herds found on the patches of grassland in the woodland zone; there most herds (33.6%) were "harems" composed of females and young with one (territorial) male. Comparison of herd size shows even more strikingly that the largest herds occur on the open plains: ¼ of all the plains herds counted numbered over 40 animals, all these being mixed herds, of which the largest Walther ever counted numbered 428 head (11). The largest herd

Table 5.2 Comparison of Grant's Gazelle Grouping Patterns in the Woodland Zone and on Open Plains in Serengeti N.P.
(F = female; M = Male)

Group type	Mean size (range)	Open plain	Woodland zone
Lone M	—	14.0%	14.2%
Male herd	4.6 (2–37)	6.8	14.5
Lone F± calf	—	5.5	5.1
FF − M	6.1 (2–37)	8.1	20.4
FF + M	9.6 (2–32)	16.3	33.6
Mixed (FF + MM)	46.5 (3–428)	49.3	12.2
TOTAL		100.0%	100.0%

SOURCE. Excerpted from Walther 1972, tables 10 and 11.

counted in the woodland zone numbered 44 animals, or below the combined average for mixed herds. Otherwise, the average number of *granti* in the different types of groups was not significantly different in the woodland and plains (table 5.2).

Herds averaged larger in the open grasslands of Ngorongoro Crater than in Serengeti N.P.: 16.4 head in a sample of 100 "harems," and 10.3 animals in 15 bachelor herds (2). In the woodland savanna of Tsavo N.P., by contrast, female groups of Grant's gazelle averaged only 3.4 (maximum of 13), and mixed herds averaged 9 head (maximum of 29). Yet here too, groups of all kinds averaged larger in the most open habitat, except for bachelor herds, which were marginally larger (3.2 vs. 2.9) in wooded habitats (8).

Apparently, really large herds of Grant's gazelle occur only when males and females aggregate on open plains. This aggregation tendency may be explained as the product of 2 attractive forces: first, the ever-present social attraction to conspecifics, and second, attraction to landmarks in a largely featureless landscape (11). The animals clearly use rock outcrops, trees, bushes, and mounds for orientation on the plains, both during migration and while resident, and these same landmarks also serve as gathering points. Migrating groups and individuals will change course to approach members of their own species, and the attraction increases with the size of the group. As a result, very large herds tend to gather near isolated landmarks (11).

In the woodland zone, limited open space and the presence of visual obstructions inhibit the formation and limit the size of mixed groups. Furthermore, territorial males are more intolerant of nonterritorial males, which, consequently, are usually forced to stay in bachelor herds in less desirable, undefended areas. On the plains, males that establish territories around landmarks frequented by large mixed herds may try to drive out bachelors but are defeated by sheer numbers and have to be satisfied with the status of alpha male. The mixed herds circulate within huge ranges of up to 20 km^2, whereas territories range from 2.5–10 km^2. A given territorial male may therefore only enjoy dominion over the herd for a few hours before it moves onto his neighbor's property. When that happens, the male usually goes along, even though he reverts to bachelor status as soon as he leaves his own territory (11).

The differences in tolerance between territorial males in plains and woodland areas may also have something to do with the availability of reference points. Walther believes that males need landmarks and/or neighbors to define the boundaries and center of each territory (11). Perhaps because territories in the woodland zone are clearly defined, they are more strictly defended, and their owners only vacate in emergencies.

Seasonal Differences in Social Organization. During the rainy season, the gazelle population is split into the smallest units, territoriality reaches its peak, and breeding herds are largest and most stable. Mixed herds are at their peak in the dry season, when territoriality is at low ebb. The migrations that occur at the beginning and end of the dry season disrupt existing herds and associations.

Walther (11) found that breeding herds of a certain size could be very sedentary and stable in the woodland zone, remaining inside the territory of a single male for up to 8 months, or for as long as the territory was maintained. In short, these were real harems for the duration. Optimal herd size appears to be 10–25 head, which is somewhat above the Serengeti N.P. average of 9.6 (11). Below about 10 head, females tend to leave the herd and the territory to seek a larger herd, possibly for greater security. Above 20–25 head, bucks have trouble preventing small groups from leaving, making large harems less manageable and stable than smaller ones (11).

Except for harem herds all the other types of social groups are fully open, the members free to come and go. Even in harem herds, there is little evidence of continuing association between individuals (other than mothers and dependent offspring) once they cease to share the same home range.

Bachelor Herds. Over ½ the Serengeti herds in Walther's sample contained only males of the same age class, including ⅓ of all the adult males. The largest herds contained all 3 age classes: adults (2 years old), subadults (1–2 years), and adolescents (7 months–1 year). Bachelor herds without adult males are rare, apparently because young males seek the company of adults and perhaps vice versa: the younger male satisfies the older one's urge for dominance (11).

In fact, dominance interactions are the main social activity in *granti* bachelor herds, in contrast to the passive social relations of females. The existence of an age-graded hierarchy in bachelor herds is evinced by aggressive interactions and also by filing order: dominants insist on walking behind inferiors. Consequently, the youngest males come first in line and the oldest last (10). Males of the same age class compete for dominance by engaging in threat and dominance displays, sparring matches, and fights. In Ngorongoro tournaments occur regularly in late afternoon, especially when bachelor herds mingle temporarily with female herds (2).

Territorial Males. When a bachelor male matures at age 3 and is ready to become territorial, he either leaves the herd in quest of a suitable, undefended location, or—if sufficiently dominant—drives away his former companions and appropriates all or part of the herd's range (11). This is easier to do in a small than in a large all-male group or a mixed herd. Although a territory may be established and maintained without females, their presence or absence influences the period of occupation, and population density influences territorial density: an influx leads to an increase in the number of territorial males. In areas with perennially resident populations such as Ngorongoro and Tsavo N.P., it could well turn out (but remains unproven) that some males hold the same territories year after year (2, 8). In Serengeti N.P., where territorial behavior reaches a low ebb during the dry season, no males were known to remain continually on territory for a full year. However, several bucks occupied and reoccupied the same territories for at least 1.5 years (11).

ACTIVITY. The activity of this species has not been studied as closely as in Thomson's gazelle (see species account), but a similar space-time system appears to be typical of resident herds. One mixed herd made a regular daily circuit of its large (5-km diameter) home range, traveling practically as far as gazelles on migration. Followed from 0600 to 2330 h, it covered a good 10 km. The animals kept moving while feeding except for relatively brief interludes, between 0900 and 1100, 1200 and 1300, and around 1500 h, during which at least part of the herd rested. They took a long rest after dark, from 1930 to 2215 h, before beginning another feeding bout (11).

POSTURES AND LOCOMOTION. See tribal account.

SOCIAL BEHAVIOR

COMMUNICATION. See tribal account.

TERRITORIAL BEHAVIOR: *linked urination/defecation, static-optic advertising, vegetation-horning* ("weaving").

Grant's gazelles do not mark with their preorbital glands but maintain dung middens where they perform *linked urination/defecation* (sometimes preceded by pawing) to mark their boundaries and declare territorial status. Another dynamic display with more aggressive overtones, sweeping or "weaving" the horns rhythmically from side to side in a clump of grass or bush, is common in this species (10). Resting bucks also advertise their presence passively by stationing themselves some distance apart from their harems, often on a slope or mound where they can see and be seen to best advantage (*static-optic advertising*). Territorial neighbors do not engage in regular border rituals or skirmishes like Thomson's gazelles, but monitor one another's movements and seem to react more quickly and strongly to a neighbor's than to a stranger's approach to the border (2).

AGONISTIC BEHAVIOR. See tribal introduction for explanations of terms.

Dominance/Threat Displays: *erect posture* with *nose-lifting* and *head-flagging, lateral presentation, looking away, parallel marching, circling, angle-horn, medial-horn presentation, head-ducking, vegetation-horning, pursuit marching.*

Defensive/Submissive Displays: *head-low/chin-out posture; head-low posture* (female defensive threat).

Fighting: *horning, twist-fighting, horn-pressing, front-pressing, fight-circling.*

Dominance and threat displays have developed to a high degree in Grant's gazelle, whereas fighting is comparatively unritualized and often rigorous, involving a wide variety of fighting styles (cf. Thomson's gazelle). *Head-flagging* is the most distinctive display and unknown in other Antilopini (fig. 5.10) (12). It occurs as an elaboration of the *erect posture* with *nose-lifting,* as the performer abruptly turns his head from side to side, often while standing in *lateral presentation.* The basis of the display, in my view, is to advertise neck development, which is a conspicuous attribute of mature males; like a man flexing his biceps, a displaying *granti* buck makes his neck swell as much as possible (2). The horns, too, are shown to advantage, and as shape and size are diagnostic of different age classes, especially in the *robertsi* race (the horntips in mature bucks bend sharply down), horns could back up neck size in identifying fully mature males. In any case, adults commonly settle the issue of dominance by displaying rather than fighting, although sometimes only after long, intensive display duels lasting up to ¼ hour and including up to 20

Fig. 5.11. Young-adult males grooming simultaneously during aggressive interaction.

head-flagging episodes, along with all the other aggressive displays in the repertoire (12).

Displaying males often circle slowly in reverse parallel as though maneuvering for just the right angle from which to *head-flag,* or they march parallel, meanwhile tilting their horns inward (*angle-horn*). During interludes between aggressive displays, *displacement grooming* (fig. 5.11), *vegetation-horning,* and *linked urination/defecation* often emphasized by pawing are commonly performed. The more prolonged and intense the dominance and threat displays, the more likely that a fight will develop. The final prelude to combat is a close confrontation in combat readiness, in which the 2 bucks stand with forelegs spread in *medial-horn presentation.* They may duck their heads down and up.

Fights may also begin spontaneously, without preliminary displays (3). Those between subadults and adults are typically strenuous, involving *front-pressing* with horns bound together, plus *twisting, turning,* and *levering* movements. Combatants may trip over and be pricked by each other's horns, and necks may be twisted until stopped by a horn tip digging into the ground (3). Occasionally bucks are unable to unlock their horns, leading them to wheel about (*fight-circling*) while making frantic efforts to jerk free (2). A contestant who is bested in a fight runs away with the victor in hot pursuit.

REPRODUCTION. In southern Kenya and Tanzania Grant's gazelle may breed through the year (8, 11). December—February and August—September are distinct birth and mating peaks in the Serengeti region, divided by breeding minima in June—July and October—November. Based on zoo records, gestation is about 27 weeks (1). The minimum breeding age is about the same as for the much larger wildebeest: 1½ years for females and 3 years for males (age of social and physical maturity).

SEXUAL BEHAVIOR: *following* in *erect posture* with *nose-lifting* and *goose-stepping, lateral presentation, sputtering* call, female defensive and evasive tactics.

An absence of hard *chasing* and *driving* makes the herding and courting behavior of Grant's gazelle seem tame in comparison with Thomson's gazelle. *Granti* bucks rarely even approach females in *lowstretch,* although they manage to direct females' movements in less hectic ways, for instance by blocking the way out of the territory with a *lateral display.* Courtship consists almost exclusively of the *mating march,* up to the point of copulation (see tribal introduction). With head and tail elevated, the buck follows the female, often making the *sputtering* noise typical of the tribe. Occasionally he lifts his nose and at the same time *goose-steps,* particularly if the female stops while he is following closely. If she trots or runs away, or makes a sudden turn and evasive jump, a suitor usually desists for the moment and very rarely gives chase. Females sometimes resist a male with incipient threat displays, including the *head-low posture* and something similar to *head-flagging,* which often seems to stimulate a male to mount (12).

Female receptiveness is signaled by holding the tail out. Many mounts may precede successful copulation, after which the male drops at once to all fours. Both partners then lower their tails and all mating activity ceases for at least ½ hour. Males may lose interest so completely that they will tolerate nearby bachelors behaving sexually toward the female, or allow her to leave the territory without intervening (10).

PARENT/OFFSPRING BEHAVIOR. A detailed protocol has been published of a birth in Ngorongoro Crater (10). The main events:

1640 h The female, in high grass 50 m from a harem herd, white amniotic sac protruding, walks restlessly with tail out, stamping nervously and scanning surroundings, pausing occasionally to nose and lick her vulva, alternately lying down and getting up.

1645. She lies on side and labors, with head raised and facing the rear. When she stands the sac bursts and she licks the fluid from the ground.

1651. The fawn's head emerges while the

female stands. She lies down and gets up repeatedly for several minutes, finally lies on side and labors, with legs stretched straight out and elevated parallel to the ground, gets up within 2 minutes, down and up, looks around, licks the ground and walks in a circle.

1705 and 1716. More 2-minute labor bouts on side.

1718. The female stands; the fawn dangles by its hindlegs, then falls to the ground. The mother turns, begins licking it and eating embryonic membranes.

1721. The fawn raises its head.

1723. It tries to rise.

1737. It stands without falling.

1739. It finds the udder, helped into the correct orientation by the mother's anal licking; suckles intermittently for 15 minutes in bouts up to 90 seconds.

1743. The fawn now holds its tail out when licked (vertically after the first day).

1743. The mother grazes briefly.

1759. The fawn moves away, followed and occasionally licked by the mother.

1825. It lies down in a patch of high grass 50 m from birthplace.

When a mother comes (from as far as 300 m) to suckle her fawn, she summons it from 50–100 m by head-bobbing and/or bleating, after making a careful approach interrupted by frequent scanning. The fawn runs to her and, after they touch noses, it dives for the udder. During the 4–6 week concealment period, mothers who have young in the same area may join in mother herds of 2–10 females which frequent zones of higher vegetation between or adjacent to territories. As the fawn's tendency to hide in isolation weakens, it becomes more gregarious, associates in peer groups with other fawns, and accompanies its mother. The mother-fawn bond persists until adolescence, even though weaning occurs by 6 months.

ANTIPREDATOR BEHAVIOR: *alert posture, alarm snort* and *stamping, flank-shuddering, stotting.*

Grant's gazelles made up less than 6% of the kills of cheetahs, wild dogs, leopards, and lions recorded in Serengeti N.P., presumably because the most preferred prey species are far more numerous and perhaps easier to catch (9). By staying away from water holes *G. granti* also makes itself less vulnerable to ambush. Probably the most important predators of this antelope are black-backed and golden jackals, which prey almost exclusively on concealed fawns. A mother can defend her offspring quite effectively against 1 jackal but not against 2, and jackals tend to hunt in pairs during their breeding seasons, which happen to coincide with gazelle birth peaks (see jackal accounts). Cooperative defense by 2 or more *granti* gazelles evens the odds. In an observed episode, 1 of a pair of black-backs managed to catch and begin throttling a fawn they had flushed several times, while 2 defending females were both chasing the other jackal. However, 1 of them always came racing back in time, and the hunters finally gave up. The fawn apparently escaped serious injury (2).

References

1. Dittrich 1968. 2. Estes 1967*a*. 3.——— 1973*a*. 4. Haltenorth and Diller 1980. 5. Hofmann 1973. 6. Kingdon 1982. 7. Lamprey 1963. 8. Leuthold and Leuthold 1975. 9. Schaller 1972*b*. 10. Walther 1965*a*. 11.——— 1972*b*. 12. Walther, Mungall, Grau 1983.

Fig. 5.12. Springbok *pronking.*

Springbok
Antidorcas marsupialis

TRAITS. A gazelle with a dorsal skin fold (a marsupium, hence the name *marsupialis*) containing erectile white hair that is displayed by *stotting* in a special way (whence the name springbok). *Height and weight:* average of *A. m. marsupialis* (the smallest race) males 73 cm, 30.6 kg, females 69 cm, 26.7 kg (1); *A. m. hofmeyri* (largest race) 77–87 cm (8), males 41 kg (33–47.6), females 37 kg (30.4–43.5) (Botswana series) (13), up to 59 kg in Namibia (1). *Teeth:* 2 instead of the usual 3 lower premolars. *Horns:* hooked tips curved inward (like a stethoscope), 35–49 cm, stoutly ringed; much thinner, very much to somewhat shorter in females (see under Distribution). Ears long, narrow, pointed. *Coloration:* rich cinnamon-brown (paler in desert areas), with extensive white areas bordered by dark markings; head white (except in young juveniles) with a thin cheek stripe and forehead patch, white underparts and backs of legs, white tail with black tip, rump patch continues over tail into fan-shaped dorsal crest with erectile hair 10–12 cm long, concealed in skin fold except when displayed. *Scent glands:* a dorsal gland (yellow, sticky secretion) underlies the erectile crest according to (11); male's preorbital glands produce yellow, oily secretion; hoof glands in all feet, no shin or inguinal glands. *Mammae:* 2.

DISTRIBUTION. South West Arid Zone and adjacent dry savanna, from western Transvaal to the Atlantic, north to southern Angola and northern Botswana. Greatly reduced in South Africa in the last century, the springbok has been reintroduced throughout its range and beyond, but nearly all those south of the Botswana border live on fenced land without their usual predators (1, 14).

Subspecies. Of the three recognized subspecies, *A. m. hofmeyri* is much the largest, especially the desert form that occurs in Namibia's Kaokoveld (8). Females of this race and *angolensis* have stronger horns that may approach males' horns in length (1, 8).

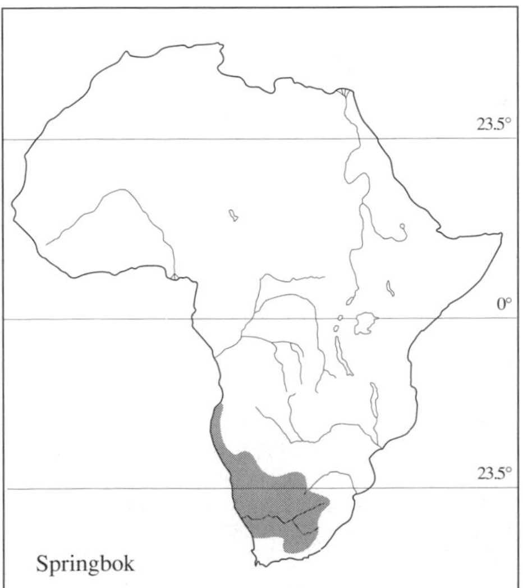

Springbok

ECOLOGY. The springbok is the most abundant plains antelope in the arid lands of southern Africa. It has a considerable habitat tolerance, from savanna with rainfall of up to 750 mm to the Namib Desert with 0–100 mm, and from the South African Highveld (2000 m) to sea level. However, it is absent from mountains and rocky hills and avoids woodland with tall, dense vegetation. In the Botswana mixed savanna it is partial to calcrete pans and fossil riverbeds where short grass and low karroid shrubs grow. It likes firm footing and does not follow the gemsbok onto the dunes of the southwestern Kalahari (1, 14).

DIET. A mixed grazer/browser, the springbok grazes as long as grasses are young and tender, then browses karroid vegetation, which includes a variety of low shrubs and succulents. After heavy rain it feeds on the ensuing lush growth of herbs and melons. In a Transvaal reserve, springboks fed on 68 different species, of which 9 grasses and 11 shrubs were the mainstay (15). Springboks near Kimberly browsed from June through August, and grazed from October through March (14). Though it drinks when water is available, and tolerates water with a mineral content too high for most other species, under experimental conditions the springbok can go indefinitely without drinking provided its feed contains at least 10% water (14).

ECOLOGICAL NICHE. As the only gazelle in the South West Arid Zone, the springbok may fill a broad niche that is partitioned between large and small gazelles of the Somali Arid Zone. The large desert form may be most comparable to Soemmering's gazelle (which incidentally has similarly shaped horns). Before South Africa's game was decimated by hide hunters, the smaller springbok, together with the black wildebeest and quagga, used to migrate between the Highveld and Karroo, just as Thomson's gazelle, white-bearded wildebeest, and plains zebra migrate today between the Eastern and Western Serengeti. Such an association still exists in parts of Botswana and Namibia's Etosha N.P. between the springbok, blue wildebeest, red hartebeest, and zebra.

TREKBOKKEN. In former times when millions of springboks inhabited the Kalahari and Karroo, vast numbers occasionally emigrated from the interior and invaded surrounding settled areas (5). Evidently these mass treks occurred during times of protracted drought, when springboks were forced outside their normal range. Many migrated southward toward the Orange River, and those that crossed eventually came into settled farm country. A green-flush brought on by an isolated thunderstorm could cause all the springboks and other game from a wide area to concentrate in one place; the invasion of Beaufort West (southern Cape Province) in 1849 is a famous example (5). A number of observers of trekbokken reported that the springboks disbanded and disappeared overnight as soon as rain began to fall in the interior.

Trekbokken still periodically occur in Botswana, though on a far smaller scale than in the past. Four eruptions involving upward of 15,000 springboks were recorded between 1946 and 1959, usually at the nominal end of the dry season (October, November) (3, 4, 7). Springboks from the Kalahari Gemsbok N.P. took a southwesterly course into Cape Province and onto farmland, where thousands were shot (4)—just like the good old days. Some got as far as 240 km south of the border. Prior to one trek an abnormal restlessness was noted as early as July and August, with herds of 100 to over 1000 springboks moving aimlessly at times when they normally fed or rested. Another characteristic of trekbokken and other huge game concentrations is the loss of normal wariness; trekking springboks marched right through towns and barely moved aside for passing vehicles (4, 6).

SOCIAL ORGANIZATION: gregarious, territorial, migratory.

The springbok has much the same kind of fluid, adaptable social organization as Thomson's gazelle, with the same types of herds: separate bachelor herds and herds of females and young during periods when males maintain a territorial network and exclude all sexually mature nonterritorial males, and mixed herds during migration and/or when springboks congregate on burns or break up into small parties in the dry season. Single females and small maternity groups separate from the "harem" herds during birth peaks. Nowadays aggregations of over 3000 springboks are considered large. Bachelor herds range up to 50 and rarely up to 300 head, and the average size of 14 mixed herds sampled one March in Kalahari Gemsbok N.P. was 58.6 (7–182) (2). In a fenced Transvaal reserve containing 55 springboks, the average herd included only 5.3 females, and ½ the herds included 1–4 females with 1 male, but 32 springboks were in the largest "harem," including 21 adult and subadult females (10).

The usual pattern of wet-season dispersal and dry-season concentration is reversed in the springbok: it tends to aggregate on short green pastures in the rains and to disperse in small groups during the dry season (14). But these wandering groups congregate when a localized storm creates a green-flush, and if the foliage of the bushes and trees that sustain them in the dry season withers and falls during protracted drought, springboks may evacuate their normal range en masse, as noted.

Territorial springboks are also comparable to Thomson's gazelle in energetic herding and chasing of females, especially during mating peaks, although their territories may be as large as those of Grant's gazelle. In the Transvaal reserve territories ranged from 25–70 ha, big enough to accommodate 3 (1 accommodated 7!) blesbok territories (10). In Kalahari Gemsbok N.P., springbok males were spaced at 1.2 km intervals along the lower Nossob River from the onset of the rut in April through the dry season, holding on to these choice locations even when there were no other springboks around (2). Although a male has been known to retain the same territory for at least a year (in the fenced reserve), this may be unusual.

Females probably do not remain for any extended period within an individual territory, but range far more widely (no home range data) and live in completely open societies. There is no regular social grooming between females. Daughters may remain attached to their mothers until calving as yearlings, but sons of 6 months and over may be relegated to bachelor herds, separation from the mother often coinciding with the birth of her next offspring.

ACTIVITY: unstudied; peaks early and late in the day, also night-active; rests in available shade in hot weather.

POSTURES AND LOCOMOTION. Fast as any gazelle, a galloping springbok has been clocked at 88 kph (3). Trotting is also an important gait in this species, including a fast, level gait and a high-stepping *style trot,* usually accompanied by head-tossing (an abrupt raising of the lowered head [16]). *Style-trotting* evidently represents a mildly excited state and is often the prelude to *pronking,* the unique springbok version of *stotting* (fig. 5.12), during which the back is bowed, the tail is clamped, the neck is lowered, and the straightened legs are bunched. Like *stotting, pronking* is a signal of medium excitation and is performed more often by fawns than by adults (1, 16). A displaying animal may bound and rebound like a ball half-a-dozen times, to a height of nearly 2 m. The dorsal crest is not automatically displayed; it must be erected by contracting the skin muscles and this is usually done before *pronking,* when a springbok is sufficiently excited (see Antipredator Behavior). The most frequent prelude to *pronking* is bending the head in a peculiarly stiff manner. Suddenly alarmed springboks run or trot away without *pronking* or, like startled impalas, take giant leaps and scatter. They can clear a road in 1 bound, yet prefer to crawl under rather than jump over a 1-m fence. The tail of an undisturbed springbok beats like a metronome, but is clamped when running away (14).

SOCIAL BEHAVIOR

COMMUNICATION. See tribal account.

TERRITORIAL BEHAVIOR: *linked urination*/defecation on dung middens; *static-optic display;* pronounced *object-horning; grunting.*

Trampled bare spots strewn with dung are the most reliable signs of an occupied territory (3). A patrolling territorial male frequently stops to urinate and defecate in sequence, often preceded by pawing, on his dung middens (14). Males also thrash or weave their horns in low bushes, but unless they rub off secretion in the process do not otherwise mark with their preorbital glands. While other springboks rest and ruminate in the shade, territorial males often lie in the open on regular resting places, thereby advertising their status (10). But calling may be the most important form of territorial advertising (16). The vibration sounds springbok produce through open nostrils (typically 3–5 calls per stanza) grade from a soft *brrl,* given during courtship, through a considerably louder *ru,* to a very loud *ru-oo* uttered while chasing another male (16). The crescendo of calling associated with the rut makes the springbok the noisiest member of the tribe.

AGONISTIC BEHAVIOR

Dominance/Threat Displays: *medial horn-presentation; head-low posture ± head-jerking and symbolic biting* (snapping); *object-horning; stabbing* movement (female threat).

Defensive/Submissive Displays: *head-low/chin-out posture.*

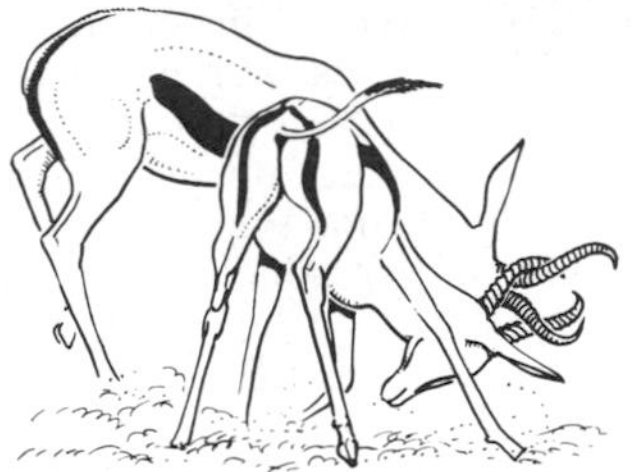

Fig. 5.13. Springbok males in twisting fight, standing parallel.

Fighting: *twist-fighting, front-pressing, fight-circling.*

Aggression is less ritualized in the springbok than in most Antilopini (16, 18). Bucks threaten simply by assuming the combat position (fig. 5.3) and only perform a dominance display when courting females. The *head-low posture* is probably defensive and so may be the accompanying *head-jerking* (forward and backward) while snapping. *Symbolic biting* (see chap. 2) by males may be unique to the springbok in the family Bovidae (16, 17).

The springbok's horns are ideally shaped for locking when 2 pairs are crossed. Fighting consists of *front-pressing,* and *lateral twisting* and *levering,* either while confronting or, more commonly, standing sideways (fig. 5.13). Springbok combat is comparatively uninhibited and dangerous. Over-the-shoulder stabbing, similar to that of the oryx and sable, has been observed, and attacks to the body are not uncommon (16). The horns sometimes become inextricably bound, leading to vehement *fight-circling* with jerking and twisting. The 2 males shown in figure 5.14—1 of which was dead, the other very much alive—were found lying in a riverbed. In a study of springbok mortality, the ratio of 50 found skulls was 5.3 males to 1 female; 83% of the males were young adults 19–30 months old which were mostly found near dung middens. Mortality in this class, attributed to territorial competition, appeared to be the main factor in a skewed adult sex ratio of 1:1.6 (6).

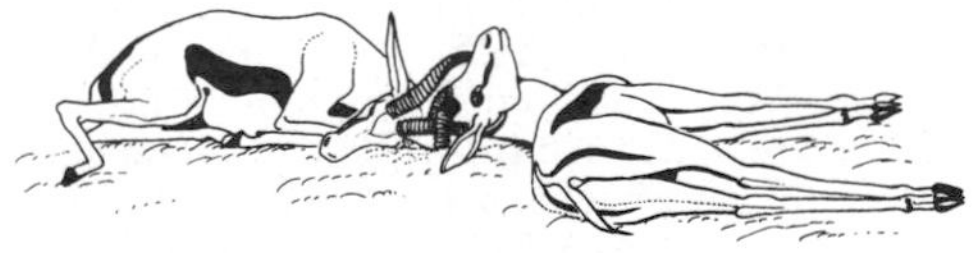

Fig. 5.14. Fatal attachment: springbok males whose horns became inextricably locked; one dead, one still alive (from photos by D. Otte).

REPRODUCTION. The springbok has been called both a seasonal and a nonseasonal breeder (1, 14). Considering that rainy seasons become increasingly unpredictable the more arid the climate, selection should favor reproductive flexibility. The birth season of springboks in northern Cape Province and Namibia varies from year to year by up to two months, indicating that such flexibility may indeed exist. The evidence suggests that the springbok at least has seasonal breeding peaks, tending to mate while in top condition early in the dry season and to calve 6.5–7 months later, in the spring near the beginning of the rainy season. When range conditions are good, females conceive at 6–7 months and calve at one year, and there may be a secondary calving peak in autumn (3). It takes twice as long for males to reach puberty (12) and they mature as two-year-olds.

SEXUAL BEHAVIOR: approach and *herding* in *lowstretch* with tail raised or curled; urination and *urine-testing; erect posture* with muzzle level, *nose-lifting,* and *head-turning; foreleg-lifting;* grunting.

Male springboks approach females in *lowstretch* made unusually conspicuous by curling the tail over the back (white upper, black lower surface and tassel). The ears are cocked vertically or horizontally, and often a male *style-trots,* brown-and-white legs flashing. The female squats low and urinates and the male then pauses to *urine-test,* with lips parted. A female in estrus may be violently pursued while the male utters surprisingly deep, bellowing grunts, until she either escapes into the next territory or slows to a walk. During the *mating march,* the male follows closely in the *erect posture,* occasionally *nose-lifting;* in *foreleg-lifting,* generally without contacting the female, he sometimes pushes her forward with his chest (9). Copulation, as in other Antilopini, involves many preliminary mounts, culminating in a single ejaculatory thrust (9).

PARENT/OFFSPRING BEHAVIOR. The tendency of springboks of the same population to breed synchronously leads to the association of fawns and their mothers in sizeable nursery herds that remain separate from harem and bachelor herds for several weeks. Male and sometimes female offspring are separated from their mothers and recruited into bachelor herds as a distinct age class (3). Little firm information is available about birth and behavior of mothers and young. The concealment stage may be shorter than usual, with fawns associating in crèches that are sometimes left alone while the mothers graze at a distance (like impala and oryx). Mother and young have been heard to exchange low grunts (1).

ANTIPREDATOR BEHAVIOR: *alert posture, alarm sneeze, style-trotting, pronking,* leaping.

The most spectacular displays of pronking are seen when a springbok aggregation is disturbed by wild (or even domestic) dogs (described under Postures and Locomotion) (fig. 5.12). Leaping and scattering in huge bounds occurs when a herd is surprised at close range, say by a stalking lion or cheetah.

The springbok has the same predators as gazelles, with the black-backed jackal probably heading the list on the basis of fawn predation. Experience with managing herds on fenced ranches for commercial hunting and meat production indicates that this species can easily withstand a mortality of 25% a year (1).

References

1. Bigalke 1963*b*. 2. ——— 1970. 3. ——— 1972. 4. Child and Le Riche 1969. 5. Cronwright-Schreiner 1925. 6. Crowe and Liversidge 1977. 7. Eloff 1961. 8. Groves 1981. 9. Hutchinson, K. 1975 (unpub. MS). 10. Mason 1976. 11. Pocock 1910. 12. Skinner and Van Zyl 1971. 13. Smithers 1971. 14. ——— 1983. 15. Van Zyl 1965. 16. Walther 1981. 17. ——— 1984. 18. Walther, Mungall, Grau 1983.

Added Reference

Hofmeyer and Louw 1987.

Fig. 5.15. Gerenuk standing and browsing.

Gerenuk

Litocranius walleri

TRAITS. A tall, thin, gazelle-like antelope with large rounded ears and a narrow head. *Height and weight:* males 89–105 cm and 45 kg (31–52), females 80–100 cm, 31 kg (28.5–

45) (4). Limbs and neck longer than in gazelles, tail hock-length (22–35 cm). *Horns*: male only, *S*-shaped, short (32–44 cm) but massive and heavily ridged. *Coloration:* similar to impala, a red-brown saddle over the back, clearly separated from the buff-colored sides and limbs by a lighter edging, white underparts and narrow rump patch; tail white near broadened base with a black terminal tuft, the underside black and hairless; head without cheek stripes, but a conspicuous white eye ring, white lips, chin, and throat patch; forehead and blaze red-brown like saddle, ears with dark backs and tips, white inside; female with black crown patch; newborn like adult but paler saddle (3). *Scent glands:* preorbitals prominent in adult males; shin and hoof glands. *Mammae:* 4.

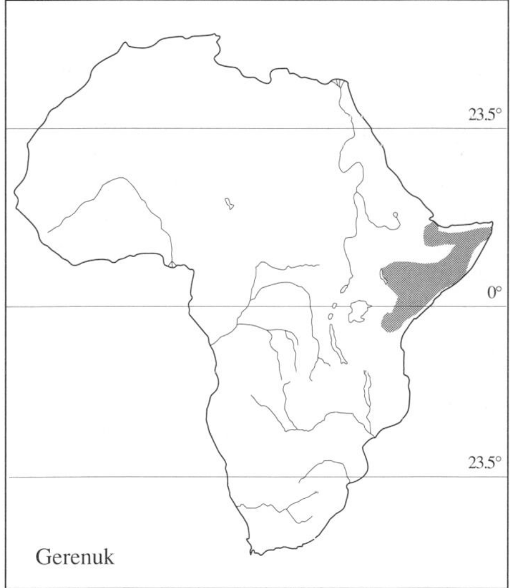

DISTRIBUTION. Somali-Masai Arid Zone from eastern and central Ethiopia through Somalia and Kenya to northwestern Tanzania, bounded on the west by the western wall of the Rift Valley. It occurs mainly in semiarid bush country below 1200 m, as in Tsavo and Amboseli N.P. and the Rift Valley. Depicted on the walls of pharaohs' tombs, the gerenuk was known to the ancient Egyptians and probably inhabited the country between the Red Sea and the Nile (4).

RELATIVES. The unstudied dibatag, found only in eastern Somalia and the Ogaden of Ethiopia, is the gerenuk's closest relative. The 2 are much alike in appearance, ecology, and behavior (observed in captive animals) but overlap in eastern Somalia, so must be ecologically separated. The question is, how?

ECOLOGY. The gerenuk is found in a variety of habitats with woody vegetation, but is commonest in dry, flat thornbush spaced widely enough so that it can freely move around and between bushes while browsing. It avoids dense bush, thickets, tall woodland, and open savanna (9). Aerial sampling of Tsavo N.P. showed that its density increases with aridity and distance from permanent water. It thereby reduces resource competition with more water-dependent browsers such as the impala (9).

The gerenuk is among the purest of browsers, not known to eat grass or even small herbs. Superbly adapted to feed selectively on small-leaved, thorny shrubs and trees, it has a muzzle proportionally as narrow and nearly as pointed as a dik-dik's, a flexible upper lip, a long tongue, slitlike nostrils and hairy muffle, low-crowned teeth, a long neck, and limbs adapted not for speed but for added height. Standing on its hindlegs, it can browse to a height of 2 m (fig. 5.15), higher than other antelopes in its weight class or even the lesser kudu, a larger associated browser that keeps all four feet on the ground. Wedge-shaped hooves, robust lower limbs, and lumbar vertebrae modified for an increased curvature of the spine while standing bipedally enable a gerenuk to stand unsupported, unlike any other known antelope. It uses its long pointed tongue and mobile lips to penetrate often thorny bushes and browse the tiniest leaflets, and uses its spoon-shaped, sharp-edged incisors to nip leaves and shoots with small upward nods. Long eyelashes and sensory hairs on the muzzle and ears serve to guide its movements and protect the eyes from scratches (4).

DIET. A typical concentrate selector, gerenuks of Tsavo N.P. were seen feeding on 84 different species of plants (7), almost exclusively on the leaves, shoots, and flowers of shrubs and trees, and when available, on the foliage of creepers and climbers. They ate the fruits of a few species. In the short growing season the animals preferred the new leaves of common deciduous woody plants. During the 8-month dry season they fed largely on evergreen shrubs and trees, which are most abundant along watercourses. So the presence of such greenery is decisive for the welfare of the gerenuk during drought, and therefore a limiting factor in its distribution.

Table 5.3 Composition and Mean and Maximum Size of 607 Gerenuk Groups (1,349 Animals) Classified in Tsavo N.P. Roadstrip Counts

Group type	% of population in groups of this type	% singles	Group size (minus singles)	
			Mean	Maximum
Males	17.7	78.6[a]	2.4	4
Females (± young)	30.2	49.1	2.5	5
Females and 1 adult male	39.5	—	3.4	9
Mixed herd (females, 1 adult male and subadult males)	12.6	—	3.4	7

SOURCE. Leuthold 1978*a*.
[a]37.6 adult, 41.0 subadult.

SOCIAL ORGANIZATION: gregarious in very small, loose groups, territorial, sedentary.

Going by dispersion pattern and group size, the gerenuk is among the least gregarious of the sociable bovids, comparable to the mountain reedbuck and sitatunga (1). For an antelope living in relatively closed habitats at low density (0.5–1/km^2 in Tsavo N.P.), small groups are predictable (table 5.3). Some 42% of all the gerenuks sighted were singles, of which nearly half were female, and mean group size (excluding singles) was only 3.2. The largest group included 12 animals and less than 5% of herds contained over 5 animals (7).

In more arid parts of northeastern Kenya and Somalia where gerenuks apparently live at higher density, herds may be slightly larger. Groups ranging from 2 to 8 have been reported as normal, with occasional herds of 10–12 and very exceptionally 25–30—these probably representing aggregations on a localized greenflush (7).

Subadult males associate with one another and, given the chance, attach themselves at least temporarily to their mothers or other adult females, with whom they may partially share a home range (7, 10). Bachelor and mixed herds include no more than 1 adult male—apparently males become permanently territorial as they mature and rarely venture off their property, which may be as large as 4 km^2 (10). The territories of 9 known Tsavo males averaged 2.1 km^2 (0.8–2.8), did not overlap to any degree, and often had no common boundaries. Female home ranges were of similar size (1.4–3.4 km^2) and in at least 1 case the home ranges of 2 females coincided exactly with a male's territory (9, 10).

Although gerenuks are generally very local and sedentary, individuals have been known to shift ranges. One female moved about a kilometer and settled in a different territory, after the proprietor of the territory where she had been living disappeared (7). Subsequently, 4 subadult males that had shared her home range left one by one and resettled 3–4 km away. Territories that have been vacated may not be reoccupied for months, and meanwhile may be divided up among territorial neighbors (7). Gerenuks with overlapping ranges may be considered as belonging to the same social unit, but the only individuals that regularly stay together are a mother and her dependent offspring. Females with or without young often associate but rarely for long (10). Even when territorial males have females resident on their property, the male may spend much time alone, for instance while patrolling and demarcating his property (see Territorial Behavior). But two known males showed opposite tendencies: one was rarely seen in company with a resident female, whereas the other was found with 1–7 females in 27 of 28 sightings. When the 2 sexes associate, it is the males that actively seek out and follow the females, perhaps mainly or entirely due to sexual attraction (10).

Male Dominance. Adult males are socially dominant over all other gerenuks, but this only becomes apparent when there is competition for the same resource. A buck can supplant a browsing female merely by approaching; if she fails to yield he may enforce his will with a threat display or a short rush during which he emits a rumbling sound (7). When moving in file, the male's customary trailing position often leads to mild threat and sexual displays by him and submission or avoidance by the female. When alarmed, the greater timidity and flight distance of females make them the first to flee. A female's flight may prompt the male to withdraw, either from the disturbance or in pursuit. Otherwise, bucks rarely chase or even actively herd females, having little need to prevent their leaving the territory or being waylaid by rivals. Adult males

Fig. 5.16. Gerenuk adult male chasing yearling male.

do sometimes chase adolescent and subadult males (fig. 5.16), especially those associated with females, but rarely persistently enough to drive them from the neighborhood. Fleeing youngsters tend to circle.

ACTIVITY. Observations of daylight activity in Tsavo N.P. have shown great variation from day to day in the duration and timing of different activities (6). Male gerenuks spent 32–64% and females (which need to eat more while lactating) spent 51–68% of the day feeding. But males spent relatively more time ruminating: the ratio of feeding to ruminating was 3:1 in females and 2.7:1 in males. The percentage of time devoted to basic maintenance activities by three different males that were observed for a total of eight days and one female observed for seven days is shown in table 5.4.

Although there were no obvious peaks of feeding activity, there was a clear trend for both sexes to ruminate between 1000 and 1300 h, and for females but not for males there were morning and evening peaks of standing/walking, doubtless reflecting their maternal activities. An increase in male standing/walking between 1530 and 1800 h correlated with increased scent-marking. Most of the time spent lying was recorded from 1000 to 1400 h, and individuals were rarely seen lying early or late in the day.

POSTURES AND LOCOMOTION. Gerenuks may remain immobile in the alert posture for many minutes, apparently as a means of escaping detection (see Antipredator Behavior). Although most bovids that live in cover or on broken ground use the *cross-walk,* the gerenuk *ambles* like other Antilopini. Trotting is normally the fastest gait and a gallop is reserved for real emergencies. When withdrawing from a disturbance, a gerenuk seems to glide effortlessly through the bushes with neck raised, ears back, and tail down (7). (A fleeing dibatag holds its tail straight up like a warthog's [11].) The gerenuk also *stots* in a special way, folding its legs at the height of each bound and keeping its tail low. Sometimes the white rump patch is flared, increasing its conspicuousness (10). Stotting has a higher threshold in this species than in gazelles and as usual, is commonest in the young.

SOCIAL BEHAVIOR

COMMUNICATION. Very little has been published about *social grooming,* dominance, leadership, and other forms of social behavior.

Tactile Communication. Although nibbling of the head and neck and especially the preorbital region has been seen between male and female captives, no similar *social grooming* has been observed in wild gerenuks. However, mothers and their offspring and females sometimes rub their heads together, usually after one solicits by putting its head under the other's chin and pressing upward (10).

TERRITORIAL BEHAVIOR: *preorbital-gland marking; linked urination/defecation.*

Systematic investigation of the preorbital-gland marks made by the same territorial male in two successive years showed that deposits of the black secretion (up to 1 cm in diameter) were concentrated in an extended narrow band, respectively 3.1 and 2.0 km long, which circumscribed an

Table 5.4 Mean Percentage of Daylight Hours Spent on Basic Maintenance Activities by Male and Female Gerenuks

Sex	Feeding	Standing/ walking	Ruminating: Standing	Ruminating: Lying	Lying without ruminating
Males	47.0	34.0	14.8	2.4	1.9
Females (lactating)	60.3	18.7	17.6	1.6	1.8

SOURCE. Leuthold, B. M., and W. Leuthold 1978*a*.

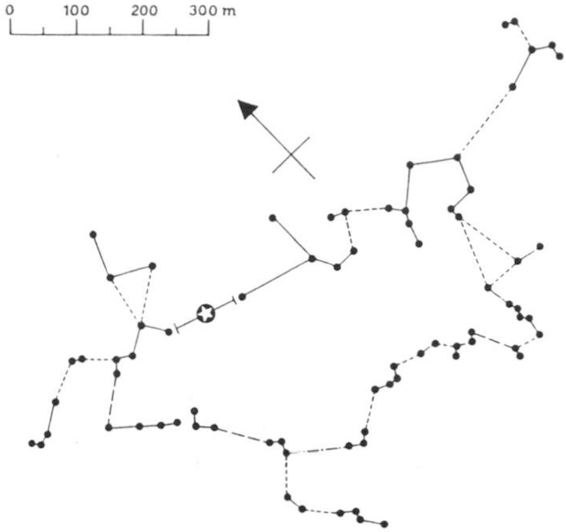

Fig. 5.17. Pattern of preorbital-gland deposits found in a gerenuk territory in 1968. Solid lines link nearest-neighbor deposits within primary clusters; dashed lines join primary clusters; dash-dot lines link secondary deposits. (From Gosling 1981.)

oval-shaped polygon of 30 and 12 ha, to which a number of radiating lines of scent marks were linked (fig. 5.17). Most marks were placed on the outermost bare, stiff twigs of preferred forage plants, often very close to or overhanging (40%) a game trail, at an average height of 116 cm (range 105–129 cm). As the radiating arms extended beyond the ovals, the male was often seen outside, and the areas demarcated were considerably smaller than the average territory, evidently only part of the range was kept marked at a given time. A hypothesis based on the costs/benefits of marking suggests that the area necessary to sustain a gerenuk plus females and young during the year is too large to keep demarcated; the oval and radiating arms represent the space a male has enough time to spare and enough secretion to post effectively (2).

Territorial males also urinate and defecate in sequence like all Antilopini, but are not known to maintain dung middens. However, interactions between adult males have rarely been witnessed; trespassing is unusual and it is still unclear whether territories normally abut and have distinct boundaries, or tend rather to be separated by neutral zones of greater or lesser extent (10).

AGONISTIC BEHAVIOR

Dominance/Threat Displays: *erect posture* ± nose-lifted; *medial-* and *low-horn presentation; head-shaking; poking* (females only); *nodding; ground-* or *vegetation-horning; charging.*

Defensive/Submissive Displays: *head-low/chin-out posture, tail-curling.*

Fighting: *front-pressing, twist-fighting, nod-butting* and *clash-fighting*(?).

Information about aggressive behavior is based mainly on observations of a captive herd (16, 17). Observations of wild gerenuks indicate that aggressive interactions and fighting (also play) are relatively uncommon under normal conditions (7, 10).

The *erect posture* and the more intense form, with nose lifted, are generally performed from frontal rather than lateral orientation, either confronting or with *head turned away* from the opponent. These same dominance displays, which show the male's muscular neck to advantage (cf. Grant's gazelle), are also performed during courtship. *Head-shaking* has been observed in both sexes; females raise their heads and flap their ears, and make empty *poking movements* either toward or away from an antagonist (typically a courting male), both being probably defensive threats. Similar behavior is seen in female dibatags and blackbucks, and in hornless females of other tribes (18).

Horning the ground may be a particularly strong threat when directed by an adult to a young male. The latter, in the captive herd, responded by dropping to his knees in the *head-low/chin-out posture,* and occasionally even lay down. But neither behavior has been observed in wild gerenuks. Submission is normally shown by arching the tail over the back while standing with the neck at shoulder level. The same young male sometimes pretended to feed during confrontations, picking up a leaf or piece of soil from the ground; this is a striking example of *displacement feeding* by a species that seldom browses at ground level (17).

Kingdon (4) describes vigorous *nodding* as a common behavior during social interactions, by both sexes beginning with juveniles, and suggests it is a dominance display or threat based on *nod-butting* and *horn-clashing* during combat. The massive horns and elongated back of the skull, taken together with the male's thickened upper neck, appear adapted to *clash-fighting,* creating a longer fulcrum that imparts extra force to forward-downward blows. However, up to now the only fighting technique that has been reported is *front-pressing* and *twisting,* though the fighting was playful and involved an adult and an adolescent male (17). In the prelude, the 2 would approach with heads low and chins out, touch noses, then carefully pull in their chins until their horns touched. One would then tilt and move his lowered head sideways as though about to jab at the other, who countered each movement exactly with his own horns. When their horns finally touched or clashed, they bound themselves together firmly by crossing or forking them, then thrust and wrestled, at the same time seeking to lever sideways and push one another out of position (cf. Grant's gazelle). Sometimes they tugged backward or stood parallel and twisted sideways (cf. springbok). During intervals between their short bouts, both contestants displayed until one lowered his head into *medial-* or *low-horn presentation,* whereupon combat was renewed. Turning the head sideways signaled the intention to break off the match, and when both did so at once, the bucks parted. Otherwise the opponent had either to continue the match or run away, at the risk of being furiously pursued by the "victor."

REPRODUCTION. A Tsavo female that was seen regularly over a period of 5½ years produced seven offspring in that time. Females usually reenter estrus within 2–4 weeks, and this one produced five offspring after intervals of only 8–9 months (9). Gestation is recorded as 6½–7 months. Thus female gerenuks, which breed the whole year, may have at least four calves within three years. The minimum recorded age of conception is 11 months and maximum known longevity is 13 years (both recorded in zoos). A wild male was known to become territorial at 2¾ years, but normally this occurs at 3–3½ years (9).

SEXUAL BEHAVIOR: *erect posture* with nose-lifted; *foreleg-lifting; scent-marking/rubbing female, humming.* Female: defensive threats (*head-shaking, poking, head-low/chin-out posture*).

As with aggression, little sexual behavior and no copulations were seen during the Tsavo study, but both have been observed in the study of a captive herd (17). Males do not approach and herd females in *lowstretch,* but simply march behind them in the *erect posture* with nose lifted. *Foreleg-lifting* is exceptionally ritualized and pronounced. Standing close behind the female, the buck slowly raises a straightened foreleg, with hoof spread apart, between or outside her hindlegs until contact is made. In this way a standing female is urged to continue the *mating march* and the buck follows uttering a humming sound. During the precopulatory stage, when the female accepts contact, the male rubs her with his preorbital glands (fig. 5.18), especially her chest and rump but also her neck, shanks, and

Fig. 5.18. Gerenuk courtship: male rubbing preorbital gland on female.

croup. This behavior, confirmed by observations of wild gerenuks, and also reported in dibatag, is otherwise unknown in the Antilopini (10, 19).

Gerenuk copulation is the most spectacular example of the unique antilopine technique: the 2-m-tall male walking bolt upright with forelegs dangling behind the moving female and managing intromission without falling over backwards. Meanwhile the female is holding her head high, at shoulder level or sometimes lower (16).

PARENT/OFFSPRING BEHAVIOR. Several births have been observed in zoos (5). The behavior of the female during parturition could just as well describe the birth of a Grant's gazelle (q.v.). The fawn's head emerged ½ hour after the protruding amniotic sac burst, and expulsion followed 55 minutes later. The fawn began struggling to rise within 15 minutes but only stood after 1 hour. It then immediately began calling, which brought the mother, uttering a similar bleating cry, to its side. The afterbirth was dropped 2½ hours post-partum and immediately eaten.

In three different births, a) the fawn was dropped while the mother stood; b) the fawn bleated as soon as it stood, and the mother answered and kept calling until the fawn followed; c) the mother refused to suckle until the afterbirth had been dropped.

In Tsavo N.P., a female that was followed while retrieving her 3-week-old fawn first left 2 other females and a male and walked in a purposeful way for 1 km. The fawn emerged from the bushes and, after a nose-to-nose greeting, began nursing at c. 1800 h (8). A fawn hand-reared in Tsavo began sampling plants and trying to stand bipedally at 2 weeks but could only do so well at 1 month. Yet it stotted and adopted the male urination stance at 5 days, made the *urine-testing grimace* at 3½ months, displayed *foreleg-lifting* and the submissive *head-low/chin-out* and *tail-curling* at 5 months, and mounted at 6 months (12). Male offspring normally part company with their mothers at 1–1½ years, afterward joining together with 1 or more other subadults until mature and ready to become territorial at 3 years (12).

ANTIPREDATOR BEHAVIOR: *immobile stance, lying-out, stotting, trotting, alarm snort, stamping.*

A disturbed gerenuk maintains the head-high *alert posture* for an indefinite period, while remaining motionless and staring constantly at the source of disturbance. The ears are either cupped forward or held back, the former possibly showing greater confidence and more typical of adult males (4). Also, an animal that flicks its ears, swishes its tail, and/or chews its cud is only mildly disturbed. But a gerenuk standing motionless in cover is very easily overlooked. Females have also been reported to *lie-out* after being frightened, behavior that is otherwise exceptional in this tribe (but characteristic of solitary antelopes and even social tragelaphines). A gerenuk alarmed to the point of flight typically goes at a flowing trot, occasionally a gallop, and sometimes *stots* (described under Postures and Locomotion).

This antelope has been hunted to extinction in parts of Somalia. On the other hand, overgrazing and the spread of subdesert bush enable the gerenuk to extend its range (13).

References

1. Estes 1974*a*. 2. Gosling 1981. 3. Haltenorth and Diller 1980. 4. Kingdon 1982. 5. Kirschshofer 1963. 6. Leuthold, B. M., and W. Leuthold 1978*a*. 7. Leuthold, W. 1971*a*. 8. ——— 1971*b*. 9. ——— 1978*a*. 10. ——— 1978*b*. 11. ——— 1984. 12. Leuthold, W., and B. M. Leuthold 1973. 13. ——— 1975. 14. Schomber 1966. 15. Walther 1958. 16. ——— 1961. 17. ——— 1968. 18. ——— 1984. 19. Walther, Mungall, Grau 1983.

Chapter 6

Reedbucks, Kob, and Waterbuck
Tribe Reduncini

Redunca arundinum, common reedbuck
R. redunca, bohor reedbuck
R. fulvorufula, mountain reedbuck

Kobus kob, kob
K. vardoni, puku
K. leche, lechwe
K. megaceros, Nile lechwe
K. ellipsiprymnus, waterbuck

This is a close-knit tribe with only 2 genera, and 6 of the 8 species are geographically paired: the bohor and common reedbuck, kob and puku, and the 2 lechwes fill similar niches in the Northern and Southern Savanna. The kob and puku are enough alike to be considered a superspecies (1). The differences between common and defassa waterbucks are at least as striking, but in their case intermediate forms exist and so they are only designated as subspecies. Even the waterbuck, lechwe, and kob remain so genetically close that they can interbreed and produce apparently viable hybrids in zoos. Nevertheless, these different species all coexist without ever making such mistakes in the wild (6).

TRIBAL TRAITS. Medium- to large-sized (65–130 cm, 20–260 kg) antelopes of sturdy build and shaggy coat (only the throat in reedbucks to all over in waterbuck). *Horns:* males only, robust and strongly ridged, short and hooked forward in reedbucks to long and lyrate (up to 92 cm) in lechwe. *Coloration:* tan to dark brown with black markings (legs, neck, shoulders, or most of body) and white underparts, rump, throat patch, and facial markings; males darker than females; tail short, wide, and bushy with white underside in reedbucks, longer and thinner in other species. *Scent glands:* large, diffuse sebaceous glands make the coat greasy and strong-smelling (especially waterbuck) (5); a pair of inguinal glands in reedbucks and kobs only; simple preorbital glands present in kob only; foot glands absent (but vestigal in reedbuck and puku hindfeet); a glandular bare spot below the ear in reedbucks. *Mammae:* 4.

DISTRIBUTION. Northern and Southern Savanna, from southern Sudan to South Africa.

ECOLOGY. Waterbuck would be an appropriate name for the whole tribe, for no other group of antelopes is so water-dependent. They are invariably found within a few miles of water (except the mountain reedbuck) and reach highest population densities on floodplains. They are all grazers, preferring grasses with high protein and low fiber. There is considerable habitat overlap between species. The lechwe, puku, waterbuck, and reedbuck coexist in some Southern Savanna floodplains, for example the Chobe River in Botswana (2). However the lechwe, the most aquatic antelope next to the sitatunga, is the only reduncine that regularly enters the water to feed; the rest prefer dry ground. Reedbucks depend on cover and lechwes depend on swamps to elude predators, while the waterbuck is a woodland/grassland-edge species. Reedbucks and the waterbuck are the most sedentary reduncines, lechwes move back and forth as the waters rise and fall on the floodplain, and the huge white-eared kob population of Southern Sudan is migratory (3, 4).

SOCIAL ORGANIZATION. The range in social organization within this tribe runs the gamut from the monogamous common reedbuck to the extremely polygynous kob and lechwe, which are among the most gregarious antelopes. Social, territorial, and reproductive organization of the Reduncini is compared in table 6.1.

The transition from solitary to gregarious can be seen in the reedbucks, for the bohor reedbuck may be either, depending

Table 6.1 Comparative Ecology and Social and Reproductive Organization of the Reduncini

Species	Habitat	Social organization	Territorial system	Age at maturation (years) M	F	Gestation (months)	Time calves hide (weeks)
Common reedbuck	Valley and upland grassland	Monogamous pairs, dry-season groups	Conventional	3	2	7.5	up to 8
Bohor reedbuck	Floodplains	Monogamous or polygynous, dry-season groups and aggregations	Conventional	3	2	7.5?	up to 8
Mountain reedbuck	Montane grassland	Polygynous, small female herds	Conventional	3	2	7.5	up to 8
Kob	Floodplains	Gregarious, female & male herds and big aggregations; white-eared kob migratory	Conventional and lek territories	3	2	c. 8	4
Puku	Floodplains	Like kob but no big aggregations	Conventional	3	2	c. 8	4
Lechwe and Nile lechwe	Swamp/flood-plain	Mobile/aggregated–sedentary/dispersed	Temporary leks + conventional	4	2.5–3	7–8	2–3
Waterbuck	Floodplain & wooded savanna near water	Female, male, and mixed herds	Territorial males tolerate bachelors	5–6	3.5	8–8.5	2–4

on ecological conditions, and the mountain reedbuck is sociable. At least 2 of the *Kobus* species have unusual forms of territoriality: the kob and lechwe at high density are lek breeders. The greatest development of sexual dimorphism is seen in these species, reflecting extreme reproductive competition. Territorial waterbucks tend to be unusually tolerant of other males, even to the extent of allowing subordinate adult males to reside on their land. Territorial size ranges from as large as 226 ha in a waterbuck to breeding arenas where the spacing between kob and lechwe males is as close as 15 m (6).

Relations between females may be characterized as peaceful and casual in the whole tribe; herds lack strong cohesion or obvious hierarchy, especially large herds of kobs and lechwes. The low level of female aggression may reflect the absence of horns, although waterbucks, for instance, occasionally snap, poke, and butt when provoked (7).

ACTIVITY. Reedbucks stay in or near cover by day and emerge to graze at night, whereas *Kobus* species are primarily day-active, and tend to withdraw into cover by night.

POSTURES AND LOCOMOTION. Reduncines have an *ambling* walk (7). Reedbucks run in a distinctive *rocking canter* that is well-adapted to flight in high grassland but inefficient on bare ground. Lechwes leap like impalas when suddenly startled, plunge through the water in flight, and skulk into hiding with lowered head. They are slow and ungainly runners on dry land and the waterbuck, too, is considered ambulatory rather than cursorial (6). The kob and puku are the swiftest members of the tribe, but they too are less cursorial than the gazelles or topi, and may seek cover when pursued (see kob account).

SOCIAL BEHAVIOR

COMMUNICATION

Olfactory Communication. The lack of hoof glands and the absence of deliberate scent-marking (see under Territorial Behavior) suggest that olfactory signaling is less important in this tribe than in others. However, the characteristically strong body odor and the presence of inguinal glands and of subauricular glands in the reedbucks clearly indicate that glandular secretions play a role—as yet unknown—in reduncine communication (see reedbuck).

Tactile Communication. *Social grooming* is unimportant in the tribe, but courting males nuzzle, rub, and lick females.

Vocal Communication. The reedbucks and kobs whistle, both in alarm and in territorial advertising, and territorial lechwes grunt. The waterbuck is relatively silent, but has an alarm snort.

Visual Communication. Definitely the most developed communication channel, visual displays are elaborate; *static-optic advertising* is employed even by the secretive reedbucks (fig. 6.1).

TERRITORIAL BEHAVIOR: *erect posture, static-optic advertising, object-horning, whistling* or *grunting.*

Reduncini do not actively scent-mark their property. Territorial status is advertised visually and vocally by the owner's presence and behavior, notably by staying apart from the herd, by standing and walking in the *erect posture,* and by aggressive displays to males and sexual displays to females. On kob and lechwe breeding arenas, territory owners are readily singled out not only by their striking appearance but also by their posturing and interactions (see figs. 6.4 and 6.12). But in a waterbuck territory, it may take some watching to determine which of several adult males is the proprietor. *Whistling,* which is usually considered an alarm signal, may be the primary form of territorial advertising in reedbucks (5).

AGONISTIC BEHAVIOR

Dominance/Threat Displays: *erect posture* in *lateral presentation* with or without *nose lifted* or *head turned away, medial-, low-, and high-horn presentation, angle-horn, head-shaking,* rhythmic *nodding, horning, penile erection, rushing.* Females only: *poking, butting, snapping.*

Defensive/Submissive Displays: *head-low posture, head and nose lifted* (waterbuck and lechwe); *champing* (mouthing).

Displacement Activities: *grazing; grooming* and *scratching.*

Fighting: *push-* and *twist-fighting, front-pressing, nod-butting, clash-fighting.*

Aggression is highly ritualized in *Kobus* and less so in reedbucks. In the waterbuck, satellite and bachelor males behave like females to appease alpha males. The significance of *object-horning* appears to differ between species, for it is performed by territorial reedbucks and kobs, but in the waterbuck by subordinate males, perhaps as redirected aggression (5). Fighting is common and consists of powerful downward blows (*nod-butting* and *clash-fighting*), pushing and *twist-fighting.* The thickness of the neck in mature males reflects the importance of a strong neck in reduncine combat (5, 6), and neck skin up to 2 cm thick (waterbuck) shields males from the sharp horntips. Bucks typically touch muzzles as a prelude to engaging horns. *Penile erection* during aggressive displays and combat is a particular reduncine trait, the significance of which remains indeterminate.

REPRODUCTION. *Kobus* species are relatively slow to mature: from 3 years in reedbucks up to 6 years in the waterbuck (table 6.1). Females mature 1 to 3 years earlier. Breeding is perennial in equatorial populations, with a more or less well-defined birth peak during the rainy season. In the Sudan the white-eared kob is a seasonal breeder, but the season extends for four months (4), and even in subtropical southern Africa reduncine breeding seasons are protracted. With a 7–8½ month gestation period, females could calve every nine months or so.

SEXUAL BEHAVIOR. Mating behavior is very similar throughout the tribe, featuring female urination on demand, with tail held out or up, when a male approaches in *low-stretch, foreleg-lifting, rubbing, nuzzling,* and/or *licking* the female's posterior, sides, and neck, and multiple preliminary mounts before copulation. Females resist male advances by *circling, symbolic biting, butting,* and *flight.* Copulating females adopt a hunched stance, with tail out or to one side (*swan-neck posture,* fig. 6.8). The male rests his chest on the female, forelegs clasping her tightly, with head high (fig. 6.9). Estrus lasts a day or less.

PARENT/OFFSPRING BEHAVIOR. Reduncine young are typical hiders. Calves remain concealed for 2 to 3 weeks (waterbuck, lechwe) up to 2 months (reedbucks). Afterward, calves of the same age class band together. The maternal bond lasts no longer than 8 or 9 months. Though hornless, female reedbucks and waterbucks have been known to defend their young against predators as formidable as wild dogs and hyenas.

ANTIPREDATOR BEHAVIOR: *alert posture, whistling* (reedbucks), *grunting* (lechwe), *snorting* (waterbuck), *skulking* into cover, *lying-out, stotting* (reedbuck) or *style-trotting* (waterbuck), leaping or bounding flight into cover or through water.

Reedbucks hide in long grass, the waterbuck also runs into cover and may then sometimes *lie-out* (6), and lechwes take to the water. Only the kob/puku flees in the open like typical plains antelopes.

References

1. Ansell 1971. 2. Child and von Richter 1969. 3. Fryxell 1980. 4. ——— 1985. 5. Kingdon 1982. 6. Spinage 1982. 7. Walther 1979.

Fig. 6.1. Territorial bohor reedbuck advertising presence/status by standing in the open in *erect posture.*

Reedbucks

Redunca Species

TRAITS. Yellow- to gray-brown, medium-sized antelopes with a round bare spot below each ear, white underparts and underside of bushy tail. *Horns:* males only, shorter than in *Kobus,* with slight to moderate forward hook. Males 10%–20% bigger with thicker necks and more defined markings than females; calves darker and longer-haired. *Scent glands:* the subauricular bare spot (hair-covered in some common reedbucks [7]) and large inguinal glands in both sexes. The measurements in table 6.2 are from (10).

DISTRIBUTION. Bohor and common reedbucks occur in suitable floodplain and drainage-line grassland throughout the Northern and Southern Savanna, respectively, overlapping only in southern Tanzania. The mountain reedbuck has a peculiar disjunct distribution, occurring only in hills and mountains of Ethiopia and East Africa, southern Africa, and the Adamana Mountains of Cameroon. Montane habitat between East and southern Africa is occupied by the common reedbuck, suggesting a greater ecological similarity between these 2 species than between the bohor and mountain reedbuck (9, 10).

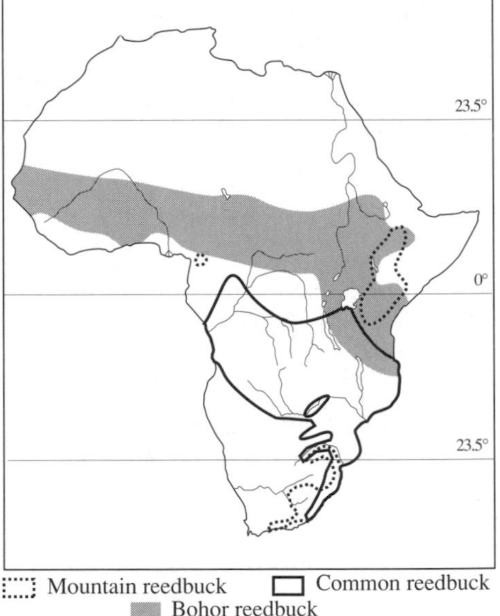

ECOLOGY. Common and bohor reedbucks frequent grassland habitats tall enough to hide them. Most common reedbucks inhabit drainage-line grassland, whereas bohor reedbucks are most abundant on wide floodplains. The mountain

Table 6.2 Reedbuck Measurements and Coloration

Weight (kg)		Height (cm)		Horns	
Male	Female	Male	Female	(cm)	Coloration
Common or southern reedbuck					
68 (51–80)	48 (39–64)	84–96.5		30–45 (curved)	Browner; black foreleg stripe; pale band below horns; white throat
Bohor reedbuck					
43–55	36–45	75–89	69–76	20–41 (hooked)	Yellower; no leg stripe, white throat patch
Mountain reedbuck					
30 (22–38)	29 (19–35)	65–76		14–38 (hooked)	Markings less developed

reedbuck frequents rolling, grassy hills and steep, rocky slopes above 1500 m. All 3 are grazers, though they eat some forbs and also browse woody vegetation to some extent in the dry season when no green grass is available (8). Studies of the common reedbuck's feeding ecology and of the mountain reedbuck's digestive system indicate that reedbucks can subsist on grasses that are either inaccessible or unpalatable to most other antelopes (3, 8). Thus, few plains species venture into the tall grassland or onto the steep slopes frequented respectively by these 2 species. Reedbucks may forgo drinking when on green pastures; the mountain reedbuck is considered comparatively water-independent, whereas the common reedbuck cannot survive more than a few days without drinking late in the dry season (7).

SOCIAL ORGANIZATION. Reedbucks bridge the gap between solitary and gregarious territorial social systems (see chap. 2). The common reedbuck lives typically in monogamous pairs, whereas the mountain reedbuck lives in small herds numbering 3 to 6 and up to 8 females and young (5, 6). If we understood why and how the transition from one form of organization to the other occurred, new insight into the evolution of gregarious habits might be gained. However, the common assumption that living in social groups is the more advanced condition and that solitary habits represent the ancestral condition has been challenged in this case, for in morphology and ecological and social adaptability the mountain reedbuck seems to be the least advanced of the three (10). Conceivably monogamy is a secondary development in the reedbucks.

The monogamy of the common reedbuck is adapted to patches and strips of grassland tall enough to provide concealment, such as those occurring along drainage lines in woodland. But the pair bond is comparatively loose: there is no mutual grooming and no territorial marking ritual. Male and female remain apart much of the time, especially during the 4 months after a calf is born (7). A female is often accompanied by 2 successive offspring. Daughters mature and leave the territory in their second year, up to a year earlier than sons, which are tolerated by their fathers almost to maturity. In most bovids males disperse first.

Dry-season drought and fires often drastically alter the habitat and force changes in reedbuck social organization. Loss of cover, forage, and watering points forces some animals to vacate their ranges and seek these necessities elsewhere, leading to concentrations near remaining waterholes, in unburned grassland, and on post-burn flushes. Reedbucks forced or tempted into the open associate in temporary herds which, however, disperse when disturbed instead of fleeing in a group like most herding species. Territorial males will tolerate displaced female immigrants and, when territories are subject to constant traffic (e.g. by reedbucks going to water), intolerance of other males and the size of the territory may both diminish. But even on neutral ground adult males avoid one another. The nearest thing to a bachelor herd of common reedbucks is an adult male accompanied by a subadult that has left its mother. Groups consist of females with young accompanied by no more than 1 adult male, and rarely number over 7 animals (7).

The bohor reedbuck appears to have similar habits in similar habitats, but on closer examination often turns out to be polygynous rather than monogamous (2). A male's territory may include the ranges of up to 4 females. As long as there is cover to hide in, the females stay separate; lacking it, they and their offspring band together in groups of up to 10 head. Territorial males of this species allegedly drive their sons away soon after their horns appear (at c. ½ year). These immature males associate in twos and threes in border areas between territories until they mature in their fourth year (10). Territorial males may tolerate and even associate with bachelors in the absence of females.

This pattern of social organization changes drastically where the bohor reedbuck reaches high population densities, as on the vast floodplains of Southern Sudan, where up to 110 reedbucks/km^2 congregate on the greenbelts bordering the rivers during the dry season, in aggregations numbering in the hundreds (1, 4). Territorial organization and perhaps even family groupings break down, and a whole aggregation seems to become a single macro-herd perhaps comparable to a wildebeest aggregation (9). But the structure and behavior of reedbuck aggregations remain to be studied. At other times and places sexual segregation is preserved: gatherings of over 100 males have been recorded in Tanzania's Rukwa Valley (13).

Clearly, the traditional view of these 2 reedbucks as asocial and monogamous needs to be revised. That it is the prevailing pattern indicates that dispersion in pairs is adapted to the conditions that prevail through much of the Southern and part of the Northern Savanna. But given conditions that can sustain a higher population density, the bohor reedbuck, at least, is capable of changing to a gregarious form of social organization. The largest known concentration of mountain reedbucks, on the other hand, was about 50, whereas a population density of 5–7 animals/km^2 is average (5).

Two measured common reedbuck territories bordering a permanent waterhole covered respectively 48 and 60 ha in the summer, but became compressed to 35 ha apiece in the winter following increased intrusions by other reedbucks to reach the water (7). In Serengeti N.P. bohor reedbuck territories were estimated at 25–60 ha, each shared by 1–5 females (2). Mountain reedbuck territories of only 10–15 ha were recorded at Kekopey Ranch, Kenya, but averaged 28 ha in the Loskop Dam Reserve, Transvaal (5). Here female ranges averaged 57 ha (36–76) and included the territories of up to several males. This emancipation of females from the territory of individual males is perhaps the most significant difference in the social organization of mountain reedbuck compared to the other species.

ACTIVITY. Although daytime activity of the common reedbuck has been studied, apparently all 3 species are primarily active at night, a period for which there is little firm information. The reedbucks studied in Kruger N.P. were most nocturnal during the rains and early dry season, when their nightly activities were extended by 1–2½ hours after dawn and an hour or less before dark (7). The animals remained inactive, lying up in high grass or the shade of a tree through the day, apart from interruptions to stand, stretch, and groom. A total of 2–3 hours of the 7½–9 hour rest period was spent chewing the cud.

As the dry season advanced and grass quality declined, the reedbucks rested less and grazed more during the day. Between August and late October, reedbucks could be seen grazing and moving at all hours. High temperatures and low humidity of the vegetation and air forced them to go to water more often (frequency uncertain). Finally these animals were feeding most of the day, from early to late morning, midday and afternoon until nearly 1600 h, and again after dark. Thus, reedbucks feeding on low-protein/high-fiber grass compensate by eating and drinking more, and this seasonal change in activity appears to apply to all 3 species (4, 9).

POSTURES AND LOCOMOTION. See tribal introduction.

SOCIAL BEHAVIOR

COMMUNICATION. Discussed in tribal introduction.

TERRITORIAL BEHAVIOR. Perhaps because reedbucks do not scent-mark their territories, the borders are not sharply defined. In the common reedbuck, family members freely trespass 100–200 m into their neighbors' land, the boundaries of which fluctuate depending on the pugnacity of the owners, environmental conditions, and local population density (7).

Advertising: posing in *erect posture, whistling, stotting* ± *popping, defecation posture* (?).

Reedbucks clearly employ visual and vocal modes of advertising territorial status, and may also use olfactory signals undetected by human observers. Bucks stand in the *erect posture* ("proud posture" of [7]) (fig. 6.1), positioning themselves for maximum visibility (e.g., on mounds during resting/ruminating periods). At the sight of another reedbuck, a buck standing in the *erect posture* may utter a shrill, reedy whistle, expelling air through his nose with such force that his whole body shakes (12). After whistling 1–3 times, he often makes a few *stotting* bounds, throwing the hindquarters high and landing on his forelegs first (cf. impala *show-jumping*). The neck is raised showing the white throat patch but the tail, contrary to most accounts, is normally held down. Both *whistling* and *stotting* are also performed in response to predators, and the distinctions between the two forms of behavior are unclear (more under Antipredator Behavior) (7).

Even though these displays are not strictly territorial, they certainly serve to advertise a reedbuck's presence. The act of defecation, too, is made conspicuous by curling up the tail to reveal the white underside and could therefore also be a display, but perhaps not of territorial status, as both sexes defecate in the same stance (7). Secretions of the subauricular and inguinal glands may also advertise status and identity, either through traces left on resting places and passageways or by scent wafting through the air, for instance during *stotting* accompanied by *popping* of the inguinal glands (explained under Antipredator Behavior) (7, 10).

Fig. 6.2. *Medial-horn presentation* by adult male common reedbuck causes subadult male to adopt the submissive *head-low/chin-out posture.*

AGONISTIC BEHAVIOR

Dominance/Threat Displays: *erect posture, low-horn presentation, head-nodding, angle-horn, head-turned-away, defecation, chasing.*

Defensive/Submissive Displays: *head-low/chin-out posture* with tail out (fig. 6.2); *lowered forequarters; lying-out.*

Fighting: *boxing, clash-fighting, pushing, twist-fighting.*

Encounters between territorial males and their neighbors or intruding strangers may include any of the above dominance and threat behaviors plus *displacement grazing* and *grooming,* and *chasing. Low-horn presentation,* which places the animal in the combat attitude, and a slow, deliberate approach are a prelude to fights. The forward hook of the horns combined with strong forward-downward blows in *clash-fighting* makes combat perhaps more dangerous than usual (10); fighting was considered responsible for ⅓ of male deaths in a Natal population where common reedbucks live at unusually high density (12).

During play fights and chases, young reedbucks have been seen to lower their forequarters in the manner of dogs inviting play (cf. klipspringer). There may be an element of submission in this posture, for young males and females approaching to sniff noses with an adult male sometimes behave in the same way (7).

REPRODUCTION. Reedbucks do not have a strict breeding season, though, as usual, there is a rainy-season breeding peak. Female mountain reedbucks may conceive at 1 year and reproduce at intervals of 9–14 months. Gestation is thought to be around 7½ months.

SEXUAL BEHAVIOR: *lowstretch* with *empty licking* and tail spread; *responsive urination* by female and *urine-testing; foreleg-lifting; preliminary mounting.*

Reedbuck reproductive behavior has not been studied closely enough to describe differences between the species. The following is a general account.

Females approached by a male in *lowstretch* adopt the *head-low posture* (fig. 6.3), then squat and urinate in the same posture adopted during defecation. Sometimes a male intent on sniffing a female's vulva flicks his tongue while approaching. An unresponsive female may run away while the male is occupied in *urine-testing,* and is seldom pursued.

Fig. 6.3. Common reedbucks in *mating march.*

During the *mating march,* the male persistently noses and licks the female's rump and makes repeated mounting attempts, but rarely *foreleg-lifts.* In rearing he rests his forequarters on the female's croup but without clasping her, thereby pushing her forward unless she stands firmly—signaling readiness to mate. A female common reedbuck courted by an outside male thwarted his mounting attempts by *circling,* moving out from under him, and even kicking. Courtship terminates with a single thrust during which the male clasps her flanks tightly. He then dismounts, the pair stand immobile for some seconds, and the male may lick her neck and rub his head on her body. The female often licks her own flanks and back, then they resume grazing (7).

PARENT/OFFSPRING BEHAVIOR. Female common reedbucks seek seclusion beginning a month before calving and during the lengthy concealment period (1½–2 months) remain close to the calf (usually within 20–30 m) (7). Last year's calf resists separation, even when chased, and its presence may be tolerated following parturition. The newborn calf is suckled only once a day and probably once or twice at night. The mother comes right up to her calf when retrieving it, apparently without signaling visually or vocally. First she licks it, from rump to head, stopping when the calf begins nursing (2½–4 minutes). Young over 4 months old are no longer licked but can still solicit grooming by approaching, touching noses, and raising the head to indicate the spot to be groomed (usually the head or ears). After an activity-play period of 10–30 minutes the calf seeks a new hiding place. Those older than 2 months graze alongside their mothers but run for cover if alarmed. The female rejoins her mate when the calf is about 4 months old; even then calves withdraw into cover to rest more often and feed less often than adults.

ANTIPREDATOR BEHAVIOR: *skulking, lying-out, alert posture, whistling, stotting ± popping, snorting.*

A disturbed reedbuck may whistle intermittently for as long as ¼–½ hour, especially at night after getting wind of, say, a stalking leopard. In the bohor reedbuck, *whistling* may precede and follow, and often accompanies and emphasizes, the typical *rocking gallop. Stotting,* in which the reedbuck bounds up and down almost in place, represents a higher state of excitation and usually involves 3–8 jumps (7). According to Jungius (7), the common reedbuck does not whistle while galloping, but instead a snorting sound is produced at each jump as the head is thrown back and air is forced through the nostrils. A number of observers have heard fleeing reedbucks make strange *popping sounds,* which they attributed to the sudden opening of the pocket-like inguinal glands. At the height of a bound, disturbed common reedbucks of both sexes have been seen to throw the hindlegs backward and outward, which coincided with a double pop at each jump (7, 10).

Lacking endurance, the reedbuck's strategy is to disappear into cover as soon as possible. In an emergency it dashes in a flat run, interspersed with long bounds. If already in cover, adult as well as young reedbucks will often lie close until a man or other predator gets within a few m. If a standing reedbuck sees a predator but thinks itself undetected, it will crouch or, if on the edge of cover, skulk into hiding. Reedbucks are more likely to be hunted successfully by leopards than other antelopes that avoid cover. They are also relatively easy and hence preferred prey for cheetahs and wild dogs during the dry season when cover is scarce.

Despite the absence of armament, mother reedbucks may vigorously defend their fawns—which are feeble runners for at least the first month. Females have been known to charge and tree a troop of baboons, and to rescue a calf about to be caught by a domestic dog by kicking out while jumping over the dog. The latter is an old and possibly unreliable story (11). Mountain reedbucks take advantage of their ability to bound rapidly up steep slopes and over stony ground, but also *lie out* like the other two species when there is enough cover for concealment.

References

1. Fryxell 1980. 2. Hendrichs 1975*b*. 3. Hofmann 1973. 4. Holsworth 1972. 5. Irby 1976. 6. ——— 1979. 7. Jungius 1971*a*. 8. ——— 1971*b*. 9. Kingdon 1981. 10. ——— 1982. 11. Millais 1899. 12. Ventner 1979. 13. Vesey-FitzGerald 1967.

Added References

Howard 1986*a,b*.

Fig. 6.4. Uganda kob territorial male standing in *erect* ("proud") *posture.*

Kob

*Kobus kob**

TRAITS. Size and general appearance similar to impala, but heavier build, different markings. *Height and weight* (10): males 90–100 cm, 94 kg (85–121); females 82–92 cm, 63 kg (60–77). *Horns:* 40–69 cm. *Coloration:* general color golden to reddish brown with white throat patch, eye ring, and inner ear, and black-fronted legs. Males darken as they mature; mature male white-eared kob, *K. k. leucotis,* are ebony-colored. *Glands:* well-developed inguinal glands in both sexes (pouches 2.5–5 cm deep, secretion yellow, waxy with pungent odor); preorbitals simple (without orifice) but functional, secreting a pleasantly scented oil (1).

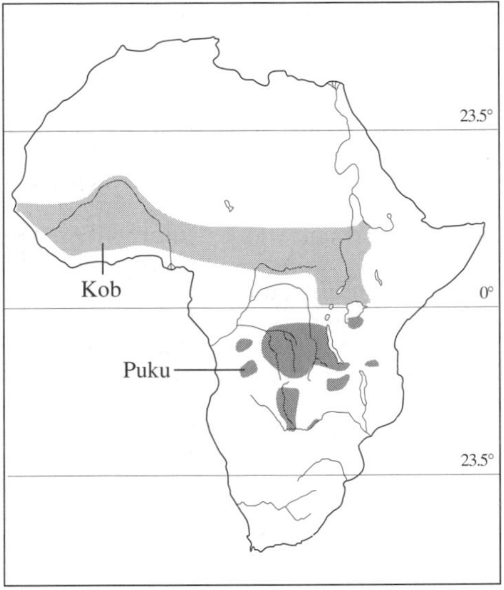

DISTRIBUTION. Northern Savanna.

RELATIVES. The puku is the Southern Savanna version of the kob, smaller (77–83 cm, 66–77 kg) and paler (tanner, without black markings) with shorter (40–54 cm), thicker horns. They are about as different as the common and defassa waterbuck, classified as separate species mainly because there are no hybrids between them, reflecting their geographical isolation. In time, separated populations become separate species; kob and puku may be considered somewhere between subspecies and full species, that is, superspecies (2). The kob has a much wider distribution and reaches very high population densities on major floodplains. An estimated 840,000 white-eared kobs range the vast floodplains of Southern Sudan, where up to 1000/km^2 concentrate near water during droughts (8, 9). The puku has a comparatively limited, patchy distribution in the Southern Savanna and fails to reach comparably high densities.

ECOLOGY. The kob probably evolved from a reedbuck-like ancestor. It remains closely tied to floodplain grasslands but is not cover-dependent (although kobs may use dense cover for shade—J. Fryxell, pers. comm.), and avoids flooded ground and steep slopes; but its preference for perennial grasses in early, palatable stages and the need to drink daily ties it to green pastures in well-watered areas (9, 10). During the rainy season, kob concentrations form in areas of short grass and higher, dry ground, keeping these pastures short while ungrazed grasslands grow tall and rank. In the dry season, they keep to the contracting greenbelts bordering waterlogged ground, becoming ever more concentrated. But dry-season fires that stimulate regrowth of low-lying pastures can completely alter kob distribution.

SOCIAL ORGANIZATION: sedentary-dispersed or -aggregated; migratory-aggregated; conventional and lek territories.

At average and low density, the kob and puku have the conventional territorial, sedentary-dispersed type of social organization (11). Territorial males spaced at least 100–200 m apart occupy the best habitat, including all areas utilized by females, which with their calves live in herds usually numbering 5–15 and up to 40 head. These herds are loosely structured, with open, changing composition and size from day to day as the animals move about their range in search of greener pastures. After weaning, young males join bachelor herds containing males of all ages, which are generally kept separate from the females by the intolerance of territory-holders. Both daily and seasonal movements are usually led by females (11).

On floodplains where the Uganda kob reaches year-round densities of 40–50/km^2 (Uganda's Semliki Valley) up to 61/km^2 (Achwa, Uganda [10]), perhaps ⅔ of the territorial males defend conventional territories; the other ⅓ cluster on traditional territorial breeding arenas or leks that may be no larger than a single conventional territory. An arena 200–400 m in diameter may be shared by 30–40 males, spaced rather evenly over the area at 15–30 m intervals (7, 9). That means conventional territories of say 50 ha become compressed on leks to much less than 1 ha (fig. 6.5)!

Arenas stand out as lawns of short grass and/or bare ground surrounded by taller grassland. The continual presence of resident males and transient females wears away the cover, leaving nothing substantial to eat. The fact that 8 or 9 of every 10

*This account is based mainly on studies of the Uganda kob, *K. k. thomasi.*

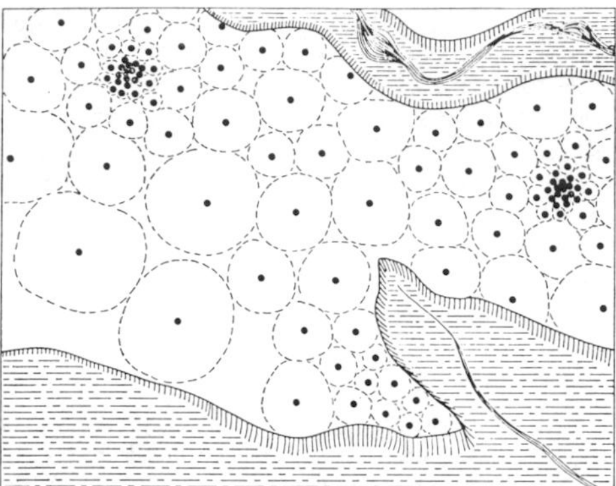

Fig. 6.5. Schematic drawing of kob territories, showing close clusters on lek breeding grounds (from Leuthold 1966).

females come to the arenas to mate makes it worthwhile for males to forgo space and food for an exceptional reproductive advantage. Females and bachelor males, often associated in amorphous large herds of up to 2000 head, circulate around an arena, which is usually located in the middle of the best pastureland, near water holes or along well-traveled trails. On conventional territories outside the arenas female herds typically number 30–50 head. But when the time comes to breed, most of these females also pay a visit to the nearest arena (see Sexual Behavior) (4, 7, 11).

SOCIAL BEHAVIOR

COMMUNICATION. Discussed in tribal introduction.

TERRITORIAL BEHAVIOR: *erect posture, prancing, whistling, grunting, object-horning.*

AGONISTIC BEHAVIOR

Dominance/Threat Displays: deliberate approach and *lateral presentation* in *erect posture, looking-away, head-nodding* and *-shaking, penile erection, object-horning, low-horn presentation.*

Defensive/Submissive Displays: *head-low posture; snapping* and *butting* (females).

Fighting: *horn-clashing, front-pressing, twist-fighting.*

Aggression between territorial Uganda kobs is highly ritualized and rarely includes fighting, either on standard territories or breeding arenas. If a neighbor trespasses, for instance while following a departing female, the owner need only walk firmly toward him in the *erect posture* (fig. 6.4) to turn him back (5). Sometimes he chases the intruder, uttering low grunts. Confrontations across a common boundary, frequent on the arenas, also feature the *erect posture:* the ears are held back or horizontally, tails wag sporadically, and attention is directed to the throat patch by abruptly *looking-away.* At higher intensity, a pair may stand confronting in *low-horn presentation,* which readily leads to a brief clash, usually preceded by touching noses. *Object-horning, head-shaking,* and *displacement grazing* and *grooming* are often part of the performance. Kobs commonly have partial erections during such encounters (5).

Fights are much more frequent between males on the leks of the white-eared kob, among which a restricted mating season makes for more intense reproductive competition (9). In the Uganda kob, severe, even fatal fights also occur, usually when an outsider attempts to take over a territory. Combatants engage in violent *clash-fighting, front-pressing,* and *twist-fighting,* meanwhile briskly wagging their tails. Apart from the considerable danger that the contestants will hurt one another, sometimes an overstimulated neighbor launches a damaging attack from the rear or the side (5, 9).

Turnovers are rare on conventional territories, whose owners commonly keep their places for a year or 2, at least, but are commonplace on the arenas. There the competition is so fierce that few males manage to hold central positions for even a week and many last only a day or 2. The spot where males and estrous females are most clustered is centrally located and there 3 or 4, at most 6 or 7, males, whose territories converge on this point like the slices of a pie, monopolize copulations for the period of their tenure (7). With literally dozens of fit have-nots in the running for a central "court," a male who vacates his place may find it occupied by a well-rested newcomer or ex-proprietor when he gets back. Yet males *must* leave to feed and drink. They reduce the risk of being supplanted by going during lulls in mating and other activity. But they get too little food and rest to make up for the energetic costs of holding a central position, and so must eventually withdraw from competition. After recuperating a week or two, a male can become fit enough to try again. Thus, in the bachelor herd nearby every kob breeding lek, there are always males in training to gain or regain central territories, and it is this reserve that makes for a brisk turnover (4, 5).

REPRODUCTION. Breeding arenas may be established wherever kob populations reach high densities, as they do on broad floodplains of the Northern Savanna. The studies of Uganda and white-eared kob have established that lekking occurs respectively in perennially breeding sedentary populations and in seasonal-breeding migratory populations (cf. lechwe lekking system). Leks have not been reported in the puku, which breeds the whole year with a rainy-season peak but does not occur in comparably large concentrations (6).

After beginning to ovulate at 13–14 months, females come into estrus every 20–26 days until bred; gestation takes c. 8 months, and in perennial breeders ovulation resumes 21–64 days after calving. Males mature at three years (3, 9).

SEXUAL BEHAVIOR: *erect posture* with *nose-lifting, prancing, foreleg-lifting*. Female response: *circling, symbolic biting, butting, swan-neck posture.*

Male courtship behavior differs on standard and arena territories. Males with large grounds try to keep females from leaving by chasing and herding them. Arena males cannot keep females on their postage-stamp courts from escaping, although they try. A male stands in the *erect posture* as a female approaches, often *prancing* around with tail raised and *nose lifted* to flash the throat patch, and *penis unsheathed* (fig. 6.6). Given the opportunity, he then proceeds to check her reproductive status by sniffing her rump and stimulating urination and, since most females who venture onto an arena are already in estrus, this leads into the *mating march,* in which the male follows closely with head up, contacting her body with raised, straightened foreleg whenever she stops (fig. 6.7). The female may try to avoid this contact by moving in a tight circle, sometimes *nipping* or *butting* the male's hindquarters. The final precopulatory stage begins with multiple *preliminary mountings,* culminating in intromission and ejaculation (fig. 6.9). The female stands with back hunched, head horizontal or up, hindlegs straddled and tail held to the side (*swan-neck posture*). The whole courtship sequence may be as brief as 2–3 minutes (average 5.5 minutes), and copulation takes only 1–2 seconds (4, 5).

Post-copulatory phase. After dismounting, the male stands still for 2–5 seconds with back arched, penis still erect and moving spasmodically, and whistles loudly 1–5 times. His neighbors and sometimes nearby females respond in kind. Follow-

Fig. 6.6. Kob courtship: male *prancing* in *erect posture* (from a photo in Buechner and Schloeth 1965).

Fig. 6.7. Kob courtship: *foreleg-lifting* during mating march (from a photo in Buechner and Schloeth 1965).

Fig. 6.8. Male licking genitalia of estrous fermale which stands in the *swan-neck posture* (from a photo in Buechner and Schloeth 1965).

Fig. 6.9. Kob copulation (from a photo in Buechner and Schloeth 1965).

ing this display, the male licks his penis, then approaches his mate again, who meanwhile has remained in the *swan-neck posture* (fig. 6.8). He licks her genitalia, and nuzzles and often licks her inguinal area as well. Then one or both graze or lie down. If the female stays around, the male may start courting again within 10–15 minutes. During the 1-day estrus a female may be served up to 20 times, by 1 or more of the central males.

PARENT/OFFSPRING BEHAVIOR. Calves lie concealed in high grass for about a month (12). They are retrieved and suckled morning and evening. Newborn calves develop the flight response within a few hours and thereafter bolt when approached closely. The innate following response is seen in the attempts of calves still too weak to stand properly to follow and suckle any moving object. But imprinting on the mother usually occurs within a day. Both mother and young use their noses to identify one another, but can also recognize individuals by voice and probably by sight. Calves past the hiding stage join crèches and rarely go into high grass, but often rest together in available shade, whereas adults regularly lie in the midday sun without apparent discomfort. At 3–4 months the young join female herds, where they continue to associate with their mothers until after weaning at 6–7 months, when males begin to join bachelor herds. Female calves may remain attached up to 9 months, when Uganda kob mothers in good condition calve again. White-eared kobs calve once a year, late in the rains (Sept.–Dec.) (9).

ANTIPREDATOR BEHAVIOR: *alert posture* with tail out or up; *alarm whistling; stotting.*

No detailed study of kob antipredator behavior has been made, although at high density it is probably the main prey of associated large carnivores. Although the kob and puku are considered slow and therefore vulnerable, it is hard for predators to approach aggregations undetected, and kobs make prodigious leaps in an emergency (Fryxell, pers. comm.). Like the lechwe, kobs may run for the nearest water or reed bed when pursued by cursorial predators (13).

References

1. Ansell 1960*b*. 2. ——— 1971. 3. Buechner, Morrison, Leuthold 1966. 4. Buechner and Roth 1974. 5. Buechner and Schloeth 1965. 6. DeVos and Dowsett 1966. 7. Floody and Arnold 1975. 8. Fryxell 1980. 9. ——— 1985. 10. Kingdon 1982. 11. Leuthold 1966. 12. ——— 1967. 13. Mitchell, Shenton, Uys 1965.

Added References

Fryxell 1987*a,b*. ——— 1988. Mühlenberg and Roth 1985.

Fig. 6.10. Territorial lechwes on a breeding arena (lek). (From a photo in Schuster 1976.)

Lechwe

Kobus leche

TRAITS. Koblike but bigger and shaggier (including neck mane), with overdeveloped hindquarters and longer horns. *Height and weight* of Kafue lechwe, *K. l. kafuensis:* males 104 cm (99–112), 103 kg (86.5–128); females 97 cm (90–106), 79 kg (61.6–97) (12). *Horns:* 70 cm (45–92), lyrate, relatively thin. *Coloration:* bright chestnut with white underparts, throat, and facial markings, and dark leg and body markings of variable extent depending on subspecies and age (males darken with age and perhaps seasonally), ranging from the black lechwe (*K. l. smithemani*) of Lake Bangweulu to the red lechwe (*K. l. leche*), with the Kafue lechwe intermediate. Hooves adapted for swampy terrain, elongated with wide splay and naked pasterns. *Scent glands:* absent or undeveloped, but coat greasy with distinct smell (especially males).

DISTRIBUTION. Floodplains bordering swamps in the Southern Savanna, centered in Zambia. The Kafue population, which possibly numbered ½ million a century ago, numbered 50,700 (± 17,500) in 1987 (7), a sharp drop from the 100,000 counted in 1971, before construction of hydroelec-

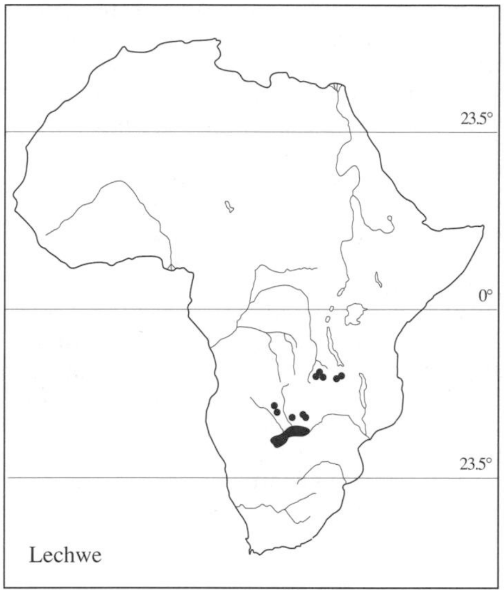

tric dams that altered the natural flooding cycle (2, 5, 10, 12). The black lechwe population, estimated at 150,000 in the 1930s, dropped to probably below 30,000 when the water level of Lake Bangweulu rose in the late 1940s, eliminating much of its habitat, but had increased to 41,000 (± 15,000) by 1983, despite poaching (5, 7). Smaller populations of the red lechwe inhabit the floodplains bordering other major rivers and swamps of Zambia, Angola, and Botswana.

RELATIVES. An estimated 30,000–40,000 Nile lechwe (*K. megaceros*) live on both sides of the White Nile in the Sudd, the most extensive papyrus swamps in Africa. Females of the 2 species look much alike but the mature male Nile lechwe is tied with the white-eared kob for gaudiest member of the tribe. It turns a dark chocolate brown with a broad white saddle on neck and shoulders as well as white neck, face, and leg markings. This species remains unstudied.

ECOLOGY. Next to the sitatunga, the lechwes are the most aquatic antelopes. They specialize in the abundant and highly nutritious grasses growing in inundated water meadows bordering major rivers and lakes, pastures out of reach for other bovids most of the year (5, 16, 17). Lechwes feed in water up to their bellies and even their shoulders. They may wade up to several km from shore, finding shelter and out-of-water resting places in the reeds. Here they overlap with sitatungas. Preference for the grasses and sedges that keep growing in shallow water makes the red lechwe literally an edge species and, since the shoreline advances and withdraws across the floodplain during the annual flooding cycle, lechwe populations tend to remain concentrated and mobile most of the time. At low water, they move into the permanent swamp, and as the waters advance they advance with them (1, 9, 12, 15). They manage thus to remain nearly the whole year on green pastures that can support 1000 lechwes/km^2 (2, 13). But very large populations occur only on the flattest plains, where extensive areas of water meadow are maintained throughout the flood cycle. Even on the Kafue Flats, once ideal habitat, *K. l. kafuensis* along with all other antelopes was forced off the floodplain into the woodland margin during years of exceptional flooding. *K. l. leche* of Botswana's Chobe River floodplain makes a regular migration which covers a distance of 80 km during years of very high water (3). During cool dry weather lechwes need not drink, but in dry hot weather they may go to water up to 3 times a day (15).

SOCIAL ORGANIZATION: highly gregarious, conventional territories and temporary leks during breeding peak; populations either mobile-aggregated or sedentary-dispersed.

The conventional arrangement of female and bachelor herds kept segregated by territorial males can be seen in small, resident populations like that of the Moremi Wildlife Reserve, but breaks down in large concentrations, although different sex and age classes are usually discernible: females and calves predominate nearest the water and males predominate inland on drier ground (4, 9, 12). The separation is caused partly by the greater dependence of females with young on proximity to a water refuge, and partly by the efforts of accompanying territorial males to exclude possible rivals. Yet in contrast to most territorial species, lechwes have minimal sexual segregation during and maximal segregation outside the breeding season (12). Most adult males only compete for territories during the rut, passing the rest of the year in bachelor herds. But approximately one-quarter of the females ovulate outside of the peak, enough to justify the optimism of males who maintain territories during other times of year. At high density, rutting Kafue lechwes become lek breeders (4, 14) (see under Territorial Behavior).

Female herds are completely open and subject to constant change, apparently with no lasting associations except between mothers and young of the year, and no particular leaders (9, 15).

ACTIVITY. Lechwes are most active from before to several hours after sunrise, and from about 2 hours before to an hour after sunset. In between, most animals rest on hot days, preferably on dry ground such as termite mounds, levees, and islands, and in the shade. Adult males appear to rest more than adult females and young males (4). Lechwes probably have at least 1 long grazing period at night (5), but published information on nocturnal activity is scant.

POSTURES AND LOCOMOTION. In becoming semiaquatic, lechwes have developed unusual ways of moving. They run in a bounding gallop, the forefeet and hindfeet moving in pairs, chin outstretched so that the male's horns lie back on his shoulders. This gait looks and is clumsy and slow on dry ground, but perhaps no other antelope can run as fast through water (11), or is as efficient in mud and shallow water. The overdeveloped hindquarters provide added thrust and the elongated hooves help to support the animal on soft bottoms. Startled lechwes also make spectacular leaps and scatter in different directions like impala, a tactic that may confuse predators (cf. kob) (9). Individuals disturbed in the open near cover adopt a *skulking posture* preparatory to sneaking or trotting toward a refuge (fig. 6.11).

Fig. 6.11. Lechwe headed for cover in *skulking attitude.*

SOCIAL BEHAVIOR

COMMUNICATION. Discussed in tribal introduction.

TERRITORIAL BEHAVIOR

Advertising: *erect posture, object-horning, grunting, penile display.*

The lechwe *erect posture* (fig. 6.10), with chin horizontal, shows the long, backswept horns of adult males (4 years and over) to advantage and, together with darker coloration, makes them stand out in a crowd of females and immature males. Males also wag their tails vigorously and often have erections during aggressive interactions. *Ground-* or *vegetation-horning,* and *whinnying grunts* (rarely faint whistles reminiscent of the kob's vocal display), stimulated by other moving or calling lechwes, complete the repertoire of common territorial advertising displays.

Leks. Lechwe breeding arenas are neither perennial nor fixed at traditional locations, at least partly because annual variation in the flooding cycle makes the locations of concentration areas unpredictable. Both the arenas and the behavior of the males who establish them are otherwise very similar to the kob's. The breeding peak comes early in the rains (December to January in Zambia), while the waters are beginning to spread across the floodplain from their lowest point. As the concentration zone moves with the advancing water, so do the territorial arenas. Because of mobility and crowding, not only territorial males have access to females; nonterritorial males, including even subadults, may be seen courting. Restriction to a conventional territory under these conditions could conceivably be a liability (9).

It happens, nevertheless, that most females visit an arena the day they enter estrus and end up mating with 1 or more of the central males (14). The lechwe rut is a time when large numbers of males, being in peak condition following months of grazing on emergent water meadows, make their bid for immortality. Yet only those that manage to win places on an arena are in a favorable position to breed and of those, the few central males monopolize most of the mating until their stamina fails and they are replaced.

Lechwe leks are located on the highest and openest available ground near water but are only recognizable as such by the presence of adult males standing around (rarely feeding or resting) looking noble (fig. 6.10). The typical Kafue arena consists of 50–100 males within a roughly circular area about ½ km across. Inter-male spacing decreases to a minimum of 15 m at the center. Here 10–20 females, presumably all in heat, may be found clustered tightly on the land controlled by 1 or 2 superstuds. All the other females remain in large mixed herds adjacent to the leks (14).

In the Moremi Reserve, where good lechwe habitat is patchy, males defend the favored waterfront areas as conventional ter-

Fig. 6.12. Territorial neighbors interacting on their border: erections during agonistic encounters are a tribal trait.

ritories. They behave like their arena counterparts, but have aggressive and sexual encounters less frequently (9).

AGONISTIC BEHAVIOR

Dominance/Threat Displays: *lateral presentation* in *erect posture, head-high-turning, head-shaking, object-horning, low-horn presentation, parallel-walk.*

Defensive/Submissive Displays: *head-low posture.*

Fighting: *air-cushion fighting, clash-fighting, front-pressing, twist-fighting, parallel-fighting.*

Much the same behavior that is used to advertise territorial status is employed and directed at adversaries during aggressive confrontations. The passage of females (or males) through territories provokes most confrontations, as the proprietor follows the female up to the boundary line and there meets his neighbor advancing to escort her onto his ground. The first male draws up abruptly in the *erect posture* and gives a vigorous *head-shake.* His neighbor follows suit, standing 2–3 m away (fig. 6.12). This display alternates with *object-horning* and *displacement grazing* and *grooming.* More intense encounters include *head-high-turning* (turning the head away while standing in the *erect posture*), *low-horn presentation, air-cushion-fighting,* and perhaps brief horn contacts (4, 8, 14). Such border engagements may last from 5 up to 20 minutes, with display bouts divided by long intervals of grazing. Erections are frequent during displays. The rare serious fight occurs, as in the kob, when an intruder seeks to oust a resident male. It usually begins without preliminaries and features violent *clash-fighting,* followed by *front-pressing,* twisting and turning with horns firmly seated (fig. 6.13). Combatants sometimes stand almost parallel and occasionally one is thrown sideways to the ground. The loser runs off suddenly, pursued by the winner who pulls up at the border and displays triumphantly. Sparring and fighting in bachelor herds and between territorial males increases in the months prior to the breeding peak (4).

Females rarely behave aggressively, but sometimes threaten in the manner of males, butt another lechwe in the flank, or butt heads (9).

REPRODUCTION. *Kobus leche* has a definite breeding season throughout its range: for ca. 2½ months during the rains, typically between November and February (12). Testes size increases before and decreases following the rut, a fact that helps to explain the more sporadic territorial behavior during the dry season. Females may breed as early as 1½ years. Males become adolescent then but only mature physically in their 5th year, when maximum testes weight is also achieved. After an estimated gestation of 7–8 months, some ⅔ of the annual calf crop is dropped during a two-month birth peak: mid-July to mid-September on the Kafue Flats (12).

Fig. 6.13. Lechwes *clash-fighting* (*top*), and *front-pressing* (*bottom*). (Redrawn from Lent 1969.)

SEXUAL BEHAVIOR: *erect posture, prancing, lowstretch, foreleg-lifting, penile display, preliminary mounting.* Female: *head-low posture, butting, swan-neck posture.*

Lechwe courtship has not been described in detail but looks very similar to the kob's. A male meets an approaching female in the *erect posture, prances* before her, goes behind her in *lowstretch,* tests her urine, and during the *mating march foreleg-lifts* (typically with the foreleg bent), tail switching and often with a partial erection. He noses the female's rump and makes many *preliminary mounts* before finally copulating in the typical reduncine posture (see fig. 6.9). In addition to adopting the submissive *head-low posture,* a female may butt suitors in the flank and show other avoidance behavior before finally accepting mounting. Receptiveness is signaled by adopting the *swan-neck posture* (see fig. 6.8). Postcopulatory behavior, if any, is undescribed.

PARENT/OFFSPRING BEHAVIOR. Females seek cover and dry ground for calving, either singly or in small groups, and calves remain concealed for 2–3 weeks (15). Mothers retrieve and suckle them early and late in the day, spending the rest of the time feeding and resting in the open. Weaning may occur at 5–6 months. Once out of hiding calves in dense populations join together in crèches of up to 50 young which are largely independent of the mothers; crèches are small or nonexistent in low-density, dispersed populations (9). Even then, reportedly only very young calves follow their mothers closely and, when fleeing, mother and offspring head in different directions without first making contact (9).

ANTIPREDATOR BEHAVIOR: *skulking* into cover (fig. 6.11); fleeing into water; scattering with leaps and bounds (confusion effect); *alarm snorting (whistling).*

Lions, leopards, spotted hyenas, and wild dogs are all major predators on adult lechwes, and even a hawk eagle (*Hieraaetus spilogaster*) has been seen killing calves (1). Drowning during stampedes into deeper water is another common source of calf mortality.

Dependence on water as a refuge from predators causes lechwes to cluster together along the shoreline before dark descends. Yet lions and leopards readily enter the swamps to hunt them. In Lake Bangweulu the lions use the numerous termite mounds as convenient places to eat their kills and rest up between hunts. But man is the lechwe's most serious predator by far. On the Kafue Flats it was formerly the practice during peak flooding to hold *chilas,* communal hunts in which hundreds of people in canoes speared as many lechwe as they could. Under these circumstances the lechwe's instinct to seek safety in the water was disastrous, and the most vulnerable members of the population were pregnant females (15). An estimated 5000 lechwes were killed on one *chila* (10).

Nevertheless, the lechwe's inability to outrun hyenas or wild dogs on dry land makes its dependence on flooded grasslands not merely nutritional but a matter of life and death.

References

1. Allen 1963. 2. Bell et al. 1973. 3. Child and von Richter 1969. 4. DeVos and Dowsett 1966. 5. Grimsdell and Bell 1975. 6. Hillman and Fryxell 1988. 7. Jeffery, Bell, Ansell 1989. 8. Joubert 1972. 9. Lent 1969. 10. Mitchell and Uys 1961. 11. Rees 1978. 12. Robinette and Child 1964. 13. Sayer and van Lavieren 1975. 14. Schuster 1976. 15. Smithers 1983. 16. Vesey-FitzGerald 1965. 17. ——— 1973.

Fig. 6.14. The waterbuck *greeting ceremony:* young-adult male (*right*) displays submission to territorial male displaying dominance.

Waterbuck

Kobus ellipsiprymnus

TRAITS. One of the heaviest antelopes, long body and neck with short, stocky legs, coarse-haired with a neck mane and ruff. *Height and weight:* (Uganda population) males 127 cm and 236 kg (198–262), females 119 cm and 186 kg (161–214) (5). *Horns:* 75 cm (55–99), curved forward, massive. *Coloration:* variable, grizzled gray to red-brown, darkening with age, lower legs black with white ring above hooves; white rump patch (*K. e. defassa*) or elliptical ring (*ellipsiprymnus*), underparts, throat patch, ear-linings, eyebrows, and snout, contrasting with bare, black muffle. *Skin glands:* none in the usual sites but numerous, diffuse sebaceous follicles emit a cloying, musky smell and make the coat oily (5).

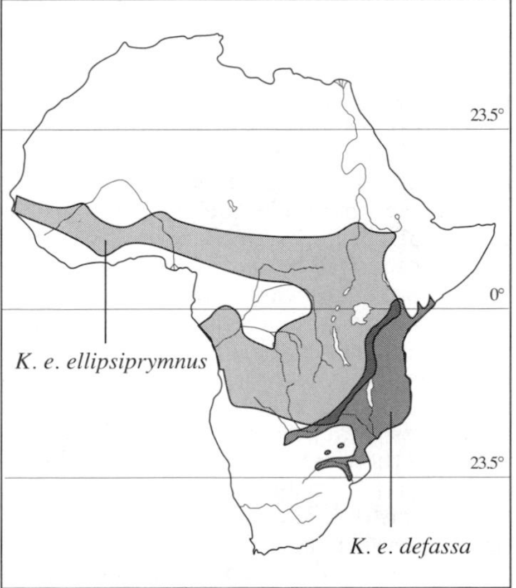

DISTRIBUTION. The common and *defassa* subspecies occur respectively east and west of the Rift Valley and its extensions, the Luangwa and Middle Zambesi valleys in Zambia and Zimbabwe. The two forms overlap and hybridize in five different places, most notably in Nairobi Park and Ngorongoro Crater (3).

ECOLOGY. The waterbuck is possibly the most water-dependent of all antelopes, with even less ability than a domestic steer to withstand dehydration in hot weather (6). Being a grazer that has to drink often, it is limited to grassland habitat within a few km of water. The combination of requirements for cover, open grassland, and water makes for a patchy ecotone distribution along drainage lines and within valleys. This antelope also prefers dry ground, although known to enter water to escape predators and sometimes to feed (5). Sedentary by nature, it is most dispersed in the wet season, when females with calves tend to frequent woodland, and spends more time in open grassland during the dry season while concentrated near water (1, 3). Low-lying bushed areas and bushy slopes are normally avoided; females like to have some but not too much cover within reach (5).

DIET. Though it feeds on a wide variety of grasses, preferably of medium and short length, the waterbuck selects and apparently requires a diet rich in protein—which accords with its high water intake and urine output (6). When green grass is in short supply, the waterbuck acquires additional protein by eating other herbage and browses such dicots as acacias and *Caparris* for up to 21% of its feeding time (5).

SOCIAL ORGANIZATION: populations distributed as semi-isolated units; territorial organization but bachelor males tolerated on territories and often in proximity to females.

The patchy distribution of preferred habitat causes waterbucks to live in often widely separated ranges that are shared by a number of females and both territorial and nonterritorial males. The sizes of home ranges and territories depend on the quality of the habitat and on population density, as well as on individual age and fitness: area decreases with increasing density and age.

In Queen Elizabeth N.P., at a density of 10.5 waterbuck/km^2, the ranges of 10 marked females, plotted for nearly 2 years, averaged 600 ha and included the territories, ranging in size from 4 to 146 ha, of 6–8 males. The oldest female, estimated at 18 years (ancient indeed), had the smallest range. The 8 bachelor males that lived in the same area (the 4.4 km^2 Mweya Peninsula) had a mean home range of 100 ha, which shifted over time and included one or more territories; bachelor males were not found in areas where there were no territories, namely in habitat avoided by other waterbucks. In an enclosed area in Kenya with a much higher density of 37–54 females/km^2, the average female home range was 33.6 ha (21–61), and bachelor-male ranges were 30 ha (24–38). Territories were the smallest on record: 13 ha (4–28) (5). Even in Lake Nakuru N.P., with 72–106 waterbucks/km^2, the highest known density for this species, territories averaged c. 32 ha (10–40) (10).

Females with overlapping home ranges associate in herds averaging 5–10 (range 1–70) (5). But females are most commonly seen alone or in pairs, and herds are in continual flux, of no particular size, without a rank order or leadership, leading Spinage (5) to conclude that groups result from largely random meetings of females whose movements are essentially independent. Waterbucks do not actively repel one another like females of solitary species; but neither do they have physical contact in the form of greeting or grooming. Thus herd formation and maximum size may depend simply on the number of resident females. Nevertheless, the same author's observation that "A doe likes the company of others" (5, p. 216) suggests a positive social attraction, and the aggressive treatment of outsiders is evidence that waterbuck herds are semiclosed societies (as in sable, roan, topi, etc.). It was noted, further, that certain pairs of females tended to associate, suggesting the persistence of bonds between mothers and daughters or sisters.

Dispersal. The patchiness of waterbuck habitat and populations could readily lead to chronic overcrowding, with accompanying habitat deterioration and poor physical condition. The finding that some female as well as male waterbucks emigrate suggests an attempt to solve this problem (cf. impala) (5). Weaned calves of both sexes are treated aggressively by females (5). Some calves continue to associate with female groups nonetheless, but others become detached and either wander alone or associate in floating spinster groups. Eventually some of these females leave during the wet season and go through a solitary wandering stage, like dispersing males, while seeking a place to settle. Two marked females were relocated 30 and 32 km from their birthplaces (5).

Bachelor Males. The separation of male offspring occurs when their horns emerge at 8–9 months, provoking territorial males to chase them out of the nursery herds. After joining bachelor herds, young males remain until mature. Although bands of 40–60 males may be encountered in high-density populations, bachelor herds are as amorphous as female herds; modal group size is 2–3 and mean herd size is comparable to that of females (5.5 in Queen Elizabeth N.P.) (5). Strangers are not welcome to join, as bachelor herds are closed to outsiders. A rank hierarchy based on seniority is maintained and probably also dominance status within age classes. Bachelors regularly spar together, beginning when one male approaches and touches foreheads with another, and matches commonly involve individuals of different age (5).

Territorial Males. Given the very limited areas of suitable waterbuck habitat, and the substantial size of territories even at high density, unusually intense male reproductive competition would be expected. Could this account for slow maturation, with bucks only beginning to compete for territories at 6, 3 years after sexual maturity, and the fact that males are past their prime by age 10? The largest, most centrally located territories visited by most females are held by males of 7–9 years; younger and older males have to settle for smaller, more peripheral properties. This is the reverse of the gradation in territorial size found in a lek system. Once mature, male waterbucks remain territorial, or at any rate solitary, to the end—there are no old bucks in bachelor herds (5). Territorial tenure ranges from a few months to several years, but averages 1½–2 years (5, 10).

Proprietors only defend their property against challenges from other mature bucks, and tolerate bachelors, including young adults, as long as they behave themselves (10). But not all territorial males are so tolerant. Some keep all sexually mature males off their grounds; others tolerate them only in the absence of females, or outside the peak breeding season. Still others accept known and reject unknown bachelors, including "satellite bulls" that assist and "understudy" the territory owner (10). (The reality of this concept and its applicability to other populations are disputed [5].)

In Nakuru N.P., about half the territorial males tolerated from 1 to 3 satellites. Satellites helped to keep out other intruding males and sneaked perhaps 10% of the copulations when the territorial male was absent or otherwise engaged. But the main advantage of being a satellite was that it improved the chances of inheriting the territory when the owner died or lost fitness. There was such a surplus of males in the Nakuru population that only 7% of the adults held territories, 9% were satellites, and the rest lived in bachelor herds (10).

ACTIVITY. Studies of daytime activity in South Africa (2), Zimbabwe (7), Zambia (1), Kenya (Wirtz, unpub. report), and Uganda (5) show considerable variation in the timing and duration of basic activities, reflecting latitudinal/seasonal differences and variation in habitat, grazing conditions, distance from water, and predation pressure. Like most grazers, the waterbuck generally has peak feeding sessions at the beginning and end of the day and spends the hours between ruminating and resting, interrupted by shorter, more individualized bouts of feeding. But in Queen Elizabeth N.P., females observed during the dry season spent only an hour ruminating (from 0900 to 1000 h) after the morning feeding peak, then fed through most of the day with time out to go to water. Territorial males followed a similar schedule, but spent less time feeding (44% vs. 62%) and more time resting (15% vs. 11%), ruminating (21% vs. 17%), and in other activities (21% vs. 7%) (5). Lactating mothers spent 10% more time feeding than other females, mainly by sacrificing resting time.

Seasonal changes in the above activity pattern appeared to be relatively minor in Queen Elizabeth N.P. In regions with less rainfall and longer dry seasons, however, waterbucks may spend progressively less time feeding and more time resting as forage quality decreases. In Zimbabwe, females who spent 60% of the day feeding during the rains cut down to only 35% in the hot dry season, with little change in rumination time (7). Except to reach distant water, waterbucks probably rarely move more than a km in a day (1). Though night activity has yet to be carefully studied, waterbucks probably have at least one feeding peak, around and after midnight, but move less and have shorter, more irregular feeding bouts than by day. The first and last hours of darkness are ruminating and resting peaks (5).

SOCIAL BEHAVIOR

COMMUNICATION. See tribal introduction.

TERRITORIAL BEHAVIOR. See also under Social Organization. Territorial neighbors rarely cross the invisible border to challenge one another, and generally seem to take one another's presence for granted. But let a strange adult male appear and a territorial male instantly reacts and moves to repel him (5). There is evidence that waterbucks can sight strangers from afar and also track them by smell (5).

Advertising: standing and moving in *erect posture.*

The absence of conspicuous territorial advertising displays and the presence of full-grown bachelor males can make it difficult to single out the rightful owner of a territory. The biggest, darkest male with the longest horns is the likely candidate.

AGONISTIC BEHAVIOR

Dominance/Threat Displays: approach in *erect posture, lateral presentation* ± *high-horn presentation, head-turned-away, angle-horn, head-shaking, parallel-walking, medial-horn presentation, object-horning, chasing.* Females: *butting, charging.*

Defensive/Submissive Displays: *champing* (symbolic biting), with tail out, *head-low/chin-out* or *head-and-nose-raised;* sniffing superior's genitalia.

Fighting: *touch noses* (preliminary); *pushing* and *twisting* (*front-pressing*).

The subordinate status of immature males and satellites is reaffirmed whenever their paths cross that of the territorial male. The dominant approaches or stands in the *erect posture* as the subordinate approaches. At high intensity the displaying male stands broadside with neck arched (*high-horn presentation*) and tail held out, curved around one flank, or up (8) (cf. sable), often with an erection. He may angle his horns toward the opponent and shake his head abruptly (fig. 6.14). A subordinate male acts like an unreceptive courted female: he extends his neck, opens his mouth, and makes chewing motions (*champing,* fig. 6.14; cf. fig. 6.15). A cautious approach to the displaying male, to sniff his nose and sometimes his penis, is typical. A sudden *headshake* by the dominant at this juncture will make a young male bolt.

Aggressive encounters between equals, whether immature or mature males, proceed from dominance to threat displays and often include mutual *lateral presentation,* leading to *parallel-walking, object-horning, displacement grazing* and *grooming* (especially *cheek/shoulder swishing*), and *sparring.* As usual, aggression is least restrained among bachelors. Battles lasting up to ½ hour and deaths from abdominal and chest puncture wounds have been recorded and may be commoner than realized (5, 9). To protect against the sharp, very dangerous horns, males are shielded by skin 2 cm thick over the neck and shoulders (5).

A territorial takeover involving a satellite bull and the next-door territorial bull was photographed in Nakuru N.P. (10). It began when the 2 males met at the border. They approached stiff-leggedly and stood about 2 m apart with heads raised and ears cocked, one *looking away.* Then they came together and sniffed noses, lowered their heads and began an *air-cushion fight* which suddenly escalated when they surged forward and met with a clash. After five minutes of *front-pressing,* the territorial male suddenly broke and ran, hotly pursued by the winner, who tried to gore his backside. The satellite proceeded to take over the territory and the loser disappeared.

REPRODUCTION. Like the male, which matures at 6 years, the female waterbuck is also relatively slow to mature, perhaps rarely conceiving before 3 years old (1). Breeding is perennial in equatorial populations, with an average of 10 months between generations (= 8–8½ month gestation + post-partum estrus of 3 weeks or more) (5). At higher latitudes, as in Kafue N.P., females calve annually during the rainy season, but males maintain territories through the year nevertheless (1).

SEXUAL BEHAVIOR: *erect posture, lowstretch, empty licking,* nosing, rubbing, licking and pushing female's posterior, *foreleg-lifting, preliminary mounting, chin-resting.* Females: *urination on demand, champing* and *butting, head-low posture, swan-neck posture.*

Males stimulate females to urinate by approaching in *lowstretch,* often with *empty-licking.* A cow nearing estrus may act skittish, toss her head and *champ,* leading to pursuit. An unreceptive female adopts the *head-low/chin-out posture* (fig. 6.15). During the *mating march,* the male keeps sniffing the female's rear, rubs his face on her, and sometimes pushes her with his forehead and horns, to which a female in estrus responds by standing in the *swan-neck posture* (see kob account, fig. 6.8). In full estrus she may sniff the base of his horns and his penis. Preparatory to mounting, the male rests his chin on the female's rump while *foreleg-lifting.* A series of mounts without erection follows, which impel the female forward unless she braces herself. At that point the male mounts with erection, grasps her loins, intromits, and thrusts while standing bolt upright (3); sometimes his feet leave the ground. Afterward the pair stands immobile for up to 4 minutes. The male may then begin courting again but usually over ½ hour elapses between copulations. Females generally copulate with 1, sometimes with 2 males on their day of estrus. Where females conceive within a few weeks of calving, resident territorial males often remain in attendance (5).

PARENT/OFFSPRING BEHAVIOR. Expectant mothers isolate in thickets a couple of days before calving, which occurs typically early in the morning. The preferred location is the place the female first calved (5). Calves often take ½ hour to gain their feet but are agile enough within a day to outrun a man, and they tend to bolt rather than freeze when discovered (5). During the

Fig. 6.15. Female leading calf past a dominant male adopts *head-low/chin-out posture, champs* and *bobs* her head (from a photo in Spinage 1969).

2–4 week concealment stage, mothers suckle them 5 minutes at a time 3 times a (24-hour) day, summoning calves with a low bleat or moo. Calves then gambol, racing in circles, *alarm-trotting* and *stotting*. Mothers go up to ½ km from the hiding place by day and stay nearby at night, often alone. They frequent woodland while calves are young. Month-old calves no longer seek hiding places spontaneously, so mothers have to sneak or run off while the calf isn't looking if they want to park them. To persuade a calf to follow, the mother holds her tail out or up, a follow-me signal unknown or unusual in other antelopes (5). Weaning occurs at 6–8 months. Waterbuck calves wander around on their own, alone or in crèches, surprisingly often. Usually but not invariably, a herd is nearby to which the calves quickly run when frightened.

ANTIPREDATOR BEHAVIOR: *alert posture, snorting, stotting, trotting,* running into cover, *lying-out,* self-defense (males).

Because waterbuck fat imparts a strong, musky odor to the meat unless carefully removed (5), the notion that lions and other predators do not readily hunt and eat this species, though untrue, has been widely accepted. Actually, in Kruger N.P. proportionally more waterbucks fall prey to lions, hyenas, and crocodiles than some other more numerous antelopes (4). Lions take more males than females, even though males who find themselves in close proximity to a lion have been known to confront it rather than flee (4). The tendency of young calves to wander about alone may make them unusually vulnerable to hyenas.

References

1. Hanks, Stanley Price, Wrangham 1969. 2. Herbert 1972. 3. Kiley-Worthington 1965. 4. Pienaar 1969*b*. 5. Spinage 1982. 6. Taylor, Spinage, Lyman 1969. 7. Tomlinson 1979. 8. ——— 1980. 9. Verheyen 1955. 10. Wirtz 1982.

Fig. 6.16. Rhebok courtship: male *foreleg-lifting* to female.

Vaal or Gray Rhebok

(Tribe Peleini)
Pelea capreolus

TRAITS. Medium-sized antelope with long neck, bulbous nose, long, narrow ears, and upstanding spikelike horns, 15–25 cm long, ringed at base, males only. *Height and weight:* 70–80 cm, 19–30 kg, male c. 10% bigger than female. *Coloration:* short, rabbitlike fur, varying shades of gray, young often browner than adults, legs with dark stripe on lower, front sides, muzzle with dark blaze; underparts lighter, inner thighs and undersurface of 15–30 cm tail white; hairless areas black, including muffle, hard palate, tongue, penis, inside of ears, and eye conjunctiva. *Scent glands:* male's preputial gland secretes copious, deep-black, pasty substance with penetrating smell; no preorbital, subauricular, or inguinal glands (4).

DISTRIBUTION. Found only in South Africa, it typically occurs in grassland habitats above 1000 meters, but its range also extends to the coastal belt in Cape Province, where it occurs almost at sea level among the dunes (2).

RELATIVES. This antelope has no close relatives and is put in a tribe by itself. It has even been suggested that the rhebok could descend from ancestors of sheep

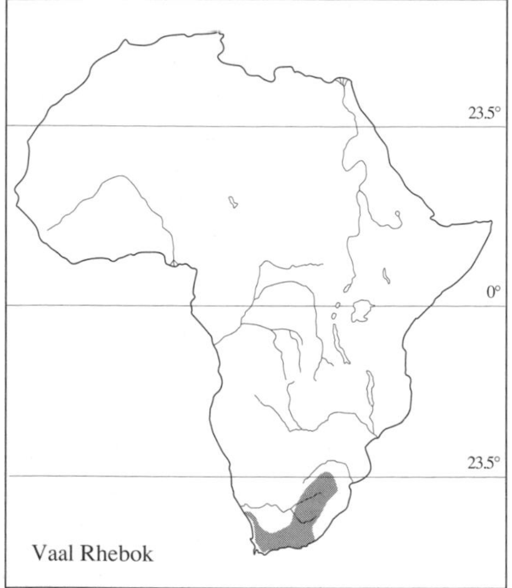

and goats that lived in Africa back in the Miocene, early in bovid history (3).

ECOLOGY. This antelope is associated with *sourveld* grassland and scrub savanna, such as that occurring on plateaus and mountain- and hillsides, where the pasturage is rich in the rainy season but becomes unnutritious after growth stops (2). A mixed feeder, the rhebok stays on the higher slopes during the rains and moves onto slope grasslands during the dry season. Here the grasses continue growing later and it browses on herbs, leaves, and green shoots. Although it overlaps with the mountain reedbuck on the slopes, it avoids the taller grasslands frequented by the latter. *Pelea* is also water-independent, able to meet its requirements by browsing.

SOCIAL ORGANIZATION: gregarious, territorial, sedentary-dispersed, closed family group or harem (2).

Rhebok herds consist of from 1 to 15 females and young and 1 adult male. Typical group size is 3–5. Females voluntarily remain within a single territory, giving chosen males an exclusive family group or harem. When the home range is larger than the territory, it extends into undefended habitat and not into another territory. In the Ohrigstad Dam Nature Reserve (Transvaal), the range of one family of 4 varied from 33 ha in summer to 78 ha in winter; another family of 6 ranged 41 and 75 ha, respectively (2). Family groups are closed to outsiders, including even strayed calves. Leadership and dominance are rarely apparent but exist, probably both vested in the senior female. The male brings up the rear during herd movements, nor does he try to control his harem when it comes to the property boundary, since this also marks the limit of their home range. The average social distance between females (3 m when resting) is about half that maintained by the male, who also goes off alone for up to several hours to mark and patrol his land. Females only quit the herd to calve or stand guard near a hidden calf.

Bachelor Males. Rheboks apparently do not form bachelor herds. Male offspring disperse as yearlings and remain peripheral until they can gain a territory. The dangers of wandering alone in unknown range are offset by early maturation at 1½–2 yrs. Acquiring a territory is only the first step in gaining a harem; the male then has to wait for females to settle on his land. The presence of a lone male may attract a yearling or adult female from a neighboring range. In one observed transfer, the female first tried out a lone male's territory during short visits, always returning to her own family. Eventually she left them and, after remaining indecisively on the border, finally settled down in the new territory. Sometimes a lone male tries to capture neighbors that come near but the females almost always run home.

ACTIVITY (daytime). Seasonal differences in the apportioning of foraging and resting were noted, but no differences between sex and age classes were detected, apart from hidden young (2). In the summer wet season, rheboks spent about equal time foraging and resting. The only clear activity peak was in late afternoon, and rest periods were irregular and spread over the whole day, apart from a rest interval of at least 1½ hr between 1200 and 1500 h. In the winter dry season, the animals had to spend more time foraging, and the resting time fell to ⅓–¼ of the day. There were long periods of foraging uninterrupted by rest pauses, and the midday rest period was also shorter.

POSTURES AND LOCOMOTION. The rhebok has an *ambling walk* on level ground, which changes to a *cross-walk* on slopes and broken ground. It also has two different forms of the gallop: a *rocking canter* interrupted by long bounds, with head and neck slightly lowered, ears erect or slightly back, and tail rolled up to show the white scut and inner thighs; and a fast gallop (up to 65 kph), body and neck low to the ground, without bounding, ears back and tail curled. *Pelea* has a *stotting* gait similar to the bounding seen during the canter, but the throwing up of the hindquarters is exaggerated. Reedbucks have similar running and stotting gaits.

SOCIAL BEHAVIOR

COMMUNICATION. Being an open-country antelope, the rhebok is primarily sight-oriented, with olfactory and auditory communication of secondary and apparently equal importance (2). Movement is detected at 400–500 m, stationary objects are noticed at 100–200 m, and standing or moving rheboks are recognized at 300–400 m. A territorial male that lost sight of an intruder he was chasing continued following it by scent, nose to the ground. Rheboks always react to unidentified sounds, but usually wait to see what caused the disturbance before fleeing.

TERRITORIAL BEHAVIOR: *urination and defecation*, standing in *erect posture, patrolling, snorting.*

Males urinate and defecate in the same distinctive crouch, with tail curled to show the white scut. Females adopt a similar but deeper crouch when urinating but do not squat to defecate. Territorial marking is thus a visual as well as an olfactory display; even the urine itself, darkened by melanin pigment produced in the preputial glands, is conspicuous (4). Patrolling males mark near landmarks and at exposed locations throughout the territory, but not at regular middens nor at any particular time of day. Standing or walking in the *erect posture* also advertises territorial status. The posture is often combined with *snorting* (or "smacking"), *stamping*, and *stotting* in reaction to a territorial neighbor. Males are more actively territorial and aggressive during summer, when calving and mating occur, than during winter, when most transfers occur.

AGONISTIC BEHAVIOR

Dominance/Threat Displays: *erect posture, high-horn presentation, looking away, angle-horn, low-horn presentation*, stabbing gestures.

Defensive/Submissive Displays: *head-low posture, lying-out* (see under Sexual Behavior).

Fighting: *air-cushion fights* (contact rare).

Territorial males interact on borders but actual fights are extremely rare, possibly because the daggerlike horns make it too risky. Fatal wounds have been reported and are most likely to occur during chases following inadvertent trespass. Male encounters feature the *erect posture* in frontal or *lateral orientation*, often while *looking away. High-horn presentation* is similar except that the head and horns are angled forward and sometimes tilted (*angle-horn*). Associated acts include *snorting, stamping, stotting, displacement grazing* and *grooming, urination* and *defecation*, and rarely *ground-horning*. At highest intensity, males rush at each other in the combat attitude (foreheads to ground) but stop 1–2 m apart and have an *air-cushion fight* with stabbing movements (cf. oribi). This may be followed by a high-speed race or chase along the border, another *air-cushion fight*, and more displays. Females employ the same intimidation and threat displays, especially *low-horn presentation* and stabbing movements; males make the same threats to females and young that violate their *individual distance*. Females sometimes butt courting males in the flank and butt heads with one another.

REPRODUCTION. Breeding is seasonal: mating occurs between January and April, and most calving occurs in November–January, after a gestation of c. 7 months (5). The rhebok is probably the only African antelope that may sometimes (but rarely [2]) produce two offspring (1).

SEXUAL BEHAVIOR: *erect posture, empty-licking, urine-testing, foreleg-lifting*, licking female's vulva, neck, and shoulder.

The male makes his approach in the *erect posture* but with an erection and *empty-licking*. He desists if the female moves off or threatens; driving and chasing are rare. A willing female adopts the *head-low posture*, urinates with tail curled, and may even approach the male to urinate while standing with rump turned looking over her shoulder. The male tests her urine with head tilted to the side, then urinates and defecates on the same spot. During this preliminary stage in an estrous cycle of 2–3 days, a courting couple spends much of the time resting and feeding, staying unusually close together. When the *mating march* begins, the female moves slowly and steadily with the male close behind, *foreleg-lifting* at practically every step (fig. 6.16). He licks his muzzle, sniffs her vulva, and begins preliminary mounting. Between attempts he sniffs her rump, *grimaces*, and may also lick her neck and shoulders. When the female stands for mounting, the male rests his brisket on her croup, clasps her haunches, and gives a single ejaculatory thrust, with neck erect. After dismounting he licks his penis and both parties may self-groom before resuming other activities.

PARENT/OFFSPRING BEHAVIOR. Beginning 2–3 weeks before calving, a female leaves the harem for several hours at a time, apparently to prospect for a birthplace affording good cover, and 2–3 days before parturition goes into seclusion. Behavior during birth has not been recorded. After 1–2 weeks a mother rejoins the harem; she may go up to 500 m from her calf's hiding place and stays at least 50 m distant except to retrieve and suckle it. The concealment period is c. 6 weeks. Preparatory to parking her offspring, the mother leads it to a suitable area 200–400 m from the last hiding place, where it then chooses its own spot. The calf is suckled early and late in the day and probably not at night. Young calves nurse ½–5 minutes in 3–7 bouts during a 1-hour activity period. Older calves are only allowed to suckle ½–¾ minute, despite begging in the *head-low posture,* and are weaned at 6–8 months. Calves up to 6–8 months immediately run to their mothers and follow closely during flight. The rest of the time they associate with any peers that may be present. Territorial males apparently do not play an active role in driving out yearling males, thus showing forbearance of their own sons (2).

ANTIPREDATOR BEHAVIOR: *alert posture, restless walking, snorting, stamping, stotting.*

Large predators are absent in most places where rheboks occur. Flight distance from dogs and caracals is 150–200 m. Even during flight a herd is reluctant to leave its territory. In hilly country they use the terrain to get out of sight as quickly as possible. Herd members cluster together at less than the usual individual distance following a disturbance. A male has been seen to threaten, attack, and chase baboons, and there are other indications that territorial males protect their families. Large eagles take young calves.

References

1. Asdell 1946. 2. Esser 1973. 3. Gentry 1978. 4. Starck and Schneider 1971. 5. Walther 1972*d*.

Chapter 7

Horse Antelopes

Tribe Hippotragini

Hippotragus equinus, roan
H. niger, sable

Addax nasomaculatus, addax

Oryx gazella, oryx
O. dammah, scimitar-horned oryx
O. leucoryx, Arabian oryx

TRIBAL TRAITS. Large antelopes with horselike conformation (Gk. *hippo* horse + Gk. *tragos* goat). *Height and weight:* from 100 cm, 75 kg Arabian oryx to 160 cm, 300 kg roan, females 10%–20% smaller than *males*. Sexes much alike (except sable), with horns nearly or quite as long in females as in males (but thinner), sickle-shaped (roan, sable, *O. dammah*, straight (other *Oryx* spp.), or corkscrew (addax), and strongly ridged. *Coloration:* off-white (*O. leucoryx, O. dammah*, addax), gray, tan, chestnut to black (sable bulls), with contrasting black and white markings on body (oryx, sable) and head (reddish in scimitar oryx); juvenile coloration unlike adult, usually tan, with only faint markings. Muffle hairy; tail hock length or longer (oryx) with a terminal tuft; short, upstanding mane (except addax); well-developed false hooves. *Scent glands:* preorbitals rudimentary, merely a thickening of the skin without a central duct, present only in *Hippotragus* and scimitar-horned oryx; hoof glands in all feet. *Mammae:* 4.

DISTRIBUTION. Members of this tribe formerly ranged throughout Africa's savanna and arid lands and beyond to the Arabian Peninsula. The addax and scimitar-horned oryx, once abundant around and in the Sahara, and on the evidence of tomb paintings even domesticated by the ancient Egyptians, have been hunted (from vehicles) to the verge of extinction. Both may have gone over the edge during the drought of the 1980s, when the last herds were forced from their desert retreats into settled areas (6). *Oryx gazella* is holding its own in the Somali-Masai and South West Arid Zones, while the roan and sable inhabit the broadleafed, deciduous woodlands of the savanna biomes. The roan, sable, and oryx overlap along the Zimbabwe-Botswana border (Wange and Chobe N.P.). The extinct bloubok (blue buck, *H. leucophaeus*), which looked like a small roan, occupied the extreme tip of South Africa (5).

ECOLOGY. The members of this tribe are ecologically diverse, each genus adapted for a different biome. The sable and roan are associated with medium to tall grassland within the savanna woodland with rainfall of 500–1200 mm, have mouths adapted to harvest the dominant, tufted perennial grasses, and must drink regularly. The oryxes and addax are adapted for subdesert and desert conditions, respectively, built for long-distance travel, and normally able to gain sufficient water from plants without drinking. The addax and caribou have a somewhat similar conformation, with level backs, short legs, and broad hooves (present also in scimitar-horned and Arabian oryxes) adapted respectively for walking on sand and snow. The addax penetrates more deeply into the Sahara than any other ungulate except the (introduced) camel, including the dunescapes of the great ergs, its last refuge from motorized hunters.

All members of the tribe are predominantly grazers. The addax and oryx take the most roughage, but all browse to some extent to gain water (oryx and addax), if not to increase protein intake. All species live at low density in regions that cannot support a high herbivore biomass.

SOCIAL ORGANIZATION: gregarious, sedentary or nomadic, more (when sedentary) or less (when nomadic) territorial and sexually segregated.

Social organization ranges from the conventional territorial system of the roan, in which females occupy traditional home ranges encompassing the territories of several males while immature males associate in separate bachelor herds, to the mixed oryx herds which wander long distances and suspend or even dispense with territoriality when conditions favor nomadism. Herds of over 100 animals are not uncommon in the sable and oryx, although the average is 15–20 and only ½ dozen for the roan. The minimal sexual dimorphism in this tribe is undoubtedly related to the tendency to form mixed herds (see chap. 2, Sexual Dimorphism). Even the penile sheath and scrotum are remarkably inconspicuous in the oryx and addax except for dominant males in breeding condition. The sable, by contrast, with brown cows and black bulls, is one of the most dimorphic antelopes (see species account), and has a pendant black-tipped penile sheath as an added masculine garnish. The comparative unimorphism of the closely related roan leads one to expect mixed herds, yet such have not been reported. There is a tendency in this tribe for dominant males to tolerate young males to a relatively advanced age (at least 3 years); even territorial sables tolerate young bulls as long as they resemble females. Bachelor herds are correspondingly small and less important (for male survival) in the Hippotragini, composed only of subadult and young-adult bulls.

Peer subgroups are in evidence in larger herds of all species, especially among seasonally breeding populations. Groups of young calves often lag behind the herd, a risky habit. Female herds are closed, with a linear rank hierarchy based, as usual, on seniority. Compared to most other tribes, female horse antelopes have aggressive temperaments to match their dangerous weapons. High-ranking female oryxes may dominate all but alpha males, but in general males become dominant over all females once they surpass them in size (8).

Female herds of sable and roan are often unattended by the territorial males on whose property they reside, sometimes for weeks at a time. A preference for more wooded habitat may explain these absences at times when no females are in estrus; also bulls regularly patrol their territories and even when accompanying a female herd may leave it for hours at a time.

ACTIVITY. The information reported here does not necessarily apply to addax and scimitar-horned oryx. Hippotragini move and forage day and night, with the usual peaks in the coolest hours of the day and the middle of the night, but the timing and duration of activity and rest periods vary with climatic conditions and forage quality. The roan and sable tend to be late risers; these species and *Oryx gazella* all make "obligatory" moves after the evening feeding peak before settling to rest (see species accounts and under Activity, chap. 2).

POSTURES AND LOCOMOTION. The oryxes and addax walk in an amble, as expected. Sables and roans are more variable, cross-walking when going slow and changing to an amble at a quicker walk. Oryxes nod their heads like topis when walking fast, especially the scimitar-horned oryx (3). The addax throws its wide-hoofed feet slightly sideways to avoid brushing against the opposite limb, but places one foot behind the other, leaving a single line of tracks. The trot is not a regular gait in this tribe but may appear as a transition between walk and gallop, and a *style-trot* is performed in situations of excitement or alarm. Oryxes have a particularly beautiful flowing trot with a suspension stage during which all feet are off the ground and the head is turned synchronously from side to side (4). Trotting scimitar-horned oryxes hold their chins raised with horns back (3). The gallop differs considerably among species. The sable and roan bound higher and flex their legs more than the rest, whereas the addax has a flat gallop with minimal flexing, appearing stiff-kneed. It and the Arabian oryx are considered to be the slowest and clumsiest runners in the tribe, perhaps reflecting their adaptations to sandy substrates. Both have great endurance when traveling on sand (3). *Oryx gazella* is probably the fastest and most enduring horse antelope, though less fleet than gazelles or topis. When running at full gallop, the chin is held out so that the horns lie back in line with the neck (7).

SOCIAL BEHAVIOR

COMMUNICATION. The visual channel is most important, as expected of open-country antelopes (see under Agonistic Behavior).

Vocal Communication. Auditory signaling is also important, especially between mothers and offspring of sable and roan, which exchange far-reaching birdlike contact and distress calls. Bulls of all species are known to roar during fights and chases, and all species snort in alarm.

Olfactory Communication. Though only the hoof glands are well-developed, olfactory communication is nevertheless important. Sable and roan bulls signpost their territories with dung (fig. 2.5), and the extreme crouch adopted by adult male oryxes and addaxes makes the defecation act a striking visual display of social status (fig. 7.1). *Urine-testing,* too, is unusually prominent in this tribe, performed for instance by both sexes and all ages of the sable and roan (fig. 7.8).

Tactile Communication. Horse antelopes are not contact species, but maintain individual distances beyond reach of their long horns. Social grooming is rare, but rubbing the head or horns on another animal's torso is a common assertion of dominance among female and immature sables and roans.

TERRITORIAL BEHAVIOR: patrolling, *ceremonial defecation, vegetation-horning, herding.*

Roan and sable bulls patrol their territories regularly, preferably following roads and paths, pausing every few hundred m to defecate (fig. 2.5). The oryx and addax are not known to patrol or demarcate territorial boundaries, yet mature males make a conspicuous display of defecating (fig. 7.1). Apparently what began as a territorial display has evolved into a conspicuous display of dominance, or it may be that some addax and oryx bulls are territorial (see oryx account). Territorial sables and roans, and alpha-male oryxes with a female herd, typically bring up the rear, but manage to direct and control herd movements by means of dominance and threat displays (see species accounts). *Vegetation-horning* is most frequent in the sable and roan, and comparatively rare in the addax (3).

Fig. 7.1. Addax bull in extreme *defecation crouch* (also seen in *Oryx*). (From a photo in Huth 1980.)

AGONISTIC BEHAVIOR

Dominance/Threat Displays: *supplanting; lateral presentation; erect posture* ± nose-lifted and head turned 45°, *high-, medial-,* and *low-horn presentation, angle-horn, symbolic butting, horn-sweeping, charging/chasing* ± *roaring* or *grunting.*

Fig. 7.2. Oryx bull in *lateral presentation* displaying *erect posture* to subordinate, which withdraws in *head-low posture* (from Walther 1984).

Defensive/Submissive Displays: *head-low posture* with chin in or out, ± kneeling; *head-throwing* (oryx); *facing away; fear gaping* (sable and roan); *distress cry; lying-out.*

Aggressive and submissive displays are highly developed in this tribe, in keeping with the long, sharp horns and relatively closed, structured social groups. The consequences of failing to respond appropriately and promptly to a superior's aggressive display are too painful to risk ignoring. Displays of dominance and submission are very common between females as well as males. Most such encounters are low-key and one-sided, involving a lower-ranking individual moving out of a higher-ranking animal's way. *Lateral display* of the *erect posture,* in which the long horns and powerful neck are shown to advantage, is the primary dominance display in the tribe (fig. 7.2). Since fighting horse antelopes deliver strong forward-downward blows, the *erect posture* may be closer to *high-horn presentation,* a strong threat (especially in oryx) rather than a dominance display (see chap. 2, Agonistic Behavior), and is often combined with tilting the head and aiming the horns toward the opponent (*angle-horn*) (fig. 7.13). The tail is held out or up by a displaying sable or roan, but usually hangs or is slightly lifted in other hippotragines. *Medial-* and *low-horn presentation,* which are the attitudes assumed during combat, are strong threats, especially when combined with *head-tossing* (symbolic stabbing), nodding (*symbolic butting*), *horn-sweeping* (fig. 7.3), or *rushing*—the last leading to a painful jab if the victim fails to escape. Finally, there

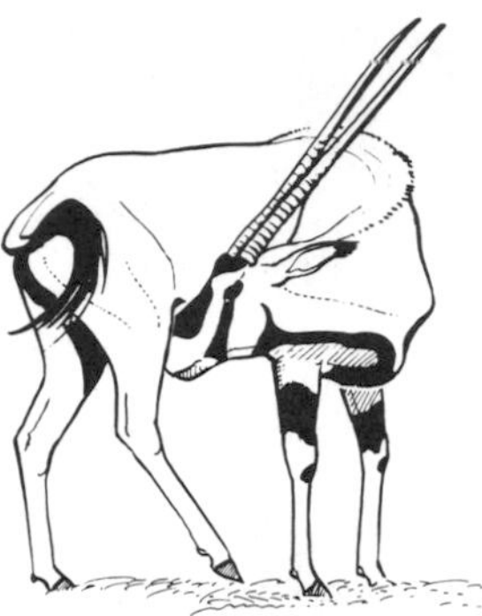

Fig. 7.3. Oryx *horn and tail sweeping* ("insect shooing"), a common gesture during agonistic interactions.

is a form of contact in sables that looks more social than aggressive but which functions as a sparring invitation between young animals and as an assertion of dominance between adult females (fig. 7.4). The challenger approaches with head at shoulder level (*medial-horn presentation*?) and presses or rubs its forehead and horns against the partner's neck, shoulder, or rump.

In addition to rushing or charging, unritualized aggression includes *chasing*, both in attack and courtship, during which the aggressor may roar (sable, roan) or grunt (oryx) and the victim may scream or bray.

Defensive and submissive displays are as similar in hippotragines as are dominance and threat displays. The *head-low posture* takes 2 forms: the first, with chin drawn in so that the horns point upward, putting the animal in position to defend itself; the second, with chin out so that the horns come into line with the neck, less defensive and more submissive (8). Oryxes go one step further, turning the head so that

Fig. 7.4. *Sparring invitation/challenge* between subadult sable bulls: the challenger approaches and touches/rubs the shoulder of bull standing side-on in *medial-horn presentation*.

the horns are screened and the throat is exposed, and also *head-throwing* as a dynamic form of the *head-low/chin-out posture*. Intimidated sables and roans meanwhile hold their tails clamped, whereas oryxes raise and wave the tail (fig. 7.2). Sables and roans also open their mouths (*fear gape*), as when *distress-calling*, so *gaping* may represent the intention to cry out (fig. 7.6). When a feared dominant such as the alpha male draws himself up in a *lateral display*, females and inferior males approach in the submissive posture and pass behind him (cf. wildebeest and waterbuck), often accelerating on the turn as if anticipating a blow or prod of the horns, especially if the bull angles, ducks, or sweeps his head (see roan account, fig. 7.7). For a persistently persecuted individual (often a sexually harassed female), the last resort—short of leaving the herd entirely—is to *lie out* like a concealed calf.

The combination of submissive actions performed by horse antelopes amounts to an elaborate appeasement ceremony, regular performance of which is probably the key to enabling males to remain in the society of females past the stage of adolescence and, in oryxes and addaxes, into adulthood.

Displacement Activities. *Scratching with hindfeet* and *incisor-grooming* during aggressive interactions are less prominent in horse antelopes than in some other tribes. However, *horn-sweeping* to the shoulder or flank (fig. 7.3), resembling the response to biting insects, is much in evidence (cf. wildebeest). As these abrupt movements are given by higher-ranking individuals when inferiors come too close, they apparently are aggressive and may convey a threat to hook or stab. In oryxes, submissive *head-throwing* may lead into *scratching the rump with the horn tips*.

Fighting: *clash-fighting; front-pressing; horn-pressing; lateral-fighting* and *sideward-stabbing*.

Despite differently shaped horns, horse antelopes have very similar and elaborate fighting styles. All species usually but do not necessarily kneel during serious combat. Most unusual is *sideward-stabbing* (usually over the shoulder) (fig. 7.5). The fact that such socially evolved antelopes are prepared to stab one another with their long, dangerous horns is evidence against the theory that natural selection has led to ritualized and relatively safe tests of fitness (2). It is simply the defensive and offensive skill of the combatants, I suspect, that makes serious fighting injuries un-

Fig. 7.5. Hippotragine *over-the-shoulder stabbing* movement performed by subadult male sable during sparring contest.

usual. That, plus thick hide (over 6 mm in the oryx, whose skin has been widely used for shields) on the head, neck, and upper torso (4). It is true, nevertheless, that *parallel fighting* and *stabbing* are less common than other techniques in Hippotragini and may be limited to sparring sessions between young males, as in the sable, or to severe fights, as in the oryx and scimitar-horned oryx, and to self-defense against predators.

REPRODUCTION. Most sable populations breed annually, whereas the roan, oryx, and addax breed perennially, ovulating within a month or so of calving (postpartum estrus), but with some synchrony between associated females and discernible breeding peaks and valleys. Females conceive as 2-year-olds and calve when nearly 3, after a gestation of 8–9½ months. Males reach maturity in their sixth year.

SEXUAL BEHAVIOR: *erect posture,* driving and chasing, *urine-testing, lowstretch* (sable, roan), *foreleg-lifting, courtship circling, resting chin on rump* (oryx, addax), sniffing/rubbing female's rump. Female: *urination on demand; head-low postures; circling;* flight; rubbing face or horns on male's rump, shoulder, or neck (aggression or solicitation).

Dominance displays are prominent in hippotragine courtship and probably necessary to convince a female to submit to a male's attentions. Females react aggressively to males unable to dominate them, but when an adult male approaches in *lowstretch,* females respond by urinating. The male holds his nose in the stream, then performs the *urine test* (fig. 7.8). In the sable and roan, females and calves regularly approach and test female urine.

During the *mating march,* the male follows closely; he sniffs and may rub his face or horns on the cow's croup. *Foreleg-lifting* (bent except in *Oryx gazella*) tests willingness to stand for mounting (fig. 2.13). Oryx and addax bulls also rest the chin on the female's croup as a mounting prelude. Female responses are as listed. *Circling* is a tribal trait which probably represents female efforts to evade the male's attempts to make bodily contact. Yet a female in full estrus, as revealed by a slightly elevated tail and willingness to stand for mounting, may rub her head and horns on the male's croup while circling, which seems to stimulate the male to mount. The male copulatory posture in Hippotragini is upright, firmly grasping the female's loins (fig. 7.16).

PARENT/OFFSPRING BEHAVIOR. Calves are concealed for up to 6 weeks; they are retrieved and suckled once or twice by day, typically early in the morning, and perhaps at night. The natal coat is tan with faint or no markings for at least 2 months and one species looks much like another. Sables may bring calves a few days old into the herd, but usually mothers isolate before and for at least a week after calving, remaining within view of the hiding place during the early part of the concealment period.

ANTIPREDATOR BEHAVIOR: *alert posture, alarm snorting, stamping, style-trotting;* flight, self-defense.

Though horse antelopes are considered by hunters to be among the most dangerous antelopes, and even lions are sometimes fatally stabbed, their dangerous horns and stabbing technique do not deter predators from hunting them and are probably rarely effectively employed (1). The tendency of all hippotragine young to lag behind or become detached from the herd should make them unusually vulnerable. Yet there is no evidence of lower-than-normal survival. The habitats frequented by these antelopes usually have relatively few predators, perhaps permitting a certain slackness in their antipredator behavior.

References

1. Eloff 1964. 2. Geist 1966. 3. Huth 1980. 4. Kingdon 1982. 5. Klein, 1974. 6. J. Newby *in litt.* 1985. 7. Walther 1965*b*. 8.——— 1978*c*.

Added Reference

Dixon and Jones 1988.

Fig. 7.6. Roan courtship: male *foreleg-lifting,* female *fear-gaping.*

Roan

Hippotragus equinus

TRAITS. The fourth-largest antelope, tall, sand-colored, with a striking black and white mask and long, tufted ears. *Height and weight:* 126–145 cm; males 280 kg (242–300), females 260 kg (223–280) (8). Rangy build, high in the shoulder, long head and wide gape. *Horns:* curving, massive, heavily ridged, relatively short, average 75 cm (55–99) (8). *Coloration:* geographically variable, pale gray to rufous, black mask and nostrils contrasting with white "eyebrows," lips, and jaw; light-colored ears; tail with dark terminal tuft. Sexes much alike but male more robust, with bigger horns, blacker mask, and pendant penile sheath often dark-tipped (as in sable); newborn calves tan without clear markings, very similar to sable calves. *Scent glands:* preorbitals (beneath "eyebrows") and pedal glands, the latter with a sharp, penetrating smell (5).

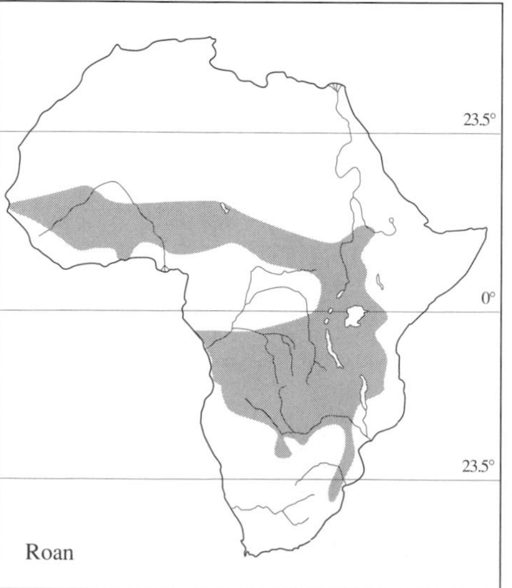

DISTRIBUTION. Formerly one of the widest-ranging antelopes, found nearly throughout the better-watered parts of the Northern and Southern Savanna, from sea level to 2400 m and penetrating into the adjacent arid zone, yet strikingly absent from the eastern part of the Southern Savanna (1). Although now greatly reduced, due both to poaching and elimination of habitat, it was never an abundant species and only a dominant herbivore in parts of its northern range, such as Waza N.P. in Cameroon (9). It is a rare species in South Africa, confined now to the eastern Transvaal, Kruger N.P. being the southern limit of its range (11), and also rare in Kenya, where it is common only in the Lambwe Valley N.P., near Lake Victoria.

ECOLOGY. Like the sable, the roan is associated with the *Brachystegia/Isoberlinia* wooded savanna, but is less of a woodland and more of a grassland/tree-savanna species, tolerating taller grass and higher elevations, including montane grasslands such as Malawi's Nyika Plateau and Zimbabwe's Chimanimani Mountains. Also like the sable, it is a selective grazer on perennial grasses that grow in leached soils of poor nutrient status which support a low herbivore biomass, offering little nourishment in the dry season except on low ground that retains enough moisture to produce growth after annual fires (3, 4). The roan also browses to some extent (up to 10–20% of rumen contents) on forbs, leaves, and pods. Like other water-dependent wildlife, the roan concentrates near water points during the dry season and disperses during the rains (7).

SOCIAL ORGANIZATION: territorial, mostly sedentary/dispersed in small herds, but aggregates seasonally in West African distribution centers.

Although the sable and roan share a very

similar social organization, the roan lives in smaller herds and maintains a greater individual distance (estimated at 7 m between lying roans vs. 3.4 m between sables [10]). Groups of females and young typically number 6–20, though herds of 35 are not unusual and temporary gatherings of 150 head have been recorded (8). In Zambia, the mean of 109 herds counted in the national parks was 9, compared to a mean of 15 in 93 sable herds (13).

A herd consists of females and young that share a traditional and exclusive home range, which in Kruger N.P. was found to cover 60–120 km^2 (7). In Serengeti N.P., a herd of 10–15 head had separate wet- and dry-season ranges some 20 km apart. The herd appeared periodically at Banagi for at least 30 years (12), before disappearing in the early 1970s (author's observ.). Ranges are smaller and herds tend to be larger in the dry season, but in any case herd composition varies from day to day. Groups composed very largely of young animals may remain separated for hours and even days from groups of adult females containing most of the mothers (cf. sable), evidence that social bonds within immature classes are stronger than between mother and young. Sex and age subgroups occur whenever young of approximately the same age are present. Observations of herds in which two females were consistently nearest neighbors suggest that bonds between calves—or possibly mother and daughter—may persist for years. A female dominance hierarchy, with the oldest cows highest and yearlings lowest, is maintained by frequent, mostly low-intensity aggression (*supplanting*, dominance displays). The rank order in a Kruger N.P. herd kept in an enclosure remained unchanged over a 3-year period (5). The possibility of outsiders being accepted into such highly structured social groups is slight.

Males are tolerated in the female herds until adolescence at around 2 years. By performing the *appeasement ceremony* (described in tribal introduction) (fig. 7.7), young males can placate the aggression of territorial males for a time, but by age 3 they are relegated to bachelor herds, where they live until mature in their sixth year (5, 6). Bachelor herds usually number under 10, rarely up to 17 head (10).

ACTIVITY. Like sables, roans tend to be late risers, especially on cool mornings when the grass is soaked with dew. In Kenya's Shimba Hills Reserve, a translocated roan herd, observed between July and September, lagged about an hour behind the indigenous sable, often resting until 0900 h, then grazing intensively between 1000 and 1100, with a resting peak from 1400 to 1500 (10). Both species had feeding peaks in the last hour of daylight and settled to rest and ruminate after dark (author's observ.). There appears to be at least one feeding peak at night. Roans go to water no less than every other day in the dry season (7). Kruger N.P. herds normally drank between 1000 and 1100, then moved away 400–800 m before settling to rest during the heat of the day.

The daily ranging of these animals was generally contained within an area of 2–4 km^2 or less. It seems to be typical for a herd to stay in the same area for up to several weeks, then abruptly move to another part of its range and settle down again (6)—just like the sable.

POSTURES AND LOCOMOTION. See tribal introduction.

SOCIAL BEHAVIOR

COMMUNICATION. See tribal introduction.

TERRITORIAL BEHAVIOR. Doubts as to whether the roan is actually territorial have been raised by a longterm study in Kruger N.P., where female herds were accompanied throughout their very large home ranges by the same herd bulls, which merely defended a zone 300–500 m wide around the herd against incursions by other bulls (5, 6). In most gregarious/territorial antelopes, including the sable, a herd's home range usually includes at least several different territories, and it seems likely that the situation in Kruger N.P. is the result of abnormally low population density. Further study of roans living at higher density is needed to determine whether males establish typical territories in which dominance is site-specific or whether, as in sable, certain master bulls are able to invade the territories of their neighbors and dominate them (see sable account).

Advertising: *patrolling, dunging ceremony, vegetation-horning.*

Territorial males frequently patrol their boundaries and deposit dung at intervals, especially along roadsides. Because of overlap in hoof and pellet size, roan and sable spoors cannot be reliably distinguished unless the ground is scored by scrape marks. The sable but not the roan scrapes before dunging. But roans, too, rub and thrash small trees, often filling the spaces between horn rings with grated bark—both a visible and audible display of territorial status.

AGONISTIC BEHAVIOR

Dominance/Threat Displays: *lateral display* of *erect posture, medial-, high-* and *low-horn presentation, angle-horn, symbolic butting* (ducking, nodding, head-throwing), *horn-sweeping, rushing, chasing, roaring.*

Defensive/Submissive Displays: *head-low posture, appeasement ceremony, fear gaping, distress-calling, lying-out.*

Fighting: *clash-fighting,* sometimes escalated to *thrust-fighting, parallel fighting* and *stabbing.*

A roan standing broadside in the *erect posture* arches its neck, holds its ears upward/outward at 45°, and lifts its tail slightly to straight out, as in the sable. The *appeasement ceremony* is also identical in the two species (fig. 7.7). Sparring roans typically stand confronting and rub heads together, then engage horns and push, one or both meanwhile dropping to their knees (6). Their tails swing from side to side and their ears twitch. Serious encounters between adults (including cows) may be settled by a few quick but violent clashes, the opponents delivering powerful forward-downward blows as they drop to their knees and then attempt to push one another backward.

Fig. 7.7. Final stage in the roan *appeasement ceremony:* a subadult male bolts in *head-low posture* as adult male lightly raps his rump.

REPRODUCTION. Under favorable conditions, apparently throughout its range, the roan reproduces every 10 months or so, mating within a few weeks of calving (6). Gestation is 9–9½ months (276–287 days) (7).

SEXUAL BEHAVIOR: dominance displays, *chasing, persistent following, urine-testing, foreleg-lifting* (fig. 7.6). Female: defensive and submissive patterns, *courtship circling, urination on demand.*

Roan courtship is typical of the tribe (see tribal introduction and sable account). *Courtship circling and foreleg-lifting* are especially frequent (2). For several known females in a captive herd, the average number of copulations during 1–2 days of estrus was 8 (6).

PARENT/OFFSPRING BEHAVIOR. Cows go into seclusion several days before calving and return to the herd about 5 days afterward, but remain within ½ km of the concealed calf without even going to water. The more dominant the cow, the shorter the period of isolation (6). The calf has a suckling/activity period early in the morning and possibly 1 or 2 at night. Mother and calf communicate with low calls and utter loud, birdlike contact calls if unable to find one another. The calls sound like sable, only higher. Calves are suckled daily up to 4 months and weaned at 6 months. Since breeding is perennial and roan herds are small, the young may not have companions of the same age. However, young of assorted ages group together and the youngest calf is the nucleus. Young roans are playful, running, chasing, *prancing, style-trotting,* bucking, jumping, *object-horning,* and play-fighting.

ANTIPREDATOR BEHAVIOR: *alert posture, alarm snort,* self-defense, flight.

The roan has been described as long-winded and sure-footed but has a top speed of only 57 kph, which is slower than many antelopes lions commonly prey upon (8). Yet roans appear to suffer relatively little predation after the first few months of life (7). Possible contributing factors are low roan and predator density, unusual alertness, large size, formidable horns, and a tendency to aggressive, effective self-defense. Although the aggressive potential is there, the evidence of self-defense is largely anecdotal (8).

References

1. Ansell 1971. 2. Backhaus 1959*a*. 3. Bell 1982. 4. East 1984. 5. Joubert 1970. 6. ——— 1974. 7. ——— 1976. 8. Kingdon 1982. 9. Poche 1974. 10. Sekulic, R. 1977, B.A. thesis, Harvard Univ. 11. Smithers 1983. 12. Walther 1972*d*. 13. Zambian Dept. of Wildlife 1957–1968.

Fig. 7.8. Sable bull testing urine voided by female.

Sable

Hippotragus niger

TRAITS. A large antelope with long, curving horns, built rather like a cob horse: short-coupled with sturdy legs, thick neck (especially males) enhanced by a brushy mane. *Height and weight:* 117–140 cm, males 235 kg (216–263), females 220 kg (204–232) (8). Long face with wide gape, ears average size without terminal tuft; a small dewlap under throat as in gemsbok and mountain zebra, function unknown. *Horns:* 50–154 cm, longer in male giant sable (the record: 164.8 cm, 64⅞ in. [1]), laterally compressed, strongly ridged, thicker and longer in males, becoming ± hooked with maturity. *Coloration:* short glossy coat, females and immatures sorrel to rich chestnut, with white facial markings, belly, and rump patch, black mane and tail tuft; brown replaced by black in males beginning in third year, except backs of ears and sometimes (typically in giant sable) hocks remain brown; calves under 2 months light tan with faint markings. *Scent glands:* preorbitals under the white "eyebrows," hoof glands.

DISTRIBUTION. Restricted to the Southern Savanna, where it is closely identified with the well-watered *Miombo* Woodland Zone that girdles Africa from eastern Tanzania and Mozambique to Angola and southern Zaire.

Subspecies. There are four well-defined sable races: *roosevelti*, occupying the coastal hinterland of Kenya and Tanzania, has the smallest horns; perhaps 100 remain in Kenya, all confined to the Shimba Hills National Reserve. *H. n. kirkii* has the largest geographical range. The giant sable (*H. n. varianii*), so-called because the males have horns averaging about a foot longer than other races, occurs only in central Angola, where possibly 1000–2000 face a clouded future in two protected areas within the war zone (3, 4). Apart from the horns, the giant sable differs only in minor ways from *kirkii* and *roosevelti*, mainly in the absence of a clearly defined cheek stripe. The most distinctively different race is *H. n. niger*, the "black, black" sable, found south of the Zambesi, in which females turn nearly as black as males. It is interesting that this race inhabits the driest savanna, undertakes seasonal movements of up to 50 km, and forms the largest herds (up to 200 head and sometimes even 300 in Zimbabwe). The reduced sexual dimorphism should enable male offspring to remain longer in female herds and would thus facilitate the formation of large, mixed herds (3).

ECOLOGY. The sable favors a mosaic arrangement of woodland and grassland. The woods have to be open enough to support an understory of grasses, which are utilized in the rainy season. Sable herds range

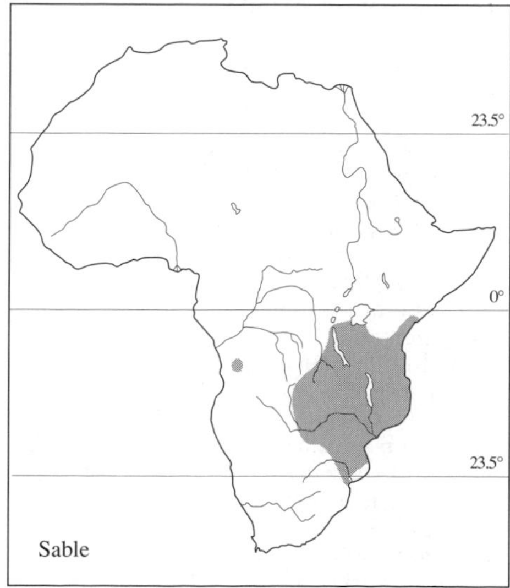

the open grasslands in the dry season in search of green plants, including the forbs and foliage that make up c. 20% of their diet. Termite mounds, which support lusher growth than the surrounding leached, ancient soil, have many of the grasses and browse plants they like best. Dry-season movements depend on the availability of water and food. Forage quality is in turn closely dependent on annual, manmade fires that burn off the tall dead grasses within a month or two after the rains end. Greenflush comes up along the drainage lines with their heavier clay soils while the droughtier woodland soils remain blackened and lifeless until woody plants put out new leaves in the *miombo* spring, a good month before the rains begin, attracting the sable back to the woods. Sables regularly visit salt licks, typically situated at the bases of termite mounds; and where soils are particularly poor, they may visit the sites of old kills to chew bones, presumably to acquire calcium and phosphorus (12).

SOCIAL ORGANIZATION: gregarious, territorial, sedentary, with a tendency to more extensive movements where wet- and dry-season ranges are separated.

Herds of 15–25 females and young are commonest, but groups of 30–75 are not uncommon, especially in the dry season. In fact, small herds are often just one part of a larger herd that has subdivided; the sections may range separately for a few hours or as many months (3). Herds tend to subdivide and disperse during the rains when food and water are not limiting, and to aggregate late in the dry season on the best available pasture within a few km of water. Like other denizens of the *miombo* woodlands, sables have a patchy distribution, often with miles of perfectly suitable-looking habitat between herds. Even when the ranges of adjacent herds overlap, meetings are infrequent and apt to be hostile, reflecting the closed, hierarchical nature of sable female society. Since females normally remain in the area where they grew up, knowledge of the home range is passed on from generation to generation, and descendants may occupy the same range indefinitely, like human clans. A stable rank hierarchy based on seniority is maintained among these resident, often related females, and the top-ranking cow is usually also the leader and the most vigilant herd member. In one regularly observed herd, the ancient lead cow usually kept 20 m or so apart from other herd members and was alert even during rest periods, when other sables gathered in clusters and lay spaced as close as 3–4 m apart (11). Rank does not always go with age, however, and a dominant animal may lose rank because of sickness or an accident such as a broken horn. Permanent separation into different herds with separate ranges may be precipitated when the alpha position is disputed by 2 senior cows (9, 11, author's observ.).

Home ranges vary greatly in size among herds and populations, and may also change over time (3, 9, 11). In the Shimba Hills Reserve, several known herds remained year-round in ranges of 10–25 km^2, which overlapped by up to 20%. The biggest was occupied by a very large herd of over 70 animals. These ranges included the territories, 4–9 km^2 in extent, of from 2 to 5 males (10, 11). In Zimbabwe's Matopos N.P., 4 herds that were studied for 27 months occupied ranges estimated at 12–25 km^2 from the range map (fig. 7 of [5]), and 2 sable herds in different Transvaal nature reserves utilized annual home ranges of 9.2 and 17.7 km^2 (13). The fact that sable herds in all three areas occupy ranges of similar size suggests they may be average for relatively sedentary populations in areas of limited available habitat. In eastern Zambia, however, sightings of one herd encompassed an area of 320 km^2 for 2 consecutive years (2), and ranges of comparable size have been reported for certain herds in Kruger N.P. (S. Joubert, pers. comm.). Such very large home ranges are plausible in populations that have separate seasonal ranges, as in Gorongoza N.P., where sables spend the dry season on a large floodplain and the wet season on the adjacent wooded plateau (K. Tinley, pers. comm.). In Angola's Luando Reserve, one herd "migrated" to a wet-season range 15 km from where it had spent the dry season, while an adjacent herd stayed put the whole year within an area of c. 12 km^2 (3).

Considering the expansiveness of sable home ranges, herds cover remarkably little ground from day to day (typically less than 1 km), and even week to week. Especially in the dry season, herds may spend days and weeks on the same pasture, leaving it only to go to water or to find shade during the hottest hours (3, 5). In Victoria Falls N.P., a herd of 125 sables remained within an area of 250 ha for at least 7 weeks (author's observ.). More typically, after spending a week or more in one place, a herd suddenly moves to another place some kilometers away, where it settles down again.

ACTIVITY. Sables have the usual morning and late-afternoon activity peaks separated by several hours during which most animals ruminate and rest, often with short intervals of renewed grazing by some. Even when pastures are green, sables visit water holes frequently, if not daily, often at midday and sometimes at night. Herds observed until after nightfall typically fed until dark, then moved in a close bunch several hundred m and proceeded to lie down and ruminate for 2 or 3 hours. A bout of grazing followed which ended irregularly sometime after midnight, followed by another general rest period. On bright-moonlight nights in the dry season herds often moved up to 1 km or more while feeding on open grassland with a flush of new grass and herbage. From well before usually to well after dawn, sables can nearly all be found lying and resting; on cool, dewy mornings Shimba Hills herds may not get up before 0800 h (3, 11).

POSTURES AND LOCOMOTION. See tribal introduction.

SOCIAL BEHAVIOR

COMMUNICATION. See tribal introduction.

TERRITORIAL BEHAVIOR

Advertising: *patrolling* with frequent *pawing* and *dunging, vegetation-horning, herding* behavior.

Territorial males spend much of their time alone, even when they have females in residence. Bulls regularly patrol along roads and pathways, moving at a stately walk with frequent stops to sniff the ground, especially the dung and urine of other sables. Every 100 m or so they deposit dung in a ceremonial manner, pawing vigorously with alternate forefeet before or while defecating with raised, waving tail (fig. 2.5). Also they often thrash bushes and sapling trees, breaking branches in the cleft of their horns with quick twists and grating the outer curve of the horns against the main stem; the accumulation of bark between annulations often makes a bull's horns look brown instead of black. During inactive periods, lone bulls often withdraw into fairly dense vegetation, becoming so inconspicuous among the shadows that they are usually overlooked; yet they are highly conspicuous in the open and less shy than other sables when with a herd.

A bull accompanying a herd generally brings up the rear and makes no effort to dictate its direction, unless it approaches the territorial boundary. He then makes dominance and threat displays (described below) and if necessary physically blocks the way. Thus sable bulls often prevent females from departing, although cows determined to leave eventually break free and the rest of the herd usually follows.

The sable territorial system is unusual in that some bulls are able to dominate others (3). In the first of several observed instances, a territorial bull without a herd penetrated over 2 km inside his neighbor's property, walked boldly into the herd residing there, and began displaying to the cows. When the resident bull, who meanwhile lay within 100 m with his back turned, eventually came forward to challenge the invader, the latter matched him display for display. A brief fight ensued, which ended, after several solid clashes and thrusts, with the resident bull's abrupt departure. Surprisingly, the usurper soon withdrew to his own territory, where he was seen next day alone as usual; the defeated bull was able to reclaim his territory and herd by default (author's observ.). Other instances occurred when bulls, having failed to prevent herds from leaving their territories, went with them and their neighbors acknowledged their subordinate status by keeping a respectful distance (3). The subdominant bulls generally looked younger than these clearly mature "master bulls." One possible explanation: the dominance of bulls that have held territories for years continues to impress junior bulls into their early territorial stage.

AGONISTIC BEHAVIOR

Dominance/Threat Displays: *lateral presentation* in *erect posture,* approach in *lowstretch, rubbing head/horns on opponent, high-, medial-,* and *low-horn presentation, angle-horn, symbolic butting, head-shaking, horn-sweeping, chasing* and *roaring.*

Defensive/Submissive Displays: *head-low posture, appeasement ceremony; fear gaping; lying-out;* braying distress call.

A territorial male herding in *lowstretch* strikingly recalls a hazing stallion. *Horn-sweeping* and loud snorting are also typical during herding. A cow or young sable approached thus will usually move in the desired direction (away from the territorial boundary) without further prompting. Failure to do so may be punished by a chase or charge with or without prior threat displays. Sable bulls clearly command a lot of respect; to be chased is so unnerving that the victim often cries out, while a pursuing bull may bellow his rage. When a bull comes up to a herd he performs the im-

Fig. 7.9. Agonistic interaction between sable cows: subordinate female in *head-low posture,* dominant female in *medial-horn presentation.*

pressive *lateral display* in *erect posture,* with chin pulled in and tail raised. Young males then come forward to pay homage, walking past him in the *head-low posture,* tail clamped, then quickly pass behind to avoid a rap of the great horns (threatened by *angle-horn* or *ducking*). Females that are the object of an adult male's aggressive displays respond in the same submissive manner.

Dominance among females is asserted by a series of actions of increasing severity, beginning with *supplanting.* Next comes rubbing the head and horns on another animal's neck, shoulder, or rump. Aiming the horns (*medial-horn presentation*) at an opponent is a rather severe threat (fig. 7.9). But most severe and painful is a charge that results in a prod to the rump or side with the horn tips, using either an upward or sideward head movement. The *lateral display* is reserved for near-equals, which may "talk back." When two sables display, standing a couple of meters apart in *reverse parallel* and sometimes circling (fig. 7.10), how high the tail and head are held is an indicator of confidence and intensity (3).

Fighting: *clash-fighting, thrust-fighting, parallel fighting, fencing* and *stabbing.*

The complete repertoire is practiced by young males, who often spar and fight with considerable abandon. Even calves go through the motions. The only form of fighting that is not directed to the opponent's head and horns is the stabbing movement practiced while two youngsters stand or kneel shoulder to shoulder *fencing* (fig. 7.5). One, usually the stronger, takes the defensive role, parrying the opponent's thrusts like a fencing master toying with a pupil. The pupil tries to get past his guard by suddenly stabbing over his own shoulder or under his opponent's neck. If performed by a bull with fully grown horns, the result could be fatal. But this technique declines with age and combat between adults is largely confined to *clash-fighting.*

Fig. 7.10. Sable cows standing in *reverse parallel* ("head-to-tail") mutually assert dominance by maintaining the *erect posture* in *lateral presentation.*

REPRODUCTION. Sables are seasonal breeders throughout most of their range, calving over a period of several months toward the end of the rains, after an 8-month gestation (240–280 days) (5, 13). The Roosevelt race has no fixed season, at least in Kenya; like roan, females have a postpartum estrus and may reproduce every 10 months or so (11). Females conceive at around 2½ and calve at 3 years.

SEXUAL BEHAVIOR: dominance displays, *lowstretch, urine-testing, foreleg-lifting,* chasing. Female: *appeasement ceremony, head-low posture, urination on demand, courtship circling, lying-out.*

Bulls accompanying a herd routinely *urine-test* all females (fig. 7.8). Other cows and even young calves also perform the urine test very frequently, unlike other ungulates. Sable courtship is conventional enough, except that bulls sometimes brutally drive cows that are sexually interesting but slow to respond (3). One female giant sable was hotly pursued around and through her herd, which she was reluctant to leave, the chase punctuated by his bellows and her drawn-out cries. Run to exhaustion, she finally lay prone like a calf, but the bull persisted, standing over her and prodding her with a foreleg until she rose. The pursuit was then renewed. This went on intermittently for 2 weeks until at last the unfortunate cow came into heat. Then she stood with head low holding her tail out almost horizontally waiting to be mounted, but moved a little when the bull reared, and circled head-to-tail with him. Meanwhile, though, she rubbed her head and horns on his rump (*soliciting*?) and once jabbed him lightly. He mounted 5 times in a few minutes, the last 2 complete mounts including ejaculatory thrusts.

PARENT/OFFSPRING BEHAVIOR. Cows usually seek seclusion before calving and remain alone for a week or more of the calf's 3-week concealment period. However, there is considerable variation. In the Shimba Hills, cows followed by brand-new calves have been seen rejoining their herd, and one even calved in the herd on bare ground (10, author's observ.). As perhaps 80% of the year's offspring are born during a 2-month birth peak in most sable populations, age classes are readily discernible (fig. 7.11). Peer groups often begin during the concealment stage, through the tendency of mothers isolated from the rest of the herd to associate in subgroups. The larger the herd the bigger the classes. Calves past the concealment stage only seek out their mothers to nurse and have still less contact after weaning at 6–8 months. After the first weeks or less, cows invariably break off suckling bouts and walk away. Calves follow closely for up to a minute before giving up; sometimes they *foreleg-lift* and attempt to mount (3). The maternal bond is so loose that even small calves sometimes end up in different sections of a split herd and may not see their mothers for several days. Such an "abandoned" calf eventually begins calling and may even leave the group, uttering ever-louder cries. Mothers that have misplaced calves or lost them to predators wander around giving the same birdlike call in a lower register until they find them or give up looking for them (within 2–3 days).

Fig. 7.11. A crèche of sable calves scrimmaging (play-fighting).

Male offspring begin to be harassed by territorial bulls when they become adolescent at c. 1½ years, but usually manage to stay with the females until darkening color and big horns override the effect of the *appeasement ceremony*, usually in the fourth year. They then join bachelor herds (generally under 10, rarely up to 25 males aged 3–5 years).

ANTIPREDATOR BEHAVIOR: *alert posture, snorting, style-trotting.*
The tendency for calves to lag behind and even to leave the herd could be expected to make sable young unusually vulnerable to predators. Leopards take a fair number of calves, and may often be the chief predator (5). However, the representation of different age classes is comparable to other antelopes, so either sable calves are not more at risk or they benefit from the typically low density of predators in the *miombo* biome (3).

References

1. Best and Best 1977. 2. Child and Wilson 1964. 3. Estes and Estes 1974. 4. Estes 1983. 5. Grobler 1974. 6. Huth 1970. 7. ——— 1980. 8. Kingdon 1982. 9. Sekulic 1976. 10. ——— 1978. 11. ——— 1983. 12. Sekulic and Estes 1977. 13. Wilson and Hirst 1977.

Fig. 7.12. Oryx bulls in a water hole confrontation. Male on left threatens with *medial-horn presentation*, other male *turning away* (yielding).

Oryx

Oryx gazella

"Above all other antelopes the gemsbuck seems to embody the spirit of the African veldt. He is at home in vast shadeless spaces under a fiery sun, reared on the pale desert grass and sheltered by the scant wait-a-bit thorn. . . . He is thoroughbred of the desert. . . . Energy, strength, endurance are the keynotes of his conformation"(1).

TRAITS. Built like a polo pony with level back, short neck, deep chest, and long limbs. Minimal sexual dimorphism: males heavier with thicker neck and horns. *Height and weight:* 115–125 cm; males 176 kg (167–209), females 162 kg (116–188) (5). Head short with blunt muzzle, small ears; long, flowing tail, and short, stiff mane. *Horns:* lancelike, often slightly curved, strongly ridged, average 105 cm (75 cm for East African races), record 120 cm, equally long in females. *Coloration:* gray to tan with black-and-white markings, most developed in gemsbok, least in fringe-eared oryx, but all races have black facial markings, sidestripe, and foreleg garters set off by adjoining areas of white. *Scent glands:* hoof glands only.

DISTRIBUTION. Somali-Masai and South West Arid Zones. The fringe-eared oryx, *O. g. callotis,* of southern Kenya and Tanzania inhabits *Acacia-Commiphora* scrub country that is less open and less arid than the regions inhabited by the beisa oryx, *O. g. beisa,* and gemsbok, *O. g. gazella;* it is browner with prominent ear tufts (cf. roan) and more subdued markings.

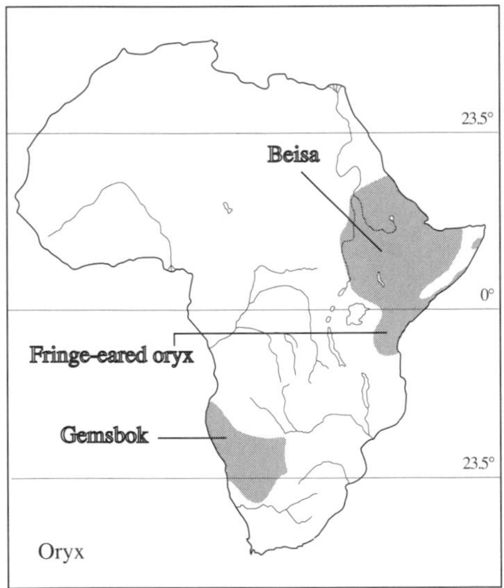

RELATIVES. Though much reduced in range and numbers, *Oryx gazella* is relatively well off compared to *O. dammah,* the scimitar-horned oryx of the Saharan region, which is nearly extinct, and the Arabian oryx, *O. leucoryx,* which became extinct in the wild but has now been reintroduced (8).

ECOLOGY. The oryx is one of the most perfectly desert-adapted large mammals, capable of subsisting in waterless wastelands where few other ungulates can survive. Perhaps most at home on level, stony plains, it also ranges over high sand dunes, climbs rocky mountains to visit hidden pools and salt licks, and goes beyond the arid biomes into the savannas of East Africa and Zimbabwe, utilizing ranges that are only used by other grazers during the rainy season. Its short face, broad, high-crowned molars, wide incisor row, and narrow gape are adapted for close cropping of coarse desert grasses such as those growing along drainage lines and around the bases of dunes. It prefers green but also eats dry grass, browses to some extent, and digs assiduously for roots, bulbs, and tubers, which, together with wild melons and cucumbers, furnish water. Three liters per 100 kg of body weight per day is all an oryx needs. Like the camel and other desert dwellers, when deprived of drinking water it uses various measures to minimize its water needs, notably by allowing body temperature to rise from a normal 35.7°C to 45° (113°F) before beginning evaporative cooling (by nasal panting and sweating) (9). It also concentrates its urine, absorbs all possible moisture from its feces, and tries to avoid exposure to the midday sun, seeking or even making shade (by excavating a hollow), moving and feeding at night and very early in the morning, when the vegetation contains the maximum amount of moisture (9).

SOCIAL ORGANIZATION: nomadic, gregarious, mixed herds and some solitary, probably territorial males.

The integration of the sexes in mixed herds sets the genus *Oryx* apart from most other antelopes. Presumably this arrangement is an adaptation to semidesert environments, where sparse and unpredictably distributed forage favors nomadism and very low population density. Oryxes have been portrayed as wandering great distances in barren wastelands, occasionally congregating in a locality where a thunderstorm has stimulated the growth of grass and ephemerals. The difficulty of finding other oryxes in vast empty spaces could well favor postponing the departure of male offspring from the comparative safety and other amenities of life in the maternal herd. But to avoid eviction by dominant males it was necessary to postpone development of male secondary characters and behavior, and this was achieved, I submit, by females mimicking their male offspring to ever-later stages, until female oryxes finally came to resemble adult males (2).

Radio-tracking of collared gemsboks in

Botswana's Central Kalahari Game Reserve has yielded evidence that *O. gazella* may be much less nomadic and wide-ranging than previously supposed (13). Two females and 6 males whose movements were monitored for 18–39 months during a severe drought resided in this waterless sandveld within ranges no larger than those of the sable and roan. The annual ranges of the females averaged 127 km^2 (66–217) (5 years of data), while 5 of 6 apparently territorial males occupied ranges of 10–16 km^2. In Namaqualand (northern Cape Province), the mean size of oryx territories was only 7.6 km^2 (4.2–9.8) (6). One of the radio-collared bulls, an old animal that may have abandoned or been displaced from his property, ranged 88 and 66 km^2 in 2 consecutive years. Furthermore, this animal and the 2 collared cows did not wander at random through their large ranges, but returned repeatedly to the same places and a number of times stayed anywhere from 3 to 10 weeks in an area of <1 km^2. All the evidence suggested that the Kalahari gemsboks circulated purposefully within accustomed home ranges even in times of drought, gaining sufficient water to meet their needs by digging for roots and tubers at depths up to 1 m (13).

Observations of fringe-eared oryxes in Serengeti N.P. suggest a more nomadic movement pattern (11). One herd of 15 traveled 17 km in the same direction in 1 day, and a lone bull walked 4 km in a straight line within an hour. The mean distance traveled by Kalahari gemsboks in 24 hours was only 2.6 km for females and 1.6 km for males, with a maximum of 9.5 km (13). However, one Serengeti herd remained localized in a 20 km^2 area for 3 weeks (11).

Oryx herds of 50 head are not uncommon and herds numbering up to 200 animals have been recorded (7). But in pooled samples of all 3 races, totaling 554 oryxes, mean herd size (not counting singles) was only 14 head, and the largest troop numbered 131 (Tsavo N.P.) (author's data). Adult females outnumbered adult males 2 to 1 and 11%–28% of the adult males were solitary. Large herds are nearly always mixed and often do not include juveniles, but small herds (say under ½ dozen) may consist of females only, females with 1 male, and males only. Cows may be seen singly while guarding concealed calves. The more conventional segregated pattern is likeliest to occur in less open habitat and/or areas with permanent water where oryx subpopulations may be resident for at least part of the year (cf. wildebeest), as in Etosha, Hwange, Tarangire, and Tsavo N.P. The presence of a network of highly dominant, resident, and probably territorial bulls could account for the segregation in these subpopulations.

Behavioral observations of mixed herds indicate that they represent lasting associations of individuals that are arranged in a stable dominance hierarchy headed by an alpha bull (7, 11). Subgroups are visible in herds containing young, but adults seem to be more integrated than the bisexual groupings of, say, wildebeests or elands. Although a high-ranking female often leads and the dominant male prefers to stay at the rear, he plays a pivotal role in directing and coordinating herd movements by aggressive displays (see below), and will leave the herd to retrieve laggards of either sex.

Observations of oryx kept in semidomesticated herds indicated that groups are semiclosed to strangers: oryxes young enough not to have been inducted into the dominance hierarchy could become integrated in time (males more readily than females), but not individuals older than c. 1½ years or heavier than 100 kg (7). However, adult females were seen joining Serengeti mixed herds with little aggression and adult males, after being subjected to aggression by herd males, were also eventually accepted. Whether these outsiders were actually strangers was not known. Also, the observer of these animals concluded that most males were dominant over all but the highest-ranking females (11), whereas adult females were dominant to all but the alpha and one beta male in the domesticated herds (7). But when gemsboks were competing for access to water holes they had excavated in dry riverbeds during a drought in the Namib Desert, adult males took precedence over all females (3, 12)

Territorial Males? The evidence of territoriality in oryx is quite convincing: the fact that known prime bulls have stayed for extended periods in particular localities (6, 13); the existence of apparent territorial networks in resident subpopulations, notably in Etosha N.P., where bulls were observed to be spaced at intervals as close as 0.5–1 km; and instances where lone bulls have approached and proceeded to dominate (or fight and defeat [11]) the males and court the females in mixed herds (author's unpub. observ.). But the presence of supposedly nonterritorial yet dominant and reproductively active males accompanying mixed herds confuses the issue. In at least some cases the dominant male could well be a territorial bull that had joined the herd before observations began. In other cases, particularly where territories are isolated and/or clustered around water points with large undefended tracts in between, a territorial male may accompany a departing herd and continue in command until and unless the herd enters another territory, as previously described in the roan (see account). The third possibility, given the same circumstances, is that a mature bull may cease to be or never become territorial, but instead accompany and attempt to defend a herd of females as his personal harem. Conceivably, all these options are exercised at different times and places.

Clearly more research on oryx socioecology is needed to reconcile conflicting interpretations of oryx social organization.

Fig. 7.13. Oryx on left makes stabbing threat to a bull displaying defensive threat. (From a photo by Walther 1984.)

ACTIVITY. Information about activity patterns of wild oryx is very limited. A Serengeti herd, watched five days at different seasons, became active and began grazing within ½ hour of daybreak, continuing with occasional pauses until c. 1000 h. Though spread out while grazing, the whole herd often moved considerable distances in the direction dictated by the alpha bull, particularly in the latter part of the activity period. A rest period ensued until 1400 or 1500, during which the oryxes stood and lay most of the time, although sporadic movement and grazing took place at variable intervals. Another general grazing period began usually after 1500, continuing until sunset. From dusk until perhaps ½ hour after dark the oryxes made an "obligatory" march of up to several kilometers (cf. sable account) before bedding down at c. 2000. Apart from individual bouts of standing or grazing, herd members lay until c. 2300, then began a midnight activity period of grazing and moving which lasted to 0100 or 0200, after which they lay down and rested until dawn (11).

POSTURES AND LOCOMOTION. Discussed in tribal introduction.

SOCIAL BEHAVIOR

COMMUNICATION. Covered in tribal introduction.

TERRITORIAL BEHAVIOR. The one oryx display that looks like territorial advertising is the ostentatious *defecation crouch* adopted only by mature—but again, not necessarily territorial—bulls (like the addax in fig. 7.1).

AGONISTIC BEHAVIOR

Dominance/Threat Displays: *erect posture* in *lateral presentation* ± *angle-horn* or head turned toward opponent or, in frontal approach, *head turned away* slightly and ear pointing at opponent; *high-horn presentation* ± slow ducking motion (threat to strike downward as in *clash-fighting*) or head turned away with horntips pointing at opponent (over-the-shoulder stab threat [fig. 7.13]); *medial-horn presentation* ± *nodding* (*symbolic butting*); *low-horn presentation* (horntips aimed at opponent) ± *head tossing; horn-sweeping; head-shaking, pawing, vegetation-horning.*

Defensive/Submissive Displays: *head-low posture, head-low/chin-out* or with *horns turned away* and throat turned to superior (purely submissive); *chin-lifting* and *head-throwing* (7), *creeping* or *knock-kneed walk* in submissive posture (11), *lying-out.*

Fighting: *boxing* (*nod-butting*), *fencing* (striking front sides of horns with oblique downward blows), *clash-fighting; horn-pressing, front-pressing* with levering and twisting, *parallel fighting* and *over-the-shoulder stabbing* (12).

The elaboration of aggressive displays and fighting styles in oryxes attests to the danger of their weapons and the frequency of aggressive interactions. Hierarchical relationships are constantly being expressed in a large herd, especially during the change from one activity to another (11). Head and horn movements, tail-swishing, and ear-flicking accompany these interactions—although it is not always easy to distinguish social signals from movements provoked by accompanying swarms of flies. Most aggressive interactions are one-sided, involving a show of dominance toward a lower-ranked animal that responds submissively. *Supplanting* is the commonest and least aggressive assertion of dominance; only if the subordinate fails to move out of the way in good time are more severe measures taken. Head-on meetings, in particular, provoke hostility and are avoided. Herd members normally maintain a social distance greater than the reach of the horns (at least 1 m), but may come closer when filing, or when resting as long as they face in different directions (7, 11).

The male rank order is maintained with more overt aggression than the female hierarchy, including frequent sparring contests that occasionally escalate into fights. However, the majority of males in mixed herds have small scrotums and are sexually inactive. Only adult males with fully developed scrotums are in contention for the top position, and these individuals also have the thickest necks and greatest bulk (12). Fights between such heavyweights are most likely when strangers meet and compete for the same resource: either an estrous female or, more often, access to water (fig. 7.12) (3, 12).

At low intensity, oryxes merely touch and press or hit their horns lightly together while staying on their feet. At high intensity, standing in *reverse parallel* with heads high, they deliver powerful forward-downward blows (*clash-fighting*), often while *circling*, then fork their horns together firmly at the bases and push, twist, and grapple, using the full force of their bodies (*front-pressing*). Stabbing fights occur only at highest intensity (4, 12). Driving sideways, the stabber pushes off with his forefeet and often lands on his knees (fig. 7.14). Fighting oryxes occasionally bellow. During pauses in long fights (up to 20 min.), oryxes graze, kneel and horn bushes or the ground, paw, defecate in a crouch, and occasionally groom themselves. The loser of a fight usually breaks and runs, pursued by the winner.

Fig. 7.14. Fighting oryx: *over-the-shoulder stabbing* attempt, blocked by bull on right. (From a photo by Walther 1980.)

Despite the oryx's dangerous weapons, its willingness to use them, and the existence of a stabbing technique, injuries from fights are remarkably rare: none of some 184 observed fights, including 30 at high intensity, resulted in even minor injury (12). It may be concluded that oryx horns are not only formidable offensive weapons but also very effective defensive weapons, and that individuals that survive to reproduce are all fencing masters. Yet, fatal body attacks have been witnessed under famine conditions, when extreme competition for access to water caused a breakdown in social organization (3). Still, even well-nourished oryxes are prone to jab one another rather freely, females no less than males, and draw blood from victims that fail to get away quickly enough. Despite the extremely thick hide over the upper body, welts and punctures are often present, and once I photographed a gemsbok with a broken horn tip imbedded in his shoulder.

REPRODUCTION. *Oryx gazella* may breed at any time of year, although the frequent presence of same-age calves suggests synchrony between associated cows. Females conceive as early as 2 years and may calve before turning 3 after a gestation of c. 8½ months (6). They mate again within a few weeks of calving.

SEXUAL BEHAVIOR: *erect-posture, urine-testing, foreleg-lifting,* chin-resting. Female: *head-low posture, head-low/chin-out, courtship circling, urination on demand, lying-out.*

Courtship circling is much in evidence in oryx courtship, recalling the whirling of fighting males, but spacing is closer and the female keeps her head low and ears back (fig. 7.15). Circling comes about through the female's efforts to move behind the bull (as in the *appeasement ceremony*), while he tries to sniff her backside. If approached frontally, a cow may respond defensively with the *head-low posture.* But sooner or later she squats and urinates, and if the *urine test* is positive the bull begins *following closely* and *foreleg-lifting.* The intent to mount seems clear in the oryx's case, as the back legs are bent in and the tail is held out as during copulation (fig. 7.16). *Nudging with the muzzle* and *resting the chin on the cow's rump* are also seen during the precopulatory phase. While and after copulating the cow stands with back

Fig. 7.15. Oryx *courtship-circling,* male displaying dominance (*erect posture*), female in *head-low posture* (from Walther 1966*b*).

legs slightly straddled and tail slightly to the side and out. In 1 recorded courtship sequence, a pair mated three times in an hour, including a number of incomplete mounts. The same pair was seen to mate again the next day after an interval of 13 hours (author's observ.).

PARENT/OFFSPRING BEHAVIOR. Like most other *hider* species, oryx cows isolate before calving and remain in the vicinity if not within sight of their concealed offspring. An example of lying-out discipline was afforded by a newborn beisa oryx (author's observ.). It was lying in plain sight beside a log in an acacia grove, where the grass had been grazed to a short sward by game and cattle. Although flies covered its face and crawled around its eyes, it never moved a muscle, except for an occasional blink, even when approached within 2 meters. Soon afterward a herd of Samburu cattle was driven right across the calf's hiding place, barely parting for the log. Certain the calf had been trampled, I approached to find it lying in exactly the same position. The mother and another cow had remained within a few hundred meters the whole time, screened behind bushes.

Fig. 7.16. Oryx copulation.

Calves remain in hiding for up to 6 weeks, although crèches of calves no older than two weeks are common, with or without any cows in attendance. Peer groups persist for at least a year and perhaps longer. Calves with horns 10–15 cm long were seen to nurse for up to 2 minutes and, after the mother broke away, to follow her another 2 minutes in a vain effort to get more, just like sable calves. A captive 3½-month calf was suckled 3 times a day (10). The horns are visible at birth as hair-covered bumps.

ANTIPREDATOR BEHAVIOR: *alert posture, snorting, style-trotting,* self-defense.

Despite anecdotes about hunters and lions that have been fatally stabbed by oryx at bay, and their undeniably dangerous weapons and aggressive tendencies, I know of no proof that predators treat oryxes with greater respect than other large herbivores. An animal ambushed by a lion or run to exhaustion by wild dogs or hyenas seldom has any chance to defend itself effectively unless the predator bungles. Like other open-country antelopes, oryxes place primary reliance on flight, and they are generally considered swift and enduring runners.

References

1. Blaine 1922. 2. Estes 1990. 3. Hamilton, Buskirk, Buskirk 1977. 4. Huth 1980. 5. Kingdon 1982. 6. Smithers 1983. 7. Stanley Price 1978*a*. 8. ——— 1988. 9. Taylor 1969. 10. Walther 1965*b*. 11. ——— 1978*c*. 12. ——— 1980. 13. Williamson and Williamson 1985.

Chapter 8

Hartebeests, Topi, Blesbok, and Wildebeests

Tribe Alcelaphini

Alcelaphus buselaphus, hartebeest
A. lichtensteinii, Lichtenstein's hartebeest

Damaliscus lunatus, topi/tsessebe
D. dorcas, blesbok/bontebok
D. hunteri, Hunter's antelope/hirola

Connochaetes taurinus, common, blue, and white-bearded wildebeest / brindled gnu
C. gnou, black wildebeest / white-tailed gnu

TRIBAL TRAITS. Large and medium-sized antelopes, with high forequarters and sloping backs, from 60 kg bontebok to 230 kg blue wildebeest, horns in both sexes, subequally developed in females. *Alcelaphus* and *Damaliscus:* elongated, narrow muzzle; horns ringed, lyrate, or complexly recurved; coat short, glossy, plain tan to chestnut ± light rump, dark and white markings; tail hock-length with toothbrush terminal tuft. *Connochaetes:* broad muzzle; horns smooth, hooked forward or sideward, with knobby bosses; coat short and glossy, with mane, beard, and facial tufts; color dark brown–black with white tail to blue-gray with black stripes, mane, and tail; tail long, flowing, horselike. Young of all species tan and similar-looking. Reduced sexual dimorphism: horns and general appearance of females like males (including fake penile tuft in wildebeest), but males more muscular, heavier, generally darker, with bigger horns. Muffle hairy. *Scent glands:* preorbitals well-developed and functional in both sexes (larger in male); hoof glands in forefeet only, exuding black, sweet-smelling secretion. *Mammae:* 2.

DISTRIBUTION. Wholly African in origin and distribution, alcelaphines reached their peak 2 million years ago when there were over 8 genera with at least 15 different species (3). Though since reduced to 7 species in 3 genera, it remains one of the most numerous and widely distributed bovid tribes. A century ago alcelaphines ranged the entire continent except for the deserts, forests, and high mountains. Today there are none north of the Sahara, several thousand black wildebeests are all that remain of the millions that once roamed South Africa's Highveld and Karroo, and the others are greatly reduced in range and abundance. The wildebeest and topi remain dominant herbivores in acacia savanna and plains but only in a few places like Serengeti N.P. and Southern Sudan do they approach their former abundance. Once the hartebeest had the widest distribution of any open-country antelope, and the hirola, found only in northeastern Kenya and adjacent Somalia, had one of the smallest. Estimated to number ca. 12,500 in Kenya and 1000–2000 in Somalia in 1976 (2), the hirola population averaged ca. 7000 in Kenya between 1977 and 1983 (4), the year of the last published counts. The hirola may be a relict population of a formerly widespread damaliscine that was later replaced by the more successful topi (9).

RELATIVES. As in the horse antelopes, there is a clear division in this tribe, with *Alcelaphus* and *Damaliscus* on one side and *Connochaetes* on the other. The first 2 are obviously closely allied and complementary in many respects (see Tribal Traits). The former has more specialized horns and skull characters, and the latter has more elaborate coloration, but otherwise they are much alike. The blesbok and hartebeest have been known to hybridize on South African farms, as have the 2 wildebeest species (1).

ECOLOGY. Alcelaphines are archetypical plains antelopes. Although Lichtenstein's hartebeest inhabits the *miombo* woodlands and the common wildebeest is associated throughout its range with acacia savanna, as a rule each species prefers the openest available landscapes with the shortest grass. The largest populations occur in the transition zones between savanna and arid biomes, where the animals move seasonally between biomes in large, mobile aggregations. Wildebeests are the most and hartebeests the least migratory/nomadic members of the tribe. Hartebeests are the most and black wildebeests the least tolerant of woodland and high grass. The hartebeest penetrates farthest into arid country and has the lowest metabolic rate and the lowest water consumption (12). The topi, which tends to dominate floodplain herbivore guilds, is the most water-dependent. However, as nearly pure grazers, no alcelaphine is desert-adapted to the extent that most gazelles, the eland, and the oryx are. As a rule, migratory alcelaphines utilize arid lands only during the rains and concentrate around permanent water in higher rainfall areas during the dry season. Yet hartebeests and wildebeests manage to live in waterless regions of the Kalahari by eating melons and digging up water-storing roots and tubers.

Wherever the 3 genera overlap, *Connochaetes* dominates *Alcelaphus* and *Damaliscus* ecologically, and *Damaliscus*, though smaller, dominates *Alcelaphus* in interspecific interactions, in my experience. The wildebeest's feeding and reproductive specializations enable it to exploit short grasslands more efficiently than other ruminants, to the point where it crowds them out when it concentrates in great armies. But in the Northern Savanna, where the wildebeest does not occur, the topi/tiang comes into its own as a dominant grazer.

SOCIAL ORGANIZATION. The wildebeest, topi, and blesbok are among the most gregarious and socially advanced ungulates, forming the largest migratory herds of all antelopes (though rivaled by the white-eared kob and lechwe). Their armies often include all the animals from a wide region which have concentrated in a limited area. But other populations and subpopulations of these same species are resident and live in the usual sedentary/dispersed pattern prevalent in the hartebeest. In resident populations, herds of 2–10 females (occasionally more) and their offspring occupy the best available range, which is partitioned into territories whose owners exclude other sexually mature males. Associated females establish dominance hierarchies and resist attempts by outsiders to join their herd. Male (and sometimes female) offspring are cut out by territorial bulls as yearlings (wildebeest and *Damaliscus*) or 2-year-olds (hartebeest) and thereafter live in bachelor herds until mature (3–5 years) and ready to become territorial. Bachelor herds in this tribe also contain mature and old males that have lost, abandoned, or temporarily left their territories.

In the mobile phase, group size is much larger and the only stable associations are between mothers and immature offspring. The similar appearance of male and female alcelaphines promotes bisexual association in migratory aggregations (6). As long as males behave like females, they neither intimidate cows and calves nor antagonize other males and can remain as anonymous members of sizeable groups.

Depending on the seasonal availability of good pasture and of water, every gradation from completely sedentary to completely nomadic habits and social organization can be found in different populations or in the same population at different times. Populations tend to be most sedentary and herds most stable in composition during the rains and most unstable in the dry season, when both sexes congregate on the best available pastures and when the territorial network partially or completely breaks down.

ACTIVITY. Like most grazers, alcelaphines feed day and night, but are generally less active when it is dark and predation risk is greatest. Areas with little vegetation to obstruct vision are favored for resting sites. See species accounts for details.

POSTURES AND LOCOMOTION. The main gaits of alcelaphine antelopes are an ambling walk and galloping, with neck horizontal and muzzle pointing downward. The canter is perhaps less graceful-looking than a gazelle's or an oryx's, but the overdeveloped forequarters may largely explain the alcelaphine's ability to maintain a tireless canter. They are also speedy. Hunters on horseback like Selous considered the hartebeest and tsessebe the hardest of all African game to ride down (11). Trotting in this tribe is usually performed as an alarm signal and in play; it is emphasized by raising the legs high and tail-lashing (*style-trot*). *Damaliscus* and *Alcelaphus* also *stot* when alarmed, a gait not seen in wildebeests.

Alcelaphines paw more than most antelopes, usually as a prelude to going to their knees, either to horn objects or simply to lie down, and territorial bulls paw before defecating. Wildebeests also roll on their backs in dirt and mud. Exaggerated head-nodding is characteristic of *Alcelaphus* and especially *Damaliscus*, whereas tail flourishes are conspicuous in wildebeest displays.

SOCIAL BEHAVIOR

COMMUNICATION

Visual Communication. The conspicuous coloration of alcelaphines indicates the outstanding importance of visual communication in this tribe of typical plains antelopes. All except the hartebeests are dark (very revealing in open grassland) and/or reverse countershaded: darker underneath and lighter on top, opposite to the way cryptically colored species are shaded. The blesbok/bontebok is about the gaudiest of all antelopes, and even the hartebeests have some races with contrasting markings, notably the red hartebeest (*A. b. caama*) of the South West Arid Zone. Also, the catalogue of alcelaphine displays is heavily weighted toward visual signals.

Vocal Communication. *Alcelaphus* and *Damaliscus* produce similar quacking and grunting calls, but are not "noisy" species. Females and young of the two wildebeests have similar lowing calls, which can become very loud when a large aggregation is disturbed, causing separations of cows and calves. But the males of these two species have completely different calls. The black wildebeest's hiccoughs carry a good 2 km, whereas most subspecies of *C. taurinus* give a series of metallic grunts (like the Hottentot *gnou* sound with the *g* aspirated). The western white-bearded wildebeest has a throatier, more sustained call, which when voiced by thousands of bulls (especially during the rut) resounds like a chorus of immense frogs.

Olfactory and Tactile Communication. Scent ranks second to sight as the most important communication channel. The use of feces, preorbital- and hoof-gland secretions, and in wildebeest the urine, in territorial behavior is described under that heading. Hoof-gland secretions help the animals keep track of one another, especially on migration. Wildebeests rub their faces and preorbital glands on one another's withers or rump as a form of social contact (fig. 8.16); otherwise the only social contact in the tribe is the sniffing-touching of noses, neck, or genitals when alcelaphines meet.

TERRITORIAL BEHAVIOR. The behavior of territorial males is extraordinarily elaborate, ritualized, and variable. Males are most active and interactive in dense populations where territorial spacing is minimal. *Connochaetes taurinus* and the topi tolerate closer spacing (20–30 m) than any species except the kob and lechwe. Territorial hartebeests, which usually defend much larger areas (but no larger than topis in patchy habitat), are perhaps less tolerant. Under conditions that allow permanent residence, males often occupy the same territories year after year. But competition for the best locations is often severe, resulting in a more rapid turnover. In migratory populations, the males that stake out territories whenever an aggregation settles down usually abandon them when the aggregation moves on. But sometimes a bull is loath to quit his claim and stays behind for days or even weeks after all other wildebeests have gone.

Advertising. Territorial status is advertised by a variety of visual displays, by olfactory marking, and also by vocal signals, especially in the wildebeests. Major displays of the whole group include *pawing, kneeling and horning/face-rubbing* the ground or vegetation, *defecation on stamping grounds* or middens, *preorbital-gland marking*, standing and cantering in the *erect posture* (*rocking canter*), *calling*, and *snorting*. Defecation is the main form of olfactory marking. *Alcelaphus* and *Damaliscus* make it a visual display as well by adopting a distinctive crouch (fig. 8.1); wildebeests barely even raise their tails; however, they compensate by vigorous preliminary pawing. The preorbital glands play a role in marking the ground, objects, the animal's self, and others (see species accounts). Stamping grounds are table-size bare places, usually near the territorial center, where the owner retires to rest and ruminate. The concentration of pawing, horning, dunging, and lying in one spot wears the grass away. The odor of dung and glan-

Fig. 8.1. Hartebeest *challenge ritual:* one male defecates while the other incisor-grooms his flank (*displacement grooming*). (After a drawing by M. Gosling.)

dular secretions then impregnates the coats of the owner and other animals that lie there. Territorial males advertise their status even while standing idle by holding their heads above shoulder level. The hartebeest and particularly topi, but wildebeest rarely, enhance their visibility and gain a good vantage point for scanning their domain by standing upon mounds (fig. 2.4).

The territorial *status quo* is maintained through daily ritualized encounters between neighbors. The closer the territorial spacing, the more often neighbors interact. The richest repertoire of displays possessed by the species is incorporated in these *challenge rituals,* including all the territorial advertising displays, displacement activities and displays derived from them, dominance and threat displays, and largely symbolic combat (5). *Challenge rituals* also include mutual, nonbelligerent contact of uncertain significance, which takes a different form in each genus: *sniffing noses* and *rubbing heads* in the topi; *sniffing noses* and *neck-sliding* in the hartebeest (fig. 8.6); *sniffing noses* and rubbing the face on the adversary's rump (usually), head, or neck in the blue wildebeest (both sexes) (fig. 8.16). Another step, unique to wildebeests, is a *urine-testing ceremony* between territorial bulls (fig. 8.2).

AGONISTIC BEHAVIOR

Dominance/Threat Displays. *Lateral presentation* in the *erect posture* (fig. 2.6) is the main alcelaphine dominance display, as it is in the horse antelopes. Tilting the head so that the horns aim toward an opponent is also common to both tribes, but has different meanings (cf. horse antelope threat displays). *Angle-horn* in the Alcelaphini, seen typically when 2 males are standing in *reverse parallel* (fig. 8.2), is apparently

Fig. 8.2. Wildebeest *challenge ritual:* as two territorial bulls stand in *reverse-parallel,* one (*left*) solicits urination and the other responds by bending away and displaying *angle-horn,* showing the intention to turn into a confronting position.

Fig. 8.3. Wildebeest bull *cavorting.*

an intention movement to turn into a confronting position. Since this is also the prelude to fighting, *angle-horn* signals readiness for combat, if not a threat to attack. Alcelaphines also nod and shake their heads, and perform vigorous *horn-sweeping* gestures which may derive from a fly-shooing gesture but have evolved into aggressive displays because of their similarity to sideward hooking (itself rare) (5). *Head-nodding* is typically performed from the normal standing position and often while animals are filing and otherwise peaceful; in this context it may simply say to the animal ahead, "Come on, get moving." *Cavorting* (fig. 8.3) is often seen during playful and aggressive interactions and combines many different actions: *leaping, bucking, spinning, style-trotting, jinking, head-shaking, head-throwing and -nodding, horn-sweeping,* and *tail-lashing.* The more spirited the performance, the more excited and aggressive the performer. In addition to all of the above, the territorial-advertising activities of adult males, basically aggressive in nature, become threat displays when addressed to a specific adversary.

Differences between species in which displays are emphasized and in other details will be brought out in the species accounts.

Defensive/Submissive Displays. Subordinates respond to dominance and threat displays with the *head-low posture* and *moving away.* When a territorial bull stands blocking the way of a passing male, the intruder alters course to pass behind him. Male and even female wildebeests often express a degree of defiance toward displaying bulls by *cavorting.* Hartebeests and topis appear to be more impressed by their bulls:

females and young assume the *"head-in" posture* (fig. 8.4). *Lying prone* is performed under continuing harassment, for instance, when a female wants to avoid a male's efforts to mount. Lying on the side, possibly the most extreme form of submission in bovids, has been recorded in the black wildebeest (see species account).

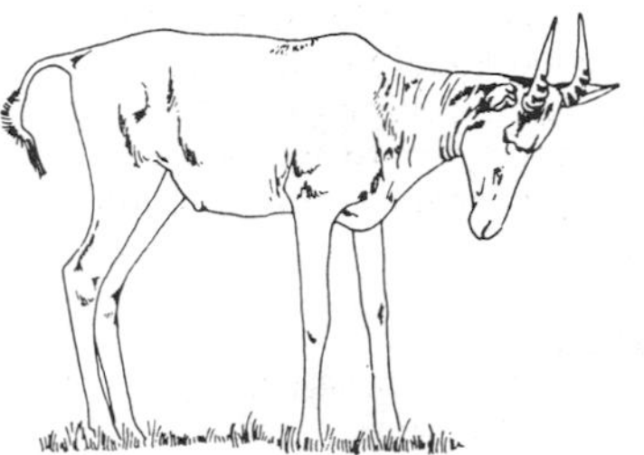

Fig. 8.4. Yearling hartebeest performing the *head-in display* (from Gosling 1974).

Displacement Activities. Ritualized or *displacement grazing* (fig. 2.11) features prominently, especially in the wildebeests, during the *challenge ritual. Horn-sweeping,* combined with *tail-swishing,* is even more developed than in the Hippotragini; in the hartebeests it has evolved into a self-marking ritual. Alarm signals, including the *alert posture, snorting,* and *stamping,* are performed during aggressive encounters (fig. 8.5). Sometimes wildebeest bulls lie down and even ruminate after *pawing, kneeling,* and *horning* during a *challenge ritual.*

Fighting. Members of this tribe all kneel during combat (figs. 8.13, 8.15). From this position they can slide forward and backward as though on runners, and they are also less likely to trip or be hurt when suddenly pushed sideways. This attitude and the alcelaphine horn configuration are adapted to head-on pushing (*front-pressing*), in which the object is to force the opponent backward and out of line. In rare hard fights, the contestants also duel with their horns, each attempting to get under the other's guard and gore or lift and throw him. Should one fall, the other would probably kill him if he could—but I've never known it to happen. Losers simply run away. But most fights are brief and inconclusive.

REPRODUCTION. A unique reproductive system, seen in no other tribe, has evolved in the wildebeests and to a lesser extent in *Damaliscus.* The key difference is that calves do not go through a lying-out stage but accompany their mothers from the moment they can stand. To survive, the calf must either be effectively protected or able to elude predators by outrunning them or somehow escaping notice. As the mother cannot defend her calf effectively against more than one predator, and the neonate cannot outrun such predators as hyenas or wild dogs, much less a cheetah, survival depends on the calf losing itself in a herd. The larger and denser the herd, the better its chances of avoiding detection and eluding pursuit. In alcelaphines other than the hartebeest, however, the tan natal coat, though well-adapted for concealment, makes calves stand out in a herd of dark adults. So unconcealed newborn calves are highly vulnerable unless there are other, slightly older calves around to give them "cover." For wildebeests, a short, sharply defined calving season insures that this cover is provided. But only *Connochaetes* is fully committed to a follower-calf strategy. *Damaliscus* calves may or may not be concealed and the calving season is longer. Hartebeests retain the hider system and breed the year round. The wildebeest's follower strategy is connected with its migratory habits, tendency to concentrate in large numbers, and preference for short

Fig. 8.5. *Displacement alarm,* performed during wildebeest challenge ritual, reduces the risk of predators approaching undetected.

grass, all of which are incompatible with a concealment strategy (7).

SEXUAL BEHAVIOR. Antelopes of the genera *Alcelaphus* and *Damaliscus* are the only bovids so far known that do not *urine-test* females (i.e., sample the urine, then curl the lip and/or open the mouth in the *flehmen* grimace). Yet they have an apparently functional vomeronasal organ (8). How do they find out when a cow is coming into heat? Hartebeest and topi cows behave just as passively as the cows of other species. Bulls approach them in *lowstretch*, which in many antelopes elicits urination, but merely sniff the vulva—as do many antelopes with functioning vomeronasal organs (see chap. 2, Communication). Presumably they use their noses to tell when a female is in heat, or perhaps the posture assumed by a cow ready to copulate, which resembles the head-in display (fig. 8.4), may be effective as a visual signal. Cow wildebeests signal estrus merely by holding the tail slightly out. However, they *urinate on demand* and bulls perform the standard *urine-test*, making a pronounced *grimace*.

Male courtship features the *erect posture* with ears out or dropped and tail raised. This posture is combined with *high-stepping* in *Damaliscus* (fig. 8.10), but *foreleg-lifting* is lacking in the whole tribe. *Lowstretch* is the main display of the hartebeest and blesbok. Copulation is similar in all the species (but see blesbok account) (fig. 8.8), and multiple copulations are the rule. Females mate dozens of times with a male, often 2 or more times in a minute. In large aggregations, cows on the move mate with a succession of territorial bulls.

ANTIPREDATOR BEHAVIOR. The common wildebeest is usually the preferred prey of the lion and spotted hyena wherever it occurs. The fact that it tends to be the dominant large herbivore in savanna ecosystems may largely explain this preference, but maybe it is also more vulnerable. Wildebeests (and other game) at high density are easier to approach than when in small groups or alone, perhaps because each individual feels and in fact is safer in a large group. The more animals present, the less likely any given individual is to be the victim. So predators have less difficulty catching wildebeests in large aggregations than in small herds. Topis and hartebeests are unlikely to be preferred prey as long as wildebeests and zebras are available. Schaller (10) considered the topi the most alert of Serengeti prey species and the hartebeest second. Yet probably throughout the Northern Savanna, the topi and hartebeest are the lion's main prey.

The wildebeest and topi give powerful alarm snorts; the hartebeest's is thinner. All species stamp, bunch together, approach or follow a lion or leopard in the *alert posture*, and perform a *style-trot*. *Alcelaphus* and *Damaliscus* also *stot*.

Wildebeest mothers defend their calves with considerable success against predators up to and including hyenas hunting singly. Jackals rarely try to attack wildebeest calves but prey on very young topi and blesbok calves.

References

1. Brand 1963. 2. Bunderson 1977. 3. Cooke 1968. 4. East 1988. 5. Estes 1969. 6. ——— 1974*a*. 7. ——— 1976. 8. Hart, Hart, Maina 1988. 9. Kingdon 1982. 10. Schaller 1972*b*. 11. Selous 1881. 12. Stanley Price 1978*b*.

Fig. 8.6. Hartebeest *challenge ritual:* two bulls *neck-sliding* (from a drawing by M. Gosling).

Hartebeest

Alcelaphus buselaphus

TRAITS. A large, high-shouldered antelope with elongated forehead and oddly shaped horns. *Height and weight: A. b. cokei* (kongoni) males 115 cm (107–120) and 142 kg (129–171); females 112 cm and 126 kg (116–148) (6); *A. b. major* (the largest race) males to 143 cm and 228 kg (10). *Horns:* 45–70 cm (6). *Coloration:* coat short with sheen, varying from pale tawny (*cokei* and *tora*) to chocolate brown (*swaynei*); rump pale or white, black facial blaze and leg markings in the most colorful races (*swaynei* and *caama*); new calves light tan and unmarked except for diagnostic brow line. *Glands:* preorbitals with a central duct, secretion black and sticky in Coke's and Lichtenstein's, colorless in lelwel and some other races; hoof glands in forefeet only.

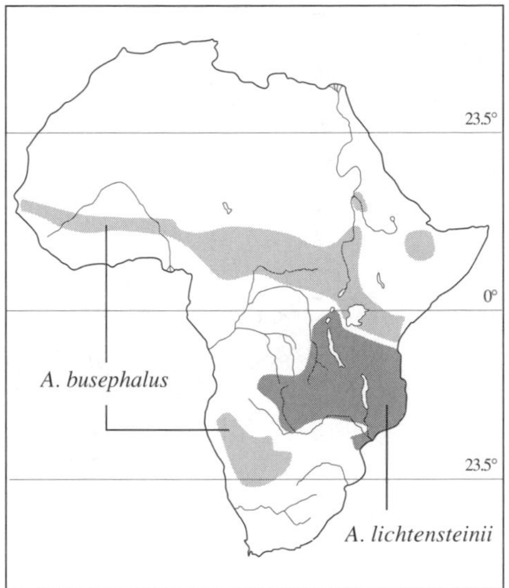

DISTRIBUTION. Historically, hartebeests occurred in savanna country from the Cape to the Mediterranean, and from Senegal to Somalia. The bubal (*A. b. buselaphus*) was one of the few antelopes of African origin which ranged north of the Sahara. The last one died in Paris in 1923 (1). The red hartebeest (*A. b. caama*) was also practically exterminated in South Africa. Yet hartebeests still occur, though in vastly reduced numbers, over the rest of their range. At least 10 populations are different enough from one another to be considered separate subspecies, but only 3 are readily accessible in parks: the kongoni or Coke's hartebeest (*A. b. cokei*) of Kenya and Tanzania; the lelwel (*A. b. lelwel*) of Uganda; and the red hartebeest of Botswana, southern Angola, Namibia, and (reintroduced) South Africa. Lichtenstein's hartebeest, seen in the *Miombo* Woodland Zone of Tanzania, Zambia, and Zimbabwe, is considered a separate species (1).

ECOLOGY. The hartebeest is a typical plains antelope, but enters open woodland and tall bush grassland more readily than the wildebeest and topi, and is found more often on the edge than far out on the plains. It grazes selectively on leafy perennial grasses and has a particular association with medium grasslands dominated by red-oat grass (*Themeda triandra*) and scrub acacia, such as the whistling thorn. In open plains the hartebeest may migrate between short (10 cm or more), well-drained pastures during the rains and long grasslands in the dry season. Where these different types occur in small patches, home ranges incorporate pastures for all seasons (5). Lichtenstein's hartebeest is still more of an "edge" species, its distribution closely associated with the drainage-line grasslands that dissect the *miombo* woodland (3). *Alcelaphus* drinks regularly where water is available, but in waterless parts of the Kalahari, it eats melons and digs up roots and tubers to meet its needs (8).

SOCIAL ORGANIZATION. Hartebeests in mobile aggregations of thousands were reported in earlier days from eastern and southern Africa (9), and an aggregation of 10,000 was recorded as recently as 1963 on the plains near Sekhuma Pan, Botswana (8). But even when plentiful, hartebeests perhaps rarely live for long in the mobile-aggregated state like wildebeests and topis. Most populations are distributed in the typical sedentary-dispersed pattern.

In Nairobi N.P., where kongonis lived at unusually high density in the 1960s and 1970s as dominant herbivore, the areas of grassland and scrub were partitioned into a mosaic of comparatively small territories (fig. 8.7), within which herds of females and young circulated (5). These herds, representing female hierarchies, semiclosed to outsiders and typically numbering 6–15 head, remained within home ranges of 3.7–5.5 km^2 which included or overlapped 20–30 territories. The time spent on a given territory ranged from a few hours to several weeks. Territorial bulls remained separate from the females except when actively herding or courting (keeping a mean distance of 86 m compared to 6 m between cows). At low density in patchy habitat, females may stay indefinitely on a single territory as harems, a pattern that may be typical for Lichtenstein's hartebeest (3).

Bachelor Males. Male hartebeests often accompany their mothers for up to 2½ years, a year longer than other alcelaphines, yet mature no later than wildebeests, at 3–4 years (5). The basis of this unusual arrangement is continuous breeding: female hartebeests calve every 9–10 months (gestation 8 months), too early for the last calf to become independent. The forbearance of territorial males for these subadult males is maintained by the *appeasement ceremony* (see tribal introduction, fig. 8.4), and by maternal protectiveness. A territorial male who chases a young male away is apt to lose its mother along with it. Males stop performing the *appeasement ceremony* once separated from the female harem;

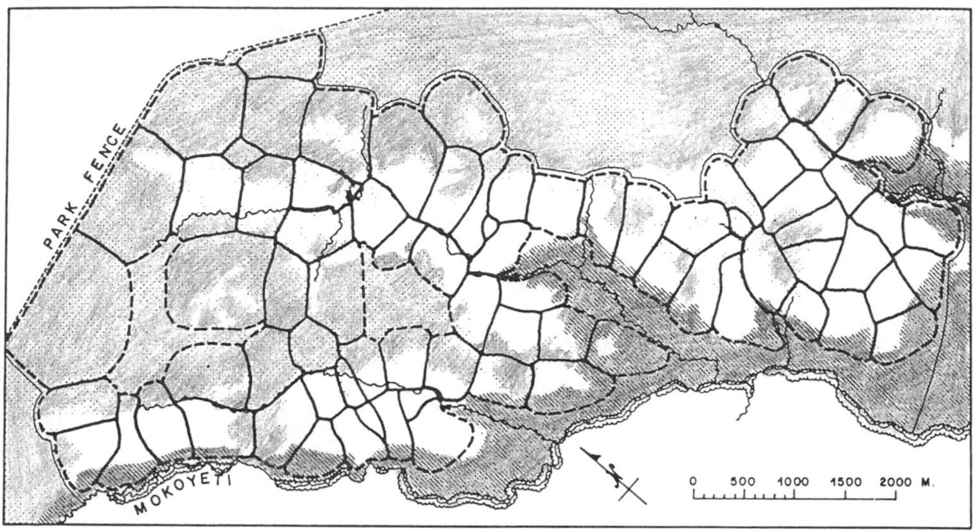

Fig. 8.7. Hartebeest territorial mosaic in Nairobi N.P. (from Gosling 1974).

they simply run away when persecuted. Between 1 and 2½ years they join bachelor herds; these are usually smaller than female herds, although male herds of up to 35 head are not unusual. In Nairobi N.P. aggregations of up to 100 males hung out near major waterholes in the dry season, these including young as well as old bachelors plus off-duty territorial bulls; the aggregations broke up into smaller parties when the animals moved off to graze. Five known bachelors had home ranges of 6.7–10.3 km^2. Between 3 and 4 years, males that have achieved high status in the bachelor herds begin the quest for territories (5).

The best territories are attractive to other hartebeests at all seasons—incorporating at least 2 different plant communities, with ready access but not too close to water, and with firm footing during the rains. Competition for such pivotal territories is fierce: among 33% of the adult males in Nairobi N.P. which held territories, only 18% held them for 6–12 months at a stretch and half for less than 3 months (5).

Prime males (4–7½ years) occupied the pivotal properties, which were passed on unchanged through successive turnovers (see fig. 8.7). Dispossessed bulls tried repeatedly, sometimes successfully, to regain the places they had held previously. But post-prime bulls had to settle for inferior locations in the scrub grassland (cf. waterbuck); here territories were bigger and rounder, and spacing between bulls averaged 700 m compared to 200 m in the best habitat. Peripheral males could hold territory indefinitely and vacated territories often remained unoccupied, in contrast to the rapid turnover of pivotal territories. Nairobi N.P. territories averaged 31 ha (ca. 16–100) compared to 290 ha (160–520) in a subpopulation of *A. lichtensteinii* in Kafue N.P. (3).

ACTIVITY. Daylight activity of 6 lelwel hartebeests was observed for 16 days (April–June) in Zaire's Garamba N.P. (2). The herd rested an average of 4 hours 25 minutes a day (more on hot and less on cool days), usually during 3 or 2 different periods from 0800 to 1630 h. Individuals slept for a few minutes at a time, as denoted by curling up or resting the chin on the ground. The territorial male spent long intervals advertising his presence by standing on a termite mound, and correspondingly less time lying than the females. The main feeding/activity peaks were early and late in the day, with lesser peaks between rest periods. Daily ranging averaged 2–3 km, the herd remaining always within the male's territory; but he sometimes went beyond the boundaries.

SOCIAL BEHAVIOR

COMMUNICATION. See tribal introduction.

TERRITORIAL BEHAVIOR

Advertising: standing in *erect posture,* preferably on elevation; defecation in crouch with tail held out; *kneeling* preceded by *pawing, face-rubbing* and *soil-horning* followed by *shoulder-wiping; rocking canter, herding* in *lowstretch,* and *grunting.*

The *defecation squat,* standing and moving in the *erect posture,* and *grunting* are the most diagnostic territorial behaviors; the other displays are also performed by nonterritorial hartebeests although less often and vigorously.

A territorial male does not defend all of his sizeable property with equal vigor. An activity center amounting to 10%–20% of the total area is actively defended; any male that comes within 100–200 m of the owner is likely to be challenged (5).

Territorial neighbors regularly meet and interact at their common border. *Neck-sliding* (fig. 8.6), "horn-tangling" (sparring), repeated *dunging* emphasized by *pawing*, and *grazing duels* with contestants parallel and 5–100 m apart (cf. Thomson's gazelle) feature prominently in encounters, which typically last over 10 minutes (5).

AGONISTIC BEHAVIOR

Dominance/Threat Displays: *lateral presentation* in *erect posture*, with head up and tail out, *neck-sliding, soil-horning, shoulder-wiping, confronting on knees* (combat attitude), *nodding, cavorting.*

Defensive/Submissive Displays: *head-in posture, head-low posture, lying-out* (rare).

Fighting: *clash-fighting, front-pressing* and *twist-fighting, horn-tangling, sparring* on feet, *chasing.*

Shoulder-wiping is a special hartebeest behavior, performed by both sexes and all ages, usually in connection with kneeling and *soil-horning*. Either while kneeling or after standing, the animal touches or wipes its face on each shoulder in turn with abrupt movements. Mud and/or preorbital-gland secretion is thereby smeared on the shoulders and flanks (5). In Lichstenstein's hartebeest and *A. buselaphus cokei*, both with pigmented secretion, these areas are typically streaked with black (3, 5).

Serious fights, though rare, can be fatal (5, 10). They are likeliest to occur when an off-duty territorial bull returns to find another bull (often a previous proprietor that lost his place in similar circumstances) has taken over. Battles royal of over an hour have been recorded (5). The loser flees, often pursued a long way by the winner.

REPRODUCTION: perennial, with 1 or 2 extended peaks (see tribal introduction).

Usually some calves are born every month in East African populations, but with 2 birth peaks in the short and long dry seasons. Further south, hartebeests have a single birth peak late in the dry season.

SEXUAL BEHAVIOR: *erect posture, lateral presentation, lowstretch* with ears lowered and tail out, *sniffing vulva* but no responsive urination and *urine-testing; chasing* and *herding, rubbing preorbital glands*

Fig. 8.8. Hartebeest copulation; the male's posture is typical of the tribe.

on female's rump. Female: evasion, *head-low posture, butting in flank.*

A territorial male greets an approaching female with head up and stands blocking her way with ears lowered, immobile. Then he wheels slowly and tries to sniff her vulva, which often causes the female to depart. A cow ready to mate stands braced with hindlegs slightly spread and tail to one side, and holds a similar but more crouched attitude after copulation. Males adopt much the same attitude prior to mounting. Multiple copulations are the rule—e.g., 4 times in 4 minutes (11) during the 1-day estrus (fig. 8.8). Male offspring tend to react aggressively and sometimes interfere, provoking a chase by the attending bull.

PARENT/OFFSPRING BEHAVIOR. Expectant females isolate, accompanied by 1 or more previous offspring. In 4 observed births calves did not begin struggling to rise before 9 minutes and it took 30 minutes postpartum before any could stand firmly (4). Within 10 minutes thereafter, calves could walk and run, and within 45 minutes would follow their mothers. The hiding phase lasts 2 weeks. In case of disturbance, a mother interrupts a calf seeking a hiding place by passing it or touching/licking its rump, which makes it stand and raise its tail for elimination.

ANTIPREDATOR BEHAVIOR: *alert posture; snorting; stamping; style-trotting; stotting* (see tribal introduction).

References

1. Ansell 1971. 2. Backhaus 1959*a*. 3. Dowsett 1966. 4. Gosling 1969. 5. ——— 1974. 6. Kingdon 1982. 7. Roosevelt and Heller 1914. 8. Smithers 1971. 9. Swayne 1894. 10. Ward 1962. 11. Wilson 1966*b*.

Added Reference

Kok 1975.

Fig. 8.9. Topi bachelor males *huddling*, often the prelude to sparring contests.

Topi and Tsessebe

Damaliscus lunatus

TRAITS. Resembles a hartebeest but smaller and darker, without elongated forehead; horns [S-shaped] not sharply angled. *Height* and *weight*: males 115 cm (104–126), 130 kg (111–147); females 113 cm (105–118), 108 kg (90–130) (Serengeti N.P.) (10). *Horns:* ringed, lunate in tsessebe, 30–40 cm; more parallel, lyrate, and longer in other races, 45–60 cm in korrigum, the record (71 cm) equal to hirola record (12). *Coloration:* short, glossy coat, tan with purple blotches increasing in hue from west to east and south to north, reverse countershading (darker below, lighter above); adult males darker than females, young calves tan without markings. *Scent glands:* preorbitals with central duct, naked, developed in both sexes, secrete clear oil; hoof glands in forefeet only.

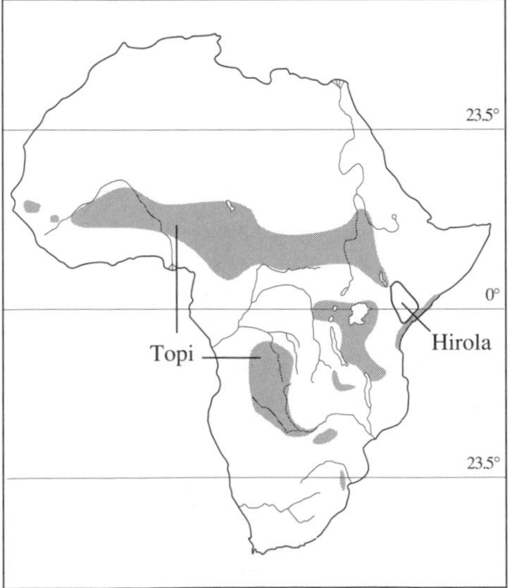

DISTRIBUTION. *D. lunatus* has a very extensive geographical range but a naturally patchy distribution reflecting specialization for certain grassland communities in the arid and savanna biomes. Hunting and habitat destruction by man have greatly accentuated the trend toward isolated populations. The tsessebe (*D. l. lunatus*) is isolated by hundreds of miles of *miombo* woodland from the nearest topi populations of East Africa. In the Northern Savanna and adjacent arid zones, *D. l. tiang* (up to ¾ million in Southern Sudan [16]) and *korrigum* fill much the same niche as the wildebeest in East and southern Africa.

ECOLOGY. The topi favors grassland habitats, ranging from vast treeless plains to lightly wooded bush and tree savanna. It prefers medium grassland (up to knee-high). Where woodland and open grassland adjoin, it frequents the edge, withdrawing into the shade during hot weather. Though occasionally found in rolling uplands (e.g., Akagera Park, Rwanda), the topi generally sticks to flat lowlands and seldom occurs above 1500 m. This species eats virtually nothing but grass. Its long, narrow muzzle and mobile lips are adapted for selective feeding; it harvests the tenderest green blades, avoiding mature leaves and stems (1, 3). Like most grazers, it relishes greenflush but is unable to harvest very short pasture as efficiently as bulk feeders like the wildebeest, waterbuck, and zebra. During the rains topis avoid very short and very mature pastures; in the dry season they prefer plant communities with the greatest abundance of grass. They do not have to drink when their fodder is green, but drink every day or 2 when subsisting on dry grass (5). The densest populations occur where green pastures persist through the dry season, notably on plains bordering lakes and rivers subject to seasonal flooding (11). Tiang and korrigum migrate between the arid and savanna zones.

SOCIAL ORGANIZATION. Topi social and reproductive organization appears to be more variable than that of any other antelope (3, 4, 6, 9, 15). Depending on the type of habitat and ecological conditions, topis can assume any dispersion pattern between perennially sedentary-dispersed and perennially mobile-aggregated. Mating strategies range from large territories with resident female herds to small temporary territories in aggregations, to breeding leks or arenas (table 8.1). And depending on habitat and social organization, calves may be either *hiders* or *followers.*

The sedentary-dispersed mode is characteristic of topis/tsessebes inhabiting patches of grassland in woodland habitat (4, 13). Territories vary from ½ to 4 km^2 and may or may not have common borders, depending on the extent of the patches. The territory contains the resources needed to sustain a herd of 2–6 (rarely over 10) females with their young of the year. Serengeti females have been known to frequent a single territory for 3 years (4). The resident male has a more or less exclusive harem. Not only are such herds closed (actually semiclosed, since persistent females may gain acceptance), but the females play an active part, as does the male, in preventing outside topis of either sex from settling on their property (3, author's observations). In the bull's absence, the dominant cow may actually behave like a territorial male, approaching intruding cows and even bulls in the *rocking canter* and performing the *high-stepping display.* The sexes look so alike that this mimicry is apparently as convincing to other topis as it is to the human observer, at least from a distance. I have seen cows from a neighboring herd which had grazed past the border turned back in this way. But once the male-impersonator's bluff was called by a feisty female intruder and another time by a passing male that saw through the disguise.

Where topis live at relatively high density (up to 47/km^2), notably on alluvial plains, the great variability in their social organization is most apparent. For example, in Uganda's Queen Elizabeth N.P. all the topis of the Ishasha Plain remain aggregated the whole time in 1 or more concentrations of up to 2000 head (6). They circulate constantly and unpredictably across

Table 8.1 Summary and Comparison of Topi Arena and Conventional Territories

Arena	Conventional
Ecotype	
Broad floodplains, especialy at intersections.	Islands of grassland surrounded by unsuitable habitat.
Population Density	
High	Low
Movement-Distribution Pattern	
Mobile-aggregated; equal access to pastures for bachelor males.	Sedentary-dispersed; bachelors relegated to marginal habitat.
Territories	
Traditional sites, individual attachment to site, seasonal occupancy (rains).	Traditional sites, individual attachment to site, year-round to seasonal occupancy.
Territorial Function	
Reproduction only; males leave territories to feed.	Resources and reproduction; full selection of preferred plant communities required in perennial territories.
Male Competition	
For central positions in arena and directly for females. Males spend much time in aggressive encounters.	For habitat most likely to attract females (= best pastures). Males spend less time in aggressive encounters.
Mating	
Females in estrus come to arena.	Females breed in conventional territories, remain in herd.

SOURCE. Adapted from Monfort-Braham, 1975.

the 80-km^2 plain, which is a patchwork of different grassland types. A temporary territorial network is established wherever an aggregation settles down for a few hours or days, with the usual result of fragmenting and segregating the concentration. Spacing between territorial males is as little as 50 m, close to the minimum in a conventional territorial system.

In Akagera N.P., Rwanda, the fittest males gather on traditional leks every year at certain locations where large open plains intersect (9). A hundred males were counted on the largest arena, spaced 100–250 m apart on the periphery down to 25 m near the center. Some known bulls managed to reclaim the same territory for 2 or 3 years in a row. Females and young remain aggregated in herds of 200–300 up to 1000 head which circulate outside an arena, accompanied by hundreds of bachelor males. Most females enter the lek singly or in small groups on their day of estrus and are bred by centrally located males. Central grounds are so small and worn that the owners have to leave in order to graze (cf. kob). They do so hurriedly during lulls in activity. As mating tapers off late in the rains, the lek network begins to break down and later all the topis migrate 30 km or so to a dry-season range. The differences between arena and normal territorial organization are summarized in table 8.1. Topi lekking behavior has also been observed in Kenya's Masai Mara Reserve (14, 15).

The topi's maternal bond lasts only a year, or until the next calf is born. Calves as young as 8 months are sometimes seen in bachelor herds, and most yearling males join by the end of the calving season, or at latest by the time of the rut 4 months later. Bachelor herds often include yearling females, too, for territorial males tend to evict them along with the males. Once they become detached from their mothers, the *head-in display* no longer protects yearlings from the bull's intolerance. The aggressive treatment of yearling females is unusual and not easily explained, unless it reflects some difficulty in distinguishing the gender of immature topis.

Topis mature about a year sooner than hartebeests. Well-fed females reach adult size in their second year and may breed at 16–18 months. Males are nearly as tall but weigh less than adults by the end of their second year and mature in their third year. However, the age when they become territorial depends on such factors as population density and sex ratio; perhaps relatively few males gain the opportunity to breed before they are 4 years old (9).

Spirited contests and fights can often be seen in bachelor herds, particularly early in the morning (fig. 8.9).

ACTIVITY. Serengeti topis typically have 2 daily grazing/activity peaks, feeding in the morning until 0800 or 0900 and in the afternoon beginning after 1600 h until dusk. The intervening hours are devoted mainly to ruminating and resting, interrupted by short, relatively unsynchronized feeding bouts, and in the dry season by going to water. This can entail a round trip of 5 km or more and meetings with other topis at water on neutral ground, or en route while passing through other territories. Rather than be subjected to harassment from territorial males and females, traveling topis tend to pass along territorial margins, although these are often overgrown areas (where risks of lion or leopard ambush are greater) that they normally avoid (author's unpub. observ.). Topis tend to feed longer and at shorter intervals during the rains than during the dry season, when the more fibrous diet requires long processing (2, 4).

Dozing topis sometimes lie with the mouth resting on the ground supporting the head, looking (with their curving horns) like a claw hammer resting on its face. Still more peculiar, topis in aggregations often rest standing in parallel ranks, with eyes apparently closed, but keep nodding to one another. This behavior is particularly frequent in male groups and seems to be a form of signaling, but the meaning has yet to be deciphered. Topi herds often, possibly even routinely, make a move at the end of their afternoon feeding peaks, just after dark before settling down to ruminate (cf. sable). Young calves then usually leave the herd one by one and go off 100 m or so to bed down for the night. The adults and older calves have 1 or 2 activity periods during the night, beginning before and lasting until well after midnight (author's observ.).

SOCIAL BEHAVIOR

COMMUNICATION. Summarized in tribal introduction.

TERRITORIAL BEHAVIOR

Advertising: standing, cantering, and walking in *erect posture, high-stepping* (courtship); defecating in crouch (as in hartebeest, fig. 8.1); *ground-horning; mud-packing; shoulder-wiping;* grunting.

Topis and termite mounds go together (fig. 2.4). Every desirable territory has elevations, some of which are regularly used as vantage points. But females and young use them too, so not every topi sentinel is a male advertising territorial status—it may be a female in the *alert posture.* Nor are *preorbital-gland marking, ground-horning,* and *mud-packing* confined to territorial males. Both sexes perform these acts in the same way; only the frequency and vigor vary. Topis insert grass stems into the preorbital duct, if necessary after preparing a stem by biting it in two at the desired height (cf. oribi) (5), and agitate it until it is well-coated with the gel-like secretion. Then they anoint their foreheads and horns by weaving head movements. This is an impressive exercise in controlled movement, but the purpose is totally obscure. It is not a major activity and plays no obvious role in territorial marking. Far commoner and very important as a display of aggressiveness, if nothing more, is the topi habit of *soil-horning* and *mud-packing.* Topis often get down on their knees, rub their faces in and plow up a patch of mud (e.g., the base of a termite mound or margin of a water hole). Next comes vigorous head-shaking and mud-slinging, followed by wiping the horns on the brisket and back (note that topis wipe themselves with their horns, hartebeests with their heads). After a good rain half the topis one sees are mud-streaked and their horns, caked with mud, look unusually impressive.

Harem males display a proprietary interest in females and young that is otherwise unusual in territorial males. While remaining vigilant against trespassers, a preoccupation that often keeps them on patrol or on sentry duty atop a mound, they play a watchdog role, warning the herd of danger (alarm snorts), leading or directing the retreat, and stationing themselves between the herd and predators. They sometimes threaten and even chase jackals and hyenas that prowl close to a herd. Males in aggregations, by contrast, are preoccupied with meeting and detaining as many females as possible. They are much more active and interactive, continually on the move and conspicuous as they canter, wheel, and turn in the *erect posture, grunting* and performing courtship displays (see below).

AGONISTIC BEHAVIOR

Dominance/Threat Displays: *lateral presentation* in *erect posture, nodding/head-casting, horn-sweeping, cavorting, ground-horning, head-shaking.*

Defensive/Submissive Displays: *head-in posture, head-low posture, lying-out.*

Fighting: *clash-fighting, ramming, front-pressing, twist-fighting.*

Dominant topis, notably territorial males, need only adopt the *erect posture* with ears cocked and tail out to intimidate and impose their will on lower-ranking individuals. The performer stands or walks very stiffly. Females and young get out of the way and respond with the *head-low posture* or the *head-in display* (like hartebeest, fig. 8.4). Territorial males block the way of intruding males by *standing broadside,* often combined with *cavorting* and emphatic nodding (*head-casting*). Tsessebes emphasize the latter more than East African topis: males compete to see which can throw its head highest, to the point of actually rearing (fig. 2.6). Other common elements of the *challenge ritual* include *touching/sniffing noses, pawing* and *defecating, kneeling* and *ground-horning, cavorting, displacement grazing* and *grooming, horn-sweeping,* confrontation in combat attitude, and brief sparring or fighting. Males with large territories interact less but often at higher intensity, usually on the common border, than closely spaced males, which usually invade one another's property to initiate an interaction (cf. wildebeest). Like hartebeests, when topi bulls fight, the winner often chases the loser a long distance.

REPRODUCTION. Topis are strictly seasonal breeders in most of their range. However, some equatorial populations have 2 calving peaks (e.g., in Virunga N.P., Zaire), and breeding may be perennial in the population north of the Tana River, Kenya (3). Elsewhere, topis habitually calve at the end of the dry season, unlike most other associated mammals (but cf. impala, warthog). Yet calf survival is equal to that of species that calve during the rains. Gestation is 8 months (4).

SEXUAL BEHAVIOR: *high-stepping* with head up, ears down, and tail out; *lowstretch; immobile stance* (in *erect posture*) behind female (pre- and postcopulation).

The bull's approach with head high, often in a *rocking canter,* may be intimidating, but dropping the ears and *high-stepping* are addressed only to females. When close, the displaying male slows down and lifts a foreleg at the knee with each stride, his movements slow and deliberate (fig. 8.10). The more intense the display, the higher the forelegs and tail are raised. The contrast between the pale lower legs and the dark purple of the upper legs and shoulders makes *high-stepping* very conspicuous. Territorial bulls returning from a challenge ritual routinely show off in this way to their cows. Males also approach in the more familiar *lowstretch,* ears cocked and nose outstretched to sniff the female's rear, meanwhile uttering a strangled cry resembling a calf's bleat (7). There is no *urine-testing* sequence. Females typically react to being sniffed by rapidly wagging the tail and run-

Fig. 8.10. The topi *high-stepping* courtship display.

Fig. 8.11. *Immobile stance* often adopted just before mounting (from a photo by B. J. Huntley).

ning a meter or so. Having detected a cow in heat, the bull follows closely in *low-stretch, bleating,* until she stands, tail raised, head up, ears up and back, in a pose similar to the *head-in display.* Prior to and after mating, the bull remains immobile in a very upright stance a meter behind the cow (fig. 8.11). Then he advances, lowers his muzzle to the base of the cow's tail and mounts. Cows mate many times, with the same or different males depending on circumstances, during their 1–1½ days in estrus (7, 8).

PARENT/OFFSPRING BEHAVIOR. The topi's calving strategy is intermediate between the ancestral *hider* system and the *follower* system of the wildebeest. Yet apart from breeding seasonally and the fact that calves may not *lie out* (at least by day), the topi has developed virtually none of the adaptations seen in the wildebeest and retains all of the characteristics of the *hider* system (cf. blesbok). Even in aggregations whose mobility and presence on open plains should select for the *follower* strategy, at least some calves leave and seek hiding places for the night, just as they do in sedentary small herds, and females in resident herds isolate to calve like typical *hider* species. Some even leave the herd's range and stay away for weeks, giving birth and remaining on guard during the calf's concealment stage in areas with more cover than usual (author's observations).

ANTIPREDATOR BEHAVIOR: *alert posture;* snorting, *style-trotting, stotting, cavorting* in flight (defiance?).

Newborn topis are vulnerable to predators as small as jackals, if found in hiding and unguarded. Cheetahs run down young up to the yearling stage, but predation on adults appears to be light where other prey species are more numerous, as in the Serengeti. Even a female topi with an injured leg who limped badly for weeks was unnoticed by the hyenas and lions, as though adult topis were simply discounted as a prey species (author's observations).

References

1. Bell 1970. 2. Child, Röbbel, Hepburn 1972. 3. Duncan 1975. 4. ——— 1976. 5. Huntley 1972. 6. Jewell 1972. 7. Joubert 1972. 8. ——— 1975. 9. Monfort-Braham 1975. 10. Sachs 1967. 11. Vesey-FitzGerald 1955. 12. Ward 1962.

Added References

13. Garstang 1982. 14. Gosling 1991. 15. Gosling, Petrie, Rainy 1987. 16. Hillman and Fryxell 1988.

Fig. 8.12. Bontebok male in *lowstretch* (from a photo in David 1975*a*).

Blesbok and Bontebok

Damaliscus dorcas

TRAITS. Smallest and gaudiest alcelaphine. *Height* and *weight* (bontebok): c. 90 cm, males 68–86 kg, females 56 kg (2). *Horns:* S-shaped, 35–37.5 cm, thicker and somewhat longer in males. *Coloration:* dark brown with white blaze, underparts, and lower legs, adult males distinguishable by darker color and white scrotum; bontebok (*D. d. dorcas*) glossier with purple-black blotches on upper limbs and sides, a small white rump patch, and facial blaze (usually) not completely bisected by a brown band; *D. d. phillipsi* lacks rump patch and blaze is bisected by band; newborn tan with dark blaze and no white markings. *Scent glands:* preorbital glands developed in both sexes, bigger in male, yellowish-black secretion (found caked between horn annuli due to habit of weaving horns through grass after marking) (8); hoof glands in forefeet.

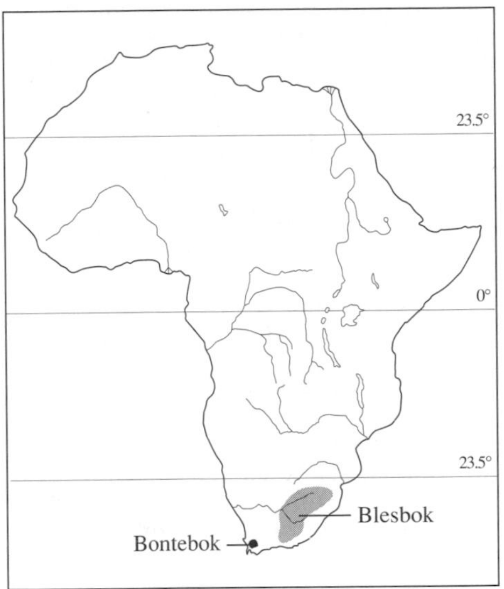

DISTRIBUTION. The blesbok is one of the 4 grazing herbivores that once thronged the South African Highveld in uncounted thousands. Although the quagga was the only Highveld species to become extinct, the springbok is the only one that is still free-ranging in natural ecosystems—outside South Africa. The blesbok and black wildebeest are confined to fenced range without predators larger than a jackal. Numbering over 50,000 head and more widely distributed now than ever, the blesbok has staged a dramatic comeback, but both its distribution and living conditions have been manipulated by man (12).

Unlike the blesbok, the bontebok had a very local distribution when European colonists first encountered it at the southernmost tip of the Cape in the seventeenth century. A blesbok population that was left isolated in the Strandveld when the Karrooveld formed and created a 300 km arid barrier between the coastal plain and the Highveld, this race evolved some differences during the ensuing millennia of isolation.

ECOLOGY. Of the 3 dominant Highveld antelopes, the blesbok is the most water-dependent, the purest grazer, and the most closely confined to the Highveld grasslands (5, 12, 13). The springbok and black wildebeest, but not the blesbok, ranged into the Karroo during the rains, and during the dry season the blesbok went farther and higher into the taller grassland, where its superior ability to graze selectively enabled it to subsist in sourveld where a black wildebeest would starve. But the original range of the blesbok indicates that it was closely associated with the short to medium-length, high-quality grasses (e.g., *Themeda triandra*), the so-called sweetveld. Both its food intake and fat reserves decrease sharply in the dry season unless veld fires or thunderstorms bring up a flush of new grass (5, 8). The observation that blesboks enter dense thickets to escape the midday heat of the Highveld summer suggests that the best natural habitat also included trees (13).

SOCIAL ORGANIZATION: territorial, formerly mobile-aggregated and migratory in the Highveld; bontebok and most remaining blesbok herds sedentary-dispersed on fenced range.

Decimation of wildlife in Cape Province and the Orange Free State in the nineteenth century, followed by fencing of the range, left one of Africa's greatest and most unusual plains ecosystems in tatters. Today blesboks are scattered in hundreds of small, separate, perforce sedentary populations. However, in certain favored locations offering perennial water and grazing, there were probably naturally sedentary subpopulations of blesboks (cf. topi and blue wildebeest). The bontebok, inhabiting the well-watered coastal plain of the southwestern Cape, may well have lived in the sedentary-dispersed mode for at least part of the year. The social organization of blesboks in the mobile-aggregated phase was undoubtedly similar to that of topis on wide floodplains; in fact, at least one Transvaal population of several hundred spends most of the time in aggregations (9).

A study of the 250 bonteboks living in Bontebok N.P. showed that this population was kept segregated into female and bachelor herds by 25–30 mature males that had divided the whole 28-km^2 area into a permanent territorial network (2, 4). Female herds averaged only 3 cows and 1.5 calves, with 9 the largest number of females seen in one group. Their home ranges typically included 2–3 territories, although sometimes a herd remained within a single territory for over a year. A territorial male almost always accompanied a herd, initiated and led their movements, and kept the members together. He would approach any that strayed in *lowstretch* and run flat out to head off escapees. The territorial bucks tolerated yearling males, contrary to the findings of another bontebok and a blesbok study (8, 11). Nevertheless, yearlings of both sexes voluntarily left their herds during or following the calving season and joined the single large bachelor herd of more than 100 bonteboks which wandered through the park without any fixed home range. Many females remained in the mobile aggregation through their second year, being recruited into the sedentary small herds apparently only after calving (3).

The demography of this population was unusual in that males of yearling class or older outnumbered females 100 to 87, reflecting the absence of predators and the impossibility of dispersing. The aggregation included males of all ages, although most were immature. With so many males in such a small area, compression of territories and a quick turnover of proprietors might be predicted. But on the contrary, territorial density was only 1.5 males/km^2, territories ranged from 10 to 40 ha, with an average spacing of ca. 300 m between males, and the minimum average tenure of 16 individually known males was 22 months (½–59) (2). One bull held the same territory continuously for 5 years. Males that did leave their territories were not driven out but apparently left voluntarily because of advanced age and decreased sexual/territorial drive. There were even 4 observed cases of abandoned territories that remained vacant, one for a year and another for at least 5 months during and after the rut. Apparently the advantages of ownership were so great that few among the 25 adult bachelors made an effort to become territorial before the age of 5, 2 years after maturation, and even prime males always deferred to territory owners.

The difference between such resource-based territories and mating territories is brought out by comparing the density and spacing of territorial blesboks in an aggregation during the rut: 169 males/km^2 spaced 77 m apart (9). In another blesbok population where rutting territories averaged only 2.3 ha, territorial behavior declined after the rut and virtually ceased during winter and spring (10). Despite the far greater size of bontebok territories, which contained pasture for all seasons, the owners only defended activity centers of ca. 175-m radius, leaving large areas where bachelor males could circulate with little or no harassment (cf. hartebeest account) (2).

ACTIVITY: morning and afternoon feeding and movement peaks, resting during middle of day, no information on night activity.

Territorial bonteboks were found to spend 55% of daylight hours grazing, 23% ruminating, 10% resting, and 13% in other activities (of which 60% were sexually or territorially related) (2). The allotment of feeding time fell to a low point of 45% during the rut and reached a maximum of 66% in midwinter. Females spent an average of 65% of the day grazing, 21% ruminating, 9.5% resting, and 5% in other activities.

SOCIAL BEHAVIOR

COMMUNICATION. See tribal introduction.

TERRITORIAL BEHAVIOR

Advertising: standing in *erect posture* (preferably on elevations [14]), defecating and lying on stamping ground, *snorting, ground-horning, preorbital-gland weaving* (by both sexes), *herding* and *chasing* in *low-stretch.*

Territories are demarcated with dung middens—an average of 4.3/ha in Bontebok N.P., where 96% of all the owner's defecations were concentrated (2). A typical marking sequence begins with *kneeling* and *ground-horning* (without preliminary pawing) on the midden. The bull then stands, walks forward, and *crouches* to defecate on the same spot.

The alcelaphine *challenge ritual* is of major importance in the relations of territorial bonteboks/blesboks. During the rut, bontebok males spent 35 minutes/day interacting with their neighbors and 20½ minutes/day at other seasons. Up to 30 different actions were performed (2) (fig. 8.13). The average ritual lasted 6½ minutes (range 1–23). The main steps in the ritual: Bull A deliberately invades the property of his neighbor, B, who stands and watches with head up and ears horizontally extended. A keeps stopping and looking over his shoulder, thereby perhaps neutralizing the aggressiveness of approaching in the *erect posture.* B also *head-flags* as he stands watching. When A gets close, B advances to meet him and they move into *reverse parallel,* a neutral position, and stand, heads high, looking about but not at each other. They sniff each other's rump in the same way that males check cows, but males react with agitated *tail-swishing* and emphatic *head-shaking. Head-shaking* is the most frequently performed action in bontebok challenge rituals, whereas *head-nodding* is infrequent between territorial males (2) (cf. topi). However, males *head-dip* from the *erect posture:* lowering the head to the ground only to raise it the next second. Eventually the males move a

Fig. 8.13. Territorial bontebok males in combat attitude (*top*), fighting (*bottom*). (From photos in David 1973.)

few meters apart and proceed to *defecate,* adopting a deeper crouch than usual. They also *horn the ground, urinate,* "weave" in the grass with their preorbital glands and horns, and *cavort.* A pair may walk or canter some distance, one following the other. The commonest displacement behaviors are *horn-sweeping, mouth-grooming,* and *grazing.* Bouts end when either A or B walks or grazes away. Even token horn contact was rare in the observed bontebok encounters (2, 8), but was commoner in territorial blesbok interactions (9).

AGONISTIC BEHAVIOR

Dominance/Threat Displays: *erect posture ± lateral presentation, ground-horning,* defecating, urinating, *nodding,* head-dipping, *head-shaking, horn-sweeping, cavorting, grazing* (during *challenge ritual*).

Defensive/Submissive Displays: *head-low posture ± approach with tail raised, looking away.*

Fighting: similar to topi.

The gaudy coloration makes the bontebok's postures and movements unusually conspicuous. Thus the white blaze and insides of the ears emphasize head and ear positions as viewed from in front, and the *erect posture with ears spread* is the primary aggressive display with or without *lateral presentation. Looking away* removes these areas from view. *Head-nodding* is a low-intensity threat when directed at another animal (e.g., a territorial male approaching a bachelor), but is also closely linked with locomotion, common even in young calves, and contagious (2).

Submissive behavior, shown primarily toward displaying territorial males, may go no further than lowering the head like unreceptive courted females. A more intimidated animal also curls its tail over its back and approaches the displaying male; the same *tail curl* is a striking feature of the *lowstretch display* (fig. 8.12). The *head-in appeasement posture* (see tribal introduction, fig. 8.4) has yet to be reported in this species, although the precursor for it is seen in the attitude of females in estrus standing for copulation.

REPRODUCTION

SEXUAL BEHAVIOR: *lowstretch,* sniffing vulva, *courtship-circling,* multiple copulations.

The blesbok *lowstretch display* is an all-purpose courtship and herding posture for approaching females, whether at a walk or on the run. At highest intensity the tail is curled over the back (fig. 8.12). Territorial males returning from seeing bachelors off or a challenge ritual routinely display to their females. Females are also checked daily to determine their reproductive status by *sniffing the vulva* (fig. 8.14). The female facilitates this check by standing with tail out and ears back. Then she moves quickly

Fig. 8.14. Bontebok sniffing female's genitalia, still in the *lowstretch posture* adopted while approaching. (From a photo in David 1975*a*.)

away, wagging her tail rapidly as do topi (5, 8). If the male tries to get behind and mount an unreceptive female, she turns toward his rear and they end up *circling,* she with head low. Unlike other alcelaphines, blesbok/bontebok males hold their heads up during copulation, meanwhile grunting (3).

PARENT/OFFSPRING BEHAVIOR. The blesbok/bontebok has committed itself to a follower-calf strategy. Most calves (80%) are born 8 months after mating, during a 1- or 2-month peak calving season (2, 5, 7). Females do not isolate, but calve either in small herds or, in the aggregated state, on calving grounds in maternity bands (9), usually in the forenoon. Calves can stand in 5–10 minutes, walk in 15–20, and do not seek hiding places or form subgroups but follow their mothers. Apart from cleaning newborn calves, mothers do not lick their rumps to stimulate elimination; afterbirths, delayed at least 3 hours postpartum, are left uneaten (3).

Although this species has taken another step beyond the topi in abandoning the hider system, it has still not gone as far as the wildebeest (q.v.) (6): calves are less precocious (some take 20–58 minutes to get up), they cannot run as soon and need over 1½ hours to locate the udder and suckle (1, 3), and they lag behind rather than run beside the mother.

Calves seeking their mothers (who may wander up to 400 m from resting calves) bleat and mothers respond (sometimes) on a lower key. A sample of bontebok calves of from 1 to 8 weeks nursed an average of 6 times for a total of 132 seconds during daylight hours (3). Calves play only briefly and infrequently.

ANTIPREDATOR BEHAVIOR. The black-backed jackal, which may prey heavily on newborn blesboks, is now the only predator that coexists with the species, apart from the leopard in some reserves (5).

References

1. Altmann 1971. 2. David 1973. 3.——— 1975*a*. 4.——— 1975*b*. 5. DuPlessis 1972. 6. Estes 1976. 7. Huntley 1972. 8. Langley and Giliomee 1974. 9. Lynch 1971. 10. Novellie 1979. 11. Rowe-Rowe and Bigalke 1972. 12. Selous 1914. 13. Smithers 1983.

Added Reference

14. Coe and Carr 1978.

Fig. 8.15. Hard impact of wildebeests *thrust-fighting.*

Common Wildebeest (Gnu)

Connochaetes taurinus

TRAITS. A large, high-shouldered antelope with broad muzzle and cowlike horns. *Height and weight:* up to 138 cm and 274 kg (*C. t. johnstoni*); western white-bearded wildebeest (*C. t. mearnsi*), the smallest race, males 123 cm and 200 kg (171–242), females 117 cm and 163 kg (141–186) (11), other races 30–60 kg heavier. *Horns:* unridged, longer (55–80 vs 45–63 cm), and much thicker in males, with a more developed boss (especially *mearnsi,* the shortest-horned race). *Coloration:* short, glossy coat, color geographically and individually variable, slate gray to dark brown, lightest on back (reverse countershading); black face, mane, and tail, dark vertical stripes of longer hair (partially shed seasonally), adult males darker than females; *mearnsi* the darkest and *albojubatus,* the eastern white-bearded wildebeest, the lightest race; the blue wildebeest (*C. t. taurinus*) slate gray with tan lower legs; beard white to tan and mane lank in *albojubatus* and *mearnsi,* beard brown to black in the other races, mane upstanding in *taurinus and johnstoni.* *Scent glands:* hoof glands in forefeet, preorbital glands larger in male, secretion a clear oil.

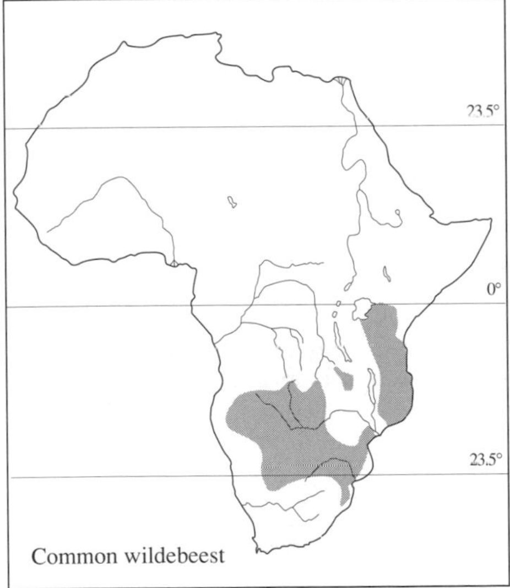

Common wildebeest

DISTRIBUTION. *C. taurinus* dominates the plains and acacia savannas of eastern Africa. Its southern distribution stops at the Orange River, coinciding with the limit of the tree savanna in South Africa, and ends just below the equator in East Africa, where its range is bounded by Lake Victoria, Mt. Kenya, and the low, arid *Acacia-Commiphora* bush country east of the high plains. Large migratory populations occur, or occurred, on the extensive short-grass plains and bordering acacia savannas of Botswana, Namibia, southeastern Angola, and southwestern Zambia, the major river valleys of Mozambique and southeastern Tanzania, and the great plains of southern Kenya and northern Tanzania. The 1.5 million wildebeests, plus a million other ungulates, in the Serengeti ecosystem represent the world's greatest remaining aggregation of large land mammals (13, 14, 18). Except for an isolated population in Zambia's Luangwa Valley (*C. t. cooksoni*), wildebeests are almost absent from the wetter parts of the Southern Savanna with their soils of low nutrient status, notably the *Miombo* Woodland Zone (2, 5).

ECOLOGY. The broad muzzle, with wide incisor row and loose lips, is adapted to close, rapid bulk feeding on short grass. Wildebeests favor plains covered by colonial grasses, which spread by rhizomes, carpet the ground, and respond to grazing, trampling, and manuring by rapid regrowth (10). Such grasses often dominate light, alkaline soils with a hardpan in semiarid environments (2, 5), and also tend to replace less tolerant species in trampled or overgrazed areas. Wildebeests at high density can maintain and even create the type of pasture they prefer, as long as the grass keeps growing. But in the dry season such shallow-rooted grasses quickly stop growing and the plains become waterless wastelands. The need to drink daily (at most every other day) then limits wildebeests to pastures within commuting distance of water, calculated at 10–15 km (3, 17). However, the perception of a heavy thunderstorm may stimulate migratory wildebeests to travel over 50 km to graze the resulting greenflush (15). The best dry-season range is around alkaline lakes and pans, where colonial and annual grasses carpet the emerging ground, and areas of higher rainfall where the soil is moist enough to keep growing or produce a greenflush after burning, and/or where thunderstorms occur during the dry season. The wildebeest is absent from temperate and montane grasslands where the hartebeest, for instance, is at home.

SOCIAL ORGANIZATION: territorial, highly gregarious in mobile aggregations or dispersed in sedentary herds.

Resident populations tend to be sedentary and dispersed, singly (territorial bulls) and in small, segregated herds. In Ngorongoro Crater, a cow herd averaging 8 females and calves may spend over 90% of its time within a range of just a few hectares, encompassing perhaps 4–5 territories and overlapping with other cow herds, as long as the grass keeps growing. Herd composition tends to be stable and transient cows are often harassed, leading to intervention and expulsion by the attendant bull. However, where herds are numerous and close together, the addition or subtraction of a few members is common from day to day, indicating a degree of interchange even during the period of maximum stability following the calving season.

This pattern begins to change when the rains end and green pastures become localized, leading wildebeests and others to congregate on them. At first, female herds return to their core areas for the night but later, herds remain aggregated on the green pastures and eventually lose their separate identities. Only mothers and their latest offspring remain together, and calves as young as 8 months leave their mothers to associate in peer subgroups. Bachelor herds are usually found attached to these aggregations. Many territorial bulls also join up during the day, returning by evening each to his own property, but others sojourn as bachelors with the aggregations until rains rejuvenate the range.

Dry-season aggregations of resident populations are like migratory populations. The latter simply remain aggregated while engaged in local or long-distance movements in search of greener pastures. How often and how far the animals move and how densely they aggregate is determined by seasonal and random environmental events: rainfall distribution and amount, fires, presence or absence of reserve dry-season pasture (e.g., marshland greenbelts), and so on. Wildebeest social organization is finely adjusted to the prevailing conditions. There are permanently resident and almost permanently nomadic populations, and all gradations in between. Both resident and migratory wildebeests coexist within the confines of Ngorongoro Crater, and resident subpopulations have been known to bud from migratory ones, as happened in Wankie N.P. in the 1930s and in Kalahari Gemsbok N.P. after wells were drilled in the dry Auob River (6).

At night, aggregated wildebeests rest in linear bedding formations containing anywhere from a few dozen to several thousand animals. A formation is 10 or perhaps 20 animals deep, with enough space between them (minimum 1–2 m, except cows and calves may lie in contact) to permit individuals to pass freely in or out. They can also take immediate flight in the event of a disturbance, such as hyenas running at the herd. If instead a crowd of wildebeests lay in a circle, those in the center would be trapped.

Males. Bachelor herds can be distinguished from nursery herds by the wider, more regular spacing between bulls. Compared to bachelor topis, hartebeests, and many other antelopes, nonterritorial wildebeests are phlegmatic. There is comparatively little play, chasing, and fighting among them. They are relegated to marginal habitat where there are few territorial bulls to harass them. However, in concentrations of females and young, territorial males may sometimes experience difficulty in distinguishing them from all the other bodies.

Males are cut out of the small herds as yearlings, beginning before the calving season, and 90% are in bachelor herds by the time the rut begins 4 months later (7). Rejection by their own mothers and by other females with new calves triggers the process, but the driving out is done by territorial bulls, who tend to react aggressively toward any wildebeest that acts ill at ease or provokes a disturbance in the herd.

When wildebeest bulls become territorial at 4–5 years (7), they undergo a behavioral metamorphosis, becoming the noisiest (*C. t. mearnsi*) and most active antelopes of the African plains. Territorial wildebeests bear out the rule that the smaller the property the more active the owner. Despite its size, this species tolerates closer territorial spacing than any known bovid except the kob and topi at the center of breeding leks. Sample quadrat counts of the Serengeti populations in rutting aggregations indicate a density of up to 270 bulls/km^2, an average of only ⅓ ha per territory (author's observations). These are of course only temporary territories, held for mere hours while the migration is passing, or for days if an aggregation settles down that long. Territories in a sedentary population are much larger (at least 1 ha in Ngorongoro Crater), but still much smaller than in hartebeest, topi, or black wildebeest populations—even the little Thomson's gazelle controls over twice as much land.

The fact that wildebeest bulls in resident populations hold their territories year-round, although 80% of all matings fall during the 3-week rut, demonstrates competition for space and the fact that it is far easier to hold on to than to acquire property. Typically only ⅓ to ½ the adult males hold property, even during the rut (7).

ACTIVITY. Detailed observations of Etosha N.P. wildebeest activity during one annual cycle revealed that the populations spent 53% of the total time resting (32% lying and 21% standing, including rumination bouts), 33% grazing, 12% moving (99% at a walk), and 1–1.5% in social interactions (4). The animals spent the same amount of time feeding on moonlit nights as by day, but moved less and rested more. There were 2 daily movement peaks, soon after dawn as the wildebeests got up and moved to day pastures, and to and from water (usually during the heat of the day when predators are least active). Seasonal differences in the times and duration of daylight activities correlated with daytime temperatures, the wildebeests spending more time feeding and less resting in cool than in hot weather, as expected. But even when temperatures reached 35°C in the afternoon and most wildebeests were resting in the shade, a minimum of 17%–18% of the population could be found grazing. Conversely, during the early-morning and late-afternoon grazing peaks, an average of over 40% of the animals would be resting, moving, and so on. Significant differences were found between different sex and age classes in the percentage of time devoted to these basic activities. In the adult classes, bulls rested and grazed less and moved more than immature males. Wildebeests in mixed herds grazed less and rested and moved more than all-male herds, reflecting access to better pastures by mixed herds (requiring less time to eat their fill) but a longer walk to water.

SOCIAL BEHAVIOR

COMMUNICATION. Summarized in tribal introduction.

TERRITORIAL BEHAVIOR

Advertising: standing in *erect posture* typically on stamping ground (rarely on elevations); *calling;* approach in *rocking canter* or *lowstretch, snorting; pawing* and *defecating, kneeling* and *ground-horning; preorbital-gland-rubbing* and *rolling* on stamping ground, also *preorbital-rubbing* and *horning* of trees.

The territorial *status quo* is maintained by daily *challenge rituals* with each neighbor, lasting c. 7 minutes (range ½–30 minutes), during which some 30 different behaviors may be performed, by either or both contestants, at any time, separately or in sequence, at low to high intensity (cf. blesbok *challenge ritual*). The more spirited interactions include a lot of *tail-swishing, horn-sweeping* (while simultaneously sweeping tail across face), *head-shaking,* marking/advertising behavior (especially *pawing, dunging, ground-horning*), *urinating* and *urine-testing,* standing and *displacement grazing* in reverse parallel (fig. 2.11), feinted attacks followed by *dropping to knees* in *combat attitude, cavorting* (fig. 8.3), *displacement scratching* and *grooming* and *alarm signals (scanning, stamping, snorting)* (fig. 8.5), and (usually brief) horn contact. Rituals are initiated by one bull walking and grazing up to his neighbor, who as often as not will be standing on his stamping ground in the heart of his territory (7).

Repeated *horn-sweeping, cavorting, alarm-snorting,* and combat (fig. 8.15) signify that the animals are aroused, whereas other encounters may be comparatively listless and uncomplicated. Sometimes a contestant even seems to forget the point of it all, lies down, and starts ruminating beside his rival!

AGONISTIC BEHAVIOR

Dominance/Threat Displays: *erect posture* ± lateral *presentation* ± *angle-horn, head-shaking, nodding, hooking, horn-sweeping, cavorting,* confronting in *combat attitude; object-horning, rubbing face* on another wildebeest's rump, head, or shoulder.

Defensive/Submissive Displays: *head-shaking/cavorting* (defiance), *head-low posture, lying-out.*

Fighting: *front-pressing* ± *twist-fighting, ramming/thrust-fighting.*

Dominance among females in small herds is expressed through *rubbing the face* on an inferior's rump or shoulder, by *hooking,* and by *chasing. Nodding* seems to be a sign of impatience, probably signaling the animal's desire to get moving, and may be a mild threat when directed at another wildebeest. The other intimidation and threat displays, though occasionally performed by females and bachelors, are deployed mainly by territorial males. Unlike other alcelaphines, wildebeests do not have an *appeasement ceremony.* An intimidated individual simply lowers its head and takes evasive action. Thus, when a passing bull finds his path blocked by a territorial male in *lateral presentation,* he alters course and passes to the rear, with head at or below shoulder level. A bolder trespasser, say a territorial male in transit, may pause and rub his head on the owner's rump and perform other behaviors seen in the *challenge ritual* (fig. 8.16).

There are 2 peaks of territorial activity when fighting becomes common: during the rut, when the greatest number of bulls compete for the space each must own to col-

Fig. 8.16. Wildebeest *challenge ritual:* one bull rubs preorbital region on rival's rump.

lect females and mate, and, in resident populations, when the rains revive pastures that had been abandoned during the dry season. Rutting bulls with females do not engage in ritualized encounters with their neighbors but simply run at and ram invaders. The impact sometimes breaks a horn, though horns are so tough and elastic that axe blows rebound harmlessly. But such contacts are usually as brief as they are violent. The longest and hardest fights I have seen involved males that were just becoming established in a permanent territorial network (7).

REPRODUCTION. Few tropical mammals have such a restricted birth season as the wildebeest. Some 80%–90% of the calves are dropped during a 3-week birth peak. (See under Reproduction in tribal introduction.)

The timing of calving and rutting is geared to the climate in such a way that both occur under favorable conditions. The rut comes at the end of the rains when the animals are in top condition, thus insuring an adult conception rate of better than 95%. First conception can occur at 16 months if yearlings have been very well nourished, but most only conceive a year later. Calving comes at the beginning of and/or about a month before the period of most reliable rainfall. Possibly a somewhat later time would be more favorable, but then rutting would occur during the dry season, on a falling nutritional plane. Given an 8–8.5 month gestation, the timing of the reproductive cycle represents a favorable compromise with the environment.

SEXUAL BEHAVIOR: *rocking canter, fast-calling, lowstretch* with ears laterally extended or lowered, *sniffing vulva, urination on demand* and *urine-testing, resting chin on female's croup, standing bipedally.*

In or out of season, territorial bulls are always ready and willing to mate. They call to and go to meet any approaching wildebeest, typically in the *rocking canter* (fig. 8.17). Depending on its identity and reproductive status, a bull will proceed to herd, mount, or chase out the intruder. Sexually interesting females elicit the *lowstretch display.* But before and during the rut, greatly increased hormone and sperm production (1, 16) lead to intensified calling, herding, and fighting (7). Sexually excited bulls call at twice the normal tempo and may become so worked up that they literally froth at the mouth. When further stimulated by the passage of many migrating cows, bachelors join territorial males in calling, producing a continuous humming and croaking like a chorus of giant frogs, which may similarly function to stimulate and synchronize female estrus. Evidence that the onset of the rut coincides with the full moon suggests that the lunar cycle serves to trigger the mating peak (cf. impala rut) (12).

Competition to gather and hold as many females as possible serves to fragment the population into pseudoherds averaging c. 16 head, each closely herded by a bull. As long as females are within his territory a rutting bull neither eats nor rests. His encounters with neighbors take the form of head-bashing as he strives to protect his investment. Yet rutting behavior is often remarkably localized in the immediate vicinity of herds containing estrous cows, and even bulls more than 2 territories removed from the spot may peacefully rest or graze, seemingly oblivious of the furious activity practically next door. In this way many bulls manage to conserve their energy and stay in the reproductive running for the duration of the rut.

Although the wildebeest reproductive

Fig. 8.17. A territorial bull advancing to meet approaching wildebeest in the *rocking canter.*

Fig. 8.18. Wildebeest *copulatory display.*

system is remarkably synchronized and elaborate, courtship is relatively simple. Apart from *lowstretch* and *urination on demand* and *urine-testing,* there are few preliminaries. Every territorial bull that an estrous female encounters tries his best to mount her (although many receptive females are overlooked by bulls preoccupied with *herding* and *chasing*)—and she may encounter dozens during her day in heat if her aggregation is on the move. A cow in full estrus holds her tail to one side, often stays near and follows a bull between copulations, and may solicit copulation by rubbing her head on his torso and sniffing his penis. As long as she will stand still, copulations may be repeated dozens of times at a rate of over 2 times per minute. If an estrous female will not stand, a highly excited bull may rear up before her with full erection in a dramatic *copulatory display* (fig. 8.18). Bulls also try to mount on the run, chin and tail outstretched, meanwhile uttering a call that sounds like a prolonged belch.

PARENT/OFFSPRING BEHAVIOR. The 3-week calving peak begins abruptly. Few calves are born ahead of time—or at any rate, few survive in areas where there are spotted hyenas, cheetahs, wild dogs, and lions. Those born later have a 2-month grace period before the natal tan coat is replaced, during which the latecomers can be overlooked among older calves. Thereafter the risks of being singled out increase sharply; of the few calves born as much as 6 months out of phase, hardly any survive (8, 9).

Wildebeests never seek isolation at parturition or afterward. In migratory populations, cows congregate in hundreds on calving grounds which are located whenever possible on short grassland (9). Serengeti wildebeests normally calve on the eastern plains, which happen to be outside the hunting territories of most of their predators; leaving predators behind is one of the advantages of migrating.

Wildebeest society is more structured at this time than at any other season. There are groupings of pregnant cows and of cows that have calved, groups of yearlings newly separated from their mothers, and bachelor herds, which are usually excluded by territorial bulls from the calving grounds. Expectant mothers forgather on these grounds every morning and drop calves by the dozens before (very few after) midday. From the first appearance of the white amniotic sac or the yellow-tipped forefeet, labor lasts a little over an hour and ½ hour, respectively. Labor can be interrupted at any time if a cow is disturbed, until the calf's head and trunk emerge, after which gravity completes the process if the cow is standing. Expulsion only proceeds while the cow is lying on her side and having visible contractions. Once the calf is out, the mother immediately turns and begins licking it. Within an average of 6 minutes (occasionally less than 3 minutes) calves gain their feet and begin seeking the udder—unless the mother moves away, in which case the calf runs beside her. The beard of the white-bearded races adds to the difficulty of spotting calves on the mothers' off-side.

Mothers with new calves associate in nursery herds. Mix-ups are a problem initially in large groups because the instinct to follow makes calves approach any moving object, including a lion, hyena, or automobile, until it learns to discriminate. It is entirely the mother's responsibility to stay with her calf for the day or 2 needed to become imprinted upon her; the imprinting process begins with the first successful suckling. Each mother recognizes her own calf, initially by scent alone, and actively rejects all others; mix-ups are largely avoided in this way, but it also means that lost calves are doomed to starvation or predation.

Because calves in the feeble stage are less likely to be spotted by predators in large groups of older calves, survival rates are higher in large aggregations than in small herds (8). But thereafter calves in sedentary small herds lead a quieter and safer existence than those in migratory armies. Wildebeests massed at water holes, river crossings, and other bottlenecks are prone to stampede. Even when none gets stuck in the mud or trampled, inevitably some cows and calves become separated. Luckily for the calves, 90% of these separations are only temporary. Separated pairs remain at the scene and begin running back and forth along the line of march, mooing loudly. When a mother and suckling calf pair up, the calf immediately dives for the udder.

ANTIPREDATOR BEHAVIOR. See tribal introduction.

References

1. Attwell 1982. 2. Bell 1982. 3. Berry and Louw 1982. 4. Berry, Siegfried, Crowe 1982. 5. East 1984. 6. Eloff 1966. 7. Estes 1969. 8. ——— 1976. 9. Estes and Estes 1979. 10. McNaughton 1985. 11. Sachs 1967. 12. Sinclair 1977a. 13. Sinclair and Norton-Griffiths 1979. 14. Sinclair, Dublin, Borner 1985. 15. Talbot and Talbot 1963. 16. Watson 1969. 17. Western 1975.

Added Reference

18. Rodgers and Swai 1988.

Fig. 8.19. Black wildebeest territorial male *chasing/herding* females; note raised tail.

Black Wildebeest (White-tailed Gnu)

Connochaetes gnou

TRAITS. A small wildebeest with level back, head adorned with tufts of long hair and dangerous horns. *Height and weight:* males 111–121 cm, 140–157 kg; females 106–116 cm, 110–122 kg (6). *Horns:* up to 78 cm long in male, bases expanded into protective shield, female horns much thinner and somewhat shorter. *Coloration:* dark brown to black, lighter in summer, males blacker than females; long, bushy mane and beard extending onto brisket; tail dirty white, long, and flowing. *Scent glands:* preorbital glands concealed under hair tuft, hoof glands in forefeet.

DISTRIBUTION. The plains where the black wildebeest, quagga, blesbok, and springbok once roamed in millions are now all fenced crop- and rangeland. These species were equivalent in the Highveld temperate grasslands and the arid Karroo, both largely treeless environments, to the migratory blue wildebeest, plains zebra, topi, and Thomson's gazelle populations of the East African plains. The South African species used the Karroo during the rains and subsisted on the cured sweet grasses of the Highveld during the winter dry season. Some wildebeests ranged into the Drakensberg montane grasslands up to 2600 m (6). During the second half of the nineteenth century, hide-hunters reduced the black wildebeest to under 600 head; these were protected on 2 farms in the Orange Free State (4). It has since increased to over 9000 head (2), but has not recovered nearly so well as the springbok and blesbok, which take more kindly to fencing, do less damage to the range, and compete less with cattle. Most herds, confined to private farms, are small and inbred.

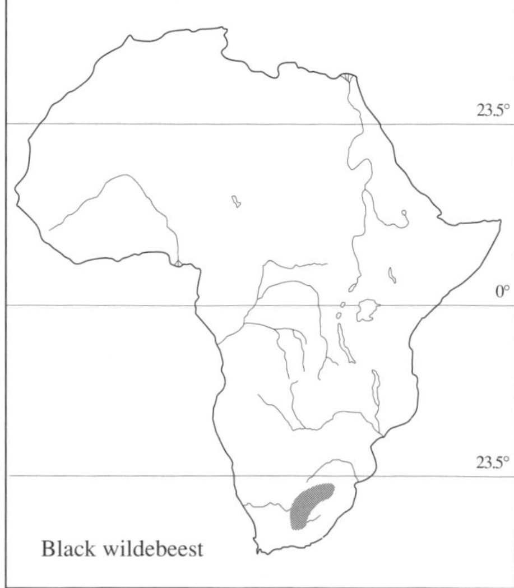

ECOLOGY. The black wildebeest, also known as the white-tailed gnu, occupied open plains similar to those the other wildebeest occupies in East Africa. But the blue wildebeest is not adapted to temperate climates and did not range south of the Orange River, where the acacia savanna ended. The black wildebeest's thick, dark coat apparently insulates it better against both cold and heat; it rarely seeks shade (1) and even in a much colder climate (Catskill, New York) needs only an open shed for protection in winter. In addition, as a mixed feeder known to take up to 37% foliage of karroid bushes and shrubs in its diet (7), the black wildebeest is adapted to subsist in the Karroo without having to drink every day (but does when pastured on standing hay).

SOCIAL ORGANIZATION: formerly migratory in large aggregations, now perforce sedentary.

The behavior and social organization of the 2 wildebeests are so similar that almost everything written in the preceding account (q.v.) probably also applies to the black wildebeest. This antelope was never studied while living in an intact ecosystem together with its natural predators. The largest existing population, maintained at c. 330 head, resides in the 120 km^2 Willem Pretorius Game Reserve, Orange Free State (2).

The mean size of female herds in 4 different populations ranged from 11 to 32 cows and young (6). Group size increased with density, being 30 (mean of 14 herds) in the Willem Pretorius Reserve, including 13 females plus calves and yearlings (4). A hierarchy exists in the cow herds, which maintain their identity during temporary aggregations, behave aggressively to strangers, and are strongly attached to home ranges averaging c. 100 ha in Willem Pretorius (larger on poor range or when space is not limited). Black wildebeests do not groom one another, or, like blue wildebeests, rub the forehead on a companion's croup. Their horns prevent it. In a captive *C. gnou* herd, females sometimes rubbed a cheek on a lying animal's neck (8).

Calves remain attached to their mothers until the next calf is dropped, then bring about their own rejection by coming between mother and newborn calf (6), just as in the blue wildebeest (see under Social Organization). In the process of driving out yearlings, bulls often separate mothers from their calves, which sometimes get hurt, even killed; this may be the main cause of calf mortality under captive conditions (4).

Rejected yearlings join bachelor herds which, as in the blue wildebeest, are relatively peaceful and inactive, and include all age classes. Bachelor herds are sometimes tolerated in company with females late in the dry season when territorial behavior is at a low ebb. As usual, the wider, more regular spacing between individuals distinguishes male from female herds.

Only 4%–11% of the Willem Pretorius population was territorial (however, surplus males were regularly removed) (5). Territorial spacing is much greater in this species, up to 1 km or more at low density, down to 180 m. In Willem Pretorius Reserve territories are clustered with practically no males in some areas that look perfectly suitable. Bulls hold on to their territories indefinitely and seldom leave them except late in the dry season, when some sojourn a few days in bachelor herds. Males mature at 3 but rarely compete for territories before 4 years (5, 6).

SOCIAL BEHAVIOR

COMMUNICATION. See tribal introduction.

TERRITORIAL BEHAVIOR

Advertising: *erect-posture; rocking canter; calling; defecation* preceded by *pawing, kneeling, ground-horning,* and *rolling* on stamping ground; *herding* and *chasing.*

The black wildebeest's territorial-advertising call sounds much like the hic, greatly magnified, in a hiccup and has great carrying power. With each hic the bull's chin jerks upward. Bachelor males rarely call and tend to produce an adolescent croak.

Territorial neighbors engage in a *challenge ritual* which resembles that of blue wildebeests with one difference: the *head-on-rump rub* is omitted (5).

AGONISTIC BEHAVIOR

Dominance/Threat Displays: *erect posture* and *lateral presentation, head-shaking, nodding (head-throwing), pawing* and *defecating, kneeling* and *ground-horning, rolling, confronting on knees* (combat attitude).

Defensive/Submissive Displays: from least to most submissive: *head-low posture* with *head turned away,* tail out to curled over back and *distress-bawling;* the same but on knees; *lying-out; lying on side,* belly exposed.

Fighting: *ramming/thrust-fighting, hooking,* and *front-pressing* with heads to ground.

The rank order in female herds is reinforced most noticeably by *head-nodding* or *-throwing* (at highest intensity) and to a

Fig. 8.20. Yearling male black wildebeest responding to *low-horn* threat of territorial male by kneeling in submission.

lesser extent by *head-shaking* (5). Bulls also nod when driving or chasing females and young. Interpreted as the intention to jab, nodding is made a strong threat by the dangerous horns of this species. Inferiors respond by trying to get behind other herd members (avoidance) and with the 4 different submissive behaviors listed above. The standing submissive attitude is very similar to the attitude of a cow in estrus. The *kneeling posture* (fig. 8.20) represents the intention to lie down, and *lying on the side,* seen only when harassment continues despite appeasement behavior, appears to be unique to the black wildebeest. While giving these displays the performer often emits a loud calflike bawl and this *distress call,* unlike the postures, stimulates other animals including the territorial bull to intervene and even assist the distressed animal (6).

The white tail is brandished, flourished, or lashed in most *gnou* displays. It is held erect by herding and chasing bulls (fig. 8.19), is held out and sometimes up by intimidated and sexually receptive individuals, and is *lashed* furiously by aggressive animals during interactions, *cavorting,* even when fleeing from a predator. *Tail-swishing,* audible at 0.5 km on a still day, may also double as an acoustic signal (3).

REPRODUCTION. The calving peak, as short and sharp as that of *taurinus,* comes 8–8½ months after the rut, in November or December, depending on locality and the timing of the rains.

SEXUAL BEHAVIOR: *lowstretch, ears down, sniffing vulva, urination on demand* and *urine-testing, resting chin on croup.*

The attitude of cows is more distinctive than that of the blue wildebeest, mainly on account of the elevated tail. When a bull approaches to sniff a receptive female she raises her tail and swishes it across his face (6). The tail is kept up during mating, sometimes vertically, while the cow stands with hindlegs straddled and back slightly bowed. The animals separate after successful mounting. Sometimes the female follows the male with tail raised and nudges his rump with her snout.

PARENT/OFFSPRING BEHAVIOR. In 5 observed births, 25–196 minutes elapsed from the emergence of the amniotic sac. Calves gained their feet in 9 minutes on the average (6–14 minutes). The afterbirth was not eaten. Calves begin to graze at 1 month and are weaned between 6 and 9 months (6).

ANTIPREDATOR BEHAVIOR: *alert posture, snorting, stamping, style-trotting, tail-swishing, cavorting.*

No current information. See tribal introduction.

References

1. R. Bigalke pers. comm. 1965. 2. East 1989. 3. Millais 1899. 4. Richter, W. von 1971. 5. ——— 1972. 6. ——— 1974. 7. Van Zyl 1965. 8. Walther 1966*a*.

Fig. 8.21. Impala territorial male in *proud posture (erect posture)* advertising territorial status.

Impala

(Tribe Aepycerotini)
Aepyceros melampus

TRAITS. The perfect antelope, comparable in size and conformation to a kob or Grant's gazelle. *Height and weight:* males 75–92 cm, 60 kg (53–76), females 70–85 cm, 45 kg (40–53) (13). *Horns:* males only,

45–91.7 cm, far larger in East than in southern Africa, S-curved, strongly ridged but comparatively thin, tips wide apart. *Coloration:* two-tone brown, a red-brown saddle sharply divided from tan lower torso and limbs; white belly, throat, lips, line over eye, inside ears, and undertail; black markings include vertical stripe down tail and both thighs, fetlocks, eartips, ± forehead patch and line from eye corners (most developed in the black-faced impala of southern Angola, *A. m. petersi*). *Scent glands:* fetlock glands on rear feet beneath black hair tufts; sebaceous glands concentrated in forehead (and dispersed on torso) of dominant males (7). *Mammae:* 4.

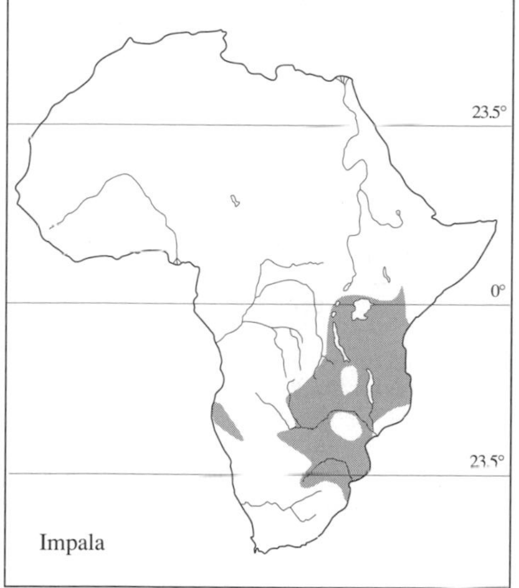

DISTRIBUTION. A Southern Savanna antelope, which achieves very high density in areas affording both grazing and a varied diet of browse within convenient distance of water.

ANCESTRY. Although efforts have been made to tidy up bovid classification by putting the impala in the same tribe as gazelles, kobs, and most recently hartebeests (13, 20, 25), the impala is so different from all other antelopes that it clearly belongs in a separate tribe. There is no evidence that more than 1 species existed at any time in the past (25).

ECOLOGY. The impala is an edge (ecotone) species, preferring light woodland with little undergrowth and grassland of low to medium height. Added to dependence on free water, soils with good drainage, firm footing, and no more than moderate slope, its special requirements produce an irregular and clumped distribution. Though rarely found more than a few kilometers from water in the dry season, impalas with access to green vegetation can go without drinking.

DIET: intermediate feeder (grazer/browser) (6).

The impala is predominantly a grazer while grasses are green and growing and a browser of foliage, forbs, shoots, and seedpods at other times. If necessary, it also eats fallen dry leaves (23). It not only changes its diet in a given area according to season, but can adapt to different habitats by being mainly a grazer in one area and a browser in another (23). The impala's ability to utilize both monocots and dicots gives it an unusually varied, abundant, and reliable food supply, enabling this antelope to lead a sedentary existence and reach densities of up to 214/km^2 in wooded savanna of Rwanda's Akagera N.P. (16). It can also thrive in areas where the natural vegetation has degenerated because of overgrazing or bush encroachment. Thus, in mopaneveld of Zimbabwe, where perennial grasses have been largely eliminated by burning and overstocking, impalas increase while pure grazers (tsessebe, hartebeest, wildebeest, zebra) disappear (4).

As in other sedentary antelopes, impala home ranges include a variety of vegetation types which are utilized at different seasons. In the *Acacia-Commiphora* woodland zone of Serengeti N.P., impalas move up and down the soil catena, staying during the main rainy seasons on the upper slopes, where visibility and forage quality are optimal, and concentrating in the drainage-line greenbelts during the dry season (7).

SOCIAL ORGANIZATION: seasonally or perennially territorial, gregarious, sedentary.

The existence of a territorial organization in impalas has been doubted by some observers (13, 16, 21, 26), affirmed by others (4, 15), and finally proven by long-term observations of known individuals (7, 19). There are several reasons for the uncertainty:

1) In southern Africa most males are only territorial for a few weeks around the time of the annual rut, and in East Africa, despite an extended breeding season, males lose territorial vigor in the dry season, when benefits in terms of mating opportunities are outweighed by the costs of herding females and excluding rival

males. Consequently, bachelor males are typically found in proximity to or actually mixed with herds of females during the dry season. 2) Even vigorously territorial males may tolerate and indeed associate with bachelors when no females are present, as long as the owners' dominance is unquestioned. 3) Although the roaring of territorial males makes the impala one of the noisiest of all bovids, and its other territorial advertising displays are conspicuous, the same displays are performed by high-ranking bachelor males (though generally at much lower frequency) and have often been interpreted as simply expressions of "activated dominance" (21). 4) Mature males alternate between bachelor and territorial status and rarely hold a territory for more than a few months at a time; the turnover of males may also have contributed to the uncertainty about whether impalas are in fact territorial.

Paradoxically, the very features of impala social organization that raise doubts as to whether it is territorial are actually the result of unusually rigorous territorial and reproductive competition. Because of its clumped distribution and locally high population density, males that win positions in the most preferred habitat during the main mating season can achieve very high reproductive success (see under Territorial Males).

Female Clans. Groupings of females and young vary greatly in size, and composition rarely remains constant from day to day (7). Herds of ½ dozen to 15 or 20 represent a minimal average (23) and herds of 50–100 are quite common. In Nairobi N.P. groups of 16–25 females and young were counted in a heavily wooded area compared to over 35 in a more open area (15). The average size of female herds in Akagera N.P. was 36, with groups of over 100 in the custody of a single male not unusual on the large plains (16).

Only after 443 impalas were caught and marked in a 12 km^2 study area in the Sengwa Research Area was it finally established that females live in discrete clans within traditional home ranges (17, 18). For the 3-year duration of the study, clans containing 30–120 impalas remained stable in composition and resident in home ranges of 80–180 ha. When the ranges of individual females were mapped, the centers were found to cluster within a radius of c. 200 m in the case of clan members, whereas the distance between centers of impalas from different clans was 800 m. Yet clan ranges overlapped by as much as 31% late in the dry season, and members of different clans sometimes mixed, though only temporarily. No more than 2% of the females per annum left their natal range and joined another clan. Most males, however, left their clan range by the age of 4 and moved an average of 1.2 km (0.4–3.2), just far enough to place them within the range of a different clan and avoid inbreeding (19).

Impala female herds are notable for their uniformity and lack of distinct peer or family subgroups. The maternal bond quickly dissolves after weaning. Juveniles often play together and may form temporary peer groups, and the Sengwa study also disclosed a tendency for young-adult females to associate (17). The only other division occurs during the calving season, when females with or without offspring tend to be segregated.

Apart from occasional episodes of *head-butting* and *reciprocal grooming* with partners apparently chosen at random, females do not come into physical contact, and there is no sign of a rank order or regular leadership (7, 17). Yet female herds are very cohesive, with an average spacing of 1 m or less between individuals (author's estimate), and herd activities are closely synchronized (16).

Territorial Males. The degree of territorial behavior and enforced sexual segregation of impala herds is closely linked to the breeding season and climatic regime. In southern Africa, vigorous territorial behavior is limited to a few months. The rest of the year males and females are free to associate in mixed herds, although a male preference for relatively dense habitats that females avoid (cf. eland and giraffe) results in a measure of separation (1, 18). The extended breeding season in East Africa leads to the maintenance of territories and sexual segregation for most of the year. But most mating occurs during the rainy season, with a peak at the end of the rains. As the dry season progresses, territorial occupancy declines, and new or continuing occupants are less vigorous about expelling juvenile males, and about keeping females in and bachelor males out of their domains (see below) (7).

Territory size varies with population density, location and habitat quality, individual prowess, and the seasons (table 8.2). Except for the surprisingly large size of impala territories in Mkuzi Reserve during the rut at high population density, the figures clearly show the tendency for terri-

Table 8.2 Relation Between Population Density, Season, and Territory Size in the Impala

Location	Population density/km²	Territory size (ha) Mean	Range	Season	Source
Sengwa Research Area	50–68	10.8	8.5–13	Peak rut	19
	49			Dry season	
Serengeti N.P.					7
Minimum average	32	17	13–>50	Peak occupancy	
Maximum average	32	58		Dry season	
	19	42	16–83	Peak occupancy	
Nairobi N.P.	15–18	51.5	20–90	Not compared	15
Mkuzi Game Reserve	80	66	50–80	Peak rut	24

tories to be small at high population density during the peak rutting season, and large under the opposite conditions.

In the case of large properties, the distinction between territory and home range is often indistinct, especially during reproductive off-seasons. Those which are occupied for an extended period include both wet- and dry-season forage; they expand during the dry season, and only the part in use is actively defended (7). In Sengwa, for instance, prime males tend to remain localized and often alone outside of the rut within ranges averaging 49 ha, where they assert dominance over other males (18). In other central and southern African populations, some 20% of the adult males are also found alone or as the only adult male with a herd of females right through the dry season (12). At least some of them maintain dung middens (see under Territorial Behavior). Even though territorial behavior may be comparatively attenuated in such males, the very fact that they remain apart from other adult males within a fixed range where they display dominance clearly sets them apart from males in bachelor herds, and it is reasonable to consider them territorial.

Attachment to a particular place is characteristic of male impalas during their prime. After roaming widely in search of a favorable place to live and reproduce, Sengwa males pick their spot and settle there for good (18). Serengeti males become imprinted on the area where they first establish a territory and come back every time they are ready to try again (known also in the kob and wildebeest). Among all the males that were observed starting a period of territoriality, an estimated 80% had held territories before, most within the immediate vicinity (7).

Although only c. ⅓ of the adult Serengeti males were territorial at any one time, nearly all those in the study area had territorial bouts during the 2-year study. In Sengwa, by contrast, there was evidence that some adult males reached old age without breeding. Only 12 of 69 males that were present in the range of 67 adult females during one rut were seen to do any mating. The same 4 prime (5½–7½ years old) males accounted for 52 (78%) and 47 (66%) of observed matings during 2 successive ruts, and 1 mated with over 30 females (19). But few if any males achieve the status of top breeders for more than 2 out of a possible 3 ruts during their prime years (19).

Actively territorial males invest up to ¼ of their time in rounding up and attending the females that enter their grounds, time that would otherwise go into feeding and ruminating (8). The more females that come and the longer they stay, the more energetically costly it becomes for the male. When a herd enters his territory, a vigorous male herds them toward the center, chases any bachelor males that may be around, and even cuts out weaned juvenile males (whose horns give them away) (fig. 8.22), while moving through the herd in search of estrous females. To forestall their depar-

Fig. 8.22. Territorial impala chasing juvenile male out of a female herd.

ture, he tries to keep the females tightly bunched and to turn back any that attempt to break free. Depending on how worked up he is, a male's herding behavior varies from approaching at a walk to violent chasing while snorting and roaring. He redoubles his efforts when the herd comes near the border, blocking the way while performing dominance and threat displays. But if the females are determined to leave, all a male's efforts are in vain. Once the first few have crossed over, the rest inevitably follow (7).

The demands of being actively territorial are such that East African males are often in visibly poorer condition than their counterparts in bachelor herds, even when food is most abundant (11). Maximum territorial tenure for 94% of the males in the Serengeti study was under 4 months, and the average tenure was 82.5 days (5–267) (7). Far more intense competition during the peak rut in southern impalas is indicated by an average territorial tenure of only 8 days (± 4.4 days) in the Sengwa population (19).

Bachelor Males. Since females outnumber males by 1.5 or 2 to 1 in the adult population, and up to half the adult males may be on territories during the breeding season, bachelor herds are typically smaller than female herds. In Nairobi N.P., male herds rarely numbered over 10 in closed or 15 in open habitats (15); the mean number in Natal's Hluhluwe Game Reserve was 3.6 (2–25) (1); and in Akagera, ¾ of bachelor herds numbered 10 or less, while the rest included 11–30 males (16).

In the Serengeti study area, male herds numbering from 5 to 35 impalas shared a range of about 6 km^2 in the wet and early dry seasons. Membership fluctuated as often as in female herds, and some bachelors left for several months in the dry season, traveling up to 10 km to reach a greenflush and returning with the onset of the rains (7). Males may be associated more often with female herds than in separate bachelor herds in southern populations, although a tendency for immature and old males to form peer subgroups is apparent, even in mixed herds (12, 18). But males are not confined like females to a clan range. Sengwa males begin moving independently of females as yearlings, and gradually increase their ranges to a mean maximum of 90 ha at 3–4 years, as they disperse from their birthplaces and prospect for suitable places to settle during their reproductive years (18).

Bachelor males maintain individual distances of 2.5–3 m (16), one indication of the greater antagonism that appears from an early age in male impalas. Nevertheless, *reciprocal grooming,* a social-bonding behavior, appears to be about as frequent in bachelor as in female herds and involves individuals of the same and different classes, including adults (28).

Observations of aggressive interactions between known Serengeti bachelors revealed the existence of a largely linear rank hierarchy (7). Males that had worked their way to the top of the hierarchy were fit to contend for positions in the territorial network, whereas those that rejoined the herd after being territorial for some months were worn out and low-ranking until their physical condition improved. Still, all currently territorial males that tolerated and interacted with bachelor males on their property always maintained dominance, even those in poor condition.

ACTIVITY. Impalas are primarily day-active, spending most of the night lying down, mainly ruminating, preferably in open terrain. By day they graze and usually stand while resting and ruminating, preferably in the shade. Feeding, standing, and lying accounted for 18.6–19.2 hours of a 24-hour day in female Serengeti herds (8). The basic activity pattern was found to be quite regular and to be closely synchronized among herd members, though less so in large than in small herds.

The impalas rose from their lying places at dawn, and after relieving and grooming themselves, fed intensively until 1000 or 1100 h, ruminated until noon or later, then began feeding again. Another well-defined rumination period often followed between 1500 and 1700, followed by another intensive feeding bout until dusk. Nearly all the impalas then lay down and ruminated. One feeding bout occurred at night, generally from c. 2300 to 0300 or 0400, but like the midafternoon feeding period was often poorly synchronized. Disturbance by a predator was often the trigger for the night feeding bout, which gave way to another lying and ruminating bout that continued until dawn. Sleeping individuals (lying with eyes closed for up to 10 minutes) were mostly seen at night.

Peaks of herd movement and of most social activity (social grooming, play, dominance interactions between bachelors) occur shortly after dawn and before dusk. From a third to a half less movement occurs at night. All kinds of social intercourse are minimal at night, except that mating activity continues at full swing by moonlight during the rut (26). Normal activity can be disrupted not only by predators but also by weather. Impalas all stop feeding when it rains, stand with their backs to the wind, bunch closely (females more than males), and begin ruminating. High winds often stimulate intense activity, including outbreaks of high-bounding, chasing, and roaring.

Activity cycles vary seasonally with the condition of the food supply. Impalas eat more, ruminate less, and move farther (up to 3 km daily vs. a minimum of 0.95 km in the wet season) within a larger home range in the dry season. The ratio of daylight feeding to ruminating time for female herds changed from 1.8:1 in April (peak rains) to 3.7:1 in July. Serengeti impalas go to water every second or third day, usually between 1000 and 1400 h, in the dry season only (8).

POSTURES AND LOCOMOTION. The impala is known for its spectacular leaps, up to 3 m high and 11 m long (13), which are triggered by sudden disturbances especially when in dense vegetation. The sudden explosion of antelopes leaping in all directions is apparently an antipredator tactic that increases the difficulty of selecting a quarry. The animals not only leap upward but to one side and then the other, and often pass or cut in front of one another (13). Impalas also engage in another form of leaping, equally spectacular, the motivation and function of which are unclear, which looks playful and is often infectious (fig. 8.23). As an animal descends from a high jump, it kicks its hindlegs nearly to the vertical and lands on its forelegs, rebounds, and brings its hindquarters down before landing again. This gait is unique to the impala (13).

Fig. 8.23. Impala *rocking high jump,* landing at a steep angle.

SOCIAL BEHAVIOR

COMMUNICATION. The roaring of adult male impalas and their repertoire of visual displays are so colorful that the importance of olfactory communication in this species via their unique scent glands has received little notice. It has been established that a male's dominance status is signaled by strong-smelling secretions of the forehead skin (7) (see Dominance/Threat Displays and Territorial Behavior). But the function of the metatarsal glands remains indeterminate. Their structure and equal development in the sexes suggest that the secretions of these glands diffuse through the air and serve some general social function, which could well be to maintain and restore contact between herd members, especially when vision is obscured (13).

TERRITORIAL BEHAVIOR. Presumably because of the higher testosterone levels associated with territorial and reproductive activity, territorial males not only are more aggressive but also have thicker necks than bachelor males, and their skin is made greasy by copious amounts of smelly sebaceous secretions. Probably as a result of rubbing their forehead secretion on branches, dark bare skin around the eyes makes their eyes appear bigger than normal (11).

Advertising: *erect posture* (=*proud posture*), *tail-raising, linked urination-defecation* on dung middens, *forehead-marking, roaring display, herding* and/or *chasing* of females and trespassing males; *prancing* or *goose-stepping* and *bipedal walking* (during herding, both uncommon) (7).

Territorial males commonly though not always urinate and defecate in sequence, in postures much like those of gazelles, and rarely excrete casually and at random as do other impalas. The tail is typically raised to show the white side, a behavior also associated with dominance displays. Territorial males regularly use dung middens, located on bare ground such as paths or roads, and often in the 15–30 m neutral zone between adjacent territories. Other impalas and also other species often use the same latrines (7). Standing in the *proud posture* (fig. 8.21), with head high but hindquarters lowered, is a static-optic advertisement of territorial status that looks similar to and may derive from the urination posture (symbolic marking) (7).

Forehead secretion is deposited both by rubbing the head up and down against objects and while *vegetation-horning;* both are performed by territorial males during all major social activities. However, horn-

Fig. 8.24. Impala *roaring display.*

ing is more closely associated with territorial defense against potential challengers, and *forehead-rubbing* is seen more often in interactions with females and inferior males. In addition to scent-marking while moving and feeding, territorial males go on regular marking patrols, often in border zones, during which they walk steadily and directionally, in an upright, stately manner, stopping intermittently to urinate-defecate, *forehead-rub,* and so on.

Roaring is the most impressive and far-reaching (2 km) of the impala's displays. It is preceded by 1–3 explosive snorts with mouth closed, and followed by 2–10 deep, guttural grunts emitted with mouth open, chin lifted, and tail gradually raised to 45° and spread (fig. 8.24). Although adult bachelor males participate in a spectacular outbreak of calling and chasing during the mating season, territorial males were responsible for 86% of the roaring episodes observed during the Serengeti study (7). Uttered only when a male is highly excited, roaring is apparently addressed specifically to other males, which pay attention and frequently react to one another's roars, whereas females appear indifferent. During the height of the rut at Sengwa, 180 *roaring displays* were observed in 1 hour on a moonlight night (26).

AGONISTIC BEHAVIOR

Dominance/Threat Displays: *erect* or *proud posture* ± *head-turned-away;* tail-raising and spreading; *yawning; tongue-flicking; high-, medial-,* and *low-horn presentation, head-tossing; head-dipping* (exaggerated nodding), *vegetation-horning; ground-horning* (rare). Females: *head-dipping; butting.*

Defensive/Submissive Displays: *head-low posture,* moving away in this attitude or while grazing, approaching and *sniffing* superior's forehead or under tail (uncommon).

Fighting: *air-cushion fighting, front-pressing* and *twist-fighting.*

Impalas assert dominance by walking stiffly in the *erect posture* while displaying their neck and horn development to maximum effect (cf. Grant's gazelle dominance displays). The ears are held back, the tail is clamped, and the performer usually faces his opponent but often with *head-turned-away.* Males appear very sensitive to even slight movements and changes of posture; for example, a barely noticeable head nod is enough to make a rank-inferior move aside (7). *High-horn presentation* invites an opponent to sniff the forehead skin during a rank-testing encounter. A subordinate responds by timidly approaching and *sniffing the forehead;* he then confirms the other's dominance by backing or walking away in the *head-low posture.*

An equal responds in kind to *high-horn presentation* and other assertions of dominance, such as raising and showing the underside of the tail, *yawning* (which intimidates inferiors), and *tongue-flicking.* The last, which is also an important courtship display, often accompanies threat displays. *Displacement grooming* (scratching with a hindfoot, scraping a shoulder with incisors) is common during confrontations; sometimes a contestant erects and mouths his penis. Often dominance displays between territorial males end with one or both moving apart while grazing.

Only 2% of observed aggressive encounters between Serengeti bachelor males ended in combat and 7% of encounters between territorial males. Yet fights are a normal risk for territorial males, as indicated by horn scars on the neck of 14 of 19 territorial males examined, versus 6 of 20 adult bachelors. Males have a dermal shield of thickened skin which protects the neck and head against deep stab wounds (10). Sometimes while pushing and pulling with interlocked horns a male trips over his opponent's horns and breaks or dislocates a foreleg (7). But apart from one broken horn, none of the 70 marked Sengwa males suffered a serious injury during a year or more of observation (19).

REPRODUCTION. Females conceive first at 1½ years, whereas males, though fertile as yearlings, only begin reproducing as they mature and gain territories in their fourth year. Gestation is 194–200 days (5).

Observations at Sengwa indicate that the 3-week peak rut among southern impalas is influenced by the lunar cycle (cf. wildebeest) (19, 26). The onset of the rut in May varied by up to 20 days in 5 years, with most mating between full moons. Males begin gearing up for the rut as early as March, when shorter days stimulate gonadal growth and hormone production, leading to increased aggressiveness and territorial behavior (2, 5, 22).

SEXUAL BEHAVIOR: *lowstretch; urine-testing; chasing; tongue-flicking; licking.*

Female impalas do not urinate on demand when approached by a male, but males respond to the urinating female's crouched posture with white tail held out by approaching and *urine-testing.* A male checks out females most energetically right after a herd comes onto his property, walking rapidly among the females in *lowstretch,* with nose and tail raised, nostrils flared, and mouth slightly open, occasionally *tongue-flicking,* turning his head to sniff at females' rumps (7, 27).

Having located a female in estrus, an excited male immediately begins an energetic courtship, of which the following is a generalized summary of what is in reality highly variable precopulatory behavior. He runs at her in *lowstretch, snorting, wheezing,* or *roaring,* and the female runs away and circles back into the herd, pursued by the male. When he relocates her he continues his suit. Presently the chase slows to a fast walk, the female keeping 3–5 m ahead of the male, which follows *flicking his tongue,* often emphasized by vigorous *nodding.* Later she allows him to close the distance and he proceeds to lick all around her vulva as she walks slowly ahead, tail now held slightly from the body (but rarely raised in response to licking). The *mating-march stage* leads into the *mounting stage.* The male runs or walks toward the female while rising to stand bipedally with body erect and head high. She usually walks forward and if he fails to penetrate he falls forward, clasping her with his forelegs and sometimes leaning his chest on her as he comes down. If his aim is true, the female usually holds still briefly. Mounting attempts are repeated at intervals of a few seconds to a minute or 2, averaging 4½ (1–14) attempts per successful copulation (fig. 8.25). In 9 of 14 observed copulations, the male roared and often ran around *chasing* bachelors and *herding* females for several minutes. But males rarely showed any further interest in females after the 1 copulation, even though females remained sexually attractive and were courted by other males (1 estrous female was seen with 4 different males) (7).

Fig. 8.25. Impala copulation.

PARENT/OFFSPRING BEHAVIOR. The female isolates in cover several hours before calving (usually in midday), though often only after strenuous attempts by a territorial male to prevent her leaving his herd (7). How long fawns remain concealed is unclear, but reportedly in as little as 1–2 days they follow their mothers back to the herd and there join a crèche of other small calves. Crèches containing a dozen or more fawns are characteristic of impalas, especially where there is a sharp annual calving season, and may be guarded by only a few females or even unguarded. Juveniles rest, move, play, and groom together, joining their mothers only to nurse, for herd movements, or when a predator is near (7).

Pushing contests between male calves begin before their horns emerge during the second month, although dominance interactions with accompanying displays appear only at 18 months and the full aggressive repertoire by 30 months (7). Impalas are weaned and able to survive without their mothers by 4½ months, which is just as well, because in East Africa juvenile males at that stage become subject to the aggression of the more active territorial males. Chasing and harassment usually continue for some months before young males accept separation from female herds and join a bachelor herd, usually by 8 months or when their horns exceed ear length (11). In South Africa, thanks to the short calving season, young males gain a reprieve from the aggression of territorial males for a whole year, since territorial organization disintegrates before their horns develop (11).

ANTIPREDATOR BEHAVIOR: *alert posture, alarm snorting, flight-intention* *movement, flight ± high-jumping.*

Where numerous, impalas are staple prey of all the larger predators, and martial eagles are big enough to carry off fawns. Their habitat preferences make impalas more vulnerable to ambush than open-country antelopes. No wonder, then, that these animals are unusually alert and quick to take flight. When walking through the undergrowth a suspicious impala will suddenly stop moving its head, while its eyes scan the landscape for movements and its ears rotate to pick up sounds. While staring at a suspicious object it cannot identify, an impala will move its head up and down and sideways, apparently to see it more "in the round." Or it begins to graze, then suddenly raises its head again and stares, a tactic well-adapted to detect a stalking lion or leopard. On the way to drink, the female that leads the file is extremely cautious, continually stops to look around with head high, and may stand motionless for minutes while her followers remain quite relaxed. The *alarm snort* is powerful and the intention to flee is signaled by a sudden upward movement of the head with neck stretched forward, which may release flight in other impalas (21). The spectacular flight and dispersal of leaping impalas is described under Postures and Locomotion.

References

1. Anderson 1972. 2. Bramley and Neaves 1972. 3. Brooks 1975. 4. Dasmann and Mossman 1962. 5. Fairall 1972. 6. Hofmann 1973. 7. Jarman, M. V. 1979. 8. Jarman, M. V., and Jarman 1973. 9. Jarman, P. J. 1972*a*. 10. ——— 1972*b*. 11. Jarman, P. J. and Jarman 1973. 12. ——— 1974. 13. Kingdon 1982. 14. Lamprey 1963. 15. Leuthold 1970. 16. Monfort-Braham 1974. 17. Murray 1981. 18. ——— 1982*a*. 19. ——— 1982*b*. 20. ——— 1984. 21. Schenkel 1966*a*. 22. Skinner 1971. 23. Smithers 1983. 24. Vincent 1979. 25. Vrba 1983. 26. Warren 1974.

Added References

27. Hart and Hart 1987. 28. ——— 1988.

Spiral-horned Antelopes and Buffalo

Subfamily Bovinae

The bushbuck, kudus, and other members of the Tragelaphini, an African tribe of antelopes with spirally twisted horns, are related to 2 tribes of Asian origin: the Bovini, which includes cattle and buffaloes, and the Boselaphini, which includes the Indian nilgai (*Boselaphus*) and 4-horned antelope (*Tetraceros*). The bushbuck and cattle may not look much alike, but the eland provides an obvious link between the 2 tribes. In fact there are a number of morphological, physiological, and behavioral traits that reveal the relatedness of these 3 tribes and set this subfamily apart from all other bovids.

SUBFAMILY TRAITS

1. Pronounced sexual dimorphism. Males mature much later than females, at 7–8 years compared to 4 years for females in the African buffalo and other very large species, and they keep growing, becoming progressively bulkier and more dimorphic (dewlap, hump, beard, mane, darker color, bigger horns) with age. Males weigh from ⅓ (African buffalo) to 2 times more than females. The largest bovids are found in this subfamily.

2. Unusual scent glands occur between the false hooves of the hindfeet in the Tragelaphini (fig. 9.1) and Boselaphini, bearing out the relationship between these 2 tribes. Otherwise there are no pedal glands in the subfamily, nor are there any preorbital glands. All species have 4 teats and a bare, moist muffle.

3. Most members of the subfamily live in closed habitats, ranging from forest (forest buffalo), to swamp (buffalo, sitatunga), to thornbush and mountains (kudu), to parkland (Asiatic wild oxen). A few have adapted to plains (eland, bison). Most are water-dependent.

SOCIAL ORGANIZATION. A key difference between the tribes in this subfamily and all other antelope tribes is the absence of territoriality (2). Male reproductive success depends on the ability to dominate other males whenever and wherever they meet and compete. Male development is prolonged and growth continues after maturation (3), resulting in an age-graded hierarchy. However, libido, aggressiveness, and endurance decline with age, so that in at least some species (e.g., buffalo), the oldest and biggest males may opt out of reproductive competition and/or be senile.

Males and females of most Bovinae remain separate except when the presence of females in estrus attracts males to them. The association of the African buffalo in mixed herds is unusual. Normally, male offspring part company from female groups as they become dimorphic. The greater the difference between the sexes, the less the social attraction, the greater their social distance, and the more divergent their habitat and food preferences (2, 3). As males get bigger and older they tend to become more solitary.

COMPARATIVE BEHAVIOR. The members of the subfamily have 3 distinctive behavioral traits in common.

1. *Lateral presentation,* by which the male's size and muscular development are shown to greatest advantage, is the primary means of asserting dominance. Many of the male secondary characters, such as great muscular development of the neck and shoulders, advertise the individual's bulk and power in lateral view, and such appendages as a dewlap (eland, cattle), beard, mane, and pendulous penile sheath contribute to the overall effect, as do color and markings that make the male stand out in a group.

2. Bovines and tragelaphines engage in *social licking,* something that is otherwise rare in antelopes. Bovines and at least some tragelaphines also use their tongues for grasping or gathering in their food; in cattle the tongue comes out and makes a sweeping movement with every bite. This is unknown in other antelopes.

3. Courtship features a *tending bond.* The most dominant of the males that gather around a female in estrus stays close to her and keeps the others away, thus ensuring exclusive mating rights.

References

1. Estes 1974*a*. 2. Estes 1990. 3. Jarman 1983.

Chapter 9

Bushbuck, Kudus, and Elands
Tribe Tragelaphini

Tragelaphus scriptus, bushbuck
T. spekii, sitatunga
T. angasii, nyala
T. buxtoni, mountain nyala
T. strepsiceros, greater kudu
T. imberbis, lesser kudu
T. (Boocercus) euryceros, bongo
T. (Taurotragus) oryx, common eland
T. derbianus, Derby or giant eland

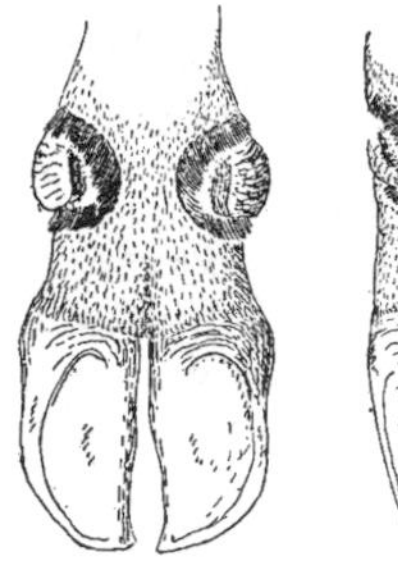
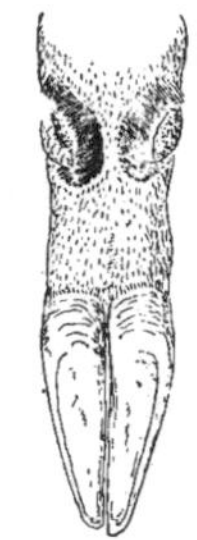

Fig. 9.1. Tragelaphine false-hoof glands (concealed beneath hair fringe) in eland (*left*) and lesser kudu (*right*). (From Pocock 1918.)

TRIBAL TRAITS. Medium-sized to very large antelopes with spiral horns, white vertical stripes, and pronounced sexual dimorphism. Females hornless (except bongo and elands). (The following descriptions apply to *Tragelaphus* species but not necessarily to the eland; see eland Traits.)

Deep- but narrow-bodied, rounded back with massive hindquarters (bushbuck type); kudus and mountain nyala more level-backed, limbs longer and nearer equal; large, rounded ears; bushy tail with white underside. Coat short, except in male nyala and sitatunga, dorsal crest of erectile white hair, better-developed in male. *Coloration:* fawn to dark brown or black, females tan to red-brown, males darken with age; stripes and spots individually and geographically variable, white chevron or bar between eyes, cheek spots, white throat patch and chest crescent (except greater kudu and nyala), upper forelegs with dark garters. *Scent glands:* a pair of inguinal glands located ahead of the teats in the bushbuck, sitatunga, lesser kudu, and mountain nyala; glands encircling false hooves in hindfeet (fig. 9.1) (absent in bushbuck, sitatunga, and bongo). *Mammae:* 4.

The markings in this tribe are concealing rather than revealing, serving to break up the animal's form (disruptive coloration) and enabling it to blend into the background. However, some markings are clearly important for social communication, notably the dorsal crest and white scut, and possibly also the white horn tips, which emphasize the horns' size and shape in most species (4).

RELATIVES. Except for the eland, the tragelaphines are much alike in appearance, preference for cover, and behavior. The bushbuck, sitatunga, and nyala are particularly close, and the bongo resembles an overgrown bushbuck or sitatunga. The kudus and mountain nyala have departed from the bushbuck form and are similar. The elands may have branched off the kudu line (4), as suggested by zoo hybrids between the greater kudu and eland. Other crosses have been reached between the lesser kudu and sitatunga, the lesser kudu and bushbuck, the sitatunga and bushbuck, and the bongo and sitatunga (4, 6).

DISTRIBUTION. Members of this exclusively African tribe are found throughout sub-Saharan Africa in virtually all kinds of wooded habitats from lowland rain forest to montane cloud forest (bongo), forest-savanna mosaic (nyala), gallery forest and thickets of all kinds from sea level to mountain heath (bushbuck), in arid lowland thornbush and stony mountains (kudus), and in savanna and subdesert (elands). The sitatunga is a swamp dweller. The common eland is the only tragelaphine that ventures onto open plains.

ECOLOGY. Tragelaphines are all browsers (except the eland and sitatunga). Like cattle, they have prehensile tongues they use for grasping food (eland excluded). No other tribe of large antelopes is so dependent on cover; in this respect tragelaphines are like the small antelopes and their disruptive markings are adapted for a concealment strategy, although they more often stand still than lie prone as do small antelopes. The forest, savanna, and swamp dwellers are water-dependent, but the kudus and elands are adapted to the arid biomes and are normally independent of surface water.

SOCIAL ORGANIZATION. An unusually wide range of social organization is found in this tribe, from the "solitary," sedentary bushbuck to the highly gregarious, nomadic eland (4). But again, it is the eland that has broken the tragelaphine mold, becoming more like an ox than an antelope. *Tragelaphus* social groups rarely exceed a dozen animals. The typical group numbers less than a half dozen, and the sitatunga is only slightly more sociable than the bushbuck. On the whole this tribe has not advanced very far along the path of sociability, leading sedentary lives in closed habitats in small, unisexual groups.

Adult males join female herds long enough to check out the cows but soon depart unless they find one in estrus. Females, too, come and go. Except for bonds between mothers and young of the year (which may persist in kudus), there appear to be no close ties between individuals and no rank hierarchy except for the usual dominance of seniors over juniors, and of larger males over smaller females. Tragelaphine society is open, loose, and notable for the low level of aggression evinced by either sex.

Although less gregarious than many bovids, females of hornless species may tolerate comparatively close individual distance. Herd members often lie touching, mostly in pairs, in a close or open star formation (7).

Males. *Tragelaphus* males remain in female herds and/or with their mothers until the appearance of secondary characters makes their gender evident. This happens usually at around 1½ years, when the horns (which only become visible at 5–6 months in species with hornless females) exceed ear length (1, 5). Adult males attracted to herds containing an estrous female tend to harass any accompanying adolescent males. Elands remain associated with females a year or 2 longer; the presence of well-developed horns in the females reduces or eliminates the impact of horns as a badge of masculinity, enabling males to remain until other secondary characters (coloration, bulk, and appendages) advertise their sex (3). Even if not driven out of the female herd, young males are drawn to seek out peers against which to vent their developing aggressiveness and test their physical prowess. In any case, males of gregarious species associate for several years between adolescence and maturity in small bachelor herds. When finally mature enough to compete for mating opportunities, questing bulls tend to wander alone.

The ranges of adult males overlap but they may avoid close proximity except when the magnet of an estrous female draws them to the same spot. The false-hoof glands, which appear to be adapted for wafting scent through the air, may play a part by surrounding the individual with an odor field that forewarns and repels adult males while attracting males to females, females to females, and so on. However, the functions of these unusual glands remain to be investigated.

ACTIVITY. Tragelaphines may be more nocturnal than most bovids, although allowance must be made for environmental influences. Thus, bushbuck and kudu populations subjected to heavy human predation may be so nocturnal and secretive as to be rarely seen, even though common. Yet the same species are often seen abroad by day in many national parks. Nevertheless, the best times to see most tragelaphines are in late afternoon and early morning, and they are likelier to venture into the open by night than by day.

POSTURES AND LOCOMOTION. The bushbuck type of build is quite like a forest duiker's and similarly adapted to bounding, rushing, and dodging through dense cover. Such a conformation is not noted for speed and stamina (4). Like most ungulates that inhabit uneven terrain or closed habitats, the bushbuck types *cross-walk.* The eland is well-adapted by size and conformation to long-distance travel, but perhaps because of its bulk can only run slowly and quickly tires. Yet elands and kudus are marvelous high-jumpers.

When calmly alert, *Tragelaphus* species hold the head up with neck curved and ears playing. When alarmed, the neck is raised as high as possible, and if the animal moves, the head moves jerkily backward and forward in rhythm with the step, in a "pecking," goatlike gait that serves to alert other animals (2, 7). It is also characteristic of *Tragelaphus* species (and various forest ungulates) that they can stop moving instantaneously when startled, often with one leg raised—something plains antelopes never do. When investigating a suspicious object, the animal's head and neck are lowered; the more unsure the animal, the lower the nose is held (also true of eland). In flight, the tail is raised to show the white scut (as in reedbucks). Tragelaphines do not stot and very rarely stand bipedally while feeding.

SOCIAL BEHAVIOR

COMMUNICATION

Olfactory and Tactile Communication. The most common form of social contact is the olfactory checking and/or inspection that occurs when 2 individuals meet. Of the places that are regularly inspected, the nose is the first and the anogenital region the second in importance. Strangers are checked far more thoroughly than familiar animals and newborn calves count as new arrivals. Males rub their horns, forehead, cheeks, and neck on objects and on females, but apparently not for purposes of scent deposition, although the skin of the cheek and the ear area is oily and attracts flies (4). Nor is dung or urine employed for scent-marking. Unlike bovines, tragelaphines do not roll or wallow.

Social grooming. Social grooming is apt to occur following *nose-to-nose contact.* It takes the form of *nibbling* and especially *licking* and may last up to ¼ hour (at least in captivity). The head, neck, shoulders, withers, and rarely the back, flanks, or hindquarters are groomed, never the chest, belly, or legs (7). *Mutual licking* of the genitalia or udder is associated with courtship behavior.

Tragelaphines solicit grooming by 1) raising the nose vertically in frontal orientation, to elicit grooming of the chin and throat, 2) raising the head with back of head turned, to elicit grooming at the base of the neck and between the ears, 3) lowering head and neck toward the grooming partner, to direct grooming to the nape of the neck, and 4) presenting the horns to be licked between the horns or on the ears. All these forms of soliciting resemble dominance and threat displays (see below), suggesting that the primary function of social grooming in the Tragelaphini is to appease the aggressive tendencies of a higher-ranking animal (7).

Vocal Communication. A loud, gruff bark is characteristic of all tragelaphines. The fact that males are generally more vociferous than females suggests that the bark functions not only as an alarm call but also to advertise sex and social status. It is heard most commonly at night and at twilight. Deep grunts have also been described in some species, as well as humming, whining, clicking, mooing, and smacking sounds, but most of these are relatively quiet calls exchanged between courting couples or mother and calf.

Visual Communication. See under Agonistic Behavior.

AGONISTIC BEHAVIOR

Dominance/Threat Displays. Males use some of the same aggressive displays to intimidate females during courtship that they use against one another—demonstrating the linkage between reproductive success and ability to dominate. The *lateral display,* performed with head high or low and hair bristling, is particularly impressive in species with dark coloration and conspicuous white dorsal crest, notably the nyala and bushbuck (figs. 9.5, 9.7). Competing males typically approach each other in a stiff, deliberate walk or trot which may be a dominance display in its own right. *Lateral presentation* is often followed by *ground-horning* (fig. 9.2). When displaying males stand confronting, they often turn their heads (*looking away*) and this movement may carry over into grooming the shoulder. *Low-horn presentation* (fig. 2.9) is the strongest threat display; the more the chin is pulled in and the horns aimed at the adversary, the readier the animal is to attack, for this is the *combat attitude.* The threat is further enhanced by symbolic thrusts, *ducking* the head down and up, or

Fig. 9.2. Nyala *soil-horning.*

feinted attack: jumping forward with the forelegs while the hindlegs remain planted. *Shaking* and *nodding* the head, performed by both sexes, are the mildest forms of threat; in fact, the *slow headshake* signals submission in elands.

Defensive/Submissive Displays. Being hornless, females have a much more limited aggressive repertoire, consisting mainly of defensive-threat displays. *Poking* with the snout is the commonest defensive threat, followed by *symbolic biting* (cf. *champing* in the waterbuck). During courtship, the *poking* is always directed toward the male's snout (7). A more intense display is raising the nose straight up (*nose-in-the-air*). This is often combined with *neck-winding* (fig. 9.6), especially during courtship when the male rests his chin on the female's neck or rump: the female stretches her neck and head forward, then bends it upward, downward, or sideward. Indicative of persistent defense, *neck-winding* could derive from the neck-fighting of hornless ancestors (7). Females may also threaten to butt.

Standing in the *head-low/chin-out posture* expresses submission. Significantly, it is the same posture adopted by tragelaphine females ready to mate. *Licking* as appeasement of a dominant but not too actively aggressive opponent has already been described.

Fighting. Rarely observed in the wild, and apparently uncommon in this tribe. The basic tragelaphine fighting technique is *front-pressing* with foreheads or horns engaged. Attack is nearly always directed at the head, although the unsociable bushbuck is possibly less inhibited about flank attack. In the kudus, the corkscrew horns go together with the unusual fighting technique of *horn-pressing* (see lesser kudu, fig. 9.9).

Female tragelaphines *neck-fight,* twining their necks and sometimes lifting from below in an effort to overthrow the adversary. They also butt and are as likely to aim blows at the opponent's body as its head. Other forms of attack are *poking* or *snapping,* with mouth open or closed. Highly excited females may bite, but the most they can do is pull out tufts of hair. When courting males are threatened or attacked by females, they show characteristic restraint and never retaliate, although they may give strong threat displays (7).

REPRODUCTION. The bushbuck, sitatunga, nyala, and lesser kudu are perennial breeders, although populations living in regions with extended dry or cold seasons will tend to have reproductive peaks during the most favorable months. Reproduction in the greater kudu and common eland is more seasonal, especially in southern Africa and in arid regions, and the bongo of Kenya's Aberdares responds to seasonal changes in its montane habitat (see species accounts for details).

Fig. 9.3. Tragelaphine courtship (eland): male *drawing alongside* female (from Walther 1972*a*).

Being nonterritorial, males have no refuge where they can sequester and court females free from the interference of rivals. A male must establish dominance over all comers in order to gain mating rights. Having found a female in estrus, a male forms a *tending bond,* staying with her and attempting to keep other males away until she is ready to mate. The prolonged estrus that goes together with this type of mating system (see chap. 2) makes it likely that the biggest and fittest male in the neighborhood will be in attendance by the time the female ovulates, and will end up siring most offspring.

SEXUAL BEHAVIOR. Male tragelaphines approaching females do so in *low-stretch.* But instead of staying behind like most courting antelopes, tragelaphines regularly *draw alongside* the female (fig. 9.3). All species drive in a straight line (courtship circling is rare). A courtship behavior unique to this tribe is *neck-pressing:* the male lays his head or neck on the female's neck and presses down (see fig. 9.12). This behavior is thought to represent the fighting technique of hornless ancestors, retained in courtship as a symbolic expression of male dominance (7). However, since the male also rests his chin and neck on the female's back preparatory to and during copulation (fig. 9.4), it is possible that *neck-pressing* is simply a mounting-intention movement. Having approached the female and sniffed her vulva, the male often rubs his cheek on her hindquarters. She responds by urinating and the male performs the *urine-test*/lip-curl (see chap. 2). *Mutual licking* and *nibbling* occur during courtship, especially by

Fig. 9.4. Greater kudu in typical tragelaphine copulation posture (from Walther 1964*b*).

the male in response to female defensive threats; males may also utter infantile feeding calls during the precopulatory stage. These appeasing behaviors are in sharp contrast to the dominance/threat displays males often perform in the prelude to courtship.

The species differ in which courtship displays are emphasized and how they are performed. Thus, *neck-pressing* is obligatory in the greater kudu and nyala, common in the lesser kudu, but unrecorded in the sitatunga or bushbuck (7). The lesser kudu investigates the female's udder (possibly smelling the inguinal glands); the greater kudu seldom does; the sitatunga and bushbuck never do. Rubbing the cheek on the female's hindquarters is seen in all species but is especially important in the lesser kudu and bushbuck (7).

PARENT/OFFSPRING BEHAVIOR. Tragelaphines are conventional *hiders.* Calves lie out for 2 weeks or so (often less in eland and much longer in bushbuck), usually in dense cover. The mother retrieves the calf for suckling by going directly toward it and can prompt it to rise with a special contact call (*smacking, clicking, mooing*). Calves eager to nurse run around in front and block the mother's path (possibly the beginning of *lateral presentation*? [7]); older calves demand feeding by vigorous jabs, butting, attempted *neck-fighting,* and mounting. Mothers suckle young calves for 5–10 minutes, but weaning tends to come early in the tragelaphines: between 4 and 6 months. Licking of the young, including cleaning, stimulation of excretion, and consumption of wastes, is well-developed. In addition, mothers and young engage in *social licking.*

PLAY. Calves begin to play in their first week. Play takes different forms: postures and displays based on the dominance/threat repertoire, fighting, and running games (7). The invitation to a running game is to run toward a potential playmate with head and neck stretched forward, ducking the head, straddling the forelegs, and swerving to one side at the last moment. Running games are contagious, and when calves start racing around the herd older juveniles and sometimes even staid cows cut capers. Running and chasing often lead into play fighting. Jumping forward and backward, *feinting attack, cavorting,* and *horning the ground* are all incorporated in running games.

ANTIPREDATOR BEHAVIOR. Except for elands, tragelaphine antipredator strategy is based first on concealment (standing still) and, failing that, on abrupt flight, often including high bounds, to elude pursuit. Both stratagems tie tragelaphines to cover tall and dense enough to camouflage the presence of a standing animal. Alert and alarm behavior is described under Postures and Locomotion, and the distinctive tragelaphine *alarm bark* was also mentioned under Vocal Communication. *Cooperative defense* of calves has evolved in the eland (see eland account).

References

1. Anderson 1980. 2. Brown 1969. 3. Estes 1990. 4. Kingdon 1982. 5. Leuthold, W. 1979. 6. Van Gelder 1977. 7. Walther 1964*b*.

Added Reference

Underwood 1984.

Fig. 9.5. Bushbuck *lateral display.*

Bushbuck

Tragelaphus scriptus

TRAITS. A medium-sized forest antelope with large ears and eyes, rounded back, hindquarters more developed than forequarters. *Height and weight:* male 70–100 cm, 40–80 kg; female 65–85 cm, 25–60 kg. *Horns:* 26–57 cm; one twist only, nearly straight; males only. *Coloration:* highly variable geographically and individually (at least 9 different races are recognized [2]); northern and western forms reddest with best-developed markings, including harness-like diagonal lines on shoulder in *T. s. scriptus* (whence the French common name *antilope harnachée*); Abyssinian and southern forms yellower with fewer markings; both sexes darken with age, but males far more than females. *Glands:* inguinals situated well ahead of mammae, no false-hoof glands.

DISTRIBUTION. One of the most ubiquitous antelopes but usually unseen owing to its secretive habits and closed habitat. It is not found on open plains or anywhere without sufficient cover to conceal it. It ranges to 3000 m in East African montane habitat and penetrates the arid biome wherever it finds cover and water to meet its needs.

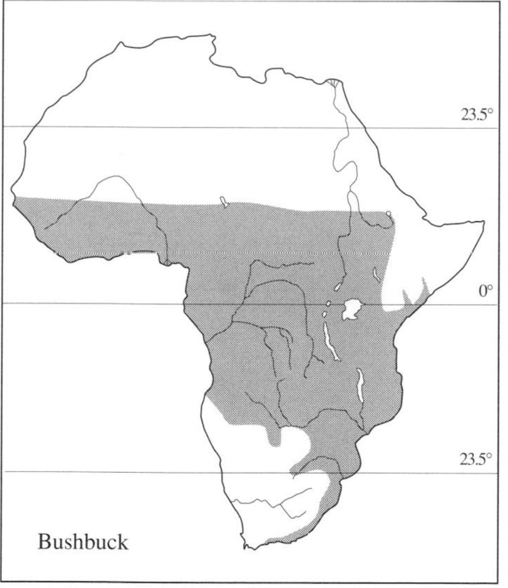

Bushbuck

ECOLOGY. The bushbuck is a forest-edge antelope, cryptically colored and disruptively marked, which only leaves cover when drawn into the open by choice food plants and perhaps the presence of other bushbucks. It eats tender new grass but is predominantly a browser on herbs and shrubby leguminous plants. It is almost invariably found near water, possibly because the dense cover it frequents by day is in ravines and along watercourses. Yet it can also obtain water by licking dew from vegetation. Fondness for garden produce can make it an agricultural pest. Bushbucks are also fond of figs, other fruits, and flowers, which would explain their frequent presence beneath feeding baboons and other monkeys (3).

SOCIAL ORGANIZATION: solitary, nonterritorial, polygynous, sedentary.

As the only strictly solitary tragelaphine, the bushbuck is the most socially primitive member of the tribe. Individuals do not herd together and the only regularly associated individuals are a female and her latest offspring. Each adult has its own exclusive lying-up place, usually in a thicket, where it retires to ruminate and rest by day when not actively foraging. On the other hand, bushbucks do not actively avoid one another, their home ranges overlap extensively and sometimes completely, and when individuals meet they often approach, interact in a friendly way, and remain in proximity for hours while feeding (6). In patchy habitat, bushbucks are often distributed in clusters which amount to discrete subpopulations; as many as 6 males and 6 females were found to share the same patch in Nairobi N.P. (1). Thus, bushbucks should not be considered antisocial but rather as loosely and casually sociable. The pattern is most likely adapted to the antipredator strategy based on concealment and a gleaner feeding role, which favor individual rather than group foraging (1, 6).

Bushbuck density may be surprisingly high: 26/km^2 in Queen Elizabeth N.P. (Mweya Peninsula), for instance, or equal to the density of the Uganda kob in the same park. Individual core areas may be as small as .25 ha for females, 0.5 ha for adult males, and 2 ha for subadult males in Nairobi N.P. (1). But in Queen Elizabeth N.P. the mean daily range was 19.6 ha (6.3–35.2), and day and night ranges were entirely different in all cases (7).

ACTIVITY. In Queen Elizabeth N.P. each bushbuck spent the day in or on the edge of a thicket, feeding 38%, resting 50% (often standing and blending into the background), and moving 12% of the time. Beginning before nightfall, the animals walked toward their night range in open habitat, where they spent 25% of the time feeding. They grazed more and browsed less at night and resting animals lay more often than stood, making them less visible in the open. With the approach of dawn each bushbuck returned to the security of its thicket (7).

POSTURES AND LOCOMOTION. See tribal introduction.

SOCIAL BEHAVIOR

COMMUNICATION. See tribal introduction.

SELF-ADVERTISING. Reproductive competition leads adult males to advertise their presence and dominance status, thereby making themselves more conspicuous than females. A male may deliberately lie out in plain sight on the edge of a cliff or stand sentinel on a hillock, especially early or late in the day (3). Males bark more frequently and in a lower tone than females. They also horn the ground and bushes (see nyala account, fig. 9.2), and both sexes rub their heads and necks on branches and such, especially the space between and under the ears, which is oily and scented, possibly as a means of social communication (3).

AGONISTIC BEHAVIOR

Dominance/Threat Displays. Male: *stiff-legged approach* and *lateral presentation* with head up and turned aside; *low-horn presentation* ± *ducking* and *feinted attack, ground-* and *object-horning.* Less common are *tossing movements, snout-thrusting,* and *snapping.* Female (offensive or defensive): *snout-thrusting,* snapping, *head-low posture, ducking.* Less common are *nose-in-the-air, symbolic butting,* and *feinted attack* (5).

Defensive/Submissive Displays: *head-low/chin-out posture; turning away, social licking.*

Fighting: *front-pressing, thrust-fighting,* charging.

Except for the dominance behavior associated with courtship, only adult males have been seen to threaten or avoid one another, although subadult males commonly horn bushes/ground when near other males (3). Two well-matched males were seen to approach and circle with tense, high steps, necks flexed, backs arched with crests erected, and tails laid over the back and fluffed, adding to the display of white (fig. 9.5). At higher intensity males spring into the air, and *circling* may lead to *chasing* interrupted by *horn-clashing.* Outright fighting is rare and potentially dangerous because of the straight, sharp horns and uninhibited stabbing attempts. Opponents stand confronting with heads lowered and feet firmly planted, forelegs straddled, engage horns, and proceed to push and twist violently (4). At highest intensity they charge into combat from a short distance (*thrust-fighting*). If one gains an opening he will stab to the body, often with fatal consequences (1). Dogs often, and sometimes their owners, have been killed by wounded bushbucks brought to bay.

REPRODUCTION. With a gestation of 6–7 months (exact period uncertain), bushbucks could and at least sometimes do produce 2 young in a bit over a year. In high rainfall areas there is no strict breeding season; in arid areas there are birth peaks in the wet season.

SEXUAL BEHAVIOR: *lowstretch, courtship call, urine-testing, genital nuzzling, licking, rubbing cheek* on female's rump, *neck-pressing, chin-resting.*

Males intent upon checking female reproductive status approach in *lowstretch* (without displaying the dorsal crest). Females respond by assuming the *urinating crouch,* with fore- and especially hindlegs flexed and tail raised. If the *urine test* is negative the male goes on his way. Before and/or after the female urinates, the male sniffs and nuzzles her vulva, which often causes the female to run forward a short distance. Such avoidance, if persistent, may provoke an aggressive response: the male pushes or drives the female in *lowstretch, snapping* and *poking* her, even hitting her flanks with his horns. He also assiduously licks her vulva, tail root, thighs, flanks, and back, and rubs his cheek and tries to rest his chin on her rump. Meanwhile he utters a *clicking, twittering call.* Until ready to stand for mounting, the female fends off the male by *poking* and *snapping (symbolic biting).*

Once the *tending bond* has been established, the male stays close to the female and keeps other males at a distance, unless a bigger male appears to supplant him.

PARENT/OFFSPRING BEHAVIOR. Calves, dropped in dense cover, do not accompany their mothers into the open for up to nearly 4 months. When retrieving the calf to suckle it, the mother not only stimulates excretion by licking it anogenitally, but the 2 also engage in *social licking* of the back of the head, the neck, and ears. The maternal bond continues until the next calf is born and sometimes longer. Offspring of both sexes reach puberty at c. 11 months. Horns appear first at c. 10 months, begin to twist at 1½ years, but only reach full size at 3 (3).

ANTIPREDATOR BEHAVIOR. A bushbuck standing in a thicket is virtually invisible as long as it remains motionless, and *freezing* is the standard reaction to the proximity of a predator. At night its disruptive white markings also help to conceal it while lying still in the open. Hyenas and lions have been seen passing unaware within 10 m of bushbucks that were plainly visible through a starlight scope (7). If detected and pursued, a bushbuck races for the nearest cover, raised tail flashing like a white-tailed deer's.

References

1. Allsopp 1971. 2. Ansell 1971. 3. Kingdon 1982. 4. Verheyen 1955. 5. Walther 1964*b*. 6. Waser 1974. 7. ——— 1975.

Added Reference

8. Smits 1987.

Fig. 9.6. Sitatunga *mating march:* male, *following closely,* rests his chin on female's croup; she responds with *neck-winding* (from Walther 1964*b*).

Sitatunga
Tragelaphus spekii

TRAITS: A swamp-dwelling antelope very similar to the bushbuck. *Height and weight:* males 100 cm (88–125), 100 kg (70–125); females 75–90 cm, 50–57 kg. Feet adapted to swampy terrain: elongated hooves with wide splay, naked, padlike pastern. *Coloration:* geographically and individually variable, adult males gray-brown to chocolate-brown, females brown to bright chestnut, calves bright rufous-red, woolly-coated, spotted, and striped; adults long-coated (especially males) and markings therefore less distinct than the bushbuck's; white neck patches but dorsal crest often brown. *Horns:* males only, 66 cm (45–90) with 2 twists, ivory-tipped. *Scent glands:* 1 pair of inguinals.

The sitatunga and bushbuck are close enough genetically to produce viable hybrids in captivity (4), and almost indistinguishable from the nyala except by pelage and hooves (1).

DISTRIBUTION. The sitatunga is so specialized that it occurs only in swamps or permanent marshes. Its center of distribution is the West African rain forest and the wetter regions of the Southern Savanna.

ECOLOGY. Although most partial to papyrus swamps, sitatungas also occur in wetlands dominated by bullrushes (*Typha*), reeds (*Echinochloa pyramidalis*), and sedges (*Cyperus*), and frequent the deepest parts of the swamp (3). They both graze and browse, often selecting plants in the flowering stage, emerging at night to graze on nearby dry land and venturing into adjacent forest to browse on foliage, creepers, and such. Sitatungas tend to concentrate their feeding activity in one small area of swamp for many days at a time, then to shift suddenly to new grounds. They often feed while immersed up to their shoulders, moving

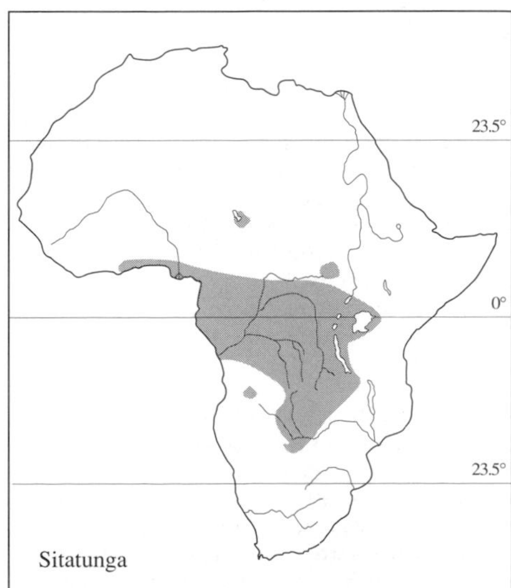

slowly and carefully through floating vegetation. They may stand with forelegs immersed and hindlegs elevated on a tussock. They also rear to reach flowers of tall reeds, sedges, grasses, and foliage, and males have been known to break branches with their horns (cf. eland and bongo) (1). When feeding on long leaves, a sitatunga wraps its tongue around a clump, pulls it into its mouth, and crops it with its incisors (3).

SOCIAL ORGANIZATION: semisocial, nonterritorial, sedentary.

The sitatunga's dependence on swamps for food and cover favors a dispersed social organization and makes it hard to observe. However, swamps are highly productive ecosystems and sitatungas can live at densities of $55/km^2$ or higher (3). Though most often seen singly, sitatunga females have a herding tendency. In 240 sightings of adult cows during a Kenya study (3), 27% were in groups with 1–3 other females, and only 33% of the 89 adult males were alone; the others were in company with females. The largest group, 9, included an adult male, 4 adult females, and 4 young. Males associate together or with females until subadult, but as adults avoid one another. Nocturnal *barking bouts* (see below) may serve a spacing function.

ACTIVITY. Sitatungas move through the swamp along established pathways. Major arteries up to 7 m wide may transect the swamp, with numerous side branches leading to feeding grounds, including exit ramps to neighboring riverine forest. In high reeds and sedges the pathways make dark tunnels through the tall vegetation. Sitatungas are active from daylight to about 1100 h and after 1700, when they may emerge to feed on open grassland or in forest under cover of darkness. Although individuals may feed out in the open at any hour in areas where they are protected, sitatungas typically lie up on platforms of vegetation each animal prepares for itself by repeated circling and trampling. A platform may be used several times. Sitatungas also often stand and ruminate in the water.

POSTURES AND LOCOMOTION. Slow, clumsy runners on land, sitatungas would be easy prey out of their element, but their plunging run is efficient in water, mud, or floating vegetation, and their long, splayed hooves keep them from sinking in as much as other ungulates. Their usual movements are slow and inconspicuous, and they are good but slow swimmers.

SOCIAL BEHAVIOR

COMMUNICATION

Vocal Communication. Like bushbucks, male sitatungas often bark, especially at night. An individual may continue barking intermittently for over 10 minutes. It may be as much of an advertisement as an alarm signal, or more; barking may be taken up serially by other males, each thereby announcing its presence and location. Females give a single, higher-pitched bark that sounds more like the conventional alarm snort of other antelopes. During or after barking the animal moves into cover. A low-pitched squeak, heard mainly during the day among feeding sitatungas, has also been described. Calves bleat sheepishly, and a male following a female in *lowstretch* may utter a suppressed roar, with slightly open mouth. A louder version is uttered during running games by both adults and young (in captivity) (5).

AGONISTIC BEHAVIOR

Dominance/Threat Displays. Male: *erect-posture, medial-horn presentation, head-turned-aside,* ducking, head-throwing, head-shaking, *ground-horning, feinted attack.* Female: *poking, snapping, ducking, butting, feinted attack.*

Defensive/Submissive Displays: *head-low posture, nose-in-the-air, neck-winding.*

Fighting: *front-pressing* and lifting, with lowered forequarters and wide-splayed legs (when pushing with maximum force).

Aggression is rare in the wild, but one fatal fight is on record (6). *Lateral presentation* is rare (hence the poorly developed dorsal crest?) and the main aggressive displays are presenting the horns and *ground-horning.*

REPRODUCTION. Reportedly most births occur during the dry season (6, 7).

SEXUAL BEHAVIOR: *lowstretch; rubbing head* on rump or flanks, *urination* and *urine-testing, mutual licking, driving, roaring.*

After approaching a female in *lowstretch* to sniff her vulva, a male often rubs his head on her flank if she will hold still (3, 5). But more often the female moves slowly away or behaves skittishly, standing still with head erect and very alert until the male comes within a few inches, then bounding away suddenly through the swamp, making considerable commotion (3). Instead of giving chase, the male follows persistently, never hurrying. However, in a zoo a courting male responded to an evasive female by running toward her in *lowstretch* while *roaring*, and the female tried to fend him off by *neck-winding, snapping,* and *poking.* During *courtship-following* or *driving,* a male stayed behind rather than *drawing alongside* the female as in some other tragelaphines (cf. kudu), laid his head and neck on her back, and lifted his forelegs off the ground in a mounting attempt. A receptive female responds by *neck-winding:* with neck angled down obliquely, the head is turned sharply up with mouth wide open, while the head is thrust forward, upward, and back (fig. 9.6). The male mounts with forelegs bent, then straightens them; his head and neck rest on the female's back, while her head and neck point forward and down.

PARENT/OFFSPRING BEHAVIOR. Calves are so well hidden that the behavior of mother and young has been recorded only in captivity. Even partially grown calves are brought into the open only when several other sitatungas are present (3). Well-used platforms have been found among reeds growing in deep water, completely secluded and quite dry, where small calves had probably been hidden. The young are unable to move slowly and deliberately through the swamp like adults, but jump from tussock to tussock and sometimes fall in the water. After learning how, the calf follows its mother closely for several months and may even remain with her after a new one is born.

Retrieval of young calves for suckling was observed in a captive herd (5). The mother would feed toward the calf's hiding place, then walk up to it, touch or lick the calf's snout, and move away. After the cow had gone 10–15 m, the calf would get up and follow and she would lead it to a protected place and proceed to suckle it. She was not heard to call, but when the calf was older she could summon it by looking intently in its direction from a distance, abruptly lowering her head to the horizontal, and repeating this movement. Two different calves sucked 2–21 minutes (average 10 minutes) on 52 occasions. One mother ceased suckling after 134 days. Up until the calf is about a half year old, the mother regularly licks it during nursing, especially anogenitally.

ANTIPREDATOR BEHAVIOR. Lions and wild dogs have been known to prey on sitatungas, and presumably leopards catch some of those that venture into riverine forest. The use of regular pathways makes sitatungas particularly vulnerable to snare-trapping. They can also be driven by beaters into nets or into deep water, where spearmen in boats easily dispatch them. Whenever possible, sitatungas strive to avoid detection, and will even submerge until only their nostrils and eyes remain above water (2, 6).

References

1. Kingdon 1982. 2. Lydekker 1908. 3. Owen 1970. 4. Van Gelder 1977. 5. Walther 1964*b*. 6. Wilhelm 1933.

Added References

6. Starin 1990. 7. Williamson 1986.

Fig. 9.7. The nyala's spectacular *lateral display.*

Nyala

Tragelaphus angasii

TRAITS. Similar to the bushbuck and sitatunga but bigger and hairier, sexes extremely dimorphic. *Height and weight:* Males, 106 cm, 114 kg (98–125); females, 92 cm, 58 kg (55–60). *Horns:* 60 cm, 1½–2½ twists, white-tipped. *Coloration:* female and young bright chestnut, smooth-coated, with contrasting stripes (8–13) and spots, chest and nose chevron; males charcoal gray (except tan lower legs) and hairy, with fringe of dark hair from throat to hindquarters and white dorsal crest longest in tribe, stripes reduced or absent. *Glands:* false-hoof glands on hindlegs, otherwise no known specialized scent glands (including forehead and cheeks [3, 4]).

DISTRIBUTION. The nyala is almost entirely confined to the Lowveld of the Southern Savanna. Its distribution closely corresponds to the area of the Mozambique coastal plain and the valleys of nearby major rivers, but its range has been greatly reduced within this century and its distribution is now very spotty. Two of the best places to see it are the Hluhluwe Reserve (Natal) and Kruger N.P.

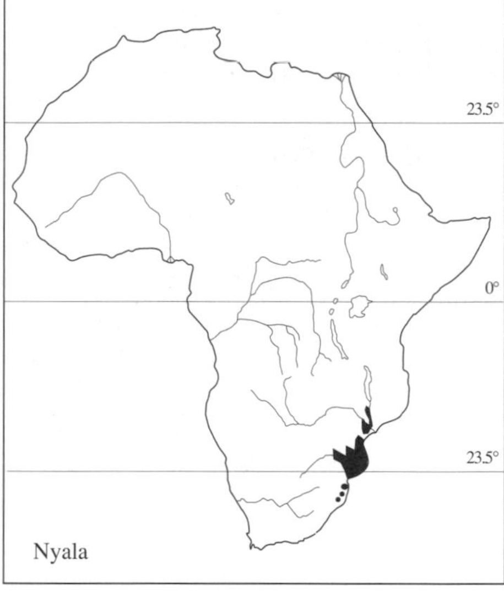

ECOLOGY. Found only in low-lying, densely wooded habitat generally near water, the nyala overlaps in habitat preferences with the greater kudu, common and red duiker, bushbuck, and bushpig. It is a mixed feeder, eating leaves and pods of acacias and other trees, various fruits, herbs, and tender young grass. It grazes more in the summer rainy season, shifting gradually to browse in the dry season, when it drinks daily (1).

SOCIAL ORGANIZATION: gregarious, nonterritorial, sedentary.

The nyala and sitatunga have a very similar social organization. The basic social unit consists of an adult female with her latest and next-to-latest offspring. Two or more units may join to form a female group, but such associations tend to be temporary and only form when each is accompanied by young. Yet only females guarding concealed young tend to be solitary, and typical group size in Hluhluwe Reserve is 5.6 nyalas (mean group size is 4) (3). There is some evidence that associated adult females are related; daughters tend to remain in their natal home range and to continue associating on a casual basis with their mothers after becoming mothers themselves. Female home ranges averaged 66 ha (33–100) and males ranged 80 ha (48–139). No exclusive core areas were detected, but at least 10% of each home range consisted of forest and up to 28% of the range was open habitat, especially floodplain and other grassland, which was utilized only at night (cf. bushbuck). Home ranges of a Mozambique population appeared to be larger: 360 ha for one known female and 550 ha for 5 adult males (5).

Like greater kudus, males do not actively avoid one another and often associate, but so casually that bachelor groups rarely remain unchanged for over 2 hours. Groups of males 1½–3 years old, typically numbering 2.7 (average 1.8, maximum 9), especially peer groups, are relatively stable, but with maturity (5½ years) males become increasingly solitary as they continue growing bigger and more dominant. The isolation may result from avoidance of superiors by smaller males, whose chances of mating in such company are nil (3).

Mixed groups of nyalas are either fortuitous or explained by the presence of an estrous female, in which case the biggest male around ends up tending her. Groups of up to 30 (occasionally up to 100) aggregate at water holes, fruiting trees, or a greenflush, but the different units ignore one another and show no cohesion. Males check out females whenever they encounter them, but failing to find any in heat depart usually within 5 minutes. When a bachelor herd meets a female group the subadult males take it in turn to *urine-test* each adult and subadult female.

ACTIVITY. Monitoring of radio-collared nyalas in Hluhluwe revealed a period of minimum activity between midnight and daylight (1). In the hot spring and summer months nyalas stay in cover between 1000 and 1400 h, typically standing motionless beside a tree and practically invisible. In winter they may rest in light shade in the open, interrupting the siesta to drink, usually before 1300.

SOCIAL BEHAVIOR
COMMUNICATION

Vocal Communication. Unlike the sitatunga and bushbuck, nyala males do not advertise their presence by ostentatious barking. (But they advertise visually by positioning themselves on the edge of clearings [author's observations].) This species is essentially silent. The bark is strictly an alarm call and any nyalas out in the open react with instant flight. Both calves and grownups bleat in distress (e.g., a male caught by a lion [1]), and females "click" to calves and during courtship.

AGONISTIC BEHAVIOR

Dominance/Threat Displays: *lateral presentation; stiff-legged approach; ground- and object-horning; medial-horn presentation.*

Defensive/Submissive Displays: *head-low/chin-out posture, turning away.*

Displacement Activities. Performed by inferior male during aggressive interactions: *horn-sweeping, shoulder-grooming; feeding; head-flagging* while *backing away* from confrontation.

Fighting: *horn-pressing, clash-fighting, front-pressing.*

Aggressive nyala bulls dig and toss the soil with their horns (fig. 9.2), preferably in soft, wet places, and horn and rub their foreheads on bushes. The spectacular *lateral presentation* display (figs. 2.7, 9.7) is nearly always decisive in intermale encounters, without recourse to other displays or combat. In fact, relations between males are notably pacific except when an estrous female is present—then they display very intensively and the biggest male wins. Rather than competing for rank-dominance status on all occasions, nyalas and perhaps other tragelaphines compete only when it counts: when access to a female in heat is at stake (3).

The male nyala's hair fringe and dorsal crest increase its apparent surface area by up to 40% during displays. Horn size is comparatively unimportant. *Lateral presentation* is performed at different intensities, depending on distance and relative status. At sight of a distant male, a bull may half-erect his crest without other overt response. Or he may approach in a special way, stiffly erect, lifting each leg high and slowly so that the orange lower legs become conspicuous, with dorsal crest fully erected. At higher intensity the bull moves slower, the head is lowered to horizontal, and the tail is half-raised (fig. 2.7). At highest intensity movement stops except to maintain the broadside orientation, and the tail is draped over the bull's rump, fully fluffed, so that the maximum area of white is displayed toward the opponent. Males that perform this all-out display always win (3). The loser lowers his crest, turns his head away, and may perform the above displacement behaviors, or even put on an alarm display.

The few fights that have been seen were notably fierce and without preliminary displays. Many adult males bear large scars on their necks and one bull was killed by a stab through the skull base (3). Sparring, by contrast, is a genteel exercise in which 2 equally matched males carefully bring their horns together and push until 1 disengages by sidestepping. Females sometimes butt each other in the side or flank.

REPRODUCTION. Nyalas breed year-round, with 2 conception peaks in autumn and spring. Gestation is 7 months and females come into estrus within a week of parturition, but usually conceive only on the third cycle (at 3-week intervals) (2).

SEXUAL BEHAVIOR: *lowstretch, sniffing tail, urine-testing, inguinal nuzzling/lifting, clicking, neck-pressing.*

Males can check female reproductive status without *urine-testing,* simply by putting the nose to the base of the tail (not the vulva) (3). *Urination* and *urine-testing* are seen only (or primarily) when a female is in estrus. A female remains in estrus for 2 days, during which males are sexually attracted, but the female will only stand for copulation in the final 6 hours.

Though protracted, courtship is simple: the male follows closely in *lowstretch,* sniffs the female's vulva, tests her urine whenever she urinates, and pushes his head between her hindlegs (*inguinal nuzzling*), and sometimes lifts her hindquarters off the ground with his muzzle. Whenever she stops, the male gets ready to mount, so the female must keep moving until ready to mate. Apart from walking away, females respond to the male's following by adopting the *head-low posture* (fig. 9.8), and may give clicks inaudible beyond 5 m (cf. bushbuck). When almost ready, the female stands until the male starts to mount, then moves. To counter this, the male

Fig. 9.8. Nyala *mating march:* female in *head-low posture,* male following nose-to-tail.

moves alongside and with his neck over her withers *presses her neck* into the submissive position. If she submits, he then aligns his hindquarters and mounts, while continuing to *neck-press* her. Standing in the submissive attitude elicits the mounting response, as revealed when estrous females are at water: when they lower their heads to drink, the attending male attempts to copulate.

PARENT/OFFSPRING BEHAVIOR. Birth has only been observed in captivity (3). The mother cleaned up all the birth fluids, even eating straw where she had lain, and spent 20 minutes licking the calf. The baby began trying to rise after 20 minutes, but only succeeded at 74 minutes, began seeking the udder at 120 minutes and succeeded at 199 minutes. Calves stay hidden 10–18 days. During estrus, females are not accompanied by their calves, suggesting that courtship may be a period when calves are unusually vulnerable. It is also a time when adolescent males finally become separated from their mothers through the agency of courting, aggressive adult males.

ANTIPREDATOR BEHAVIOR. See tribal introduction.

References

1. Anderson 1978. 2. ——— 1979. 3. ——— 1980. 4. Pocock 1918. 5. Tello and Van Gelder 1975.

Added Reference

Anderson 1985.

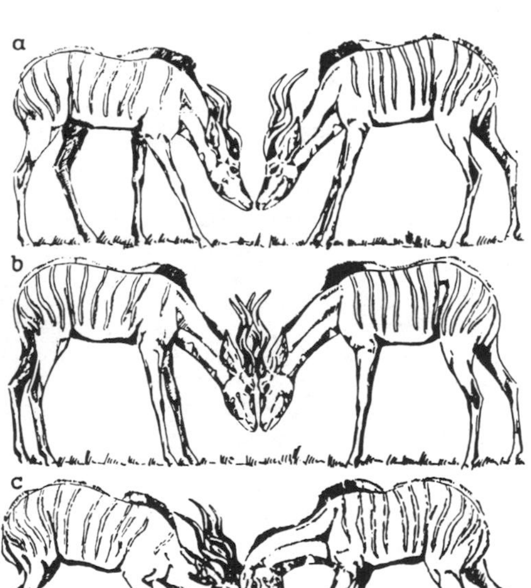

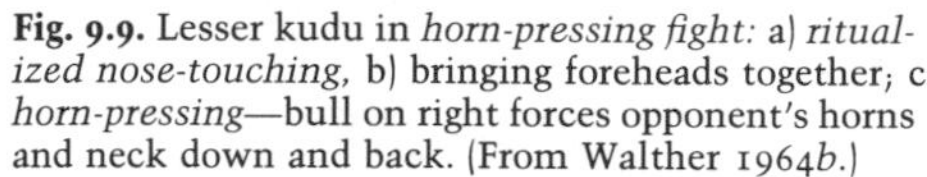

Fig. 9.9. Lesser kudu in *horn-pressing fight:* a) *ritualized nose-touching,* b) bringing foreheads together; c) *horn-pressing*—bull on right forces opponent's horns and neck down and back. (From Walther 1964*b*.)

Lesser Kudu

Tragelaphus imberbis

TRAITS. Conformation and corkscrew horns similar to the greater kudu's. *Height and weight:* males 95–105 cm, 92–108 kg; females 90–100 cm, 56–70 kg. *Coloration:* females and young bright red-brown, males blue-gray, darkening with age; well-defined white stripes (11–15) but few if any torso spots; large, distinct throat and chest patches, incomplete nose chevron and 2 cheek spots; bushy tail with black tip and white underside; legs tawny with black and white patches, males with short white dorsal crest but no throat beard. *Horns:* males only, 50–70 cm in straight line (60–90 along outer curve), 2½ (rarely 3) spirals. *Scent glands:* false-hoof and inguinal glands.

DISTRIBUTION. Endemic to the Somali-Masai Arid Zone of Ethiopia, Somalia, Kenya, and Tanzania, lesser kudus are

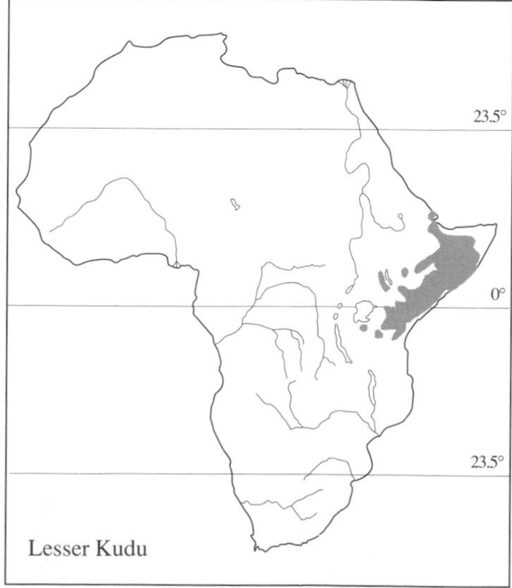

found below 1200 m, their distribution corresponding with the level, monotonous *Acacia-Commiphora* bush country (*nyika*).

ECOLOGY. Lesser kudus depend on thickets for security and are seldom found in open or scattered bush. They are almost as pure browsers as the associated gerenuk, feeding on leaves of dominant trees and shrubs, also on pods, seeds, fruits, vines, and sometimes a little green grass. The kudus of Tsavo N.P. were found to eat 118 different species (1). Unlike gerenuks, lesser kudus seldom browse acacias and do not stand on their hindlegs to feed. They are water-independent, eat succulents such as wild sisal in the dry season, but sometimes drink when water is available (1, 2).

SOCIAL ORGANIZATION: gregarious (bonds between adult females), nonterritorial, sedentary.

The basic social units are composed of 1–3 females with their offspring, which in Tsavo N.P. reside within home ranges of 210 ha (58–505) (3). Female associations tend to be exclusive and long-lasting, which is unusual in tragelaphines. Three known females stayed together for at least 4–5 years. Individual bonds may be based on kinship (mothers and daughters). However, there are no indications of consistent leadership, no rank hierarchy, and no mutual grooming (2).

Males. Male offspring remain in the maternal unit until 1½ years old, when developing secondary characters (horns longer than ears, darker color, dorsal crest) cause them to separate; whether willingly or because of harassment by adult males is unclear (2). Subadult males often associate in pairs, but casually, with rotating companions; as they mature at 4–5, they become solitary, avoiding one another even though 4–5 bulls may share much the same home range and have no exclusive core areas. No evidence of an established rank hierarchy has been noted in associated males. Male home ranges are also relatively small, averaging 230 ha (55–540), whereas subadult males, the most unsettled class, range up to 670 ha (average 300) (2). Males do not advertise themselves by barking.

Where the range includes riverine thickets with evergreen vegetation, lesser kudus concentrate there in the dry season, and disperse somewhat into deciduous scrub in the rains. However, group size is larger in the rainy season. Population density is rarely more than 1 lesser kudu/km^2.

Subadult and adult males associate with females only for reproduction, or when numbers of kudus are attracted to the same place. (The record in Tsavo N.P. was 24 head.) Female groups do not avoid one another and in fact often join, but sooner or later separate again. This antelope rarely associates with other animals, except when drawn to the same food source (e.g., fallen fruits).

ACTIVITY. Lesser kudus spend about 35% of the daylight hours feeding, 36% standing and lying, and 29% moving (3). As usual, activity peaks occur early and late in the day, and the warmest hours are spent resting and ruminating. Resting kudus spend hours standing in thickets, blending so well that only the flick of an ear or tail betrays their presence. They are probably active at night, especially in the dry season, but data are lacking.

POSTURES AND LOCOMOTION. Thinnest of the tragelaphines, *T. imberbis* is adapted to move easily through its dense thornbush habitat. Like the greater kudu, it runs at a bounding gallop and is an accomplished high-jumper, soaring over bushes up to 2 m high (see under Antipredator Behavior). (More on tragelaphine postures and locomotion in tribal introduction [2].)

SOCIAL BEHAVIOR

COMMUNICATION. See tribal introduction and below under Agonistic, Sexual, and Antipredator Behavior.

AGONISTIC BEHAVIOR

Dominance/Threat Displays: *bristling* of dorsal crest and tail, *lateral presentation* with head up; *stiff-legged approach; feinted attack; chasing; vegetation horning; medial-horn presentation.* Female: *symbolic poking; nose-in-the-air; butting; feinted attack; chasing.*

Defensive/Submissive Displays: *head-low/chin-out posture; looking away; social licking.*

Fighting: *horn-pressing, front-pressing/* wrestling with interlocked horns.

Very few aggressive interactions and no serious fights were observed between adult males (2). Bulls sometimes react aggressively toward subadult males, especially in the presence of females. The reaction ranges from *bristling* the dorsal crest and raising the tail to show the white scut to *stiff-legged approach, horning bushes* (very common), *feinted attack,* and *chasing.* The subadults respond by evasion (sidestepping, dodging behind obstacles) or running away. *Lateral presentation* and the other listed behaviors have been seen in captive animals (4).

In a *horn-pressing* bout between a sparring captive adult and subadult male, the adult forced the smaller animal's horns back onto his shoulders and his forequarters to the ground (fig. 9.9) (4). The corkscrew horns can become inextricably locked together in serious fights, dooming the combatants to perish miserably (2) (see greater kudu account, fig. 9.11).

REPRODUCTION. Assuming the Tsavo N.P. population is representative of the species, there is no set breeding season (2). Births may occur in any month, despite a higher mortality of calves born during the dry season. Gestation is estimated at 7½–8 months, and birth intervals of 8–10 months are normal for females in good condition.

SEXUAL BEHAVIOR: approaching in *lowstretch, sniffing vulva, cheek-on-rump rubbing, urination/urine-testing, cheek-on-neck rubbing, inguinal nuzzling, neck pressing, whining.*

Males check every female they encounter in the usual way: approaching in *lowstretch, sniffing the vulva,* and *rubbing the cheek* on the female's rump. Females usually respond by *urinating,* whereupon males *urine-test.* Having located a female in estrus, the male follows closely, rubs his cheek on her rump, and attempts to draw alongside and rub his cheek on her chest, neck, and head. The female responds by assuming the *head-low posture,* or she may *poke* and even *butt* his shoulder. Meanwhile the male *whines* and appears to be gasping for breath. Preparatory to mounting, he lays his neck on her haunches.

PARENT/OFFSPRING BEHAVIOR. Females are known to isolate before calving and to remain alone for some days afterward. The calf (birth weight 4–7½ kg [2]) lies out for probably 2 weeks, during which the mother retrieves it only for suckling, cleaning, and short activity bouts. She probably calls it from a short distance, and the calf may bleat to the mother. The calf's identity is checked by sniffing its rump or neck before allowing it to nurse; suckling bouts may last up to 8 minutes during the first month. Mothers sometimes butt calves that persistently try to resume nursing after the mother breaks the contact.

ANTIPREDATOR BEHAVIOR: *freezing* in *alert posture, alarm bark, distress bleat,* flight.

A lesser kudu standing in cover reacts to an approaching predator by becoming motionless (*freezing*), facing the danger with ears cocked in the *alert posture.* If detected, the kudu takes sudden flight, often after venting a single sharp bark like a bushbuck's. Tail curled over the back, it gallops with frequent high bounds and often leaps over bushes, showing remarkable ability to avoid obstacles and to disappear. Captured animals bleat loudly.

References

1. Leuthold, W. 1971c. 2. —— 1979.
3. Mitchell 1977. 4. Walther 1964b.

Fig. 9.10. Greater kudus fighting.

Greater Kudu

Tragelaphus strepsiceros

TRAITS. The tallest antelope after the eland, with the longest horns. *Height and weight:* males 135 cm (122–150), 257 kg (190–315); females 121 cm (100–140), 170 kg (120–215) (1). *Horns:* largest of any antelope, up to 180 cm, average 120 cm along outer curve. Huge, cupped ears. *Coloration:* reddish brown to blue-gray (darkest in South Africa); males darken with age; 6–10 stripes, prominent nose chevron and small cheek spots, no throat or chest patches; dark leg garters, black-tipped tail with white underside; dorsal crest in both sexes, beard in males only. *Scent glands:* false-hoof glands in hindfeet.

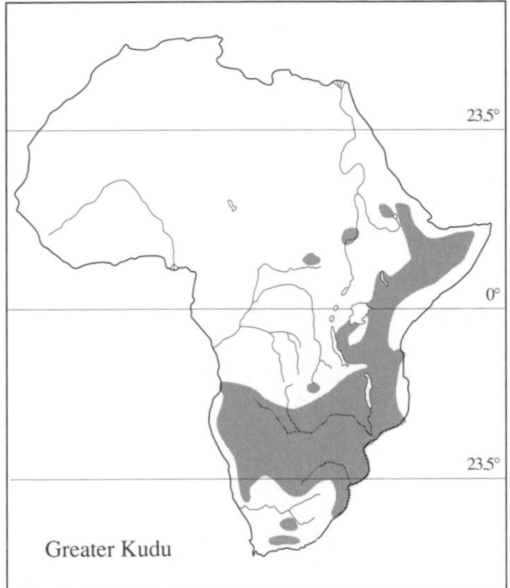

DISTRIBUTION. Wide but patchy distribution in southern and eastern Africa, in different biomes that afford bush and thicket habitat. The greater kudu is one of the few large mammals that are capable of surviving in settled country. So long as adequate cover remains, it is wonderfully adept at concealment, emerging to feed (often in cultivation) only at night. It has actually reclaimed much of its former South African range after being decimated in the great rinderpest epidemics early in the century. But in East Africa the greater kudu's lowland habitat has been largely eliminated and it survives only as isolated populations on some mountains, ranging to elevations of 2450 m (6).

ECOLOGY. A nearly pure browser, the greater kudu feeds on leaves of many kinds, herbs, fallen fruits, succulents, vines, tubers, flowers, and a little new grass (9). It drinks in the dry season but can also subsist in waterless regions. In some areas this antelope makes wider seasonal movements than other *Tragelaphus* species, dispersing in deciduous woodland in the rains and concentrating in ecotones (riverine, hillside base) where the richest, most varied vegetation is found in the dry season (2).

SOCIAL ORGANIZATION: gregarious (lasting social bonds between females), nonterritorial, sedentary.

The typical herd includes 1–3 females and their offspring. In the Zimbabwe Lowveld, mean herd size varied seasonally, being smaller (less than 2) early and late in the dry season, and larger (3–4) early in the rainy season (9). Similar seasonal variation was found in the Loskop Dam Nature Reserve, Tranvaal, where mean herd size in a population of c. 200 kudus was 4 head (7). In Kruger N.P., cow herds in two study areas averaged 5–6 (range 2–15), and seemed to be based on continuing associations between the same cows. There was no obvious rank hierarchy. Two herds had ranges of 360 and 520 ha. Mergers of herds were common but transitory; the largest groups numbered 20–30 animals (1, 2).

Males associate in transient bachelor groups of 2–10 head which may include more than one adult after the annual mating peak, suggesting that greater kudu bulls are more tolerant of one another than are lesser kudus, at least part of the year. Males with overlapping home ranges associate often enough to know one another, but the existence of an established male rank hierarchy remains to be confirmed. Bulls scatter widely and often singly during the rains but some at least return to the same core area every year after a 4–5 month absence (1, 2). A bull's home range (11 km^2 in the case of two radio-collared individuals) may include the ranges of 2–3 female herds, but during the rains bulls disperse away from cow groups, with which they associate only during the breeding season.

ACTIVITY. Two groups of adult female kudus studied in Kruger N.P. spent 50%–58% of a 24-hour day foraging (year-round average), 45% of which was done at night. Around 73% of active periods was spent feeding, 14% moving, 11% standing alert, and 1% grooming, excreting, and in social activities (2). Even though food abundance declined by 75% in the dry season, the kudu's rate of food intake fell only 6% (from 70% to 64%). The animals spent more time moving (22% versus 15%) but compensated by reducing from 15% to 9% the time spent standing alert and in comfort and social activities.

SOCIAL BEHAVIOR
COMMUNICATION

Vocal Communication. The kudu's gruff bark is one of the loudest noises antelopes make (3). Males also grunt loudly when rutting and fighting (5), and make a suppressed whine during courtship (8). A hum that becomes a loud moo when the mouth is opened has been noted in captive cows and a young male. Mothers summon calves with a smacking sound, and calves utter a *u-u-u* distress call.

Fig. 9.11. Locked horns of two kudus that died in combat (from a photo of the author's).

AGONISTIC BEHAVIOR

Dominance/Threat Displays: *lateral presentation* with head up (sometimes *turned away*) or down, back arched, with horns presented; *head-tossing, ground-* and *vegetation horning.* Females: *butting, poking, symbolic biting, neck-twining, feinted attack* (8).

Defensive/Submissive Displays: *head-low/chin-out posture, neck twining, nose-in-the-air.*

Fighting: *horn-pressing, front-pressing, parallel-fighting*(?).

Most of the listed aggressive patterns have been seen only in captivity, where greater kudus are forced into closer association and with different companions than usual in the wild, where aggressive interactions occur at very low frequency. Even in the presence of an estrous female the chances of 2 equally matched, mature bulls meeting and fighting are very low. What usually happens is that 2 bulls of unequal size meet, the bigger bull displays dominance, and the smaller one withdraws. Should he attempt to join a defended herd, the bigger defender chases him out. In short, unless equals are involved, there is no contest. Sometimes a displaying bull tosses his head and flashes his tongue (1, 2).

The fact that kudus do fight is proven by finding sets of interlocked horns. Those in figure 9.11 were found and photographed in Victoria Falls N.P. Conceivably, horns so ideally designed to mesh and bind could become wedged during sparring, either by force or if fitted together just so—like a ring puzzle. However, if most "wedged sets" proved to be adult, like those shown, it would suggest jamming usually happens during hard fights. *Parallel-fighting,* suspected from zoo observations, could readily lead to inextricable locking (fig. 9.11) (8). Forking or crossing horns and wrestling head-on would also seem riskier for kudus than for smaller-horned tragelaphines.

REPRODUCTION. The greater kudu is a seasonal breeder in southern Africa; in equatorial regions it tends to calve during the rainy season and to mate near or after the end of the rains. Well-nourished females can breed as early as 2 years while still immature; they mature at 3, males at 5 years (1, 4, 7, 9).

SEXUAL BEHAVIOR: approaching in *lowstretch; sniffing vulva, urination/urine-testing;* dominance displays, *cheek-on-rump* or *-shoulder rubbing, mutual grooming, head-resting, neck-pressing, whining, clucking,* or *grunting.* Female: aggressive and defensive/submissive displays.

In greater kudu courtship, the tragelaphine patterns that are emphasized include *moving alongside, neck-pressing* (fig. 9.12), and (assuming captive behavior is representative) *lateral presentation* to stop and/or impress unreceptive females, female defensive threats, and *mutual licking/nibbling.* Despite the male's performance of aggressive displays, bulls are inhibited from actually abusing females, even when cows butt them (8).

PARENT/OFFSPRING BEHAVIOR. Calves are born (after 9-month gestation) when the grass is high. They stay hidden for at least 2 weeks before joining the herd, and continue to lie out, at least by night, for 4–5 weeks. A newborn calf stood first at 45 minutes and found the udder soon afterward. The mother stayed with it, licked it all over, and suckled upon demand. Later she would suckle 6.8 minutes (range 1–18). She stopped licking the calf anogenitally and consuming its wastes at 18 days, when the calf's lying-out habit began to weaken.

Fig. 9.12. Greater kudu courtship: male *neck-pressing* female (from Walther 1966*b*).

The mother made a smacking sound to summon it and could also retrieve it by lowering and raising her neck several times (1, 8). Cows other than the mother also lick calves, but only in the areas reserved for social licking; calves respond by empty licking or by licking themselves. Young kudus are nutritionally independent and weaned by 6 months but remain in the maternal herd up to at least 1½–2 years (males) or longer (females) (1, 2).

ANTIPREDATOR BEHAVIOR. A kudu standing in a thicket will rely on crypsis to the point of letting a person come within 12 m before fleeing (4). Alternatively, a kudu may try to sneak away from a disturbance. During flight, when obstacles up to 2½ m high are easily cleared, the tail is rolled up to display the white scut, and bulls raise their chins high so that their horns lie at shoulder level.

References
1. Kingdon 1982. 2. Owen-Smith 1979. 3. Shortridge 1934. 4. Simpson 1972. 5. Stevenson-Hamilton 1947. 6. Stewart and Stewart 1963. 7. Underwood 1978. 8. Walther 1964*b*. 9. Wilson 1965.

Added Reference
Owen-Smith 1984*a*.

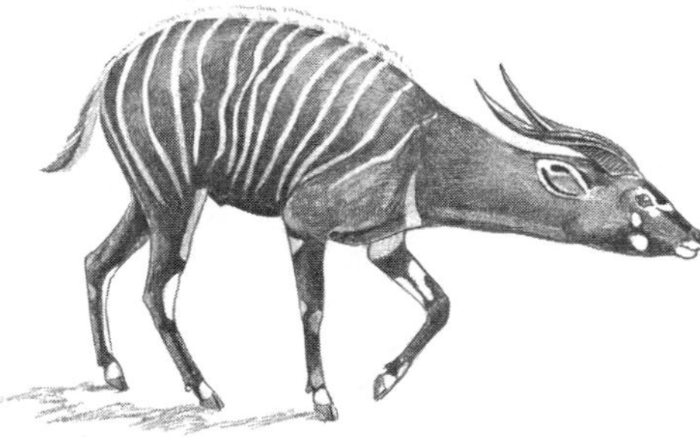

Fig. 9.13. Bongo male in *overstretched posture* (redrawn from Hamann 1979).

Bongo
Tragelaphus (Boocercus) euryceros

TRAITS. Built like an oversized bushbuck; the largest, most colorful of forest antelopes. *Height and weight:* 122–128 cm; males 300 kg (240–405), females 240 kg (210–253) (3). *Horns:* in both sexes, 1 spiral, average 75 cm up to 99 cm, ivory-tipped, male horns massive, female's often as long but thinner, more parallel. Ears enormous, tail hock-length, thin with terminal tuft. *Coloration:* bright chestnut, darker on head, neck, chest, belly, and legs; old males nearly black. Markings clearly defined, 12–14 stripes continuing into black-and-white dorsal crest (cf. zebra mane), large white chest crescent and cheek patches, ears edged with white, a broad nose chevron, legs banded black and white. Calves similar but lighter colored with dark areas reduced and bushy tail. *Scent glands:* none described.

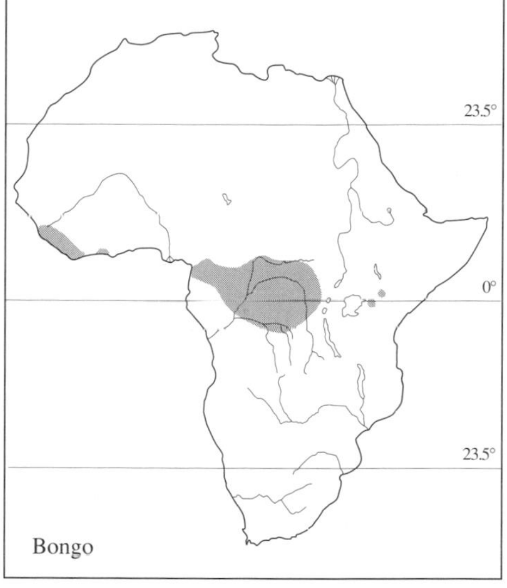

DISTRIBUTION. The bongo has a discontinuous distribution in the Lowland Rain Forest of West Africa and the Congo Basin to the Central African Republic and Southern Sudan. There are also several isolated populations in the montane forests of Kenya, of which that of Mt. Kenya's Aberdare Forest, estimated at over 500, is by far the largest; the others have drastically declined (3). Overnight visitors to the Aberdare's game lodges often see bongos visiting the salt licks.

ECOLOGY. Canopy rain forest has too little vegetation at ground level to be good bongo habitat (2). This antelope (along with other large forest ungulates such as the buffalo, forest hog, okapi) depends on openings in the forest which let the sunlight in and support dense growths of bushes, herbs, creepers, and bamboo. These provide both food and cover for the bongo. Accordingly, this species can thrive in areas where the forest is regenerating following logging, cultivation (2), or heavy elephant damage. Bongos are selective browsers on high-protein vegetation, including leaves, flowers, and twigs of various shrubs and vines, thistles, and flowering tops of rank succulents. They also raid plantations and visit deserted gardens to feed on coco, yams, cassava, and sweet-potato leaves. The long, mobile tongue is used as a feeding tool and the horns are sometimes employed to break high branches (3). Kenya's bongos range between 2100 and 3100 m, from the *Hagenia/Podocarpus* cloud forest through the bamboo to the moorland zone. The Aberdare population roams quite widelv (home ranges undetermined but may be over 100 km^2 [3]), moving up and down the mountain seasonally: up in the dry seasons (January–March, June–August), and down to the lower cloud forest in the rainy seasons (April–May, September–December).

SOCIAL ORGANIZATION: gregarious, nonterritorial, seasonal movements in montane habitat.

Next to the eland, the bongo forms the largest herds of any tragelaphine. Herds of up to 50 have been recorded in Kenya's Aberdares (3), and up to 44 in southwestern Sudan (2). Yet the mean group size of 3500 Aberdare bongos seen at The Ark in 13 years was only 2, and groups of a dozen or more are considered big (3). In the Sudan, herds of over 5 or 6 bongos almost invariably contained young of the year, whereas smaller herds usually contained only adults (2). The average of 23 groups containing calves was 8 in the Aberdares; the largest groups form several months after the peak calving period (July–September), when cows with calves past the hiding stage band together in herds of 9 or more (cf. eland). Climatic conditions also affect bongo grouping patterns. The bongos in montane habitat concentrate at higher levels during the February/March dry-season peak, then descend to lower levels and disperse during the long wet season of cold rain and mist (3).

In contrast to female bongos, adult males are often solitary. However, 6 of 18 Sudan herds that contained adults of both sexes included 2 or more adult males (2). During the October–January mating peak in the Aberdares, nursery herds are often accompanied or trailed by a black bull. Here adult males are thought to avoid one another (3).

Female herds appear to be led by senior cows. Thus, before a herd emerges from cover to visit The Ark's salt-lick glade, a female often scouts the scene from the edge, withdraws, then reappears leading a file of cows and young. Aggressive interactions and *social grooming* between cows have been observed, an indication of a rank order and that individuals associate together for some time. In the Sudan study, 3 known females were always seen together but in groups of variable composition, suggesting that groups are short-term aggregations of smaller, more stable units, as in the eland and buffalo (2).

ACTIVITY. The fact that bongos visit the Aberdare lodges' salt licks almost exclusively at night proves that these animals are night-active but not that they are nocturnal, since most members of the genus prefer to emerge from cover at night. Though they stay in cover during the day, bongos move about and feed early and late, resting and ruminating between about 1030 and 1600 h in thickets. In bad weather they remain inactive in dense cover, but emerge to feed once it stops raining (3). Most activity among the bongos observed in Sudan was recorded from dusk to 2 hours after dawn. The available evidence suggests the bongo is mainly nocturnal, with activity peaks around dawn and dusk.

POSTURES AND LOCOMOTION. The bongo's rather massive build and short legs are adapted to movement through dense vegetation. It might be a clumsy, short-winded runner in open terrain, but it has the ability to disappear like magic in the forest. It skulks away silently from a disturbance, or bolts suddenly from hiding, gallops into dense cover, nose out and horns laid back, and stands still again. Bongos in zoos have been seen to *style-trot,* and to jump gates or other animals 1½ m high (4), although wild bongos are noted for going under rather than over obstacles (5).

SOCIAL BEHAVIOR
COMMUNICATION

Vocal Communication. A very quiet animal, the wild bongo gives very few calls: a bleat of variable intensity, given by calves and courting bulls, and a loud distress bleat or moo. Bulls have also been known to grunt and snort, and in captivity they click their tongues, with mouths closed, during courtship (1).

AGONISTIC BEHAVIOR

Dominance/Threat Displays: *lateral presentation* with head up, *slow walk, medial-horn presentation, ground-* and *object-horning*/rubbing; *branch-breaking.* Calves: *neck-fighting, neck-rubbing, pawing.*

Defensive/Submissive Displays: *head-shaking/throwing, nose-in-the-air, neck-twining, head-low posture, kneeling, lying, social-licking, symbolic poking* and *snapping.*

Fighting: *neck-fighting, front-pressing.* Few observations of bongo aggression have been reported. The main dominance display is *lateral presentation* with head up, often combined with the *slow, stiff-legged walk. Medial-horn presentation* may be commoner between females than males (since adult males avoid contact). Males and sometimes females have the ground-horning/digging habit; they also thrash and rub trees, and break branches, as displays of aggression and not only to reach food (1).

Fighting is rare and probably dangerous: a male died from a punctured lung at Treetops Hotel, and a captured adult male was found to have broken ribs and old horn wounds in the shoulder and chest (3).

REPRODUCTION. Whether the calving peak described in the Aberdare Mountains (see under Social Organization) applies to other, lowland populations remains to be seen. A zoo-reared female first came into heat at 20 months and cycled thereafter at 3-week intervals, staying in estrus for 3 days. Gestation takes 9 months (4, 7).

SEXUAL BEHAVIOR: dominance displays; *lowstretch, stiffly erect posture; head-to-side rub* (both sexes).

Estrous females in a zoo herd often reacted aggressively to the male's initial approach (fig. 9.13), followed by flight (1). A cow about to flee opened her mouth, presented her horns, then abruptly raised her nose and twisted her neck while leaping to one side. The male kept following, pausing occasionally to display his side, with *head turned away.* After several approaches, the cow's escape tactics diminished and the *mating march* began. The bull followed closely, sometimes *drawing alongside,* maintaining *lowstretch,* meanwhile *clicking his tongue* and *licking his lips.* Rather than running, unreceptive females sometimes lay down to escape the bull's attentions. The bull rubbed his cheek and face on the croup of a cow that would stand still, tried to rest his head on her rump, or else stood stiffly erect behind her as both the eland and topi do preparatory to mounting. Females may also buy time by urinating, then moving off while the bull is *urine-testing;* a cow ready to mate stands by while the test is carried out. Following copulation, both partners may rub their heads on each other's sides (7).

PARENT/OFFSPRING BEHAVIOR. The fact that the largest herds contain calves suggests that young bongos, like eland calves, form tight crèches, thereby forcing their mothers to associate. Calves lick one another and play together beginning at a young age (in zoos), practicing *neck-fighting, head-shaking/throwing, nose-in-the-air,* and so on (1).

ANTIPREDATOR BEHAVIOR: concealment, *alert posture, head-forward posture,* laying tail over rump, running away, *style-trotting, distress bleating.*

Spotted hyenas and leopards are the bongo's main predators other than man, and lions have been known to kill a bongo on the moorland of Mt. Kenya (3). Pythons have been recorded as calf-predators in West Africa (7). Bongos hunted by dogs come to bay in water, *grunting* and *stamping* defiance. A mother has been seen charging a hyena and another counterattacked a charging forest hog (3). Bongos in herds are said to lose much of the timidity that is characteristic of bongos alone or in pairs (3).

References

1. Hamann 1979. 2. Hillman 1986. 3. Kingdon 1982. 4. Reuther 1964. 5. Roosevelt and Heller 1914. 6. Verschuren 1958. 7. Xanten, Collins, Connery 1973.

Fig. 9.14. Eland high-jumping.

Eland

Tragelaphus (Taurotragus) oryx

TRAITS. The largest, most oxlike antelope. *Height and weight:* males 163 cm (151–183), average 500–600 kg (450–942), females 142 cm (125–153), 340–445 kg (317–470). *Horns:* both sexes, 1–2 tight spirals, males' far thicker but usually shorter, 54 cm (43–67), females 60.5 cm (51–69.6). Head short and comparatively small, ears small and narrow; a prominent dewlap. Tail hock-length, cowlike with black terminal tuft. *Coloration:* individually and geographically variable, general color tawny, darkening with age, the young reddish brown. Markings (more pronounced in northern race, *T. o. pattersonianus,* fading going south, faint or absent in the South African race, *T. o. oryx*): 10–16 white stripes, a dark dorsal crest, white markings on legs, and black garters on upper forelegs and around hooves. *Scent glands:* false hoof glands (fig. 9.1) and possible apocrine glands under forehead tuft (1).

Sexual Dimorphism. Adult males develop a tuft of dark hair on the forehead and nose, a massive neck, an enlarged dewlap, and a short, curly neck mane. They turn blue-gray with age, as the dark skin shows through a thinning coat, but may also change seasonally (1).

DISTRIBUTION. The common eland ranges the Somali-Masai Arid Zone from Southern Sudan to Tanzania, the Southern Savanna, and most of the Southwest Arid Zone. It was eliminated long ago from most of South Africa. Because it is highly susceptible to rinderpest and intolerant of human settlement, its range has been greatly reduced in other parts of Africa as well (5, 6).

RELATIVES. *Tragelaphus derbianus,* found west of the White Nile in the Northern Savanna, has longer, more massive horns (regularly 1 m and up to 1.2 m), because of which (and not greater body size) it is also called the giant eland (cf. giant sable). It is also a richer color with more pronounced markings, and has a prominent neck ruff, large, round ears, and a dewlap that begins on the chin instead of at the throat. The paler coloration, less conspicuous markings, lengthened snout, and broader hooves of the common eland are probably adaptations for a more arid environment, making this the most specialized of all the tragelaphines. The Derby eland is adapted to the broad-leafed, deciduous *Isoberlinia* woodlands.

Though the eland is so bovinesque that the Masai and other pastoralists consider it fit to eat, the common ancestors of elands and cattle separated millions of years ago. Hybridization attempts have always proved infertile, though mating has been accomplished on several occasions (7).

ECOLOGY. The eland is one of the most adaptable of antelopes, equally at home

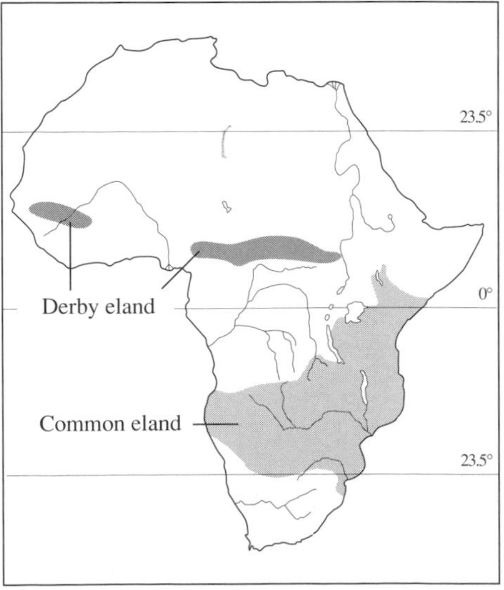

in subdesert, acacia savanna, *miombo* woodland, floodplain, and mountains up to 4600 m (1). It does not venture into the desert and avoids dense forest. This broad habitat tolerance is reflected in an extremely varied diet consisting primarily of browse, but also including fruits, pods, seeds, herbs, tubers, and grasses when green and tender (50%–80% of wet-season diet [2]). Food is gathered by the lips, not with the tongue. Feeding elands sometimes use their horns to pull down and break branches whose leaves and fruits would otherwise be out of reach (2, 5).

Although the eland is less desert-adapted than the addax or oryx and some of the gazelles, it can go indefinitely without drinking. Physiological studies have shown that water-deprived elands allow their body temperature to rise as much as 7°C during the day, letting the cooler nighttime temperature cool them down again (8, 9, 10). Elands thereby avoid sacrificing water for evaporative cooling. Their great bulk keeps their temperature from rising as quickly as a gazelle's or even an oryx's, and they can store more heat: a 500-kg animal would need to evaporate c. 5 liters of water to dissipate the number of kilocalories it can store with a 7° rise in body temperature (8, 10). Water loss is further reduced by concentrating the urine, excreting dry feces, lowering the metabolic rate, breathing slower and deeper, and by seeking shade in the heat of the day. In all these respects the eland is less efficient than the desert antelopes. Nevertheless, by feeding at night and early in the morning, an eland can theoretically gain all the water it needs even during a severe drought (9).

SOCIAL ORGANIZATION: gregarious, nonterritorial, nomadic.

The eland is one of the most mobile antelopes. Females of Nairobi's Athi-Kapiti Plains ecosystem had home ranges of at least 174 up to 422 km^2 (2). This species also forms larger herds than most bovids: I have counted groups of up to 500 on the Serengeti Plain, in a still-uncensused population estimated at up to 60,000, and herds of over 100 are common in many areas. Since eland density overall is usually less than 1 per km^2 (2), their distribution is unusually clumped. However, grouping pattern and herd size vary according to habitat and season. Large herds form on the plains during the rainy season and invariably contain numerous calves and juveniles (2). It is not unusual for young to outnumber adults and one may even see herds composed entirely of calves and juveniles. Conversely, unisexual or bisexual groups consisting of adults only are common. This variability reveals the open, fluid nature of eland society (2, 11). Groups and singles meet, mingle, and separate casually—as do other tragelaphines. Herds are entirely open and the only regular associations of individuals are between mothers and their calves.

On the Athi Plains, unisexual adult groups typically numbered 3–5, bisexual groups 10–12, whereas herds containing juveniles and calves averaged 48 (range 2–427) (2). The observed tendency of calves to coalesce into close subgroups whenever separate nursery herds meet may create and maintain such large nursery herds. The mutual attraction between calves forces lactating females to remain together, in attendance, as it were, on their offspring (4).

Young calves may in fact be more attached to one another than to their own mothers (1, 2). They maintain the least social distance (under 1 m), play, often lick one another, and lie in contact. Social distance increases with age, to 4½ m in juveniles and 11 m in adults, indicating that sociability declines (1, 2) (with horn development?). Although members of the same age and sex class continue to associate (except older males), approaches within 2 m usually provoke aggression. Consequently eland herds tend to be spread out except for clusters of young animals.

Adult males are, as usual, the least sociable class. They often travel alone and range a comparatively small area averaging 50 km^2 (14–60) in the Athi Plains population and less with age. Bulls frequent denser cover, and browse more than females. Nursery herds, conversely, frequent the most open habitat where there is the least danger of predator ambush (2).

Perhaps the most unusual feature of eland social organization is cooperative defense of the young (1, 2). Adult females will jointly defend calves against large predators, including lions (more under Antipredator Behavior). This trait makes elands still more like cattle and less like other antelopes (but see gazelle tribal account).

ACTIVITY. The daily activity of the eland is extremely variable, as this highly selective mixed feeder forages widely in search of food to sustain its bulk. In a cool climate, elands alternately eat and ruminate all day in roughly 2-hour bouts. They continue feeding at night until ca. 0200, then rest to 0600. But in a hot climate such as Tsavo N.P. in the dry season, elands may rest in the shade all day and feed the whole night (2). In South Africa, elands introduced to Loskop Dam Nature Reserve exhibited 4 main phases of daylight activity in winter: 1) intensive feeding, ruminating, and walking in the morning; 2) feeding and ruminating for c. 2 hours in the afternoon; 3) walking and browsing followed by a brief period of inactivity; after which 4) they turned around and walked/browsed back the way they had come. In summer, these animals moved and fed in the morning, rested at midday, and fed intensively in late afternoon (11).

POSTURES AND LOCOMOTION. Theodore Roosevelt found the eland the easiest of all big game to ride down on horseback. "We have rounded up a herd quite as easily as we round up old-style Texas cattle" (6). Elands only gallop when badly frightened (or playing) and if pursued quickly tire. A bull chased by Land Rover to get within darting distance slowed to a trot within 1 km (author's observation). This is the eland's fastest gait under usual conditions; it can trot at a rate of 35 kph for several kilometers, or much further at a slower rate (11). Is it the eland's bulk that makes it slow, or is it simply a tragelaphine trait the eland has been unable to change in adapting to open habitats? Cows are not particularly bulky and certainly calves not at all, yet both are slower than other plains antelopes. The fact that elands, like their closest relation, the kudu, are incredible high-jumpers is also against the bulk argument. Fleeing elands often display their prowess (fig. 9.14) by jumping effortlessly right over a neighbor, and youngsters can sail over a 3 m fence from a standing jump (2).

SOCIAL BEHAVIOR

COMMUNICATION

Tactile Communication. The role of the young as binders of eland society is seen, not only in their cohesiveness and contact-seeking, but also in the attraction of other elands to them. Young calves elicit *social licking* from all other classes, too. Otherwise, social licking is limited, as in other tragelaphines, to situations in which a lower-ranking individual attempts to appease a higher-ranking one.

Vocal Communication. Most of the sounds elands make are either inaudible at any distance or infrequent. There is a mother-calf contact call that resembles a creaking door slowly opened, and a faint bleat that is given by calves and/or mothers when together, a sound also made by courting bulls trying to mount. Elands have a typical tragelaphine *alarm bark.* The most unusual sound elands make is the castanet-like *clicking* produced by mature bulls (rarely and irregularly by old cows). Exactly what it is that clicks remains to be explained, but probably it is a tendon slipping over the carpal (knee) joint or another foot bone (2). The sound can be heard for hundreds of meters on a still night. By advertising the presence of a moving gray male, clicking helps regulate eland social relations. For instance, an immature bull that was following adult females by night moved away at once when clicking heralded the approach of a mature male (1).

Olfactory and Visual Communication. Gray bulls (6 years and up) regularly rub their foreheads and thrash the ground and objects with their horns. While demonstrating to the world at large the performer's power and ferocity this behavior also enhances his odor and changes his appearance. Bulls like to plow up soft earth, mud, and piles of vegetation, and to thrash low bushes and weed patches. They often end up with a crown of hay or weeds (fig. 9.15), or with mud-caked horns. They also use their horns to snap branches and sapling trees up to 4 cm thick as an expression of aggression distinct from the branch-breaking associated with feeding. *Horning* is linked with *rubbing the forehead/nasal brush* in odorous substances. Wet earth elicits this

Fig. 9.15. Mature male a) horning pile of dead grass, b) walking off with a haystack on his horns.

behavior and urine is the preferred wetting agent: their own or another eland's, and even the urine of other mammals (elephant, buffalo, etc.). First the bull noses the ground and usually *lip-curls* in response to urine, then grinds his forehead into the earth, while the horntips rake and dig into the soil. At high intensity he pivots on his head as when fighting and may even lift his hindquarters off the ground. The brushes of shot eland bulls reek of urine and vegetation (they single out aromatic species for thrashing). Young males sometimes approach to sniff and lick the brush of a resting senior bull, exercising great care in the process (cf. impala, waterbuck) (1, 2).

AGONISTIC BEHAVIOR. See tribal introduction and ref. 3.

Dominance/Threat Displays: *glance-threat, head-toss/horn-wipe, medial-* and *low-horn presentation, feinted attack, ground-horning/forehead-rubbing* (see Olfactory Communication). Challenge: *approach with head up ± looking away,* blocking path in *lateral presentation, mounting* by dominant.

Defensive/Submissive Displays: *slow head-shaking, head-low posture, displacement grazing.*

Fighting: *front-pressing, ramming.* Sparring: *neck-wrestling, horn-tangling.*

Most aggression between elands is low-key and one-sided, therefore easily overlooked. A bigger animal supplants a smaller one (e.g., from the shade of a bush) simply by moving toward it; the other moves out of the way. If not, the dominant may resort to a *glance-threat:* raising its head slightly from ground level and looking at the inferior, meanwhile stopping any other activity such as chewing or following a female (2, 3). The intimidated animal then moves away, often giving the *slow head-shake. Mutual head-shaking* may be seen between elands anxious to share a limited resource such as a salt lick without fighting. *Grazing attitude* is a less obvious expression of inoffensiveness, for instance, by an adult bull when a superior bull tending a cow *threat-glances* at him. *Head-tossing ± wiping the horntips* across the withers is a stronger threat. Stronger still is *low-horn presentation,* with horns aimed at the adversary (fig. 2.9); this display can be further intensified by lunging (*feinted attack*). *Chasing* of a defeated adversary is rare in the eland (2). *Ground-horning* is often seen during aggressive interactions; for instance, a big bull urinated, stepped back, and rubbed his forehead in his urine while displaying to a younger bull, ending a sequence that began with *horn-wiping* (the shoulder) and *ground-horning* (author's observations).

The low incidence of fighting between elands, even in the presence of estrous females, has impressed observers (2, 5, 11). The development of male secondary characters and the behavior patterns that advertise status enable elands to assess one another's combat potential quite precisely and to avoid unequal contests. The single most important character is neck development, for the power to lift and to twist an opponent's neck is decisive in eland combat. Horn length is apparently unimportant, but shape may count because wide-apart tips that point forward can pierce the back of an opponent's neck during sparring or combat. The direction of the horntips depends on the stage of horn rotation; prior to the stage when they point forward, the tips face backward, then inward. However, the final shape of eland horns is highly variable, wear reduces their length and sharpness with age, and anyway the skin over the neck is about 15 mm thick (2)!

Neck-wrestling, the supposed holdover from hornless ancestors, can be seen during play fights between juvenile elands. Sparring, including horn contact, is common between subadult and young adult males, especially in early morning. It is initiated by deliberately infringing the social distance, approaching either in *grazing attitude* or in the *alert posture,* or by *mounting.* Surprisingly, lower-ranking males usually take the initiative, even in mounting, which is normally an assertion of dominance (2). Having approached within touching distance, the contestants tuck in their chins and delicately entangle their horns, then push and twist as in real combat. In contests between equals, neither gives ground and the bout ends with the bulls backing away. In unequal contests, the weaker bull is pushed back and withdraws by leaping to one side (1).

Real fights, which can occur when 2 equally matched gray males meet while checking out a herd, are brief but violent. The antagonists charge from 1–2 m apart and ram heads, followed by vigorous pushing and wrestling, during which they attempt to throw each other off balance (2).

REPRODUCTION. Mating and births occur most of the year but with definite birth peaks late in the dry and early in the rainy season. Mating peaks also fall within wet periods, especially when elands aggregate in large nursery herds on the plains.

Fig. 9.16. Male eland *testing* female's urine (redrawn from D. Altmann 1969).

Estrus lasts up to 3 days (2). Most females conceive first at 2½ years (some a year earlier) and gestation is 8–9 months.

SEXUAL BEHAVIOR: *urination/urine-testing; lowstretch* ± *tongue-flicking, head-on-rump resting, head-up posture.*

All elands perform the *urine test* beginning c. 3 months. Adult males, the main practitioners, stimulate females to urinate by *close-following,* and will approach from a distance any cow seen to adopt the distinctive *urinating squat* (fig. 9.16). Males also routinely test one another's urine (2) (cf. wildebeest).

Having located an estrous female, the dominant male on the scene maintains a close *tending bond.* During the precopulatory phase, mounting attempts are preceded and followed by the bull's attempts to rest his head on the cow's rump. If she stops moving, the bull draws himself up in the *erect posture,* meanwhile bleating like a calf. He may also make short runs at the cow in this attitude or in *lowstretch,* with rapid *tongue-flicking* accompanied by copious salivation (2). The cow may respond by *horn-wiping/hooking, ritualized biting, low-horn presentation,* and even *feinted attack.* Instead of retaliating in kind or displaying dominance, the bull behaves submissively: turning away and *slow head-shaking.* The cow may also bolt, in which case the bull tries to stay close, seeking to *draw alongside* (fig. 9.3) and rest his head on her withers. This chasing attracts the attention of other males and often leads to the replacement of a smaller by a bigger bull. When finally ready to mate, the cow is attended by the most dominant male.

PARENT/OFFSPRING BEHAVIOR. Calf crèches are joined by calves within a few days of birth, whereas newborns without access to crèches may lie out for up to 2 weeks. A calf will only go to its mother when summoned to suckle (by calling) and will go no further than 50 m from its peers (2). Juvenile groups often lag behind the herd and sometimes end up entirely alone. Mothers readily go off on their own or in splinter groups for hours, even days at a time (cf. sable). Weaning occurs between 4 and 6 months; thereafter the mother-calf bond is still weaker, helping to explain the detached herds of juveniles containing few if any adults. Juvenile groups of mixed sex persist for 2 years, after which elands form unisexual peer groups (2, 11).

ANTIPREDATOR BEHAVIOR. Eland cows defend their calves against large predators either alone or in concert. Although whole herds do not engage in mobbing, adult elands will not (cannot?) flee even from lions, and females in groups will go on the offensive to defend their calves (2). Whether a lone female will attack a lion, and whether only females with calves cooperate, remains unclear. Cows in herds have been seen to advance on lions and to chase cheetahs (and be chased by them), but they more often ignore predators unless their calves are threatened. However, any creature that startles them by suddenly appearing out of the undergrowth is likely to be attacked including warthogs, jackals, baboons, cattle, and even birds (1).

Apparently the eland has substituted size and cooperative maternal defense for speed to protect itself and its offspring against predation. Elands fear only man and have the longest flight distance (300–500 m) of all African game. Might this be a habit acquired since the advent of firearms, or could it be that hunting peoples are able to run down an eland?

References

1. Hillman 1974. 2. ——— 1979. 3. Kiley-Worthington 1978. 4. Moss 1975. 5. Posselt 1963. 6. Roosevelt and Heller 1914. 7. Skinner 1966. 8. Taylor 1969. 9. ——— 1970*a*. 10. ——— 1970*b*. 11. Underwood 1975.

Added References

Hillman 1987. ——— 1988.

Chapter 10

Buffaloes and Cattle

Tribe Bovini

African species
Syncerus caffer, buffalo

TRIBAL TRAITS. The largest, most massive bovids (up to 1000 kg), heavy-bodied with short, thick legs and neck, broad muzzle with naked, moist muffle. *Horns:* present in both sexes, never ringed, round in cattle (*Bos* and *Bison*), flattened in buffaloes, growing sideways and curved or hooked inward at the tips. *Coloration:* coat generally short and often scant in adults (except bison and yak), brown to black, uniform or with white stockings. Tail long with terminal tuft. Hooves broad with well-developed false hooves and no interdigital cleft. *Sexual dimorphism:* males up to twice as big as females, their bulk further exaggerated by thick neck, hump, dewlap, or hair, males of brown species darker. *Scent glands:* no specialized scent glands but cow-barn smell characteristic of tribe, presumably produced by diffuse oil or sweat glands in the skin. *Mammae:* 4. Dung in the form of cowpats, unlike pellets of other bovids.

ANCESTRY. The latest and most advanced ruminants to evolve, bovines originated in Asia in the lower Pliocene, as an offshoot of the nilgai (Boselaphini) lineage (see family introduction). Most of the 5 genera and 11 existing species are distributed in tropical and subtropical Asia. The African buffalo, which is not closely related to the Asiatic or water buffalo (*Bubalus*), is the only wild representative south of the Sahara. Its origins remain a mystery (4, 7).

ECOLOGY. Bovini are all water-dependent roughage feeders, most of which inhabit wooded or forested country interspersed with glades. Only the American bison was a dweller on open plains—and significantly it was the most numerous of all the bovines, with probably the greatest biomass of any bovid. The African buffalo ranks second among Bovini in range, abundance, and density, although of course greatly outnumbered by domestic cattle, even in Africa. But the latter are still poorly adapted to African conditions and are only able to compete with wild bovids through man's intervention.

A wide incisor row and massive molars enable bovines to harvest and grind coarse grasses in the quantity needed to sustain their bulk. A prehensile tongue assists ingestion by gathering the grass in a bunch, which is then gripped between the incisors and upper palate and ripped off. When grass is long enough to bundle (15 cm or longer), bovines can take extra-big bites. But they are less efficient than smaller herbivores on shorter pastures and cannot feed selectively or effectively on really short grass because of their stiff, relatively immobile lips (2).

Lacking efficient mechanisms for conserving moisture, bovines have to drink regularly. Most species readily enter and buffaloes often live in swamps, conditions for which the broad hooves without interdigital cleft and the prominent false hooves are well-suited. Some bovines have the wallowing habit, though only the water buffalo habitually spends its day submerged.

SOCIAL ORGANIZATION. Bovines are highly gregarious and nonterritorial. Dominance among males is based on size, which increases with age. Males continue growing after maturation, leading to pronounced size dimorphism and separate rankings of the 2 sexes, with males dominant over all females. Mixed herds are formed during the breeding season, but at other seasons the sexes tend to range separately. The African buffalo, which lives in mixed herds, is unusual. It is also less dimorphic

Table 10.1 Comparison of Bovine Agonistic Behaviors

	Buffalo		Bison (*Bison*)		Cattle (*Bos*)				
Behavior	*Syncerus*	*Bubalus*	*bison*	*bonasus*	*grunniens*	*gaurus*	*banteng*	*sauveli*	*taurus*
High-horn presentation (threat)	+	+	+	+	+	+	+	+	+
Lateral display	+	+	+	+	+	+	+	+	+
Rubbing face in earth	+	+	+	+	+	+	+	+	+
Earth tossing	+	+	+	+	+	+	+	+	+
Horning bushes	+	+	+	+	+	+	+	+	+
Pawing	−	+	+	+	−	+	+	+	+
Pushing fight	−[a]	+	+	+	?	+	+	+	+
Aggressive roaring/calling	−	−	+	?	+	+	+	+	+
Rolling in mud/dust	+	+	+	+	+	+	−	−	−
Circling	+	?	−	?	?	+	?	?	−
Wallowing in deep mud	+	+	−	−	−	−	−	−	−
Submissive bellow	+	−	−	−	?	−	−	−	−
Nose-up threat	−	+	−	−	−	−	−	−	−

SOURCE. Sinclair 1977*b*, table 6.
[a]Observed by Sambraus (1969).

than other bovines, with males only 30% heavier than females.

Bovini have evolved a group defense which serves to protect both young and adults (cf. eland). A mobbing attack can be triggered by a distinctive distress call (see buffalo Antipredator Behavior). Because of the group defense, calves need not be hidden or particularly precocious.

POSTURES AND LOCOMOTION. A group defense may be related to the size and formidable horns of bovines: their bulk increases their defensive capability while reducing their ability to outrun predators. Bovines can travel long distances at a walk or slow canter but are not designed to run fast or far.

Rolling (in mud or dust, onto the side and back but usually not completely over, as in equids) is a habit peculiar to the Bovini and to the wildebeest. It appears to be an expression of aggression that is largely limited to mature males (7).

SOCIAL BEHAVIOR

COMMUNICATION. Olfaction is the best developed sense in the Bovini, employed for finding food, for detecting enemies, and in social communication. Vision and hearing are less acute, but visual and vocal signals are nonetheless important in bovine communication (see Agonistic Behavior and buffalo account).

Social Grooming. Social licking is a bovine trait that presumably promotes maternal-offspring bonds and social bonds among herd members. It should not be confused with the *maternal licking* that is designed to stimulate hider calves to eliminate wastes; bovine *social licking* is directed mainly to the head and neck (6). Studies of *social licking* in domestic cattle suggest that it is a basic drive, satisfaction of which is essential to well-being (5). Though not based on any need for actual skin care, the urge to lick and be licked grows stronger the longer it is denied.

AGONISTIC BEHAVIOR

Dominance/Threat Displays. Bovine aggressive displays are listed and compared in table 10.1. The *lateral display,* typically with head high and chin pulled in, is the primary means of asserting dominance in the whole tribe (see fig. 10.4, buffalo account) (7). From this position the displaying animal's height and bulk are shown to maximum effect, enabling opponents to assess each other's combat potential. In addition to the usual *low-* and *medial-horn* threats, aggressive actions include *head-tossing* and *hooking, rubbing the face and chin on the ground* while kneeling (fig. 10.1), *soil-horning/tossing, vegetation-horning, pawing, roaring,* and *rolling* in mud or dust.

Defensive/Submissive Displays. Submission is expressed by lowering the head with chin outstretched (*head-low/chin-out posture*) so that the horns lie back at or below shoulder level, thus eliminating them as part of the animal's silhouette. A bull intimidated by a *lateral display* can withdraw gracefully by *grazing away.* Buffaloes also signify submission by bellowing when threatened or attacked by a dominant animal.

Fighting. The bovine fighting style is

simple, consisting of *front-pressing* and *ramming*, in which weight and brute force count more than technique. There is also some dueling and attempted hooking in efforts to get past an opponent's guard and gore him. The loser of a fight must make good his escape by wheeling and running away quickly enough to avoid being gored in the flank or rump.

REPRODUCTION. See buffalo account.

SEXUAL BEHAVIOR. Courtship and mating are also simple and direct. The account of the buffalo is representative of the tribe.

PARENT/OFFSPRING BEHAVIOR. A 9-month gestation period is characteristic of most Bovini (but cf. the African buffalo). Females of most species are mature by 4 years, males not before 7 years. The group defense and follower calves of the buffalo are representative of the tribe (see species account).

ANTIPREDATOR BEHAVIOR. When disturbed, all bovines assume the same distinctive *alert posture* seen in cattle: the head is raised high, the back is arched (hollowed), and the tail base is lifted in an arc (4, 6) (see fig. 10.5). If the disturbance is visible, the chin is lifted; the animal(s) stares fixedly and may approach for a closer look. At this point the observer—and perhaps the animals—cannot tell whether they will charge or flee. But unless something happens to trigger the mobbing response, bovines almost always run away from danger, with tails raised high, frequently accompanied by spontaneous defecation. Bovines hold their tails in the same way during running games. When put to flight, most species take refuge in dense cover.

References

1. Hafez and Schein 1962. 2. Hofmann 1973. 3. Lott 1974. 4. Mloszewski 1983. 5. Sambraus 1969. 6. Schloeth 1958. 7. Sinclair 1977*b*.

Fig. 10.1. Buffalo bull kneeling and *neck-rubbing*, an aggressive display.

Buffalo

Syncerus caffer

The African buffalo comes in 2 forms, different enough to be classified as separate species if they did not freely hybridize wherever their ranges overlap. The largely unstudied red buffalo, *S. c. nanus*, is a small form adapted to West African forests; the other, more familiar black buffalo, *S. c. caffer*, is the savanna form.

TRAITS. *S. c. caffer*—the bulkiest and most formidable African bovid. *Height and weight:* 135–170 cm (4); 425–870 kg (5), males 686 kg, females 576 kg (10). Head broad with wide muzzle and bare, moist muffle; ears large and round, drooping, fringed with long hair. *Coloration:* adults black or seal brown, coats of old bulls scant with dark brown hide showing through, often with white patterns of grizzled hair on their heads; young calves black or dark brown, smooth-coated, often changing to dirty yellow-brown after several months, then gradually darkening to reddish or chocolate brown. *Horns:* highly variable in size and shape, width up to 100 cm, length along the curve up to 141 cm, female horns much thinner and usually shorter; a massive shield develops across the forehead in males only, the 2 boss halves meeting at maturity (about 7 years). *Scent glands:* none described. *Mammae:* 4.

S. c. nanus—height 97–108 cm and no more than 300 kg (1). Horns average only 30–40 cm in males, curving backward rather like a water buffalo's to make a closed crescent, with massive horn bases but no complete horny shield. *Coloration:* red-brown to brown, males darker (sometimes black) than cows and calves.

DISTRIBUTION. The buffalo is one of the most successful African mammals, in terms of geographical range, abundance, and biomass. It is or was ubiquitous throughout the Northern and Southern Savanna, and also inhabits areas of second growth and clearings within the Lowland Rain Forest. Altitudinally, it ranges from sea level to the limits of forest on the highest mountains. Buffaloes also penetrate the arid biomes wherever there are rivers, lakes, and swamps.

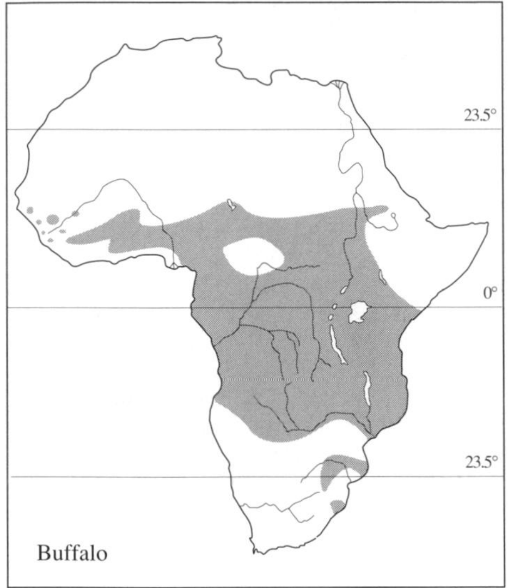

During the great rinderpest epizootic of the 1890s, no animal was harder hit than *Syncerus*. By some estimates, 10,000 died for every 1 that survived (7). After the disease had swept the continent from Ethiopia to South Africa, the survivors proceeded to recolonize their range. Buffaloes are so productive that herds protected in national parks often need culling to avoid habitat degradation caused by overpopulation (6).

ECOLOGY. The buffalo reaches maximum density in the well-watered savannas of western Uganda and eastern Zaire, in the Serengeti ecosystem, in swamps and floodplains, and in the montane grasslands and forests of major mountains. Good buffalo habitat usually affords dense cover such as thickets or reeds, although herds may also live in very open woodland. Herds of several hundred are commonplace and groupings of over 1000 animals are not uncommon (2, 10).

Syncerus is a bulk grazer able to subsist on grasses too tall and coarse for most ruminants, and is less partial to young tender shoots than most grazers. It also browses to some extent, perforce when grass is scarce or of very poor quality, but does some selective browsing (up to 5% of diet) even on good pasture (5). Although it cannot crop as closely and selectively as the zebra, warthog, or wildebeest, for example, its wide incisor row and use of the tongue enable it to consume more grass more quickly. In exercising its preference for tall pasture, the buffalo also plays a pioneering role in the savanna grazing succession, reducing grassland to the height preferred by more selective feeders. Being highly mobile, buffaloes rarely linger on trampled or depleted pasture as long as good stands of grass are still available within their home range (10).

SOCIAL ORGANIZATION: highly gregarious, nonterritorial, large mixed herds with male dominance hierarchy.

Buffaloes are distributed in discrete population units which remain in separate, traditional home ranges. There appears to be minimal interchange between units, for strangers that attempt to join a herd are subjected to continuing harassment and kept on the periphery (10). The nature of the habitat and the primary productivity of the ecosystem largely determine the size of these population units. As usual, larger herds occur in more open habitat. In the *Acacia-Commiphora* woodlands of Serengeti N.P., herds number from about 50 to over 1500 animals, averaging 350; on a broad floodplain in Kafue N.P., the average was 450 (19–2075) buffaloes. In the forests and glades of Mt. Meru (Arusha N.P.), by contrast, mean herd size is 50, and in Zaire's rain forest, 20 or less (5, 10).

Herds probably rarely include all the buffaloes that share the same home range. Sedentary old bulls remain in separate herds the whole year, and bachelor herds of subadult and prime males separate from the main herds during the dry season when no breeding is taking place. When preferred longer grassland is reduced to isolated patches close to water, females and young may also divide into smaller units each based in a different part of the range. Herds tend to be largest during the rains, especially near the end during the peak mating season (10).

The basic units of buffalo herds are stable groups or clans of presumably related cows that are arranged in a nearly linear dominance hierarchy, to which are attached a number of adult and subadult bulls that are also ranked according to age and dominance status (5). Each cow is accompanied by successive offspring up to perhaps 3 years of age, although the subadults tend to associate in subgroups. For instance, a smallish herd of 139 buffaloes included 3 subherds containing 11, 18, and 15 cows plus their offspring, each accompanied by 4–5 adult bulls. In addition there were 3 subgroups of 6–9 subadult and adult bulls, 1 of which included the most dominant bull; these bulls were not closely

Fig. 10.2. Close individual distance in a resting buffalo herd, including contact.

attached to any of the subherds.

The social distance between buffaloes is often zero within resting subherds, whose members may lie touching (fig. 10.2). Certain individuals, which need not be dominant or even adult, regularly take the lead when a herd is moving, serving in turn as trusted "pathfinders" while the buffaloes traverse specific parts of their home range, the members of subherds clustered in close columns. Each subherd also has pathfinders, which assume leadership only when the main herd splits into smaller units (5) (fig. 10.3).

Female buffaloes presumably stay within their clans as they mature, though thcy cease following their own dams closely once they become mothers themselves at age 3 or 4. The maternal bond with males, in contrast, ends with adolescence (3 years), after which males associate together in bachelor subgroups. Immature males tend to keep their distance from adult bulls (10).

Home range size, herd size, resource distribution, and grassland productivity are all interrelated. Home range size also differs among sex and age classes. Bachelor and solitary males are the most sedentary. Groups of old males, typically 5–10 but numbering up to 50 head, may keep within a surprisingly small area: some well-known Serengeti bulls stayed within a range of 3–4 km² for at least 4 years (10). A breeding herd of 138 buffaloes ranged an area of 10.5 km² in Queen Elizabeth N.P. and a herd of 1500 had a range of 296 km² (13). During daily ranging, too, buffaloes in breeding herds move 2 to 3 times as far in a 24-hour day as bachelor herds: a mean of 5.5 km compared to no more than 2 km (10). During the dry season, herds travel up to 17 km while commuting between pasture and water (5).

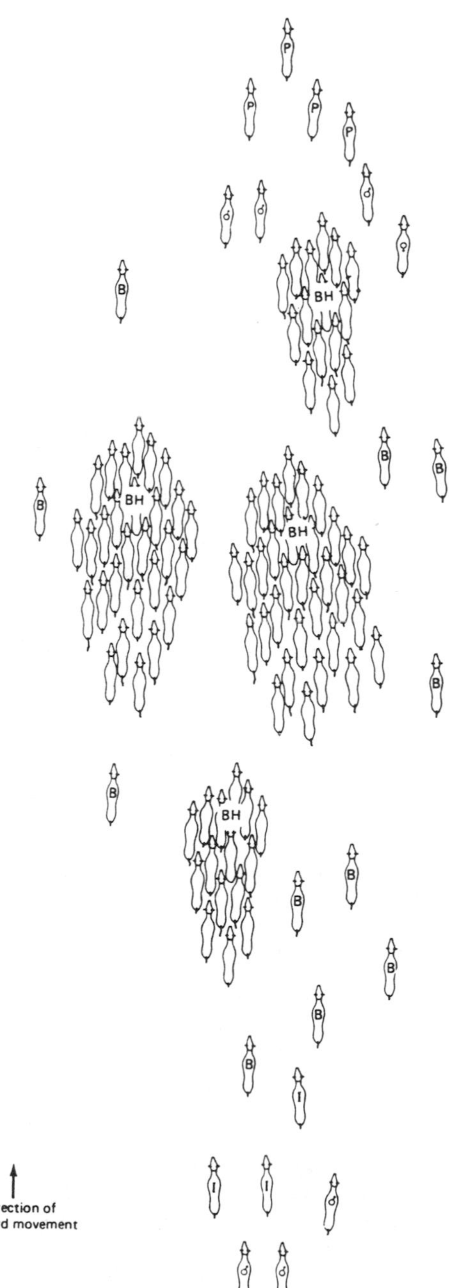

Fig. 10.3. Buffalo column formation typical during passage through wooded country. *BH*, basic herd. *P*, pathfinder. Male symbol, high-status male. *B*, bachelor male. Female symbol, high-status female. *I*, invalid or aged animal. (From Mloszewski 1983.)

ACTIVITY. Although the buffalo is often considered mainly night-active (10), herds not subjected to human predation tend to spend about as much time grazing and resting by day as by night (5). Buffaloes studied in western Zambia grazed from 8½ to 10⅓ hours of a 24-hour day, 51% by night and 49% by day. The buffaloes fed a little more by daylight in the dry season (52%) and more at night during the wet season (56%), but the basic pattern remained unchanged. The main diurnal grazing peak fell between 1500 and 1800 h in both seasons, although grazing intensity was much greater in the dry season. A secondary grazing peak was in the morning from 0800 to 0900 in the wet season and to 1000 in the dry season. At night, most grazing occurred between 2000 and 0330 in both seasons. Between then and sunrise was the peak night resting/ruminating period, and from noon to 1600 was the main rest period by day. Other shorter and less predictable rest periods were scattered over the 24 hours, occurring at intervals of ½–3 hours. These and other observed buffaloes preferred to rest in the open, and only entered woodland or thickets when it was extremely hot or where subjected to hunting by man (5).

Buffaloes go to water at least once every 24 hours, mainly in daylight and most often between noon and dusk; in the dry season 42% of all drinking by the Zambian animals occurred between 1600 and 1900 h. It takes an adult about 6 minutes (4.4–9.6) to drink its fill of c. 34 liters (5, 6).

Wallowing. Although individuals of both sexes and different status occasionally take mudbaths, neither wallowing nor entering the water is a regular habit of the African buffalo (5). Dominant males wallow most frequently, and in the Serengeti population bulls regularly spend a couple of hours a day in mudholes, especially during the hottest hours. But their high status and associated aggressive actions, notably *digging* and *tossing mud* with the horns, *kneeling and neck-rubbing* (fig. 10.1), *rolling*, and *urinating*, suggest that wallowing has a primarily social significance. Furthermore, access to wallows, which are seldom big enough for 2, depends on rank: a subordinate bull quickly withdraws at a superior's approach (10).

SOCIAL BEHAVIOR

COMMUNICATION. The high degree of coordination within herds containing hundreds of animals that are grouped into subherds, and the close individual distance in this species, would indicate that communication is continuous and efficient under all conditions of light and cover. The role of olfaction is unclear but is very probably important for individual recognition as well as for detection of predators. To the human nose, buffaloes and their excrement smell just like cattle.

Vocal Communication. As the following catalogue of vocal signals demonstrates, this channel is very important in buffalo communication (5). Many of the calls bear a resemblance to the lowing of domestic cattle (*Bos taurus*), but are generally lower-pitched. Buffaloes are also far less vociferous; they rarely call except when in a group and even then usually less than half the members of a herd join in calling, which is usually initiated by a few individuals of high status.

1. Signal to move: a low-pitched 2–4 second call repeated several times at 3–6 second intervals. Coordinates herd movements.
2. Direction-giving signal: a "gritty," "creaking-gate" sound given intermittently by leaders in the early stages of a herd movement.
3. Water signal: an extended *maaa* call emitted by one or a few individuals up to 20 times a minute before and during movement to a drinking place.
4. Position signal: call emitted by high-ranking individuals in a herd that announces their presence and location.
5. Warning signal: a more intense form of the same call sounded as a warning to an encroaching inferior.
6. Aggressive signal: an explosive grunt which may be extended into a sequence or become a rumbling growl; also given by dominant males after a stampede.
7. Mother-to-calf call: a croaking call emitted by mothers seeking their calves; heard in chorus after a stampede and as a prelude to mobbing attacks (10).
8. Calf distress call: a higher-pitched version of no. 7.
9. Danger signal: a drawn out *waaaa* sound, heard only 3 times, by daylight, when hunting lions were detected at a distance by individuals in a resting herd.
10. Grazing vocalizations: a variety of sounds, including brief bellows, grunts, honks, and croaks, often heard as a herd is grazing, which may help to keep the herd moving in the same direction, signal that all is well, and so on.

AGONISTIC BEHAVIOR

Dominance/Threat Displays: *lateral presentation, low-* and *high-horn presentation, threat-circling, stiff-legged walk,* grunting/growling, *head-tossing* and *hooking, vegetation-horning, ground-horning* with *dirt-tossing, rubbing neck on ground, rolling.*

Defensive/Submissive Displays: *head-low/chin-out posture,* bellowing; *flight-intention posture.*

Fighting: *charging* with chin raised, *ramming, front-pressing.*

During the *lateral presentation* and when approaching an opponent, the head is held at or above shoulder level with chin pulled in (fig. 10.4). Tossing and hooking movements accompany the display. A lower-ranking male responds by adopting the *head-low/chin-out posture* and often approaches and places his nose under the dominant's belly, sometimes under his neck or between his hindlegs. Thus, behavior associated with suckling has seemingly evolved into an *appeasement ceremony.* Immediately afterward, the subordinate wheels and runs away, meanwhile bellowing, during or even before which the mouth is opened and the tongue is curled. A subordinate that is suddenly subjected to attack bolts a few steps, gives a short grunt, and stops with nose raised high, back sloping, and tail clamped in an exaggerated version of the *alarm posture* (fig. 10.5) seen in buffaloes on the point of flight.

When one bull approaches and threatens another that, instead of giving way, responds in kind, the full repertoire of aggressive displays may be enacted. Standing 20 m or so apart and often circling slowly, they perform the *lateral display,* meanwhile tossing their heads and pausing to punish nearby shrubbery or toss clods in the air. Such contests usually end within 10–15 minutes with one bull giving up and walking away.

The few fights that have been witnessed were violent but usually brief. Typically,

Fig. 10.4. Bull in *medial-horn presentation head-tossing.*

Fig. 10.5. Buffalo wheeling and fleeing.

2 bulls charge from up to 30 m apart, making the deep rumbling call, and ram each other just once (5, 10). While charging, the head and neck are stretched forward and at the last moment the chin is pulled in so that the impact is taken on the horn bosses. The bull with the greater mass wins. The loser turns immediately and flees, with the winner in hot pursuit for up to 100 m. Sparring, which is common between subadult bulls, is a much milder exercise, involving no charge and only a little horn-tangling and halfhearted pushing. It begins with one approaching another slowly, then waiting with head down for the other to come forward. *Cavorting* and *frisking* are often associated, suggesting an element of play (10).

REPRODUCTION. Females calve first at about 5 years, 3–4 years before males are ready to enter reproductive competition. Where wet and dry seasons are clearly defined, buffaloes are seasonal breeders. Since this species has much the longest gestation period in the tribe (11½ months compared to 9 months for cattle), both mating and calving are largely confined to the rainy time(s) of year. The birth peak comes early and the mating peak late in the season (10). The interval between births ranges from 15 months, under good conditions, to 2 years.

SEXUAL BEHAVIOR. Bulls regularly monitor female reproductive status by *urine-testing,* and can prompt cows to urinate by *licking the vulva.* A cow coming into heat is closely guarded by a bull who attempts to keep other bulls at a distance (cf. eland). But cows often prove evasive, thereby attracting the attention of other bulls and leading to a series of displacements by males of successively higher rank. By the time a cow comes into full estrus after 2 or 3 days in proestrus, the most dominant bull in the herd or subherd is usually in attendance.

A tending bull tests the cow's readiness

to stand and be served by licking and resting his chin on her rump; if she doesn't bolt, he mounts and a receptive cow stands still with tail arched. A minimum of 2 copulations within ½ hour were observed in the mating of 6 different females (10). Cows in heat have been seen to mount and be mounted by other cows, and one may solicit a bull by resting her head on his rump or pushing it under his belly. The sight of a copulating pair excites other bulls, which may gather around and then begin fighting among themselves (10).

PARENT/OFFSPRING BEHAVIOR. Birth weight averages 45 kg (35–50 kg) and calves have been known to gain their feet within 10 minutes of birth (12). Yet buffalo calves are too feeble to follow for several hours and remain poorly coordinated and slow runners for several weeks (10). A group defense is clearly adaptive. Nevertheless, it is not unusual for a mother, after giving birth in the middle of the herd (typically during the morning or afternoon rest period), to be left alone to defend her calf when the herd moves off to drink or feed. She may then try to lead it into the nearest cover and hide with it until it is strong enough to follow her back to the herd, a slow and halting procedure (10). Or she may urge it to follow by *croaking,* whereas calves only call if they lose contact with their mothers, for instance when left behind by a fleeing herd. The mother responds by dropping out of the herd, but may shuttle back and forth between herd and offspring, obviously pulled in both directions. Sometimes the pull of the herd is stronger and young calves are abandoned (3).

Unlike cattle, buffalo calves nurse between the mother's hindlegs, sucking for 3–10 minutes at a time at irregular intervals, and may be suckled right up until the next calf is born (10). However, cows normally stop lactating during the seventh month of pregnancy, when the current calf is as young as 10 months old. Possibly going through the motions lessens the emotional pangs of weaning and helps to maintain the maternal bond, which lasts considerably longer in the buffalo than in most bovids. The state of grace ends abruptly with the next birth, whereafter the mother keeps her yearling offspring out of the way with unloving horn jabs. Nevertheless, yearlings continue to tag along for another year or longer (5, 10).

ANTIPREDATOR BEHAVIOR: *alert posture* ± advancing to investigate, *head-tossing,* wheeling and flight, stampeding, individual and group attack.

Indirect evidence of the relative safety enjoyed by buffaloes in herds is the occasional presence of blind and crippled individuals in good health, and the fact that the adult sex ratio of males remains almost equal to that of females through middle age. The ratio becomes skewed toward females in older classes, marking the departure of most males from the breeding herds at 10–13 years. As formidable as the old bulls may look, they are more vulnerable to lion predation than buffaloes in mixed herds. Some of the bachelors and loners are decrepit and diseased, and they are certainly far less alert and active than the mostly younger, still vigorous bulls seen in herds. In addition, lions that pick on animals in the mixed herds run the sometimes fatal risk of a mobbing attack. A number of such incidents have been recorded. In Manyara N.P., for instance, 4 lionesses and a lion were charged and kept treed for hours by 200 buffaloes after the lions killed a herd member. Another time a lagging cub was overtaken and trampled before it could get to a tree (8).

A mobbing response is most often triggered by a calf's distress call. It evokes the same croaking call given by a mother seeking a lagging calf, and both sexes make it while approaching a disturbance in a tight phalanx in the *alarm posture.* The distress cry of an adult can evoke the same protective response, even in bachelor herds. Herds fleeing from a predator (including hunters on horseback) often crowd close together and run at much less than their top speed, making it very difficult to single out a quarry (5, 9). However, with an estimated top speed of 48–56 kph (5) healthy buffaloes are capable of outrunning lions, their only major predator apart from man.

References

1. Blancou 1958. 2. Kingdon 1982. 3. Ludbrook 1963. 4. Macdonald 1984*a*. 5. Mloszewski 1983. 6. Pienaar 1969*a*. 7. Roosevelt and Heller 1914. 8. Schaller 1972*b*. 9. Selous 1908. 10. Sinclair 1977*b*. 11. Verheyen 1954*a*. 12. Vidler et al. 1963.

Added References

13. Grimsdell 1973. 14. Prins 1987.

Chapter 11

Giraffe and Okapi
Family Giraffidae

Giraffa camelopardalis, giraffe
Okapia johnstoni, okapi

FAMILY TRAITS. Large ruminants with high shoulders and sloping hindquarters; no upper incisors or canine teeth; 2 skin-covered "horns" present in male only (okapi) or both sexes (giraffe); tongue very long and prehensile; nostrils slitlike and closable, muffle hairy; large eyes, tail hock-length with a terminal tuft; no false hooves. *Scent glands:* present in okapi hooves, none described in giraffe. *Mammae:* 4.

Senses: smell and hearing most important for the okapi; the giraffe depends on eyesight and may have color vision (1).

ANCESTRY. Giraffes and deer are considered distantly related, largely because antlers are thought to have begun as bony projections and giraffe horns somewhat resemble antlers "in the velvet." It is more likely, however, that they and the bovids all descend from traguloid ancestors of the chevrotains, the most primitive living ruminants (see chap. 1, Ancestry), which persist in the rain forests of Africa and Southeast Asia. Giraffids of varied forms once ranged Eurasia. The immediate ancestor of the modern giraffe was *Paleotragus,* which arose in the African Pliocene. When the okapi was finally discovered early in this century, lo and behold, it was as if the ancestral form had survived for 12 million years in the hidden recesses of the Congo Basin, the okapi is so like it (fig. 11.1) (3, 4, 20, 22).

SPECIES COMPARISON. Not enough is known about okapi behavior to make a detailed comparison with the giraffe. The 2 species do share the following traits, at the very least. Both have an *ambling walk.* The tongue is adapted for selective browsing, and also for grooming. They have to straddle or bend the forelegs to reach the ground with their mouths, indicating that this limitation existed before the giraffe's neck and limbs became elongated (fig. 11.3). The horns are used only for intraspecific fighting, by males only; both species defend themselves against predators by kicking. Like many antelopes, males *foreleg-lift* during courtship (a widespread ruminant trait). Both species have a long gestation period (14.5 months). Young calves hide, though giraffes are less committed to a hider strategy.

There are also, of course, many dissimilarities, apart from the obvious differences in size, conformation, and coloration. The okapi has hoof glands and the giraffe does not. Scent and voice are important in okapi communication, whereas visual signals are paramount in the giraffe. The okapi is unsociable whereas the giraffe is gregarious.

Fig. 11.1. Okapi dominance display: male (*right*) and female *nose-lifting* (from Walther 1960).

Fig. 11.2. A juvenile giraffe galloping; the curled tail is characteristic.

Giraffe

Giraffa camelopardalis

TRAITS. The tallest animal. *Height and weight:* males up to 5.5 m, females 4–4.5 m to top of head; males 1100 kg (800–1930), females 700 kg (550–1180) (9). Neck elongated, with a short, erect mane, shoulders much higher than croup but limbs of nearly equal length. Tail hock-length with long, black terminal tuft. *Horns:* the main pair up to 13.5 cm, borne by both sexes, the ends knobbed and hairless in adult males, thin and tufted in females and young; a median, lumpy horn and 4 or more smaller bumps in males only. *Coloration:* ground color brown to dark chestnut (sometimes black), broken up into patches and blotches by a network of light-colored hair, the pattern individually unique; males darken with age. *Scent glands:* possible apocrine glands on eyelids, nose, lips (18); adult males have a pungent odor (2, 6, 9, 16).

Special Features of the Horns and Skull. Unlike all other mammals, the giraffe has the main pair of horns present even in the embryo, formed of cartilage from a layer of skin and unattached to the skull (19). They lie flat and present no obstacle to birth, but within a few days stand erect and appear prominent, mainly because of terminal tufts of long hair. While growth proceeds from the base, bone deposition begins at the tips and proceeds downward. Fusion with the skull occurs at 4–4½ years in males and after 7 years in females. Rather than attaching to the frontal bones, as in deer and antelopes, the giraffe's main pair take root further back, in the parietals. Male horn growth continues through a unique process whereby bone of dermal (skin) origin is deposited over the whole surface of the skull except where muscles attach. Apart from the parietal and median horns, bone accumulates as knobs at the base of the skull, over the eyes, and on the nose. Gradually the male's head becomes a massive knobbed club which is used to gain the dominance that spells reproductive success. The skull of a 15-year-old bull may be 7 kg heavier than that of a bull half his age, and assuming body growth also continues after maturation, older bulls also have a size advantage (19). Maximum age in the wild is about 25 years (16).

DISTRIBUTION. Formerly throughout the arid zones and drier regions of the Northern and Southern Savanna, wherever trees grow; marginal in the *Miombo* Woodland Zone except for the isolated population in the Luangwa Valley (*G. c. thorncrofti*) (2, 6). Though eliminated from most of its former range in West Africa and no longer present in the southern Kalahari, it is still reasonably common outside parks and reserves.

Of 8 different races, the most distinctively different is the reticulated giraffe (*G. c. reticulata*) of northeastern Kenya, whose coat with its latticework of thin lines separating dark patches with regular edges is also most unlike any other mammal's (9). The most irregular, jagged patches occur in the Masai giraffe (*tippelskirchi*), which

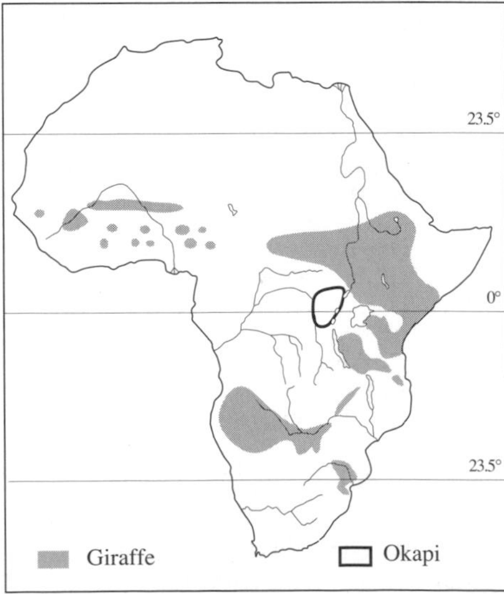

overlaps and interbreeds with *reticulata* south of the Tana River. Reticular patterns predominate in the northern races and blotchy patterns in the eastern and southern races, but the spots become less jagged in *angolensis* and *giraffa.* However, there is so much individual variation that patterns supposed to distinguish different races can often be found within a single herd.

ECOLOGY. Growing tall gave the giraffe access to a 2-m band of foliage beyond reach of all other large browsers but the elephant. Aided by its 45-cm tongue and a modified atlas-axis joint that enables the head to tilt to the vertical, a giraffe can feed on the crowns of small trees. Big bulls can reach up to 5.8 m (19 ft.) (16), nearly a meter higher than cows. Where a choice exists between high and low browse, there is a clear ecological separation between the sexes, the bulls browsing high while females concentrate on regenerating trees and shrubs below 2 m. The sexes of distant giraffes can usually be predicted by whether the animals are feeding high or bending low (17). Differences in feeding ecology as well as lower vulnerability to predators (based on size and absence of parental responsibility) allow males to enter taller and denser woodland more readily than females, leading also to a measure of spatial separation of the sexes.

Browsing by giraffes can have profound effects on their food trees: pruning of young trees keeps them short for years longer than usual and makes waistlines around trees whose tops are beyond reach. Over 100 different plants have been recorded in the giraffe's diet, but various species of acacias and *Combretum* form the bulk of their diet in most areas (6, 12, 17). They tend to feed on deciduous foliage in the rainy season and rely on evergreen species at other times. Like many savanna herbivores, they disperse widely during the rains and concentrate along watercourses in the dry season, where high-quality foliage continues to be available (13, 17). Horny papillae protect the lips and tongue against thorns. The narrow muzzle, extremely flexible upper lip, and long, prehensile tongue enable the giraffe to strip leaves off branches or select individual leaflets from between sharp thorns; thus it can both feed selectively and consume the quantity of foliage (up to 34 kg a day) needed to sustain its bulk. Giraffes may drink at intervals of 3 days or less when water is available, but can also fill much of their need from green leaves, especially when covered by dew (fig. 11.3) (8).

Fig. 11.3. A giraffe straddling its forelegs to drink; the other way is to bend the forelegs without straddling.

SOCIAL ORGANIZATION: gregarious, nonterritorial; loose, open herds.

The giraffe is not only physically but also socially aloof, forming no lasting bonds with its fellows and associating in the most casual way with other individuals whose ranges overlap its own. Giraffes rarely group closely, except when browsing the same tree, when the approach of a predator makes them nervous, or in large numbers in open tree grassland. Even at rest they stay over 20 m apart. A dozen feeding giraffes may be spread out over an area greater than a kilometer—can such a collection of individuals be called a herd? There is no leader and a minimum of coordination in their movements. Yet even widely dispersed giraffes may well maintain visual contact, thanks to their lofty vantage point. Rival bulls recognize and react to one another from a long way off, demonstrating the capability of long-distance communication.

Except when guarding a newborn offspring, females are rarely sighted all alone; usually there are other females somewhere within view. Mothers with small calves are most likely to associate. The tendency of calves to cluster together in crèches anchors mothers to the same locality; in addition, females return faithfully to particular calving grounds (see Parent/Offspring Behavior) (16). Even so, giraffe herds are so fluid that a Serengeti cow who was seen on 800 consecutive days was only twice found with a group whose composition remained unchanged over a 24-hour period (16). The fact that a few adult individuals are regularly found associated can probably be explained as the occasional continuation of social bonds between mothers and daughters, which typically remain within their natal home range. Though behavioral interactions between females are rarely observed, the existence of some sort of hierarchy is indicated when one cow yields its feeding place at the approach of another (16).

Males tend to associate in bachelor herds beginning at puberty (3 years) and to emigrate outside their natal ranges (16). As males mature they become increasingly solitary, dividing their time between feeding and monitoring the reproductive status of the females within their core ranges. These tendencies, like the males' preference for more heavily wooded habitat, promote sexual segregation of adults. However, males in herds or alone frequently mingle with females and young. Consequently, giraffe herds may contain almost any possible combination of sexes and ages at any given moment—and will almost certainly not remain in the same configuration for many hours. Variability is the only rule.

Home Range. In Tsavo N.P. the overall mean home range of 60 male and 50 female adults and subadults was found to be virtually equal: 164 km^2 and 162 km^2, respectively. Ranges varied greatly, from as little as 5 up to 654 km^2. The giraffes spent more than half the year in relatively small (up to 100 km^2) dry-season ranges centered on riverine vegetation and wandered most widely during the rains. Immature males had the largest ranges (>500 km^2 in 3 cases), reflecting their tendency to disperse before settling down, after which males occupy smaller ranges than females (13). Similarly, in the Serengeti population, cow ranges averaged 120 km^2, although 75% of all sightings were concentrated within the central 30% of the ranges (16, 17).

ACTIVITY. A giraffe spends approximately the same amount of time feeding as smaller browsers such as the impala or Thomson's gazelle; a mean of 53.2% and 43.2% of a 24-hour day for females and males, respectively (17). As browse declines in quality and abundance in the dry season, more time is spent feeding, but the increase is significant only in bulls (48% vs. 40%); cows take longer and select more nutritious foliage at all seasons. They spend less time ruminating (27% vs. 30%) and walking (12.7% vs. 21%) than bulls. The male's extra walking time (during which a giraffe may also ruminate) goes into searching for females in estrus.

The first and last 3 hours of daylight are peak feeding periods, whereas the hottest hours in between are spent mainly standing and ruminating. Females spend 72.4% and males 43.2% of the daylight hours feeding. The main nocturnal activity is ruminating, but giraffes feed more and ruminate less on bright than on dark nights: ruminating 48.6% and browsing 22.4% on dark nights compared to 39.8% and 33.6% on moonlit nights. Most ruminating by night occurs while lying. All giraffes lie down part of the night, beginning right after sunset; lying frequency increases through the night, reaching a peak in early morning, the time when fewest giraffes do any foraging. Occasionally a lying giraffe curls up and goes to sleep for 5 minutes or so. "Other" activities, including social and sexual interactions, take up less than 2% of a female's and 5% of a male's time (17).

POSTURES AND LOCOMOTION. The giraffe has only 2 gaits, an *ambling walk* and a gallop (fig. 11.2) When walking, the entire weight is supported first on the left legs, then on the right legs, as in a camel. Otherwise the long limbs on the short trunk would interfere with one another. The neck moves in synchrony with the legs and helps the giraffe maintain its balance. At a gallop forefeet and hindfeet work together in pairs as in a running rabbit, the hindfeet landing outside and ahead of the forefeet. Even when traveling at 50–60 kph the giraffe is so long-legged that it appears to be moving in slow motion; meanwhile the tail is held curled up on the rump and switched at regular intervals (fig. 11.2) (5).

SOCIAL BEHAVIOR

COMMUNICATION

Vocal Communication. Giraffes are silent but not mute. Apart from an alarm snort, calves bleat and make a mooing/mewing call (15). Cows seeking strayed calves may give a roaring bellow, and courting bulls have been heard to utter a raucous cough. Moaning, snoring, hissing, and flutelike sounds have also been reported (6, 8, 19).

AGONISTIC BEHAVIOR

Dominance/Threat Displays: *erect posture* with chin and head high, neck vertical, *angle-horn, neck extended with chin raised, stiff-legged approach, lateral presentation.*

Defensive/Submissive Displays: *head and ears lowered with chin in, jumping aside, retreating.*

Fighting: head-to-body slamming, *necking* (sparring).

Position in the male dominance hierarchy is largely a function of seniority and between peers is decided usually before maturity through contests in the bachelor herds.

Fig. 11.4. Males sparring ("necking").

The challenge to a duel begins with an apparently nonchalant approach. When close, the challenger raises his head and stands in the *erect posture* facing his opponent. If the other responds in kind, they then stalk forward stiff-leggedly and stand parallel, or they may march in step with *necks extended* horizontally, looking straight ahead. At low intensity, they proceed to rub heads and necks gently together, and may lean heavily against each other with ears flapping and rub shoulders or flanks—probably assessing their comparative weight and strength. Sometimes they pause and gaze into the distance. The prospective winner can be predicted by noting which male holds himself more erect (21). At higher intensity, the contestants aim blows at rump, flanks, or neck, often standing in *reverse-parallel*. Preparatory to giving or receiving blows, a giraffe braces himself by straddling his forelegs; then he draws his neck sideways and swings upward and backward over his shoulder to strike his opponent with the parietal horns, thereby concentrating the blow in a small area (19). But blows seldom land solidly, for each does his best to avoid being hit by moving his neck away at the last moment, meanwhile getting ready to return the blow. Movement and countermovement appear rhythmical and synchronized, imparting the sinuous grace of a stylized dance (fig. 11.4). There may be long pauses between blows when the animals stand motionless. Sometimes they lift their noses as if trying to look taller or to snatch a mouthful of leaves (*displacement feeding*).

The force of a solid blow is literally staggering. The heavier the skull and the wider the arc of the swing, the harder a giraffe can hit—hence the advantage of age, height, and weight. Mature bulls know their place in the hierarchy and normally avoid confrontations, even when an estrous female is at stake (11). A serious fight is apt to be brief but violent and involve a stranger with the top male in the local hierarchy. After a few heavy blows, one gives up and runs away (15, 16). In one case a bull was knocked senseless and lay stretched on the ground for 20 minutes before recovering (6).

REPRODUCTION. Reproduction is typical of nonterritorial ungulates (discussed in chap. 10, Bovini introduction). Lone mature males go from herd to herd checking female reproductive status by *urine-testing*. Breeding is perennial but with most conceptions during the rains (7, 10) gestation is long (14–14½ months), and cows conceive again several months later (minimum of c. 16 months between births [11]). Females become pregnant in their fourth year, whereas males are not mature enough to compete before age 7.

SEXUAL BEHAVIOR: *urine-testing, tending bond, mating march, licking* and *head-on-rump rubbing, circling, foreleg-lifting.*

Having located an estrous cow, a bull attempts to maintain a *tending bond* and keep rivals away (11). He follows the female

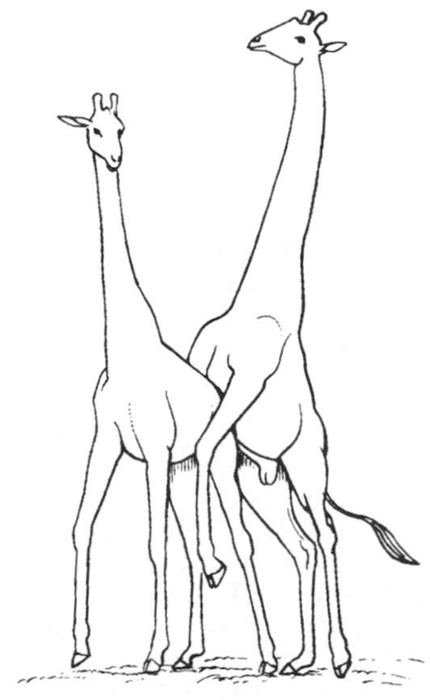

Fig. 11.5. Copulation.

closely, licking her tail and sometimes taking it between his lips, standing and *lip-curling* when she urinates. He may rest his head on her rump, nudge her with his muzzle, butt her gently with his horns, or try to rest his neck on her back. Moving in still closer, he *foreleg-lifts*, which causes her to move forward. When she is ready to stand for copulation, he stands immobile with head high (cf. topi), then mounts by sliding his forelegs loosely onto her flanks, and stands bolt upright while giving an ejaculatory thrust that propels the cow forward (fig. 11.5) and ends the courtship bout.

PARENT/OFFSPRING BEHAVIOR. Serengeti giraffes have preferred calving areas within their home range, to which each cow returns, alone, to drop successive calves (16). As mothers are usually standing at the time, calves literally drop—about 2 m. A calf may gain its feet within 15 minutes and suckle in the first hour. For the first week or more the mother remains in isolation, avoiding or warding off approaches by other giraffes and very much on guard against predators (18) (see Antipredator Behavior). The calf spends half the day and most of the night lying, preferably beside a bush or fallen tree. It lowers its head when alarmed and its color and blotchy markings help to camouflage it (10). The mother often stays farther than 100 m from a concealed calf and may leave it alone for periods of up to 4½ hours to travel 4 km or more to water (18).

From the second week through the second month calves spend most of the day just standing and looking (42%–43% vs. 20%–27% lying). The sight of apparently unguarded small calves, some with umbilical strings still attached, has contributed to the widespread belief that giraffes are careless mothers (6). But detailed observations of maternal behavior have shown that in fact the mother-calf bond is strong and enduring (up to 22 months), although subject to great individual variation (10, 18). The modal distance between Serengeti mothers and calves for the first 2 weeks is only 11–25 m. The distance increases to over 25 m within 8–30 days, after mothers allow their offspring to begin associating with other calves. Up to 9 young giraffes may associate in crèches, over half the time at distances of under 10 m. Mothers of young calves continue to be watchful and quick to come to the aid of their offspring. However, the enhanced security within a maternity group frees mothers to browse 100–200 m away or even to absent themselves for hours at a time. Nevertheless, mothers of calves 1 month or older were present in 44%–98% of all sightings (18). Calf groups are rarely left entirely unattended. Absent mothers usually return before dark, suckle their offspring, and remain with them overnight, then suckle them again before going off to spend the day browsing (15). Beginning at 1 nursing bout per hour during the first week, suckling frequency declines to once every 3 hours after a fortnight, and from 4 months on to once every 5 hours. The mean duration of 415 bouts was 66 seconds (4–226) (18). Calves begin ruminating between 3 and 4 months and thereafter spend 53%–64% of the day browsing. They accompany and feed with the maternal herd starting at 5–6 months and are weaned as yearlings; most become independent at 12–16 months (10, 11).

Even as calves, males spend less time browsing and more time standing (also running) than females, and move much farther from their mothers, particularly in the second year (most over 100 m compared to less than 50 m for most yearling females) (18).

PLAY. Newborn calves often frolic around their mothers and may run off 50 m or more and return. Calves in maternity groups often gambol, especially early and late in the day, but do not pursue one another or make bodily contact (but greet one another by touching noses and foreheads [18]). Older juvenile males practice play-fighting (9, 11, 13).

ANTIPREDATOR BEHAVIOR. Great size, superior vision (day and night), speed, and formidable hooves make grown giraffes largely invulnerable to predators—although lions are able to kill even bulls if they can get them down and secure a throat or nose hold. But small calves are very vulnerable and ½–¾ of those that are subject to lion and spotted hyena predation die within the first month or so (6, 15, 16, 18). The first lines of defense are concealment on the part of the calf and the mother's vigilance. Mothers of newborn calves spend much of the night scanning in the alert posture, and prefer either open ground with an uninterrupted view or bushy areas where the calf is well-hidden and the mother cannot be seen silhouetted against the sky (18). The mother may browse but keeps within 40 m of the calf's hiding place. Upon spotting a stalking lion, a mother usually leads her calf away but can and will stand over and defend it with her hooves against lions as well as hyenas. Young calves are faster than adults over a short course and may well outdistance a lion, but lack the endurance to outrun a hyena.

An integral part of the giraffe's antipredator strategy may be to grow big as quickly as possible. The large amount of time calves spend lying and standing, and the relatively small amount of active movement they engage in, enables them to put most of their food intake toward growth. A calf grows as much as a meter in the first ½ year and nearly doubles its height within a year (18).

References

1. Backhaus 1959*b*. 2. Berry 1973. 3. Churcher 1978. 4. Cooke 1968. 5. Dagg 1962. 6. Dagg and Foster 1976. 7. Hall-Martin, Skinner, Van Dyk 1975. 8. Innis 1958. 9. Kingdon 1979. 10. Langman 1977. 11. Leuthold, B. M. 1979. 12. Leuthold, B. M., and W. Leuthold 1972. 13. ——— 1978*b*. 14. MacClintock 1973. 15. Moss 1975. 16. Pellew 1984*a*. 17. ——— 1984*b*. 18. Pratt and Anderson 1982. 19. Spinage 1968. 20. Thenius 1969. 21. Walther 1984. 22. Young 1962.

Part II

Hoofed Mammals: Nonruminants

Other Even-Toed Ungulates,
Order Artiodactyla
Family Suidae, swine
Family Hippopotamidae, hippos

Odd-Toed Ungulates,
Order Perissodactyla
Family Rhinocerotidae, rhinos
Family Equidae, zebras, asses, horses

Near-Ungulates,
Superorder Paenungulata

Order Hyracoidea
Family Procaviidae, hyraxes

Order Probiscidea
Family Elephantidae, elephants

Order Sirenia, dugongs

Part I dealt with a comparatively homogeneous group of ungulates, all of which belong to the same order and suborder (Ruminantia), of which all but two species are in the same family. Part II lumps together a diverse array of mammals from 4 different orders and 6 different families.

The nonruminant ungulates have all been around much longer than the ruminants. They had their heyday before ruminants appeared on the scene and became the dominant large herbivores in the Miocene (see table 1.1). Membership in an ancient group does not mean that a species is necessarily primitive or poorly adapted for existing conditions. Rhinos may look like archaic relics, and they have not been dominant members of any ecosystem in recent times, but the African black rhino was doing fine wherever suitable wooded habitat existed until the demand for its horns made killing it nearly as profitable as dealing drugs. Looked at another way, animals that have been around for ages represent designs that have stood the test of time. Some of Africa's most successful mammals are nonruminant ungulates, notably the common or plains zebra; the warthog; the large hippo, modern representative of a family that started out as a smallish, solitary rain forest denizen quite like the pygmy hippo (*really* a living relic); and most of all the elephant, only recently replaced by man as Africa's true king of beasts.

Where the ruminants have succeeded by specializing, partitioning African ecosystems into narrow segments that enabled this group to produce a large number of species, most of the more successful nonruminants have persisted as generalists, able to utilize a very broad range of vegetation and habitat types, the elephant being the prime example. The broadness of their niches has kept the number of species small. Although the hippo is specialized as a grazer that spends its days submerged and its nights feeding on shore, this very widespread, productive niche is monopolized by just this one species. (But it was not always so—see the hippo family introduction.)

Other Even-toed Ungulates
Order Artiodactyla

Chapter 12

Swine
Family Suidae

African Species
Potamochoerus porcus, bushpig *Hylochoerus meinertzhageni*, giant forest hog *Phacochoerus aethiopicus*, warthog *Sus scrofa*, wild boar (North Africa only)

FAMILY TRAITS. The smallest nonruminant ungulates. Thickset, short legs and neck, proportionally large head, and nose a cartilaginous disk with closable nostrils. *Teeth:* canines enlarged as persistently growing tusks, the lower pair honed to a knife-edge against upper pair; molars simple, low-crowned (bunodont, brachydont), not serrated as in other ungulates. Coat coarse and sparse in most species. *Coloration:* drab (except bushpig), piglets horizontally striped (except warthog and giant hog). *Feet:* elongated false hooves (trotters) nearly reaching ground. *Scent glands:* the most diverse of all ungulates (fig. 12.1). *Mammae:* 4 (warthog) to 12 (domestic pig).

Senses: eyesight generally poor, hearing good, and sense of smell outstanding.

ANCESTRY. Although living genera are comparatively recent (Pliocene to Pleistocene), swine arose early (mid-Oligocene) and, proving well-adapted to the role of rooting omnivore, survived with only minor change and specialization. Wild swine occur throughout most Old World wooded habitats but failed to cross the Saharan savanna. The bushpig fills the equivalent niche in the African woodlands; this species and the giant forest hog evolved later from the same parent stock (subfamily Suinae). Only warthogs, a divergent line that arose in Africa in the Miocene, emerged to exploit savanna grassland. The 8 living pigs are relicts of a formerly rich and diverse array, in-

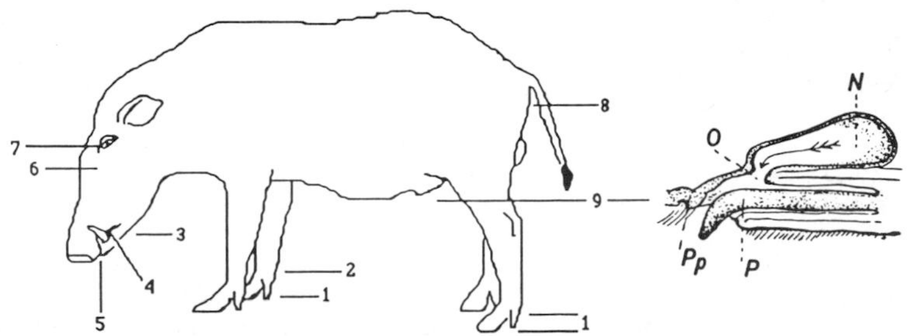

Fig. 12.1. Outline drawing of a pig showing scent-gland locations. 1, hoof (digital) glands (bushpig). 2, carpal gland (bushpig). 3, chin (mandibular or mental) gland. 4, tusk gland (warthog, bushpig, *Sus*). 5, salivary gland (all suids?). 6, preorbital glands (warthog, forest hog). 7, Harderian gland (warthog). 8, anal glands. 9, preputial gland. Inset drawing of domestic boar's gland: *P*, penis; *Pp*, prepuce; *O*, gland orifice; *N*, preputial sac (from Mohr 1960).

cluding at least a dozen African species in a half dozen additional genera that became extinct only in the late Pleistocene (2, 3).

SOCIAL ORGANIZATION: gregarious, nonterritorial, one-male harem/family group and matriarchal groups usually unassociated with adult males.

Pigs are highly social, contact animals that live in sounders within definite, overlapping home ranges. A home range includes a number of fixed points interconnected by a regular network of paths: a resting area with nest or den, and drinking, wallowing, and rubbing places, often close together. The basic social unit consists of a sow with her litter, which remains closely attached until and sometimes after the next litter is born. Several sows with their young may band together, but meetings between different sounders tend to be hostile and are usually avoided. Adult males are either solitary (warthog and wild boar) or protect sounders as 1-male harems (bushpig and forest hog). Male dominance is based on seniority and size, which are advertised by marking trees at maximum reach with tusks and tusk-gland secretions (1, 4, 5, 12). In the warthog and bushpig immature males sometimes associate in bachelor herds.

ACTIVITY. Most pigs are largely nocturnal, at least where hunted by man, but the warthog is strictly and the forest hog is partially diurnal (6, 9, 10). Like other nonruminants, pigs sleep soundly for extended periods, followed by long activity periods.

FORAGING BEHAVIOR. Pigs use their snouts to locate and uncover food. The nose disk is moved over the ground or in a circular motion through the subsoil (excepting the herbivorous giant hog and the Asian *Babyrousa*). The tusks may also be employed on roots and such. When digging deeper, all species use the back of the nose for shoveling earth out of the hole.

SHELTER AND COMFORT. Pigs build communal nests of litter or lie up in dense thickets. The warthog utilizes burrows. Such places not only provide refuges from predators, but also shelters from the elements, an important consideration for animals with poor thermoregulation, especially young pigs and the nearly naked warthog. Similarly, the suid wallowing habit helps pigs cool off in hot weather and may protect the skin against insects. Nevertheless, the warthog and bushpig are preferred prey of tsetse flies. Pigs scratch and rub themselves on objects after wallowing, enabling them to groom places like the rump, flanks, back, and head that they cannot reach otherwise. They also groom and massage each other, using their noses and incisors.

POSTURES AND LOCOMOTION. Pigs move in a *cross-walk.* Galloping is the prevailing rapid gait, but warthogs trot except in emergencies (fig. 12.2). All species are competent swimmers. Preparatory to lying, pigs sit doglike on their hams, then lower their forelegs; they rise in reverse order. They lie on their sides with legs extended, or on the stomach with legs gathered and chin on the ground.

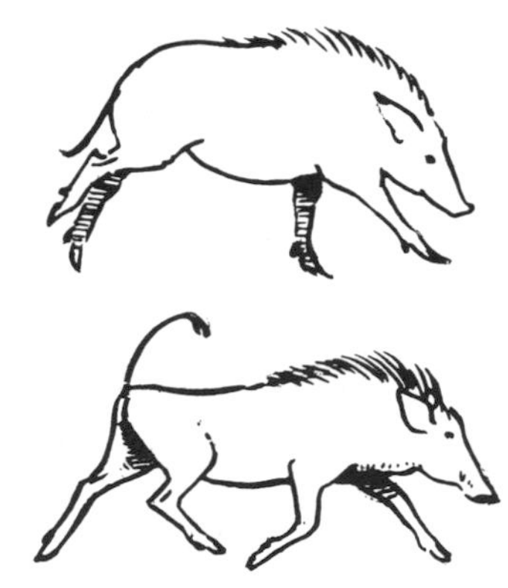

Fig. 12.2. Suid locomotion: a) running bushpig, b) trotting warthog (redrawn from Kingdon 1979).

SOCIAL BEHAVIOR

COMMUNICATION

Vocal Communication. Pigs share the same basic repertoire of grunts, squeals, snarls, and snorts, expressing warning, pain and fear, rage, sexual arousal, and well-being. Males also champ and grind their jaws, churning their saliva to foam, during sexual and aggressive behavior. Females and young tend to be most vociferous (6).

Olfactory Communication. The functions of most of the glands remain unclear, even in the domestic pig. Best known is the steroid pheromone in the boar's saliva, secreted by the submaxillary salivary glands, which induces estrous females to stand for copulation (5, 11). The preputial gland, which imparts a strong smell to male urine, has the same effect. The tusk gland, recently discovered in the warthog, then in the bushpig (fig. 12.1) and *Sus,* is employed for marking objects and the secre-

Fig. 12.3. Bushpig tusking tree and depositing tusk-gland secretion.

tion is probably also present in the saliva (5) (fig. 12.3). Estrus is detected by *urine-testing* but pigs, unlike most other ungulates, do not grimace in the process.

Tactile Communication. Pigs frequently lie in contact, and in greeting make nose-to-nose and nose-to-mouth contact. Nursing young *poke* and *massage* the mother's udder, and boars *massage* and *poke* estrous females. *Mutual grooming* is important for social bonding.

Visual Communication. Blimpish shapes, unexpressive faces, weak eyesight, and nocturnal habits give pigs comparatively little scope for elaboration of visual displays (see Agonistic and Sexual Behavior).

AGONISTIC BEHAVIOR

Dominance/Threat Displays: *lateral presentation, strutting, bristling* (especially of dorsal crest), *ears cocked, head-up posture,* sideways and upward head movements (*boxing* and *slashing intention*); *grunting, growling, gnashing tusks* and *grinding molars; pawing, object aggression, feinted attack.*

Defensive/Submissive Displays: *ears back, head-low posture, backing away, bolting, squealing.*

Fighting: *jostling, snout-boxing* and *lifting, sideswiping, slashing, biting, frontal pushing* and *ramming.*

Suids have 2 basically different fighting styles: *lateral* and *frontal combat,* the former being the more primitive and damaging (7, 9). Fighting style is reflected in morphology. Pigs with a long, narrow face, short tusks, and no warts (wild boar and other *Sus* spp.) fight laterally. Those with a broad head and thick skull, bony shield, thickened pads or warts, and long tusks (forest hog and warthog) fight head to head. Bushpigs do both. *Lateral combat* begins with boars *strutting* shoulder to shoulder and *circling.* They *jostle shoulders, sideswipe* with their heads, and may try to overturn each other by putting the snout in the groin and *lifting.* At high intensity they *grind their teeth, gnash* and *snap* their jaws, and utter deep, barking grunts. Finally, they begin *slashing* at neck and shoulders with the lower tusks and try to bite a front leg, neck, or ears (1) (fig. 12.4).

The tusks cut deep vertical gashes, but the hides are shielded by thick connective tissue that forms during the mating season (14). *Frontal fighting* is more ritualized and seldom damaging (see warthog and forest hog accounts). Combatants push and box with blunt upper tusks and snout, do not bite, and rarely slash (fig. 12.13). There is no specific submissive posture, but squealing serves to inhibit attack, giving a loser the chance to disengage and run for it (6).

REPRODUCTION. Pigs are the only truly multiparous ungulates (4–5 in a typical litter, up to 14 in domestic pig). Gestation is correspondingly short (114–175 days) and birth weight is only about 1% of the mother's weight. Females may conceive at 18–20 months, whereas boars mature at 4–5 years (8).

SEXUAL BEHAVIOR. Estrus is marked by swelling of the labia and a mucous discharge, frequent urination, general restlessness, loss of appetite, and mounting

Fig. 12.4. *Lateral fighting* in wild boar, the primitive form of suid combat (redrawn from Grzimek 1972).

of and by other females. Interest in males is shown by nuzzling, playful biting and skittishness, and smelling and licking male genitalia. Boars intently smell the female's urine, *test* it, and may spray their own urine (and preputial-gland secretion) over it. Courtship combines visual, vocal, olfactory, and tactile stimulation: *strutting* to establish dominance, raised tail, *rhythmic grunting* (the *chant de coeur,* "lovesong"), *champing* and *foaming, nuzzling, massaging,* and *scent-marking* sides and genitalia with the snout. Experiments have shown that the lovesong and preputial- and salivary-gland odors will each induce an estrous sow to adopt the rigid mating stance. Suid copulation is a lengthy affair, lasting from 3 to 20 minutes in domestic breeds, during which the boar stands erect with feet resting on the sow's back and makes repeated pelvic thrusts. The fibro-elastic penis has a corkscrew shape. The huge testes produce 150–600 ml of semen containing perhaps 290 billion sperm per ejaculation, and a boar copulates 7–11 times at minimal intervals of 12 minutes (8). Equivalent data are unavailable for wild suids, but the figures would presumably be scaled down.

PARENT/OFFSPRING BEHAVIOR. Shortly before farrowing, all pigs (except warthog) prepare nests in which the young are born and spend their first days. If domestic and captive wild pigs are typical, births occur while the sow lies on her side and may take 3–4 hours for a numerous litter. Being small and rotund, the young can come out either head or tail first. Suids neither cut the umbilicus, lick their young, nor help free them from enveloping fetal membranes. Sows are careful not to lie on babies, however, and keep them together and warm in a clean dry nest, care that is essential because of poor thermoregulation in the first weeks (13). The young follow the mother closely, kept together by her short, rhythmical grunts. Frequent *nose-to-nose contact* is seen and is apparently vital for maternal and social bonding; sows will not accept other piglets after imprinting has occurred. The mother summons piglets to suckle by grunting and lying on her side. *Sus* piglets each occupy a particular teat from the first, which they defend by biting and slashing with well-developed milk canines. This behavior helps insure survival of the hardiest individuals and prevents weakening of a whole litter should there be more young than teats or too little milk to go around (anterior teats are more productive than posterior pairs). After prompting milk letdown by vigorous *nosing* and *poking*, piglets lie quietly sucking for all they are worth, because flow ceases within ½ minute. Older young can actually force a sow to lie and deliver by violently *massaging* her sides and belly until the suckling reflex is triggered (8). Probably weaning occurs by 6 months in most species.

PLAY. Piglets often engage in playfighting and running games, or in solitary play such as whirling (tail-chasing) and make-believe scratching, digging, and wallowing. Sometimes piglets toss and play with objects, much like puppies (6).

ANTIPREDATOR BEHAVIOR. Powerful build, razor-sharp tusks, and willingness to use them make pigs unusually dangerous to predators. Loud grunts and squealing roars, *bristling,* and *mock charges* are intimidating, and if unheeded may be backed up: disemboweled dogs and leopards, and fatal attacks on man have been recorded. The essence of the suid matriarchal organization is courageous defense of the young, including *mobbing attacks* by sounders containing more than 1 adult, in response to piglets' distress squeals. A warning grunt causes piglets to *freeze,* relying on camouflaging stripes to escape notice (6).

References

1. Beuerle 1975. 2. Cooke 1968. 3. Cooke and Wilkinson 1978. 4. Cumming 1975. 5. Estes, Cumming, Hearn 1982. 6. Frädrich 1974. 7. Geist 1966. 8. Hafez and Signoret 1969. 9. Kingdon 1979. 10. Mohr 1960. 11. Signoret et al. 1975. 12. Skinner, Braytenbach, Maberly 1976. 13. Slijper 1960. 14. Snethlage 1967.

Fig. 12.5. *Snout-boxing* fighting technique of the bushpig (redrawn from Kingdon 1979).

Bushpig

Potamochoerus porcus

TRAITS. Big head, body short with rounded back. *Height and weight:* 55–80 cm, 70 kg (54–115) (6). *Coloration:* the most colorful pig; western races red, southern races dark with light head, individuals darken with age; white eye ring, cheek ruff, ear tassels, and dorsal crest; natal coat dark brown with rows of light spots. *Teeth:* 42, tusks hardly visible but lower pair 7 cm long and razor sharp. Males with projecting bony ridges and calluses on muzzle. *Scent glands:* tusk gland a large pouch (males only), pedal and carpal glands, and glands under neck mane (6). *Mammae:* 6.

DISTRIBUTION. Widespread in wooded habitats from lowland forest up to 4000 m in montane forest, except where the giant hog is common (*Hylochoerus* hogs the best sleeping places [6]), throughout the savanna biome and marginally in the Somali-Masai Arid Zone (1). A serious agricultural pest that increases in farming areas because of reduced predators and increased food supply.

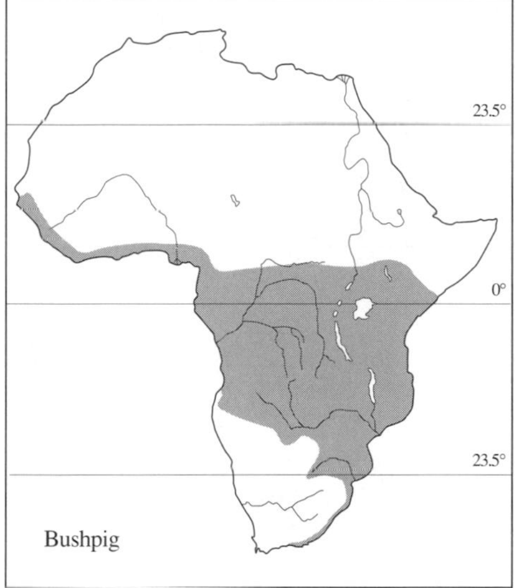

ECOLOGY. Bushpigs can subsist in almost any kind of habitat that affords concealment and food, including marshes and swamps. More or less nocturnal (less when protected), sounders lie up in dense thickets by day, often in litter shaped into rough nests which they leave only if driven. The bushpig is omnivorous, but roots, bulbs, and fallen fruit are the mainstays. Bushpig rooting leaves sizeable areas literally plowed up and stripped of all standing plants. The tusks are used to dig deeper holes. Bushpigs readily eat animal food, including carrion and excrement, and also kill small mammals and birds (6).

SOCIAL ORGANIZATION: gregarious, nonterritorial, 1-male harems or family groups (male + 1 female and young).

Highly social, bushpigs live in sounders of up to 15 head containing several adults and young of different ages. They sometimes aggregate in scores. Sounders forage widely within sizeable home ranges of 0.2–10 km² (7). Ranges overlap but 2 sounders that came regularly to a feeder were antagonistic (7). A family is normally led and protected by a boar. Offspring remain attached for at least a year and some females perhaps permanently. But older juvenile males may be bullied by their father if they approach his feeding or wallowing place too closely. He displays, rushes, and may buffet them with his heavy muzzle (fig. 12.6). Sows compete similarly with younger females over food. Bushpigs on the move communicate by soft grunts given by the dominant animal. Boars communicate their presence and status by rubbing and tusking trees (fig. 12.3). Tusk-gland secretions are enhanced by bacterial decay of vegetation that accumulates in the cheek pouches behind the upper tusks (3, 5). Greeting bushpigs blow their breath at each other (2).

SOCIAL BEHAVIOR. Bushpig behavior is poorly studied but would seem very similar to wild boar, including vocalizations, feeding and other maintenance behavior, and displays (see family account).

COMMUNICATION. Discussed in family introduction.

AGONISTIC BEHAVIOR

Dominance/Threat Displays: *strutting* in *lateral presentation, erecting crest* and *bristling, grunting, champing, pawing, rolling* followed by shaking, *tusking* trees or other objects.

Fig. 12.6. Adult male bushpig chasing yearling males (redrawn from Maberly, 1960).

Defensive/Submissive Displays: *ears flattened, backing away, bolting, squealing.*

Fighting: *frontal* and *lateral, sideswiping, slashing,* biting.

The bushpig's *lateral display* is spectacular (4). Standing or circling 1–2 m apart with dorsal crest erected, coat bristling, and tessellated ears held forward, displaying males *strut* with mincing steps, tails lashing. In addition to the primitive shoulder-to-shoulder combat, bushpigs *box* and *push frontally* by crossing and engaging the bony protuberances and gristly cheek pads (fig. 12.5). This armament also protects the muscles and tendons against slashing. Fighting is accompanied by low-register squealing which can change into a sort of roar (6).

REPRODUCTION. Bushpigs breed perennially, probably with a rainy season peak. Females conceive at c. 21 months and produce litters of up to 6, but typically 2–4.

SEXUAL BEHAVIOR. Displays are undescribed.

PARENT/OFFSPRING BEHAVIOR. Females isolate and prepare nests in which they farrow. For up to 2 months the young often stay hidden while the mother is foraging. Dams accompanied by offspring may join forces; large sounders (over 20 head) include several adults and numerous ½–¾ grown young. *Mutual sniffing* and *grooming* are frequent, and litter mates are inseparable until well-grown.

ANTIPREDATOR BEHAVIOR: *alarm grunts,* flight, concealment, *distress squeals, threat grunts, feinted* and actual (including *mobbing*) *attack.*

A broad range of predators, from lions to eagles, prey on bushpigs at one stage or another, with leopards and spotted hyenas probably most important, particularly where other natural game has disappeared. Being nocturnal may make this species more vulnerable than the warthog, for which the bushpig's savage self-defense may partially compensate. Family boars and sows defend the young, often cooperatively. An aroused boar gives deep, ominous grunts. A fleeing sounder crashes through the undergrowth giving loud grunts and snorts—only to stop and turn after a short distance. Until their spots fade (before 6 months), piglets respond to the *alarm grunt* by *crouching* and *freezing.*

References

1. Ansell 1971. 2. Cumming 1975. 3. Estes, Cumming, Hearn 1982. 4. Frädrich 1974. 5. Jones 1978. 6. Kingdon 1979. 7. Skinner, Braytenbach, Maberly 1976.

Fig. 12.7. Male giant forest hogs *ramming* heads (based on a drawing in Kingdon 1979).

Giant Forest Hog

Hylochoerus meinertzhageni

TRAITS. Resembles an oversized bushpig. *Height and weight:* 96 cm (86–101), males 230 kg (145–275), females 180 kg (130–204) (2). Massive head with broad snout, inflated, naked cheek pads bordering the eyes of males. *Teeth:* reduced to 32–34 and modified for grazing/browsing, tusks smaller and straighter than warthog's (up to 30 cm), flaring outward. Broad, pointed ears and long, tasseled tail. *Coloration:* bristly coat of long black hair, slate-gray skin, inconspicuous patches of lighter hair on face; newborn straw-colored soon darkening to brown or black, unstriped (1). *Scent glands:* preorbitals within male's cheek pads, and preputial gland. *Mammae:* 4 (5).

DISTRIBUTION. Across equatorial Africa in a variety of forested habitats, including Lowland Rain Forest, forest/savanna mosaic and postcultivation thickets, and montane forests into the bamboo and subalpine zone.

ECOLOGY. The common denominator of giant hog habitat is dense cover and a

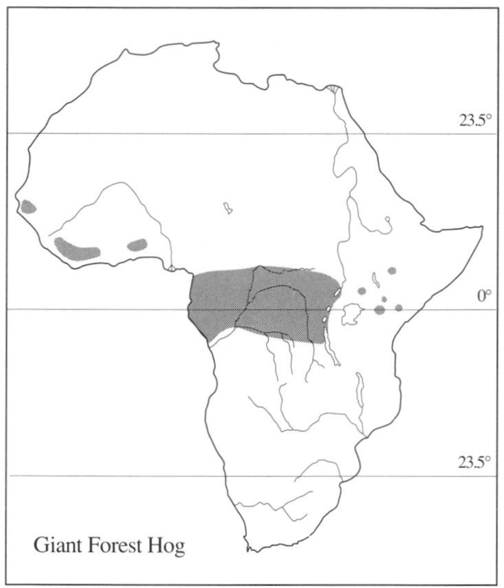

Giant Forest Hog

yearlong supply of green fodder (2). It emerges into clearings to feed on the grasses, sedges, and other herbage that require sunlight for growth, and retires into dense thickets between feeding bouts. In acquiring the apparatus for grazing and browsing (a wide muzzle, lips and incisors adapted for plucking grass, and molars adapted for sideways grinding action), *Hylochoerus* lost the ability to root, except in soft soils. To excavate the mineral earth it eats in quantity, it uses its lower incisors (2).

SOCIAL ORGANIZATION: family group accompanied by adult male, overlapping home ranges.

A sow with offspring of up to 3 successive litters is the basic social unit, each family normally accompanied and defended by an adult male. A home range (area undetermined) includes pastures (openings with ground-level herbage), sleeping areas, water holes, wallows, rubbing places, and mineral licks, linked by a network of paths (4). A bedding ground consists of shallow scrapes under logs or roots in thick cover. Sounders sleep here part of the day and after midnight, huddling together for warmth (2). They change sites often and use communal latrines at the bases of trees. Up to 4–5 sows may associate regularly in a sounder, and sounders often mingle when, for instance, they meet at water or mineral licks. However, boars are mutually intolerant and no doubt usually acquainted; when 2 sounders meet, 1 of the boars usually turns away and goes off, with or without his family. Otherwise they have a confrontation (see Fighting). Sounders of over 20 giant hogs have been seen but groups of over a dozen are usually temporary mergers of 2 or more families. Solitary hogs are usually adult males and pairs are courting couples. The occurrence of bachelor herds is uncertain (1).

ACTIVITY. At the Ark in Kenya's Aberdare N.P., giant hogs regularly emerge to feed in a large glade with salt lick and pond, at times dependent on the weather. They tolerate both low and high temperatures (in equatorial rain forest) but stay in bed in cold, wet weather and avoid strong sunlight. They are most likely to be seen in the open between dusk and midnight, when hogs feed steadily for up to 4 hours.

SOCIAL BEHAVIOR

COMMUNICATION. See family account.

AGONISTIC BEHAVIOR

Dominance/Threat Displays: *strutting* with coat *bristling* and *head raised* to horizontal, *champing* and *foaming, grunting.*

Defensive/Submissive Displays: giving way and flight (no details).

Fighting: *snout-pushing* and *-knocking; forehead-pushing* and *-ramming.*

Forest hogs engage in 2 different forms of *frontal combat.* In *snout-pushing,* 2 males approach with heads horizontal, sniff noses, then engage the broad nose disks and push. This is a safe test of weight and strength and between adults usually ends quickly with one running away. During sparring matches between young males, they also deliver sideward blows (*snout-knocking*) and try to push the opponent's snout to the side. Adult males may alternatively or in addition press their wide, hollowed foreheads together and push until one gives ground and bolts (*forehead-pushing*). At high intensity, boars charge from a distance (up to 30 m) and *ram foreheads* on the run (fig. 12.7). Although their skulls are reinforced, with air sinuses in the cranium and massive temporal and brow ridges, and overlain with thick protective tissue, the impact is so great that skulls quite often show healed breaks (2). When the forehead hollows meet exactly, the resulting compression of air makes a loud report. Evenly matched boars may *ram foreheads* for ½ hour before one gives up. Confrontations between sounders sometimes involve skirmishes between the young as well as the 2 boars, and adding insult to injury, the defeated boar is harried by the small fry (2).

REPRODUCTION. There is no strict breeding season, but most mating occurs late in a rainy period (2).

SEXUAL BEHAVIOR: *close-following, grunting,* and *butting* (details lacking).

A courting pair leaves the sounder, the male following the female closely, grunting loudly and frequently urinating. He pushes and butts her hindquarters roughly with snout and forehead, until she stands for copulation, which may take 10 minutes (2).

PARENT/OFFSPRING BEHAVIOR. Sows farrow in a dry nest. The 2–11 piglets accompany the dam very soon but remain in thick cover a week, and in the open walk beneath the mother. The young playfight by the hour. Females may conceive and leave the family as yearlings. Males stay until secondary characters (cheek pads and glands) develop (2).

ANTIPREDATOR BEHAVIOR: *alarm grunts, snorts, feinted* and real *attack,* including *mobbing, defensive ring, standing at bay.*

Piglets are vigorously defended both by sows and the family boar against hyenas (the main predator) and leopards. Piglets *freeze* when they hear the single alarm grunt. Adults surround piglets when threatened in the open. Hogs fall easy victims to men hunting with dogs, because of their habit of coming to bay with backs to cover and defending themselves (2).

References

1. Dorst and Dandelot 1970. 2. Kingdon 1979. 3. Maberly 1962. 4. Rahm and Christiaensen 1963. 5. Roosevelt and Heller 1914.

Fig. 12.8. Warthog suckling young.

Warthog

Phacochoerus aethiopicus

TRAITS. Level-backed, nearly hairless swine with large, flattened heads; prominent tusks and facial "warts" in adult males. *Height and weight:* 65–84 cm, males 82 kg (68–100), females 65 kg (45–71) (7). Skin gray with scattered bristles, dark mane of long, coarse hair, tail tufted, and white cheek whiskers. *Teeth:* reduced to 32–34, molars and jaw articulation modified for grinding grass, upper tusks to 60 cm, lower tusks to 13 cm in males. "Warts" made of thickened skin and gristle, larger pair below eyes up to 15 cm, another pair close to the tusks; tusks smaller and warts much less developed in females. *Scent glands:* tusk and preorbital glands in both sexes but more developed in males (4); dark stains around eyes of males represent secretions of Harderian glands; preputial glands. *Mammae:* 4 (1).

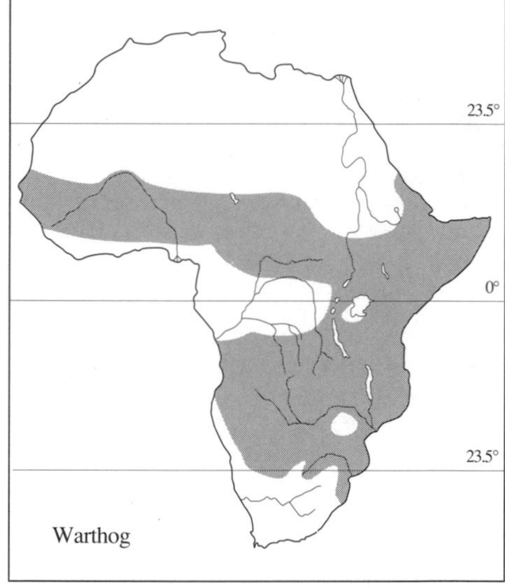

DISTRIBUTION. Northern and Southern Savanna, extending into adjacent arid zones.

ECOLOGY. The only pig adapted for grazing and savanna habitats, the warthog avoids forest and dense undergrowth. It drinks and on hot days wallows daily. In the wet season it grazes; in the dry season it specializes on the underground rhizomes of perennial grasses and sedges, also bulbs and tubers, which it unearths, even in hard ground, with its tough snout. Unlike other pigs, it grazes and roots while resting and walking on its calloused "knees" (carpal joints or wrists) (3).

Fig. 12.9. Warthog burrow (redrawn from Mohr 1960).

SOCIAL ORGANIZATION: matriarchal, traditional home ranges shared by filial sounders; bachelor and solitary males.

The basic social unit consists of females and young, usually 1 mother with young of the year, but also 2 or more sows with assorted offspring up to 2 years old. Sounders may include up to 16 hogs but typically number 5 or less. The bonds between mature females and between mothers and female offspring can be stable and continue through successive breeding cycles. There are also bisexual, usually transitory, groups of immature warthogs. Subadult males often associate in bachelor groups of brothers or unrelated males. Individual males remain associated for only a few months and become solitary as adults (over 4 years). Boars only accompany sounders containing estrous females. A 6-year study (3) revealed that both males and females tend to remain in the natal home range (average area 174 ha [64–374]); so sounders belong to clans of related individuals. Yet sounders seldom meet or exchange members. They use the same network of holes, but on a first-come, first-served basis: a sounder finding a hole occupied seeks another. Each sounder uses up to 10 holes on a rotating basis. Large burrows with multiple entrances (aardvark diggings, erosion-gulley holes) are preferred over simple chambers (fig. 12.9).

Fig. 12.10. Warthog piglets greeting mother.

Group members greet after a separation with explosive grunts and *nose-to-nose contact* (fig. 12.10). They also *social-groom,* which may include stripping the long mane hair through the lips or incisors. To solicit grooming, one warthog lies prone before another.

ACTIVITY. Highly diurnal, females and young go underground before dark, but males often stay up an hour or 2 later. Daytime activity varies according to season and weather. Hogs emerge later on cold or rainy days, seek deep shade earlier and spend more time wallowing on hot days. They have early-morning and especially late-afternoon feeding peaks, but also graze between irregular resting/sleeping periods of roughly an hour. Distance traveled per day averages c. 7 km (2).

SOCIAL BEHAVIOR

COMMUNICATION

Olfactory Communication. Both sexes mark objects and one another with the tusk and preorbital glands, but males more frequently (fig. 12.11). Boars but not sows urinate in their wallows.

AGONISTIC BEHAVIOR

Dominance/Threat Displays: *lateral presentation* with *strutting* (fig. 12.12), *grunting, growling, woooomph warning, head jerk, mock attack, chasing.*

Defensive/Submissive Displays: head-low posture with mane lowered and ears flattened, *squealing* and *squealing growl, bolting.*

Fighting: *frontal pushing* and *lateral swiping* with upper tusks and snout.

Warthog combat is a highly ritualized pushing/boxing match, the object of

Fig. 12.11. Warthog rubbing preorbital-gland secretion on tree (redrawn from Cumming 1975).

which is to strike the opponent in the head with the tusks and muzzle and push him off balance while holding him off with tusks and snout (fig. 12.13). There is no attempt to slash or stab with the lower tusks. The warts absorb the blows and the big pair helps shield the eyes (3).

REPRODUCTION. Wherever there are marked seasonal climatic changes, warthogs are seasonal breeders, rutting at the end of the rains or early in the dry season and farrowing near the beginning of the rains. In East and southern Africa, for instance, the yearly crop of young usually appears in October or November (3, 9). But in Queen Elizabeth Park and probably other populations living in higher-rainfall equatorial regions, there is no strict breeding season (1). Gestation is 160–170 days, and the average litter is 2–3 (range 1–8). Both sexes reach puberty at 1.5 years.

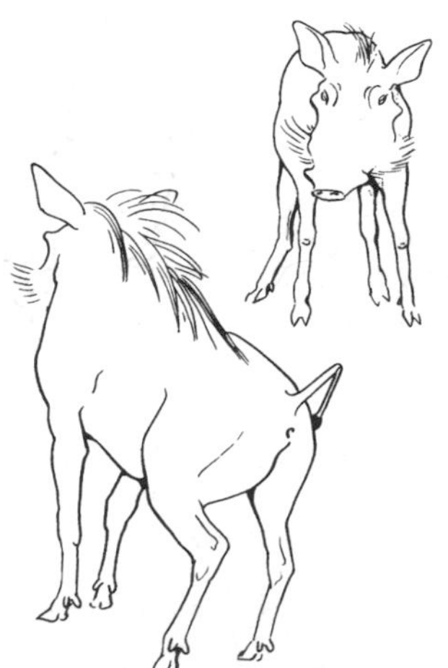

Fig. 12.12. Warthog *lateral display* (*left*).

SEXUAL BEHAVIOR: *strutting, close-following, champing.*

During the rut, boars locate estrous females by visiting burrows, and either wait for them to emerge or track them (3). Estrous females urinate often and a discharge from the swollen vulva discolors the rear end. Males intently check the urine and may deposit their own. A boar first *struts* to demonstrate his dominance, then follows the female closely in a springy, hip-rolling gait, tail out and bent, and tries to rest his chin on her, meanwhile making clacking noises that sound like a 2-stroke engine. Copulation may take less than a minute; estrus lasts up to 3 days (1).

PARENT/OFFSPRING BEHAVIOR. Sows separate from their families to farrow in a hole where the young remain for 6–7 weeks except for brief excursions or to change burrows (fig. 12.9). The mother rarely leaves the hole in the first week and eats less than usual for several weeks. Between 3 and 6 weeks the young are suckled 12–17 times a day at about 40-minute intervals (fig. 12.8). Piglets start grazing within 2–3 weeks and are weaned by 6 months. All sounder activities are closely coordinated and juveniles file in a fixed order (6). Newborn piglets express discomfort with a birdlike squeak followed by a churring *eeek-chrrr.* Separated juveniles rush around giving high squeaks, and mothers answer with a brief series of grunts. Soft, low grunts summon young from the hole, and a single, snortlike grunt is an alarm that causes piglets to freeze, then race for the nearest hole (3).

Fig. 12.13. Warthog fighting consists of *frontal-pushing.*

ANTIPREDATOR BEHAVIOR: *trotting with tail vertical* (fig. 12.2), *alarm grunt, distress squeal, growl-grunt* and *woooomph call, mock* and real *attack,* flight to underground refuge and defense thereof.

Whether trotting with the tail erected like an antenna is really an alarm signal is debatable, since the tail seems to be automatically erected whenever a warthog trots. Warthogs are slower with less endurance than most savanna antelopes and when pursued head for the nearest hole. Juveniles pile in headfirst—and sometimes come to grief: 5 piglets ran into a culvert and 4 immediately ran out again, the fifth was snatched by a resting hyena (author's observation). Mortality in the first year may be as low as 10% but can be 64% or more (3). Adults reverse direction at the last second, enabling them to use their tusks against an enemy that tries to enter the hole. Lions sometimes take the trouble to dig warthogs out (8).

References

1. Clough 1969. 2. Clough and Hassam 1970. 3. Cumming 1975. 4. Estes, Cumming, Hearn 1982. 5. Kingdon 1979. 6. Monfort 1974. 7. Robinette 1963. 8. Schaller 1972*b*.

Added Reference

9. Mason 1986.

Chapter 13

Hippopotamuses

Family Hippopotamidae

Hippopotamus amphibius, common hippopotamus
Choeropsis liberiensis, pigmy hippopotamus

FAMILY TRAITS. Bulky, piglike conformation, amphibious habits, naked skin with mucous glands, 4-toed feet. *Teeth:* lower canines enlarged as tusks, simple, low-crowned molars, jaws with extreme (150°) gape. Laterally flattened tail fringed with modified hairs (18), internal testes and recurved (backward) penis; short gestation period (6–8 months), single young; long-lived (35–50 years).

Senses: Smell, hearing, and night vision all important; diurnal vision also good in *Hippopotamus.*

BEHAVIORAL COMPARISION. The main behavioral traits shared by the two species are:

- Nocturnal activity
- Marking with dung and urine, during which the excrement is scattered by side-to-side beating of the tail
- The female lies down during copulation
- The young can nurse underwater

ANCESTRY. Pigs and hippos had common ancestors which diverged in the late Eocene, but the early history of hippos is unknown. They appear in the fossil record only in the lower Miocene (see table 1.1), in a form already more advanced than the living pygmy hippo. There is no record of *Choeropsis's* ancestry nor links between the latter and *Hippopotamus.* However, the pigmy hippo is living evidence that hippos started out as solitary, forest-dwelling herbivores with semiaquatic habits. Whether the line arose in Africa or Eurasia is unknown, but during the Pleistocene a number of different species lived on both continents. In Kenya's Lake Turkana, 4 different hippos, at least 1 more specialized than *H. amphibius,* formerly shared the ecological niche now occupied by this 1 species (2, 7).

Fig. 13.1. Hippo *yawning threat display* (redrawn from Kingdon 1979).

Common Hippopotamus

Hippopotamus amphibius

TRAITS. Barrel-shaped and short-legged; head with enormously expanded muzzle, noticeably bigger in males, eyes, nostrils, and ears all placed on top. *Height and weight:* 140 (up to 165) cm, males 1600–3200 kg, females 1400 kg (655–2344) (9, 12). *Coloration:* brown to gray-purple with pink underparts and creases, naked and smooth except for short bristles on head, back, and tail. *Teeth:* canines and lower middle incisors enlarged, especially in males, continuously growing, incisors round, 30–40 cm, tusks up to 50 cm, kept sharp by honing against short upper canines. *Skin glands:* no scent or sweat glands described, but mucous glands secrete a viscous red fluid that dries like lacquer and serves to protect the thin epidermis against water loss, sunburn, and perhaps infection (11). *Mammae:* 2.

DISTRIBUTION. Formerly everywhere south of the Sahara in suitable habitat. Now extinct or rare in most populated areas.

RELATIVES. The pygmy hippo, standing 70–80 cm and weighing 250–273 kg, occurs only in the lowland forests of West

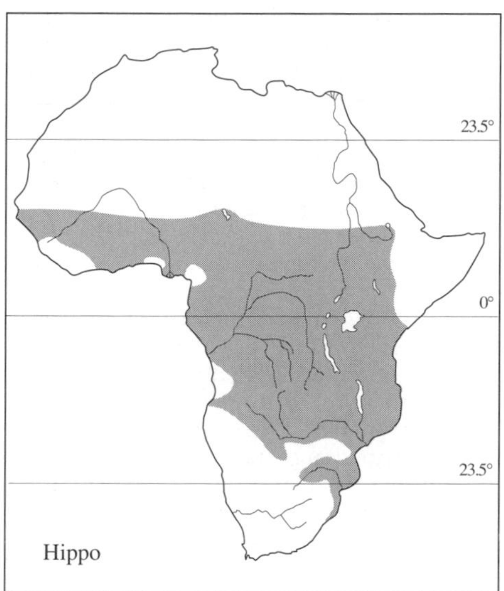

Africa from Guinea to the Ivory Coast. Found on the edges of swamps and rivers, it spends the day on land, in a hole, a wallow, or dense undergrowth. At night it enters the water to feed on aquatic plants. It also forages in the forest but stays near water, which serves as a refuge from predators (4, 19).

ECOLOGY. The 2 essential requirements are water deep enough to submerge in and nearby grassland. Because it has a unique skin (thin epidermis, no sweat glands) which loses water at several times the rate of other mammals, a hippo out of water in hot weather risks rapid dehydration and overheating. A gently sloping, firm bottom, where herds can rest half-submerged and calves can nurse without swimming, is preferred and rapids are avoided (12). Muscular lips up to 50 cm wide enable hippos to graze a broad swath and harvest short grasses. They consume c. 40 kg of grass a night, or only 1–1.5% of body weight, feeding rapidly, intently, and noisily. Lawns kept short by repeated grazing are preferred over longer, coarser grassland. At high density (up to 31/km^2), hippos can have a devastating impact on the vegetation and soils (3, 7, 9).

SOCIAL ORGANIZATION: highly gregarious, contact species in water, but solitary when foraging; males territorial according to (9), nonterritorial with a dominance hierarchy according to (12).

The hippopotamus is socially schizophrenic. In the water or resting ashore, it tolerates closer contact than most other ungulates but nevertheless—or perhaps therefore—is highly aggressive. When foraging, hippos are unsociable: each animal becomes an independent unit, except for females with dependent offspring. Being largely immune from predators, adults need not herd together for safety on land. In the water, it is hard to tell whether clustering has evolved for protection of calves against crocodiles, or simply to accommodate the maximum number of bodies. Either way, aggression or trampling by other hippos, especially bulls, is now the greatest danger to calves (see below).

Herds typically number 10–15, but range from 2 to 50 and, at very high density in large bodies of water, sometimes up to 150 hippos. Average density in lakes is 7 hippos/100 m of shoreline and 33/100 m in rivers (9). Crowding is greatest during the dry season; hippos disperse more widely during the rains, especially males, and may occupy temporary pools and even wallows many kilometers from permanent water (7, 9, 15, 16).

Mature bulls (20 years and up) control 50–100 m sections of a river or 250–500 m of lakeshore and shallows as exclusive mating territories (9). Some individuals have been known to keep the same territories for at least 8 years in lakes and 4 years in rivers. But under more variable climatic conditions which cause major changes in grouping patterns and density, territorial turnovers may occur every few months (9). Territorial (or if nonterritorial, alpha) bulls usually tolerate bachelor males within their domain and even in cow herds, so long as they behave submissively and refrain from sexual activity. But at times they drive out potential rivals with great ferocity, and bachelor herds are also found living apart from females in the least favorable habitat; lone hippos may be either outcasts or territorial bulls without herds (9).

The infrastructure of female herds remains to be explained. Although herd composition may remain fairly consistent for several months at a time, there appear to be no close ties between cows. However, maternal bonds with daughters persist at least to the subadult stage. Accordingly, cows may be followed by up to 4 successive offspring (7).

ACTIVITY. Hippos feed at night on land, rest and digest by day in the water. Shortly before dark they begin to commute to inland pastures along branching paths up to 2.8 km long. After grazing for up to 5 hours, they return before dawn, having traveled 3–5 km (maximum 10 km).

POSTURES AND LOCOMOTION. The nostrils close and the ears fold into recesses when a hippo dives. Resurfacing, the nostrils open wide releasing the pent-up breath, and the ears spring erect throwing showers of droplets. Mature hippos can stay under up to 5 minutes, but on average surface every 104 seconds, compared to 20–40 seconds for a 2-month calf. Hippos rise to breathe even in their sleep, the act of surfacing being as involuntary as breathing itself (7). Swimming hippos in effect gallop underwater; they also walk on the bottom while submerged, as lightly as astronauts on the moon. On land they look clumsy but can gallop up to 30 kph in an emergency, although a jouncy trot is normally the fastest gait. They can climb steep banks, yet cannot jump and are reluctant even to step over obstacles. They lie down and get up like pigs, sitting down on their haunches before reclining and rising front legs first (fig. 13.2).

Fig. 13.2. Seated hippo (from a sketch in Kingdon 1979).

SOCIAL BEHAVIOR

COMMUNICATION. Since hippos spend their days submerged and come ashore at night, show very little sexual dimorphism, have inconspicuous appendages and coloration, and have no facial expression apart from an open mouth, auditory, olfactory, and possibly tactile communication should be particularly important in this species. The resonant honking call made by submerged hippos is one of the most familiar and impressive African wildlife sounds (but on land hippos are notably silent). The way hippos use their exhaled breath to express threat and alarm is described under Agonistic and Antipredator Behaviors. *Dung- and urine-showering* in the water and on land (described below) is clearly of central importance in hippo social life (4, 7, 12). In addition to normal olfactory reception, *urine-testing* with the vomeronasal organ presumably functions to communicate the reproductive status of females and possibly of males as well (urine-testing, *flehmen,* and the vomeronasal organ are described under Courtship Displays, p. 25). Although hippos do not make the *flehmen* grimace that signifies urine-testing in most other ungulates, the hippo vomeronasal organ appears to be designed to function under water, operating like a syringe bulb to draw in a sample of voided urine (large incisive ducts lead to the vomeronasal organ from the mouth but do not penetrate, as usual, into the nasal passageway; a thick layer of muscle surrounds the ducts and organs [author's observations]). This possibility gives added significance to the habit of females and subordinate males of urinating when approached by bulls.

Visual signals obviously do play a role in the daytime interactions of submerged hippos, notably in threat displays such as *yawning* and *charging; dung-showering* is also conspicuous and the performance varies according to social rank (see below). The greater development of male teeth, the hippo's most obviously dimorphic character, is doubtless observed by individuals addressed with a *yawning display.*

TERRITORIAL BEHAVIOR: *wheeze-honking, dung-showering* (fig. 13.3).

Territorial bulls have frequent ritualized encounters. After approaching the common boundary, they stop and stare at each other, then turn tail, elevate their rumps, and shower dung and urine over each other with rapidly paddling tails, following which they withdraw (9). The territorial significance of the conspicuous dung middens that are found along hippo paths has been questioned (9, 12, 15). Various passing hippos add their excrement, which may well assist in orientation and communication at night. However, a number of observers believe that only bulls use dung middens (1, 12, 15). Paths start at the water's edge and therefore lead out of and into territories, whose owners take a proprietary interest in them. Middens are most frequent near shore, and are renewed nightly by bulls on the way to pasture. They also emerge by day to defecate on shoreline dung heaps, and on rocks or islands in the water (12). Preparatory to marking, a bull smells the existing deposit, then backs up and urinates and defecates to the rear, broadcasting his excre-

Fig. 13.3. Bull hippo *dung/urine-showering* on established dung midden.

ment by vigorous tail-paddling (fig. 13.3). Bull excrement is particularly smelly and interesting to other hippos; a juvenile often follows a bull on the way to a midden, intently smelling or licking his anal area, then spends minutes nosing and even eating his excrement (12).

AGONISTIC BEHAVIOR

Dominance/Threat Displays: *yawning* (fig. 13.1) ± *water-scooping* and *head-shaking, rearing, lunging, roaring* and *grunting, chasing, explosive exhalation* above or below the surface, *dung-showering.*

Defensive/Submissive Displays: *facing aggressor with mouth open,* in self-defense, *turning tail, slow tail-paddling while urinating, lying prone, flight.*

Fighting: *tusk clashing, rearing and pushing* with lower jaws engaged, *slashing* and biting.

The above display inventory bears out the quip about hippos that "all social interactions revolve around the 2 opposite poles of the body" (7, p. 260). Territorial bulls are mutually intolerant, are frequently beastly to younger males, and surprisingly often attack and even kill calves (15). Mothers in turn may attack bulls that disturb a nursery herd (see under Sexual Behavior).

The full significance of showering excrement by tail-paddling and differences in this behavior between males and females have yet to be clearly explained. In males it is apparently an assertion of dominance, if not also a form of territorial advertising. When females and subordinate males are approached by a bull, they respond by turning their rears and tail-paddling, but slowly, splashing the water and probably urinating but not necessarily defecating (4, 12, 15). This behavior appears very similar to the *urination-on-demand/urine-testing* sequence so common among antelopes. And like them, perhaps hippos also employ this tactic to appease or divert aggression. Likewise, *lying prone* in submission may gain added appeasement value through resemblance to the posture of estrous females during copulation.

Males begin testing themselves by the time they become adolescent at 7 years (4–11), engaging with their peers in *jaw-to-jaw sparring* and *yawning* contests. Any disturbance in a hippo pool will set off a wave of *wheeze-honking, yawning,* and sometimes chasing and fighting. Usually fights are undamaging, because the combatants can fend each other off with their jaws. But disengaging is a problem for the loser, who thereby exposes his body to the winner's tusks. Deep gashes are cut during hot pursuits, but often look worse than they are, for skin up to 6 cm thick protects the hindquarters and sides. The most serious injuries result from crushing bites on the legs, head, and neck (7).

Aggression is most frequent and intense during the dry season, when living conditions are most crowded. Serious injuries and deaths from fighting are not uncommon, and this is also the time when most attacks on calves occur (12).

REPRODUCTION. Breeding is not strictly seasonal but most conceptions probably occur in dry-season concentrations, and the rainy season is the time of peak births. Females conceive at c. 9 years (7–15) and calve at 2-year intervals. Gestation is 8 months (9, 12, 17).

SEXUAL BEHAVIOR. Bulls questing for mating opportunities may wander through basking nursery herds sniffing at cows' backsides, at the risk of being mobbed should the cows become disturbed. To avoid that, a bull moves very carefully and at the first sign of trouble, lies down—a display of submission that is unusual in a dominant male animal (15).

Having located an estrous female, a bull hippo wastes no time on courtship displays. He pursues her into the water until she turns and clashes jaws with him, then forces her into prostrate submission, whereupon he mounts. The female's head is often forced underwater, and when she raises it to breathe, the bull may snap at her (7). Courtship is often punctuated by *wheeze-honking.*

PARENT/OFFSPRING BEHAVIOR. Cows isolate prior to calving, on land or in shallow water, rejoining the herd after 10–14 days. Baby hippos (birth weight 25–55 kg) are adapted for nursing underwater. Even out of water, their ears fold and their nostrils close while sucking (9). Every few seconds a submerged sucking calf pops to the surface, breathes, and goes back to the nipple, gripping it between the tongue and the roof of the mouth. Calves graze a little by 1 month, and in earnest by 5 months; they are weaned at c. 8 months (9, 12).

Expectant and new mothers tend to be savagely protective, keeping other hippos at a distance. The bond between mother and calf is close. The mother licks, nuzzles, and scrapes the calf with lower incisors, and calves reciprocate. However, small calves may be left in crèches guarded by 1 or a few hippos while their mothers are away at pasture. When accompanied by young on land, a mother punishes any tendency to stray by *nudging, sideswiping,* or even biting the calf, which responds by prostrating itself (15). Calves in crèches often engage in playfights and chasing games.

ANTIPREDATOR BEHAVIOR. Baby hippos would be easy prey for lions, hyenas, or crocodiles if not protected by their mothers. A hippo's jaws are wide and powerful enough to bite a 3-meter crocodile in two (14). A bull hippo that was set upon by 3 famished lions dragged and carried them along until he reached the river, when he entered and submerged, little the worse for wear, leaving the lions to swim back to shore (7). Nevertheless, lions are capable of killing a full-grown hippo if they can get it down on its back, with its throat and chest exposed to their jaws (5).

References

1. Bourlière and Verschuren 1960. 2. Coryndon 1978. 3. Field 1970. 4. Frädrich 1967. 5. Guggisberg 1961. 6. Hediger 1951. 7. Kingdon 1979. 8. Laws 1968. 9. ——— 1984. 10. ——— Laws and Clough 1966. 11. Luck and Wright 1964. 12. Olivier and Laurie 1974. 13. Rode 1943–44. 14. Stevenson-Hamilton 1947. 15. Verheyen 1954*b*.

Added References

16. Karstad and Hudson 1986. 17. Kayanja 1989. 18. Kranz 1982. 19. Lang 1975. 20. O'Connor and Campbell 1986. 21. Owen-Smith 1988.

Odd-toed Ungulates

Order Perissodactyla

Rhinocerotidae, Rhinoceros Family
Equidae, Horse Family
Tapiridae, Tapir Family (Asia, South America)

The members of this order preceded the even-toed ungulates (Artiodactyla) as dominant herbivores in the Eocene and Oligocene, a period of some 20 million years (table 1.1). Of this incredibly diverse group, which included 12 different families, with 350 species in the Equidae alone, today only 15 species in 3 different families survive. Africa is home to 6 of them: 2 rhinos, the 3 zebras, and the wild ass. In this order the middle (third) digit came to bear the main weight of the limb. Evolutionary refinement caused the other digits to be discarded until only the third toe was left in modern horses (6). The intermediate stage is seen in the 3-toed rhinos.

Comparing communication in the 2 orders, perissodactyls have few specialized scent glands such as the foot and preorbital glands which are so common in the artiodactyls. Nevertheless, they demarcate trails and property boundaries with urine and dung, and olfactory communication is important. The mouth and lips are more mobile in perissodactyls, and the teeth are weapons in tapirs and equids; facial expressions are accordingly important. Perissodactyls are also highly vocal.

Chapter 14

Rhinoceroses

Family Rhinocerotidae

African Species

Diceros bicornis, black rhinoceros
Ceratotherium simum, white rhinoceros

FAMILY TRAITS. Graviportal form typical of massive (and mostly extinct) mammals: vertebral column a girder balanced on the forelegs, with the head counterbalancing the body weight and the hindlimbs providing the main propelling force, the wide-set, pillarlike limbs and 3-toed feet making a firm foundation (6, 7). Head rather turtle-like, ending in beaklike snout (all except white rhino); small, puffy eyes low on the cheeks, and wrinkled skin. *Teeth:* massive molars and premolars adapted for grinding coarse vegetation, no incisors or canines in African species. *Horns:* not composed of modified hairlike fibers but of keratin, like true horn, though lacking a bony core and unattached to the skull, resting on nasal- and frontal-bone pedicels, continually growing (7). *Skin:* thick and hairless except for ear fringes and stiff bristles terminating the long, thin tail. *Coloration:* uniform gray to dark brown. Testes internal and penis recurved to rear (like hippos). *Scent glands:* none described in African species except possible preputial gland in white rhino (1). *Mammae:* 2. Brain comparatively small; longevity 35–50 years.

Senses: weak eyesight, hearing and smell both acute.

ANCESTRY. The ponderous, armor-plated rhinoceroses look like relics of a bygone age. Of a diverse group once found throughout Eurasia, North America, and Africa, with over 30 different genera in the fossil record, only 3 genera and 5 species survived into modern times. All except the well-protected white rhino now face extinction because rhino horn is literally worth its weight in gold (5). Stemming from small tapirlike ancestors, the first true rhinos appear in the Oligocene as already large but still hornless terrestrial browsers (2, 3, 4, 6). They reached their peak in the Miocene and Pliocene (table 1.1). The African rhinos belong in a separate subfamily of 2-horned rhinos, along with the Sumatran rhino (*Didermocerus sumatrensis*), the smallest and most primitive of the 3 Asian species. *Diceros,* the direct forerunner of the black rhino, is known from the early Pliocene of Europe, but the white rhino, *Ceratotherium,* the last and most specialized species, is wholly African and known only from the Pleistocene.

BEHAVIORAL SUMMARY. Like other hairless mammals, rhinos are confirmed wallowers. They also dustbathe, rub and scratch themselves against objects, regularly visit salt licks, and readily enter swamps. They can cool off by sweating, too, thereby losing moisture which they have to make up by drinking or by browsing water-rich plants (4). The lips are used to gather browse and grass, taking the place of the incisors. Rhinos sleep soundly, lying on their briskets with legs gathered, sometimes on their sides, or while standing. They alternately feed and rest day and night, being least active and wallowing most often when it is hot. They are generally sedentary and solitary, but the white and also the black rhino are at least semisocial when young, and even territorial bulls tolerate subordinate males. Rhino home ranges are dissected by regularly used trails and scent-posted with dung middens used by both sexes. Territorial bulls also mark by spraying urine (fig. 14.5). African rhinos have much the longest horns, which are employed as stabbing weapons against predators but usually only as staves in encounters with other rhinos (4). The white rhino is far less excitable and dangerous to man than the black rhino. Rhinos make a variety of sounds, from loud puffing snorts of alarm to high-pitched squeals quite absurd for such behemoths. Both courtship and mating are remarkably prolonged. Gestation lasts 15–16 months and the single offspring remains with the mother 2–4 years.

References

1. Cave 1966. 2. Colbert 1969. 3. Hooijer 1978. 4. Kingdon 1979. 5. Martin and Martin 1987. 6. Romer 1955. 7. Young 1962.

Added Reference

8. Owen-Smith 1988.

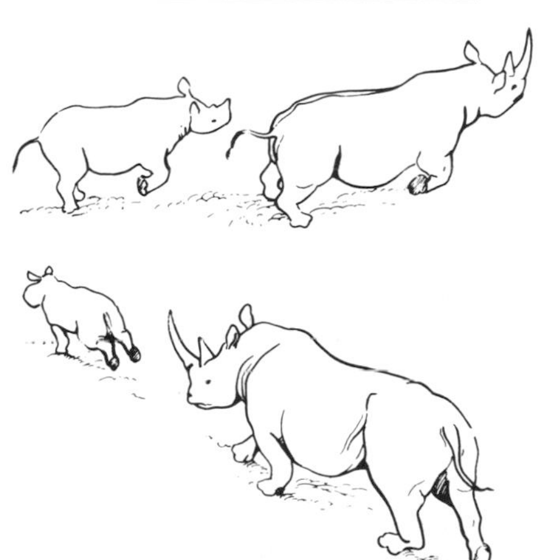

Fig. 14.1. Black rhino calves follow and white rhino calves lead their mothers during flight.

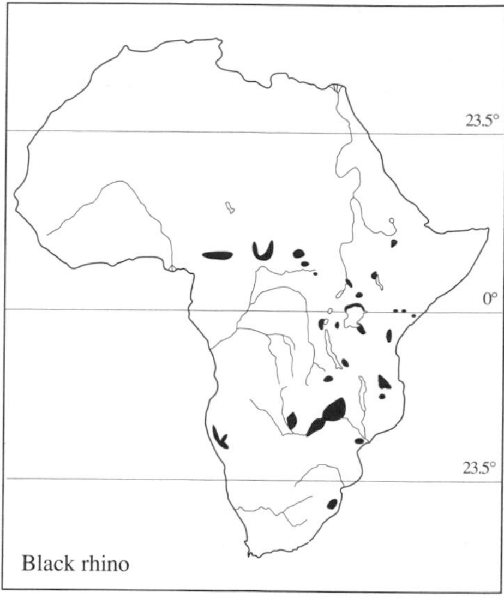

Black Rhinoceros

Diceros bicornis

TRAITS. No taller than a buffalo but twice as heavy: 1.7 m (1.4–1.8), 996–1362 kg (7); females as big but ±100 kg lighter. *Coloration:* dark gray. *Horns:* extremely variable in shape and length, front horn up to 132 cm, rear horn shorter, more flattened sideways. Upper lip triangular and flexible, used as grasping organ.

DISTRIBUTION. Formerly the most widespread and numerous rhino and a notably successful member of the herbivore community. The range and number of *Diceros* dwindled steadily with the increasing human population. Its decline has been tremendously accelerated in the last decade by poaching to meet demand for rhino horn from Arab and Asian countries (8), and it is rapidly approaching extinction. Yet all it needs is protection: it is actually increasing in the same Zululand reserves where the white rhino lives.

ECOLOGY. A nearly pure browser with a marked preference for leguminous herbs and shrubs, the black rhino eats over 200 plants in 50 different families. Although they can go 4–5 days without drinking by chewing succulents (7, 9), rhinos may travel 8–25 km to water nearly every day (5), and dig for water in sandy riverbeds during droughts (7). When wetlands are available, rhinos frequent them in the dry season. The prehensile upper lip enables *Diceros* to browse selectively, but it uses its molars to take big bites and to eat woody or fibrous plants. The sound of munching carries up to 400 m. The horns are sometimes used as tools to remove bark, break off high branches, and excavate soil at salt licks.

SOCIAL ORGANIZATION: sedentary, overlapping home ranges, females rarely alone, adult males usually solitary and possibly territorial.

This animal is not as solitary and antisocial as commonly portrayed, but instead rather similar to the white rhino. Thus, bulls may tolerate male acquaintances as long as they remain submissive (6). Although cows with calves tend to stay alone, they sometimes allow an unrelated immature male or female to attach itself following rejection by its own mother. Such an attachment may continue to maturity (4). In Ngorongoro, 2 cows associated together for 13 months, and after mating, an adult male and female, her calf, and a subadult male

Fig. 14.2. Male black rhinos confronting horn to horn.

stayed together for 4 months (4). Rhinos often associate briefly but amicably when they meet at water holes and salt licks. Up to 13 have been seen in a group (4).

Yet displays of bluff and bluster are typical of black rhino meetings, indicating a degree of tension. An encounter between a bull and a cow is usually marked by *puffing snorts* from either or both. The bull then approaches with cautious steps, head lowered and ears forward, thrashing his head from side to side or rooting the air with his horns. This is pure bluff, for if the cow advances in turn or lunges at him, he wheels and gallops off, only to circle and make another cautious approach. As the preliminary to courtship, this performance may continue for hours and lead to the formation of a tending bond if the female is nearing estrus. Females also approach each other cautiously though less aggressively. They usually end up horn to horn (fig. 14.2) and gingerly nudge heads or joust with their spears, finally separating with evident indifference. Male encounters may be aggressive or peaceful. At least some bulls have overlapping home ranges; meetings on their common ground tend to be peaceful, including gentle head and horn nudging and resting side by side. More often bulls avoid one another after establishing identity through an approach that may include dominance and threat displays. But a resident bull "invariably attacks" when he meets a stranger in his home range (4)—or is it a territory? The question is still unresolved. Black rhinos studied in Natal seemed to be very similar to white rhinos (q.v.): radio-tracked bulls had territories of 3.9–4.7 km², and female ranges were 5.8–7.7 km² (6).

Home range size varies greatly, depending on the habitat and to some extent on sex and age. The average in Ngorongoro was 16 km² (2.6–44) for both sexes (4). But Serengeti rhinos had ranges of 43–133 km² (2). Subadult males wander most widely.

ACTIVITY. Most of the time rhinos do little more than sleep and eat. Like other wildlife, they tend to be most active early and late in the day and least active at midday. They move about and feed more but also spend time sleeping at night (4, 10). Although they will lie out in hot sunlight, they tend to wallow during the hottest part of the day, smearing mud on both sides and sometimes rolling on their backs.

POSTURES AND LOCOMOTION. See Antipredator Behavior.

SOCIAL BEHAVIOR

COMMUNICATION

Olfactory Communication: scent-marking by *dung-scattering, urine-spraying.*

Rhino dung middens are scattered seemingly at random and used by all passersby. The depositor sniffs the pile of dung intently, and may sweep and root it with the forehorn (especially males), then *shuffles* through it stiff-legged before defecating on the same spot. After or less often during the act, the animal scatters its dung by kicking backward in slow motion. Calves have been known to defecate in imitation of their mothers (12). Dung middens undoubtedly serve as message stations, and the odor of the individual's dung is also dispersed on its feet (5). In addition, adult males perform a *urine-marking ceremony.* Using a bush or shrub as a target, the bull backs up and sprays it 2–4 times (fig. 14.5). Sometimes one attacks and demolishes bushes after *urine-spraying* or *dunging* (4, 10).

Vocal Communication: *puffing snorts, squealing, shrieking, grunting, groaning, mewing.*

Black rhinos produce *puffing snorts* during tense encounters with conspecifics and other species. They also *shriek* like pigs and *grunt* and *groan* when fighting; and females *squeal* while mating. A cow separated from her calf calls it with a high, thin *mewing.* The calf's distress cry, a throaty *squeal,* summons not only the mother but other rhinos on the double.

AGONISTIC BEHAVIOR

Dominance/Threat Displays: *stiff-legged walk* with *lateral presentation,* *dunging ceremony* with *horn-sweeping, approach with head lowered, ears flattened, eyes rolling,* and *tail raised,* giving *puffing shrieks, feinted attack* with *puffing shrieks* and *empty horn thrusts.*

Defensive/Submissive Displays: confronting aggressor silently in *head-low* (combat) *posture, backing away,* slow or rapid *retreat.*

Fighting: *jousting* (horn-clubbing) or *stabbing* (rare), with vocalizations (see Vocal Communication).

A resident bull approaching a stranger produces a *screaming groan* with the upper lip curled (3). Meanwhile the intruder faces him silently with head lowered to defend himself should the other attempt to club or gore him. A bull that turns tail and runs for it is immediately pursued, often for over 2 km. Usually the outsider withdraws slowly and deliberately. The other dominance and threat behaviors appear during encounters between equals, which may lead to or include *jousting.* Fights are likeliest to occur in the presence of an estrous female, but seem rarely to result in serious injury (6).

REPRODUCTION. The black rhino has no strict breeding season. Females mature at c. 7 years, but individuals of 3.8–5.7 years have been known to conceive; males are mature probably by 10, but one 4.3 years of age was seen mating. Gestation is 15–16 months, lactation continues for 1–2 years, and calving intervals range from 2¼–4 years (7, 9).

SEXUAL BEHAVIOR. Male: cautious approach and *urine-testing, retreat* and *circling,* dominance and threat displays (*rushing, jabbing, horn-clubbing, puffing, dunging ceremony*), *nudging* with horn and head, *rubbing muzzle on female's sides and shoulders, resting chin on croup.* Estrous female: threat and feinted or real attack, frequent *spray-urination, mouthing, squealing.*

A bull that detects a female approaching estrus—by sampling her urine with a pronounced *grimace* (fig. 14.3)—becomes her consort and begins the time-consuming process of conditioning her to accept contact and eventually copulation (see white rhino account, fig. 14.7). When a cow is finally in full estrus, the bull begins a series of *preliminary mounts without erection* which may be repeated for several hours, interspersed with bouts of feeding and walking. Copulation lasts for about ½ hour, during which the cow stands quietly, sometimes emitting a *low-pitched squeal* and

Fig. 14.3. Male black rhino *testing* female's urine.

making *mouthing expressions.* A female may mate with several different bulls during a 3-day estrus.

PARENT/OFFSPRING BEHAVIOR. In the 1 published observation of a wild birth, the calf was dropped within 10 minutes of becoming visible, while the mother stood (7). She then removed the birth sac and the calf stood within 10 minutes. Before and after calving, females are particularly irritable, and at this stage the previous offspring, by now 2½–3½ years old, is driven away and may seek a substitute companion. Calves follow very closely (fig. 14.1) (cf. white rhino) and in case of alarm companions press their rumps together and face outward. Calves rarely have peers to play with, but have been seen tossing vegetation and picking up sticks (1).

ANTIPREDATOR BEHAVIOR: *puffing snorts, distress squealing,* mock and real charges with stabbing, *flight with tail raised.*

A mother has been seen to kill a lion that tried to take her small calf, but mothers may also ignore attacks on their offspring until alerted by the calf's squeals. Despite their bulk, rhinos can gallop at 50 kph and turn in their own length (7). When given advance warning of danger—by the chirring of attendant oxpeckers, scent, or noisy movement—rhinos will usually make off at a fast trot, tail curled over the back and head high.

References

1. Frame 1971. 2. ——— 1980. 3. Goddard 1966. 4. ——— 1967. 5. ——— 1970. 6. Hitchins 1968, 1969. 7. Kingdon 1979. 8. Martin and Martin 1987. 9. Owen-Smith 1984*b*. 10. Schenkel 1966*b*. 11. Schenkel and Schenkel-Hulliger 1969. 12. Spinage 1962.

Fig. 14.4. Territorial white rhinos in confrontation.

White Rhinoceros

Ceratotherium simum

TRAITS. The biggest land mammal after the elephant (but outweighed by hippo). *Height and weight:* 171–183 cm; males 2040–2260 kg, females average 1600 kg (4). Twice the bulk of the black rhino with proportionally longer, larger head, wide, square mouth, large ears, and pronounced hump. *Horns:* average c. 60, maximum 150 cm. *Coloration:* slate gray to yellow brown. *Scent glands:* preputial papillae with possible olfactory function (1).

DISTRIBUTION. Widely distributed in the Northern and Southern Savanna within human memory (cave paintings), this species has been in decline for centuries because of failure to adapt to human predation and space competition (see black rhino, map). The southern population was wiped out by early in this century except for a few hundred in Natal. The northern population (*C. s. cottoni*), which ranged west of the White Nile in Uganda, Zaire, Central African Republic, and southwestern Sudan, has been virtually exterminated within the last few years, following civil wars and general chaos in these countries. Fortunately the approximately 2000 white rhinos in Natal's Umfulozi and Hluhluwe Game Reserves have produced a surplus, which has gone to restock parks within (and outside) the species' former range (7).

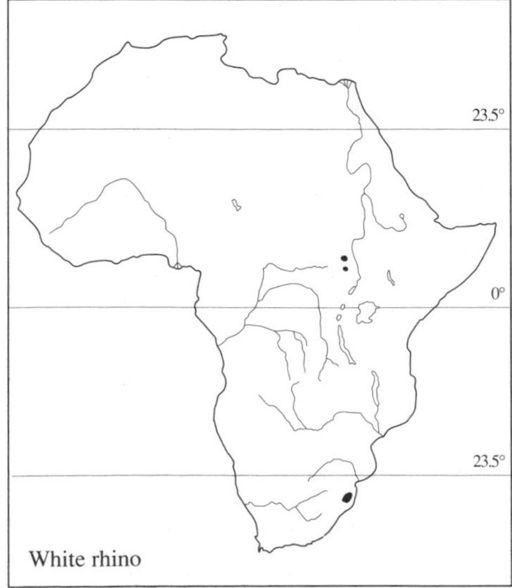

ECOLOGY. Possibly the largest pure grazer that ever lived, this bulk feeder has a wide, square mouth which enables it to graze most efficiently in dense swards of short, green grass (cf. hippo) (4, 8). Good habitat includes trees and ample water as well as open grassland. It often drinks twice daily but can go 2–4 days without drinking if water is distant. White rhinos near the White Nile used to concentrate in wetlands during the dry season and migrate at least 10 km inland to higher ground during the rains (2).

SOCIAL ORGANIZATION: sedentary, semisocial, territorial and satellite males.

Adult females live in overlapping home ranges which encompass 6–7 territories averaging 1.65 km^2 (0.8–2.6). Females and subadults are rarely solitary. They associate in pairs, typically a female with her most recent offspring. A juvenile rejected at 2–3 years when the mother calves again seeks another companion, preferably of the same age and sex, but may attach itself transiently to another cow with calf. A calfless cow will tolerate 1 or more juvenile substitutes, and 2 calfless cows may join forces. Stable herds of up to 6 head can be formed in this way; larger groups, of up to a dozen, represent temporary aggregations, especially during midday heat when white rhinos sleep on breezy ridges. Adult females rarely permit any individual except their partners to come closer than several meters, but the young are less standoffish and often engage in horn-wrestling.

Unlike cows and adolescents, adult bulls are basically solitary. Except to check the females' urine, they associate only with those in estrus. In Natal, all available range is partitioned into territories held by only ⅔ of the mature bulls, whose average tenure is 3 years. Perhaps because dispersal is prevented by fences, the other ⅓ live as satellites on the territories. One or occasionally 2–3 satellite bulls reside within a particular territory whose owner becomes conditioned to their presence to the point of ignoring them, as long as they behave submissively (cf. black rhino, waterbuck). The fact that territorial bulls treat nonresident intruders far more aggressively promotes the residency arrangement.

SOCIAL BEHAVIOR

COMMUNICATION

Olfactory Communication: *dunging ceremony, urine-spraying, scraping.*

A territorial male maintains 20–30 dung middens where he always defecates, scattering and excavating the pile with slow, ritualized kicks before and after. Females and young, and nonterritorial males, add their deposits but without kicking. Only territorial males spray urine and only while on territory. First the bull *horn-wipes* a bush or the ground, then *scrapes* with all 4 feet, then sprays this spot 3–5 times (fig. 14.5). Patrolling bulls *urine-mark* about 10 times an hour. They also pause to *scrape-mark* their paths every 30 m or so. Marking is most concentrated along territorial boundaries (4, 5, 6).

Vocal Communication. These are sounds commonly made by the white rhino.

Panting: a contact call common in groups, for example, by a calf that has lost visual contact with its mother.

Whining: juvenile begging call.

Squeaking: calf in distress.

Gruff squealing: chasing.

Snarling: defensive/submissive, grades into a shriek at highest intensity.

Chirping: a peculiar sound made by a rhino fleeing from an aggressor.

Hic-throb: male lovesong.

Loud wailing: by a courting male trying to prevent a female's departure.

Gasp-puffing: fright (alarm).

AGONISTIC BEHAVIOR

Dominance/Threat Displays: *silent approach with head raised and ears cocked, standing horn to horn* and *staring at opponent, horn-to-horn blow, feinted attack, wiping horn on ground.*

Fig. 14.5. Territorial male white rhino *spraying urine.*

Defensive/Submissive Displays: *head down and forward with ears back, snarl-threat/shrieking, feinted attack, backing away,* flight.

Fighting: *horn-fencing, shouldering, horn-thrusting* (very rare).

Aggression is rare and mild except on the part of territorial males. The owner examines the credentials of every rhino he meets on his land, approaching in the dominant manner described until within olfactory and visual range. Except for small calves, all—including passing territorial males forced to go to distant water holes—respond with defensive/submissive behavior (fig. 14.6). Females and adolescent males are seldom molested, but older males, particularly nonsatellites, often have to sue loudly for peace. The territorial bull continues his approach until he is *standing horn to horn* with the interloper and may then simply *stand staring* at him. Or he may *bang horns,* which the latter parries to the accompaniment of *trumpeting shrieks,* while *backing away.* Sometimes subadult and young adult bulls will make a break for it, but slower adults thereby invite pursuit and expose their rears to a stab from the extremely dangerous front horn. It is safer to face the danger and even to bluff an attack, but meanwhile screaming for mercy.

Fig. 14.6. Submissive posture (with ears back) of a male white rhino responding to threat behavior of a territorial male.

When territorial bulls meet at the border, both assert dominance (fig. 14.4). One may fake a charge by trotting up with lowered head, checking at the last moment, but usually they both walk up with heads raised and *stand horn to horn staring at each other.* Next they *back away* and *ground-wipe* with the front horn, readvance, *back up,* and so on. Sometimes they *clash horns* briefly. After interacting for anywhere from a few minutes to over an hour, both bulls *back off* one last time, turn, and go their separate ways, often after scraping and urine-spraying. A territorial bull that is caught trespassing lets himself be walked backwards by the owner until he reaches the boundary, whereupon he asserts his rights in the usual way. One of 2 serious fights that were seen in a 3-year study occurred when the trespassing neighbor refused to kowtow but chose to fight his way home across the owner's whole territory. Most of the time the combatants fenced slowly with raised heads, but occasionally one tried to get under the other's guard by suddenly dropping his head and *stabbing* upward. The resident bull succeeded several times in goring his rival in the head, shoulders, and body (5). Attempts to take over a territory or to head off an estrous female as she is about to enter a rival's land are likeliest to provoke serious fights. However, 2 of 3 observed changeovers came about without a long fight. Interestingly enough, each of the defeated bulls remained on his property, simply changing to the status of satellite male.

REPRODUCTION. Females calve for the first time at 6½–7 years, after a 16-month gestation. There is a conception peak early and a calving peak late in the rains in Natal (4). Estrus cycles occur at monthly intervals.

SEXUAL BEHAVIOR: *hic-throbbing* call, *urine-testing, following, chin-on-rump resting, preliminary mounting.* Estrous female: frequent urination, *mock charges* and *threat-snarls, curling tail and standing* for copulation.

Bulls make *hic-throbbing* sounds when approaching cows, only to be driven backward by *mock attacks.* When the urine test reveals a cow approaching estrus, the territorial bull begins a protracted courtship lasting from 5 to 20 days, during which he shows remarkable restraint. Apart from heading off any attempt to quit the premises (blocking the way and *chasing,* while *squealing*), he keeps the distance set by the female (5–30 m) until she comes into full estrus. Stimulated by her urine, he comes closer, crooning his lovesong, but retreats each time she threatens and even puts up with interference by her calf or juvenile companion. Eventually the cow allows him to *rest his chin on her rump,* followed a little later by *attempted mounting.* Finally, after 15–20 hours of persistent approaches, the female *stands with tail curled* during the half-hour copulation (fig. 14.7). The consortship usually continues another 2–5 days, even though repeat copulations are the exception (5).

Fig. 14.7. Male white rhino mounting female.

PARENT/OFFSPRING BEHAVIOR. Cows seek seclusion in dense cover before and for several weeks after calving. Calves can stand within an hour but remain wobbly for a couple of days. Within a month the cow resumes her normal routine and range, always closely accompanied by the calf, which moves in front but responds immediately to the mother's changes of direction. During flight the calf gallops ahead and the mother follows close on its heels (up to 40 kph) (fig. 14.1). The calf suckles on demand for 2–3 minutes or until satisfied, and begins grazing at 2 months. Weaning occurs at 1 year (4, 7).

ANTIPREDATOR BEHAVIOR. There are few reports of predation on the white rhino. Only calves are vulnerable to hyenas and lions, and the close maternal bond insures that the young are well-protected. When alarmed, companions press their hindquarters together and face in different directions. The white rhino has only 1 enemy: man. Unfortunately that is 1 too many.

References

1. Cave 1966. 2. Foster 1967. 3. Kingdon 1979. 4. Owen-Smith 1973. 5. ——— 1974. 6. ——— 1975. 7. ——— 1984*b*. 8. ——— 1988.

Chapter 15

Zebras, Asses, and Horses

Family Equidae

African Species
Equus burchelli, plains or Burchell's zebra
E. zebra, mountain zebra
E. grevyi, Grevy's zebra
E. africanus, wild ass

FAMILY TRAITS. Large, single-hoofed ungulates built for speed and long-distance movement. Males c. 10% bigger than females, otherwise little sexual dimorphism. *Teeth:* adapted for grazing, strong upper and lower incisors and large, high-crowned molars; males have spade-shaped canines used in fighting. *Coloration:* striped or solid-colored (asses and true wild horse dun-colored). Brushlike mane and long tail with terminal tuft; penis vascular, without bone. *Scent glands:* none described. *Mammae:* 2.

Senses: all good, especially eyesight.

ANCESTRY. The 6 surviving *Equus* species (including the horse, *E. przewalskii,* and the Asian wild ass or onager, *E. hemionus*) are the end products of 60 million years of evolution which started with *Hyracotherium* (formerly *Eohippus,* the "dawn-horse"), a small, solitary forest browser. Its descendants emerged onto the plains as grasses spread in the Oligocene and Miocene, and in the process became bigger and increasingly horselike: the teeth adapted to grazing, the limbs lengthened, and the digits, from 4 on front and 3 on rear feet, became reduced to just 1, the third. Some 20 genera came and went, lastly *Equus,* which appeared only about 2 million years ago, in the early Pleistocene, in North America. Although ruminants had by then replaced other ungulates as dominant herbivores, *Equus* was so superbly adapted that zebras, asses, and horses spread across the prairies, plains, and steppes of all the continents, evolving into the 6 existing species (8, 16).

AFRICAN DISTRIBUTION. Not counting domesticated forms, 4 of the 6 *Equus* species are confined to Africa: the 3 zebras and the African wild ass, the source of all domestic donkeys and asses. The horse, adapted to temperate grasslands, never penetrated the Sahara prior to domestication. (The wild form is now extinct.) The ass, adapted to more arid conditions, did penetrate but only in late Pleistocene and early Paleolithic time, along the Red Sea and Somali coastal zones, conceivably with the aid of man (i.e. already domesticated). In North Africa it replaced the plains zebra, which persisted in northern Algeria and Tunisia until the early Neolithic (10,000 years ago) (3, 8). Grevy's zebra is the relict of a form that was formerly widespread (even outside Africa). Except for one last stronghold (map), its place has been taken by *E burchelli,* presumably because the latter is better adapted to the savanna biome.

Three of the 4 African equids are rare and/or very restricted in distribution. The fourth, the plains or Burchell's zebra, rivals the horse as the most successful member of the family. Indeed, it occupies the tropical equivalent of the plains and steppes the wild horse occupied in North America and Eurasia. The other 3 African species are all confined to the arid zones (see maps). The Somali ass, which looks like a large, strong donkey (fig. 15.1), is one of the most endangered large mammals. A few scattered bands still survive in the remotest reaches of Somalia and Ethiopia. But it interbreeds so readily with donkeys, its domestic counterpart, that it is hard to know whether the wild herds are composed of feral stock, hybrids, or the true wild ass. It is going the way of the wild horse and wild cattle.

ECOLOGY. Although their digestive system is less efficient than a ruminant's (as attested by the coarser dung), equids compensate by eating more, including vegetation too fibrous and low in protein for ruminants to digest. The equids' strong incisors enable them to crop tough and tall, as well as short, grasses. The mobile, sensitive lips aid in gathering and pushing herbage between the incisors (4). Although the breakdown of cellulose in the large equid cecum is less thorough than in the rumen, processing is also quicker: throughput is 30–45 hours in the horse compared to 70–100 hours in the cow. A zebra that is only ⅔ as efficient as a wildebeest at extracting protein from its food ends up extracting more protein from less nutritious grasses because of the faster rate of digestion and assimilation (1, 6). However, to consume an adequate amount of herbage equids have to spend 60% of their time eating, night and day, under the best conditions, and over 80% under poor conditions. When even coarse grass is unavailable, in times of famine, equids resort to browsing and even consume bark and roots. They are consummate survivors.

SOCIAL ORGANIZATION. Two different types of grouping/mating system occur: a conventional territorial type, represented by the asses and Grevy's zebra, and a nonterritorial, 1-male harem system, represented by the other two zebras and the horse, an uncommon type found in only a small number of mammals (cf. hamadryas and gelada baboons, gorilla, bushpig and giant hog). The differences between the two systems are brought out in the species accounts.

The equid territorial system is adapted to arid environments in which resources are widely scattered and defendable. Males with territories to or through which females must come to reach safe drinking places or good pasture enjoy a reproductive advantage and may stay put even when the rest of the population migrates to distant parts of the range (11, 14). Females and bachelor males live in unstable groups in which there is no discernible rank order. There appear to be no lasting bonds between adults, although individuals of similar social and reproductive status tend to associate, and related females may share the same range (as in gregarious, territorial antelopes). Grevy's stallions and jackasses maintain territories as large as 10 to 15 km^2, respectively, but these are small compared to home ranges measured in hundreds of km^2 which females and bachelor males utilize in the course of a year.

The equid harem system probably developed later, as an adaptation to environments in which better, more evenly distributed resources favored nomadism with no restraints—other than the need to drink at least every few days, which keeps all equids from ranging more than 20–30 km from water (10). In acquiring a harem, males substitute movable for fixed property: a stallion herds and defends 2–5 mares and his offspring by them against rivals and against predators. Though unrelated, harem females spend their adult lives together, and continue to associate even if the herd stallion is replaced. In horses and plains zebras—but not necessarily in mountain zebras—a rank hierarchy based on time of joining the harem is revealed by a consistent filing order: the top mare comes first, followed by her offspring, the number 2 mare second, and so on. The stallion usually brings up the rear, but can exert control over his herd's movements and when necessary takes the lead. The mares in these semi-exclusive groups react aggressively to intrusions by any outsider, including foals. Their offspring depart in adolescence, females by abduction to other harems, males voluntarily to join bachelor herds until they mature at 5–6. Harem and bachelor units may assemble into bands or go their separate ways.

The dichotomy between these two systems has proven less absolute than earlier studies indicated, and more responsive to environmental variables. Thus, under certain conditions feral horses may be territorial and feral asses may keep 1-male harems; Grevy's zebra mares have been known to stay together for months, and some plains zebra herd stallions form alliances, maintaining proximity and cooperating in pairs to keep herds of bachelors from invading their harems (14).

ACTIVITY

Sleep. Equids may spend as much as 7 hours out of 24 asleep (4). During the midday heat they sleep usually while standing, and at night lie with legs gathered or—especially foals—flat on their sides. Sometimes individuals sleep so soundly that predators can sneak up and grab them (15). But almost always at least 1 member of a resting herd remains standing and alert; a herd member that fails to react to alarm snorts may be urged to rise by a nudge.

Rubbing and Dustbathing. Equids habitually rub their bodies and heads against objects such as trees, rocks, and termite mounds, and roll in the dust. There is often a waiting line, and the order depends on social rank.

Coordination of Activities. As would be expected of animals living in cohesive social groups, the activities within herds of Equidae are closely synchronized. Individual members rest, feed, move, groom, dustbathe, suckle, and excrete on much the same schedule, as though these activities were all infectious. This is less true of resting and grazing than of movements and the other specific activities; thus, foals sleep more, and mothers graze longer, than other classes. Furthermore, the activities of one group infect other groups, leading them to join in mass movements to and from pasture and water (see plains zebra account, Activity) or on migration. Even the passage of other species (wildebeest, oryx, springbok, etc.) may stimulate equids to join processions (7, 8, 10, 12).

Fig. 15.1. *Mutual head-resting* by two wild asses.

POSTURES AND LOCOMOTION.

The gaits of equids are the same as the natural gaits of horses. The walking gait is a single-foot or 4-beat sequence: left hind, left fore, right hind, right fore, at about equal intervals (5). The trot is a diagonal 2-beat stride; the pacing gait (lateral stride) is found only in certain horses bred for pacing or a running walk. At a gallop, either the right or left foreleg "leads," depending on which foreleg reaches out further, changing as the animal changes direction. Though zebras have a top speed of at least 55 kph, they are no match for a good horse (13). Larger equids trot more readily and with a freer action than the smaller species (e.g., the Grevy's compared to the plains zebra [13]).

Equids lie down and get up quite differently from ruminants. In lying, they gather the legs beneath the body, bend the knees and hocks, then settle on one side with legs folded on the opposite side and neck arched over the forelegs. The sharpness of the sternum precludes resting on the breastbone in the manner of ruminants (4). In getting up, the forelegs are extended to lever up the front end, after which a thrust of the powerful hindlegs brings the animal into a standing position. Rolling is an equine habit, especially in connection with dustbathing, and most species can roll completely over (not mountain zebra), unlike ruminants (but see Bovini introduction).

SOCIAL BEHAVIOR

COMMUNICATION

Tactile Communication: *social grooming*, standing and *head-resting* in pairs, *challenge* or *greeting ritual*.

Social grooming consists of scraping and nibbling with the incisors and lips, is confined to the neck, shoulders, and back, and is performed simultaneously by 2 individuals that are on friendly terms. In the plains zebra and horse, all members of the same social group groom each other, although at different frequencies reflecting individual preferences. *Social grooming* is thought to play a role in social bonding; yet grooming between zebra mares is least frequent and was observed only once in a long-term study of the Cape mountain zebra (12). Mothers and foals are the main practitioners and siblings also groom each other, suggesting that maternal and kinship bonding was the original and remains the main function of equid *social grooming*. It commonly involves individuals of unequal rank and the initiative is taken by the inferior one; the other may refuse by turning away or threatening, or may respond in kind. Thus *social grooming* plays a role in appeasing aggression and confirming social status (10).

Another characteristic and amicable equid contact behavior is *standing in pairs*, either head-to-tail or looking over each other's shoulder (fig. 15.1). In this position with heads down, a pair can swish flies off each other's faces, and with heads up, can see in all directions.

Greeting/challenge ritual. Individuals meeting for the first time or after a separation perform a ritual that begins with *nasal contact and sniffing* (fig. 15.2), followed by *rubbing cheeks*, then *moving alongside* and *touching/sniffing the genitalia*. This is typically followed by *pressing and rubbing shoulders* and sides together, and/or *resting the head on the partner* (fig. 15.11). The ritual is performed primarily by territorial and herd stallions, and by bachelor males more or less in play (10). Dominance and aggressive behavior is included and overall the so-called greeting ritual is more challenging than friendly (7, 12). Yet the cere-

Fig. 15.2. The *greeting/challenge ritual* between plains zebra stallions.

mony usually proceeds without threat displays or fighting, and the initial approach resembles the invitation to *social grooming.* Thus, a plains-zebra herd stallion comes out to meet and challenge any potential rival that ventures within 50–100 m of his harem, approaching with head outstretched and ears cocked (fig. 15.2). If the other stallion is of equal status, he either stands with head high as the challenger approaches, or advances in the same manner, and the ceremony runs its course. But a social inferior, such as an immature or bachelor male, assumes the posture and facial expression of an estrous female (fig. 15.3; cf. fig. 15.9), in which case the interaction goes no further than nasal contact. The challenge ritual differs slightly among species (see species accounts) (10).

Fig. 15.3. *Greeting* between family stallion (*left*) and bachelor male displaying submission (*ears back and chewing* with open mouth).

Olfactory Communication. The *nasonasal* and *nasogenital contact* of the *greeting ceremony* is surely olfactory as well as tactile. In addition, territorial males demarcate their property with dung middens (fig. 15.7). Possibly as a carryover of this habit, herd stallions of the harem-forming species perform a conspicuous *dunging ritual* near the herd, often covering the dung of their mares or that of rivals. Stallions also routinely check the urine of females, approaching at the sight of a urinating mare and

Fig. 15.4. The spectacular equid *urine-testing posture* (plains zebra).

then performing an unusually conspicuous *lip curl* (*flehmen*) (fig. 15.4).

Vocal Communication. *Equus* species share a repertoire of a half dozen calls. There are 2 alarm calls: a *loud snort* and a quiet, *hoarse gasp, i-hah,* with open mouth. A *short squeal* is given when an animal is kicked or bitten, but also during playfights and often when stallions greet; and equids emit a *drawn-out squeal* when caught or injured. *Blowing* with loose lips, often heard in a grazing herd, is a sound of contentment. The only specifically different vocalization is the *contact call:* the *whinny* and *nicker* of the horse, the *barking* call of the plains zebra, and the *braying* calls of the ass and Grevy's zebra (10).

Visual Communication. The mobility of the lips and the significance of teeth in fighting and social grooming make facial expressions important in equid communication. A considerable range of emotions can be signaled simply by different combinations of ear position and facial mime; general posture and the positions and movements of the head and tail are also incorporated in visual displays. Thus, when a horse lays back its ears and at the same time lashes its tail, a kick is likely to follow. A hazing stallion not only lays his ears back but holds his head low and weaves it from side to side, the strongest form of threat (see Dominance/Threat Displays) (fig. 15.5). The facial expression, the ear position, and particularly the posture of a mare in heat (fig. 15.9) is a very conspicuous and effective visual display (see Sexual Behavior).

AGONISTIC BEHAVIOR

Dominance/Threat Displays: *erect posture* with head high and ears pricked, *high-stepping gait, tail arched, lateral presentation* (barring the way), *threat face* (teeth bared and ears back), *biting threat* (biting-intention movements ranging

Fig. 15.5. *Hazing posture* of a plains zebra stallion herding an orphaned foal.

from moving head only, through rushing), *kick threat* (intention movements to kick with fore- or hindfeet: *rearing, lifting hindquarters); head-low threat* (hazing posture, fig. 15.5); *challenge* (greeting) *ritual, stamping, tail lashing, squealing.*

Defensive/Submissive Displays: kicking with hindfeet, grooming initiative, *submissive display* (neck lowered, tail extended, *chewing* with lips retracted, ears back).

Fighting: *biting* (ears, throat, neck, and particularly legs), *rearing and biting* or *striking with forefeet, kicking with hindfeet.*

Adopting the posture and facial expression of an estrous female is a *submissive display* (fig. 15.3) comparable to social presenting in monkeys, which is known to dampen male aggression by deploying a sexual stimulus. In essence, the posture of a mare in heat (fig. 15.9) is merely the *female urination posture,* adoption of which stimulates males to approach and perform the *urine test.* The so-called *mare-in-heat face* (Ger. *Rossigkeitsgesicht*) resembles a *biting threat,* with only the *chewing movements* to suggest grooming intention (17). In the mountain zebra, the display includes a high-pitched submissive call (12). Interestingly enough, only immature animals and adult bachelor stallions perform ritualized appeasement; it is seen in adult females only if they really are in heat and is much less conspicuous and prolonged in mares than in fillies (10, 12). Apparently the danger of being kicked by the hindfeet is enough to deter aggressors from persecuting adults of lower rank, thus obviating the need for elaborate submissive signals; simply moving out of the way of a threatening individual normally suffices to appease aggression (12).

Fights between stallions over estrous fillies (fig. 15.10) are a stern test of endurance and determination (see plains zebra account). Fighting is relatively unritualized and carries the risk of serious injury if a kick connects with the jaw or a leg. Although bites can be painful, especially when one contestant gets a grip on the other's hock or carpal joint, the canine teeth are the wrong shape to slash or stab.

REPRODUCTION. Although gestation lasts a year in this family, mares may come into estrus within a few days to a month of foaling and thus can reproduce yearly under favorable conditions (14).

SEXUAL BEHAVIOR. Estrus lasts 5–10 days and is marked by frequent urination, mucous discharge, swollen, everted labia, and the conspicuous *estrous display* (fig. 15.9). A female in full estrus stands with hindlegs spread, tail raised, and mouth open, and may invite mounting by backing into the stallion. Male courtship tends to be rather violent, especially in asses and Grevy's zebras, including *hard chasing, nipping,* and frequent *mounting attempts.* Mares resist by lashing out with their hindfeet (fig. 15.6). But more gentle courtship, including *nibbling, mutual grooming,* and *resting the chin on back and croup,* are also characteristic, especially when courting fillies. However, a recent study of feral horses showed that stallions that take over an existing harem actually rape the mares, causing those less than 6 months pregnant to abort. They then quickly reenter estrus, enabling the new owner to implant his own genes (2).

PARENT/OFFSPRING BEHAVIOR. A birth is described in the plains zebra account. Foals follow at their mothers' heels from the time they can stand. A mother makes all other zebras keep their distance for several days until the foal becomes imprinted, recognizing the mother's smell, appearance (stripe pattern in zebras), and voice. The bond with the previous offspring is broken or weakened at this stage. However, it is quite common for zebras, for instance, to foal only every other year. In the territorial species, mothers of new foals

Fig. 15.6. Female Grevy's zebra threatening to kick a stallion attempting to mount.

tend to associate in nursery herds (see Grevy's zebra account). In the harem species, the young stay in the natal herd until adolescence. Males are not driven out by their fathers but leave of their own accord in search of peers with which to compete and play. Young males interact far more than females, making bachelor herds a much livelier society than female groups. In the absence of other playmates, foals try to play with their mothers, but mares tend to be singularly unplayful (10).

ANTIPREDATOR BEHAVIOR: *alert posture* (head high, ears pricked, staring), *i-hah* and *alarm snort,* calling (after a disturbance), slow flight in tight group, defense by stallion, flight.

Perhaps the most important advantage of the harem over the territorial system is *group defense* of the young. Instead of running away from pursuing hyenas or wild dogs, a herd of plains zebras containing a small foal surrounds it and moves off in a tight bunch, while the stallion guards the rear and attacks the predators. The herd will also adjust its pace to that of weak or injured members, and the stallion actively searches for missing individuals (10).

References.

1. Bell 1971. 2. Berger 1983. 3. Churcher and Richardson 1978. 4. Hafez, Williams, Wierzbowski 1962. 5. Hildebrand 1965. 6. Janis 1976. 7. E. Joubert 1972*c*. 8. Kingdon 1979. 9. Klingel 1969. 10. ——— 1972. 11. ——— 1974*a*. 12. Penzhorn 1984. 13. Roosevelt and Heller 1914. 14. Rubenstein 1986. 15. Schaller 1972*b*. 16. Simpson 1951. 17. Trumler 1959.

Fig. 15.7. *Dunging ceremony* during border encounter between territorial Grevy's zebras.

Grevy's Zebra

Equus grevyi

TRAITS. A striped ass with a massive head and very large ears, the largest equid. *Height and weight:* 145 cm (140–160); males 430 kg (380–450), females 386 kg (352–450) (2). *Stripe pattern:* close narrow stripes, especially the bulls-eye pattern on the croup, bolder stripes on neck and chest, extending through mane (making neck appear thicker), belly unstriped; striking black ear markings and stripe along spine (2).

DISTRIBUTION. Now reduced to several thousands, almost entirely confined to northern Kenya.

ECOLOGY. Grevy's zebra fills a narrow niche between the more water-dependent

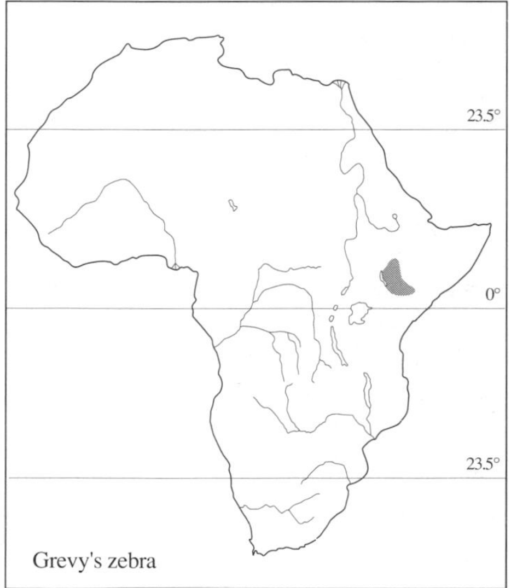

Grevy's zebra

plains zebra and the more desert-adapted wild ass (2). It inhabits *Acacia-Commiphora* thornbush country and barren plains in the Somali Arid Zone, but depends on better-watered highlands and drinking water to carry it through the dry seasons and longer droughts when the rains (July–August, October–November) fail. However, where and when water and grazing remain adequate, Grevy's zebra does not migrate. It can subsist on grasses too tough for cattle to eat or digest, takes browse when the grass is gone, and digs and defends water holes in dry streambeds (cf. mountain zebra) (1, 3).

SOCIAL ORGANIZATION: territorial, migratory, open herds.

The basic social unit is a mare with her latest 1 or 2 offspring. These units often combine, but herds are unstable, without established rank hierarchies. Female herds typically include fewer than 10 mares, averaging smaller in the dry than in the wet seasons. However, mixed aggregations of 100 to 200, up to a maximum of 450 zebras, are not uncommon during migration and around water points in the dry season, and may include up to 40 mares. Within these bands, animals of the same class tend to associate in subgroups, especially mares with young foals. Most bachelor herds number 2–6, rarely over 10 stallions (4). In the overlap zone between Grevy's and plains zebras, it is not unusual to find both species in the same herd.

Mature males (over 6 years) maintain territories of 6 km^2 (2.7–10.5) in the wet-season range and, provided local water sources last, some stay put during the dry season, even when the rest of the population sojourns many kilometers away in cool montane grassland. This fidelity is rewarded when the migrants return at the beginning of the rains, followed soon by a peak in breeding, nearly all of which takes place within the territories (5). Territorial stallions tolerate other, subordinate males, and when no females are present often mix with the bachelors; they even socialize with territorial neighbors (3, author's observations). When an estrous female is present, they will not tolerate the invasion of another territorial male, but merely keep bachelors at a distance rather than trying to drive them off the territory.

ACTIVITY. See family introduction. On a normal day while a population is resident, female herds range 10–15 km (5).

POSTURES AND LOCOMOTION. See family introduction.

SOCIAL BEHAVIOR

COMMUNICATION. See family introduction.

TERRITORIAL BEHAVIOR

Advertising: dung middens, *demonstrative defecation and urination, braying,* dominance displays, *chasing.*

Stallions maintain large dung middens on their territorial boundaries, which commonly follow such topographic features as streambeds (2, 3). A territorial male defecates in a distinctive posture (fig. 15.7) and also urinates frequently, possibly as part of the marking procedure. *Braying* also advertises territorial status. Grevy's zebra produces an extraordinary cry that sounds something like a hippo's grunt combined with a donkey's wheeze. While uttering this drawn-out, brazen cry, a stallion often breaks into a clattering gallop, cutting a somewhat ridiculous figure.

Challenge ritual. Territorial stallions approach each other, often with heads high and necks arched, touch noses, rub and press against each other, and examine genitalia. Often one rests his chin on the other's croup or rubs his head on his shoulder, and the other jumps as if to kick. They may also take turns defecating on a dung midden and urinating during these border encounters (3).

AGONISTIC BEHAVIOR

Dominance Threat Displays: *approaching in erect posture ± braying* (see Territorial Behavior), *hazing, biting threat* (ears back, teeth bared, tail arched in air and rapidly swished).

Defensive/Submissive Displays: turning back and raising hindleg (*kicking intention*), *kicking, submissive display* (see fig. 15.3), *nuzzling male's chest or groin.*

Fighting: biting, neck-wrestling, rearing and foreleg flailing, kicking.

Bachelor stallions appease the aggression of territorial males that approach them in the *erect posture* by mimicking estrous behavior, including nuzzling the stallion's chest or groin.

REPRODUCTION. Most breeding occurs early in the rains (5). Females breed at 3, males not before 6. (See family introduction for more information on equid reproduction.)

SEXUAL BEHAVIOR: male dominance displays, *braying, chasing* and *hazing, urination/urine-testing, estrous display.*

The sight of a urinating female excites male interest. A territorial stallion approaches in the *erect posture* with slow, measured tread, and sniffs at her rump. The mare may sniff or nudge his groin before turning sharply away. If she is in heat the stallion often attempts to mount at once, even while she is running away and kicking up her heels at him (fig. 15.8). Courtship and copulation proceed only after the stallion has subdued the mare by performing dominance displays, including frequent loud braying. If an estrous mare approaches or crosses a territorial boundary during courtship, 2 territorial neighbors may come to blows.

Fig. 15.8. Grevy's stallion attempting to mount a mare on the run.

PARENT/OFFSPRING BEHAVIOR. Young foals are so long-legged that they have to splay their limbs to nose the ground (2). The natal coat is russet-colored with a long hair crest down the back and belly. The mother often rubs her chin and neck over the foal's back and sometimes nibbles the dorsal crest (2). The pair often engages in mouth-to-mouth nuzzling, and the foal tries to play with her, including mock attacks and kicking (especially colts). The foal follows the mother very closely and often nurses between the hindlegs. The young become more independent after ½ year but continue to follow their mothers for up to 3 years. During droughts, when forced to travel long distances to water, mothers may leave young foals unguarded all alone or in a crèche. Lacking any instinct to hide, the foals just stand around, fair game for any passing predator (4).

References

1. Ginsberg 1987. 2. Kingdon 1979. 3. Klingel 1974*b*. 4. ——— 1975. 5. Rubenstein 1986.

Added Reference

6. Ginsberg 1989.

Fig. 15.9. Plains zebra courtship: female displaying the *mare-in-heat face.*

Plains Zebra
Equus burchelli

TRAITS. Broad torso stripes, no "gridiron" on rump and no dewlap (cf. mountain zebra). *Height and weight:* 127–140 cm; males 250 kg (220–322), females 220 kg (175–250) (4). *Coloration:* white to buff background, no gender difference, geographically variable, the northernmost race (*E. b. boehmi*) most completely and boldly striped, decreasing southward; natal coat longer and softer with brown stripes.

DISTRIBUTION. Somali-Masai Arid Zone through the Southern Savanna and marginally in the South West Arid Zone, from southeastern Sudan to South Africa and Angola. The southernmost races, *antiquorum* and *burchelli,* have shadow stripes, with unstriped bellies and lower legs. The South African quagga, exterminated last century, had stripes only on the head, neck, and back. It was probably not

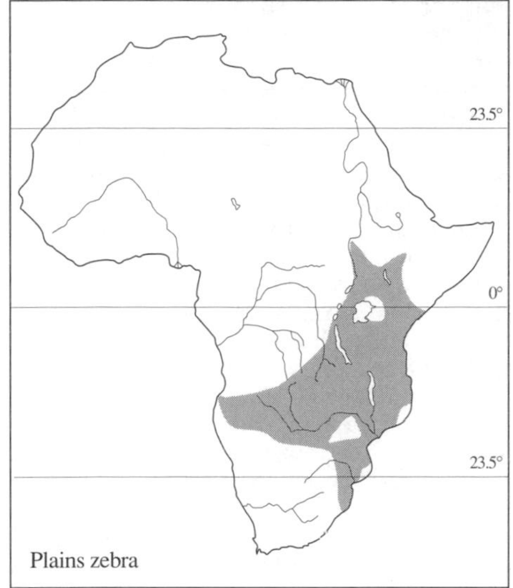

a separate species, but a subspecies of the plains zebra which was adapted to the temperate grasslands of the Highveld (4, 10).

ECOLOGY. One of Africa's most adaptable and successful grazers, the plains zebra utilizes a broad range of savanna habitats, from treeless short grassland to tall grassland and open woodland. Equipped to deal both with long, tough stems and the early stages of a flush, the zebra is often the pioneer that leads the way into taller, more wooded, or wetter pastures and prepares it for the wildebeests, gazelles, and other associated antelopes. It is also among the most water-dependent of the plains game (1).

SOCIAL ORGANIZATION: nonterritorial, nomadic, 1-male harems and bachelor herds.

Field studies of the plains zebra provided insights into the social organization of the horse, which had ceased to exist as a wild animal before anyone thought of undertaking naturalistic behavior studies (2, 5, 6). The most striking features of the equine harem system are the absence of overt male sexual competition over mares and the permanence of harems. Ownership of a harem is hardly ever disputed as long as the stallion is fit, and when he is replaced the herd remains intact.

To found a harem, a stallion must first abduct a filly from her herd. Then he must fend off other rival stallions every time she comes into heat, which happens at monthly intervals, lasts up to 5 days, and may recur for a year before the filly conceives (6). The filly stays with the stallion that impregnates her. In Ngorongoro, up to 18 stallions have been counted around a herd containing an estrous filly. Given such fierce competition, it is unlikely that the first abductor will still have her when she finally ovulates. Normally each new member of the harem (up to 5–6 females, rarely more) is also acquired through conquest. The hostility of the mares already in the harem to outsiders takes several weeks to overcome, during which a newcomer is kept at a distance, herded and protected by her sponsor.

A mare assumes the telltale *estrous display* (fig. 15.9) only briefly during full estrus while the herd stallion is courting her. Attenuation of the display, which is so conspicuous and protracted in fillies, helps to explain why harems are not besieged by rivals each time a mare comes into heat. But the herd stallion's presence and vigorous reaction to potential rivals are also essential, for any passing stallion will court any undefended estrous female. If the herd stallion's defense lacks vigor, it may be taken as a sign that he is infirm and ripe for replacement. The *challenge ritual* between stallions functions as a test of fitness: as long as the herd stallion conveys the message, "This herd has an owner that is fully capable of defending it," he won't be put to the test (3). The way in which injured, sick, or very old herd stallions are replaced sounds surprisingly peaceful (7). A bachelor begins shadowing the herd and gradually takes control, pushing the original stallion out without a fight. The same procedure has been documented in mountain zebras (3, 9). But it may take weeks before the mares accept and acknowledge the new overlord.

Home ranges vary from a minimum of 30 km^2 in a small rich ecosystem like Ngorongoro Crater, up to 600 km^2 in the migratory Serengeti population (6).

Bachelor Herds. From 2 to 15 zebras (average 3–6) live in quite stable all-male herds usually led by a young adult. Rank is based simply on age. Bachelors pass much of their time horsing around, rehearsing their adult roles by engaging in play fights and *greeting/challenge rituals.* Bachelors ready to start harems (at 5 years) often leave their companions and wander alone in search of fillies in heat. Plains zebras are otherwise rarely solitary.

ACTIVITY. The pattern of daily activity varies according to the nature of the habitat, grazing, and weather conditions. In general, plains zebras spend most of the daylight hours grazing, with time out for dustbathing, rubbing, drinking, and irregular, rather brief rest periods. In Ngorongoro Crater the zebras typically move between bedding grounds on the shortest, openest grasslands and pastures in taller grasslands. A herd followed throughout the day traveled some 17 km (6). On warm, dry days the herds get up at sunrise and trek single-file across the Crater floor on well-worn game trails to their current pastures. Frequently they join or are joined by wildebeests filing in the same direction. After reaching a pasture, zebra groups then spread out and move somewhat independently the rest of the day, though movements to water are often coordinated. In late afternoon the animals begin another mass movement back to the short-grass areas, during which social activity reaches a peak, with the running, play, and fighting of bachelor males most conspicuous (author's observations). In cold, rainy weather, the zebras become active 2–3 hours later, may not trek to distant pastures, and return earlier to night-bedding grounds.

At night the zebras move very little, except when disturbed by hunting lions or hyenas. Most rest and sleep lying down, apart from at least 1 animal per herd which stands guard (6). Three more or less distinct rest periods were noted during observations on moonlit nights: 2000–2100, 0100–0200, and 0445–0615 h. Between these periods most zebras grazed in the immediate neighborhood, while some continued resting (6). Zebras also graze on dark nights (author's observations).

POSTURES AND LOCOMOTION. Described in family introduction.

SOCIAL BEHAVIOR

COMMUNICATION. Discussed in family introduction.

AGONISTIC BEHAVIOR. The equid display repertory is given in the family introduction. The *challenge ritual* of the plains zebra differs from that of other equids in 2 ways. During nasal contact, herd stallions make *chewing movements* with open mouths (not shown in fig. 15.2), and interactions end with a jump that looks like an intention movement to rear and bite; the performer may lift both forelegs, lift one and stamp, or merely lift his chin abruptly. The interactions of bachelors are usually less aggressive and may end with *chin-resting* on each other's backs. If one of the stallions is inferior, it behaves submissively (fig. 15.3). The dominant male then appears to lose interest in continuing the ritual.

Fighting. Plains zebra fights consist mainly of *biting* and *wrestling*, rarely *flailing* with the forefeet, and only occasional *kicking*. Combatants rear, *jostle*, and *neck-wrestle* while trying for an ear or mane hold (fig. 15.10*a,b*). Seeking a grip on a fore- or hindleg, they drop to the ground and *circle on their knees* (fig. 15.10*b,c*). Having secured a grip on the hock or carpal joint, a stallion holds on and grinds away. In figure 15.10*d* the victim, a herd stallion, gave up and lay still, possibly because of pain, exhaustion, or both. But the moment he was released, he got up and continued fighting to keep his daughter from being abducted (2).

REPRODUCTION. Breeding is not strictly seasonal, but an annual birth and mating peak comes in the first months of the rainy season. Mares come into "foal heat" within a week of giving birth, but if not in good condition may not conceive again for a year. (Equid reproduction is discussed in the family introduction.)

SEXUAL BEHAVIOR. A female entering estrus urinates more often and the urine has a milky color. The buildup of female hormones in the urine arouses the stallion's sexual interest 3–4 days before the mare becomes receptive. He begins to follow and court her, *sniffing her vulva*, *nipping* her, and *rubbing* his muzzle on her neck, back, and croup. The mare tries to discourage him by kicking him in the chest and withdrawing, sometimes running away with the stallion in pursuit (as in Grevy's zebra, fig. 15.8). However, the kicks seem only to stimulate him further and he may begin periodic *mounting attempts* (without erection). Whenever the mare defecates, the waiting stallion smells the spot and either urinates or defecates on it in turn. He also covers the mare's urine with his own after *urine-testing* (fig. 15.4). This marking behavior may be stimulated by the excrement of other stallions, too.

Copulation begins only when the mare is in full estrus. She stands with hindlegs planted and may back into the stallion when he mounts, meanwhile giving the full *estrous display* with mouth gaping open (fig. 15.9). A mare often turns her head during copulation as though to bite. Ejaculation follows a few thrusts and may be detected by the rhythmic clamping or pinching of the stallion's tail, which initially is

Fig. 15.10. Battle between a herd stallion and an interloper attempting to abduct his daughter: a, b) head- and neck-biting; c, d) combatants drop to ground to counter leg-biting attempts; e) interloper gets a grip on herd stallion's hock; f) lashing out with back feet, the most dangerous form of combat.

held out. Soon afterward he slides off, moves away, and begins to graze. Copulation is repeated at intervals of 1–3 hours for about 1 day (6).

Sometimes other mares react jealously, particularly if the estrous female is of low rank. They threaten her with ears laid back, and bite and chase her out of the herd.

PARENT/OFFSPRING BEHAVIOR. One of the few observed births occurred while the mare lay on her side, watched over from nearby by the herd stallion, with the rest of the herd grazing 50 m away (8). The foal, completely enclosed in a tough caul, extricated itself unassisted and got up after 11 minutes, could walk after 19 and walk well in 32 minutes, and broke into a canter after 44 minutes. It nursed at 67 minutes. The mare licked its rear and back briefly and chewed but did not eat the remnants of the fetal membranes. Whenever the stallion or other herd members came near the foal for the first few days, the mother laid back her ears and chased them away. The previous offspring is also rejected, but later may establish bonds with its sibling through *mutual grooming.*

ANTIPREDATOR BEHAVIOR. When hyenas approach a herd that has a young colt, the mother hides with it behind other family members. By staying close together instead of bolting and leaving behind a weak or sick animal, the whole herd cooperates to protect any threatened member. However, it is the stallion that actively defends his group and does not hesitate to attack hyenas and wild dogs. This defense is so effective that only hyenas hunting in sizeable packs have much luck with zebras. And only the stallion goes looking for missing mares or older offspring (5).

References

1. Bell 1971. 2. Estes 1974*b*. 3. Joubert 1972*b*. 4. Kingdon 1979. 5. Klingel 1967. 6. ——— 1969. 7. ——— 1974*a*. 8. Klingel and Klingel 1966. 9. Penzhorn 1984 10. Rau 1983.

Fig. 15.11. Mountain zebra stallions in *challenge ritual;* one rests/presses his chin on the other's croup.

Mountain Zebra

Equus zebra

TRAITS. No shadow stripes, torso stripes close-set, a "gridiron" of narrow stripes across the croup, fully striped legs, white belly, ears larger than in *E. burchelli,* and a small dewlap under the chin. *Height and weight: E. z. hartmannae* c. 150 cm, males average 298 kg (336 kg for males over 7 years), females 276 kg; *E. z. zebra* c. 50 kg lighter (10), with smaller, narrower stripes.

Although plains and mountain zebras look and act very much alike, the former has 44 chromosomes, the latter 32 (3).

DISTRIBUTION. Formerly widespread in the arid mountain ranges that parallel the coast from southern Angola to the Transvaal. Around 215 members of the South African race, *E. z. zebra,* live in the small (65 km^2), fenced Mountain Zebra N.P. near Cradock (1980 census); otherwise only small groups survive in 4–5 other mountain locations in southern Cape Province

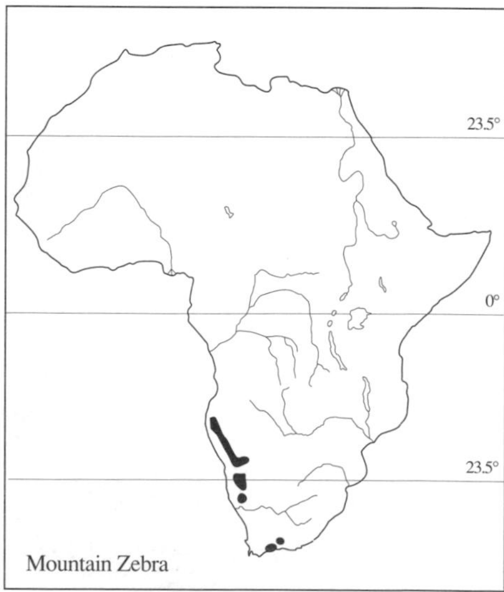

(10). Of the few thousand Hartmann's zebras inhabiting the mountains bordering the Namib Desert, the Angola population has presumably declined drastically after many years of guerilla and civil war.

ECOLOGY. Adapted to subdesert plains and barren, rocky uplands, the mountain zebra spends the rainy season on the plains, where it comes into contact with *E. burchelli antiquorum* (e.g., in Angola's Iona N.P.), and also penetrates deep into the desert, even to the coastal dunes, after rainfall (2). Where the range remains unfenced, herds still move distances of 100 km or more (5). In the dry months the mountain zebra ranges the uplands, following ancient paths that lead to springs and rainwater pools in the mountain fastnesses. In the hot dry season it drinks once or even twice daily (4). Mountain zebras dig for subsurface water in dry streambeds, excavating pits that are defended against strangers and herd members alike (cf. Grevy's zebra) (2, 6). Other game, such as the oryx, springbok, plains zebra, kudu, giraffe, hyena, and lion, subsequently drink from these holes.

SOCIAL ORGANIZATION: 1-male harems and bachelor herds.

Mountain zebra harems are slightly smaller than those of the plains zebra, averaging 4.7 (2–8), with 2.4 (1–4) mares per herd in both subspecies (6, 8, 9). Compared to the plains zebra, the mountain zebra is more inclined to remain dispersed in separate herds and rarely forms large aggregations; it is also much less vociferous, even lacking the distinctive *qua-ha* contact call (but see under Vocal Communication). Stallions fight less, usually take the lead, and exercise vigilance when a harem goes to water (as do plains zebra and wild horse stallions); they also initiate and lead ⅓ of other herd movements (9). Herd members maintain a greater individual distance, particularly while grazing, and *mutual grooming* is infrequent, although mares have brief rubbing contact with the herd stallion on a daily basis (4).

The extent to which these differences are species-specific or attributable to ecological circumstances remains to be determined. The harsher environment, especially a sparser food supply, and comparatively small populations could account for more dispersed social groups, which consequently interact less. Yet the tendencies to a larger individual distance and silence persist in mountain zebras kept in zoos (1).

Both male and female offspring leave the herds at a mean age of 22 months (13–27), often about 4 months after the birth of a sibling (5, 9). Mother Hartmann's zebras are reported to become so intolerant of last year's foals, before and after foaling, that they try to drive them away from the herd, and up to ⅓ leave between 14 and 16 months of age (5). Fillies that leave before adolescence join bachelor herds rather than stay alone, where they may spend a year or more before being taken into a harem. Again, harsh conditions might account for such early separation, for no such intolerance was observed among Cape mountain zebras. In fact colts had virtually to escape from their family units before they could join bachelor herds, as their fathers tried to prevent contacts with other young males and went looking for missing sons (8, 9). One colt was still in the paternal harem at 37 months (9).

Bachelor herds include not only young males and occasionally yearling females but also old or otherwise unfit stallions that have lost their harems. Yet stallions that remain healthy may retain harems for 15 years, possibly more (10). Male groupings are continually changing as 2-year-olds join and young adults depart to found harems; there is no regular hierarchy, although younger males are outranked by older, bigger stallions.

Evidence from known individuals that remained together for at least 10½ years indicates that mares from the same harem generally remain associated their whole adult lives (9). A few herds in Mountain Zebra N.P. have been known to split after being taken over by a new stallion, however, with 1 or more females separating and being taken into a different harem (9). Yearling and 2-year-old males generally depart after a takeover, and often end up in the same bachelor herd as their sires (9).

Home Range. Mountain zebras are naturally migratory; even in fenced reserves they have separate summer and winter ranges (5). The grounds of free-ranging Hartmann's zebras observed in Namibia were 120 km apart. But while in residence the zebras were notably sedentary. Harem but not bachelor herds showed preferences for particular core areas averaging only 10–20 km^2 in the winter range, and smaller still in the summer range. Daily movements seldom exceeded 1–3 km while foraging and 5 km to drink (4).

ACTIVITY. Nocturnal activity has not been studied. The zebras in Mountain Zebra N.P. had 3 main grazing periods by day:

from dawn to 0830 h in winter (stopping later in summer), 1000–1230 (more regularly in winter than summer), and 1400 to dusk (from 1600 in summer). At both seasons the zebras spent over half the time grazing, but spent significantly more time grazing in summer than winter. Since the animals burn up more calories keeping warm in winter, when the grass is least nourishing, they would be expected to spend more time grazing, not less. It was therefore concluded that they must be feeding more by night (9). Cape mountain zebras were never seen to seek shade, whereas Hartmann's zebras regularly rest in the shade, even in winter (4). However, both populations withdraw into sheltered places when it is cold and windy. They dustbathe (in regularly used dust bowls) and drink daily in the dry season, but can go 2 or more days without water during the rains (4). Mountain zebras rarely lie down to rest by day, but sleep on their feet, heads hanging low and ears drooping at right angles.

Rolling. In addition to dustbathing, mountain zebra stallions (rarely mares or young) roll in wet sand near their water holes, but apparently cannot roll over completely (young up to 1½ years can) (6).

POSTURES AND LOCOMOTION. Described in family introduction.

SOCIAL BEHAVIOR

COMMUNICATION. See family introduction.

Vocal Communication. A call similar to the *barking contact call* of the plains zebra is uttered only rarely by mountain zebras (mainly stallions), apparently as an alarm signal (5). The only other vocalization that has been described in wild populations is a *high-pitched squeal,* seemingly a distress cry, uttered by foals rejected by mothers, estrous females during courtship, and bachelor males displaying submission to herd stallions (5, 9). The loud *alarm snort* and the loose-lipped *blowing* "sound of contentment" complete the list of sounds.

AGONISTIC BEHAVIOR. Repertoire given in family introduction.

Challenge Ritual. During nasal contact, mountain zebras chew but keep their mouths closed (7). The parting jump seen in plains zebras is not performed, but mountain zebra stallions press and rub against each other more forcefully, meanwhile circling; they also rest and press down their chins on each other's croup (fig. 15.11), sometimes to the extent of unweighting their forelegs (9). At the end of the encounter, stallions part abruptly but then graze slowly apart.

REPRODUCTION. Very little information; apparently irregular, adapted to the erratic rainfall of the arid environment (10). (Equid reproduction is discussed in the family introduction.)

SEXUAL BEHAVIOR. The description of plains zebra courtship and mating applies equally to *E. zebra.*

PARENT/OFFSPRING BEHAVIOR. As in plains zebra, except as noted above under Social Organization.

ANTIPREDATOR BEHAVIOR. No information that would indicate any difference from the behavior of plains zebras.

References

1. Antonius 1951. 2. Blaine 1922. 3. Heinichen 1969. 4. E. Joubert 1972*b*. 5. ——— 1972*c*. 6. Klingel 1968. 7. ——— 1972. 8. Penzhorn 1979. 9. ——— 1984. 10. Smithers 1983.

Near-Ungulates
Superorder Paenungulata

Order Hyraciodea, hyraxes
Order Proboscidea, elephants
Order Sirenia, dugongs

This superorder contains both the largest and the smallest ungulate-mammals. The common ancestors of hyraxes, elephants, and sea cows separated early in the Age of Mammals, so they are not closely related. In fact, recent detailed anatomical studies of hyraxes strongly suggest that they are more closely allied to odd-toed ungulates, and should be classified with the Perissodactyla in the superorder Mesaxonia. Nevertheless, they also share certain anatomical pecularities with elephants and dugongs; accordingly, until their taxonomic status is settled, keeping them in the Paenungulata serves to emphasize their rather striking common traits.

The members of all three orders have teats between the forelegs (pectoral mammae), the same type of placenta and womb, internal testes, and a vascular penis which recurves to the rear except when fully erect. The upper incisors are reduced in number and modified as tusks or altogether absent (Sirenia), and the molars have transverse ridges. Hyraxes and elephants both retain relatively unmodified limbs with several digits that end in primitive claws more like toenails than hooves. They walk on the whole foot (plantigrade) instead of on their toes (digitigrade) like more progressive ungulates and carnivores.

References
Anderson and Jones 1967. Fischer 1986. Maglio and Cooke 1978. Romer 1959. Sale 1966*b*. Young 1962.

Chapter 16

Hyraxes

Order Hyracoidea, Family Procaviidae

Procavia, rock hyraxes
- *P. habessinica*, Abyssinian hyrax
- *P. capensis*, Cape hyrax
- *P. johnstoni*, Johnston's hyrax
- *P. welwitschii*, Kaokoveld hyrax
- *P. ruficeps*, western hyrax

Heterohyrax, bush hyraxes
- *H. antinae*, Ahaggar hyrax
- *H. brucei*, yellow-spotted hyrax
- *H. chapini*, Matadi hyrax

Dendrohyrax, tree hyraxes
- *D. arboreus*, southern tree hyrax
- *D. validus*, eastern tree hyrax
- *D. dorsalis*, western tree hyrax

FAMILY TRAITS. The smallest ungulate-type mammals, size and general appearance of a woodchuck/marmot. Sexes alike except mature males have thicker necks and blunter features, also stronger, sharper tusks and bigger larynx and guttural pouches to amplify territorial call. *Teeth:* 2 upper incisors, modified as tusks; lower incisors comblike, adapted for grooming; cheek teeth rasplike, high-crowned (hypsodont) in *Procavia*, low-crowned (brachydont) in other genera (6). *Feet:* rubbery soles kept moist by sweat glands, 4 front and 3 rear toes equipped with rounded, hooflike nails except for a curved grooming claw on the inside rear toe. Tail rudimentary. *Coat:* soft in bush and tree hyraxes, harsh in rock hyrax, sprinkled with long, sensory hairs; color gray, yellow-brown to dark brown. *Scent glands:* dorsal gland, a bare dorsal spot 20–75 mm long with c. 8 secretory pores leading to sebaceous follicles, surrounded by contrasting erectile hair and conspicuous when opened (see fig. 16.3); secretion an aromatic fluid (12). *Mammae:* 2 pectoral, 4 inguinal (some tree hyrax populations lack pectoral or have only 2 inguinal mammae) (2). *Longevity:* up to 12 years.
Senses: all acute (including tactile), but near-vision possibly weak (11).

DISTRIBUTION. In suitable habitat over most of the continent and the Arabian Peninsula. *Procavia*, the biblical coney, is the most arid-adapted, being found in the mountains of the Namib, Sahara, and Arabian deserts. The main distribution of *Heterohyrax* is the Southern Savanna, but it also ranges northward through eastern African semidesert to the Mediterranean and the Sinai Peninsula. Tree hyraxes occur in most types of forest and all 3 genera also occur in the Afromontane Zone. In South Africa, where hyraxes are called dassies, *Procavia* has increased and extended its range into plains, following the destruction of natural predators, and become an agricultural pest (6).

Whereas rock, bush, and tree hyraxes are different enough from one another to justify placing them in separate genera, the differences within genera would seem to nonspecialists too minor to justify designating the geographical forms as different species.

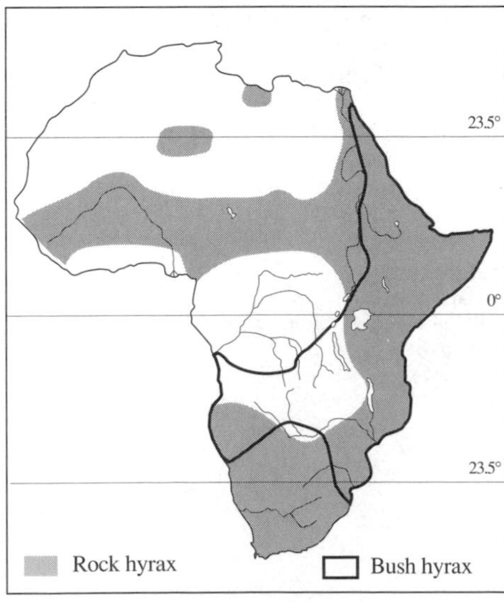

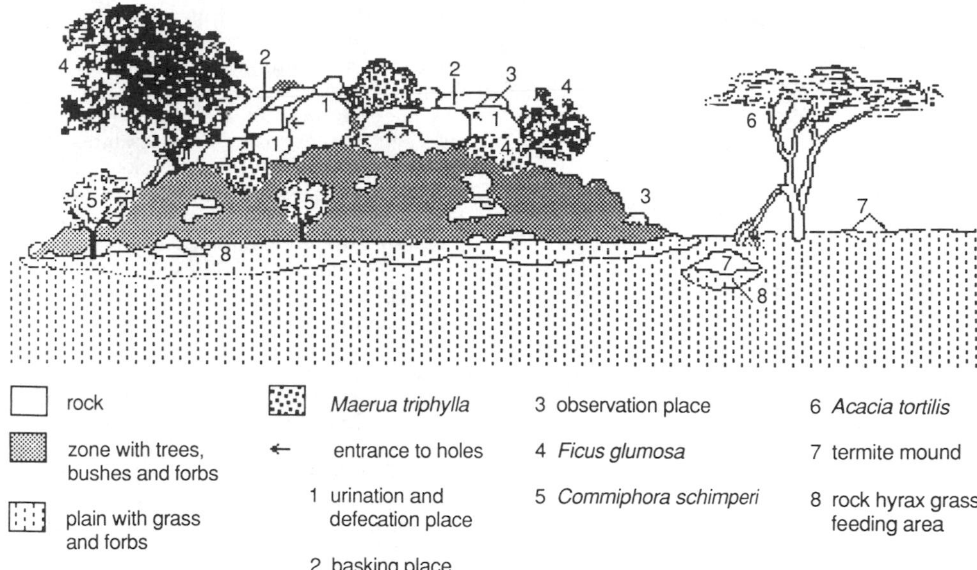

Fig. 16.1. Kopje habitat frequented by rock and bush hyraxes in Serengeti N.P. (from Hoeck 1975).

ANCESTRY. Originally terrestrial, hyraxes were dominant medium-sized herbivores some 40 million years ago. During the Oligocene, they were both numerous and diverse, ranging in size from current hyraxes to giants the size of forest hogs (6). Then came the ruminant radiation of the Miocene. Only hyraxes agile enough to negotiate terrain too steep for other ungulates, small enough to find shelter in holes, cracks, and crannies, and able to feed in trees and other places beyond the reach of ruminants survived.

To turn animals with clawless feet not unlike an elephant's into agile climbers is an impressive feat of adaptive evolution (3, 6, 13). *Procavia,* the most widespread genus, is thought to retain the most primitive features and the tree hyrax probably evolved last, from the same line as *Heterohyrax,* which is intermediate. This is the opposite of the usual evolutionary scenario, since *Procavia* is a diurnal, gregarious grazer and *Dendrohyrax* is a nocturnal, "solitary" browser. However, by acquiring a thick fur coat and becoming nocturnal, *Dendrohyrax* freed itself from dependence on rocky habitats and acquired a whole new kingdom. Nevertheless, it has also succeeded in colonizing rocky habitats in the Afromontane Zone that were unoccupied by rock or bush hyraxes; these tree hyraxes are gregarious and partly diurnal (see species account) (6).

ECOLOGY. Hyraxes are the dominant herbivores in rocky habitats, reaching biomasses comparable to successful plains antelopes (fig. 16.1) (4). Hundreds of tree hyraxes/km^2 also occur in montane forests, as on Kilimanjaro and Meru. Key hyrax adaptations include the ability to economize on energy and resources, the adaptability to eat virtually all kinds of vegetation, and tolerance of wide differences in altitude, temperature, and rainfall. This despite having so little control of their body temperature that rock and bush hyraxes cannot exist without shelter from cold and heat. Thus, the deep temperature in *Heterohyrax* fluctuates up to 7°C in response to a 42°C change in air temperature, and individuals exposed to direct solar radiation in moderate air temperature (22.5°–26.5°) suffer severe heat stress in less than 2 hours (1). Since their clawless feet are also useless for digging, hyraxes are utterly dependent on natural cavities. Like reptiles, hyraxes cleverly use the environment to avoid extremes of temperature and humidity. *Procavia* and *Heterohyrax* sunbathe to warm up in the morning (fig. 16.2), escape midday heat and

Fig. 16.2. Rock and bush hyraxes *huddling* in early morning sunlight.

Fig. 16.3. Rock hyrax opening wide to bite off vegetation with its molar teeth.

desiccation by seeking shade, and shelter at night in places where temperature and humidity fluctuate half as much as outside.

Hyraxes do not manipulate or carry food. However, by using their cheek teeth and wide gape to bite off mouthfuls of food, they eat very rapidly (fig. 16.3). The rock hyrax can take as big a bite as a sheep, enabling it to fill its far smaller stomach within an hour or 2 (10). The rest of the time rock and bush hyraxes do very little, and by thus conserving energy can subsist on resources too sparse to support more active animals of their size.

SOCIAL ORGANIZATION: territorial, sedentary, gregarious/colonial, solitary or family units (mated pair + offspring).

Rock and bush hyraxes have an almost identical social organization and often associate (see species account). Colonies may include up to 35 animals, but typically consist of 1 territorial male with a harem of several related females plus their young. Adolescent males but not females are forced to disperse (5). Colony members huddle together and at night in the dens even stack together for warmth. Otherwise they are not particularly sociable: there is no *social grooming* and low-level aggression is frequent (11). Though generally less sociable, tree hyraxes in alpine habitat have much the same social organization (6).

ACTIVITY. See Ecology and Postures and Locomotion, and rock/bush hyrax account.

POSTURES AND LOCOMOTION. Hyraxes are too long and low to sit erect or walk bipedally for more than a few steps. They crouch with all feet on the ground and back hunched. They often lie prone, forelegs extended and chin resting between them; when they are totally relaxed—or heat-stressed—the hindlegs extend backward with the soles up, facilitating evaporative cooling (11). Hyraxes move in a creeping walk and are especially sluggish in early morn, indicating low body temperature. They need an hour or 2 to warm up before becoming active and starting to feed. At maximum speed they can run at about 5 m/second for up to 120 m (4). Yet they are extremely agile climbers (fig 16.6) and good jumpers, the only ungulates that can climb smooth trees and rocks.

Comfort Behavior. In addition to basking, huddling, stacking, and lying prone, hyraxes dustbathe and bathe in and drink water when available. All species deposit dung pellets in regularly used latrines near their sleeping quarters. Rock and bush hyraxes also urinate on vertical rock faces; crystallized calcium carbonate in the urine makes a highly visible white stain that advertises colony sites.

SOCIAL BEHAVIOR

COMMUNICATION

Vocal Communication. A variety of different calls, of which the territorial advertising calls, distinctly different in each genus (see species accounts), and predator-alarm calls are most striking. The young make mewing lost and begging calls, there are twittering or whinnying contact calls, and threatening hyraxes growl and grind their molars.

Olfactory and Visual Communication. The importance of latrines and urine for olfactory communication is unknown, but the dorsal gland and erectile hair play a central role in hyrax social and reproductive behavior (11, 12). The dorsal spot opens whenever a hyrax is at all excited, from slightly to fully depending on the stimulus (fig. 16.4). The hairs of the neck are often erected at the same time, indicating an aggressive component. Inactive in young hyraxes, the gland develops fully only in older, dominant animals, and secretes most copiously in the mating season. Males assume a special posture while giving their territorial call, typically from favorite lookout posts, crouching with head raised and jaws parted.

AGONISTIC BEHAVIOR

Dominance/Threat Displays: *opening dorsal gland, raising head and shoulders, approach/stare, showing tusks, growling, grinding molars, snapping, chasing.*

Fig. 16.4. Rock hyrax with hair over dorsal gland fanned open.

Defensive/Submissive Displays: *presenting rump, backing up, closing gland, flattening ears.*

Face-to-face meetings between hyraxes are apt to provoke low-level threat, as signaled by raising the hair around the dorsal gland and retracting the upper lip (*showing tusks*) (10). To avoid confrontations, feeding or *huddling* hyraxes face slightly away in a fan pattern (fig. 16.2) or align themselves in *reverse parallel.* Approach or even a *direct stare* thus amounts to threat. At increasing intensity, a threatening animal erects the gland fringe fully, raises its head and shoulders, growls and grinds its teeth, attacks and chases, snaps and bites. The tusks can inflict fatal wounds.

Fighting. Hyraxes defend themselves by *presenting their rumps.* When joining a *huddle* and often when entering a hole a hyrax backs into it. An intimidated hyrax keeps its gland closed and ears back. Fighting hyraxes back into each other and maneuver for position to stab and bite the opponent's head, neck, and shoulders, with dorsal glands wide open and exuding secretion. Most aggression is directed toward trespassing subadult and adult males by territorial males (5, 11).

REPRODUCTION. Hyraxes are seasonal breeders. Testis size, glandular activity, and territorial/aggressive behavior all increase during the mating season, and births within the same colony are synchronized. Litter size ranges from 1 to 6, averaging 2–3 (3, 6).

SEXUAL BEHAVIOR. Females show no visible signs of estrus, but quantity and quality of dorsal-gland secretion may change (see tree hyrax account). A female may attack a courting male until ready to copulate, then backs into the male, which grasps her firmly during copulation (fig. 16.7). Anatomical differences in the penis reflect and contribute to the reproductive isolation of the genera. The organ of the male bush hyrax is over 6 cm long with a file-shaped appendage at the tip within a cuplike glans, and opens on the belly (fig. 16.7). The other 2 have much smaller and simpler organs, and the tree hyrax's has a left-hand bend (3).

PARENT/OFFSPRING BEHAVIOR. The gestation period of 7–8 months is surprisingly long for such small animals, possibly a legacy from larger ancestors (9). Consequently the young are remarkably large and well-developed at birth, have their eyes open, and look like miniature adults. Expulsion takes only a few minutes and after being licked clean the baby climbs onto the mother's back, where it perches on the dorsal gland (fig. 16.5). Apart from keeping warm and out from under a cluster of adults, babies may thereby acquire and become imprinted on the mother's scent. Hyraxes become agile and can jump 40–50 cm within 2 days; they begin sampling vegetation within 2–4 days and serious feeding within 2 weeks. Weaning usually occurs within 3 months (8).

ANTIPREDATOR BEHAVIOR. Hyraxes have many predators, including snakes, eagles, owls, jackals, and cats from servals up to lions; even mongooses may take babies. The ranking predators on rock and bush hyraxes are Verraux's and martial eagles, cobras, and leopards. Hyraxes counter predation by being alert, remaining close to cover as much of the time as possible, and responding immediately to alarm calls. They live in crevices too small to admit larger predators. Rock hyraxes may have a cooperative defense, judging by reports of adults in the open converging on a young jackal, and of growling and gnashing their teeth at dogs from the safety of a cranny (6).

References

1. Bartholomew and Rainy 1971. 2. Bothma 1971. 3. Hoeck 1978. 4. ——— 1982. 5. Hoeck, Klein, Hoeck 1982. 6. Kingdon 1971. 7. Meyer 1978. 8. Sale 1965*b*. 9. ——— 1966*a*. 10. ——— 1966*b*. 11. ——— 1970*a*. 12. ——— 1970*b*.

Added Reference

13. Fischer 1986.

Fig. 16.5. Rock hyrax with two infants resting on her back.

Rock Hyrax, *Procavia johnstoni*
Bush Hyrax, *Heterohyrax brucei*

ASSOCIATION BETWEEN GENERA. Where their distributions overlap, rock and bush hyraxes often share the same rocks and associate more closely than any 2 known mammals except for monkeys that live in multispecies troops (see chap. 27) (6). They sleep in the same holes, huddle together in the morning sun (fig. 16.2), use the same latrines, have much the same vocal repertoire, forage together, and in the dry season have much the same diet, all virtually without aggression. The young of both genera play together in mixed nursery groups. Accordingly, both hyraxes are treated here in the same account. The main differences between the 2 genera are in size, feeding specialization, reproductive organs, and territorial advertising calls. If behavioral differences exist between species within the genera, they have yet to be reported.

TRAITS. See table 16.1.

DISTRIBUTION. See family introduction.

ECOLOGY. Wherever there are rocky cliffs, boulder screes, or rock outcrops (kopjes) with cavities sufficient to shelter a colony, hyraxes are likely to occur. In addition, the site must have some tree or brush cover, have multiple cavities (holes, cracks, crannies, overhangs) too small to admit most predators and facing away from any strong prevailing winds, and offer good vantage points where hyraxes can catch the early morning rays (fig. 16.2) (10, 11). Finally, the site must not be more than a few kilometers away from another cliff or kopje, unless temporary refuges occur along the way, because crossings between outcrops are perilous, and whole colonies can be wiped out by disease (e.g., sarcoptic mange) (5). Good but remote sites may remain unoccupied indefinitely.

The rock hyrax, being a grazer, has to leave its sanctuaries and venture up to 60 m to reach a pasture, with consequent exposure to predators. However, the ability to eat rapidly enables *Procavia* to spend less than

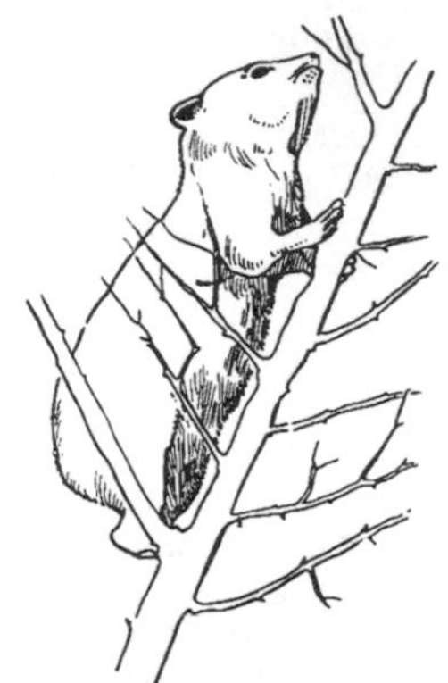

Fig. 16.6. Bush hyrax climbing a thorn tree.

an hour a day feeding (10). Led by mature individuals, the whole colony descends to pasture in the morning and again in the afternoon, using well-worn, direct pathways. They graze in a group, moving forward and fanning out, partially erected dorsal spots revealing nervousness and/or maintaining individual spacing. The territorial male often stands guard on the nearest elevation. If he, a bush hyrax, or a bird sounds an alarm, the whole herd scurries for the rocks.

The bush hyrax obtains most of its food by climbing into trees and bushes, concentrating on foliage. Though catholic in its tastes, *Heterohyrax* is partial to acacias and the way it clambers among the thorns with impunity is extraordinary (fig. 16.6). When grass is scarce or unpalatable, rock hyraxes browse many of the same plants, but cannot climb as well as the smaller bush hyrax or reach terminal branches accessible to the latter (8, author's observations).

SOCIAL ORGANIZATION: gregarious/colonial, several related females living and breeding with a territorial male.

The number of hyraxes in a colony depends on the size and resources of their home

Table 16.1 Traits of Rock and Bush Hyraxes

Traits	*Procavia johnstoni*	*Heterohyrax brucei*
Appearance	Bigger with blunter snout	Smaller with sharper snout
Length[a]	49 cm (39.5–58)	43 cm (32.5–47)
Weight[a]	3 kg (1.8–5.4)	1.8 kg (1.3–2.4)
Teeth	High-crowned, adapted for grazing	Low-crowned, adapted for browsing
Coloration	Yellow-brown, underparts a bit paler; fur coarse	Gray to gray-brown, white underparts, spot pale yellow; fur soft

[a]Serengeti samples from (5).

range. On the Serengeti kopjes, rock hyrax colonies number from 2 to 26, bush hyraxes from 5 to 34 (6). Kopjes with an area of less than 4000 m^2 are monopolized by 1 territorial male with a harem of 3–7 females—or shared by a colony of each genus. Larger outcrops can support several colonies of the same species, each occupying a different traditional range. Although female home ranges overlap, core areas averaging 2100 m^2 (1080–4050) for *Heterohyrax* and 4250 m^2 for *Procavia,* wherein colony members spend over 60% of their time, are more or less exclusive. The core areas correspond closely to male territories, averaging 2950 and 4800 m^2, respectively (7).

Associations among colony females are lasting and probably most of those that share the same home range are related (5). However, in rare cases of emigration, females have been accepted in another colony after the initial hostility of the resident females waned. There is no obvious rank hierarchy among females, but older individuals are usually the leaders, the most dominant and vigilant. The territorial male is dominant over all other members of the colony, and plays a classic watchdog role against both rivals and predators (e.g., see social mongoose and primate accounts).

Dispersing and Peripheral Males. Male offspring are forced to disperse after becoming adolescent, usually between 17 and 24 months, at latest by 30 months. Although their own fathers may remain comparatively tolerant, away from home the tusks of every territorial male are turned against them. The minimal distance traversed by dispersing Serengeti rock and bush hyraxes is 2 km (6). Several sightings of hyraxes moving overland at night suggest that dispersing males may choose to move under cover of darkness (author's observations). Some of them succeed in finding undefended spots on the larger kopjes, but for most there is no accessible, unoccupied refuge. The fact that females remain in the colony largely explains a skewed adult male:female ratio of 1 : 1.5–2 and 1 : 1.5–3.2 in the Serengeti rock and bush hyrax populations. Whole colonies may also change residence periodically, but these are usually local moves between dens and/or rock outcrops within the home range, made in daylight (7, author's observations).

Males that find places on the periphery of a colony do not associate in bachelor groups or become territorial. However, each one defends his own sleeping hole and peripheral males have aggressive interactions among themselves which help to determine dominance. When a territorial male disappears or loses fitness, the top peripheral male takes his place (7).

The remarkably peaceful association between rock and bush hyraxes, despite very similar needs, can be partly explained by the fact that *Procavia* always takes precedence in cases of direct competition, for instance over access to dusting and drinking/bathing spots (author's observations). Relations between the 2 genera might be very different if they were more evenly matched in size.

ACTIVITY. Except in very warm weather, hyraxes only emerge from their lairs at sunup and spend an hour or so basking before beginning to forage (fig. 16.2) (1). Feeding is largely limited to the cooler hours of the day, from 0730 to 0930 and 1530 to 1830 h, with maximum feeding activity always in the evening. Strong sunlight and low humidity are avoided by taking shelter during the hottest hours. No differences were found in the feeding activities of the 2 hyraxes in the Serengeti study, except that the rock hyrax typically started feeding at 0830 in the dry season, an hour later than the bush hyrax. In the wet season, both species began feeding earlier in the afternoon, *Heterohyrax* as early as 1330. Both genera call, and may move about and feed on moonlight nights. Rain causes hyraxes to stop feeding and take cover; on cold, rainy days they never come out at all (1). Territorial males move around more and appear to feed less than other colony members (1).

SOCIAL BEHAVIOR

COMMUNICATION. See family introduction.

TERRITORIAL BEHAVIOR. The call of the rock hyrax is one of the more arresting and entertaining sounds of the African wilds. A series of far-carrying yips and yelps, individually distinctive, it continues intermittently for minutes, builds to a climax, and ends with a series of guttural grunts. The hyrax equivalent of the lion's roar, the grunts convey a sense of power and menace even to a human listener; one can imagine the call putting the "fear of God" into subordinate males. Other territorial males are impelled to respond in kind, leading to outbreaks of calling. The bush hyrax's call, a thin, high-pitched mewing, is completely different and much less impressive, although far-carrying, too, and repeated for up to 5 minutes in a bout (7).

AGONISTIC BEHAVIOR. Described in family introduction.

REPRODUCTION. Both genera breed seasonally. Near the equator the bush hyrax responds to the 2 rainy seasons with 2 reproductive peaks, each lasting about 3 weeks. However, each colony reproduces only once a year; thus, 2 adjacent Serengeti colonies may give birth in March–May and December–January, respectively. Annual breeding is the rule over their whole range, with births occurring during the main rainy season (7, 8). The increased calling and aggression by adult males can be explained by an increase in testis size during the rut, which in territorial males is 20 times the off-season size (6)!

SEXUAL BEHAVIOR. Differences in the copulary behavior of rock and bush hyraxes may reflect the anatomical differences in their genitalia (see family introduction). In the prelude to mating, a *Procavia* male usually calls, then approaches the estrous female while *weaving his head,* with penis and dorsal spot both erected. Mounting, he embraces her firmly, makes several thrusts terminating with a short jerk, then jumps down. Copulation lasts a few seconds. A *Heterohyrax* male gives a sharp, shrill cry as he approaches, and the pair performs a sort of dance during which he sniffs her vulva. Preparatory to mounting, he often rests his chin on her rump, then sliding on and grasping her flanks makes thrusting movements while *swinging his head from side to side,* sometimes with mouth open (probably calling). The dorsal spot is apparently not erected by either party. The long, erect penis presses against the labia but without intromission, slackens, and ½ minute later stiffens as he thrusts again (fig. 16.7). After 3–5 minutes of this behavior, he suddenly thrusts home. The female jumps, turns and bites, then chases the male (4).

The females that peripheral males approach and occasionally succeed in mating during the rut are generally under 28 months, either subadults or young adults. Territorial males prefer to mate with mature females (6).

PARENT/OFFSPRING BEHAVIOR. Bush hyrax litters average 1.6 (1–3) compared to 2.4 (1–4) for rock hyraxes in the Serengeti (5). In a captive *Procavia* colony, females sought darkness, made no effort to isolate, but were aggressive toward other hyraxes shortly before parturition. A hoarse squeaking noise was given during contractions while the female stood still with her rear elevated, staring straight ahead with head somewhat raised. Parturition may take less than 10 minutes and the young may be expelled within a minute of the appearance of the head (9). The baby begins moving about almost immediately, and once the mother has licked away the enveloping membranes it begins trying to climb on her back, often before seeking the udder. The tendency to rest on the backs of the mother or other adults (see family introduction and fig. 16.5) persists for about 5 months, especially when colony members are heaped or huddled together. Adults do not sit on one another.

Fig. 16.7. Bush hyrax copulation: differences in the genitalia may actually prevent interbreeding.

In both hyrax genera, a newborn makes a 2-syllable *twittering distress call,* to which the mother responds with the deeper, *guttural twitter* (the adult contact call) before coming to its aid. Suckling attempts are often preceded by intense *twittering* and may serve to keep the mother in place. Nuzzling for the teats causes the mother to raise up. Suckling mothers emit a *whinnying call* with the upper lip drawn up exposing the tusks. They suckle only their own offspring, and the young partition the teats among them, maintaining the same ones throughout the 1–5 month suckling stage (6) (fig. 16.8). Babies older than a few days seldom suck for more than 3 minutes at a time (9).

ANTIPREDATOR BEHAVIOR: *alarm calls, flight to refuge,* threat and possible *mobbing* of smaller predators, *playing dead* (more under family introduction).

Procavia has an alarm call of variable intensity: a sharp, *low-pitched squeak* with mouth slightly open alerts nearby animals. A loud *squeak,* with mouth open and lips drawn back, sends the whole colony scurrying for cover. The most dominant animal generally brings up the rear, and may turn to threaten the enemy with its tusks, growling, until the last possible moment (10). The bush hyrax utters a *piping whistle.* The two species respond equally to each other's signals (1).

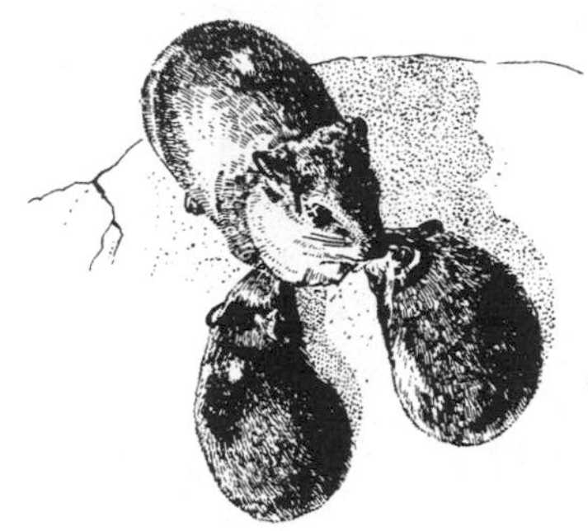

Fig. 16.8. Bush hyraxes sucking on forward pair of mammaries.

Hyraxes that have been caught by predators may escape by *playing dead,* as the following example illustrates (author's observations). At dusk, a subadult male leopard swarmed up a 10-m acacia tree beneath which his mother was waiting. A bush hyrax leaped from the crown and the mother caught it the moment it hit the ground. She tossed it about playfully for a few minutes, then carried it to the adjacent kopje and let her cub play with it. It showed no signs of life. But when he left it lying for a minute on a ledge, it leapt suddenly into a crevice choked with sansevieria. The leopard leapt after it, landing spread-eagled on the daggerlike leaves.

References

1. Hoeck 1975. 2. ——— 1976. 3. ——— 1977.
4. ——— 1978. 5. ——— 1982.
6. ——— 1984. 7. Hoeck, Klein, Hoeck 1982.
8. Kingdon 1971. 9. Sale 1965*a*.
10. ——— 1965*b*. 11. ——— 1966*a*.
12. ——— 1970*a*. 13. ——— 1970*b*.

Added References

Carr and McDonald 1986. Hoeck 1989.

Fig. 16.9. Tree hyrax aggressive display (redrawn from Kingdon 1971).

Tree Hyrax
Dendrohyrax arboreus

TRAITS. About the size of a rock hyrax, but with a larger head. *Weight and length:* 1.7–4.5 kg, 32–60 cm (2). Fur dense, velvety, color variable, usually darker in areas of high rainfall: from dark to light gray-brown to pale gray, underparts and dorsal spot usually lighter (3).

DISTRIBUTION. See family introduction. *D. validus,* found only in the mountains and offshore islands of Tanzania, may be a relic of the prototypical tree hyrax, which has been replaced by *arboreus* in eastern Africa, whereas *dorsalis* occupies the lowland rain forests (3).

ECOLOGY. To gain an arboreal-herbivore niche in African forests, in competition with rodents and primates, the ancestral bush hyrax had to become independent of rocks for shelter, improve its thermoregulation and climbing ability, and become nocturnal. The improvisations succeeded so well that the tree hyrax is one of the dominant herbivores in both lowland and montane forests. Furthermore, it retained its ability to live in the manner of rock and bush hyraxes, as demonstrated by colonies of rock-dwelling, semi-diurnal tree hyraxes that live in the Afromontane Zone of the Ruwenzori and other high mountains up to 4500 m (3). The usual diet of forest-dwelling tree hyraxes is foliage, fruits, vines, and twigs, but it can also exploit

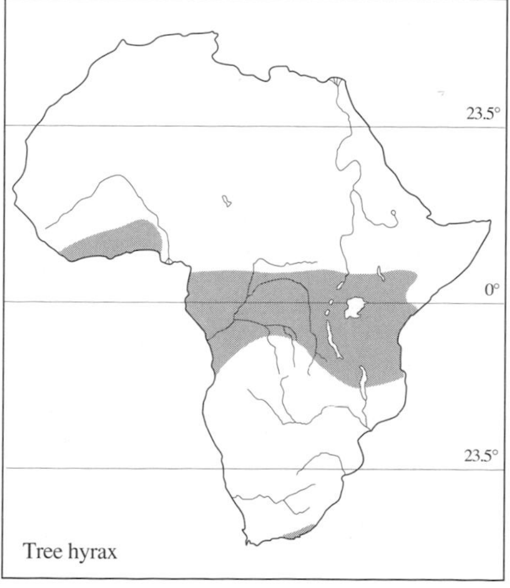

grasses, sedges, forbs, tree lichens, and such. Like other hyraxes, *Dendrohyrax* eats rapidly and it consumes up to ⅓ its own body weight per day (3).

SOCIAL ORGANIZATION. Though the tree hyrax is generally considered solitary and nocturnal, its social organization and activity patterns are clearly flexible and can be adapted to suit widely different conditions. A definitive study has yet to be made. In general, apparently, it is solitary, or lives in pairs or small families. A tree with a hole wherein the animals sleep and where they sometimes sunbathe during the day is the center of the small home range, often identified by the droppings accumulated around the base. They use regular arboreal pathways and descend to the ground regularly enough so that they can be trapped with foot nooses set at the base of their den trees (1). When necessary, say, to reach isolated trees or patches of woodland, tree hyraxes cross open country. Individuals have even been found sheltering in termitaria in comparatively open woodland (3).

SOCIAL BEHAVIOR

COMMUNICATION

Vocal Communication: *territorial cry, neighing, squealing, "flapping" alarm call.*

The presence of tree hyraxes would usually go undetected were it not for the weird, electrifying calls of the territorial males (3). Females are also known to call but, having a smaller larynx and guttural pouches, sound feeble by comparison. The call starts with a series of spaced cracking sounds, likened to the rusted hinge of a huge gate slowly opening, followed by a series of expiring screams suggesting a soul in torment. The times vary but there are typically 2 calling periods in a night—often 2 or 3 hours after dark, probably following an intensive feeding bout, and in the hours after midnight. Males seem to have fixed calling perches and cries are often answered by neighbors. A male may repeat his calls every few minutes for up to an hour, then abruptly fall silent. Calling is particularly common in the dry season and on clear, moonlight nights (3). Preparatory to calling, a captive male spread its dorsal spot fully and opened its jaws slowly until its teeth were bared; afterward it invariably erected its penis (1).

An infantile distress call develops into a *neighing call* given by agitated adults with raised head and open mouth. It has been heard as a precopulatory call by males, and by individuals exerting themselves as when trying to escape. The alarm call is a piercing cry. Mothers respond to the infant's distress cry with sounds like a piglet, and mothers with infants make *flapping sounds* when disturbed (3).

AGONISTIC BEHAVIOR. Males become particularly aggressive during the mating season. A captured *D. dorsalis* male gave *wheezing growls, ground its teeth,* showed the nictitating membrane over the eyes, *snapped its jaws,* and made short charges. Meanwhile its dorsal spot was fanned and secretion could be seen and smelled (fig. 16.9) (4).

A probably fatal fight was witnessed in Ngorongoro Crater. Two hyraxes that had called from adjacent fig trees during the night fought furiously next day when one invaded the other's tree. Either the invader or the defender was so badly hurt it could only crawl away and take refuge in a hole at the base of a tree trunk.

At 2200 h it came out and dragged itself back to the other fig tree. On subsequent trips to this place, a popular campsite, only 1 hyrax called (author's observations).

REPRODUCTION. See family introduction.

SEXUAL BEHAVIOR. A captive female was noted to be secreting a cinnamon-scented oil from its dorsal gland several days before mating (4).

PARENT/OFFSPRING BEHAVIOR. Tree hyraxes bear 1–2, rarely 3 young (3).

ANTIPREDATOR BEHAVIOR. The crowned hawk eagle, *Stephanoaetus coronatus,* is generally the ranking predator on tree hyraxes.

References

1. Fey 1960. 2. Hoeck 1984. 3. Kingdon 1971. 4. Richard 1964.

Chapter 17

Elephants

Order Proboscidea, Family Elephantidae

Fig. 17.1. Elephant female herd in defensive formation.

African Elephant

Loxodonta africana

TRAITS. The largest land mammal, nose prolonged as a trunk with 2 fingerlike projections at the tip. *Height and weight:* geographically and individually variable; fully mature bulls average over 3 m (up to 3.3, record 4 m) and 5000 kg (maximum 6000), cows 2.5 m and 3000 (maximum 3500) (14). *Teeth:* 1 functional cheek tooth in each quarter, a total of 6 sets of progressively larger size; tusks (upper incisors) grow continuously, size genetically controlled but also age-dependent, averaging 61 kg at 60 years in bulls and 9.2 kg in cows (record 106 vs. 25 kg), record length 355 cm. *Feet:* 4 nails on fore-, 3 on hindfeet. *Skin:* gray or brown, wrinkled, up to 3 cm thick but pliable and sensitive, sprinkled with sharp, coarse bristles; trunk, ears, and forehead fuzzy with long sensory hairs. *Ears:* up to 2 m high and 1.5 m wide. *Genitalia:* opening downward through skin flap between rear legs in both sexes; testes internal, extended penis recurved to rear except when fully erect. *Scent glands:* unique temporal glands weighing up to 3 kg in males, rarely over 1 kg in females. *Mammae:* 2 (between forelegs) (14). *Longevity:* up to 60 years.

Senses: hearing and smell excellent, eyesight moderate, best in dim light.

DISTRIBUTION. The elephant was formerly ubiquitous south of the Sahara wherever water and trees occurred, but its range and numbers have shrunk as human population, development, and poaching have increased. As recently as the early 1980s an estimated 1.3 million elephants survived, an indication that sizeable blocks of wilderness remained (especially in Zaire) remote and sparsely settled, for parks and reserves sheltered only a small fraction of the total. But a greatly accelerated slaughter had already begun, fueled by a rise in the price of ivory to $100/kg (from $7.44/kg in 1970); by the end of the decade elephant populations over much of eastern Africa had declined by up to 80%, both outside and inside the parks. In Douglas-Hamilton's words, "The premature elimination of elephants over much of their range represents one of the most wasteful mammalian tragedies of the century" (9, p. 22).

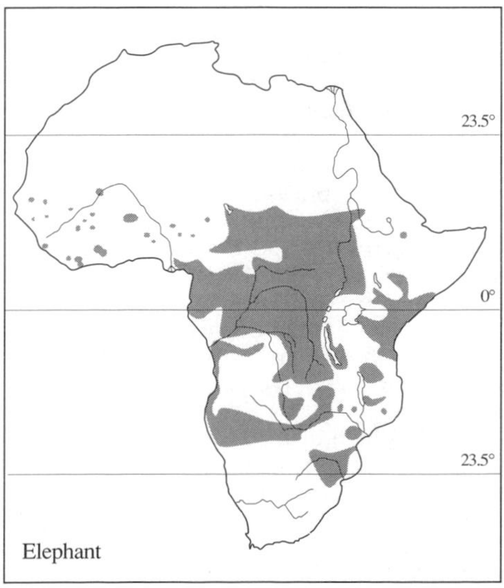

Elephant population in 1984 (redrawn from Douglas-Hamilton 1987)

Neolithic rock paintings show that African elephants inhabited the Sahara before it became desert. Indeed the elephants used by Hannibal came from North Africa, and the species persisted in the Atlas Mountains, along the Red Sea coast, and in Nubia well into the Christian era (5, 6).

Subspecies. Forest elephants of the Congo Basin and West Africa are so unlike other African elephants that they would be considered a separate species if the 2 types did not hybridize (e.g., in Uganda's Budongo Forest (6, 26). *L. a. cyclotis* stands 2.4–2.8 m and weighs 1800–3200 kg (11). The tusks are straight, parallel, and directed downward, the ears are more oval, and the feet have 5 nails in front and 4 behind.

ANCESTRY. Formerly one of the most widespread and successful groups of large mammals, which included mastodons and mammoths, the probiscideans are now reduced to just 2 species. The order arose in Africa in the middle Eocene and the Elephantidae, which first appeared during the late Miocene (see table 1.1) and in which both living elephants belong, is only 1 of perhaps 5 different families. What they all had in common was a tendency to be large, ponderous creatures with pillarlike legs, tusks, and, of course, an extended flexible nose (5, 6).

ECOLOGY. Elephants can subsist in virtually any habitat that provides adequate quantities of food and water. The forest elephant lives in the West African rain forest. The savanna race inhabits lowland and montane forests, upland moors and low-lying swamps, floodplains, and all types of woodland down to scattered tree savannas. It can even live in subdesert where gallery forest or neighboring mountains provide shade, browse, and water (e.g., in Namibia's Kaokoveld). Optimum habitat affords both grass and browse: the forest edge, where there is a variety of small trees, vines, and herbaceous plants; woodland with open or continuous canopy undergrown by shrubs and interspersed with glades; and the bush country of more arid regions (14).

DIET. The proportions and kinds of forage eaten vary seasonally and according to availability. Generally, elephants select the most nutritious and palatable of the plants available in quantity. Like other mixed feeders, they tend to concentrate on grasses and herbs in the rainy season and on woody plants in the dry season. Elephants therefore wander most widely across the savanna in the rains, and spend more time in forest and concentrate near water points at other times.

Elephants have a more catholic diet than practically any other herbivore. Thanks to their size, the milling action of the 2 pairs of long, rasplike molars (which slide forward and backward rather than side to side as in most herbivores), and the incredibly versatile trunk, they can feed from ground level up to 6 m (rarely rearing), pick up nuts, strip off leaves and bark, break off branches, and uproot shrubs and small trees. Some bulls master the technique of pushing over mature trees (12). An elephant consumes about 4% to 6% of its body weight daily: a 5000-kg bull needs c. 300 kg (75 kg dry weight) and a 2800-kg cow needs 170 kg (42 kg d.w.) of food per day. This is surprisingly little for such a huge animal. Nursing mothers eat proportionally more than other classes (14). The fibrous dung shows that digestion is very incomplete. Only 44% of the food is assimilated, compared with the ruminant's 66%. But elephants compensate by processing much coarser forage at a rapid rate (c. 12 hours) (21).

ENVIRONMENTAL IMPACT. Elephants have more environmental impact than any mammal but man. The beneficial effects (trails, wells, bringing food within reach of smaller browsers) are overshadowed by their destruction of trees, which has been spectacular in many African parks (7, 15, 16, 17). In Murchison and Tsavo, for instance, woodland was transformed into grassland, with great detriment to the elephants and many other species (especially black rhino). The root of the problem is the increasing confinement and concentration of elephants within parks. Tree destruction is a normal elephant activity and when spread over a wide area contributes to habitat diversity and soil turnover (7).

SOCIAL ORGANIZATION: matriarchal with herds, bond groups, and clans of related females; males separate in herds or alone, their rank order determined by seniority and reproductive condition; populations nomadic/migratory within large home ranges.

The basic social unit is a group of related females consisting of a mother and young with her grown daughters and their offspring (8, 20, 30). Herds range in size from 2 to 24 or so, with 9–11 typical. The members of a family unit keep together, individuals rarely venturing more than 50 m from the nearest neighbor. Activity, direction, and rate of movement are set by the matriarch,

Fig. 17.2. Elephants supporting and attempting to raise a wounded companion.

recognizable as the largest cow. When she stops to feed, the herd spreads out and forages, too; when she moves on, the rest soon cease feeding and accompany her. If disturbed, the group immediately clusters around the matriarch and follows her lead (fig. 17.1). First-year calves get under their mothers, and older calves press close to the maternal side. A family and even large herds often reveal anxiety when crossing an open area in daylight by bunching together and moving rapidly until they regain cover. Typically the matriarch goes in front and another big female brings up the rear (8, 12).

Leadership and experience play such a crucial role in elephant social organization that the female life span extends past the age of reproduction (15). Postreproductive survival is also true of man, but otherwise very unusual. The sudden loss of the matriarch, say, by shooting, completely disrupts and disorients her followers. They often mill around and let themselves be shot rather than abandon her. Cows try to lift her to her feet and if she can walk they support her, one on either side (fig. 17.2). Such behavior demonstrates the remarkable development of altruism in elephants. Yet matriarchs too old and feeble to lead any more (between 50 and 60 years old) either leave or are abandoned by their families (15). The next-oldest cow then assumes leadership.

When the number of elephants in a family increases beyond 10 or so, it tends to split in 2. Although the new herd has its own matriarch, the 2 families continue to associate closely at least half the time, remaining on nearly as friendly terms as ever, as evinced by the intense greetings between members after even a short separation. Unrelated families never greet each other so warmly (see *greeting ceremony*, under Communication) (20). Associated family units compose a "bond group" of related individuals. In Manyara N.P., bond groups included an average of 2.5 families, with a mean total of 28 (14–48) elephants, that stayed within a kilometer of one another (8, 9). Depending on the productivity of the area, home ranges may be as small as 14 km^2, as in Manyara, to over 3500 km^2 in arid country (8, 16, 17). Bond groups that share the same range make up clans of probably related elephants (8, 9, 30).

Elephants stressed by range compression and hunting pressure tend to band together in larger groups, often composed of reunited clans. Temporary aggregations of up to 200 (formerly up to 1000) elephants also occur, and these usually include bulls. The animals feed and move along together in a coordinated manner, the movements of one group having a contagious effect upon others. Aggregations are formed mainly during the rains and are associated with peak mating activity. In times when elephant movements were unrestricted, such large herds were often migratory (1).

Bull Society. Male elephants leave the maternal herd in adolescence (12–13, up to 20 years) and thereafter never stay for long with any cow herd. Separation is a gradual process and adolescent bulls may become peripheral, following the family unit at a distance, long before becoming independent. A decisive factor is female intolerance of the boisterous and sexually precocious behavior of pubescent males, which arouses maternal wrath especially in mothers with young calves (8). After becoming independent, bulls alternately associate in bachelor herds and wander alone. Herds of over 35 and up to 144 bulls are sometimes encountered, but 2–14 is more typical (15). Immature bulls, in particular, tend to associate with peers, but bachelor herds often include a wide age range.

Although bulls wander more widely than cows, they frequent particular small core areas for months at a time during periods of sexual inactivity, alone or in company (12, 20). In Amboseli N.P. the same individuals could be found year after year in the same "retirement" areas. Moreover, some of the bulls had 1 or more companions with whom they regularly associated in the bull areas; several pairs were together over 30% of the time they associated with other bulls (20). During periods of musth, mature bulls leave the retirement areas and wander alone in search of mating opportunities. Musth periods are neither synchronized nor seasonal (more under Sexual Behavior) (20, 32, 33).

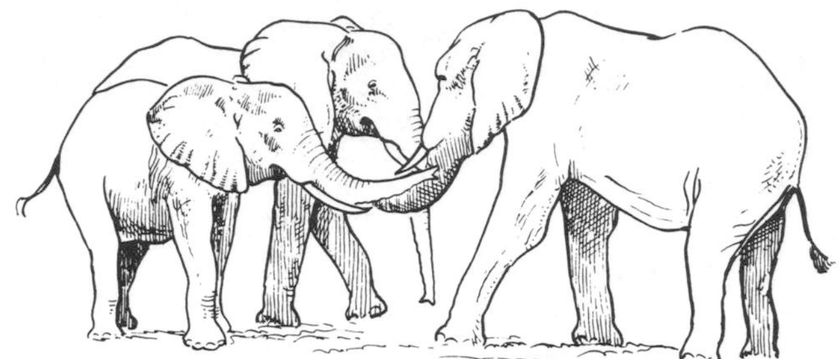

Fig. 17.3. Elephant *greeting ceremony.*

Very old bulls are the most sedentary of all elephants. They stay near swamps where they find soft vegetation they can still chew with their worn-out teeth.

ACTIVITY. Elephants spend 16 hours a day feeding, with peaks in the morning, in the afternoon, and around midnight, and eliminate c. 155 kg of wastes at the rate of 3–5 dung boluses every 1.4 hours (29). They sleep 4–5 hours in 24, sometimes lying down.

Drinking and Bathing. Elephants can go several days without drinking, during which they range up to 80 km from water, but drink and bathe daily by choice. A large bull may drink 100 liters at a time and up to 227 liters a day. Bathing follows drinking. They roll and wallow in shallows and may submerge completely in deep water. At small holes, an elephant uses its trunk as a hose to wet itself, then to splatter itself with mud or dust, followed by rubbing against objects. A need for extra sodium in the diet is met by visiting mineral licks, where elephants excavate pits and even caves with their tusks, or by drinking saline water (28).

POSTURES AND LOCOMOTION. The remarkable movements and postures that circus (mostly Asian) elephants perform are also performed by wild African elephants. They roll, kneel, squat, or sit up on their back legs like monstrous canines, and have even been seen seated on logs browsing meditatively (26). Preparatory to lying down or getting up, they first sit on their haunches. They can stand semierect on their hindlegs to reach some food item (and of course when copulating), and can negotiate precipitous slopes that people also have to negotiate on all fours. They cannot run or jump like other animals and have only one basic gait: an *ambling walk.* The normal walking rate is 6–8 kph, which may be increased to 10–13 kph by taking longer, quicker strides. An elephant can reach 30 kph at top speed—fast enough to overtake a running man.

SOCIAL BEHAVIOR

COMMUNICATION. With their highly mobile trunks, keen sense of smell, glandular secretions, great ears, and varied vocalizations, elephants are notably well equipped to express themselves.

Tactile and Olfactory Communication: *greeting, caressing, twining, slapping,* checking reproductive status, temporal gland secretion.

Elephants are very much contact animals. Family members often stand touching while resting or drinking. They lean and rub their bodies together, and often touch one another with their trunks in various contexts. A *greeting ceremony,* in which the lower-ranking animal inserts its trunk tip into the other's mouth, enables elephants of different rank and relationships to come closer together amicably (8) (fig. 17.3). The performance may derive from the calf's habit of putting its trunk into its mother's mouth, both to sample food and for reassurance. Holding the trunk out toward an approaching elephant is a *greeting-intention movement.* Mothers often guide a calf by gripping its tail (fig. 17.4), and following older calves sometimes hold the mother's tail. A touch, an embrace, or a rub with a foot reassures and a slap disciplines a calf. Courting elephants may caress each other

Fig. 17.4. Mother "steering" her calf with a grip on its tail.

and *twine their trunks;* playing and fighting elephants *trunk-wrestle.*

As an olfactory organ, the trunk is used for tracking and to check female reproductive status—this by putting the tip to the genital opening, afterward inserting it into the mouth to waft scent or secretions into the vomeronasal organ. The temporal gland, which in the Indian elephant is apparently associated only with the male's sexual cycle (musth, discussed under Sexual Behavior), is active in all African elephants beginning in infancy. Its full significance is undetermined, but the watery, variably faint-to-musky smelling secretion flows whenever African elephants become excited or anxious, staining the cheeks below the orifice of the gland (3, 35). The apparent association with stress suggests the gland is under autonomic control, and may quickly respond to massive releases of adrenaline in the circulation.

Vocal Communication. The elephant's vocal repertoire is limited to 4 different sounds, but gradations in pitch, duration, and volume enable elephants to express a wide range of emotional states.

Rumbling, a deep growling sound believed by generations of hunters to emanate from the stomach, is actually a vocalization and the main form of distance communication among elephants. The discovery that rumbling calls cover a broad range of frequencies, most of which are below human hearing thresholds (31), has stimulated research to determine whether elephants communicate over distances of several kilometers, as circumstantial evidence has long suggested. Preliminary results indicate that very loud infrasounds (as low as 14 Hz and as loud as 103 decibels) probably carry even further and play an important role in social and sexual behavior, enabling elephants to meet (clan members, females in estrus with males in musth) or avoid (musth males) each other (34).

Quiet *rumbles* audible to human ears are often uttered as a herd feeds. When a member has strayed some distance from the rest, it may give a louder *growl,* evoking a response in kind usually from the matriarch. Elephants often *growl* when greeting and there is evidence that their voices are individually recognizable (8). When the mouth is opened, the growl changes character. It becomes a *bellow* of pain or fear when an elephant trips, bumps, falls, or is suddenly attacked by another elephant. A drawn-out growl becomes a *moan.* As an expression of anger, *growling* increases in volume with the degree of arousal, becoming a terrifying *roar* when elephants threaten man or other predators.

Trumpeting, a sound of excitement, is produced by blowing through the nostrils hard enough to make the trunk resonate, meanwhile usually holding it straight down or curved slightly backward. The sound can be modulated, from a short blast, given by a startled animal, to a prolonged reverberating cry of rage. It is combined with *growling* and *screaming* in threat displays. *Trumpeting* can also sound an alarm or a cry for help; it is also voiced during intense *greeting ceremonies* (8, 20). The trumpeting of small elephants sounds very much like that of adults.

Squealing is a juvenile distress call that elicits an immediate response from the mother and other females.

Screaming is the adult equivalent, which is used along with trumpeting to intimidate opponents.

AGONISTIC BEHAVIOR.

Dominance/Threat Displays: *turning toward, spreading ears, standing tall* (or the *erect posture), head-nodding, -jerking, -shaking,* and *-tossing; forward-trunk-swish, demonstration* and real *charges.*

Defensive/Submissive Displays (mostly signs of fear and indecision): avoidance (*turning away, backing up, running away*), *flattening ears, arching back, raising tail, agitated trunk movements, touching temporal gland, throwing dust, pawing, foot-swinging, swaying, exaggerated feeding behavior.*

Fighting: *trunk-wrestling, pushing, ramming, tusking.*

Some of the postures and movements that express threat are also seen in other contexts, making for a certain ambiguity. Thus, *spreading the ears* and holding them tensely forward or flapping them is a characteristic sign of arousal. But elephants spread and also flap their ears to regulate their temperature, or simply to hear better, actions that are normally ignored by other elephants. The trunk may be raised or extended by an angry elephant, or by one simply sniffing the air; and the tail may be raised in either anger or fear. Such actions mean little by themselves and are ignored by other elephants. But when an elephant flaps its ears and lifts its head high at the same time, nearby herd members react immediately to this *alert posture* (fig. 17.5). Such combinations of postures and movements, together with sounds and perhaps odors, usually make an elephant's feelings, if not its intentions, perfectly plain.

Fig. 17.5. Bull elephant in the *alert posture* testing the wind.

Standing tall (fig. 17.6), a dominance display, resembles the *alert posture,* but the head is raised higher and the elephant peers over its tusks, ears cocked and trunk hanging at an acute angle.

Up-and-down and side-to-side head movements are a feature of most threat displays. Walking toward an antagonist while nodding with ears half-spread is a common display between cows. Merely turning toward a potential antagonist is also a mild threat which often serves to reinforce dominance relations in a herd (8). A threatened individual generally gives way without responding in kind. *Head-jerking* is a single, abrupt upward movement, followed by a slower return. *Head-tossing* is a pronounced version of the jerk; the head is lowered and then lifted sharply so that the tusks describe an arc. In *head-shaking,* the head is first twisted to one side, then quickly rotated from side to side, causing the ears to slap against the face with the sound of a snapping towel (fig. 17.7) (2).

Fig. 17.6. *Standing tall.*

The *forward-trunk-swish,* typically accompanied by trumpeting or an air blast, is a threat usually addressed to a smaller antagonist. The trunk is rolled up then abruptly unfurled and swished toward the opponent. Using this same gesture, sometimes elephants pick up or rip up and throw bushes, grass, or other objects toward an enemy.

Since elephants grow bigger with age, dominance is largely a matter of seniority. Males learn their relative strength and status, within and between age classes, as they grow up, through frequent playfights and semiserious fights. Dominance is usually determined and acknowledged the moment 2 bulls meet, simply on the basis of which stands taller. Where the size difference is obvious, the smaller animal *flattens its ears,* keeps its *head lowered, moves backward and sideways,* and makes *writhing trunk movements.* When contact is made, it puts its trunk in the other's mouth (*greeting*). The dominant animal asserts superiority by holding its head high, trunk hanging and ears more or less spread. During contact, he may hold his trunk tip to the inferior bull's temporal gland (8).

Fig. 17.7. Immature bull makes a *head-shake threat* (*left*) during agonistic encounter.

Fig. 17.8. Bulls *taking measure* of each other's height and tusk length.

When 2 equal-size bulls meet, they take each other's measure by coming together with heads raised until their trunk bases and/or tusks are firmly engaged (fig. 17.8). This is also the position during combat. Each tries to be taller, to press downward on the other's trunk. Or, if contact is made on the underside of their raised trunks, they may grapple for a hold. Tusk length is important, but so are height and weight, for these determine who can push whom around. The elephant that holds his head higher, whether because he is bigger or more confident, usually wins. Aggressive interactions between mature bulls are rare and usually low key, although serious and even fatal fights sometimes occur when 2 bulls of equal size compete over a cow in heat and/or are simultaneously in musth (32, 33). Fighting technique is best seen between young bulls: for instance, when families are gathered at water holes. Threat displays are issued as a challenge, or one youngster may simply ram another in the side. The 2 then approach with ears spread and engage tusks and trunks with heads raised. They entwine trunks, then try to twist or push each other out of line (fig. 17.9), while using the tusks as levers to force the opponent's head down or sideways. When they break apart, the first to turn may be prodded gently in the flank by the victor. In a serious fight, this is the time and the manner in which fatal goring may occur (1). Fighting elephants also ram each other head-on (amazing photos in 32).

Fig. 17.9. *Trunk-wrestling.*

REPRODUCTION. First conception typically occurs at 10–11 years (8–20), and the interval between calves ranges from 4 to 9 years, depending on nutrition and population density. Gestation is close to 22 months (mean of 656 ± 4 days [19]) and twins are rare (13, 20, 25). Mating and births are most frequent during the rains.

SEXUAL BEHAVIOR. Estrus is detectable by a female's behavior in the presence of adult males (19). Normally cows allow bulls to approach and check their reproductive status (described under Tactile and Olfactory Communication). But a female approaching estrus is wary of bulls: she carries her head higher than normal, her eyes are wide open and watchful instead of downcast, and she moves quickly out of the way of an approaching adult male. If persistently approached and followed, she leaves the herd and walks rapidly away, with head high and/or turned to one side as she looks back at her escort(s), and tail raised. She maintains a lead of 10–75 m and eventually circles back to rejoin her family. If the bull is aroused to the point of having an erection and attempts to overtake her, the "estrous walk" turns into a chase, during which the female may go a kilometer or more from the herd over a period of several hours. The cow stops fleeing when and if a bull gets within touching distance, whereupon he attempts to mount her. First he lays his trunk along her neck and head, then rests his own tusks or chin

Fig. 17.10. Copulation.

on her rump and levers himself onto his bent hindlegs, resting his forefeet on her pelvis. A cooperative cow holds still or backs into him. As his penis grows fully erect (weighing then over 27 kg), it curves forward and upward, probing for the entrance to the vagina. After achieving intromission, the bull straightens his hindlegs and assumes a normal mounting posture (fig. 17.10), pulling the female's vulva upward in the process (cf. spotted hyena copulation). Mounting and copulation are completed within c. 45 seconds, without thrusting or apparent ejaculatory pause. Sometimes one or both rumble, groan, or scream. After the bull dismounts the cow may run off, but most stand still (19).

The distraught behavior of elephant cows during the 2–6 day estrus creates a disturbance and often excites other elephants. Chasing and particularly copulation may provoke *ear-flapping, head-shaking,* and loud vocalizing. While a pair is copulating, other herd members may *back toward them,* urinating and defecating. One result of the commotion is to bring other bulls to the scene, which then proceed to compete for mating rights—supplanting, chasing, threatening, shoving, tusk-poking, and sometimes fighting one another. On 154 occasions when estrous females were observed in Amboseli N.P., an average of 10 adult bulls (1–67) (> 25 years old) were present. Not only did older and bigger bulls dominate young animals but cows also clearly preferred to mate with them. Out of 134 chases, cows eluded all but 30.6% of their pursuers; medium and large bulls accounted for 83% of the chases that ended in copulation. Furthermore, the 19 largest bulls in the population of 160 adult males performed 65% of the observed copulations. And among them, the 1 or 2 bulls that happened to be in musth at any given time were continually in quest of mating opportunities, dominated all other elephants, and were preferred above all others by estrous cows. Apart from heightened aggression associated with elevated testosterone levels, musth is revealed by the copious and viscous secretions of the temporal glands, and a swollen and partially extended penis which drips a green, strong-smelling secretion (20, 23, 33). When a large bull in musth stays with a herd, it usually indicates he is in consortship with an estrous female. Not only does the bull follow her but the cow stays close to him (and sometimes solicits mating), especially when other bulls attempt to court her. Because other males give a bull in musth a wide berth, females consorting with them thereby escape much of the harassment that can make estrus stressful and exhausting.

PARENT/OFFSPRING BEHAVIOR. Very few births have been observed in the wild (20). The cow stays close to her family. The calf has to pass a sharp bend in the vagina to emerge between the hindlegs, and is dropped, either head or tail first, while the mother stands (24, 27). She turns and examines it with her trunk, then touches it gently with a forefoot. One baby raised its trunk and began struggling to rise within 4 minutes, gained its feet unassisted after 20 minutes, but then fell down repeatedly. The calf also has to locate the teats between the forelegs on its own; while it is sucking, the trunk, which only gradually becomes useful, dangles on one side. Calves are feeble for several days and remain shaky for weeks.

The bond between mothers and offspring is very close and can endure for 50 years. Small calves remain in almost constant contact. If one wanders more than 15–20 m away, the mother goes after it. She pushes it under her to protect it from danger or the hot sun, boosts it up steep places by crooking her trunk around its posterior, and lifts it down embankments or over fallen trees and out of wallows. She assists it to water and washes it by gently squirting water over it and scrubbing it with her trunk. In time of drought a mother will regurgitate water from her stomach and spray it to cool her calf (24). As the calf develops, the bond gradually changes to a leader-follower relationship in which the burden of staying close shifts to the calf. Even at 9 years a calf may spend over half the time less than 5 m from its mother. Calves may be weaned as early as 1–2 years, but often nurse 4 years or more. Related cows suckle one another's calves and some cows may continue lactating through their childbearing years (22).

ANTIPREDATOR BEHAVIOR: *group defense,* threat displays, *demonstration* and real *charges* (see Dominance/Threat Displays under Agonistic Behavior).

Predation on young calves when the opportunity arises would account for the aggressive manner in which families containing young calves often react to lions and hyenas (1). The *group defense* is employed against all potential predators, man included. This is the one animal against which a person is not safe inside a vehicle.

Disturbed elephants undecided whether to stand their ground, attack, or flee perform displacement behavior in between threat displays. The trunk coils and uncoils, writhes and swings. An animal may touch its ear, mouth, or temple glands, or rub an eye. *Displacement feeding* may include breaking off branches, or pulling up tufts of grass and slapping them against a foot with extra emphasis.

People who know elephants well can usually judge the mental state of an aroused individual or herd and tell the difference between a *demonstration* and a real *charge.* Signs of indecision are evidence that the elephant is not prepared to press home an attack (fig. 17.11). But experts have been killed by elephants. The fact is, the elephant itself may not know what it is going to do when it starts a charge. Thus, it is generally true that cow herds with small calves are most excitable and dangerous, whereas bull herds and single bulls (or cows) rarely make real charges. But a bull suffering from an old gunshot wound or an abscess may become a rogue.

A serious charge may come without warning, presumably because the elephant is unafraid and therefore not constrained to make threat displays. According to one expert, "When an elephant means business, it nearly always comes immediately and without warning except for a momentary rocking motion of its body; sometimes it comes without a sound. The tail stiffens, the tusks are held high and the trunk is drawn in against the chest. The ears are usually, but not always, spread out to their full extent giving the head a most awesome frontage of 12 feet or more" (1, p. 82). The trunk may also be rolled back, rolled under to one side, and sometimes extended. The head may be raised or lowered.

Fig. 17.11. Bull in *demonstration charge;* the winding trunk reveals ambivalence.

A *demonstration charge* ends in retreat, with head high, back arched, and tail raised. The animal makes off at an angle, looking back with head turned.

In contrast to their occasional ferocity, elephants can be timid to the point of absurdity. A large bird landing nearby, the presence of a squirrel, or a mongoose blocking the path is often enough to alarm them, and sometimes a whole herd will bustle away from some purely imaginary danger (1). Most elephants react to human disturbance by fleeing, silently or in noisy disarray, depending on how badly they are frightened.

References

1. Bere 1966. 2. Buss and Estes 1971. 3. Buss, Rasmussen, Smith 1976. 4. Buss and Smith 1966. 5. Carrington 1958. 6. Coppens et al. 1978. 7. Croze, Hillman, Lang 1981. 8. Douglas-Hamilton 1972. 9. ——— 1987. 10. ——— and Douglas-Hamilton 1975. 11. Haltenorth and Diller 1980. 12. Hendrichs 1971. 13. Laws 1969. 14. ——— 1970*a*. 15. ——— 1970*b*. 16. Laws, Parker, Johnstone 1975. 17. W. Leuthold 1977*b*. 18. Leuthold and Sale 1973. 19. Moss 1983. 20. Moss and Poole 1983. 21. Napier-Bax and Sheldrick 1963. 22. Perry 1953. 23. Poole and Moss 1981. 24. Poppleton 1957. 25. Short 1966. 26. Sikes 1971. 27. Slijper 1960. 28. Weir 1972. 29. Wing and Buss 1970.

Added References

30. Moss 1988. 31. Payne, Langbauer, Thomas 1986. 32. Poole 1987. 33. ——— 1989. 34. Poole et al. 1988. 35. Gorman 1986. Hall-Martin 1987. Lee 1987.

Part III

Carnivores

Families represented in Africa	Genera	Species
Viverridae, genets, civets, mongooses	18	34
Hyaenidae, hyenas	3	4
Felidae, cats	3	10
Canidae, foxes, jackals, wolf, wild dog	4	10
Mustelidae, weasels, polecats, skunks, badgers, otters	6	7

Chapter 18

Introduction to Carnivores

Order Carnivora

ANCESTRY. Herbivores and carnivores both arose from insectivore ancestors soon after the Age of Dinosaurs ended some 65 million years ago. They have evolved and differentiated together from the beginning, inextricably joined by the predator's nutritional dependence on its prey and the selection predation exerts in favor of the fittest prey.

The earliest carnivorous mammals were long-bodied, short-legged, and small-brained predators belonging to an archaic order, the Creodonta ("flesh-tooth").

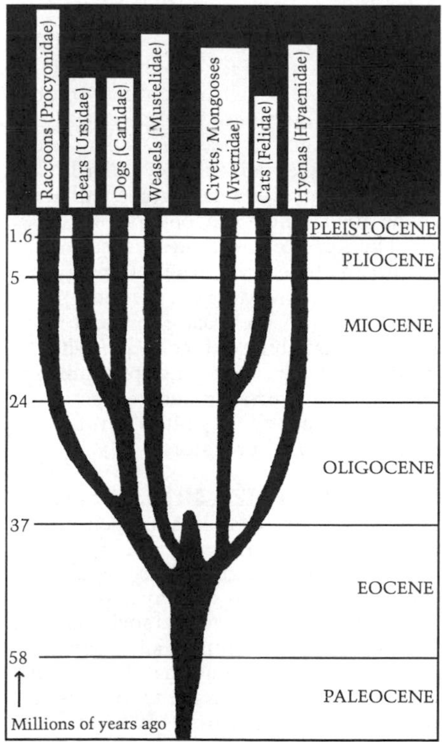

Fig. 18.1. Carnivore family tree (adapted from Macdonald 1984).

They differentiated and thrived as long as herbivores remained clumsy and slow-witted. But as speedier and brainier herbivores developed, only 1 family of creodonts was able to keep pace: the miacids, small, mongoose- or weasel-like creatures with a proportionally larger, more advanced brain, which hunted game in dense undergrowth or in trees. All recent carnivores descend from the miacid line (2, 3, 8, 11, 15).

Between the late Eocene and the end of the Oligocene (40–24 million years ago), all but 2 of the existing families of carnivores arose, represented by species very like recent dogs, cats, weasels, and mongooses (fig. 18.1). The carnivore adaptive radiation was paced by the rapid speciation of other mammals during the same period, notably the odd- and even-toed ungulates and the rodents.

The Carnivora divided very early into 2 separate lines, one of which culminated in dogs and the other in cats. Dogs, mustelids, bears, and raccoons are grouped in the superfamily Canoidea, whereas cats, civets, mongooses, and hyenas are allied in the Feloidea. (Seals, walruses, etc., considered carnivores in some classifications, are generally treated as a separate order, the Pinnipedia, and are not included in this *Guide*.) Judged by similarity to the ancestral miacids, viverrids are the least-changed of modern carnivores (3).

CARNIVORE TRAITS. For all their diversity in size and life-style, carnivores share a number of physical and behavioral traits that show their common ancestry.

(*a*) Their teeth are specialized for killing and eating other vertebrates: long, sharp canines for gripping prey, stabbing or slashing victims or enemies of other species, and attacking rivals of their own kind; cheek teeth reduced in number and modified for cutting meat and cracking bones, especially the carnassials (the fourth upper premolar and first lower molar), which function like shears to slice meat; and chisel-like incisors useful for gripping and stripping meat from bones. The jaws are correspondingly powerful and the skull is reinforced to anchor the jaw muscles.

(*b*) The body is typically long and lithe, the limbs well-muscled and movable in all directions, with 4 or 5 toes, usually equipped with strong claws, which, however, only retract fully in cats and a few other species (including genets and palm civets).

(*c*) Some carnivores walk on the whole foot (plantigrade), but most walk on their toes (digitigrade) with the heel well off the ground. Many are fast sprinters (cats), long-distance runners (dogs), or adept climbers.

(*d*) Coloration and markings are cryptic, except for species with warning coloration and the African wild dog.

(*e*) Carnivores tend to be intelligent, possessing the mental alertness and ability for coordinated action needed to outwit, capture, and kill other animals.

(*f*) Predatory and fighting skills are rehearsed and developed in play, a major activity of most young carnivores.

(*g*) The senses are well-developed, especially smell and hearing; also the sense of touch, aided by whiskers and other sensory hairs (vibrissae) on cheeks, head, ears, torso, legs, and tail. The cats, which hunt largely by sight, have exceptional binocular vision; they and some mongooses may also have color vision (see viverrid family introduction, ichneumon and suricate accounts).

(*h*) Although males are generally somewhat larger with bigger canines, the only carnivore that is highly sexually dimorphic is the lion (3, 12). Gender differences even in size are minimal among species that live in monogamous pairs: social mongooses, most canids, ratels, and some otters (6).

ECOLOGY. Of all carnivores, 85% weigh less than 11 kg and range in size between a weasel and a fox. Hardly 2 dozen of the 231 different species are heavier than 20 kg and only 11 are as big or bigger than an average man (70 kg): the tiger, the lion, the jaguar, most bears, and the giant panda. Of the 66 African carnivores, half measure less than 50 cm long (head and body length) and 90% are under 1 m. Weight ranges from the 300 g dwarf mongoose to the 172 kg lion (3, 10, 14).

DIET. Carnivores generally select prey smaller than themselves. The great majority depend on the host of prolific small creatures that are accessible to them out of the vast array of vertebrate and invertebrate organisms. Few carnivores are strictly carnivorous, as are cats and weasels. Most are omnivorous to some degree, supplementing their diet with a variety of plant foods, especially fruits. Some canids and mustelids, and many viverrids, subsist largely on insects. The presence of molar teeth adapted to grinding and crushing hard-shelled foods, instead of just slicing up meat, reflects the omnivory of many viverrids, mustelids, and canids.

Some carnivores are quite specialized or otherwise limited: otters for aquatic existence, aardwolf and bat-eared fox as insectivores, the wild dog as a pack hunter of medium-sized antelopes, the palm civet as a frugivore. Others take a wide variety of prey, from tidbits the size of insects up to animals of their own size or even larger. The more versatile the predator, the less dependent it is on particular habitats, and the wider its geographical range. The leopard and lion, cheetah, caracal, wildcat, golden or Asiatic jackal, ratel, and striped hyena are examples of carnivores that are or were widely distributed in Asia as well as Africa.

Surprisingly, there is no niche for a pure mammalian scavenger in African ecosystems (12). The supply of carrion is unpredictable and seldom adequate for more than a few scavengers. Vultures, with their network of aerial spotters, dominate the scavenging scene by day. Even hyenas and jackals are primarily hunters and only secondarily scavengers. Many other carnivores will also scavenge food when the opportunity arises. The keys to carnivore success are opportunism and conservation of energy; the opportunist always takes the food that is obtainable with the least effort and risk. The exceptional ability of jackals and hyenas to capitalize on scavenging opportunities while remaining versatile and competent hunters explains why they often outnumber other associated predators.

SOCIAL ORGANIZATION. The main types of carnivore social organization are shown in table 18.1. The relatively few species that have become gregarious, notably the lion, wolf and wild dog, hyenas, and some mongooses, have evolved societies that are among the most complex and cooperative to be found among mammals (see species accounts). As indicated in the table (reading from top to bottom), carnivore social groups consist primarily of related individuals; a pack is formed when offspring remain with the parent(s) instead of dispers-

Table 18.1 The Main Types of Carnivore Social Organization

Type (examples)	Dispersion pattern and mating system	Territorial behavior	Offspring	Care and feeding of young	Foraging
Solitary (most carnivores: palm civet, slender mongoose, serval, leopard)	Male and female live singly, associate only for mating; male range typically includes up to several female ranges (polygyny)	Defense against outsiders of same sex	Disperse in adolescence; females may settle in mother's range	By mother only	Solitary
Monogamy (bat-eared fox, ratel)	Male and female live in pair terriotry	Joint marking, defense against outsiders of same sex	Both sexes usually disperse in adolescence	Male shares parental duties; provisions offspring and/or mate	In pairs or alone
Family (jackals)	Pair plus yearling offspring and young of year	Yearlings help parents defend against own sex	One/both sexes defer emigration and breeding	Offspring help care for and provision younger siblings	In pairs or alone, rarely in packs
Clan or pack	Group may include multiple adults of both sexes:				
(Dwarf mongoose)	A. Dominant pair monopolizes reproduction	Scent-marking mainly by dominant pair	Some transfer to other packs	Communal care	Individual-foraging in packs; no food-sharing
(Wild dog)	B. Dominant pair monopolizes reproduction; males related	Scent-marking mainly by dominant pair	Extra females emigrate	Communal care	Pack hunters; food-sharing
(Lion)	C. Several different females and males may breed; females with kinship ties	Individual and group defense against same sex; males form coalitions to win and hold female ranges	Males and some females emigrate	Communal care	Individual and group hunting; food-sharing
(Spotted hyena)	D. Small to large clans, females dominant over males; large clans include different matrilines	Group and individual marking and defense	Males emigrate	No communal care, mothers rear own cubs unassisted	Individual and pack hunting; food-sharing

ing. Generally daughters remain and sons disperse, making for matrilineal kinship units. But in the dog family, offspring of both sexes postpone emigration and breeding for a year or more and participate in rearing their younger siblings (see jackal accounts). The reverse is true of the African wild dog: sons remain in the natal pack and daughters disperse. When they have the same father and mother, helpers are as closely related to their younger siblings as they would be to their own offspring; such is the case in the Canidae and social mongooses, where monogamy is the rule. Even among species that live in sizeable clans containing several adults of both sexes, reproduction is usually monopolized by a dominant pair (but see lion, banded mongoose, and spotted hyena accounts).

The great majority of carnivores are apparently unsociable. They range singly and probably in most species the sexes associate only long enough to mate. But to label all these animals "solitary" is misleading if it implies that their social organization is simple or primitive, unvarying, or the same from one species to the next. For whenever a solitary species is studied in depth, social organization is found to be variable and often complex (e.g., see palm civet, slender mongoose, and leopard accounts). Although the adult members of a population may be dispersed singly, it does not mean they are socially isolated. On the contrary, the residents of such a community in most cases know their neighbors individually and stay in communication even while avoiding meeting, through chemical and acoustic signals.

The commonest spacing arrangement is for females to defend territories large enough to meet the food requirements of themselves and their offspring, while males undertake to control a much larger area encompassing the ranges of several females (7, 9, 10). The biggest and fittest males in the population are the most polygynous. This pattern has been found in most of the unsocial carnivores studied up to now. But unexpected variability in a species's social organization has also been found, under different ecological conditions. Where food is unusually abundant and concentrated, a normally "solitary" species may show sociable tendencies. For instance, feral housecats, belonging to a prototypical solitary species, live in kinship groups that have a social organization remarkably like the lion's (see cat family introduction) (9, 10). Other examples of supposedly solitary species that may become sociable at high density include the ichneumon, striped hyena, and spotted-necked otter. There are also other species that were assumed to be solitary because they forage alone, but which turn out to den communally and be sociable: notably the brown hyena and yellow mongoose. Conversely, there are normally social carnivores which, at low density, may live in a solitary mode: the lion, spotted hyena, and wolf, for instance.

Home Range and Group Size. Carnivore home ranges/territories tend to be larger than those of herbivores of comparable size, for obvious reasons: food plants are much more numerous and concentrated than animal prey. A pair of golden jackals weighing 10 kg apiece defend a range of at least 2–4 km^2, whereas a whole herd of 100-kg topi can subsist in an area of less than 1 km^2. Yet home ranges may vary enormously in size within the same species: from 10 to over 60 km^2 in the leopard, from 40 to 1000 km^2 in the cheetah, from 30 to 2000 km^2 in the spotted hyena. Studies of various different carnivores indicate that home range size depends on the distribution pattern of prey and other food resources. Group size in sociable species, however, depends on the abundance of food. If there is only enough for an individual or pair for part of the year, offspring will have to disperse. If there is a surplus, even normally solitary species may become sociable (9).

PREDATORY BEHAVIOR. Carnivores are born hunter/killers, endowed with the instinct to hunt. Predatory behavior can be readily stimulated even in a satiated predator. The normal hunting and killing techniques of the species appear spontaneously during development, but some practice is needed to acquire the necessary skill. The ability to discriminate between prey and nonprey species also has to be learned, in response to such factors as relative abundance, size, effort involved in capture, and parental example. Carnivores have only the innate tendency to flee from any aerial object or an approaching ground object above a certain size, to pursue any moving object below a certain size, and to hesitate to attack unfamiliar animals. An inhibition of predatory response, especially of the killing bite (see below), may be necessary to prevent mistakes with one's own species during mating or while transporting young (3).

Techniques for locating and capturing prey vary with the type of prey and predator. As the species accounts make clear, few carnivores are limited to just 1 or 2 basic techniques and some, such as foxes, jackals, and leopards, are very versatile indeed. Yet most carnivores are specialized at least to the extent of being terrestrial or arboreal, diurnal or nocturnal (more under Ecology).

The most familiar predatory techniques are cursorial and ambush hunting, as typified by dogs and cats, respectively. These also happen to be the most specialized techniques. The cursorial hunters have developed the speed and/or endurance to run down the fleetest prey, whereas the ambush hunters combine power and meathook claws with cunning, stealth, and quick reactions to capture and subdue animals that are faster and in some cases bigger than themselves. Big cats kill big game by strangling or suffocating it, whereas cursorial hunters, having run their quarry to exhaustion, have to "worry" it to death by biting and tearing at its vitals.

Predators tend to be more successful and can kill larger prey when hunting together than when hunting alone, an advantage of undoubted importance in the evolution of social carnivores (3, 9). However, this would not apply to social mongooses, in which selection for increased security against a host of potential predators is the likely explanation.

The most primitive method of killing prey, seen in the handling of small animals, is to grab, bite, and toss or throw it. A precise killing bite to the head or neck, such as the felid *neck bite* and the canid *death shake,* are later, more refined developments. Similarly, grasping the prey with the paws is more specialized than grasping with the jaws (3).

SOCIAL BEHAVIOR

COMMUNICATION. Although carnivores have a rich repertoire of signals that utilizes all the senses, the development of scent glands and the prominence of scent-marking throughout the order would indicate that this is the most important channel of communication.

Olfactory Communication. The secretions of various skin glands, urine, and feces are all employed in olfactory communication. It has already been demonstrated in a number of species (e.g., 2 mongooses, badger, red fox, wolf) that these secretions

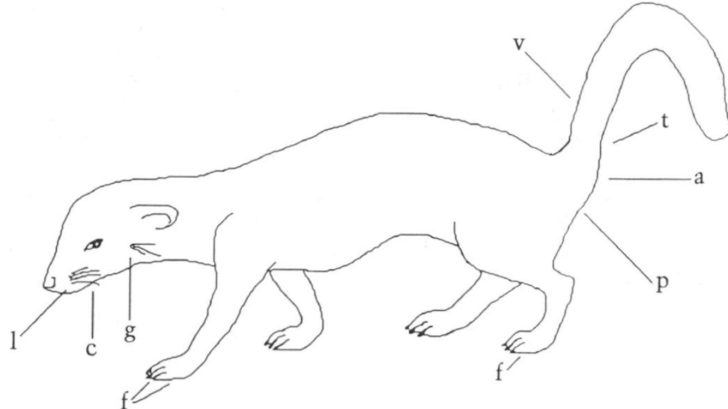

Fig. 18.2. Carnivore scent glands: *a*, anal glands, including paired anal sacs (present and variably developed in most carnivores, including antipredator "stink glands"), circumanal glands, anal pouch (mongooses [fig. 19.1*b*], hyenas [fig. 20.1], ratel [fig. 23.1], proctodial glands, etc.; *p*, perineal ("perfume") gland (civets, genets [fig. 19.1*a*]); *c*, chin gland (cat); *l*, lip glands (foxes, cats?); *g*, cheek (genal) glands (mongooses?); *f*, foot (digital) glands (canids, cats?); *t*, tail root (otter, cat); *v*, "violet" gland (canids).

convey information about the sex, age, status, and identity of each individual (see viverrid introduction, chap. 19). Scent-marking behavior differs among species but is most common along the paths an animal regularly follows, on territorial boundaries, and around dens. Probably most species mark with greatest frequency during the mating season (8). Figure 18.2 shows the location and names of scent glands that are known to occur in African carnivores. Invisible beneath the fur, many of these glands were discovered only because animals were observed rubbing the part against objects. The cat's chin gland is an example. Very likely some of these glands will be found to occur in species not presently known to possess them; it is also possible that entirely new scent glands will be discovered.

The most specialized and important scent glands in carnivores are found near the anus and genitalia (3). Accordingly, sniffing the partner's anal region features in meetings of most species. Nearly all carnivores possess anal sacs, though these structures are reduced in otters and relatively small in the dog and cat families. The paired sacs lie beneath the skin and connect to the outside via a short canal on either side or just inside the anus (see fig. 19.1). Packed secretory cells, often of more than 1 type, drain into the sac or into the canals. In species that have developed an obnoxious, liquid secretion for self-defense, the whole structure is surrounded by a muscular coat which can squeeze the sac like a syringe bulb to squirt fluid up to several meters. But in most carnivores anal-sac secretions serve a scent-marking function.

In mongooses, hyenas, and the ratel, the sacs and the anus are surrounded by a pouch (fig. 19.1*b*) which can be everted to bring the pouch lining, covered with secretion, into contact with the substrate or to waft the odor into the air (see figs. 19.4, 19.6, 20.2). In these and some other carnivores, the feces and also the urine and vaginal secretions may be mixed with anal-gland secretions. In addition to the anal sacs, diffuse skin glands of uncertain function open directly onto the skin surrounding or just inside the anus. Civets and genets possess a unique perineal gland, the product of which is used as a perfume base (see viverrid introduction, chap. 19).

Vocal Communication. Having less to fear from predators than most other mammals, many carnivores can afford to be noisy. Howling, wailing, barking, yapping, roaring, screaming, and whistling make effective long-distance contact calls. The more sociable the species, the larger and more varied the vocal repertoire, and the more "conversational." Social mongooses, for instance, maintain contact and coordinate the movements of pack members through an almost continuous churring and murmuring. In contrast, adults of solitary species may require only 1 sound, with the opposite meaning of socially negative threat calls, to signify readiness for social contact when the time comes to mate (3).

Visual Communication

Facial expressions. Carnivore facial expressions are basically alike. The similarity extends even to primates: a comparison of the repertoires of several canids with those of monkeys and apes indicates that all but 2 of the expressions and the situations evoking them are similar (4). In both orders, the importance of the canines as weapons makes the face the main focus of attention during social interactions; positions of the jaws, ears, and eyes which are associated with aggression, self-defense, and fear are virtually the same across families and species (1). As these expressions and movements are automatically assumed by animals experiencing aggressive and fearful emotions, they also function as visual displays (see carnivore and primate family introductions). However, the development of facial musculature and the degree of facial expression is far greater in cats and canids than in, say, mongooses, civets, and weasels, which have so little ability to alter their facial expressions that the difference between baring the teeth in anger and baring them in play may be imperceptible. Even in Canidae, the ability to show the canines by raising the upper lip is developed in only the most sociable species (3, 5).

Postures. An alert and confident carnivore moves with an erect bearing and holds its ears cocked as it actively investigates its surroundings. An individual that is ill at ease or fearful has a tendency to crouch with ears back, to glance about nervously as though it wanted to make itself smaller or hide.

From these and other signs that show which individual is more confident, an observer can predict the probable winner of an aggressive interaction. Confidence spells the difference between offensive and defensive threat. Both types tend to be impressive in carnivores, because of their lethal capability. Threat displays follow 3 basic principles: (1) display your weapons, (2) look as big as possible, and (3) attempt to startle and disconcert your opponent. These objectives are accomplished by opening the jaws; by erecting the hair and standing broadside (*lateral presentation*), extending the legs for maximum height, humping the back (*threat arch*), and raising the tail; and by abrupt movements, including *feigned attack,* combined with short, explosive sounds such as spitting, barking, growling, and roaring (3, 7).

One in 4 African carnivores has a dorsal crest or cape of longer hair that it erects to increase its apparent size (see figs. 19.8, 20.5) (14). First prize goes to the aardwolf, whose cape enlarges its silhouette by up to 74%. Enlarging the silhouette is most developed among ground-dwelling carnivores of medium size which are either slow runners or poor climbers, or else live in places where trees are scarce and come into contact with a variety of somewhat bigger carnivores. In short, erectile capes and crests probably evolved for use against other species, specifically against competitors and potential predators, as an alternative to flight.

Markings. Markings serve to accentuate parts of the animal that are important for communication. A black gum line emphasizes open-mouthed threat in many carnivores, through contrast with the white teeth and light adjacent hair. The black markings on the backs of a lion's ears and the white-tipped tail of the cheetah, wild dog, and side-striped jackal may function both as follow-me signals and as accentuations of ear and tail positions and movements. The large ears of hyenas and wild dogs, and the caracal's ear tufts, also increase ear visibility. The black mask of the bat-eared fox may function as a target for social grooming, and the white cheek ruff in some other canids and the black neck markings in civets may serve to direct bites to relatively well protected areas during ritualized fights (3, 4). In many species a white belly emphasizes the submissive attitude of lying on the back (13). Bold black-and-white markings, visible even in the dark, are warning signals to potential predators that the bearer is obnoxious. Although skunks have the most effective chemical defense, a number of African mustelids and several viverrids can also emit nauseating chemicals, and they too display warning coloration.

REPRODUCTION. Since most carnivores are solitary, wide-ranging, and widely spaced, finding mates can take time. Even after a meeting, courtship between 2 dangerously armed individuals that normally avoid each other or interact aggressively takes time. The fact that females typically stay in heat for several days or longer provides the necessary leeway. Extended estrus may, in turn, help to account for several other aspects of carnivore reproduction: (*a*) the possibility that most carnivores are induced ovulators, (*b*) the presence of a sizeable penile bone (a baculum occurs in all families but the hyenas), and (*c*) protracted (up to 3 hours in some mustelids) or oft-repeated copulation. Having ovulation depend on the stimulus of copulation insures that eggs and sperm will meet while both are viable. Aspects *b* and *c* may help to provide sufficient stimulation to induce ovulation (3).

PARENT/OFFSPRING BEHAVIOR. Carnivores produce multiple young (2–16) after a relatively short gestation period (3–16 weeks). With few exceptions (e.g, spotted hyena), the young are born blind and deaf, and locomotion is limited to crawling on the front limbs. But unlike many rodents, newborn carnivores are not naked. The young require extended care, and many have to learn hunting techniques before they can become independent. The participation of the father, and in social species of siblings, has evolved in various families to meet the demands of rearing young predators.

ANTIPREDATOR BEHAVIOR. The saying, "eat or be eaten," hardly applicable to ungulates or other herbivores, could serve as the carnivore motto. Smaller carnivores are "fair game" for larger carnivores, and the young of even the largest species are vulnerable to adults of the smaller species. Consequently carnivores have evolved a rich repertoire of antipredator stratagems, some of which are discussed under Communication (see also family and species accounts).

References

1. Andrew 1963. 2. Colbert 1969. 3. Ewer 1973. 4. Fox 1970. 5. Kleiman 1967. 6. ——— 1977. 7. Leyhausen 1979. 8. Macdonald 1980. 9. ——— 1983. 10. ——— 1984*b*. 11. Romer 1959. 12. Schaller 1972*a*. 13. Schenkel 1947. 14. Wemmer and Wilson 1983. 15. Young 1962.

Added References

Gittleman 1989. Gorman and Trowbridge 1989. Sandell 1989. Taylor, M. E. 1989.

Chapter 19

Genets, Civets, and Mongooses

Family Viverridae

	Subfamilies	Genera	Species
African viverrids	3	18	34
Total[a]	8	37	66

[a] Includes Madagascar and Southeast Asia.

African Representatives

Subfamily Viverrinae, civets and genets[a]
- **Genetta genetta*, common genet
- *G. felina*, feline genet
- **G. tigrina*, large-spotted genet
- *G. servalina*, servaline genet
- *G. maculata*, forest genet
- *G. abyssinica*, Abyssinian genet
- *G. angolensis*, Angolan genet
- *G. thierryi*, false genet
- *G. victoriae*, giant forest genet
- *G. johnstoni*, Johnston's genet
- *Osbornictis piscivora*, fishing genet
- *Poiana richardsoni*, African linsang
- **Civettictis civetta*, African civet

Subfamily Paradoxurinae, palm civets
- **Nandinia binotata*, African palm civet

Subfamily Herpestinae, mongooses[b]
- **Herpestes sanguineus*, slender mongoose
- **H. ichneumon*, ichneumon or Egyptian mongoose
- *H. naso*, long-nosed mongoose
- *H. pulverulentus*, Cape gray mongoose
- **Atilax paludinosus*, marsh mongoose
- **Ichneumia albicauda*, white-tailed mongoose
- *Rhynchogale melleri*, Meller's mongoose
- *Bdeogale nigripes*, black-legged mongoose
- *B. crassicauda*, bushy-tailed mongoose
- *Paracynictis selousi*, Selous' mongoose
- **Cynictis penicillata*, yellow mongoose
- *Dologale dybowskii*, Pousargues's mongoose
- **Helogale parvula*, dwarf mongoose
- **Mungos mungo*, banded mongoose
- *M. gambianus*, Gambian mongoose
- *Crossarchus obscurus*, kusimanse
- *C. alexandri*, Alexander's mongoose
- *C. ansorgei*, Angolan or Ansorge's mongoose
- *Liberiictis kuhni*, Liberian mongoose
- **Suricata suricatta*, suricate or meerkat

* Species treated in separate accounts in this chapter
[a] Reference 14
[b] Reference 8

The largest family of carnivores, with 37 genera and 66 species (12, 14), the viverrids are a diverse array of small carnivores widely distributed in the Old World tropics. Of the 8 different subfamilies, only the 2 containing mongooses, genets, and civets are at all familiar to most people. The others live in Madagascar and Southeast Asia. The arboreal palm civet represents the only other subfamily that occurs in Africa. Yet Africa, with 34 species, is rich in viverrids, especially mongooses. Eleven of the 13 Herpestinae genera and 17 of the 27 species occur only in Africa, making this continent and not Asia the Land of the Mongoose (4).

ANCESTRY. Genets and civets are considered the most generalized and primitive of living carnivores (see carnivore introduction). Viverrids of this type have been around since the Lower Oligocene, and the earliest fossils are nearly indistinguishable from the miacid ancestors of the Carnivora (fig. 18.1). Viverrid teeth and skele-

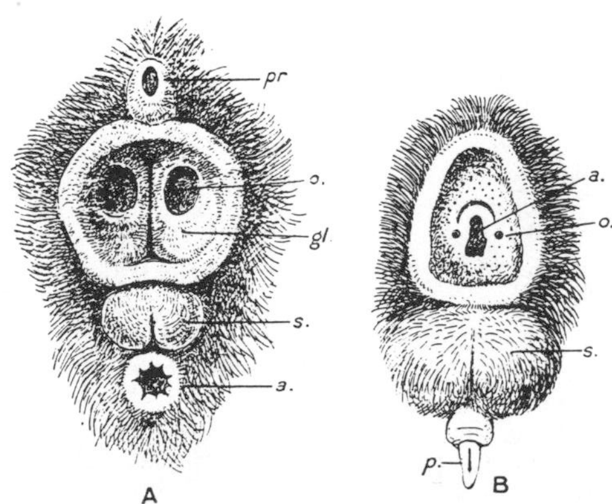

Fig. 19.1. A, perineal ("perfume") gland of African civet. B, anal pouch of marsh mongoose, shown spread open. Key: *pr*, prepuce; *o*, orifice (of pouch in A, of anal gland in B); *gl*, interglandular space; *s*, scrotum; *a*, anus; *p*, penis. (From Pocock 1915, 1916.)

Table 19.1 Comparison of Mongoose and Genet/Civet Traits

Trait	Genet/civets: Viverrinae	Mongooses: Herpestinae
Skull	Cranium long and narrow, muzzle narrow and low (except *Poiana*)	Variable, from long, narrow cranium and short, high muzzle (*Herpestes*), to wide cranium and short jaws, with widened palate (social mongooses)
Ears	Typically catlike (oval), high on head, conspicuous	Semicircular, small, set low on head, inconspicuous, closable
Teeth	Relatively primitive, shearing carnassial teeth	From general carnassial type, but with smaller canines (*Herpestes* type), to shortened and widened molars with sharp cusps (insectivores)
Body	Long and narrow, back rounded, neck and tail typically long to very long, legs short	From weasel-like (slender mongoose) to broad and squat with short neck, legs, and tail (social and marsh mongooses)
Feet	Wide, subdigitigrade, 5-toed, claws short, curved and retractile (except civet), adapted to climbing or walking	Narrow, fifth digit often reduced or absent, claws long and nonretractile, feet adapted for walking and digging
Scent glands	Well-developed perineal glands between the scrotum and penis/vagina (fig. 19.1*a*), anal gland	Anal glands opening into an anal pouch (fig. 19.1*b*); cheek glands present in some species
Penis	Located below the scrotum between the hindlegs, urethral opening directed backward; bone (baculum) present	Ditto, urethral opening an elongated slit on underside of short, smooth penis; baculum present
Coat	Typically soft, furry, and short with longer hair on spine and tail; spots or tabby markings and ringed tail	Typically coarse and long, especially on hindquarters and upper tail; not spotted; hairs banded, giving grizzled or pepper-and-salt look

tons have barely changed in the last 40–50 million years (12). Mongooses go back at least to the early Miocene, and a 30-million-year-old skull makes *Herpestes* the oldest existing carnivore genus (7).

FAMILY TRAITS. Mongooses are so different from other viverrids that some systematists consider them a separate family (8). Rather than give a short list of the traits all viverrids share, it seems more useful to compare the 2 subfamilies to which all but 1 of the African viverrids belong (table 19.1). Their anatomical differences reflect differences in postures, locomotion, displays, and other behavior, and in ecological niche. Broadly speaking, all the viverrids except the mongooses are somewhat catlike, stalking predators, nocturnal, more or less arboreal, and omnivorous. Most mongooses are terrestrial, insectivorous, and diurnal (see table 19.2) (2).

ECOLOGY. The majority of viverrids are associated with well-watered, closed habitats, especially forest and woodland. Within these broad limits, however, the family has radiated into a variety of niches second only to the weasel family. Arboreal, terrestrial, fossorial, and aquatic types with carnivorous, insectivorous, piscivorous, omnivorous, and even frugivorous diets are all represented. But as a family, viverrids tend to be more omnivorous than most carnivores. Perhaps the prior occupation of carnivorous niches by more efficient predators, notably cats and canids, forced African viverrids to diversify (8). Comparing subfamilies, the mongooses have departed most radically and recently from the ancestral prototype. The ecological niches occupied by African viverrids, as presently understood, are summarized in table 19.2. Because their habitat and habits make most species hard to observe, very few have been studied in the wild.

SOCIAL ORGANIZATION. Most viverrids are solitary. This life-style is linked with predation and opportunistic foraging by small animals that depend on stealth to capture living prey and/or to avoid becoming prey themselves (see chap. 18, Social Organization). The combination of hunting small vertebrates at night in dense vegetation almost guarantees a solitary mode of life. Most civets and genets fall into this category, but other viverrids that do not fit the profile are also solitary, including a nocturnal frugivore (palm civet) and, among the mongooses, the slender mongoose, a diurnal hunter, along with most other members of the genus *Herpestes* and probably all the nocturnal insectivores—*Ichneumia albicauda, Paracynictis selousi, Rhynchogale melleri, Bdeogale nigripes,* and *B. crassicauda.*

Table 19.2 Ecological Niches of Some African Viverrids

Species	Weight Range (kg)	Habitat (Closed——Open)	Diurnal/ nocturnal (D/N)	Foraging behavior	Main diet	Social
Genets	0.8–3.5	rain forest–arid bush	N	Ground/arboreal stalk–pounce	Carnivorous	No
Civet	7–20	most types	N	Terrestrial, opportunistic	Omnivorous	No
Palm civet	1–5	rain forest–gallery forest	N	Arboreal fruit seeker	Frugivorous	No
Slender mongoose	0.5–0.7	ubiquitous in forest–bush	D	Ground/climbing predator on small vertebrates	Carnivorous	No
Ichneumon	2.4–4	Undergrowth near water	D	Ground predator on vertebrates and invertebrates	Carnivorous	±
White-tailed mongoose	3–5	savanna	N	Forager on surface/subsurface insects	Insectivorous	No
Marsh mongoose	2.5–4	widespread near water	N	Raccoonlike probing in water and ashore	Invertebrates	No
Selous' mongoose	1.5–1.6	forest–moist savanna	N	Surface forager, especially ants	Insectivorous	No
Yellow mongoose	0.3–0.8	dry savanna, sandveld	D	Forages individually on surface insects	Insectivorous	Yes
Dwarf mongoose	0.2–0.4	savanna with termite mounds	D	Pack forager, scratches and probes for insects	Insectivorous	Very
Banded mongoose	1–2	savanna with termite mounds	D	Pack forager, scratches and digs up insects	Insectivorous	Very
Kusimanse	1–2	rain forest–savanna, near water	D	Pack forager, digger, probes for invertebrates	Insectivorous	Very
Suricate (meerkat)	0.6–0.9	dry savanna, calcareous	D	Pack forager, digs for buried insects	Insectivorous	Very

A family and clan organization has developed in about half the genera of mongooses, comprising some 10 species. With the exception of the ichneumon, all are diurnal, terrestrial, and insectivorous. The most gregarious mongooses, *Helogale, Mungos, Crossarchus, Suricata,* are obligate members of social groups that are among the most complex and cooperative in mammals (see dwarf and banded mongoose and suricate accounts) (10). The social mongooses exploit extremely abundant but patchily distributed insects that live on and beneath the ground, often in open locations where the mongooses are exposed to predation. Foraging in packs provides necessary security for the members, which however forage independently without sharing food that is mostly too small to be divided.

How did the complex social systems of the gregarious mongooses evolve from the primitive condition represented by a solitary carnivore like the genet? Different stages in the progression from solitary to gregarious may be seen within the family.

1. A pattern that is widespread among solitary mammals has been found so far in genets, the palm civet, and the slender mongoose. Adult females live and reproduce within exclusive home ranges which they defend as territories. The largest males have much larger ranges that include up to several female ranges. Although male palm civets and slender mongooses exclude rival males, both may tolerate other smaller, nonbreeding adult males on their grounds (1).

2. Female offspring may remain in the maternal range instead of dispersing before maturation. Sometimes 2 or more adult female white-tailed mongooses, which are normally intolerant of female neighbors, share and reproduce in the same range, and even sleep in the same den. The condition necessary for such female clans to form is a rapidly renewed supply of insect prey (9, 10).

3. In the ichneumon, a diurnal predator on vertebrates and invertebrates which would accordingly be expected to be solitary, females become sociable when food is unusually abundant. Presumed relatives breed and care for offspring communally. However, family members forage individually, and the single adult male makes a minor parental investment and drives subadult sons away.

4. The yellow mongoose is a social species that lives in clans but forages individual-

ly (cf. brown hyena). Though a number of subadult and young-adult males may live in a clan territory, breeding is monopolized by a dominant pair.

5. The most gregarious mongooses are all insectivores which live and forage in mixed packs, often in open habitats where they are vulnerable to a wide range of predators. In the dwarf mongoose, usually only 1 dominant pair breeds, while other pack members, including unrelated immigrants from other packs, assist in guarding and provisioning offspring. In the banded mongoose, and perhaps also in the suricate and kusimanse, several females and males may breed (see species accounts).

How the many unstudied mongooses fit into this scheme is something that remains to be seen.

POSTURES AND LOCOMOTION. A comparison of mongooses with genets and palm civets (table 19.3) brings out fundamental differences between viverrids that are arboreally adapted, with short legs, retractile claws, and long bodies and tails, and those that have become terrestrial insectivores.

SOCIAL BEHAVIOR

COMMUNICATION. The sense of smell is preeminent in this family, both

Table 19.3 Postures and Locomotion in Mongooses and Genets/Civets

Posture or movement	Genets/Civet, palm civet	Mongooses
Sitting, standing	Bipedal sitting and standing: genets only (fig. 19.2)	Bipedal sitting (*low-sit*) (fig. 19.30) and standing (*high-sit*), braced by tail (fig. 19.17)
Crouching	Present	Present
Lying		
Resting	Side or back	Side or back, also prone and spread-eagled
Sleeping	*Side-curl* (lying on side, nose in crotch, tail wrapping head and body); circle before lying	*Ventral-curl* (seated, curled forward with forehead to ground); lie without circling
Grooming	Self-licking all over, clean face with tongue-moistened paws, singly or both at once (genets, palm civet); also nibble and scratch; genets and palm civet "sharpen" claws	Minimal, by *nibble-grooming,* lick genital area only; brush muzzle and clean teeth with (unlicked) forepaws and claws; also scratch; do not sharpen claws
Stretching	Catlike: (*a*) hump back, extend forelegs, back arched, then stretch hindlegs; (*b*) walk-stretch, extend foreleg and hindleg of opposite sides simultaneously	Extend fore- and hindlegs simultaneously
Gaits		
Walk	Slinking (genet and palm civet)	Slinking only in the more carnivorous species (*Herpestes*)
Run and fast trot	Diagonal stride	± similar
Jump-run	Forelegs and hindlegs move in pairs, body flexing and extending	Same
Jumping	Good jumpers (fig. 19.3)	Mostly poor jumpers
Climbing	Genets and palm civets agile climbers, traverse upside down, descend headfirst (fig. 19.16)	Poor climbers (except slender mongoose), descend tailfirst
Digging	Very poor	Mostly good, but few dig own burrows
Throwing	Not in the repertoire (?)	Forelegs propel eggs, millipedes, snails, and such through backlegs (as in digging) (fig. 19.27)
Manipulation	Genets and palm civets hold prey and each other with claws; civet usually eats using mouth only	Feet sometimes used to hold food down while pulling with mouth; a few species with raccoonlike dexterity (e.g., marsh mongoose) (fig. 19.21)

Fig. 19.2. Genet standing bipedally.

Fig. 19.3. Genet jumping sequence (from photos in Wemmer 1977).

for locating food and for communication. No doubt most viverrids also have acute hearing, whereas vision ranks third, even for some diurnal mongooses (see ichneumon account, Senses). However, the suricate and the dwarf and banded mongooses spot avian predators at great distances, and some species are suspected of having color vision (suricate, ichneumon, Indian mongoose, *Herpestes edwardsi*) (2).

Olfactory Communication. Most viverrids defecate at regular, often communal dung middens/latrines. The practice of placing feces along paths, as seen in various carnivores, is not a common viverrid practice, although genets have been known to leave scats on branches. A few viverrids are known to urine-mark: genets, male civets, and male suricates. Civet and palm civet urine usually has a pungent odor (8). But the main form of scent-marking in this family is with anal-gland secretions (4, 8). The civet perineal gland and the mongoose anal pouch (fig. 19.1) are specially adapted for depositing long-lasting scent on objects and are developed in both sexes. Depending on the position of the object, marking may be accomplished by *anal dragging* (fig. 19.4), by backing up to and depositing secretion at nose level against a tree or other vertical object (fig. 19.5), and by doing a *handstand* and wiping the gland against the underside of an overhead object (fig. 19.6). In palm civets the perineal gland is conveniently located for marking branches, being placed ahead of or surrounding the genitalia. Many viverrids mark any novel object, including food and also members of their own social unit.

Anal-sac secretions of the small Indian mongoose, *Herpestes auropunctatus*, have been analyzed and found to include a series of volatile short-chain carboxylic acids that are produced by bacterial breakdown of sebaceous and apocrine secretions. Each of the 6 different acids has a unique odor and their relative concentration differs individually and consistently, giving every mongoose a unique odor profile. Further investigation established that mongooses can in fact recognize one another by their anal-gland secretions (6).

Another Indian mongoose, *Herpestes edwardsi*, has enlarged sebaceous glands at the bases of the cheek whiskers which produce a honeylike odor. It deposits this secretion by rubbing its cheeks on objects, followed by anal-gland marking. The latter odor persists for 2 weeks, the former for 2 days. Dwarf, banded, and marsh mongooses and the kusimanse cheek-rub, too (4). Genets and the African civet rub their necks against objects, suggesting the possible presence of additional sebaceous glands. Finally, anal-sac secretions of greater or lesser potency are sprayed or dribbled in self-defense by some species as a response to fear or pain (see carnivore introduction).

Tactile Communication. *Social grooming* (nibbling and licking) is important in maintaining bonds between members of the same social group, including parent(s) and offspring, mated pairs, and other pack members (e.g., see ichneumon and dwarf mongoose accounts). Scent-marking of group members by dominant adults (yellow and dwarf mongooses) or mutual marking (banded mongoose) involves contact, but presumably functions to give group members a common, familiar smell.

Vocal Communication. The viverrid vocal repertoire averages 5–6 different kinds of calls per species, most of which are graded, that is, variable in intensity, pitch, or tempo (12). Most viverrids growl (threat), but palm civets and members of some other non-African subfamilies hum instead. Whining, an infantile response to various kinds of discomfort or distress, tends to disappear during development; adults scream in response to specific stimuli such as pain or extreme fear (11). *Ichneumia* barks loudly. Fighting is the commonest eliciting situation. Curiously enough, some species (e.g., palm civet, large-spotted genet, dwarf mongoose, suricate) lack a cry of pain. The palm civets all utter brief, loud, repetitive hoots (11), which are probably territorial advertising/mating calls (see palm civet account). The absence of such calls in other viverrids may be partly explained by the danger to small terrestrial carnivores of advertising their presence by means of easily located vocalizations; scent-marking is a safer, more durable form of advertising in a 2-dimensional environment (11). However, the social mongooses produce repetitive, loud whistles, chirps, barks, and the like as antipredator alarm signals. Packs also maintain contact through continuous murmuring or peeping sounds derived most likely from the nest-chirp, a juvenile contact call that is widespread in the family (5).

Visual Communication. Facial expressions are poorly developed in this family. Ear positions and gaping jaws have the usual meanings (see carnivore introduction), but the main elements in visual displays are

Fig. 19.4. Yellow mongoose *squat-marking* with anal gland (redrawn from Earle 1981).

Fig. 19.5. Civet marking tree with perineal-gland secretion.

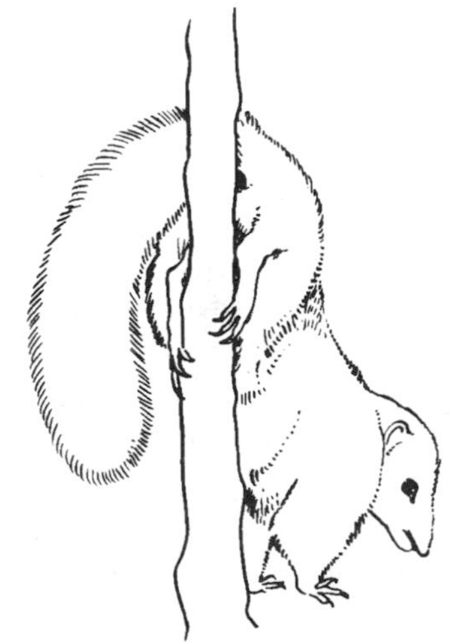

Fig. 19.6. Kusimanse assuming the *handstand posture* to deposit anal-sac secretion (from a photo in Ewer 1973).

Table 19.4 Aggressive Displays of Some African Viverrids

					Tail Position			Body Orientation		
Species	Leg extension	Stiff-legged gait	Arched back	Depressed head and shoulders	Down	Level	Up	Lateral, angled	Frontal	Hair erection
Large-spotted genet	+	+	+	+/−	+	+	−	+	+	+
Common genet	+	+	+	+/−	+	+	−	+	+	+
Civet	−	−	−	+	+	+	−	+	+	+
Palm civet	+	+	+	−	−	+	+	+	+	+
White-tailed mongoose	+	−	+	+/−	−	+	+	+	+	+
Ichneumon	?	+	+	+	−	+	−	−	+	+
Banded mongoose	+	+	+	+?	+?	+	+?	+	+	+
Dwarf mongoose	+	+	+	−	?	?	+	?	+	+
Suricate (Meerkat)	+	+?	+	−	−	+	+	−	+	+
Kusimanse	+	+	+	+	+	−	−	−	+	+

SOURCE. Based on Dücker 1965, Wemmer and Wilson 1983.

body and tail attitudes which, in combination with erectile hair, increase or decrease apparent size, and movements.

AGONISTIC BEHAVIOR

Dominance/Threat Displays. The aggressive displays of some African viverrids are compared in table 19.4. Nearly all species make themselves appear bigger by straightening their legs, arching their backs, and *bristling.* Species with crests of longer hair (civet, palm civet, genets, some mongooses) tend to display laterally (fig. 19.8). Other mongooses (e.g., ichneumon, suricate, banded mongoose) *bristle* all over and seem to swell when seen from any angle (13). *Staring* is a common reaction when members of solitary species encounter one another (e.g., genet, palm civet). Whether *staring* is offensive or defensive is revealed by posture, orientation, and movement: crouching and advancing while staring is offensive; arching the back and gaping is more defensive (11).

Defensive/Submissive Displays. *Head-darting* is a common response to close approach by a stranger, with mouth opening wider and including threat vocalizations at increasing intensity (fig. 19.9). *Snapping* is a strong defensive threat. As a last resort, genets, social mongooses, and perhaps other viverrids lie on their backs, where they can use claws and jaws to defend themselves while the back of the neck is protected. Mongooses enhance their threat displays against potential predators not only by vocalizing, *head-darting,* and *snapping,* but also by *mock attacks.* The banded mongoose rocks from side to side and the suricate rocks up and down to create the illusion of forward motion while actually going nowhere. The *backing attack* is an even more specialized defensive tactic that the ichneumon, suricate, and banded and dwarf mongooses use against their own kind. The elevated posterior is presented to the antagonist, while the animal bites over its shoulder or between its hindlegs (fig. 19.7) (cf. hyrax).

Submissive behavior remains to be described in most viverrids. The African civet and social mongooses lie on the side when intimidated. However, submissive signals are less important in social mongooses than in most other social carnivores, perhaps because pack members are generally on friendly terms, frequently in contact, and rarely in food competition (4).

Fighting. Genets, palm civets, and others with sharp retractile claws may use them to hold an opponent while biting it (so do mongooses—see suricate account, fig. 19.28), but apparently do not strike with their forepaws or rake with their hindclaws in the manner of cats. But fighting is so seldom observed that information on the subject is very limited. Fighting techniques are most frequently seen in play (see suricate account). Fights between palm civets, though rare, may have fatal consequences (see spe-

Fig. 19.7. Mongoose *backing attack* (defensive threat). (From a photo in Dücker 1965.)

cies account, under Social Organization). In the social mongooses, encounters between rival packs tend to be hostile and may end up in brawls reminiscent of battles between baboon troops (see banded and dwarf mongoose accounts).

REPRODUCTION. For their size, viverrids seem to be relatively slow to mature and reproduce. Although adolescence is reached within a year or less in probably most species, seemingly few individuals breed before the age of 2, including the dwarf mongoose, the smallest species. The largest viverrids take up to 5 years to mature (e.g., the fossa, *Cryptoferox fossa,* of Madagascar). Gestation ranges from 2 to 3 months. Litters of 2–4 (range 1–6) are typical (4). The civet and some mongooses reproduce 2 or 3 times a year, but others do so annually (see ichneumon and suricate accounts).

SEXUAL BEHAVIOR. Common elements in viverrid mating behavior include courtship vocalizations (especially in civet and palm civet subfamilies), *close-following* by the male, *licking of the female's genitalia, mutual rubbing and marking with anal or perineal glands, elevation of the hindquarters* and *deflection of the tail* when ready to copulate, and a real or symbolic *neck grip* by the male during copulation.

PARENT/OFFSPRING BEHAVIOR. Most viverrids are born blind and deaf, barely able to crawl, and sparsely haired with the adult markings only faintly indicated. They stay 1–5 weeks in hollow trees, termite mounds, or holes in the ground, sometimes in leaf or grass nests, or simply in tangled vines or a thicket. Very few, even among the best diggers, dig their own holes (perhaps excepting suricate and yellow mongoose) (9). Development seems to proceed in the same general way as in other carnivores (2, 5, 11). A male parental role is known only in the social mongooses, in which communal care has developed even further than in other carnivores: dwarf mongoose helpers include unrelated immigrants (9). Mongooses suckle newborn young in the seated sleeping posture (*ventral curl*); later they lie on the side or back (3). Suckling genets and probably most other viverrids lie curled on one side and warm the offspring by blanketing them with the tail, later stretching on the back or side (2). *Milk-treading* has been recorded in genets and less definite treading in suricates (2, 3). The newborn have to be stimulated by the mother's licking to void wastes, which she consumes while the young remain in the nest. Infants are transported by the back, neck, or head.

PLAY. The young of all species but generally not the adults are playful, engaging in ambushing, chasing, and fighting games typical of carnivores. Ichneumon, slender mongoose, and suricate young are unusually playful and inventive, and sometimes play with the young of other species (see species accounts).

ANTIPREDATOR BEHAVIOR. See under Vocal Communication and Dominance/Threat Displays.

The evolution of social groups in mongooses is plausibly explained as an antipredator adaptation that enabled small, vulnerable carnivores to exploit an abundant food resource (insects) of the African savanna where individual risk is reduced by group membership (9).

References

1. Charles-Dominique 1978*a*. 2. Dücker 1965. 3. Ewer 1963. 4. ——— 1973. 5. Ewer and Wemmer 1974. 6. Gorman 1976. 7. Hinton and Dunn 1967. 8. Kingdon 1977. 9. Rood 1986. 10. Waser and Waser 1985. 11. Wemmer 1977. 12. ——— 1984. 13. Wemmer and Wilson 1983. 14. Wemmer and Wozencraft 1984.

Fig. 19.8. Genet displaying offensive threat, its apparent size increased by *bristling* of tail and mane (from a photo in Gangloff and Ropartz 1972).

Genets

Genetta Species

G. felina (formerly *genetta),* the feline genet, *G. genetta,* the common or European genet, and *G. tigrina,* the large-spotted, blotched, or tigrine genet, are the only species that have been observed in any detail. Behavioral differences between species have yet to be distinguished.

TRAITS. Small, long-bodied, short-legged, semi-arboreal carnivores with spots in long rows, or stripes, and an equally long, ringed tail. *Weight and length* (head and body): *G. felina/genetta* males 2 kg and 49 cm, females 1.8 kg and 48 cm; *G. tigrina* 1.8–3.2 kg and 50–55 cm (Botswana series [11]). Pointed muzzle with long whiskers; eyes frontally placed (good binocular vision) with vertically oval pupil; ears prominent, oval-shape; feet all with 5 toes, curved, sharp, retractile claws. *Fur:* dense undercoat with outer coat of varying length, tail and dorsal-crest hair longer. *Coloration:* extremely variable with individually unique markings, rust-brown to black spots or stripes on a pale gray to red-brown background, markings rust-brown to black, forest species generally darkest and arid forms lightest; pale marks around mouth and eyes divided by dark chevron, tail tip black in *tigrina,* usually white in *felina. Scent glands:* well-developed anal sacs and small perineal gland, secretions of flank and perhaps cheek skin also employed for marking.

DISTRIBUTION. One of the most widespread carnivores, genets are found virtually throughout Africa except for the Sahara, and they range into southwestern Europe and the Arabian Peninsula. They can subsist virtually wherever there is sufficient cover to hide in and food to eat. Genets probably originated in the rain forest but very early radiated and dispersed into woodland and increasingly arid environments, adapting and changing in the process to suit local conditions (5). The different species are remarkably similar in appearance and habits, and presumably in behavior, although present knowledge is based primarily on observations of captive European and large-spotted genets (7, 8, 9, 13). Most of the visible differences between species are in markings and coloration, which in some cases seem barely enough to distinguish subspecies, let alone species; without their skins, most species are virtually indistinguishable (10). Yet *felina* and *tigrina,* which are by far the most widespread and familiar genets, overlap in many areas without interbreeding (5, 11). These and a few less well known forms with sympatric distributions are demonstrably different species. So is Johnston's genet, which has greatly reduced teeth apparently adapted for insectivory (15). The almost unknown fishing genet (unspotted, with teeth adapted for holding slippery fish and amphibians) and the African linsang, both West African rain-forest species, are different enough from other genets to be placed in separate genera.

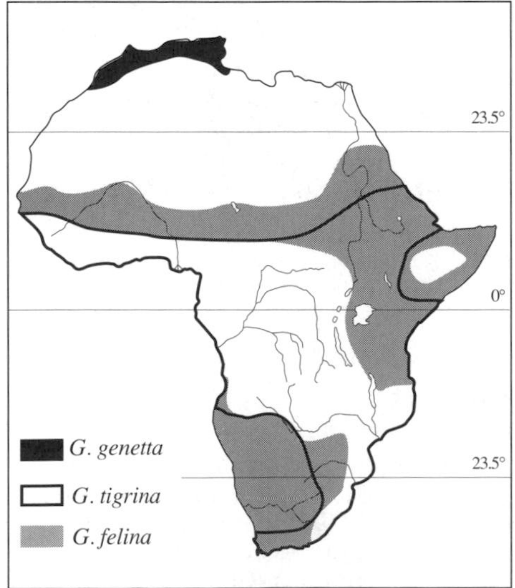

ECOLOGY. Genets live on a wide variety of foods, including small vertebrates (rodents, bats, bush babies, birds up to the size of chickens, lizards, snakes, frogs), invertebrates (grasshoppers, crickets, beetles, spiders, centipedes, scorpions, moths), fruits, and nectar (e.g., from trees that produce copious nectar at night to attract bat pollinators [6]). The habitat tolerances of the different species and subspecies effect ecological separation. *G. victoriae, servalina, angolensis,* and *johnstoni* are forest dwellers. *Maculata* occurs in forest-savanna mosaic, *thierryi* in forest and Guinea Savanna; *tigrina, abyssinica, felina,* and *genetta* range drier, unforested regions. The last 2 tolerate desert conditions and the openest habitat, whereas *tigrina* is seldom found far from water but is most adaptable to human settlement (5, 11). Larger size and largely terrestrial habits make *victoriae* a link of sorts between genets and civets.

SOCIAL ORGANIZATION: solitary, polygynous, dominant males defend ranges of several females.

Although pairs and family groups are sometimes seen, genets are considered solitary. A mother with young seems to be the most complex social unit, persisting only until weaning at 4½–6 months, although maturation takes 2 years (9, 13). However, in a sample of marked genets, at least 4 adults used the same area (P. Waser, pers. comm.). Whether genets are territorial is unknown. Retrapping records of poultry-thieving African genets and radio-tracking of common genets in Spain indicate that males wander widely within ranges as large as 5 km^2, whereas females may remain permanently within exclusive ranges as small as 25 ha. Females released as much as 35 km away from capture sites have returned home within a few days (1, 3). Males may move at the rate of 3 km/hour while actively ranging.

ACTIVITY: nocturnal (see under Predatory Behavior).

PREDATORY BEHAVIOR. Genets are active, versatile predators, equally at home in trees and on the ground (figs. 19.2, 19.3). They will alternately lie in ambush and dash at their prey. Like other small nocturnal carnivores, feline genets in Serengeti N.P. tend to be most active in the first part of the night, but are also unusually active by bright moonlight (cf. white-tailed mongoose and bat-eared fox [12]). A female *G. tigrina* that was radio-collared in Ethiopia's Omo N.P. was also most active between sunset and midnight (4).

SOCIAL BEHAVIOR

COMMUNICATION

Olfactory Communication: mutual sniffing during meetings (nose, anogenital region, back), *urine-testing,* scent-marking with urine, feces, perineal gland, and skin glands in flanks and cheeks.

Studies of scent-marking in captive common genets indicate that olfactory communication in this group is complex, and that genets can discriminate between familiar and unfamiliar individuals by their urine, feces, and perineal and flank secretions (7, 8). Feces, with anal-gland secretions probably added, are deposited in communal latrines. Urine and perineal-gland secretions are placed on various objects. Observed males made 90% of their marks with urine and only 10% with the perineal gland, at the rate of 8–10 times per hour; whereas 80% of the females' marks were perineal and 20% urine, and they marked 18–64 times per hour (8). But as the mating season approached, male marking activity increased, while female marking decreased. When 2 males were kept together with females, the dominant male marked most frequently and the subordinate male not at all. After the males were separated and each was caged with a female, the second male began marking, but marking frequency declined in all individuals, indicating reduced social stress (7).

The perineal gland is normally closed, showing only a vertical slit between the anus and vulva or scrotum (13). Preparatory to marking, the surrounding muscles open and evert the lips, exposing an underlying pad covered with fine white hair and a clear, oily emulsion. This pad is pressed against the object to be marked, either by lowering the hindquarters, by backing up to a raised surface, or by *handstanding.* Genets often rub the pad rhythmically from side to side to spread secretion on the object. A mildly sweet-scented musk, the deposit turns dark brown upon coagulation; wood permeated to several centimeters after prolonged use may smell of genet for at least 4 years. In a colony of captive large-spotted genets, overhead (*handstand*) marking was mainly a male and *squat marking* was mainly a female prerogative (13). They did not demarcate aerial pathways.

Rubbing the flanks on objects is usually associated with aggression (8). In addition, genets rub their cheeks, head, neck, and back on objects and, especially when wet, they lie down, rub, and roll like dogs. Finally, genets scuff the ground with their hindfeet while crouching, but without urinating. Whether these behaviors are associated with scent-marking remains uncertain (13).

Vocal Communication: growling, *coughing,* hissing, *spitting, panting-hiss,* screaming, purring, whining, meowing.

Coughing, singly or in volleys, is a contact call, heard first in newborn kittens (nest chirp?), which continues throughout life as the most common adult call (13); courting males also *cough* repeatedly while following females. Hissing and *spitting* (explosive hissing) are keep-away calls associated with *gaping* and *head-darting,* the intensity of which increases as the mouth opens wider. The *panting-hiss* is similar to *spitting* but results from accelerated breathing while *gaping* with retracted lips showing the lower teeth. Both types of hissing may occur with growling and screaming, the latter a response to pain. Screaming replaces whining with open mouth, a juvenile distress call that ceases after the animal becomes independent. Purring, another sound uttered only by juveniles, is produced with mouth closed or while nursing.

AGONISTIC BEHAVIOR

Dominance/Threat Displays: crouching and staring, *head-darting,* rushing.

Fig. 19.9. *Head-darting* sequence: genet on left extending head, genet on right recoiling. Shaded images represent last frame in cine sequence. (From Wemmer 1977.)

Defensive Displays: *arching back, bristling, gaping, head-darting,* accompanied by above calls; discharging anal-sac secretions.

Submissive Displays: *turning away, lying down.*

Fighting: *head-sparring, wrestling, grappling* with *pawing* and *clasping,* biting.

Social contact between genets includes *anogenital-* and *back-sniffing* as an identity check (8) and, in aggressive encounters, *head-darting,* during which the head is rapidly extended and (often) immediately retracted (fig. 19.9). The threat to bite may be defensive or offensive and *jaw-gaping,* growling, spitting, and screaming represent increasing levels of intensity. The recipient of *head-darting* typically responds submissively by lying down, turned sideways or at an angle to the aggressor. The outcome of encounters is often predictable: the one that stares more is likely to crouch and rush more, that is, take the offensive and win. The animal that *jaw-gapes* and growls more is apt to lose. *Tail-flaring* is common during encounters, with or without contact, and is associated with *moving away.* Standing with arched back (fig. 19.10) is the usual response to an approaching and staring genet. The more the hair is erected, the more the light tail rings show up. The tail rings and contrasting facial markings may serve as orientation clues between genets, for example by helping offspring follow their mother in the dark (5). But in seeming contradiction, the genet's coloration is also believed to serve as camouflage, especially while hunting on moonlit nights (3).

Fig. 19.10. Genet defensive threat: arched back and flared tail (redrawn from Wemmer 1977).

Mutual *head-darting* (fig. 19.9) leads to sparring, during which opponents swing and toss their heads to avoid being struck. Playing or fighting genets stand up and wrestle, paw, and clasp each other while *head-darting.* Evenly matched animals fighting head-to-head direct bites to the head, neck, and breast and may pull out tufts of hair, but do not inflict wounds. The loser screams, urinates, and emits the evil-smelling contents of the anal sacs.

REPRODUCTION. Females may breed twice a year, producing 2 to 3 kittens (1–4) after a 70–77 day gestation. Some populations have birth peaks, especially at higher latitudes.

SEXUAL BEHAVIOR. After sniffing a female's vulva, or her urine or musk deposits, a male genet grimaces, indicating that the odors are entering the vomeronasal organ (13). A courting male follows a female closely, giving *grumbling* and *coughing* calls. She keeps *turning away* from him, tail and hindquarters low, and may flee, but eventually she answers his calls and allows him to come within touching distance. They sniff faces and genitalia, and rub cheeks. The female holds her tail up and finally invites copulation by *crouching with raised hindquarters* and *tail deflected* sideways. The male clasps the female about the groin with chest and belly resting on her lower back, and intromits with pelvic thrusting as the female curves her spine. He sometimes bites the female's neck hair during the final seconds of the 5-minute copulation. Coupling pairs often *meow.* Afterward the female may *anal-drag* and roll on her back, and both lick their genitals (2, 5, 13).

PARENT/OFFSPRING BEHAVIOR. Birth takes place in a hole or a nest of leaves. The mother spends long periods with the kittens, licks them anogenitally, and consumes their excrement. She grips kittens across the back when transporting them. Kittens meow, and hiss or spit when alarmed. Their fur is gray, the markings indistinct. Their eyes and ears open about the tenth day (5–18). They begin to take solid food at about 6 weeks, a fortnight after the molars erupt, and gradually become competent predators in stages that vary individually in length. First the kittens sample bits of entrail dropped by the feeding mother, then take food from her mouth (8–18 weeks), which the mother tolerates after initial resistance. Attempts to pursue prey begin at 11–18 weeks, but almost all prey eaten by the young are killed by the mother until 18–24 weeks, when competence at catching and killing prey develops through maturation, without active participation or instruction by the mother (9). At about the same time, young genets begin scent-marking with the perineal glands.

ANTIPREDATOR BEHAVIOR: offensive and defensive threat displays with accompanying vocalizations (see Dominance/Threat and Defensive displays, and Vocal Communication).

References

1. Carpenter 1970. 2. Dücker 1965. 3. H.J.H. 1984. 4. Ikeda et al. 1982. 5. Kingdon 1977. 6. Lack 1977. 7. Roeder 1978. 8. ——— 1980. 9. Roeder and Pallaud 1980. 10. Rosevear 1974. 11. Smithers 1971. 12. Waser 1980. 13. Wemmer 1977. 14. ——— 1984. 15. Wemmer and Wozencraft 1984.

Fig. 19.11. African civet's submissive posture: *lying with head deflected* (from a photo in Wemmer 1977).

African Civet
Civettictis civetta

TRAITS. The largest African viverrid, the size and shape of a raccoon. *Weight* 7–20 kg, *length* (head and body) 68–89 cm (3), *height* 35–40 cm, males larger than females. Heavy-set, relatively long-legged terrestrial carnivore, hindquarters higher and more powerful than forequarters, feet 5-toed with blunt, nonretractile claws; tail ⅓ of total length. Blunt features, teeth large and doglike with broad molars, oval ears. *Coloration:* geographically and individually variable, buff-colored with irregular black spots, blotches, and lines; distinctive black-and-white head and neck pattern, including raccoonlike mask, white lips, banded ears, and neck bands; dorsal crest of long (up to 11 cm), black-tipped hairs; young dark brown with indistinct markings. *Scent glands:* large perineal gland (fig. 19.1) and anal-sac glands.

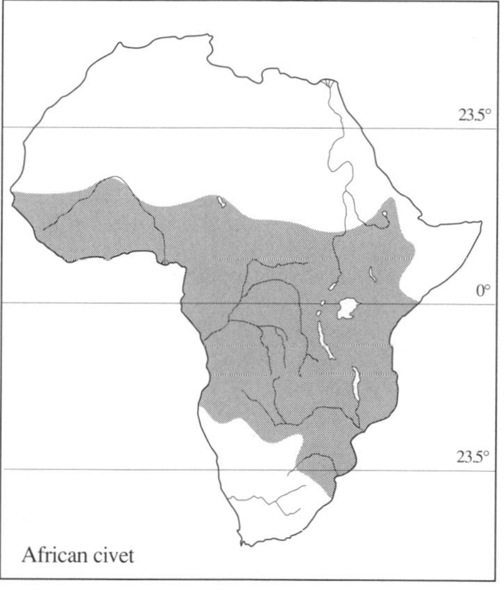

DISTRIBUTION. Throughout tropical Africa except the most arid parts, wherever there is adequate cover, usually close to water.

ECOLOGY. As omnivorous as a raccoon, but without the raccoon's manual dexterity and climbing ability, the civet eats a variety of vegetables and fruits as well as insects and other invertebrates, preys on vertebrates up to the size of a newborn antelope, and readily eats carrion.

SOCIAL ORGANIZATION: solitary, nocturnal, territorial(?).

Civets are nocturnal, sheltering by day in holes or dense undergrowth. *Civettictis* is sedentary and very regular in its habits, using pathways (especially roads) which it demarcates at intervals (see under Olfactory Communication). Despite the obvious importance of scent-marking and evidence that dung middens, called "civetries," are located at the boundaries of adjacent ranges, the species is too poorly known to say definitely whether either or both sexes are territorial. In captivity, aggression during courtship and a tendency for playfights between cubs to turn serious suggest that adults are quite unsociable and that the sexes associate only for purposes of procreation. The distinctive face mask may help civets recognize their own kind at close range in the dark (3).

ACTIVITY: nocturnal, rarely seen abroad in daylight.

FORAGING AND PREDATORY BEHAVIOR. The civet forages at night, moving quietly with head close to the ground, using its acute senses of smell and hearing to locate prey in the undergrowth. It does not stalk or chase prey like a genet, but grabs from their hiding places or pounces upon lizards, rodents, ground-nesting birds, tortoises, large insects, and such. Although civets rely entirely on their jaws to capture prey, their method of attack varies according to the size and defensive ability of the quarry (2). In dealing with prey that may retaliate, a civet first nips at it and immediately retreats. Next it grabs and throws the prey to one side with a quick head jerk, often leaping off the ground right afterward. Finally it delivers the *death-shake*: gripping the prey while shaking its head so violently that the backs of rodents, snakes, etc. are broken, often in several places. The victim is then dropped or thrown, and in the case of snakes, the civet usually leaps aside to avoid any possible retaliation. It finishes off the victim with a hard killing bite which, in the case of larger animals, is deliberately aimed at the skull.

Civets eat rapidly, bolting their food with minimal chewing. The lack of shearing carnassials prevents their cutting meat effectively, so they have to tear off mouthfuls while holding food down with the paws. The smell of carrion and entrails, or any unfamiliar animal food, often stimulates civets to rub the chin, neck, and shoulders upon food before eating it (fig. 19.12) (5).

POSTURES AND LOCOMOTION. The slow digitigrade walk of a foraging civet can be speeded up to a brisk pace if it chooses. It gallops when alarmed, dodging and jinking if pursued, and can bound over obstacles ½ m high without breaking stride. But more often it slinks into the undergrowth (more under Visual Communication). Trotting appears only as a transition between the other 2 gaits (2, 3).

Fig. 19.12. Civet *neck-sliding* on odorous substance (from a photo in Wemmer 1977).

SOCIAL BEHAVIOR

COMMUNICATION

Olfactory Communication: scent-marking with dung, urine, perineal gland.

Perineal-gland marks appear to be concentrated on trees fronting roads and pathways, especially trees that produce fruit eaten by civets. A passing civet pauses every 85 m or so to press the everted gland against a trunk (fig. 19.5) (4). The secretion is a thick, yellowish grease that hardens and turns dark brown and more visible with age, while the powerful and disagreeable scent remains detectable for at least 4 months (6). The musk scraped periodically from the perineal gland of captive African civets is refined into civetone, which "exalts" the fragrances of expensive perfumes. Apparently both sexes leave their strong-smelling dung in civetries (latrines or middens), whereas only adult males urine-mark, squirting backwards like cats against shrubs, grass, and other vertical surfaces (1).

Visual Communication. With relatively immobile lips and nose, and ears that cannot be folded back, postures are more important than facial expressions in conveying a civet's mood (2). A relaxed, alert animal moves with its head above shoulder level and legs well-extended. Anxiety or fear causes a civet to lower its head, narrow its eyes, flex its legs, and slink, with its dorsal crest flattened more than usual. Similarly, a startled civet reacts by sitting down while staring for perhaps 2 seconds at the disturbance, after which it lowers its head and disappears from view (fig. 19.13). An alerted civet may also reveal indecision by holding 1 foreleg raised while staring toward the stimulus. Fear causes the civet to erect its dorsal crest, adding a band of

Fig. 19.13. Civet in *startled attitude* (redrawn from Ewer and Wemmer 1974).

white hair edged with black which increases its apparent size by ⅓ in side view (7). It is mainly a defensive threat and its effectiveness is enhanced by creating a contour the opposite of the slinking posture of pure anxiety or fear (2). Not only confrontations with other predators but also attacks on prey that can retaliate elicit the display.

Vocal Communication: growl, cough-spit, scream, contact call, *meow.*

The first 3 calls are associated with aggression (see Agonistic Behavior). The contact call, a short *ha-ha-ha* sound repeated 3–4 times in succession, low- or high-pitched depending on the size of the animal, is given most often by kittens that find themselves alone, and by mothers to summon kittens (see Parent/Offspring Behavior). *Meowing* takes 3 forms: (*a*) a distress *meow* given by young kittens, very like a domestic kitten's *meow;* (*b*) a sex call given by estrous females, ranging from a *moaning meow* to a *hoarse meow* at high intensity; (*c*) a catlike *meow* given by the female during copulation, usually before trying to move out from under the male (2).

AGONISTIC BEHAVIOR

Offensive and Defensive Threat Displays: *head-darting,* snapping, growling, *cough-spitting.*

Submissive Displays: *lying on shoulder, lying on side, turning away.*

Fighting: *ritualized biting* without *grappling.*

A deep, menacing growl is part of the civet's defensive repertoire. A low-pitched, explosive sound, *cough-spitting,* is produced by a cornered animal, often combined with leaping in the air and maximum piloerection. By all accounts, civets do not discharge anal-sac secretion in defense; in any case the glands do not open directly to the outside, but into the rectum (1). Fights between civets tend to be formalized, consisting of *head-darting* and snapping/biting at the neck markings, whose function may be to target bites to an area where little damage is likely (2, 3). A losing combatant may scream. Unlike genets, civets have too little movement of wrists and ankles to clasp and restrain each other during wrestling or fighting (5). *Turning away* and *lying on the side,* first on one shoulder (fig. 19.11), is an appeasement display that effectively cuts off aggression (see under Sexual Behavior).

REPRODUCTION. Although capable of breeding 3 times a year in captivity, civets may be seasonal breeders in southern Africa, with young born early in the rains (3). Females are known to mate as yearlings and may reenter estrus within 3½ months after having young. Gestation is c. 80 days (2, 3).

SEXUAL BEHAVIOR: contact call, *urogenital sniffing* and *testing, courtship pursuit; mating crouch, copulatory meow,* and postcopulation *rolling* (female).

Sexually active civets may range farther and be more restless than usual (1, 5). The male gives the *ha-ha-ha* contact call, sniffs at female perineal secretions and urine, and apparently employs the vomeronasal organ to test reproductive status, adopting a characteristic stance with head raised and mouth slightly open (*flehmen*). Initially females may react aggressively to approaches, whereupon the male reclines submissively (fig. 19.11). Mounting attempts can be thwarted by remaining upright. A female in full estrus may incite pursuit by running past the male, and crouches with rear end slightly elevated during copulation. The male straddles her while keeping his forefeet on the ground, treading with the hindfeet and pelvic-thrusting. If the female meows and tries to move forward, the male may hold her by the neck hair. Following the 1-minute copulation, both partners lick their genitals; 1 female somersaulted and rolled excitedly (2).

PARENT/OFFSPRING BEHAVIOR. Genets give birth to 1–4 kittens in a hole (3). Kittens are unusually precocious; their eyes open at birth or within a few days, they can crawl, and within 5 days they can walk. However, active exploration outside begins only in the third week (2). Kittens begin to play at 2 weeks, to run and jump in the fourth week, and to react to danger by taking cover and *freezing* in the third week. When a strayed kitten gives the *coughing* contact call, its littermates automatically move toward and join it. The mother makes use of this clustering response by answering rather than going to a calling kitten, but goes to the aid of kittens meowing in distress (2). Each kitten owns its own teat and, while nursing, young civets *tread* like cats and *tug* like dogs. Weaning begins at 1 month, soon after the young begin taking solid food, and is complete after 14–20 weeks. The musky civet smell and smelly feces develop during the fifth month; at the same time males begin to test urine, to mark, and to develop larger testes (2).

ANTIPREDATOR BEHAVIOR. Displays discussed under Visual Communication.

References

1. Ewer 1973. 2. Ewer and Wemmer 1974. 3. Kingdon 1977. 4. Randall 1979. 5. Wemmer 1977. 6. ——— 1984. 7. Wemmer and Wilson 1983.

Fig. 19.14. Palm civet walking along a branch.

Palm Civet

Nandinia binotata

TRAITS. A long, low-slung arboreal carnivore with muscular, ringed tail longer than body. *Weight:* females average 2.3 kg, males 3 kg (1.2–4.7). *Length*: head and body 49 cm (42–54), tail 56 cm (49–60). Short powerful jaws with flattened molars and large canines, male teeth and skull more massive. Thick legs modified for climbing: highly flexible joints, wide paws with extensive, moist, naked pads (traction), 5 toes with retractile claws, the first (inside) digit divergent. *Coloration:* cryptic, olive-brown with indistinct spots and blotches and yellow-green eyes (daytime camouflage) (2), 2 pale shoulder spots (hence *binotata*). *Scent glands:* perineal gland situated ahead of genital opening, a glandular pocket between toes 3 and 4 overlain by yellow fur, a possible mammary scent-gland in lactating females (see Parent/Offspring Behavior), and chin or salivary glands (2).

DISTRIBUTION. Rain forests of West and Central Africa and relict forests of East Africa, extending into forest-savanna mosaic and via gallery forest into the savanna biome.

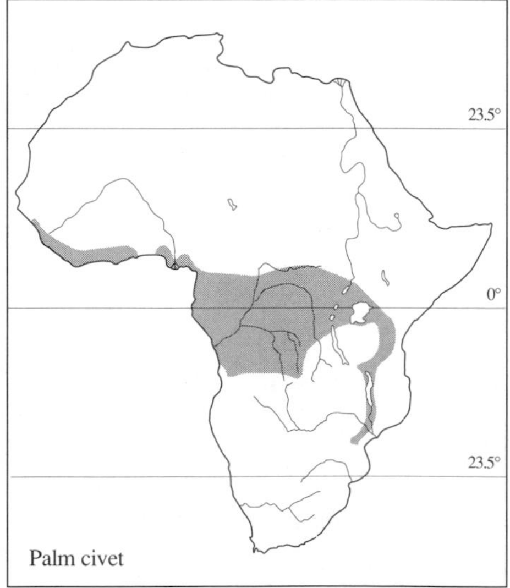

ECOLOGY. The palm civet is no more carnivorous than the other 14 nocturnal and arboreal fruit-eating mammals with which it associates. All frugivores need small quantities of animal or vegetable protein to gain the nitrogen needed for cell construction and most get it by eating insects, the most available protein source (1). The ecological separation between these arboreal mammals is accomplished mainly through size differences. Being largest and least vulnerable to predation, *Nandinia* frequents lower, more open levels of the canopy, traveling along bigger, more horizontal branches within 10–30 m of the ground. It takes the largest prey, with invertebrates, birds, and rodents usually no bigger than squirrels making up about 10% of its diet. To gain fruit in the needed quantity (stomach capacity averages 75 g), it depends on trees that produce abundant fruit (e.g., *Musanga, Uapaca, Ficus*). Since such trees are usually widely scattered and do not ripen synchronously, palm civets have to range widely but, having located a tree in fruit, may stay as long as the bonanza lasts. In primary rain forest they rarely descend to the ground, but where the forest has been reduced to patches *Nandinia* is adaptable enough to cross between them. Marauding palm civets will travel a kilometer or more overland to raid banana and papaya plantations or a chicken coop (2).

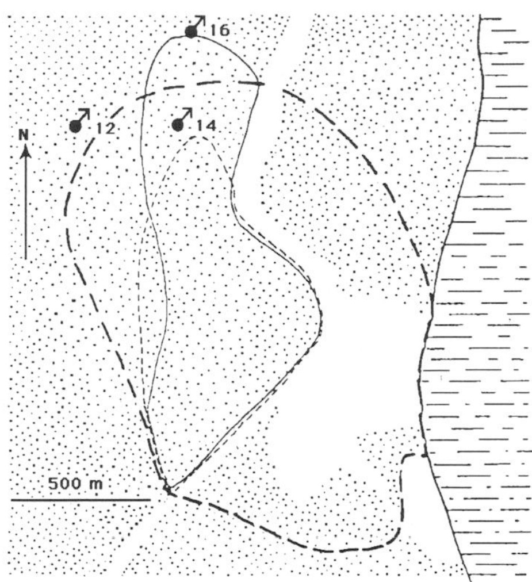

Fig. 19.15. Diagram of a male palm civet's territory (male 12) in which two subordinate adult males (14 and 16) were tolerated (from Charles-Dominique 1978*a*).

SOCIAL ORGANIZATION: unsocial, both sexes territorial, male territories encompassing 2 or more female ranges.

This type of organization is now known to be commonplace among carnivores (see chap. 18) and other "solitary" mammals, e.g. greater bush baby, bohor reedbuck. The palm civet, with an abundant and dispersed food supply, exists at a higher density than most carnivores: 5/km^2 in Gabon (up to 8/km^2 in groundwater and riparian forest) compared to 1 genet/km^2 in the same habitat. Up to ½ dozen or even more individuals may assemble at the same spot when all the area residents happen to be drawn to a particular fruiting tree. The ranges of adult females studied in Gabon averaged 45 ha. A territorial female will tolerate her own immature female offspring and even unrelated subadults on her property, but not another adult female. A common border is demarcated with perineal-gland marks which females renew during periodic visits. They also scent-mark their fruit trees.

Male territories averaged 100 ha and contained up to 5 males, including several adults. However, only 1 was territorial and had access to the resident females: the biggest, which also had the largest testes and perineal gland and a swollen neck and body. The alpha male circuited his range within a period of 5–10 days, during which he visited the territory of each female and spent several hours at points where his property contacted the territory of another dominant male. As long as social conditions remain stable, territorial neighbors are unlikely to meet except in such a contact zone, where both would rarely be present at the same time. Subordinate males remain within smaller sectors of the territory (fig. 19.15). A young adult may behave territorially and even approach females when the big male is absent, but keeps at least 300 m, and more often 900–1800 m, away when he is present. Palm civets gathered in the same fruit tree keep 10–20 m apart and threaten one another with long *meows* of variable intensity and pitch. Females answer males and are then allowed access to food. In any case no fights were seen during such gatherings (1).

When palm civets do fight, changes in social status are the usual cause or effect; changes are of course inevitable over time as young animals mature and mature individuals pass their prime. Fights, between females as well as between males, tend to be violent and losers are often so badly wounded that they die on the forest floor from starvation and infected bites. Some subordinate adult males are losers that recovered; they remain thin, with small testes and perineal glands, and often have broken teeth.

ACTIVITY: strictly nocturnal.

FORAGING BEHAVIOR. During its nightly foraging, the palm civet indulges in repeated 5–10 minute feeding bouts, after which it moves off 10–20 m for a siesta while digesting the fruits it has consumed (1). Digestion is rapid (2–3 hours for banana and papaya) but inefficient (about 78% absorption compared to 91% for a potto). Although none of the 28 radio-collared palm civets was observed hunting or capturing prey, resting animals must have opportunities to ambush unsuspecting bush babies, bats, arboreal mice, tree frogs, and such. Captive palm civets are quick to snap up presented prey (without using their paws) (1). Yet on the whole this animal is less active than most carnivores, and in daytime is downright sluggish.

POSTURES AND LOCOMOTION. Although *Nandinia* lacks the rapid movement of primates, or even its South American counterpart, the kinkajou (*Pottos flavus*), it is still a superb climber. It can shinny up and down the smoothest trunks (fig. 19.16), hang by its hindfeet while manipulating food with its forefeet, and crawl headfirst down a vertical or walk across a horizontal string, assisted by the traction pads on its hindfeet, the hooks on its inside toes, the flexibility of its limbs, and its muscular tail (balance) (fig. 19.14). It can also stand erect and even walk on its hindlegs. In an emergency it can move rapidly, leap between trees, and drop many meters to the ground without harm (1, 2). It sleeps by day in a tree crotch or vine curled up in a ball, beautifully camouflaged even in the open, and during night siestas lies full-length along a branch.

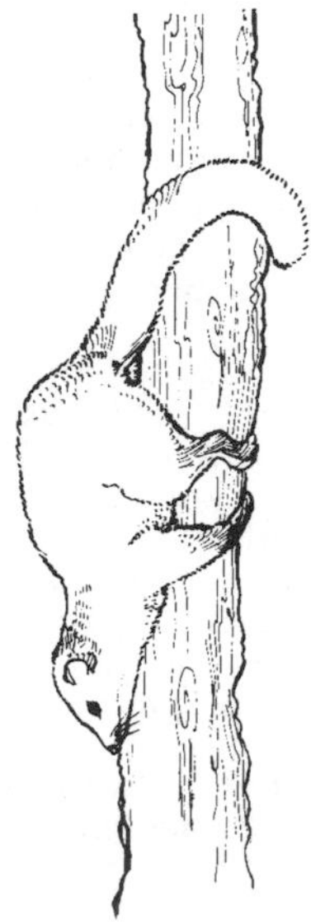

Fig. 19.16. Palm civet descending a tree headfirst.

SOCIAL BEHAVIOR

COMMUNICATION

Olfactory Communication: scent-marking with perineal glands (see Social Organization).

Vocal Communication: *humming, meowing,* hissing, snorting, screaming, *hooting,* purring (juveniles only), and *bleating* (see Social Organization, Sexual Behavior).

The owl-like *hooting* of the palm civet is unique and no repetitive sound of equivalent intensity has been heard in other viverrids (3). Far-carrying (1 km) and easily located, *Nandinia*'s hoots are well suited for advertising an individual's presence and social/reproductive status. *Humming* may be a specialized derivative of growling. It sounds like a siren and lasts from a few seconds to over ½ minute. Often 2 palm civets hum in duet. *Bleating,* a possible variant of whining given by juveniles in response to hunger, cold, and other sources of discomfort or pain, is retained in adults as a response to handling and mild pain (3).

AGONISTIC BEHAVIOR. See Social Organization and family introduction.

REPRODUCTION. In Gabon palm civets have a birth peak that coincides with the main rainy season (Sept.–Jan) (1). Gestation is c. 64 days and there are up to 4 young in a litter.

SEXUAL BEHAVIOR. Although alpha males continually monitor the reproductive status of resident females while cruising their territories, they associate only with females in heat. The sexes establish vocal contact by *hooting.* In Gabon, calling was only heard during periods of sexual activity, which come several months before the main rainy season. A male meeting an unreceptive female accompanied by 2 juvenile daughters approached and sniffed branches she had traversed. Her repeated meows kept him some meters away. After approaching the juveniles in the same way, he fed on the fruits that had attracted them to the same tree and left. A male may follow a female nearing estrus for days, keeping 20–40 m away and sleeping near her by day until she becomes receptive.

PARENT/OFFSPRING BEHAVIOR. Kittens, weighing c. 50 grams apiece, are born in a tree hole or leaf tangle, with eyes and ears closed. They purr while sucking and give a lost call that is probably the precursor of the adult contact hoot. During lactation females are said to secrete a bright orange substance which stains the kittens' fur (2). As soon as they are mobile, juveniles accompany their mother until they reach the first fruiting tree, where they stay and play. But the tendency to be solitary can already be seen in juveniles. Males separate at weaning and spend some months wandering before settling in some male's territory. Females may accompany their mothers until nearly mature in their third year, but sleep at least several meters apart after weaning, though always in the same tree as the mother.

ANTIPREDATOR BEHAVIOR. No information.

References

1. Charles-Dominique 1978*a*. 2. Kingdon 1977. 3. Wemmer 1977.

Fig. 19.17. Slender mongoose in *low sit* posture.

Slender Mongoose
Herpestes sanguineus

TRAITS. The most weasel-like mongoose: long, sinuous body with short legs and equally long, tapering tail with (usually) a black tip. *Weight and length:* males 637 g (523–789), head and body 32 cm, tail 28 cm; females 459 g (373–565), head and body 29 cm, tail 26 cm (Zimbabwe series [5]). Sharp face with rounded, very broad, low ears; carnassial teeth relatively large and sharp; feet narrow with first digit much reduced and short, curved claws. Coat sleeker, silkier, and shorter than in most mongooses, with abundant guard hairs little longer than the dense underfur. *Coloration:* variable, generally speckled, from brindled gray to dark brown (darkest in forest habitat), red eyes (greenish gray at birth). *Scent glands:* in anal sac, glandular skin on chest, and cheek/eyebrow area (2). *Mammae:* 4.

DISTRIBUTION. One of the most widespread and successful African carnivores, found south of the Sahara virtually wherever there is cover, except in deserts.

ECOLOGY. The most predaceous of the mongooses, this species is an efficient killer of small vertebrates (chiefly rodents, lizards, snakes, birds). It also eats some insects (grasshoppers, fly pupae, termites), but rarely fruit. Diurnal and very active when foraging, it is often glimpsed darting across roads. It sleeps and takes refuge in termite mounds, rock piles and crevices, holes in the ground, and tree bases.

SOCIAL ORGANIZATION: solitary, diurnal, overlapping male-female territories as in palm civet.

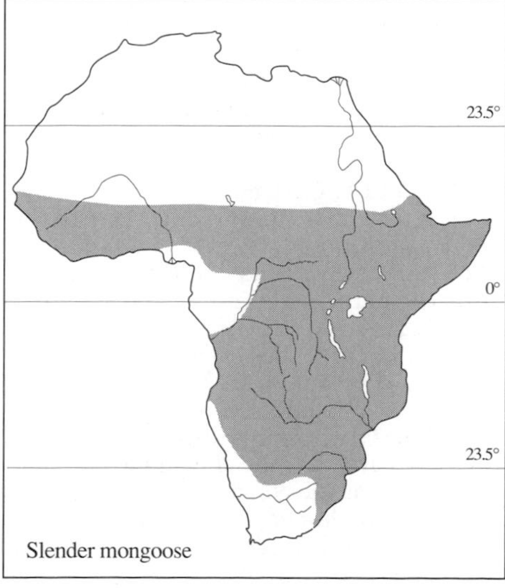

Although this mongoose has yet to be closely studied in the wild, opportunistic observations of individuals captured and marked during a study of other mongooses in Serengeti N.P. suggest a territorial and polygynous spacing/mating system. There, mature males maintain territories of at least 50 ha which typically encompass the territories of 2 females (4). The ranges of adult females are mutually exclusive, but 2 adult males often forage in the same area, usually at different times but occasionally within sight of each other. When 2 such pairs were captured and weighed, 1 proved to be much heavier (739 vs. 579 and 758 vs. 657 g), suggesting that, as in palm civets, a dominant male may tolerate subordinate males on his property as long as they don't compete reproductively.

In one (unusual) case, 2 males and a female came daily for 4 days to a giraffe kill, where they fed exclusively on carrion-fly larvae that they dug up from beneath the carcass (6). Each was independent but at times the 2 males foraged simultaneously. Normally they kept at least 2 m apart but once, when approaching from different directions, they met up in dense brush. With tails elevated and backs partly arched, they confronted each other, noses almost touching. Then they lunged and grasped each other with their forefeet, and rolled over and over giving high-pitched shrieks. After 5 seconds one broke away and ran into a nearby thicket. As usual except during estrus, the female did not interact with either male. However, one male was known to share a drainpipe for 4 months with a female and her young and even to go hunting with them (3). Male tolerance of the young has also been observed during consortship with estrous females, the male greeting kittens nose-to-nose and even allowing them to climb on his neck in play (4).

ACTIVITY. Daytime activity varies according to climate and weather conditions. Slender mongooses observed in South Africa became active before full daylight and spent the noon hours resting and sunning (2). In East Africa it is often a late riser and may sun near its shelter on warm mornings, stretched out on its belly, curled on its side, or standing broadside with hair erected and belly skin pulled around to catch the early rays, head held forward and low. Individuals have been seen feeding in sunlight on hot days (shade temperature of 35°C) without evident distress. On cold overcast days they may remain in cover and in any case they retire for the night before dark.

POSTURES AND LOCOMOTION. Endowed with quick reflexes and sinuous grace, a foraging slender mongoose slips silently between grass tussocks and through narrow openings (2). It travels at a fast trot along regular pathways, tail low with just the tip raised, slowing to a walk as it casts about for scent or sound of prey, nose to ground and back slightly humped. A slender mongoose runs with a bounding gallop and when pursued or excited changes to a flat, undulating run, flipping its tail vertically like a squirrel just before disappearing into cover. It can change direction in full flight, often by leaping at and caroming off objects, and can spring over 1 m vertically with ease. It also moves backward surprisingly fast, as when playing with prey or retreating from a suspicious object. Though primarily terrestrial, *sanguineus* is by far the best climber in the mongoose subfamily. Its short, curved claws (poorly adapted for digging) enable *sanguineus* to scamper up trees and along branches like a squirrel, and to descend headfirst. It shinnies up smooth tree trunks, embracing them with the forelegs and thrusting with the hindlegs. When startled, it *freezes* motionless, and when viewing something strange, it stands staring intently (fig. 19.17), head craned forward and moved up and down while it sways from side to side. Like other mongooses, *sanguineus* hurls eggs, nuts, millipedes, and (in play) other hard objects between its legs against rocks (2, 5).

PREDATORY BEHAVIOR. Slender mongooses employ all their faculties to locate and capture prey, typically pouncing upon or chasing after small vertebrates or insects flushed from hiding during exploratory foraging (2). Grasshoppers may be followed and put to flight repeatedly and are sometimes caught in midair. A well-fed pet female frequently played with her catch, particularly mice and other vertebrates. She would push a prey with her nose or tap it lightly with a paw to make it move, then chase and recapture it. When there was no further response, the mongoose would end the game with a single killing bite to the head/neck area. No *death shake* was observed, although a dead quarry was often thrown about or shaken. After killing its prey, a mongoose retires to a safe place to eat it. Vertebrates are eaten from the head downward, without skinning or plucking; the mongoose eats into the carcass, leaving only the skin, tail, and feet (2).

Prey Reactions. Birds and small mammals (hyraxes, dwarf mongooses, squirrels) react to the presence of a slender mongoose with alarm calls. Small birds mob and scold this predator and cats in the same way, whereas dwarf and banded mongooses go on the offensive if a slender mongoose comes near a pack with young (4).

SOCIAL BEHAVIOR

COMMUNICATION

Olfactory Communication: scent-marking with urine and dung, cheek, chest, and anal glands.

Both sexes urinate and defecate at latrines that are regularly visited (2). A tame pair also urine-marked any strange object. Marking with the anal glands, by the male, was observed only during the mating season, although the female squirted or dribbled pungent fluid from the anal sac in fright and self-defense. The slender mongoose also rubs its cheeks on objects, and during estrus the female may grasp another mongoose and rub her chest on it.

Vocal Communication: growling, *spitting,* caterwauling, shrieking, *tschaarrr* alarm call, *haah haah* sound, *whoo* contact call, *tsherr* distress call, purring.

The *whoo* contact call can be soft or loud and vary in pitch, becoming louder and higher with increasing excitement. The young begin responding in kind to the mother's *whoos* at 4 weeks, and give a harsh chirp (*tsherr*) of distress. Playing slender mongooses utter a breathy *haah haah* sound with open mouth followed by biting. A low, *droning growl* often ending in explosive spitting is given by mongooses defending their food or offspring; 2 strangers may caterwaul like cats upon meeting, and shriek during fights. A badly frightened female uttered a growling *tschaarrr* that was audible for 50 m. Newborn kittens emitted a faint purr while nursing and sleeping but were otherwise silent (2).

AGONISTIC BEHAVIOR

Dominance/Threat Displays: *crouching, bristling tail and body hair, open-mouth threat* and *snarling, head-darting,* snapping, with vocal accompaniment.

Defensive/Submissive Behavior: defensive posture lying on back, *curling up in a ball and biting between hindlegs, turning head away, submissive grimace, submissive approach.*

Fighting: *grappling, rolling, biting throat.*

In low-intensity threat, only the tail bristles, while the mongoose emits a *droning growl*; at high intensity the body hair is also erected and the animal utters a nearly inaudible *haah* (2). *Threat-gaping* was observed between 2 mongooses kept in the same cage when the dominant individual attempted to intercept the other, approaching its shoulder with slightly opened mouth turned in its direction and lips vertically retracted in a *snarl* that wrinkled the snout and fully exposed its teeth (1). *Snapping* and *spitting* often accompanied threat displays. The subordinate mongoose usually responded by *turning its head away,* by lowering its head with mouth parted and lips retracted horizontally, showing the pink lips and gums (*submissive grimace*), or by approaching slowly and cautiously with head and body low, sometimes dragging on the ground (*submissive approach*). When attacked, a slender mongoose may defend itself by falling on its back and using its claws and teeth. Two males fighting over an estrous female concentrated their biting in the neck and shoulder area, resulting in only superficial wounds (more on fighting under Social Organization) (2).

REPRODUCTION. Females may reproduce several times a year in equatorial Africa: in Serengeti N.P. one gave birth in March and again in July (4). Two to 4 young per litter is usual. In South Africa, births take place during the summer rains, mostly from October through December (5). Gestation is 58–62 days.

SEXUAL BEHAVIOR. Males accompany females during their estrus period, which may last for over a week (2). As observed in a captive couple, the male followed the female closely and attempted to mount whenever she stopped. The first copulations were brief, lasting only 30 seconds; 6 days later the duration was 2½ minutes. A copulating female kept pushing at the male with her nose and he gaped at her but did not bite her neck. In another courtship, however, the male grabbed the female by her ear and tried to pull her along, then mated with her while clasping her middle with his forearms.

PARENT/OFFSPRING BEHAVIOR. Mothers are fiercely protective and will attack animals and even people that come too close to their babies. Under captive conditions, males also behaved protectively and carried babies (which curl up and become passive when picked up by the head or neck) (2). Suckling mothers crouch over their kittens or lie on one side.

The young develop rapidly, reaching ⅔ adult weight within 50 days. Their eyes are fully open at 3 weeks; they may leave the nest at 4 weeks and begin eating solid food, brought by the mother, about the same time. They are weaned between 7 and 9 weeks, and at 24 weeks have all their second teeth (2). Separation occurs usually by 10 weeks. Females mature and become pregnant early in their second year, males probably considerably later. However, one female was still with her 3 offspring 2 months after they were full-grown, and another was sighted in company with her son 5 months after the 2 had separated. The daughter of a female that disappeared inherited the territory (2).

PLAY. This mongoose is known for its playfulness, a tendency that persists even in adults (2). As usual, the actions associated with predation, fighting, and sex (mounting) are performed in play, especially stalking, pouncing, flight and pursuit, capture, feinted attack, sparring, wrestling, biting, and play with captured prey, dead or alive. Playful interactions between *sanguineus* and other species have been observed, including hyraxes, antelopes, and monkeys. Ground squirrels sometimes invite chases by approaching very closely and then running away (3).

ANTIPREDATOR BEHAVIOR. Defensive threats: *bristling, wide-mouth threat* (also seen in play) (5), *mock attack,* alarm call, and growling.

Like other small mammals, this species quickly takes alarm at the appearance of eagles, but it appears to have little or no fear of snakes (2). It has even been known to attack large, poisonous kinds that are major predators on mongooses: for instance, a marked male took on a spitting cobra, which retreated into a rock crevice (4).

References

1. Baker 1981. 2. Jacobsen 1982. 3. Kingdon 1977. 4. Rood and Waser 1978. 5. Smithers 1983. 6. Vaughan 1976.

Added Reference

Rood 1989.

Fig. 19.18. Ichneumon eating a rat.

Ichneumon, Gray or Egyptian Mongoose

Herpestes ichneumon

TRAITS. Like an enlarged version of the slender mongoose. *Weight, total length, and tail length:* males 3.4 kg (2.6–4.1), 108.4 cm (100–117), 52 cm (45–58); females 3.1 kg (2.4–4.1), 109 cm (105–112.5), 53 cm (48.5–56) (Zimbabwe population) (10). Head narrow with blunt snout and low, wide ears; legs short and muscular, feet with 5 long digits and long, slightly curved claws; tail long and tapered. Guard hairs wiry and long (up to 10 cm), especially on hindquarters and upper tail, partly obscuring legs; dense undercoat. *Coloration:* guard hairs banded light and dark, making coat speckled gray and brown (to dark reddish brown in West African high forest [8]), underfur and underparts buff-colored, head and limbs darker, muzzle black and tail with a conspicuous black terminal tassel. *Scent glands:* anal pouch. *Mammae:* 6.

DISTRIBUTION. Pan-African, from Mediterranean to Cape, common throughout the savanna and part of the Somali-Masai and Sudanese Arid Zones, ab-

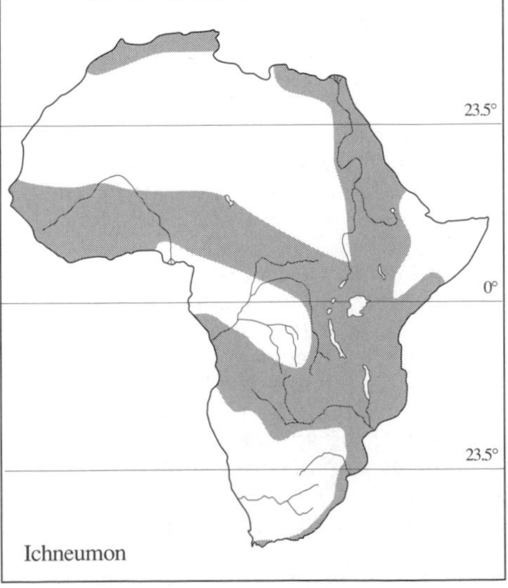

sent from deserts. Its range extends through the Arabian Peninsula to Israel, and into southern Spain and Portugal; it is the only mongoose that occurs in Europe.

ECOLOGY. One of the more carnivorous and active mongooses, the ichneumon (Greek for "tracker") frequents dense vegetation, such as riverine forest, reedbeds, and thickets bordering rivers, lakes, and ponds. Though diurnal, it is usually seen only fleetingly as it moves along the edge of or crosses an opening in the undergrowth. Its diet is varied, including insects, freshwater crabs, crayfish, and other invertebrates, but consists largely of small vertebrates, especially rodents, reptiles, birds, frogs, and fish (1, 9, 10).

SOCIAL ORGANIZATION: diurnal, polygynous, and sociable at high density, including communal care of offspring.

Although the ichneumon seems in many ways to be a larger version of the slender mongoose, it turns out to be much more sociable than its solitary foraging/hunting behavior suggests. In fact many of the characteristics of social mongooses are seen in this species, which may stand just a step below the yellow mongoose in the progression from solitary to social mongooses.

Apart from occasional sightings of ichneumons in groups (10), the first indication that this species might be less solitary than commonly supposed came from sightings in Spain, where pairs and groups of 3–7 accounted for ½ of all the mongooses that were counted (3). But the first firm information about its social organization comes from a study conducted on the Mediterranean coastal plain of northern Israel, in an area of artificial fish ponds where food was superabundant (1). Here 4 families lived in permanent home ranges within an area of 3 km^2. Whether these ranges were defended as territories or not is unclear.

Each family consisted of a male and 2–3 females plus their offspring. Family members used and maintained the same network of paths, latrine and anal-gland sites, and eating and sleeping places. In a family that was observed off and on for 2½ years, the same male was present the whole time, but was mostly seen alone; he was found with other family members in only 7 of 18 observations. The females, in contrast, regularly associated and cooperatively reared young, sometimes assisted by yearling offspring of either sex which failed to disperse before the next litter was born. When 2 or more females had offspring, they suckled and brought food to one another's babies (more under Parent/Offspring Behavior). Youngsters groomed and played with all females equally. One or more adults served as baby-sitters and also accompanied them on excursions. Changing of shifts was often observed: the one with the juveniles going off after another adult arrived. A yearling and even the adult male also baby-sat, but the father never played with his offspring.

Intolerance of adolescent sons may be the usual prelude to dispersal of male offspring, whereas tolerance of daughters may be the basis of ichneumon polygyny. A 6-month-old male disappeared after being chased and harassed by his father, and in a captive family the father persistently chased and bit his yearling son while sparing 2 yearling daughters.

The possibility that the ichneumon is less sociable and more solitary in Africa, where its niche in a numerous-predator guild may be much narrower, should be kept in mind (5).

ACTIVITY. Ichneumons may hunt most actively in early morning and late afternoon. The Israeli mongooses withdrew into the shrubbery when the temperature went above 27° or below 11°C, and during strong winds or rain. After a storm they often came out to sun. These animals habitually followed the same routes, thereby creating a network of permanent trails through the family home range. A mongoose that went off the trail for any reason always came straight back to the nearest pathway, even if it meant a longer trip to its destination.

POSTURES AND LOCOMOTION. A very active and agile animal, the ichneumon is strictly terrestrial, moving habitually at a trot while foraging, with head low to the ground and long tail extended with the tip curled forward. It cannot run very fast, however, and takes refuge in the nearest hole if pursued. It takes readily to water and swims well: this mongoose has been caught in fish traps while trying to steal the catch (5). It sits and stands upright to monitor its surroundings, and sleeps in a *ventral curl* like the suricate and the banded mongoose.

PREDATORY BEHAVIOR. When a foraging ichneumon comes upon prey suddenly at close quarters or flushes it from hiding, it attempts to seize it in its jaws with a sudden lunge, and if it misses may continue its restless quartering (5). But more often it either lies tense on the ground or stands immobile for a few seconds before rushing and grabbing its quarry (1). Large prey is caught and dispatched with a neck bite; small animals are killed instantly by biting the head (fig. 19.18). When dealing with snakes, an ichneumon attacks from the rear and retreats before the victim can retaliate (5). This species breaks eggs with the standard backward throw (see banded mongoose account, fig. 19.27), and has been rumored to dig out crocodile eggs—at any rate it is efficient at digging out invertebrate prey (10).

To obtain birds, which apart from ground-nesters would normally be invulnerable, the ichneumon allegedly lures them by performing bizarre antics (cf. canid predatory behavior). Unlikely as it sounds, individuals have been repeatedly seen to chase their tails, and one was even caught while totally engrossed in this pursuit (5). Such antics seem to lure birds closer by stimulating their curiosity and their mobbing/scolding response. There is one report of a mongoose killing 4 guinea fowl after luring them by rearing up and falling from side to side, then rolling around—nearer and nearer to the gawking birds (6). Pied crows tried to peck the black tail tassel of a captive ichneumon, and monkeys were stimulated to chase one that was running in circles (5).

In the Israeli study, members of the best-known family had 2 preferred eating places, both inside dense shrubbery, where they withdrew to eat their prey, some of which was scavenged carrion (fish, chicks, garbage) (1).

SENSES. As the name "tracker" suggests, the ichneumon relies most on its sense of smell, followed by hearing, and least on sight. Smell and hearing are used to locate distant objects, smell alone for nearby or static objects, and moving objects are detected visually. This species may have limited color vision, reacting to red and yellow (1).

An ichneumon responds to distant, even faint voices, but then points its nose and sniffs to acquire further information. Or it may walk toward a sound and, having come within a couple of meters, ignore the voice and make sweeping movements of the snout. Chemical analysis of the anal gland product suggests the ability to recognize each other by individual differences in secretions (4). Yet despite its acute senses of smell and hearing, the ichneumon has a limited ability to localize nearby objects with either sense.

SOCIAL BEHAVIOR

COMMUNICATION

Olfactory Communication. The 2 Israeli-study families each had only 1 known spot where all the members left anal-gland marks (1). Large stones or the corners of rocks on or near trails were the chosen sites, and were marked in the usual *anal-drag* (as in fig. 19.4). Whenever an ichneumon passed near the family marking place it paused to sniff and in almost ½ the observations of one family the mongoose(s) proceeded to scent-mark. No differences were detected based on sex or age, but marking frequency increased under stress. Each family also used permanent communal latrines, always located in half-open areas, some of which were in continuous use for at least 2 years. In 2 ranges occupied by newcomers after the owners were killed by dogs, the old latrines were deserted and new ones were established.

The Israeli study makes no mention of the peculiar way ichneumons are said to move in groups: single file, each individual with its nose practically glued to the anal pouch of the one ahead, making a procession resembling a long snake or giant caterpillar (5). There is even a report of 5 well-grown offspring following one another blindly round and round in a circle after their mother took flight and left them (3).

Tactile Communication. *Mutual sniffing, licking,* and *grooming* between adults, including male and female, and of the young have been reported but not described in any detail (1).

Vocal Communication: short-range *o-o-o* or *pip*(?) contact call, *clucking* or *cackling, chattering, growling, screaming,* and explosive *spitting* (threat); *ha-ha-ha* call of sexual excitement.

Ichneumon calls have not been catalogued and the above list probably includes different authors' descriptions of some of the same calls; for instance, clucking, cackling, and chattering may represent just 1 rather variable alarm or defensive-threat call. The *pip* call of reference 1, emitted by the young and heard during courtship, may be basically the same as the continuous *o-o-o* contact call reported in moving groups by reference 2.

AGONISTIC BEHAVIOR

Dominance/Threat Displays: *bristling, backing attack, gaping, lunging,* with vocal accompaniment (growling, explosive *spitting, cackling*).

Defensive/Submissive Displays. No specific information; see family introduction.

Fighting. Techniques described under Play.

REPRODUCTION. This species typically reproduces annually, but has been known to reproduce more than once after losing a litter or when prey species become exceptionally abundant (1). One to 4 young (average of 10 litters = 3.3) are born after a gestation of c. 60 days (1).

SEXUAL BEHAVIOR. Descriptions of mating in both tame and wild ichneumons are in close agreement (1, 2). More frequent licking of her vulva than usual by the female signals the onset of estrus (2). The male, *following closely,* also sniffs and licks her vulva, and the 2 emit the *o-o-o* contact call (2) or "*pip* tones" (1) similar to but lower than those emitted by the young. They engage in prolonged *social grooming,* often initiated by the female (1). The tame female would *crouch* as the male approached, then run away with the male in pursuit. The coquettishness of this behavior became apparent in full estrus, when it was always the prelude to assuming the *copulatory attitude,* with hindquarters raised, head lowered, and tail deflected. If she failed to *crouch,* the male would try to press her down with his chin, but could not copulate without her cooperation. Mounting attempts at decreasing intervals usually preceded mating; the male licked his genitals after each attempt, which lasted ½–2 minutes. During copulation he embraced the female's midsection and thrust with his hindlegs pressed against hers. He repeatedly poked and mouthed her neck but without gripping her, and the pair often dropped to their sides, without uncoupling, during the 4–5 minute coitus. In the wild couples, several short copulations without ejaculation were seen, lasting ½–1 minute, after each of which the female escaped from the male, which chased her violently and even bit her. Once 10 such couplings took place in 35 minutes. Complete copulations took 6–7 minutes. Both partners *cackled* (2) or uttered *pip* tones (1) while mating. Postcopulatory behavior was marked by unrest and by defensive threats (biting intention, loud screams) from the female, which also licked her vulva, *rolled* on the ground, and cleaned herself all over (2).

PARENT/OFFSPRING BEHAVIOR. Because females litter down in dense thickets, wild births have not been observed (1). A pet female in labor stood with hindlegs straddled and slightly bent. In carrying the 2 babies to the nestbox, she held them around the middle (2). Ichneumon young are born in a very immature condition, with eyes and ears sealed (1). After opening, the eyes of 2 hand-reared mongooses remained milky-colored until they were 21 days old, when they turned shiny black, and the irises only turned the normal gray color on day 45. The babies began reacting to voices at 25 days and appeared to sniff objects beginning on day 39. They tried to stand and walk at 21 days but fell over after a few steps. At 4 weeks they could walk normally and they began jumping at 37 days. At 2 months they stood bipedally to look and sniff and at 72 days the hand-reared ichneumons performed complete hunting behavior.

In the Israeli study area, young were born in spring and first appeared outside the natal thicket at c. 6 weeks. They began eating solid food at 1 month, when the mother and other family members brought food to the nursery thicket, and were weaned after another month or so. An ichneumon lies on its side while suckling. The wild youngsters began going on excursions at 6 weeks, always accompanied by an adult, which would leave them in a nearby bush to go off hunting alone, returning with a gift of food for them. Adults helped provision any offspring from their own family (more under Social Organization). Although the young were capable of hunting for themselves and began foraging independently from 4 months onward, they continued to be provisioned until they were 1 year old or even older: one adult female gave up her food to a 1½-year male and then stayed around to guard him while he ate (1). And the family females, never far away, would run to help older juveniles in distress.

But juveniles older than 3½ months were forced to chase after adults and compete for the food they brought (cf. banded mongoose and suricate). When a female caught something, she would run away with it, but only after the juveniles became aware that she had food and began trying to get it from her. This behavior has been compared to the way birds reward the nestlings that beg most vigorously: the ichneumon system also promotes survival of the fittest offspring, ensuring that some recruitment will occur even in years of food scarcity (1).

PLAY. In captivity, young and even adult ichneumons are exceptionally playful and exploratory. Whether it is natural or a way of compensating for enforced inactivity remains to be seen. Several different kinds of play have been described (7).

Running games. The mongoose flees in a straight line, then turns suddenly at an acute angle and either stops and sits up or makes short, vertical bounds. A pet male sometimes jumped at a wall and rebounded with all 4 feet toward the opposite wall or turned a backward somersault.

"Mole play." Like most mongooses, the ichneumon is irresistibly attracted to dark holes and tunnels. Pets crawl under carpets, blankets, newspapers, and such, scurry about playfully, and sham-attack a moving hand or tapping fingers while making the *clucking* threat call. They run blindly with a paper bag over the head, banging into furniture with gay abandon.

Playfighting. The commonest kind of play between ichneumons, playfights give an idea of what real fights are like. Jumping and sometimes mounting is the invitation. The attacker rears and comes down on the other, delivering a bite to the neck. The opponent counters by rolling on its back, defending itself with quick neck bites and kicking hindfeet. Wriggling free, the defender may now become the aggressor, while the other assumes the defensive position; or both lie on their sides clasping each other with their forelegs while raking with the hindfeet. Sometimes they crouch eyeball to eyeball, with heads turned flat to the ground, preparing to spring, or they stand facing in tense attitudes, tails lashing horizontally or vertically. Often they *grapple* and *jaw-fence* in the *low-sit attitude,* from which they may fall and roll over in a wrestling bout. The *backing attack* (see fig. 19.7) is also performed. Playfights often end in flight and hiding games.

Games with prey. Captive ichneumons often play with living or dead prey. The duration and intensity depend on whether the mongoose is hungry or not, or in a playful mood, and on the activity of the prey. Incomplete segments of predatory motor patterns are enacted: lying in ambush (often with lashing tail), chasing, grabbing, and throwing. Often the victim is not injured.

Experimental games. In playing with objects other than prey, ichneumons perform spontaneous, variable, and inventive behavior, such as running around with bags over their heads. They also drop pieces of food and small objects in their water dish, then grope for them (7).

ANTIPREDATOR BEHAVIOR: defensive threat (bristling, lunging, *backing attack, cackling,* screaming, hissing, *spitting*); concealment.

Active defense of offspring by all adult family members, including the father, has been observed (1). An ichneumon surprised in the open and cut off from cover may try to escape notice by *lying prone,* when its outline may be effectively concealed by its long hair (10).

References

1. Ben-Yaacov and Yom-Tov 1983. 2. Dücker 1960. 3. Fuente 1972. 4. Hefetz et al. 1982. 5. Kingdon 1977. 6. Preston 1950. 7. Rensch and Dücker 1959. 8. Rosevear 1974. 9. Shortridge 1934. 10. Smithers 1983.

Fig. 19.19. White-tailed mongoose defensive threat display (redrawn from Kingdon 1977).

White-tailed Mongoose

Ichneumia albicauda

TRAITS. A large mongoose with a conspicuous white tail and dark legs. *Weight and total length:* maximum of 5 kg, males 3.6 kg (2.9–4.1) and 103 cm (92–112), females 3.4 kg (3.1–4.2) and 103 cm (96–111) (Serengeti sample [10]). Long-legged, back sloping forward from higher, well-developed hindquarters; tail 35–46 cm (8); feet 5-toed with long, curving claws; head moderately broad with pointed snout, broad molar teeth adapted for crushing; ears large, rounded, set low on head. Fur long (up to 65 mm) and loose, especially on tail and hindquarters, with a very thick undercoat. *Coloration:* grizzled gray or gray-brown with a light-colored undercoat, guard hairs banded black and white, terminal ⅔ of tail usually pure white (often black in West Africa); infants pale with few bristles (6). *Scent* *glands:* anal pouch and nauseating anal-sac secretions (2, 3). *Mammae:* 6.

DISTRIBUTION. Widespread over most of sub-Saharan Africa, except high forest, mountains above 2500 m, and the driest parts of southern Africa and Somalia; also ranges into southern Arabia (1).

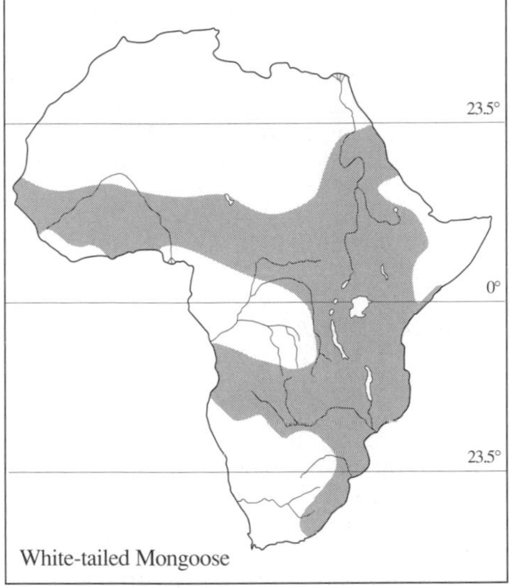

White-tailed Mongoose

ECOLOGY. Found in habitats ranging from desert to savanna woodland, this mongoose forages actively in the open at night and is accordingly one of the more commonly seen nocturnal mammals. It occasionally takes small vertebrates but subsists primarily on comparatively large insects and other invertebrates, feeding opportunistically on whichever kinds are most available at the surface (9, 10). In the Serengeti ecosystem, harvester termites and ants are its mainstay during the mid–dry season; when the rains begin, *Ichneumia* concentrates on the dung beetles and their larvae that abound wherever ungulates concentrate. Extensive overlap in habitat, diet, and activity exists among this mongoose, the bat-eared fox, and the aardwolf. Although these and other small carnivores rarely interact, it seems that the fox and aardwolf are commonest where white-tailed mongooses are scarce, and vice versa (9). The main limiting factor on all 3 species may be denning sites rather than food, for all rely on secure refuges where they can sleep by day and rear young.

SOCIAL ORGANIZATION: solitary, polygynous, nocturnal, territorial.

Where ungulates are numerous and resident, as in western Serengeti N.P., up to 4 white-tailed mongooses/km^2 have been recorded (9), but even there their density was found to vary locally according to the availability of suitable dens. Most dens were located in inactive termite mounds (*Odontotermes*), or in hollows and crannies at the bases of *Acacia* and *Balanites* trees (probably also rooted in ancient mounds). Although *Ichneumia* may enlarge ventilation shafts to reach underground chambers, it is not known to dig its own dens from scratch, and sometimes lies up in holes so shallow that it is partially exposed (10). The home ranges of 7 females included an average of 3 den sites (2.5–4.3), not all necessarily usable. Solitary mongooses used 2–3 different sites a month, residing in the same den for 5–15 days before moving on. After a season of exceptional rains, when 5 of 7 known den sites were washed out, the population crashed to less than 1/km^2 (9).

In some 390 sightings of white-tailed mongooses in the Serengeti study, 87% were of solitary animals (no other mongoose within 100 m) (9), making this species even more solitary than the slender mongoose (72% of sightings) (4). Although as many as 9 adult-sized individuals have been seen foraging and moving together in the same direction (1), such gatherings occur only on rare occasions when prey can be found in patches, for example, a termite swarm or a heavily manured pasture during the rainy season. Weaned juveniles begin foraging alone for several hours each night and are on their own by the age of 9 months and/or a weight of 2 kg. However, the young continue to forage within the maternal home range for at least 4 months after weaning and, where mongoose density is comparatively high and food is sufficiently abundant, some female offspring fail to disperse.

In part of the Serengeti study area several ranges were shared by 2–3 adult females plus 1–2 juveniles (up to 5 in the same range). The adult and subadult members of these clans foraged and often slept alone. They rarely came within 100 m of one another and when 2 did meet they usually paused only long enough to sniff noses or genitals (except twice when adult females *social-groomed*), tails bristling, before going their separate ways. But they did not avoid and never behaved aggressively toward one another, whereas only aggressive interactions (chasing) were seen with outsiders (10). In 14 of 38 times when the occupants of dens could be identified (visually or by radio transmitters), there were 2 mongooses, usually a female with young but on 5 occasions 2 particular adult females. There was very little overlap between home ranges (maximum of 250 m), and the boundaries were scent-marked (see below); clan members used the same marking sites.

Males, too, maintained exclusive home ranges. Theirs overlapped the ranges of females completely but did not share common boundaries; consequently, a male's range typically included parts of 2–3 female ranges, and a female's range often lay partly in the ranges of 2 or more males. The home ranges of 9 females averaged 64 ha (39–118), whereas the average of 4 males was 97 ha (80–123). Range size did not appear to change seasonally or according to population density in the study area, but evidently does vary in different habitats, as ranges of up to 8 km^2 have been recorded in Kenya (8).

Ichneumia may reach its greatest abundance in suburbia, where this species, like skunks, readily takes to garbage-can raiding. In Entebbe, 45 mongooses were killed in 1 year by a pack of 3 dogs (1).

ACTIVITY. Strictly nocturnal, white-tailed mongooses emerge from their dens only c. 25 minutes after sunset, when it is just too dark to see them even through binoculars, and go to ground before dawn (10). More individuals were seen in the first ⅓ then in the latter ⅔ of the night, and 50% more mongooses were abroad on overcast, moonless nights than on clear, moonlit nights. They continued foraging during light rain and brief thunderstorms. Some individuals returned to dens after only a few hours on the surface, but most remained active the whole night (up to 10 hours), with only short spells of inactivity and negligible time spent in social interactions.

FORAGING BEHAVIOR. Almost all the time spent above ground is devoted to food-seeking, during which a mongoose zigzags at a walk or trot, stopping frequently to nose or bite at the ground. Occasionally one will lick and bite repeatedly for a longer period. Moving at an average rate of 650 m/hour, a mongoose covers its entire home range each night and consumes c. 240 g of insects, equal to 3200 harvester termites (9, 10). Unlike the aardwolf and bat-eared fox, which rely largely on hearing, *Ichneumia* seems to rely mainly on its nose. Though it sometimes leaps and snaps at emerging dung beetles, it rarely captures flying insects, or stalks and pounces (e.g., on grasshoppers), or digs, except for beetles or winged termites that are close to the surface. Mostly it uncovers beetles in dung and picks insects off the vegetation (9).

This mongoose occasionally kills and eats poisonous snakes. It also smashes eggs in typical herpestine fashion (fig. 19.27), and like genets, bats, and others, harvests insects that are attracted to outdoor lights.

POSTURES AND LOCOMOTION. *Ichneumia* moves much like a dog, walking on its toes (digitigrade) with an *ambling* gait at c. 4.2 kph, head low and croup higher than the shoulders, tail down and trailing. It often moves at a brisk trot and in an emergency gallops, though too slow to outrun most predators (see Antipredator Behavior) (8).

SOCIAL BEHAVIOR

COMMUNICATION

Olfactory Communication: scent-marking with anal glands, urine, and dung, by adults and subadults of both sexes.

White-tailed mongooses deposit their sizeable scats in latrines close to occupied dens and also in large, centrally located middens, of which 1 or 2 occur in each home range/territory. When urinating, both males and females *arch their tails* in a characteristic way that tends to call attention to the act. Whether this is a form of scent-marking is unclear, but mongooses traveling together often *tail-arch* in the same locations, whereas lone animals *tail-arch* infrequently (<twice/hour). Secretions of the well-developed anal pouch, delivered in the stereotyped *anal-drag*, are deposited with greatest frequency, at the rate of once every 2 minutes by one mongoose on a border patrol. Marking rates tend to be much lower during normal foraging but increase when another mongoose is near. Thus, on 2 out of only 3 occasions when males were seen interacting, they did not approach each other but each engaged in intense scent-marking (10).

Vocal Communication: *muttering, whimpering, purring, growling, screaming/shrieking, barking.*

Perhaps unusually vociferous for a solitary mongoose, *Ichneumia* sometimes makes a *muttering* sound while digging for insects. Individuals exchange whimpering calls, and pets make a guttural, vibrant purr when stroked or groomed. A trapped animal growls, shrieks if approached, screams loudly when hurt, and, most unusual, emits an explosive nasal bark or yap (1).

Visual Communication: *tail-arching* and *-waving, bristling.*

Ichneumia's visual displays have not been described in detail, but clearly the white tail that advertises this species in the dark is employed in encounters both with its own and with other species (see also under Play, Olfactory Communication, and Antipredatory Behavior). Tail position and movement would be most effective in signaling an individual's state of mind and intentions in the dark, and the more the hair bristles the more conspicuous the appendage becomes. *Tail-arching* and *bristling* are seen when mongooses interact and when they urinate; also juveniles following their mother closely hold their tails conspicuously arched (10).

REPRODUCTION. Very little information has been published about this species. One or 2 young per litter (rarely 3 or 4) is the rule. Probably most births occur during the rains. Females within the same range may rear young simultaneously, but without cooperating or the help of older siblings as in social mongooses (10).

SEXUAL BEHAVIOR. When a foraging male and female meet, they pass without interacting or the male may approach and briefly examine the female's genitalia, while she *crouches* or *arches her tail.* But during estrus, a male and female pair up for at least several hours, remaining within 25 m of each other while foraging. Six observed copulation sequences each lasted about ½ hour, during which the male mounted 15–20 times. After achieving intromission the male sometimes rolled back and sat holding the female in his lap. Mounts lasted 5–20 seconds before the female broke away; usually she then *circled* and the male *followed closely,* mounting again when she stopped. In 4 cases the male and female remained together after copulation, and in 1 case the female went on to mate with another male within the hour (10).

PARENT/OFFSPRING BEHAVIOR. Other than the fact that young are born in a den, blind and thinly haired, very little is known about mothers and newborn (see family introduction).

PLAY. Hand-reared young are playful. They chase, drag cloths and other objects to invite pursuit, turn somersaults, and cavort with tails raised, waved, and often switched against the play companion (1). Once juveniles begin foraging independently, they rarely interact with their mothers away from the den. But when a well-grown male rejoined his mother at the den they still shared, they chased each other playfully before retiring (10).

ANTIPREDATOR BEHAVIOR. When a foraging white-tailed mongoose is disturbed, its initial reaction is to *freeze,* perhaps stand up to assess the danger, then to run for cover. It can run surprisingly fast for a mongoose and reverses direction suddenly, but when overtaken in the open, it puts on an impressive threat display (fig. 19.19). The long hair is erected and the tail is raised to increase the animal's apparent size and visibility. It emits growls, explosive grunts, and a clear short bark (7). According to references 2 and 3, this display is backed up, if need be, by discharging the evil-smelling and persistent secretions of its anal sacs, but Waser (pers. comm. 1986) was unable to provoke this defense in the animals he trapped and handled. What, then, deters predators and gives this mongoose the confidence and warning coloration of a skunk? Adults forage unconcerned near hyenas, and one mongoose was seen casually investigating a large python (10). However, wild dogs ran down and killed one flushed from a thicket by daylight (9).

References

1. Kingdon 1977. 2. Maberly 1962. 3. Pitman 1954. 4. Rood and Waser 1978. 5. Roosevelt 1910. 6. Rosevear 1974. 7. Smithers 1971. 8. Taylor 1972. 9. Waser 1980. 10. Waser and Waser 1985.

Fig. 19.20. Marsh mongoose hurling an egg to the ground (downward throwing technique; cf. the more usual mongoose technique, fig. 19.27).

Marsh Mongoose

Atilax paludinosus

TRAITS. A large, very dark mongoose. *Weight and length:* 2.5–5 kg, head and body 62–64 cm, tail c. 40 cm. Broad head with pink nose and lips (1) (possibly black in West Africa [4]), and low, broad, rounded ears; sharp-edged upper canines and long lower canines, stout premolars adapted for crushing shellfish; short legs with 5 long toes, unwebbed (unlike all other mongooses [4]), palms soft and naked, lacking thick pads; thick tail tapering to point. Shaggy coat with dense underfur. *Coloration:* usually very dark brown, but varies to medium-, red-, and speckled-brown. *Scent glands:* large anal pouch, secretion brown with a heavy musk (5) or foul smell (2). *Mammae:* 4 (2–6) (1).

DISTRIBUTION. Common in the vicinity of water and cover throughout sub-Saharan Africa, up to 2500 m, including papyrus and mangrove swamps (2, 4).

ECOLOGY. One of the more specialized mongooses, *Atilax* is rather raccoonlike, but less versatile, being strictly terrestrial, more closely tied to water, and less omnivorous. Both have very similar hands (fig. 19.21) (cf. also clawless otter) adapted for feeling delicately and dextrously in cracks and crannies, under rocks, and in mud and sand for concealed prey. The marsh mongoose requires both water and cover; otherwise it can adapt to a broad range of climatic conditions. Considered nocturnal, it is seen abroad by day fairly often. Its diet consists primarily of snails, freshwater crabs and mussels, frogs, and aquatic insects, also fish, marsh birds and eggs, reptiles, worms, and, according to reference 6, crocodile eggs.

SOCIAL ORGANIZATION: solitary, mainly nocturnal (3), territorial(?).

Since marsh mongooses appear to be linearly spaced along the waterfront, they scent-mark, and one has been seen threatening and fighting another that approached its burrow, the species is presumed to be territorial (2). However, *Atilax* remains unstudied. It is only certain that individuals are regular in their habits, following surprisingly smooth and well-beaten pathways (their soft feet are vulnerable to thorns) (2). They *handstand* to place anal-gland secretions on the undersides of logs, rocks, and such (especially males), rub their cheeks against objects, and deposit their musky-smelling scats in latrines situated in open areas, typically bordering water.

ACTIVITY: mainly nocturnal but also active by day (see Ecology).

FORAGING BEHAVIOR (includes Postures and Locomotion). *Atilax* is rumored to swim almost as well as an otter, by undulating movements of its body, despite the absence of webbing between the toes and the fact that its underfur becomes waterlogged (7). Normally, though, it forages like a raccoon, wading and groping in shallow, usually muddy water. In deeper water it holds its head high while reaching with its hands. In shallow pools it will sift the bottom sediments rapidly, methodically working the whole pool over and over again. Small food items are chomped between the back teeth. Struggling prey like frogs or fish are put out of action by chewing bites with the canines and premolars (2). Marsh mongooses open mus-

23.5°
0°
23.5°
Marsh Mongoose

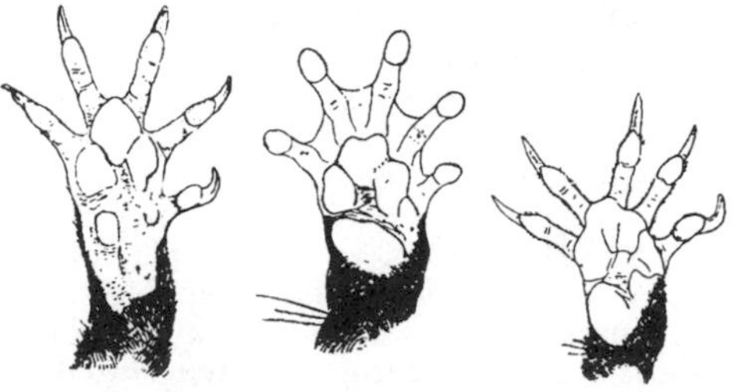

Fig. 19.21. From left to right, hands of marsh mongoose, clawless otter, and raccoon (from Pocock 1916, 1921*a*, 1921*b*).

sels and crabs by hurling them down against rocks from a standing position (fig. 19.20). When moving between the daytime lair and foraging grounds they travel at a steady trot, with the tail outstretched or slightly curled up.

SOCIAL BEHAVIOR
COMMUNICATION

Vocal Communication. Quarreling mongooses utter explosive bark-growls, presumably similar to the defensive-threat barks given toward potential predators: a low growl reinforced by a sudden barking growl in a lower register (2). A high-pitched cry, an open-mouthed bleat (excitement), a peculiar moan, and purring (pleasure) have also been described (2).

AGONISTIC BEHAVIOR. See Vocal Communication and Antipredator Behavior.

REPRODUCTION. Up to 3 young are born, at no particular season in West Africa, probably early in the rains in southern Africa (e.g., November/December in Zimbabwe) (4, 5).

SEXUAL BEHAVIOR. Undescribed.

PARENT/OFFSPRING BEHAVIOR. Undescribed.

ANTIPREDATOR BEHAVIOR. A formidable foe, the marsh mongoose has been seen to bite a dog in the throat and nearly drown it while its own leg was being chewed—again, recalling the raccoon's defensive tactics. When cornered and terrified, it urinates, exudes anal-gland secretion, and tries to curl itself into a ball, turning round and round. It escapes danger by taking to water, may swim away with head and back emergent, dive, or hide with just its nose out of water (2). It stands bipedally to scan its surroundings.

References

1. Ewer 1973. 2. Kingdon 1977. 3. Lombard 1958. 4. Rosevear 1974. 5. Smithers 1971. 6. Stevenson-Hamilton 1947. 7. Taylor, M. E. 1970.

Fig. 19.22. Yellow mongoose alerting posture (note tail position). (Redrawn from Earle 1981.)

Yellow Mongoose

Cynictis penicillata

TRAITS. A small mongoose with a bushy, white-tipped tail, short, pointed muzzle, and large ears that extend above the head. *Weight and length:* males 589 g (478–797), 496 cm (412–582); females 553 g (440–797), 506 cm (447–580); tail 211 cm (180–250) (2). Long guard hairs and thick, woolly undercoat. *Coloration:* tawny yellow or orange in South Africa to gray in northern Botswana (2), head grayer with white chin, underparts and forelegs paler. *Scent glands:* anal pouch and cheek glands (1). *Mammae:* 6.

DISTRIBUTION. Southern Africa in open, sandy terrain.

ECOLOGY. This species occurs in a variety of open habitats with sandy soils, including acacia scrub woodland, savanna, parkland, floodplains, and the fringes of alkaline pans. Although it is diurnal, at least some individuals may be active on warm nights; the possibility that these are usually peripheral colony members that become partly nocturnal in order to avoid harassment has been suggested (1). Mainly insectivorous, the yellow mongoose relies on harvester termites, beetles, grasshoppers, and the like, but also takes some small vertebrates (mice, lizards, snakes) and eats fruit in winter when insects are scarce (1, 2, 3).

SOCIAL ORGANIZATION: diurnal, territorial, colonial (also in families and pairs).

This mongoose lives in small colonies averaging c. 8, up to 20 animals (1, 2). A study of the species was conducted on a 20-ha island near Johannesburg, where the space limitation, inability to disperse, and so on suggest that the social organization of this population might not be typical.

The island was divided among 5 colonies, each consisting of a breeding male and female, their immature offspring, 1–2 young adults or adults (usually males), and sometimes 3–4 other, loosely associated individuals. Several old mongooses (including a deposed alpha female) lived separately in nearby springhare holes, but each maintained contact with its colony. There was a clear rank hierarchy in the order listed. The young were not only dominant over other adults, but even dared to take food from the alpha male (cf. spotted hyena). Very old mongooses were lowest in rank.

The ranges of the different colonies were exclusive, the boundaries demarcated and defended by the alpha male, which regularly patrolled and marked the borders, sometimes assisted by the alpha female. Roads or a row of trees often served as territorial boundaries (more under Olfactory Communication). Each colony occupied a den consisting of a network of tunnels with multiple entrances; 21 out of 66 holes were in active use in the largest colony. Dens were located on rising ground and the mongooses either dug their own tunnels (in sandy soil) or enlarged existing springhare holes. Colony members all deposited their dung at latrines located close to the entrances, either in hollows or the openings of unused holes. *Cynictis* often shares warrens with other species, especially the suricate and the ground squirrel (*Xerus inauris*) (1, 2, 3) (fig. 19.23).

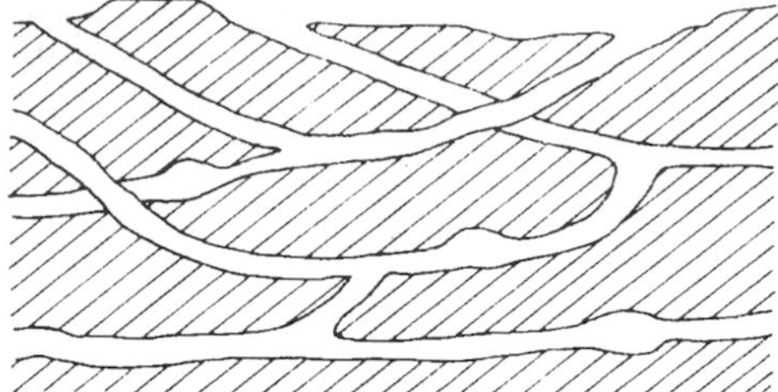

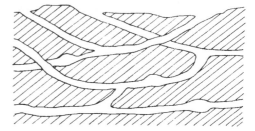

Fig. 19.23. Schematic drawing of a den shared by yellow mongooses, suricates, and ground squirrels (from Lynch 1980).

ACTIVITY AND FORAGING BEHAVIOR. Weather conditions, especially temperature, greatly affect the daily activity cycle. In summer, colony members may emerge by 0600 h and, after using the latrines, busy themselves digging and cleaning out hole entrances before starting to forage. They spend the hottest hours inside or lying on loose sand near the entrances. Then they go foraging again, returning and retiring by 1830, shortly before dark. On cold mornings, yellow mongooses come up to visit the latrines between 0630 and 0700 (sunrise), then withdraw until it becomes warm enough to sunbathe. They devote up to 1½ hours a day to this pursuit, adopting standing, sitting, and lying postures in a regular sequence as they soak up the rays with hair bristling to let in the warmth (fig. 19.24). Lying is accompanied by much stretching and yawning. Around 0900 the mongooses set off foraging. Although several animals may be in sight of one another, and even close together at a concentrated food source like a garbage dump, colony members forage individually and not in packs within their 5–6 ha territory. In April and May, foraging areas are strewn with innumerable shallow holes where the mongooses have been digging for beetle larvae. Although members of one colony scavenged dead chicks from an ibis/heron colony, they showed no interest in the eggs. The mongooses returned from foraging in late afternoon (1730) to sun and socialize before retiring for the night (1).

Fig. 19.24. Yellow mongoose sunbathing (redrawn from Earle 1981).

SOCIAL BEHAVIOR

COMMUNICATION

Olfactory Communication: *anal-gland* and *cheek-gland marking, back-rubbing,* dung middens; individual recognition.

The alpha male of each *Cynictis* colony marked each member of his group daily with anal-pouch secretion. First he would jump over the animal and, on landing, mark while standing on all fours, tail raised. He also spent 15–30 minutes making a daily marking round, usually before starting to forage (1). Once he started, there was no stopping him: even though chased and given no pause to mark, he would complete his regular route on the run. A few drops of anal-sac secretion (or urine?) were deposited at each marking site in a squatting attitude, without *anal-dragging* or leg-cocking. In addition, he rubbed his cheeks on grass stems, rubbed his back against grass tufts and stumps, and defecated on boundary latrines shared with the neighboring territorial male. Males marked throughout the territory, but most frequently and intensively on clan boundaries. Marking every 5–10 m was normal, and where contacts between neighbors were particularly frequent or tense, males stopped to mark practically every meter. A rival's fresh sign could even stimulate an alpha male to mark the same spot 3–4 times.

All colony members left cheek-gland marks, especially around the warrens. But only the alpha male *back-rubbed* and only on the border, turning almost on his side in the process. Accumulated loose hair and boluses at these sites added a visual component to the presumed scent marks. Alpha females rarely marked, but did defend the territory against dominant adults from neighboring colonies (see under Dominance/Threat Displays). When 2 mongooses unsure of each other's identity meet, they *approach in a crouch* and *sniff cheek glands.*

Vocal Communication: warning growl, usually accompanied by hissing (food defense), high-pitched scream (by frightened inferior and both combatants during fights), short bark (danger signal), soft purr (males during copulation) (1).

Visual Communication. Tail positions are important, both the white tip and degree of *bristling* conveying information. The normal tail position is lower than the back with the tip close to or touching the ground. When raised in an S-curve and spread, it is an alarm signal (fig. 19.22) (more under Dominance/Threat and Submissive Displays).

AGONISTIC BEHAVIOR

Dominance/Threat Displays: *scent-marking another mongoose, holding tail tip higher than back; rising above, licking,* and *biting adversary.*

Submissive Displays: *crouching, lying on side,* vocal signals.

The social standing of yellow mongooses can be told by dominance/submissive behavior. A submissive animal lies on its side, and if frightened screams loudly. After 2 mongooses complete an identity check (see Olfactory Communication), the dominant then *rises higher* and *licks* or *bites the inferior's neck* until it assumes the submissive posture. All colony members submit to the alpha male and female, but other adults do not behave dominantly or submissively to one another, and members of neighboring colonies do not dispute rank when they meet on neutral ground. A subtler indicator of dominance is holding the tail out with the white tip displayed above the back; an unsure animal holds the tip down.

Fighting. Border skirmishes between territorial males were common. They only lasted 3–4 seconds, ending with each antagonist running back into his own domain. The frequent presence of scars on males' faces suggests that occasional severe fights occur. Alpha males were never seen off their own territories and they even avoided neutral areas (1).

REPRODUCTION. Breeding is seasonal, with births concentrated early in the rains (October/November).

SEXUAL BEHAVIOR. A dominant pair was never observed mating. But young-adult females were seen coupling with alpha and other males, sometimes in quick succession (1). Within 4 hours, 1 receptive female approached 3 males in a submissive manner, *crouching* and then *lying down* when sniffed anogenitally. Two of the males lost interest in the first minute. The third began to purr, whereupon the female got to her feet and the male mounted, clasping her loins. He thrust for 8–12 seconds, then paused for 10–30 seconds before thrusting again. During pauses the male's weight pulled the female back into his lap; to regain the copulatory crouch they had first to roll onto their sides. Two observed copulations lasted 37 and 45 minutes. The male kept up his soft purring while the female continuously bit or licked his ears and neck. Although yearling females mate, pregnancies have not been found in this age class (4). The mating of subordinate females with different males and without resulting pregnancy sounds very like the dwarf mongoose (q.v.).

PARENT/OFFSPRING BEHAVIOR. Information is fragmentary. The privileged status of the juveniles in the study colony changed as they approached adult size at c. 10 months. After they became more independent, some immigrated to other colonies, to which they gained admission by displaying submission.

References

1. Earle 1981. 2. Lynch 1980. 3. Smithers 1971. 4. Zumpf 1968.

Fig. 19.25. Dwarf mongooses *social-grooming* (from a photo in Rood 1983*a*).

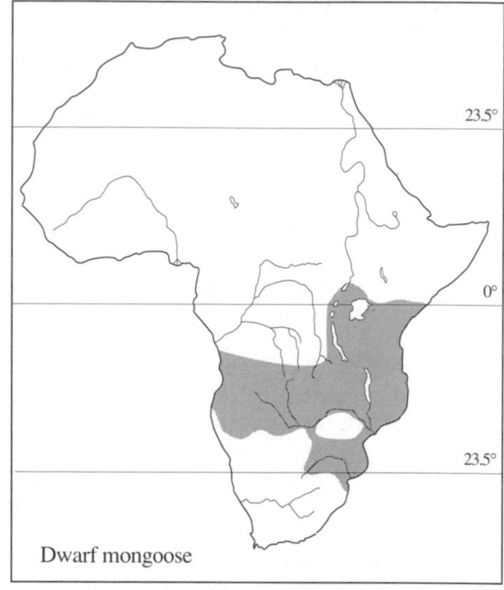

Dwarf Mongoose

Helogale parvula

TRAITS. Size and color of a red squirrel. *Weight and length:* males 267 g (227–341) and 20 cm (16–22.7), females 269 g (213–341) and 21 cm (18.5–23), tail 14–18 cm (Botswana series [8]); Serengeti males 326 g (265–415), nonpregnant females 315.5 g (221–395) (6). Domed forehead, rounded ears, short muzzle, cheek teeth adapted for insect diet; tail tapering to point; short-legged, 5 toes with long digging claws. Comparatively smooth-coated. *Coloration:* uniform, from reddish brown to grizzled gray and even glossy black (especially in Namibia) (7). *Scent glands:* well-developed anal pouch, cheek glands. *Mammae:* 4.

DISTRIBUTION. Southern Savanna and adjacent parts of South West and Somali-Masai Arid Zones, up to 2000 m.

ECOLOGY. In its preferred savanna habitat with numerous termite mounds for refuge, the dwarf mongoose may often be the most numerous mammalian carnivore, reaching a density of 31/km^2 in Serengeti N.P. (6). *Helogale* forages in thickets, around logs and trees, and in grass litter for adult and larval beetles (a mainstay), termites, crickets, and grasshoppers, also spiders, scorpions, some small vertebrates (lizards, snakes, rodents, birds), fruits, and vegetable foods. Although it drinks available water, it also inhabits waterless areas (8).

SOCIAL ORGANIZATION: diurnal, gregarious, territorial, communal care with related and unrelated pack members assisting dominant breeding pair.

Just before the birth season, the mean size of 66 Serengeti packs was 8.4 mongooses (2–18), including 2.4 adult males, 3.3 adult females, and 2.7 juveniles (6). One mated pair, usually the oldest male and female, are co-dominant and produce all the offspring. Packs include both related individuals and unrelated immigrants, all of which help care for and feed the young.

Such altruism toward unrelated young is unexpected. However, in this species immigrants of both sexes can usually become dominant and reproduce earlier than stay-at-homes, despite the higher risks of dispersing. The secret is to find another pack with fewer same-sex adults than the natal pack. The more conventional way to attain breeding status, pairing off and founding a new pack, is less fruitful for dwarf mongooses: suitable habitat is fully occupied by resident packs and there are few transients (mostly single males and occasional groups of 1–3).

Out of 31 individually known mongooses that achieved alpha status, 25 (12 males and 13 females) were immigrants, compared to only 6 (2 males, 4 females) that stayed home. Seven young females that transferred to other packs all became alpha and produced viable offspring at around 2 years of age, whereas only 3 that remained in the natal pack reproduced by the age of 4. The average tenure for mongooses that achieve alpha status is several years. Almost twice as many males as females leave home and they probably disperse farther (although the current record of 5 km is held by a female). Males suffer higher mortality on both counts, leading to a female-biased (3:2) adult sex ratio, especially among mongooses in their second and third year. Most mongooses transfer at that age, predominantly at the start of the rainy season. The information encoded in pack scent marks may help prospective immigrants decide where to try their luck.

Intrapack Relations. Dwarf-mongoose packs are highly cooperative societies with well-developed social roles and remarkably amicable intrapack relations (6). Pack members *nibble-groom* (fig. 19.25) and *scent-mark* one another, and jointly mark the home range, especially grass stems and bushes close to dens. The dominance of the alpha pair is rarely displayed except in reproductive activities (see under Sexual Behavior). However, there is a sinister side to *Helogale* reproduction: other adult females may mate and become pregnant at the same time as the alpha female, but either abort or lose their young after parturition. They then help suckle the alpha female's litter. It could be that alpha females practice infanticide (cf. wild dog), if only occasionally, for communal suckling is comparatively uncommon (see Parent/Offspring Behavior).

Other pack members help to feed the young by catching and bringing insects back to the den. They also guard, warm, groom, play with, and transport babies between dens, almost from the day the litter is born. Indeed, the mother spends less time tending her offspring than most other pack members (at least by day) and more time eating, enabling her to bear the added nutritional strain of lactation (cf. suricate).

The alpha female is also the pack leader, being usually the first to emerge and the first to set off foraging each morning; the alpha male is the most vigilant and protective pack member, backed up by subadult males (3). Most of his time at the den is spent atop the termite mound scanning for possible danger while the others are playing and grooming. In small packs without other adult helpers, he also fills the baby-sitting role. Alpha males have been known to make heroic efforts to rescue pack members: for instance, by tunneling to free individuals trapped by a cave-in (see also banded mongoose account) (6).

Interpack Relations. Each pack has its own territory. Of 7 packs that ranged a 2.2 km^2 study area, 2 were still using the same core areas and dens after 5 years. Territories averaged 34.6 ha (28–48) and had 10–30 termite mounds that served as dens. In tree-savanna habitat, territories can be as large as 160 ha. The ranges of adjacent packs may overlap slightly where mounds are close to the border and here packs sometimes meet and display hostility. Smaller packs avoid larger ones. If a large pack arrives to find a smaller pack already in possession of a mound, members of the superior force may rush at the holes with tails *bristling,* lunging and trying to bite their hidden antagonists. This is followed by intense *mutual gland-rubbing* and marking around the den. Interpack encounters generally end with each pack withdrawing toward the center of its own territory.

Association with Other Species. Despite similarity in their distribution and diet, and frequent presence in the same area, dwarf and banded mongooses do not jointly occupy burrows like suricates and yellow mongooses. Occasionally *Helogale* is forced to share an overnight refuge when a pack of banded mongooses moves into the same burrows too late for the smaller species to depart. But in general, *Helogale* can use tunnels much too small for *Mungos,* giving it a far larger choice of hiding places (Rood, pers. comm. 1986).

Dwarf mongooses form foraging communities with a number of insectivorous birds, such as the drongo, the white-headed shrike, lilac-breasted rollers, and 2 bush hornbills, *Tockus flavirostris* and *T. erythrorhynchus.* Although the first 3—and other—species also exploit other mammals as beaters to put up prey for them to catch, a true mutualism has been found between the dwarf mongoose and the 2 hornbills (3). In the arid thornbush scrub of eastern Kenya, hornbills regularly go foraging with mongoose packs, hopping along with them and eating much the same things, sometimes even competing with though subordinate to the mongooses. The mongooses benefit from the warning calls the hornbills give when they spot avian predators, which are numerous in the area and a menace to juvenile dwarf mongooses in particular. The pack posts fewer guards which give less frequent warning calls than usual when hornbills are present; evidently the mongooses are more secure and can devote more time to foraging when they have hornbills with them. The birds even give alarm calls for raptors that are predators on the mongooses but not on the hornbills. Although they catch and eat rodents of equal or larger size, the birds refrain from preying on baby mongooses.

The most tangible evidence of the mutualism between mammal and bird is the fact that each waits for the other before beginning the daily foraging. Every morning probably the same hornbills arrive at their pack's overnight lodging and wait over ½ hour for the mongooses to emerge, sunbathe, socialize, and move out. Sometimes when the mongooses are later than usual 1 or 2 birds actually call down their holes to wake them, or walk among sunbathers and try to lead them away—and these tactics work. For their part, if no birds are present, the mongooses pace back and forth nervously, delay departure, and often come running back to shelter.

ACTIVITY. Very like the banded mongoose (q.v. for details), the dwarf mongoose is perhaps even more influenced by weather conditions. Social and sexual activity occur mainly while relaxing at the den before and after foraging expeditions. Foraging mongooses move together in a spread and maintain vocal contact (see under Vocal Communication), but make more noise scratching and digging in dry leaves and litter (8). On cold nights they stack together front to back to share body heat (9).

FORAGING BEHAVIOR. *Helogale* gains most of its food by uncovering concealed invertebrates and does not stalk prey, but will pursue a moving quarry at a quick trot interspersed with little jumps, then bite or pin it with its forepaws (9). Prey that defends itself, say a grass rat standing erect, is pounced upon from the side and put out of action with head or neck bites and *death-shakes.* Larger food items are often carried to a protected spot before being devoured.

POSTURES AND LOCOMOTION. *Helogale* is an exceptionally active and agile little creature, possessed of the restless energy that goes with diminutive size and a rapid metabolism. It often moves at a fast trot, gallops in the typical *jump-run,* can bounce as high as 90 cm, can change direction in full flight, and climbs well for a mongoose. It is narrow and agile enough to get into the holes of small rodents (5, 7).

SENSES. Vision, hearing, and smell are all acute. The eye pupil is horizontally elongated, giving the dwarf mongoose an enlarged visual field, the better to spot aerial and other predators.

SOCIAL BEHAVIOR

COMMUNICATION

Vocal Communication: *peep* calls (contact and play), *twitter* (excitement), *tchee* and *tchrr* alarm calls.

The importance of vocal communication for maintaining contact and coordinating movements of pack members foraging over ground where visual contact is often obstructed has led to an extensive and complex repertoire of calls in this smallest mongoose (1). Most activities have a vocal accompaniment, and specific calls are associated with particular behaviors; in addition, calls intergrade, so that when one activity gradually changes to another, the vocal accompaniment changes along with it. In general, calling rate increases with the degree of activity. Thus the contact call exchanged by members of a foraging pack, a *peep* emitted at the rate of 1–3 times/second, increases and fuses into a *chur* when the pack begins a general movement, becoming the *"moving-out call"* (1). A variation of this is the *"panic twitter,"* which is uttered during a dash for cover: the call is repeated very rapidly (8–12 times/second) in short bursts in a harsher tone. The *"excitement twitter,"* which signals the discovery of an unusual stimulus (strange object or animal, another pack), is quite different, a rapid chittering noise with abrupt changes in frequency. When the stimulus is suspicious or potentially dangerous, the longer, frequency modulated *tchee* call is emitted, which changes to the harsher, noisier *tchrr* when a predator is sighted. Further variations in the warning call even signal the type and proximity of the predator: the nearer, the higher the noise component and the shorter the call (3). Three different calls are associated with play, 2 derived from the contact *peep* by accelerating the call rate, and 1 from the *excitement twitter.*

Babies keep up a continual *nest-chirp* and give high peeping calls when deserted. At 3 weeks they begin making adultlike contact calls, which change to purring interspersed with chirping when in body contact (contentment sounds); these juvenile sounds cease around 5 months (9).

Fig. 19.26. Dwarf mongoose *dominance mounting* (redrawn from Rood 1979).

Olfactory Communication: scent-marking with anal and cheek glands, communal latrines.

Dwarf mongooses rub both cheek and anal glands on objects, sex partners, and offspring. Marking with the everted anal pouch typically follows cheek-rubbing, and is done either in the usual *anal-drag* (often preceded by lifting 1 hindleg), or in a *handstand* with treading motions, depending on the object to be marked. Feces and urine are deposited in communal latrines.

AGONISTIC BEHAVIOR

Dominance/Threat Displays: *rearing,* growling, *hip-slamming,* mounting (fig. 19.26), inhibited and real biting, chasing with bristling tail.

Defensive Displays: growling, hissing, snapping and *head-darting, backing attack* with hair bristling, head and sometimes forequarters facing backward; lying on back, using claws and teeth (9).

Submissive Displays: *turning head during greeting, crouching* low during approach, often with one leg raised (intention to roll on side), *rolling on side.*

Fighting: *lunging,* biting, *chasing, backing attack.*

Other than by estrus, aggression is most commonly provoked by immigration, and is intrasexual. Joining may take as little as 2 days or as long as a month, and one large pack took in no transients. The beta male is the most actively antagonistic to male immigrants. Sometimes males that try to transfer are seriously wounded or killed (6). Subordinate males display submission most often, primarily to the alpha male and to the alpha female during estrus.

REPRODUCTION. Serengeti alpha females produce 2–3 litters yearly during the rainy season (October–May), reentering estrus within 2–4 weeks after giving birth. Litters averaged 2.9 young (1–6) (in 68 litters aged less than 6 weeks old [5]). The earliest known conception was at 15 months and the earliest intromission was by a yearling male.

SEXUAL BEHAVIOR. The labia become slightly red and swollen with the onset of estrus. Males sniff, lick, and mark the female and objects more often, and the alpha male's testes grow larger (5). The rank order becomes apparent as the alpha male keeps close to his mate and chases other males away. The pair engages in *mutual grooming* while *twittering* softly. When the female moves away, the male *follows nose-to-rump* and repeatedly places a foot on her hip as if to mount. She runs and he pursues. She stops and *crouches,* only to dart away as he comes up, or snaps at him to ward off mounting attempts, holding her tail out straight. When ready to copulate, she *crouches with hindquarters raised and tail deflected.* During a typical 4-day estrus, the male was seen to mount 119 times, of which 24 were complete copulations—assuming that *genital licking* by both partners after dismounting signifies ejaculation occurred (2). The median duration of mounts was only 21 seconds, whereas the longest copulation lasted nearly 11 minutes.

Often subordinate males also gain mating opportunities. Either the alpha male loses interest in the alpha female, or he consorts with subordinate females that have come into heat a day or 2 after the alpha female. Beta males are free to copulate with such females until then; however, these females may try to entice the alpha male by coming between him and his mate, then lifting a hindleg to invite anogenital licking or by mounting him. If the alpha male responds by mounting, the alpha female does not attack her rival but mounts him in turn, and while still in full estrus can make him dismount and follow her simply by moving away. Thereafter, presumably already inseminated, she will accept subordinate males (5).

PARENT/OFFSPRING BEHAVIOR. Only a small number of subordinate females that mate become visibly pregnant, that is, carry fetuses into the final fortnight of the 5-week gestation period. It happened in only 12 out of 63 litters born to the alpha females of 7 packs, in 5 of these packs. Meanwhile there were 42 females that were old enough to conceive (over 2 years). In addition to the 12 subordinate females that became pregnant, another 8 lactated without becoming visibly pregnant and helped suckle 6 litters. All 6 of the 8 lactating females whose ancestry was known were either daughters or sisters of the alpha female. In one case a younger sister successfully reared a litter orphaned at 2 weeks (5).

Although pack members sometimes carry outside and extensively groom day-old infants, the young first emerge voluntarily at about 3 weeks, and thereafter spend much of the time playing and exploring around the den. Litters are occasionally left unguarded for up to several hours, but 88%–92% of the time at least 1 pack member stands guard until the young are mature enough to leave the den at 5–6 weeks. In 32 days spent observing one litter, 3 or more mongooses were present 37% of the time, and half the time 1–3 baby-sitters (usually 1) remained on duty while the rest of the pack was off foraging (4). Pack members took turns, so that all had the chance to forage. Three different dens were used during this time, 310 and 160 m apart, and the young were transported (by the nape) between them on days 14 and 27. The moves were made for no apparent reason, seemed to be haphazard and confused, and exposed the young to danger. Once they were transported to all 3 dens on the same day. Subordinate males began the transfers but all pack members took part, sometimes in relays. In late afternoon the alpha female carried her babies back to the original den. Moves were exciting for the mongooses, as evinced by increased marking of one another and the new den sites.

On average, the alpha female spent only 67 minutes (8–217) per day with her young. Both related and unrelated mongooses spent more time baby-sitting, provisioning, and grooming the babies than the parents did. The mother suckled her offspring for 4 minutes 38 seconds (15 seconds–20 minutes) between days 17 and 25, usually in the (ventral) sleeping attitude, sometimes while lying on the side with one hindleg lifted. Weaning occurs between 6 and 7 weeks and no mongooses older than 55 days were observed nursing (4.)

ANTIPREDATOR BEHAVIOR: *low-sit* and *high-sit* alert attitudes, alerting and alarm calls, defense of young, flight to cover, warning signals of associated birds (see Association with Other Species).

Dwarf mongooses do not employ a *bunching strategy* (cf. banded mongoose) when a pack finds itself in danger out in the open. Being too small to bluff most predators, they simply run and hide. In defense of young at the den, however, pack members have been known to cooperate in chasing away potential predators such as slender mongooses and other, larger dwarf-mongoose packs (6).

References

1. Maier, Rasa, Scheich 1983. 2. Rasa 1977. 3. ——— 1983. 4. Rood 1978. 5. ——— 1980. 6. ——— 1983*a*. 7. Shortridge 1934. 8. Smithers 1971. 9. Zannier 1965.

Added References

Rasa 1986. Rood 1990.

Fig. 19.27. Banded mongoose breaking millipede against rocks (redrawn from Eisner and Davis 1967).

Banded Mongoose
Mungos mungo

TRAITS. A medium-sized, squat mongoose with stripes. *Weight and length:* 1–2 kg, head and body 33–40.6 cm, tail 15–28.5 cm, males slightly heavier than females (6). Head broad with pointed snout, wide, sharp-pointed cheek teeth adapted to eating insects. Short legs, feet 5-toed with long, slightly curved claws. Coat coarse and wiry with hardly any undercoat. *Coloration:* geographically variable, grizzled gray, back with distinct transverse banding, lower limbs and tail tip dark, underparts lighter. *Scent glands:* anal pouch and cheek glands (1). *Mammae:* 6.

DISTRIBUTION. Northern and Southern Savanna, extending into Somali-Masai and South West Arid Zones, replaced by suricate in drier parts of Botswana and Namibia.

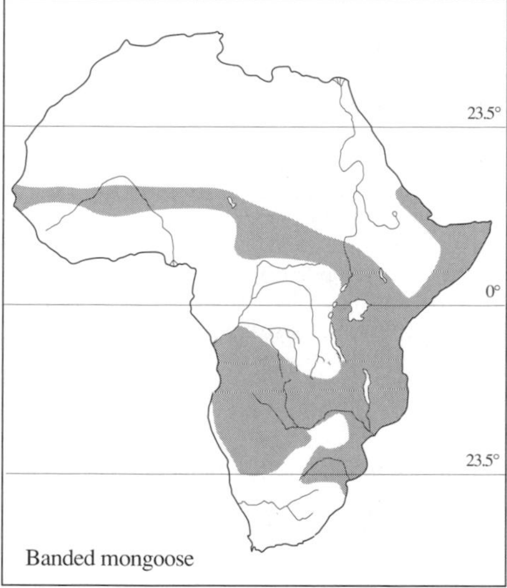

Banded mongoose

RELATIVES. The Gambian mongoose, *M. gambianus,* inhabits the Guinea Savanna of West Africa. It probably has a similar social organization and food habits. Though unbanded, this mongoose has perhaps the most conspicuous markings of any African mongoose: black-and-white diagonal stripes on the sides of the neck (1, 9).

ECOLOGY. The banded mongoose is associated with wooded savanna and termite mounds usually near (but sometimes distant from) water. It avoids forests but likes undergrowth, and rarely ventures far into open country, although packs range onto the Serengeti short-grass plains (3, 6, 8, 9). The diet consists mainly of beetles and their larvae, millipedes, earwigs, ants, crickets, termites, spiders, and the like. Vertebrate prey (mice, toads, nestling birds and eggs, lizards, snakes) accounts for probably less than 12% of the diet. Banded mongooses drink irregularly and sparingly (by lapping or by licking wetted forepaws (6).

SOCIAL ORGANIZATION: diurnal, highly gregarious, living in ± closed, territorial packs with several breeding adults.

This mongoose forms the largest groups, with up to 35 adults and subadults in a pack (6), and in some areas (Kruger N.P.) possibly twice that many (4). Instead of a single breeding pair, there are typically 3–4 breeding females and perhaps as many males. This holds true in large as well as small packs, suggesting that the dominance of breeding individuals somehow inhibits reproduction in low-ranking adults (see dwarf mongoose account). Leadership and dominance may be vested in the most senior pair (8). Female offspring may often remain in their natal pack, but males usually emigrate and attempt to join or take over another pack (8).

The mean number in 5 known packs living on Queen Elizabeth N.P.'s Mweya Peninsula was 14–17 (range 6–35) (6). Here there were up to 19 mongooses/km^2. The adult sex ratio was approximately equal, and the mean home range was 80 ha (38–130). The 2 biggest packs occupied the largest (>1 km^2), openest ranges.

Over half of 144 investigated den sites were in thickets and mostly in termite mounds, 21% were in erosion gullies, 15% were in bare termite mounds (but with cover nearby), 11% were holes in the open, and 3% were man-made (e.g., a bone pile). Favored den sites were situated on slopes and had 1–9 entrances. A den excavated after being used for breeding had 3 entrances, with tunnels of c. 9 cm diameter leading 135–210 cm to central chambers 150 by 90 cm and up to 50 cm high. Two side tunnels led to chambers big enough for a few mongooses. Latrine areas were found in a side chamber and 2 tunnels, evidently supplementing the latrines just outside the den (3). Dens are often used just once for only a few days, even when there are small offspring, but customs vary in different packs. A favored site may be revisited often, or continuously occupied for up to 2 months (3, 6).

Although females have been known to dominate males, at least in captivity (8, 11), there is no clear indication that either sex dominates. Rather, rank seems to be determined by age and by individual traits such as temperament (2). Males are more active in scent-marking (see below) and more aggressive toward other packs (6). Each pack is a closed and highly integrated social unit. In the Uganda study, no immigration was seen in over 3 years; all recruitment of members came from within. However, some of the packs split, apparently voluntarily, and males singly and in small groups were occasionally seen following and trying to join other packs, suggesting that males, anyway, sometimes transfer.

Group cohesion is maintained through *mutual scent-marking,* social grooming, and vocalizations. Banded mongooses not only mark the ground and novel objects (especially around dens) with their anal glands, but also one another (cf. yellow and dwarf mongooses). The frequency and intensity of marking reflect the degree of excitement. Typically, one pack member invites another to mark by lying on its side and showing its striped back, which the other then rubs with its anal pouch; this behavior may be or derive from submissive-dominance behavior. At highest intensity, as when confronting a predator, pack members mark one another feverishly while writhing and squirming in a mass (see Antipredator Behavior) (2). Also, pack members may take turns rubbing their cheeks, throats, and bellies on objects such as small stones or their own feces that other pack members have marked, thereby anointing themselves with a strong communal odor, which may serve both to identify and to give confidence to the group.

Packs tend to be mutually antagonistic, even sections of a pack that has split into 2. Since each pack stays within its own home range, meetings are infrequent, and the usual reaction when neighboring packs see each other is mutual withdrawal from the border. However, there is often some range overlap, typically of overnight refuges located near a common boundary (cf. dwarf mongoose). Meetings occur by chance when 2 groups head for the same den on the same evening. The bigger group then chases the smaller group. Equally matched packs may fight (6). Encounters feature loud screeching and angry churring, frequent scent-marking and chasing, and occasional grappling and biting. After approaching in close formation like rival gangs, pack members skirmish individually and in small groups. They may become scattered over a wide area and so preoccupied as to be oblivious to possible danger from other animals. Two encounters between large packs of 28–29 only ended with darkness, after 1½ and 2 hours. Several estrous females were present in one pack and males from the other attempted copulation, at least once successfully despite the determined defense of their pack males (6). (Unanswered question: Did the females resist?)

ACTIVITY. After spending the night in an underground burrow huddling together for warmth, a pack emerges typically about an hour after dawn. First one, then another pokes its head out of a hole, looks about, and sniffs the air. If satisfied, they come out, relieve themselves at a communal latrine, and gather in groups to scratch, *nibble-groom* themselves and one another, scent-mark, play, and socialize for c. 15 minutes (1–95) before beginning the day's foraging (6). After 2–3 hours of intense feeding, the pack takes a rest break. On hot days the pack seeks deep shade in a thicket or gully and may rest, with brief periods of individual local foraging, until late afternoon (1600 h). Rain may delay departure from the den, and afterward send the pack down the nearest holes. The mongooses return to the den usually within ½ hour of sunset and they socialize again before retiring at dusk.

The activity pattern of banded mongooses with young of less than 5 weeks old is different. The pack returns to the den after foraging 2–4 hours and thereafter the animals forage only intermittently in small groups. Usually 1 adult male remains at the den. Adult males were the guards in 42 of 57 observed cases, and 77% of the times only 1 adult guarded. Occasionally nonlactating females took the morning shift, and sometimes a subadult kept an adult company.

FORAGING BEHAVIOR. A traveling pack goes quickly in an undulating file or column, bunching and hurrying most when forced to cross open spaces (cf. elephants). Following a zigzag course, The Ugandan packs covered 2–3 km (1–4) a day; in the more open savanna of the Serengeti and Kenya Mara, packs range up to 10 km a day (6, 8). A senior female is usually the leader. Juveniles, which play more and forage less, often lag in the rear (10). Foraging mongooses spread out over an area usually less than 60 by 40 m and advance slowly to the accompaniment of continuous *churrs, chirps,* and *twitters.* They scratch up litter like a flock of chickens, poking their noses into every opening, turning over stones and dung. The front claws are used to extract prey from openings too narrow for their snouts and to dig up millipedes, beetle larvae, and such sniffed out underground. Earwigs, caterpillars, millipedes, big spiders, toads, scorpions, slugs, and snails are given a shake, then repeatedly pawed and rolled with the front feet to remove stings, bristles, or noxious secretions. Large millipedes and beetles, dung balls, snails, and large eggs may be smashed by catapulting them between the hindlegs (fig. 19.27). Each mongoose resists attempts to share treasures such as an ant nest or a large beetle, jumping at thieves with angry squeals and growls. But since any individual that uncovers a bonanza can't refrain from uttering excited *twitters* and *churrs,* food sharing is unconsciously promoted (as in hyenas). Thus, all 15 members of a pack homed in on a pile of elephant dung loaded with dung beetles after the mongoose that found them blurted the news (6).

POSTURES AND LOCOMOTION. See family introduction.

SOCIAL BEHAVIOR

COMMUNICATION

Vocal Communication: churring, grading from a low *grunt* to a high *twitter* (contact call); a strident *churr* or *chitter* (alarm signal); a low growl, explosive *chattering* or *squealing* (threat and anger, cf. suricate *clucking*); *whining* (a version of the contact call, uttered during courtship); low *purring* (contentment).

The repertoire of young mongooses is similar but higher-pitched (3, 12).

Olfactory Communication. Banded mongooses mark the ground and objects in an *anal-drag.* At highest intensity, the rear legs are raised and the full weight rests on the rump. The whitish musk is deposited in small quantities, primarily around dens and also at water holes, where mongooses like to rub and roll in damp soil (possibly to enhance their odor, especially in dry weather) (2). Males reportedly squirt a yellowish secretion (urine?) over small young (12), and individuals also squirt objects while *treading* (2).

AGONISTIC BEHAVIOR

Dominance/Threat Displays: standing with foreleg over subordinate's shoulder and *biting the nape* (which may lead into friendly *nibble-grooming*); *threat gaping; sidling stiff-legged around opponent* with weight on hindlegs, hair *bristling; ritualized chasing: following* fleeing opponent in a smooth, *scurrying run, tail dragging* along ground and rustling in the litter (9).

Defensive/Submissive Displays: sudden *short rushes,* ending with *explosive spitting, backing attack; crouching, lying on side* (infrequent).

Displacement Activities: digging and scent-marking (2, 10).

REPRODUCTION. Reproduction is synchronized within but not between packs. Nevertheless, most births occur during the rains. Several females come into estrus at once, usually within 1 week after littering, and mate with several different males during a 6-day estrus (2, 5). Females may begin breeding at c. 11 months and have up to 4 young. Gestation is about 2 months and a pack may produce up to 4 litters a year (5). Probably under half of all juveniles survive the first 3 months.

SEXUAL BEHAVIOR. Vigorous chasing and play mark the early stages of courtship. The males' anal glands secrete copiously and may become enlarged. A courting male holds his tail raised and coats the female with secretion between mounting attempts, while *circling, nuzzling,* and *pushing* with his head. The female also marks her suitors. During a 10-minute copulation, the male clasps the female's waist and touches her neck with open mouth (3).

PARENT/OFFSPRING BEHAVIOR. The mean number of lactating females in the 5 Mweya Peninsula packs was 2.4; 4 females with a total of 12 pups was the maximum (5). As in lions, the young are suckled communally, and all pack members play with, groom, and transport (with head or skin grip) the babies. Baby-sitting frees mothers to go foraging, which they invariably do in the mornings. Judging from captive animals, males may also play a major role in socializing and training the young to forage. During the second month, one father played twice as often with the juveniles as the mother, actively soliciting play by lying down and head-rolling or scratching at objects; later he stimulated the kittens to hunt by scratching at a food source, or by holding food in his mouth while *rolling around squeaking* and *churring* (11).

Banded mongooses' eyes open by the ninth day, and beginning at 3–4 weeks they leave the den on short afternoon foraging excursions. By 5 weeks, pups accompany the pack when it leaves the den each morning, and by 6 weeks they have acquired their adult coats.

ANTIPREDATOR BEHAVIOR: *low-sit* and *high-sit* alert posture, hair sleeked; *alarm chirping,* rushing to cover, *mobbing attack.*

The mobbing attacks of banded mongooses are so impressive that even animals as formidable as servals and large dogs may be intimidated, not to mention lesser foes such as jackals, eagles, vultures, and herons. Packs have been known to mob bushbucks, geese, and other harmless creatures somehow perceived as potentially dangerous. The combination of close *clustering* and *writhing* while advancing with rearing heads (as different mongooses stand up and peer or threaten), and while *growling, churring, spitting,* and *snapping,* can suggest the approach of a single large and relentless adversary (2). When alarmed away from cover by a member's *churring,* a bird's alarm call, or an antelope's snort, the pack immediately clusters, with the young in the center. Mutual defense extends even to the rescue of members captured by predators. Thus, an alpha male climbed a tree where a martial eagle had alighted to eat an adult male it had caught and made the eagle drop its prey, which survived the ordeal (7).

References

1. Ewer 1973. 2. Kingdon 1977. 3. Neal 1970. 4. Pienaar 1964. 5. Rood 1974. 6. ——— 1975. 7. ——— 1983*b*. 8. ——— 1986. 9. Rosevear 1974. 10. Simpson 1964. 11. Smithers 1971. 12. Viljoen 1980.

Fig. 19.28. Suricates *grappling* during a playfight.

Suricate or Meerkat

Suricata suricatta

TRAITS. A distinctive-looking mongoose with dark eye patches. *Weight and length:* males 731 g (626–797), head and body 28 cm (24.5–29), tail 22 cm (20.5–24); females 720 g (620–969), head and body 26.4 cm (26–28.5), tail 22.7 cm (19–23) (7). Slender build with long thin legs, 4 toes with long claws; thin, pointed tail; rounded forehead and small, crescent-shaped ears, pointed face, broad molar teeth with sharp cusps. Coarse coat with long guard hairs. *Coloration:* grizzled gray or tan with buff to yellowish underparts, black eye patches, ears, and tail tip, and dark interrupted banding on back. *Scent glands:* anal pouch. *Mammae:* 6.

DISTRIBUTION. South West Arid Zone and adjacent Southern Savanna, the Highveld and Karroo.

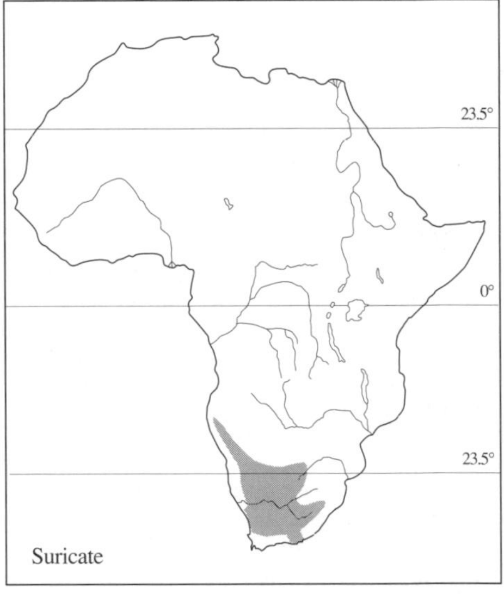

ECOLOGY. Found in savanna and open plains, *Suricata* inhabits the driest and openest country of all mongooses. It is particularly associated with firm-to-hard calcareous ground such as that occurring around alkaline pans, also the stony banks of dry watercourses, and rarely with sandy soil (3, 7). Insectivorous diggers, suricates concentrate on beetle larvae, the larvae and pupae of moths, butterflies, and flies, and harvester termites, crickets, spiders, scorpions, and other invertebrates that are buried or hidden. Occasionally they eat lizards, small snakes, birds, and mice. In waterless areas they may obtain water by chewing tsama melons and digging up roots and tubers (2). Suricate colonies often share quarters with the social ground squirrel (*Xerus inaurus*) and the yellow mongoose (2).

SOCIAL ORGANIZATION: diurnal, highly gregarious, territorial(?).

Although suricate behavior in captivity is well-known (1, 2, 9), the full results of a definitive study in the wild of this highly social and visible mongoose have yet to be published (4, 5). As in the banded mongoose, reproduction is not monopolized by an alpha pair; a pack may have several breeding females and males (6). Pack size ranges from a few up to 30 suricates, with selection (by predators and interpack competition) favoring larger over smaller packs (5). The average of 10 Botswana packs was 15 mongooses (8–30) (4), and of 16 packs on farms in the Orange Free State (OFS) was 10.6 (2–17) (3). The mean composition of 8 OFS packs averaging 11.3 suricates (9–17), whose burrows were completely excavated, was 5.4 adult males (2–9) and 4 females (1–7), 0.75 juveniles (0–2), and 1.1 subadults (0–4).

In the Kalahari Gemsbok N.P. an intensively studied pack of 12 suricates ranged an area of 15½ km^2 (5). Although the most muscular male played a prominent role in aggressive interactions, and another male was most vigilant, no indication of a linear rank hierarchy was noted; rather, the pack members seemed singularly amicable with one another while being ferociously hostile to other packs. Nevertheless, a female outsider managed to gain acceptance and was immediately pressed into baby-sitting by the mother of new babies (5). Males may routinely emigrate from their natal pack and attempt to join/take over another pack (6) (cf. banded mongoose).

A home range typically includes some five warrens spaced 50–100 m apart, which are occupied in rotation for months, even years, at a time (8). Suricate dens tend to be extensive and complex. The average warren is about 5 m across with ±15 entrances, ranging from simple burrows with a few holes up to labyrinths 25 by 32 m across with 90 entrances (3, 5). The sites are slightly elevated, even on level plains, due to the accumulation of excavated soil. Suricates can and do dig and extend tunnels when the ground is soft after rain, but usually they occupy existing burrows, especially ones prepared by ground squirrels. Tunnels run at two or three different levels, as deep as 1.5–2 m. Most interconnect and have chambers c. 30 cm high by 15 cm long at intervals (see fig. 19.23). Temperature recordings from the surface to 1.2 m revealed a daily variation in the deeper tunnels of less than 1°C. The average daily minimum in winter was 10° and the average maximum in summer was 23°, compared to −4° and 38.5° outside, with daily fluctuations of 29° and 17°, respectively. Thus suricates, like other small desert animals, use underground shelters to avoid climatic extremes.

ACTIVITY. Suricates emerge from the den well after sunrise. One by one, first just the nose followed by the head and shoulders, each checks carefully before coming out and sitting up to continue scanning. Finally the whole colony is clustered around the entrances sunbathing, *social-grooming*, playing, and so on. Eventually one drops to all fours and begins moving about, whereupon the rest fall in and set off for the day's foraging (7).

Like most social mongooses, suricates spread out and forage individually while maintaining visual and vocal contact. An irresistible urge to follow and catch up with any pack member that moves away, first noted in pet suricates, functions to keep individuals from straggling (2). Having caught up, laggards then resume foraging. This "catch-up pattern" replaces the juvenile following response (see under Parent/Offspring Behavior).

FORAGING BEHAVIOR. A pack thoroughly and systematically forages within its home range, taking a different route every day and usually allowing at least a week for an area to renew its food supply between visits (5). Suricates continually use their olfactory sense to locate concealed prey, then dig it out with rapid movements of their long-clawed forefeet. The pointed snout is thrust into the narrow trenches it excavates to grasp unearthed beetle larvae and the like. Even under artificial conditions when there is no reward, scratching, digging, and probing are performed automatically, for their own sake (2). If large or unfamiliar prey is encountered during foraging, the suricate batters it with the foreclaws before biting it tentatively. Then it bites hard and gives the *death-shake*. Birds, lizards, and mice are pulled to pieces by clutching the prey with the claws close to the suricate's mouth and tearing downward while holding the head up. Only three innate responses to active prey have been noted: a tendency to chase any small fleeing object, to bite it at the most actively moving part, and to eat hairy prey (i.e. mammals) starting at the head. The throwing response, whereby hard-shelled prey and eggs are broken, is apparently lacking (2).

Pet suricates exhibit "food envy" to a marked degree, perhaps as an artifact of the type of food and unnatural proximity while feeding. Not only will they not share, but each tries to snatch food from the other, even when not hungry (2). In the wild, however, all adults readily share food with juvenile pack members (5).

POSTURES AND LOCOMOTION. *Suricata* is well-designed for digging, for moving through tunnels, and for sitting and standing erect, but not for climbing (5) or running—even a man can outrun one. It walks on its fingers and toes (digitigrade) and has two usual gaits, a walk and the *jump-run.* The profile of a walking suricate, with head low, tail trailing, and hindquarters higher than forequarters, is distinctive (7). In addition to the *low-sit* and *high-sit* postures, this species adopts a comical-looking *lazy-sit posture,* slumping on its spine with tail forward and forelegs dangling. This converts to the sleeping attitude when the suricate curls forward with its nose between its hindlegs (*ventral-curl*).

Most of the seated and lying postures adopted by the suricate concern the relation of its thinly haired underside to the environment (cf. yellow mongoose) (2). Early on cold mornings a pack *high-sits* outside the burrow looking like little old men, exposing their fronts to the sun's rays. Later they bask while sprawled on their backs or twisted to one side. To absorb the heat of warm rock—or to contact a cool surface on hot days—they lie spread-eagled. At night pack members sleep in seated groups, curled over one another, making a pile of bodies.

SENSES. Vision is acute enough to identify distant specks as hawks, but poor in dim light, suggesting a pure cone eye and possible color vision (1). *Suricata* also has a phenomenal sense of smell, but its hearing and ability to pinpoint sounds are no better than ours (2).

SOCIAL BEHAVIOR

COMMUNICATION

Vocal Communication. Almost any activity is accompanied by noises that sound rather like murmured human conversation. Females murmur most; the adult male of the captive group usually kept quiet (2). Other, more distinctive sounds are specific to different circumstances.

(*a*) Growling and explosive *spitting:* offensive threat.

(*b*) *Clucking,* a scolding, "cross" noise: defensive threat, given, for example, by a mother resisting attempts to nurse, by young picked up against their will, or by one adult disturbed by another.

(*c*) A clear, drawn-out note of varying intensity: alerting call given in response to avian predators. A mongoose on guard duty gives uneasy *peeping* calls (5).

(*d*) *Waauk-waauk:* a gruffer, more abrupt warning call given to ground predators, possibly expressing threat plus alarm.

(*e*) *Alarm bark:* a repetitive, short, sharp call given in response to various disturbances, especially noises, expressing defiance and given perhaps primarily by dominant males. A high-pitched version is a juvenile distress call.

(*f*) *Wurruck-wurruck:* a soft, trilling sound expressing contentment.

(*g*) Bird-like chirps: *nest-chirp* of newborn young sustained while awake, diminishing as adult calls develop in the first months.

(*h*) Cry of pain: lacking in this species (2).

Olfactory Communication: scent-marking with anal pouch and urine, communal latrines.

Males often (females seldom?) mark vertical and flat surfaces with strong-smelling anal-gland secretion, especially around den entrances. The animals usually sniff before marking and afterward may rub the body along the same spot. One hindleg is often raised slightly during the *anal-drag.* A pregnant female has been observed marking in the same way (and the heaviest female anal glands were collected from pregnant individuals) (1). The use of communal latrines may originate in the tendency of littermates, beginning in the third month, to use places where other suricates have excreted. Like dogs, males urinate against vertical objects with the near leg cocked.

AGONISTIC BEHAVIOR

Threat Displays: *bristling, bounding in place,* charging, calling (see also Antipredator Behavior).

Fighting: *backing attack, inhibited biting* (see also Play).

Except when a suricate kills prey, the attack tendency is always inhibited to some extent, even against enemies of different species. Fighting between troop members is exceptionally ritualized and harmless (2) but is often fierce during encounters between different packs (5). The *backing attack* (fig. 19.7), used only defensively by the banded mongoose and *Herpestes* species, is used offensively, too, by the suricate.

REPRODUCTION. Suricates protected from climatic influences produce litter after litter: one female reproduced 11 times in 31 months (3). The season is also extended in the wild, but nearly all young are born during the warmer, rainier part of the year. Females may have up to 3 litters, averaging 3 pups, after an 11-week gestation.

SEXUAL BEHAVIOR. Undescribed in the wild. Sexual activity of observed pet

Fig. 19.29. Suricate mounting posture.

suricates usually began with bouts of semi-serious fighting, when one gripped its partner firmly by the muzzle (2). Sometimes the female provoked it by nipping at the male's cheeks, which led him to attempt mounting (fig. 19.29). If the female resisted mounting, the male would grip her by the nape, which induced passivity just as in babies being carried. During copulation the male maintains his position by clasping the female's middle, without the *neck grip.*

PARENT/OFFSPRING BEHAVIOR. As in the banded mongoose, nonbreeding helpers of both sexes guard, and also provision, suricate young. A mother does neither until after weaning but keeps foraging in order to sustain an adequate milk supply (4). Her interactions with her offspring are not competitive or strenuous (2, 7). She has to lick the perineal region to stimulate elimination of wastes in small babies, and also cleans them by licking the belly and face. She suckles in the *ventral curl* until the young grow large, after which she lies flat on her back. The young suck all 6 teats at random, occasionally *milk-treading* and purring. Babies are carried by the nape or by the back.

During weaning, beginning at 3–4 weeks and ending at 6–9 weeks, mothers exploit their babies' newly awakened food envy to teach them to forage for themselves (part of the paternal role in the banded mongoose). Taking food in her mouth, one mother would run back and forth with it before her young ones (2). They responded by snatching it from her and devouring it. If the young were unresponsive, the mother would eventually lay it down uneaten. This behavior, which persisted to the end of the weaning period, coincided with a strong following tendency in the juveniles and a distinctive contact call given only by young suricates, high-pitched and repetitive, which functioned to keep them grouped closely (cf. banded mongoose). These behaviors waned, along with the mother's generosity, as the youngsters became nutritionally self-sufficient. However, alarm calls by either parent would send the juveniles running to their mother and they would remain pressed close to her (fig. 19.30) as she made her way cautiously toward safety (2).

Fig. 19.30. Suricate mother guarding young after predator alarm (redrawn from Ewer 1973).

PLAY. All pack members often engage in acrobatic sparring during rest periods (5). Young animals of the same size practice fighting techniques: *grappling* (fig. 19.28) (standing and clasping each other with forelimbs), *wrestling* (one animal lying on its side or back [fig. 19.31], the other standing over it with the forefeet on it or on the ground), biting, pawing, leaning against the opponent, and so on (9). Attack from ambush, "mole play" (crawling into dark openings), *displaced digging,* and *stiff-legged rocking* (antipredator threat) are also seen in play (2).

ANTIPREDATOR BEHAVIOR: alarm calls, *high-sit* alert stance, flight to cover, defensive threat, *mobbing attack,* self-defense, covering young.

Fig. 19.31. *Back defense* adopted by juvenile suricate in playfight.

In keeping with their poorly developed running, jumping, and climbing ability and preference for open habitat, suricates have evolved elaborate intimidation/threat displays against predators. The suricate's antipredator and rival-suricate-pack defense turns it from a long, low animal into an almost spherical object of far greater size with an alarming demeanor (2, 5). The hair bristles, the legs are extended and back arched, the tail is stiffly erected, and the head slightly lowered. Next the suricate approaches or seems to approach the enemy in stiff-legged high bounds, but in fact rocks back and forth in nearly the same place. Meanwhile it utters menacing growls while *head-darting* and *spitting* like a cat, with increasing intensity if the enemy comes closer. If the suricate is attacked despite this bluff, it lies on its back with all weapons presented and the nape of its neck protected.

Toward hawks and eagles, which are the suricate's main enemies, it maintains constant vigilance, sounding the alarm when one is sighted and fleeing for cover if an attack seems imminent (2). A pack surprised in the open by a goshawk covered their juvenile offspring with their own bodies (5).

References

1. Dücker 1962. 2. Ewer 1963. 3. Lynch 1980. 4. Macdonald 1984*b*. 5. ——— 1986. 6. Rood 1986. 7. Smithers 1971. 8. Synman 1940. 9. Wemmer and Fleming 1974.

Chapter 20

Hyenas and Aardwolf

Family Hyaenidae

Hyaena hyaena, striped hyena
H. brunnea, brown hyena
Crocuta crocuta, spotted hyena
Proteles cristatus, aardwolf

TRAITS. Doglike carnivores of medium to large size, 8–85 kg, sexes approximately equal except spotted hyena (females bigger). Skulls massive with prominent sagittal crest for attachment of exceptionally powerful jaw muscles, hyenas with 32–34 robust teeth, the conical, blunt premolars specialized for cracking bones; cheek teeth reduced to pegs in aardwolf. Ears large and pointed or round (spotted hyena), tail ¼–½ body length; long neck and high shoulders (45–90 cm) sloping to hindquarters, legs long and muscular, feet with 4 toes (5 in aardwolf forefeet). Coat short and sparse (spotted hyena) to long and shaggy with erectile mane and bushy tail. *Coloration:* buff to dark brown, with dark stripes or spots, cubs black (spotted hyena) or like adults. *Genitals:* boneless penis similar in all 3 genera (see fig. 20.10), 11–20 cm long, the basal portion of the glans covered with backwardly pointed spines; females with conventional genitalia except for spotted hyena, which has a phallus exactly like the male's. *Scent glands:* a large anal pouch containing a copious sebaceous paste secreted by the anal sacs (fig. 20.1). *Mammae:* 2–6.

Senses: all well-developed, with the sense of smell preeminent.

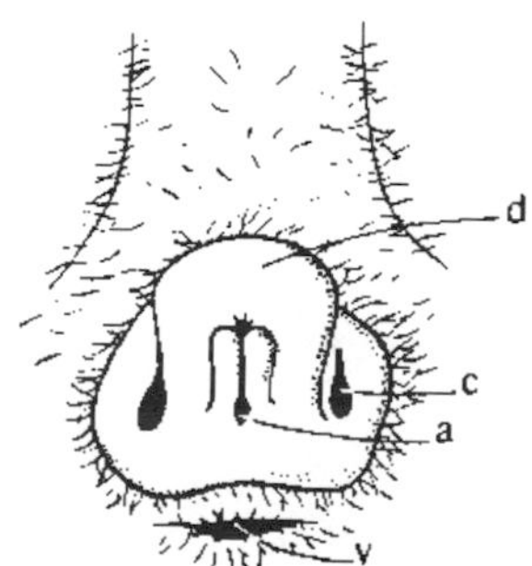

Fig. 20.1. Anal pouch of a striped hyena, shown protruded: *d*, protrusion of the everted pouch; *c*, crypt containing opening of anal sac; *a*, anus; *v*, vulva. (From Ewer 1973, fig. 3.11.)

DISTRIBUTION. Arid and savanna zones of sub-Saharan Africa; the striped hyena also ranges North Africa and Asia. The brown hyena of southern Africa has the most limited distribution. The spotted hyena is less adapted to arid conditions, but inhabits the greatest variety of habitats and is numerically Africa's dominant large carnivore.

ANCESTRY. The latest and smallest carnivore family, hyenas probably arose from civets (*Progenetta*) and the earliest known fossils (from Eurasia) date from the late Miocene (10 million years ago) (9). Spotted and striped hyenas divided into 2 separate lines at the outset, then evolved so rapidly that hyenas much like recent species appeared by the early Pliocene (fig. 18.1). The opportunity to scavenge the kills of saber-toothed cats, which were poorly equipped to eat the skin and bones of their large prey, may have fostered the hyenas' outstanding ability to eat and digest these parts (2). Whether the family arose in Africa or Eurasia is now in doubt, but continental interchange of hyaenids occurred both in the Miocene and succeeding epochs. Maximum diversity was achieved early in the Pleistocene with 4 genera and 9 species, including the 3 surviving hyenas. In Eurasia, *Crocuta* diversified into many different species before becoming extinct in the late Pleistocene; in Africa there was and is only 1 spotted hyena (1, 9, 10). Despite an absence of fossil clues to the origin of the aardwolf, its appearance, anatomy, and behavior all testify that it is a hyena transformed into an insectivore (11).

ECOLOGY. The exceptionally powerful jaws and teeth of hyenas, combined with large size and a digestive system capable of dissolving bones and even teeth within hours, give these animals the ability to utilize the remains of large vertebrates more completely than other carnivores. Hair is practically the only animal product they cannot digest; alone among carnivores, hyenas have solved the problem by disgorging hairballs (like owls), along with hooves, pieces of horns and bones, and grass. As predators, more hyenas can subsist on fewer prey animals than their competitors can, and as scavengers they can gain nourishment from what other predators leave uneaten. The high proportion of bone in the diet (especially *Crocuta*) may largely explain their ability to milkfeed their young for a year or longer. Striped and brown hyenas cache surplus food in thickets, whereas spotted hyenas sometimes cache parts of a kill underwater (5).

The spotted hyena is the biggest, most carnivorous, and most advanced living hyena, with the greatest ability to kill and consume large prey. The striped hyena is the least specialized and most omnivorous, retaining a relatively primitive dentition, and the brown hyena is intermediate (2, 4). The aardwolf is one of the most specialized carnivores. The ecology of the different species is compared in table 20.1.

SOCIAL ORGANIZATION. The spotted hyena was long considered the only gregarious member of this family. But longtime studies of the brown hyena revealed that it, too, is sociable (7, 8), and preliminary evidence suggests the other *Hyaena* has a similar organization. That leaves the aardwolf as the only truly solitary hyaenid. The social systems of the 4 species are compared in table 20.2.

But hyena social organization is notable for its flexibility, varying within species according to habitat and the type, abundance, and dispersion of the main food resource. The spotted hyena is often a solitary forager and lives in small groups (3–12) where there are few large mammals. Wherever possible, though, it actively hunts medium and large prey, often in packs, and where ungulates abound, it forms the largest social groups of any terrestrial carnivore. Clan members feed together but do not cooperate in caring for offspring. Cubs are totally dependent on their mother's milk

Table 20.1 Ecological Separation of the Hyenas
– absent (no), + present (yes), ++ highly developed

	Striped hyena	Brown hyena	Spotted hyena	Aardwolf
Climatic/vegetation tolerance				
Desert–semidesert	+	+	–	+
Semidesert–savanna	+	+	+	+
Montane forest	–	–	+	–
Forest-savanna mosaic	–	–	+	–
Habitat				
Open plains–tree grassland	–	–	+	+
Bush savanna	+	+	+	–
Semidesert scrub	+	+	+	–
Woodlands	–	–	+	–
Diet [(s)cavenger; (p)redator]				
Large mammals (topi and bigger)	s	s	p/s	–
Medium mammals (steenbok–impala)	s	s	p/s	–
Small vertebrates (smaller than steenbok)	s/p	s/p	s	–
Insects	+	+	–	++
Fruits and vegetable matter	+	+	–	–
Caches surplus food or carries to den	++	++	+	–
Activity				
Nocturnal	+	+	+	+
Rests between 2 peaks	+	+	+	+
Diurnal				
Late afternoon	+	+	+	+
Early morning	+	+	+	–
Other times	–	–	+	–
Use of dens				
As refuge for cubs	+	+	+	+
By adults	often	often	rarely	always

Table 20.2 Hyena Social Organization
− no, + yes, ? unknown or uncertain

	Striped hyena	Brown hyena	Spotted hyena	Aardwolf
Solitary	−(+?)	−	−	+
Family: male, female plus cubs and subadult offspring	+	+	−	−
Parents and subadults bring food to cubs	+	+	−	−
Clan: multimale, multifemale plus cubs of different ages	?	+	+	−
[(*a*) small clans only (<15) (*b*) large clans]	(*a*?)	*a*	*b*	−
Communal denning (with cubs of more than 1 mother)	(*a*?)	+	+	−
Communal suckling	?	+	−	−
Communal feeding	?	+	−	−
Food-sharing (large carcasses only)	+	+	+	−
Among a few only	+	+	−	−
Among many	−	−	+	−
Foraging/hunting				
Solitary	+	+	+	+
In packs	−	−	+	−
Territorial (at high, not necessarily low, density)	+(?)	+	+	+
Group defense (patrols, marking, aggression)	−	−	+	−

until developed enough to travel to kills. It is basically a competitive society; competition accounts for female dominance over males and the accompanying masculinization of the genitals that sets *Crocuta* apart from other carnivores.

The other 2 hyenas are strictly solitary foragers and scavenge but rarely hunt medium and large ungulates. No more than 2 or 3 animals eat from the same carcass at once. However, clan members, most of which are closely related, help provision cubs by carrying food back to a central den, to which cubs remain attached for more than a year. Female brown hyenas may have and indiscriminately suckle young of different ages (8). In the spotted hyena, females stay home while males emigrate during adolescence—the typical mammalian pattern. Resident adult males are unrelated immigrants and the dominant one sires most of the offspring (see species account). In the brown hyena, male as well as female offspring often remain resident as adults, and the males do not attempt to mate with clan females but defer to nomadic males (7).

Territorial defense against neighbors of the same sex has been observed in all species but the striped hyena, which remains to be closely studied. Only *Crocuta* scent-marks and skirmishes in groups. However, intense and frequent territorial behavior is practiced only at high density where ranges have a common border (see spotted and brown hyena accounts).

ACTIVITY. Three of the 4 species are quite strictly nocturnal. The spotted hyena, though primarily nocturnal, is often seen and sometimes active in the daytime.

POSTURES AND LOCOMOTION. Sleeping postures are canidlike: curled to one side nose to tail, lying prone with muzzle on forepaws and hindlegs both to one side or (especially *Crocuta*) stretched backward. Hyenas squat to urinate and defecate like most carnivores, with tail straight up (and often bristling). A similar posture is adopted while scent-marking (*pasting*) (fig. 20.2).

Hyaenids lick, shake, scratch, and stretch in typical carnivore fashion. A hyena licking its genitals sits like a cat, with 1 hindleg raised high. The long neck, massive, high shoulders, and rounded hindquarters make a walking or seated spotted hyena look bearlike.

Despite these and other resemblances to different carnivores, hyenas are unique in ways that are readily seen but hard to describe—for instance, the look of a fearful hyena glancing over its shoulder or peering intently with raised head, lowered hindquarters, and clamped tail. Kneeling and

Fig. 20.2. Brown hyena *pasting* anal-pouch secretion on grass stem.

crawling on the wrists (*carpal-crawling*), and lying with forepaws folded under are other distinctive family traits. The lower hindlegs look peculiarly loose-jointed in a galloping hyena, as the hindfeet are thrown out sideways during the recovery phase (fig. 20.3) (4). Anatomical differences of the spotted hyena translate into different gait preferences: the other hyenas often move at a trot, whereas spotted hyenas prefer to lope and seldom trot.

PREDATORY BEHAVIOR

Capture and Killing Techniques. Hyenas use the *death-shake* to kill small prey they can pick up by the back or neck, but have no special killing bite. The spotted hyena simply grabs and bites a large prey animal from the rear until it drops, and kills by disemboweling. The more hyenas, the quicker the victim dies. The 2 *Hyaena* species rarely attack large animals and are inefficient even at capturing small game (see brown hyena account). Although hyenas do not carry out concealed stalks, a stealthy approach may be employed to get close enough to grab a naive pet or a young wild ungulate (see striped hyena account).

SOCIAL BEHAVIOR

COMMUNICATION

Olfactory Communication. All species paste a greasy, smelly anal-gland secretion on grass stalks and other objects, in a way that recalls the mongoose's *anal-drag* (cf. figs. 20.2 and 19.4). The large anal pouch (fig. 20.1) is everted both while marking and during social interactions. Striped and brown hyenas present the everted gland to superiors as a submissive gesture associated with greeting behavior, whereas the spotted hyena everts the pouch when excited and aggressive. This species has a unique greeting ceremony, featuring mutual olfactory examination of the extended phallus. Both *Hyaena* species and the aardwolf scent-mark their whole territory and the brown hyena even deposits 2 different anal-sac secretions. *Crocuta* scent-posts territorial boundaries, usually after defecating at a regular dung midden, and in all sorts of socially stressful situations (3). All 4 species maintain latrines, those of hyenas made conspicuous by the chalk-white color of dried feces containing bone residues. In ancient times the dung was refined into *album graecum*, a face powder (6). The aardwolf buries its potent-smelling scats in latrines. Brown and spotted hyenas both scratch the ground as an aggressive and courtship display, thereby scent-marking with an interdigital gland. Urine seems to be voided randomly by all species.

Tactile Communication. *Social-licking* and *-nibbling* is seen in the hyenas, but mainly between mothers and offspring, with lower-ranking individuals taking the initiative. It thus appears to denote submission, although mothers of course lick and groom small cubs, especially while suckling.

Vocal Communication. The spotted hyena is one of the most vociferous carnivores. The others are comparatively quiet. If they have distance calls comparable to *Crocuta*'s whoop, few people have heard them. The calls that have been heard sound like fainter echoes of the spotted hyena's cries (5). The aardwolf is the most silent, but growls, roars, and barks impressively as part of its antipredator defense.

Visual Communication. The facial mimic is comparable to that of other carnivores that cannot snarl (see chap. 18, Communication). Hyenas show only their lower teeth in threat, although they also show the upper teeth while yawning and when grimacing in fear (fig. 20.4). Nor is the mouth puckered or pursed in offensive threat. As indicated under Agonistic Behavior, similar postures, ear, and tail positions express basically the same emotions in hyenas as in other carnivores. *Crocuta* has the most developed visual patterns in the family, using a greater variety of head movements,

Fig. 20.3. Spotted hyena running gait.

Fig. 20.4. Intimidated brown hyena making *fear grimace.*

and the tail position is emphasized by the contrastingly colored tip.

AGONISTIC BEHAVIOR

Dominance/Threat Displays: *erect posture, bristling, ears cocked, tail raised, stiff-legged walk.*

Defensive Displays: lips retracted, teeth bared; erection of dorsal crest and tail to enlarge lateral silhouette (especially against other species; absent in spotted hyena) (fig. 20.5).

Submissive Displays: *cringing* posture with hindquarters, tail, and ears lowered, *fear grimace* (fig. 20.4), *turning head away* or *showing neck* (striped and brown hyena), *carpal-crawling.*

Fighting. Spotted hyenas are sometimes mauled and even killed while trespassing in another clan's territory. But fighting within clans or families is rare, owing to well-established dominance hierarchies. Aggression tends to be highly ritualized, taking the form of *jaw-wrestling* or *neck* and *shoulder biting.* A dominant animal may bite and/or shake an inferior hard enough to draw blood, but the usual target areas are protected by massively thickened skin (3, 8).

REPRODUCTION. The aardwolf may breed seasonally but young of the other species can be born at any time of year. Even associated females do not synchronize their reproductive cycles. Spotted hyenas usually have 2, the others 3–4 young. Gestation is 3–4 months. Courtship and copulation, unique in the spotted hyena, are conventional in the other species, featuring *genital and anal-gland sniffing* and *licking, close-following,* and *preliminary mounting* attempts. Copulation may last up to 10 minutes and be repeated during estrus lasting a week or longer. In *Crocuta* the resident alpha male sires most offspring. In the brown hyena resident males are generally sons or brothers of clan females, which breed with nomadic males (7).

PARENT/OFFSPRING BEHAVIOR. See table 20.2. The fact that brown and perhaps striped hyena females suckle one another's offspring and other family/clan members help provision them, whereas mother spotted hyenas do neither, suggests close genetic relationship in the former 2 species, with cooperation based on kinship.

Hyenas are suckled for 8 months to a year or more, compared to 2 months in most canids. Assuming that sustained lactation is linked to the amount of bone in their diet, the question arises as to whether bone-eating evolved first or might have developed through selection for increased milk production. Either way, sustained milk production and carrying food to the den are alternatives to the canid habit of carrying food in the stomach and disgorging it. Milk-treading has been observed in hyena cubs of both genera.

RELATIONS WITH OTHER PREDATORS. Next to the lion the hyenas are the biggest African predators, and since dominance is usually a function of size even striped and brown hyenas outrank the leopard (see species accounts). However, the wild dog is another matter: by virtue of its superior pack organization and cooperation, wild dogs can usually rout spotted hyenas, though only ⅓ their size. Spotted hyenas take much greater liberties with lions than do other hyenas and sometimes pay with their lives. But occasionally they manage to take over or take back kills by intimidating lions with loud calls and threatening behavior.

There is some sort of mutual attraction between *Crocuta* and *Hyaena,* the former dominating and the latter playing a submissive role. The aardwolf avoids close encounters with all its relatives.

References

1. Colbert 1969. 2. Ewer 1973. 3. Frank *in lit.* 1986. 4. Kingdon 1977. 5. Kruuk 1976. 6. Matthews 1939*b*. 7. Mills 1983. 8. Owens and Owens 1978. 9. Savage 1978. 10. Thenius 1969. 11. Ulbrich and Schmitt 1968.

Added Reference

Mills 1989. ——— 1990.

Fig. 20.5. Striped hyena with mane erected in defensive threat.

Striped Hyena

Hyaena hyaena

TRAITS. A long-haired hyena with stripes and big, pointed ears. *Height and weight:* 65–80 cm and 25–45 kg, average 26–27 kg (4), males slightly heavier than females. Rangy build with long limbs and neck, pronounced slope from shoulders to hindquarters, large head with robust teeth. Coat 7 cm and mane up to 20 cm long. *Coloration:* gray to straw-colored with bold black markings: muzzle, 2 cheek stripes, throat, body, and leg stripes, torso stripes partly obscured by the mane; cubs like adults but maneless with black spinal stripe. *Scent glands:* anal pouch equally developed in both sexes. *Mammae:* 6.

DISTRIBUTION. From central Tanzania through drier parts of eastern and northern Africa, the Middle East, and Asia north to the Caucasus and southern Siberia; occurs up to 3300 m in Pakistan (8).

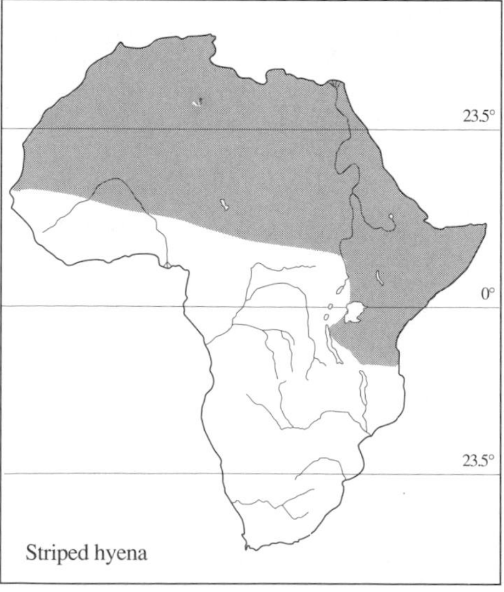

ECOLOGY. A denizen of arid, often mountainous wastelands through most of its range, in East Africa the striped hyena frequents *Acacia/Commiphora* scrub woodland and thornbush, avoiding the more open grassland habitats preferred by the spotted hyena (4). It dens in rocky hills, kopjes, and ravines. Like the brown hyena, it scavenges but rarely hunts large or medium-sized species, and kills small animals opportunistically, supplementing its diet with fruits (especially *Balanites,* the desert-date) and insects (4). It drinks regularly when it can but can live in waterless areas.

SOCIAL ORGANIZATION: largely unstudied, but probably similar to the brown hyena (see species account); solitary forager with family social organization, young provisioned at communal den, with help from older offspring.

Striped hyenas forage individually and are rarely seen together (4). The species has therefore been considered solitary. However, it is known to associate in family groups that may include offspring of different ages (4, 5). Subadults participate in feeding younger siblings by bringing surplus food back to a central den (4, 6, 8). Five Israeli dens that were investigated all had accumulated animal remains. A boneyard of 40 m^2 covered the floor of one maternity cave; a 2 m^2 sample contained 267 pieces from 57 different individuals (9).

Thus the striped hyena is actually a social carnivore whose dependence on dispersed, small food items favors solitary foraging. But whether it ever lives in clans including several adults of both sexes like the brown hyena remains to be seen. In captivity, a male and female peacefully share the same enclosure, lie in contact, and lick one another, but females older than 15 months are mutually intolerant (8).

The home ranges of one radio-collared Serengeti female and male were 44 and 72 km^2, respectively (4). No evidence of territorial marking or defense was noted, but in Israel latrines and anal-gland marks were found near feeding sites and along well-used pathways, preferably on sandy or dusty spots (5). In the Serengeti no tendency to use latrines, apart from the habit of moving some meters from the den entrance before defecating, was noted, nor did the hyenas use regular pathways (see Foraging Behavior).

ACTIVITY. Striped hyenas are quite strictly nocturnal and unlike the spotted kind are rarely seen in daytime, which they spend under a rock overhang or a fallen tree or in caves and other dark recesses. Radio-tracked Serengeti animals rarely used the same lair 2 days in a row. Beginning at dusk, they foraged until 2200–2400 h, rested for several hours, then foraged again during the last hour or 2 before daybreak (4). The mean duration of activity was 6¼ hours, or for about ¼ (26%) of each 24-hour day (compared to 16% for Serengeti spotted hyenas). A subadult male observed for 10 full days traveled 7–27 km a night, averaging 19 km.

FORAGING BEHAVIOR. The Serengeti hyenas seldom traveled the same route twice, even when returning to a particular foraging site, such as a fruiting *Balanites* or a household garbage pit (4). Walking at 2–4 kph or trotting at 8 kph, a foraging hyena always quartered ground it had not visited in the recent past, keeping its head at or below shoulder level with tail hanging or slightly raised. Apparently in search of sleeping birds, concealed small mammals, reptiles, or insects, it stopped to investigate every bush or rock, poking its snout into dense shrubs, clumps of grass, and holes, and sniffed the bases of trees. The direction of travel bore no relation to wind direction, but individuals were seen to respond abruptly to the scent of carrion, making a sharp turn upwind to pick up pieces of bone or skin or a dead bird. Herds of antelopes and zebras were often encountered, usually without any show of interest. The prey most actively hunted was insects: grasshoppers, beetles, and especially flying moths, termites, and beetles. Hyenas ran as far as 30 m, made acrobatic leaps, and snapped loudly while trying to grab flying insects. A few chases after larger prey were seen, over distances of 100 m or less at speeds up to 50 kph: hares, bat-eared foxes, cheetah cubs, birds, gazelle fawns, dik-diks, and once a reedbuck. Only a bat-eared fox, taken unaware, and a hare were actually caught. Curiously enough, the hyenas had their manes erected during chases, which might be thought to impede rather than assist their efforts (4).

Striped hyenas often cache bones, pieces of skin, and such, using the snout to push the food deep into a dense shrub or grass clump (4). They also carry surplus meat back to their lair or den.

SOCIAL BEHAVIOR

COMMUNICATION

Vocal Communication: *whining, giggling, yelling, growling, lowing,* and *lowing-growl.*

According to reference 4, these calls closely resemble the calls of spotted hyenas (q.v.), but are much softer and uttered only during intraspecific encounters. Striped and brown hyena calls should be still more alike, but remain to be compared. *Whining* was heard from cubs trying to nurse, *giggling* when one hyena chased another carrying food, *growling* and *lowing-growling* during foodfighting and playfighting, and *drawn-out lowing* by a cornered or trapped animal. A long-distance call, described as a *cackling howl,* has been reported (3), but is rarely heard, and captive pairs may exchange quiet *"hoos."*

Olfactory Communication: *pasting, treading,* dung deposits, greeting, and *presenting the anal gland.*

Foraging striped hyenas leave scent-marks throughout their range, preferably on tall grass stems. The *pasting* procedure is highly stereotyped and very similar if not identical in the 3 hyenas (see fig. 20.2). If the spot has been previously marked, the hyena first sniffs the stem intently. Then it walks forward straddling and bending the stalks, typically with 1 foreleg lifted and bent around the grass and the hindlegs flexed. Meanwhile the anal pouch is fully extruded and everted, and a coating of yellowish pomade is deposited on a 2–5 cm length of grass as the hyena moves forward. A more conventional method is used to mark bushes and other unbending objects: the hyena backs into the bush with tail erect and smears a twig with pomade by waggling its rump from side to side (5). In both cases the scent-marks are located at convenient nose height 30–60 cm above the ground. The secretion looks rather like zinc ointment (2) and smells like essence of spotted hyena to the human nose (4); and the 2 species react to each other's deposits (see Relations with Other Predators). *Treading* or trampling with the hindfeet for up to 10 seconds has been observed both after *pasting* and after defecating at latrines (latrines discussed under Social Organization) (5).

Greeting and presenting the anal gland. Striped hyenas *sniff noses* at every meeting, even after a brief separation, often followed by *anogenital sniffing.* Subsequent behavior depends on the social standing and comportment of the participants. Immature young display submission by *pres-*

enting the anal gland: while lying on the brisket or *carpal-crawling* around the adult with tail and sometimes a hindleg raised, bringing the everted anal pouch under the adult's nose; while lying on the side or even rolling on the back while pawing at the dominant hyena's chin (3, 4, 5). A hand-reared cub spontaneously everted the anal gland whenever handled, beginning at 1 month (1).

Visual Communication. Visual displays were infrequently seen during interactions between Serengeti striped hyenas but as noted below (Agonistic Behavior), tail and ear positions, facial expressions, and the erectile mane make aggressive and submissive displays visually distinctive. Practically any situation that makes a striped hyena feel aggressive or fearful causes its hair to rise (4). The tail is also expressive, held horizontally while hunting and playfighting, vertically during meetings with another hyena, and curled over the back during a dispute over food (4).

AGONISTIC BEHAVIOR

Dominance/Threat Displays: *bristling*, ears and tail erect, *hunched posture*, growling.

Submissive Displays: *presenting anal gland* (see under Olfactory Communication), *crest lowered, ears flattened, fear grimace, upright stance, turning head from side to side, licking lips, rolling eyes.*

Fighting: ritualized *muzzle-wrestling, biting neck and rump, leg-biting.*

During an aggressive interaction, the dominant animal may assume a low, *hunched posture* (fig. 20.6), while the subordinate lowers its crest but *stands upright* and may back away *displaying* its *black throat*, meanwhile *turning its head from side to side, licking its lips*, and *rolling its eyes.* Lowering the head may express the intention to snap at an opponent's legs, which has been observed in a real attack. To protect their long, vulnerable limbs, both contestants drop to their knees. But contests typically consist of ritualized *wrestling matches* as in the brown hyena, with biting directed to the cheeks, neck, and rump (3).

REPRODUCTION: nonseasonal breeding, beginning at 2–3 years.

One to 6 cubs (average 2–4) are born after a 3-month (88–92 day) gestation.

SEXUAL BEHAVIOR. No detailed descriptions of sexual behavior have been published. Based on captivity observations, females mate several times at minimal intervals of ¼–⅓ hour during a 1-day estrus (8).

PARENT/OFFSPRING BEHAVIOR. The cubs are born in a den, with eyes and ears closed and barely able to crawl, weighing 660–675 g (8). Intense digging behavior prior to whelping has been noted in captive females. The cubs' eyes open at 5–9 days and they may venture aboveground at 10–14 days. A Serengeti female visited her cubs regularly, usually at nightfall and often around midnight. One yearling female left a whole wildebeest leg at the den after carrying it 2 km (4). Cubs may start eating tidbits offered by the mother at 1 month, and begin accompanying her on foraging excursions at 6 months, but are suckled for a year or more.

RELATIONS WITH OTHER PREDATORS. Striped hyenas keep a good 50 m away from lions (and tigers [7]). Like the brown hyena, they are capable of driving or keeping leopards from their kills—though sometimes leopards also chase hyenas (3, 4, 7). Normally the cheetah is also subordinate, but a hyena that tried to catch cheetah cubs was repeatedly charged and finally

Fig. 20.6. Striped hyenas fighting (redrawn from Leyhausen 1979).

driven off by the mother (4). Jackals, foxes, aardwolves, servals, and other small carnivores are mostly ignored, but play it safe by giving the striped hyena a wide berth.

Relations between striped and spotted hyenas appear ambivalent. There seems to be a mutual attraction based probably on similarities in their appearance and smell. Upon meeting they approach and circle each other at 6–8 m. The striped hyena behaves fearfully and submissively, allows the larger hyena to steal its food, and is chased away when it attempts to scavenge from spotted hyenas (4). Apparently its ability to appear 38% bigger with its mane erected (10) fails to impress its heavyweight cousin. Striped hyenas have been seen to sniff and paste grass stalks previously marked by spotted hyenas and to use their latrines (4). Visible proofs of interspecific dominance are only trivial signs of the spotted hyena's ecological dominance, however. In the Serengeti ecosystem, *Crocuta* monopolizes the open habitats and the large ungulates that live there, consigning its lesser relative to substandard habitats where the game is smaller and resources are more dispersed (4).

References

1. Fox 1971c. 2. Holzapfel 1939. 3. Kingdon 1977. 4. Kruuk 1976. 5. Macdonald 1978. 6. Mills 1978. 7. Prater 1966. 8. Rieger 1979. 9. Skinner and Ilani 1979. 10. Wemmer and Wilson 1983.

Fig. 20.7. Brown hyena carrying springhare (from a photo in Mills 1978).

Brown Hyena

Hyaena brunnea

TRAITS. The southern equivalent of the striped hyena. *Height and weight:* 78.7 cm (70.6–86.8) and 39 kg (35–50), no significant difference between the sexes (1, 6). Muzzle broad and short, robust teeth adapted for cracking bones; long, pointed ears; long legs. Coat shaggy with tail and mantle of much longer hair. *Coloration:* mostly dark brown, partly obscured by mantle of straw-colored hair, legs dark yellow-brown with black stripes; cubs gray, maneless, with more stripes on body and legs. *Scent glands:* large anal pouch that produces 2 secretions (white and black). *Mammae:* 4.

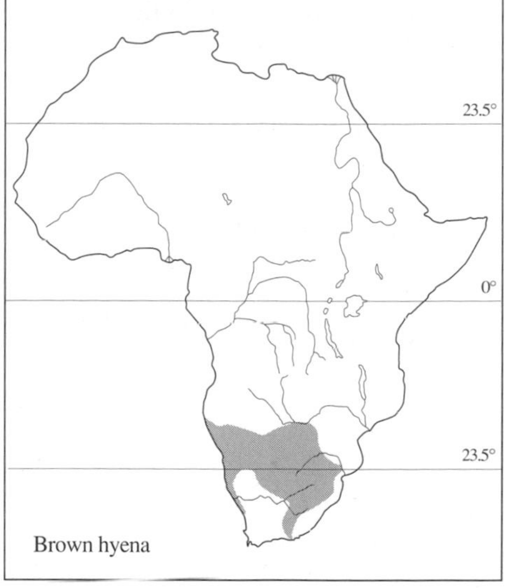

DISTRIBUTION. South West Arid Zone and adjacent dry savanna south of the Zambezi River. It has been exterminated in South Africa except in northern districts of the Transvaal and Cape Province.

ECOLOGY. The dominant large carnivore in the South West Arid Zone, the brown hyena is an opportunistic forager that manages to survive in country where game and water are both very scarce by eating almost everything with any food value except herbage: some 58 different kinds of food have been identified in its droppings (10). It kills only c. 6% (by weight) of the food it consumes, and is purely a scavenger of large mammals, the richest but least dependable food. Shortages of preferred vertebrate food are made up by eating insects and various fruits and vegetables (2, 8, 9, 10). In Namibia the brown hyena penetrates the Namib Desert to the Atlantic and scavenges a living on the Skeleton Coast, where it is called the *Strandloper.*

Since most of the large mammals that live in the arid zone are migratory, whereas the brown hyena is resident, its diet may change radically between the wet and dry season. In Botswana's Central Kalahari Game Reserve, springboks, red hartebeests, and wildebeests immigrate during the rains (November—April), joining the resident gemsboks (10). Lions and cheetahs and occasional spotted hyenas come with them, and the brown hyena population lives well on the remains of their kills. In the dry season the proportion of large-mammal remains in brown hyena scats falls to only 17%, while the percentage of fruit and vegetable foods increases, especially tsama and other wild melons, the main source of moisture during 8 waterless months. Presumably these cucurbid fruits have nutritional value, too, for brown hyenas also eat them while water is available, when they may drink 1–3 times a night (8, 10).

RELATIONS WITH OTHER CARNIVORES. Though it is seldom a predator on big game, the brown hyena is large and aggressive enough to rank fourth in the carnivore hierarchy, and effectively second in the most arid parts of its range, where spotted hyenas are few and wild dogs even rarer than usual. Brown hyenas show no more respect for leopards than for cheetahs: a lone female robbed a male leopard of the springbok it had just killed and when the leopard tried to reclaim it, chased it up a tree (10). Yet the same species keeps at least 200 m from lions on kills and allows ½ hour after the lions leave before moving in (10).

SOCIAL ORGANIZATION: kinship groups including sons and daughters; communal suckling and provisioning at the den; foraging alone, clan members mark and defend territory against same-sex neighbors; females mate with nomadic males.

Long considered a solitary species, the brown hyena has one of the more advanced and elaborate carnivore social systems, as shown by an 8-year study in the Kalahari Gemsbok N.P. (KGNP) (refs. 1–8) and a 5-year study in the Central Kalahari Game Reserve (CKGR) (9, 10, 11). From 4 to 14 hyenas, including 1–4 adult females, 0–3 adult males, 0–5 subadults, and up to 4 cubs, live in very extensive home ranges of 235–480 km^2 (330 km^2, mean of 6 clan ranges) in the KGNP (2, 3). In the less-arid central Kalahari, the main study clan (average of 13 hyenas, including 5 females) had a comparatively small range of 102 km^2 (10). Range size is apparently determined by the distribution of food, whereas clan size is correlated with food quality and abundance (3).

Not only female but also male offspring may remain in the maternal group after maturing at 2½ years, but whether this is the usual arrangement remains uncertain. In a sample of 9 male offspring born into 2 KGNP clans, 4 disappeared as subadults (beginning at 22 months) and 5 remained as adults. One of these was still living at home 6 years later, but 3 others left within a year. However, only 1 of 9 adult males of known origin was observed to join a group of unrelated hyenas. In any case, resident males generally show no sexual interest in females of their own or even other groups, and show little antagonism toward nomadic males. These, representing 33% of the adult males and about 8% of the total population, wander through the clan territories and mate with any estrous females they encounter (4).

But both resident males and females behave intolerantly toward trespassers of the same sex from neighboring territories, often chasing and *neck-biting* those they catch (see Fighting). Clan members post their territory and at the same time communicate with one another by scent-marking throughout their range (see Olfactory Communication) (1, 7).

Apart from the submissive behavior shown by the young toward adults, no evidence of a rank hierarchy was noted in the KGNP study (5), whereas a definite rank hierarchy was described in the CKGR study group, which was reinforced at every meeting through ritualized displays of aggression and submission (9). A male was the alpha animal but otherwise dominance was not sex-linked. The presence in the group of 3 unrelated immigrants, 2 females and a male, could have caused intragroup relations to be unusually competitive. The greater seasonal abundance of medium and large carcasses, combined with a smaller range, would also increase food competition and frequency of meetings between individuals. Two or more hyenas were present at over half of 87 carcasses of springbok or larger size, and occasionally up to 6 gathered and socialized between bouts of leisurely feeding. Even adults that were unrelated and unmated often (at 23 carcasses) fed together; this would seem to reflect the same tolerance of nomads and individuals of opposite sex shown by resident KGNP hyenas. However, no more than 3 hyenas fed together at the same time; the rest had to wait, sometimes in vain (10).

Communal Suckling and Provisioning of Offspring. Several months after whelping, clan females bring their young to a centrally located den where the cubs are reared communally. Four females in the CKGR clan had 5 cubs between them, ranging in age from 4 to 20 months. They all suckled one another's young and brought food back to the den (a springhare, birds, a jackal carcass, carrion, tsama melons, etc.). Even a cubless female brought food and tried to suckle cubs. Related subadults of both sexes also brought food, but sometimes scavenged others' offerings themselves. Females and subadults would often spend several hours a night sleeping and socializing around the den, but no adult remained on guard; they spent the days in favorite resting spots up to several hundred meters distant. (More under Parent/Offspring Behavior.) The adult males did not provision the young and none was seen at the dens.

In the southern Kalahari population, there was rarely more than 1 litter of cubs in a den at a time and even when 2 females had offspring they only occasionally suckled each other's cubs. Adult males helped provision cubs to which they were related (5).

ACTIVITY. In KGNP, brown hyenas were active 80% of the time from 1800 to 0600 h and, walking at the rate of 4 kph, traveled a distance of 31 km (1.2–54.5) a night in search of mostly small food items (6). They invariably foraged alone and found vertebrate food items or the nutritional equivalent (10 wild fruits) every 9.2 km. Resident spotted hyenas, in contrast, foraged mostly in groups averaging 3 animals and had to move a mean distance of 32.7 km to make a kill or find a nearly intact carcass (6). Brown hyenas of the CKGR traveled as far as the more southerly population in the dry season (20–30 km a night), but only 10–20 km during the rains, when they had 2 activity peaks from c. 1930 to midnight and 0230 to dawn, with an intervening rest period (10).

FORAGING BEHAVIOR. A foraging individual quarters the ground in a zigzag course with head raised and ears erect, relying mainly on its sense of smell, especially in dense bush and tall grass. Often a hyena will stop and turn abruptly into the wind, head and muzzle raised and turning from side to side. A female winded a calving oryx 0.6 km upwind, ran to the spot through thick bush, and tried to approach, but desisted after the oryx's third charge (10). Hearing—and comprehension—are also impressive: one foraging hyena turned and loped 400 m downwind and appropriated a kori bustard that it had apparently heard 2 jackals kill. Within 15 minutes 2 other hyenas arrived.

Although brown hyenas sometimes stalk a bird or hare, they usually make no effort to move quietly and will pursue small game disturbed by their passage for only a short distance. Occasionally one will undertake long, zigzag chases after springhares, which cannot hop very fast but dodge superbly; sometimes a hyena manages to dig one from its burrow. But only 9.6% (6) to 14% (10) of observed hunts were successful.

Food Caching. Brown hyenas cache surplus food. Some 70% of the times hyenas were seen at a large carcass, individuals carried a piece, preferably a hindleg, 100–600 m and cached it in a thicket, usually returning to eat it early the next night. One animal may remove up to 3 legs before any competitors discover the kill. Other hyenas and jackals seldom find a cache, even though the owner scent-marks bushes and grass within 15–20 m of the site (9). Hyenas may also carry scavenged items, mostly small game, back to the communal den instead of caching them. In the KGNP study, 26 out of 40 times that a hyena was seen to find suitable food, all or part of it was carried back to the den, an average distance of 6.4 km, and equally often by males and females (8).

SOCIAL BEHAVIOR

COMMUNICATION

Olfactory Communication: scent-marking with anal-sac secretions and dung; *presenting anal gland* during greeting ceremony (described under Agonistic Behavior).

Foraging brown hyenas pause 2–3 times/km to paste grass stems, bushes, and other objects with anal-sac secretion. The locations of scent-marks 3 adult males and 2 adult females were seen to deposit within their territory in 1 year are shown in figure 20.8. This clan left an estimated 145,000 marks throughout their property, of which at least 20,000 would be potent at any given time (the odor being detectable to the human nose for 30 days minimum). The highest concentration of marks was in the central, most-used part of the range, especially along trails and near water holes, dens,

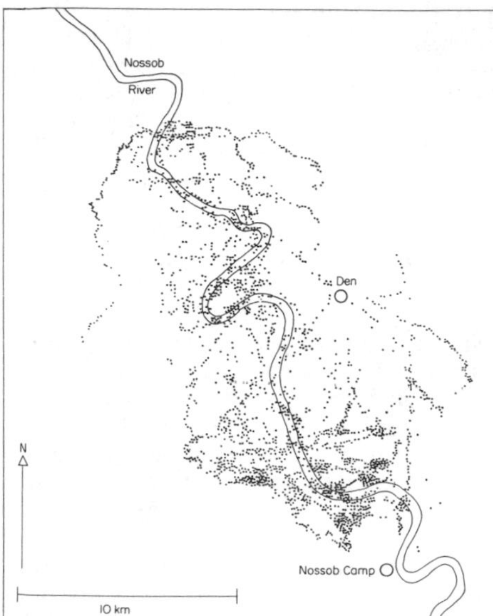

Fig. 20.8. Map of anal-gland scent marks found in a brown hyena clan territory (from Gorman and Mills 1984).

latrines, large carcasses, food caches, and other focal points. Marking density decreased in the outer part of the range, although the rate of pasting increased when hyenas visited the borders, as it did during and after social interactions. Through computer modeling, it could be shown that intruders into a territory would be likely to encounter an active scent-mark within the first 250 m (1).

In addition, brown hyenas maintain latrines where up to 15 defecations may accumulate. The distribution of some 85 latrines that were found in the above territory was more clumped than random, and groupings of latrines occurred mainly near the border. The majority were situated next to a tree or bush (75%), or other landmark such as the side of a road (15%) or pool (2%). Furthermore, 24 out of 34 randomly chosen latrines beside *Boscia* trees were situated on the south side. Thus latrines were specifically associated with landmarks and those on the border were visited more frequently by clan members than interior latrines (1).

Hyaena brunnea deposits 2 different anal-sac secretions in turn, first the usual white pomade, then a black secretion (see fig. 20.2 and description of *pasting* in striped hyena account). The white secretion is produced by sebaceous glands and its odor is long-lasting; the rather watery dark secretion is produced by apocrine glands and its smell fades faster. Gas-liquid chromatographs of the secretions from different hyenas show consistent differences in the concentrations of the constituents, which could serve for individual identification. There is also behavioral evidence that brown hyenas can distinguish between the marks of clan members and outsiders. The secretions probably function not only to establish territorial rights but also as chemical messages among pack members which reveal both who made the mark and how long ago. One important benefit may be to avoid foraging in areas already covered by other hyenas (1, 7).

Tactile Communication. Members of the same social group occasionally lick or nibble one another, concentrating on the neck region. Performed most often between subadults and adults, *mutual grooming* probably has a social as well as a sanitary function (reaching an inaccessible spot), serving to reinforce social and kinship bonds (5).

Vocal Communication: squeal, squeak, scream, yell, and growl or grunt, according to reference 10; 2 whines, 4 growls, yell, and hoot, according to reference 6.

a. Squeal: a shrill, sharp cry emitted by juveniles or other subordinates while approaching to greet or beg food of a dominant hyena.

b. Squeak: a hoarse or rasping cry associated with *carpal-crawling* (abject submission).

c. Scream: high-pitched, cackling shriek, given by a hyena whose neck is being bitten.

d. Yell: a loud, abrupt, high-pitched call associated with defensive threat.

e. Growl or grunt: low-pitched, breathless, throaty sound given while muzzle-wrestling.

All these calls except the yell are short-range signals audible at no more than a few hundred meters (6).

AGONISTIC BEHAVIOR

Dominance/Threat Displays: *bristling, erect posture* with ears pointed, *scratching ground with forefeet,* growl/grunt.

Submissive Displays: *fear grimace* (fig. 20.4), *ears flattened, presenting anal gland, carpal-crawling, lying prostrate,* plus vocal accompaniment (above).

Fighting: *muzzle-wrestling, neck-biting.* When 2 hyenas meet, they may either perform the *greeting ceremony,* ignore or actively avoid each other, or behave aggressively and sometimes fight (5).

Greeting between equals begins with *mutual sniffing* of head, neck, flanks, and anus as 2 hyenas stand in *reverse parallel,* with tails over their backs. An inferior individual *presents the anal gland: cringing, ears flattened, grinning, cackling,* it cuts in front of the other and passes its everted anal pouch under the other's nose, often repeatedly. Greetings are exchanged mainly between group members and *presenting* is most commonly seen when cubs and subadults greet adults; adults were not seen to present to younger hyenas (5).

Behaving submissively has some tangible benefits. A hyena feeding at a kill may tolerate another that *carpal-crawls* up and lies low while eating. And a hyena carrying food can usually be forced to surrender it to another that runs up with ears flattened, body lowered, tail raised or outstretched, and squeals while latching onto the meat. This behavior derives from and is mostly seen in juveniles begging from adults bringing food to the den (9).

Encounters between members of neighboring clans of the same sex were almost always hostile and accounted for all but 2 of the fights observed in KGNP. However, only 16 fights were seen in 2500 hours of hyena-watching (5), and these involved highly *ritualized neck-biting bouts* lasting from a few seconds to a few minutes. They occurred mainly in border areas and the behavior of the contestants foretold the outcome from the start: one was always the aggressor, which bit, pulled, and shook the other by the neck; the other yelled and growled but failed to bite back. The defeated animal always left the area while the winner usually remained. Several far more severe bouts were observed in the CKGR: one between 2 unrelated females that lasted 2¼ hours, during which the inferior's neck was continuously gripped for up to 18 minutes at a time, and an engagement of over 3 hours between 2 unrelated males (10).

Members of the same group often engage in *muzzle-wrestling,* which may represent the contest version of the one-sided *neck-biting* encounters between residents and intruders. *Standing tall* with *hackles raised* and facing at about a 135-degree angle, 2 contestants throw their heads wildly while biting and pushing at each other's muzzle and neck, with mouths wide open and teeth showing, meanwhile growling or grunting (10). Most bouts involve young animals and appear playful, especially those between younger and older animals. Females may dominate males and vice versa, but as usual greater ferocity is shown toward members of the same sex. Subadults try, often successfully, to engage adults, which rarely jaw-wrestle with each other. Engagements between adults and subadults are sometimes serious, the adult becoming increasingly aggressive while the subadult behaves submissively and whines pitifully. Subadults of both sexes which were treated in this way subsequently emigrated. Thus, dispersal of 4 male and 2 female offspring of the CKGR study clan was preceded by progressively more severe conflicts with same-sex adults over a 3–7 month period beginning at c. 22 months (9).

REPRODUCTION. Breeding is unsynchronized and perennial. Litter size is 1–4. The modal number in 15 KGNP litters was 3 (6).

SEXUAL BEHAVIOR. Nearly ½ (46%) of observed encounters between clan females and outside males included sexual behavior, compared to only 6% of encounters with group-living males (*sniffing vulva* or *urine,* only). Two females were known to mate with at least 4 different nomadic males during an estrus period of several days. Mating is conventional and doglike, featuring some chasing and defensive aggression, *mutual anogenital sniffing,* and *close-following.* During copulation, which may last 2½–10 minutes, the female stands with head low, back slightly arched, tail raised and deflected. The male rests his whole body on her (see spotted hyena, fig. 20.14) and may lightly nip and pull her nape hair. A captive pair copulated at the rate of 15 times in 9 hours (maximum 5 times in 1 hour) during the second part of a 2-week estrus period (12). In the wild, too, copulations with different males were seen 9 days apart, between which the female was often alone (4).

A combination of factors is thought to account for the strategy of mating with nomadic males: erratic, often long intervals between estrus cycles (12–23, up to 41 months between litters), very large territories, and solitary foraging all select against attempts by resident males to monopolize mating with a group of females.

PARENT/OFFSPRING BEHAVIOR. Communal dens may remain in continuous use for several years. In the CKGR, dens are typically situated on bush-covered dunes preferably overlooking a river bed. The one occupied by the study clan had 2 entrances 183 cm apart, which narrowed from 152-cm diameter to 50 and 80 cm in the first ½ m, then to 30–35 cm (9). Adults entered the vestibule to rest a few hours or leave food, but only younger cubs could use the small tunnels, excavated by themselves. Around the entrances discarded soil made an earthwork where cubs and adult females spent a lot of time resting while nursing and eating, monitoring their surroundings, and socializing.

Brown hyenas give birth in a separate den, transferring their cubs to the communal den when they are c. 3 months old. Cubs begin to eat meat at 3 months but continue to nurse until at least 10 months old and may only be fully weaned at 15 months (6). They suck for 25–30 minutes at a time, apparently on a first-come, first-served basis. The mother of a 4-month juvenile suckled a 10-month youngster without her own offspring even being present. Females with small cubs spend up to 5 hours a night attending them, often paying 2 visits. But after parking them in the communal den, mothers pay only 1 visit a night and may leave juveniles older than 8 months unattended for 2–3 nights. Thanks to the communal nursing and provisioning arrangement, these youngsters need not fast while waiting (cf. spotted hyena) (9).

With tunnels for safety, the presence of older along with younger cubs makes adult baby-sitter/guards unnecessary. The smallest cubs make running dives into the den at the slightest disturbance. Cubs that have outgrown the tunnels put on a defensive-threat display (see striped hyena account, fig. 20.5). Those over 14 months old may sleep in or away from the den, returning each evening. Though they go off foraging beginning at 14 months, the den remains a focal point through the second year (9).

PLAY. Cubs of different ages lie in contact, *social-groom,* and play together. They hold free-for-alls in which older cubs treat younger ones so roughly that all cubs acquire neck scars during jaw-wrestling and neck-biting playfights. Yet the younger juveniles often ask for it by biting the ears, legs, or tails of resting older cubs, by dancing to and fro before them or flopping onto their heads. By the time they become adults, brown hyenas are played out (9).

References
1. Gorman and Mills 1984. 2. Mills 1978. 3. ——— 1982*a*. 4. ——— 1982*b*. 5. ——— 1983. 6. ——— 1984. 7. Mills, Gorman, Mills 1980. 8. Mills and Mills 1978. 9. Owens, D. D., and M. J. Owens 1979. 10. Owens, M. J., and D. D. Owens 1978. 11. ——— 1984. 12. Yost 1980.

Added References
Mills 1989. ——— 1990.

Fig. 20.9. Spotted hyena carrying young cub (from a photo in Van Lawick and Van Lawick-Goodall 1970).

Spotted Hyena
Crocuta crocuta

TRAITS. The largest, commonest hyena. *Height and weight:* 70–91.5 cm and 50–86 kg; males 84 cm (79–85.5) and 59.3 kg (55.8–62.5), females 86 cm (84–88.5) and 71 kg (67–75) (Kalahari Gemsbok N.P. [19]); males average 48.7 kg, females 55.3 kg in Serengeti N.P. (12). Powerfully built, especially forequarters, sloping back, legs long and muscular, tail short (25–35 cm) and bushy; massive skull, teeth robust, outer incisors like small accessory canines; ears broad and round. *Coloration:* varies with age and individual; solid seal-gray for 1.5–2 months, head lightening first; yearlings heavily spotted with grayish ground color and legs still unspotted; spots fade and color lightens with age, from reddish brown to tan, but muzzle and tail tip remain dark. *Scent glands:* anal pouch equally developed in both sexes, secreting white pomade, and presumed glands between front toes (12). *Genitalia:* identically masculine in both sexes (fig. 20.10) until puberty, when the female urogenital opening and the 2 mammae enlarge and the male's scrotum becomes hairless and black (8, 25).

DISTRIBUTION. Africa south of the Sahara except rain forest, up to 4000 m. Exterminated in most of South Africa and greatly reduced, along with wild herbivores, in many savanna areas. In undisturbed savanna and plains it tends to be the most abundant large carnivore (6).

Fig. 20.10. Comparison of male and female external genitalia in the spotted hyena, illustrating male-looking genitalia of the unmated female. From left to right, adult male; subadult female approaching first estrus; anestrous female that has borne young (note enlarged nipples). (From Matthews 1939*a*.)

ECOLOGY. The most carnivorous member of the family, *Crocuta* rarely eats vegetables or fruits; its diet consists very largely of vertebrates and especially ungulates (1, 3, 12, 15). Other carnivores waste up to 40% of their kills; the spotted hyena eats virtually everything but the rumen contents and the horn bosses of the biggest antelopes like the wildebeest. Disgorged castings up to 20 by 8 cm, containing hair and other undigested material (grass, bone and horn fragments, entire hooves, stones, wood, etc.), are often found around dens (1, 12). Bone slivers up to 9 cm long are swallowed and digested within hours; teeth take a little longer. The desiccated corpses of wildebeests that died months earlier are consumed and yield protein, fat, calcium, phosphorus, and other minerals that help sustain milk production during lean times. Probably no other carnivore utilizes vertebrate prey so efficiently (1, 12).

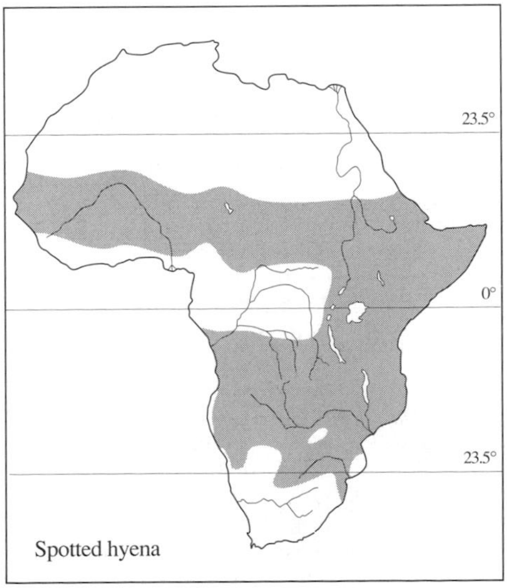

23.5°
0°
23.5°
Spotted hyena

SOCIAL ORGANIZATION: highly gregarious and polygynous, territorial clans led and dominated by females, independent and cooperative hunting, scramble competition at kills, communal dens but no cooperative suckling or provisioning of offspring.

The social organization of this hyena is as variable as its diet (1, 7, 12, 15, 19). In many areas it is known primarily as a solitary forager and was traditionally regarded as a skulking scavenger too craven to prey on any but the infirm, the old, and the helpless young of large mammals. But in areas where wild ungulates are numerous, the spotted hyena forms the largest social groups and outnumbers all other large carnivores. In Ngorongoro Crater and Serengeti N.P., where the first long-term field study of the species was made, *Crocuta* turned out to be a formidable predator which hunted cooperatively as well as singly; in fact the Crater lions scavenged the hyenas' kills more often than the other way around at that time (12). Ngorongoro's 25,000 resident ungulates support a population of some 470 hyenas, which live in 7 clans numbering 35–80 adults (mean number 55). The 265 km^2 crater floor is partitioned into exclusive territories averaging 30 km^2.

The spotted hyena's social system is unlike that of other social carnivores, including the other 2 hyenas. The communal care of offspring, characteristic of societies based on kinship (e.g., brown hyena, lion, jackals, wild dog, social mongooses), is conspicuously absent in *Crocuta*. Instead each female provides for her own cubs only, suckling them for as long as 1½ years (12).

In place of cooperation, the spotted hyena has an openly competitive system, in which access to kills, mating opportunities, and time of emigration for males depend on the ability to dominate the other members of an often numerous clan (6, 12, 19).

Reliance on milk to nurture young spotted hyenas enhances food competition between mothers and with all other hyenas, for how long and how much milk a mother can produce depends on how well she eats. The volume and duration of milk production (see Parent/Offspring Behavior) is undoubtedly linked to the high percentage of bone in the diet (1, 12, 15). The famous "hermaphroditism" of the spotted hyena can be understood as the outcome of selection for female dominance. Apparently the nutritional advantage of being dominant led to increased female aggressiveness and size (12, 21).

Aggressiveness is mediated by male sex hormones, especially testosterone. To compete successfully with males for food, *Crocuta* females not only had to equal or surpass them in weight but also in testosterone production. Preliminary evidence of this (13, 14) was borne out in a study indicating that females and males have equal concentrations of male hormone in their circulation (16). Furthermore, testosterone levels measured in twin female fetuses were similar to the mean for adult females, supporting the hypothesis that females become masculinized during fetal development because of high androgen concentrations (23). Thus, the presence of a male phallus in female spotted hyenas may be an incidental by-product of the process whereby females achieve dominance.

Female dominance is asserted on almost all occasions when there is any question of precedence. When 2 hyenas of opposite sex meet, it is the male that gives way. If a male is lying asleep in a puddle only big enough for 1 and a female wants to lie there, the male surrenders his place. And it is females that lead packs on a marking expedition or into battle (12).

An ongoing 11-year study in Kenya's Masai Mara Reserve of a single clan (mean composition 16.7 adult males, 22 adult females, 13.7 subadults, and 18 cubs) has provided a clearer picture and modified earlier concepts (e.g., ref. 2) of hyena society (6, 7). Mean testosterone levels were found to be significantly higher in the males, except for the biggest, most dominant female: her testosterone level was 6 times higher than in other females and considerably higher than in most males. All adult females were nevertheless dominant over all males, with the notable exception of sons of the alpha female: they outranked all other hyenas except her, and the younger took precedence over the older (21). (In another study, cubs of alpha females took precedence over other males but not over adult females [19].) However, males born in the clan abstained from reproductive competition. Only immigrants were in contention and among these the male that dominated the rest also monopolized matings with clan females (more under Sexual Behavior).

The Mara study bears out the conclusion of earlier observers that the offspring of high-ranking females eat better, grow faster and bigger, and achieve high rank as their birthright (cf. savanna baboon) (12, 20). Not only is the alpha female able to claim the largest share of any kill, but her cubs gain privileges denied to the lower classes (more under Parent/Offspring Behavior). Bonds between mothers and daughters persist, as female offspring remain in the clan, whereas males emigrate in adolescence (6, 15). The mean dispersal age in the Mara clan was 25½ months (17–34); but the sons of the alpha female stayed home until they were 41 months old and nearly full-grown.

Thus, the odds that sons of high-status females will become dominant in another clan are stacked in their favor. Added to the nutritional advantages they have enjoyed, highborn males tend to have the self-confidence that is lacking in sons of low-ranking females, which have been conditioned to subservience. The genetic payoff for males that achieve alpha status is remarkably high: given a 3-year tenure in a sizeable clan (over 20 females), a male could sire about 50 surviving offspring, compared to a mean of 0.78 cubs successfully reared per female per year. An alpha female that produced 4 alpha males could have over 200 grandchildren. The fact that the top female in the Mara clan produced 4 and perhaps 5 surviving male offspring in a row, against the odds favoring a 50:50 natal sex ratio, suggests that female spotted hyenas might even control the sex of the offspring they conceive or allow to survive (6).

Dens. The focus of activity within a clan territory is a communal den, used by up to 10 females, each with 1–2 cubs of various ages. A mound of bare earth, beaten paths, bits of skin, bone, horn, and hairballs, but rarely sizeable bone middens, identify occupied dens. Holes ½–1 m wide narrow within ½–3 m to tunnels of 15 cm or less, through which only cubs can crawl. Good denning sites are used generation after generation, often by different species, but spotted hyenas rarely stay at one den for more than a few months before moving (12). Some dens include large tunnels dug by aardvarks and warthogs, underground waterways, caves, or crevices in a hillside. Big hyenas may rest and take refuge in these, and mothers of small cubs may spend the day at the den. But generally adults lie up at a distance, returning in late afternoon to suckle their offspring and socialize. Males play no parental role and only a few privileged ones are allowed anywhere near the den. In the Mara clan, the alpha male was the only one that was seen to interact with cubs (6). Cubs sometimes join in bullying males (6, 20).

Variations in Social Organization. Territorial behavior is much in evidence in the Ngorongoro Crater population, including boundary patrolling and marking and aggressive encounters between neighboring clans (see Agonistic Behavior and Olfactory Communication) (12, 20). But territorial behavior has been less in evidence or even unseen in other studies where prey and hyena densities are lower.

1. Because of patchier distribution of open grassland in the Mara Reserve, clan ranges, centered on plains utilized by ungulates, are separated by wide buffer zones of scrub woodland. There is accordingly much less contact between hyenas of neighboring clans, and where contact occurs, group aggression is rare (6).

2. Where the ungulate population is mainly migratory, the movements and social organization of hyenas that depend on ungulate prey vary seasonally. Thus, while the migration is concentrated on the Serengeti Plain, hundreds of hyenas live in its midst and set up clan territories. But when the rains end and the grazers migrate into the woodland zone, the plains hyenas can only follow by intruding into the domain of the woodland clans (12). From dens on the nearest border of their territories, they usually have to travel at least 30 km to reach the nearest concentrations of wildebeests, their staple prey (12). Spotted hyenas of southern Botswana have been known to travel 80 km between the den and their food supply (3). Cubs may have to wait 5 days or even longer before their mothers return, and many die.

In the southern Kalahari and Namib deserts, where conditions are marginal for spotted hyenas, small groups (3–12) range enormous areas of 500 to 2000 km^2 which would be impossible to defend (15, 19, 26).

ACTIVITY. The first of 2 nightly activity peaks runs from an hour before dark to about 2000 h, starting with socializing at the den followed by foraging. A shorter peak begins before and continues for perhaps 2 hours after dawn, when feeding, hunting, and social activities can often be seen by daylight. A radio-collared Ngorongoro female followed for 12 days ranged a bit over 10 km per 24-hour day and spent 84% of the time lying down. Relatively little time was spent foraging and eating; she simply shared kills made by other clan members (12). In the Kalahari Gemsbok N.P., in contrast, spotted hyenas traveled 26.5 km (0.5–69) and were active 31% of the time (15) (cf. brown hyena of the same park).

POSTURES AND LOCOMOTION. *Crocuta* walks with a long, *ambling* stride. It can lope tirelessly 10 kph mile after mile, gallop over 60 kph at top speed, and run 40–50 kph for several kilometers in pursuit of prey (12, 19) (fig. 20.3).

FORAGING, HUNTING, AND EATING BEHAVIOR. The spotted hyena's foraging habits can be summed up in a sentence: it will scavenge whenever possible and as a predator will always select the most easily captured prey. Yet a lone hyena is capable of running down and killing healthy prey the size of a bull wildebeest. Such an all-out effort is made only as a last resort. Hyenas usually go foraging alone or in pairs. Where ungulates are common, they try to locate very young or incapacitated individuals, relying mainly on their eyesight and sense of smell. A very common technique, practiced on game concentrations of all kinds from wildebeests to flamingos, is to make them run. This is done by suddenly loping toward the concentration. The animals, which hardly move out of the way of a walking hyena, flee in the opposite direction for 50–100 m, enabling the hunter(s) to detect any stragglers. Why prey species allow hyenas to get closer than other large carnivores remains unclear. (This statement does not apply to mothers with newborn young.)

Although most prey species that rely on open flight can run faster than a hyena, victims often fail to go at top speed early in the race and, unmindful of the hyena's staying power, end up being run to exhaustion within 1.5—5 km. Where hyenas are numerous, a chase initiated by a lone hunter attracts other clan members which join in pulling down, killing, and eating the victim. Once overtaken, the quarry is brought to a stop by hyenas biting at its legs and belly or gripping it by the tail. There is no special killing bite, but the victim is usually disemboweled and dies within 5 minutes or so. Prey the size of a newborn wildebeest calf may be grabbed by the neck and shaken. Few victims put up any resistance after a hard chase. They appear to be in a state of shock and exhaustion; some stand or recline on their briskets as if in a trance while being eaten alive, until their vital organs are torn out and the heart finally stops beating (4). The more hyenas, the speedier the end.

Most cooperative hunting by hyenas comes about thus spontaneously, when individuals foraging for themselves grab the opportunity to be in on the killing and eating of a quarry singled out by one of their number. But spotted hyenas also deliberately stage pack hunts: up to 25 Ngorongoro hyenas (average number 11) forgather and deliberately set off to hunt zebras, walking for miles through thousands of wildebeests until they encounter a zebra concentration (12). Their concerted effort is needed to circumvent the determined defense of herd stallions and break up zebra families. A female hyena usually leads such pack hunts, but typically more males than females participate.

Hyenas also operate as packs during interclan disputes over kills made near a territorial boundary, and in attempts to intimidate lions (fig. 20.13). But cooperation is far less developed in spotted hyenas than in wild dogs. When these 2 species compete the wild dogs come to one another's aid, whereas it is every hyena for itself; singly they are helpless against the concerted attacks of the dogs and consequently a small pack of dogs can stand off a large number of hyenas (4, 5).

Similarly, whereas wild dogs share kills amicably, hyenas engage in scramble competition. Ironically, their disputes promote food-sharing. One or 2 hyenas may feed silently. But let several come together on a kill and they begin a noisy quarrel that serves to attract all other clan members within hearing. I have seen over 65 hyenas around a wildebeest kill in Ngorongoro. Only on such occasions do most or all clan members come together (12). The more abundant and the larger the prey, the greater the number of hyenas that can live and eat together. The average number at a Ngorongoro wildebeest kill is c. 26, compared to 14 in the Serengeti, where overall prey and hyena density is much lower (12).

Instead of fighting one another, spotted hyenas compete by eating as quickly and as much as possible. A hyena can consume up to ⅓ its own weight, compared to ¼ for a lion. An adult Serengeti wildebeest will fully feed 12–13 hyenas, yielding c. 13.6 kg per animal (17). The daily food intake of this population has been estimated at 2–3 kg (12), but individuals often feed only every other day and may go up to 5 days without eating. In struggling to reach the kill, hyenas climb over, jump on top of, and wriggle underneath one another until the carcass disappears beneath a seething mass of bodies, to the accompaniment of crunching bones, tearing flesh, and sporadic outbursts of calling. Twenty hyenas will polish off a 100-kg yearling wildebeest in 13 minutes, and 35 can consume an adult zebra (220 kg) in as many minutes (12).

SENSES. Sight, smell, and hearing are all acute and important to the hyena for foraging, hunting, and communication. Its ability to detect something wrong with an animal that seems perfectly normal to us is uncanny. A hyena may see about as well as man by day but sees very much better in the dark, for instance reacting to a small disturbance in a wildebeest herd miles away and downwind when a person could barely discern the herd at all. As for hearing, a hyena can hear other hyenas calling when people cannot, and its sense of smell is incomparably better (1, 12).

SOCIAL BEHAVIOR

COMMUNICATION

Olfactory and Tactile Communication: *greeting ceremony* (phallic inspection), *pasting, pawing, social defecation* at latrines, *social grooming*.

Greeting ceremony (fig. 20.11). Meetings between spotted hyenas feature mutual sniffing of the erected phallus. As 2 individuals come together, they first sniff each other's mouth, head, and neck briefly, then

Fig. 20.11. Spotted hyena *greeting ceremony*, featuring phallic inspection.

typically move into *reverse parallel* and proceed to nose each other's phallus, which meanwhile has been extended to its full length of 14.5–19.5 cm. The hindleg is cocked during an inspection lasting up to ½ minute. The ceremony is performed by hyenas of both sexes, beginning in the first month. The lower-ranking of 2 individuals usually takes the initiative, often after showing submission by rubbing against, nosing, and licking the mouth of the superior while bobbing its head and wagging its tail vigorously with crouched hindquarters. The dominant hyena often yawns. The meaning of this behavior remains unclear. According to Frank (8), "Greeting only occurs between animals that know each other well . . .; it is clear that they are able to recognize each other long before they get close enough to greet . . . It is likely that, among other functions, greeting serves to reinforce knowledge of each other's scent . . . Adult males virtually never greet with females; it just isn't done."

Pasting. Spotted hyenas routinely mark tall grass stems with their anal glands (described in striped hyena account; see fig. 20.2) (22). The behavior and secretion appear as early as the second month (8). While *pasting* its soapy-smelling white pomade along 2–5 cm of the stalks, the tail is curled over the back and the phallus is extended. Ngorongoro hyenas regularly go in groups to *paste* along territorial borders. Aggressive behavior is usually displayed by these marking parties, suggesting that *social pasting*, at least, is associated with aggression and territoriality. In the southern Kalahari region, where range perimeters are too extensive to be scent-marked effectively, spotted hyenas *paste* throughout the clan's range (cf. brown hyena) (21, 24).

Social defecation. Latrines are also located at territorial boundaries and may be used by hyenas of one or both neighboring clans, often in conjunction with *pasting* and *pawing* during border patrols. A green color when fresh, the bone residues in hyena scats cause them to turn chalky white when dry, making latrines conspicuous. The strong-smelling urine is either voided at random or while the animal is resting, in which case urine permeates the coat. Having noted that males regularly sniff at a female's flanks rather than her genitalia, Frank (8) suspects they thereby test the urine in the fur.

Pawing. Hyenas often scrape vigorously, with one forepaw at a time, after *pasting* or after defecating at a border latrine. Like pasting, *pawing* is stimulated by the presence of hostile hyenas or lions on a kill, or an estrous female who rebuffs courtship attempts. It seems then to express frustration and possibly aggression, but may also convey an olfactory message, since the scratch marks have a distinctive smell (12).

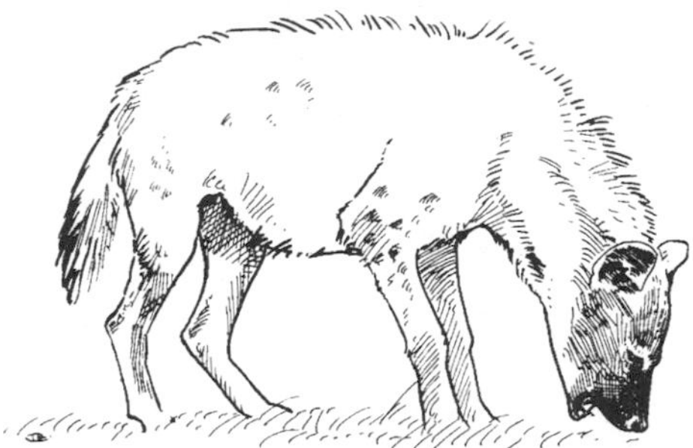

Fig. 20.12. Spotted hyena *whooping.*

Social grooming. Mothers and cubs lick and *nibble-groom* one another, but *social grooming* between adults is uncommon. However, a mother and grown daughter may lie touching, the mother with one paw draped over her offspring (20).

Vocal Communication. *Crocuta* is one of the noisiest African animals. Its calls are both highly distinctive and remarkably varied, with 11 recognizably different but intergrading calls. Like lions, spotted hyenas use distance calls to establish or maintain contact (fig. 20.12), and a variety of other vocal signals for close-range communication. They can presumably recognize one another individually by their calls (20).

1. Whoop. The characteristic distance signal, a rising *o-o-o* call, repeated up to 15 times in a series and audible up to 5 km. Usually given while walking with head low to the ground and jaws parted (fig. 20.12), the vast majority of whoops are emitted by males. The calls of males are apparently ignored by other clan members, whereas on the rare occasions when a female whoops, her offspring and relatives come running (8).

2. Fast Whoop. Accelerated, higher-pitched, given by excited hyenas, for instance when competing with lions or another clan at kills, or by a youngster subjected to aggression (8). Clan members (or the youngster's mother) rally to this call, often answering in kind on the run.

3. Low. A drawn-out *o-o-o* sound, low-pitched, rising and falling. Commonly heard among hyenas waiting at a kill, it is often followed by fast-whooping and may express rising impatience and/or aggressive intentions.

4. Giggle. The high, cackling laugh which gives *Crocuta* its other common name, laughing hyena, is typically made by a fleeing individual while being attacked or chased, and indicates intense anxiety or fear.

5. Yell. The loudest, most intense call, starting as a scream and changing to a roar, emitted when a hyena is being attacked and trying to escape.

6. Growl. A deep, loud rumble, often with a staccato vibration. Emitted by an animal crouched defensively while threatening to bite or actually biting an attacker, this call often alternates with yelling.

7. Rattling-Growl. Soft, low-pitched staccato grunts given in rapid succession. An expression of alarm and possibly threat, this call is given by hyenas surprised at a kill or den as they run away from an approaching lion or person. A louder version is voiced during encounters between clans or as hyenas try to displace lions at a kill.

8. Grunt. A quiet, very low growl given with mouth closed by a hyena behaving aggressively, for example, a female with cubs toward an approaching male.

9. Groan. Similar to the grunt but drawn out, higher and variable in pitch, often uttered before or during the *greeting ceremony,* usually between relatives; also voiced by females and subadults interacting with cubs (8).

10. Whine or Whinny. Loud, high-pitched squeals and chattering noises made by begging hyenas. Cubs keep it up for minutes on end during weaning.

11. Soft Squeal. Like the whine but quieter and without chattering, squealing accompanies friendly meetings and submissive behavior after long separations.

Visual Communication. See Agonistic Behavior.

AGONISTIC BEHAVIOR

Dominance/Threat Displays

Attack posture (fig. 20.13): tail raised, hair bristling, head high, ears cocked, mouth closed, combined with a stiff-legged approach. The anal pouch may be rhythmically opened and closed at the same time (12). Two or more hyenas advancing shoulder-to-shoulder this way in the *parallel walk* is frequently the prelude to a mobbing attack on another hyena or an attempt to hurry lions off a kill. Attacks on another hyena normally focus on the shoulder region, which is protected by thick skin (8).

Defensive/Submissive Displays

Defense posture: ears flattened, *fear grimace,* hindquarters lowered and tail clamped. Commonly seen when a hyena is attacked by other hyenas or by wild dogs.

Fig. 20.13. Spotted hyena aggressive intent: advancing in *parallel walk* with tails erect (from a photo in Kruuk 1972).

Fleeing posture: ears flattened, hair sleeked and tail clamped.

Carpal-crawling or *groveling* (8): legs flexed, carpal joints bent under, ears flattened, *fear grimace,* tail often straight up or bent over back (cf. brown hyena).

Gradations and combinations of the above postures express varying intensities of the tendencies to approach or withdraw. A bold hyena motivated by curiosity may approach in the attack posture but keep its tail down. A timid one makes a hesitant approach with head high, but with ears back and tail low and flicking from side to side. An extremely fearful hyena driven by hunger may grovel while approaching, say, a feeding or resting pack of wild dogs, but without raising the tail. The first 3 postures are usually accompanied by vocalizations (see under Vocal Communication).

Behaviors associated with scent-marking, notably *pawing, pasting,* and *social defecation,* are also visually distinctive.

REPRODUCTION: nonseasonal, 2-week estrus cycle, 16-week gestation (98–132 days), 2 young (1–4), maturation at 3 years, females later than males (12).

Reproduction in the spotted hyena is not only unique but wildly improbable: copulation and birth both occur through the phallus (see fig. 20.10). The female's labia have been replaced by a false scrotum complete with testes-like fatty tissue. The vagina has merged with the urethral duct to form a common urogenital tract that makes a sharp bend and exits through the phallus (technically a peniform clitoris) (25). In immature females, the 2–3 mm slit in the glans is just like the male urethral opening. Copulation only becomes possible after this opening splits down the middle and enlarges to about 15 mm with sexual maturity; at the same time the black, wrinkled skin of the prepuce grows slack and baggy (12, 16).

SEXUAL BEHAVIOR. Copulation is rarely witnessed, because of the brief periods of estrus at intervals of 1–1½ years during which females are carrying or caring for dependent young, the lack of a definite breeding season, and the hyena's nocturnal habits. Because of the female's dominance, courting males are unusually diffident.

Frequent olfactory monitoring (see under Olfactory Communication) and persistent following of a female are reliable signs of approaching estrus (or sometimes postestrus) (7). But depending on the female's receptivity, males take anywhere from a week to over a month to work up the nerve to attempt mounting. During the active-courtship stage, fear of the female's aggressive response results in the rather comical *bowing display*. A male standing several meters away from a reclining female, with penis extended, suddenly lowers his muzzle almost to the ground, moves forward quickly, bows again, and *paws the ground* close behind her (20). If the female responds by growling or lunging, or merely raises her head and stares at him, the male instantly retreats. Yet groups of males have been known to gang up on and bait a female a day or two before she became sexually attractive to them (20).

Active courtship may continue for a week or more, during which males of successively higher rank succeed one another and move nearer the interesting female (6, 12). When at last she is ready to copulate, the alpha male follows her nose-to-tail awaiting opportunities to mount, meanwhile threatening other males that try to cut in. To gain entrance, he first grasps the female's loins, then slides back on his haunches and partly under her while his penis, which curves forward and upward when fully erect (cf. elephant copulation), probes for the opening. The female's organ, in contrast, is fully retracted, leaving the urogenital opening surrounded by the flaccid foreskin. She holds still with tail slightly raised. Having achieved intromission, the male stands up again and assumes a conventional doglike copulatory posture (fig. 20.14), thereby pulling the urogenital canal backward and upward and allowing deeper penetration. Toward the end of the 4–12 minute copulation, the male rests his head and body heavily on his partner.

Licking of the genitals is common following copulation, which may be repeated at intervals of ¼ hour or less for several hours (7).

PARENT/OFFSPRING BEHAVIOR. Females whelp in the entrance to a burrow, or in a hollow or thicket at some distance from the communal den. Again, the phallus is contracted during birth and the cubs appear at the greatly enlarged opening of the prepuce (12). Nevertheless, their passage through the phallus stretches it and sometimes tears or lacerates the urogenital opening. Weighing 1½ kg, the cubs are born with their eyes open but blind (8), with

Fig. 20.14. Spotted hyena copulation; a rival male (*left*) tries to interfere.

Fig. 20.15. Adolescent spotted hyena in *weaning tantrum* (from a photo in Van Lawick and Van Lawick-Goodall 1970).

canines and incisors already erupted, and able to crawl surprisingly well on their forelegs.

Considering the tortuous and constricted passageway spotted hyenas must negotiate to enter the world, the fact that they are larger and better-developed than other hyenas at birth seems still more bizarre. But the production of just 2 precocious young at a time is simply another manifestation of the unusually severe nutritional competition among lactating females. They and their offspring are in a race for dominance, which correlates with size (21). The more precocious the cubs and the faster they develop, the sooner they can begin traveling to kills and eating meat. Cubs are ready to eat meat by 2½ months, but only a privileged few get the chance to eat regularly before they are 7–8 months old. Despite one published record (10), mother spotted hyenas rarely bring meat back to the den.

After 10 days or so in a separate burrow, either the mother carries her cubs (by the back, neck, or nape) to the communal den (fig. 20.9), or else older cubs, which are strongly attracted to infants, move in on them. Juveniles only a few months old wander between burrows on their own initiative and their mothers may have to go looking for them to suckle them (12, 20). Dens are left entirely unattended by adults for long intervals, but it is then rare to see any small cubs aboveground. Babies are suckled half-a-dozen times a day and half as often at 3 months. A feeding bout lasts 1–1½ hours (!), during which the cub lies next to its mother in *reverse parallel,* while the second cub lies at right angles or between the hindlegs.

Though mothers continue suckling offspring for up to a year, weaning is a prolonged and traumatic experience. Yearlings refused milk frequently throw *weaning tantrums* during which they circle their mothers whining and whinnying, teeth exposed in a fearful grimace (fig. 20.15) Sometimes mother and young get mad enough to bite one another (20). Cubs may tag along on hunts at 1 year, but only begin to hunt effectively on their own at 1½ years.

PLAY. Young hyenas are playful and participate in the usual carnivore chasing, tug-of-war, and fighting games during evening and morning social hours at the den. Most of all, cubs enjoy reducing objects to the smallest possible size (8). Females play with cubs and sometimes adults romp in the water or around kills after eating their fill (12, 20).

References

1. Bearder 1977. 2. Bertram 1979. 3. Eloff 1964. 4. Estes 1967*b*. 5. Estes and Goddard 1967. 6. Frank 1986*a*. 7. ——— 1986*b*. 8. ——— *in lit.* 1986. 9. Frank, Davidson, Smith 1985. 10. Hill 1980. 11. Holzapfel 1939. 12. Kruuk 1972. 13. Matthews 1939*a*. 14. ——— 1939*b*. 15. Mills 1990. 16. Racey and Skinner 1979. 17. Schaller 1972*b*. 18. Smithers 1983. 19. Tilson and Hamilton 1984. 20. Van Lawick and Van Lawick-Goodall 1970.

Added References

21. Frank, Glickman, Zabel 1989. 22. Gorman and Mills 1984. 23. Lindeque and Skinner 1982. 24. Mills 1987. 25. Neaves, Griffin, Wilson 1980. 26. Tilson and Henschel 1986.

Fig. 20.16. Foraging aardwolf (from a photo in Kruuk and Sands 1972).

Aardwolf

Proteles cristatus

TRAITS. Like a miniature striped hyena. *Height and weight:* 40–50 cm and 8–12 kg, sexes equal (9). Slender build, sloping back, long legs, forefeet with 5 toes, hindfeet with 4. Head with pointed, hairless snout, wide palate, cheek teeth reduced to pegs, incisors normal and canines thin but long and sharp. Coat c. 2.5 cm long with thick, woolly underfur, long (16–20 cm) mane and bushy tail 19–28 cm long (11). *Coloration:* buff to reddish brown, paler underparts, with vertical brown torso stripes and black horizontal stripes on upper limbs; muzzle, ear backs, dorsal stripe, outer tail, and feet black; young resemble adults, but darker. *Scent glands:* anal pouch as in hyenas. *Mammae:* 4.

DISTRIBUTION. Discontinuous: grassland and tree savannas of the Somali-Masai Arid Zone from Ethiopia to central Tanzania; Southern Savanna and South West Arid Zone, absent from intervening *Miombo* Woodland Zone.

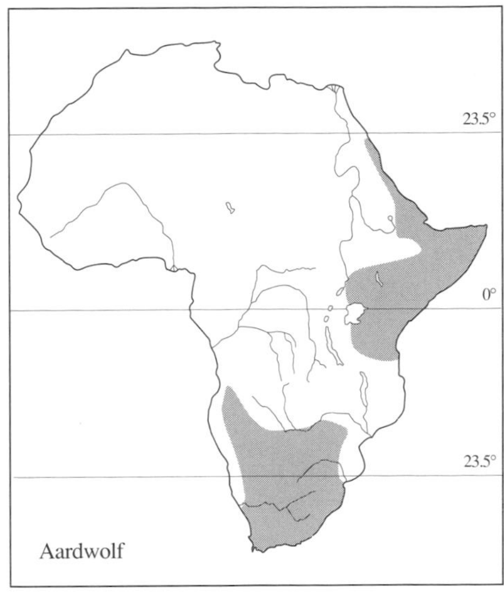

Aardwolf

ECOLOGY. One of the most specialized carnivores, the aardwolf dines almost exclusively on harvester termites of the genus *Trinervitermes* (2, 7, 11, 13). Unlike most termites, *Trinervitermes* leave their subterranean nests or low mounds at night by the thousands and crawl around on the ground to harvest dried grass. Some species stay underground and live off their harvest for part of the year. Aardwolves concentrate on certain species in which the storing habit is least developed; these kinds forage in the open at all seasons (7). The distribution of aardwolves is thus largely dictated by that of the termites they prefer: open, dry country with short grass, especially areas that have been overgrazed by ungulates (6).

Chemical analysis of harvester termites showed that 1 kg is equivalent to ¾ kg of lean meat (5). Thus the aardwolf has specialized on a protein-rich diet that it can obtain with minimum effort or competition from other predators. A foraging individual may consume up to a kilogram or 200,000 harvester termites in a night (7, 9). Other insects account for at most 10% of the diet, notably termites of different types, especially *Hodotermes,* the other, more diurnal harvester termite, plus ants, moths, beetles, and such. The diet is more varied in the rainy than in the dry season. In addition, there are a few reliable reports of *Proteles* having eaten small rodents, carrion, eggs, birds, and baby tortoises (2).

SOCIAL ORGANIZATION: nocturnal, solitary forager, probably monogamous (9, 14).

Because aardwolves forage and usually sleep alone, this species appears to be solitary, and indeed it is clearly not a social carnivore that happens to forage singly like the brown and striped hyenas. However, there is mounting evidence that male and female *Proteles* share a 1–2 km² territory (larger in subdesert conditions [8, 11, 14]) together with their most recent offspring, which males have been known to guard (see Parent/Offspring Behavior). Whether aardwolves generally live in pairs, and if so during all or only part of the year, and whether they are truly monogamous or not are still open questions. Apart from occasional presence in the same burrow, no evidence for the existence of pair bonds (*mutual grooming,* coordinated movements) has been reported; in fact, even nonhostile meetings with aardwolves tend to be brief and undemonstrative, with nothing resembling the elaborate greeting ritual between hyenas (9). Unconfirmed reports of several females and their offspring sharing the same den (10) also raise the possibility that daughters may sometimes settle and breed in the mother's range, as in the spotted hyena.

The evidence of territoriality rests on observations of scent-marking along home-range boundaries (see Olfactory Communication), on the mutual avoidance of 2 individuals that met exactly on the boundary of their respective ranges (7), and on aggression toward intruders, including long chases and serious fights if the intruder is overtaken (9). In the Serengeti, I once witnessed a chase of over ½ km in broad daylight, which ended with growling and threat displays after the pursued animal came to bay in a hole. However, the territorial system has been known to break down during periods of food shortage. In Namibia's subdesert Namib-Naukluft N.P., several aardwolves from at least 3 different territories foraged simultaneously for harvester termites, sometimes less than 100 m apart (2, 8). Social interactions among them were limited to staring or one running toward another.

Aardwolves are considered weak diggers but may burrow extensively when the ground is soft, extending or enlarging the holes of aardvarks, bat-eared foxes, springhares, and others. An excavated aardwolf burrow ended in a chamber after 2.3 m (2). A territory may include up to 10 different burrows, 1 or 2 of which may be occupied more or less regularly for 4–6 weeks before moving (9). But any burrow may be used as a temporary resting place or refuge; 1 female rested in 7 different burrows in 7½ hours (2), and in an emergency an aardwolf bolts unerringly for the nearest hole (11).

ACTIVITY. Aardwolves of southern Africa, where night temperatures regularly go below freezing in midwinter, usually emerge 1½ hours before sunset, and may occasionally sun in the den entrance at other times of day. In summer, dusk is the usual time of emergence (2). They generally return to their burrows by daybreak at all seasons. An aardwolf typically comes out of its earth abruptly and may begin foraging at once, but more often begins by scent-marking or proceeds straight (up to 450 m) to the nearest latrine to relieve itself (more under Olfactory Communication). Most animals take time out from foraging for rest breaks averaging 36 minutes (2–137), usually at or near a den (2).

FORAGING BEHAVIOR. A foraging aardwolf (fig. 20.16) walks quietly across the grassland, with head somewhat below shoulder level, covering perhaps 1 km in an hour at a walking pace of 3–4 kph between feeding and scent-marking stops. Its progress is erratic because of frequent, often abrupt changes in direction, usually into rather than away from the wind. Evidently patches of termites are located primarily by hearing, since the large ears are constantly moving, from sideways to forward (cf. bat-eared fox), while the muzzle remains still instead of moving from side to side as in animals questing by scent. People can detect the noise of termites cutting brittle grass from up to several meters, and the aardwolf, judged by the enlarged bony capsule enclosing the inner ear and the large external ears, has unusually acute hearing (7). It was also noted that aardwolves stopped hunting the moment it began to rain, suggesting that the patter of raindrops masks the sounds of termites.

Having located a patch of termites, the aardwolf proceeds to lick them up with its broad tongue, coated with sticky saliva, as rapidly as a cat lapping milk, accompanied by pushing and rotating head movements. The ears, cocked as the aardwolf homes in on a patch, are laid back. Inevitably, other items stick to its tongue along with the 5–6 mm termites: sand, dirt, straws carried by the termites, and the like amount to 15%–66% of the feces (2, 7). A female twice closely observed while foraging by daylight stopped and fed, on the average, every 88 seconds for 22 seconds (3–70). The same pattern was recorded at night, indicating that it took only a little over a minute to find patches of termites in the Serengeti savanna. In the drier Namib grassland aardwolves took over 2 minutes to locate food, and they spent much more time at each patch: 9.2 minutes (3–15) before midnight and 4.6 minutes (1–15) after midnight. Patches of termites averaging 16 m (2–40) across were spaced 53 m (10–100) apart one year and 26 m (2–80) apart the next, wetter year. All individuals that were closely monitored emerged and fed nightly (2).

Kruuk and Sands (7) noticed that aardwolves left food patches while termites were still milling around in thousands. These turned out to be mostly soldiers, which make up perhaps 20%–30% of the population. Soldiers protect the defenseless workers by spraying noxious terpenoids that most predators probably find at least mildly distasteful. When a colony is attacked, the workers immediately run for their holes, while the soldiers come out to cover their retreat. Apparently, an aardwolf stops feeding when only soldiers remain. However, the approximately equal representation of soldiers and workers found in sampled aardwolf scats (2, 7) would indicate they continue feeding for some time after soldiers come to outnumber workers.

SOCIAL BEHAVIOR

COMMUNICATION

Olfactory Communication: latrines, *pasting* with anal-sac secretion.

The ammoniacal smell of termites pervades the aardwolf's feces, which are exceptionally large and numerous. Scats 5 cm in diameter and equal to 8% of the animal's weight may be passed in a single defecation. The feces are deposited in latrines, of which a territory may contain up to 10 in active

Fig. 20.17. Aardwolf *pasting* (redrawn from Kingdon 1977).

use, though serving no apparent territorial function: the feces are buried, perhaps to minimize contamination of foraging areas (2, 7). Latrines are oval-shaped areas of bare earth averaging 4–6 by 3–4 m in sandy soil, with a pitted surface and exposed pieces of old scats. Location does not appear to be related to landmarks but latrines are often clustered; for instance, 8 were found in an area of 21 ha which at least 3 aardwolves were using.

An aardwolf prepares to defecate by digging a narrow trench with slow, alternating strokes of its forepaws, then crouches close to the ground while depositing its dung. Afterward it scratches dirt over the feces to a depth of 5 cm, and often urinates on the same spot, in a similar deep squat, followed by perfunctory scraping (8).

Like other hyaenids, the aardwolf scent-marks by extruding its anal pouch and smearing secretion on grass stems, bushes, or other objects (a hand-reared cub insisted on marking its bottle before feeding [11]). But instead of straddling and dragging grass stalks between its hindlegs, it swings its rear around to align the everted pouch with the stem, then coats a 3–6 cm section by waggling from side to side, causing its vertically erected tail to swing in a 90-degree arc (fig. 20.17). If long stalks are unavailable, an aardwolf will squat to mark short stalks or a shrub (8). Oddly enough, the secretion of aardwolves from Namibia is dark brown or black, but is yellow-ochre or orange in Botswana, eastern South Africa, and East Africa (6, 7, 8, 11). But regardless of color, it has a strong aromatic smell with an overtone of sweat (7), which persisted for over 3 years on a tainted skin (6).

Scent-marking is concentrated around dens and latrines, where aardwolves commonly paste nearby plants after defecating. As many as 71 fresh or nearly fresh blobs of secretion have been found around a den, with up to 5 deposits on a single stalk of grass, at heights of 10.5–37 cm (8). An average of 5 deposits (0–20) were found around the periphery of latrines. In general, marking frequency is highest at the beginning of nightly activity.

In addition to these visible deposits, aardwolves leave hundreds of nearly invisible scent-marks on grass stems wherever they forage, pausing for barely a second to wipe the anal sac sideways over the stem. The animal crouches with tail straight up but not waved. The marks may be left in rapid succession often less than 5 m apart. Conceivably, these more ephemeral marks are used to scent-map areas an aardwolf has already foraged, so that it or other family members will not waste time searching for depleted food patches (cf. brown hyena) (8).

Serengeti aardwolves of both sexes were also observed on 5 different occasions in the act of boundary marking, but by which of the 2 methods is unclear (7). The animals marked in prolonged bouts of 5–22 minutes, walking at normal speed and zigzagging as when foraging but without feeding, crouching and marking every minute and every 50 m or so. In South Africa, scent-marking reaches a peak during the mating season (9).

Vocal Communication: growl, grunt, roar, bark, click(?), howl(?), whistle(?).

The aardwolf is a very quiet animal except when defending itself or fighting. Then it emits growls, barks, and roars of surprising depth and volume for such a small animal (11). In addition to barking and growling softly, cubs are said to make a clicking noise very like the warning clicks of termites (1). A whistling call between mates and a howl like a striped hyena have also been reported (9). These last 3 calls need confirmation.

AGONISTIC BEHAVIOR

Offensive and Defensive Threat Displays. The aardwolf has one of the most impressive defensive-threat displays of any African carnivore. It can enlarge its silhouette by up to 74% when it stands side-on with mane and tail bristling (as in fig. 20.5) (12). But meanwhile it keeps its mouth closed—so that enemies won't realize what weak teeth it has, one scientist speculated (3). Threatening aardwolves sometimes drop to their knees like striped hyenas, and the neck mane can be erected independently of the dorsal crest, emphasizing the head

(6). These and other behaviors may be common during encounters between aardwolves, which remain to be described in detail.

Fighting. The contestants *bristle,* bark, and roar in more or less the same way as when confronting enemies of another species. Males may have 1 or 2 fights per week during the breeding season (9).

REPRODUCTION. Most young are born during the rainy months, and there is a defined mating season in South Africa (9, 11).

SEXUAL BEHAVIOR. Territorial boundaries are often violated during the mating peak. Despite the spirited defense of resident males, intruders sometimes succeed in mating with resident females (9, 14).

PARENT/OFFSPRING BEHAVIOR. After a 3-month gestation (90–100 days), 2 or 3 (1–5) offspring are born, often in an old aardvark burrow, with eyes open but otherwise helpless. They spend 6–8 weeks underground before venturing outside. During the 3 months they remain attached to the den, males have been known to spend up to 6 hours a night on guard while the mother foraged (9). The cubs begin foraging for termites at this stage, accompanied at first by at least 1 parent. At 4 months they often forage alone but usually sleep in the same burrow as the mother. Aardwolves reach adult size by 9 months (6) and at yearling stage already venture far beyond the boundaries of the parental territory. They disperse before the next litter begins foraging (9).

Cubs taken from the mother when less than ⅓ grown almost always develop a nervous disorder and die, despite administration of vitamins (11). In play, cubs dart about waving their tails like jackals.

ANTIPREDATOR BEHAVIOR. Recurrent speculation that the aardwolf deliberately mimics the striped hyena is belied by the fact that aardwolves look the same regardless of whether they are associated with brown or striped hyenas (4). The aardwolf's threat display helps to compensate for its inability to outrun larger predators (other than lions and leopards), its small size, and its weak teeth. However, like the bat-eared fox, aardwolves are very shifty runners, using the raised tail as a rudder and perhaps as a teaser (to deflect bites from the body) (5, 10).

References

1. Bartlett and Bartlett 1967. 2. Bothma and Nel 1980. 3. Ewer 1968. 4. Goodhart 1975. 5. Ketelhodt 1966. 6. Kingdon 1977. 7. Kruuk and Sands 1972. 8. Nel and Bothma 1983. 9. Richardson and Bearder 1984. 10. Shortridge 1934. 11. Smithers 1983. 12. Wemmer and Wilson 1983.

Added References

13. Richardson 1987*a*. 14. ——— 1987*b*. 15. Skinner and Van Aarde 1986.

Chapter 21

Cats

Family Felidae

African Species

Felis lybica, African wildcat
F. margarita, sand cat
F. nigripes, black-footed cat
F. aurata, golden cat
F. chaus, swamp cat
F. serval, serval
F. caracal, caracal

Panthera pardus, leopard
P. leo, lion

Acinonyx jubatus, cheetah

FAMILY TRAITS. Carnivores resembling the domestic cat, with long, lithe body and long tail, or shorter body and bobtail (lynx type), males somewhat larger and more muscular than females, otherwise minimal sexual dimorphism (except lion). Legs short (sand cat) to long (serval, cheetah); big, soft-padded feet with 5 front toes and 4 rear toes armed with hooked, sharp claws normally retracted in sheaths (except cheetah). Head rounded, with large cranium and short jaws; big, frontally placed eyes; ears triangular to rounded, ± prominent; conspicuous whiskers on upper lip. Teeth typically reduced to 30, specialized for eating meat: large, shearing carnassials and only 1 small upper molar, small, chisel-like incisors, canines long and laterally flattened; tongue rough, covered with horny papillae. *Coloration:* cryptic, pale gray to brown, sometimes black (especially leopard and serval), with pale or white underparts, most species spotted, striped, or blotched, some plain-colored (lion, caracal); many with black and/or white markings on ear backs, face, and tail. *Scent glands:* paired anal glands, scent glands in foot pads, chin, and cheeks of at least some *Felis* species. *Genitalia:* short, barbed penis situated just below testes with urethral opening facing backward, bone present but reduced.

Cats are so much alike in nearly every way except size that all but the cheetah could arguably be placed in the same genus, *Felis*. They are so close genetically that species within the same size range can interbreed and even produce fertile hybrids (e.g., tiger × lion, leopard × lion, domestic cat × various wildcats). But most taxonomists separate the 7 big cats, lumping 5 in the genus *Panthera*. This group shares an anatomical specialization that enables them to roar: the hyoid apparatus (voice box) is suspended by an elastic ligament that permits vibration and sound magnification (3, 5).

Senses. Vision is highly developed in the Felidae. Apart from good binocular vision of the frontally placed eyes and cone receptors providing visual acuity (and possible color perception) equal to the human eye by daylight, cats can see at least 6 times better in the dark, partly because of the light-reflecting layer (the tapetum lucidum) behind the retina, and the very large pupil (3). The iris varies in color from orange and yellow to green, gray, or brown; under bright light the pupil contracts to a small dot in the big cats and caracal, a vertical slit (wildcat, black-footed cat), or spindle shape (serval). Hearing is also acute in most felids, especially the smaller species that use their large mobile pinnae like dish antennae to locate hidden prey precisely (see serval account). The sense of smell is least important for predation but preeminent in felid social interactions (see under Social Organization and Olfactory Communication), and urinalysis with the vomeronasal organ enables males to monitor female reproductive condition. The presence of prominent whiskers and other sensory vibrissae indicates the importance of the tactile sense in this family (6).

DISTRIBUTION. Worldwide except Australia, the Antarctic, Madagascar, and other oceanic islands. *Felis* is one of the most widespread mammal genera, and the leopard is one of the widest-ranging land mammals in the Old World.

ANCESTRY. The first recognizable felids date back 40 million years to the upper Eocene, and representatives of modern cats, subfamily Felinae, have existed since the Miocene, a good 24 million years ago. However, from then until just a couple of million years ago the family was dominated by saber-toothed cats. True cats (*Felis*), the size of a large wildcat, date from the late Miocene of Eurasia and the Pliocene of South Africa (table 1.1, fig. 18.1). The modern big cats arose (probably from a leopard-like prototype) as predators on modern ungulates, which the saber-tooths, apparently specialized for grappling and killing ponderous, primitive mammals, were poorly adapted to capture (3).

The seldom-seen but common golden cat of Africa's rain forests may be close to the prototype from which modern cats evolved as they moved into more open habitats. The golden cat and the caracal are considered to be survivors of an earlier radiation (3, 5). The cheetah has been running down gazelles since the late Pliocene, but the other big cats only became dominant in the early Pleistocene, during the Golden Age of Mammals.

ECOLOGY. Found in all major habitat types from desert (sand cat) to equatorial rain forest (golden cat, leopard), to swamp and marsh (swamp cat, fishing cat), to high mountains (leopard). Apart from the cheetah and the adult tiger and lion, probably all cats are good climbers; yet Africa, unlike Asia and South America, has no specialized arboreal hunters (7). Most cats climb trees for their own safety; those that live on treeless plains take refuge underground.

DIET. Of all the carnivores, cats are the most committed meat-eaters. Although reptiles, amphibians, fish, shellfish, and even insects are included in the diet, rodents, rabbits, and hares are no doubt the mainstay of most small cats. The bigger ones prefer ungulates, but take whatever is most readily available.

SOCIAL ORGANIZATION. Because of their secretive and nocturnal habits, few cats have been closely studied in the wild. But judging from those few and observations of others in captivity, social organization and behavior are as similar as felid morphology. The familiar assertion that a housecat is simply a miniature tiger is essentially true. Knowledge of this domestic feline provides insights into the psyche of all cats.

With few exceptions, cats are truly solitary animals which avoid one another except when sexually attracted. In closed habitats, prey species are usually dispersed and best hunted alone, and mostly too small to be shared by a group. Relying mainly on scent-marks, which carry information about when the mark was made and the sex, reproductive status, and identity of the marker, individuals can share a hunting range without meeting unless they choose to do so (1, 6). Adult males of probably all species defend at least part of their ranges against other males. The ranges of females are generally smaller and several may be included within a male's territory. The extent to which females are territorial is unclear in most cases, but even when their ranges overlap, adult females usually avoid both one another and males. The African lion is the only social cat and has evolved an elaborate clan-territory system based on close kinship bonds among females (9). The pride owns an extensive range frequented by large, open-country ungulates, which pride members hunt and eat cooperatively. But latent sociability may be seen in the fact that tigers, normally solitary, summon one another to share large kills (8), and adult male cheetahs, like lions, form coalitions. Furthermore, feral housecats subsisting on an abundant food supply such as a garbage dump develop a social organization comparable to the lion's, including baby-sitting by subadult males and communal guarding, suckling, and feeding of kittens (7).

ACTIVITY. Although many cats can be active in daytime, all but the cheetah are primarily nocturnal, night being the time most favorable for stalking and ambushing predators to go hunting.

POSTURES AND LOCOMOTION. Although few cats are truly arboreal, they have retained much of the freedom of movement associated with climbing, which terrestrial carnivores have sacrificed in becoming more cursorial (4). The use of the claws in capturing prey and in fighting also requires limb mobility, as well as a muscular build. In acquiring the strength to overpower large prey, cats sacrificed speed and endurance. Most prey of rabbit-size and up can outrun their felid predators, which accordingly have to rely on stealth to get close enough to capture them. If their quick rush or pounce misses, they usually quit, for cats (including even the cheetah) are notoriously short-winded. A comparatively small heart (e.g., amounting to only 0.36% of a leopard's body weight) partly explains the felid's lack of endurance (cf. canids, Postures and Locomotion). But they cover long distances at a walk or trot, and they are also good swimmers and jumpers. The tail is important as a balance organ, especially in the more arboreal species.

PREDATORY BEHAVIOR. All cats capture and kill prey in basically the same way as the domestic cat (6). The different steps involved in hunting, capturing, killing, and eating prey are traits shared by the whole family which appear at the appropriate time in the development of each individual. Yet practice is needed to perfect these innate motor patterns, and experience leads to individual variations and improvements on the basic techniques. Interestingly enough, each link in the predatory chain exists as an independent drive: the urges to stalk, to catch, and to kill can each stand alone and separate from the urge to eat (6). Once this is understood, some of the inconsistencies in the predatory behavior of cats become less puzzling: for instance, why gorged lions can't resist the chance to make another easy kill, why cats often abandon a quarry after killing it and even when hungry usually don't begin eating right after killing their prey, and why cats play with and may not even kill the animals they catch.

Considering that the usual prey of most cats is small, why do they need to be so powerful? The capability of killing animals their own size or bigger increases the food resources that are available to the predator and the areas where it can subsist. There is always the risk of injury, though, particularly from well-armed prey. The cat's strength and killing techniques minimize the risks of bringing down and immobilizing large animals while administering the coup de grace. Cats hardly ever jump out of trees upon formidable prey. They spring or rush forward and grab the quarry by the rump, back, or shoulder with a forepaw; both paws are used when pouncing on small prey and to grapple large prey. The back feet remain on the ground, providing purchase to pull the quarry off its feet. The unsheathed claws sink through the skin like meathooks when the cat pulls the prey toward itself. But while chasing at full speed behind a fleeing animal, the claws remain sheathed when the cat adroitly trips it up with an extended foreleg—to be attached to a large animal as it falls would be dangerous. This technique is practiced by cheetahs and lions as well as by other cats (6).

Fig. 21.1. Leopard strangling gazelle (redrawn from Leyhausen 1979).

Cats use a very precise bite to kill small prey: a canine tooth is deftly inserted between 2 cervical vertebrae, wedging them apart and partially or completely severing the spinal cord, bringing instant death. But the big cats dispatch large prey by suffocation, holding the throat or muzzle with jaws designed to clench with maximum sustained force (fig. 21.1) (3, 5). Small prey is always eaten beginning at the head. Large prey, with skull too large to crack, is eaten beginning at the belly or chest, although there are minor differences among species (9). Meat is sheared from the bones with the carnassials or pulled off with the incisors and swallowed without chewing. *Felis* species eat in a crouched position and rarely use their paws to hold their food, whereas *Panthera* species normally lie down and hold their meat between or beneath their paws (6). Most cats partially pluck the feathers before eating birds.

Cats tend to be frugal with their food, caching the remains of a kill by covering it with leaves or grass and returning to feed on it as long as it lasts. Leopards and sometimes caracals take their kills up trees. Many species will eat carrion, and many can go for days between meals, making up for a fast by gorging up to ¼ their own weight. Their catholic choice of prey species extends to all smaller cats, sometimes including members of their own species.

SOCIAL BEHAVIOR

COMMUNICATION. Felids have the most mobile lips and changeable facial expressions of all carnivores. Why should cats need a more complex communication system than other solitary animals? Perhaps because of the unusually dangerous consequences of fighting between animals armed both with long fangs and sharp claws (9, 10). The ability and readiness of cats to defend themselves with these weapons tend to inhibit outright attack. Instead, cats employ discrete distance signals (scent-marks and loud calls) to avoid surprise meetings, and variable, graded visual and vo-

Fig. 21.2. Felid offensive and defensive body postures: A_0B_0 denotes a normal, relaxed posture, A_3B_0 a posture of offense unmodified by fear, A_0B_3 a purely defensive posture. Other illustrations show postures assumed when offensive and defensive emotions are both present, in varying degrees, culminating in A_3B_3 when a cat is strongly and equally motivated by offensive and defensive tendencies. (From Leyhausen 1979.)

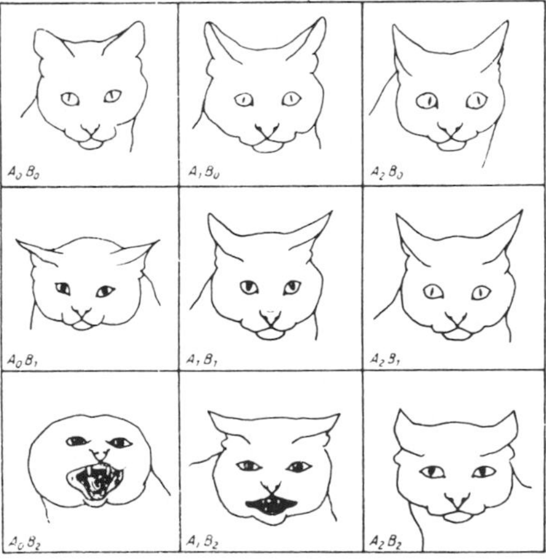

Fig. 21.3. Offensive and defensive facial expressions: same arrangement as in fig. 21.2, but to a less extreme degree. Here A_0B_0 denotes a relaxed expression, A_2B_0 an offensive mood unaffected by fear, A_0B_2 a purely defensive threat, and A_2B_2 equally strong offensive and defensive moods. (From Leyhausen 1979.)

cal signals, plus olfactory and tactile messages, to communicate their feelings and intentions at close range under any and all conditions (9, 11). The domestic cat, a typical small feline virtually identical to the African and European wildcats, uses at least 9 facial expressions and 16 different tail and body postures for a total of 25 visual patterns, 16 of which occur in combination. It has 8 distinct calls, 3 kinds of scent-marks, and about 7 contact patterns (rubbing, clasping, patting, etc.) (2, 11). The lion's communication system employs at least 17 visual patterns, 13 calls, 5 olfactory signals, and 7 contact patterns. To make the positions and movements of the ears, tail, eyes, and lips as plain as possible, these organs have contrasting markings in most cats.

Visual Communication

Offensive and Defensive Facial Expressions and Postures. Some of the possible variations in feline visual signals are shown diagrammatically in figures 21.2 and 21.3, where facial expressions and postures associated with offensive and defensive moods are illustrated in the domestic cat. The upper left-hand drawings represent the face and body in the relaxed, neutral state; the upper right figures represent the strongest threat and the lower left figures the greatest readiness for defense. Note that the ears are erect and turned outward, presenting a triangular shape, in the most intense offensive threat, and the eyes are wide open with small pupils (fig. 21.3 A_2, B_0); whereas the ears are flattened, presenting no silhouette, and the eyes are narrowed with dilated pupils in a purely defensive mood (A_0, B_2). The other drawings show gradations between the 2 extremes when a cat is activated by both emotions to varying degrees and intensities. The lower right drawing in figure 21.2 (A_3, B_3) shows what happens when motor patterns associated with flight, defense, and attack are all superimposed, at high intensity. This is the classic posture of a cat confronting a dog. Only at highest intensity are facial expressions and body postures completely synchronized. Otherwise the face registers changes of mood quicker and more subtly than does the body (6). Consequently it is necessary to illustrate expressions and postures separately.

The mildest expression of aggression is a bland unwavering look (fig. 21.3 A_1, B_0), as if the cat was looking through or past its opponent (9). Conversely, the mildest expression of an intimidated cat is *looking away* or *looking around.* When a cat in strange surroundings is approached by an-

Fig. 21.4. *Consummatory expression* of fully fed lion licking its bloody paws.

other cat, it sits or crouches and calmly surveys its surroundings in every direction but that of the other cat. The stranger often appears more at ease than the resident. In like manner, a stalking lion sits up and looks around indifferently when it is spotted by its intended victim. In both situations cats are seen to be disconcerted by a direct stare, and indeed a cat on the defensive can also halt the approach of an aggressive opponent by looking directly toward it (6).

Nonaggressive Facial Expressions

Consummatory face (fig. 21.4). Associated with certain activities apparently pleasurable to the animal, notably eating contentedly, urinating and defecating, sharpening the claws, rolling in carrion, catnip, or dung. The eyes may be closed, or open but "dreaming" or "staring into space," the ears partially flattened.

Alert face. An expression of intentness as a cat surveys its surroundings, particularly after sighting potential prey; mouth closed or slightly open, ears cocked, and eyes wide open.

Play face. Mouth slightly open, corners drawn back and slightly raised without showing teeth; ears and eyes as in relaxed face or fairly alert.

Urine-testing (flehmen). Wrinkling up nose with mouth open and eyes closed, after smelling urine, estrous females, and sometimes carrion. Given mostly by males.

The *flehmen* grimace is much more pronounced in *Panthera* than in small cats. Otherwise facial expressions and ear movements are very similar in all cats. So are the postures and behaviors associated with defense, although the big cats do not arch their backs. The greatest differences are seen in the postures and movements associated with offensive threat and attack. For instance, threatening lions keep their heads low, backs stretched, and rib cage lowered so that the shoulder blades stand out prominently (6).

Vocal Communication. A wide range of different calls is represented in the family, including the cheetah's birdlike contact chirp. Other monosyllabic short sounds include panting, inspiratory gasp (leopard only), expiratory bursts, "knocking," and guttural hissing. Examples of variable, longer calls are yowling, meowing, roaring, cooing, snorting, rumbling, growling, grunting, gargling, snarling, and purring. Mothers use a special call to summon kittens when bringing home live prey. Call intensity is clearly different when the prey is small and harmless (mouse) or large and potentially dangerous (rat), and the young respond accordingly (6).

Most small cats lack distance calls comparable to the roars of *Panthera* species, although some *Felis* females announce the onset of estrus with loud calls, and males of seasonal-breeding species may call more often then or utter special mating calls. As in visual signals, most vocal signals are concerned with offensive and defensive behavior (11). Thus, most small cats have 6 calls in common: hissing, spitting, growling, screaming, purring, and meowing, of which all but the last 2 are hostile (11). Several of these calls grade into one another and express rising excitement by increasing noise and pitch: growling—yowling—screaming (the highest, noisiest vocalization), followed by a volley of spitting.

To demonstrate how calls are often linked with visual, olfactory, or tactile signals as part of a complex display, consider growling and snarling, the one offensive and the other defensive in character. It is basically the same vocalization, but when given with the mouth closed during offensive threat it comes out as a growl, and when the mouth is opened with lips retracted in a defensive threat it becomes a snarl (9). In this way, the vocalizations of an aroused, angry cat express every shade of emotion from pure aggression to pure fear while precisely tracking the visual signals.

Meowing signals mild distress, especially in kittens, helping to remind mothers of their whereabouts, whereas purring, as everyone knows, expresses contentment, as when kittens are suckled or licked, or when adults greet or groom one another. Felines have no special alarm calls; they simply growl (9).

Fig. 21.5. Lion patrolling and scent-marking defended part of pride territory by *urine-spraying.*

Olfactory Communication. The main form of scent-marking is *urine-spraying* directed backward against objects at nose level or higher. The placement and form of the penis seem specifically adapted for this purpose (6), and indeed adult males are the main practitioners. But adult females also spray, in some species more frequently than males, particularly during estrus (6). The basic marking technique is as follows: first the cat sniffs the place to be marked intently, especially if it has been marked before, in which case it often *grimaces.* Next, or after urinating, it rubs its head against the object, thereby anointing itself with its own or another's urine, meanwhile wearing the *consummatory face* (fig. 21.4). Finally, it backs up and sprays its own urine on the spot (figs. 21.5, 21.13). Feline urine has a potent smell, especially the male's, but there is no evidence that anal-gland secretions are sprayed or otherwise mixed with the urine (11), as sometimes asserted (9). Indeed, the function of the cat's anal glands is still a mystery.

Scuffing. A rhythmic treading and forward-kicking motion of the back feet with claws extended is performed by many of the bigger cats (fig. 21.6). If, as commonly occurs, the cat also urinates, the urine becomes mixed with soil and impregnates

Fig. 21.6. Lion *scuff-marking* (from Schaller 1972*b*, drawing by R. Keane).

the hindfeet. Scuffing visibly disturbs the ground.

Rubbing and rolling. Head-rubbing is stimulated by odors other than urine (e.g., catnip) and may serve to deposit glandular secretions as well as to anoint the animal with scent. After sniffing, or licking and biting the place, and often after *grimacing,* the cat proceeds to rub its lips, cheeks, head, and neck on the spot. Copious amounts of saliva often appear on the lips during a vigorous rubbing bout, soaking the spot and the rubbed areas of the cat (11). Cats are also stimulated to rub their cheeks and neck on smelly, decomposing substances such as carrion and dung, followed sometimes by rolling and writhing of the whole body on the spot. The parts of the body which are most commonly rubbed, namely the forequarters, are also the places most commonly sniffed and rubbed during social encounters between cats (more under Tactile Communication).

Defecation. The act of defecation is similar among species and sexes, but other associated acts vary. Feces are apparently unimportant for scent-marking, at least in many smaller cats that rake soil or litter over their scats. *Panthera* species make no attempt to cover theirs and a few deposit dung at specific or conspicuous sites (see leopard account).

Display claw-sharpening. Probably all cats rake the claws of their forepaws on trees. Aside from removing loose claw sheaths, clawing leaves visual and possibly scent marks (assuming glands described in the feet of domestic cats exist in other cats [11]). The act of clawing may also be a visual display.

There are few clues to the significance of these different marking behaviors, either for the sender or the receiver. Urine-spraying is often considered territorial behavior, but if so there is no evidence that these marks intimidate or deter other individuals from using the same areas, although cheetahs have been seen to alter course after checking fresh scent-marks (1, 6). Such marks may well be signposts from which cats find out who is ahead and how far. *Spraying* and *scuffing* may also be assertions of self-confidence if not space claims, since cats often scent-mark during or following aggressive encounters, around kills, and so on (6, 9, 11).

Tactile Communication. Nonaggressive meetings between cats begin with *mutual sniffing* of noses, head and neck stretched forward and ears erect in friendly curiosity (6). They proceed to sniff and touch light-

Fig. 21.7. Lion *greeting ceremony.*

ly with their whiskers the nape and flanks, and finally the anogenital area. A friendly cat raises its tail to allow this inspection, but often one is more fearful, keeps its tail down, and sidesteps, so the pair ends up *circling.*

Greeting ceremony. When family members meet they go through a ritual that begins with sniffing noses, followed by rubbing their heads and often their sides together (fig. 21.7). The tail is erected with the tip bent toward or draped over the partner. Small kittens rub against the mother's chest and throat. The similarity between greeting and the sexual presenting of an estrous female is striking, and comparable to social presenting in primates (6). The ceremony is more frequent and pronounced in some of the larger cats, above all in the lion.

AGONISTIC BEHAVIOR. See also Vocal and Visual Communication. The following account of aggressive interactions, based on studies of feral housecats (6), applies with minor variations to probably most small and at least some of the big cats.

The intensity and outcome of hostile encounters between cats vary according to circumstances and the sex and motivation of the participants. Ranking among males whose ranges overlap or adjoin is usually settled by fighting. *Rival fights* are serious combats with the underlying intent of inflicting damage by delivering a neck bite. A single fight between adult males usually settles the issue of dominance, but young adults may go through many fights before their standing is established. Dominance determines which of the males attracted by the calls and scent of a female in heat ends up mating with her.

The approach of an aggressive male is made in the posture shown in figure 21.2 A_3, B_0. The animal moves very slowly and stiffly and when close, raises and moves its head from side to side, meanwhile growling deeply and yowling. The tail tip twitches rapidly and spasmodically. The defensive cat gets ready to roll onto its back, first by turning its forequarters, freeing a forepaw, and raising it to strike the enemy on the nose if it continues to advance. The attacker can only press home a biting attack by disregarding the risk of injury to itself. Usually the aggressor stops, draws in its own neck, and hits out with a forepaw, indicating that it is on guard and ready to repel a counterattack. The defender immediately responds by rolling onto its back, with all 4 paws ready for action (fig. 21.8). Should the aggressor follow through, the defender clutches and pulls him toward his open jaws while his hindfeet trample and rake his exposed underside. The aggressor has no choice but to follow suit and they now roll on the ground, the aggressor yowling, the defender hissing,

Fig. 21.8. In a dispute over food, a lion (*right*) resorts to the characteristic felid *rollover self-defense,* which often inhibits an aggressor's attack.

spitting, and growling, culminating in hair-raising shrieks. After breaking off, the winner typically sniffs the ground (*displacement sniffing*) and slowly moves away. The loser remains crouched until its rival is out of sight (6).

Territorial and maternal defense. Both types feature a sudden, swift attack, chasing, and hitting without preliminary displays. If the trespasser stops and defends itself, both stand or rear with paws raised to strike, showing that territorial and maternal attacks are defensive and not purely offensive. Mothers of small cubs are the most determined and fearless.

Long chases are associated only with territorial defense and *mock pursuit* during mating. In *rival fights,* the stiffness associated with offensive threat display actually prevents the aggressor from pursuing a rival that elects to flee (6).

REPRODUCTION. Gestation ranges from about 2 months in small cats to over 3 months in the lion and tiger. Litters include 1–8 offspring, with 2–4 the usual number. The smaller species may reproduce yearly beginning at a year or younger; the larger ones begin a year later (males later still), and reproduce at intervals of no less than 1½ years. But when a litter is lost, females reenter estrus within a few weeks (see lion account), and keep cycling every few weeks until fertilization occurs, either during a regular mating season or, particularly in the tropics, at any time of year. Estrus may last anywhere from a day to 2–3 weeks (6).

SEXUAL BEHAVIOR. Frequent copulation over a period of days is a felid characteristic (see lion account). Cats are induced ovulators and repeated copulation is necessary before ova are produced and conception can occur (3, 9).

A female coming into heat advertises her approaching receptivity by become hyperactive, calling, scent-marking, *rubbing against objects,* and *rolling* sinuously on the ground. This behavior disrupts the usually solitary existence of mature but sexually inactive cats, attracting the local males who proceed to compete for the right to mate. They engage in a lot of display and some combat (6). The female's courtship behavior is alternately coquettish and defensive. She *rubs* against the male, *presenting her posterior with tail raised* invitingly, but when he makes bold to approach, she turns upon him *spitting,* scratching, and yowling. The next moment she writhes on the ground at his feet or invites pursuit. A female in

Fig. 21.9. Felid copulation (lion), with female crouching, male making biting movements at her neck.

full estrus adopts the *copulatory crouch* (fig. 21.9), head low, eyes narrowed, ears slightly back, rump elevated, tail raised and deflected. Sometimes a female *treads* with the hindfeet.

Male courtship behavior features frequent *urine-spraying, rubbing heads* with the female, frequent licking of the erect penis, long pursuits, and patient waiting. When the female invites copulation, the male tries to grip her neck before mounting. This may induce passivity in the female, in the same way that kittens go limp when so gripped. But quite often a female will try to escape, twisting and scratching. Once the male straddles the female, instead of grasping her with the forelegs as in most carnivores, copulation is usually accomplished with a few pelvic thrusts. Small cats commonly maintain the *neck grip* the whole time, whereas in *Panthera* species the bite has been reduced to a mere baring of the teeth or a *symbolic bite* a moment after ejaculation (6). Immediately afterward the female screams or snarls, and often twists and hits out at the male, which jumps back, growling. The female then licks her vulva and/or rolls. Within a few minutes courtship is resumed. After several days of this, the initiative shifts increasingly to the female, which uses *rubbing, rolling,* and *presenting* to renew the male's flagging interest. Not infrequently males lose interest while the female is still going strong, and another male takes over (see lion account). This could explain the observation that males of solitary species tend to be more tolerant of one another than are females. Waiting may be a more viable strategy than all-out fighting to gain mating opportunities (6).

PARENT/OFFSPRING BEHAVIOR. The young are born blind, deaf, and barely able to crawl. In probably all species but the lion, the mother bears sole responsibility for feeding and rearing them. Males play no paternal role. Cubs remain hidden in a den or nest for several weeks until they become mobile. During this time the mother's normal flight response may be blocked, leading her to attack even against impossible odds.

Older cubs are warned to take cover by the mother's hissing and slapping at them; in this way they learn which animals are to be feared (6). The mother readily changes hiding places if much disturbed, carrying 1 kitten at a time by the head, nape, or back skin. The mother retrieves and vigorously licks strayed babies—but only if they meow loud enough to be heard.

Felid mothers train their young to fend for themselves, prior to separating from them shortly before the next litter is born. While her offspring are still very young, the mother begins bringing prey back to them, first dead, later still alive. At first the kittens are more frightened than attracted, but the escape attempts of live prey soon stimulate their hunting urge. All the motor patterns of prey catching and eating are practiced long before the young become independent. The neck bite is the last link to be perfected. Seemingly 1 or 2 experiences in delivering the bite are necessary before the pattern "clicks" into place and becomes automatic (6).

References

1. Eaton 1970*a*. 2. Eisenberg 1973. 3. Ewer 1973. 4. Jenkins and Camazine 1977. 5. Kingdon 1977. 6. Leyhausen 1979. 7. Macdonald 1983. 8. Schaller 1967. 9. ——— 1972*b*. 10. Tembrock 1968. 11. Wemmer and Scow 1977.

Added Reference

Packer 1986.

The only 2 strictly African cats are the serval and black-footed cat. The others also occur in Eurasia or did until very recently. Because of their extremely secretive habits, too little is known about the behavior and ecology of these small cats to present comprehensive species accounts. Dissimilarities in their form and distribution clearly indicate still unstudied differences in their ecology. However, observations in captivity reveal relatively minor differences between species in displays and other basic behavior patterns. Descriptions in the family introduction should be consulted to fill gaps in the species accounts.

Fig. 21.10. African wildcat, resting in a disused hole, its refuge in open grassland.

African Wildcat
Felis lybica

TRAITS. A wild tabby cat. *Weight and length* (head and body): (Botswana population) males 5 kg (3.7–6.4) and 58 cm (53–63), females 4 kg (3.2–5.4) and 55 cm (51–58), tail 31–37 cm (7). *Coloration:* geographically and individually variable, lighter in dry and darker in wet regions, usually gray or tan but varies from orange to black; dark garters on upper limbs, outer tail ringed, ending typically in black tip; striped forehead and cheeks, fainter stripes on torso and indistinct spots on chest; underparts lighter. Distinguished from domestic tabby by rufous, unmarked ears and longer legs, resulting in more upright seated posture (8). *Scent glands:* anal glands, also chin, cheek, and possibly foot-pad glands. *Mammae:* 8.

DISTRIBUTION. Pan-African outside the Lowland Rain Forest and the Sahara, it is the commonest African cat. The wildcat of Europe and Asia, *F. silvestris,* is considered by some authorities to be the same species, despite a heavier build, shorter legs, and long hair (3). Embalmed cats retrieved by thousands from Egyptian tombs show that the wildcat was domesticated probably as early as 4000 B.C. The Egyptians worshiped Bast, the cat goddess of the hunt, love, and pleasure (3). Egyptian drawings and figures of cats depict the distinctively upright seated posture of *F. lybica,* an indication that domestic cats descend from the African wildcat. But the long legs that account for the posture disappear, along with the translucent, unspotted ears, when the wild form crosses with modern-day housecats from African villages and towns (8).

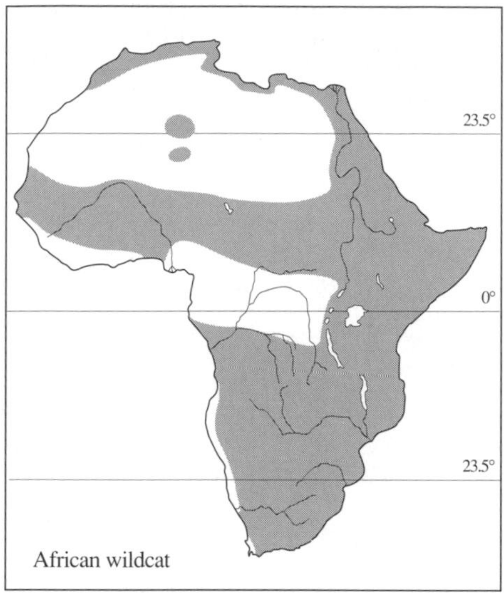

African wildcat

ECOLOGY. The wildcat occurs in virtually all places where rats and mice are plentiful, including the outskirts of villages and towns. Attraction to human habitation may have led to domestication in the first place. Wildcats even live in treeless open grassland, as on the floor of Ngorongoro Crater. Here they depend on holes dug by other species for refuges against a host of bigger, faster predators (fig. 21.10). One orange-colored cat with greenish eyes, seen occasionally sunning, lived in a burrow complex that also housed hyenas, porcupines, and bat-eared foxes (author's observation).

The food habits of the wildcat differ little from those of free-ranging housecats. Rodents are the mainstay, with small birds, lizards, snakes, frogs, and large insects (beetles, grasshoppers, winged termites, centipedes, spiders) taken as opportunity affords. The springhare and the African hare are normally the largest prey of this species.

SOCIAL ORGANIZATION: solitary, territorial.

Since the African wildcat and the housecat are one and the same species, interbreeding as readily as wolf and domestic dog, it should be one of the most familiar of Africa's wild animals. But no proper study of a wild population has been made, so in this case knowledge of a wild animal rests on studies of the domesticated form, and on observations of the closely related (if not conspecific) European wildcat. The variability in social organization found in feral housecats, notably the formation of clans and the cooperative care of offspring where there is a food surplus (see family introduction), has yet to be seen in African wildcats. They appear to be quite strictly solitary and territorial (8). Yet a tendency toward cooperative care can be seen in the behavior of 2 hand-reared wild females. After littering, each became territorial, marking and fighting over the same range. The only way to keep the dominant one from attacking and driving the other away was to let them out by turns. However, despite their unrelenting antagonism, each brought back and left food offerings (mice, birds) for the other's kittens.

Since the African wildcat is dispersed in much the same way as its European and rural domestic counterparts, it probably has a similar social organization (see family introduction). The following information on the dispersion pattern of wild and feral cats in Europe gives some indication of what to expect of *F. lybica*. Home ranges in good habitat may be 50–100 ha (4, 5). A tomcat's territory encompasses the ranges of up to 3 females, reaching maximum extent in the mating season. Females are more intolerant of one another than are males, but only actively defend a core area within the home range. Here cats that are subordinate elsewhere (normally) become dominant and may fiercely attack even much larger individuals. The rest of the range is shared and all members of the local community have the right to use it on a first-come, first-served basis, regardless of rank (4). Cats entering a common hunting range check carefully to see and smell whether another cat is already on the scene, often going to a vantage point to scan, sniffing vegetation where others have sprayed urine, scraped, or scratched, and leaving their own marks in turn. If another cat is spotted, the latecomer waits for it to move off, although occasionally 2 cats may hunt in the same field while completely ignoring each other. African wildcats have been seen hunting in pairs or family groups, the members moving in the same general direction and spaced 3–30 m apart (3).

ACTIVITY: nocturnal.

POSTURES AND LOCOMOTION. As in domestic cat.

PREDATORY BEHAVIOR. Described in family introduction.

SOCIAL BEHAVIOR

COMMUNICATION. Discussed and illustrated in family introduction.

Olfactory Communication: *urine-spraying* by both sexes, also *chin-* and *cheek-rubbing* and *claw-sharpening.* The feces are buried as in housecats.

AGONISTIC BEHAVIOR. Described in family introduction.

REPRODUCTION. Housecats remain in estrus about 4 days and ovulation occurs a day after copulation (1). Several males have been seen accompanying a *F. lybica* female in estrus (8). (More in family introduction.)

PARENT/OFFSPRING BEHAVIOR. Both in southern Africa and East Africa young tend to be born during the rains, the time when most prey species are at peak abundance. There is no evidence of more than 1 litter being born in a year in the wild state, whereas a hand-reared, free-ranging female littered 8 times in 3⅔ years, twice producing 2 and once 3 litters in 1 year, for a total of 20 offspring (8). Gestation is 56–60 days and typically 3 kittens (2–5) are born in a hollow tree, hole in the ground, rocky crevice, or nest in dense grass. Their eyes open at 10–14 days, and they become mobile within a month, accompany the mother when less than 3 months, and are independent by 5 months. They are fully grown and begin breeding at 1 year, but males probably rarely have a chance until fully mature at 2–3 years (3, 4).

References

1. Asdell 1955. 2. Ewer 1973. 3. Kingdon 1977. 4. Leyhausen 1979. 5. Lindemann 1955. 6. Maberly 1962. 7. Smithers 1971. 8. ——— 1983.

Fig. 21.11. Black-footed cat in defensive theat (from a photo by A. Bannister).

Black-footed Cat

Felis nigripes

TRAITS. The smallest cat, distinguished from the wildcat by size, bold markings, and ears the color of the body (never reddish) (4). *Weight and length* (head and body): males 1.6 kg (1.5–1.7) and 40 cm (38–43),

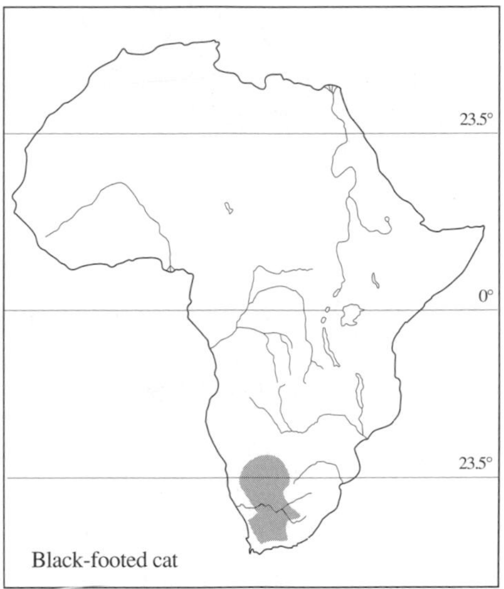

females 1.1 kg (1–1.4) and 27–36 cm, tail less than ½ head and body (13–20 cm) (4). Legs relatively long, wide head with large, rounded ears. *Coloration:* from cinnamon-buff (southern) to tawny (northern population), heavily spotted and barred: 3 broad transverse bars on upper limbs and narrow, shorter bars on lower limbs, undersides of feet black (hence *nigripes*), 4 black bands from head or nape extending onto the back and flanks, tail spotted with black tip, underparts pale with white chin, chest, and insides of thighs. *Scent glands:* undescribed, presumably as in other *Felis* species (see family introduction). *Mammae:* 6.

DISTRIBUTION. Restricted to the arid central parts of the South West Arid Zone in Botswana, Namibia, and the Karroo south of the Orange River (4). Though it is often considered rare and endangered, its nocturnal habits and extreme shyness make its exact status hard to determine.

ECOLOGY. This cat inhabits open country with some cover in the form of scrub, bushes, and clumps of grass. By day it lies up in disused burrows (springhare, jackal, fox, etc.) or in termite-mound ventilation shafts. The arid country it inhabits, with rainfall of 100–500 mm, is mostly waterless, but it will drink occasionally when rainwater pools are present. Its food consists of small rodents such as gerbils and mice (present in 4 of 7 collected stomachs), spiders and solifugids (present in 3 stomachs), beetles and other insects, plus small ground

Fig. 21.12. Black-footed cat in *skulking attitude* (from a photo by A. Bannister).

birds (coursers) and lizards (each present in one stomach) (4).

SOCIAL ORGANIZATION: solitary, territorial.

Virtually nothing is known about the social behavior of the black-footed cat, which is not only nocturnal but so shy that it hides at the slightest disturbance (fig. 21.12) (3). These and other aspects of its biology, such as the abbreviated mating period, prolonged gestation, and relatively precocious young (see below), may be adaptations that enable a very small cat to live in a treeless environment. In the Kalahari, they become active only 2–3 hours after sunset (4). Black-footed cats have the reputation of being more intractable and antisocial in captivity than any other cat (2, 4). Even kittens taken before their eyes open and hand-reared make defensive threats (fig. 21.11) and retreat to dark corners, or attempt to attack their owners (4).

ACTIVITY: nocturnal.

POSTURES AND LOCOMOTION. See Predatory Behavior.

PREDATORY BEHAVIOR. It seems likely that black-footed cats forage widely, since captives habitually walk and trot around for hours during nightly activity periods (2). Captives go off their food unless provided with plenty of grass. Skill and persistence in digging suggest that they dig for insects, spiders, and other small subterranean prey. To secure wriggling prey, one cat was seen to straddle and clamp it with its forearms, anchoring with the dewclaws while waiting, with head back, for the right moment to deliver a killing bite to the nape (1).

REPRODUCTION: gestation 63–68 days, 1–3 young, born mainly during the summer.

SEXUAL BEHAVIOR. Estrous behavior lasts no longer than 36 hours in this species, compared to at least 6 days in the housecat and some other small felines. A female will actually accept copulation for a period of only 5–10 hours, during which a pair mates c. half-a-dozen times at intervals of 20–50 minutes, the intervals being longest near the beginning and end (2). The male's readiness to copulate may match the female's period of receptiveness, as one that mated with 2 domestic cats lost interest long before they did. The same male would no longer mate with housecats once he was provided with a female of his own species; since these 2 species overlap in southern Africa without interbreeding, there clearly is some sort of reproductive barrier. Hybrids produced in captivity have not been successfully back-crossed (2).

PARENT/OFFSPRING BEHAVIOR. With a gestation period nearly a week longer than in the considerably larger domestic cat (2), black-footed kittens are more mature at birth and at first develop faster, perfecting coordinated movements a whole week earlier than domestic kittens. Their eyes open at 6–8 days; they leave the nest at 4 weeks, start eating solid food at 5 weeks, and capture prey (mice) at 6 weeks. A captive mother kept her kittens from leaving the nest until they were 3 weeks old, but later actually hauled kittens reluctant to leave out by the nape. She tried to move the nest site every 6–10 days. Once the kittens could run well, the nest ceased to be a refuge from danger, and the mother made no effort to drive them toward it, unlike housecats and a tame wildcat (cf. wildcat account). Instead, the kittens scattered and took cover, lying motionless until the mother sounded the "all clear": a peculiar, staccato but almost inaudible "*ah-ah-ah*," accompanied by synchronous up-and-down movements of the half-flattened ears (2). No similar sound has been heard in other cats. The kittens responded by relaxing and typically ran toward the mother.

Some days before her offspring were ready to eat meat, the mother began bringing dead prey back to the nest, and brought live prey when the kittens were c. 5 weeks. She neither stimulated the kittens' interest by coaxing or competing with them nor disabled the prey. She simply released it near them, sat back, and watched. If the prey was on the point of escaping, she skillfully drove it back toward the kittens with her paw rather than catching it. Consequently the kittens never saw how prey was killed before they did it themselves (1).

After about the fifth week, the growth and development of black-footed cats slows down, and this species matures considerably later than the wildcat (2). A female came into heat for the first time at 21 months, twice the normal age for a housecat. Another female of 15 months was considerably smaller than adult females and showed no signs of being sexually mature.

References
1. Leyhausen 1979. 2. Leyhausen and Tonkin 1966. 3. Smithers 1971. 4. ——— 1983.

Fig. 21.13. Serval male *urine-spraying.*

Serval

Felis serval

TRAITS. Tallest of Africa's small cats, with long legs and very prominent ears. *Length* (head and body) 67–100 cm; *height* 54–62 cm; *weight* males 13 kg (10–18), females 11 kg (8.7–12.5) (3). Slender build with long neck and medium-length tail (24–35 cm). *Coloration:* variable, tawny to russet, darker forms in moister regions but bolder markings in drier savanna; neck, shoulders, legs, and tail barred and blotched, large solid spots on sides and legs, ear backs black with white blotches. A freckled form (perhaps representing the ancestral condition) occurs together with the common type in western Guinea Savanna, sometimes in the same litter (3); newborns grayer than adults, with suffuse markings. *Scent glands:* undescribed (see family Traits). *Mammae:* 6 (6).

DISTRIBUTION. Occurs in all kinds of African savanna where grass grows, especially along the edges of gallery forest and marshland and in montane grassland and forest glades, usually near water. It also adapts readily to abandoned cultivation and second growth in rural areas (3). It does not occur in rain forest or in subdesert regions (6).

ECOLOGY. The serval is a specialist in catching rats, mice, and other rodents and birds that live in tall grass. Its ears function like dish antennae to pick up and pinpoint the noises made by such prey, even underground (2). Its long legs are not designed for speed but for added height, the better to hear and see over and into the grass, and to make a high leap as it pounces (see Predatory Behavior). The commonest rodents are the mainstay of the serval's diet, especially grass rats (such as *Arvicanthis, Otomys*), mole rats, and such. Thus, in Botswana 30 of 31 serval stomachs contained rodent remains and 20 contained only rodents (5). But the diet may also include birds up to the size of flamingos and storks, snakes, lizards, frogs, fish, insects, and concealed young of small and medium-sized antelopes (3).

SOCIAL ORGANIZATION: solitary, territorial.

Evidence from a study made in the Ngorongoro Conservation Arca indicates that servals of the same as well as opposite sex may have overlapping home ranges, but normally avoid meeting one another (1, 2). In Ngorongoro Crater, a male and a female ranged a 7.5 by 4 km area of open grassland bordering a swamp, parts of which were also used by up to 5 other servals (2). A total of 12 other servals were seen in this area during 2½ years, including juveniles, subadults, 4 adult females, and 1 adult male; but the last occupied a different range and

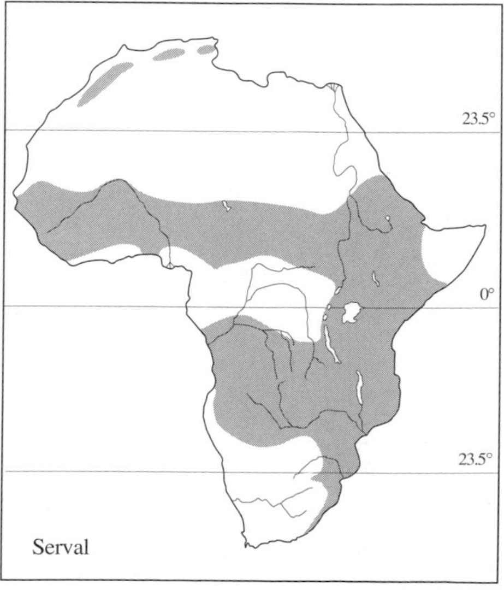

was seen only on the border of the study area (but see next paragraph). Probably at least some of the adult females were related, for female offspring are tolerated much more than males, which the mother and the resident male force to emigrate during adolescence.

Most of the individually known servals in the study disappeared for up to 3 months at a time, but whether they stayed in the neighborhood or dispersed was never determined. In the acacia woodland around Lake Ndutu, bordering the Serengeti Plain, 2 known males changed their ranges from one year to the next but moved no more than 2½ km (1). These 2 had overlapping hunting ranges of 1–2 km^2 which they shared amicably; once they were seen hunting 300 m apart. However, each male had an exclusive though not sharply defined core area.

ACTIVITY. Servals are more diurnal than most cats. They are very active in early morning, often up to 1000 or 1100 h, and again after 1600–1700. On cool and overcast days they stay active longer. At night servals spend 3–4 hours resting and, as usual, have 2 activity peaks, beginning at 2200–2300 and 0400–0500, the latter often carrying through into the morning (2). The average distance covered in a day by Ngorongoro servals was c. 2 km; one traveled 6.7 km. During the heat of the day, servals doze and groom themselves in the shade of a bush or grass clump, sitting up frequently to look around. They are less alert when there are no larger predators or game concentrations around.

POSTURES AND LOCOMOTION. See Ecology and Predatory Behavior.

PREDATORY BEHAVIOR. Servals hunt in typical cat fashion (see family introduction). Having located and stalked a rat concealed in the grass, a serval leaps into the air and comes down with both front feet on its victim, in the manner of a pouncing jackal or fox (fig 21.14). Sometimes one will make a whole series of spectacular jumps and pounces, but it usually moves on after a miss. A serval not only can jump high and far but has a wonderful ability to swerve and change direction at full speed. Its actions and reactions are so quick that it can pluck birds from the air, sometimes jumping high and batting with powerful downward paw strokes (caracals bat sideways) (4). One male, while stalking something in a bush, suddenly looked back at a tree 18 m away, turned and ran to it, jumped 2 m high, and snagged an agama lizard (1). Servals also reach deep into holes to catch mole rats and hole-nesting birds, aided by their elongated forearms. They readily enter water in pursuit of wading birds and waterfowl, and one spent 1½ hours in a pool after a heavy storm stalking frogs and toads (2).

Fig. 21.14. Serval jumping to pounce on concealed prey (from a photo in Van Lawick and Van Lawick-Goodall 1970).

Having caught a rat or other quarry, a serval may play with it for up to 10 minutes, throwing it 2–3 m in the air, catching it as it falls, and so on. The cat may eat its prey in the open or carry it to a protected spot. It eats in the same manner as a housecat, crouching down and beginning at the head end, bolting small rodents whole after a little chewing, but often leaves the viscera (1). Judging by the hunting-success rate of Ngorongoro servals, this feline is an efficient predator: 40% of its pounces by day and 59% by night were successful. One big male caught 20 rodents in 31 pounces (2).

SOCIAL BEHAVIOR

COMMUNICATION

Olfactory Communication: *urine-spraying, rubbing face on objects.*

Both sexes mark by spraying (males) and squirting (females) urine on the grass and sometimes against conspicuous clumps, bushes, or reeds. They also rub their faces against grass stems or on the ground, meanwhile salivating freely. In the Ngorongoro study area, the male spent much more time than the female in patrolling, visiting different parts of his range and pausing to urinate frequently: 556 times in one 12-hour period (2). Although the cats seemed to take certain preferred routes, for example, along the edge of the swamp, no individual was seen marking the same bush twice. Scats are deposited at random along paths or roads, preferably in a depression or patch of short grass. Beyond a few cursory scrapes with the hindfeet, no effort is made to cover the feces (6).

AGONISTIC BEHAVIOR. See family introduction. The aggressive displays of servals differ from other cats' in 2 respects: during offensive threat the head is thrown up and down instead of sideways, and an aggressor may prod its opponent with an outstretched foreleg (4). An encounter between 2 male neighbors began when A intruded into B's core area (1). B was just becoming active after a rest when he noticed A approaching. He sat down and watched as A paused to spray a bush. When he was within about 30 m of B, A suddenly started racing and bounding around, tail arched over his back and with all 4 feet sometimes off the ground. B ran toward him and joined in this activity until suddenly they stopped and stood facing only 30 cm apart. With tails raised, hair bristling, and backs somewhat arched, they threw their heads up and down while yowling in a low key and showing their teeth. Onc and then the other raised a forepaw only to lower it again. This went on for 5 minutes, when the intruder backed away and around B, then slowly moved off, glaring back with tail still fluffed. When he got back to his own valley, he sniffed and sprayed bushes (fig. 21.13).

REPRODUCTION: gestation 65–75 days, 2–3 young/litter (1–5).

SEXUAL BEHAVIOR: unrecorded.

During estrus a male and female may consort for several days during which they hunt and rest together (1).

PARENT/OFFSPRING BEHAVIOR. Kittens are hidden in dense vegetation and subject to frequent moves. A Ngorongoro female that had 2 litters in 2½ years hid hers in tall sedges bordering the marsh. Her daily routine was altered radically while they were young: she spent most of the day hunting and bringing food back to the kittens, only resting in late afternoon. Returning over a kilometer with a mouse she had caught, she began calling as she came within 50 m of the marsh, stopped, listened, then called again. Three pairs of ears appeared behind a clump of reeds. When the mother called again, 1 kitten galloped to her, followed by the others, all meowing. One kitten picked up the mouse, while the mother licked, then suckled the others, meanwhile purring loudly (2).

When the kittens were older, she had trouble making them go into hiding so that she could resume hunting; it took up to ½ hour, with increasingly severe growling and spitting on her part, to get rid of them. The males are driven away soon after becoming competent at hunting for themselves, whereas daughters are tolerated much longer (2).

ANTIPREDATOR BEHAVIOR. Ngorongoro servals behave very fearfully when they encounter spotted hyenas, immediately crouching or ducking into cover and waiting. If a hyena comes close, the cat flees with tail raised in leaps and bounds with sudden changes of direction. Without high grass for cover, servals would be highly vulnerable, in particular to this predator and to wild dogs.

References

1. Geertsema 1976. 2. ——— 1981. 3. Kingdon 1977. 4. Leyhausen 1979. 5. Smithers 1971. 6. ——— 1983.

Fig. 21.15. Caracal carrying prey.

Caracal
Felis caracal

TRAITS. The African version of a lynx. *Weight:* up to 20 kg (6), heaviest of Africa's small cats; Botswana series: males 14 kg (12–18), females 10 kg (8–13). *Length* (head and body): males 78 cm (62–91), females 73 cm (61–81), tail 26–34 cm. *Height:* 40–45 cm (8). Long legs with hindquarters higher and more developed than forequarters, and big feet; short face with powerful jaws and long, tufted ears. *Coloration:* plain tawny to rufous coat with spots faintly indicated in whitish underparts; lips, ear backs and tufts black, dark facial markings on cheeks and over the eyes, bordered by white fur; newborn light yellow with relatively distinct spots and markings already formed. *Scent glands:* anal sacs, other glands undescribed (see family introduction).

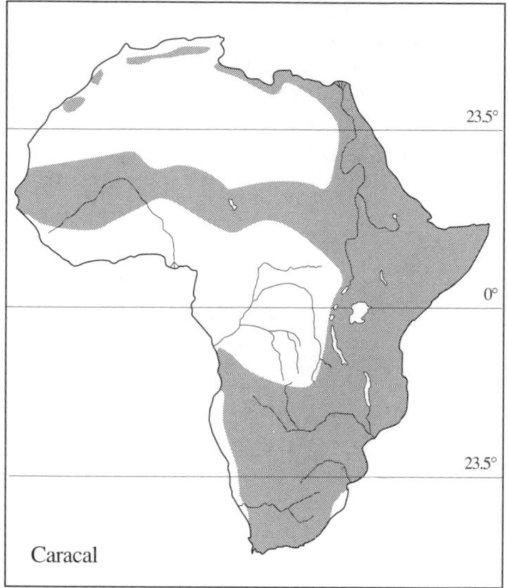

DISTRIBUTION. Arid zones and dry savannas of Africa and North Africa, through the Near East to India and Russia's Karakum Desert.

ECOLOGY. Though equally at home on plains, rocky hills, and mountains, probably most caracals live in arid bush country. They may venture into open grassland at night to hunt but seem to require woody vegetation for cover, while avoiding dense evergreen forest (8). This cat is replaced by the serval in better-watered areas where grass is dominant for much of the year (5).

In South Africa's Mountain Zebra N.P., where the caracal is the largest predator, mammals made up 94% of the prey species identified in collected scats (4), with rock hyraxes predominating (53% of the kills). In terms of biomass, however, adult mountain reedbucks (average weight c. 30 kg) accounted for 70% of the prey harvest. Other prey included springboks, steenboks, gray duikers, hares, rabbits, springhares, small rodents (5%), and Cape gray mongooses. Birds represented 5% and reptiles 1% of the identified remains. The caracal also preys on sheep in South Africa. Like jackals, perhaps it tends to take bigger animals in areas where larger carnivores have been eliminated or where the choice of prey is limited. Records of caracals taking adult impalas, a sitting ostrich, and young kudus attest its predatory prowess, but the mainstay appears to be hyraxes and other mammals in that size range, including jackals, monkeys, and game and other birds in addition to the kinds already listed. The caracal's predatory niche is thus very different from that of the serval, a specialist on rodents.

SOCIAL ORGANIZATION: solitary, territorial.

During the aforementioned study of caracal food habits, the cats were sighted a total of 57 times, mainly early and late in the day (4). Forty-one of the sightings were of single adults, large juveniles and adult pairs were seen 11 times, and the rest were family groups consisting of a female with 1–4 kittens. The estimated population in the 6536 ha park was 15 adult and 10 juvenile caracals, giving a density of 1 caracal/261 ha.

ACTIVITY: mainly nocturnal (9) but may be active at twilight and also by day.

POSTURES AND LOCOMOTION
The caracal is fast over a short distance, propelled in long bounds by its powerful hindquarters, which also give it remarkable jumping powers: a housepet, startled as it lay asleep, bounded against a wall, its front feet reaching the height of 3.4 m (9). It is also an agile climber, using its strong dewclaws to scale smooth tree trunks (9). The combination of speed, strength, and agility makes this a very formidable predator for its size.

PREDATORY BEHAVIOR (fig. 21.15). The caracal kills medium-sized antelopes over twice its own weight by employing the same technique as the big cats: suffocation with a throat bite. It uses its jumping and climbing ability to catch hyraxes in the rocks. It also takes sleeping birds from their perches, including martial and tawny eagles, and ground birds up to the size of a kori bustard. Sandgrouse, doves, pigeons, and the like are vulnerable to caracals when they come to drink at waterholes with nearby cover (5). A caracal is so quick that it sometimes gets 2 birds from a flying flock by leaping high and batting sideways with its paws, but more often it uses both paws together to snag a bird from the air. The dewclaws also serve to hold prey (or an opponent) securely. A pet female killed hares with a bite to the nape. If the hare continued thrashing, she would throw herself on the ground and hold the hare while raking it with her hindfeet.

The caracals studied in Mountain Zebra N.P. ate small birds completely except for a few feathers, and partially plucked larger birds (4). The viscera, primaries with attached flesh, portions of skull, and occasionally a lower leg were left uneaten. The viscera of all mammal prey were left untouched, and the fur of mammals of hyrax size and up was partially plucked with the incisors. Caracals also avoided eating hair by shearing meat neatly away from the skin. Access to antelope kills was gained through the anus. The meat on the hindquarters was consumed first, then the forequarters. They rarely ate more of the larger antelopes before the meat became putrid, though they returned to all but 2 of 21 kills. Grass and leaves were raked over 1 carcass; in other areas, small kills may be cached in tree forks (5).

A tame but free-ranging subadult caracal weighing 11 kg ate 796 grams of meat a day. Assuming that an adult would eat 1 kg a day (up to 2 kg at a time), the 25 caracals in Mountain Zebra N.P. consume about 7200 kg of meat a year, thereby accounting for 2995 hyraxes and about 190 mountain reedbucks (4).

SOCIAL BEHAVIOR

COMMUNICATION

Visual Communication. Caracals are cryptically colored except for the head, which is boldly marked and surmounted by remarkably conspicuous ears, white on the inside and black on the outside, with black tufts, so that every twist, turn, and flick is clear to see (5). White fur bordering the black lips repeats the same basic pattern, emphasizing facial expressions and bared teeth. Taken together, these markings make an effective semaphore system. Thus, when 2 caracals spot each other and, in typical cat fashion, sit and *look around,* the movement is exaggerated and ritualized as a side-to-side head-flagging, during which the ears flicker rapidly (5). But slightly flattening and pulling back the ears causes a narrow wedge of sandy fur along the leading edge to mask the contrasting black and white pattern from the prey of a stalking caracal. Thick fur over the eyelids also masks the eyes (possibly against glaring sunlight), except when a caracal opens them wide (5).

Vocal Communication: meowing, growling, spitting, hissing, purring, screaming, loud coughing calls (in mating season) (5).

Olfactory Communication. Caracals have been seen urine-marking grass tufts, rocks, and trees (2). A free-ranging pet sometimes buried its dung but at other times left it exposed (4).

REPRODUCTION. Though births have been recorded through the year, there is an extended birth peak during the summer, at least in South Africa (1). The average litter numbers 2.2 kittens (1–4) and gestation varies from 62 to 81 days. Both sexes reach puberty at 7–10 months. Apparently females do not reenter estrus until after lactation ceases at c. 4 months.

SEXUAL BEHAVIOR. Estrus lasts 3 to 6 days and recurs every 2 weeks until conception occurs. Copulation occurs during a period of 1 to 3 days, initiated by the female assuming the *copulatory crouch,* and averages 3.8 minutes (1.5–8). Immediately afterward both partners groom themselves (1).

PARENT/OFFSPRING BEHAVIOR. In zoo observations, the expectant mother prepared a nest of hair and feathers, and stopped eating a day before littering (6). In the wild, caracals may litter in burrows; for instance, 2 kittens were found in an old aardvark hole (9). In one observed birth, the mother behaved restlessly, licking her genitals and teats and scratching at the earth floor for 35 minutes before visible labor contractions of the abdominal muscles began. The emergence 110 minutes later of the kitten was followed shortly by the placenta, which the mother ate after licking the baby.

A rich rufous color like the adults, with the black facial markings already pronounced, the offspring of a pet caracal opened their eyes at about 9 days, but avoided the light the first 2 weeks and were very sensitive to noise (9). The ears remained folded back on the head, only becoming fully erect after 30 days. Kittens stop crawling and begin walking at 9–17 days, and begin cleaning themselves at 11 days, licking their front paws and legs. At 12 days a hand-reared pair began uttering twittering food calls. Kittens may only begin eating meat at 1–1½ months and are weaned at 4 to 6 months (1). Caracals as young as 3 weeks may leave the nest and begin to chase moving prey (3, 6, 9).

References

1. Bernard and Stuart 1987. 2. Fuente 1972. 3. Gowda 1967. 4. Grobler 1981. 5. Kingdon 1977. 6. Kralik 1967. 7. Pringle and Pringle 1979. 8. Smithers 1971. 9. ——— 1983.

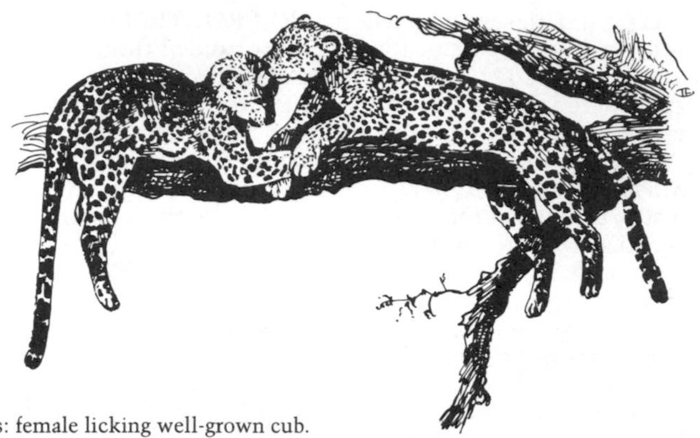

Fig. 21.16. Leopards: female licking well-grown cub.

Leopard

Panthera pardus

TRAITS. A long, low, spotted (sometimes black) big cat. *Height and weight:* males 60–70 cm and 35–65 kg, females 57–64 cm and 28–58 kg (12, 15, 16). Limbs short and massive, head wide with short, powerful jaws and long canines. *Coloration:* ground color tan, highly variable, generally darker and more closely spotted in forests, spots grouped in rosettes on torso and upper limbs. Cubs woolly and dark with close-set, indistinct spots. *Scent glands:* anal and probably other glands (see family Traits). *Mammae:* 4.

DISTRIBUTION. The most ubiquitous of all cats (except the domestic cat), the leopard occurs throughout Africa wherever there is sufficient cover for concealment, and from the Arabian Peninsula through Asia to Manchuria and Korea. Though eliminated in many densely settled areas and heavily poached for its valuable pelt, this cat is so stealthy and so bold and versatile as a predator that it manages to survive in places other, smaller cats cannot. Where its natural prey has been eliminated, it may subsist largely or entirely on domestic animals and sometimes (in Southeast Asia, very rarely in Africa) becomes a man-eater. It is the only large predator in Africa's rain forests, and it also ranges the montane forests and moors of the highest mountains; on Kilimanjaro a leopard carcass was found frozen in the ice at 5692 m (9). Despite dire predictions of extermination, it turns out that leopards are far commoner over much of Africa than conservationists had supposed (10). The least specialized of the big cats, the leopard is successful wherever diversified habitats afford a variety of small to medium-sized mammals.

ECOLOGY. A leopard will consume protein in almost any form, from beetles up to antelopes twice its own weight. It readily eats carrion, and often caches sizeable kills in trees, returning nightly to feed on them. In Serengeti N.P. a sample of 150 leopard kills included over 30 different species, compared to about 12 regularly taken by lions. Medium-sized antelopes (impala, Thomson's gazelle, reedbuck) and the young of larger species (topi, hartebeest, wildebeest, zebra) predominated, with hares, birds, and several small carnivores rounding out the list (2, 3, 14). In Zimbabwe's Matopos N.P., an area with a variety of mammals and no great preponderance of ungulates, 72% of 200 leopard scats contained the remains of small mammals (< 5 kg) and 19%, mammals of medium size (5–45 kg) (7). Birds, reptiles, and invertebrates (in-

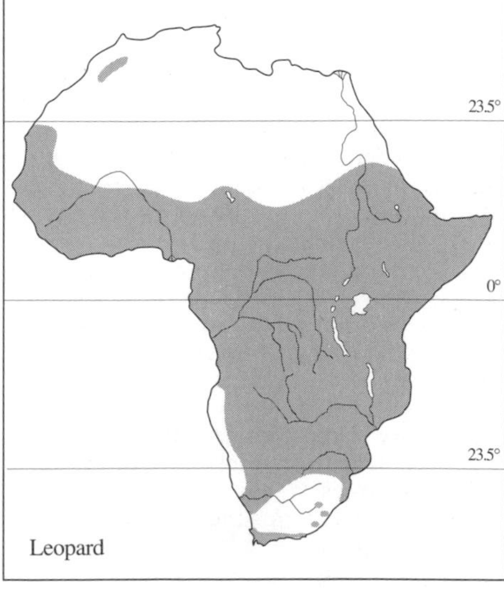

sects, etc.) accounted for 10% of the animal remains. Although leopards prey on baboons in some areas (e.g., Kruger N.P. [11]), the common notion that this primate is a mainstay of their diet is false (2); clever leopards don't risk their own hides by attacking baboon troops on the ground, and clever baboons sleep in safe roosts at night. Unclever ones may indeed turn up missing in the morning—as do even large, savage dogs a prowling leopard finds sleeping.

SOCIAL ORGANIZATION: solitary, territorial.

The leopard is a typical cat. Adults associate only long enough to mate and the young become independent as subadults, at around 22 months. The ranges of females often overlap (see below), but authorities differ about whether male ranges also overlap. The practice may well vary in this highly adaptable cat, just as home-range size is widely variable.

In Sri Lanka's Wilpattu N.P., home ranges are as small as 8–10½ km^2 and males apparently defend territories, by calling and by demarcating the boundaries with urine, dung, and tree-scratching (5). In Tsavo N.P. the home ranges of 10 leopards that were radio-tracked for 3 years varied from 9 to 63 km^2 (3). Density was estimated at 1 leopard/13 km^2. Each leopard used only about half its range at a time and overlapped with others by as much as 70%. The core areas of 2 males were only about 0.8 km apart, and 1 female had sporadic contact with at least 4 different males. Female ranges in Serengeti and Kruger N.P. were estimated at 10–25 km^2 (3, 11). Several may be included within a male's range. One adult male and no less than 3 females have been known to hunt independently along the same 5 km stretch of river (14).

The maternal bond is strong and enduring (fig. 21.16). Mother and offspring often have reunions after separating, and the mother may continue to share kills until her offspring become fully self-sufficient. For instance, a radio-collared 2½-year male who stayed within a 5 km^2 range during the 4 months he was observed often joined and went hunting with his presumed mother at night (2). A similar association is described under Predatory Behavior. Continuing tolerance of adult female offspring probably explains overlapping female ranges (cf. cheetah). A 1½-year female, hunting alone, spent her time stalking hyraxes, hares, and small birds, and was unsuccessful at capturing larger prey. Her mother killed an impala or Thomson's gazelle every few days, stashed it in a tree, then went to fetch her daughter. They would feed 1–2 days together intermittently until it was gone, then separate again. Six weeks later the mother came into heat and spent several days consorting with a male. The 2 females, both radio-collared, were never found together again; however the daughter stayed in the same range (2).

ACTIVITY. See also under Parent/Offspring Behavior. Leopards spend the day and part of the night inactive, draped over a tree limb or lying up in a dense thicket. Beginning perhaps ½ hour before dark, they move around intermittently all night and for a couple of hours after dawn. The radio-tracked Tsavo leopards regularly wandered 25 km a night, covering much of the used part of the range within a few days. They seldom rested in the same spot 2 nights in a row (8).

POSTURES AND LOCOMOTION. See Predatory Behavior.

PREDATORY BEHAVIOR. The leopard is the quintessential ambush and stalking predator. A lion rushes upon its quarry as it is trying to escape; a leopard seeks to pounce *before* its quarry can react (2), stealing to within at most 20 m and preferably within 5 m before pouncing. It seldom chases if it misses and then for no further than 50 m, although its estimated top speed is a respectable 60 kph (fig. 21.17) (3). Hunting by day is hardly worth a leopard's while: 61 of 64 daylight attempts by Serengeti leopards failed. Even at night its success rate per attempted stalk is thought to be comparatively low (2).

Some individuals develop a particular fondness for dogmeat and are bold enough to come onto a veranda or even through open windows to carry off sleeping pets. And yet a barking terrier can easily tree a leopard by daylight, and one even gave way to a yapping jackal (2). But another leopard, which visited my camp in Ngorongoro Crater almost nightly for several weeks, brought

Fig. 21.17. Leopard running gait.

Fig. 21.18. Leopard eating springbok it carried up a tree (from a photo by J. Dominis, *Life* magazine).

back 11 jackals in that time and proceeded to eat them on a tree platform beside my cabin, sometimes after cat-and-mouse play with them (6). She also caught 2 large male Grant's gazelles, members of a bachelor herd that often ventured dangerously close to tall streamside vegetation to feed at night. At 70–75 kg apiece, they were too heavy for her to carry up a tree, so she left each in turn on the ground after feeding on the carcass, first scratching leaves and a little dirt over it.

Once in the Serengeti I witnessed apparently deliberate teamwork between a female and a subadult male in capturing a bush hyrax, shortly before dusk. The male ran swiftly up the trunk and into the crown of a 10-m umbrella acacia, while the female waited below. A hyrax jumped out and the female caught it as it landed. After playing with it briefly (tossing it in the air and catching it), she surrendered it to her son, who carried it onto a nearby rock outcrop. He laid it down, then ignored it to survey his surroundings. The hyrax suddenly came to life and jumped into a patch of sansevieria (Spanish bayonet). The leopard jumped after it, but too late; he landed right on the stiletto-tipped plants, yet showed no pain.

RELATIONS WITH OTHER PREDATORS. Of the 7 large African carnivores, the leopard only outranks the cheetah. Not only the lion but all 3 hyenas outweigh a leopard, and wild dogs, though smaller, operate in packs. But that an animal as well armed and powerful as a leopard would surrender its own kills to a single hyena is somewhat puzzling (see brown hyena account). Not that the outcome of an encounter is invariable: a leopard with young cubs is far more likely to take the offensive than to run from a hyena, for instance. However, the practice of eating and caching kills in trees (fig. 21.18), so common in Africa, is not observed where leopards are the only large predator (5).

SOCIAL BEHAVIOR

COMMUNICATION

Olfactory Communication. Leopards use all the usual feline scent-marks plus feces to communicate with one another (14). Large trees with inclined trunks or big branches 2–3 m from the ground are preferred scent posts. Here a leopard pauses to sniff at previous scratch marks, stretches out along the branch or trunk, and "sharpens" its own foreclaws. (Sometimes it scratches with its hindclaws.) The marking ritual is completed by spraying urine at the base of the tree. Leopards (especially males) also spray bushes and other objects at frequent intervals during their nightly wanderings, after rubbing the head and cheeks on the spot (5, 14). In addition, feces are deposited along paths and roads, where other leopards are likely to encounter them.

Vocal Communication: "sawing," grunting, growling, snarling, hissing, caterwauling, purring; cubs: meowing, *urr-urr, wa-wa-wa.*

The distance call of a leopard, given both while inhaling and exhaling, sounds a lot like someone sawing wood. A typical series has 13–16 "strokes" in a period of a dozen seconds. Sawing is repeated at intervals of 6 minutes or so during peak calling bouts, which tend to occur early in the evening and shortly before dawn, often while the leopard is moving (5). Occasionally 2 leopards duet, for instance when 2 males "shout" at each other from near their common territorial boundary. In addition, leopards grunt when alarmed, growl, snarl, and hiss in fear/rage, and sometimes caterwaul when treed by dogs (16). Small cubs give a soft *urr-urr* call to their mother, a *wa-wa-wa* when comfortable, and a high-pitched meow of distress (1, 14). A mother summons her cubs with loud, abrupt purrs (8).

Visual and Tactile Communication: as in other cats (see family introduction).

The white tail tip, which looks like a flitting moth when flicked in the darkness, no doubt functions as a follow-me signal between mother and cubs (8).

REPRODUCTION. Leopards are capable of breeding at two years. From 1 to 3 cubs are born at intervals of c. 25 months, after a 90–100 day gestation (4, 13). Estrus is known from zoo records to last about 7 days (4–12) and to recur every 46 days (25–58) until conception occurs (13).

SEXUAL BEHAVIOR. Behavior during estrus is typically feline (see family account). A precopulatory sequence involving wild leopards began when the female came up to the male that had been following 10 m behind her and *rubbed cheeks* with him as he stopped to scent-mark a tree. She then *presented* in the *copulatory crouch,* whereupon he mounted, only to leap off the next moment as she whirled about with a snarl (15). Another time a female removed a gazelle kill that her male consort had cached in a tree that morning, and dragged it to a patch of bush where the 2 proceeded to share it. Such unwonted sociability was a sign that the female was nearing estrus.

Leopards may copulate as often as lions and the male emits harsh, growling cries (*copulation call*) (5, 17). One captive male broadcast every copulation by growling, snarling, and teeth-gnashing as he symbolically bit the female's nape. Judging by the calls, copulation recurred night and day at roughly ¼-hour intervals (author's observation).

PARENT/OFFSPRING BEHAVIOR. Weighing only 400–600 g at birth, cubs are concealed in dense thickets, hollow trees, or caves. Their eyes open at 6–10 days, but 6 weeks pass before they venture from hiding and make short excursions with their mother. At the same time they start to eat meat. Weaning may occur as early as 3 months, when cubs weigh about 3 kg, but more than a year passes before they are ready to fend for themselves. A radio-collared mother spent 44% of her time with and 56% of her time away from her 2 cubs when they were 3 weeks old (15). During 168 hours of continuous recording, she stayed with them 62% of the nighttime and 30% of the daytime. After spending up to 33 hours with them, she would leave them for up to 36 hours at a time, but went no farther than ¾–2 km (average 1½) from their hiding place (which she changed from time to time). When the cubs were 2 months old she spent about half her time with them, staying for up to 39 hours at a stretch, and left them for up to 25 hours while feeding on a sizeable kill and guarding it against vultures and jackals. During a further 127 hours of continuous monitoring, she spent 67% of the time with them and 33% away. Throughout this period the female used an 8 km² range. Her movements and activities were shaped by the time required to find and kill prey, and the need to protect it from scavengers and consume it before it decomposed in the tropical heat. Since leopard milk is not markedly richer than housecat milk, it is clear that leopards become adapted from the very beginning to long fasts relieved by intervals of gorging (15).

References

1. Adamson 1969. 2. Bertram 1974. 3. ——— 1976. 4. Crandall 1964. 5. Eisenberg and Lockhart 1972. 6. Estes 1967*b*. 7. Grobler and Wilson 1972. 8. Kingdon 1977. 9. Moreau 1944. 10. Myers 1976. 11. Pienaar 1969*b*. 12. Robinette 1963. 13. Sadleir 1966. 14. Schaller 1972*a*. 15. Seidensticker 1977. 16. Turnbull-Kemp 1967. 17. Ulmer 1966. 18. Wilson 1968.

Added References

Bothma and LeRiche 1986. Norton and Henley 1987. Scott 1985.

Fig. 21.19. Lion consort pair.

Lion

Panthera leo

TRAITS. The largest African carnivore. *Weight:* adult males 189 kg, females 126 kg (Serengeti N.P. sample) (12); up to 260 kg (1). *Average height:* males 120 cm, females 110 cm (19). Coat short; mane in males only, variable in length, extent, and color—just visible in 2-year-olds, maximum development in early prime, declining in late prime (16), but sometimes undeveloped. *Coloration:* tawny with white underparts; tail tuft, ear backs, and lips black; mane tawny to red-brown to black; cubs spotted with woolly, grayish coats, changing to adult color beginning at 3 months, but faint spotting may persist in adults (especially in East Africa). *Scent glands:* anal glands and possible foot glands. *Mammae:* 4.

DISTRIBUTION. Formerly from the Cape to the Mediterranean wherever suitable prey existed, except in desert and rain forest, and in biblical times or later throughout the Near and Middle East as far west as Greece and as far east as India, where a few hundred lions still persist in the Gir Forest (19). Lions were eliminated from North and South Africa by the end of the nineteenth century, except for the area that is now Kruger N.P. Since then, this cat has

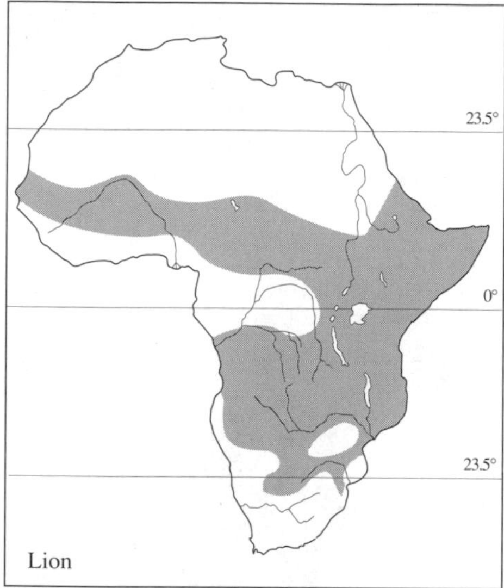

been shot out of or greatly reduced over much of Africa along with other big game. But in savanna and plains habitats where large herbivores still abound, it is the most numerous large carnivore next to the spotted hyena, reaching a density of c. 38 lions/100 km² in Ngorongoro Crater, 26/100 km² in Nairobi N.P., 12.7 in Kruger N.P., and 12 in Serengeti N.P. (13, 14, 17).

ECOLOGY. Lions eat a great variety of animal food, and will scavenge a meal sooner than hunt for one. But their size and gregarious habits are specifically adapted to predation on the larger ungulates. In most areas of abundant wildlife, only around 5 of the most numerous ungulates weighing between 50 and 300 kg make up ¾ of their prey (14). Lions that hunt cooperatively often succeed and they feed communally on large kills, whereas smaller and weaker individuals often lose out when the carcass is small. Single lions readily bring down animals up to twice their own weight, wildebeests and zebras being particular favorites, and some lions regularly kill buffaloes 4 times their weight; even animals weighing well over 1000 kilos (e.g., bull giraffe) are sometimes killed. Usually a cooperative effort is used to overcome such large prey; males, being much bigger and stronger, though slower, tackle large prey more readily than do females.

SOCIAL ORGANIZATION: gregarious, territorial, matriarchal society, communal care, male coalitions.

Most lions live in resident prides that occupy hunting ranges large enough to support a certain number of animals during occasional periods of game scarcity. The home range may be as small as 20 km² in the best habitat, and over 400 km² where prey density is low. The average area of 9 Serengeti prides was c. 200 km² (14). There may be as many as 40 lions in a pride, including from 2 to 18 adult females and up to 7 adult males, but it is rare that all assemble in one spot. Pride members are usually scattered singly and in small groups that change from day to day as individuals come and go on their own initiative. Groups typically number 3–5 lions, and any given pair of females may only be found together 25%–50% of the time (10). Mere presence in a pride's territory is no proof of membership, as some lions are transients or "lodgers" (squatters). Pride membership can only be determined by observing which individuals interact amicably over a period of time; that is, one has to recognize the lions individually. Both in the Serengeti and Kruger N.P. the typical pride numbers about 13 lions. In Kruger, the average composition of 14 prides totaling 181 lions was 1.7 adult males (1–4), 4.5 adult females (2–9), 3.8 subadults, and 2.8 cubs (including yearlings). In another sample of 60 Kruger prides, the mean number of pride males was 2.1 (15).

The part of the range currently used by the pride is defended against outsiders of the same sex. Adult females are normally all related, descending from females that have lived generation after generation in the same range. However, if the number of lionesses should fall below the capacity for the range and there are no female offspring to fill the gap, subadult immigrants may be permitted to join (16, 17). Conversely, when the roster is filled female cubs have to emigrate as 2-year-olds. The existence of a "female capacity" for each territory is illustrated by the fact that the number of lionesses in 6 neighboring Kruger prides remained constant for 2½ years, despite a turnover in membership (17).

Because of the clumping of females in discrete territories, male lions can monopolize mating opportunities to an unusual degree: in the Kruger sample, there were 2.6 adult females per pride male. The degree of male competition in this species has led to innovations in social behavior unlike those of any other carnivore, but similar in some ways to baboon society. The development of greater sexual dimorphism than in other carnivores is only the most obvious outcome of this competition

(14). The advantages of size and a showier mane have caused males to become so beefy and conspicuous that their hunting ability is actually impaired. It takes them 5 years to mature and they only reach puberty at 2½ years (range 26–34 months) (16). By 3 years at the latest, male offspring are forced to emigrate, at least partly because of the intolerance of adult males (5). From then until they mature and succeed in taking over a pride range, these males and also emigrant females lead an unsettled existence, either as nomads following migratory game, as in the Serengeti ecosystem, or within the ranges of established prides as unwelcome lodgers which have to avoid the proprietors (16).

Being sociable, outcasts associate, preferably with members of the same sex. This is just the reverse of the primarily intrasexual aggressiveness displayed between resident and nonresident lions. Within prides, too, females associate and behave much more affectionately with one another than with males, and vice versa for males. Companionships may last a lifetime. In fact this social bonding, usually between closely related individuals that have grown up together, and particularly between mothers and daughters, is the very foundation of lion society.

Male littermates often stay together after leaving their pride, and when ready to compete for a territory they operate as a team. A single territorial male that finds himself confronting 2 challengers is clearly at a disadvantage. Two established males are at a disadvantage against 3 challengers, and so on. But the advantage of an additional male decreases as coalition size increases and 9 is the largest number I have heard about. A Serengeti septet dubbed the Seven Samurai took over and divided their time among 3 or 4 different prides. But this arrangement did not work well, for other males invaded territories left unguarded and kept most of the females from rearing cubs (6).

Wherever lions are plentiful, single males can rarely take over or keep a territory. Their mating opportunities are accordingly limited to nomadic females, which seldom manage to rear young, or to pride females in estrus left unattended by the resident males. The reproductive advantage enjoyed by coalitions forces even unrelated males to band together (2, 6, 12). In the Serengeti population, 42% of the coalitions composed of known individuals included 1 or more nonrelatives (7). However, out of 40 sightings of transient males seen passing through Serengeti territories, all but 3 were pairs and singles, indicating that most coalitions are small and that some males fail to team up at all (8).

Even males in coalitions have a remarkably brief reproductive career. Pride tenure for most lasts no more than 2 years, 4 at the outside, and very few ever get a second chance (2, 16). The prime years are between 5 and 9, and males are biggest and fittest between 5 and 6. After 8 they lose weight and mane hair. Only 2 of 25 known-age Kruger males were older than 9 and both were alone: a 10-year-old and a decrepit 16-year-old that had survived several years beyond the usual life span (16, 17). Large coalitions are likelier to have a longer tenure than small ones: Serengeti coalitions of 1–2 reigned for less and coalitions of 3 reigned for more than 25 months, whereas coalitions of 4–6 often lasted longer than 46½ months (8). As we shall see, the effect of these differences in tenure on reproduction is the difference between success and failure for both sexes.

Often one of the first acts of males after taking over a pride is to kill any cubs under a year old which they can catch. Older juveniles may escape with their lives but cannot survive unless their mothers leave with them—as sometimes happens (9). Lionesses often attack males in defense of their cubs and may be wounded and sometimes killed. They are most effective when they gang up on the infanticidal males. Females with very young, concealed cubs may be kept from going to them until the cubs starve, or they may simply abandon them. The more males in the coalition, the likelier it is that they will do in all the cubs, including any born in the months following the takeover (16). Such behavior used to be considered aberrant, but it has now become apparent that lions are one of a number of highly social mammals in which infanticide is a vital part of male reproductive strategy (2, 11). Normally, lionesses produce cubs at intervals of no less than 2 years, and only come into estrus when their offspring are c. 1½ years old. The average interval between birth and the next estrus is 530 days. However, the loss of a litter causes a lioness to reenter estrus and mate within a few days or weeks (5). With an average tenure of only about 2 years in which to propagate their genes, males that delay breeding will leave few progeny and their cubs will be liable to fall victim to their successors.

And what about the females? Is there no way they can prevent this waste of their far greater maternal investment? Lionesses mate without becoming pregnant for an average of 134 days after losing their cubs in a takeover (9). During this interval they engage in an unusual amount of mating activity and display heightened sexuality, initiating more copulations and seeking more partners than in normal times. Even pregnant females may go through apparent estrus (16). Other lionesses come into heat every few weeks but apparently fail to ovulate. The period of infertility may well be an adaptation that (*a*) protects the females against the consequences of desertion by the new males, (*b*) helps to bond the new males to the pride and reduces the likelihood of their deserting, and (*c*) allows time for the strongest coalition around to come onto the scene and take over. Serengeti females often mated with males of more than 1 coalition during their infertile stage and in 5 observed cases the larger coalition ended up staying in the territory (9).

Although lions have no special breeding season, reproductive synchrony is very common within prides. Several adult females conceive and give birth at about the same time, then care for the cubs communally. It now appears that synchronization commonly follows a takeover: once the infertility period ends, the chances that pride females will ovulate and conceive at nearly the same time are greater than at any other time; there is no need for—and no evidence of—any more complicated synchronizing mechanism (8, 17).

Perhaps the most extraordinary aspect of lion social organization, considering the degree of male sexual competition, is the gentleman's agreement governing mating rights: it is a matter of first-come, first-served, literally. Instead of fighting over a lioness in heat, pride males race to be first at her side; the winner becomes her consort, and the losers keep their distance. Only if one male is clearly stronger and fitter than another will he preempt mating rights. Competition to be first leads to the practice of guarding a female some days before she comes into heat (7, 10). The extended period of estrus, marked by hundreds of copulations, and the frequent synchrony of female cycles help make the system work. Consort males often lose interest in females toward the end of their estrus period, giving another male a chance to mate, and more than 1 lioness was in heat during 43% of the times that Serengeti females were observed in estrus (7). As over a third of these matings were fertile, it would mean that second-ranking males could sire up to 20% of the offspring, thus making it reproductively worthwhile to belong to a coalition (6). Half a loaf . . .

ACTIVITY. Lions typically spend 20–21 hours a day resting (14). To see lions doing anything other than conserving energy requires luck or persistence. The times when they are most likely to become active are late afternoon, when females often suckle their young followed by play and other social activities, early and late at night, when hunting activity shows slight peaks (14), and the early hours of daylight. However, lions will seize opportunities to capture prey at whatever time and even if they are too gorged to eat (see Predatory Behavior, family introduction). Prides living in woodland habitats with plenty of cover hunt in daylight more often than those living on open plains. Lions are obviously also well aware of the advantage of darkness for hunting: hungry animals will lounge on moonlight nights until the moon disappears, then suddenly become active (14).

POSTURES AND LOCOMOTION. See Predatory Behavior and family introduction.

PREDATORY BEHAVIOR. With a maximum speed of 48–59 kph which they can maintain for hardly more than 100 m, lions need skill, patience, and judgment to capture such fleet animals as antelopes and zebras. An experienced lion will rarely charge unless it is within 30 m and/or the quarry is turned away so that the lion can start its rush before the victim becomes aware of the danger and can get up speed.

To get within range, lions rely primarily on stalking. They use cover with great skill and show a remarkable ability to anticipate and take advantage of opportunities to get close (even using moving vehicles to screen their approach). Lions frequently attempt a running approach toward an animal that is inattentive or distracted, freezing instantly and sinking to the ground when the quarry faces in their direction. Lions of the Serengeti Plain approached by stalking and running in 88% of over 1300 observed hunts, and seldom tried to ambush prey (3%) (14). But since ambushes commonly take place in daylight, typically at water holes with convenient cover, this is the kind of hunt park visitors are most likely to see.

Some 48% of the 1300 hunts involved 1 lion, 20% involved 2, and the rest were communal hunts, with 3–8 (occasionally up to 14) participants (14). Lionesses were the hunters 85%–90% of the time. On group hunts, males usually trailed behind without taking part, and as soon as a quarry was caught ran up and claimed "the lion's share." The success rate of solitary hunters was only 17%–19%, compared to 30% when 2 or more lions hunted together. Schaller found that hunting lions respond to one another's postures and movements, not to facial expressions or sounds (they maintain silence). On communal hunts they usually move on a broad front. Sometimes prey is encircled by the wings while those in the center lie low, evidently expecting the game to flee in their direction. When game concentrates on strips of green grass beside streams or marshes in the dry season, lions may advance on a front quite openly, causing the animals to panic as they see their escape route blocked and giving the hunters chances to intercept animals as they dash through the line. There was no indication that lions deliberately roar to drive prey into an ambush (in my experience prey species normally ignore lion roars), and in 300 hunts where wind direction could be determined, the hunters approached downwind as often as upwind. Even though upwind stalking (of gazelles) was 3 times as successful, lions apparently do not learn to take this factor into account (14). But on the whole, African ungulates are not very reactive to the scent of predators, perhaps because their scent-marks are so pervasive and persist in their absence.

Killing Technique. Once within catching distance of its prey, a lion has to overpower and kill it without getting hurt in the process. Game the size of an impala or reedbuck is brought down with a slap on the haunch, tripped, or clutched with both paws and simply dragged down, then quickly killed with a bite to the neck or throat. To bring down large game, a lion usually comes in at an oblique angle, rears and throws 1 paw over the shoulder or rump, and using its full weight and strength pulls the quarry down backwards and sideways. To gain additional leverage a lion may grip the quarry's neck, shoulders, or back in its jaws (fig. 21.20). Once the victim is down, the lion lunges for its throat or nose and proceeds to kill it by strangulation, maintaining a firm grip until all movement ceases, for up to 13 minutes (see leopard, fig. 21.1). Assertions that lions kill big game by twisting the head so that the victim breaks its own neck in falling are largely anecdotal, although lions in the Kalahari Gemsbok N.P. may possibly

Fig. 21.20. Lioness pulling down buffalo (from Schaller 1972*b*, drawing by R. Keane).

have developed some such tactic for killing gemsboks (4). None of several hundred lion kills Schaller examined had broken necks (14). The stranglehold protects a lion from the horns and hooves of the struggling victim and makes it easy to hold it down. Very few victims ever get a chance to defend themselves before being overpowered.

Food Sharing. Lions share food but often grudgingly, depending on how hungry they are and the size of the prey. Pride males are the most irascible and piggish, regularly monopolizing small antelopes killed by the lionesses. When game is scarce, cubs are the first to suffer; even their own mothers will not share until they have eaten their fill (10). At best, feeding lions often growl, snarl, and slap at one another. Schaller attributes this scramble competition to the lack of an accepted rank hierarchy in lion society. If weaker lions simply deferred to stronger ones without putting up a fight, they would be far likelier to starve during periods of food scarcity, for lions may gorge until they have eaten ¼ of their own weight (up to 50 kg), whereas the average daily consumption is only 5–7 kg. In normal times there is more sound than fury among feeding lions and all members of a pride get enough to eat. Even crippled and old animals may survive on the leftovers. The best evidence that sharing is a lion trait is that nomadic strangers will feed on the same kill (14). If there is anything left over, lions stay and guard carcasses against other scavengers. They themselves scavenge kills whenever possible rather than hunt, keeping a watchful eye out for vultures and responding to the rallying calls of hyenas as readily as other hyenas. Large packs of hyenas can often intimidate lionesses into leaving kills before they are ready, but get nowhere with males in their prime.

Fig. 21.21. Lion roaring (from Schaller 1972*b*, drawing by R. Keane).

SOCIAL BEHAVIOR

COMMUNICATION

Vocal Communication: roaring, grunting, moaning, growling, snarling, hissing, spitting, meowing, purring, humming, puffing, woofing.

The lion's roar is rightly considered one of the most impressive natural sounds. Loud roaring typically begins with a few moans, followed by a series of 4–18 full-throated, thunderous roars, ending with a series of grunts (average c. 15, up to 57) (14). Males begin roaring at 1 year, females a couple of months later (3). The male's roar is deeper and louder, but sex and distance are both hard to judge because of a ventriloquial quality and gradations in volume: the same call grades from a soft "*huh*" or moan barely audible at 100 m to full-throated roars audible up to 8 km. Lions can roar from any position and even while running, but usually stand. The facial expression is distinctive (fig. 21.21). Active animals roar most frequently, hence night and shortly before dawn are peak times, but how often pride members call in a night is also highly variable. One pride monitored continuously for 16 days averaged 6.5 roaring episodes (range 0–16) a night. The stimulus is often indeterminate. Lions definitely respond to other lions' roars, especially when near, but they also roar spontaneously (14). Probable functions of roaring, depending on who is sending and receiving, include (*a*) territorial advertising, (*b*) location of pride members, (*c*) intimidating rivals during aggressive interactions, and (*d*) strengthening social bonds (roaring in chorus).

Like roaring, meowing and growling/snarling are graded calls that enable lions to express a wide range of emotions by changes in the volume, intensity, tempo, and tone of the call. The meow of cubs, a signal of light distress, varies from short and yippy when greeting an adult, being licked, and moving close to the group, to harsh and snarly when competing for a nipple with teeth bared; adults also meow while snarling. The cough is a short, explosive growl often given during a charge. The snarl is simply a slurred growl given with open mouth, and hissing is emitted with the mouth open as if to snarl. An abrupt hiss becomes spitting, given at sight of a strange lion (or when a vehicle approaches too closely).

Purring and humming are sounds of contentment, heard during affectionate interactions and when cubs suckle. Puffing, a faint "*pfff-pfff*" through closed lips given repeatedly as lions approach each other, signifies peaceful intentions (comparable to *prusten* in tigers), and *woofing,* an abrupt sound given by a startled lion, expresses alarm.

Grunting (which grades into roaring) is used at low intensity by mothers to summon cubs and as a contact call between close-by adults. It may grade into moaning or soft roaring. Cubs also grunt and moan while trailing behind a moving group. In fact, starting with a hiss at 9–10 days, cubs acquire the full vocal repertoire in rudimentary form by the age of 1 month (14).

Olfactory Communication: *urine-spraying, scuffing, clawing, urine-testing.*

Lions may pay little attention to the odors of other species but display intense interest in the odors of other lions. Anal sniffing is common when lions meet, and males very often sniff females in heat and make a spectacular *grimace.* They can follow the scent trace left by another lion's feet, and pay close attention to one another's scent-marks. Pride males spend a lot of their time patrolling and marking the currently defended part of the range, stopping to spray bushes and other landmarks at frequent intervals. Although females occasionally squirt urine against vertical objects, urine-marking is mainly a male behavior. But both sexes perform the *scuffing ceremony,* beginning at about 2 years, raking the ground 2–30 times with the backfeet, with or without urinating (fig. 21.6). Prominent objects bordering roads, paths, kopjes, water holes, and solitary trees tend to become regular scent posts that reek of lion urine. Individuals and groups often scuff- and spray-mark after aggressive interactions with other lions and competitors such as hyenas, before leaving a kill, and (males) beside estrous females. It seems to be an assertion of the right to be at that place, if not a claim to ownership. Lions of all ages also rake trees, and on treeless plains claw the earth like cats clawing a carpet (10).

Tactile Communication: *greeting ceremony, social licking,* lying in contact.

The kind of close, affectionate contact seen between mothers and young in other cats carries over to the adult stage in lions. Pride members have to go through the *greeting ceremony* whenever they meet, as a proof of membership in the pride and of peaceful intentions (fig. 21.7). Two lions approach each other, often moaning softly, rub heads then sides together, with tails raised and frequently draped over the partner's back. They lean against each other, so hard that a standing lion may roll over on top of a lying one; and when females roll on cubs in greeting, you wonder why the little ones are not crushed. But cubs more often rub lionesses than the other way around, and females greet each other more often than they greet cubs. Both females and cubs often seek to rub pride males, which may allow it but reserve their greetings for other males of their coalition. The fact that weaker animals tend to take the initiative indicates that greeting serves an appeasing function (14).

Social-licking is less prevalent than *head-rubbing* but also plays a part in reinforcing social bonds. Any pride member can initiate a licking session and it may be one-sided or reciprocal. Cubs are licked all over, whereas licking is otherwise largely confined to the head, neck, and shoulders.

Visual Communication. See family introduction. The black ear backs, lips, and tail tip accentuate the expressions and movements of these parts.

AGONISTIC BEHAVIOR

Dominance/Threat Displays: *strutting, head-low threat,* accompanying vocalizations.

Defensive Displays: ears flattened, teeth bared, head turned, eyes narrowed, crouching, lying on back, with accompanying vocalizations.

Fighting: slapping, grappling, biting, ganging up.

The *strutting display* is performed only by adult males (and by cubs in play), primarily to females, as a show of dominance (fig. 21.22). The lion makes himself as tall as possible and presents his side to or walks stiffly around the female with erected tail looped over his back.

The offensive threat is unconventional in that a lion stands with head low, forelegs straddled, and shoulders higher than normal, gazing steadily at its opponent. The tail is often lashed up and down—perhaps the only unambiguous tail movement in this species (14). If the other indicators of feline aggressive threat are also present (mouth slightly open with lips set in a straight line covering the teeth, eyes wide and pupils small, ears twisted so that the black marks face forward), accompanied by growls or coughs, a charge is imminent. The defensive threat is typically feline (see family introduction, Visual Communication and Agonistic Behavior).

Fig. 21.22. Lion *strutting* (dominance) *display* (from Schaller 1972*b*, drawing by R. Keane).

The efficiency of the lion's scent and sound communication system enables nomads, lodgers, and trespassing neighbors to avoid encounters with territorial proprietors. For their part, pride lions show less inclination to rush upon outsiders in blind fury than to intimidate and drive them away without coming to blows. Once strangers have been put to flight, residents may escort them for miles at a fast walk or trot while staying at least 10 m behind (14).

Fighting is commonest between pride members competing over meat, but rarely goes beyond slapping, which is just as well since lions show little restraint when biting and may inadvertently administer the killing neck bite. However, severe, often fatal fights are usual during takeover attempts, females may fiercely attack infanticidal males in concert, and occasionally members of male coalitions fight over females in estrus. These battles are usually provoked by a female moving close to a male that has been keeping his distance, but sometimes 2 males get to a female at exactly the same time and neither is willing to defer (7, 10).

REPRODUCTION. Breeding is nonseasonal but often synchronized within prides (see Social Organization). Females begin breeding at four (43–54 months), only a year earlier than males (12). Gestation is 3½ months (14–15 weeks), the typical litter numbers 3 cubs (1–4), and the birth interval is 20–30 months if cubs survive (as a rule half die in the first year). A preponderance of male cubs in synchronized births that occurred within 300 days of a pride takeover (129 males:96 females) in the Serengeti lion study suggests that mothers are somehow able to bias the sex ratio of their offspring (possibly through division of male zygotes to produce identical brothers) to take advantage of the higher lifetime reproductive success enjoyed by the members of larger male coalitions (6, 11).

SEXUAL BEHAVIOR. It has been estimated that lions copulate 3000 times for every cub that survives to the yearling stage (2). The discovery that females may cycle for months after a takeover without actually ovulating helps to explain why only 1 mating cycle in 5 results in progeny (9). However, even in normal times, many copulations are apparently necessary to induce ovulation. Once ovulation occurs, the fertility rate is very high: 95%.

Estrus typically lasts about 4 days (2–6), during which couples couple at the rate of 2.2 times per hour (8), each copulation lasting c. 21 seconds (8–68). One male copulated 157 times in 55 hours, with 2 different females (14). The male's rate increases to 3.5 copulations per hour when serving 2 females—evidence against the notion that females but not males are inexhaustible. Giving other males a chance late in the estrus cycle may be of more social than reproductive significance (8).

Mere proximity of a male and female is usually a reliable sign of a mating pair (fig. 21.19), and the way the male follows, practically tripping on the female's feet, is characteristic. The lioness rebuffs mounting attempts with snarls and slaps at first, but later often initiates mating by crouching and presenting before the lion. Instead of biting her nape during the climax, the male makes at most a few quick, light biting movements (fig. 21.9). The female keeps up an ominous rumbling the whole time, possibly from resentment at being straddled. The male may react by meowing in distress, leading into loud, harsh yowls at the climax. Immediately afterward he either dismounts or jumps aside, sometimes with a snarl or half-roar, as the lioness twists her head with an explosive snarl, prepared to cuff him. But once unburdened, she then languidly rolls onto her side or back (14). During estrus, a female will accept any adult male.

PARENT/OFFSPRING BEHAVIOR. Lion cubs are born in a well-hidden lair, tiny (1–2 kg) and completely helpless. Their eyes open at 3–11 days; they can walk at 10–15 days and run at 25–30 days. They can scratch the first day but bite only after the milk teeth erupt at 21–30 days (3). At 4–8 weeks mothers begin leading their cubs to kills; they can keep up with the pride by 7 weeks and are weaned at 7–9 months (at latest by 10 months), but are unable to fend for themselves before 16 months. Cubs have the best chance of surviving when the births of several females are synchronized and there are no older cubs around to hog most of the milk of females suckling one another's cubs communally (fig. 21.23) (8). Suckling bouts last 1–10 minutes and at 4 months cubs suck about 15 minutes a day (14).

Cubs are often left alone for over 24 hours, while their mothers consort with other pride members. They stay closely hidden and if disturbed crawl into nooks and crannies to escape detection. They are quick to take cover even when grown-ups are present. Starvation and abandonment are most likely to occur when large prey is scarce. A mother will not slow her pace for cubs more than 5–7 months old; those too weak to keep up and single offspring are often abandoned. The risk of starvation recedes only when the young become capable of killing for themselves.

Already at 2½ months cubs observe prey movements, and as usual play enacts the stalking, ambushing, grappling, and killing motor patterns used by adults to capture prey. Females never lose their playfulness, whereas males older than 3 rarely join in. Cubs as old as 7 months are still unresponsive to the stalking postures of their elders and often manage to spoil hunts by blithely running ahead, playing, meowing, and so on. A little later they stay back and observe intently, but become participants only when c. 11 months. Unlike various other cats, lionesses rarely school their young in killing by bringing them young animals

Fig. 21.23. Lioness suckling her own and other cubs (communal suckling).

to chase. But lion cubs stay in the maternal group much longer than any other cat.

ANTIPREDATOR BEHAVIOR. Adults in good health have little to fear from any animal but man, who has been a major competitor from the days when our kind began eating meat. Small cubs left alone in hiding are very vulnerable to predation by spotted hyenas, which probably also prevent most starving, diseased, and old lions from dying of natural causes.

References

1. Berry 1983. 2. Bygott, Bertram, Hanby 1979. 3. Cooper 1942. 4. Eloff 1973*a*. 5. Hanby and Bygott 1987. 6. Packer, Herbst et al. 1988. 7. Packer and Pusey 1982. 8.——— 1983*a*. 9.——— 1983*b*. 10.——— 1983*c*. 11.——— 1984. 12. Pusey and Packer 1987. 13. Rudnai 1973. 14. Schaller 1972*b*. 15. Smuts 1978. 16. Smuts, Hanks, Whyte 1978. 17. Starfield, Furniss, Smuts 1981. 18. Stevenson-Hamilton 1947. 19. Van Orsdol 1984.

Added References

Hanby 1983. Packer 1986. Van Orsdol, Hanby, Bygott 1986.

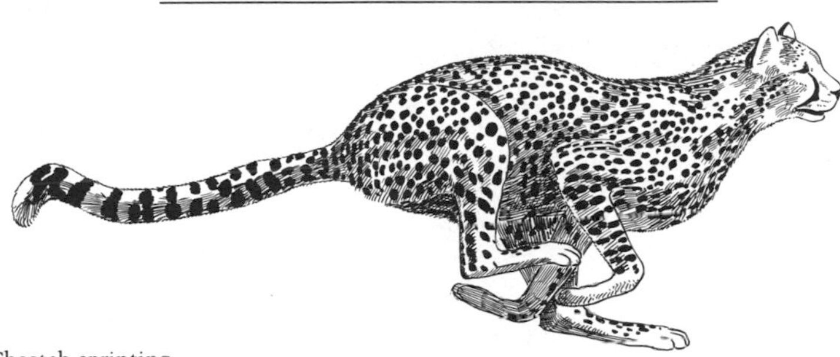

Fig. 21.24. Cheetah sprinting.

Cheetah

Acinonyx jubatus

TRAITS. The felid version of a greyhound, the most specialized cat. *Height and weight:* 78 cm (70–90), and 50 kg (35–65), males more robust and c. 10 kg heavier than females. Lightly built with long, comparatively thin legs and small feet with blunt (except dewclaw), unsheathed claws; swayback, short neck, small, rounded head with foreshortened face, broad but low ears, teeth relatively small, especially canines. *Coloration:* tawny, with white underparts, a short ruff, more developed in males, and fluffy hair on belly and chest; spots small and solid, outer tail ringed black and white, tip usually white; black ear backs, lips, nose, and "tear stains" from eyes to mouth; juvenile coat black with faint spots and a cape of long, blue-gray hair. *Scent glands:* presumably anal glands. *Mammae:* 12.

DISTRIBUTION. Similar to the lion's: the cheetah formerly ranged through the Near East to south India and throughout Africa except in equatorial forests and true desert. The cheetah became extinct in India in the early 1950s, and is now extremely rare elsewhere in Asia and North Africa (14). It is still widely but sparsely distributed south of the Sahara in the savannas and arid zones, but greatly reduced in range and numbers. This species is far less adaptable to man than the leopard, but in Namibia has benefited from the removal of lions and hyenas and the reintroduction of antelopes on ranches (13). However, an extraordinary degree of genetic uniformity in this species raises the possibility that a disease could devastate wild populations (15).

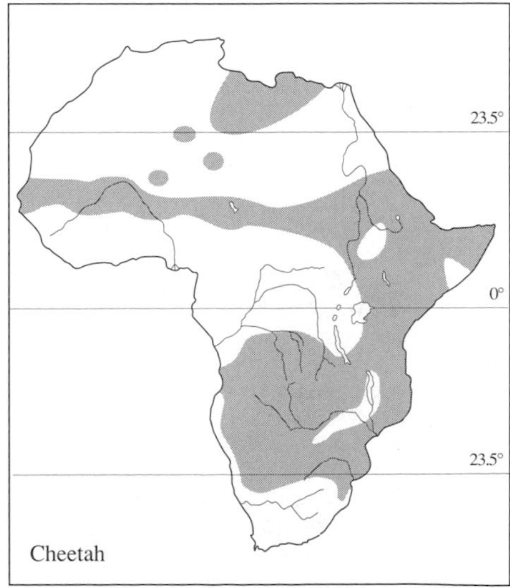

ECOLOGY. The cheetah is specialized to prey on the fleetest antelopes, especially the gazelles and their close relations, the blackbuck and springbok, which dominate(d) the arid wastelands of Asia and Africa. That the cheetah is built for speed (fig. 21.24) is demonstrated by the fact that the fastest dogs—greyhound, whippet, Afghan hound—produced after centuries of selective breeding, have a similar conformation. No other mammal is as fast as a cheetah, which has a top speed of 90–112 kph (60–70 mph), but dogs have something the cheetah lacks: stamina (19) (more under Predatory Behavior). To capture its prey, the cheetah has to overtake it within 300 m. It has a hard time getting within sprinting range on plains with no cover; optimum cheetah habitat therefore includes cover in the form of bushes, medium-length (not tall) grass, trees, broken ground, and such. In the bush and savanna woodlands, the impala takes the place of gazelles as the cheetah's mainstay, accounting for 68% of 1092 cheetah kills in Kruger N.P., for instance (16). This cat also occurs in the *Miombo* Woodland Zone, where its main prey is probably the bush duiker, reedbuck, lechwe, oribi, juvenile sable and roan, and warthog. Single cheetahs seldom kill antelopes heavier than themselves, and often take smaller game; in Serengeti N.P. hares are second (12%) in importance to Thomson's gazelles (62%), especially during times when gazelles are scarce (8). Serengeti cheetahs regularly go 4 days between drinks, after which they will travel 5–10 km to water if necessary. In the Kalahari, where they eat melons, cheetahs easily go 10 days without drinking (8).

SOCIAL ORGANIZATION: diurnal, territorial, females solitary, males in coalitions or alone.

Being an inhabitant of open country, diurnal, and often seen in groups, the cheetah could well be supposed to be a social carnivore. However, these groups almost always turn out to be mothers with subadult cubs, young adult siblings recently separated from their mother, or coalitions of males. Adult females are as solitary and shy as other cats (7, 17). In acquiring the speed to catch the fleetest antelopes, the cheetah sacrificed the strength and weapons needed to defend its kills and offspring against competitors (5, 7). Perhaps its delicate build is enough to explain the timidity that characterizes this cat. It is afraid of lions, surrenders its kills to hyenas, and can even be intimidated by vultures—maybe because it realizes that vultures attract other predators. To avoid being victimized, it makes itself as inconspicuous as possible, aided by its cryptic coloration.

Home Range and Territory. Where their main prey is migratory, cheetahs cover a wide area: the annual range of Serengeti females that subsist largely on Thomson's gazelles is c. 1000 km^2 (7). But several females whose histories were recorded for up to 11 years had narrow ranges 50–65 km long which included habitat utilized by gazelles both during wet and dry seasons. The cheetahs were able to move with the gazelles in a regular cycle that took them from end to end of their home ranges. When offspring reached 17–23 months and separated from their mother, daughters remained within the maternal home range but each stayed alone. Once a mother and daughter were seen only 20 m apart, but parted without coming any closer. However, despite minor adjustments of range during ensuing years, a broad overlap between the ranges of the mother and 2 daughters persisted. The daughters' ranges overlapped much less.

Male offspring emigrate and typically wander huge distances while maturing and seeking to establish territories. Nine males marked in Namibia were retrapped at distances of over 200 km from the marking site (13). Among other dangers, transient males run the risk of injury, even death, if caught trespassing on the territories of established males (example under Agonistic Behavior); consequently females may be twice as numerous as males in the adult population (7). Males compete for the best hunting grounds, and defend areas much smaller than the females' ranges—the reverse of the arrangement seen in most carnivores. However, males may range outside their territories (3). The areas defended by Serengeti males were found to be 39–78 km^2; they had no overlap and sharp boundaries. Territorial males did not trespass, whereas females passed through a number of territories during the circuit of their ranges.

In the Serengeti study, 41% of the adult males were solitary, 40% lived in pairs, and 19% lived in trios (3). The benefits to male cheetahs of forming coalitions are far less obvious than in the case of lions. Female cheetahs are solitary, and as single males apparently meet as many as do males in coalitions, singletons would not have to share or compete for copulations with companions. However, there is evidence that single males have a harder time acquiring and keeping a territory. Only 4% of the

Fig. 21.25. A male coalition of 3 cheetahs, drinking.

observed single males ever held a territory, whereas all the single males that joined coalitions acquired territories. Furthermore, estimated territorial tenure increased with coalition size: the median tenure for singletons was 4 months (3½–9), it was 7½ months (2¾–16) for pairs, and 22 months (18½–24½) for trios. One trio held a territory for 6 years (3).

How Coalitions Are Formed. Littermates often stay together for several months after separating from their mother. One by one the females drop out before they reach 2 years of age, presumably at the onset of estrus. Some males also separate, whereas others stay together as permanent companions (fig. 21.25). Of 6 Serengeti litters containing more than 1 male, brothers of 4 litters went their separate ways (7). A single male joined a pair of brothers after separation while they were still living in their mothers' adjacent ranges. Then they moved 20 km and established a 36 km^2 territory on the Serengeti Plain. Fourteen months later they extended their range by 22 km^2 during a time when gazelles were scarce, but within a month were back in their usual area.

Social relations between cheetah males are more restrained and less affectionate than relations within lion coalitions. Cheetahs seldom lie in contact and their greeting ceremony proceeds no further than *cheek-rubbing.* During 38 hours of observation over a five-day period, males of the above trio exchanged cheek rubs only five times (7). The unrelated male, which proved to be dominant, administered 4 of the 5 cheek rubs, two to each brother, and he issued 5 of 9 play invitations, all declined (the 2 brothers played once for a minute). No cheek rubs were initiated by the lowest-ranking member of the trio. In addition to cooperative defense of their territory, the males jointly marked their property. They sniffed existing marks 55 times and urine-marked 59 times in 34 hours. Number 3 male marked most (27 times) and number 2 marked least (13 times), although he sniffed as often as the alpha male (20 times).

The largest number of cheetahs seen in the Serengeti totaled 9 and consisted of 2 females with their offspring. Associations between adult females were seen about once in every 500 cheetah sightings and lasted only a few hours. Of 1260 cheetah sightings, 35% were lone adults, 40% were females with offspring, 7% were littermates on their own, 7% were male coalitions, and 3% were female-male consorts (7). In Namibia, however, groups numbering from 10 to 14 are not uncommon; 16% of 102 adult females were observed in groups of 2 or more, and 28% of all litters were accompanied by more than 1 adult. This population also kills bigger game than usual, including adult kudus and wildebeests (13). These differences may be explained by the fact that the cheetah has become the dominant predator on fenced range, where other large carnivores have been exterminated. The implication is that the cheetah's social organization and predatory role are shaped elsewhere by the presence of hyenas and lions (13).

ACTIVITY. The most diurnal cat, cheetahs do almost all of their hunting by daylight, but usually rest during the hottest hours. Rarely, cheetahs will engage in brief chases by moonlight or even on dark nights. In Nairobi N.P., females and cubs moved about 3.7 km per day, compared to 7 km for males (12).

POSTURES AND LOCOMOTION. See Predatory Behavior and family introduction.

PREDATORY BEHAVIOR. Of 493 observed hunts by Serengeti cheetahs, 203 were successful (41%) (7). Only 40% of the stalks of Thomson's gazelles, which made up nearly 60% of the kills, ended in chases (fig. 21.24), but half of those chased were caught. Male coalitions and females with subadult young use their combined might to pull down game the size of yearling and 2-year-old wildebeests. Four males in Nairobi Park even killed a zebra and a waterbuck (6).

An actively hunting cheetah walks along alertly, utilizing termite mounds and trees with low branches as vantage points from which to spot potential prey. To get within sprinting range (50 m or less), a cheetah uses several different techniques, depending on the terrain and the type, dispersion pattern, and behavior of the prey. It may simply wait at a vantage point if it sees that a group of gazelles is moving in its direction. Or it may approach slowly and openly an alert herd on the open plain, whose members stand and watch or even trot closer for a better look. If it can get within 60–70 meters before the animals take flight, the cheetah may gallop at them, but will accelerate to full speed only after selecting a particular quarry. Alternatively, if all the animals are grazing unsuspectingly, the cheetah may rush at them from over 100 m (rushing) and try to get close enough to select a quarry before being detected. But whenever cover is available cheetahs stalk as close as possible, walking semicrouched with head lowered to shoulder level, trotting, freezing in midstride when the game looks up, dropping to the ground, lying crouched, and sitting (8, 22). In a sample of 129 complete hunts, stalking was employed 85 times, taking anywhere from a few minutes to more than an hour. In ⅓ of the hunts the cheetah first walked closer, and then began stalking or tried rushing. Forty-six of the hunts ended in failure, either because the prey spotted the hunter before it could get within range (28 times), the quarry wandered away during the stalk (5 times), or the cover ran out before the cheetah could get near enough. Five times cubs alerted the quarry. Lone animals at a little distance from a group and individuals near cover are most likely to be singled out (22), but the cheetah does not look for individuals in poor condition and takes no account of wind direction. However, it will single out and pursue small fawns that leave their hiding places, coming on the run from as far away as 500–600 m. Hunting success with fawns is 100%—a small but sure meal (17).

Once a cheetah gets within range, often flight seems to trigger pursuit. An antelope or warthog that stands its ground may well not be chased (11). Although a cheetah can accelerate to about 112 kph (70 mph), the average speed during a chase is less than 64 kph (40 mph). If it fails to overtake its quarry within 300 m the cheetah's breathing rate goes up to 150 a minute, its temperature soars, and it has to cool down for half an hour before trying again (11, 19). Having overtaken its quarry the cheetah may yet fail to catch it, for the tommy, in particular, reacts by turning sharply and the cheetah is hard put to follow closely, the more so the greater the difference in speed. A tommy seldom makes more than 3–4 such turns, but by then the cheetah is often winded (11).

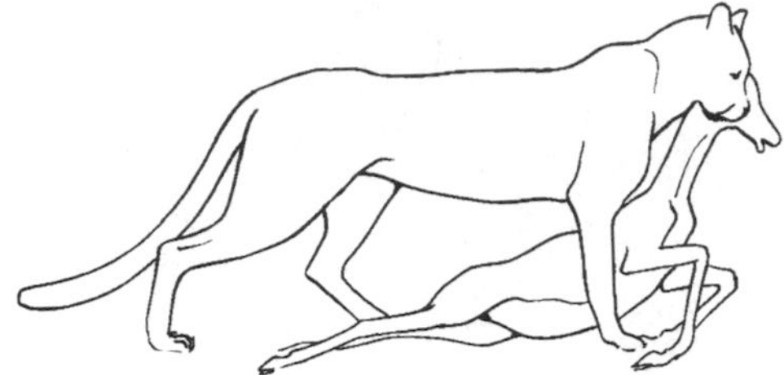

Fig. 21.26. Cheetah dragging prey to cover.

The harder its quarry is running, the easier it is for a cheetah to unbalance it by striking rump, thigh, or hindleg with a sideward or downward stroke of a forepaw, or simply by tripping it. The victim crashes on its side or even flips end over end, sometimes breaking a leg (11). If the prey is going slower or standing, the cheetah rears, hooks into the flank or back with a dewclaw, and yanks backward, causing it to fall on its side. Using one or both legs and its chest to hold the struggling animal down, the cheetah lunges for the throat and secures a grip on the windpipe, usually from behind, out of the reach of flailing hooves. The canines are so small that they only penetrate a short distance, but the skull is designed to give the cheetah a viselike grip (11).

After the victim's struggles have ceased (typically within 4½ minutes), the cheetah proceeds to drag it into nearby cover (fig. 21.26), if any, and there settles down to eat, while keeping a wary eye out for other predators and scavengers. It opens the carcass by shearing the belly skin with its cheek teeth, then eats the muscles of the limbs, back, and neck first. It may eat up to 14 kg at a sitting, then not kill again for 2–5 days (9). A female with cubs is kept much busier. One with youngsters 3–4 months old that was followed for 35 days captured 31 gazelles and a hare during that time (17). An estimated 4 kg of meat per day is available in a cheetah's kills. Most bones, the skin, and the digestive tract are left uneaten. Unlike other big cats, cheetahs do not hold their meat between their paws, but gnaw or tear off large chunks. They do not return to their kills, and only occasionally rake dirt over a carcass before leaving it (1).

Relations with Other Predators. Serengeti cheetahs lose about 10% of their kills to other predators, mainly spotted hyenas and lions, half the time before they have begun feeding (9). Sometimes a cheetah, especially a mother with cubs, will resist giving up a kill to a hyena or wild dogs, growling, moaning, spitting, and even making mock charges, but will not hold its ground if a larger carnivore continues to advance. Lions catch and kill cheetahs whenever possible; cheetahs will often flee even from a lion's voice. Leopards and wild dogs have also been known to kill cheetahs. Man is another one of the predators which have long exploited the cheetah's hunting skill: a silver vase from the Caucasus dated to 2300 B.C. depicts a cheetah wearing a collar (11).

SOCIAL BEHAVIOR

COMMUNICATION

Vocal Communication: *chirping* or *yelping, churring,* growling, snarling, moaning, *bleating,* purring. Maternal and juvenile: *whirring, nyam nyam, ihn ihn* (mother calling cubs), *prr* (maternal "follow me" call).

Many of the cheetah's calls are unlike the sounds of other cats, particularly the 2 discrete contact cries, *chirping* and *churring,* which are often given alternately and repeatedly, at varying intensity. The birdlike chirp, which sounds like a yelp or a dog's yip at high intensity and may be audible for 2 km, is the usual call given by females to summon hidden or lost cubs, by greeting or courting adults, and by cubs around a kill, the intensity reflecting the degree of excitement. *Churring* is a staccato, high-pitched growling sound that is less far-carrying. These 2 calls have been compared to the lion's loud and soft roars (11). Cheetahs also growl, snarl, hiss, and cough in anger or fright, but less frequently than other big cats. When forced to surrender its prey to another predator, a cheetah may hiss and sometimes moan loudly. *Bleating,* equivalent to meowing, is a sound of distress, as in the lion; for example, a female circling a lion that had stolen her kill uttered a growling bleat (11). Cubs squabbling at a kill made a whirring sound possibly equivalent to growling in other cats, which rose to a ferocious squeal at peak intensity and subsided to a rasp. A sound like *nyam, nyam, nyam* is also associated with eating, and captives anticipating food sometimes utter a curious humming (2, 11). Contented and friendly cheetahs purr like huge housecats, especially while greeting or licking each other. Other calls heard between mothers and young include *ihn, ihn, ihn,* which, like chirping, is used to summon young; a sharp *prr, prr* which elicits close-following when the mother is moving; and a short, low-pitched sound that makes the young stay still. Small cubs disturbed in hiding sometimes make sounds like breaking sticks (11).

Olfactory Communication: scent-marking by *urine-spraying, scuffing* (± urination), defecating on landmarks, clawing (rare).

Scent is apparently the main communication channel among cheetahs. They spend much time searching for, smelling, and depositing their own scent on previously marked places. Elevations and other landmarks used as observation points are preferred marking places, thus serving as regular stations where virtually every cheetah receives and leaves olfactory information. When a cheetah comes to a marking place, it crouches on its forelegs and sniffs long and carefully. In a group, the posture stimulates other cheetahs to approach and do likewise. A male then sprays urine on the spot and (unlike lions) often sniffs his own mark before continuing on his way (17), which may zigzag between established scent posts (4). Females also urine-mark, with increasing frequency as they come into estrus, but less actively than males; in Botswana, ranchers have trapped up to 40 cheetahs in a decade—all males—at traditional marking trees situated at junctions between several territories (13). Both the urine and feces of estrous females also attract males from far and wide. It is not known whether feces have any territorial significance, but both sexes often defecate on the mounds, boulders, and trees they use as observation and scent posts.

Tactile Communication. The cheetah *greeting ceremony* features much mutual sniffing (oral and anogenital), face-licking, and cheek-rubbing, but no body-leaning or side-rubbing (11, 17).

Visual Communication. The "tear stains" and black lips edged with white fur direct attention to the face and clearly signal changes of expression at close range. A cheetah sitting and *looking away* (mildly intimidated) keeps its lips pursed so that the mouth line is concealed, whereas the lips and tear stripes merge in an emphatic geometric figure in cheetahs with confident and aggressive expressions (fig. 21.27) (11). But otherwise the cheetah's coloration and markings are cryptic except at close range. The continuation of spots onto the face makes the head as hard to see as the body from a distance. The black ear-

Fig. 21.27. Cheetah charging (offensive threat). (From a photo in Eaton 1971.)

marks only show up well from behind or when a cheetah has its head lowered in threat. However, the tail with its black and white rings is conspicuous, functioning perhaps mainly as a follow-me signal for the young.

AGONISTIC BEHAVIOR

Dominance/Threat Displays: *stiff-legged approch with head below shoulder level,* often in *lateral presentation,* charging (fig. 21.27), and aggressive facial expression (see fig. 21.3).

Defensive/Submissive Displays: *looking away, lying on side, rolling on back, crouching with wide-mouthed snarl, lunge or mock charge, slapping* and *snapping.*

In defensive threat, a cheetah crouches low, snarling with mouth wide open and ears flattened, eyes glaring upward. At highest intensity the animal makes sudden lunges and thumps the ground sharply with downward strokes of its forepaws, meanwhile moaning and hissing, but rarely growling. Two cheetahs performed this display convincingly enough to discourage 8 hyenas from moving in on their kill (11).

Fighting. Rare as long as unassociated cheetahs avoid meeting, fights are most likely to occur when males collect around an estrous female or catch other males on their territory.

Fatal fights are occasionally reported, including one gruesome case in which the coalition of 3 Serengeti males mentioned earlier repeatedly attacked and finally killed 1 member of another trio of known males which intruded on their territory (7, 10). The territorial trio chased the intruders, overtook and proceeded to attack one viciously while the others withdrew to a distance of several hundred meters and merely looked on. All 3 resident males participated, tearing out fur and biting the intruder hundreds of times, but the unrelated, alpha male was the most persistent and vicious. However, the number 2 male administered the coup de grace, by securing a throat grip and strangling the intruder like a gazelle. Strangely enough, the victim made almost no effort to defend himself, but just lay there and took it.

REPRODUCTION: minimum age at first conception 21–22 months; interval between birth and next conception 18 months; perennial breeding with possible mating peaks after short and/or long rains; gestation 90–95 days; average litter 3–4 (1–8) (8, 17).

SEXUAL BEHAVIOR. Courtship may be extended or brief, stormy or calm, depending on the female's receptiveness and individual temperament and the number of competing males.

A lone female that revealed she was nearing estrus by stopping to sniff virtually every tree, bush, and grass clump she came to, and urine-marked at the high rate of once every 10 minutes, was observed for 8 days (10). Early on the sixth day a male came across her tracks and broke into a fast walk, alternately *yelping* and *staccato-purring.* When the female heard him, she immediately turned and trotted toward him, then lay down as soon as she saw him. The male mated her practically without preliminary, while maintaining a hold on her nape. Afterward the female rolled, groomed her face and legs, and studiously ignored the male. He growled and hissed whenever she moved, then followed and sniffed the grass where she had lain. After resting the whole day, the pair mated again at dusk, and stayed together until the next afternoon. The female then stole away while the male was sleeping.

In another, more typically feline courtship sequence, the female alternately tempted and resisted her suitor (17). Both uttered *churring* and *chirping* calls (18).

PARENT/OFFSPRING BEHAVIOR. Since females withdraw into cover to give birth and carefully hide their cubs, cheetahs less than a month old are rarely seen. A litter of 3 cubs born in captivity emerged at 20–25 minute intervals; the female broke the fetal membrane of each in turn with her teeth (18). Blind newborn cubs weighing 150–300 g can crawl, turn their heads, give soft *churring* calls, and spit explosively. Four wild-born cubs only 3 days old were found in a patch of high grass when the mother was seen transporting them (by the back, once by a foreleg). After depositing them in a thicket, she returned twice to the former site as if to make sure none had been overlooked (17). A mother with 10-day cubs moved them at least every other day. Another female kept her cubs within a 1 km^2 area and hunted a 10 km^2 area for the first month (8).

The zoo-born cubs' eyes opened at 10 days; they could walk on day 16 and got their first teeth at 20 days (18). A litter of wild cubs was first led to a kill at c. 5½ weeks and thereafter followed wherever the mother led, except when she chased prey. However, they often ignored her stalking movements and sometimes spoiled hunts by playing or trotting ahead of her. She countered this by sitting and patiently waiting until they came back. After 3 months, when the cubs were probably already weaned, they stayed behind, following slowly or waiting for the kill. They formed a close-knit family with remarkably little friction, even when feeding together on small prey. After eating, the mother would lick their faces clean, to the accompaniment of purring (17).

Cheetah cubs begin practicing catching and killing for themselves well before they become independent. The mother brings back gazelle fawns and the like and lets the cubs try catching them before they are ½ year old. Between 9 and 12 months, cubs may hunt and capture hares and fawns for themselves while the mother remains on the sidelines, but they are rarely able to make the kill. Even at 15 months, 3 cheetahs took turns swatting and bowling over a gazelle fawn they flushed 10 different times; finally the mother rushed in and bit it in the neck (17).

PLAY. Joy Adamson described the spirited, incredibly athletic games played by young cheetahs in great detail (1, 2): mainly stalking, pouncing, chasing, boxing, wrestling, and tug-of-war. Playing animals run with tail raised in typical cat fashion; they also climb trees and play king-of-the-mound. The commonest form of play, beginning at about 3 months, is chasing and swatting at one another's hindquarters, the typical way of bringing down prey (17).

ANTIPREDATOR BEHAVIOR. See Relations with Other Predators. Probably less than half of Serengeti cheetahs survive the first 3 months. They are fair game for a whole range of predators down to the size of the larger eagles. Whether the resemblance of the juvenile coat to a ratel's color pattern has any deterrent effect on predators (eagles, for instance?), a recurrent suggestion, remains to be tested. Mothers may courageously defend their offspring. One with 3 one-third grown cubs went out of her way to charge and tree a leopard, normally a dominant competitor which could kill a cheetah (author's observation).

References

1. Adamson 1969. 2. ——— 1972. 3. Caro and Collins 1987*a*. 4. Eaton 1970*a*. 5. ——— 1978. 6. Foster and Kearney 1967. 7. Frame 1980. 8. Frame, G. W., and L. H. Frame 1976. 9. ——— 1977. 10. ——— 1980. 11. Kingdon 1977. 12. McLaughlin 1970. 13. McVittie 1979. 14. Myers 1975. 15. O'Brien et al. 1985. 16. Pienaar 1969*b*. 17. Schaller 1972*b*. 18. Spinelli and Spinelli 1968. 19. Taylor and Rowntree 1973.

Added References

20. Caro and Collins 1987*b*. 21. Caro, FitzGibbon, Holt 1989. 22. FitzGibbon 1989. 23. Frame, G. W., and L. H. Frame 1981.

Chapter 22

Foxes, Jackals, and Dogs
Family Canidae

African Species

Vulpes (Fennecus) zerda, fennec
V. pallida, sand fox
V. ruppelli, Ruppell's fox
V. chama, Cape fox
Otocyon megalotis, bat-eared fox

Canis aureus, golden jackal
C. mesomelas, black-backed jackal
C. adustus, side-striped jackal
C. simensis, Simien jackal*

Lycaon pictus, wild dog

*Also misleadingly called the Simien fox or Abyssinian wolf

FAMILY TRAITS. Carnivores of foxlike or wolflike appearance, ranging in size from the 1.5 kg fennec to an 80 kg timber wolf. Slender build, long, thin legs with nonretractile claws; forefeet with 5 digits, counting the extra dewclaw (absent in *Lycaon*), hindfeet with 4; bushy tail hock-length or longer. Head with elongated, pointed muzzle, and upstanding, prominent ears. *Teeth:* 38–50, canines long, sharp, recurved, and laterally flattened, well-developed carnassials and sharp premolars, molars adapted for crushing, incisors small and chisel-like. *Pelt:* rough-coated with dense underfur, longer, erectile hair on neck, shoulders, back, and tail. *Coloration:* from white (Arctic fox in winter) to black (melanistic fox or wolf), but generally inconspicuous gray, tan, or brown, often grizzled because of banding of the guard hairs, with contrasting markings on ears, face, shoulders, and tail; natal coat dark brown in most species (4). *Scent glands:* various glands of the anogenital and head regions (see fig 18.2) described under Olfactory Communication; also glands in foot pads or between toes (7, 18). *Mammae:* 6–14. Large penis with bone.

Despite a considerable range in size and shape, canids all conform to a common structural plan. There is, in fact, far less variation in all the wild forms put together than in the dog with all the scores of different breeds that man has created from descendants of the wolf (*Canis lupus*). The relationship between the different *Canis* species is close enough so that dogs can cross with jackals and the coyote (although possibly with reduced fertility) as well as with the wolf (20).

Senses. The sense of smell is particularly well developed in the Canidae, being important for both locating food and social communication. Canids also have acute hearing. The vision of dogs is usually considered inferior to ours, yet visual displays are clearly important in this family. Wild dogs and jackals, at least, sight prey, enemies, descending vultures, and such at impressive distances (author's observations).

DISTRIBUTION. All continents except Antarctica and Australia, from the Arctic through the tropics to the tips of the southern continents, but largely absent from tropical rain forests. The red fox (*Vulpes vulpes*) and the wolf (both found in North Africa) probably have the widest ranges of all wild animals.

ANCESTRY. Although Canidae have been around for some 55 million years, the genus *Canis* appeared only within the last 2 million years (fig. 18.1) (3, 9, 20). By then, several dozen genera and at least 3 subfamilies had come and gone. Today only 10 genera and 37 species survive (17), of which all but the bat-eared fox may be closely related enough to belong in the same subfamily, Caninae (11). The earliest African canids appear in Miocene deposits and are closely allied to forms found in Europe, whence they probably came (3).

ECOLOGY. Although canids are highly adaptable, their tendency to prey mainly on abundant ungulates, rodents, hares, and rabbits ties them to grassland habitats, from tundra and prairie to steppe and savanna, and all sorts of woodlands. African species differ in their habitat preferences, adaptability, and geographic range. The wild dog has the broadest habitat tolerance and is most widespread, and the long-legged jackal that lives on Ethiopia's Simien Plateau is the most specialized and limited. The other 3 jackals replace one another geographically: the golden is a plains dweller, the side-striped a woodland species, and the black-backed intermediate. The fennec, sand fox, and Ruppell's fox are Saharan species, specialized for desert conditions but still unstudied, as is the Cape fox of South Africa. The bat-eared fox is a denizen of the Somali-Masai and South West Arid Zones.

Canids are among the most intelligent, adaptable, and opportunistic carnivores. All are capable predators but most, following the law of minimum effort, will scavenge sooner than hunt. Most are at least partly omnivorous and less strictly nocturnal than most cats (the wild dog is diurnal). Foxes and jackals like fruits and berries, and domestic dogs can exist on cereal alone. Insects are the major food of the bat-eared fox and very important at least seasonally in the diets of jackals and the other foxes. The wild dog, in contrast, is one of the purest carnivores, and also the most dependent on cooperative hunting, specializing on medium-sized antelopes. Most other canids, including pack-forming species, prey heavily on small game of all kinds, especially rodents, and the young and unfit adults of larger mammals.

SOCIAL ORGANIZATION. The basic canid social unit is a monogamous pair. A tendency to polygyny (one male with up to several related females) has been observed in foxes (see bat-eared fox account), but as a rule canids pair for life (11, 15). The comparatively small amount of sexual dimorphism in this family is typical of monogamous species (12). The couple jointly marks and defends a territory, and may forage and hunt together. The male makes a major contribution to rearing the young, and also feeds his mate while she is confined to the den. In the foxes, a reduced male parental investment goes together with polygyny (14).

Although food may be carried back to the den by mouth, canids have developed the much safer and easier way of transporting meat in their stomachs and disgorging it on demand (fig. 22.1) Participation of both parents enables canids to rear comparatively large litters (4–8, up to 19). However, the task can be so demanding that additional help substantially increases reproductive success. In the jackals and coyote, offspring from the previous litter often stay with their parents and help rear the next litter, investing in the survival of younger siblings while deferring parenthood a year (16).

Fig. 22.1. Wild dog disgorging the meat it ate at a kill, then carried safely concealed in its stomach to feed pups at the den.

It would seem but a short step from such a family group to the pack organization of the wild dog, dhole, and wolf. In these social species, too, one breeding pair is the rule and other pack members, which are either their offspring or siblings, stay on as helpers for an extra year or longer. However, more is involved in a pack system than simply prolonging the term served by helpers. Along with increasing body and group size, there is an increase in litter size and a decrease in birth weight of pups relative to the mother's weight; an increase in litter size with maternal weight has not been recorded in any other family of mammals (17). The production of more offspring born at a more immature stage means that more helpers are needed over a longer period, which the pack organization is adapted to provide. In the wild dog account we see that large packs are more successful in rearing young than are small packs.

Apart from larger adult size, bigger litters, and slower maturation, another important difference in the pack-forming canids is that they hunt in groups and bring down large prey, whereas the less social species normally forage and hunt singly or in pairs for small game. Sometimes jackals and (more often) coyotes also form packs and hunt larger prey. These medium-sized canids, and even the smaller red and arctic foxes, are also known to form larger social groups where a concentrated food supply (carrion, garbage, seabird nesting colony, etc.) provides surplus food and obviates the need for early offspring dispersal (see golden jackal account) (14, 15, 16).

In these larger social groups sometimes 2 or 3 females breed at the same time, then proceed to suckle and rear their pups communally like other social carnivores (cf. lion, brown hyena, and banded mongoose). Yet dominance by an alpha pair still prevails in canid society, and subordinate females often lose their litters if competition with the alpha female becomes severe (see wild dog account).

The social canids have evolved special signals for coordinating group activities, notably the *mobbing response,* which underlies pack hunting, killing, and competition with rival packs and other species (see wild dog account). Contagious behavior and social-facilitation signals are undeveloped in solitary species (11). Sociability has been achieved by perpetuating the dependency and strong social bonds that exist between parents and offspring and between littermates in all canids. The same begging and submissive behavior that pups display toward their elders serves to maintain social bonds between adults, forming the basis of the canid *greeting ceremony* (see under Communication) (4). *Social grooming,* which is of prime importance in maintaining pair bonds, is reduced in the most social species, except between adults and young and during courtship.

The dominance of the breeding pair is asserted with varying degrees of severity. Aggression and submission are both very evident in wolf packs, whereas submissive behavior is the hallmark of wild dog society. Yet in both species, adults are ranked in separate male and female dominance hierarchies wherein the inhibiting effects of being subordinate normally suppress reproductive hormone production and behavior (16).

If the Canidae were arranged according to degree of sociability and complexity of their social behavior, foxes would rank at one end and the wolf/dog at the other, with jackals and the coyote intermediate. Yet it is clear that canids' adaptability extends to their social organization, which can vary as much within as between species, depending on the ecological circumstances (14, 15).

ACTIVITY. See Ecology.

POSTURES AND LOCOMOTION. In becoming the most cursorial carnivores, canids sacrificed freedom of limb movement to achieve more efficient forward propulsion (8). They walk on their toes (digitigrade) and their long limbs are made more rigid by fusion of the wrist bones (scaphoid and lunar) and locking of the 2 forearm bones (radius and ulna), which prevents rotation of the forelimbs. The normal gait of a traveling canid is a trot or a lope (slow gallop) of 12–16 kph. When chasing prey, canids run in a bounding gallop, forelimbs and hindlimbs working in pairs, flexing and extending the spine to achieve maximum stride length. The lean build, with thin legs (compared to cats) and deep but narrow chest containing large lungs and a heart up to 1% of body weight (22), gives canids far greater stamina than cats (cf. cat family Postures and Locomotion). Healthy adult ungulates can outrun wolves, whereas African wild dogs regularly run the fleetest antelopes to exhaustion (see Thomson's gazelle Antipredator Behavior). Canids are poorly equipped for climbing, but their blunt-clawed feet are well-adapted for digging and they are good jumpers.

PREDATORY BEHAVIOR. All canids employ a stalk-and-pounce technique to capture rodents and other small prey hidden in the grass. The hunter moves carefully, head low and ears pricked as it uses all its senses to detect a quarry. When a mouse is detected and its position pinpointed, the hunter springs high in the air and lands with all 4 feet together on the spot, pinning the quarry to the ground (fig. 22.2). It is then grabbed and given a quick shake that either stuns or kills it. The *deathshake* is considered less specialized than the felid *neck bite* but seems nearly as effective. The canid technique for killing large game, though, is indeed unspecialized compared to the felid suffocation method, since dogs are unequipped to trip or wrestle their prey to the ground. They can only bite, usually directing their attack to the hindquarters, especially the thin skin over the groin. But wild dogs and wolves may use a nose grip to immobilize a large animal like a zebra or moose while the rest of the pack worries it from the rear.

Fig. 22.2. The way canids catch rodents: a golden jackal pouncing on a concealed grass rat (from a photo in Van Lawick and Van Lawick-Goodall 1970).

The young of pack-hunting species take longer to become self-sufficient than do solitary foragers. First they must learn by experience how to select, tire out, and kill susceptible large prey, and how to cooperate with their companions. Hand-reared individuals that have grown up without having this experience tend to be afraid of large animals (4).

Whenever possible, canid predators prey on injured, sick, young, or otherwise vulnerable animals, which are the easiest to capture and kill. Carrion represents the least effort of all. Most canids bury surplus food in well-hidden caches, programmed by the same genes that lead our dogs to bury bones.

The most unusual predatory technique employed by canids is the *"fascination display,"* whereby the wily hunter lures ducks on a pond, a rabbit, a sheep in a field, or other prey that cannot readily be run down or approached under cover, within catching distance by performing strange antics such as dancing in a circle, somersaulting, rolling on the back, and squirming (4). The foolish quarry is meant to come closer to get a better look. Meanwhile the hunter may also contrive to close the distance without the prey noticing. The red fox, coyote, wolf, and black-backed jackal are all credited with using this ploy, but proof that it is a deliberate stratagem is lacking.

SOCIAL BEHAVIOR

COMMUNICATION

Olfactory Communication. The importance of scent in canid communication is self-evident, but the kinds of information transmitted by the different canid scent glands and urine are very difficult for humans, with little ability to detect and discriminate odors, to comprehend. We can only make deductions and inferences about the information canids may be exchanging through social sniffing and scent-marking.

A large array of different scent glands is located in the 2 areas that are most important in social sniffing. The secretions of at least some of these glands are doubtless influenced by the levels of sex hormones in the circulation and can thus contain information about social and reproductive status.

Anogenital region: a pair of anal pouches and proctodeal glands inside and circumanal glands outside the anus; at least 3 kinds of glands in the penis (hair-follicle sebaceous glands, alveolar preputial or smegma glands, and tubular preputial glands); glands in the vulva plus vaginal secretions.

Head region: circumoral glands in the mouth corners, possibly other glands at the bases of the sensory vibrissae on the nose, cheeks, forehead, and neck (18).

The wealth of olfactory information concentrated in the canid rear end makes the intense interest in this region more understandable, and indicates that tail positions are correlated not only with visual (see below) but also with chemical signals. *Mutual sniffing,* first the head and then the anogenital region, is characteristic of meetings between canids. A self-confident animal raises its tail to facilitate the inspection, whereas one that keeps its tail lowered, thereby damping its own scent, reveals timidity and inferiority (4). Bitches in heat carry their tails constantly raised and turned to one side, in effect broadcasting their condition.

The violet gland (so called for its smell in a fox) on the upper tail surface is supposedly present in all canids except the dog and the wild dog (17, 18). The gland secretes most actively during the denning season, and is more developed in females than in males (in some foxes, at least); presumably the secretions are brushed against den entrances as the animals come and go. The gland is also checked during social sniffing.

Scent-marking. All species use urine for scent-marking. Feces are rarely used, even though remarkably smelly in many canids (probably because of anal-gland secretions) and evidently of olfactory interest. Based on observations of 10 different species, male canids selectively urinate on familiar landmarks which, when located on territorial boundaries, become regular scent posts examined and marked in turn by every passing conspecific (10). New objects, particularly if odorous or eye-catching, are also marked, thereby making them part of the owner's property. Adults of both sexes urine-mark but in different ways. The cocked hindleg, well-adapted for squirting vertical objects, is the typical attitude for males. This behavior is innate and under hormonal control: it can be induced in dogs at ½ the usual minimal age of 19 weeks with testosterone injections (2). In social species the top male marks most often and emphatically, whereas low-ranking males may squat and urinate like puppies (4). As in dogs, urine-marking is often followed by vigorous raking movements with the feet, sometimes accompanied by growling and other aggressive behavior.

Females squat or crouch to urinate, and when deliberately scent-marking typically hold 1 hindleg forward (fig. 22.3). Males

Fig. 22.3. The female *urine-marking posture* (golden jackal).

also do this (but crouch less) when marking low objects. Females may also mark vertical objects, by backing up to them, deflecting the tail and raising 1 hindleg under the body (10). Male wild dogs and female golden jackals have (rarely) been seen to urinate in a handstand (fig. 22.22). Both sexes urinate more frequently during the mating and denning seasons, typically in tandem, thereby demarcating territories and/or an area around the den.

The canid habit of rubbing and rolling in smelly substances, including carrion, feces, and the like (cf. civet), hardly differs among species. The animal lowers its forequarters and rubs its temple and sides of the neck on the spot several times, then turns on its back and proceeds to writhe and wriggle so that its shoulders and back are thoroughly impregnated. No one really knows the meaning of this behavior, though one plausible idea is that the animal makes itself more socially "interesting" to its fellows (4).

Tactile Communication. Scent communication involves contact during social and sexual sniffing and licking. In addition, *social grooming* is important in canid social relations. Licking of infants is necessary to stimulate voiding wastes and to clean them, whereas *nibble-grooming* functions to establish and maintain bonds between a mated pair and among the members of a family or pack (fig. 22.4). Intention movements to lick and nibble are seen during aggressive encounters (see below), where they serve an appeasing function. Adults are begged to disgorge food by licking and poking at their mouths, and this is also the basis of the *greeting ceremony* (see below).

The distance that separates resting animals is one measure of sociability. Wild dogs and bat-eared foxes lie touching, whereas other foxes, jackals, and wolves/dogs tend to maintain a wider social distance as they grow older.

Vocal Communication. Canids possess the richest repertoire of calls of all carnivores. The short-distance vocalizations of the social species are most elaborate, featuring graded signals and the simultaneous combination of different calls (e.g., whining and growling as an expression of ambivalent emotions). Distance signals, which simultaneously maintain spacing from strangers and rivals and establish or maintain contact with mate or family members, are particularly developed in the unsocial species. But thankfully no wild canid is as vociferous as the domestic dog.

Short-range calls. Whining, growling, snarling, and yelping are familiar from the calls of dogs. Sounds such as growling and snarling, designed to increase social distance, tend to be emitted in rhythm with an increased rate of respiration, then checked or choked off. At high intensity these sounds have an explosive or roaring quality (21). Such infantile sounds as whining or mewing, which elicit parental care, are also used by adults to reduce social distance and express submission during courtship, greeting, and aggressive encounters. Mothers use a faint, rhythmic "luring call" to summon pups from the den for feeding, and repeat this call during courtship (21). Other calls are described in the species accounts.

Distance signals. Barking is a universal warning signal in the Canidae. The barking of dogs in a variety of situations, often without apparent cause, is an artifact of domestication. But barking may also be used in territorial advertising, as in some foxes.

Fig. 22.4. *Social grooming* in golden jackal family (from a photo in Van Lawick and Van Lawick-Goodall 1970).

A bark may identify not only the species but also the sex, emotional state, and identity of the caller, through differences in the frequency, duration, rhythm, amplitude, and tonal quality of the call. The main contact call in the wolf and other *Canis* species is howling (fig. 22.12), which is thought to have evolved from barking by fusion of rhythmic barking strophes into a continuous string of sounds (21). In fact, howling is often preceded by a series of yipping barks (see golden jackal account). Howling has an infectious effect on members of the same family or pack, which regularly join in choruses that are obviously important in maintaining social bonds. A chorus is commonly preceded or followed by other social interactions such as *mutual sniffing* and greeting (11).

Visual Communication. As usual, appendages that are important in visual signaling are made more visible by increasing their size, color contrast, or texture. The ears are light inside and dark outside, and a white or black tip emphasizes tail position and movement. Light-and-dark facial markings, whiskers, and other sensory hairs accentuate facial expressions. Aggressive emotion causes the hair to bristle, increasing apparent size, and longer hair, often banded and/or black-tipped, along the whole spinal column and tail amplifies the effect. Conversely, the paler, often white, belly, inner legs, chest, and throat accentuate submissive postures, like showing the white flag of surrender (4). Alternatively, marking patterns on the face, cheeks, neck, and shoulders may serve as targets for attack, these being the areas, relatively well protected by thick skin and hair, that are most often bitten by canids. Facial markings such as the black mask and muzzle of the bat-eared fox are thought to define targets of social grooming, whereas a dark tail patch marks the location of the violet gland (10).

The richest repertoire of visual signals is dedicated to displays of dominance and submission (see Agonistic Behavior).

AGONISTIC BEHAVIOR

Dominance/Threat Displays. Offensive threat combines these elements: rigid body, stiff-legged walk, raised hackles, mouth puckered and closed or slightly open, ears erect and directed forward, a direct stare with eyes wide open, tail raised level with or above the back (arched in bat-eared fox) and held stiffly (wild dog, wolf) or lashed from side to side (jackals). At high intensity the wolf/dog, coyote, and golden jackal retract the upper lip to expose the canines in a snarl (usually with growling); other canids merely open their mouths slightly, even when about to attack.

Fig. 22.5. Dominance display between a pair of golden jackals: the T-*formation*.

Partners in a dominance interaction typically stand in a T-*formation* (fig. 22.5), one presenting its side, making the bar of the T, while the other approaches stiffly and stands close to its shoulder, forming the base of the T. A self-confident animal holds itself stiffly while presenting its side, back rounded, hackles raised, nose pointing down, ears forward and tail up, as if daring the approaching animal to attack (4, 6). The aggressor may press its chest against the other's shoulder or—most assertive short of actually biting the scruff or *hip-slamming*—stand up with stiff forelegs braced on the other's shoulder. Mounting may also be seen as an assertion of dominance during T-sequences.

Defensive Displays. The mixture of fear and aggression felt by animals that would rather flee but are prepared to fight if necessary can be read from their postures and expressions, in which elements of threat and submission are combined (fig. 22.11). Or first one emotion and then the other may visibly alternate, as the animal responds to internal and external stimuli. Thus, a dog may simultaneously grimace in fear and snarl, and alternately growl and whine. If hard-pressed, a defensive animal snaps viciously—but at empty space. The hair is erected, the back humped, but at the same time the animal crouches more or less, with head low and muzzle raised, keeps its tail and ears down, and avoids staring directly at its opponent.

Submissive Displays. Postures and expressions are the antithesis of offensive threat: tail between the legs or to one side and wagging weakly, body crouched, hackles lowered, ears flattened or spread sideways opening downward, eyes narrowed and head turned away (fig. 22.14), lips retracted in the *submissive grin* or *fear grimace*.

Fig. 22.6. Ritualized begging (*left*) in wild dog, basis of the canid *greeting ceremony* (from Estes and Goddard 1967).

Looking away, the antithesis of the aggressive stare, is an easily overlooked but important submissive/appeasing behavior (cf. *looking away* in cats), which may be exaggerated in some species as ritualized avoidance of eye contact (4). At the same time, the submissive animal may raise a paw, thereby flashing the white chest patch, and show its tongue (licking intention). This behavior is often effective as a cutoff to aggression. A dominant animal may also look away and lick its lips to break off an encounter or reassure a subordinate, and exaggerated *looking away* is seen again in play, often alternating with a fixed (insolent) stare (4).

A useful distinction can be made between active and passive submission (19). An actively inferior animal approaches a superior in the *submissive attitude* (fig. 22.6), licks its muzzle, and pokes the corners of its mouth, often whining like a puppy. If the dominant continues to be assertive, threatens, or actually bites, the subordinate becomes passively submissive, rolling on its side and back with a hindleg raised to expose the groin area (fig. 22.15). Immature individuals may even urinate in this attitude (*submissive urination*). Turning up the head and raising a foreleg are intention movements to roll over. Both active and passive submission are derived from infantile behavior. Licking and poking at the mouth is *ritualized begging,* which may derive from nursing behavior (nuzzling, licking and biting the nipples). Passive submission derives from the response of young puppies when nosed or licked in the groin area by a parent: they become immobile, and are stimulated by licking to urinate and defecate.

Greeting ceremony. Actively submissive behavior is also enacted between members of a family or pack practically any time they meet, as a ritualized greeting that promotes amicable group and individual relations (fig. 22.6). The adult response to a puppy's overly persistent begging is to seize it by the muzzle (*inhibited bite*). Similarly, dominant individuals may react in this way toward actively submissive inferiors.

Fighting. In most dogfights the amount of damage is trivial considering the blood-chilling growls, roars, and shrieks, the baring and snapping of wicked fangs. This is because even fighting is largely ritualized; all-out combat with intent to kill is rare among canines, mainly because losers can cut off aggression by surrendering (see wild dog account, Fighting). Fighting between canids is generally over status and ends once dominance is conceded. As usual, serious fights are likely to occur only in situations where dominance is not clearly established or is deliberately challenged. Fights ending in death have been recorded in jackals after one member of a pair died and another pair moved in on the survivor, and in wolf packs when a deposed leader was mobbed by other pack members. But in wild dogs, deposed leaders commonly stay on as nonbreeding senior citizens.

When a dog attacks, it flattens its ears or turns them sideways and downward, and almost closes its eyes for self-protection (4). Depending on the distance and reactions of its opponent, it may lunge at, stalk toward, or chase after it, and it aims either at the back (scruff) of the neck, shoulder, cheek, or muzzle. The target area varies according to species. Throat biting has a high threshold. How hard the combatants bite also varies and often depends on the degree of resistance: the more an opponent struggles, the harder the bite. The most severe bites are inflicted by head-shaking, a technique usually reserved for killing prey. In maneuvering for position, fighting canids slam each other with shoulder or hip, kick out with foreleg or hindleg against the opponent's torso, and rear up to *jaw-* or *cheek-wrestle* or try for a throat grip (fig. 22.24). Turning the hip and flourishing the "brush" in an opponent's face is frequently used to fend off an attack.

REPRODUCTION. Most wild canids are seasonal breeders that whelp once a year, beginning at 1–2 years of age (1, 5, 15). Litters of 2–6 are more usual than larger ones, but 16 or more in a litter have been recorded in the wild dog, domestic dog, and Arctic fox. Gestation is 53–72 days. Fetal growth is slow during the first half of pregnancy; consequently female canids are only briefly heavy with young.

SEXUAL BEHAVIOR. Promiscuous mating is probably only common in the dog.

Not only do most unsocial species live in permanent pairs, but so do most social species, and even some dog breeds show definite mating preferences (11). Members of a pair behave aggressively toward interlopers of the same sex.

Pair formation can be a drawn-out and elaborate process and may take place months before a female comes into heat (see golden jackal and wild dog accounts) (6). During the week or more that a bitch is in heat, she holds her tail out and turned aside in *sexual presentation*; the vulva is swollen and there is a bloody discharge, making the area visually conspicuous. The male follows closely, constantly smelling and licking the vulva, mounting and pelvic-thrusting while firmly clasping the female's flanks. Following intromission and more thrusting, the bulbourethral process at the base of the penis swells inside the vagina, effectively locking the pair together for anywhere from a few minutes up to 1/2 hour (the interval varies among species). The adaptive significance of the canid *copulatory tie*, which occurs in no other carnivore, remains unexplained. It certainly exposes mating canids to predators if not to rival attack, although the male dismounts and the pair stands rear-to-rear, a position from which they can defend themselves fairly well until the male detumesces sufficiently to pull free (fig. 22.7). As repeated ejaculations occur during the extended intromission, the tie may be found to play a vital role in the fertilization process.

The main significance of the exclusive mating bond in canids is to guarantee the paternity of the male, so that his parental care will be invested in his own rather than another male's genes (16).

PARENT/OFFSPRING BEHAVIOR. The production of large litters requires canids to be very small and immature at birth. They are blind and deaf, able to crawl on their forelimbs just enough to seek warmth, and can turn their heads from side to side in search of the udder, when aroused by touch. They require constant care and attention for the first few weeks. Meanwhile development is rapid: many adult motor patterns appear in at least rudimentary form between the second and third weeks (5). Pups start eating solid food at about 3 weeks, when their eyes open; they are weaned by the end of their second month and most species become independent between 4 and 9 months.

Fig. 22.7. The canid *copulatory tie* (golden jackals).

Initially the pups have to be stimulated to void wastes by the mother's licking; by consuming these, she keeps the nest clean. When old enough to venture out of the den at 3–4 weeks, the pups use a latrine area. Probably mothers of most species fast for the first day or 2 while remaining in constant attendance on the pups, usually in a subterranean den excavated by prior tenants or by the mother, beginning some days or weeks before whelping. She continues to spend much time underground for a week or more, rarely going hunting. Meanwhile, her mate and other helpers feed her with regurgitated food. But the main effort goes into feeding the pups during the 2 months or so until they are mature enough to go foraging with the adults (2, 5, 11, 15).

The socialization process begins when the pups' eyes open. Facial expressions appear in place of the startle or threat gape of neonates, coinciding with loss of the dark natal coat and acquisition of the species-typical markings. Maternal care becomes less intense and the young are left increasingly alone or with helpers, leading to the formation of strong social bonds among littermates. These bonds lay the foundation for future pack formation in social species (5, 17). But in unsocial canids intolerance of littermates increases as pups mature, and they eventually go separate ways.

PLAY. All young canids are playful, but adults of the unsocial species seldom play except during courtship. In social species, though, the playful spirit is prolonged, along with other juvenile traits. As in other carnivores, play behavior is mostly derived from motor patterns associated in adults with fighting and predation. There is a typical progression from the clumsy chewing, rolling, and tail biting of young pups to tossing, chasing, pouncing, chewing, and tearing at toys, to stalking, chasing, and increasingly unrestrained playfighting with siblings, culminating in actual fights.

Playful intent is signaled in all species by a *"play face"* (4): lips drawn back horizontally, an open-mouthed grin, ears ± erect, eyes partially closed and slanted. In the less expressive canids the play face may be confused with defensive threat. But the essence of the play face is a relaxed expression and demeanor. Often the tongue hangs out and the playful animal pants. Panting is one of a number of *play invitations* which may follow one another in rapid succession: dancing and tapping with the forefeet, pawing, boxing, poking with the snout, sidling, approach and withdrawal, an insolent stare followed by exaggerated *looking away* or violent head-shaking, and especially lowering the forequarters with hindquarters and tail elevated, the *play bow*. Whining and growling (and barking in dogs) are a frequent vocal accompaniment (4).

ANTIPREDATOR BEHAVIOR. See species accounts.

References

1. Bekoff, Diamond, Mitton 1981. 2. Berg 1944. 3. Cooke 1968. 4. Fox 1971*a*. 5. ——— 1971*b*. 6. Golani and Mendelssohn 1971. 7. Hildebrand 1952. 8. Jenkins and Camazine 1977. 9. Kingdon 1977. 10. Kleiman 1966. 11. ——— 1967. 12. ——— 1977. 13. Kleiman and Eisenberg 1973. 14. Macdonald 1983. 15. ——— 1984*b*. 16. Macdonald and Moehlman 1983. 17. Moehlman 1986. 18. Schaffer 1940. 19. Schenkel 1967. 20. Simpson 1945. 21. Tembrock 1968.

Added Reference

22. Miller 1955.

Fig. 22.8. A pair of bat-eared foxes, apparently asleep but in reality stressed by the presence of the observer's vehicle.

Bat-eared Fox
Otocyon megalotis

TRAITS. A sizeable fox with huge, cupped ears and a face mask. *Height and weight:* c. 30 cm, males 4 kg (3.4–4.9), females 4.1 kg (3.2–5.4) in Botswana (14) compared to 2.5 kg average in Kenya's Masai Mara population (8). Pointed muzzle and broad forehead, back rounded, hindquarters more developed than forequarters, forefeet equipped with 20-cm claws very efficient for digging; tail bushy but short (23–34 cm). *Teeth:* 46–50, canines well-developed but other teeth small and numerous (4–8 extra molars), adapted for mincing insects (2). *Coloration:* grizzled gray (owing to black guard hairs with white tips); dark brown or black face mask, including markings on muzzle and lips and across eyes; and black earbacks, lower limbs, upper tail and tip; white to buff-gray underparts, insides and edges of ears, and forehead band bordering mask (in adults); newborn sparsely covered with gray underfur, changing to adult color by 4–5 weeks (10, 11). *Scent glands:* see Family Traits; no *dorsal* gland (5). *Mammae:* 4.

DISTRIBUTION. Discontinuous as in other arid-zone species (cf. black-backed jackal, aardwolf): Somali-Masai and South West Arid Zones, absent from *Miombo* Woodland Zone.

ECOLOGY. The bat-eared fox inhabits open grassland, light acacia woodland, and overgrazed rangeland, preferring short grass and/or extensive bare ground like that occurring on volcanic-ash and calcareous soils with an underlying hardpan, and sandy soils. Its habitat requirements and geographic range are almost the same as those of the harvester termite *Hodotermes mossambicus*, a species that comes out of the ground both night and day to cut and collect grass and in the process often creates bare patches (cf. aardwolf) (14).

Seventy-two bat-eared foxes collected in Botswana (14) had eaten mainly insects (in 88% of the stomach samples): harvester termites had a 56% occurrence, followed by

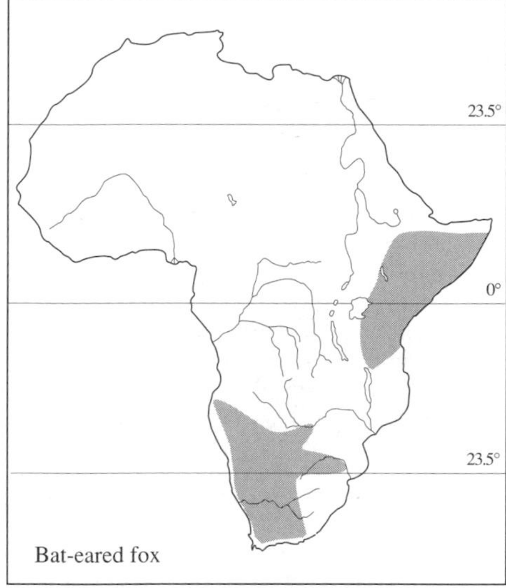

beetles (44% adult and 33% larval Coleoptera) and grasshoppers (22%, including some crickets). Scorpions were the next-most-common food item (22%), followed by rodents (mainly mice and gerbils, 17%), reptiles (14%, 7 lizard species and 2 little snakes), wild fruits (14%), hunting spiders (11%), and millipedes (7%). Seasonal changes in the abundance of different foods were reflected in the diet. Late in the rains, insects predominated in every stomach examined, with wild fruits, scorpions, and mice following in that order. Millipedes were eaten during the first week or 2 after the rains began, before they started secreting a noxious fluid that made them unpalatable. With the onset of the dry season (May) insects became scarce and were replaced by mice as the most important food; fruits too were poorly represented. In Serengeti N.P., foxes were observed eating eggs and chicks of ground-nesting birds encountered during foraging. The predominantly insectivorous diet was otherwise very similar to that of Kalahari populations (5, 10). In the Mara Reserve, Doryline (safari) ants and *Odontotermes* termites are the dry-season mainstay (8).

SOCIAL ORGANIZATION: monogamous, nonterritorial pairs and occasional trios (male + 2 females); father (and sometimes helpers) guards but rarely provisions young.

Whether bat-eared foxes normally mate for life remains to be established, but they are normally seen in pairs or family groups; only 1/4 of 457 Serengeti sightings were singles (5), and the pair bond appears to be as close as in any canid. Pairs are seldom separated except while small cubs are confined to the den, when the male often remains on guard while his mate forages (8, 9). Couples sleep in the same burrow, forage and rest together, often lying in contact (fig. 22.8), *social-groom* and play with one another, and protect and assist each other. In Zimbabwe, 2 hand-reared foxes stayed paired for 12 years, during which they raised 8 litters (more under Parent/Offspring Behavior) (14).

Where harvester termites, beetles, and other insect prey are abundant, there may be as many as 19–28 bat-eared foxes/km^2, a density higher than reported for any other canid, though 0.5–3/km^2 is more typical (8). Couples inhabit comparatively small and temporary home ranges. The average of 8 Serengeti groups was 0.79 km^2 (0.43–1.57) (5), and of 3 Masai Mara groups was 3.53 km^2 (2.5–4.9) (8). Ranges overlap to a large extent and foraging family groups intermingle freely and with little antagonism (8, 11). The suggestion that Serengeti foxes are territorial (5) is contradicted by evidence that resident and transient pairs may occur in the same area, that dens of different pairs may be clustered within a few hundred meters, and that pairs apparently do not keep the same range from year to year, but seek out good foraging areas (which change from year to year) and denning sites before each breeding season (5, 8).

Burrows at ground level with several entrances and overgrown with vegetation are preferred as dens. *Otocyon* rarely needs to use its superior digging power to create a new den, but merely modifies holes excavated by warthogs, aardvarks, springhares, etc. Excavation of a Botswana burrow system some 3.5 m in diameter revealed 4 entrances and 3 chambers interconnected by tunnels. The first chamber was 1.2 m from the entrance, 33 cm below the surface, and 18 cm in diameter. From there a 3-m tunnel rose gradually to the surface. The deepest chamber was c. 70 cm underground, and was 30 cm wide by 18 cm high. Except for juvenile scats near the main entrance, which was located beneath a thornbush, all the tunnels and chambers were perfectly clean, without nesting material (14).

Despite the prevalence of pairs, several instances of a male living with 2 females are known (8, 16). In one triad the females had 5 cubs between them which they suckled and otherwise cared for communally; the young would nurse first one, then the other (16). In another case, 2 females with 10 half-grown cubs were repeatedly seen near the Serengeti Research Institute in grassland laid bare by harvester termites (author's observ.). Recent observations indicate that the extra female is usually a daughter (6).

The differences in social organization that distinguish *Otocyon* from other canids may all derive from becoming specialized as an insectivore (5):

1) Cubs are not provisioned at the den, but begin foraging with their parents when approximately ½ grown; 2) adults rarely regurgitate food; 3) cubs are weaned later and mature earlier than most canids, reaching adult size at 4–6 months (more under Parent/Offspring Behavior) (7, 13).

Although their invertebrate food resource is abundant, it takes many hours for bat-eared foxes to eat their fill (see under Activity and Foraging Behavior). As Malcolm (9) points out, "Bat-eared foxes rarely have 'excess' food, unlike larger canids with a kill. It is more efficient to produce milk than to regurgitate largely indigestible insects." In this connection, the male's extended guard duty at the den may be necessary to enable a mother to eat the extra insects necessary to sustain lactation. Nevertheless, if and when parents happen to catch vertebrate prey, they often do bring it back to the cubs (9).

For small cubs to venture away from the den is obviously dangerous, so the advantages of foraging in family units must outweigh the risks. It might seem particularly dangerous by night, but in fact diurnal birds of prey generally represent the greatest hazard for young bat-eared foxes. The need for cubs to go foraging beginning at an early age could explain the unusually prolonged lactation and early maturation in this species (5). Because harvester termites run for their holes as soon as a predator begins feeding on them, a fox can only consume a limited number before the rest escape. When several animals are present, each can consume as many termites from a dense patch as one fox alone, and they can actually harvest more termites by foraging in a group than if they hunted separately over the same ground at the same time. There are also the other, usual advantages of being in a group such as enhanced ability to detect predators, protection by both parents, and the opportunity for offspring to learn by imitation what to eat and how to get it (5). Interestingly enough, foxes in the Masai Mara Reserve, where *Hodotermes* termites do not occur, generally forage singly (8).

ACTIVITY. In the Serengeti and probably elsewhere in tropical Africa, bat-eared foxes are primarily night-active (85% vs. 15% by day), but may often be seen resting outside in the daytime (5). Beginning at 1630–1700 h, foxes leave the holes or thickets where they passed the hottest hours and lie in the open until nightfall. They then become active and, after socializing a while (*mutual grooming,* play, etc.), the pair or family begins foraging. The foxes continue feeding with hardly a break until midnight, followed by 1 or more rest periods from 0100 to 0400. They then forage some more, returning to the den at dawn, where they bask and socialize until the heat makes them withdraw for the main rest period. Sometimes, especially in cool weather, foxes will forage again before retiring and emerge in early afternoon to loaf around the den or do a little feeding. In the southern Kalahari, bat-eared foxes are mainly diurnal in winter, when freezing temperatures keep harvester termites and other insects underground at night (10).

POSTURES AND LOCOMOTION. See Antipredator Behavior and Play.

FORAGING BEHAVIOR. Bat-eared foxes rely mainly on their ears and noses, and only secondarily on their eyes, to locate food (14). When feeding on harvester termites, they move about with heads low, licking up the insects much in the manner of an aardwolf. Pursuing grasshoppers and winged termites, the foxes leap like acrobats, and they capture lizards, gerbils, and young birds with a quick dash (14). They pounce on mice with both front feet in the typical canid manner. But most impressive are the bat-eared fox's highly developed ears and digging ability, which enable it to locate and unearth beetle larvae and other subterranean invertebrates more efficiently than most other carnivores. A searching fox moves erratically, its head turning from side to side, ears cocked to pick up the faintest sound. It then stops and moves its ears back and forth like twin dish antennae to spot the source. Having done so, it approaches and stands with ears cupped just above the spot where, say, a beetle larva lies buried some centimeters deep in the soil (fig. 22.9). Now it begins digging, rapidly excavating a narrow trench with paws striking alternately in the same place, pausing often to

Fig. 22.9. Bat-eared fox taking a binaural fix on a subterranean insect (from Smithers 1983).

Fig. 22.10. Bat-eared fox offensive threat: note inverted U-tail (from Lamprecht 1979).

Fig. 22.11. Defensive threat (from Lamprecht 1979).

listen. It can dig through calcrete hardpans—or the stoutest carpet (13, 14).

Foxes crunch beetles, millipedes, and other hard-bodied prey with very rapid bites, drop them, pick them up, and chew them again until they stop moving, then break them into small fragments while eating them. Scorpions and spiders are broken up but not thoroughly chewed. Small mice may be swallowed with only the skull crushed, and small birds are swallowed whole, feathers and all, after thorough chewing. Reptiles and large rodents are cut or pulled to pieces and swallowed after much chewing (14).

Foraging is an individual activity and bat-eared foxes are unwilling to share food except with inexperienced cubs. Attempts to approach a digging or eating individual may be discouraged with growls and flattened ears (11, 14). Like foraging mongooses, adults and older offspring remain spaced out while seeking individual food items, but keep the same direction and pace.

RELATIONS WITH OTHER SPECIES. Foxes often meet other species while foraging, including other insectivorous carnivores such as the white-tailed mongoose, aardwolf, and zorilla. They usually pay little or no attention to one another (14, 17). However, different observers have seen foxes provoke baboons, guinea fowl, chickens, birds of prey on the ground, gazelles, and so on by chasing after, fleeing from, and snapping at them. Probably in most cases the foxes were guarding dens or offspring. Parents are fiercely protective and will harass jackals and even hyenas that prowl near the den, threatening with all hairs standing on end, as in figures 22.10 and 22.11.

SOCIAL BEHAVIOR

COMMUNICATION

Vocal Communication. With just 2 loud calls, voiced only in emergencies, and no group call, the bat-eared fox is one of the most silent canids. The 9 calls that have been distinguished mostly function for within-group signaling and are inaudible at a distance (11).

Contact calls:

Soft whine/mew (birdlike *"who-who"* of ref. 13). Elicits approach or following by cubs or mate. An adult that has lost contact with its mate may give a long quavering mew, very similar to the juvenile "searching whine."

Chirping call (sibilant whistle of ref. 14, searching whine of ref. 5). A very distinctive, birdlike call given by cubs when separated or deserted, and in response to parental soft whine/mew. It can become loud and piercing, lasting for 2–4 seconds and descending in pitch (5). According to references 5 and 14, this call develops into the adult contact call.

Woof/growl/hiss. Soft warning calls which cause cubs to run for cover.

Short whistle. One of the commonest calls, uttered during *social grooming,* by either partner.

Agonistic and alarm calls:

Growl/snarl/hiss. Defensive response when threatened or chased. A rattling growl is also emitted in play (14).

Snarl/yap. High-pitched sounds of fighting, heard also during play.

Scream/distress cry. The loudest sound, given, for example, by a bitten fox.

Short bark. Warning/alarm, a loud cry given when surprised or threatened.

High-pitched bark. A loud call given, for example, when predators approach a den with young pups; it stimulates other bat-eared foxes to join a mobbing attack (11). This is a rare call never heard by Lamprecht (5).

Olfactory Communication: scent-marking (urination), urine- and genital-sniffing.

Urination postures are similar to those of other canids (see family introduction). Bat-eared foxes project urine onto grass tufts and low bushes, rarely against tree trunks or other uprights, and never scratch the ground before or afterward (3, 5, 11). The male holds his tail horizontally. Lifting the leg and/or prior sniffing of the spot both distinguish purposeful scent-marking from simply voiding urine. The primary purpose of scent-marking in this nonterritorial canid may be in establishing pair bonds. It is seen most commonly during the period of pair formation by foxes unaccompanied by young, and often takes the form of double-marking: the female urinates first, then the male covers her mark after first sniffing it. The initial stimulus for the double-marking ritual is frequently the male's anogenital sniffing and/or licking (11). Females rarely overmark their mate's urine (9 of 89 observed urinations), and deliberately scent-mark only about ¼ of the times they urinate (i.e., after first sniffing the spot).

Bat-eared foxes may mark anywhere in their home range, but particularly along trails and borders and after investigating possible den sites (5). No single foxes were seen urine-marking by Nel and Bester (11). However, this species generally urine-marks at much lower frequencies than jackals and other social canids. Foxes of the southern Kalahari average less than 1 urination an hour and in the summer heat, while caring for offspring, they are further constrained from urinating as a water-conservation measure (11).

There is no evidence that feces are used for scent-marking. Family members often defecate near dens and other resting places upon becoming active, but do not maintain regular latrines and elsewhere defecate seemingly at random (8).

Tactile Communication: *social grooming, nose-to-nose greeting, huddling.*

Pairs and young often lie touching and in cold weather huddle together for warmth. The predominant social activity, both in frequency and duration, is mutual, *simultaneous nibble-grooming* (rarely licking). It occurs between parents and offspring, and year-round within pairs, interspersed with bouts of self-grooming (scratching, nibbling, licking) (5). A hand-reared female invited grooming by approaching submissively or by throwing herself flat on her stomach (*belly-flopping*), tail out and chin on the ground, before her mate (a position also identified with defensive threat—see Agonistic Behavior). Sometimes both solicited at once, lying facing a few inches apart. During grooming of the cheeks and ear bases, the groomee elevates its snout (13).

The typical canine *greeting ceremony,* featuring poking and nuzzling the muzzle, is greatly deemphasized in *Otocyon,* apparently reflecting the rarity of regurgitation/begging in this species (8). However, bat-eared foxes touch and sniff noses upon meeting, and cubs regularly lick and nibble a parent's muzzle after a separation (5).

Visual Communication. Although *Otocyon's* facial mobility is limited, its huge ears and the tail, both made conspicuous by contrasting markings, are particularly expressive and in combination with positions of the head and body signal a wide range of emotional states. The one striking difference between the displays of this species and those of other canids is the inverted U (arched) position of the tail (fig. 22.10), but this attitude is assumed in too many different situations to be diagnostic all by itself: threat, fear, flight, play, defecation, and sexual arousal (11). *Otocyon* also does not wag its tail, except sometimes slowly in play (4, 11). The basic visual displays associated with aggression, submission, sex, play, and predator response are described under the appropriate headings.

AGONISTIC BEHAVIOR

Dominance/Threat Displays. *Tail arching with head up: erect posture* with tail, back, and neck arched, coat bristling, head raised, eyes open, ears erect and directed forward (the closer together, the more intent and alert the fox), no aggressive pucker (fig. 22.10) (5, 11).

Defensive Displays. *Tail arching with head down:* head lowered and ears pulled back (fig. 22.11) (tail elevation diminishes with aggressiveness). *Belly flop:* standing confronting, one or both opponents suddenly flop to the ground and lie flat with tail straight back (never clamped between legs), chin pressed to the ground and ears sidewards, mouth slightly open ± soft growling (5, 11).

Submissive Displays: *belly-crawl* with

ears and tail as in belly flop; *falling on back with hindleg raised + fear grimace; head lowered* or *turned sideways* and *ears maximally flattened.*

Fighting: sudden charge accompanied by snarling and the victim's screeching, followed by a short chase.

Parks visitors who manage to drive close to bat-eared foxes usually see them lying curled up with ears flattened (fig. 22.8). Though the animals are apparently resting calmly, this behavior actually signals submission or avoidance (*looking away*) induced by fear. Other, more fearful individuals simply flee when approached.

Eleven of 16 encounters Lamprecht (5) observed between Serengeti foxes from different groups were hostile and included offensive threats by all participants; all but 1 also involved biting and chasing. Significantly, 10 of the encounters occurred during the prelude to the mating season (May–July) when pairs are prospecting for suitable breeding grounds. Four of the 6 encounters that took place later, at the beginning of the whelping period, involved resident neighbors and were peaceful or even friendly. The sex of opponents is often hard to determine, but it is known that both males and females fight, usually against members of the same sex. Aggression within families rarely goes further than a female snarling and snapping when a mate's sexual advances become too persistent. However, in the case of a triad, a gravid female was seen to attack and bite the other vixen, after first driving her out of the burrow (5).

REPRODUCTION. Although *Otocyon* could potentially breed twice a year, it breeds annually, whelping near the beginning of the rainy season during a period of peak insect abundance, after a 60–70 day gestation (5, 14). *Otocyon* reaches adult weight by 6 months at latest, but females apparently first come into estrus at c. 18 months (11, 14).

The tame pair from Zimbabwe had 7 of 8 litters in October and November. The number of cubs ranged from 4 to 6, and the average number of fetuses carried by 4 wild females was 5. Yet the average number of young in 13 Botswana and 32 Serengeti families was 2.4 and 2.9, respectively. A high juvenile mortality, especially during the 3 weeks the cubs are confined to the den, appears likely. Some of it may be due to infanticide: the tame vixen ate the extra cub(s) in 4 litters that initially numbered over 4 cubs. A pronounced difference in the birth weights of litter mates (99–142 g) has also been noted, and persists at 1 month: 2 cubs weighed nearly 2 kg and the other 2 weighed 1/4 less. Such an inequality could be adaptive, enabling *Otocyon* to raise large litters during bountiful years and small litters even in bad years, when the smaller, weaker pups will lose out to their more robust siblings (14).

SEXUAL BEHAVIOR. Mating is undescribed in wild foxes. In a zoo, a pair mated up to 10 times a day for a week. The female showed no estrous swelling and the male became interested in her only the day before copulation began (12). But before and during estrus females marked with unusual frequency. The male followed closely, licking her vulva and periodically mounting. After achieving intromission, he dismounted, and the pair remained tied (duration unspecified).

PARENT/OFFSPRING BEHAVIOR. The hand-reared pair began excavating burrows several months before whelping, indeed even before mating, and they persisted in doing so even after a cement den was provided and accepted for breeding.

Pups' eyes open at 9 days and they emerge from the den at c. 17 days. Weaning of the tame pair's offspring began within a month, when the mother started leaving small, thoroughly chewed birds at the burrow entrance. Such provisioning has not been observed in the wild (5). Exploratory excursions from the den were closely watched by the parents, who would growl softly in warning when the young were approached. If they failed to react, either adult would pick them up and deposit them in the den. But the cubs rarely needed prompting, reacting to practically any outside stimulus by bolting for safety. Often all 4 would jam together at the entrance, resulting in much scuffling and growling before all could get inside. Until the cubs were 4 months old, the female would always emerge from the burrow first and scrutinize the surroundings with great care before summoning them with the "*who-who*" call. Both adults were remarkably sensitive to any changes and would investigate anything that was out of place or new from all angles, with nose and front legs extended, tail down, and ears cocked (13, 14).

The same pattern of parental care has been observed in wild families, and the father's participation in guarding, warning, carrying (by the back), grooming, playing with, and leading the young appears equal to and sometimes greater than the mother's. However, the young are suckled for up to 15 weeks, the mother either standing or sitting. A male was twice seen regurgitating food for 9½-week cubs (5).

PLAY. Young bat-eared foxes play by the hour and some of this playfulness is never outgrown. For instance, a wild adult was seen playing by itself with a stick (13). The tame pair played together year after year. To solicit play one would prance and bob from side to side before the other, ears up and tail arched or, more submissively, with ears back and head low. If the other was resting and unresponsive, the playful one would *belly-flop* facing it and whine, or try to prod it into action by digging at it with the forefeet. As a last resort, the active fox would nip the lazy one's haunches or tail tip, then prance some more, until it finally became aroused. Vigorous play was accompanied by rattling growls. Often one would race past the other, bite, then leap nimbly aside before the victim could retaliate, or jump backward and upward in a twisting motion with all feet off the ground. During a chase, the pursued fox sometimes turned suddenly and jumped over the pursuer. In playfighting, the pair would take turns rushing in with head lowered, ears back, and tail arched, which was countered by presenting the rump with raised tail or by leaping out of the way. Occasionally one (usually the female) would get a grip on the side of the other's throat and hold it down until it gave in by whining (13).

ANTIPREDATOR BEHAVIOR. This small carnivore is vulnerable to predators down to the size of eagles and jackals, even as an adult. To elude pursuit it relies not on speed but on its extraordinary dodging ability, which earned it the name of "turning jackal" (*Draaijakkal*) in South Africa (a title also applied to the Cape fox). With a flick of its plumed tail, it can reverse direction at a flat run without losing speed. These sudden changes of direction have been attributed to the practice of pivoting on 1 foreleg (16), but clearly the tail also functions as a rudder. In addition it can be used as a shield and flaunted in the face of an adversary, possibly serving to deflect or decoy the grasp of talons or jaws. A fox fleeing from an aerial predator zigzagged furiously to the nearest hole with its brush flailing over its back. One that was caught by a martial eagle escaped by lying on its back with erect tail flailing, while snapping at the bird's upper legs. The eagle cried out in pain and quit (2).

References

1. Hendrichs 1972. 2. Kingdon 1977. 3. Kleiman 1966. 4. ——— 1967. 5. Lamprecht 1979. 6. B. Maas pers. comm. 1987. 7. Malcolm 1984*a*. 8. ——— 1986. 9. ——— in lit. 1986. 10. Nel 1978. 11. Nel and Bester 1983. 12. Rosenberg 1971. 13. Smithers 1966. 14. ——— 1983. 15. Van der Merwe 1953. 16. Van Lawick and Van Lawick-Goodall 1970. 17. Waser 1980.

Fig. 22.12. Golden jackal family howl (from a photo in Van Lawick and Van Lawick-Goodall 1970).

Golden Jackal

Canis aureus

TRAITS. A small canine without prominent markings. *Height and weight:* 40 cm (38–50) and 11 kg (7–15), sexes equal (11). Long, pointed snout and ears; short (20–30 cm), bushy tail. *Coloration:* individually, geographically, and seasonally variable, also changing with age (6); generally yellowish to silvery gray, limbs redder than torso, longer shoulder and tail hair black-tipped in adult but fading in dry season; tail tip, nose, and mouth usually black, insides of legs, throat, chest, and markings on muzzle and cheeks white; eyes amber-colored. *Scent glands:* listed under Family Traits. *Mammae:* 4.

DISTRIBUTION. The main range of this jackal is Asian; it is found throughout the Near and Middle East to Burma, the Balkans, and southern Russia. Its African distribution extends barely beyond the limits of the Sudanese and Somali-Masai Arid Zones. In temperate climes a luxuriant winter coat makes *C. aureus* appear larger and handsomer than its sometimes scruffy-looking African counterpart (6).

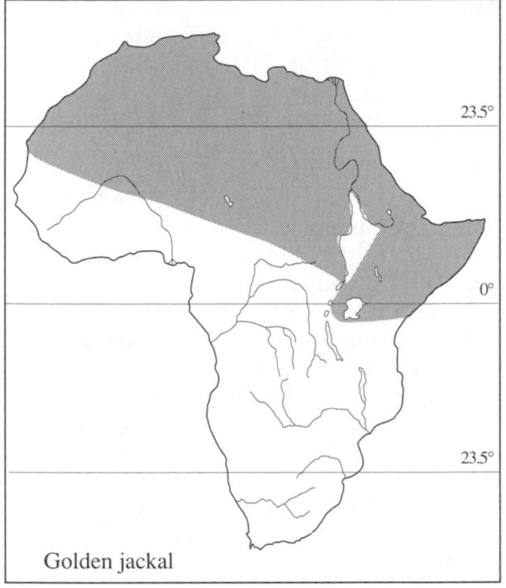

Golden jackal

ECOLOGY. The most desert-adapted jackal, *C. aureus* is a plains and steppe dweller that can subsist in waterless country. It is often the most numerous carnivore in the East African grasslands. Although it scavenges carrion, the kills of other carnivores, and man's refuse, it is primarily a predator on invertebrates and vertebrates up to the size of a gazelle fawn, and sometimes larger prey. It also eats fruits such as figs, desert dates (*Balanites*), grapes, and various berries.

SOCIAL ORGANIZATION: monogamous, territorial, yearling offspring serve as helpers, larger social groups under special conditions.

On the Serengeti Plain, where golden-jackal behavior has been studied for over a decade, pairs defend permanent territories of 2–4 km^2, the boundaries of which may change very little over a couple's lifetime (11). Here food is abundant during the rains but so scarce during the dry season that resident jackals may subsist principally on invertebrates such as excavated dung-beetle larvae, spiders, solifugids, and ants, supplemented with an occasional lizard or small bird. Yet golden-jackal pairs only go outside their territory to drink or when attracted to a large carcass.

In a sample of 8 litters, yearling offspring helped rear 6. The 12 cubs that were known to survive to 14 weeks all remained with their parents until c. 20 months old, although both sexes become adolescent by 11 months (10, 12). However, some offspring of ½ year or older left home during the dry season and apparently sojourned in the nearest woodland, returning when the food supply increased after the onset of the rains. Unlike mated pairs, helpers often trespass on neighboring territories. One of the probable advantages of serving as a helper is the opportunity to prospect for undefended or inadequately defended ranges from a secure home base. The possibility of inheriting part of the parental territory is another (11).

Golden jackals have comparatively close and amicable family relationships. Although the young remain subordinate to their parents, with endocrine function and sexual behavior suppressed as long as they remain at home (yearling males have small testes and abstain from territorial marking), the parents are less assertive than black-back pairs and have never been known to drive their offspring away (9, 11). They rarely regurgitate to helpers but share kills with them. Family members rest closer together and *social-groom* more than black-backed jackals (fig. 22.4). Partly because siblings quarrel less and because the parents break up fights, severe fighting between siblings is rare. And yearlings participate in territorial defense against invaders of the same sex.

The golden jackal is capable of forming different and larger social groups under conditions—possibly only man-made—in which an abundant, localized source of food is available. At a garbage dump in Israel, 2 stable groups of 20 and 10 jackals were observed, the larger of which included several breeding females and several adult males, 1 of which was dominant (8). All 20 lived within an area of only 11 ha which they defended as a communal territory. The border was clearly demarcated by dung middens—the only reported instance in which jackals have used feces for territorial marking. Another, 2-year study of 17 jackals inhabiting coastal sand dunes near a kibbutz showed that mated pairs were regularly followed by unrelated individuals (5). The followers were usually unmated females or "engaged" pairs less than 2 years old. They all behaved submissively and continued to follow even when subjected to threats and *hip-slamming* by the established pair. It could be supposed that these young adults were seeking surrogate parents. In fact, one male follower persistently fed an established pair's pups for several months. Different adult pairs, in contrast, almost never interacted with one another and males showed sexual interest only in their own mates (5).

Pair Formation. The process of pairing up may begin before leaving home. A young-adult female began to associate with a strange male on the boundary of her family's territory. At sight of him, she lay down and curled up, ears back and nose pointed at him as he approached with hair bristling. After circling behind her, his hackles relaxed and he sniffed her rump, then withdrew 10 m, curled up, and lay down, too. From time to time he glanced at her, and she in turn stole looks at him (13). In 3 other observed pairings, the female took the initiative by approaching with raised, waving tail and standing side-on to the male in the T-*formation* (fig. 22.5) (5). T-*sequences* are repeated with great frequency by pairs during the long buildup to the mating season (more under Sexual Behavior; also see family introduction, Dominance/Threat Displays). At the end of a sequence the female followed the male and urine-marked when and where he did. Once, 4 females were seen to converge on a male stranger after he uttered a series of short, peculiar calls between a bark and a howl; 1 of the females then initiated a T-*sequence.* After pairing up, each couple limited its movements to a particular area, followed certain paths, and rested under the same bushes. Once established, a pair was never known to separate (5, 11).

ACTIVITY. Where unpersecuted by man, golden jackals may be nearly as active by day as by night, especially early and late in the day. In Ngorongoro Crater, goldens are more in evidence by day than are black-backs, which appear to predominate at lion and hyena kills at night (2).

POSTURES AND LOCOMOTION. See under next two headings.

PREDATORY BEHAVIOR. In foraging for the insects that make up part of their diet at all seasons, jackals nose and turn over dung piles to find dung beetles, and in the dry season excavate dung balls containing the fat larvae. They pounce upon or leap into the air to grab grasshoppers, flying termites, and such. Where rodents are present, they hunt them in the usual canid manner (fig. 22.2). They catch lizards and small snakes, and forage beneath trees containing bird colonies (weavers, ibises, cormorants, etc.) for fallen eggs and nestlings. When hatchling tree ducks flutter to earth to begin the hazardous journey to water, Ngorongoro jackals are often waiting with open mouths. Jackals also forage under trees for fallen fruits.

The largest prey normally hunted by jackals are young gazelle fawns, hares, and birds up to the size of a flamingo (flamingos become vulnerable to jackals when they mass in very shallow water). Hares are faster than jackals and gazelles defend their fawns effectively against single hunters, often in teams of 2 or 3 mothers (see gazelle accounts). Jackals have better luck hunting fawns in pairs: in a sample of 12 chases all 5 undertaken by pairs were successful, whereas only 1 of the 7 attempts by singles resulted in a kill (7, 14). Jackals methodically search for concealed fawns, approaching groups of female gazelles and then quartering patches of taller grass, bushes, and other likely hiding places. Finding nothing, the jackal(s) scans the horizon and then trots straight toward another gazelle herd, there resuming the quest (7).

Although golden jackals have been known to kill animals 2 to 3 times their own weight, most such cases probably involve disabled or newborn individuals. Otherwise jackals rarely attempt to hunt healthy game even as large as themselves. Sometimes families with helpers or large pups may hunt in packs, but apparently goldens are generally less predatory on mammals than are black-backed jackals (7).

SCAVENGING AND RELATIONS WITH OTHER SPECIES. Although golden jackals take whatever food can be obtained with least effort, in natural ecosystems such as the Serengeti and Ngorongoro probably less than 20% of their diet comes from scavenging (12). Territorial intolerance largely restricts each family to what is available within its own range. There are rarely more than a few goldens around a kill (10, 11). Within their territory, however, goldens compete very aggressively with other predators and scavengers for a share of the spoils. A jackal can dominate even the largest vultures, and may hold dozens at bay while it monopolizes the remains of a kill, threatening, snapping, and lunging with ears flattened and tail whipping from side to side (see black-backed jackal, fig. 22.16), sometimes jumping into the air to bite at a bird that tries to land too close. They give way to larger carnivores, and seem to be warier of lions than are black-backed jackals, but are bolder with wild dogs and possibly hyenas (2, 4). Ngorongoro golden jackals sometimes stole meat from wild dogs and, when a dog tried to take it back, the jackal would stand its ground, head down and snarling, snapping viciously when the dog continued to advance. This bluff worked surprisingly often. Superior agility enables jackals to take liberties with hyenas, too. One may dart in and snatch a bone or chunk of meat from under a hyena's nose, or play tug-of-war with a piece of gut. Once I saw a golden run with several hyenas and help capture a newborn wildebeest, prey that jackals normally pass up unless something is wrong with it. A pair of jackals may outwit and steal small items from a single hyena by employing the "yo-yo gambit." Figure 22.13 shows a pair moving in on a hyena that is feeding on a flamingo it caught. When the jackals were nearly within biting distance, the hyena couldn't stand it any longer and snapped at one, whereupon the other darted in and grabbed at the bird. After several attempts, one jackal got away with most of the flamingo (3).

Hyenas, of course, also rob jackals, relieving them of gazelle fawns they have captured or scavenged food they may be carrying back to the den. Perhaps to minimize losses to larger competitors, pairs regularly dismember their larger kills, then separate 30 m or so while each eats its portion. If one jackal loses its meal to a hyena, they then share the other's (7). They also regularly cache surplus food in holes and under bushes, returning usually within 12 hours to retrieve it (7). But the best insurance against loss is to eat as much as possible in the shortest possible time: food in the stomach is safe until a jackal voluntarily surrenders it to its mate or offspring. Accordingly, jackals bolt as much as they can hold with minimal chewing.

SOCIAL BEHAVIOR

COMMUNICATION

Vocal Communication: howling, barking, growling, whining, cackling.

The golden jackal's high, keening wail is one of the characteristic sounds of the African plains. Actually, 3 variations of the call can be distinguished: a long, continuous howl on one tone, a wail that rises and falls, and a series of short, staccato howls. Adults perform all 3 in succession, whereas jackals of less than 2 years are said to voice only the staccato howls (6). Howls differ individually and, apart from advertising presence to outsiders, serve as contact calls between pair and other family members, which generally answer one another. Pups respond by running to their parents and greeting them. Howling in chorus (fig. 22.12) is thought to reinforce pair and family bonds, while at the same time serving to advertise territorial status (11). Howling may also be contagious between groups and, especially near dusk, the calls of one family are often taken up by others until the plains re-echo with their wailing.

Tactile Communication: social grooming, greeting.

Reciprocal grooming (fig. 22.4) is most frequent and prolonged during courtship, when sessions may last up to ½ hour. A male started nibbling his fiancée above a front leg, proceeded to her waist, then the throat, and when she rolled over, did her underside as well (5). In a long-established pair the female usually took the initiative. Beginning on her mate's ear, she proceeded in the way described, while the male sat and then lay with a *consummatory facial expression.* She also groomed her pups regularly and intensively; they in turn groomed one another and their parents, increasingly as they grew up. Grooming can be solicit-

Fig. 22.13. A pair of golden jackals cooperating in an attempt to steal meat from a hyena.

Fig. 22.14. Submission by a trespassing female golden jackal (*left*) in response to the aggressive approach of the female proprietor.

ed by falling over on the side in front of another animal. Brief nibbling of the face and neck is also common in the *greeting ceremony* (described in family introduction).

Olfactory Communication. Both the boundaries and the interior of the territory are regularly *urine-marked,* preferably at nose height, in the usual canine manner. The urine has a potent smell. Pairs typically patrol together, moving at a springy trot and stopping to mark tufts of grass, bushes, rocks, and so on every few hundred meters or less. During 25 hours of focal samples of Serengeti pairs, 77% of the urinations during foraging were done in tandem (10). Usually the male leads and is the first to mark, but the practice varies and in some pairs the female may be the more active marker (fig. 22.3) (6). Like feisty dogs, jackals may vigorously scratch the ground with fore- and hindfeet afterward, meanwhile bristling and snarling with puckered lips and narrowed eyes. Occasionally both sexes also place their feces in the middle of a tuft of grass, reminiscent of fox behavior (13). But using dung middens as boundary markers has only been recorded in one group-defended territory (noted under Social Organization) (8).

TERRITORIAL BEHAVIOR. The fact that jackal couples patrol and mark in tandem and howl in chorus is an indication that their joint efforts are needed to maintain a territory against pressure from other jackal pairs (see black-backed jackal account). Both partners and also helpers behave aggressively toward trespassers (fig. 22.14), but reserve their greatest ferocity for members of the same sex: pair members do not assist each other against intruders of the opposite sex (11). In guarding against invaders of the same sex, each pair member also safeguards the state of monogamy.

AGONISTIC BEHAVIOR. Displays described in family introduction.

Fighting: *hip-slamming,* rearing, gripping muzzle, *jaw-wrestling,* biting and shaking neck or shoulder (more under family introduction).

The facial expressions and postures of the golden jackal are typically canine. This species even lifts its lip to expose the canine teeth like a dog; the other jackals have less facial mobility. Interactions between approximately equal partners typically take place in the T-*formation* (fig. 22.5). The most frequent, involved, and lasting T-*sequences* occur during pair formation (see Sexual Behavior) (5, 6).

REPRODUCTION: annual, up to 6 pups (maximum 9) born during rainy season, after 60-day gestation. Pairs have been known to produce at least 8 litters (12).

SEXUAL BEHAVIOR. Courtship is a remarkably long and involved process in jackals, during which a couple remains almost constantly together. Once 2 jackals have paired off and acquired a territory, courtship proceeds through 3 intergrading stages, each characterized by certain behaviors (6).

1. *Joint patrolling* and scent-marking of the territory.

2. T-*sequences* (fig. 22.5), in which the following patterns are prominent: (*a*) the female holds her tail out and angled to one side, exposing a distinctive genital pattern that bitches develop during the mating season; (*b*) the partners approach each other while whimpering and lifting their tails, with fur bristling, and stand in the T-*formation,* displaying varying intensities of offensive or defensive threat; (*c*) the female sniffs/licks the male's genitals and he nuzzles her fur. They may circle head-to-tail or head-to-head, and sometimes they may *hip-slam* and fight briefly. T-*sequences* are often stimulated by howling in chorus and by aggressive interactions with other jackals.

3. *Copulatory stage,* in which T-*sequences* end with the male licking the female's

vulva, with increasing frequency as she approaches estrus, while aggressive behavior declines. The male repeatedly mounts the bitch but without erection or pelvic-thrusting. Several days later he mounts, makes genital contact, and thrusts, but still without intromission. Finally, after days of increasingly intense following and licking, actual mating occurs and is repeated frequently for about a week. The *copulatory tie* lasts c. 4 minutes (fig. 22.7).

Toward the end of estrus, marking behavior and T-*sequences* reappear sporadically, the two stay farther apart and sometimes separate altogether. The female often approaches and greets her mate submissively. He responds by disgorging food or surrendering prey, if he has any. Thus the pattern of feeding the mother and young at the den appears long before the litter is whelped. The presumed point of having such a prolonged and involved engagement is to promote the state of monogamy, and assure the male that all the offspring will be his and the female that her mate will stick around and share parental duties.

PARENT/OFFSPRING BEHAVIOR. Serengeti golden jackals whelp in December/January during the season of abundance. The main birth peak of the gazelles comes then, followed by the annual wildebeest calving season, during which mammalian and avian scavengers get to share ½ million afterbirths. Food is so readily available that a pair of goldens can feed pups at about the same rate as a pair of black-backed jackals with 2 helpers (cf. black-back account). In *C. aureus,* the mother contributes significantly more food than the father or the helpers. Nevertheless, the food she receives from them helps reduce the nutritional stress of lactating, contributing to her future reproductive success. In addition, their presence at the den frees the pair to hunt as a team, thereby improving their ability to kill larger prey and to defend it against competitors (see Scavenging and Predatory Behavior). Pups at dens without helpers were left alone 29% of the time, compared to under 6% for those with helpers (11). The latter were thus better protected against predators and enjoyed a more active social life. Even so, the presence of helpers did not significantly increase reproductive success in the study area, for here flooding of the dens during heavy thunderstorms was the main, indiscriminate source of pup mortality.

The mother spends 90% of her time with her cubs during the 3 weeks they remain in the den. She suckles the cubs perhaps 5 times a day and twice at night for the first month, summoning them from the den with a soft whine. The pups' eyes open at c. 10 days (13). At 2 weeks they begin exploring and playing around the den while grown-ups are resting nearby. The mother often licks them thoroughly, pinning 1 at a time with a forepaw. The father and helpers occasionally *nibble-groom* pups, more often as they get older (11).

Puppies begin taking disgorged food after a month. When a grown-up returns to the den, the pups surround it, then jump up and lick its face, with ears back, grinning and wagging their tails in the *begging/greeting ceremony.* Sometimes the adult has to take evasive action before it can lower its head enough to disgorge. At about this stage the pups begin eating any insects they encounter around the den. Their nearly black natal coat has turned yellow.

Weaning begins at 2 months and is complete by 4 months. Meanwhile the pups become increasingly independent, venturing up to 50 m from the den and trailing the parents when they set off foraging. They spend more time aboveground and often sleep in the open. Their play often turns into disputes as they compete for rank dominance, which becomes established by the age of 6 months. Young-adult helpers are more sociable with each other than are half-grown pups, which interact less and less, forage independently, and often rest separately. When a father brought in the hindquarters of a gazelle fawn and presented it to 4 10-week-old pups, the biggest one monopolized it, afterward burying what he couldn't eat (13). Such are the perks of rank.

When adults are at the den, they warn pups of danger (e.g., an approaching eagle) by growling or barking, whereupon the pups dive for the burrow. If a hyena comes nosing around, the guard distracts it by nipping its hocks. In one case, the mother then howled to her mate, which rushed to the scene and nipped the hyena again, and when the hyena turned to him, the bitch attacked, and so on, in another example of the "yo-yo gambit." Very soon the hyena shuffled off, squatting low to protect its backside from its tormentors.

Jackals may change dens from time to time: one pair moved 5 times in 3 months; the longest move, of over a kilometer, was made by daylight. Each time the pups followed under their own power, even when only a few weeks old (13).

PLAY. In solitary play a cub tosses a stick, a feather, pebbles, dry dung, or whatever in the air, pouncing as it hits the ground. It tugs at grass tufts and chases its own tail. Pups invite one another to play with a shake of the head. They often play with objects for up to ½ hour, rolling, tugging, and chasing it and each other. Small pups gently bite and tug at one another, tumble over, paw, and sometimes mount one another. Play becomes progressively more vigorous after fighting patterns develop. Adults sometimes play with pups, especially helpers, but seldom with one another (13).

ANTIPREDATOR BEHAVIOR. See Parent/Offspring Behavior and black-backed jackal account.

References

1. Estes 1967*a*. 2. ——— 1967*b*. 3. ——— 1973*b*. 4. Estes and Goddard 1967. 5. Golani and Keller 1975. 6. Golani and Mendelssohn 1971. 7. Lamprecht 1978. 8. Macdonald 1979. 9. Moehlman 1978. 10. ——— 1979. 11. ——— 1983. 12. ——— 1984. 13. Van Lawick and Van Lawick-Goodall 1970. 14. Wyman 1967.

Fig. 22.15. Aggression in black-backed jackals: the subdominant individual, displaying passive submission, is kicked after being *hip-slammed* (from a photo in Van Lawick and Van Lawick-Goodall 1970).

Black-backed or Silver-backed Jackal

Canis mesomelas

TRAITS. An elegant small canine distinguished from other jackals by a dark saddle in bold contrast to the tan coat. *Height and weight:* (East Africa) 38–48 cm, 8.5 kg (7–13.5) (4); (Zimbabwe series) males 7.9 kg (6.8–9.5), females 6.6 kg (5.4–10) (8). Pointed, foxy face with long, pointed ears and bushy tail. *Coloration:* general color reddish brown to tan, redder on flanks and legs, males typically brighter-colored, particularly the winter coat (9), saddle black with intermixed silvery hair, black-tipped tail, white throat and underparts; young lead-gray with indistinct saddle. *Scent glands:* see Family Traits. *Mammae:* 6–8 (10).

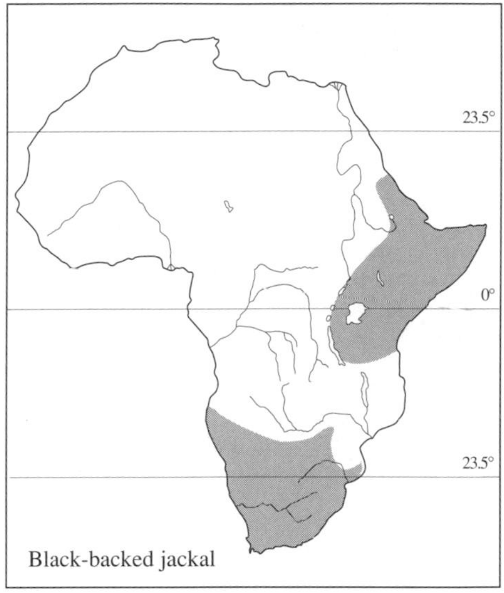

DISTRIBUTION. Somali-Masai and South West Arid Zones, absent from the intervening *Miombo* Woodland Zone (cf. bat-eared fox).

ECOLOGY. In northeast Africa this jackal occupies habitats intermediate between the plains where golden jackals live and the broad-leafed, deciduous woodlands inhabited by the side-striped jackal. Thus, in the Serengeti region *mesomelas* predominates in the *Acacia/Commiphora* woodlands whereas nearly all the jackals seen on the short-grass plains are *aureus.* The 2 species are about equally common in the longer grasslands along the woodland edge (5). In Ngorongoro Crater all 3 species live in treeless grassland, but preferences for short or longer grassland and bushland effect a degree of separation. In southern Africa, *mesomelas* inhabits a broader range of habitats, including the more open and arid country that *aureus* occupies in its range, although it still prefers areas with scattered bush (8).

DIET. Like the golden jackal, the black-back is omnivorous and follows the path of least resistance to gain its living. Indeed, there are no clear differences in the food habits and foraging or predatory behavior of the 3 species; diet seems to be determined by the type of habitat and what kinds of foods occur there rather than by preferences for any certain kinds. Invertebrates (beetles, grasshoppers, crickets, termites, millipedes, spiders, scorpions, etc.), rodents, hares, concealed young of antelopes up to the size of a topi calf, carrion of all kinds, lizards and snakes, and various fruits and berries are the mainstay in most areas (2, 4, 9, 10). In the Serengeti woodland zone the single most important food is a small, diurnal grass rat, *Arvicanthus nilotica*, which can reach densities of up to 32,083/km², a biomass of c. 1925 kg/km², in years of good rainfall, but may fall to 1/10 that amount in years with little ground cover (7). Grass rats reach peak abundance in the dry season, when black-backs whelp. However, *mesomelas* also breeds at this season (August to October) in southern Africa, so presumably it is not dependent on such a narrow spectrum of prey species. Many animals succumb to starvation and disease in the late dry season, making it a favorable time for scavenging, and certain abundant ungulates have their young during or right after the dry season, notably gazelles in East Africa (semiannual peak), the impala, the topi/tsessebe, and the warthog. On southern-African ranches, black-backed jackals may take a considerable toll of young sheep and goats and some individuals even kill adult animals. Consequently war is waged against the jackal, in the same way as in the United States against coyotes (10).

SOCIAL ORGANIZATION: monogamous, territorial, some offspring serve as helpers.

The social organization and behavior of golden and black-backed species are alike in most respects, but long-term comparative studies of the 2 species in the Serengeti region revealed some interesting differences (cf. golden jackal) (6, 7). In 17 of 18 black-backed litters, the assistance of helpers had a direct bearing on pup survival. Starting with litters averaging 3–4 (2–6) pups, pairs without helpers had only 1.3 surviving offspring after 14 weeks, equivalent to 1 offspring per year per couple. Pairs with 1 helper had 3.3 survivors (8 litters), those with 2 helpers had 4 (2 litters), and in 1 family with 3 helpers all 6 pups survived (7). Prey density was not critical, for even during years when grass rats were abundant, if offspring outnumbered the adults the pups tended to be scruffy-looking and listless. One pair without helpers lost all 5 cubs; the year before, with 1 helper, they had reared 4. Unassisted parents have to work much harder to feed their pups and themselves, going 8–10 km often without catching many rats. Meanwhile the cubs may be left unguarded: 5 litters observed for 267 hours were left alone 40% of the time.

Not only did helpers contribute up to 30% of the regurgitations to pups and lactating mothers but litters with 1 helper were alone only 15% of the time. Helpers spent 60% of their time at the den, slightly more than the mother and slightly less than the father. With 2 or more helpers, the mother spent more time resting at the den, while the father spent more time foraging. The pups were guarded over 90% of the time, during which they were free to explore, play, and socialize outside the den.

Guarding may be more important for this species than for golden jackals, because the presence of cover makes it easier for predators to approach undetected. Adults warn the cubs of danger by barking or rumble-growling, and drive hyenas away by nipping their haunches. Parents with helpers can spend more time hunting together, enhancing their chances of killing gazelle fawns and hares (more under Predatory Behavior).

Considering the difference helpers make to reproductive success, it is surprising that only 12 of 19 litters were attended by helpers, and that only ¼ of all survivors (10 male, 10 female) stayed and served their apprenticeship (7). Because full siblings share as many genes as parents and offspring, yearlings can promote their own genes by caring for their parents' offspring as much as by caring for their own progeny (see Social Organization in family account). The maturity and experience gained as a helper also contribute to their own competitive and parental success. Certainly it is in the parents' best interest to have help. The fact that all observed golden-jackal offspring in the area served time as helpers (cf. species account) makes the dispersal of black-back young all the more surprising.

A difference in the gregariousness of the 2 species may have a bearing (7). Black-backed siblings become increasingly quarrelsome and unsociable as they develop; they maintain a wider individual distance and establish a more rigid rank hierarchy. Rank is asserted more rigorously both among littermates and between parents and offspring of the same sex. Dominance may determine which cubs stay and which ones have to leave. Dominant cubs hog food, and also tend to be more independent and better foragers. They would be most likely to emigrate under favorable conditions, but when the costs of leaving are high, they should stay and force subordinate siblings to leave. Single surviving cubs almost all stay on as helpers.

The Pair Bond. None of the known pairs in the Serengeti study changed partners and one pair stayed together for at least 8 years (7). After 4 years, 2 of 5 pairs were still together. The other 3 couples had disappeared. Presumably one or both partners had died, as Serengeti black-backs hold permanent territories. Territory size changes over time, however, depending on food availability and resulting jackal density—not on the number of jackals in a family. Thus, in a defined area near Lake Ndutu where 3 pairs held territories in an average year, there were 5 pairs after 2 years of increased rainfall resulted in more grass and rats. Mean territorial size overall was c. 2.5 km^2.

The activities of pairs are closely synchronized and cooperative, including marking and defending the territory, hunting, sharing food, and providing for offspring. Two parents whose daytime activity was observed for 121 hours spent 46% of the time within 100 m of each other (36% of this interval while resting and 10% while foraging, which is done mostly alone in the absence of helpers). *Urine-marking* was carried out in tandem (male, then female) 76% of the time, and the marking rate of jackals patrolling in pairs was twice that of single animals. The pair hunted as a team and shared all food but bite-sized items. They called and answered each other when separated, called in chorus with their pups, and engaged in *social grooming.* The bitch usually took the initiative, but could also elicit grooming by lying or sitting before her mate (7). Another study reported equality in social grooming and no signs of dominance within pairs (3).

Territorial Defense. That territorial defense really is a joint venture becomes apparent when something happens to one partner (7). Within a week after the male disappeared, a pair of jackals trespassed and scent-marked on a territory despite the female's attempts to drive them away. They ate and perhaps had also killed the month-old pups, and the new female attacked the mother. Barely able to walk after 3 days of conflict, she disappeared and the new pair took over the territory. In another case a strange male trespassed near the den occupied by a widow and her 13-week-old pup. Next day she had serious wounds and subsequently disappeared. The third day the new male and a female scent-marked near the den; then the cub disappeared and 4 days later the new couple took over the territory.

Even in pairs, black-backed jackals are unable to prevent trespassing when a large animal dies in their territory, for up to 6 pairs may converge on the carcass.

ACTIVITY. Mainly nocturnal but often day-active as well.

PREDATORY BEHAVIOR. Black-backed jackals may be particularly efficient predators of young Thomson's gazelle fawns, since ¾ of observed chases resulted in kills, compared to ½ the chases of golden jackals (5). Moreover, black-backs were almost as successful hunting singly as in pairs or trios. However, these were small samples (12 chases by each species). Success rates were found to be similar for both species in another study: 16% for singles and 67% for pairs (12). In my experience, the gazelle mother's spirited defense makes it hard for single jackals of either species to kill fawns (1, 2). When a jackal foraging near a tommy herd comes upon a concealed fawn, it tries to grab the fawn before it can run. Older fawns spring up at the last moment and go stotting away. The mother then rushes to the defense, coming between her offspring and the pursuer, sometimes charging and actually hitting the jackal (see Grant's gazelle Antipredator Behavior). When there are 2 hunters, however, one can often dispatch the fawn while the mother is pursuing the other. Even when 2 or 3 female gazelles cooperate, the jackals often succeed through superior teamwork.

Mesomelas has occasionally been reported to hunt in packs and kill prey the size of adult Thomson's gazelles (5, 10, 11). Such packs presumably consist of a pair with adult offspring, or possibly of unmated young adults. In South Africa, a jackal was seen to perform a *fascination display* (see family introduction, Predatory Behavior) before a flock of curious sheep, rolling and squirming on the ground until a sheep came close enough to be grabbed (10).

SCAVENGING AND RELATIONS WITH OTHER SPECIES. As scavengers, black-backs seem to be on better terms with lions and hyenas than are goldens. They often follow hunting lions and dart in with remarkable boldness to sneak tidbits while lions or hyenas are feeding. In

Ngorongoro, up to 15 or more black-backs may be found around kills, and only a few goldens, although goldens are more numerous in the Crater. It may be that the black-back is more nocturnal (2). Like other jackals, *mesomelas* caches surplus meat, burying a piece at a time in a dispersed pattern, usually near some landmark. Probably most caches are retrieved within 24 hours, although many may be found by hyenas (11, 12).

SOCIAL BEHAVIOR

COMMUNICATION. See golden jackal and family Communication.

Vocal Communication: yelling, yelping, woofing, whining, growling, cackling.

The Swahili name for *C. mesomelas, bweya,* is descriptive of this animal's distance call, an abrupt yell followed by several shorter yelps. Family members respond to one another's calls and usually ignore those of other individuals (7). The yell is often voiced as an alerting or alarm signal, for instance at kills or when a predator prowls near a den with pups. In southern Africa this species also howls, sounding a lot like golden jackals in East Africa (4, author's observ.). Startled jackals give a woof of alarm and when terrified (e.g., when caught in a trap) cackle like a fox.

TERRITORIAL BEHAVIOR. See under Social Organization.

AGONISTIC BEHAVIOR. See family introduction and golden jackal account.

Two differences in threat behavior should be noted: (1) this species cannot raise its lip to show its fangs like golden jackals; (2) the back is less humped in the defensive threat posture, and the tail is waved slowly but vigorously from side to side while the jackal growls menacingly (fig. 22.16). In figure 22.15, a male is shown kicking a prostrated opponent following a *hip-slam.*

Fig. 22.16. Black-backed jackal defending kill against vultures.

In a sample of 116 social interactions, 38% included aggression. Hostilities were particularly common in the presence of food; only pairs shared kills amicably (3).

REPRODUCTION. Three to 4 (up to 6) pups are born in the dry season (July–October), after a gestation of 60–65 days (3, 9, 10).

SEXUAL BEHAVIOR. The protracted courtship period and other behaviors associated with mating described in the golden jackal account (q.v.) seem to apply equally to this species. Courting couples discourage offspring that persist in following them at this period by growling and even biting at them. But cubs may continue to follow at a distance and share or scavenge from their parents' larger kills.

PARENT/OFFSPRING BEHAVIOR. Jackals usually den in holes (often in termite mounds) dug by other species but can also dig their own in soft earth, in which case the female excavates a 1–2 m tunnel with 1 entrance (10). For up to 3 weeks after whelping, the mother spends 90% of the time in the den, suckling and keeping the pups warm. Meanwhile her mate and any helpers provision her with food transported in their stomachs (7). The pups begin coming outside at 3 weeks, after which the mother spends about ⅓ of the day at the den and the rest of the time foraging.

Between weeks 4 and 13 the parents spend about 40% of their time within 100 m of the den and 60% of the time foraging. The presence of helpers makes a difference (see Social Organization). A family may change dens every couple of weeks, always on the female's initiative; she lures the pups into following her by whimpering to them to come out and nurse, then leads them to the new den. Sometimes a pup stays behind, in which case the father or a helper stays with it until the mother returns. At 3½ months, pups sleep under bushes rather than in the den, and begin following the adults on foraging expeditions. They slowly learn the territorial limits and how to hunt. By 6 months they hunt on their own with some success but the parents continue to groom, play with, and feed them. Family members typically rest separately by day, but assemble in late afternoon in response to yipping calls. After becoming self-sufficient at 6–8 months, most offspring leave the territory (7).

PREDATORS. Small pups are vulnerable to any predator big enough to carry them away. Eagles are probably their main threat. Even the smallish bateleur has been known to catch and carry a 10-week pup, and martial eagles can carry off subadults (11). The main predator on adults may well be the leopard (see leopard account).

References
1. Estes 1967*a*. 2. ——— 1967*b*. 3. Ferguson 1978. 4. Kingdon 1977. 5. Lamprecht 1978. 6. Moehlman 1978. 7. ——— 1983. 8. Smithers 1971. 9. ——— 1983. 10. Van der Merwe 1953. 11. Van Lawick and Van Lawick-Goodall 1970. 12. Wyman 1967.

Added References
13. Avery et al 1987. 14. Moehlman 1986. 15. Rowe-Rowe 1983.

Fig. 22.17. Side-striped jackal chased by a Grant's gazelle defending her concealed fawn.

Side-striped Jackal

Canis adustus

TRAITS. Drab-colored jackal with white-tipped tail and more or less distinct side stripe. *Height and weight:* (Zimbabwe series) c. 38 cm, males 9.4 kg (7.3–12), females 8.3 kg (7.3–10) (2); (East Africa) 41–50 cm, 6.5–14 kg (1). Muzzle blunter and ears shorter and more rounded than in other jackals. *Coloration:* looks overall light gray or buff at a distance, markings subdued except the diagnostic dark tail with the white tip (not invariably present); the white or pale-colored lateral band edged with black running from shoulder to the tail base is variably developed; underparts pale. *Scent glands:* see Family Traits; pups reportedly strong-smelling (1). *Mammae:* 4.

DISTRIBUTION. This jackal dominates woodlands in the more humid regions of the Northern and Southern Savanna. It overlaps with both the black-backed and golden jackal in East Africa, replacing them going northwest from Uganda and from cen tral Tanzania south to Botswana. It has a wider distribution than the more familiar black-backed jackal.

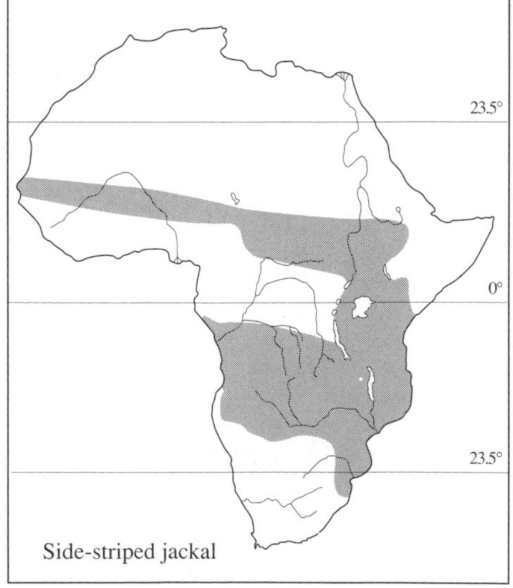

ECOLOGY. The only jackal in the Guinea Savanna and *Miombo* Woodland Zone, this woodland species occurs right through the forest-savanna mosaic to the edge of the equatorial forest, and in areas where forest and woodland have been transformed by cultivation. It also inhabits bush, grassland, and marshes offering good ground cover, and montane habitats up to 2700 m (1).

The stomachs of 71 jackals trapped in a cultivated area of Zimbabwe contained wild fruits (48% occurrence), small mammals (35%, including mice, rats, and a hare—the largest prey), and insects (31%, mainly grasshoppers, crickets, beetles, and termites) (2). The occurrence of carrion (11%) is probably low compared to areas in which larger predators and prey are plentiful. In jackals collected in Uganda, also from a settled area, insects, fruits, and village offal predominated (1).

SOCIAL ORGANIZATION: monogamous, territorial pairs.

This species remains unstudied, but presumably has a social system very similar to that of the golden and black-backed jackals. It definitely occurs in well-spaced pairs and family units of up to 6, and from 8 to 12 have been counted at kills or scavenging offal outside towns (1). Whether offspring serve as helpers is unknown.

ACTIVITY. Unstudied; definitely night-active but also seen in daytime.

FORAGING AND PREDATORY BEHAVIOR. According to most authors, this jackal is more sluggish, nocturnal, and omnivorous and less predatory than *mesomelas* and *aureus.* It is rarely heard or seen by day, at least outside parks, and seems to be mainly active early at night and from before until an hour or 2 after dawn (2). Foraging individuals move at a walk or slow trot. A free-ranging pet was seen to dig up earthworms, millipedes, and crickets (1). It actively pursued and snapped up or pounced on grasshoppers and winged termites. Caterpillars were left alone. It tried to catch small birds by suddenly dashing at them and sometimes dashed after rodents, but usually pounced. It was never seen to run down prey. The same animal ate all kinds of fallen fruit. It almost always took its food into cover before eating it. Instances of side-striped jackals eating safari ants have been reported: one that was collected had consumed 1.5 kg of white larvae mixed with adults, and many hundreds of soldiers were attached to its lips, gums, feet, and pelt (1).

The belief that side-striped jackals are less predatory than other species may be based simply on the fact that game is generally sparse in the *miombo* woodlands. There is evidence that in fact *adustus* is just as predatory and in the same way as *mesomelas* and *aureus* under similar conditions. In Ruwenzori N.P., it forages on the Ishasha Plain and cleans up topi afterbirths (1). In Ngorongoro Crater, where it is vastly outnumbered by the other 2 species, I have seen it try to catch Grant's gazelle fawns, vying with black-backed jackals (fig. 22.17). There are also reports of side-striped jackals trailing other predators when they go hunting (1).

SOCIAL BEHAVIOR

COMMUNICATION. See family introduction.

Vocal Communication: yapping, *"bwaa,"* hooting, whining, growling, screaming, cackling, *croaking.*

The commonest call is an explosive *"bwaa,"* usually uttered by a single animal at night, while standing (1). The *adustus* version of howling is an owl-like hoot. Sometimes an excited yapping is heard. An animal brought to bay may scream, and terrified animals cackle (2). A fatally wounded young female made a croaking noise that drew an adult male and 4 other jackals despite the presence of people (1).

Olfactory Communication. Trapped individuals give off a very strong smell (probably the discharge of the anal sacs), which so contaminates the traps that other species will not approach them (2).

AGONISTIC BEHAVIOR. See family introduction and golden jackal account.

REPRODUCTION. Three to 6 pups are whelped during the rains or just before (August–January in southern Africa, June/July and September in western Uganda) (1, 2). Gestation is reported to vary from 57 to 70 days (1).

PARENT/OFFSPRING BEHAVIOR. Dens are located in termitaria, aardvark holes, and hillsides, which the female may modify, for instance, by digging a second tunnel off the main chamber to serve as an emergency exit. The breeding chambers in 2 different dens were located ¾–1 m below the surface and 2–3 m in from the entrance. At one of these dens the mother jackal could be seen to look around carefully from both entrances before coming out. Pups still too small to venture outside may crawl to the entrance and sunbathe, while the mother, resting in nearby cover, keeps watch (2). For the first few weeks after whelping, the male forages alone, presumably bringing back food to provision his mate. Later they both bring back food, carrying it either in their mouths or stomachs, to feed the pups. Small ones are fed inside and larger ones outside the den entrance. Mothers of small pups readily move them to another den if much disturbed (2).

References

1. Kingdon 1977. 2. Smithers 1983.

Fig. 22.18. Rough play between male and female wild dogs in the process of pair formation (from a photo in Frame et al. 1976).

Wild Dog or Cape Hunting Dog

Lycaon pictus

TRAITS. A medium-sized, pack-hunting dog, the African equivalent of a wolf. *Height and weight:* 60–75 cm, 20–25 kg in East Africa, up to 30 kg in southern Africa, males very slightly larger than females. Head rather hyenalike (the German name is *Hyänenhund*) with heavy, blunt muzzle, dark eyes, and very large, rounded ears; slender-bodied with deep chest and long legs, 4 toes (no dewclaw) on all feet. Coat short, coarse, with little or no underfur. *Coloration:* tricolored, black and white at birth, tan patches developing from black areas beginning second month, the pattern unique in each individual (but recognizably similar in close relatives); the only consistent markings are the white tail tip and the facial pattern (tan forehead, black muzzle). *Scent glands:* see Family Traits; a disagreeable body odor is characteristic. *Mammae:* 12–14.

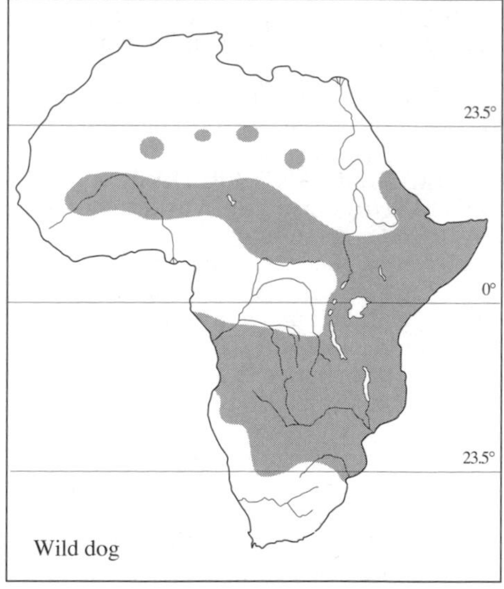

DISTRIBUTION. Widespread in the savanna and arid zones, wild dogs avoid forest but range through dense scrub, woodland, and montane habitats wherever suitable prey occurs. A pack of 5 dogs was seen on the summit of Kilimanjaro (5894 m) (19). Wild dogs of the *Miombo* Woodland Zone and southern Africa are larger and generally have more white and less black in their coats, making them appear lighter and handsomer than the dogs of East Africa.

ECOLOGY. This carnivore is specialized as a pack hunter of the commonest medium-sized antelopes: the Thomson's gazelle on the East African plains; the impala, reedbuck, kob, lechwe, and springbok in central and southern Africa. Although prey as small as hares and gazelle fawns and as large as adult zebras are taken, only 11% of 298 Serengeti kills weighed over 65 kg (15). At any given time and place, wild dogs tend to select just a few of the commonest antelopes within the 14–45 kg size range. In Kruger N.P., where impalas are about as numerous as all other herbivores together, this antelope made up 87% of 2745 wild dog kills (13). In the Serengeti ecosystem Thomson's gazelle is the mainstay (75% of kills July–December), but the dogs concentrate on wildebeest calves for some months following the annual February birth peak (3, 15). If no one species in the preferred size range is particularly numerous, wild dogs are perforce less selective, as in Kafue N.P., where gray duikers and reedbucks made up ½ of 96 kills, hartebeests 15%, and 12 other species the balance (12).

Lycaon is one of the most exclusively carnivorous carnivores. Apart from small amounts of grass, it eats no plant food or insects, and rarely eats carrion (15). Although it occasionally scavenges kills made by other predators, it kills most of what it eats on pack hunts (see Predatory Behavior). While a pack is traveling, dogs may forage individually, grabbing concealed prey such as gazelle fawns and hares, and on moonlit nights, springhares (3). But single dogs would have a hard time killing their normal prey and a still harder time defending it against hyenas, and might be unable to cope with animals as big and strong as a yearling wildebeest. Wild dogs rarely return to a kill or cache surplus food. However, 4 different dogs have been seen caching food (mostly regurgitated meat) down holes, in each case while a pack was tied to a den with pups. The dogs retrieved and ate the meat themselves within 7–25 hours (8). Wild dogs drink regularly and like to rest in wet spots on hot days, but can go without drinking for long intervals. They avoid evaporative water loss while chasing prey by letting their temperature rise (18).

SOCIAL ORGANIZATION: diurnal cooperative hunters, packs consisting of a breeding pair and nonbreeding adults that assist in provisioning lactating mother and pups; related males remain in natal pack, females emigrate.

In the wild dog, the canid traits of cooperative hunting and provisioning of the young produced by a single breeding pair have developed to the point of no return. The species has specialized on an abundant food resource that it can only exploit efficiently by hunting in packs, and the breeding pair requires the assistance of other adults to provision large litters during an extended (12–14 month) period of dependence. The details of its social and reproductive system, worked out during a decade-long study of Serengeti wild dogs, reveal an array of adaptations that should make the wild dog one of the most successful African carnivores (5, 6, 8, 9).

Wild dogs live in packs that often number 20 or more dogs, sometimes 40 and perhaps even 60 or more animals (7). But the mean, both in East Africa and Kruger N.P., is about 10 (6, 13). Serengeti packs (up to 1975) typically contained 6 adults (1–18): 4 males (0–10) and 2 females (0–7) (6). Hunting ranges are often huge: 1500–2000 km^2 in the Serengeti. But where prey is resident and numerous, ranges may be much smaller. A pack that spent a year in Ngorongoro Crater stayed mostly within an area of 265 km^2—but that was only part of its total range (3). Except for the vicinity of occupied dens, wild dogs do not defend their ranges. Different Serengeti packs had range overlaps of 10%–80%. But when 2 packs meet, the larger usually chases the smaller one away (6).

Separate rank hierarchies exist among male and female pack members, headed by the breeding pair. The role of leader is usually filled by one or the other; if not, they can still dictate pack movements by refusing to follow the leader. There is normally so little overt aggression among pack members that the existence of a rank hierarchy was overlooked by early observers (including the author). Instead of the bared teeth, growls, and snarls of most canids, the wild dog asserts dominance simply by approaching in much the same stalking posture seen during hunting (fig. 22.19). One reliable way of recognizing the alpha pair is by observing which ones *urine-mark* (see

Fig. 22.19. Wild dog *stalking posture,* used to approach game and sometimes other wild dogs, when it appears to express a tendency to attack (author's observations).

under Olfactory Communication), as this is the prerogative of the dominant male and female.

Instead of aggression, submissive behavior has been emphasized in wild dog social relations. The essence of their social and reproductive system is cooperative hunting and food-sharing, and these tendencies are continually reinforced by behavior patterns derived from infantile begging. Whenever a pack becomes active after a rest and invariably after a period of separation, the members greet intensely and display active submission (fig. 22.6). Dogs that have recently eaten may even disgorge food during the *greeting ceremony*. The same kind of begging behavior persuades dogs to share kills. With lips retracted in a toothy grin, ears flattened, forequarters lowered, and tail curled over the back, 2 dogs contending for the same chunk of meat twitter excitedly and try to burrow beneath each other—contending as it were to be underdog instead of top dog (fig. 22.20). Through this behavior young wild dogs succeed in displacing adults from kills until they are fully a year old, whereupon they lose their privileged status and are absorbed into the rank hierarchy. But decrepit or disabled individuals unable to keep up with the pack can use this technique to receive food, either first- or secondhand.

Litter size is greater in wild dogs than in any other canid: the average of 26 litters was 10 pups (6–16). This means that 1 female is capable of producing an average-sized pack every year. If 2 females bred in the average pack, the other adults would be unable to provide enough food for all the cubs plus the lactating females. Normally only the alpha female and male breed (20 of 26 dens with pups). When a subordinate female does breed, it is usually at the same time as the alpha female and intense competition between the mothers is the result. The alpha female's insistence on controlling access to the pups (see ref. 20) may effectively prevent other adults and even the mother from feeding the subordinate's pups; or the alpha female may kill them one by one (20). In only 1 of 6 cases did any of the subordinate female's litter survive (6).

In wild dogs, unlike most sociable mammals, males remain in the natal pack and females emigrate (cf. red colobus, gorilla, and chimpanzee) (5). Still more unusual, the sex ratio is skewed in favor of males. Not only are there 2 or 3 males for every female in the adult population, but in a sample of 96 pups in their first month, 59% were males (6). In the Serengeti population, female offspring stayed until 2½ years at latest, then transferred to a pack of males, mostly in company with one or more sisters (example below). Subordinate females eventually emigrated again, leaving only the dominant, breeding female in a pack composed of males that were related to one another and unrelated to her. This arrangement is so effective in preventing inbreeding that none occurred even in a subpopulation numbering fewer than 100 dogs. However, in Kruger N.P. a father and daughter formed a breeding pair after the daughter displaced her mother as alpha female (14).

This is the only known case in which a subordinate replaced a dominant breeding bitch. An alpha female normally remains with the group into which she first transfers, and has a tenure of up to 8 years. The tenure of alpha males ranges from 1 to 8 years. Thus alpha status may last most of a lifetime, for wild dogs rarely live beyond 9–10 years. However, one Serengeti male bred at 1¾ years, and four others became dominant at 3–5 years. A young adult may replace his father or grandfather, who then stays on as a subordinate. In 10 challenges to the alpha male, the challenger was successful 7 times (6).

Whereas females have to emigrate from their natal pack and keep transferring until they find the opportunity to breed, males do better to stick together. Other packs

Fig. 22.20. Two wild dogs competing for meat through aggressive begging.

containing males will not accept immigrant males, and singly their chances of finding a female and successfully rearing young would be slight, even if there were enough females to go around.

Are Wild Dogs in Decline? Wild dogs seem to have an exceptionally efficient reproductive system, yet this is the least common of Africa's large predators. Furthermore, the species may be undergoing a general decline, though hopefully less drastic than that of the Serengeti population. There, the number of packs that ranged the 5200 km^2 study area fell from 12 with 95 adults to 7 with 26 adults between 1970 and 1978; that is, density dropped from the 1960s level of 1 adult/35 km^2 to only 1/200 km^2 (6). By the early 1980's it appeared that only 1 pack lived in the middle of the richest game country on earth. However, two other small packs were located late in the decade (4), offering hope that this population is finally on the increase.

The cause of this dramatic decline and of the apparently general decline of the species remains a mystery. The immediate problem in the Serengeti was that 4 of every 5 pups died, most between 3 and 12 months old (10). Between mid-1974 and early 1976 no pups survived to yearling stage (6). Disease and competition with hyenas have been suggested as major factors in high pup mortality, also flooding of dens during the rains and starvation during the dry season when dogs denned far away from the nearest game concentrations. Wild dogs are known to be highly susceptible to various canine diseases, especially to diseases spread by domestic dogs such as distemper, which was introduced to East Africa at the beginning of the century (7). If disease is responsible for the general decline of wild dogs, however, it remains undiagnosed. Even when a blood sample was obtained from a dying Serengeti dog and sent to a laboratory in England, the disease that killed its whole pack could not be determined. (4).

Persecution by man should also be mentioned as a factor in the wild dog's decline. Astonishing as it may seem, it was the widespread practice in African national parks (including the Serengeti) to shoot wild dogs on sight as late as the mid-1960's, due to a prevailing hatred of this carnivore and the gamekeeper mentality of wardens who believed that herbivores should be protected against carnivores.

At least the decline of the Serengeti population may increase our insight into the close relationship between social organization and population dynamics in this species. Large packs seem to be more efficient and viable than small packs. Packs containing less than the typical 5–6 adults may suffer far greater pup mortality than bigger packs. Even in the late 1960's, when some 153 wild dogs ranged the Serengeti Plains, the 55 large young dogs were all in the biggest packs; packs numbering under 11 dogs had none (15). Between 1975 and 1977 the mean number of adults per pack fell to 3.6, meaning that most breeding pairs had fewer than two helpers. At that point the decline of this subpopulation may have been irreversible. So we begin to see how dependent the wild dog system is on the recruitment of large litters consisting mainly of males:

(*a*) Large packs often fission, typically when a subordinate female or 2 leaves with 2 or more males, usually when the female is in estrus (6). For example, 1 female and 8 males separated from a pack of 20 and settled for a year in Ngorongoro Crater, where the female produced a litter of 9 pups. Although she died within a month thereafter, the males successfully reared the whole litter (3).

(*b*) Emigration from the natal pack by males occurs only if there are 3 or more littermates. Male emigrants travel much farther than females: only 3 of 21 Serengeti males that disappeared from their natal packs as presumed emigrants were ever seen again, whereas 13 of 36 female emigrants were resighted in other resident packs (6).

(*c*) The risks to emigrating females decrease with numbers and increase with time and distance traveled before joining a male group. One or 2 females cannot hunt or defend their kills as successfully as 4 or 5. When males are scarce, females have fewer opportunities to found packs, and may have to travel farther and search longer. Normally, females stay in their pack until an opportunity to transfer comes along, and spend little time on their own. However, it took one group of 4 sisters 3½ months to make a connection after emigrating (6).

(*d*) Small packs run the risk of being raided by marauding males which, instead of waiting for unattached females, may forcibly abduct a breeding female. Three young Serengeti emigrants disrupted 2 packs in succession that had only 2 males each and killed 2 of them.

Pack Formation. The process of pack formation, including the establishment of a dominant breeding pair, has been observed several times (6). For example, 2 sisters of about 2 years left their pack, including 3 male littermates, their parents, and an old male, and joined up with 3 males that had lost their breeding female 10 months earlier. The transfer occurred after the 2 packs came within 5 km or less of each other, and probably had been in olfactory, vocal, and/or visual contact. Instead of following their own pack north, the sisters repeatedly broke away and trotted south, intently sniffing the ground, and one (A) frequently *hoo*-called. The rest of the pack kept coming back to them and for the next 2 days the dogs remained on the southern edge of their range, while the sisters continued their restless questing behavior. On the fifth day after the 2 packs came close, the sisters left for good, trotting southeast with heads low as though scent-tracking, female B leading. They traveled over 70 km in the first 24 hours, most of it outside their former home range, then for the next 2 days hung around on the plains, feasting on wildebeest calves and making short excursions seemingly at random. Female B had become hoarse from *hoo*-calling. On the morning of the seventh day the females encountered the 3 males while both groups were chasing wildebeest calves. The sisters ran toward them and the 3 males initially fled. Then the 2 younger males circled around and met the 2 females nose to nose, while the third, old male, stayed in the background. Standing stiffly with head high and ears cocked, the meeting dogs aggressively thrust their noses against one another's muzzles and necks but kept their mouths closed. The females behaved playfully and were actively submissive. One would roughly paw at a male's head or flanks, rear up with forepaws braced on the male's shoulder (fig. 22.18), or press her nose into his groin, while the males responded defensively and aggressively but never actually attacked. They kept trying to get into position to sniff the females' behinds.

Within a minute of meeting, the dominant, 3-year-old male threatened the second, younger male away from both sisters, and thereafter the other 2 males kept their distance. Then female B suddenly asserted her dominance over A by threatening and attacking her whenever she came near any of the males. But the alpha male appeared equally interested in both females. Beginning within ½ hour of meeting, he tried to mount A 12 times and B 17 times within 16 hours. He began *urine-marking* within 1½ hours and presently female B followed suit, while A was intently sniffing the spot. Afterward A rolled on the marks but was constrained by her subordinate status from further marking activity. The 2 dominant dogs marked together 7 more times, the female seconding the male, often after first rolling on his urine.

The alpha male continued to interact with both females twice a minute through the first day and about once a minute by the next morning. Meanwhile the 5 dogs hunted and killed wildebeest calves in the evening and morning, the 2 younger males selecting and capturing them. The females joined in killing and eating the prey, but the 2 subordinate males could not join without coming close to the females and making the alpha male jealous. Five months later female B littered. By then the old male had disappeared, presumably dead at age 8. Female A had not bred.

This appears to be a typical example of the way new packs and new pairs are formed. The sisters' departure was probably stimulated by the proximity of unattached males. After meeting them, the females "broke the ice" by a combination of playful (juvenile) and sexual solicitation that effectively appeased the males' aggressiveness and led to rapid integration of the group. The dominant male and female asserted their status and pair-bonded months before the female came into heat and mated. Out of 8 such cases, 4 females conceived no sooner than 10–12 weeks after joining new males. Immediate conception occurred only when the couple was already acquainted, as when a pack fissioned. Emigrations and transfers occurred only during the rainy season, when 2 dogs or even 1 dog could obtain food without difficulty, and when the concentration of game on the plains attracted wild dogs to the same general area, increasing the chances of meeting (5, 6).

ACTIVITY. *Lycaon* shares with the cheetah the distinction of being the most diurnal of large African carnivores. Since it runs down its prey in a fair chase, the wild dog does not need cover or darkness to conceal its approach, and it can afford to be conspicuously colored. Hunting by daylight makes it easier to select and keep a quarry in sight. The dogs are inactive on dark nights, but often travel and hunt as usual on moonlit nights. Otherwise they have 2 well-defined activity periods early and late in the day, each lasting 1 to 2 hours (3). Sometimes the dogs hunt at other times, especially on cool, overcast days. A pack moves an average distance of c. 10 km/day. The rest of the time is spent resting or, particularly if game is scarce, in travel. Then a pack commonly traverses its home range in 2–3 days, and often covers 40 km in a day, trotting at 9–11 kph. One pack, tied to a den after the game had migrated, made daily round trips of 70 km before the pups starved to death (5, 6).

POSTURES AND LOCOMOTION. See Predatory Behavior.

PREDATORY BEHAVIOR. Wild dogs hunt so effectively that in good game country a pack has regular breakfast and dinner times (fig. 22.21). In Ngorongoro, a pack averaged 2 kills per day, and caught 85% of the animals it chased, usually within 25 minutes of becoming active (3). A Serengeti pack followed for 20 consecutive days averaged 1.5 kills per day, and captured 70% of the game it pursued, usually in less than an hour of hunting (15). A success rate of only 39% was recorded in another series of 91 chases (20), but that still exceeds the hunting success of all the other large predators except the cheetah.

What makes the wild dog such an efficient predator? One popular theory is that a pack chases in relays, fresh dogs replacing tired leaders until the quarry is run to exhaustion. It is true that the quarry is run to exhaustion, but without relay-running. The dogs simply have greater endurance (18). Although their top speed of c. 64 kph is slower than a gazelle's, they can maintain a pace of 56 kph (35 mph) for several kilometers and at least 48 kph for up to 5 km (3, 15, 20). Most prey is overtaken within 1½ to 3 km and the dogs rarely continue a pursuit for over 5 km. As a rule, 1 dog remains in the lead and the rest of the pack streams out behind it, with immature and infirm animals straggling a kilometer or more in the rear. Often the quarry runs in a big circle or zigzags, giving other pack members the chance to cut across the curve and narrow the gap. Followers may thus become leaders; it is this tactic that has been misconstrued as running in relay (3).

The moment a quarry the size of a gazelle or impala is overtaken it is grabbed and thrown down, or simply bowled over, and speedily disemboweled. A group of dogs literally tears a Thomson's gazelle or wildebeest calf apart in the first minute by yanking from all directions (3). It is very much harder for an unassisted dog to kill such prey. Even a large pack is inefficient at dispatching big herbivores and it becomes evident that a wild dog is far less formidable than a hyena or wolf—one tends to forget how much smaller the dogs are. It took a pack of 21 dogs over 5 minutes to pull down a cow hartebeest weighing c. 126 kg, from the time the pack came up to where the leader was holding her by a hindleg (3). They mobbed her from the rear, never from in front, tearing at her belly and hindlegs until she finally sank down from sheer exhaustion. When a rent had been made in the belly skin, dogs took turns putting their heads inside and yanking out the entrails, while others ate their way in through the anus. This is the way wild dogs usually eat, and it guarantees a speedy demise for their normal prey. An average pack will consume a tommy in less than 10 minutes, leaving only the head with attached backbone and skin, and dogs feeding on larger kills usually become gorged within 15 minutes. But considering that it took a pack 1½ hours to kill a bull wildebeest, while another harassed a zebra for 1¾ hours without killing it, it seems surprising that wild dogs would attempt to kill anything that big, except in desperation (15). Yet one Serengeti pack made a habit of hunting zebras and would pass thousands of gazelles while searching for a herd (11, 20). The same technique horsemen use to control wild mustangs, by holding onto and twisting the upper lip (called the nose-twitch), was used to restrain a zebra while the pack tore at its hindquarters. Once the zebra was immobilized, it took the dogs only 5½ minutes to kill it (average of 18 kills [11]).

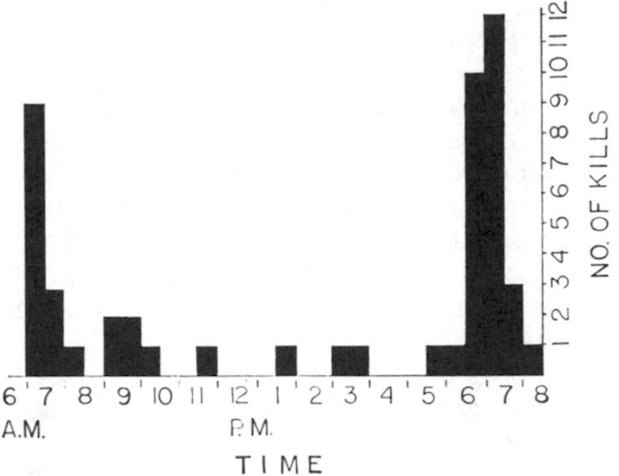

Fig. 22.21. Graph of wild dog activity, showing morning and late-afternoon hunting peaks (from Estes and Goddard 1967).

When wild dogs hunt large prey they apparently try to locate infirm individuals, and they also single out unfit and young animals when hunting their regular prey. Like hunting hyenas, wild dogs do this by running at and scattering game concentrations, often testing many groups before undertaking an all-out pursuit. A pack may try to get as close as possible to a concentration by approaching at a slow walk in the stalking attitude (fig. 22.19), and when the animals take flight they run after them on a front, chasing one and another until an individual that may be a little slower than the others is selected as the quarry. Usually the other dogs then abandon their own chases and concentrate on the same animal. But when game is scattered, a quarry is more often selected before the dogs begin to run, and the pack follows the leader, ignoring all other animals. Sometimes wild dogs run full speed over a rise of ground apparently as a tactic designed to take any game on the other side by surprise. In wooded habitats, wild dogs typically advance on a front like beaters and flush game from hiding (11, 15). Dogs moving through high grass often stand up to look for a quarry or a nearby enemy.

Prey Reactions. Confirmed wild dog haters often justify their prejudice by claiming that all the wildlife abandons an area in which the dogs have been hunting. This assertion is contradicted by the observations of trained biologists. It is true that repeated disturbance by a hunting pack can make game more flighty than usual, which is probably a main reason why the dogs tend to keep moving where game is abundant. Nevertheless, after a pack had been hunting daily for a year in Ngorongoro, the 25,000 herbivores that live crowded together on the Crater floor had not emigrated, acted disturbed only when and where the dogs were actually hunting, and were as numerous as ever (3).

RELATIONS WITH OTHER SPECIES. The spotted hyena tries to capitalize on the dogs' hunting prowess for its own benefit. Hyenas in force can take over kills from small packs and larger packs whose members fail to keep together. One or more hyenas may actually wait all day for a pack to go hunting, tag along when the dogs become active, stay in the vanguard during the chase, then run in and snatch the quarry away from the leader right after the capture. Sometimes the hyena gets away with it. More often, the dogs mob the hyena(s) and take back their kill (3). Hyenas also represent a threat to wild dog pups, perhaps less at the den, where they are usually guarded, than afterward, when still-small dogs often trail far behind the adults. Yet against all odds, a Serengeti pup with a badly broken leg was alive and reasonably well-fed after a year (author's observ.).

SOCIAL BEHAVIOR

COMMUNICATION

Olfactory Communication. A strong body odor might have evolved to meet the unusually strong dependence of wild dogs on group living, helping separated individuals to track the pack, to stay in contact on dark nights, and so on. The *urine-marking* behavior that is so conspicuous in most canids is infrequent and incomplete in this species, reflecting the absence of territorial behavior. However, the alpha female may often *urine-mark* around an occupied den, which is defended both against predators and other dogs, squatting with 1 leg half-cocked to sprinkle a little urine on a tuft of grass. Newly formed couples urine-mark in tandem and females in heat, as usual, urinate with increased frequency. The alpha male assiduously adds his own urine each time, as though claiming ownership. Usually the male also squats with leg only partly cocked, but sometimes one eager to hit the exact same spot at the same time resorts to a handstand (fig. 22.22)

Vocal Communication: *hoo,* bark, growl, whine, *twitter,* ultrasonic sounds (pups).

All *Lycaon's* calls are graded to some extent (see carnivore Vocal Communication). The contact call is a bell-like hoot or cooing that carries for several kilometers. A deep, gruff bark is an alarm signal, and growling is another, low-intensity expression

Fig. 22.22. A male handstands to *urine-mark* the spot where his mate (in estrus) is urinating (from a photo in Van Lawick and Van Lawick-Goodall 1970).

Fig. 22.23. A heap of resting wild dogs.

of alarm. Females have also been heard making a growly noise while excavating dens (10). Whining is a graded call emitted in at least 4 different situations: (*a*) a soft, plaintive whine of distress (pups and sometimes adults); (*b*) an intense whine or squeal given by a subordinate attacked by a dominant dog; (*c*) an abrupt, loud whine which summons pups from the den; and (*d*) a prolonged nasal whine given by pups begging food, and by adults in conjunction with twittering (15). *Twittering,* a birdlike, intense sound, is uttered during prehunt rallies, during the chase as the quarry is overtaken, while mobbing a hyena, and when competing over food. Finally, pups at play may utter ultrasonic calls (10).

Tactile Communication. Social grooming is comparatively rare in wild dogs. But they are otherwise very much of a contact species, habitually resting in heaps, both to keep warm on cold days and perhaps to share the flies and one another's shade on hot days (fig. 22.23).

AGONISTIC BEHAVIOR

Aggressive/Submissive Displays. Behavior designed to appease aggressive tendencies and/or solicit feeding is so much in evidence in wild dog society (see under Social Organization)—and so effective—that overt aggressive behavior is comparatively rare. Every type and degree of submissive behavior is expressed, from licking and nibbling the superior's lips to lying on the back in total surrender (more in family introduction). Only when a subordinate forgets or disputes its status does a dominant dog need to threaten with erect ears and raised tail. Approaching in the stalking attitude may signal the intention to mob another dog. As the dogs all rush about, greeting, chasing, biting, and wrestling playfully, a bunch of dogs will often gang up on 1, get it down, and bite its exposed underparts the same way they treat prey or a hyena—except that they almost never bite in earnest. Even so, the prospect of becoming the target of a mobbing attack may be frightening enough to make the stalking posture a potent threat display.

Fighting. An all-out fight between second- and fourth-ranked females consisted mainly of biting at the head and neck while the animals reared with forelegs supported against each other (20). Wild dogs often playfight in the same way (fig. 22.24), but the aim of the combatants was to secure a throat grip and throttle the opponent. Number 4 got a death grip on number 2 and only relented when number 2 collapsed with a shriek and afterward lay absolutely still. When she was eventually allowed to rise, number 2 acknowledged the reversal of status by behaving submissively from then on (20). Wild dogs caught by hounds have also been known to play dead (16, 17). Perhaps, like foxes, they feign death as a survival tactic when confronted with superior enemies.

Fig. 22.24. Wild dogs rearing in playfight (from a photo in Van Lawick and Van Lawick-Goodall 1970).

REPRODUCTION. (More under Social Organization.) Serengeti wild dogs have been known to whelp in every month but September. However, 60% of all births (35 litters) fell in the second half of the rainy season (March–June). Similarly, in Kruger N.P. whelping occurs mainly in May and June (14). The youngest age at first whelping was 22 months; the youngest breeding male was 1¾ years and the oldest was 7½ (6). Gestation takes 69–73 days (1, 2) and the interval between litters is typically 12–14 months.

SEXUAL BEHAVIOR. When a bitch comes into heat she is closely attended by only 1 dog, which keeps other males away with threats or—if he is not the dominant male—by getting between her and them and pushing her side or head away with his muzzle (20). When the bitch moves, he follows closely, frequently resting his head on her rump or shoulder. When she lies down, he settles beside and often leans against her. Every time she urinates, he quickly covers her urine with his (except in the case of a subordinate suitor) (20). A *copulatory tie* lasting 50 seconds to 6 minutes has been recorded following intromission (6).

PARENT/OFFSPRING BEHAVIOR. See also Social Organization.

A pack may return to the same general area and sometimes to the same den every year. The expectant mother selects the burrow and readies it with a minimum amount of cleaning out and digging. After whelping, the mother spends the first days in the nest chamber, and afterward remains on guard while the rest of the pack goes hunting. She is inclined to keep other dogs away from the entrance until the pups are old enough to eat solid food (3–4 weeks), and she may change dens, carrying the pups clumsily by the back, the head, or 1 leg.

Pups venture out of the den at c. 3 weeks, and thereafter the mother usually suckles them outside, summoning them with a whine. They are allowed to nurse for 2–3 minutes at a time, usually while the mother stands, but often she withdraws or snaps at them and the pups are weaned as early as 5 weeks. By this time they are taking the meat provided by returning pack members, which disgorge upon demand of the pups, the mother, and any other adults that have stayed at home (fig. 22.1). Arriving adults run up to the pups with twittering cries, flip them on their backs with a toss of the nose, and lick them eagerly. But the pups are intent only on begging food, and they proceed from polite poking to biting at the adults' lips and legs if they fail to disgorge at once. Adults with nothing more to donate sometimes snap at them or grip their noses.

Beginning at around 7 weeks, pups swiftly change from clumsy babies into rangy adult shapes (15). Muzzle, ears, and legs all lengthen, and the coat changes from black and white to tricolor. They wander up to 2 km from the den behind the pack, until an adult leads them back. Like all puppies they are playful, their games consisting of chasing, mock-fighting, and tug-of-war. When offspring are 8–10 weeks old the pack abandons the den, often after moving from one den to another in the days preceding departure. Thereafter the pups get most of their food directly from kills, but take no part in chasing or killing for many more months. They just tag along. The adults gorge at a tremendous rate until the pups arrive, then obligingly surrender their places. Otherwise, the pups chase them away by employing the same aggressive begging seen at the den. This privilege ceases when pups finally reach adult status as yearlings (10).

References

1. Cade 1967. 2. Dekker 1968. 3. Estes and Goddard 1967. 4. T. Fanshaw pers. comm. 1987. 5. Frame, L. H., and G. W. Frame 1976. 6. Frame et al. 1979. 7. Kingdon 1977. 8. Malcolm 1980. 9. ——— 1984*b*. 10. ——— in lit. 1986. 11. Malcolm and Van Lawick 1975. 12. Mitchell, Shenton, Uys 1965. 13. Pienaar 1969*b*. 14. Reich 1978. 15. Schaller 1972*b*. 16. Selous 1899. 17. Stevenson-Hamilton 1947. 18. Taylor et al. 1971. 19. Thesiger 1970. 20. Van Lawick and Van Lawick-Goodall 1970.

Added References

Frame, G. W., and L. H. Frame 1981. Reich 1981.

Chapter 23

Weasels, Otters, Zorilla, and Ratel
Family Mustelidae

This most diverse family of carnivores includes some 64 to 67 different species in 26 genera and 5 subfamilies:

Subfamily	Animals included	Genera	Species	African species
Mustelinae	Weasels, ferrets, mink, marten, wolverine, polecats	10	c. 33	*Poecilogale albinucha*, striped weasel *Ictonyx striatus*, zorilla *Poecilictis libyca*, Libyan "weasel"
Mellivorinae	Ratel	1	1	1
Melinae	Badgers	6	8	0
Mephitinae	Skunks	3	13	0
Lutrinae	Otters	6	12	*Lutra maculicollis*, spotted-necked otter *Aonyx capensis*, clawless otter *A. congica*, Congo clawless otter

Next to the Viverridae, it is also the largest carnivore family, and next to the Canidae the most widespread, occurring on all the major land masses except Australia, Antarctica, and Madagascar (4). Mustelids are adapted to just about every biome and habitat, but are essentially Holarctic in distribution. Only 7 species occur in Africa south of the Sahara, the stronghold of the civet-mongoose family, with which the mustelids have much in common (see carnivore introduction). The 2 weasels, *Mustela nivalis* and *M. erminea,* the ferret, *M. putorius,* and the common otter, *Lutra lutra,* which occur in North Africa are Eurasian colonists and are not treated separately in this book.

FAMILY TRAITS. Small to medium-sized carnivores, ranging from the dwarf weasel (*Mustela rixosa*), at 35 g the smallest carnivore, up to the 35 kg sea otter (*Enhydra*); minimal sexual dimorphism except in size, males typically 1/4 bigger but up to twice as big as females. Weasel and otter types, the most carnivorous and predatory, are long and lithe with very short legs; badgers are broad and squat (fossorial build); skunks are short-bodied; the wolverine and ratel are powerfully built and have relatively long legs. Feet digitigrade to semiplantigrade, all with 5 digits, nonretractile claws shaped according to use: short, curved, and sharp in climbers; long, curved, and blunt in diggers; lacking or shaped like fingernails in clawless otters. Head typically tubular with long posterior section, short, blunt snout and powerful jaws, with 28–38 teeth, adapted according to predatory role: long, stabbing canines in most species; shearing carnassials in weasels, ferrets, and other specialists on warm-blooded prey; becoming broad and molariform in ratels, otters, skunks, and the like which are more omnivorous and/or have to cope with hard-shelled, invertebrate prey. Ears set low, rounded or triangular, closable in most mustelids (including all African species). Tail long and muscular (otters), bushy (skunks, striped polecats), to short (ratel, badgers). Fur soft and luxurious (ermine, marten, sable, otter) to coarse and sparse (ratel, badger). *Coloration:* black, through varying shades of brown to white (e.g., ermine in winter); uniform (mink) to spotted or striped; contrasting black-and-white markings in species with potent chemical defenses (skunks, African mustelines, badgers). Penis with a well-developed bone (bacculum), prepuce always situated on the belly well ahead of the scrotum (as in dog family and unlike civet-mongoose family).

Scent glands. Many mustelids are protected by stink glands that discharge an obnoxious fluid, which in skunks has both a nauseating smell and irritates mucous membranes, temporarily blinding an enemy unwise enough to come within squirting range. Although all mustelids have anal

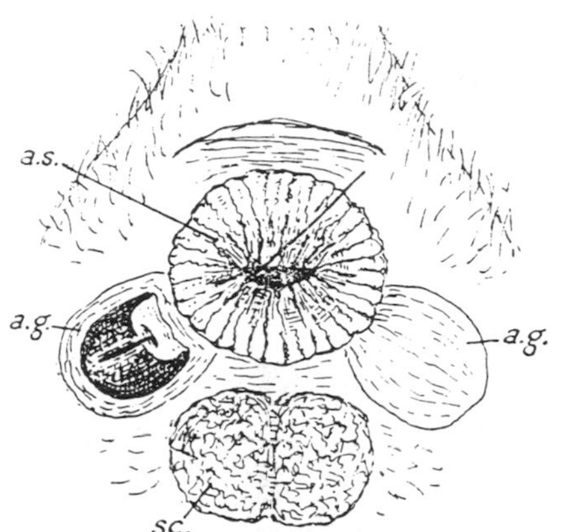

Fig. 23.1. The ratel's anal pouch: *sc.*, scrotum; *a.s.*, anal sac, spread open; *a.g.*, anal glands, left side dissected to show secretion reservoir and the end of a bristle inserted through the duct from the pouch. (From Pocock 1920.)

glands, the degree of development, the potency, consistency, and quantity of the secretion, and the distance and accuracy of squirting all vary between species. Those that are best protected by—and most dependent upon—a chemical defense carry conspicuous black-and-white warning coloration (see striped weasel and zorilla accounts). The anal glands not only are involved in the production and storage of the defense chemical but also secrete entirely different musky secretions employed in scent-marking. In fact the anal glands are compound, consisting of different layers and types of secretory cells which encircle and empty into the ducts of the anal sacs (fig. 23.1). Much remains to be learned about their functions and chemistry. Other glands have been described in one or another species (e.g., cheek, preputial, and subcaudal glands), but only anal glands are known in the African species (10, 11, 17).

Senses. The sense of smell is most important in this family, both for finding food and for social communication. Hearing is also acute, but the ability to close the external ears suggests this sense is at least partly shut down while mustelids move through subterranean passageways. Those species may rely on the sense of touch, especially mustelids that pursue prey through labyrinths or underwater. Otters have especially long, stiff whiskers and other sensory vibrissae on chin and elbows which are sensitive to water turbulence (1); in addition, clawless otters feel for crabs and other prey with sensitive fingers. Eyesight is clearly important at close range, especially in social communication, but many species appear not to notice prey or enemies at a distance. Otters, for instance, have frontally placed, round, and protruding eyes that give them good binocular vision and enable them to find and pursue fish visually, but out of water they look decidedly myopic (see spotted-necked account).

ANCESTRY. The mustelid branch separated from the carnivore main stem in the Oligocene, beginning with a small, long-bodied, generalized predator resembling the weasels (table 1.1, fig. 18.1). A major adaptive radiation took place during the middle and late Tertiary Period, but the earliest fossils known from Africa date only from the middle Pliocene. Weasels and other mustelines are closest to the prototype, being highly carnivorous, extremely active ground and tree-climbing predators. The members of the other subfamilies are more specialized in form, yet apart from the otters, the most specialized of all mustelids (18), they are more generalized in their diet and less carnivorous. In Africa, the niches for omnivorous and insectivorous carnivores were filled by genets, civets, and mongooses, leaving few openings for mustelids (16).

SOCIAL ORGANIZATION. Although only a few European and North American species have been closely studied, it looks as if most mustelids are as unsocial as civets, genets, and cats; that is, males and females are distributed singly and associate only to breed (19). However, gregarious tendencies have developed to some degree and in different directions within the family, ranging from the winter aggregations of skunks (*Mephites*) in cold climates (presumably for warmth), to large herds of sea otters, the most sociable mustelid. The otters, ratel, and badgers may live in monogamous pairs for at least part of the year, and 2 or more succeeding litters of giant otters may remain with their parents in close-knit family groups (1). European badgers share communal dens (6). The steppe polecat (*Mustela eversmanni*) is said to live in colonies and extra males associate in bachelor groups; the latter are also reported in the spotted-necked otter (see species account).

Home ranges can be as small as 1 ha in a weasel up to 500 km^2 in a male wolverine or otter. Male otters, martens, fishers, weasels, and others make a regular circuit around the perimeter of their ranges, often using an established network of trails along which they leave scent-marks at frequent intervals. The home range usually includes a den or lair which is the focus of the individual's activity. Diggers like the badgers and ratel can easily excavate their own, but most species prefer to use or adapt holes dug by other animals or natural shelters such as crevices, piles of stone, tree hollows, and caves under roots. A den may be used for long periods: for example, a winter lair and dens used for rearing young. These "permanent" dens may be lined with fur, feathers, grass, and such carried by mouth. In addition, home ranges include obligatory temporary shelters.

There is considerable evidence that all or part of the home range is defended as a territory, and that the circuits made around the perimeter are primarily scenting expeditions for the purpose of posting the territorial boundary. Studies of European and North American mustelines and otters suggest that the prevailing spacing pattern among unsocial mustelids is intrasexual territoriality: males defend their ranges against males and females against females (14). Male territories are much larger and typically encompass the ranges of 2 or more females. Female territories may not even abut, as proved to be the rule among stoats (*Mustela erminea*) and weasels (*M. nivalis*) studied in Scotland (8). Accordingly, resident females had little contact but kept out transients and, during late pregnancy and for the several months required to rear a litter, also fiercely resisted invasion by the territorial male. This, too, seems to be a common mustelid trait. At other times females were subordinate to the males, which in these 2 species are considerably larger, reflecting the enhanced male competition that is inherent in polygynous mating systems. In some other mustelids such as otters and the ratel, the sexes are nearly equal in size—a characteristic of mammals with monogamous mating systems (5). In the giant otter, the female is the dominant sex (1).

ACTIVITY. The most active of all carnivores belong to this family, *Mustela* species and otters in particular. A very rapid metabolism is characteristic; digestion of their protein-rich food is quick, providing abundant energy but at the same time impelling them to eat at frequent intervals (1). Many mustelids seem to be active day and night, but probably nocturnal habits predominate, certainly in skunks, badgers, and other species with conspicuous black-and-white coloration, this being perhaps the one pattern that is conspicuous by night as well as by day. Many mustelids remain active through the northern winter, and those that go underground (skunks and badgers) are not true hibernators (their metabolism and body temperature do not change) (4).

POSTURES AND LOCOMOTION. The family includes both extremely quick, lithe movers such as weasels and martens, and slow-moving skunks and badgers. The characteristic running gait in long-bodied mustelines is the *jump-gallop,* in which the forelegs and hindlegs move in pairs; the back is humped when the legs are gathered and concave while all feet are extended and off the ground. Weasels also glide and weave in a serpentine manner and have a scampering run. The ratel and wolverine move in a bearlike (plantigrade) shuffle and gallop in awkward-looking but tireless bounds. These and other specialized mustelids like badgers, skunks, and otters are slow runners but compensate through chemical and/or other defense mechanisms (see Antipredator Behavior). Many mustelids show the ability to move backward with agility when frightened or transporting young or nesting material, a legacy of their dependence on burrows, and weasels can go into and turn around in any tunnel big enough to admit their heads. The habit of creeping into and exploring holes and crannies is a trait mustelids share with viverrids, and is most developed in those species (weasels, ferrets) that pursue rodents through their tunnels. Most mustelids can climb well, even some predominantly terrestrial types such as the ratel, but badgers and most otters are poor climbers. Jumping prowess is greatest in the agile climbers (fisher, marten) and least in the slow runners. Most species also swim well, the otters being among the most agile of all aquatic mammals (fig. 23.7).

Alert postures are similar to those of viverrids: sitting up on the haunches (*low-sit*) and standing bipedally with forelegs dangling and steadied by the tail (*high-sit*). Long-bodied, short-legged species cannot sit like dogs or cats, but may assume an S-posture (clawless otter) or remain crouched on all fours (spotted-necked otter). These 2 species also illustrate the full range of manipulative ability in this family (see species accounts and below under Predatory Behavior). Sleeping mustelids adopt the *lateral curl* with nose to root of tail.

Some mustelids are not merely agile but acrobatic. Otters are best-known in this respect because of their habit of sliding down muddy or icy slopes. Somersaulting—presumably also for amusement or as a stereotyped behavior in captivity—has also been observed in otters and the ratel.

PREDATORY BEHAVIOR. Mustelids fill an exceptionally wide range of predatory niches, from pure carnivores specializing on warm-blooded vertebrates to partly vegetarian omnivores. Weasels and their kind, the specialists on warm-blooded prey (mainly rodents and insectivores), are the most avid and bloodthirsty hunters—but do not in fact suck their quarries' blood, as often asserted (4, 10). Relying mainly on the sense of smell, weasels, stoats, and ferrets tirelessly track and pursue rodents through their tunnels. Martens and fishers outclimb and capture squirrels in open pursuit. Mustelines grab prey with the mouth, not the paws, and dispatch it with a precise bite to the back of the skull, holding it down with the forepaws or hugging and meanwhile kicking with the hindlegs. Weasels sometimes also attack and kill animals much larger than themselves, for example, hares and chickens. When their capturing and killing responses are continually restimulated, say by hysterical chickens in a coop, they have been known to kill dozens at a time.

Otters have the speed and agility to overtake and catch fish in an open chase. However, only *Lutra* species deserve the name of fish otters, these being the most aquatic and specialized for the purpose (18). The clawless otters and the sea otter have specialized on shellfish, developing crushing jaws and handlike forefeet (see fig. 19.21).

The least carnivorous mustelids belong to the skunk, badger, and ratel subfamilies. All are more or less specialized for digging, and most eat some plant food (fruits and berries) and have flattened molars to implement their omnivorous habits. However, badgers can dig so rapidly that some manage to prey largely on rodents and insectivores, whereas skunks are formidable mousers that pounce on their quarry like foxes and cats, after locating it by sound and scent. But skunks, the zorilla, and the ratel are also insectivorous, probing in crevices and turning over stones in their foraging. In addition, badgers and the ratel take every opportunity to raid beehives; their fondness for honey is recognized in their scientific names, *Meles* and *Mellivora*, derived from the Latin for "honey."

Most mustelids eat vertebrate prey beginning at the head. Caching of surplus food for future use is a widespread mustelid trait, especially developed in northern climes. At least 2 species (wolverine and European badger) scent-mark their caches (4).

SOCIAL BEHAVIOR

COMMUNICATION

Vocal Communication. Eight different kinds of calls have been identified in the family (see species accounts). Since few species have been studied in depth, the list may well be incomplete. The smaller species tend to be the most vociferous in their threat behavior (4).

Olfactory Communication. Scent communication may be as important in mustelids as in any group of mammals, yet the functions, the chemical composition of the secretions, and even which glands are involved remain poorly known. Feces (together with urine in otters) and anal-gland secretions are both used to mark property. Many species leave their feces in exposed locations (atop rocks, mounds, logs, etc.), and also maintain middens or latrines, especially near dens. The feces tend to be black, shiny, and smelly and may well carry additives of the proctodeal if not the anal glands (17). Both sexes wipe their anal glands against prominent and novel objects within their territories (see otter accounts), "musking" most frequently during the mating season. A musky odor, sweet-smelling in some species, strong in others, is discernible to humans; it is definitely not the stench produced in self-defense. Clearly the mix of glandular secretions varies, but the mechanism remains a mystery. The stink glands typically become functional and can be used in self-defense even before the eyes open, whereas scent-marking behavior develops much later.

When the existence and uses of other glands (cheek, preputial, subcaudal, etc.—see Family Traits) are taken into account, in addition to the employment of anal glands, feces, and urine, the range of marking substances, techniques, and functions within the Mustelidae is impressive.

Visual Communication. An inventory of mustelid displays has yet to be made, and it is uncertain which of those described up to now are broadly representative. Speaking here only of behavior associated with aggression and submission, mustelids share with other carnivores bristling of the hair, the aggressive stare, stalking, ambushing, snapping, neck-biting, defensive attack, and lying on the back as a defensive and submissive response to aggression.

AGONISTIC BEHAVIOR. The offensive and defensive behavior of the ferret or polecat, perhaps the most thoroughly studied mustelid, is representative at least of the subfamily Mustelinae (12, 13).

Dominance/Threat Displays. Listed in order of increasing intensity:

1. *Dancing.* Springing in the air and snakelike weaving from side to side.
2. *Inhibited biting.*
3. *Neck grip.*
4. Mounting with *neck grip* (as in copulation).
5. *Dragging* (as of prey or estrous female), often with vigorous shaking.
6. *Sideways attack,* following *sideward approach with head turned away,* given as a counter to an opponent's defensive threat.
7. *Backing attack* (like no. 6 except aggressor backs into opponent; cf. mongoose *backing attack*).
8. *Sustained neck-biting,* usually leading to retaliation and fighting, in which both contestants roll over and over until the grip is broken (the neck skin is very thick in many mustelids).

Defensive Displays

9. *Inhibited biting* or snapping without touching opponent.
10. *Rolling over* onto back and inhibited bite to aggressor's chin.
11. Defensive threat A. Crouching or standing with back humped, neck raised, and head parallel to ground, facing opponent and backing away, meanwhile hissing or screaming with teeth bared. (fig. 23.5).
12. Defensive threat B. Body, neck, and head close to ground, teeth bared, the animal stays put but flees at first chance.
13. *Flank shielding.* Side of neck and flank presented to and leaning toward the aggressor, with head averted, while standing still or moving away (same as no. 6 but in opposite direction).
14. *Lunge-hiss* or *scream attack.* Rapid approach to opponent in defensive threat B posture, snapping at neck (usually without contact), followed by immediate retreat.

REPRODUCTION. Some aspects of mustelids' reproduction set them apart from other carnivores: prolonged copulation, delayed implantation, young born very immature (nearly hairless in some species) after a short gestation. Delayed implantation occurs only in species of northern latitudes, in which selection has decreed that mating and birth should both occur during the short season of abundance. Probably most species start breeding as yearlings, but larger mustelids mature at 2 (fishing otters) up to 4 years (wolverine) (15). At the other extreme, ferrets have been known to reach adult weight at 3–4 months and to conceive before their eyes open at 5–6 weeks (10). Ovulation is induced by copulation in at least some species.

SEXUAL BEHAVIOR. Detailed observations of courtship have been made on only a few species and usually on captive animals. But as far as present knowledge extends, the following traits are widespread:

There is usually a definite mating season, during which marking activity increases in both sexes and the male's testes increase in size and activity. Females in estrus may also utter an enticing or begging call that attracts males. Courtship may be involved and lengthy, including frequent body contact, jostling, exchanges of caresses, chasing, playing, genital licking, and marking. Or it may be simple, as in ferrets, in which the male simply pursues the female, grips her by the scruff, and drags her about until she stops resisting his attempts to mount (13). In any case, *dragging the female* around by the neck is typical of musteline courtship (see striped weasel account) and the *neck grip* during copulation is a family trait, derived from transporting the young. The grip induces a passive, limp state in both young and adults.

The copulatory attitude of long, low-slung mustelids is lying on the belly, neck extended, with tail raised and deflected (see ref. 13, fig. 6). The male mounts and firmly grips the female with his forelegs and maintains the *neck grip.* Copulation lasts from as little as 5–20 minutes in a skunk (*Mephitis*) up to 2–3 hours in the mouse weasel. However, the former copulates up to 10 times in a row, compared to 3–5 copulations within 3–4 days by the latter (4). *Mustela* couples often lie on their sides during their long copulatory bouts.

PARENT/OFFSPRING BEHAVIOR. Gestation ranges from as little as 4–5 weeks in weasels up to 2 months in otters, the latter more typical of other carnivores. The number of young ranges from 1 in the sea otter to 13 in the stoat and ferret; 2–4 offspring is average for most species. The young are more immature at birth than any other carnivores except bears, weighing as little as 1/100 of the adult weight, blind, deaf, nearly naked, and virtually immobile. The sea otter, in contrast, is one of the most precocial carnivores, born with eyes open and able to swim (3). The eyes of many mustelids only open at 5–6 weeks, compared to 2–3 weeks in dogs and cats, and it takes up to 2 months for the young to develop enough to accompany the mother.

Although the males of some species may provide food and protection for the young (reportedly otters, badgers, mink, and common weasel), typically the mother seems to be in sole charge. The following outline of behavior of mother and development of the young probably applies to most species:

About a week before birth the female prepares a nesting den, lined with grass, fur, or feathers. She becomes very aggressive toward intruders, including conspecific males, and remains fiercely protective as long as the young are in the nest. Seemingly unprovoked attacks on people by "brave" weasels probably involve females defending nesting territories (4, 8). For the first couple of weeks a mother may only leave the nest long enough to hunt and visit the nearest latrines, and through the whole nesting period she remains in the vicinity. Weasels and martens are known to close the den entrance with nesting material on departure (4).

During parturition, at least some mustelines squat or lie on their backs and draw the young forth by the head, bite the umbilical cord, and eat the placenta. The tiny, helpless babies huddle together for warmth and give a continuous nest-chirp when abandoned or otherwise unhappy. While suckling, the mother encircles and warms them; later on, she stretches out on her back or side. She licks the young to stimulate excretion and deposits their feces in a corner of the den. At least some species *nibble-groom* their offspring frequently. Most species readily transport the young to different dens in response to disturbance. Young weasels are gripped by the belly or flank, older ones by the scruff, which is provided with a pad of longer hair and thickened skin (see striped weasel account). Big juveniles may be dragged (like estrous females) or straddled like prey.

Weaning may occur between 5 weeks (mink) and 8–10 weeks (some weasels, otters). Otter and ferret mothers begin bringing food to the nest in the first month, before the offspring's eyes have opened. Though blind, they respond to the smell of meat, and many species begin to play and use a latrine area before they can see. The nose, ears, and eyes become functional in that order. Soon after their eyes open the young finally venture outside, but return immediately if the mother gives a warning call. Eventually they accompany her on foraging expeditions, characteristically trailing closely in single file. It takes time for the young to become self-sufficient; in otters and other larger mustelids the young may stay with their mother for a year or more, and most species reproduce only annually (4).

Play. Probably all mustelids are playful when young; otters remain playful as adults, which together with their acrobatic skill and inventiveness makes them particularly engaging to watch (see otter accounts). Aggressive play, exploratory play (creeping into holes), ambushing, and chasing are most characteristic of the family. (See species accounts for details.)

ANTIPREDATOR BEHAVIOR. More or less the identical aggressive behaviors are deployed against enemies of the same or other species. The defensive threat displays of mustelids, including mock and real attack, often with little provocation, have earned many species reputations for fearlessness and ferocity (see ratel account). The ultimate defense, firing the stink glands, is also deployed during one-sided intraspecific fights.

Even the largest members of the family are vulnerable to predators such as lions, crocodiles, and bears, whereas species of decreasing size face an increasing array of predators, down to foxes, housecats, hawks, and owls in the case of weasels and the young of larger species. Any adaptations that help reduce vulnerability to predators would be strongly selected for. Given anal glands secreting a strong-smelling musk as basic mustelid equipment, selection for an increasingly potent repellent and the ability to project it to a distance could lead to the rapid evolution of chemical defenses. In addition, means of warning a potential predator before it goes ahead and catches an obnoxious quarry would be favored, leading to the evolution of clear and unmistakable warning displays that enable predators to learn, usually through a single unpleasant experience, to give all similar looking, smelling, sounding, and/or behaving creatures a wide berth.

Strong selection for special defenses would not be expected for arboreal, aquatic, or fleet terrestrial species, and in fact conspicuous coloration should be selected against in predators whose hunting success depends on stealth or concealment. It is the slow-moving terrestrial forms which need protection, notably species that have exchanged mobility and climbing prowess for digging ability. Furthermore, conspicuous coloration is not a serious disadvantage for carnivores with omnivorous habits, for those whose prey is hidden, helpless, or immobile, and for those that hunt strictly at night. Thus potent stink glands, warning coloration, fossorial habits, slow locomotion, nocturnal habits, and concealed or slow-moving prey go together.

At least 16 mustelids, distributed in all the subfamilies, have warning coloration. The prevailing black-and-white pattern is further enhanced by reverse countershading (darker below and lighter above). Mustelid markings range from the modest black-and-white mask of the ferret to the gaudy stripes of skunks, the zorilla, and the striped weasel. Stripes may have evolved in stages, beginning with a plain gray back like the ratel's. A bushy, all-white or all-black tail is an additional ornament of the real stinkers and typically is erected or curled over the back as part of the warning display, along with general pilo-erection. The warning may be further emphasized by waving the tail, shuffling forward and backward, drumming, growling, or screaming (fig. 23.5).

But warning coloration is not limited to species with the most potent chemical defense. The ratel, badgers, and some others advertise their physical invulnerability and ferocity. All have nearly impenetrable skins, so loose-fitting that they can turn and bite while being held, devastatingly powerful jaws, and a disposition to match (see ratel account).

References

1. Duplaix 1984. 2. Erlinge 1967. 3. Ewer 1973. 4. Goethe 1964. 5. Kleiman 1977. 6. Kruuk 1989. 7. Lindemann 1957. 8. Lockie 1966. 9. Macdonald 1983. 10. Muller 1970. 11. Pocock 1921*a*. 12. Poole 1966. 13. ——— 1967. 14. Powell 1979. 15. Rue 1967. 16. Savage 1978. 17. Stubbe 1972.

Added References

18. J. A. Estes 1989. 19. Sandell 1989.

Fig. 23.2. Yawning striped weasel, showing its wide gape.

Striped or White-naped Weasel
Poecilogale albinucha

TRAITS. A weasel with warning coloration, the smallest African mustelid. *Length and weight:* males 44.5 cm (41–51), including 15.5 cm (13–16) tail, 263 g (218–355); females 42.5 cm (40–46.5), including 13–17 cm tail, 173 g (116–257) (Zimbabwe series) (11); males 49 cm (41–52) inc. 17.5 cm (14–20) tail, 310 g (215–355); females 45 cm (42–48) inc. 16 cm (14–18) tail, 240 g (230–250) (Natal series) (8). Typical weasel build, very long, slender body and long neck with very short legs (making this proportionally the longest African mammal) (4); feet with moderately long claws; fluffy tail; head with elongated cranium, blunt nose and short, powerful jaws, ears small and simple compared to true weasels (*Mustela* spp.), closable. *Teeth:* 28–30, large, sharp carnassials and long, stabbing canines. Coat short and sleek with sparse underfur. *Coloration:* jet black with white tail, head, ears, and nape (hence *albinucha*), torso with four white to buff or yellow stripes (variable among and within individuals), young like adults. *Scent glands:* potent stink glands, used in self-defense; other glands undescribed. *Mammae:* 4–6.

DISTRIBUTION. Southern Savanna, from eastern Cape Province to southwestern Uganda, barely reaching equator. Although seen most often in open savanna, it is probably commoner in more closed habitats in higher rainfall areas than the zorilla, with which it overlaps in most of its range (4).

ECOLOGY. As a specialized predator on mice and rats, the striped weasel is probably only common in habitats with perennially dense rodent populations, that is, areas having good grass cover which are not burned off every year, such as highlands, moist areas along drainage lines, forest-savanna mosaic, and agricultural land (4). Although its natural diet is largely unknown, food preferences of captive specimens, dentition, and behavior all indicate that *Poecilogale* is pretty strictly carnivorous, and would be unlikely to take prey larger than a rat—hardly adult springhares, chickens, or guinea fowl as asserted in the older literature. In addition to mice and rats, captive weasels accept birds and lean meat, and some but not others eat eggs. Reptiles, toads, insects, fat, fruits, and vegetables were not eaten (7, 9).

SOCIAL ORGANIZATION. In 37 sightings of striped weasels, 23 were singles, 6 were pairs, 3 were trios, and 5 were quartets. Two of the quartets included 2 adults and 2 young, and the other sightings of 3–4 animals were of 1 adult and youngsters (9). The possibility that males and females form monogamous pairs is suggested by the fact that they live together amicably in captivity, whereas adult males caged together fight (although not invariably) (7, 8). Pairs share the same nest box and food, and sometimes cooperate in excavating tunnels (7). One male and one female were kept together for nearly 2 years without any signs of aggression. When first introduced, the female, then subadult, behaved submissively, nibbling the male's cheek and repeatedly crawling under his belly. But otherwise *social grooming,* prevalent in pair-bonded species, was not observed. When a second female was introduced, all 3 shared the same sleeping box after the new female stopped responding aggressively to the residents. She then replaced the first female as the male's nearest neighbor but was found dead on the fourth morning, of unknown causes (7). The tolerance, especially toward same-sex individuals, is untypical of solitary, usually territorial carnivores, but still there is no clear evidence of monogamous pairing.

The size of a home range, territorial behavior, and foraging distance and behavior all remain unknown. If the fact that captive individuals always bring food back to the nest before eating (see under Predatory Behavior) is an indication that striped weasels use a regular home den, then presumably they either have small home ranges or use different dens in rotation.

ACTIVITY. Although sometimes seen sunning or moving early and late in the day, striped weasels studied in captivity proved to be almost entirely nocturnal, remaining totally inactive inside their nests during daylight hours (7). The weasels usually emerged c. 2 hours after sunset and remained active until an hour or so before sunrise. This animal is thus more strictly nocturnal than most true weasels, perhaps at least partly as the result of competition with such diurnal carnivores as slender and gray mongooses, and partly to avoid the large number of diurnal birds of prey (4).

POSTURES AND LOCOMOTION. *Poecilogale* moves in much the same way as other weasels: sinuously, quickly, and restlessly (see family introduction). In walking, the head is held low with the nose, the main sensory system, near the ground, tail in line with the back or slightly drooping, and back slightly humped (4). It cannot outrun bigger, longer-legged predators. Adapted to pursuing its rodent prey into and out of tight places, it is so flexible that it can turn around in any tunnel wide enough to admit its head (11). It is also an avid and accomplished digger (7). The front feet are used alternately for excavating, then together to scoop accumulated dirt between the hindlegs. Meanwhile the forehead is pressed to the ground, to support the body or to keep dirt from falling back into the hole. Finally, all 4 feet are used to fling dirt out of the hole with a backward movement of the body.

Although the striped weasel can climb to some extent, it is almost wholly terrestrial. It sleeps in the typical *side curl,* but on hot days may lie on its back or stomach with legs outstretched. On emerging from its den after dark, it yawns repeatedly, revealing a remarkably wide gape (fig. 23.2). Yawning was also noted as a displacement activity (7). The weasel is too long-bodied to maintain a seated attitude, but crouches like a genet, with forelegs under its shoulders and back hunched. When investigating a strange object, a weasel may sit up on its haunches, but typically explores on all fours, head weaving from side to side and body often rocking rhythmically from one foot to the other. This movement of the rather flat, snakelike head and long neck makes the Afrikaans name of *slangmuishond* ("snake-mongoose") quite descriptive (1).

Grooming: licking, scratching, nibble-grooming.

Licking is the primary form of grooming. Striped weasels typically have clean, shiny fur, and it has been noted in a captive colony that those allowed to tunnel in the dirt had the shiniest coats (7). A pet couple licked themselves regularly and thoroughly in their den before going to sleep (1), but members of a captive colony were seldom seen grooming (7).

PREDATORY BEHAVIOR. This weasel kills and probably hunts its prey in much the same manner as other weasels. It relies on scent to locate its quarry, and may pass within 5 cm of a mouse without seeing it (2). Having located a mouse, the weasel immediately darts at it and tries to seize it by the back of the neck. Then, maintaining the grip of its powerful jaws, it pulls the victim to its chest and rolls over and over, sometimes kicking it with its backfeet. The forefeet are not used to capture prey or to manipulate food. Kicks with the hindfeet, powered by the long back, may serve to dislocate the vertebral column and immobilize the victim (8). All food was carried or dragged into the den before being eaten, and surplus food was hoarded in the den (1, 7, 8). The weasels never teased or played with their prey, but up to 6 mice were killed and carried to the nest box before beginning to eat. A male and female pair often preyed simultaneously, but whichever one made the catch, the other turned away and made no attempt to appropriate it.

Poecilogale has certain fixed feeding habits. When eating mice and small birds, first the head is consumed, then the entire animal is swallowed with minimal chewing. The wings, feathers, and legs of dove-sized birds are left uneaten. To eat a rat, a weasel gains access through the belly and chest. When finished, the head, most of the skin, the legs, and the tail are left, all attached. Weasels are capable of eating up to 3–4 rats in a night (7).

SOCIAL BEHAVIOR

COMMUNICATION

Vocal Communication. The vocal repertoire has been recorded and sonographed (3). Striped weasels are generally silent, and being unsocial, the variety of their calls is not great. Offensive and defensive calls are most developed.

Offensive Calls

1. Warning call. A quiet growl, a simple, continuous sound with a duration of about a second. A low-intensity threat, the commonest sound heard in captivity.

2. Aggression call. Half bark, half scream, a sudden high-intensity sound that accompanies the threat display, and may precede attack and stink-gland discharge.

3. Warning-aggression transition call. A sound structurally intermediate between nos. 1 and 2 in duration and dominant harmonics, infrequently used.

Defensive/Submissive Calls

4. Release call. A simple sound of high intensity and amplitude with full expression of harmonics, usually accompanied by stink-gland discharge. Longer duration and otherwise unlike no. 2 (the sonograms look completely different). This call is part of a desperate defense, given by a male losing an unequal fight or to intimidate an enemy.

5. Submission call. A high-frequency sound with complex tones, similar to the juvenile contact call but louder with higher harmonics. It is used by sexually unreceptive females to pacify persistent males, by a subordinate male to a dominant male, and to terminate playfights between young weasels.

6. Greeting call. A quiet churring sound lasting up to 5 seconds, this call is a modified form of the adult submission call, faster and emphasizing only the lower harmonics. Used by young to the mother, and by a male prior to mating or in the presence of a strange female, it may signal unaggressive intentions.

Juvenile Calls. These are characteristically rhythmic and repetitive sounds. However, the calls of blind young sound like primitive versions of adult calls.

7. Distress call type A. A simple, rhythmic sound with no harmonics suppressed, given by blind infants during the first month, when the mother is away from the nest or when a baby becomes separated.

8. Distress call type B. A number of notes separated by constant intervals, making an on-off/on-off sound. This call replaces type A after the eyes open, up until 12 weeks.

9. Contact call. A repetitious sound uttered rapidly (10 times/second) by blind infants when the mother enters the nest prior to suckling.

Olfactory Communication. Striped weasels apparently do not mark objects with their anal glands but employ them purely in self-defense. The yellowish fluid of the anal pouches, present even in juveniles, can be ejected to a distance of about 1 m. The smell, though described as foul, rather sweet, very heavy, and pungent, apparently lacks the nauseating characteristics of zorilla and skunk secretions and is less persistent (1, 7). Feces may be used for scent-marking. The weasel adopts a distinctive posture with tail straight in the air (also when sniffing an interesting scent) (2), backs up to a vertical surface, and plasters its dung against it (cf. otters). The feces are black and strongly scented, but nothing like the anal-gland secretion (1). Captives always urinate and defecate, usually simultaneously, at a regular latrine (1, 7).

Young weasels have been seen following behind their mother in a line, each with its nose to the other's anus. Whether this *processioning* has to do with maintaining contact by scent, as postulated for *processioning* in the Egyptian mongoose (q.v.), is unknown (4).

Visual Communication. An account of *Poecilogale* visual displays has yet to be published. In offensive and defensive threat, the hair along the back is erected but the coat is so short that the animal hardly looks any bigger, although the tail looks like a bottle brush. However, the effect is heightened by *bark-screaming* and *feigned attack*, during which the weasel bounds forward with a rocking-horse motion, bouncing on both front feet at once with stiff legs, stopping just short of the target. The climax of this display is to fire the stink glands, during which the tail is held vertically (1, 7).

Greeting and submissive displays have been observed between young and adult weasels (see under Parent/Offspring Behavior).

REPRODUCTION. *Poecilogale* breeds during spring and summer in South Africa. Males have enlarged testes at this time of year. One to 3 young are born after a 31–33 day gestation (10).

SEXUAL BEHAVIOR. When a male and female meet the male makes a quiet churring call (no. 6). A receptive female crouches and allows the male to nibble her cheek, smell her vulva, and grasp her by the nape. One female in full estrus danced playfully around the male when the 2 were put together. Eventually the male drags her by the nape under cover and copulation ensues, lasting 62–78 minutes. Meanwhile the male maintains his hold on her nape skin and clasps her pelvic region with his forepaws. He makes 3–5 pelvic thrusts, pauses 6–8 seconds, thrusts again, and so on. The female keeps quiet apart from occasional yaps (10).

PARENT/OFFSPRING BEHAVIOR. A tame female whelped first at 19 months, although striped weasels reach adult weight at 21–25 weeks. All observed births occurred in daytime (10). Mothers ate nothing the first night, sustained by the afterbirths they consumed, remaining continually in the nest with the pink-skinned, practically hairless babies for the first 36 hours. The newborn were c. 70 mm long, including a 15 mm tail, weighed c. 4 g, and bore a raised hump of thickened skin covered with a distinct mane on the neck—a convenient carrying handle while it lasts (7 weeks). Unable to crawl the first week, separated babies could only hold up their heads and move them jerkily side to side, up and down, while squeaking without interruption. At 10–14 days babies could move in a circle on their forelegs and by 19 days could crawl feebly, but after a month they were only half as mobile as zorillas of the same age (cf. *Ictonyx* account). They could walk, albeit unsteadily, after 56 days, soon after their eyes opened (51–54 days), could trot and follow their mother after 11 weeks, and could *jump–run* a week later.

The young lived on milk alone until their canine teeth erupted at 35 days, nursing at all hours to 6–7 weeks and often falling asleep while attached to a nipple (no teat preference noted). The transition to carnivory began with licking the fluids and eating the viscera of partially eaten prey the mother brought to the nest. They were eating meat and small bones by the time their eyes began to open and were weaned by 11 weeks.

After 12 weeks the mother stopped killing and bringing mice to the nest, but left them kicking outside. She would then hurriedly enter the nest, nose the babies and trot back to the mouse, repeating the process until they followed. At first they were timid and merely smelled it and the spot where it had been killed. But after several days they began to bite at a kicking mouse, sometimes tugging at it and growling. Both members of a litter killed their first mouse during week 14, and the following day killed 6 mice between them, performing all the elements of the adult killing pattern. Their interest in and ability to kill prey efficiently coincided with replacement of the milk canines by the permanent canines.

Greeting Ceremony. Beginning about 2 weeks after their eyes open, juveniles approach and nuzzle the mother's groin when she enters the nest, tumbling over each other in their haste and excitement, meanwhile giving the greeting call (no. 6). The mother responds in kind.

Submissive Behavior. When 14-week weasels were introduced to adult males, they lowered their forequarters and laid their heads sideways against the ground. The males responded with the greeting call (no. 6).

PLAY. The young began playing as their eyes opened, at first quietly in the nest box, more energetically as their locomotor ability improved. Play consisted mainly of fighting games: wrestling, grasping the cheek skin, rolling laterally and kicking while grasping the nape or lumbar skin in the jaws. Running and aggressive displays were also featured: running and darting in all directions, tails raised and bristling, and bouncing on the forefeet. Playfighting simulates all the elements of real combat except for sustained biting and screaming, and also includes elements of prey-killing: kicking and lateral rolling. Play outside the nest begins in the eleventh week but playfulness declines after 17 weeks, although hand-reared animals may play a bit even as adults (10).

References

1. Alexander and Ewer 1959. 2. Ansell 1960c. 3. Channing and Rowe-Rowe 1977. 4. Kingdon 1977. 5. Lockie 1966. 6. Pocock 1927. 7. Rowe-Rowe 1972. 8. ——— 1978a. 9. ——— 1978b. 10. ——— 1978c. 11. Smithers 1983.

Fig. 23.3. Zorilla killing a rat, with coat bristling as during defensive threat.

Zorilla or Striped Polecat

Ictonyx striatus

TRAITS. The African version of a skunk, though zorilla is diminutive of *zorro*, Spanish for "fox." *Length and weight:* males 63 cm (57–67), including 26 cm (24–30) tail, 0.97 kg (0.68–1.5); females 60 cm (56–63), including 26 cm tail, 0.7 kg (0.6–0.9) (8). Long and low (c. 10–13 cm high), but much broader than a weasel with longer, more muscular legs; narrow feet with curved foreclaws up to 18 mm. Coat long and silky with fine underfur and bushy tail. *Coloration:* geographically and individually variable (especially on head) (2); jet black legs and underparts, alternating with pure white markings on upper parts; 4 white dorsal stripes, tail usually all white but varying to all black; forehead with white spot, white stripe or spot on each cheek, and white-backed ears. *Teeth:* 34, canines short but strong, front part of carnassials adapted for shearing, broad molars adapted for crushing/grinding. *Scent glands:* well-developed stink glands (described under Family Traits). *Mammae:* 4.

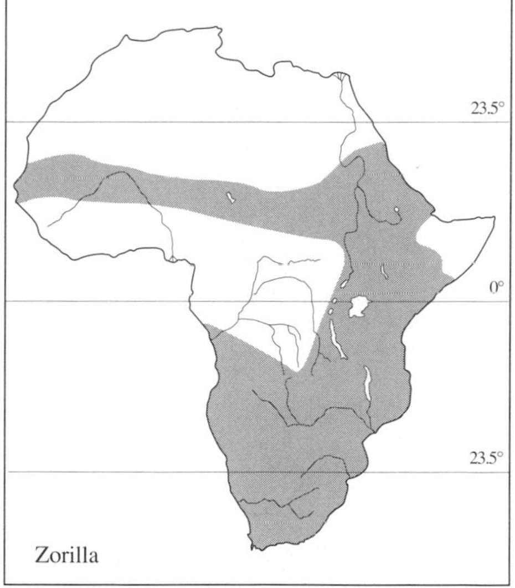

DISTRIBUTION. Widespread south of the Sahara in savanna and arid zones, absent from forested areas.

RELATIVES. The so-called Libyan weasel, *Poecilictis libyca,* which occurs in northern Africa south to about 12° N, looks like a miniature zorilla (3). Since there is considerable range overlap between the two, *Poecilictis* is properly considered a separate species, but could belong in the same genus as *Ictonyx.*

ECOLOGY. Present but scarce in dense woodland and savanna with thick cover, and able to subsist in the coastal sand dunes of the Namib Desert, the zorilla is commonest on relatively open rangeland where domestic or wild ungulates keep the grass short, especially in high and dry uplands (2). Rodents and dung beetles are both abundant in grassland habitats protected from burning, and *Ictonyx* has greater freedom to forage in the open by night than most other small carnivores because of its chemical protection.

The zorilla is more carnivorous than most mongooses and unlike genets eats no fruit. It preys on virtually all kinds of small vertebrates up to the size of a small hare, catch-as-catch-can, while devoting most of its foraging time to hunting insects. In preference trials conducted on tame zorillas, the only proffered items that were refused were toads, crabs, snails, slugs, millipedes, and stinkbugs (4). Mammals (mainly mice and rats) made up 46% by volume of the stomach contents of 21 zorillas collected in South Africa, compared to 21% insects; but insects occurred in 62% of the stomachs and mammals in only 38%. Insects were also prevalent in a sample of 36 zorillas collected in Zimbabwe and Botswana, occurring in 61% of the stomachs, followed by murid rodents (17%), lizards and snakes (17%), and scorpions (11%). Dung beetles and their grubs, grasshoppers, and crickets, in that order, made up the bulk of the insect prey in both samples (7). Zorillas can subsist on water derived from their food, enabling them to live in regions without free water, but will drink sparingly (4).

SOCIAL ORGANIZATION: solitary, nocturnal.

Most sightings are of single individuals or groups consisting of a female with young. Field study of individually known animals is needed to answer such basic questions as whether the species is territorial. In captivity, by one account, adult males are mutually intolerant, and females avoid males except during estrus (6). In apparent contradiction, another account states that different families have been kept together, not only without strife but on such friendly terms that individuals often engage in *mutual grooming,* which is solicited by flopping over and presenting the pure-black underside (2). When molested, zorillas put on an impressive threat display (see under Antipredator Behavior) and, if necessary, follow through by spraying the foul-smelling, caustic fluid from their anal sacs (2). But the first resort of a zorilla disturbed while foraging is to run for the nearest refuge; apparently all the bolt holes within the home range are well-known. One is used for an unknown length of time as the home burrow (7). Zorillas have the habit of following well-defined paths on their nightly foraging expeditions, which makes them peculiarly vulnerable to automobiles when their paths happen to cross highways. Whole families may be wiped out, as the other members often stay around if 1 is run over (2).

ACTIVITY. One of the more strictly nocturnal carnivores, *Ictonyx* is rarely seen before 2200 h and retires before dawn.

POSTURES AND LOCOMOTION. *Ictonyx* is a very active animal, capable of turning and reversing direction instantly. When foraging it typically moves at a buoyant trot, with back arched and tail horizontal, and it can gallop at a good (but unspecified) pace (8). It can swim well and, under stress, even climb. But its claws and general build are primarily adapted for digging. Unlike the striped weasel, it can sit up on its haunches but is a poor jumper (see under Foraging and Predatory Behavior).

FORAGING AND PREDATORY BEHAVIOR. A foraging zorilla quarters at a trot, nose to the ground, with sudden stops to poke its nose into litter or dung for insect prey (cf. white-tailed mongoose). Senses of smell and sight are both well-developed, with primary reliance on the former (audible sniffing) to locate insects in dung or litter, worms, and other subterranean invertebrate prey. Dung-beetle larvae buried down to 30 cm are dug out with the long foreclaws (cf. bat-eared fox). More mobile insects (moths, beetles, crickets, grasshoppers, mantises, and winged termites) are snapped up or first held down with a forefoot. No attempt is made to capture insects on the wing, but a zorilla may follow and capture them after they land.

Vertebrate prey is usually stalked, then either grabbed with the mouth or pinned with 1 or both forefeet and bitten in the neck or head. In the killing bite to the neck, the canines penetrate and force apart the vertebrae, breaking the spinal cord, whereas the head bite penetrates the skull below the orbits and bases of the ears (4). During feeding experiments, a zorilla often grabbed prey at some other point, then, after pinning it with a forefoot, it would deliver a killing bite (fig. 23.3). Very large rats were killed with a throat bite. Rolling while biting was seen only 3 times (cf. striped weasel), again while grappling with large rats; twice the zorilla straddled a rat, clasped it with its forefeet, tucked in its head, then somersaulted forward while biting the back or side of the neck. It may also make sure a formidable prey is dead by poking and biting it some more after killing it, especially if the corpse makes reflex movements (4). *Ictonyx* attacks snakes in much the same manner as does a mongoose: the zorilla slunk up slowly, body low and hair bristling, bit the snake near the middle or near the tail, then retreated rapidly. After several bites, the zorilla delivered a brief but vigorous shake. Four or 5 such attacks usually caused the snake to uncoil and retreat, whereupon the hunter moved in for the kill, pinning the snake with its forefeet, then biting it 10–15 cm behind the head, sometimes with more shaking, until the victim stopped struggling. The still-writhing body was then subjected to further bites at different locations.

Ictonyx crouches while eating small rodents, holding the food between its forepaws with head uppermost while cutting pieces off with its carnassials, chewing first on one side then the other (4). A different technique is used on prey with large, hard skulls. The zorilla chews through the shoulder skin, then eats from that point downward, holding the skin with its forefeet while pulling off pieces of flesh with its teeth. The prey is thereby skinned out, leaving the head, large bones, feet, and skin uneaten. Because of the zorilla's short gut length (3.2–4.3 times head and body length), the food eaten in 1 night is completely digested by the time the animal becomes active the next night (4).

SOCIAL BEHAVIOR

COMMUNICATION

Vocal Communication (cf. striped weasel) (1).

1. Warning. A quiet growling which, together with bristling hair, is the common response to any disturbance, by its own or other species.

2. Aggression. A high-amplitude, sudden utterance with 2 different components, equivalent to but different from the weasel's *bark-scream.* Often combined with the threat display and may precede fighting or spraying.

3. Release call. A high-intensity scream with descending pitch given in conjunction with squirting as a desperate defensive threat, intended to stop further attack by a victorious rival or to intimidate an enemy.

4. Submission. A complex, high-frequency call that rises and falls, consisting of up to 7 discrete subunits. It is voiced by a submissive individual during encounters between males, by an unreceptive female in response to sexual harassment, and by young polecats during playfights, in each case to appease aggression.

5. Greeting. A low-amplitude call with a duration of 1/10–1/5 second, similar to the submission call but faster and emphasizing only the lower harmonics. Signifying friendly intentions, it is uttered by males during courtship or on meeting a female or young zorilla, and by the young to the mother.

6. Mating calls. Females vocalize almost continuously while mating, uttering calls of 3 recognizably different types, 1 similar to the submission call and 2 to juvenile calls, with an overall duration of 2–15 seconds.

Juvenile Calls (similar to striped weasel)

7. Distress type A. A continuous nest-chirp or squeak with rising and falling frequency, given by young up to 2½ weeks old, usually when the mother is away from the nest.

8. Distress type B. A complex biphasic mewing call heard from the second to the eighth week.

9. Contact. A loud, rapidly repeated (10 notes/second) yap decreasing in frequency and amplitude, uttered when the mother enters the nest during the first 2½ weeks. This is the precursor of the adult submission and greeting calls.

Olfactory and Tactile Communication. Only the use of the anal stink glands for self-defense (see under Visual Communication and Antipredator Behavior) has been described. Other secretions of these compound glands (see Family Traits) probably play a role in social communication.

A juvenile greets its mother by nuzzling her groin, then rolling on its side and cocking a hindleg. The mother then sniffs its groin while the juvenile continues to sniff hers (6).

Visual Communication. Only a few displays have been described. The zorilla is arguably the most conspicuous African mammal for its size, particularly at night. The long mane and tail hair, automatically erected whenever a zorilla is fearful or angry, increases its visibility and apparent size. The back hair radiates from a point in the center to produce a long, diamond shape when viewed from above. Instead of the black-and-white pattern they show their enemies, zorillas present their black underparts during sociable interactions with each other. The individuality of the spot pattern on the head should make it easy for them to recognize one another visually (2).

AGONISTIC BEHAVIOR

Defensive Threat Displays. An aroused zorilla keeps its rear end turned to its enemy, with bristling tail sticking straight up or curled over its hunched back, head drawn in, and hindquarters elevated. It may spring forward and backward from all 4 feet in mock attack, and if cornered will growl or scream loudly (7). When squirting it swings its hindquarters so that its spray covers an arc, increasing its chances of hitting a target it cannot see well with its back turned (2).

Submissive Display: lowered forequarters, mouth open, and head turned with cheek resting on ground.

This display was observed in estrous females responding to courting males, and in 9-week youngsters when put into the cages of adult males. The males responded by *nibble-grooming* them or by briefly licking the inside of their mouths (6).

Fighting. See under Play.

REPRODUCTION. Breeding, at least in South Africa, is limited to the wetter, warmer half of the year (September – April), with a birth peak in October/November (6). Males' testes reflect this seasonal breeding, being smaller the other half of the year. Females are seasonally polyestrous but, even given a 36-day gestation, the period of postnatal care is so extended that polecats whelp only once a year (unless a litter is lost soon after birth). A hand-raised female mated at 9 and whelped at 10 months. One male bred first at 22 months.

SEXUAL BEHAVIOR. Mating has been observed in a captive colony (6). Unreceptive females yapped and snapped when males tried to sniff or mount them. A fully receptive female crouched and allowed the male to sniff her vulva, and sometimes adopted the submissive posture. The male then grasped her by the neck and dragged her under cover. He kept up the quiet, reassuring greeting chirp throughout these exchanges. Eventually he mounted, gripping the nape skin in his teeth and clasping her waist with his forefeet, while the female maintained a crouch with tail to one side. During the long copulation (106 minutes in the only one timed from start to finish), the male alternately gripped her nape and stretched his throat parallel to and touching hers. He also alternated pelvic thrusts of c. 10-second duration with pauses of c. 35 seconds. After ½ hour he was thrusting for 3–5 seconds then pausing for c. 6 seconds. Meanwhile the female yapped with increasing frequency and volume for 5–15 seconds at a time.

PARENT/OFFSPRING BEHAVIOR. One to 3 young are born in a nest, weighing c. 15 g and 110–115 mm long, pink and nearly hairless (6). Pigment is already present in the zone of dark stripes and by 10 days, when babies are covered with white hair, the future black stripes appear gray. Distinct black-and-white stripes develop by 19–21 days. The young are unable even to crawl for the first week. However, zorilla babies develop more quickly than striped weasels. Within 7 days a zorilla can move in a circle on its forelegs; it can crawl a little by day 10 and as well by day 15 as month-old weasels. The eyes begin to open at 35 days and are completely open by day 42. A zorilla can walk unsteadily at 39 days, and trot and follow its mother at 7–8 weeks. By then a juvenile can curl its tail over its back and within another week it can perform the complete threat display, including reversing direction. Mothers transport newborn young by gripping them across the shoulders and hold older ones by the nape.

The lower canines erupt at 32 days, several days before the eyes begin opening, and at that point the young start eating solid food: the soft parts of prey the mother puts into the nest. The offspring lick the juices and eat the viscera. From then on they eat more meat and drink less milk, but continue to nurse day and night until weaning, which is complete by 8 weeks. By then young zorillas were foraging actively for insects, scratching crickets from under stones and pinning them with the forefeet while eating. Within another week a male and a female juvenile each killed its first rat (6).

PLAY. About a week after their eyes open the young begin playing, more vigorously and aggressively as motor coordination improves. From 8 weeks, when the young first play outside the burrow, until c. 15 weeks, play is the major activity, decreasing thereafter in intensity and duration in wild zorillas (up to 24 weeks in pets). Playfighting is most frequent: youngsters wrestle, rolling and kicking while holding each other by the cheek or neck skin with their teeth. Wrestling and biting alternate with chasing games, in which the roles of pursuer and pursued change. Pets made play threats, advancing and reversing rapidly from different angles, sometimes performing lateral rolls, on alternate sides, and landing on the feet again to resume the threat display. The same behavior patterns performed in play are later employed in real fights: nipping the base of the tail, flight, rolling, biting the cheeks and side of the neck (6). The skin around the throat is thickened as a shield against neck-biting (cf. ratel) (8).

ANTIPREDATOR BEHAVIOR: threat display, discharging stink glands, playing dead.

Although the zorilla's anal-sac secretion is not only nauseating and persistent but also produces a burning sensation if it contacts the eyes (3), it is not always enough to discourage attack. In that case the zorilla has a second line of defense: it plays dead. However, a limply lying zorilla will suddenly flip over if its enemy circles it, in order to keep its back turned (2).

References

1. Channing and Rowe-Rowe 1977. 2. Kingdon 1977. 3. Rosevear 1974. 4. Rowe-Rowe 1978*a*. 5.——— 1978*b*. 6.——— 1978*c*. 7. Smithers 1971. 8.——— 1983.

Fig. 23.4. Ratel with honeycomb robbed from bees (from a painting by R. Orr, as printed in *International Wildlife*).

Ratel

Mellivora capensis

TRAITS. A badgerlike carnivore with conspicuous white or gray upper parts and black lower parts. *Height:* 23 – 28 cm; *length and weight:* 95 cm (90–102), including 22 cm (18–22) tail, and 12 kg (8–14.5) (12). Broadly and powerfully built with stout legs and broad feet, foreclaws like curved knives, 3.5 cm long with sharp edges, backclaws straighter, 1.5 cm long without sharp edges (12). Head wide with short muzzle, massive skull with powerful jaw muscles and robust teeth; stout canines, cheek teeth reduced to 4 per side, adapted for crushing more than shearing; external ear reduced to thick ridges of skin which close purselike when digging or raiding nests of biting insects; eyes small and dark. Skin very loose and up to 6 mm thick around neck; coat coarse (hoglike), from 1 cm long on head to 7 cm on hindquarters and tail, with sparse underfur. *Coloration:* jet black except for the white to iron-gray (sometimes black) mantle extending from crown to tail, which may darken with age, remaining whitest on head and neck; anal and genital area scantily haired or naked. *Scent glands:* anal glands present in a very large anal pouch resembling that of mongooses and hyenas (fig. 23.1); other glands undescribed. *Mammae:* 4.

DISTRIBUTION. The ratel occurs both in Asia and Africa. Though nowhere common, it is practically ubiquitous south of the Sahara, being absent only from deserts and rare in lowland rain forest. In northern Africa, the species occurs in Río de Oro, Mauritania, and Morocco.

ECOLOGY. The ratel tolerates a wide range of conditions, from very dry to very wet (100–2000 mm of rain), and from sea level to 1700 m in montane forest. Its ecological role could be compared to that of a European or American badger, although it is somewhat less fossorial and decidedly more mobile. Like the striped polecat, it is an opportunistic insectivore and

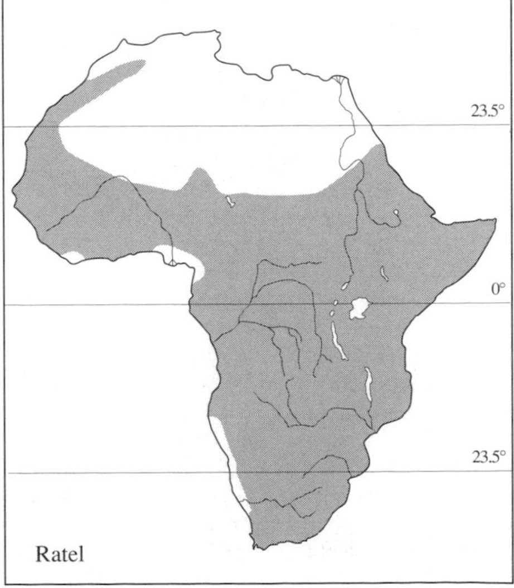

carnivore, but takes an even wider range of invertebrate and vertebrate prey, from insects to the young of large mammals, and it eats carrion. It also eats berries and fruits. Its digging ability, perhaps second only to the aardvark's, with whose holes its holes are often confused, enables it to extract buried food that is inaccessible to less-accomplished excavators (see under Foraging and Predatory Behavior). In addition, the ratel is one of the few mammalian predators on bees, feeding both on larvae and honey, hence the scientific name *Mellivora*, "honey eater," and the common name, honey badger. Its technique and association with the honey guide, considered below, make the ratel altogether one of the most extraordinary African animals. At least in the *Miombo* Woodland Zone, colonial insects may be a major source of food for this animal: termites, ants (also beetles and other insects) in the wet season, and bees in the dry season (4).

Seven ratels that were collected in dry regions of western Zimbabwe and adjacent Botswana had eaten mainly scorpions (71% occurrence), mice (57%), spiders (57%), lizards (43%), other insects (29%), centipedes (29%), small birds, snakes, and bee larvae and honey (14% each). Since both scorpions and spiders were particularly common and accessible at the time, the sample is not generally representative (12).

SOCIAL ORGANIZATION. Frequent sightings of 2 adults, the lack of any pronounced sexual dimorphism, and the affectionate behavior of pets suggest that ratels may live in monogamous pairs. However, the social organization and behavior of this animal remain unstudied. In a sample of 24 ratel sightings, 20 were singles, the others duos and trios (8). One presumed pair caught together proved to be adult females. A maximum of 12 ratels have been seen in one place: a Masai cattle kraal where the animals came to dig for dung-beetle larvae (4).

The ratel has the reputation of being a remarkably tough and courageous animal best left alone. Although it is known to use its anal-sac secretion in self-defense, apparently the fluid is dribbled rather than squirted and stinks less than striped weasel and zorilla secretions. The ratel's main defense is a good offense: when molested it attacks, no matter how big and dangerous its adversary (fig. 23.5). People and even motor vehicles that come too close are not exempt. An all-black ratel that was followed first by a hyena, then by a jackal, and lastly

Fig. 23.5. Ratel defensive-threat display.

by me as it galloped across the floor of Ngorongoro Crater early one morning, rounded on the car and bit the tires. The hyena and jackal seemed merely curious, and were diverted, each in turn, after the ratel stopped and rubbed its bottom on the ground or grass; it looked as if the scent held them spellbound. In another instance, a man who got out of his car to film a pair of ratels was promptly chased by the male, which proceeded to scratch at the car door and growl for 5 minutes before returning to its mate. When the photographer tried again, the performance was repeated, while the female waited patiently at a distance (12). The ratel's courage is backed up by powerful jaws and limbs, sharp claws, and a nearly impenetrable skin, which teeth and even buckshot (except from very close) will not penetrate (12). At the same time, the looseness of its skin enables it to twist about and grab an assailant in its own vise-like jaws. According to folklore, backed up by some circumstantial evidence, the ratel goes for the scrotum when it attacks large animals (bull buffalo, wildebeest, waterbuck, kudu, man) that have offered some real or imagined provocation (13). Whether this last is fact or fancy, it is clear that the ratel fully deserves its warning coloration.

ACTIVITY. In settled areas the ratel seems to be more or less completely nocturnal, but in parks and in remote areas where bees are plentiful it must normally be active by day to have an association with the strictly diurnal honey guide. Tame ratels are active for long periods day and night. They move around, eat, and play, with relatively short intervals of rest, during which they withdraw to their sleeping quarters (caves, burrows dug by itself or other species in the wild, sometimes empty beehives). Ratels have also been seen moving about during heavy downpours (4).

POSTURES AND LOCOMOTION. Its squat fossorial form, plantigrade feet, and long foreclaws make the ratel a slow and clumsy runner, whose gallop has been compared to a dachshund's canter (10). It often moves in a lumbering, rather pigeon-toed jog-trot which, however, it can maintain indefinitely, thereby covering possibly up to 35 km in a night. While sniffing out food it moves at a rolling walk. Badgerlike, it can move backward with considerable agility, and its body and limbs are remarkably supple (10). Though certainly not an agile climber, it regularly ascends rough-barked trees to reach beehives. Instead of climbing down, a ratel may simply drop, not bothered by hard landings. It reportedly swims very well (it has webbed feet), even chasing turtles and other creatures underwater (4). A peculiar motor pattern it shares with otters and the wolverine is somersaulting down slopes, presumably in play. In captivity somersaulting may become a stereotyped activity (5).

FORAGING AND PREDATORY BEHAVIOR. There are not many animals a ratel can outrun, but it is adept at finding—especially unearthing—concealed invertebrates and vertebrates, such as dung-beetle larvae, scorpions, spiders, estivating tortoises, turtles, frogs, fish, rodents in their burrows, termites, and the creatures that shelter in the ventilation shafts of termite mounds, including snakes, lizards, mongooses, and such (4). Every hole and cavity is explored. The sense of smell is most important but the ratel also uses the aardvark trick of blowing vigorously into a hole and then cocking its head to listen for a response. A tame one used to blow under manhole covers, which, after listening to the echo, it would flip off to investigate the underlying hole (4). Ratels also turn over heavy stones and tear bark off dead trees in search of prey (10). Where elephants and other large herbivores are abundant, ratels busily forage for dung-beetle larvae, digging innumerable vertical shafts to reach the large dung balls encasing the grubs. A ratel observed as it foraged on a Botswana pan was hunting baboon spiders, which live in silk-lined holes. It moved at a slow walk, nose to the ground, pausing now and again to dig vigorously and effortlessly a hole 15–25 cm deep through the hard, calcareous soil and extract a spider with its teeth (12).

Whether the ratel's skin is impervious to snake's fangs is uncertain, but it is reported to catch and eat even the deadliest kinds. One was seen to follow a mamba into an aardvark hole, drag it out, and devour it with complete unconcern (13). Another fought and killed, then ate, a 3-meter python, but shrieked and puffed during the battle, indicating some fear. The snake was as mangled as if it had been run over by a train (10, 13).

By far the most fascinating aspect of *Mellivora*'s food habits is its predation on bees and its association with the greater honey guide, *Indicator indicator* (fig. 23.4). Of the 10 species of honey guides, a family related to the woodpeckers, this is the most widespread savanna species and the only one that habitually solicits people and other animals to follow it to a beehive. The honey badger is the only animal other than man that regularly accepts the invitation (3). The basis of the partnership is the honey guide's craving for wax and the ratel's fondness for bee larvae and honey. Neither is dependent upon the other for its survival, or even to find and gain admittance to beehives: the crops of all the nonguiding honey guides are nearly always found to contain beeswax (a substance other birds cannot digest). These birds are basically insectivorous and only start eating wax after fledging, for they are nest parasites whose young are reared by other species. Nevertheless, cooperation between the ratel and the greater honey guide must pay off or the arrangement would never have evolved in the first place. Probably the ratel finds more nests with less effort when guided, whereas many hives are inaccessible to the birds until opened up by larger animals.

When a greater honey guide sees a potential follower (mostly people), usually it approaches to within 5–15 m or else remains perched, calling. The call has been compared to the sound made by a small box of matches shaken rapidly (3). Churring constantly, the drab bird fans its tail so that the white outer feathers are displayed. As soon as it is followed it turns and flies off a little way with an initial conspicuous downward dip, tail widespread, and alights in a tree, often out of sight, where it continues calling until the follower appears. Then the process is repeated.

According to eyewitness reports, a willing ratel follower answers its guide with a grunting, growling sound or a "slight sibilant hissing and chuckling" (13, p. 243). Hunters of wild bees often imitate the ratel's call when following a honey guide and hammer against trees to simulate chopping noises, but the simple act of following is sufficient encouragement for the bird.

Indicator may lead a honey hunter anywhere from a few meters up to 2 km and the journey may take up to ½ hour. Those who bother to retrace their steps have often found that they were guided in a roundabout and erratic manner, not in a straight line. Furthermore, it appears likely that the honey guide may often have no specific hive in mind at the start of the expedition (3). Beehives are very plentiful in many woodlands; the bird can easily locate one simply by observing the flights of bees, or, more likely, it already knows the locations of the hives within its range.

Having guided its follower to the vicinity of a beehive, the honey guide then falls silent. It does not indicate the exact location, but goes and sits unobtrusively and patiently in a nearby tree, for over 1½ hours if necessary, and waits for its guest to open up the nest and eat its fill. Only after the guest has departed does it claim its share of the spoils. Should its partner go away having failed to locate the hive (which may be in the ground, a rocky cleft, or a termite mound as well as in a tree), the honey guide may try to lead it to another hive.

How frequently honey badgers follow honey guides is unknown. There is evidence that the impulse to guide other species is completely innate in the greater honey guide, but that the bird has to learn by trial and error which species respond and which do not (3). The honey guide learns to single out the ratel, the baboon, and man because these species all raid beehives on their own anyway. Occasionally approaches to mongooses and monkeys are made, but receive no encouragement. Where no suitable followers can be found, the guiding habit may be lost. In fact, it has been lost in many urban and suburban areas where the people are no longer interested in collecting wild honey (3, 4).

The ratel's method of dealing with a beehive, if true (it may be just a folk tale), is truly extraordinary. According to Kingdon (4), quoting African honey hunters, the ratel uses its anal glands to fumigate bees and other biting insects before attacking their nests, in the same way that honey hunters use smoke and with the same effect. Backing up to the opening of the hive, the ratel is said to rub its everted anal pouch all around, meanwhile swirling its tail. Sometimes it performs a handstand in the process of releasing the copious secretion, the odor of which has been described as "suffocating" (6). The bees either flee or become moribund. After a ratel had attacked a hive hung in a baobab tree, the owners noted a pervasive sharp smell and found bees collected at one end of the hive and inactive. Other beekeepers have found many dead bees after a ratel depredation. Tanzanian beekeepers estimate that over 10% of their hives (typically log or bark cylinders hung in trees) are damaged yearly; hunters of wild bees estimate that ratels find and open a much higher percentage of natural hives (4). How much does the guidance of *Indicator indicator* contribute to the ratel's success?

Sometimes ratels are found dead inside or near a beehive, apparently stung to death. Kingdon suggests that the ratel's hide is not proof against bee stings and other insect bites and that ratels may be stung to death if their anal glands fail to function properly: one was seen rolling and rubbing on the ground to dislodge clinging soldiers while raiding and gassing a nest of safari ants.

Eating. The ratel's long claws rule out the kind of delicate manipulation seen in otters, yet *Mellivora* displays not only incredible strength but also delicate skill in uncovering, peeling away, and extracting its chosen foods from inedible outer coverings. It can extract individual bee larvae from a honeycomb with its incisors while holding the comb between its claws, separate dung beetle larvae from their dungballs, lungfish from their clay capsules, and tortoises from their shells, skin out a rodent or hedgehog, eviscerate an antelope through a small hole, and so on. Ratels have been known to cache surplus honeycomb and other food (4).

SOCIAL BEHAVIOR

COMMUNICATION

Vocal Communication: growling, grunting, hissing, barking, screaming, squeaking, whining.

Ratel calls have not been properly catalogued or analyzed, and the same calls may be described differently by different authors. Calls may also vary in frequency and pitch (9). Deep growls, a *whistling hiss*, and a high-pitched *bark-scream* or *rattling roar* are associated with the warning/intimidation display (11, 13). Foraging, possibly disturbed ratels have been heard uttering a breathy "*hrrr-hrrr*" sound, and the call of one following a honey guide has been variously described (as noted above) (4, 9). Pairs have been heard grunting loudly to each other, and a female that seemed to be nest-building made fussy squeaking sounds (10). Cubs are said to give plaintive whines and a hiccuping distress call.

Olfactory Communication. Apart from the fumigating and self-defense uses of the anal-sac secretion, the ratel very frequently scent-marks with its anal pouch, presumably using different secretions. It *anal-drags, handstands,* and backs up against objects such as tree trunks, buttress roots, and rocks, especially around dens, swirling its tail in the process (cf. the above description of "fumigating" bees). The anal pouch is said to be everted whenever a ratel becomes excited (4). A pet regularly anointed its owner but not other people (9). Feces may also play a role in ratel communication. This species digs a hole and defecates into it but leaves the site uncovered. Because ratels investigate every hole they encounter, the theory is that other ratels will therefore be more likely to find the feces (4).

Visual Communication. Undescribed apart from the warning/threat display (see under Social Organization and Antipredator Behavior). Although ratels actively attack their enemies, the basically defensive character of the performance is indicated by screaming, *gaping,* and *bristling.*

REPRODUCTION. Very little is known about ratel reproductive behavior. From 1 to 4 young are born in a leaf- or grass-lined chamber. The gestation period, given as 6 months in most reference books, is very much longer than in other mustelids apart from those with delayed implantation (see family introduction), and requires substantiation. Judging from a zoo specimen that lived for 26 years, the ratel is a very long-lived animal (6).

ANTIPREDATOR BEHAVIOR. There appear to be no natural predators on adult ratels, which itself is evidence of how formidable this animal is, for it weighs hardly more than a jackal. A lion was recorded as having killed one, but there were signs of a terrific struggle (13). In another encounter between these species, 3 ratels took a kill away from 3 subadult and 4 half-grown lions (1). Another indication of the ratel's strength and truculence is the fact that one that entered a steel live trap escaped by tearing the end to pieces with its claws, then continued to mangle the trap with its teeth (11). Wounded ratels have been reported to play dead (4).

References

1. Cowie 1966. 2. Ewer 1973. 3. Friedmann 1955. 4. Kingdon 1977. 5. Lydekker 1896. 6. Nowak and Paradiso 1983. 7. Pocock 1908. 8. Rowe-Rowe 1978*a*. 9. Sikes 1963. 10. ——— 1964. 11. Smithers 1971. 12. ——— 1983. 13. Stevenson-Hamilton 1947.

Added Reference

Dean, Siegfried, Macdonald 1990.

Fig. 23.6. Spotted-neck otter eating a frog.

Spotted-necked Otter

Lutra maculicollis

TRAITS. A small otter with a light, freckled throat. *Length and weight:* overall length about 1 m (85–105 cm), tail long and thick, c. ⅔ of head and body length, and 4.5 kg; females slightly shorter and c. 1 kg lighter than males (4, 6, 11); old males distinctly larger and more muscular than females and younger males, with more prominent testes (7). Feet webbed to the ends of the digit pads, each digit with a well-developed claw. *Teeth:* total 36, adapted for shearing more than for crushing. *Coloration:* variable, from reddish to chocolate brown, the white underfur tipped with brown; white blotches ranging from nil to most of underparts; chin usually dark but upper lip white or dark; white patches may increase with age; young usually darker; old otters have grayish muzzles and old males are often lighter overall (7). *Scent glands:* anal glands (see Family Traits). *Mammae:* 4.

DISTRIBUTION. Africa south of 10° N in lakes, larger rivers, and swamps, but unaccountably absent from many seemingly suitable bodies of water, notably most of the rivers and lakes of eastern Africa (4, 11). It is most abundant in Lake Victoria and in Zambia, where, however, it is absent from the Luangwa Valley and the Zambezi below Victoria Falls. There are no records of this species living in estuaries or the ocean (cf. clawless otter). Its distribution may be restricted by competition with *Aonyx.*

ECOLOGY. Far more aquatic than the clawless otters, this small otter leaves the water only to eat its food, sleep, sunbathe, scent-mark, and whelp. Limited ability to travel overland confines it to perennial bodies of water and usually to areas of open water. Yet it is sometimes common at higher elevations in mountain streams (4).

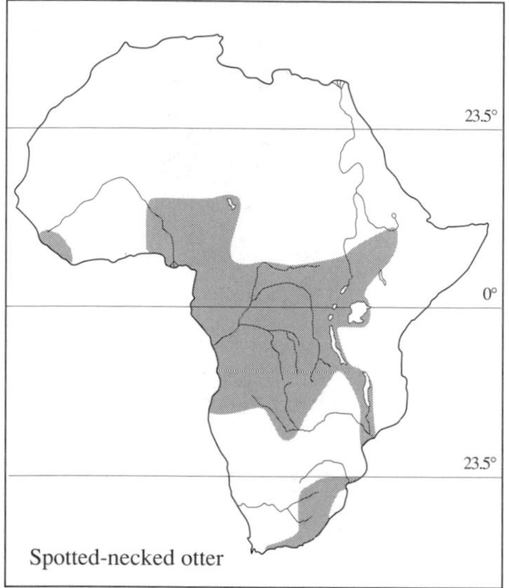

Spotted-necked otter

Though it is more of a fish-eater than the clawless otters, frogs and invertebrates may make up as much as half its diet, so presumably there is considerable dietary overlap with *Aonyx*. In collected scats of otters living on 2 Natal streams, crabs occurred in 57% and 50%, frogs in 29% and 45%, compared to 56% and 42% occurrence of fish (8, 9). But the diet varies in different waters and probably also seasonally. In Lake Victoria, the slowest-moving fishes 10–20 cm long are the mainstay of this otter (7).

SOCIAL ORGANIZATION: uncertain; semisocial under certain conditions, otherwise probably like other river otters (*Lutra* species), male and female separated at least part of the year.

Studies of river or fish-eating otters have almost all been done in Europe and North America (1, 2, 3, 5). As in other unsocial carnivores, *Lutra* males maintain large home-range territories that often include more than 1 female territory. Although the sexes may associate for part of the year, they do not live in monogamous pairs (cf. clawless otter). Females rear offspring unassisted for at least the first few months. Pups remain with the mother for a year until the birth of the next litter, then disperse and go through a wandering stage until they find ranges of their own. The size and shape of an otter's home range depend on the kind of water (lake, stream, swamp, etc.) and the available resources. Otters living on a large lake may defend only a stretch of the shoreline while sharing the same fishing grounds, whereas otters living on a stream may control a long section as an exclusive hunting ground. In any case, otters tend to use their home range in a consistent pattern, their regular passage along the same routes creating worn paths. One dog otter was known to travel within the same range for 5 years, with only small variations in the routes he followed (1).

Most sightings of spotted-necked otters are of singles or a mother with (typically) 2 offspring. The largest groups seen in most areas number no more than 5 (11). However, at Bukoba on Lake Victoria, where spotted-necked otters reached densities of up to 1/1.5 km of shoreline, loose schools of 4–6 were seen most often (13 times), with 1–3 otters the next commonest group size (11 sightings). Schools of 7–9 and 10–12 otters were sighted 6 and 5 times, respectively. The two largest schools numbered 19–21 and 16–18 (7). The numbers of otters remained fairly constant during the period of observation (June–March). The larger groups were usually composed of passing bachelor males that interrupted trips across the bay to rest a few minutes on the rocks at the entrance of Bukoba Port. The more resident otters were seldom seen in groups of over 5, of which most usually turned out to be females. It appeared that young of the same sex banded together in schools after the breakup of families during the dry season and the beginning of a new breeding cycle. Prior to that time, family parties typically consisting of 2 adults and 3 nearly grown offspring were in evidence. Assuming the adults were male and female, it appears that spotted-necked otters at least sometimes live as mated pairs for some part of the year, and that males may participate in caring for the young.

ACTIVITY. Where not subjected to heavy human predation, this otter has daytime activity peaks for 2–3 hours after dawn and before dark (6, 7, 8). In Botswana waters it may be mainly crepuscular (11). Otters may also be active at night, particularly by moonlight. The otters of Bukoba emerged from their holts at first light, cruised around aimlessly for ½ hour or so until sunup, then embarked on 2–3 hours of intense fishing, after which they either emerged from the water to bask and groom on the rocks or withdrew again into their holts. Most traveling was also seen in the morning, although individuals fished and traveled at irregular hours, too (7).

POSTURES AND LOCOMOTION. The spotted-necked otter is at home in the roughest waves and rapids and freely dives into the water from heights of 2 m leaving barely a ripple—unless alarmed, when it hits with a loud splash. It is as lithe and powerful as other otters, "all movements being performed with agility, grace and control," including surprising dexterity with the forefeet considering its completely webbed fingers (6). Yet it seems to be a comparatively slow swimmer and short-winded diver. Its swimming speed while chasing fish (fig. 23.7) has been recorded at 4 kph. The main propulsive force is provided by the hindfeet thrusting together, and the forefeet come into play when making sharp turns (7). An average of 15 seconds was recorded between breaths and 45 seconds was the maximum time spent underwater. The first dive of a pet female was invariably the longest of a series (6). A diving animal descends almost vertically (usually no deeper than c. 2 m), kicking powerfully with both fore- and hindlegs to gain depth quickly. A traveling individual typically swims underwater for 3–4 m, then rises to breathe and swim on the surface for 1–2 m before submerging again, sometimes keeping this up for 5–10 km (7). If an otter becomes tired far from shore, it can rest by floating on its back.

On land, the spotted-necked otter is mobile enough to scramble up rocks and leap a 1-m gap, and can climb to some extent (e.g., a wire fence), but may become overheated and distressed after only a short walk. It walks and runs with a diagonal, bouncy stride (*cross-walk*), body raised a few centimeters off the ground and head up. Normally it goes no faster than 6–7 kph, but it can gallop a short distance nearly as fast as a man can run, humping its body far more than a clawless otter (4, 6). When resting ashore and also while eating, *maculicollis* lies flat on its belly, tail outstretched, head lowered or slightly raised. It sleeps in the usual *sideward curl.* Two other characteristic postures are crouching with the head up while peering around, and standing bipedally and braced by the tail to see over high grass or other obstruction.

Grooming. As it emerges from the water, the spotted-necked otter always shakes its head, causing the neck hairs to stand up stiffly, but does not shake itself all over. The coat dries by evaporation, assisted by rubbing the flanks and belly against the ground, rocks, and such. Much time is devoted to scratching and *nibble-grooming* the fur. Procter (7) saw *mutual-grooming* in groups at Bukoba but could not tell whether the individuals were mates or otherwise related. Licking plays no part in grooming, but captives of all ages often suck their own paws or belly fur while composing themselves to rest (4, 6).

PREDATORY AND FEEDING BEHAVIOR. Spotted-necked otters apparently rely entirely on sight to locate prey. They capture and transport fish and all other prey by mouth and, except for small items like snails, minnows, and the like which may be consumed while treading water or lying on the back, leave the water to eat their catch, holding it under or between the forepaws (fig. 23.6). In these and other respects their predatory and feeding behavior is very similar to other *Lutra* species and quite unlike clawless otters. A pet was observed groping on the bottom beneath stones for frogs and invertebrates (5), but that is considered unusual. Another pet otter used to poke its nose under stones and bring out crabs attached to its face, which it then pushed into its mouth. Perhaps that would explain another otter's reluctance to pick up crabs longer than 50 mm (8).

A hunting otter makes repeated short dives, swimming rapidly and randomly while searching the surrounding water. When a fish is sighted, the otter accelerates and overtakes it in a straight chase (fig. 23.7). Hungry otters pursue fish with great vigor, grab and bite a quarry usually in the middle or near the tail, and then land and consume it at once. A nearly full otter will often continue pursuing fish and toy with a victim after crippling it, throwing it out of the water, butting it with its nose, and so on (6). Cooperation in driving or catching fish, sometimes reported in other *Lutra* species, has not been documented and is of doubtful occurrence.

Fig. 23.7. Fish otter pursuing its prey (redrawn from Maxwell 1960).

All but the smallest fish are eaten tail first and eaten completely, but the heads of large fish are discarded. Frogs too are eaten from the rear, after biting off the legs (fig. 23.6). When eating crabs, the carapace is usually discarded (cf. *Aonyx*) (8, 11).

SENSES. Like other otters, *L. maculicollis* presumably has good binocular vision, especially underwater. Nonetheless, it is myopic out of water, failing to distinguish a motionless person as near as 3–6 m (7, 11). The iris of adults is orange-red. A pet otter could smell and locate a glass of water. Hearing is also considered acute (6).

SOCIAL BEHAVIOR
COMMUNICATION

Vocal Communication: chattering, mewing (*"ya-a-a-a"* or *"yea-ea-ea-ea"*), squealing, whistling, *squeak-trilling.*

A chattering or rattling call very similar to the call note of the pied kingfisher, but fuller and less piercing, is a warning or scolding call, usually given while on land. The mewing call seemed to Mortimer (6) to be a challenge, also given while on land and usually stationary. A squealing whistle was given by excited otters during play-fighting. The commonest and most variable (graded) call given by hand-reared pets was a squeak or twitter (6). A high-pitched squeak was given as a begging or distress call. The tempo increased when the otter was frightened and changed to a high-pitched trill when the animal was hurt, whereas a lower, harsher trill produced far back in the throat expressed rage (6). The nest-chirp of *maculicollis* infants remains undescribed.

Olfactory Communication: *musking, sprainting* (defecating), discharging anal-sac secretion.

This species is known to maintain dung middens and to leave its dung on elevations near water, preferring rocks with flat, moderately sloping tops. The animals come ashore for the express purpose of *sprainting,* and at the same time urinate, crouching with bellies nearly touching the ground. Males erect their tails vertically; females hold theirs nearly horizontal (7). Regular rolling/musking places have also been reported (10). When suddenly alarmed or hurt, these otters discharge the milky secretion of their anal sacs. The smell is pungent but not comparable to the stench of a zorilla or weasel (6).

AGONISTIC BEHAVIOR: uncatalogued.

In the prelude to an attack, which consists of a sudden, vicious bite, a spotted-necked otter weaves its head in the snake-like manner common to many mustelids, and voices its trilling battle cry. The defensive/submissive attitude is lying on the back with feet waving and sharp claws spread, meanwhile squeaking (6). An otter stranger has been seen to hold its neck out to a bolder, investigating animal, after which it was ignored (7).

REPRODUCTION. So little is known about reproduction that it is even unclear whether breeding is seasonal or perennial. It is thought to be seasonal, with typically 2 young (1–3) born late in the dry season or early in the rains (e.g., September on Lake Victoria [7]). Although a pet female reached her adult weight by 7 months, the presence of schools of nonbreeding yearlings in Lake Victoria would indicate that there, at least, females do not breed before their second year, and thereafter reproduce once a year. Nothing has been published about sexual behavior of this species (see family introduction).

PARENT/OFFSPRING BEHAVIOR. Following a 2-month gestation, the otter bitch whelps in a burrow, hole in the rocks, hollow tree trunk or, lacking a proper holt, a dense thicket or reedbed. She prepares a nest of grasses and other plants. Expectant and new mothers are extremely intolerant of other otters, at least until the cubs are old enough to leave the den and swim, at probably 10–12 weeks. The father may then join the family and help bring up the cubs, but there is still too little evidence to decide whether or not this is usual. Parental training in fishing has been observed in the spotted-necked otter: a bitch that repeatedly dropped live and flapping fish on a rock beside her 3–4 month pup. The pup was still unable to dive properly or to catch fish; apparently it had trouble submerging. But a hand-reared pup caught fish without prior instruction or assistance. The mother of this youngster died defending her 3 offspring against a man (6).

PLAY. Apart from the tendency of satiated animals to toy with their prey, spotted-necked otters often play with pebbles, shells, sticks, and such. When playing on land, a pet otter patted rounded objects with either hand and chased after them, tossed objects held in the mouth backwards over its shoulder, and pushed with its nose others too big to be thrown or carried. Social play includes wrestling, tag, ducking, rolling, biting, and the like. Grappling animals hold each other in a strong bear hug (6). Up to 5 wild subadults have been seen play-fighting in the water, in bouts lasting several minutes (7).

References

1. Erlinge 1967. 2. ——— 1968. 3. Harris 1968. 4. Kingdon 1977. 5. Macdonald and Mason 1980. 6. Mortimer 1963. 7. Procter 1963. 8. Rowe-Rowe 1977*a*. 9. ——— 1977*c*. 10. ——— 1978*a*. 11. Smithers 1983.

Added Reference

12. Gorman, Jenkins, Harper 1978.

Fig. 23.8. Clawless otter walking on three legs while carrying a fish.

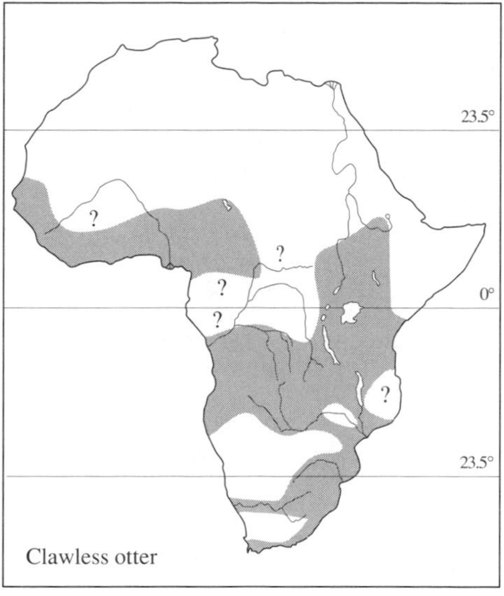

Clawless Otter

Aonyx capensis

TRAITS. A very large otter with unspotted white markings. *Length and weight:* males 128.5 cm (111–138), including 51 cm (44–57) tail, and 13 kg (10–18); females 139 cm (114–160), including 51 cm (48–53) tail, and 13 kg (12–16) (southern Africa) (16). Forefeet specialized for probing and manipulation, unwebbed, clawless, naked underside, opposable thumb (4); hindfeet webbed to outer joints (14). Massive skull and robust teeth (36 total), with long canines and very broad molars adapted for crushing. Whiskers particularly long and prominent (all white or black and white). *Coloration:* from tan to chocolate brown with variable white or off-white markings, including chin, cheeks, throat, chest, and belly, sometimes a frosted mantle of white-tipped hair from forehead to shoulders. *Scent glands:* anal glands (see Family Traits). *Mammae:* 4.

DISTRIBUTION. Widespread south of the Sahara in suitable habitat, the clawless otter occurs in small and large streams and in lakes, ponds, and swamps, but is frequently absent from large rivers, for example from the Zambezi between Victoria Falls and its delta in Mozambique and from parts of the Limpopo. It is also at home in salt water, living both in estuaries and in the ocean, but only on rocky coasts or in mangrove swamps in the vicinity of fresh water (13).

RELATIVES. In the Lowland Rain Forest the smaller, less-specialized Congo clawless otter, *A. congica,* replaces *A. capensis* (7), but in parts of West Africa, the southeastern Congo Basin, and Uganda the 2 species coexist (16).

ECOLOGY. *A. capensis* is the third largest otter; the maximum recorded weight of c. 34 kg (5) is close to the mass of the Brazilian giant otter. It is more amphibious and less aquatic than the spotted-necked otter, with fish making up a small part of the usual diet. In a scat analysis of Natal otters, the percentage occurrence of crabs was 65%, followed by frogs (23%), with only 4% fish. Other prey, including insects, snakes, waterbirds, and mollusks, occurred in 1%–2% of the samples (12). In another scat sample of a coastal Cape Province population, the diet was varied, including some 35 different crabs, fishes, and octopuses; 2 crab species led (38% occurrence), followed by 2 species of sedentary fish (21%) and an octopus (7%). Here otters foraged mainly along vertical rock faces just below the surface, but also explored crevices and tidal pools (18). In a small sample of 10 Zimbabwe otters, in contrast, the dominant food was fish (78%), especially tilapia and bream (17). Although bivalves and snails are abundant in many waters, the available evidence indicates that they are rarely eaten by clawless otters.

The clawless otter's ability to subsist on crabs, frogs, and other invertebrate foods enables it to exploit a wide variety of watery habitats, right up to the headwaters of river systems. Its mobility on land is adaptive for traveling overland in search of new food resources. The type of vegetation matters little: it occurs in rain forest with up to 2000 mm rainfall (Liberia) and in subdesert with 100 mm of rain (e.g., lower Orange River), and in mountains to around 3000 m (10). It is only necessary that the surrounding terrain offer some form of shelter (4).

SOCIAL ORGANIZATION. A definitive study of known individuals is needed to clarify the presently vague picture of *Aonyx* social organization. Otters inhabiting the south coast of Cape Province (South Africa), studied for 17 months, appear to have a clan-type organization, in which a number of adult males share a range and even forage together (1). But this maritime population, linearly distributed along a rocky coastline, lives in an ecosystem that is hardly representative of the species; the dispersion and grouping patterns of populations living in swamps, rivers, and lakes may be quite different, reflecting contrasting ecological conditions.

Population density and home ranges of otters living along the 58.5 km of coastline included in Tsitsikama N.P. were estimated by capturing, marking, and recapturing operations, which included radio-collaring 5 individuals and injecting several others with a radioactive isotope that could be detected in the feces (spraints) (1). The total population was estimated at 30 or 31 otters, with a mean density of 1 animal/1.9 km of coastline. The known ranges of 6 adult males extended along 3.3 to 19.5 km of coastline, but only the latter figure, based on 73 days of radio contact, was considered to be a reliable estimate of the full range; it is the longest range that has been recorded in any coastal otter population (1). However, this long linear distance may reflect the fact that on this steep coast only a 100 m band of inshore water is shallow enough for the otters to reach bottom; the surface area of this range was therefore only 195 ha. Furthermore, 90% of the times it was located the male was in a 12 km core section, which it shared with 3 other adult males.

Range overlap is apparently usual among coastal populations of otters, but there was preliminary evidence to suggest that these clawless otters were sharing and possibly defending a clan territory. (*a*) They shared the same holts, near which (*b*) they deposited their spraints, and (*c*) 3 of the males were known to forage together in pairs some of the time.

The position of females in this population remained uncertain. Arden-Clarke suggests that the two sexes might occupy different core areas. One female whose range was plotted on the basis of radioactive scats usually stayed within a 7.5 km core area, but had a minimum home range covering 14.3 km of the coast (1).

Aonyx home ranges may be considerably smaller in inland waters. In southwestern Tanzania, three distinct groups (clans or families?) were known to inhabit a 5 km wooded stretch of river for at least 25 years (7). A well-worn slide down a mud bank and into a pond was located near one holt.

Holts on the Cape Province coast are only shallow scrapes located in dense vegetation within 15 m of a perennial source of fresh water, used as refuges during periods of inactivity on shore. The average density was 2 holts/km of coastline. Elsewhere, otters use holes in banks, under rocks, between tree roots, and the like as holts. Despite clawless hands poorly adapted for the purpose, *Aonyx* is also known to dig burrows, up to 2.8 meters deep with 1–2 entrances above water, in sandy, alluvial soil (13). Low water during a drought revealed an underwater entrance to a riverbank den (15).

ACTIVITY. Clawless otters can be active day and/or night. The observed activity peaks of Natal otters were late afternoon and early evening, and sometimes early morning (13). Coastal otters were active all night up until c. 0700 h, with a peak between 2000 and 2200, then remained inactive until 1500 (18). Occasionally, however, clawless otters come out of their holts in the middle of even the hottest days (15).

POSTURES AND LOCOMOTION. *Aonyx* resting and eating ashore assumes the S-shaped crouch typical of long-bodied mustelids (cf. spotted-necked otter). Walking is a leisurely gait, but more often this species bounds along at a good pace, and it gallops in the usual inchworm fashion but with less humping than *Lutra* (7). In swimming, propulsion comes mainly from the hindfeet, the tail acting as a rudder. While swimming on the surface, the head and sometimes part of the back shows. Preparatory to diving, an otter raises up, then arches forward and dives vertically, making hardly any disturbance.

Fig. 23.9. Clawless otter sleeping (redrawn from Maxwell 1960).

Drying Out and Grooming. On emerging from the water clawless otters throw off water by shaking first the head, then the whole body. They dry off by rolling and rubbing on the ground in sand, litter, or short grass (16). Probably they also mark with the anal glands during this performance (3). Observation of a captive female, which dried herself on the furniture and carpets, disproved the idea that an otter's coat is waterproof: the underfur was sodden, its appression to the body contributing to the sleekness of the coat (16). After rubbing and rolling, otters like to bask in the sun, lying on their backs, bellies, or sides often in comical attitudes (fig. 23.9).

PREDATORY BEHAVIOR AND FEEDING. The specialization of the clawless otter's hands for probing and grasping, and of the teeth for crushing, are reflected in its prey preferences and feeding behavior. Given a choice of crabs, frogs, and fish, a wild-caught female always chose crabs first and frogs second, taking these animals in the same proportions as represented in the scats of wild otters (11). A series of experiments with this animal showed that it was only slightly harder for her to catch crabs and frogs in muddy water or at night than in clear water by day, whereas it took 4 times as much effort to capture fish under conditions of poor visibility. The simple explanation: crabs and frogs hide and are located mainly by touch, whereas fish are usually detected and pursued by sight (12).

A female and 2 well-grown cubs were observed as they hunted in an oxbow pool in late afternoon (11). Each foraging independently, they crisscrossed the 150-m length of the pool from side to side, returning and reemerging close to their starting point after 35 minutes. The otters were diving and coming up with crabs and frogs, which they ate while treading water. The female secured prey in 23 of 37 dives. When the otters later reentered the pool they swam with heads out of the water, apparently no longer hungry.

The captive female dove in the same manner, and while submerged probed under rocks for crabs and frogs, body angled downward at 45°. When fishing, she swam below the surface, looking from side to side, and on sighting a fish, accelerated and tried to grab it. In the case of a school of fish swimming overhead, she would pursue the first one that broke ranks and fled—a typical predator response. The average duration of 389 dives was 18 seconds, with a typical variation of between 14 and 24 seconds (minimum 6, maximum 49) (11). The otter made 2–5 dives for each fish caught, the effort required depending on the fish's swimming ability (12). Barbels took only 32 seconds and 2.3 dives, tilapia 91 seconds and 5.4 dives. But the fastest species, *Barbus natalensis,* took an average of 12.8 dives and 218 seconds to catch (12).

A clawless otter wading in a narrow mountain stream kept its head submerged, looking left and right as it felt under rocks. Two otters moving down a shallow river with a bottom of rounded stones groped under them and even moved stones to catch an occasional crab (11).

With rare exceptions, all prey is captured with the hands. Rough skin on the palms and fingers enables a clawless otter to grasp slippery fish and frogs securely. The capture of a fish is followed by a bite to or behind the head. By using its hands to hold its food, the clawless otter can eat without taking the time to come ashore (cf. spotted-necked otter) (fig. 23.10) It can even eat with its head underwater, but usually tilts the head back while chewing. Food items longer than 8 cm are generally taken into shallow water, where the otter sits or stands on the bottom while holding its catch in one or both hands and eating. The captive female picked up small items with her left hand and carried them in her right

Fig. 23.10. Clawless otter eating a fish.

hand. When going ashore, she held prey in her mouth while climbing out, then hobbled on 3 legs while holding the prey against her chest with her right hand (fig. 23.8). She also fed prey into her mouth with the right hand. Another individual was left-handed (2).

Crabs are held in both hands, bitten across the shell, and then crunched up; the claws and legs of larger ones are chewed off first. Pieces which fall out of the mouth are caught against the chest with the hands. Sometimes crabs bite an otter's cheek, lip, or even tongue. The otter merely pulls them off, without any indication of pain. When eating a frog, the limbs are held in one hand while the otter tears off pieces with its jaws. Fish and frogs are always eaten starting at the head end, except for barbel (*Claria*) and other fish with very big or hard heads, which may be eaten tail first and the head left untouched (11). Freshwater mussels with shells too tough for an otter's jaws (e.g., *Aspatharia wahlbergi*) are opened by smashing them on nearby rocks, tree trunks, and such. Middens of broken shells accumulate around these anvils (2).

When full, the tame otter sometimes captured several fish and then carried them to her sleeping place. Later she would take them back to the water and eat them. Whether this form of caching occurs in the wild is unclear. The sight of terrestrial prey such as rats (ashore or in the water) did not arouse predatory behavior (11). She did capture and kill a fledgling cattle egret by holding it down with a forefoot and biting its head, then ate all but a few feathers. But grasshoppers, beetles, and eggs were all discarded after being sniffed.

After completing a meal, the clawless otter cleans its face and feet in the water in an elaborate ritual (16).

SOCIAL BEHAVIOR

COMMUNICATION

Vocal Communication: *"hah!,"* growling, snarling, screaming wail, humming, *"wow-wow-wow," "whack-o,"* chirping, squealing, moaning, mewing, snuffling, *"whee."*

Although all of the above sounds have been mentioned in accounts of the clawless otter, the calls have not been properly catalogued and it is very possible that there is duplication in those listed. However, this species has quite an elaborate and expressive vocal repertoire which has more elements in common with the other large otters than with the spotted-necked and other fishing otters (3).

Hah! is an explosive alarm snort that is given by all otters studied up to now (3). In addition to growls and snarls uttered as threats, a screaming wail uttered at highest intensity impressed Maxwell (8, p. 32) as "of all the animal cries I have heard the most vindictive." Humming, a sound similar to a cat's defensive threat, is a begging call in *Aonyx* (9). A wailing *wow-wow-wow* or *ow-ow-ow* is an expression of apprehension or displeasure which often alternates with snarling while the otter lies on its back waving its paws (cf. high-pitched squeak of spotted-necked otter). Maxwell's pets greeted him with a loud, repeated 2-note chirp that sounded like *"whack-o, whack-o,"* and also uttered a wide range of "conversational" noises: affectionate squeals comparable to those of human infants, moaning, mewing, and snuffling noises (the last comparable to a cat's purr), and a *hah!* begging call. Cubs utter a squeaky *whee-whee* call grading from soft to loud, short to long, to express different emotions. This may be the juvenile version of the adult conversational calls. The basic juvenile contact call is a high-pitched whistling *whee* (8).

Olfactory Communication: latrines (sprainting sites); anal-gland marking.

Rolling/rubbing (and presumably *musking*) places have been found in grass and on sandbanks, earth ledges, and rocks, often near regularly used latrines. In the study of the Cape Province coastal population, 85% of all collected spraints were found within 50 m of a holt, mostly close to water (1). In swamps mounds are often latrine sites (15). Clawless-otter feces can usually be told apart from those of *Lutra* by the prevalence of ground-up crab shells. Despite their coarseness, the scats are sticky enough to adhere to vertical surfaces, against which clawless otters like to plaster their dung while at the same time urinating, like other otters (5).

Visual Communication. Mostly undescribed.

AGONISTIC BEHAVIOR

Threat/Attack Displays. The body posture and facial expressions of a threatening clawless otter have not been exactly described or illustrated, but the whole demeanor of an angry otter, coupled with the noises it makes, is not only unmistakable but downright frightening. Here is an animal with the same kind of loose, impenetrable skin as a ratel, the same incredible strength and viselike grip, the same courage, only bigger and more flexible. Pound for pound, it is arguably the most formidable African carnivore. Females with cubs readily attack people as well as other enemies in their defense. When brought to bay, clawless otters have been known to drag big hounds underwater and drown them (6). Delightful as it may be to have one as a pet, the tendency to fall into sudden jealous rages adds more than a touch of danger: Gavin Maxwell's pet female bit the first joint off a finger on each hand of a teenage boy who was helping to look after her during a sudden, vicious attack (9).

Defensive/Submissive Displays. Lying on the back with paws waving can apparently be either defensive or submissive, depending on the accompanying sounds and movements (9).

REPRODUCTION. It is unclear whether breeding is seasonal or perennial. Births (of 2–3 young) have been reported in most months, with a possible peak in the dry season south of the equator (13, 16). Gestation is 2 months.

SEXUAL BEHAVIOR. Undescribed (see family introduction).

PARENT/OFFSPRING BEHAVIOR. Two cubs about a week old weighed 260 g apiece (13). The young stay in a holt at least until their eyes open at c. 1 month, after which they begin to venture outside when the mother is present. Small cubs keep up a constant chirping *whee-whee* (see under Vocal Communication). In transporting a cub, a mother sometimes holds it in both arms while staggering along on her hindlegs (4). The pale color of the woolly underfur, which persists to the yearling stage, appears to attract grooming by adults when a cub lies on its back in the submissive attitude. It has been suggested that the adult's white bib and underparts may serve to inhibit aggression when similarly exposed, based upon association with the juvenile coloration (7).

PLAY. In Botswana, a family group of 4 clawless otters regularly leaped and slid down a grassy river bank, one after the other, for up to ¼ hour at a time (15). Diving for pebbles or shells is a favorite solitary game. The otter carries a pebble out into the water and drops it, then dives and recovers it before it reaches the bottom. It may catch the pebble on its flat head, then go through a whole series of gyrations while keeping it balanced (6). Clawless otters have also been seen diving with floating objects, which were released underwater and allowed to float to the surface (15). Like other otters, *Aonyx* also plays "cat and mouse" with caught fish in much the same way it plays with inanimate objects. Perhaps more unusual, *Aonyx* uses its tail to scoop objects within reach of its hands (8, 15). The dexterity of its hands, as demonstrated in play with minute objects, definitely is outstanding.

ANTIPREDATOR BEHAVIOR. See under Agonistic Behavior.

References

1. Arden-Clarke 1986. 2. Donnelly and Grobler 1976. 3. Duplaix 1984. 4. Ewer 1973. 5. Eyre 1963. 6. Harris 1968. 7. Kingdon 1977. 8. Maxwell 1960. 9. ——— 1963. 10. Rosevear 1974. 11. Rowe-Rowe 1977*a*. 12. ——— 1977*b*. 13. ——— 1977*c*. 14. ——— 1978*a*. 15. Smithers 1971. 16. ——— 1983. 17. Smithers and Wilson 1979. 18. Van der Zee 1982.

Part IV

Primates

Chapter 24

Introduction to Primates

African Primates
Suborder Strepsirhini: prosimians
Superfamily Lorisoidea
Family Lorisidae: lorises, pottos, bush babies
Subfamily Lorisinae: pottos (2 species)
Subfamily Galaginae: bush babies (6 species)
Suborder Haplorhini: monkeys, apes, humans (formerly Catarrhini: Old World monkeys and apes)
Superfamily Cercopithecoidea: Old World monkeys
Family Cercopithecidae: monkeys and baboons
Subfamily Cercopithecinae: guenons, mangabeys, baboons* (33 species)
Subfamily Colobinae: leaf-eating monkeys (7 species)
Superfamily Hominoidea: apes and humans
Family Pongidae: great apes (3 species)

*Classification follows Jolly 1985 except for *Papio* and Colobinae species

The primate order, to which we belong, includes some 185 species, divided among 11 families and 56 genera (10). Our living relations include creatures as unlike us as the nocturnal potto (fig. 25.1) and as like us as the chimpanzee, as small as a 60-g bush baby and as big as a 200-kg gorilla.

There are 4 major branches of the primate tree: prosimians ("almost-monkeys"), New World monkeys, Old World monkeys, and hominoids (apes and humans). Monkeys and apes all share obvious simian characteristics, whereas the connection with prosimians is far less apparent. Yet the evidence indicates that the first primates were also small, nocturnal, solitary-foraging, arboreal, vertical clingers and leapers, from which diurnal, gregarious monkeys and apes evolved (7, 15, 17, 24).

Tangible evidence is skimpy, because of the generally rapid decay of tropical-forest organisms, but Madagascar contains living proof that such an evolutionary scenario has occurred. Here, isolated from competition with higher primates, the prosimians underwent an adaptive radiation, in the course of which diurnal, social lemurs evolved to fill niches occupied on the continents by monkeys (9, 10). The 39 living prosimians are placed in 4 different families, of which all but 11 species and 1 family are native to Madagascar. The small nocturnal lemurs are very like bush babies, which are found only in Africa. There is also 1 oddball family in Southeast Asia, the tarsiers, with 1 genus (*Tarsius*) and 3 species. They look a lot like bush babies but their internal anatomy, biochemistry, chromosomes, and reproduction prove they belong with the higher primates (10). Because of this single nonconformist genus, the traditional primate classification had to be rearranged. The suborders of Prosimii and Anthropoidea have been replaced by the Strepsirhini (Greek "turned nose"), including all the prosimians except tarsiers, and the Haplorhini, (Gr. "simple nose"), containing the tarsiers and all other primates (10, 11).

Africa and Asia have 51 and 52 primate species, respectively. Seventy-eight of these belong in 1 family, the Cercopithecidae. The African branch is dominated by the forest guenons (20 *Cercopithecus* species), whereas Asia has most of the leaf-eaters (20 species, mainly langurs, subfamily Colobinae) and the baboonlike macaques (15 *Macaca* species). Except for the presence of the hamadryas baboon in southern Arabia, and the Barbary macaque (*M. sylvanus*) on Gibraltar, the primate faunas of the 2 continents are completely separate.

In this section of the *Guide,* a chapter is devoted to each of the primate families that occur in Africa, followed by separate accounts of the relatively few species that are commonly seen (or heard in the case of bush babies) in the national parks of eastern and southern Africa. The differences between leaf-eating colobus monkeys and the rest of the family Cercopithecidae are recognized by giving the subfamily Colobinae a chapter to themselves.

TRAITS. Because of the gulf that divides prosimians and simians, the list of traits shared by all primates is not long and for the most part may represent anatomical characters inherited from the earliest true mammals, with further development of features adapted for living in trees (24):

Generalized skeleton with limbs and body highly mobile, capable of varied movements, locomotor patterns, and postures; all primates capable of sitting and standing, but not necessarily walking upright. Hands used for both feeding and locomotion.

Five fingers and toes, equipped with flat nails instead of claws (for protection of sensitive fingertips), at least 1 pair of digits (usually thumb and big toe plus second digit) opposable and able to grasp. Exceptions: thumb reduced in colobus, foot unable to grasp in humans, and a few New World monkeys have claws.

Clavicle (collar bone) large and specially developed to increase forelimb mobility.

Eyes surrounded by compete bony ring, frontally placed with overlapping fields giving binocular vision needed for accurate depth perception; diurnal primates with color vision and a central depression (fovea) of sharp focus.

One pair of pectoral mammae (some bush babies with 1–2 additional pairs).

Testes in a naked or thinly haired scrotum, suspended penis (sheath not tied to abdominal wall) often with bone.

A well-developed cecum (reduced to the appendix in humans).

The following traits show progressive development in the order: large, complex brain, comparatively long gestation period, slow maturation, and long life span.

ANCESTRY. Mammals with the above skeletal characters had evolved as long ago as 65 million years, earlier than any other order of true mammals except insectivores and carnivores (22, 24). They evolved in the tropical forests, from arboreal insectivores probably similar to tree shrews. Monkeylike, large-brained, possibly group-living primates have been around for at least 30 million years, and bipedal hominids date back c. 4 million years (16).

DISTRIBUTION. With the exception of a few macaques and langurs that have adapted to temperate Japan and China, primates are confined to the tropics and 80% of them live in rain forests, where they tend to be the dominant mammals. As many as 17 or 18 different species coexist in Zaire's Semliki Forest and on Mt. Cameroon (10).

More than half of all primates are adaptable enough to be able to live in either moist or dry forests. Yet few indeed—and only Old World species—have managed to leave the trees and venture into savanna and subdesert: 3 baboons, the patas and vervet monkeys, the chimpanzee, and a few Asian langurs and macaques among living species. All still depend on trees or cliffs for security at night. Only man and the gorilla sleep on the ground, and the gorilla fashions a nest. Humans too need shelter, for comfort as well as for safety. Without caves, roofed nests, clothing, and/or fire to ward off the chill, could our species have afforded to become naked? And why did we, anyway?

ECOLOGY. Over a dozen different primates can share a forest by partitioning its resources and their time in such a way that no 2 species have precisely the same diet or regularly compete for the same foods at the same time and place.

The forest is a 3-dimensional habitat, differing in character from the sunlit canopy, where most of the foliage and photosynthesis occur, to the sparse growth of the dim-lit floor. Some primates range from top to bottom, but many live and forage within just 1 or 2 tiers. Most of the species that live in the lowest layers depend on the secondary growth that occurs when sunlight reaches the ground through openings created by watercourses and fallen (or felled) trees.

SEPARATION BY SIZE AND DIET. The ecological separation of primates is effected first of all by differences in size. All primates need protein for growth and tissue replacement, carbohydrates and fat for energy, vitamins (e.g., B_{12}), and trace elements. But size largely predetermines the animal's primary food sources. Because small animals expose a relatively larger surface area (to volume) than big ones, they have to burn proportionally more fuel to maintain a constant body temperature (5). Protein and energy are both more concentrated in animal than in plant tissues. The smallest bush babies can gain all the nourishment they need from insects, supplemented by gum exudates from certain trees (4). Bigger primates cannot locate and capture enough of such small, elusive food items, so have to rely on less nourishing vegetable foods to sustain their greater bulk. But insects are still an important protein supplement for many monkeys. Small monkeys like the redtail or vervet are quick enough to catch jumping and flying insects (though less adept than a bush baby); larger ones (blue monkey, baboons, chimpanzees) probe in holes for slower types of invertebrate and vertebrate prey (5, 20).

New leaves, buds, shoots, seedpods, and such are an alternative source of protein for many primates. Edible gums consisting of long-chain sugars are also important for some monkeys (see vervet account) and apparently were a mainstay of the early prosimians, whose lower comblike incisors probably evolved as a tool for scraping gum (4). In fact, the needle-clawed bush baby subsists almost entirely on gum (see chap. 25). But the staple food and main source of energy for most primates is fruit. Among 45 species that have been studied for more than a year, including prosimians and Old World and New World monkeys, fruit amounts to at least ¼ of the diet of all but 5 species (10). Fruit contains very little protein, however, so even the purest frugivores (e.g., potto, gray-cheeked mangabey, and crowned, DeBrazza's, and moustached guenons) have to eat some foliage or invertebrates. Animal food is also the main source of vitamin B_{12}, which helps to explain why most primates eat insects and even small vertebrates at least occasionally.

Some primates subsist mainly on foliage, and these same species eat virtually no animal food. Colobine monkeys have a ruminantlike digestive system that enables them to utilize the most abundant forest product, while also gaining protein from the bacteria and protozoans they host (see subfamily introduction). Since fermentation is more efficient in large containers (either the stomach or the enlarged cecum and colon of the gorilla and other hindgut fermenters), leaf-eaters tend to have pot bellies and larger frames (though no more muscle) than other monkeys. They also tend to be more sluggish (see colobus account). But even folivores eat some fruit, flowers, and the like. Of the above-mentioned 45 primates, only 2 colobus monkeys and the gorilla consume foliage in amounts exceeding ¾ of their diet (10).

SEPARATION BY SPACE AND TIME. Different species can share the same space and many of the same foods by keeping different hours. Pottos and bush babies, together with other strictly nocturnal animals such as the palm civet, genets, various rodents, and owls, have African forests to themselves at night while monkeys and birds are asleep. There is no equivalent day shift among African mammals, for none of the diurnal Old World primates is as small as the forest bush babies, perhaps because niches for small diurnal insectivores and frugivores were preoccupied by birds. That could explain the sharp difference in size between nocturnal and diurnal primates: monkeys average 10 times heaver than nocturnal prosimians (6, 10).

Even diurnal primates can share feeding sites by adopting different schedules. Smaller monkeys are chased out of fruiting trees by larger species, but can still get their share by coming to feed earlier and later than their competitors (examples in chap. 26 and blue monkey account).

The least specialized primates are the most adaptable, being omnivores that can exploit various different habitats. The king of the "tramp species" is the savanna baboon, which includes grass in its diet and ranges from the southern edge of the Sahara to the tip of South Africa. But within their own narrow spheres the specialists are better-adapted than such generalists. Although the savanna baboon seems perfectly at home in trees, apparently there is no niche room for this species among the array of rain-forest monkeys (17).

SOCIAL ORGANIZATION. The different levels and types of primate social organization are mostly exemplified in the species accounts. Though primate society ranges from comparatively simple to most complex (ourselves) and can be extremely variable within species, one starts from the fact that all monkeys and apes are sociable. The great majority live in female-bonded groups (23): that is, females never leave the group and range wherein they were born, whereas male offspring emigrate and have to join or take over another group before they can reproduce. The greater risks attendant upon dispersal and male sexual competition make an adult sex ratio skewed toward females a corollary of this arrangement (6). The great majority of monkeys and apes live in groups composed primarily of related females and their offspring, accompanied by 1 or more adult males.

How did higher primates all happen to become sociable? Did each primate line evolve social groups independently, or do they share a common heritage? Probably it all began with nocturnal prosimian ancestors, for prototypes of the basic primate grouping patterns can be found among their living representatives. The earliest identified higher primate fossils, which are similar in anatomy and presumably ecology to Eocene prosimians, show sexual dimorphism, an indication that primates have been polygynous from the beginning (10, 16). The major types of primate social organization are outlined here, proceeding from simple to increasingly complex, with examples of each type. Some of the variability within systems and species is brought out in the species accounts (26).

Nocturnal Primates

1. *Solitary, territorial.* The potto and other members of the lorisine subfamily are apparently the only primates that can be properly called solitary (see bush baby/potto introduction). Unrelated female pottos maintain separate territories, several of which may be included in the range of a mature male, a pattern very like that of solitary carnivores (cf. palm civet). Female offspring commonly settle nearby and a mother may donate part of her own range to successive offspring, resulting in a cluster of related females with overlapping home ranges (4).

2. *Monogamous pairs.* Two nocturnal lemurs and possibly one guenon (*Cercopithecus neglectus*) are examples of what is a rare arrangement among Old World primates (10). Obligatory pair-bonds distinguish animals in which male parenting or protection is essential for offspring survival (discussed in dog family introduction).

3. *Female kinship group* (matriline) with 1 reproductively dominant male. Adult bush babies are solitary while foraging at night, but females and young, joined sometimes by the male, associate in sleeping groups by day. Bush babies from different matrilines repel one another by scent-marking territorial overlap zones. The resident male may control several female ranges and tolerate smaller adult males as long as they remain sexually inactive (4).

Diurnal Primates. As a rule, all individuals except peripheral males travel, forage, and sleep together in groups. Separate male groups equivalent to ungulate bachelor herds are rare in this order (gelada and hamadryas baboons, the patas monkey, and some langur and black-and-white colobus populations are known exceptions). The change from solitary to group foraging may be linked with a shift to diurnal habits, as clearly happened in lemurs (9). Bush babies emerge from concealment together at nightfall before going their separate ways. If such animals needed to come out of hiding and become day-active, it would seem only natural for sleeping groups to stick together for mutual reassurance and increased safety—the danger of being singled out and captured decreases for each individual as group size increases.

Another plausible theory is that intraspecific food competition forces individuals to forage in groups in order to gain access to food that is distributed in small, defensible patches (23). Two cooperating individuals can displace 1, 3 can prevail over 2, and so on. The ability of females to bear and rear offspring being directly dependent on an adequate food supply, food competition is a more immediate concern for them than for males, whose reproductive success depends first and foremost on access to females. The tendency of female offspring to remain in their natal group/home range, and kinship selection favoring assistance to closely related individuals, would both promote social groups built around a core of related females. And the clustering of females in groups to which males can control access leads inevitably to a polygynous mating system.

Other things being equal, selection should favor large groups over small ones. This is apparently true when different troops range over the same area (see example no. 6, below), but does not apply to a territorial system. In any case, upper limits on groups size are set by the distance a troop must go to find enough food. The more mouths to feed, the more ground must be covered and the greater the food competition within the troop, until the point is reached where low-ranking members have to starve or emigrate. This is not to say that group size is infinitely variable and flexible. In fact, each species has norms for troop size and composition that are adapted, like its anatomy and physiology, to its own particular way of life, and the degree of variability in the system is also a species trait (5, 10, 20).

4. *One-male territorial troops.* This is the prevailing system among higher primates. The blue monkey and black-and-white colobus accounts represent omnivorous forest guenons and leaf-eating colobine monkeys, respectively. Forests are generally productive enough to support from 10 to 25 or 30 females and young within ranges small enough to be defended as exclusive territories, under the supervision of 1 resident adult male (10, 23). Succeeding generations of females can subsist in the same territory by keeping their numbers within its carrying capacity. Males compete directly for mastery of a female group and incidentally, it seems, become territorial to defend their mating rights. This arrangement is comparable to a lion taking over a pride of females and unlike the commoner form of territoriality represented by gregarious antelopes (see chap. 2). Nevertheless, the resident-male primate takes the lead in repelling invaders (at least other males) and part of his watchdog role includes de-

fending his troop against predators, using special displays (backed by formidable fangs) for the purpose. Even in age-graded troops that include subadult and young-adult males, the watchdog role is the prerogative of the 1 "resident male" (see black-and-white colobus account).

To become troop overlord, a male can either take over an existing troop by ousting the resident male (unless perchance he finds a troop whose leader has died), or he can found a new troop by joining emigrating females that leave their natal band. Fissioning occurs when a troop becomes too big for its territory and usually involves the emigration of younger females. They may be accompanied by male siblings or perhaps by a lurking solitary male that has been shadowing the troop.

Recent studies indicate that resident males can be cuckolded even while nominally in charge. As described in the blue monkey account, troops containing females in estrus may be subject to simultaneous invasions of up to 6 males, which manage to avoid the resident male and sneak copulations while he is off chasing one or another interloper (21). Among langurs, such invasions are the prelude to a takeover, often followed by infanticide of the former overlord's nursing offspring (8). Infanticide has now been observed in both the blue and red-tailed guenons and the red colobus (see colobine introduction) (3, 19).

5. *Multimale societies.* This grouping pattern, typified by the savanna baboon, generally goes together with terrestrial foraging in large troops within extensive, undefended home ranges, increased size, and pronounced sexual dimorphism: males very large and dangerously armed, estrous females of most species with conspicuous genital swellings (10, 20, 25). Relatively high levels of aggression and steep dominance hierarchies are normal (6). Social relations are extremely complex and variable, reflecting kinship bonds, male coalitions, and male-female alliances. Female offspring assume their mother's social rank, leading to a hierarchical arrangement of families, in which adult females of low rank have to kowtow even to juveniles of a highborn family (1). Relations between troops are generally competitive, particularly when a fruiting tree or other prize is at stake, and larger troops with more males usually win over smaller troops with fewer males (23). Males take the initiative in intertroop hostilities. They also confront and even attack predators when necessary to protect the troop (see baboon account).

A number of primates that form multimale troops do not fit the typical pattern.

a) The vervet, a guenon that left the forest, represents a possible transitional stage between 1-male and multimale organization. It is too small to venture far from safe refuges, only slightly dimorphic, and territorial. In more closed habitats with relatively abundant food supply and high population density, vervets tend to live in typical 1-male troops of standard size, but in acacia savanna they usually live in larger troops that include a number of adult males. Both sexes are actively territorial, and in fact female vervets cooperatively attack and sometimes kill male strangers, perhaps incited by a perceived threat of infanticide. In the more dimorphic species females seldom dare to attack males (10).

b) There are a few arboreal monkeys that live in large, multimale troops, notably the red colobus and the little talapoin (see family and subfamily introductions).

c) Chimpanzee social organization is unusual in at least 3 respects: (1) Chimps wander widely in small parties and alone, rather than staying together in their troop, in search of ripe fruit; (2) troop males are fiercely territorial; and (3) females rather than males emigrate. The lack of kinship bonds between adult females is reflected in lower levels of social grooming and other affiliative behavior, whereas strong social bonds exist between males (23). The red colobus is one of the few other mammalian examples of male-bonded societies (see wild dog account).

To derive a multimale troop from the 1-male troop of forest guenons, it is easy to imagine that a resident male might be forced to accept the presence of several persistent rivals near and finally in his troop, if the effort of driving them away was exhausting enough to make him vulnerable to overthrow. (The problem is illustrated in the colobus and blue monkey accounts.) The resident male's only viable option might then be to conserve his energy and settle for dominance. How hard it would be to exclude rivals would depend partly on troop size, openness of the habitat, and the presence of sexually receptive females.

Speaking of sexually receptive females, it has been suggested that the development of colorful and swollen genitalia during prolonged estrus, promiscuous mating, and sexual receptivity outside of estrus are female strategies for recruiting males, whose presence enables them to dominate competing troops, thereby ensuring their food supply and reproductive success (23).

6. *One-male harems.* In a true harem system, a male literally owns 2 or more females. The females are usually unrelated and may behave less aggressively or affinitively with one another than is usual in female-bonded societies. Troop movements are initiated and led by the male. The hamadryas baboon and the gorilla are 2 examples of this unusual system. The hamadryas and the plains zebra have a rather similar social organization. In both cases large troops consist of harems and groups of bachelor males in independent units (13). A hamadryas male acquires a harem by abducting juvenile females one at a time and keeping them until they grow up, or by attaching to a harem owned by an old male and gradually replacing him at stud while leaving leadership to the patriarch. Gorilla females are not forcibly kept in bondage as in the hamadryas. In both species, encounters between troops tend to be prolonged and intense, as males compete to lure females from each other's harems; most transfers occur during or shortly after such encounters (23).

7. *One-male, nonterritorial system.* The patas monkey is a guenon that has adapted to terrestrial life by developing a greyhound build and speed, ranging widely, and being nonterritorial. Unattached males associate in bachelor groups. Some observations suggest that the male who accompanies the troop may sometimes be merely a caretaker or figurehead, who is promptly overthrown by bachelor males when females come into estrus (18). Pronounced sexual dimorphism in this species is an indicator of intense male reproductive competition.

ACTIVITY. See Ecology.

POSTURES AND LOCOMOTION. Attempts to classify primate locomotion introduce artificial divisions in what is actually a continuum. Primates have the flexibility and limb mobility to move any way they choose. However, the ratio of leg to arm length gives a rough idea of the preferred form of locomotion. The following classification is adapted from reference 15 (table 1, p. 385), limited to the locomotion types that apply to African primates.

1. *Vertical clinging and leaping:* most prosimians, from the dwarf bush baby to the 1-m-tall Madagascan *Indri.* Clinging to vertical supports with body upright and froglike jumps propelled by elongated legs. Limb proportions and jumping technique vary considerably (illustrated in bush baby introduction).

2. *Four-footed walking and running* (quadrupedalism). Hands and feet grasp supports while climbing, providing stability.

- Slow-climbing type: the Lorisinae (potto). No leaping or running, no more than 1 hand or foot moved at a time.
- Branch-running and -walking: the guenons. Walk on palms (plantigrade), hands and feet usually employed to grasp supports. Includes climbing, jumping, and dog-like leaps.
- Ground-running and -walking: baboons, macaques, patas. Walk on fingers (digitigrade) on ground and often along branches.
- Old World semibrachiation: most colobine monkeys. During typically downward leaps between trees, arms held out to grasp foliage and break fall when landing on all fours. Otherwise walk and run along branches like guenons (see colobus account).

3. *Ape locomotion* (excluding brachiation). Terrestrial knuckle-walking: chimpanzee and gorilla. Occasional arm-swinging (brachiation) by chimpanzees and juvenile gorillas.

4. *Bipedalism:* man.

A seated posture with upright trunk is characteristic of primates while feeding, resting, and sleeping.

The Primate Hand. All primates have prehensile hands which they employ for climbing and for eating. But the ability to pick up and manipulate objects increases progressively from prosimians to the apes, improving with opposability of the thumb and independent mobility of the index finger. Not all primates have a completely opposable thumb, or even 5 fingers. In New World monkeys the thumb is aligned with the other fingers. Colobus monkeys are thumbless, whereas the second digit is greatly reduced in the loris subfamily.

Hand use in each major prosimian line is governed by a single motor pattern that is conserved through all variants of feeding and locomotion (2). Lorisoids (bush babies, pottos) reach for a branch or an insect by spreading the digits wide and then closing them after the object contacts the palm. Lemurs reach with digits nearly parallel and first touch objects with the fingertips. Higher primates have gone beyond this stage to develop a precision grip, as distinct from a power grip, in which the thumb and forefinger can be separately controlled to manipulate small food items or pick tiny particles during grooming sessions. New World monkeys can only use a sideways scissor grip, whereas Old World monkeys and apes have developed a tweezer grip between thumb and forefinger. But maximum dexterity is only realized in species with hands in which the tips of thumb and forefinger meet at right angles. This ability is seen in ground-living species, notably baboons and macaques, and in the apes, which are only a step below humans in dexterity despite the necessity of walking on their hands (knuckles), too (15).

SOCIAL BEHAVIOR

SENSES AND COMMUNICATION. The common vocal, visual, and olfactory signals employed in primate social communication are catalogued in chapter 26. Just a few general points about primate signaling systems will be mentioned here.

Olfactory communication is of primary importance for nocturnal prosimians, which retain highly developed olfactory and accessory olfactory (vomeronasal) systems, followed by vocal communication. Visual signals are comparatively undeveloped.

In diurnal primates, 3-dimensional color vision has replaced the olfactory system as the predominant sense organ and communication channel.

Stereotyped and Graded Signals. The forms and relative importance of visual and vocal communication depend primarily on the form of social organization, and secondly on where a species lives. For rainforest monkeys living in small groups whose members are typically dispersed and not in visual contact while foraging, vocal signaling may be at least as important as visual. Both visual and vocal signals designed for communication between groups are highly stereotyped and distinctly different from species to species. (See "How Monkeys Communicate," chap. 26.)

The most advanced primate societies have the least stereotyped and most variable visual and vocal signals. One signal grades imperceptibly into another, elements of different displays are combined in varying combinations, and so on, making it possible to communicate very subtle changes of mood and intention (14). The ability to change facial expression progresses up the social scale, from nil in the prosimians (apart from opening the mouth) to the extraordinary expressiveness of a chimpanzee's face. Similarly, the features of great apes, but not of monkeys, are as individualistic as our own, presumably because individuality is equally important in their societies.

Social Grooming. A universal primate trait, social grooming is for most species the main form of tactile communication; its primary significance is as a social-bonding mechanism. Prosimians groom each other by mouth, especially by licking, whereas monkeys and apes use their hands (15).

REPRODUCTION. It is hard to make generalizations about primate reproduction which apply across the different suborders and superfamilies. The following traits are, however, widespread.

Primates tend to mature late, live long, and reproduce slowly. Many have a definite breeding season, including lorisoids, guenons, and baboons, during which Old World monkeys cycle 3–4 times at 20–35 day intervals (10, 17). Monkeys and apes menstruate but blood is rarely visible, especially in guenons. The genitalia of most primates swell and flush slightly during estrus. New World monkeys also have conspicuous swellings (10), but conspicuous sexual skin is confined to Old World primates that live in multimale troops: baboons, some macaques, mangabeys, red, olive, and black colobus monkeys, the talapoin, and the chimpanzee (25, 27) (fig. 26.5).

SEXUAL BEHAVIOR. In various multimale societies, females copulate outside of estrus (e.g, during pregnancy, rarely during lactation), but in 1-male troops copulation is typically confined to estrus (except following a male takeover, when females may simulate estrous behavior [10]). Since females generally spend most of their adult lives pregnant and lactating, sexual activity is correspondingly infrequent.

Except for humans and pygmy chimpanzees, ventral/dorsal copulation is the rule. The usual position in monkeys is with the female standing planted on all fours while the male mounts her, feet grasping her ankles and hands holding her hips (fig. 27.13). The number of copulations and ejaculations varies widely, even between closely related species: some are single, others are multiple mounters/ejaculators (10).

PARENT/OFFSPRING BEHAVIOR. Gestation lasts much longer in primates than in most other mammals of the same size. It is 4 months even in bush babies and mouse lemurs no bigger than a mouse (which has a 3–4 week gestation) and 5–6 months in most monkeys, compared to c. 2 months in carnivores of matching size (10, 17). Pregnancy in the great apes is about as long as in humans, and the stages of parturition are the same in all primates, though less drawn out. A single offspring is the rule except in bush babies (2 or even 3) and South American tamarins (2 sets of twins a year). Although no newborn primate is precocious enough to follow its mother, baby monkeys and apes are more developed than human infants. They are protected from the elements by a fur coat, able to climb enough to reach the mother's teats and to cling to her while being transported. Judging from bush babies and some lemurs, the first primates were born nearly naked and immobile in nests (see bush baby chapter). The fact that humans have reverted to nearly the same condition may be explained by our enlarged brain: babies must be born in an earlier fetal stage in order for the head to pass through the mother's pelvis (10).

Bush baby young develop rapidly and become independent often within 6 months, enabling greater and lesser bush babies to reproduce twice a year. A few guenons may breed yearly, including the vervet (see species account), but most breed only every other year, and apes only every 4–5 years or longer (10, 22). The duration of infancy increases progressively from prosimians to man, correlating with brain size and how much time is needed for the primate to learn what it has to know to survive, become socialized, and reproduce. The social environment differs greatly in different primate societies, and even between troops of the same species (e.g., in availability of peers as playmates). In species with a rank hierarchy, the mother's social standing affects the treatment her offspring receive—from other troop members and the mother—and how long they remain dependent: anxious mothers curb their babies' exploratory attempts (1). Socialization in some species (e.g., vervet, black-and-white colobus) begins in the first few days with other females handling and carrying the baby; mothers of other species (blue monkey, baboon) will not hand over their infants until much later, if at all.

Male parental behavior is very limited in most primates, except those that live in obligate monogamy (many New World species and gibbons). However, males of many species behave protectively, particularly toward younger infants, and male baboons, for instance, carry and may even adopt orphaned offspring of their female "friends." In general, male protectiveness of the young is seen in species where the protector is likely to be the father (10).

ANTIPREDATOR BEHAVIOR. The strategies primates use against predators are varied and depend on size, habitat, and the type of predator. Almost no species is limited to only 1 of the following tactics and some may deploy all, depending on the circumstances.

1. *Hiding.* Nocturnal prosimians depend on immobility and cryptic coloration to escape predators by day, and on agile leaps by night. Many monkeys also hide themselves in the trees, especially lone males (see blue monkey).

2. *Flight.* Bush babies escape by rapid leaps and bounds. Concealed infants drop to the ground and stay immobile. Colobus monkeys and some others also escape aerial predators by plummeting into the undergrowth. Terrestrial primates jump out of trees and run overland to escape predators such as man, whereas arboreal ones race away through the trees.

3. *Scolding/mobbing.* Many monkeys and even bush babies will surround and scold ground predators from the safety of the trees, like so many—indeed often in company with—mobbing birds.

4. *Threat behavior.* Mobbing and scolding contain an element of aggression, which is particularly clear in the calls and visual displays of adult males. Special alarm calls, branch-shaking, and leaping displays are also given during aggressive encounters between troops (see monkey accounts).

5. *Attack.* In terrestrial species, notably baboons and the common chimpanzee, males actively defend the troop against predators even as formidable as a leopard, and have canines dangerous enough to make even lions think twice, especially when males make a united front.

References

1. Altmann 1980. 2. Bishop 1964. 3. Butynski 1982. 4. Charles-Dominique and Martin 1979. 5. Clutton-Brock and Harvey 1977. 6. Clutton-Brock, Harvey, Rudder 1977. 7. Fiedler 1972. 8. Hrdy 1977. 9. Jolly 1966. 10. ——— 1985. 11. Kavanagh 1984. 12. Kingdon 1980. 13. Kummer 1968. 14. Marler 1976. 15. Napier and Napier 1967. 16. Pilbeam 1984. 17. Rowell 1972. 18. Rowell and Hartwell 1978. 19. Struhsaker 1977. 20. Struhsaker and Leland 1979. 21. Tsingalia and Rowell 1984. 22. Washburn and Hamburg 1965. 23. Wrangham 1980. 24. Young 1962.

Added References

25. Dixson 1983. 26. Dunbar 1988. 27. Hrdy and Whitten 1986. 28. Martin 1990. Lee, Thornback, Bennett 1988. Zucker 1987.

Chapter 25

Bush Babies and Pottos

Family Lorisidae

African Representatives

Subfamily Galaginae, bush babies
Galago (subgenus *Galago*)
G. senegalensis, lesser bush baby
G. crassicaudatus, greater bush baby
G. alleni, Allen's bush baby
G. inustus, small needle-clawed bush baby
G. demidovii, Demidoff's dwarf bush baby
G. (Euoticus) elegantulus, needle-clawed bush baby

Subfamily Lorisinae, pottos
Perodicticus potto, potto
Arctocebus calabarensis, angwantibo or golden potto

The place to see prosimians is Madagascar, where there are some 28 different kinds, including 23 lemurs (family Lemuridae), many of them diurnal and conspicuous (6). Asia and Africa have only 11 species between them, and all of them are nocturnal, arboreal, and inconspicuous (11).

The Lorisidae includes 2 very different types of prosimians: the bush babies, an exclusively African group of highly active little creatures that spring through the trees like miniature kangaroos, and the pottos and lorises, somewhat larger creatures of radically different build which move hand-over-hand in slow-motion (fig. 25.1).

Of the 8 African Lorisidae, only the lesser and greater bush babies are widespread outside the Congo and Guinea rain-forest blocks; they may be heard—though rarely seen—not only in most parks but also on the outskirts of towns and cities. Very few people have ever seen the 4 bush babies and 2 pottos that live in the rain forest; yet a remarkable study made in Gabon has yielded a wealth of information about 5 of the 6 species (2). The remaining unstudied species, *G. inustus*, which occurs in forests of Rwanda and western Uganda, may be a connecting link between the lesser and needle-clawed bush babies (8).

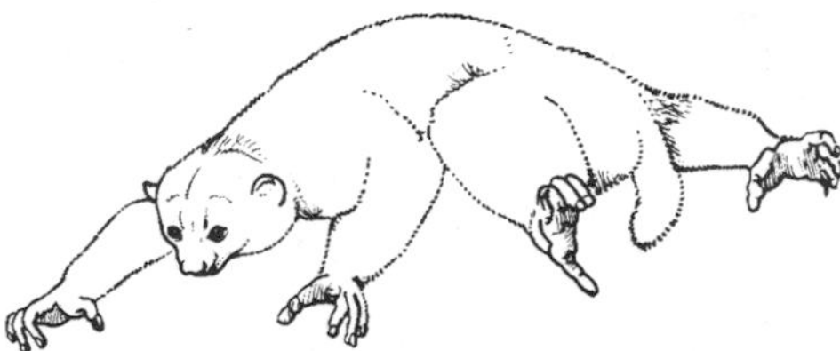

Fig. 25.1. Potto climbing technique; note the powerful hands with reduced second digit equipped with grooming claw (redrawn from Kingdon 1971).

BUSH BABY TRAITS. Small nocturnal primates with large (greater bush baby) to enormous (needle-clawed bush baby), frontally placed eyes; short, pointed snout; long, batlike ears (membranous, ribbed, foldable); slender body and long limbs, the hind pair greatly elongated; and tail longer than body and head. *Length and weight* (head and body): from the 12 cm, 60 g dwarf bush baby, the smallest primate, to the 37 cm, 2 kg greater bush baby (2); males slightly heavier than females but no obvious sexual dimorphism. Digits tipped with flat nails except for a long grooming claw on the second toe, the thumb and big toe opposed to other digits. Coat soft, thick, and somewhat woolly, tail more (greater bush baby) or less (dwarf bush baby) bushy, face hairy. *Coloration:* cryptic, gray, rufous, or brown with lighter underparts and subdued markings, especially on head (dark eye patches, white nose bridge), with white-tipped tail in needle-clawed and dwarf bush babies; eyes reddish with black pupils closing to a vertical slit in daylight, immobile in sockets. *Teeth:* 34–36, lower incisors and canines modified as a scraper/grooming comb, from which a special "second tongue" equipped with denticles that fit between the comb's grooves removes the hairs. Penis with bone, clitoris as large as an unerect penis in some species (e.g., greater bush baby). *Scent glands:* scrotal/labial gland in all species, salivary and lip glands, and chest gland (greater bush baby). *Mammae:* 2–3 pairs (pectoral) (3, 12). *Longevity:* up to 14 years in captivity (15).

LORISINAE TRAITS. Head rounded, eyes frontally placed but not particularly large, ears inconspicuous, tail reduced (potto) or absent (angwantibo), limbs subequal, hands and feet modified as pincers, with enlarged thumb and big toe opposed to other digits, index finger reduced to stub.

Despite their pronounced differences, members of the 2 subfamilies share a number of similarities, such as the specialized tooth scraper/comb and "second tongue," flattened fingernails and grooming claw, and similar scrotal/labial glands (2). They also share many behavioral traits.

ANCESTRY. Bush babies are considered the closest living representatives of the earliest primates. The dentition of the dwarf galago is hardly distinguishable from fossil forms dating back to the Paleocene (11). The vertical clinging and leaping form of locomotion has been retained and remarkably developed in some of the larger lemurs, such as the *Indri* and *Propethicus.* The loris/potto form of creeping locomotion appears to be later and more specialized, mainly as an antipredator tactic; in any case the 2 lorisid subfamilies have been separate at least since the Miocene (11).

ECOLOGY. Three bush babies and 2 pottos are able to coexist within the same rain forest on the same nocturnal schedule with minimal competition by staying at different levels, using different supports, moving in different ways, and specializing on somewhat different kinds of food. Spatial separation is illustrated schematically in figure 25.2. Knowledge of the ecological differences between these species has led to a new understanding of the adaptive significance of their morphological and behavioral traits (2, 5).

As in other forest mammals, the amount of insect food in the diet of prosimians correlates with body size (2). The smallest species can subsist almost entirely on insects. The larger ones cannot catch enough insects to meet their daily needs, but have to rely on fruits and/or gum as a staple diet, supplemented by insects and occasional small vertebrates to furnish the protein needed to compensate for nitrogen loss (5).

Allen's bush baby (fig. 25.7) (250 g) and the potto are primarily frugivorous (73% fruit, 25% insects, and 65% fruit, 21% gums, 10% insects, respectively), but their spatial separation precludes any competition: the former eats fruits that have fallen to the ground and those produced in the undergrowth layer, whereas the potto only eats fruit growing in the trees. Swallowing fruits up to cherry size whole, this bush baby eats a lot (up to 8% of its body weight) rapidly, thereby minimizing the time it spends foraging on the forest floor. Its

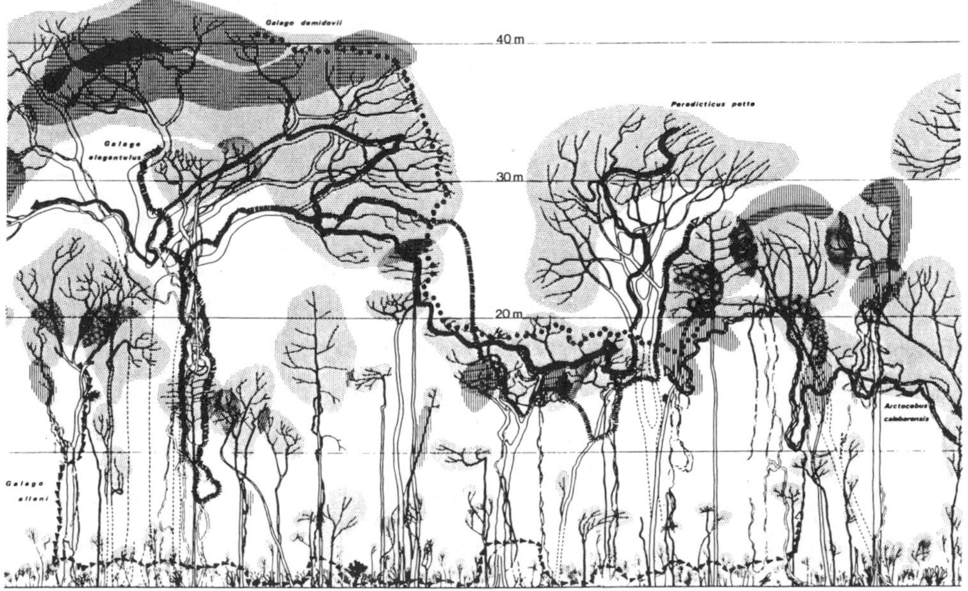

Fig. 25.2. The pathways taken by five species of prosimians in Gabon primary rain forest (passing from left to right). Dotted areas represent tree foliage and hatched areas represent lianas. (From Charles-Dominique 1974.)

main insect prey, beetles and moths, are similar to those selected, in the forest canopy, by Demidoff's dwarf bush baby (fig. 25.6). This tiny creature climbs among small-diameter branches and lianas, is mainly insectivorous (70% insects, 19% fruits, 10% gums), and is the most energetic and acrobatic of all the galagos (5).

Largest of the forest bush babies at 300 g, the needle-clawed bush baby (fig. 25.5) is a "gummivore" (cf. lesser bush baby). Fresh gums from 3 tree and 2 liana species, all in the mimosa family, made up 75% (by weight) of its stomach contents, followed by insects (20%) and fruits (5%). Its keeled and pointed nails enable it to gain purchase on trunks and limbs too smooth to be gripped and too large to be embraced, thereby giving this bush baby access to sources of gum beyond reach of competing species. It moves up and down as well as between trees, keeping to the main stems and branches. Individuals follow a set itinerary night after night, each stopping c. 100 times an hour to harvest droplets of exuding sap with its long tongue or robust tooth scraper, depending on the exudate's consistency. Although gummivory is considered a primitive prosimian habit, *G. elegantulus* has become a highly efficient specialist. Nevertheless, the availability of gum depends largely upon wounds to trees made by wood-boring insects (2).

The potto ranges the forest canopy, traveling at a snail's pace between fruit trees, preferably along large branches. To reduce exposure to predation while eating in often sparsely leafed fruiting trees, it too has developed the ability to eat a lot quickly: it can put away the equivalent of 2 bananas a minute in an expandable stomach able to hold up to 8% of its body weight. Accordingly, pottos are rarely seen in fruiting trees but often rest and digest in an adjacent, leafier tree. Whereas galagos eat only soft, sweet fruits, the potto's jaws are big and strong enough to cope with larger, tougher kinds. It also eats gum, but moves too slowly to go around collecting it in small droplets, so it takes large lumps that are too stale and hard to interest bush babies. Nor can it capture most of the more active and palatable insects that galagos eat, but takes slow-moving and unpalatable kinds, primarily ants supplemented by slugs, caterpillars, spiders, centipedes, and large beetles.

The angwantibo or golden potto is radically different both in form and ecological niche, a slender, much smaller creature (200 g compared to the 1100 g potto) which creeps about in the lowest, densest vegetation and is insectivorous: 85% animal food and 14% fruit. Caterpillars, including kinds protected by venomous hair, ants, beetles, and crickets, especially of kinds unpalatable to other animals, account for the bulk of its protein.

Population densities of rain-forest prosimians range from 8 to 80 animals/km^2, relatively low compared to densities of lesser and greater bush babies (see species accounts) (2). However, the combined biomass of the 5 forest prosimians is comparable.

SOCIAL ORGANIZATION: pottos unsocial; bush babies associate in kinship groups (sleeping groups by day), forage alone by night; both sexes defend core areas against own sex.

The following generalized sketch of lorisid social organization is based mainly on Charles-Dominique's studies in the Gabon rain forest (2). The trouble with pointing out what bush babies have in common is that the differences, which take much longer to discover and too much space to describe, are neglected. The lesser and greater bush baby accounts give some indication of how these savanna species differ from rain-forest galagos and from each other. However, important intraspecific differences also exist within different populations of species that live under different ecological conditions. As A. Jolly (7) points out, the social structure of almost any species is likely to be tighter and more territorial in rich stable habitats than under drier, more seasonal conditions. The social tolerance and lax territoriality shown by South African greater bush babies, compared to rainforest species, may be viewed from this perspective; but whether equatorial populations of greater bush babies are more territorial is a question requiring further study.

Despite the surprisingly complex social organization of bush babies, all lorisines forage individually by night instead of in coordinated groups like diurnal lemurs and monkeys. By day, bush babies that share the same home range huddle together in nests or other regularly used sleeping sites. Though pottos sleep alone, individuals that live in the same area have frequent social communication and regulate their spacing by scent.

Bush baby sleeping groups are matriarchal, consisting of a mother with offspring of assorted ages. Adult males generally sleep alone, but occasionally (most species) or regularly (needle-clawed bush baby) 1 or even 2 will sleep with a group.

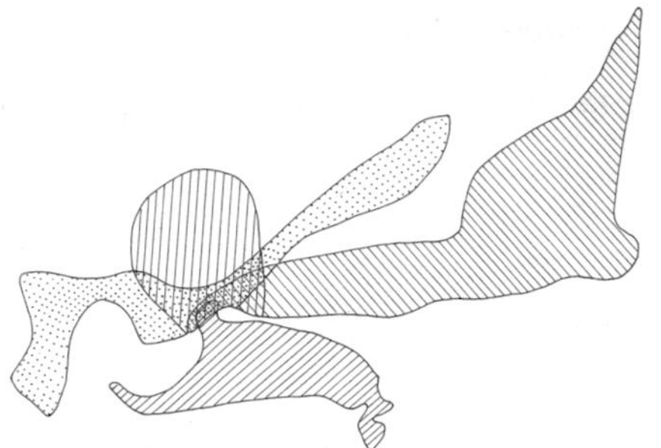

Fig. 25.3. The territories of 4 neighboring male Demidoff's bush babies, in which there is a central overlap zone (from Charles-Dominique and Bearder 1979).

The ranges of related females typically overlap partly or completely, and they associate together with their young in sleeping groups numbering up to 10 (dwarf bush baby). Females from different ranges, in contrast, treat one another with the hostility typical of territorial neighbors. Adult female pottos do not associate, but mothers have been known to donate part or all of their home range to daughters (2). In all species females tend to be sedentary, whereas male offspring leave their mothers' ranges at puberty.

Territorial Males. The ranges of mature adult males overlap only a little along common borders (fig. 25.3). Neighbors regularly come there to check on one another and assert their territorial claims by scent-marking and calling. The biggest, fittest "central" or "resident" males hold territories that extend through the ranges of several females, with whom they maintain social bonds regardless of female reproductive condition. Male Allen's bush babies, unusually polygynous, defend mating rights with as many as 8 or more females, each with a range of 8–16 ha, and manage to visit up to 5 a night. Their territories of 30–50 ha are huge, especially compared to the 0.5–2.7 ha territories of male dwarf bush babies, which transect the 0.6–1.4 ha ranges of 1 to several females (fig. 25.4).

Despite their slow cruising rate, central-male pottos also control sizeable areas: 9–40 ha, compared to female ranges of 6–9 ha. A marked male visited each of his 2 females every 2–4 nights for ½–3 hours at a time, always within 2½ hours of becoming active. He did not make actual contact, but came close enough to exchange olfactory signals (urine and glandular secretions) (2). Meetings between bush babies, in contrast, entail at least facial contact and often *social grooming*.

Central males may tolerate other adult males within their ranges: young, much smaller individuals whose ranges are very small and which have no contact with the females (cf. palm civet). Outside the territories of the central males are peripheral males of intermediate size. They occupy sizeable ranges adjacent to but not including female ranges and such males may have cordial relations when they meet. These are mostly individuals waiting for vacancies to open up in the central network—where the male turnover rate is high because of greater activity, competition, and exposure to predators. Finally, there are nomadic or "vagabond" males: mostly adolescents in the process of dispersing from their natal range (2).

Sleeping Sites. Bush baby sleeping sites may be tangles or clumps of foliage, the fork of a tree, a hole, or a nest of green leaves constructed usually by the senior female of a group (see species accounts). Preference for one or another kind of site varies among and also within species. Allen's bush babies use tree-hollow nests, whereas needle-clawed galagos always sleep in contact groups in clumps of foliage. Groups of dwarf bush babies either huddle together in the foliage or sleep in leaf nests. Pottos sleep in dense foliage and do not build nests (2).

SENSES. The prosimian eye is adapted to making precise leaps at night in forests where only one-hundredth of the ambient light filters down through the canopy (2). The frontal location of the eyes, their great size, and a brilliant reflecting layer (the *tapetum lucidum*) enable prosimians to see well and judge distances accurately under most conditions. Although bush babies' (but not pottos') eyes are immobile, their roundness enables a bush baby to take in an arc of some 250° (more than 50° greater than man can), and the head can be rotated 180°, as in owls (9).

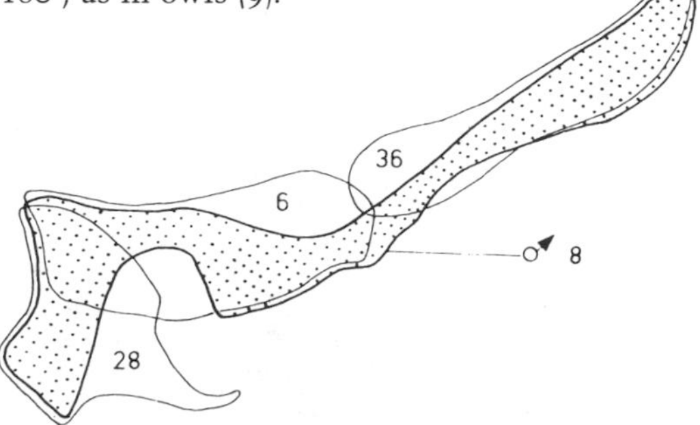

Fig. 25.4. Three female Demidoff's bush babies' home ranges overlapped by the territory of a male (one of those shown in fig. 25.3). (From Charles-Dominique 1977.)

Apart from its role in locomotion, capture of moving prey, and detection of predators, vision is less important than the senses of smell and hearing in finding food and in social communication. However, all 3 senses are well-developed in most nocturnal prosimians, which have not developed one system at the expense of the others (12). Only the angwantibo, unlike the related potto, has weak visual powers and relies mainly on smell to locate its prey (2).

ACTIVITY. Ambient light intensity is the primary factor regulating the onset and cessation of activity in nocturnal prosimians (2). The pattern is very similar in all the observed African species. About ½ hour before dark they begin to stir, stretching and yawning, followed by protracted bouts of self-grooming. Bush babies leave their sleeping places 5–10 minutes before dark, whereas both pottos wait for full darkness. Allen's bush baby, the only 1 of the 5 associated rain-forest lorisids to sleep in hollow trees, checks the time by coming to the entrance and emerges when the light reaching the entrance falls to 50–20 lux (2). All 5 species rest, usually when full, and are most active early and late at night. Light rain is ignored, whereas all but the potto seek shelter in heavy rain (but see lesser bush baby account, Activity). Activity is arrested at the first hint of dawn, which nocturnal prosimians, with eyes fully accommodated to darkness, detect well before man or a light meter. All the African prosimians remain inactive during daylight hours, although capable of seeing and moving perfectly well if disturbed in their sleeping places. Very occasionally one may be seen peering down from its hiding place in full daylight: for instance, a lesser bush baby in its den tree (author's observation).

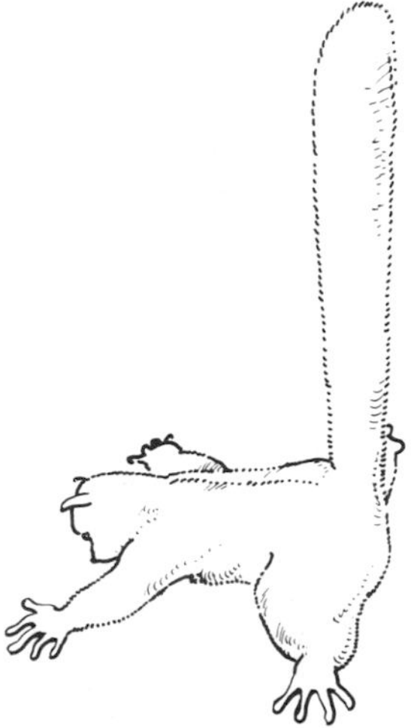

Fig. 25.5. Needle-clawed galago "parachuting"; this bush baby free-falls up to 8 m in moving between trees (redrawn from Charles-Dominique 1977).

POSTURES AND LOCOMOTION. Adaptations for leaping and running tend to be incompatible, since elongating the legs for leaping reduces rapid limb retraction during running. Galagos have partially solved the problem through elongation of the foot (tarsus), which permits rapid forward leg movement while running (2). However, running and jumping technique and proficiency differ from species to species, reflecting differences in size, relative development, and proportions of the limbs that correlate with their habitat preferences. The tail serves a significant balancing function but its use also varies with jumping technique.

The greater bush baby is more of a runner than a jumper and comparatively lethargic (see species account). The needle-clawed bush baby is the most accomplished jumper; a leap measured in the field covered a horizontal distance of 5½ m and a drop of over 3 m (2). Sometimes it drops as much as 8 m when crossing between trees, spread-eagled limbs and bushy tail slowing its descent (fig. 25.5). But normally it jumps in an upward arc with its body vertical. To gain purchase on surfaces too broad to be grasped with the hands, it has unusually large hands and broad terminal pads on fingers and toes. The pointed nails assist in climbing, but are not used for takeoffs or landings.

Demidoff's bush baby, with the most-developed tarsus for its size, is adapted for running through interlaced, mostly fine branches of the canopy that go in all directions. It runs along horizontal supports (fig. 25.6) and leaps gaps in virtually a straight line, using its forelimbs to absorb the shock of landing whether on vertical or horizontal branches. The tail serves to change body attitude from horizontal in flight to vertical in landing.

Allen's bush baby has the most specialized form of locomotion. Most of the supports in the undergrowth are thin uprights and it leaps from one to the next with incredible agility and speed, clearing distances of 2½ m without losing height. It can make 5–6 jumps and travel a dozen meters within 5 seconds, pausing only 1/25 second between landing, pivoting around the trunk, and taking off again (fig. 25.7) (2). On the

Fig. 25.6. Demidoff's dwarf bush baby branch-running (redrawn from Charles-Dominique 1977).

ground it both runs and hops, unlike the lesser bush baby, which prefers to hop whether on the ground or traveling along horizontal branches (14) (see species account).

Members of the subfamily Lorisinae move in a totally different manner, aptly termed "cryptic locomotion" (2). They are exclusively climbers and never leap between supports. The tail is greatly reduced, the limbs are considerably lengthened, and all the joints move with great freedom, permitting all sorts of contortions. The hands and feet are long and broad and close around branches with a powerful pincer action; a potto, moving hand over hand (usually above but also below a branch) always holds on with at least 2 of its extremities, on opposite sides of the body, and proceeds so slowly, silently, and steadily that predators' eyes and ears, adapted for detecting fleeting movements and sounds, receive no stimulus (fig. 25.1). Stealthy movement may also enable these animals to get within grabbing distance of prey. The slightest disturbance makes a potto freeze and remain immobile while looking toward the source of danger. Yet when forced to use exposed passageways, as when traversing a vine between 2 trees, it moves with surprising alacrity, only slowing to its normal pace when again under cover (2).

The slender angwantibo, no heavier than most bush babies, moves like a potto and has anatomically similar hands and feet, but so much smaller that it is only capable of gripping relatively thin supports. It is thus well-adapted to conditions near the forest floor. However, in primary forest the undergrowth is often so sparse that the only way to get across gaps is on the ground. As angwantibos do this only as a last resort, they are largely confined to openings created by fallen trees and such where sunlight supports dense undergrowth and a network of lianas.

The Bush Baby Hand. The long slender fingers of bush babies with widened, flattened tips, and the way bush babies capture insects, suggest that the morphology and neuromuscular control of the prosimian hand evolved as an adaptation for catching the quicker kinds of flying, jumping, and running insects (1, 2). Bush babies seize such prey with a rapid, stereotyped movement of one or both hands, in which all the fingers converge on the palm at once. Plucking insects out of the air in darkness requires wonderful eye-hand coordination and quickness; the highly mobile, antenna-like ears are also adapted to locating insects by sound. Yet prosimians lack the neuromus-

Fig. 25.7. Allen's bush baby clinging and leaping sequence (from Charles-Dominique 1977).

cular control to move their fingers separately. They cannot, therefore, groom themselves or each other in the simian manner, but employ the grooming claw (second toe) and incisor scraper instead; with these a prosimian can reach everywhere.

SOCIAL BEHAVIOR

COMMUNICATION. The primary channel of communication is by scent-marks; however, vocal signals are also important in social and territorial relationships (2).

Vocal Communication. The vocal repertoire of the loris/potto subfamily is limited to weak or moderately developed calls used during infrequent encounters at close range; for solitary animals whose antipredator strategy depends on concealment, the emission of frequent loud calls would be self-defeating. Bush babies have evolved a large variety of vocal signals, but compared to the graded vocalizations of gregarious higher primates, their calls are generally more stereotyped and discrete.

The calls of African lorisids have been analyzed and compared according to the situations with which they are associated, and may be conveniently divided (although with some overlap) between distance (mainly loud) and short-range (mainly soft) vocalizations (2, 10).

Distance Calls:

Territorial. The loud cry of the greater bush baby, which sounds something like a crying human infant, gave this group its common name. The same call functions for communication within as well as between groups (see species accounts). Each of the other species has its own distinctive, though less strident and powerful, cry: *G. alleni* croaks, *elegantulus* emits a birdlike "*quee*" call, *demidovii* makes a "plaintive squeak," and *senegalensis* barks. The call of Allen's bush baby, in common with the calls of most birds and mammals that live in dense undergrowth, uses frequencies between 1000 and 1500 Hz, which carry up to 250 m; relatives that live in more open habitats utilize a much broader frequency range. Apparently all bush babies utter these calls during territorial encounters between rival males or females, but they are not limited to territorial advertising. The croaking call of Allen's bush baby elicits a ± automatic response from most other conspecifics. Males use the call to announce their presence and locate their wives when making the rounds of their large territories, and females use a faint croak to communicate with their young.

Alarm. Bush babies have comparatively elaborate graded calls to express varying degrees of anxiety caused by any kind of disturbance. Alarm calls are also uttered during aggressive interactions and spontaneously after waking and during other activities. The "*tee-ya*" of the needle-clawed and the "*kiou-kiou-kiou*" of Allen's bush baby are more stereotyped and unvarying, but other bush babies have intergrading calls which build up from weak to very powerful and sustained (see species accounts). The calls of *alleni* and the "chip calls" of dwarf bush babies are individually different enough to indicate the identity as well as the presence and location of the caller. Bush baby alarm calls, unlike those of primates that stay together in groups, are not infectious.

Short-Range Calls (social contact and contact-seeking):

Infant. Infants of all species emit similar click and "*tsic*" calls, singly or in series, in response to any excitation. Mothers respond maternally. The clicks grade through intermediate sounds (see lesser bush baby account) into a squeak (or "*tsic*") (2) at highest intensity.

Maternal. Used by mothers to evoke an answer which helps to localize a "parked" infant, these calls vary between species, from the soft croak of Allen's bush baby to the faint "rolling call" of the lesser bush baby. Pottos and needle-clawed galagos use *tsic* calls similar to but more powerful than the infant call.

Gathering. More or less powerful calls, typically derived from infant and maternal calls, which bring the members of bush baby sleeping groups together preparatory to going to their resting place. These sounds, which may also be voiced on awakening and sometimes at night, are only infectious when uttered near dawn. Distinct individual differences probably serve to identify each caller. Specific gathering calls appear to be absent in Allen's, lesser, and greater bush babies.

Courtship. The male potto utters a *tsic* call very like that of mothers and babies when following an estrous female, and male greater bush babies emit a "low, grating call" (3).

Short-Range Calls (distance increasing):

Aggressive. Staccato calls graded in intensity and pitch, which normally serve to arrest approach of a conspecific. "*Ki-ki-ki*" (*elegantulus*), "*quee-quee-quee*" (*alleni*), "sob, spit" (*senegalensis*), "hack" and "spit" (*crassicaudatus*), and "*hee*" (potto) calls have been described.

Fright. A 2-phase grunt (uttered while

breathing out and in), which comes out as a hoarse growl in the 3 forest bush babies and a groan in the potto, is added to the aggressive call when an animal is badly frightened. The "scream" of the greater and the "fighting chatter" of the lesser bush baby are equivalent. This call is associated with the defensive-threat posture, and sometimes with the voiding of urine and feces.

Distress ("weet"). An expression of extreme fear and/or pain, fairly constant among species. The call has been seen to arrest fighting in dwarf bush babies. It brings other bush babies—and predators—to the scene.

Olfactory Communication. The main form of scent-marking in this family utilizes urine, in ways adapted to arboreal habits. Scent-gland secretions are also employed for marking. These substances are deposited in strategic locations (especially borders) where other individuals will be likeliest to encounter them. In addition, social partners may be marked with urine and glandular secretions (2, 12).

Urine-washing. This involves transferring small quantities of urine to the hands and feet and then onto the substrate. It is the most stereotyped of all forms of urine-marking and is characteristic of lemurs as well as bush babies. The description of *urine-washing* in the lesser bush baby account is typical.

Direct urine deposition. G. elegantulus does not urine-wash, but deposits urine directly onto a support, first lowering its rear slightly with tail raised until the penis or clitoris touches the surface. Captive animals direct fine urine jets at water bowls and even people from as far as 1 m. The other bush babies also deposit urine directly onto supports, often in bouts termed *rhythmic urination* (see lesser bush baby account). The potto urinates in similar fashion, lowering its rear with tail raised until sensory hairs projecting beyond the penis or clitoris brush the surface, then depositing a urine trail c. 1 m long, preferably on a broad surface. The angwantibo, however, voids large quantities of urine, most of which simply falls to the ground. Urine is also voided in a normal stream by the other species.

Marking with scent glands. In all members of the lorisid and lemur families, a bare patch of skin on the male's scrotum, corrugated like a waffle iron, bears enlarged sebaceous and apocrine glands (12). Similar glands are found surrounding the female's labia. These structures are developed to a varying degree and are of varying importance (unstudied in most cases) in different species. They are best developed in the potto. While grooming an estrous female, the male scratches his scrotal gland with one hand and then grasps the female, thereby marking her. The angwantibo male marks a female by direct contact with the gland while passing over her during social grooming (2). Bush babies brush the gland against supports, perhaps in conjunction with rhythmic urination.

Bush babies also rub their muzzles, lips, and cheeks on supports, during which Allen's and lesser bush babies leave a saliva trail (12). The skin in these areas is known to contain large apocrine glands. Male and female greater bush babies rely on the secretion of a chest gland to regulate traffic and social contact within their ranges (see species account).

Lastly, the potto and its Asian counterpart, the slow loris, emit a strong, currylike body odor that is detectable from 20–30 m. Whether the smell serves for individual recognition and/or to repel predators is unknown.

Tactile Communication. *Nose-to-nose sniffing* followed by *nose-to-face contact* is characteristic of meetings between individuals of all species, and the prelude to other social interactions. The anogenital region is the second most frequent target of sniffing and contact. *Social grooming* is the primary means of maintaining bonds between mother and offspring, among members of bush baby sleeping groups, and within mated pairs (more under Sexual Behavior). Grooming individuals lick each other in hard-to-reach places (e.g., head, ears) and only use the tooth scraper if the partner's fur is matted. Grooming can be solicited by presenting the neck and armpit in an attitude that is typical of all lorisids and many higher primates. Bush babies in sleeping groups often maintain contact; for example, needle-clawed bush babies sleep in a tight ball (2).

Visual Communication. Postural signals and facial expressions are comparatively undeveloped in nocturnal prosimians. The facial mime is limited to opening the mouth and furrowing the brow. Ear and tail positions are most expressive in bush babies, but mostly linked with particular vocalizations (2).

AGONISTIC BEHAVIOR

Dominance/Threat Displays. An extended posture of the body and tail, especially standing erect, widely spread ears and opened mouth denote aggressiveness in the whole family. At higher intensity, threat calls are emitted and finally one hand is raised ready to strike. This is usually sufficient to cause an opponent to retreat.

Defensive/Submissive Displays. If prevented from retreating, a threatened animal crouches with mouth open and ears widespread, often with hand held before its body defensively (fig. 25.14), meanwhile uttering the 2-phase growling call. The potto and angwantibo tuck the head between the forelegs and under the armpit, respectively (fig. 25.9).

Submission is shown by folding the ears, lowering the head, and passing beneath the aerial runway while withdrawing, and by flight. Juvenile dwarf bush babies have a special posture for appeasing aggression in adults, which features curling up the tail like a corkscrew.

Fighting: eyes partially closed, ears fully retracted, cuffing, wrestling, hands always used to grasp fur prior to biting.

Fighting is virtually never witnessed. Yet Charles-Dominique (2) found evidence that males occasionally engage in serious fights, beginning after puberty. In a sample of trapped individuals, over 20% of adults of all Gabon species but the dwarf bush baby showed signs of healed wounds, with an incidence in males 2–4 times commoner than in females (but not statistically significant, according to ref. 3). None of 147 prepubertal animals showed signs of wounds. Pottos and greater bush babies have been known to fall to the ground while locked in combat (4).

REPRODUCTION. The bush babies and pottos of the rain forest are reproductively distinct from the 2 savanna bush babies. The latter are strictly seasonal breeders and regularly bear twins or triplets—twice a year in the case of the lesser bush baby. The rule among the forest species is one young, once a year. Apart from the potto, whose young are born from August to January, the others breed the whole year, with a peak during the rainiest months (2, 13). Such low fecundity is characteristic of many inhabitants of rain forest and other stable, climax communities (2). Gestation ranges from 112 days in the dwarf bush baby to 193 days in the potto.

SEXUAL BEHAVIOR. Sexual relations in prosimians are limited to the time of estrus (in contrast to higher primates), indeed precluded by a membrane covering the vagina (absent in greater bush baby). Also, the testes only descend into the scrotum in the mating season, in sexually active males. Yet males stay in touch with their females between matings. A bond is established in the first place through courtship, which males initiate with any unattached female they encounter regardless of age and reproductive status. The pattern is much the same in all species. The male tries to approach and *greet* the female, which at first refuses any contact and runs away, with the male in pursuit. Eventually she tolerates *nose-to-nose sniffing,* then *nose-to-face contact* and *facial-licking,* and finally *mutual grooming.* The male marks frequently with urine and glandular secretions and rubs his head and forequarters on the female's urine marks. A period of intensive *social grooming* ensues, lasting several days and for up to several hours in a bout, during which the couple typically hang suspended below a branch by their feet, face-to-face, and lick each other all over. The bond now established, grooming gradually subsides to a maintenance level of ½–15 minutes during male visits. Potto couples apparently have no physical contact except to mate, the male's nose telling him all he needs to know during regular visits to the female's range.

Copulation. The precopulatory stage begins when the female's vagina opens in proestrus, releasing attractant secretions (see species accounts for details). Females remain receptive for only 1–2 days. Intromission lasts for a few minutes (incomplete copulation) up to several hours with multiple ejaculations, during which spines on the shaft lock the penis to the vaginal wall. The standard mounting position is as in higher primates, with the male grasping the female's heels (fig. 25.8). Dwarf bush baby and potto females hang suspended below a

Fig. 25.8. Galago courtship and copulation (lesser bush baby). (Redrawn from Doyle et al. 1967.)

branch by hands and feet, whereas lesser, greater, and needle-clawed bush babies remain right side up (2).

PARENT/OFFSPRING BEHAVIOR. Compared to higher primates, the young are precocious, especially lorisines, which are born while the mother clings to a branch and, without maternal manipulation, have to climb over her and cling to her belly (2). Both galagos and lorisines are born with eyes open. The latter are better protected against the elements by thicker fur. A potto mother continued foraging soon after giving birth and the baby clung to her without trying to nurse; infant pottos suck only by day while the mother sleeps.

Bush babies are generally born in leaf nests built by the mother, although Allen's and sometimes lesser bush babies use tree hollows. The mother licks and manipulates her newborn infant and devours the fetal membranes. Female dwarf bush babies withdraw from their group and remain isolated for 1–2 weeks after parturition. If disturbed in the nest, a mother abandons it at once, carrying her baby by the flank skin (fig. 25.10). But after 3–7 days the nest is used only for sleeping. The mother emerges at dusk holding the baby and parks it on a convenient branch where it remains motionless while she forages nearby. Before moving to another tree she retrieves it, and parks it again, and so on until approaching dawn sends her back to the nest. The immobile baby is almost impossible to detect except by eyeshine under artificial light.

The above account applies with minor variations to nearly all bush babies (cf. greater bush baby). Newborn babies are able to grip with their hands only, but can crawl around the nest and burrow under the mother to reach the teats. At 2–3 weeks a baby begins to move more actively and independently; at the same time it acquires a thicker coat. A month-old dwarf bush baby eats alongside its mother and snatches food fragments from her mouth; by 45 days it is weaned. During night siestas, mothers and infants engage in prolonged social grooming and the mother tolerates her baby's playful assaults. More details on development are given in the species accounts.

Adult weight is reached in 5–6 months by dwarf, 6–8 months by Allen's, and 8–10 months by needle-clawed bush babies. Angwantibos and pottos reach adult weight at 8–9 months and 8–14 months, respectively. Pottos stop sleeping with their mothers at 6–8 months, by which time males but not females leave the maternal range (2).

ANTIPREDATOR BEHAVIOR. Forest lorisids have little to fear from predators, except perhaps snakes, as long as they stay off the ground (2). The only truly arboreal African carnivores are the palm civet (primarily a frugivore) and genets. Eagles and hawks, being diurnal, are not a problem. Allen's bush babies and angwantibos are vulnerable to the leopard and golden cat when they forage on the ground. Bush babies can escape by jumping away from almost any predator unless taken by surprise. They react to the presence of a dangerous or strange animal by giving intense alarm calls, and often follow and harass it. The calls of one species alarm other species as well. Parked infants respond to movement of their perch by dropping to the ground, take several hops after landing, then lie immobile again. Often they give a distress call while falling, which brings not only the mother but other bush babies to the scene. A predator or person is scolded until it moves away far enough for the mother to retrieve her infant.

The potto and angwantibo rely on crypsis, including slow locomotion, to escape notice. If confronted by a predator, a potto clamps its hands and feet securely to its support and maintains a facing position while assuming the defensive attitude (fig. 25.9). With head below its shoulders, it presents a neck shielded by thick skin and a row of 4–6 low tubercles. These knobs cap specially elongated spines of the neck vertebrae which project nearly to the surface. A potto that was being molested by a palm civet chattered and made sudden, short rushes during which it struck the branch violently and noisily with its bared teeth. If grabbed, the potto surges forward in a surprise move that may dislodge its attacker. An immature palm civet was thus knocked out of a tree (2). The angwantibo has an even more peculiar defensive posture, adapted for

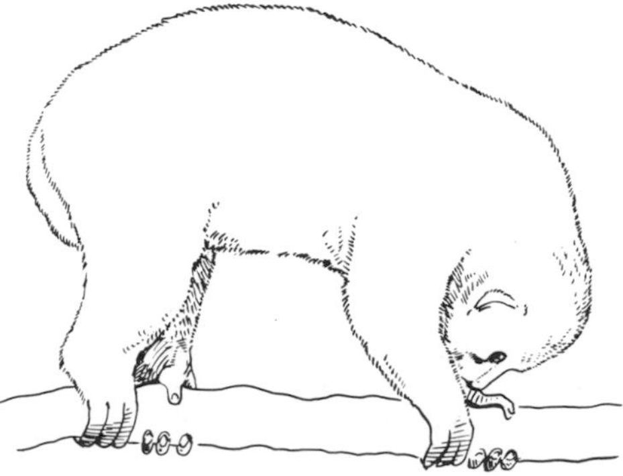

Fig. 25.9. Potto defensive posture.

use on the ground when attacked from above. It rolls up so that its head is completely protected but with opened mouth presented beneath its armpit. If grabbed it bites the attacker on its snout and holds on; the violent efforts of the predator to dislodge it may fling the angwantibo to a distance, giving it a chance to escape (2).

References

1. Bishop 1964. 2. Charles-Dominique 1977. 3. Clark 1985. 4. Epps 1974. 5. Hladik 1979. 6. Jolly 1985. 7. ——— in lit. 1986. 8. Kingdon 1971. 9. Pariente 1979. 10. Petter and Charles-Dominique 1979. 11. Petter and Petter-Rousseaux 1979. 12. Schilling 1979. 13. Van Horn and Eaton 1979. 14. Walker 1979.

Added Reference

15. Charles-Dominique 1984.

Fig. 25.10. Female lesser bush baby carrying baby and jumping (from Doyle et al. 1969).

Lesser Bush Baby

Galago senegalensis

TRAITS. Smaller than a squirrel, with large, somewhat pointed ears. *Weight:* 150–200 g (up to 300); *length* (head and body): c. 170 mm, tail 230 mm (5, 6). Coat short, soft, woolly; tail thin-haired, bushier toward tip. *Coloration:* light to bluish gray, sometimes suffused with brown or yellow on crown and neck, with yellowish arms and thighs. Face with brown to black cheek and eye patches and white nose bridge, underparts white or cream, throat and breast sometimes tan, hands and feet white to gray. *Scent glands:* scrotal/labial glands, salivary glands, head and neck glands (?), males with bare but nonglandular chest patch (14). *Mammae:* 6 (2 pectoral, 4 inguinal).

DISTRIBUTION. The commonest and most widespread of the bush babies, found in all types of wooded habitat outside of the main forest blocs, from sea level to about 1500 m (12).

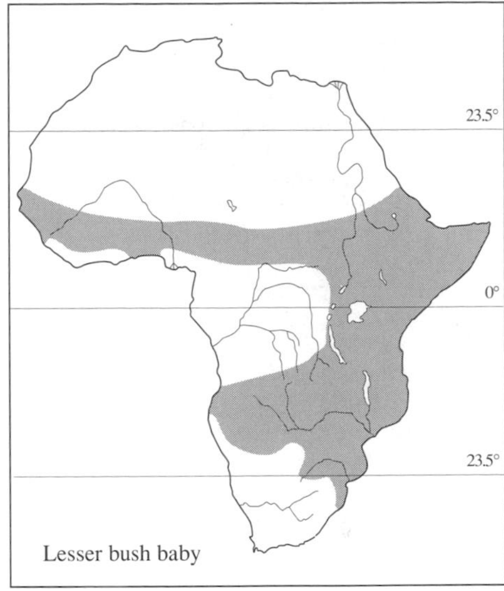

ECOLOGY. The lesser bush baby reaches peak abundance in savanna habitats, especially acacia savanna and *miombo* woodland. This small creature is perfectly at home in the thorniest thickets, where few other vertebrates dare venture (cf. bush hyrax). Motorists driving at night along tree-lined roads often see its eyes reflecting bright orange in the headlights.

DIET. *G. senegalensis* lives mainly on tree gum and on various insects. Bush babies of the northern Transvaal acacia savanna were found to subsist entirely on acacia gum and arthropods, especially the former when insects were scarce (6). *Senegalensis* in the wild shows no interest in vegetation or small vertebrates, nor has it been seen drinking water or licking dew, even in dry weather (7), whereas captive specimens have been known to eat both and to drink (by lapping rather than by dipping and licking the hands).

SOCIAL ORGANIZATION: strictly nocturnal, territorial (both sexes), polygynous, solitary foraging, social sleeping groups.

Lesser bush babies may reach surprisingly high population densities: 95/km^2 in riparian bush, 200/km^2 in acacia savanna, up to 500/km^2 in thickets of *Acacia karoo* in the northern Transvaal (3). In uniform habitat bush babies may be fairly evenly distributed. But in patchy habitat they may occur in isolated subpopulations (demes), a trend that is accelerated by cutting and clearing operations.

A family group consisting of an adult male and female with offspring of 1 to several generations is the basic social unit (3, 7, 8). However, the largest males often control the ranges of several females. Females with overlapping ranges often associate, together with their offspring, in sleeping groups numbering up to 7 bush babies. Such groups usually include an adult male: 29 of 42 sleeping groups had 1 adult male, 3 contained 2, and 10 groups contained none (10). In another sample, the average of 119 groups was only 2 animals, 75% of which were in groups of 2–4, usually containing a pair with young, a female with young, or a pair alone (7). Mean home range size of 15 bush baby groups was 2.8 ha (1.25–3.75) (cf. Demidoff's bush baby, fig. 25.6). Roads, open spaces, and the like often serve as range boundaries. But the size, shape, and position of home ranges tend to change over time.

The tendency of female offspring to be sedentary fosters continuing association between the mother and female siblings. The number of different females that associate thus reflects the mother's fecundity, the mortality rate of her young, and her tolerance of grown-up daughters (which varies individually) (5). Observations of a captive breeding colony showed that females associate on the basis of mutual familiarity (7). Unrelated females kept in the same cage slept and bred in the same nest, and behaved aggressively to outsiders. Adult males could not be kept together, but generally tolerated subadults. In the wild, the ranges of mature males overlap only slightly if at all, and changes in their population may be frequent (11).

Sleeping Sites. Lesser bush babies sleep with nearly equal regularity in nests or on branches, usually 4–6 m above ground level in dense vegetation such as a bush acacia tree. In a sample of 168 animals, 52% were found in nests, 40% huddled on branches or tree forks, and 6% in hollows (3). In Angola, a hollow tree beside my house housed 7 bush babies, with occasional absences of a few days up to a couple of months, during the whole year I lived there. Nests are merely platforms constructed of broad, soft, fresh leaves (in Namibia, nests may be domed). Sometimes old birds' nests provide the foundation. Nest size corresponds to the number of occupants. Observations in a laboratory colony indicate that females use nests as delivery tables and thereafter to provide resting places and camouflage for the babies (7).

Up to 13 different sleeping places may be used during the year in any 1 home range. The same trees may be used for generations, with new nests being constructed near old ones. Preferred sites offer concealment and protection from predators, and shelter from temperature extremes. Bush babies seem unaffected by weather conditions at night but try to avoid temperature extremes by day. Sleeping groups are more likely to withdraw into nests during hot weather, in search of coolness, than in cold weather, when they often sleep exposed and sometimes sunbathe. In East Africa, apparently, nests are rarely used (10). Changes in leaf cover and temperature may cause a group to change quarters. Different sites may also be used from one day to the next, for no obvious reason, and sometimes 1 or more members of a group sleep in different places. Males are more apt than females to sleep alone, but in a Kenya population mature males slept nearly every day with 1 or 2 associated females and their offspring (11).

As in Demidoff's bush baby, probably the senior female usually takes the initiative in choosing a sleeping site and building a nest (5). Younger females may bring a leaf or 2 in their mouths without actually adding it to the nest structure. Otherwise there is no evidence of a dominance hierarchy and there is no direct food competition, since each bush baby forages separately.

Social relations within a matriarchal group can be quite complex, but lesser bush babies are essentially solitary most of their waking lives, spending 70% of each night alone (cf. greater bush baby) (3). Although they sleep in contact, group members spend little time in *social grooming* and this even applies to mothers of young infants. On waking, bush babies yawn, stretch, and lick themselves and each other. They leave the nest one by one, bouncing along the same pathways within and between trees, then disperse and begin foraging. Should one follow another, it is apt to be aggressively discouraged. For the first hour after dispersing, group members often maintain vocal contact. Thereafter they usually keep quiet until the group reassembles near the sleeping site, where they again engage in mutual licking and other social activities until bedtime (more under Activity).

ACTIVITY AND FORAGING BEHAVIOR. Like other nocturnal prosimians, lesser bush babies remain inactive until twilight and return to their sleeping sites before human observers can detect a hint of dawn (5). The period of maximum activity comes during the first 2 night hours, with a secondary peak in the hour before dawn. The rest of the night bouts of feeding, exploration, marking, rapid movement, grooming, and sleep alternate without any pattern or set rhythm; each bush baby is an independent entity. Weather seems to have little effect on activity: lesser bush babies move around actively even during heavy thunderstorms (12). However, time devoted to foraging, which averages 6 hours a night, varies seasonally, reflecting changes in the availability of insects and gum. Thus, bush babies in the northern Transvaal, subsisting on gum in the late dry season, typically fed 12 minutes per hour in 21 separate feeding bouts, and spent up to 15 minutes an hour on the ground, apparently searching for scarce grasshoppers and other terrestrial arthropods (6). Once the rains started they spent much less time licking gum and on the ground, and relied more on arboreal insect prey. In midsummer (February–March), feeding time averaged 6 minutes per hour in 1-minute bouts, with a maximum of 20 minutes and 43 bouts/hour.

Lesser bush babies spend most of their foraging time seeking out, licking, and eating gum that exudes from bark penetrated by boring insects. The same sites may be revisited on successive nights to obtain the exudate while it is still fresh and soft (6, 7). On an average night, a bush baby may travel 2 km and visit 500 different trees, repeatedly using the same crossing points to jump between trees or traverse open spaces. However, the route and movements are often varied, indicating an intimate knowledge of the entire home range. Areas of heaviest use also vary seasonally, but usually all parts of the range are visited at least once a week. How high bush babies range and how often they descend to the ground depend on the type of habitat. Lesser bush babies living in primary forest are extremely reluctant and seldom need to descend to earth, in contrast to those living in open country with scattered trees and bushes (6).

Insects are caught either by leaping and grabbing them, often as they are taking flight, or by creeping to within grabbing distance. Prey is always taken by hand and not in the mouth, in a highly stereotyped movement which is too quick for the eye to follow (4). At the moment of capture the bush baby shuts its eyes and flattens its ears, reflexively protecting its face from prey such as locusts whose flapping wings or spurred legs could cause injury. Immediately afterward it bites the quarry's head, then its eyes open and ears come forward. Items too big to be held in one hand are grasped or held down with both. The lower incisors (the "toothcomb") are not used in eating, but only for detaching pieces when the food item is too big to be chewed by the molars and canines.

POSTURES AND LOCOMOTION. In preparation for a long (4–5 m) horizontal jump, a bush baby rotates its head up to 180° as it checks in all directions from all angles, ears meanwhile each moving independently. During the leap, the arms are held forward and above the head for added momentum and the tail helps maintain balance and attitude (vertical clinging and leaping techniques of other bush babies are compared in the family introduction; see fig. 25.7). Depth perception of 1 cm at a distance of a meter has been demonstrated in this species. One bush baby was seen traveling along a fence by hopping between posts spaced at 10-ft. (3 m) intervals; it also walked 70 m along the top wire strand. If chased and prevented from taking a careful fix on its landing point, a bush baby can still jump again and again with complete accuracy, under all light conditions except extreme darkness (6). It can see well for probably 20–30 m by starlight. On the ground, a bush baby jumps like a kangaroo. In climbing, *senegalensis* is equally at home on the largest branches and trunks and on the finest and thorniest twigs. It can climb in any position, upside down on horizontal branches, head- or tailfirst descending vertical surfaces (16).

Sleeping postures vary greatly, especially in groups. A bush baby may sleep on its side, head covered by its hands and tail; crouched and curled forward with bent head resting on its hands and covered by the tail; lying on the back; or in a seated position. During sleep the ears are folded close to the head. *G. senegalensis* stretches like a cat, lowering the forequarters and elevating the hindquarters. One of many variations in stretching attitudes is hanging by the hindfeet with forelimbs extended (4).

SOCIAL BEHAVIOR

COMMUNICATION

Vocal Communication. The lesser bush baby has no raucous calls to compare with the greater bush baby's, and even its strongest vocalizations usually go unnoticed by people. Yet this species emits 10 or 11 basic, discrete sounds which can be combined to produce some 25 recognizably different calls (also see Vocal Communication in family introduction) (13).

Distance-increasing Calls:

Advertising: bark.

Alarm: grunt, sneeze, shivering stutter, "*gerwhit,*" cluck, whistle, yap, wuff and wail, caw, explosive cough. A hoarse grunt marks the onset of excitation and grades into a sneezing noise; as the animal(s) becomes worked up, the calls increase in pitch and intensity, changing in quality to produce the above list of different sounds. Excited bush babies may voice alarm calls for ¼–1 hour and, when the members of a group are gathered near their sleeping place, all may join in what appears to be a typical mobbing response (author's observations). Bush babies sit still while alarm-calling, only changing orientation if necessary to keep the disturbing object in view. At other times intense alarm calls, warning of a predator's presence, will cause bush babies within hearing to move higher in the trees and remain absolutely still (7, 13).

Aggressive: sob, spit, spit-chatter, spit-grunt, rasp. These calls are also a graded series of increasing intensity and pitch which express different degrees of antagonism and serve to regulate individual relationships.

Fighting: fighting chatter. When attacked by another bush baby or predator, caught in a trap, or otherwise frightened, a lesser bush baby emits a 2-phase grunt which is superimposed on the aggressive calls to produce the so-called fighting chatter (13).

Distress: scream. A high-pitched, plaintive scream expresses intense fear and/or pain. It attracts other bush babies and stimulates mobbing of predators; small predators also respond, evidently in hopes of finding an injured bush baby.

Physical Contact (Friendly) Calls:

Mother-Infant: clicks, crackles, squeak, coo, soft hoot. Babies seeking contact with their mothers emit high-pitched clicks ("*tsic*" calls) and crackles, grading into a squeak at highest intensity. The mother responds with maternal care and retrieves the infant if it is isolated, answering with coos or soft hoots. She may also call to elicit a squeak response by which to locate a hidden offspring.

Olfactory Communication: *urine-washing, rhythmic urination, salivary deposition, genital rubbing, chest-patch rubbing.*

Urine-washing, the primary form of scent-marking, is performed by lesser bush babies of both sexes beginning as juveniles. The hand and foot on one side are raised simultaneously, and the cupped hand is held under the urogenital opening, wherein a few drops of urine are deposited; then the foot sole is rubbed or firmly grasped by the hand 1 or more times (7). Next the opposite hand and foot are wetted in turn, the whole stereotyped procedure taking only a few seconds. The bush baby's hands and feet now automatically leave scent traces as it moves along. However, marking is most frequent in the sleeping tree between waking and leaving to begin foraging; thereafter a bush baby usually marks only once or twice an hour, during which it may move through 45–50 trees. It does not leave or follow a scent trail or mark any particular spots (7). Urine-washing is stimulated by: presence in a strange area, gazing at a strange or disturbing object, aggressive interactions, and genital-sniffing or grooming another bush baby (2). Males and females *urine-wash* in the same way but males do so with greater frequency, and dominant individuals most of all. During estrus, the male's *urine-washing* rate increases by nearly a third and much of his urine is deposited directly on the female (fig. 25.11) (7).

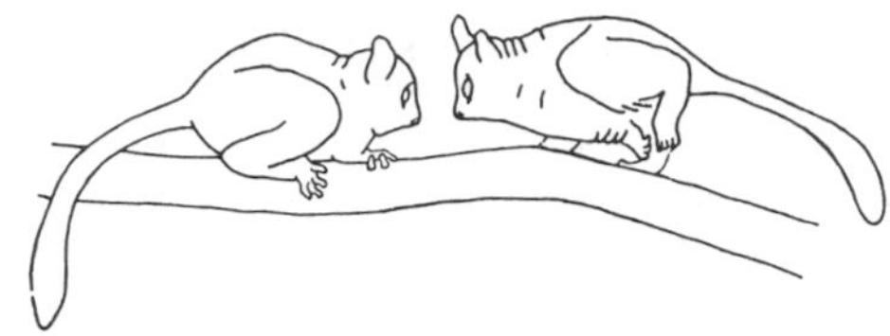

Fig. 25.11. Lesser bush baby courtship: male (right) frequently *urine-washes* when in contact with the female (from Doyle et al. 1967).

Rhythmic urination, conversely, is more frequent in females. The hindquarters are lowered and a few drops of urine are released while the animal wriggles forward (7). The performance may be repeated every few steps as a bush baby explores a strange area (e.g., in a laboratory colony) (2).

Secretions of the scrotal and vulval glands (see family introduction) may also be rubbed onto supports while urinating or in a posture that resembles *rhythmic urination.*

Adult males rub the bare spot on their chest against branches, stubs, and the like, often followed or preceded by rubbing the chin, lips, and armpits. While *chin-rubbing,* the lower lip is dragged back and a trail of saliva is deposited. Males also rub their chests vigorously on a female's flanks and back after genital sniffing, meanwhile adopting a mountlike position. Since there are supposedly no scent glands on the bush baby's chest, the meaning of this behavior is obscure; one suggestion is that the male thereby collects rather than leaves odors and may also transfer odors he has already collected (14).

Tactile Communication: *nose-to-nose sniffing* and *nose-to-face contact* (greeting behavior), genital examination, *social-licking* and *-combing, huddling* in sleeping group (see under family account).

Apart from courtship, mating, social grooming, and very occasional play, which take place mainly in the intervals before and after individual foraging, adult lesser bush babies have comparatively little physical contact, except while sleeping in daytime. In self-grooming, a bush baby employs tongue, tooth scraper, and grooming claw, whereas *social grooming* is largely confined to *reciprocal licking.* Grooming partners thereby exchange salivary secretions, and urine on the hands is transferred while a groomer grasps its partner's fur.

Visual Communication. Postures and expressions are relatively undeveloped and usually associated with vocal signals (see family account) (5).

AGONISTIC BEHAVIOR

Dominance/Threat Displays. These are the signs of a strong tendency to attack: ears upright, eyes wide open, round, and staring, mouth wide with tongue and upper canines partially exposed.

Defensive/Threat Displays: eyes at first wide and ears fully extended as the feared antagonist approaches, becoming flatter while the mouth becomes rounded; lips not tense but covering the teeth, only the tongue visible, and head raised. If threatened further or cornered, a bush baby assumes the *defensive attack posture:* standing like a boxer with hands upraised at head level ready to cuff or to pounce and bite (see fig. 25.14). A bush baby that sways while making this display betrays a strong tendency to flee.

Submissive Displays: cringing attitude, body very low and tense, head retracted between hunched shoulders, eyes narrowed, and ears flattened (7).

Fighting: cuffing, wrestling, biting, hair-pulling.

Serious fighting, only likely when 2 dominant males meet, is normally avoided through efficient olfactory and vocal communication (7). Most aggressive encounters are one-sided and take the form of chasing: subadult male by adult male, female by female. The retreating animal emits the full range of submissive calls. Disputes between equals, say, between juveniles or between adults of opposite sex, take the form of sparring matches: the opponents stand bipedally, cuff each other, and try to grab a hand or leg. If one gets a grip they begin wrestling, holding and kicking each other while rolling over together. They do not bite and no damage is done.

REPRODUCTION. Lesser bush babies reproduce twice a year during the wet season: October and early November, and from late January to March, in South Africa. Gestation is 121 to 123 days. Females reenter estrus soon after the first but not after the second birth; this enables them to reproduce twice during the time when food and foliage (for concealment and nest-building) are abundant. Both sexes become adolescent their first year and females may conceive at c. 200 days. After the first birth, often of 1 offspring, twins are usual (13 sets of twins and 7 singles in 20 births) (3).

SEXUAL BEHAVIOR. At the approach of estrus the membrane that normally seals the female's vagina opens, the labia and clitoris become red and swollen for 1 or 2 days, and the vaginal lining develops a shining white appearance. Since there is no increase in the female's urination rate, the discharge of a whitish vaginal secretion probably serves to signal the onset of estrus. Signs of estrus last up to 7 days, but the female is only sexually receptive 1–3 days. Estimates of estrous cycle length range from 30 to 43½ days; 31 cycles of 4 females averaged 31.7 days (15).

The male's routine for checking female reproductive status (frontal approach, greeting followed by anogenital sniffing) increases during estrus, along with *urine-washing* (fig. 25.11). The female is more inclined than usual to let the male make contact before jumping away. He pursues, clucking quietly. After 3–4 attempts to make contact, he is allowed to mount. Approaching from behind, he may lick the female's genitals before grasping her waist with his arms, thighs spread wide and feet flat on the support (occasionally the male grasps the female's ankles with his feet in the more usual primate mount). His back is fully arched and he presses his chin into the female's nape (fig. 25.8), but does not bite or grip the fur. During the mount he may emit a loud call ending in a whistle, and his eyes remain half-closed. The female crouches low, usually looks straight ahead with eyes wide open, and keeps silent (9).

Copulation is prolonged and females in the early stages of estrus often break free and bolt after 10–50 seconds. One female was mounted 22 times on the first day of estrus, of which only 5 mounts lasted more than ½ minute. The second day she was mounted thrice, each time for 4.6–6.3 minutes, and remained passive throughout. Both partners rested in the final 2 minutes, during which the male's hold relaxed and he remained motionless. After he dismounted, both groomed themselves. Neither showed sexual interest in the other during the long intervals between copulations.

During estrous periods both the male and female drove away subordinate males that tried to approach, whereas anestrous females were tolerated and sometimes even groomed one or both partners during copulation (9).

PARENT/OFFSPRING BEHAVIOR. Infants are born on a nest platform with their eyes open, can crawl within an hour, and cling tightly to the mother's fur. By 4 days they can climb on wire mesh and otherwise use their hands like adults (4). They begin grooming themselves within a week and twins have been known to groom each other the second day. Development is very rapid, the 12 g (8.5–15.5) birth weight doubling in the first week and increasing fivefold within a month. By day 10, infants are very active in the nest. They climb over adults, wrestle, jump up to 15 cm, stand bipedally, and spend a good deal of time peering out of the nest. They are mature enough to venture forth by 2 weeks of age. Within another week they can run fast horizontally and their vertical leaping ability is improving rapidly; they often lose their balance but not their grip. However, at 3 weeks the mother still has to carry her infants between trees (by mouth), 1 at a time (fig. 25.10). She parks them in a tree, together or up to 40 m apart, then goes off to forage, returning at shorter or longer intervals, but always retrieves them in time to reach the sleeping site before dawn. Mothers recognize their offspring by scent (and later no doubt by sound as well) and accept only their own.

Lesser bush babies start eating gum within about 3 weeks, and they capture and eat insects within a month. Six-week juveniles leave the sleeping site of their own accord and spend the night feeding, resting, and moving over increasing distances. They can negotiate gaps of up to 1 m between trees, and descend to the ground for the first time, proceeding in short (30 cm) hops. Though approaching nutritional self-sufficiency, the young may still nurse on rejoining the family back at the sleeping tree (fig. 25.12). But from 6 weeks onward there is little maternal care of any kind and weaning is probably complete by 10–11 weeks (8).

Fig. 25.12. Mother lesser bush baby grooming baby as it begins nursing; second infant clings to her back (from Doyle et al. 1969).

Social behavior patterns also appear early: the threat posture before 2 weeks, *urine-washing* (often following anogenital sniffing and grooming) by 3 weeks. At 1 year males display the full adult aggressive pattern, emit male calls, and begin to urine-mark with adult frequency. By then many females are already mothers (7).

PLAY. Twin infants begin playing together by the sixth day. Juveniles engaged in solitary play perform complex locomotor patterns (manipulation of objects, hanging by hindfeet, etc.). Social play is often associated with mutual grooming, and features fighting but not sexual behavior. Galagos often wrestle while hanging upside down facing, and kick while hanging by their hands. When adults play it is usually with juveniles, rarely with one another (7).

ANTIPREDATOR BEHAVIOR. Though vulnerable to virtually the full array of avian and arboreal predators, bush babies reduce the risks by remaining well-hidden by day, and by being alert and hard to catch because of their extraordinary jumping powers. However, they are easily captured while on the ground. Alarm calls, mobbing of predators, and retrieval and transport of young in the mother's mouth are also effective antipredator measures (4).

References

1. Anderson 1969. 2. Andrew and Klopman 1974. 3. Bearder and Doyle 1974. 4. Bishop 1964. 5. Charles-Dominique 1978*b*. 6. Charles-Dominique and Bearder 1979. 7. Doyle 1974. 8. ——— 1979. 9. Doyle, Pelletier, Bekker 1967. 10. Haddow and Ellice 1964. 11. Harcourt and Nash 1986. 12. Kingdon 1971. 13. Petter and Charles-Dominique 1979. 14. Schilling 1979. 15. Van Horn and Eaton 1979. 16. Walker 1979.

Fig. 25.13. Greater bush baby at rest in tree.

Greater Bush Baby

Galago crassicaudatus

TRAITS. The largest (cat-sized) bush baby. Size varies with subspecies. *Weight:* females average 1.2 kg and males 1.4 kg in the Transvaal (4). *Length:* head and body 27–47 cm, tail 33–52 cm (6). Robust build, large rounded ears. Fur thick, soft, woolly, especially tail (*crassicaudatus* means "thick tail"). *Coloration:* geographically variable, from silver gray with nearly white tail to dark brown with black-tipped tail, underparts lighter, varying from white to pale yellow to gray; face same color as body or lighter with dark marks on muzzle and around eyes, and light nose. *Scent glands:* scrotal/labial glands, salivary glands and possibly other glandular secretions from mouth or lips, apocrine glands in longitudinal bare chest patch of adults (larger in male) (9). *Mammae:* 1–3 pairs: either pectoral, or pectoral and inguinal, or pectoral, axillary (armpits), and inguinal.

DISTRIBUTION. Apart from the lesser bush baby, this is the only other member of the family that is widespread outside equatorial rain forest, being found in eastern Africa from southern Sudan to southern Natal and Angola in forested and wooded habitats. Though more limited in distribution and numbers, the greater bush baby is far more familiar because of its raucous advertising call, from which the name bush baby originated.

ECOLOGY. *G. crassicaudatus* is largely confined to dense evergreen forest and riparian bush, from sea level to 1800 m (1). It is adaptable enough to live in different man-made habitats, such as pine, blue gum, eucalyptus, and coffee plantations, suburban gardens, and the like, and tolerates temperatures down to freezing. It subsists mainly on fruits and gums, supplemented by nectar, seeds, and insects of easily caught types. The availability of fruit may be a main limiting factor on this galago's distribution: it is only found in regions where fruits are abundant for at least half the year (2). The annual diet of Transvaal bush babies observed during 281 feeding bouts was gum (62%), fruit (21%), flower secretions (8%), seeds (4%), and insects (5%) (including a few unidentified items) (2). In another population, gum deposits of *Acacia karoo* were the mainstay (4, 5). Despite numerous reports of predation on birds, small reptiles, and mammals, not a single instance was observed during the above studies; in fact, bush babies were seen to ignore roosting birds they could have caught. Apparently meat-eating is extremely rare under natural conditions, although individuals may sometimes become habitual chicken killers (2).

There can be considerable dietary and some habitat overlap with the lesser bush baby, especially in the winter/dry season when fruits and insects are scarce and both species depend on gum. *Senegalensis* densities tend to be lower than usual where the 2 species overlap (1). However,

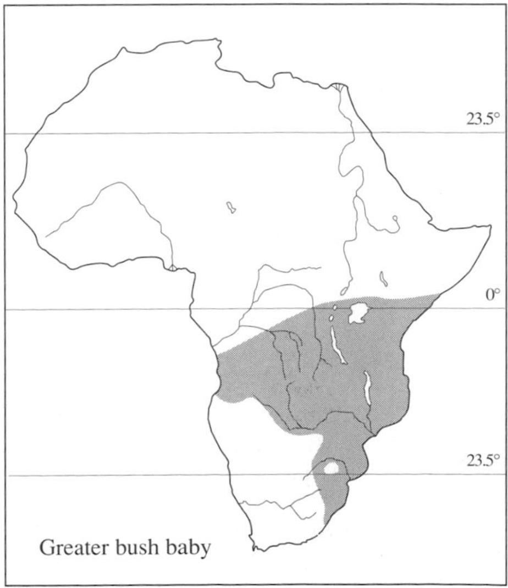

the greater bush baby stays in denser, moister habitats and the lesser bush baby keeps to the edge of the forest and ranges into open savanna and bush. Thus overlap occurs mainly along the edge.

The greater bush baby sometimes licks dew but does not drink regularly.

Samples of *G. crassicaudatus* populations in the southern part of its range indicate densities from 72/km^2 in lowveld riparian bush and savanna, to 110/km^2 in mixed woodland and plantations, up to 125/km^2 in Zululand coastal dune forest (1).

SOCIAL ORGANIZATION: nocturnal, semisocial, females occupy and defend ± separate ranges, several of which may be controlled by a dominant male.

In a study of individually known greater bush babies inhabiting a northern Transvaal *Acacia-Combretum* gallery forest, the oldest and largest male in the study area was found to control the ranges of 5 females, 4 of them adult and 1 subadult (7). The male tolerated other, subordinate males within his range, which changed seasonally. Two of the adult females shared the same range, and the subadult female lived with her mother. Females from different ranges behaved territorially, marking and fighting one another in border overlap zones, where however they also engaged in occasional *social grooming* sessions. Even the females that shared a home range apparently kept to separate core areas and came separately to collect gum from the same trees (but at least sometimes socialized between feeding bouts, according to an earlier study of the same population) (3, 4).

The adults were permanently resident, whereas younger animals either left the area or died (7). Males but not females returned sporadically after departing, but avoided meeting the big resident male. During the summer they were rarely aggressive to one another, sometimes foraged together, and often engaged in *social grooming.* They became less sociable and more aggressive toward one another in winter, particularly during the mating season. All the bush babies that lived in or visited the study area seemed to utilize many of the same gum and fruit trees and to follow the same aerial routes night after night. Yet individuals and maternal groups were temporally separated, and experiments established that the scent-marks deposited at strategic locations by the different bush babies effectively regulated bush baby traffic (more under Communication).

Another Transvaal study focused on a single isolated group composed of a female with her offspring and 1 male (1). The female's range of 7 ha made up only part of the male's range, which also varied seasonally along with his feeding grounds. The same pair associated for a number of years, often on a daily and nightly basis: the male was found in company with the female and young 20% of the time (411 observation hours), and was sleeping with the group (numbering 2–6 bush babies) or in a nearby tree on 32% of the days the animals were found. The female always slept with at least 1 of her offspring and at night was accompanied by her infants 90% of the time.

Sleeping Sites. A home range contains certain favored sleeping, feeding, and resting places which may be located at various unrelated points.

Up to 12 different sleeping sites may be used but typically several are clustered in particular trees. Although this bush baby has been known to use tree hollows, all but 4 of 35 sleeping sites were situated in forks or on branches of trees, usually in dense tangled vines or interlaced limbs 5–12 m above ground. Greater bush babies only occasionally sleep on exposed branches in daylight. Nests, consisting of inaccessible leafy platforms with a central depression, are made only when females have small infants.

The greater bush baby's ranging behavior tends to be more stereotyped than that of *senegalensis,* reflecting the importance of fruit and gum in its diet compared to insects, along with its larger size and slower locomotion. Another major difference is the mother greater bush baby's habit of taking infants along while foraging. Infants too small to follow are carried in her mouth, 1 at a time, or clinging to her back (never her front). Often 2 and sometimes 3 babies ride at once. This form of transport is otherwise known only in *G. elegantulus.* The adaptation may be favored by the greater bush baby's slow, quadrupedal gait and ample size (the infants are comparatively tiny) (2).

ACTIVITY. Similar to lesser bush baby (q.v.): awaken in twilight, depart sleeping place for individual foraging after dark, peak activity and calling early and late at night, return to sleeping site before dawn.

FORAGING BEHAVIOR. Greater bush babies may spend several hours in a fruiting tree, and keep returning as long as ripe fruits last. *Ficus* species, *Diospyrus* and other trees that keep producing for a long time, and acacias and other trees that regularly supply gum are mainstays. Apparently most food items are located by scent, including even gums as well as fruiting and flowering trees. Small fruits such as figs are plucked directly by mouth with or without a guiding hand and then carried to a convenient perch where the fruit is eaten while being held in one hand (2). To deal with large fruits, a bush baby chews away and discards the peel, then licks or scrapes out the soft insides with its tooth comb.

In searching for gum exudates or nectar, a bush baby carefully seeks out likely sources and eats those that are productive. Freshly exuded gum is licked up, whereas older deposits are chewed. Sometimes bark is gnawed away with the cheek teeth to expose a source of gum. To reach gum in inaccessible spots, bush babies assume remarkably contorted postures. Dry fruits, seeds, and insects are grabbed deliberately with one hand and put in the mouth. This species lacks the stereotyped, lightning hand movement of the insectivorous galagos (2). Nor does it pursue insects actively, or take stationary insects in preference to fruit (2).

POSTURES AND LOCOMOTION. This species moves very differently from other galagos, walking or running along branches like a monkey and making short jumps only when moving rapidly. An alarmed individual can make longer jumps but not in the usual way: it generally jumps in a downward direction with comparatively little forward momentum. Greater bush babies also have a slow, stealthy way of moving, sometimes while hanging below a branch, which recalls the creeping gait of a potto and may be similarly adapted to avoiding predator detection. On the ground, this bush baby either hops like a kangaroo, gallops with both forefeet and both hindfeet together, or walks and runs as when traveling along branches, but with hindquarters and tail elevated.

SOCIAL BEHAVIOR

COMMUNICATION

Vocal Communication. A repertoire of 10–11 basic, discrete calls, with intergradation and combinations of some calls, closely matches that of the lesser bush baby. However, the distance call of the greater bush baby is much louder, and the close-range calls are more graded, reflecting the greater amount of social contact during activity periods (2).

Distance (loud) Calls:

Raucous Cry. The greater bush baby's famous cry is a cawing or barking noise produced by adults with mouth slightly open, often "broadcast" from tall trees while turning the head. Meanwhile the body shakes violently (6, 10). Adult males call most, especially during the mating season, and their voices are distinctive enough for people to recognize different individuals.

Alarm: sniff, knock, creak, squawk, whistle, whistle-yap, chirp, chatter, moan, rattle. These calls represent a graded series, beginning with the weak sniff and progressing through knocking, creaking, and so on, becoming more powerful and sustained with increasing excitement. Thoroughly aroused bush babies may sound off for up to an hour.

Contact-rejection Calls: hack, spit, scream. Staccato sounds that normally function to arrest approach, these calls can be modulated through varying intensity and pitch, according to the animal's emotions. The scream is superimposed on the other 2 calls when a bush baby is frightened. The defensive-threat posture, urination, and defecation are often associated with intense screams.

Distress Call: yell. A high-pitched, plaintive call given in response to extreme fear or pain, as when fighting, captured, or wounded. It evokes a mobbing response in other bush babies and also attracts predators (see lesser bush baby account).

Contact-seeking Calls:

Mother-Infant: buzz, click, squeak, cluck. The infant's clicks and squeaks and the mother's clucks sound and function like those of other bush babies (see lesser bush baby account and family introduction). In addition, older offspring emit a buzzing sound while accompanying their mother during nocturnal movements.

Adult-Adult. The raucous cry, noted above, doubles as a gathering call.

Sex: crack calls. Low-frequency, grating sounds given only by males, usually to females, for instance while following and when grooming, but not confined to courtship (10).

Olfactory Communication: *urine-washing, foot-rubbing, rhythmic urination, anogenital- and chest-gland marking.*

As scent-marking behaviors of greater and lesser bush babies are indistinguishable in most respects, only the differences will be mentioned here (refer to lesser bush baby account).

Foot-rubbing, given in conjunction with urination, appears to be unique to this species. Adult males (occasionally juveniles) rapidly rub first one foot and then the other against a branch or other object, producing a loud scraping noise because of the presence of wartlike spines. This behavior, the meaning of which remains unclear, is common during social encounters and also when confronting a predator, often in conjunction with *urine-washing, rhythmic urination,* and *chest-gland marking.*

Both males and females rub their bare chest patches against tree trunks and branches. The patch contains apocrine glands, which secrete a yellow oily substance beginning at about 12 weeks (4). Particular spots along their regular arboreal pathways are repeatedly rubbed, usually after prolonged sniffing and often in conjunction with biting the bark and *foot-* and *anogenital-rubbing.* The whole performance is highly stereotyped and the bush baby assumes odd postures to bring its chest up against hard-to-reach spots such as the underside of a branch, followed by an upward sliding movement to make its mark. This appears to be the most important means of marking and dominant males are the main practitioners. When tending an estrous female, males also *chest-rub* as a threat to rivals (7). Peripheral-male intruders were seen to turn around and retreat after encountering the fresh marks of the resident male. They left after prolonged sniffing and "a peculiar display" during which the galago surveyed its surroundings, then sniffed some more. Marks more than an hour old were disregarded, as were marks away from commonly used pathways (3, 7). Chemical analysis showed that 1 of 3 main compounds (benzyl cyanide) dispersed within the first hour, whereas the other 2 persisted for days, evaporating very slowly (cf. mongoose introduction, chap. 19). When the 3 compounds were presented separately and together as artificial scent-marks, only the mixture stimulated bush babies to sniff. Experiments indicate that chest-gland secretions encode the time of deposit in addition to the sex, age, individual identity, and even the subspecies of the marker, thereby enabling bush babies to share a restricted habitat such as narrow gallery forest while avoiding potentially hostile encounters (3, 4, 7).

Tactile Communication: *nose-to-nose greeting, anogenital checking, social grooming* (see lesser bush baby account for details).

Social grooming is more frequent and protracted in the greater than in the lesser bush baby, reflecting the longer-lasting and closer maternal bond. Associated individuals

Fig. 25.14. Defensive threat of greater bush baby.

of all sexes and ages engage in *reciprocal licking/combing* when they gather at night between feeding bouts. However, males more than 2½ years old do not groom one another (4). No scent-marking of partners, as described in lesser bush babies, has been observed (4).

Visual Communication. See Agonistic Behavior.

AGONISTIC BEHAVIOR. The description of lesser bush baby aggressive and submissive behaviors applies equally to *G. crassicaudatus* (see also under family introduction). Figure 25.14 illustrates defensive threat.

REPRODUCTION. With a gestation period of c. 18 weeks, this galago could reproduce twice a year and may do so in parts of equatorial Africa (8), but in regions with a single rainy season, and also in coastal Kenya, it has an annual breeding season (November in southern Africa, August–October in Kenya) (11, 12). Twins and triplets are usual for the multiparous wild females of the Transvaal: 12 of 16 bush babies were twins, three were triplets, and only one was a singleton (1). In another study, all multiparous females seen within two weeks after giving birth had triplets. But in Kenya a single offspring is the rule (12). Data from zoo and wild births indicate that this is one of the few mammals with a male-biased natal sex ratio, of c. 3:2 (4) (see under Social Organization: cf. wild dog).

The genitals of both sexes undergo changes during the mating season. The estrous female's become red and swollen. The male's scrotum, normally fur-covered and inconspicuous, enlarges and a patch of pink, granular skin appears in the midline groove. These changes presumably reflect increased activity of the labial and scrotal glands, respectively (see family introduction). Females become sexually receptive c. 5 days (2–16) later and stay in heat c. 6 days (2–10). The mean cycle length is 44 days (35–49) (11). According to Clark (4), the vagina is not sealed between breeding cycles as in other galagos.

SEXUAL BEHAVIOR. During a rut that is as brief as a fortnight in South Africa, young-adult males often cluster around estrous females and harass the resident male, provoking growls, grabs, and spats (5). A courting male gives the clucking call when approaching to greet a female, after which he tries to sniff her vulva. A receptive female crouches and deflects her tail; an unreceptive one withdraws and threatens. Attempts to check the female's genitalia continue, interspersed with mutual grooming bouts, as long as the female remains unresponsive. When she stops threatening, the male tries to mount, sometimes gently biting the female's back. Having grasped her ankles with his feet and her waist with his arms, he holds still a few seconds and then either dismounts or begins rapid "searching" thrusts which may continue in repeated bouts of 30–80 seconds before intromission is achieved. He then makes slower, deeper thrusts for 4–8 minutes, followed by ejaculation. But that is only the beginning of a copulation that lasts several hours, during which the male thrusts 5–6 times every 20 minutes or so (11).

PARENT/OFFSPRING BEHAVIOR. Babies spend their first 2 weeks in a leaf nest or tangle of vines. The mother may move them during this period, carrying them by mouth 1 at a time. Later she takes them with her during her nightly foraging, the infants often riding on her back (discussed under Social Organization). They are coordinated enough to begin following her short distances by c. 25 days, and begin licking gum by the fifth week, but continue to be carried and suckled up until c. 10 weeks. This close association continues at least until the young are subadult (at c. 300 days), when they begin to forage independently (2).

ANTIPREDATOR BEHAVIOR. Adult greater bush babies have little reason to fear arboreal predators such as genets or palm civets and any but the largest birds of prey. The young are well-concealed by day and under the mother's protection by night. The slow movements of this galago (see under Postures and Locomotion) while foraging probably function to avoid attracting the attention of predators (2). In habitats where bush babies spend time on the ground, they are vulnerable to carnivores such as the leopard and hyena. They are accordingly very cautious and alert when leaving the trees and quickly return if startled. Larger snakes elicit intense alarm calls and withdrawal. On discovering a predator, the typical response is to retreat to a safe vantage point, from which the bush baby stares fixedly, or to approach hesitantly while emitting alarm calls.

References

1. Bearder and Doyle 1974. 2. Charles-Dominique and Bearder 1979. 3. Clark 1978*a*. 4. ——— 1978*b*. 5. ——— 1985. 6. Haltenorth and Diller 1980. 7. Katsir and Crewe 1980. 8. Kingdon 1971. 9. Schilling 1979. 10. Tandy 1974. 11. Van Horn and Eaton 1979.

Added Reference

12. Nash and Harcourt 1986.

Chapter 26

Old World Monkeys

Communication in the Family Cercopithecidae

This family includes all the monkeys of the Old World, totaling some 74 species in 2 subfamilies, the Cercopithecinae and the Colobinae. Tropical Asia is the stronghold of colobine monkeys and macaques, and tropical Africa is the stronghold of guenons and baboons. The split between the two continents is approximately half-and-half: 35 Asian and 39 African species.

FAMILY TRAITS. Dog-shaped (cynomorph) primates which stand and move horizontally on four legs with head directed forward and downward, form and movement particularly doglike in the more terrestrial species (baboons, macaques, patas). Buttocks padded with ischial callosities (bare and often colorful in baboons and macaques, small and hidden in other monkeys). Face partially naked. *Size range:* 1.2 kg talapoin to 40 kg chacma baboon and mandrill. *Sexual dimorphism:* size dimorphism ranges from minimal (female c. 80% of male weight) in talapoin and vervet to great in most large, terrestrial species with multimale organization (females 50% of male weight in baboons); *teeth:* long, sharp canines in adult males only (baboons, mangabeys, macaques), or both sexes (guenons); dimorphism in coloration and markings mainly limited to the genitalia of some guenons, macaques, and baboons, also manes and facial hair in baboons, macaques, and a few guenons (e.g., Hamlyn's). *Coloration:* family includes the gaudiest mammals, notably the mandrill, with blue, violet, and red hues on muzzle and posterior, such pure primary colors confined to areas of bare skin; fur colors range from white to black, yellow to orange or reddish, often arranged in striking patterns but no more saturated than in other mammals; coloration of newborn contrasts with adult coloration in most species (see species accounts). *Scent glands:* known so far in few species, e.g., chest gland in vervet and mandrill, armpit glands in gray-cheeked mangabey (more under How Monkeys Communicate). A pair of capacious cheek pouches extending beneath the neck skin holds food gathered by hands in all members of the subfamily Cercopithecinae, (absent in Colobinae). *Mammae:* one pair, pectoral.

The great majority of African monkeys live in the Lowland Rain Forest of the Congo Basin and West Africa. Like forest birds, monkeys are hard to see, not only the ones that are cryptically colored and deliberately hide, but also the most brilliantly colored and active ones, because of the screening foliage and dim light. But primatologists are a determined lot and, despite the manifest difficulties and discomforts of working in the forest, the behavioral ecology of over a dozen forest monkeys is known in considerable depth.

The members of the two different subfamilies are treated separately (chaps. 27 and 28). A general discussion of the distribution, ecology, social organization, activity, postures and locomotion, reproduction, and so on of the family is given in the primate introduction (chap. 24). This chapter focuses on monkey communication.

HOW MONKEYS COMMUNICATE. The most remarkable fact about communication among monkeys (and apes) is the sameness of their signal systems (4, 5,). This applies particularly to visual, tactile, and olfactory signals, but also to many vocalizations, especially those expressing alarm and distress, which sound virtually identical in various species (9). Summing up their survey of visual signals in Old World monkeys, Gautier and Gautier-Hion (4, p. 950) wrote, "Ultimately it seems conceivable that no truly unique facial display or posture exists . . ."

Given a common repertoire of gestures and facial expressions, what is there to prevent interbreeding of closely related

Fig. 26.1. Mona monkey *head-bobbing* (from Kingdon 1980).

monkeys like forest guenons, most of which are anatomically and genetically enough alike to hybridize (see chap. 24 and blue monkey account)? Differences in coloration and markings are of primary importance, followed by differences in male loud calls (discussed under Vocal Signals). Very similar displays are performed by players wearing very different uniforms. The bold facial markings of many guenons are in effect masks that actually obscure their facial expressions at long or medium range (6). Selection has favored the development of striking patterns that will convey both the message and the identity of the sender to distant receivers in dim light. To be effective, such signals have to be highly stereotyped and unambiguous. In a perceptive analysis of face patterns and visual signals in guenons, the artist/zoologist Kingdon found that minor variations in the way different species perform the same displays serve to translate their particular coloration/marking patterns into highly specific visual signals (6).

Take threat displays, for example. The repertoire of all guenons includes head-bobbing and sudden threatening stares accompanied by loud calls and interrupted by *cut-off gestures* (turning away, bowing the head, and other movements that obscure the visual signal). However, the rate, amplitude, and/or direction of the movements are consistently different, making the most distinctive elements of the species' color pattern flash like semaphore signals (see fig. 27.3). Thus, the head-bobs of the crowned guenon have a pronounced lateral component, emphasizing facial markings that are most visible in side view, whereas the mona monkey bobs more vertically, displaying its white front bordered by black lower limbs and its black robber's mask set in a pale facial disk (fig. 26.1). The white beard of DeBrazza's monkey (fig. 27.3) waves from side to side to side like a flag as the male slowly wags his head and reverses direction while emitting his loud call (6).

In the baboon-mangabey line, the emphasis has been almost the reverse, on communication within rather than between social groups. Eye-catching markings are unnecessary to identify a species at close range and, instead of highly stereotyped, discrete displays, the more complex social relations within large, multimale troops have selected for graded visual and vocal signals capable of expressing subtle shadings of emotions and intentions (8, 9). Baboons and their kind have developed both greater mobility of their features (especially lips and brows) and more visible expressions, by baring more of the face, by lightening skin tones, and by the use of contrasting color on the eyelids to emphasize brow movements associated with staring (more under Facial Expressions). So, even though the basic repertoires remain alike, baboons are capable of sending and receiving much more complicated messages than are guenons, and the frequency and intensity of their exchanges are also far greater (4).

VISUAL SIGNALS

Species, Sex, and Age Recognition. In addition to brilliant colors and striking mark-

Fig. 26.2. *Penile display* of a sentinel baboon (from Wickler 1967).

ings on head and body, the length, color, and bearing of the tail play a major role in species recognition, especially in mangabeys (fig. 26.3). It is noteworthy that the "riding-whip" tail carriage and doglike head of the savanna baboon are more distinctive than its coloration, which is drab indeed by comparison with its forest counterpart, the mandrill, whose multihued face and posterior are the most colorful of any mammal.

Infants of many species are a different color than adults: black-and-white colobus are born white, newborn baboons are black with bright pink faces and behinds, and so on. Brightly colored genitalia (typically a blue scrotum and red penis) distinguish sexually and socially mature males from immature males in many guenons. In the gelada, olive, and hamadryas baboons, capes and tufts of facial hair identify mature males (the cape turns lighter with age in the hamadryas). Equally important, the male's canine teeth come in only at maturity; yawning displays the fangs and adult status. Female reproductive state is advertised by cyclic change in color and swelling of the anogenital area in most monkeys that live in multimale societies (11), including the baboons, mangabeys, talapoin, Allen's monkey, and red, black, and olive colobus (see chap. 24).

Body (and Tail) Language

1. *Postures and social status.* The postures and movements of monkeys often reflect their social status (4). A confident animal appears relaxed, walks with its limbs extended and back level, looks around casually, and appears equally at ease when resting. A subordinate walks with back hunched, limbs somewhat bent, and tail low or curved downward, and sits hunched over, lowers its head, and peers about nervously (no. 15 below). Dominant males may exaggerate their status by walking with a swagger and even sit down with ostentation. When trying to intimidate a rival, a monkey stands as tall as possible and bristles, which can make species with capes and face ruffs of long hair (especially baboons) appear much bigger. Adult males of species with colorful genitals display them to impress other males (see vervet *red-white-and-blue display*) (fig. 27.7). Dominant males are also recognizable by their surveillance behavior ("watchdog role"): they sit apart on high vantage points facing away from the troop and scan, often with legs spread to expose the partially erect penis (fig. 26.2) (see species accounts) (4, 15). Dominant males of various forest monkeys, especially mangabeys, elevate their tails in a conspicuous and species-specific manner. The crested mangabey stands with tail arched over its back and touching its head (fig. 26.3), the gray-cheeked mangabey brings its tail to the vertical, and the moustached guenon curls its tail into a question mark. These poses are termed *stopping postures* (4).

AGONISTIC BEHAVIOR

Dominance/Threat Displays

2. *Supplanting or displacing.* The commonest and least conspicuous way of asserting dominance. One monkey simply approaches (in a confident manner) and displaces another from its resting, feeding, or watering place, position next to an estrous female, or other desirable spot.

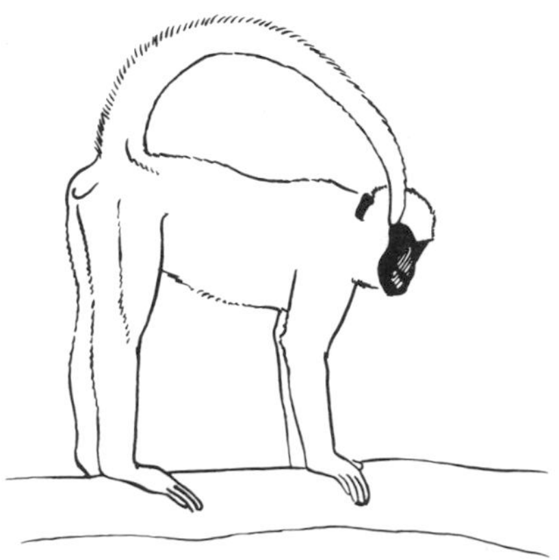

Fig. 26.3. *Stopping posture* of the crested mangabey, with tail tip touching head.

3. *Head-bobbing, arm-pumping (push-ups), jumping in place, and branch-shaking.* Aggressive displays addressed to predators (including human observers) and distant rivals, typically in combination with staring and open-mouth threat. In forest guenons movements of head and body are most stereotyped and conspicuous, serving to maximize the transmission distance and clarity of their specific color patterns by creating striking semaphore patterns (fig. 27.3) (6).

4. *Leaping display.* Violent, noisy movement through the trees, often involving spectacular free-falls and deliberate branch-breaking, this display is the prerogative of dominant males and, like no. 3, is addressed to predators and rival troops. It frequently accompanies type 2 loud calls (see colobus and blue monkey accounts).

5. *Slapping and scrubbing the ground, striking toward opponent, and feinted attack.* Strong threats made by most monkeys during aggressive encounters at close range.

6. *False pursuit and false flight.* Distinguished from real pursuit by failure to overtake and a leisurely pace. False pursuits typically involve a lower-ranking monkey retaliating for a real pursuit or attack by a superior. The pursued monkey is clearly not intimidated and may at any moment reverse the roles.

Submissive, Appeasing, and Friendly Displays

7. *Social presenting.* Presenting consists of standing or crouching with the rear end pointing toward the receiver and tail diverted to display the genitalia (fig. 26.4). Sexual and social presenting look so similar that telling one from the other may hinge on the identity and condition of the presenter

Fig. 26.4. *Social presenting* by baboons: juvenile female to infant.

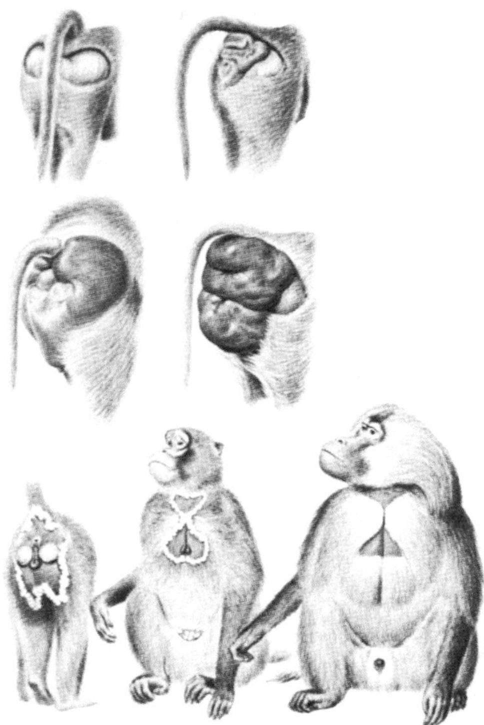

Fig. 26.5. Baboon posteriors. *Top:* savanna baboon male callosities (*left*) and female in estrous. *Middle:* hamadryas female estrous swelling (*right*) and mimicry thereof in the male (*left*). *Bottom:* sexual skin of female gelada (*left*), and representation thereof on the chests of the female and the male (*right*). (From Wickler 1967.)

and the response of the receiver. When a male or a female whose sexual skin indicates nonestrus presents and the receiver fails to copulate (although he/she may mount and thrust in response to social presenting), it is clearly *social presenting.* In its commonest form, presenting is a brief greeting given by one monkey as it passes before another, usually higher-ranking one, which may respond (if at all) by reaching out and touching the presenter's rump, by anogenital inspection, or the like.

Social presenting is the most obvious and important display of peaceable intentions among primates and probably the most effective in appeasing aggression, since aggression and sex are closely linked, both being mediated by male sex hormones (3). The nearer social presenting remains to sexual presenting the more effective it should be. Mimicry of the estrous female's sexual skin color or swelling by males, as seen in some of the most sociable and aggressive multimale societies (e.g., hamadryas and gelada baboons, red and olive colobus) may have evolved as a means of enhancing the effectiveness of social presenting (15) (fig. 26.5). It seems equally plausible that nonestrous females, too, mimic the condition of estrus for the same purpose: to appease the aggression of dominant individuals. The colorful sexual skin worn by females of some of these same species when not in estrus could have evolved in this way. One might go further and suggest that the huge sexual swellings seen in baboons and macaques have evolved through competition between social and sexual presenting, as females in heat were driven to ever larger, more colorful swellings to advertise their condition unmistakably (author's speculation).

8. *Grooming invitation.* Most monkeys have specific postures that signal their desire to be groomed or (less often) to groom (e.g., *lip-smacking,* grabbing a partner's forelock and pulling its head down). Grooming may be directed to the desired area, say the chest, back, or genitals, by lying, sitting, or standing in front of the intended groomer (figs. 27.4, 27.10).

9. *Play invitation.* A lurching, "drunken" run, swinging from a branch, jumping, somersaulting, mouthing, and other antic behavior performed usually by the young while displaying the "play face" (see below).

Facial Expressions

Dominance/Threat Expressions:

10. *Staring.* A fixed intense look, the eyes only slightly widened, eyelids barely raised, and the lips with perhaps a slight pout (fig. 26.6). Staring contrasts with the glancing and vague looks of monkeys during peaceful interactions. In baboons, mangabeys,

Fig. 26.6. *Staring* display of a spot-nosed guenon (from a photo in Gautier and Gautier-Hion 1977).

Fig. 26.7. *Tension yawning* of a male mandrill.

and the talapoin, the stare is intensified by raising the eyebrows and retracting the scalp to display contrastingly colored eyelids, and the face skin is stretched by pulling the ears back. In the guenons, movements of the ears, scalp, and brows are concealed by hair.

11. *Staring with open mouth.* Mouth opened more or less, frequently in an O-shape, teeth covered. This display is often associated with head-lowering (baboons, macaques) or head-bobbing (guenons, mangabeys, hamadryas), and with both bobbing and arm-pumping in forest guenons.

12. *Bared-teeth threat.* Canines exposed (fig. 27.11). Also linked with staring, it can be either offensive or defensive.

13. *Yawning.* Displays the canines of adult males, sometimes in directed threat (e.g., a baboon harassing a rival that is in consortship with an estrous female), but more often as an expression of tension (fig. 26.7), evoked, for example, by the presence of a rival group or predator. Its main function may be to advertise status ("here is an adult male with long, sharp teeth").

Ambivalent Expressions or Head Movements:

14. *Head-shaking.* In conflict situations, as when undecided whether to approach or withdraw, many monkeys shake their heads. Like lip-smacking, presenting, and grimacing, head-shaking indicates a sociable tendency and is not to be confused with the head-bobbing that accompanies staring, branch-shaking, and such (no. 3). DeBrazza's monkey head-shakes while making closed-mouth chewing movements. The gray-cheeked mangabey and mandrill both shake their heads vigorously, the one while lip-smacking and staring, the other while grimacing (4).

Submissive, Appeasing, and Friendly Expressions:

15. *Rapid-glancing and looking away.* A threatened monkey may respond to a stare or other threat with fearful rapid-glancing (alternately looking toward and away from the opponent), by turning away and deliberately looking in the opposite direction (*visual cutoff*; cf. *looking away* in cat family), by feigned indifference (pretending not to notice a stare), or by feigned preoccupation (with grooming or with some object).

16. *Fear grimace.* Lips retracted and turned up, showing clenched teeth. Displayed by virtually all Old World monkeys, the grimace is both an expression of fear, often accompanied by fearful calls ("gecker" scream, shriek), and an appeasing or reassuring signal that functions to reduce aggressive and flight tendencies. It is comparable to and often alternates with *lip-smacking* and *presenting.* The most exaggerated grimace is given by the gelada baboon, which actually folds back the upper lip ("lip flip") to reveal a large expanse of pink—and in males their fearsome canines (fig. 26.8).

17. *Lip-smacking.* Common in most Old World monkeys, *lip-smacking* may derive from teat sucking (1) but has evolved into a social display that generally signifies peaceable intentions. As such it facilitates approach to or by individuals that may otherwise feel insecure, and is the common prelude to *social grooming,* infant-handling, and copulation (4). The degree of lip movement and tongue protrusion varies among species. This positive social role is particularly clear in baboons and mangabeys. In the guenons, however, *lip-smacking* is rarely seen as a preliminary to social encounters or as a friendly interaction. In these less socially evolved monkeys, it is mainly confined to grooming. It also appears during aggressive encounters, often in combination with staring, head-bobbing, and other threat behaviors, suggesting the performer is beset by ambivalent emotions. In the mandrill, lip-smacking appears only during *social grooming.*

18. *Teeth-chattering.* Similar to and perhaps derived from *lip-smacking,* but less common. For example, male chacma and hamadryas baboons, and white-collared mangabeys, may lip-smack or teeth-chatter as a sign of friendly intentions when approaching a presenting female (4). But when a patas monkey gnashes its teeth it denotes a combination of threat and fear (5).

19. *Play face.* Also known as the "relaxed open-mouth face" (12). The teeth are bared as in bared-teeth threat but there is an obvious difference in the eyes, which are often slitted or partially closed, never staring (see fig. 29.5*c*).

Fig. 26.8. The gelada *lip flip* (from a photo in Spivak 1971).

Sexual Expression:

20. *Pouting.* Protruding, pouting lips are often a sign of sexual receptiveness in a female, and may or may not be combined with copulatory calls in baboons, mangabeys, macaques, guenons, and patas. Pouting is also seen during copulation in most species.

VOCAL SIGNALS. Considering that Old World monkeys share a common ancestor, it should not be surprising if the calls of closely related species sound alike. Differences develop over time because of genetic drift in isolated populations, or in response to selection. There seems to have been little selection for the close-range calls exchanged between family and group members to become different, and in the case of alerting and alarm signals, notably the chirps of females and young, positive advantages of having the same alarm calls may have selected against divergence (5, 8, 9).

The acoustical properties of different sounds in different environments are also important in determining call characteristics. In the case of alerting and scolding calls, selection dictates a call structure that permits rapid maximal diffusion without pinpointing the position or identity of the caller (4). Conversely, male loud calls, which function to rally scattered group members and keep rival groups at a distance, are designed to carry far and to indicate the location and identity of the caller. Wide-bandwidth noisy calls such as barks and roars are best suited for vertical and horizontal localization in open habitats. But in forests the foliage, restricted line-of-sight distance, and background noise interfere with sound propagation. Calls that carry farthest with least attenuation through the canopy are low-pitched, frequency-modulated pure sounds delivered in sequence, such as the *boom* calls of the blue monkey (see type 1 loud calls, below) (2, 10, 13, 14). Some high-pitched, clear calls, such as the red colobus's scream and the talapoin's whistle, are also vertically locatable and relatively unimpaired by masking noise.

Loud calls are most discrete and stereotyped in forest monkeys; not only the species but probably also the identity of the calling male is encoded. The *whoop-gobbles* of mangabeys are perhaps most specialized in this respect: the whoop serves as a cue that attracts the attention of all mangabeys within hearing, and the gobble then identifies the calling male through a set temporal pattern of broadband pulses. Unlike most calls, in which nuances of pitch and quality are rapidly lost through attenuation and reverberation, pulse-repetition rate is not degraded by distance (13). In terrestrial species such as baboons, harsh calls are often emitted at close range and are typically accompanied by facial displays (2). Such calls may function at least partly to direct attention to facial expressions, resulting in compound visual-vocal signals.

The Vocal Repertoire. The watchdog role adult males play in most monkey societies has led to the evolution of calls peculiar to this class, whereas "conversational" noises are uttered primarily or only by females and young, at least in 1-male systems where the resident male is in effect attached to the group only through the adult females (5, 9, 10).

Male Loud Calls:

Males of most species emit loud calls of 2 different types: (1) a new addition to the male repertoire that is acquired with maturity, and (2) calls given in response to stimuli that may disrupt the troop (4). The latter derive from alarm and aggression calls.

Type 1. Examples are the *boom* calls of the blue, spot-nosed, and mona guenons, the bark of various other guenons, the mangabeys' *whoop-gobble,* the *grunt-roars* of baboons, and the croaking roar of the black-and-white colobus (2, 5, 9). These calls tend to be the most distinctive and far-

Fig. 26.9. Male spot-nosed monkey giving boom call (from a photo in Gautier and Gautier-Hion 1977).

reaching in the species' repertoire, and are made more resonant by the presence of vocal sacs in some guenons (e.g., blue, spot-nosed, and DeBrazza's) and an enlarged larynx in black-and-white colobus. The calls are uttered typically at the beginning and sometimes the end of the day in a regular ritual, either responsively or spontaneously (fig. 26.9). In some forest guenons, types 1 and 2 loud calls are linked in the same stereotyped sequence (e.g., the crowned guenon).

In 1-male social systems, type 1 calls are emitted only by the dominant animal, even when a troop contains other adult males (see blue monkey account).

Type 2. Examples are the baboon's 2-phase bark, the blue monkey's *"pyow"* and *"ka-train"* calls, and the redtail's *"hack."* Such calls are the immediate response to any perceived disturbance to the troop, notably the proximity of another troop or a predator. Type 2 calls are less species-specific and stereotyped than type 1 calls, becoming less stable and more variable with increasing excitation. For instance, the bark of a spot-nose and *pyow* of a blue monkey sound very similar (4). Type 2 calls develop in subadults and are not necessarily inhibited in the presence of a dominant male. Calling regularly goes together with branch-shaking and leaping displays, whereas type 1 calls require concentrated effort and are usually given while the male is sitting still (fig. 26.9). Type 2 calls function to restore intertroop spacing after it has been disturbed, to warn the troop of danger, and to threaten the source of the disturbance. They express a higher level of excitation than type 1 calls (4).

Alerting and Alarm Calls Emitted by Females and Young:

3. *Alerting signals.* Derived from progression calls (no. 5), alerting signals are provoked by any slight change in the environment such as sudden clouding over, a gust of wind, or a small animal that rustles the

leaves. The tone of the progression calls intensifies as call energy frequencies become progressively concentrated. The change in call quality alerts other troop members. The chirps of spot-nosed, moustached, red-eared, blue, redtail, and most other guenons can hardly be told apart. Unlike male loud calls, which tend to inhibit calling by other troop members, the alerting calls given mainly by adult females are highly contagious.

4. *Alarm calls.* High-frequency sounds often covering a broad frequency range and of short duration, which may be uttered in rapid, sometimes long series. The chirps of guenons, shrill barks of baboons and macaques, "*chuckles*" of mangabeys, "*chittering*" of patas, and "*ists*" of the talapoin are examples. Sudden, relatively intense disturbances, such as the appearance of a predator, a loud noise, or a large animal bolting in the undergrowth, provoke alarm calls which vary in rate, duration, and intensity according to the duration and severity of the disturbance. Alarm calls are structurally close to and probably derive from calls linked with aggression. They are the high-pitched response of females and juveniles to the same stimuli that elicit type 2 alarm calls in adult males. The infectious alarm calls of females and young serve to bring the whole group to the same state of alertness. If the stimulus is strong, the troop flees; if weak, the animals will soon return to other activities. Should the stimulus continue but present no immediate danger, the troop may come close and scold like mobbing birds.

Most forest or semiwoodland monkeys have 2 different alarm signals, one for avian and the other for ground predators, for example, the "*ka-train*" and "*pyow*" calls of blue monkeys, and the "*rraup*" and "*nyow*" of vervets, which have yet a third call specifically for snake predators (see species accounts). In young monkeys, a leaf or any other falling object will provoke the avian alarm call. Adults usually reserve the call for the 2 really dangerous raptors, marital and crowned hawk eagles. Troop members respond appropriately: running for the trees if on the ground *freezing* or dropping into dense foliage if exposed in a tree.

Group-Cohesion Calls:

With the possible exception of colobus monkeys, seemingly all Old World monkeys have calls that serve to maintain cohesion of the social unit by repeated vocal exchanges between members (cf. social mongooses) (4). This function is particularly important for forest monkeys and other species that live in habitats where visibility is obstructed. Apart from the usual sounds (no. 5) that monkeys exchange during troop progressions, calls uttered by troop members that have lost, are searching for, or are soliciting contact with their group also function to preserve social cohesion. The different types of calls may intergrade completely, as in the talapoin, or be at least partially discrete and structurally more diverse as in many guenons (see vervet account, Vocal Communication). The most variable cohesion calls are given by the most sociable species (mangabeys, baboons) and are generally combined with visual and tactile signals during interindividual exchanges (4).

5. *Progression calls.* Usually nasal grunts of short range (mangabeys, baboons, guenons) uttered without change of facial expression. A group at rest is generally silent, but as soon as the members begin stirring they start vocalizing. The calls are not addressed to anyone in particular but evoke a response in kind from all group members except adult males. The call changes in pitch, emission frequency, and so on with increasing size and age, in such a way that troop members are probably kept informed of the sex and age of their nearest neighbors through their calls alone. In males, the change occurs abruptly at adolescence, when the progression grunt drops from shrill to low pitch, just as a boy's voice changes. In many guenons, though, progression calls disappear from the male's repertoire to be replaced by type 1 loud calls, which also play an important role in group cohesion and mobilization (4, 5). Call rate and volume vary with the activity and external conditions, increasing with the potential risk of losing contact while in dense vegetation or dim light, as when troop members are dispersed or moving rapidly. Progression calls increase in rate and excitement following a disturbance or when, for instance, a troop arrives to feed at a tree in fruit.

6. *Isolation calls.* Derived from progression calls in some species (e.g., DeBrazza's and crowned guenon), structurally new and pure (moustached guenon), or "noisy" (spot-nosed guenon). Infants and juveniles are the main callers, but adults that become separated from the troop also utter isolation calls. Thus, a separated group of baboons gives shrill barks to reestablish contact with the troop.

Calls Associated with Peaceful Individual Relations:

Apparently closely related to and derived from cohesion calls, these quiet calls are voiced during peaceful close interactions between individuals. Visual signals such as lip-smacking and presenting are often associated with exchanges of close-range calls.

7. *Calls exchanged between dominant and subordinate individuals.* In vervets a subordinate male gives "*woof-woof*" and "*wa*" calls to an approaching dominant male, females say "*wa-waa*" and juveniles "*rraugh*" to a dominant female (see vervet account). "Greeting grunts" of faster or slower tempo than progression grunts are emitted by male baboons and mangabeys to reassure approaching juveniles.

8. *Infant-mother exchanges.* The calls of infant guenons may be structurally similar to cohesion calls in some species, or shrill calls of an entirely different, specific structure. Blue, redtail, spot-nosed, and moustached guenons emit trills that are lacking in the repertoires of crowned and DeBrazza's guenons. Through frequency modulation and varying intensity, infant calls acquire tones that sound plaintive, sometimes interrogatory and peremptory, and even violent, given in response to increasingly frustrating situations. For instance, talapoin infants *gecker* when contact with their mother is slightly disturbed, whistle if contact is broken, and break into screams when seriously distressed. The mother responds to her infant's calls in kind or with progression calls (4).

Aggression and Flight Signals. Signals associated with aggression and flight make an almost perfectly graded system, in which visual displays are accompanied by a variety of vocal and nonvocal sounds (4, 12). In general, calls that accompany threat behavior are low-pitched, whereas those associated with fear and flight are shrill. Monkeys beset by conflicting emotions emit composite calls in which low-pitched and shrill components are blended.

9. *Aggressive calls.* Short, low-pitched calls with rolling sonority which derive from cohesion and peaceful-interaction calls. Examples are the baboon's *2-phase grunt* and the grunts of various guenons that have cohesion grunts of the vervet type. With increasing intensity, the structure of threat calls diversifies. They remain low-pitched but their tempo is modified by the effects of movements, breathing rate, and other physiological factors. *Two-phase grunts* develop into roars in baboons and macaques, and grunts of increasing duration and intensity turn into barks in guenons, mangabeys, and patas. Adult males are more likely to give roars and barks than grunts.

Low-pitched rhythmic calls are still more commonly associated with threat, such as the repeated "*uh-uh*" of baboons, "*huh-huh*" of the patas, "*chutters*" of the vervet, and similar emissions of other guenons, the talapoin, and mangabeys. Such emissions make a bridge between aggression and flight calls. There is a distinction between calls associated with alarm and those specifically linked to aggressive interactions among monkeys. Adult females and immature subordinates of probably most guenons and mangabeys often begin aggressive interactions with noisy, comparatively low-pitched, rhythmic calls which apparently derive from the *gecker* of infants and juveniles. Similarly, the "yakking" of baboons derives from chirplike clicking sounds emitted by infants.

10. *Distress and defensive calls.* As the intensity of an interaction increases, rhythmic vocalizations become shriller and develop into high-frequency calls, either pure-toned whistles and squeals or harsher screeches and screams, in practically all species. Such calls, indicating a high level of arousal, function as defensive signals which inhibit aggression and evoke very strong protective responses in other troop members, especially mothers and adult males.

Copulation Calls. Sometimes males but more often females emit copulation calls. Patas and mangabeys, for instance, emit particular calls that indicate receptivity and may encourage males to copulate, and the same calls may occur during copulation. Emission rate may vary with breathing rhythm, call length, and intensity depending on the female's state of excitation.

OLFACTORY AND TACTILE SIGNALS. Although olfaction is considered generally the least important communication channel in higher primates, there are indications that odors play a significant role in the social and reproductive behavior of at least some primates, including humans. The vervet monkey and mandrill both have chest glands. An odor characteristic of guenons is localized in a skin fold behind the head and helps to identify different species and individuals (4).

A number of highly stereotyped behavior patterns involve contact with genital and other areas that release odors. The distinction between tactile and olfactory signals is completely unclear in such cases. But when female gray-cheeked mangabeys sniff

Fig. 26.10. *Mutual anogenital sniffing* by mangabeys (redrawn from Gautier and Gautier-Hion 1977).

one another's armpits, and raise an arm when approached, a connection with the strong odor of flour that emanates from the armpits seems quite obvious. The back is sniffed by black-and-white colobus, the side and back of the neck by talapoins, and juvenile male spot-nosed monkeys approach and sniff the muzzle of an adult male that has given loud calls (4). That sniffing may also be employed to learn or confirm the identity of an individual is suggested by the fact that the presence of a newcomer, including a new baby, stimulates olfactory and tactile exchanges. So do serious disturbances, and competitive interactions between adult males.

1. *Nose-to-nose contact.* Described in most species, but relatively rare in baboons and mangabeys, in which it may retain its original function of finding out what another animal ate. It is the usual greeting in guenons, and the prelude to other contact such as grooming, anogenital sniffing, play, heterosexual mounting, and so on.

2. *"Kissing" (mouth-to-mouth contact).* Perhaps equally widespread and more tactile than olfactory, *kissing* is often seen when 2 monkeys are holding each other, face-to-face. *Kissing,* accompanied by *grinning, lip-smacking, tongue-clicking,* and the like, is particularly common between adult males.

3. *Genital-sniffing.* Widespread or universal, especially in response to female presenting. Males also routinely check females by sniffing the anogenital area directly or by probing and afterward sniffing their fingers. Such sniffing is uncommon in most guenons compared to baboons and mangabeys. But stereotyped *mutual genital sniffing* is practiced by some guenons (blue, spot-nosed, crowned, moustached) as males circle each other head-to-tail. Gray-cheeked mangabeys sniff one another in a variety of unlikely-looking attitudes (fig. 26.10). Sniffing the newborn's anogenital area, with or without manipulation, is also very widespread. Olfactory interest is particularly clear when adults turn an infant upside down and hold its genitals below their nose. In vervets, only females join in these examinations; in mangabeys and talapoins the whole group takes an interest.

4. *Social grooming.* Already discussed in chapter 24 and identified as the most important kind of tactile behavior in primates. Virtually the same motor patterns are shared by all species and all parts of the body are groomed, but frequency and duration of grooming and the participation of adult males vary between species. Male baboons participate in *social grooming* (though less than females), whereas adult-male guenons and mangabeys receive most of the grooming without reciprocating (or pay back by rapidly and even brutally hitting a partner's coat instead of grooming properly). Mother-infant grooming is important in the baboon lineage, whereas mangabey and guenon mothers groom small infants rarely and perfunctorily, but do groom juvenile offspring. Being groomed is clearly a pleasurable experience and even if not reciprocated can be socially rewarding for the groomer. Thus, it offers a way for lower-ranking animals to approach and make friendly contact with dominant individuals. Juvenile and adult females that want to handle a small infant attempt to gain access by grooming its mother. Mothers help soothe and distract offspring during weaning by grooming them, and may likewise divert a male that is behaving aggressively toward another troop member.

5. *Contact postures.* Sitting together, *huddling, tail-twining.* Affinities between individuals are identifiable from the composition of sleeping and grooming groups. The same animals often continue

to associate in such subgroups year after year. Baboons do not huddle closely, although hamadryas harems may huddle on the ground during daytime rest periods (7). Long-tailed monkeys, including various guenons and the gray-cheeked mangabey, often twine their tails together while huddling on a branch.

6. *Other tactile signals.* Mounting without copulation, generally considered an expression of dominance, is often seen before or after grooming, in play, and during aggressive interactions. It is rare between adult females, although mangabeys, baboons, and talapoin mothers will mount females that present themselves. Touching as a gesture of reassurance or appeasement is common in baboons, but does not exist in forest guenons. Touching and manipulating the genitals, especially of males, is seen in patas, vervet, and spot-nosed monkeys, the savanna baboon, and at least 1 mangabey (4).

References

1. Anthoney 1968. 2. Brown 1982. 3. Davis 1964. 4. Gautier and Gautier-Hion 1977. 5. ——— 1982. 6. Kingdon 1980. 7. Kummer 1968. 8. Marler 1965. 9. ——— 1978*a*. 10. Quris 1980. 11. Struhsaker and Leland 1979. 12. Van Hoof 1967. 13. Waser 1982*a*. 14. Waser and Waser 1977. 15. Wickler 1967.

Chapter 27

Guenons, Mangabeys, and Baboons

Subfamily Cercopithecinae

African Species

Genus Cercopithecus, guenons
C. nictitans, spot-nosed
C. mitis, blue, Sykes, or diademed
C. l'hoesti, l'Hoest's
C. preussi, Preuss's
C. cephus, moustached
C. erythrotis, red-eared
**C. sclateri*, Sclater's
C. ascanius, redtail
C. petaurista, lesser spot-nosed
C. erythrogaster, red-bellied
C. dryas, dryas
C. mona, mona
C. campbelli, Campbell's
C. denti, Dent's
C. wolfi, Wolf's
C. pogonias, crowned
C. neglectus, DeBrazza's
C. hamlyni, Hamlyn's or owl-faced
C. diana, diana monkey
C. salongo, salongo
C. aethiops, vervet, grivet, or green monkey
Miopithecus talapoin, talapoin
Allenopithecus nigroviridis, Allen's swamp monkey
Erythrocebus patas, patas monkey

Genus Cercocebus, mangabeys
C. torquatus, white-collared or sooty
C. galeritus, crested or agile
C. albigena, gray-cheeked
C. aterrimus, black

Baboons
Papio cynocephalus, savanna baboon
P. hamadryas, hamadryas
Mandrillus sphinx, mandrill
M. leucophaeus, drill
Theropithecus gelada, gelada

Macaques
Macaca sylvana, Barbary macaque

*Species listed by reference 13, not by 10 or 24

TRAITS. See Family Traits, chapter 26. All but 6 of Africa's 39 monkeys belong to this subfamily. In the following annotated list, related species are grouped together.

GUENONS

Genus *Cercopithecus*. The main source of bewilderment, this genus includes 17 (ref. 24) to 20 (refs. 10, 12) species, or half of Africa's complement of Old World monkeys. They are so closely related that probably most could hybridize—although it rarely happens in the wild (but see blue monkey account).

According to Jonathan Kingdon's analysis of this genus, the guenons divided early into two different lines of large monkeys (fig. 27.1) (13). In each line small, highly arboreal monkeys subsequently evolved and radiated into a variety of forest niches.

1. The *C. nictitans-cephus* line (fig. 27.2). The members of this lineage are adapted to various types of dense secondary growth both within and outside the main forest blocks.

Superspecies *Cercopithecus (nictitans)*
C. nictitans, spot-nosed guenon
C. mitis, blue or Sykes monkey

Apart from the white nose spot of *nictitans* and the white diadem and bib of some *mitis* races, these monkeys are drab-colored, relatively slow, and retiring in

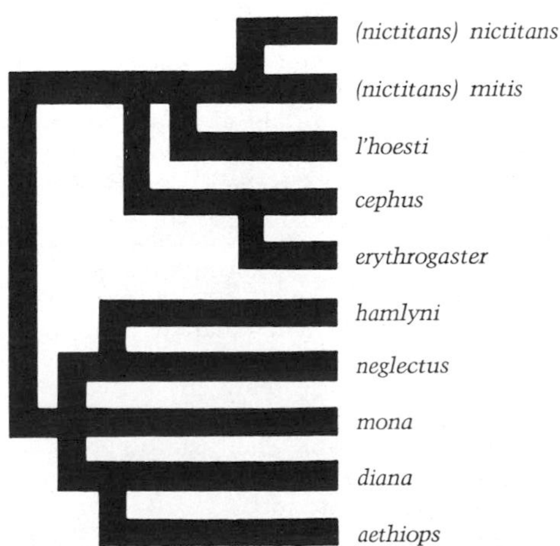

Fig. 27.1. Dendrogram showing possible affinities between guenon species groups (from Kingdon 1980).

Fig. 27.2. Representatives of the *C. nictitans/ C. cephus* group. *Upper row:* from left to right, *nictitans, mitis, cephus. Lower row: erythrogaster, ascanius, petaurista.* (Paintings by J. Kingdon.)

their habits, relying more on vocal than on visual advertising. However, the spot-nosed monkey lives in the canopy of West Africa's high primary forest, whereas the blue monkey stays in the middle layers of rain forest and is also at home in other types of forest—it is in fact the most widespread guenon next to the vervet (see species account).

Superspecies *Cercopithecus (l'hoesti)*
C. l'hoesti, l'Hoest's guenon
C. preussi, Preuss's guenon

Adapted to montane or secondary forest with discontinuous canopy and dense undergrowth, two widely different populations of this monkey exist in Cameroon and western Uganda/eastern Zaire. The former, *C. preussi,* distinctive enough to be regarded by some as a separate species (10, 13), probably shares common ancestry with the blue monkey; it has a facial mask (black nose with paler margins and black mouth) which is the antithesis of *nictitans'* pattern. This could represent a mechanism for achieving reproductive isolation from *nictitans* (12). The eastern population of l'Hoest's guenon is distinguished by a prominent white bib.

Superspecies *Cercopithecus (cephus)*
C. cephus, moustached guenon
C. erythrotis, red-eared guenon
C. sclateri, Sclater's guenon
C. ascanius, redtail guenon
C. petaurista, lesser spot-nosed guenon
C. erythrogaster, red-bellied guenon

Small (2–6 kg), agile monkeys much alike in size and proportions, the members of this group include some of the most colorful, active, and successful Old World monkeys. They are more abundant than most associated species, reaching densities of up to 228 redtails/km^2 (18), and are almost continuously distributed in the forest block. The members of the (*cephus*) superspecies replace one another geographically, with narrow zones of overlap (typically along divides between major watersheds) where some hybridization may occur. Although differences in color and pattern, especially of face and tail, make the members of this group look as unalike as more distantly related guenons, Kingdon (13) found that all the different facial patterns could be derived from a relatively undifferentiated monkey similar to but smaller than *C. nictitans.*

2. The *C. diana-mona* line. The species in this lineage are more diverse than the above line, implying an earlier radiation (fig. 27.3). Ecological adaptations range from the semiterrestrial DeBrazza's monkey to the completely arboreal *diana* and *mona.*

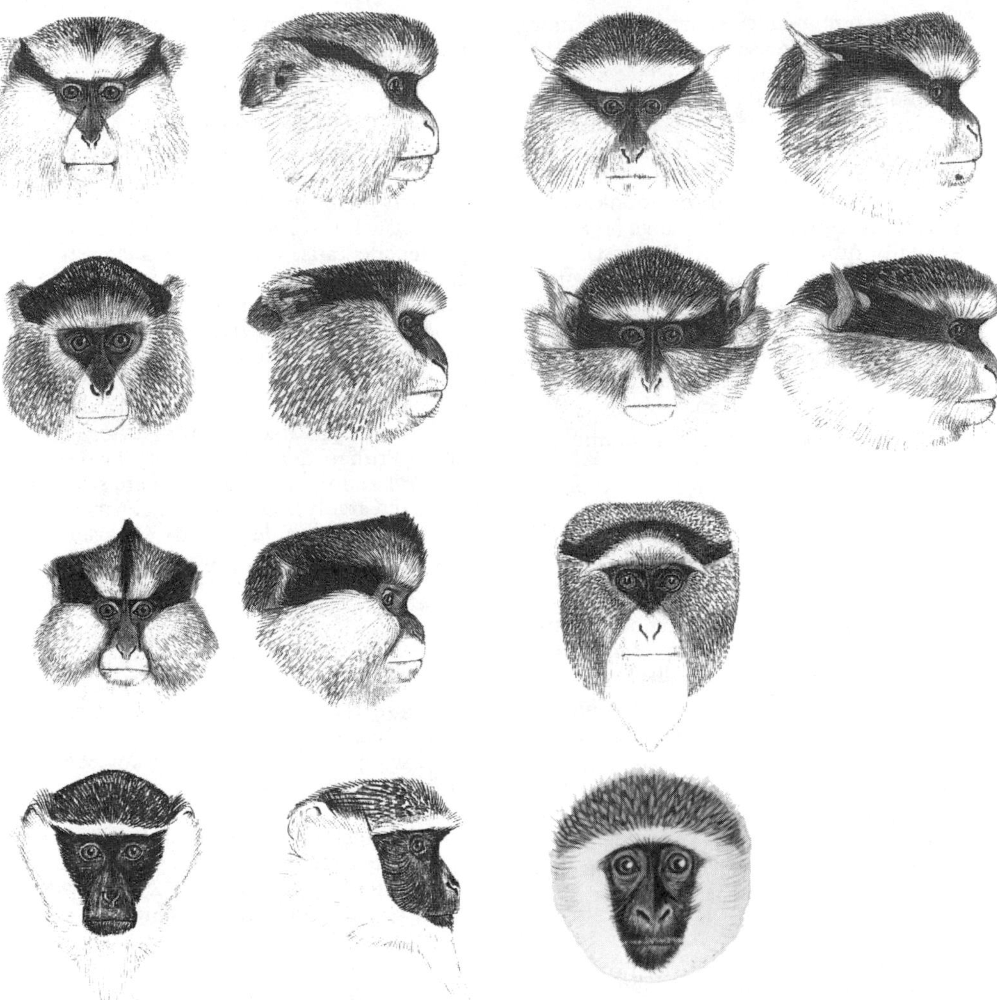

Fig. 27.3. Facial patterns in the *C. diana/mona* species group. *Top row:* from left to right, *mona, denti. Second row: campbelli, wolfi. Third row: pogonias, neglectus. Bottom row: diana, aethiops.* (Paintings by J. Kingdon.)

Superspecies *Cercopithecus (mona)*
- *C. mona,* mona monkey
- *C. denti,* Dent's guenon
- *C. campbelli,* Campbell's guenon
- *C. wolfi,* Wolf's guenon
- *C. pogonias,* crowned guenon
- *C. hamlyni,* Hamlyn's or owl-faced guenon
- *C. neglectus,* DeBrazza's monkey

Like the (*cephus*) species group, the members of the *mona* group replace one another geographically, except for an overlap zone between the mona and crowned guenons in Cameroon and eastern Nigeria (where hybridization occurs [16]). Members of this group are enough alike that *campbelli* and/or *denti* and *wolfi* are considered by some to be merely *mona* subspecies (8, 24) (fig. 27.3). As in the diana monkey, the limbs and body are most conspicuously marked; the face is naked with a dark blue or gray mask which ends at the nose, bordering a flesh-pink muzzle, making expressions of the mouth unusually visible. The more brightly colored species have conspicuous ear tufts. *Mona* species have nearly identical sneeze-like alarm calls, and adult males give resonant *boom* calls, typically followed by a series of *hack* calls

(16). The crowned guenon, smaller, more active, and far wider-ranging than the mona monkey, may have displaced it from the Lower Congo Basin. *C. mona* is most successful in mangrove and secondary forest, whereas *pogonias* is a dominant monkey in high, open forest (13).

DeBrazza's and Hamlyn's monkeys are similar in being relatively large (4–8 kg), sexually dimorphic, semi-terrestrial, and shy. They live in small 1-male units often containing only 1 adult female, and hide from danger, relying on their drab, grizzled (agouti) coat color. Resident males use similar deep booming calls to advertise territorial status and position.

Superspecies *Cercopithecus (diana)*
C. diana, diana monkey
C. salongo, salongo

One of the most colorful and active of all guenons, the diana monkey has a black face framed by a white beard and wide bib, like a bull's eye (fig. 27.3). It lives in the upper canopy of mature forests west of Dahomey. Males give *hack* calls often combined with a spectacular leaping display.

Superspecies *Cercopithecus (aethiops)*
C. aethiops, vervet or green monkey

The only member of the genus that is adapted to savanna, the vervet may be most closely related to the diana monkey, which it resembles in having a black face framed by a white ruff (see species account).

Species Related to the Guenons

Miopithecus talapoin, talapoin

The smallest member of the family at 1–1.5 kg, drab-colored and very slightly dimorphic, this monkey lives in large, multimale troops often numbering 60–80 monkeys or more, and is more insectivorous and less frugivorous than guenons. It is very active and forages at various levels in various types of forest, but is never found far from water since it sleeps in swamps or trees overhanging water (5).

Allenopithecus nigroviridis, Allen's swamp monkey

An unstudied, small (2.5–5 kg), sturdy monkey with nondescript dark olive-green coloring and pink face and ears. It inhabits swamp forests in the Congo Basin.

Erythrocebus patas, patas monkey

The nearest thing to a greyhound in the monkey world, the patas is a guenon that has been remodeled for savanna life by acquiring the build and speed to outrun most predators—at least as far as the nearest trees. Males are much bigger than females, weighing over 10 kg to 7 kg for females.

MANGABEYS. Mangabeys (*Cercocebus* species) may not look much like baboons, but they move, behave, and sound a lot like them (fig. 26.3). Sizeable (7–11 kg), leggy forest monkeys with long, semiprehensile tails, coarse, rather shaggy coats, and generally drab coloration, they move more slowly and deliberately than most guenons. They live in small- to medium-sized multimale troops (average 15–17, range 6–36 [19]) and have extensive (400 ha), overlapping home ranges. Estrous females develop large sexual swellings (see under Primate Sexual Behavior). Two species are mainly arboreal and 2 are semiterrestrial. Largely frugivorous, mangabeys have massive incisors (especially *albigena*) which enable them to eat fruits and seeds that are too hard or tough for guenons to open. Mangabeys are among the more vociferous monkeys, emitting along with the usual shrieks and screams chuckles, barks, grunts, and roars. Most distinctive and far-carrying are the *whoop-gobble* spacing calls of adult males (19).

At one point, when the rain forest was more extensive, mangabeys ranged from the Atlantic to the Indian Ocean, but they are now confined to the forests of West and Central Africa except for 2 relict populations of the crested mangabey, 1 on Kenya's Tana River and the other in montane forests of Tanzania's Uzungwa Mountains (11).

C. albigena, gray-cheeked mangabey
C. aterrimus, black mangabey

These 2 species are very much alike in behavior and ecology, down to a nearly identical *whoop-gobble.* There is evidence that they diverged from the baboon line more recently than the other 2 mangabeys. Both are largely arboreal and come to the ground for only short periods. The black mangabey lives only in central Zaire and remains unstudied.

C. torquatus, white-collared mangabey
C. galeritus, crested or agile mangabey

Smaller with less robust incisors, these 2 are semiterrestrial and closely associated with gallery and seasonally flooded forests.

BABOONS. Strictly speaking, only *Papio* species are called baboons. However, the drill and mandrill and the gelada all have the long, doglike head and relatively short tail, the large size (18–40 kg), and the pronounced sexual dimorphism that distinguish this kind of terrestrially adapted monkey (figs. 26.7, 26.8, 27.11).

Papio cynocephalus, savanna baboon
P. hamadryas, hamadryas baboon

Many zoologists treat the 5 most distinctive geographical forms of the savanna baboon as separate species (10), but there are no clear-cut distinctions in their behavior or ecology to compare with the differences between them and the hamadryas, which lives in the semidesert Horn of Africa (with an Asian connection across the Red Sea on the Arabian Peninsula). This small baboon (adult males weigh only 18–19 kg) forms enormous bands (up to 350, average 50–100) which sleep at night on high cliffs and forage by day for grasses, roots, flowers, and occasional fruits. The hamadryas's unique harem system is outlined in chapter 24 (see Social Organization, type 5).

Theropithecus gelada, gelada

Slightly smaller than the hamadryas (males 15–20 kg), the gelada is the sole surviving species of a genus that 50,000 years ago ranged the African savanna south all the way to the Cape. A relatively slow monkey with very little climbing ability, the savanna gelada may have been exterminated by man and/or displaced by the savanna baboon (9). The surviving gelada inhabits Ethiopian montane grassland in the vicinity of towering cliffs where it sleeps and takes refuge. It subsists almost exclusively on grass (all different parts), which it is adapted to consume more quickly than other baboons (15). Geladas forage intensively within a very small area, remaining seated most of the time or shuffling along on their well-padded bottoms (3, 9). Possibly because its backside is usually invisible, the gelada has evolved a patch of red, beaded sexual skin on the chest which charts the female's reproductive cycle and is mimicked by males (fig. 26.5) (2, 20). Like the hamadryas, geladas live in large bands (numbering up to 600 but generally from 50 to 250) which are composed of small 1-male harems (2–8 females) and bachelor groups of 5–10 subadult and young-adult males. Females are not kidnapped as infants but lured away as adolescents by single young-adult males that attach themselves as followers of a harem. Or an old male may be challenged and replaced as harem master. Female members associate closely with their relatives (especially mother and oldest daughter), the dominant female decides where the harem moves (forcing the male to follow), and female coalitions may chastise overly aggressive males (2, 3, 21).

Mandrillus leucophaeus, drill
M. sphinx, mandrill

These dark-coated, powerfully built, nearly tailless inhabitants of the West African rain forest are considered to represent baboons that reinvaded the rain forest (fig. 26.7). They occur only on the Atlantic drainage between the Niger and Congo river basins, the drill north and the mandrill south of the Sanaga River, Cameroon (7). Both species forage mainly on the forest floor in multimale troops sometimes numbering in the hundreds but averaging 9–55. Home ranges may be as large as 50 km^2. It is unclear whether there is any overlap in the distribution of these 2 least-known baboons. Males are over twice the size of females and stand out in a crowd: their red, pink, blue, and lilac color patterns appear luminous in the blue-green gloom of the forest floor. The ridged muzzle of a mature male mandrill bears a fair likeness of its penis (red) and scrotum (lilac), set off by a saffron-colored beard (20).

ASSOCIATIONS AMONG SPECIES. With so many species of monkeys inhabiting the African rain forest, one would expect to encounter quite a variety during a jungle walk or drive, just as one sees a variety of antelopes on the plains. In fact, the number of different monkeys at any one place is less than might be supposed, because so many species are allopatric: closely allied forms that replace one another geographically and fill the same ecological niche (1, 13). The maximum number of sympatric monkeys is 6–8 species. For instance, the red colobus, black-and-white colobus, blue, redtail, and l'Hoest's guenons, and gray-cheeked mangabey inhabit Uganda's Kibale Forest, along with the chimpanzee and in lumbered areas the savanna baboon (17). The red and Angola colobus (*C. angolensis*), mona, crowned, and l'Hoest's guenons, white-collared mangabey, and drill, plus the chimpanzee, occur at Idenau, Cameroon (16). The black-and-white colobus, spot-nosed, moustached, crowned, and DeBrazza's guenons, talapoin, and gray-cheeked and crested mangabeys, plus the chimpanzee and the gorilla, share the rain forest of northeastern Gabon (5). Outside the rain forest, up to 4 monkeys may be found at places where different habitat types converge, for example the savanna baboon, vervet, blue monkey, and black-and-white colobus on the Kenya coast (14).

Forest monkeys are often found in mixed troops containing up to 5 or even 6 different species, though associations of more than 2 or 3 species are unusual and ephemeral (6, 18). The tendency to associate with other species varies, from nil to slight in most ground-foragers such as baboons and drills, l'Hoest's, Hamlyn's, and DeBrazza's monkeys, and crested and white-collared mangabeys (also the apes), to constant in the case of certain arboreal species that live in semipermanent mixed troops. Thus, spot-nosed monkeys regularly associate with crowned and moustached guenons and, in different areas, with other members of the (*mona*) and (*cephus*) superspecies. The blue monkeys of Kibale Forest often associate with redtails. Other monkeys that are frequently found associated are: red colobus and redtail guenon; gray-cheeked mangabey and redtail or moustached guenon; mangabey and crowned or mona guenon; mangabey and spot-nosed or red-eared guenon.

Mixed troops are much commoner among African monkeys than among Asian or South American monkeys (1). The greater danger of avian predation, owing to the ubiquity of the crowned hawk eagle throughout African rain forests, may largely account for the difference (5, 17). But what would account for the fact that associations are more prevalent and permanent in West Africa than in Central and East Africa? A study in Gabon of a mixed troop of spot-nosed, crowned, and moustached guenons yielded many insights into the mutual benefits that may accrue from polyspecific associations (6) (see also [25]).

A wide-ranging species such as the crowned guenon or the gray-cheeked mangabey can benefit from the more detailed local knowledge possessed by species with small home ranges, like moustached and redtail guenons, to locate fruiting trees (6, 18). But on balance, it looks as if the smaller species, which are most vulnerable to eagle predation, have most to gain from the associations. The alternative for them, as illustrated by unmixed troops of moustached guenons, is to stick close to cover in less-productive habitats. The chance to gain access to more-productive open forest as inconspicuous members of polyspecific troops would explain why members of the *cephus* group (red-eared guenon in Cameroon, redtail in Uganda, lesser spot-nosed guenon in Ivory Coast, etc.) have among the highest association indexes (4, 6, 18). In Gabon, moustached guenons gain the same advantages by dispersing themselves in large troops of the smaller but wider-ranging and more active talapoin (5).

Other studies suggest that the species that associate most regularly and amicably have minimal overlap in their diets, larger species such as the spot-nosed and blue monkeys being more folivorous and the smaller species more insectivorous. There is virtually no diet overlap between red colobus and redtails, the monkeys that are most often associated in the Kibale Forest, and most interactions take the form of grooming, with colobus grooming redtails. A clear correlation was found between aggression and the degree of dietary overlap; thus most interactions of redtails with blues and mangabeys were competitive, occurred at feeding sites, and involved displacement of redtails by the larger species. Interspecific play was the least common form of interaction and involved only immature monkeys. In general, smaller and less numerous species joined larger, more numerous species (18).

Mixed troops of monkeys are comparable to mixed herds of antelopes and other ungulates (see antelope introduction).

References

1. Bourlière 1985. 2. Crook and Aldrich-Blake 1968. 3. Dunbar 1978*b*. 4. Galat-Luong and Galat 1979. 5. Gautier-Hion 1978. 6. Gautier-Hion, Quris, Gautier 1983. 7. Grubb 1973. 8. Haltenorth and Diller 1980. 9. Jolly, C. J. 1970. 10. Jolly, A. 1985. 11. Kavanagh 1984. 12. Kingdon 1971. 13. ——— 1980. 14. Moreno-Black and Maples 1977. 15. Napier and Napier 1967. 16. Struhsaker 1970. 17. ——— 1975. 18. ——— 1981. 19. Waser 1982*a*. 20. Wickler 1967.

Added References

21. Dunbar 1984. 22. ——— 1988. 23. Gautier-Hion and Gautier 1986. 24. Smuts et al. 1986. 25. Cords 1987. Gautier-Hion et al. 1988. Norris 1988. Zucker 1987.

Fig. 27.4. Blue monkeys grooming (from a photo by L. Leland).

Blue or Sykes Monkey

Cercopithecus mitis

TRAITS. A dark, stoutly built forest guenon with round facial disk and no beard. *Length:* (head and body) males 48–67 cm, females 44–52 cm, tail 55–109 cm. *Weight:* up to 12 kg (9), males 5–7 kg, females 3.5–4.5 kg (8). Fur thick and quite long, with a brow patch of bristling hair. *Coloration:* too geographically variable for a general description, ranging from drab to colorful (the Sykes monkey (*C. m. kolbi*) of Mt. Kenya being most colorful) (6, 9, 10); general color bluish to reddish, with long grizzled facial hair, forelimbs and tail usually black and always darker than body, face skin purple-black, crown hair pure black to grizzled gray, bordered in some populations by a white diadem, a showy white throat east of the Rift Valley; underparts vary from light gray to black. Genitalia uncolored.

DISTRIBUTION. *C. mitis* is the only forest guenon that is widespread outside the rain forest, occurring in eastern and southern Africa, where visitors to parks containing forest often see it. The spot-nosed guenon (*C. nictitans*), a relative close enough to be placed in the same superspecies, fills a similar ecological niche in West African forests (see catalogue of species in subfamily introduction) (9).

ECOLOGY. The blue monkey occurs in various evergreen forests, including rain forest, montane forest types up into the bamboo zone at above 3000 m, and coastal mangrove forest, and through the Southern Savanna wherever there are patches of forest. It is, nevertheless, a highly arboreal monkey that frequents the middle zone (12–20 m) in primary forest (1) and is intolerant of strong sunlight; on both counts it contrasts with the vervet.

DIET. *C. mitis* is an omnivore but more folivorous than most forest guenons, a difference that correlates with its larger size, slower locomotion, greater inactivity, broad habitat tolerance, and wide geographical range. Two troops studied in Uganda's Kibale Forest fed on 93 different plants (14). Fruits were taken most (43%), followed by foliage (21%), invertebrates (20%), and flowers (12%), with seeds, nectar, galls, and unidentified plant items making up the balance. The monkeys fed selectively, rely-

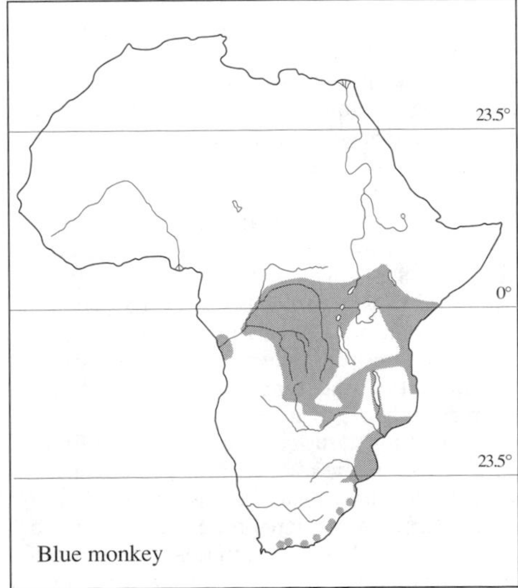

ing on only 10 different species for nearly ⅔ of the fruits and foliage they ate, and in each month tended to concentrate on just 1 or 2 specific foods that were eaten much less or not at all in other months. Thus, invertebrates were eaten more than anything else in May, September, and March; flowers were preferred in July and February; and foliage made up 41%–64% of the feeding observations from March to May. Slow-moving (e.g., caterpillars) and immobile insects were thought to make up the bulk of the invertebrate foods (see under Activity and Ranging). Blue monkeys may drink from tree holes after rain, by sucking or by dipping a hand in and then licking the fur, but succulent fruits and foliage usually provide adequate water.

By eating the most abundant available foods, blue monkeys can remain within a relatively small home range. The associated gray-cheeked mangabey must travel far more widely in Kibale Forest to find enough fruit and insects to meet its needs. The same study revealed feeding and dietary differences among different classes of blue monkeys. Adult females and subadult males fed more than predicted. Adult males devoted least time to feeding, concentrating more on fruit, whereas invertebrates were most important in the diets of immature classes. The diets of 2 neighboring troops were also very different. Among 74 different species of food plants that grew in both ranges, they shared only 33 and in general utilized them at different frequencies.

SOCIAL ORGANIZATION: territorial, female-bonded, 1-male units.

The blue monkey has a social organization that is archetypical of forest guenons, in which dominion over a group of females and dominion over their home range/territory go together (see Social Organization, chap. 24). Theoretically only such a resident male has the privilege of reproducing and the stakes are high: 5 Kibale Forest troops totaling 89 blue monkeys, averaging 17.8 animals (11–24) to a troop, included 6–18 females (average 10) (3). Such a high breeding sex ratio presupposes intense male competition, helping to explain the fact that males are a good 50% bigger than females.

According to most studies, the harem organization of the blue monkey is remarkably stable and peaceful, with resident males firmly in control for 2 years or more (1, 3, 5, 14). Related females assisted by their offspring actively defended their boundaries against neighboring troops, maintaining permanent, spacious territories averaging 61 ha in Kibale Forest (14), 16–26 ha in Uganda's Budongo Forest (1), and 13–14 ha in the Kenya Highlands (5). In a 2-year study of the 5 Kibale troops, composition changed only slightly, as a result of births, disappearance and probable mortality of a few infants and females, recruitment of the young into successive age classes, and the emigration of subadult males (14). Solitary males were notable for their absence. Only 8 were seen, which were evicted within a few weeks at most by the resident male, who was quick to chase any outsider that tried to approach the troop. Only resident males gave loud calls. Each troop reacted to its neighbors' calls about half the time. The resident male often responded in kind to *ka-train* alarm calls, but *pyows,* which evoked chirps and sometimes *phrased grunts* or trills from females and young, never elicited countercalling (it does in the Budongo population [1]) (see Vocal Communication). There was just 1 male turnover in 2 years; it occurred when an outsider joined a troop and took leadership from the resident male, which afterward hung around for several months in an apparent 2-male troop.

During a further, 20-month study involving apparently the same 5 troops (3), no solitary males were seen in 1994 hours of observation and all but 1 troop remained stable. But that troop experienced both turnovers and infanticides. Unusual circumstances surrounding this case will be discussed after first considering another study, the results of which seem to undermine the very concept of the 1-male harem.

A troop that was intensively studied in Kenya's Kakamega Forest for 6 months had not 1, not 2, but 13 different adult males in attendance during that period (17)! Five of them were present in each of 6 months and the other 8 were present in 1 to 4 different months. The troop occupied a territory of c. 24 ha, shared borders with 5 other troops, and included 18 females, of which 4–6 came into estrus and mated during the 6-month period of intensive study (the others had small offspring and remained sexually unreceptive). One of the 13 males was clearly dominant. He was known to have been the only resident male for the previous 2 years, during which he was seen to copulate with the females and to chase other males that attempted to join the troop. And he was again the only male when the troop was revisited 3 and

7 months after the study ended. During the 6-month study he was seen on all but 8 of the 113 days when the troop was followed and probably never actually left it. The other males, though, came and went; they spent part of the time alone and also visited other troops. The 4 "regulars" were seen in the troop 37–71 days; the shortest stay of an individually known animal was 5 days in 2 different months.

Despite his almost undisputed dominance, the alpha male's sexual performance made him seem more like the harem eunuch than the harem master: he performed fewer copulations than any of the other males that were regularly present. Of 9 males that were seen copulating, most copulated at least once every day they were present; no doubt it was they that sired the several offspring born during and after the 6–10 month multimale occupation. It was not as if the alpha male was impotent or uninterested in sex; his cuckoldry was apparently caused by a combination of events beyond his control:

1. The troop was large and dispersed while actively foraging.

2. Dense foliage kept many troop members out of sight much of the time.

3. The presence of 4–6 sexually receptive females attracted a number of single males to the territory.

4. Males behave cryptically when alone, sitting still and even actively hiding in dense foliage.

5. The continuing presence of various rivals wore down the resident male, making him less intolerant than if invaded by 1 male at a time. In fact, he often chased rivals that showed themselves, but they merely went into seclusion without quitting the area.

6. The receptive females chose to mate with the outside males, on the sly. Three-quarters of observed approaches that led to pair formation were made by the females (see under Sexual Behavior). However, they only approached males that were regularly present, and did not respond to rare visitors or complete strangers.

Clearly this last was the most crucial factor: there was no way for the alpha male to prevent the females from sneaking copulations. Had they been unwilling to mate with the outsiders, they could have brought their overlord to the rescue by squealing (literally) on males that tried to approach them. Females did in fact expose strangers and sometimes even chased one themselves.

The regular hangers-on made noises like territorial males, including *booms, katrains* and *pyows.* Whether or not that influenced the females, their choice of mates was made regardless of the existing rank order among the outsiders. Ironically enough, though, mothers of small infants—including the "bastards"—all looked to the alpha male to be their protector and leader, staying close to and often grooming him. Nor did he discriminate in his treatment of females and young.

How did the resident male finally manage to regain control of his troop? After all the females had been bred, The Uninvited left of their own accord (17).

Clearly the reproductive advantage that goes with dominion over a troop and its territory does not hold in all cases. Was this a highly unusual, isolated event, or does the presence of sexually receptive females in a troop often lead to invasions and disruptions of the status quo? Similar invasions have been seen in other monkeys with a 1-male, territorial system, notably in the red-tail monkey (4) and black-and-white colobus (see species account); in the Indian langur, invasion by male groups is the typical prelude to a takeover (see chap. 24).

Probably most takeovers occur through the initiative of prospecting lone males that roam the forest covertly monitoring the performance of different resident males. If the prospects look encouraging, a solitary may proceed to test the overlord by approaching females and/or emitting male loud calls within his territory. A resident that makes less than an all-out effort to answer the challenge may be ripe for a takeover. But other males may also be lurking in the background and one turnover can lead to another. In the case of the Kibale Forest turnover cited above as unusual, 5 males replaced one another in the space of a year (3). The average tenure was 2.2 months, the maximum 4.4 months. Usurper number 1, which took the troop from a male that had been resident at least 9 months, lasted less than a month. During the first few days he killed 3 of 5 infants (3–6 months old), the first reported case of infanticide in this species. The next 2 males were apparently not infanticidal but the fourth, an immature animal that moved in when no other male was present, made 13 attempts in the first 2 days. However, he was unable to overcome the determined defense of the mothers assisted by other females and subadult young. He, too, was replaced within a month.

The troop in question was atypical in a number of respects. (*a*) Its home range abutted on a large area of primary forest where no other troops occurred and at 350 ha was the largest home range ever recorded for a forest monkey with a 1-male social system (3). (*b*) On the periphery of the population, the ratio of males to females in the empty quarter was about 7 times higher than in central ranges. Apparently these extra males were also sex-starved: so much so that lone blues tried and sometimes succeeded in mating with female redtails while living with troops of this smaller guenon (see below). At least 2 female and 1 male hybrid resulted, of which the 2 hybrid females bred twice with a redtail and once with a blue monkey (3, 11, 15). (*c*) The birthrate among this troop of blue monkeys was unusually high: 0.54 per female per year, compared to only 0.11 in the 4 other, stable troops (3). Summing up, it seems that 2 key factors may largely determine the degree of male competition and the turnover rate: the number of have-not adult males and the availability of sexually receptive females.

Individual Distance and Troop Cohesion. Blue monkeys are not a close-contact species and tend to disperse widely while foraging (14). The average spread of a 24-monkey Kibale Forest troop was estimated to be some 50 m. Yet the troop remained a cohesive unit and did not break up into separate parties, as recorded in Budongo Forest (1). Mean distances between monkeys of the same sex and age class ranged from 1 m in the case of small juveniles to 2.6 m between adult females. The adult male kept closer to females than to other troop members, a mean distance of 2.1 m (i.e., closer than the interfemale distance), and juveniles kept furthest from him.

Leadership. Group progressions are coordinated and are normally initiated by several monkeys moving in the same direction. Individuals of any age or sex may initiate movements. Sometimes the troop will follow the resident male when he gives the *pyow* rallying call; but at other times the male follows behind the troop (14).

Association with Other Monkeys. Blue monkeys are frequently seen together with other species. In savanna habitats they may be found on the woodland edge with vervets, the only other monkey around except the baboon, and sometimes with black-and-white colobus in gallery and montane forest. In the rain forest, blue and redtail monkeys associate closely with each other and with red colobus—up to 6 or 7 hours a day in Kibale Forest (see subfamily introduction, Associations among Species). Here mixed troops containing up to 5 species (the others being black-and-white colobus and gray-cheeked mangabey) occur. But 85% of all blue monkey associations were with one other species, usually redtails (64%). Blues use redtails to locate ripening fruit, benefiting from the superior local knowledge this more abundant species has of its comparatively small home range. The essentially exploitative nature of this association is demonstrated when blues proceed to oust redtails from fruiting trees. However, the redtail probably derives some predator protection from the presence of its larger, less vulnerable congener. In mixed troops, redtail males *hacked* in response to about half the *pyows* of male blue monkeys, whereas blues did not respond to *hacks* (14).

Occasional friendly interspecific contacts were observed in Kibale Forest, including play between young monkeys and grooming (not reciprocal but one-way). However, 88% of the 201 observed interactions with other species were aggressive (14). The great majority (79%) involved food competition with redtails, with blues the aggressors in 92% of the encounters. Blues also dominated black-and-white colobus during rare encounters but were subordinate to red colobus. In disputes with mangabeys, the largest monkey in the forest, blues were able to win by forming coalitions of up to 4 monkeys. The adult male often lent his support but subadults and juveniles were the main protagonists.

ACTIVITY AND RANGING. Kibale Forest blue monkeys typically spend a third of the day eating, 9% foraging for invertebrate prey, 16.5% moving, 29% resting, 6% grooming, and 6% in other social and nonsocial activities (14). A troop whose ranging was monitored for 67 days covered an area of 4–10 ha daily and traveled 1000–1100 meters (445–1961) (14). Foraging areas usually varied from day to day. The troop moved most between 0700 and 0800 and between 1700 and 1800, and least between 1300 and 1400 h. Rainfall inhibited movement, the monkeys sitting hunched until it stopped. But if rain continued more than 35–40 minutes, the monkeys usually resumed activity, albeit more slowly than usual. After a heavy rain they engaged in long grooming bouts. The troop preferred to roost at night in tall trees (>30 m) near the center of its range, and often began and ended the day in the same general area. After feeding until dusk in a fruiting tree, the troop would move to a different tree when it was almost completely dark, returning in the morning to the one in fruit.

SOCIAL BEHAVIOR

COMMUNICATION

Vocal Communication. The blue monkey has a repertoire of 7 different, largely discrete calls, of which only 2 (phrased grunt and growl) seem to intergrade (12).

Male Loud Calls:

Boom. The distinctive *boom* of the male blue monkey is a short tonal call with nearly all its energy concentrated at a very low frequency (c. 125 Hz). Experiments have shown that sounds in this frequency band penetrate the forest canopy most effectively with least interference from background noises (more under Vocal Communication, chap. 26). The resonance and carrying power of *boom* calls are enhanced through activation of air sacs (7). Furthermore, blue monkeys are particularly sensitive to low-frequency tones (2). These two specializations greatly enhance the species' ability to communicate through the forest canopy.

The *pyow* is a loud, resounding call, audible for several hundred meters, delivered singly in a slow, regular rhythm, at average intervals of 8 seconds (3.5–25), with a mean of 5 calls (1–18) in a series. The calls of different males are easily locatable, consistent, and individually distinct. Primary function: to rally troop members and warn neighboring resident males and solitaries to keep their distance (1, 12, 14).

Ka-train calls. Given singly or in series, these calls consist of an initial burst of sound followed by a burst of pulses, at an average rate of 11 calls/2.4 seconds. *Ka*-calls given by adult male blue and redtail monkeys are nearly identical; they are uttered typically when a crowned hawk eagle is spotted, and during aggressive interactions.

Growls (12) or snarls and nasal screams are emitted during chases and fights involving adult and subadult males.

Calls of Females and Young:

Phrased grunt. Soft, rhythmical sound given during and often as a prelude to group progressions. Function: to maintain group cohesion (12).

Chirps. Short, shrill, birdlike alarm calls given by females and subadults. Function: to bring troop members to a state of alertness, evoke flight from open areas, elicit calling in kind, and maybe stimulate scolding/mobbing action.

Trills. Soft, oscillating calls, usually descending in pitch, given by subadults approached closely by an adult. Apparently signifies submissiveness.

Visual Communication. Facial expressions, postures, tail positions, and movements as in other guenons (see chap. 26, Visual Communication). The lack of strongly contrasting color patterns correlates with the absence of striking visual displays (10).

AGONISTIC BEHAVIOR

Threat/Advertising Displays: leaping, bouncing, branch-shaking, staring ± open mouth and brow movements, sometimes while standing (3); sitting with pale blue-gray scrotum exposed (9).

Defensive/Submissive Displays: *presenting, fear grimace, displacement yawning and eyebrow flicking.*

Fighting. Within troops, the prevalent form of aggression is harassment of subadult and large juvenile males by the adult male. In a Kibale troop, 61% of aggressive encounters were over food and 67% of these were high-intensity, involving prolonged chasing and physical contact including slapping, grabbing, grappling, and biting (14). Females defending infants ganged up on an infanticidal male, and repeatedly chased and several times cornered him (3).

During the multimale occupation of the Kakamega troop recounted under Social Organization, occasional fights were often preceded by mutual glaring and lunging, interspersed with *displacement yawning* and *scratching,* until one rushed at the other. Sometimes one male would stalk another from a distance and suddenly pounce upon him, taking him by surprise. The attacked monkey countered by dropping below his opponent, while emitting a loud nasal screaming and *geckering.* The aggressor would then wait, glaring, until the other male either faced him or fled. Chases quite often led to counterchasing, with the fleeing male becoming the attacker (17). No more than 2 males at a time were seen fighting; other males have been seen to turn their backs on fights, and a male may avoid an encounter by moving away or withdrawing into cover and sitting hunched with face down until a rival has passed. Adult males often bear signs of bite wounds in the tail, right hand, and face.

Aggression between troops is comparatively rare: only 13 interactions were seen during the Kibale study (14). Four of the encounters occurred within 5 days in an overlap zone between 2 troop territories; adults, subadults, and juveniles of both sexes were involved and a number of them were injured. Seven out of 10 encounters between groups occurred at fruiting fig trees.

REPRODUCTION. Breeding tends to be seasonal: in Kibale Forest births occurred from December to May, and in a colony kept at Kampala, where there are 2 wet and dry seasons, conceptions occurred in the dry seasons. However, blue monkeys studied in the Kenya highlands mostly conceived during the "small rains" (August–December) and birthed before the long rains of April (13). The minimum interval between birth and the next conception is normally 15 months. But a female that lost her infant copulated 45 days later (13). It seems, therefore, that the waiting time for a new resident male can be cut anywhere from 7 to 13 months by killing small infants (3).

SEXUAL BEHAVIOR. Blue monkeys give no physical indications of their reproductive state (13). During a 30-day menstrual cycle, females tend to mate in a series of bouts each lasting several days, more often toward the middle than the end of the cycle. A receptive female typically takes the initiative in approaching a male (see under Social Organization), with tail raised, sometimes croaking (= *phrased grunts*) repeatedly (17). Coming close she pouts, raises her chin, and shakes her head from side to side, then stops a meter from the male and *presents*. The male usually mounts at once; if not, the female may look into his face, put her muzzle to his, or nuzzle his genitals. During copulation, which usually consists of a single mount, the female may walk slowly forward and look back over her shoulder, *pouting*. A hesitant male may first inspect and sniff the presenting female's vulva. One that has recently copulated may solicit grooming instead of mounting a presenting female, lying down before her with neck presented, and even hit at her if she persists in *presenting*.

These observations of sexual behavior were made in the Kakamega troop during the multimale invasion. Consortships lasted several hours, during which one partner followed the other as they foraged—70% of the time the female followed the male. Females were also seen to change consorts, sometimes following an attack in which the former consort was defeated. An example of a sneak copulation was afforded by a female that had been following the resident male all morning but apparently without copulating (no sign of ejaculate on the vulva). At 1130 h he spotted and proceeded to chase a strange male through the forest canopy. The moment he left, the female raced to a male that was 150 m away on the edge of the troop, copulated with him, and ran back. When the alpha male returned 20 minutes later, she was sitting just where he had left her, and she continued to follow him, without copulating, for the rest of the day (17).

In the case of the Kibale troop with 2 males, the second one being the former resident male, the females would only mate with the new overlord (14). Pregnant blue monkeys rarely copulate (cf. vervet monkey) (17).

PARENT/OFFSPRING BEHAVIOR. The gestation period of c. 140 days is comparatively short (cf. vervet), and may be related to slow development of infant blue monkeys, which remain in continual contact with their mothers for the first month (13). The infant's plain gray coat becomes grizzled like an adult's after about 2 months, at which time it begins to venture away from its mother (9). Later she may leave it briefly with other females and juveniles while feeding alone. She encourages her offspring to follow the troop's aerial pathways and leap gaps by jumping first and waiting for it to follow. Juveniles play in the usual way, biting, grappling, pouncing, and tail-pulling, especially during midday rest periods.

ANTIPREDATOR BEHAVIOR: alarm calls, threat displays, mobbing, hiding.

The blue monkey seems to be subject to little predation except by man, although sight of a crowned hawk eagle elicits intense alarm calls: male *ka-trains*, chirps from females and young, which withdraw into dense foliage if an eagle comes close (12, 13).

References

1. Aldrich-Blake 1970. 2. Brown and Waser 1984. 3. Butynski 1982. 4. Cords 1984. 5. DeVos and Omar 1971. 6. Dorst and Dandelot 1970. 7. Gautier and Gautier-Hion 1977. 8. Haltenorth and Diller 1980. 9. Kingdon 1971. 10. ——— 1980. 11. Leland and Struhsaker 1983. 12. Marler 1973. 13. Rowell 1970. 14. Rudran 1978. 15. Struhsaker 1984. 16. Struhsaker and Leland 1979. 17. Tsingalia and Rowell 1984.

Added Reference

Henzi and Lawes 1987.

Fig. 27.5. Vervet female and nursing infant.

Vervet Monkey

Cercopithecus aethiops

Other common names: grivet, green monkey

TRAITS. The common monkey of the African savanna, light-colored with a black face. *Length and weight:* (head and body) 38–62 cm, tail 42–72 cm, 5–9 kg (14). *Size dimorphism:* (head and body) males 49 cm and 5.5 kg, females 45 cm and 4.1 kg (Botswana series) (19). *Teeth:* both sexes with long, sharp canines. *Coloration:* geographically variable, from silvery gray to olive, yellow, or reddish green, underparts white to yellowish; black face framed with white ruff and brow band (variably developed in different races), eyelids pale pink; hands, feet, and tail tip conspicuously black in some races; adult males with pale blue scrotum, red penis, and white perineal skin, similar to colors of estrous swelling in some females (25); natal coat dark and silky, face skin pink. *Scent glands:* possibly present on cheeks, jaw, and chest.

DISTRIBUTION. The most widespread African guenon, occurring throughout the Northern and Southern Savanna, from Senegal to Sudan and south to the tip of South Africa (11). Up to 20 subspecies are recognized, of which several are different enough to be considered full species by some systematists (9, 17).

ECOLOGY. The vervet is adapted to practically all wooded habitats outside of the equatorial rain forest. It is one of the few African primates to escape the rain forest and come down from the trees. However, being far smaller than baboons and lacking the fleetness of the patas monkey, the vervet cannot afford to venture far from the safety of trees and is also more dependent on trees for food. It is essentially an edge species and typically associated with riverine vegetation; in the dry savanna, fever trees (*Acacia xanthophloea*) and vervets go together.

DIET. The vervet is an opportunistic omnivore which takes what is most abundant and available: fruits, seeds, seedpods, leaves, buds, sap, flowers, various herbs and grasses, invertebrates, and occasionally vertebrates (lizards, eggs, nestling birds) (20). In habitats dominated by fever trees, as in Amboseli N.P., this 1 species can be the vervet's mainstay, providing some form of forage year-round in the form of new leaves and still-tender thorns, flowers, pods and peas, gum, and even powdery bark and wood. Herbs, grasses, and insects are less important than tree food and are taken mainly during the rainy season. On Lolui Island in Lake Victoria, a botanically richer habitat, vervets help propagate their favorite fruit trees by leaving their dung atop termite mounds in the grassland: the seeds germinate and grow into thickets of the species they have eaten (11).

While foraging on the ground, vervets commonly turn over sticks, small logs, and the dung of large herbivores in search

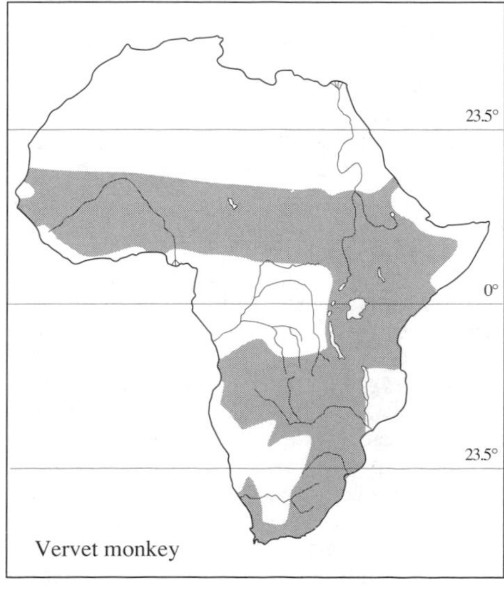

of beetles and other invertebrates, and undigested seeds. Most plant foods are eaten without manipulation other than to pick the items and bring them to the mouth. Vervets bite into and chew acacias and other trees to stimulate the flow of sap, sometimes causing considerable damage in the process. They also raid crops and among primates rank second only to baboons as agricultural pests.

ASSOCIATION WITH OTHER MONKEYS. Vervets are often found in the same areas as baboons, with whom they share many of the same foods, water holes, and sleeping trees. Whenever the 2 species compete for the same resources, *Papio* supplants *C. aethiops*; only the small terminal twigs of trees are more accessible to the smaller monkey. The baboon is also an occasional predator of young vervets. But during the seasons of food scarcity, baboons forage in the grassland much more than vervets, thereby reducing competition for the same resources (20). Vervets also associate and compete with the blue monkey at the forest edge (see species account).

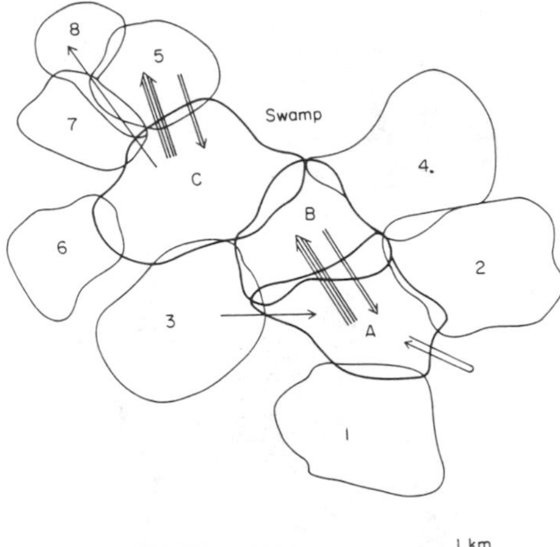

Fig. 27.6. Territories of vervet troops in Amboseli N.P. Arrows show incidence and direction of migrations by natal and young-adult males of three study groups (territories A, B, and C) in a 5-year period; each line represents one male. (From Cheney and Seyfarth 1983.)

SOCIAL ORGANIZATION: multimale troops, both sexes territorial, males emigrate.

Successive, long-term vervet studies in Amboseli N.P. have revealed a complex social structure perhaps closer to the baboon's than to other guenons (4–8, 26, 27; 15, 16; 20–24). Troops average c. 25 monkeys, but vary in size from 8 to 50, with anywhere from 1 to 8 males, just in this one population. In such a widespread, adaptable species there is bound to be great variability in troop size and social organization (see 10, 11, 13). On Lolui Island, for instance, troops average c. 11 (6–21), typically including 2 adult males, 4 adult females, and 5 young (10).

Amboseli troops live in a typical mosaic of stable territories varying in size from c. 18 to 76 ha (20), 2–10 times the size of Lolui vervet territories (10, 11). The mean area occupied by 3 Amboseli troops with contiguous boundaries was 40 ha (fig. 27.6) (8).

A vervet troop is a hierarchy of families whose members sleep, forage, and rest together. Males emigrate as they near maturity beginning at 5 years, whereas females stay home and take their places in a female-bonded society wherein the mother's rank predetermines the daughter's. Beginning in infancy, vervets are accorded a social position just below their mother's even by adult monkeys, perhaps mainly because the mother takes the part of her offspring in disputes with other troop members. Since high-ranking females gain priority to food, water, and other amenities, they and their offspring are likeliest to survive in times of famine. Social inferiors vie with one another to groom these dominant individuals, to handle their infants, and to enlist their support in their own disputes (21). Social interactions among adult females mostly involve individuals of adjacent rank which are also close relatives (7).

Playback experiments to test individuals' ability to identify other vervets by their calls alone have revealed an unexpectedly high degree of social sophistication. (*a*) Mothers identify the offspring of other females as well as their own by the infants' squeals and screams. (*b*) Troop members not only all know one another as individuals but also classify one another according to rank and degree of relatedness. (*c*) Both males and females recognize the voices of individuals in neighboring troops (shown by playbacks of the "long *aarr*" call), and react very strongly if a voice known to belong in one troop is played from an unaccustomed location (4).

Although adult males are able to dominate females in one-on-one competition, females and juveniles need only *chutter*, *rapid-glance*, squeal, or scream to bring assistance from other females. Even coalitions of 2 juveniles are usually sufficient to make a male back off, knowing that reinforcements can quickly be summoned (4).

Territorial Hostilities. Females and young tend to be more unremittingly hostile to the neighbors than are males, and during border disputes behave aggressively toward outsiders regardless of sex. The difference can be sought in their respective reproductive strategies: females defend the limited resources of their small, ancestral range, whereas the females themselves are the resource that adult males defend (see chap. 24). Since most breeding is monopolized by dominant males, they have the greatest stake in repelling invading males and should be most vigilant and vigorously aggressive, as indeed they are.

Male Transfers. Low-ranking males, by contrast, may try to improve their lot by transferring to another troop. During an intertroop fracas, such individuals sometimes try to cultivate friendly relations with members of the enemy troop (8). Juvenile males in Amboseli were found to have significantly more wounds than would be predicted if all the monkeys in a troop were equally subject to fighting injuries (22). The evidence suggests that the onset of puberty between 2 and 3 years exposes young males to the aggression of adult males, coinciding with the appearance of colorful genitals. Despite harassment, however, males usually transfer only when nearly adult.

In addition to such usual dangers as predation during the dispersal stage, immigrant vervets also face possible serious injury from adult females. Full-grown females are a match for 5-year-old males and, singly or in coalitions, may attack and chase immigrants as vigorously as do resident males. Amboseli males even the odds by emigrating in the company of brothers or other male peers (fig. 27.6), or by transferring to troops containing an older brother (8). Although they could have chosen any of 5 neighboring troops, 14 of 16 individually known migrants transferred to groups that contained previous associates, and in 9 cases brothers or peers emigrated together. No males have been known to lead solitary lives for more than 2 successive months. All 6 males that emigrated before age 5 did so in company. Evidence that troops which exchanged males were less aggressive toward each other than those that did not suggested it was easier for additional males to transfer. In contrast, only 6 of 12 fully mature males transferred to troops containing members of their former troop; 5 of these males were known to be transferring for at least a second time. Nearly all transfers (88%) occurred during the mating season.

Nonrandom transfers of males would restrict gene flow and increase the risks of inbreeding if regularly practiced. As it turns out, however, males that transfer together actually reduce the risk of breeding with one another's offspring—provided they do not remain longer than the 4–5 years it takes females to begin reproducing. The fact that over half the Amboseli males transferring a second or third time went to other troops suggests that vervets may have hit on the best compromise solution. The mature males that transfer are usually low-ranking; they transfer randomly and alone. Nine of 11 achieved higher rank after transferring (8). (See 28–30.)

ACTIVITY AND DAILY RANGING.

Daily ranging and activity patterns of vervet troops vary with the type of habitat, predation risks, and the distribution and nature of their food, watering places, and sleeping trees. Lolui Island vervets represent the general activity pattern, although they enjoy virtual freedom from predators (12). Activity begins well after dawn, with the monkeys socializing and feeding near their sleeping trees up until c. 0800 h (an hour or 2 later on rainy days) before setting off for the day's foraging. The usual morning and late-afternoon feeding peaks, divided by a midday rest period, are observed, but often some monkeys do the opposite of the others. Also, different troops may have consistently different routines. One had morning and afternoon movement peaks and rested between 1200 and 1400, whereas a neighboring troop had 3 movement peaks divided by 2 brief rest periods around 1130 and 1530. In Amboseli N.P., troop progressions are commonest from 0700 to 1100 and 1600 to 1900. Amboseli vervets usually go to water every other day, mainly in early afternoon (20). Lolui vervets rarely drink, but usually come back to their sleeping trees by 1700, where they socialize and feed until dusk (12).

Although vervets are more limited in their overland movements than patas or baboons, Amboseli vervets may range as much as 2560 m or as little as 330 m in a day. The average of 1 group was 945 m, compared to 1430 m by a neighboring troop with a less-productive and larger home range (20). The longest recorded daily movement was 18 km by a Zimbabwe troop barely subsisting in dry woodland on mopane seeds and baobab fruits (2).

A troop foraging on the ground disperses over a sizeable area and progresses slowly on a broad front, taking up to 20 minutes to travel 50 m. In unproductive habitat, the spread may be 200–500 m (11), whereas a troop about to go to roost will typically crowd within a space of 100 m or less. During group progressions, for instance when passing overland between groves of trees, the troop usually proceeds purposefully in close single file. There is no regular leader but senior females generally initiate and lead progressions (11).

POSTURES AND LOCOMOTION.

The vervet is considered anatomically generalized, intermediate in form between the arboreal guenons and the terrestrial patas monkey (11). Its conformation and limb proportions are not specialized for any particular habitat—nor is it limited to any; adaptability to a wide variety of habitats is one secret of the vervet's success.

C. aethiops cross-walks and its fastest gait is a bounding gallop. The same gaits are employed to travel overland or through the trees. The tail is carried horizontally or arched over the back while walking (more under Visual Communication), from horizontal to vertically erect when galloping. Excursions through grass too tall to see over are punctuated by standing erect, by *spy-hopping* off the hindlegs while running, and by climbing mounds and other elevations to take a good look around while standing erect with tail hanging. Agile climbers, vervets ascend and descend vertical trunks and limbs headfirst, and often use the tail as a balance or as a brace. They rarely free-fall more than a meter but when displaying or diving for cover may plummet 10–15 m. Vervets are also good swimmers and sometimes take to water during flight (14).

SOCIAL BEHAVIOR

COMMUNICATION. The vervet's behavioral repertoire includes at least 60 physically distinct gestures and 36 physically distinct sounds (22, 23). Most play a role in communication similar to the visual and vocal signals of other guenons (see chap. 26, Communication). Only displays of special significance in vervet society are singled out here.

Visual Communication (Agonistic Behavior)

Threat/territorial advertising displays: staring, eyelid exposure, head-bob, forequarter jerk (sometimes sideward jerk and *lateral display*), *penile display,* jumping around and branch-shaking, attack-chase, noncontact grabbing and slapping.

Staring and *eyelid exposure* can also be defensive; the associated posture and other behaviors must be taken into account. The pale eyelids become very conspicuous when the brows are raised. Variations in the amplitude and vigor of the forequarter jerk (performed while staring) express different degrees of aggressiveness. The red color of the penis in contrast to the white stomach makes penile erection conspicuous, especially when enhanced by a jerking action. Adult males engaged in vigilance scanning may sit with both penis and scrotum exposed to the view of rival groups (fig. 27.7), and even stand up the better to be seen during intertroop encounters (14, 22) (cf. baboon penile display). This may be a territorial version of the *red-white-and-blue* dominance display described below. Chasing often starts with an attack launched at high speed. The victim flees, glancing back and uttering a variety of shrill screams and squeals. Ostentatious jumping around and branch-shaking are typically seen during intertroop encounters, performed by adult and subadult males, and by females in abbreviated form (cf. black-and-white colobus leaping display).

Fig. 27.7. *Penile display* of a sentinel vervet (from Wickler 1967).

Dominance displays: confident walk, *red-white-and-blue display.*

Supplanting of a subordinate by a dominant vervet is done in a confident walk and manner, and the tail position is an indicator of the degree of assertiveness: the more it is looped forward over the back, the more confident the displayer and the surer the outcome (fig. 27.8) (1). The *red-white-and-blue* is a relatively stereotyped display of dominance between males of the same troop (22). The displayer approaches the receiver in the confident walk and exhibits his red penis and blue scrotum by walking back and forth or sometimes encircling the

Fig. 27.8. Vervet tail signals, ranging from confident (*left*) to fearful (*right*). (Redrawn from Bernstein et al. 1978.)

receiver, standing before the receiver in the presenting attitude, or by rearing up bipedally while walking, circling, or standing side-on. Sometimes the performer has an erection. The receiver responds with submissive signals.

Defensive/submissive displays: rapid-glancing, crouching, gaping, grimacing, false-chase, hop backward, head-shake, lip-smack, teeth-chatter.

Rapid-glancing (alternately looking toward and away from an aggressor) conveys subordination and also solicits support from other vervets. *Crouching* is interpreted as defensive threat and is combined with *gaping* and *staring,* whereas *grimacing,* accompanied by *staring* and squealing/screaming, signals subordination and inhibits attack. *False-chasing* is recognizable by the relaxed, confident movements of the chased monkey and the jerky, hesitant movements of the pursuer, which also utters fearful *woof, woof* calls.

Vocal Communication. The vervet has a considerable vocal repertoire, composed mainly of discrete calls (table 27.1). Only squeals, *chutter,* screams, *rraugh,* and *aarr* calls are continuous or graded signals. At least 21–23 different messages and 18–22 different responses may be conveyed in 21 different situations (23). The absence of male loud calls adapted to maintaining intertroop spacing is noteworthy (cf. blue and colobus monkey accounts), although male predator-alarm calls may also serve a spacing function.

Tactile Communication: *nose-to-nose greeting, social grooming.*

When 2 vervets approach each other with friendly intent, to groom, play, mount, or whatever, they usually touch muzzles first, after approaching cautiously with heads extended and lips pursed but faces otherwise relaxed (11). *Social grooming,* very similar in form and function to grooming in other guenons (chap. 26), is frequent and important in maintaining kinship and other bonds (more under Social Organization).

Olfactory Communication. On Lolui Island, adults of both sexes have been seen rubbing their chests, and less often their cheeks and lower jaws, against branches and other objects in places known to be territorial boundaries and not elsewhere (11). The behavior is performed in a characteristic and stereotyped manner: with arms wrapped around the branch, the monkey sniffs it intently, then rubs its chest on the spot. The performance may take 2–3 minutes, during which the monkey appears totally absorbed and inattentive to outside stimuli. The presence of scent glands in these areas remains unconfirmed (see chap. 26, Olfactory Communication).

REPRODUCTION. Reproduction is seasonal. In Amboseli, mating takes place in the dry season and births fall during the months after the short rains (October–January) (22). It is the reverse on Lolui Island, births beginning with the onset of the long rains (April) and continuing through the dry season (10). Apparently most females breed yearly, beginning at 3½ to 4 years (24). Males mature a year or more later (8).

SEXUAL BEHAVIOR. Vervets do not form consort pairs, nor do females copulate with a succession of males; 1 successful copulation in a day is usual. There is evidence that dominant males do most of the mating (7, 22). At any rate, no harassment was seen between adult males, even when one copulated within full view of another.

In most vervet populations, there are no external indications of estrus, nor is copulation limited to a particular period in the

Table 27.1 Vervet Monkey Calls

Call description	Performer	Context, message, and response
Dominance/threat		
1. *Chutter.* Low-pitched, monotonal, staccato; mouth closed, teeth covered.	Adult females and juveniles	Expresses aggressive threat and solicits support. Response: lend aid or flee.
2. Bark. Low-pitched, gruff call.	Adult and subadult males	Generally given while running toward quarreling monkeys. The message: "Knock it off!" Response: stop fighting.
3. Intertroop grunt. Similar to *progression grunt.*	Males	Response to sighting vervets from another troop by sentinel males.
Defensive/submissive		
4. Squeals and screams. High-pitched, piercing sounds that vary with mouth position, from open and teeth covered to fear grimace.	Females and juveniles	A call for help and defensive threat in response to threats or attack. Response: inhibit aggressor and bring help.
"Woof," "wa," and *"woof-wa"* calls.	Subordinate males	Express submission, probably increasing in intensity from *woof* to *woof-wa.*
5. *"Woof-woof."* Non-tonal, deep, guttural call, with mouth closed or slightly open.		
6. *"Wa."* A sustained, tonal exhalation, combined with a grimace.		
7. *"Woof-wa"* Combines (5) and (6) but somewhat changed.		
"Aar" and *"rraugh"* calls. Very similar sounds, but given in different situations. Each has a long and a short form.		
8. "Long *aar.*" Given with mouth slightly open and puckered, teeth covered.	Females and juveniles (subadult males)	Trespass by outsiders. Rallies other troop members to repel invaders.
9. *"Rraugh."* Mouth closed or partly open, teeth covered.	Mainly yearlings	Signal of nonaggression or appeasement, during approach to older dominant monkeys.
c. *"AArr-rraugh"*	Young juvenile females only	Ambivalent situations (e.g., approach to groom an adult, proximity of rival troop).
Contact-seeking and cohesion sounds		
10. *Teeth-chattering.*	Adult and subadult males only	Usually while grooming. Sometimes in response to *red-white-and-blue* display.
11. *Lip-smacking*	All classes	Prelude to grooming, etc.
12. *Progression grunt.* Deep, guttural call of short duration, given with mouth closed or slightly open.	All vervets older than 4.5 months	Contagious calls preparatory to a troop progression, usually within 10–15 minutes.
13. *Purring.* A very quiet sound, seldom heard.	Juveniles	Play-wrestling.
Predator-alarm calls		
14. *"Uh"*	All but infants	Low-intensity call response to a minor predator.
15. *"Nyow"*	All but infants	Moderate-intensity response to sudden movement or appearance of minor predators.
16. Chirp. Short, sharp call of low frequency, given with mouth wide open and teeth exposed, far-carrying and easily localized.	Females and juveniles	Reaction to major mammalian predator.
17. *"Rraup."* Rough, short call seldom repeated.	Females and juveniles	Sighting of avian predator. Response: descend from treetops or run into thickets if surprised in the open.
18. Threat-alarm bark. Male equivalent of the *rraup,* uttered repeatedly.	Adult and subadult males	A high-intensity alarm call and aggressive threat; tempo and volume increase with proximity of disturbance.

Table 27.1 ***(continued)***

Call description	Performer	Context, message, and response
Infant calls		
19. *Lost "rrr."* Combined with the *pout-face* expression. Very distinctive and stereotyped, individually recognizable (true also of 20 & 21).	Infants and juveniles	Signals distress and identity to mother and other troop members.
20. *Lost "eee."* Given with lips parted, mouth corners drawn back, teeth covered. (See note under 19.)	Infants and juveniles	
21. *Lost "rrah."* A less-distressed version of the lost *rrr* (see note under 19).	Infants and juveniles	
22. *"Eh, eh."* A quiet, short, non-tonal call, most like lost *"rrah."*	Infants and juveniles	Often given after reunion with mother. Message and response undetermined.
23. Screams and squeals. Similar to but higher-pitched than distress calls of females and older juveniles (see 4).	Infants and juveniles	See no. 4.

SOURCE: Struhsaker 1967*d*. For a recent study of vocal development see Seyfarth and Cheney 1986.

menstrual cycle; in fact, females commonly continue to copulate during pregnancy, at least in captivity (18). However, in Serengeti N.P. 2 females that came into heat during a brief study developed bright blue, swollen labia and a red clitoris, very similar in overall effect to the male genitalia (25). The vulva was flesh-colored in other females. Red and blue sexual skin has also been noted in Ngorongoro vervets (author's observations). Periods of sexual receptivity range from as little as 7 days to as long as 66 days (10, 22). As usual, *presenting* or stopping instead of moving away when approached by a male indicates that a female is receptive. Estrus is also indicated when a male is seen nuzzling a female's genitalia, grabbing her hips, or embracing a standing female from behind. Unreceptive females avoid male attempts to approach or touch them, employing various means from merely remaining seated, lying down, crouching low, and walking or running away, to squeal-screaming, grabbing, slapping, and lunging at overly persistent suitors (22).

The copulatory stance is typical of Cercopithecine monkeys (see fig. 27.13). The male makes short pelvic thrusts until intromission, then an average of 7 (1–25) long strokes to ejaculation, followed immediately by dismounting. Meanwhile the female stands with limbs slightly flexed and tail hanging or held aside, glances over her shoulder, sometimes reaches back and grabs one of the male's legs, or walks forward a few feet. In Amboseli, 3 out of 4 mounts occurred on the ground, the remainder in trees. Definite peaks of sexual activity were recorded approximately every other hour between 0800 and 1900 h (22).

In 1 out of 10 mountings, juvenile males threatened, grabbed, slapped, pulled, or leaped over the male partner. Such harassment was occasionally enough to make the male dismount. Sometimes he and/or the female retaliated, but usually they ignored the heckler.

PARENT/OFFSPRING BEHAVIOR.

The average gestation of 165 days (18) is considerably longer than most guenons (cf. blue monkey). Vervets are born at a correspondingly advanced stage and grow up faster than most other monkeys. For the first week after birth, the baby spends at least ¾ of the day attached to its mother's front, its fingers and toes securely gripping her fur, the slightly prehensile tail often curled around her leg, tail, or back (24). During the first hours the mother secures the baby with at least one arm around its neck or shoulders and sits forward in a crouch. Infants a few days old sit between the mother's legs and nurse, and the vervet's nipples are so close together that a baby sucks both at once. The amount of time spent clinging (always ventrally, never jockey fashion as in baboons) declines to only ¼ of the day within 3 months.

Handling by Other Monkeys. Vervet mothers typically allow other troop members to handle babies beginning at a remarkably early age (15, 24) (fig. 27.9). Females, especially juveniles of 2–3 years, are strongly attracted to babies, and spend much time touching, grooming, cuddling,

Fig. 27.9. Handling of newborn infants is often tolerated by vervet mothers.

playing with, and carrying them until they lose the natal black fur and pink face at about 3 months (24). Handling may begin within hours of birth: a baby less than 14 hours old was carried about for 9 consecutive hours by several young females, up to 10 m distant from the mother (24). Such relaxed and friendly contacts with infants are characteristic of vervet society. Babies that are handled most are also those that are groomed most by juvenile and adult females. Vervets prefer to handle the offspring of high-ranking females, but different low-ranking individuals may spend more time grooming mothers than infants. Firstborn infants receive most handling regardless of the mother's rank (16).

Grown-up males show practically no interest in babies, but juvenile males also contact and care for babies of under 3 months, especially siblings. Even unrelated males have been known to respond to the lost calls of infants the troop has left behind by picking them up and protecting them (16).

Maternal bonds with older offspring change with the birth of the next baby. Yearlings or 2-year-olds suddenly barred from their previous privileged status often act depressed, or try to assert their rights, whereas older juveniles, attracted to their new sibling, tend to interact more with their mothers than before the birth. Other family members have the readiest access to infants, which may thereby come to identify close relatives (15).

After only a few weeks, vervets show an increasing tendency to extend their network of playmates and baby-sitters, until it includes all the young monkeys in the troop. If there are numerous peers, the infants tend to form a cohesive feeding and play group; if not, older juveniles are their main companions, with which they may groom more and play less. But for at least the first 3 months the baby remains strongly dependent on its mother for food, protection, warmth, and comfort. Gradually it begins to eat what other vervets eat, learning by watching and sampling the foods its mother selects. By 6 months it is foraging and eating like an adult, returning to its mother only when danger threatens or if bullied by another vervet (15).

Weaning may begin as early as 3 months, the mother pushing the infant away from her nipples and nipping it about the head or arms if it persists (24). It is forced, under protest, to relinquish its warm, secure place on her front (fig. 27.5) by 4 months and walk on its own. Still, vervet mothers tend to be gentle, try to distract juveniles bent on nursing by grooming them instead, and may suckle them occasionally until the next baby is born. Once that happens, the juvenile takes its place as an established, well-known member of its troop, while continuing to associate most closely with its immediate family (15).

PLAY. Play is nearly always social rather than solitary (with objects) in vervets, involving from 2 to as many as 6 monkeys in any combination of sexes. Infants may start to play by day 15 and adults may briefly play with juveniles, but vervets older than 4 are seldom playful (24). Juvenile pairs spend many minutes at a time roughhousing, biting, chasing, wrestling, or tumbling in the undergrowth. Several may play king-of-the-castle on a termite mound or log, and swing on and slide down vines (15). Grabbing, mouthing, embracing, hopping, jumping, and chasing are also common play activities (24).

ANTIPREDATOR BEHAVIOR. Because of their small size and terrestrial habits, vervets are vulnerable to more predators than probably any other African primate. Amboseli vervets have to worry about at least 16 potential predators, but only 4–5 are dangerous under normal conditions: the 2 largest eagles, the leopard, and the serval (20). These, along with lions and cheetahs, are the predators to which vervets react most strongly, using different alarm calls for avian and ground predators, and another one for snakes (see under Vocal Communication). Baboons sometimes kill and eat young vervets and represent a threat on the ground, but are not agile enough to catch vervets in trees. Vervets seldom venture more than a few hundred meters away from a refuge. Like other guenons, they also hide from danger by flattening against branches and keeping still. If surprised in the open, adults may even counterattack small predators (20). Apart from their own alarm calls, vervets also react to the warning signals of associated ungulates, primates, and several birds (francolin, guinea fowl).

References

1. Bernstein et al. 1978. 2. Brain 1965*a*. 3.——— 1965*b*. 4. Cheney 1983. 5. Cheney, Lee, Seyfarth 1981. 6. Cheney and Seyfarth 1980. 7.——— 1982. 8.——— 1983. 9. Dorst and Dandelot 1970. 10. Gartlan 1969. 11. Gartlan and Brain 1968. 12. Hall and Gartlan 1965. 13. Harrison 1983. 14. Kingdon 1971. 15. Lee 1979. 16.——— 1983. 17. Napier and Napier 1967. 18. Rowell 1970. 19. Smithers 1983. 20. Struhsaker 1967*a*. 21.——— 1967*b*. 22.——— 1967*c*. 23.——— 1967*d*. 24.——— 1971. 25. Wickler 1967.

Added References

26. Cheney and Seyfarth 1986. 27. Cheney et al. 1988. 28. Moore and Ali 1984. 29. Packer 1979. 30.——— 1985.

Fig. 27.10. Baboon social grooming: female and juvenile work on adult male.

Savanna Baboon

Papio cynocephalus

Subspecies

P. c. papio, Guinea baboon
P. c. anubis, olive baboon
P. c. cynocephalus, yellow baboon
P. c. ursinus, chacma baboon

TRAITS. A large, terrestrial monkey with a doglike head (*cynocephalus* means "dog-headed"). *Weight:* up to 50 kg (19); average of *Papio* races and species, adult males 23 kg and adult females 11–12 kg (1); *P. c. ursinus* from Botswana (32), adult males 32 kg (27–44), adult females 15.4 kg (14–17). *Length:* head and body, *ursinus* males 72.5 cm (72–73), tail 60–84 cm; females 60 cm (52–65), tail 56–61 cm (32). *Teeth:* razor-sharp 5-cm canines in males only. Eyes small, deep- and close-set beneath prominent brow ridge; ears sizeable and nearly naked. Limbs sturdy, nearly equal in length, hands and feet short and wide with stubby digits. Coat: coarse, hairs brindled, length varying from long, males with well-defined cape in Guinea and olive baboons, to short in yellow and chacma baboons. *Coloration:* reddish brown (*papio*), olive-brown (*anubis*), yellow-brown (*cynocephalus*), to greenish-brown with blackish lower limbs (*ursinus*); newborn black with red face and rump, juveniles distinctly lighter than adults; males darken and females lighten with age. Bare skin of nose, lips, ears, and extremities black, eyelids whitish. Callosities and adjacent rump skin same color as face in adults, shiny and often with purplish tinge (pinkish in *P. c. papio*), female sexual swelling the largest of all monkeys, most extreme in captive animals.

DISTRIBUTION. *Papio's* adaptation to terrestrial life in grassland habitats has enabled it to become the most widespread African primate, found virtually throughout the savanna and arid biomes wherever water and trees or cliffs occur. Forest re-

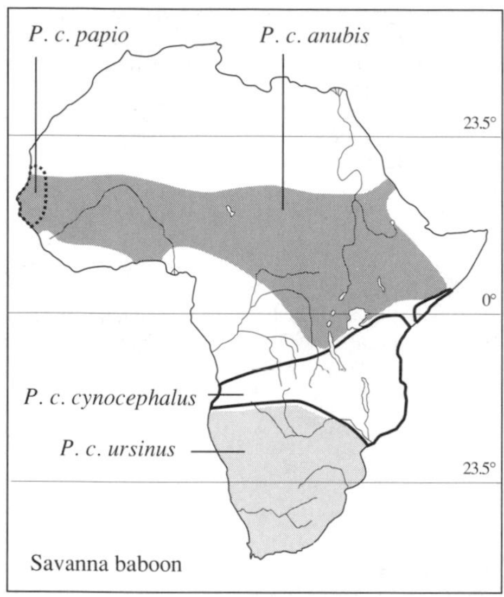

Savanna baboon

placement by man-made savanna, agricultural expansion, and local extinction of large predators have enabled the savanna baboon to extend its range (1).

ECOLOGY. The savanna baboon, in forms distinct enough to be considered different species by many primatologists, virtually monopolizes the broad niche open to a monkey able to forage equally well in trees and on the ground, whose size and social organization allow it to venture far from refuges during its daily ranging. The closely related hamadryas baboon and the gelada are still more terrestrial, subsisting largely on grasses, but are only able to dominate in, respectively, the most arid, treeless regions of northeast Africa and Ethiopia's high plateaus. The savanna baboon dominates everywhere in between there and the rain forest, and in the South West Arid Zone manages to live in habitats similar to those the hamadryas occupies in the Somali Arid Zone (14, 36).

DIET. Savanna baboons utilize virtually all the accessible edible plants within their range, supplemented by animal foods. Detailed foraging studies in different populations all agree that the baboon menu features the most nutritious foods available (1, 38). Grasses, flowers, fruits, seeds, buds, leaves, shoots, twigs, bark, sap, roots, tubers, bulbs, aquatic plants, mushrooms, and lichens are all included. An incomplete list of the plants eaten by baboons near Cape Town included 94 species (7). In typical acacia savanna habitat, grasses tend to be much the most important food source, accounting for over ½ a troop's feeding time in the course of a year (15), with leaves, stems, flowering heads, and underground portions taken in different seasons and growth stages. Early in the rains a troop may eat sprouting grass almost exclusively. Late in the dry season Nairobi Park baboons depend almost entirely on the still-nutritious rhizomes and corms of dry grasses, for which their only mammalian competitors are rooting warthogs (7). Among the variety of fruit and vegetable foods baboons find in trees, figs and acacia pods and seeds are particular favorites.

Pronounced size differences between different sex and age classes lead to some ecological separation within the troop (28). Young juveniles venture onto branches too small to support older animals, and older juveniles pluck fruits and flowers from the outermost twigs of tall forest trees. Females with small infants often sit on the ground, eat a higher proportion of grass-seed heads than do other classes, and feed more on low bushes.

Grasshoppers, spiders, scorpions, and other invertebrates may be important dietary supplements, especially in more arid areas. Baboons living by the sea forage in tidal pools and caves, where they feed on mussels, limpets, crabs, small fish, and the like (9). Vertebrate prey is seldom deliberately sought but is taken fortuitously, including fish, frogs, lizards, terrapins, the young of ground-nesting birds, bird and crocodile eggs, small rodents, hares, and the offspring of antelopes in the concealment stage (7, 15). Usually only adult males catch prey the size of hares and fawns, which they rarely share. But cooperative hunting and food-sharing have developed into a tradition in one long-studied Kenya troop, with females and juvenile offspring participating (16) (more under Foraging and Predatory Behavior). In South Africa, adult males can become major predators on young sheep and goats (36).

LIMITING FACTORS. The availability of water and of secure sleeping sites, in the form of tall trees or cliffs, is of primary importance in baboon distribution. Although baboons can gain most of the water they need from their food or in the form of dew, and in some areas do not regularly go to water (9), probably most troops do drink regularly. During droughts when surface water is unobtainable, baboons dig in streambeds to reach subterranean sources (4).

SOCIAL ORGANIZATION: terrestrial, nonterritorial, multimale troops composed of resident females and young organized in kinship groups, each associated with specific males.

Baboons live in troops ranging from 8 to over 200 animals but typically numbering 30–40, in which adult females outnumber adult males 2–3:1 and immature classes make up half or more of the total. Very small troops may represent offshoots of large troops. Home ranges vary in size from as small as 400 ha to as large as 4000 ha, depending on the habitat and without regard to troop size (15). The ranges of different troops usually overlap widely. Troops tend to avoid one another, although different bands may share the same sleeping sites if refuges are scarce; competition over roosting space, water, or fruiting trees can lead to intertroop aggression (11, 36).

Baboon social organization is highly complex and variable, with a communication system capable of signaling subtle gradations of emotion and motivation (21). As usual, females make up the stable core of the group. They remain their whole lives in the same troop (or an offshoot, should fission occur), whereas males emigrate during adolescence and often transfer repeatedly between troops (1, 23–26, 38). A strict and stable rank hierarchy exists between the adult females and their offspring. Each mother heads up a kinship group within which her offspring are ranked according to age in descending order, beginning with the youngest. In social relations between matrilines, the young take the rank of the matriarch; a lower-ranking adult female has to defer even to the infant of a superior (cf. vervet) (38). Initially, deference is enforced, if need be, by the active intervention of the baby's mother or siblings, but by the age of 2½ years a juvenile female's rank is fixed for life.

Male offspring remain in the kinship groups and subordinate to their mothers until about 4, when they become adolescent and experience a growth spurt during which they develop long, sharp canines and become bigger than and dominant over all females. Females begin reproducing in their fifth year, before they are quite fully grown, whereas males have to mature for another 3–5 years before they can compete on equal terms with other males for reproductive opportunities. It is this quest that motivates males to transfer between troops, staying for as little as 1 month or as long as 10 years (38).

Male dominance status and reproductive success are of course linked. Females tend to be receptive to dominant males, especially to those that have recently joined the troop (17, 23). Accordingly, low-ranking males would have most to gain by emigrating. However, dominance alone does not ensure mating rights. Females prefer to mate with particular males which are regularly associated with them at other seasons, with whom they are in effect pair-bonded (1, 27, 33, 38) (more under Sexual Behavior). Observations of male-female relationships in several different populations revealed that each adult female had 1 and usually 2 (sometimes 3) favorites among the adult males; they were most often near her during daily ranging, roosted in the same sleeping subgroup, and were regular grooming partners. Proximity was maintained primarily on the initiative of the females, which preferred to stay within 15 m of a favorite male while foraging and more than 15 m from other males (33). When females came into estrus, their favorite males were usually their consorts, on the male's initiative but with the female's cooperation (27). In an Amboseli N.P. troop, fewer than half the 10 adult males had special relationships with the 17 adult females. All of them were fully mature and in the top half of the hierarchy. The first- and fifth-ranked males, allied with 6 and 5 females, respectively, were associated with the most mothers and infants and were the most successful reproductively (1).

Godfathers. Males that associate with females play "godfather" roles to their offspring, beginning in infancy and lasting at least into the second year, when juvenile baboons become largely independent of their mothers. Almost all friendly contacts between males and infants, including holding, carrying, grooming, food-sharing (e.g., meat scraps), and so on, reflect such relationships. When adult males come to the aid of juveniles or females that are being bullied by other troop members, it is on behalf of the individuals with whom they are socially bonded (1, 34). Not only do young baboons receive increased protection against the dangers of injury by other baboons and increased access to resources, but godfathers have even been known to act as foster parents of juveniles whose mothers died (1, 5).

Apart from the advantages derived by juveniles, low-ranking females may gain greatly from the presence of male protectors. Mothers of infants up to 5 months old often suffer considerable interference and distress at the hands of dominant females that are strongly attracted to babies in the black phase. Most baboons (especially juveniles and adult females) find such infants attractive, but only higher-ranking individuals have the nerve to handle infants clinging to the mother's front, despite her and her baby's distress. Babies may be roughly pulled and even kidnapped from lowly females. An infant's cries will bring a godfather to the rescue. But the mere presence of an adult male is usually enough to inhibit other females from approaching a mother.

In many cases godfathers must be the real fathers of the infants they sponsor. According to one study, though, males that were probably not the father were as likely as actual fathers to play godfather roles (34). The inference is that males also derive benefits from godfathering infants other than their own. These may include (*a*) establishing bonds with a female that will culminate in the fathering of future offspring; (*b*) a means for male immigrants to gain acceptance into a new troop more quickly; and (*c*) opportunities to use infant, juvenile, and female associates as buffers in aggressive encounters with male rivals. In sum, there is increasing evidence that the reproductive success of male baboons is affected to a large extent by their nonsexual relationships with females.

Although sexual competition may have been the driving force that caused male baboons to become twice as big as females and develop lethal weapons, these same attributes enable troops of baboons to venture away from the safety of trees and cliffs, despite the presence of large carnivores. Indeed, a leopard is more than a match for a single male, but the readiness of males to join together in defending their troop (2, 31; N. Monfort pers. comm.) could well account for the fact that even lions seldom molest baboons (J. Popp pers. comm., author's observations). A group defense represents cooperation rather than competition, but there is a precedent; males form coalitions in order to gain an advantage over more dominant males. By joining together to protect the troop against predators, they are at the same time safeguarding their own reproductive investment.

ACTIVITY. The baboon is a diurnal monkey which usually becomes active well after dawn and retires to a secure sleeping place before dark. The troop is a cohesive unit which keeps together during foraging and the activities of the members are highly coordinated—90% do the same thing at the same time (1, 15). Social (primarily grooming) and sexual activity are most frequent before and after the day's foraging. Juveniles remain more or less continually active; their antics start before and end after all other baboons have begun and ceased stirring (2, 9).

Unlike most other African wildlife, baboons do not have feeding peaks early and late in the day divided by a midday inactivity period. Although they may drowse in the shade when it is very hot, what sets this species apart is the lack of any set pattern that holds for different troops and populations. A given troop may be active at any time of day. Baboons have been known to get started before dawn, to retire ½ hour after dark, and even to leave their roosts on moonlight nights to feed in fruiting trees or to raid crops (2, 9, 31, 36). Distribution and availability of water, food, and sleeping sites largely determine how much time a troop has to spend foraging and traveling in a day, and how much will be left over for resting and socializing (7). In Amboseli, cumulative rainfall was found to be a good predictor of the length of a troop's day journey and of the time the baboons spent feeding (1). Day ranging was longest (6 km) in dry months and shortest during the rains (4–4.5 km). More than 25% of the day was spent walking and 52%–55% feeding during the dry season, compared to 17%–19% walking and 40%–42% feeding in the rains. Daily ranging of 5–6 km is about average for baboons living in acacia savanna; those living in richer habitats such as semideciduous woodland and gallery forest travel only half as far (2, 7, 15, 28). A troop living in the semiarid northern Transvaal made day journeys of 10.5 km (36). But the record may belong to hamadryas baboons living in the Ethiopian semidesert lowland: one troop averaged 13.2 (9.2–19.2) km per day (20).

POSTURES AND LOCOMOTION. Adult males and females move and carry themselves differently enough that a knowledgeable observer can sex baboons without other clues. More subtle differences between same-sex baboons reflect an individual's self-confidence and social rank. However, tail carriage is not a reliable indicator of rank, as sometimes asserted. It *is* an indicator of age: the lower part of the tail stands more erect and the bend becomes more acute with age (18). The "riding-whip position" is diagnostic of the genus *Papio,* but there are also differences between species and subspecies in the amount of bend. The hamadryas baboon and the Guinea baboon carry their tails in a smooth curve, whereas the tails of other races of savanna baboon have more kink (9).

The way a baboon sits depends on how alert it is, ranging from the hunched squat of a relaxed animal, hands resting loosely on the knees, to the straight back with head raised and turning of a vigilant baboon (fig. 26.2). Lying positions vary from flat on the stomach on a branch with arms and legs dangling, to flat on the back with limbs in the air. The most abandoned poses are seen during grooming sessions (fig. 27.10).

Baboons move in the usual *cross-walk.* Running gaits are comparable to the canter and gallop of a horse (9), hindlegs and forelegs moving in pairs. Although the limbs are somewhat modified for terrestrial life, baboons still possess impressive climbing skill, seeming indeed hardly less competent than forest monkeys after allowing for the difference in size. Baboons associated with redtail and colobus monkeys in a Uganda forest could jump as far and were equally agile (28). However, the baboons jumped to the ground when seriously alarmed and fleeing, whereas the monkeys fled through the trees. Equally surefooted on cliffs, baboons ascend steep places with short leaps and grope for handholds on sheer surfaces. Falls are very rare, except during play, and baboons land catlike on all fours; one female dropped 10 m and landed undamaged on a bush-covered ledge (9). In Namibia playing juveniles somersault down sand dunes.

Despite their stubby, calloused fingers, baboons have exceptional manipulative skill. They pluck grass and pick up small objects between thumb and index finger, and use their fingernails to dig through hard soil and extract roots or beetle larvae. Actually, the index finger and thumb are the only digits that can be moved independently, although the fingers are flexible enough to bend back at right angles to the palm. The hand can be bent under more but the forearm rotates less than ours (4).

FORAGING AND PREDATORY BEHAVIOR. Baboons feed from hand to mouth and often sit while eating. There is no evidence of tool use; seashore baboons crush shellfish with their teeth rather than break them open against rocks (9). Similarly, Serengeti males use their canines to pry open river oysters imbedded in rocks exposed during extreme low water; the bivalves burst open with a loud pop (A. Root pers. comm.). But when digging up grasses to eat the rhizomes, baboons brush or knock off adhering dirt against a forearm. Invertebrate prey is sought by turning over rocks, logs, and dung, tearing off bark, groping in holes, and digging up the ground. When eating sizeable vertebrate prey such as hares or antelope fawns, baboons use their incisors to strip meat from bone and begin eating at the underside. Such prey is seldom deliberately hunted, but in a troop where hunting has become a tradition (discussed under Ecology), males cooperate in stalking and chasing game, and even share meat (16).

SOCIAL BEHAVIOR

COMMUNICATION. Baboon communication is very similar to that of other cercopithecine monkeys (see chap. 26). However, the displays of baboons and the closely related macaques are more variable than those of other monkeys. They have more expressive faces and highly graded vocal signals. To understand a given behavior, it is necessary to recognize seemingly trivial differences in posture, movements, facial expression, and accompanying vocalizations, for these register subtle changes in the animal's emotional state and intentions. To present a list of displays is therefore to make artificial subdivisions of what is actually a continuum of visual and vocal signals (21). The more subtle visual signals are not even listed. For instance, small changes in normal locomotion (change of pace, halting, looking around) by high-ranking troop members give the signal for a troop to set off on the day's foraging or to change direction (12).

Visual Communication (Agonistic Behavior)

Dominance/threat displays: staring, with *eyebrow-raising* and *rapid-blinking, tooth-grinding*, canine display* (tension yawning), penile display*, bristling, push-ups,* slapping ground, rubbing ground, branch-shaking, rearing on hindlegs, lunging, and charging.

The asterisks identify displays performed only or primarily by adult males. *Canine display* (fig. 27.11) or *threat-yawning* is often seen when a lower-ranking male harasses a superior that is preoccupied with an estrous female or meat. Unlike *tension-* or *displacement-yawning* (which is seen in all classes), *threat-yawning* is accompanied by *eyebrow-raising,* ear-flattening, and oth-

Fig. 27.11. *Canine-threat display* of male olive baboon.

er aggressive gestures. *Tooth-grinding* can be heard when 2 males are threatening each other at close quarters (12, 24). Advance warning of aggressive intent or mood can be read from an attitude of tension, as when a dominant male assumes an alert posture in a context other than danger, especially if he stares at someone and slaps the ground; or if a seated animal gets to his feet and stands with hindlegs planted, arms rigidly straight, hair on end, and stares with raised muzzle. Adult males positioned on vantage points and keeping a lookout while the rest of the troop is foraging often sit with thighs spread and flesh-colored penis prominently displayed (fig. 26.2). This behavior may be a signal to outside males that an adult male is present and guarding the troop females (39).

The commonest and mildest form of intimidation is *supplanting,* when a dominant baboon approaches and occupies the place of a subordinate.

Submissive displays: avoidance, *social presenting, rapid-glancing, fear grimace,* undirected staring with eyes wide, *erecting tail, crouching, lying prone.*

When a subordinate baboon passes close to a feared dominant, averting the gaze or undirected staring, the *fear grimace,* and *ear-flattening* (as during threat) are often combined. *Grimacing* with open mouth, combined with *yakking* (*gecker*), is a more intense expression of fear and submission. A baboon that is cornered during a chase crouches or lies flat on the ground, often rigid with fear, while *grimacing* and *churring.*

Social presenting (fig. 26.4) is seen in several different situations, each with some postural variations. (*a*) Brief presenting by nonestrous females and by juveniles, typically to a dominant male; females approached by males other than regular companions present and make submissive gestures, which may elicit genital inspection and mounting by the male (33). (*b*) Adult female to a female with a black infant, with hindquarters sometimes lowered, looking over the shoulder and *lip-smacking.* (*c*) Invitation to a juvenile to be carried. (*d*) Lowering the hindquarters by one adult male to another—typically subordinate—male, which usually grasps the presenter's rump but rarely mounts. The hindquarters are lowered more in (*b*) and (*d*) than during *sexual presenting* (12).

Tension, uncertainty, or displacement activities: yawning, scratching, twitching (head, arms, shoulders), shoulder-shrugging, muzzle-wiping with hand.

These and other motor patterns are commonly classified as displacement activities when performed out of context and often in a jerky, desultory way. Shoulder-shrugging and muzzle-wiping may appear after a baboon has been startled, as by the sudden appearance of a snake. The male of a consort pair which is being harassed by rivals may make rapid, sometimes frantic grooming movements and hurry copulation. Mounting and even the act of copulation may be performed as a displaced activity by a disturbed male (12).

Friendly or reassuring displays: grooming solicitation, *lip-smacking, greeting,* standing bipedally before another baboon.

Behavior in this category usually goes together with or leads to body contact (more under Tactile Communication; see also under Vocal Communication, e.g., *lip-smacking*).

Tactile Communication. See also under Sexual and Parent/Offspring Behavior: *social grooming, nose-to-nose greeting, social mounting,* embracing, touching side.

The greatest part of baboon social interactions is invested in *social grooming.* The complicated web of relationships, affiliations, and rank order within a troop can be deciphered by observing who grooms whom and how (relaxed or tense). The baboon's day typically begins and ends with long grooming sessions; sleeping subgroups become grooming subgroups as soon as they come down from their roosts. Mothers groom their own infants more than they groom any other baboon and receive more

Table 27.2 Savanna Baboon Vocal Signals

Call description	Performer	Context and message
Dominance/threat		
1. *Two-phase bark.* Deep, loud "*wahoo*," often repeated at 2–5 second intervals.	Adult male	(a) External disturbance, especially feline predators. (b) Intra- and inter-troop aggression between males. Advertises male presence and arousal: a distance-increasing signal.
2. Grunting (sometimes 2-phase "*uh-huh*"). Soft prelude to no. 3.	Adult male	Low-intensity threat, occasional prelude to no. 1.
3. Roaring. Crescendo of *2-phase grunts.*	Adult male (*anubis* only?)	Voiced during olive-baboon fights; apparently not given by chacma baboons (13).
4. Grating roar. Deep, resonant call. Follows a *2-phase grunt* as a series of single grunts with descending volume and pitch.	Adult male (*anubis* only?)	(a) Heard a few times by (12) after a fight, from dominant male. (b) Frequently heard at night from males in roosting troops of olive baboons, usually following some disturbance (24, author's observations).
Fear and alarm		
5. Screeching. Repeated, high-pitched screams, sometimes changing to churring noise when caught.	All classes	Expresses fear, the usual response to aggression by a dominant individual, as when chased by adult male. Functions to inhibit aggression.
6. *Yakking* Short, sharp "*yak*" accompanying open-mouth fear grimace.	Subadult and adult	Withdrawal from threatening animal.
7. Chirplike clicking. Probably same as "*ikk*" of ref. 1 and more often combined with "*ooer*" in no. 8.	Infant-juvenile	Equivalent of yakking in young baboons (see no. 9).
8. "*Ick-ooer*." Two-phase call: "*ick*" followed by cooing, given with pursed lips (hard to locate [1]).	Infants	Given, for example, in response to maternal rebuff, expresses low-level fear and/or moderate distress. Cooing expresses distress (e.g., during weaning) rather than fear.
9. Shrill bark. Single, sharp explosive sound.	All classes—except adult males?	Alarm signal, given especially in response to sudden disturbance, usually evoking flight of other troop members.
Friendly (distance-decreasing)		
10. Grunting. Low, soft, rhythmic grunts.	All classes except infants	Close-range contact and progression call, the commonest vocalization. Also given during approach, signaling friendly intentions.
11. Doglike bark. Higher-pitched, less staccato than shrill bark, with more quaver.	Subadult-adult	Contact call given by individuals or subgroups separated from troop.
12. Chattering. Nasal, rapid, gruntlike sounds.	Juvenile chacma baboons	Heard during play.
Sexual		
13. Muffled growl. Sound given with mouth nearly closed; cheeks puff in and out in chacma (perhaps not other) baboons.	Estrous females	Associated with copulation: normal in chacma baboons (10) but unusual and less audible in olive baboons.

SOURCE:. Modified from Hall and DeVore 1965.

grooming from other females than they give while their infants retain the natal coat (see under Social Organization). Otherwise, the direction of grooming is usually from lower to higher rank, males receiving much more grooming than they give (except to estrous females) (fig. 27.10). *Social mounting,* embracing, and touching the side are frequent responses to *social presenting* and probably signal friendly reassurance. They are also seen during aggressive interactions, between coalition partners, and by individuals attempting to enlist support (by *rapid-glancing,* screeching, etc.; see under Fighting).

Vocal Communication. Most of the calls

Fig. 27.12. Male coalitions: a triadic interaction among yellow baboons (from Hinde 1983).

listed in table 27.2 grade into one another, for instance grunting into barking and roaring, screeching into *yakking* (repetitive *churring*) (12, 21).

AGONISTIC BEHAVIOR

Fighting and Overt Aggression. The frequency and intensity of aggression seen in a baboon troop vary greatly, largely as a result of troop composition and relationships between individual males, in particular. If the rank hierarchy is well established, dominance is asserted and acknowledged with little or no overt aggression. But troops with recent male immigrants or newly matured resident males tend to be unstable. Dominant newcomers may be unusually tyrannical and sexually active (33). Rivalry for dominance may include frequent threat and counterthreat, the formation of male coalitions (2 males joining forces to defeat a third, more dominant one) (fig. 27.12), the use of infants to buffer aggression of a rival, and sometimes fighting (12, 25, 35, 36). The male's slashing canines make fighting extremely dangerous and correspondingly infrequent. Yet most adult males bear scars, especially about the head.

Even in stable troops, practically any disturbance, internal or external, is apt to provoke disciplinary action by dominant males. Teasing or bullying of small baboons by older juveniles and squabbles between females frequently trigger chases punctuated by shrieks, barks, and roars. If a male catches his victim, he may seize it, beat it with his hands, rub it on the ground, bite it in the neck, and even pick it up in his jaws. Despite their apparent ferocity and the disparity in size, such attacks rarely last more than a few seconds or result in visible injury. Attempts of dominant males to herd the troop away from an external disturbance (predator, another troop) may result in generalized aggression as the males canter, double back, and race from side to side to move the females and young and round up strays (13, 36).

Just as the incidence of aggression within troops is highly variable, so do relations between troops vary, apparently depending on the degree of competition for the same resources (e.g., roosts, fruit trees, water) and the existence or otherwise of social ties: baboons from troops that have recently fissioned may mingle and socialize peacefully when they meet (29). The most spectacular displays of baboon aggression occur during hostile encounters between troops, when nearly all the assembled baboons may join in a scene of pure bedlam (13, 36). Normally hostilities proceed no further than noisy squabbling, in which males take the lead: jumping about, branch-shaking, barking, herding their own feisty females and young away from the skirmish line. But sometimes from 3 to 10 males of one troop make a mass charge into the ranks of the rival troop, followed by withdrawal with or without a countercharge by males of the other troop. There is rarely any physical contact, but the troop with fewer adult males is usually the first to withdraw (13).

REPRODUCTION. Females begin menstruating at 4½–5 years, conceive about a year later, and thereafter produce young at roughly 1½–2 year intervals. Mothers do not normally resume cycling for 10–12 months after giving birth (minimum 5 months) and typically experience 4–5 cycles before conceiving again. Given an average cycle length of 36 days, a 6-month gestation, and a life span of 20–30 years in the wild, it can be calculated that female baboons spend half their adult lives taking care of dependent offspring, another third pregnant, and the rest menstrual-cycling (1).

Female reproductive condition is plain to see in *Papio* from the coloration and degree of swelling of the sexual skin (fig. 26.5). During proestrus (7 days), the perineum is black with a pink tinge and beginning to inflate. Estrus follows, lasting c. 10 days during which the sexual skin is bright pink and fully swollen. Fading and deflation begin 2–3 days after ovulation and end in 7 days with the sexual skin black and flat, a condition (anestrus) that lasts for 9 days, when proestrus begins anew. In the interim, menstruation occurs but seldom with any external signs of bleeding. In the event of pregnancy the naked skin over the hips and around the callosities turns from black to scarlet within a few weeks; the sexual skin is not involved. The sexual skin of lactating females is as in anestrus (black and flat). The size of the estrous swelling is small in adolescent females and tends to increase with age up to 5½–6 years. But the enormous swellings sometimes seen in captivity, perhaps resulting from repeated cycling without pregnancy, are not seen in wild baboons (30).

SEXUAL BEHAVIOR. Females in proestrus and estrus can be singled out of a troop of foraging baboons by behavioral clues even before their posteriors become visible (29). While other females remain seated except to move from one spot to the next, an estrous female moves about restlessly, and either approaches and solicits different males or has an adult-male consort. Females in proestrus often approach and *present* to several males within a few minutes, standing with tail bent to one side and looking back over the shoulder, sometimes *lip-smacking* and showing the eyelids (fig. 27.13). Females in this stage may allow subadults and even juveniles to mount them. It is the female, then, which takes the initiative prior to entering consortship with an adult male, after which either partner may take the initiative (12).

Although inflating and swollen females *present* equally often, males are more likely to copulate with those in estrus than in proestrus, and rarely mount adolescent females (which have small swellings). Even 4 out of 5 presentations by a tumescent female may only elicit a touch, *lip-smacking*, or brief grooming, or no response at all (29). Males become more discriminating with age; mature individuals rarely mount and virtually never ejaculate into any but fully swollen females (12, 36). Conversely, mating opportunities for subadult and large-juvenile males are limited to females in the inflating stage. Although a keen interest in a copulating pair is evinced by the presence of onlookers, there is surprisingly little evidence of sexual jealousy among adult males. The existence of pair bonds may be the cause or the effect of this tolerance. In any case, it has become apparent (from observations by female observers) that female preference influences the behavior of consort pairs more than was previously assumed (from observations by male observers). Females consorting with highly preferred partners cooperate with efforts to herd them away from other males, and groom and present to their favorites significantly more often than to less-preferred partners. For their part, preferred males follow and herd favorite females more closely, and mount females that strongly prefer them more than females with no preference for them. In short, the level of a female's receptivity closely reflects her preference, which in turn affects the male's sexual performance (27).

Copulation. The copulatory positions are the same as in other cercopithecine monkeys, the female standing as during sexual presenting, with tail deflected to the side (fig. 27.13). However, there appear to be some differences between subspecies. Chacma baboon copulation consists of a series of

Fig. 27.13. Copulation of olive baboons; note the female's facial expression, with white eyelids displayed.

mounts, ranging from 3 in 11 minutes to 9 in 191 minutes (36), culminating in a mount with an ejaculatory pause after an average of 10 pelvic thrusts (4–20) (10). Females give the grunting copulatory call after a series of thrusts, then as the male dismounts typically leap away while continuing to call. Grooming occurs after about half the copulations. Olive and yellow baboons, by contrast, usually ejaculate during a single copulation lasting 8–20 seconds, after an average of 6 thrusts, with intervals of at least ½ hour between copulations. Females rarely call.

PARENT/OFFSPRING BEHAVIOR. No strict birth season has been noted, but conception peaks tend to occur during periods of food abundance (rainy season/summer). Most births occur at night while baboons are roosting and therefore, although wet infants carried by females with bloody behinds, hands, and mouths may be seen in early morning, parturition has rarely been observed in the wild. One female was seen to go into labor after the rest of her troop had gone to roost. She alternately jerked her limbs and lay on her side on the ground, but gave birth only after ascending the trees at dusk (1). A captive female labored nearly 7½ hours and delivered in a squatting attitude while gently pulling on the infant's head. Afterward she licked it clean and ate the placenta. It was nursing within 23 minutes of the appearance of the head (8).

Selection may favor night births because of the difficulty exhausted mothers have in keeping up with the troop, especially while supporting a newborn infant with one arm (clamping its mouth on a nipple gives the baby an additional anchor). The bright pink skin and black hair of the neonate contrasts strongly with adult coloration; the male's penis is very conspicuous (adult handlers almost invariably turn infants upside down and closely examine the genitalia [1]). This coloration begins to change in the third month and adult pelage is acquired by the sixth month, although males retain a pink scrotum for at least another year.

From a completely dependent and helpless start, an infant baboon gains locomotion skill by the end of a month and during the second month develops enough climbing skill to clamber over small logs and adult baboons. Riding in the jockey position (fig. 27.14) begins at 6–12 weeks, but some baboons never change from the ventral position (1). During the second month babies

Fig. 27.14. Juvenile baboon riding in the jockey position (from a photo in Altmann 1980).

develop the skill to pick up plants and other objects by hand that they had previously picked up in their mouths. In months 3 and 4 baboons begin interacting with other infants that are their companions in the maternity subgroups formed by mothers with small young. Although still dependent on their mothers for virtually all food and transportation, they have begun to feed on the more accessible foods and can climb about within a tree fairly well. Yearling baboons are quite independent and could theoretically survive if their mothers should die. Yet the youngest surviving orphans in the Amboseli study (from which the above observations are drawn) were 17, 20, and 21 months old (1). Evidently the continuing protection and guidance of the mother are essential for some time after nutritional independence, despite the care provided by godfathers.

ANTIPREDATOR BEHAVIOR. Considering that baboons take extra risks by venturing into open grassland far from refuges, a correspondingly higher predation rate would be predicted. Yet nearly all baboonologists have commented on how little predation they have observed. Some observers maintain that males do not protect females and young, but only worry about their own skins (28, 30; R. Harding pers. comm.). If females and young were really unprotected and vulnerable whenever they left the shelter of the trees, surely the larger predators would take advantage of the situation? The fact that they do not is indirect evidence that the presence of adult males deters attack. There is also direct evidence that not only adult but also subadult males, and even adult females, will confront predators as large as leopards when there is no alternative (2, 19, 31, 36).

The way baboons respond to a predator depends very much on its identity and on the circumstances. The response of a troop to the presence of a distant lion or leopard depends on whether they are far from or near a secure refuge. An attempt to come between a troop and the safety of a grove of trees can create panic (30). In contrast, baboons are more relaxed while in short grassland than when moving through cover that could conceal a predator, and are most cautious when going to water, approaching, a few at a time, only after reconnoitering, while adult males often stand guard in trees or other vantage points (2).

When a foraging troop is alerted to danger by an alarm bark (often given by some juvenile or subadult foraging on the troop periphery), adult males move in the direction of the disturbance and proceed to demonstrate typical vigilance behavior. A sudden disturbance nearby provokes a barrage of loud barking and a stampede for cover, which would probably confuse a lurking predator's attempt to single out a victim if its initial rush had failed. The whole troop then makes good its escape while males and sometimes adult females form a rearguard and, if necessary, go on the offensive. At least ½ dozen cases are known in which leopards have been mobbed and severely wounded by baboons (19, 31; N. Monfort, pers. comm.).

Whereas the leopard is probably correctly regarded as the most effective baboon predator, lions, spotted and striped hyenas, chimpanzees, pythons, large eagles, and humans are also known to have killed and/or fed on baboons. In the presence of man, baboons rarely put up any defense (but they have been known to threaten people holding screaming infant baboons—J. Popp pers. comm.). They run away silently and will jump out of their sleeping trees up until dark. At night, however, baboons will usually not budge, but manage to discourage intruding humans very effectively with a rain of stinking excrement (author's observations).

References

1. Altmann 1980. 2. Altmann and Altmann 1970. 3. Bolwig 1959. 4. ——— 1963. 5. Busse and Hamilton 1981. 6. Crook and Aldrich-Blake 1968. 7. DeVore and Hall 1965. 8. Gillman and Gilbert 1946. 9. Hall 1962*a*. 10. ——— 1962*b*. 11. ——— 1966. 12. Hall and DeVore 1965. 13. Hamilton, Buskirk, Buskirk 1975. 14. ——— 1978. 15. Harding 1976. 16. Harding and Strum 1976. 17. Hausfater 1975. 18. ——— 1977. 19. Kingdon 1971. 20. Kummer 1968. 21. Marler 1965. 22. Napier and Napier 1967. 23. Packer 1979. 24. Ransom 1981. 25. Ransom and Ransom 1971. 26. Rasmussen, D. R. 1979. 27. Rasmussen, K. L. R. 1983. 28. Rowell 1966. 29. ——— 1967. 30. ——— 1972. 31. Saayman 1971. 32. Smithers 1971. 33. Smuts 1983*a*. 34. ——— 1983*b*. 35. Stein 1984. 36. Stolz and Saayman 1970. 37. Strum 1975. 38. ——— 1981. 39. Wickler 1967.

Added References

40. Smuts 1985. 41. Strum 1987.

Chapter 28

Colobus Monkeys
Subfamily Colobinae

Colobus guereza, Eastern black-and-white colobus or guereza
C. polykomos, Western black-and-white colobus
C. angolensis, Angolan black-and-white colobus
C. satanas, black colobus
C. badius, red colobus
C. (Procolobus) verus, olive colobus

Colobinae are a subfamily of monkeys with ruminantlike stomachs specialized for efficient digestion of leaves. Most species are found in Asia (see chap. 24), but colobus monkeys occur only in Africa.

TRAITS. Long-tailed and mostly long-haired monkeys, ranging in size from the 5 kg olive colobus to the 23 kg (maximum) black-and-white colobus. The same 2 species span the range of coloration and markings, from one of the most nondescript to one of the most conspicuous of all monkeys. Infants are distinct in color and pattern from adults (except olive colobus). The trait that distinguishes colobus from all other monkeys is the reduction of the thumb to a small stub.

DISTRIBUTION. Tropical Africa, from The Gambia of West Africa south to Angola and across the continent to Zanzibar. Olive and red colobus are essentially rainforest species, their range largely confined to the equatorial forest blocks, whereas the more adaptable black-and-white colobus inhabit many kinds of forest and woodland over a broader geographical range (14).

ECOLOGY

THE COLOBINE DIGESTIVE SYSTEM. Leaf-eating monkeys have a ruminantlike digestive system (see chap. 1). The stomach is greatly enlarged and divided into chambers wherein ingested vegetation undergoes fermentation and predigestion by anaerobic bacteria before entering the acid pyloric part of the stomach (1). The large stomach capacity (which makes colobine monkeys pot-bellied) permits the accumulation and slow passage of food necessary for thorough fermentation. Fermentation rates similar to those of small ruminants and comparable concentrations of volatile fatty acids have been found. Large numbers of cellulose-digesting bacteria indicate that plant structural carbohydrates are an important energy source for colobine monkeys. Perhaps equally important, bacterial digestion apparently detoxifies leaves, seeds, and fruits containing concentrations of secondary compounds that would be poisonous to other monkeys with conventional hindgut fermentation. The ability both to break down cell walls and to cope with secondary compounds would explain the ability of the black-and-white colobus to subsist on an unvaried diet of mature leaves for months on end (see species account). Speaking of this subfamily as a whole, the evolution of a ruminantlike digestive system has enabled the leaf-eaters to occupy ecological niches that are inaccessible to other primates.

Forest folivores exploit a superabundant resource which requires far less active and extensive searching than is necessary for an omnivore. Consequently, colobines achieve far higher population densities than other associated monkeys. In Uganda's Kibale Forest, for instance, the red colobus is the most abundant of the 6 monkeys that live there, estimated at 300 *C. badius*/km² (11).

ECOLOGICAL SEPARATION. The black-and-white colobus and the black colobus, which some authorities consider only subspecies rather than separate species, are equally similar ecologically. However, the red and olive colobus are very different from them and each other. Little is known about the olive colobus, other than that it apparently frequents dense undergrowth in the rain forest, regularly associates as an inconspicuous member of mixed-species troops, and eats foliage (2). But the differences between the red and the black-and-white species have been closely studied (9, 10, 12, 14, 16). *C. badius* feeds mainly on young growth, and travels comparatively widely to find trees coming into leaf and flower, utilizing a great variety of different species at all levels—but particularly the upper strata—of the forest. It also eats the mature leaf stalks of a wide variety of plants. Black-and-white colobus, in contrast, have a comparatively simple and unvaried diet, with up to half their food coming from a single common tree species. They, too, prefer young over mature growth stages, but if necessary can subsist for months on fibrous mature leaves. In Kibale Forest there is only a 7% overlap in the food species utilized by these 2 monkeys. Even when there is overlap, preferences for different parts or growth stages may effect separation: thus, both monkeys ate *Markhamia* leaves, but *badius* ate only the stalks and *guereza* ate only the blades (11).

The red colobus eats and apparently needs a wide variety of plants throughout the year, which largely limits its distribution to rain-forest habitat, whereas the black-and-white colobus's ability to get by on a coarse, unvaried diet of mature leaves makes it much more adaptable (9, 10, 16).

SOCIAL ORGANIZATION. The black-and-white colobus represents the prevailing form of social organization in the subfamily (and in forest guenons, for that matter): groups consisting of several related females with assorted offspring which live in a small home range defended by a territorial male. Some groups contain 2 or more large males, but as a rule only 1 is dominant, has access to the females, and behaves territorially (8). Typically, female offspring remain in the natal troop and males are forced to disperse, usually as subadults. Relations with the neighbors are hostile but within the troop are peaceful and close, without a steep dominance hierarchy, reflecting the more placid dispositions of leaf-eating monkeys compared to most other members of the family. In black-and-white colobus (and some langurs), mothers permit other group members to handle babies less than a week old. Few other primates, including red and olive colobus, are that permissive (8).

The red and the olive colobus are nonterritorial and live in multimale troops. *C. badius* lives in large troops, commonly numbering 40–80 monkeys in the best habitat (half as many in poorer habitat), which share home ranges of c. 32 ha (11). From 2 to 10 males live in a ± stable rank hierarchy, with the alpha male siring most progeny. Relations between troops are hostile and larger ones usually supplant smaller ones when they meet. But troops are far less cohesive than black-and-white colobus groups, with wider individual distance, and in at least some areas troops often break up into subgroups (10). Females and their juvenile offspring may range in different groups sometimes for days (cf. sable antelope). Further proof of weak maternal bonds is the transfer of females from their natal troop, generally before maturity, sometimes repeatedly and with little or no difficulty. So far, fewer than a dozen primates are known in which female transfers are the norm (see gorilla and chimpanzee accounts). Male red colobus also leave their natal troop, usually as a result of increasing harassment by older males beginning at adolescence. But their efforts to join another troop meet with much more resistance, not only from resident males but also from females, which have been known to participate in fatal mobbing attacks against recent immigrants (10), possibly triggered by maternal defensiveness against suspected infanticidal males. Infanticide following male takeovers, now known to occur in various mammals (see lion account), was first described in langurs (3, 15). It has been recorded in red colobus (6) and may also occur in black-and-white colobus (see species account).

The social organization of red colobus (large, loose troops, communal home range, nonterritorial) seems to be adapted to its foraging strategy, which keeps troops moving in search of leaf flush and flowers. Food distribution is clumped in time and space; a single clump may feed many monkeys for a few days but a large area is necessary to find sufficient food throughout the year—too large to be monopolized as an exclusive territory. But the abundance of food within undisturbed rain forest supports a phenomenal biomass of red colobus, estimated at 1760 kg/km^2 in Kibale Forest, compared to only 162 kg/km^2 of redtail monkeys, the second most abundant species (11).

POSTURES AND LOCOMOTION. Squirrel-like quadrupedal galloping and leaping (see black-and-white colobus account). Struhsaker (11, p. 4) considers the red colobus one of the clumsiest climbers, whose "general mode of locomotion can best be described as suicidal."

SOCIAL BEHAVIOR

COMMUNICATION. In black-and-white colobus, emphasis is on communication between troops, with loud calls, conspicuous coloration, and leaping displays designed to maintain territorial spacing and discourage transgressions (7). In the red and presumably the olive colobus, vocalizations and visual displays are designed for interpersonal communication within large, multimale troops, and interactions at close range with other troops competing for the same resources. One would accordingly expect these species to have a richer, more graded communication system.

Vocal Communication. The black-and-white colobus has the largest and the red colobus has nearly the smallest larynx in the subfamily. Male red and olive colobus sound like alto and soprano choirboys compared to the rolling bass of the black-and-white colobus. Yet a repertoire of 16 different sounds has been attributed to the red colobus compared to only 6 so far described in the black-and-white colobus (11). Furthermore, red colobus calls intergrade, whereas most black-and-white colobus calls are discrete. Male red colobus use 3 different calls (bark, *chist, wheet*) in response to a great number of different stimuli, as expressions of alarm, threat, and so on. Although the functional differences among these calls remain unclear, intertroop communication and spacing are functions they all have in common. Red colobus calls may also be more individually distinct than those of black-and-white colobus (11).

Visual Communication. Red colobus are decidedly more individualistic than any of the other species in appearance, notably in facial features, hair color, postures, and movements (10). Likewise, age classes are better defined and sexual dimorphism among adults is more obvious than in other colobus. However, gender may be said to be "deliberately" obscured in juveniles of both red and olive colobus by the presence of a unique "perineal organ," which mimics the bright red and swollen genitalia that develop in adult females during estrus (5, 11). Mimicry of the female structure is so faithfully copied in young males, and their own genitals are so inconspicuous, that even juveniles in hand are hard to sex (fig. 28.1) (5). The probable function of this mimicry is to reduce the intolerance of adult males toward young males (see chap. 26, Agonistic Behavior, *social presenting*). Harassment often culminating in eviction begins in adolescence as the perineal organ fades to the gray color seen in adult males and nonestrous females (11). Interestingly enough, there are 2 forms of *social presenting* in this species: submissive presenting directed by juveniles and adult females mainly to adult males, and a slightly different form in which dominant males show their backsides mainly to subordinate males; this second form apparently functions to reinforce the dominance hierarchy (11).

There are no external signs of estrus and no perineal organ in other colobine monkeys, although the black and Angolan colobus, like the red and the olive colobus, live in multimale troops (13).

REPRODUCTION. Strict birth seasons have yet to be described, but birth peaks during rainy months have been noted in *C. badius* (13) and suspected in *C. guereza.*

SEXUAL BEHAVIOR. The typical primate copulatory position, with the male's feet grasping the female's ankles, is not

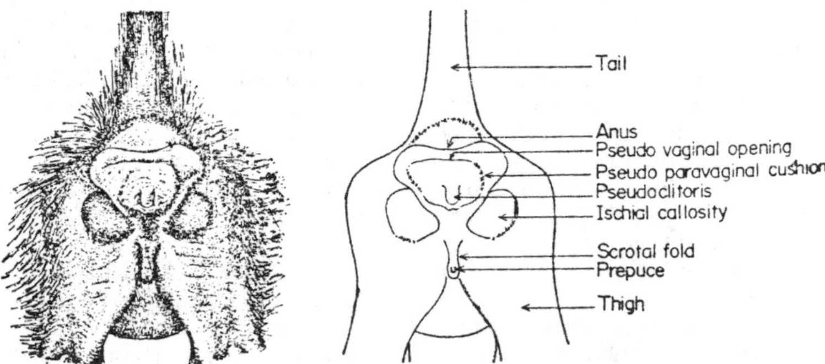

Fig. 28.1. The red colobus perineal organ, which in juveniles of both sexes mimics female's estrous swelling (from Kuhn 1972).

seen in all colobines. In some cases the male's feet remain on the substrate; both positions are employed by red and by black-and-white colobus (8, 11).

PARENT/OFFSPRING BEHAVIOR. The red and perhaps the olive colobus seem to be the only colobines that do not engage in communal handling of small infants. There is probably a connection with the looser female-female bonds and female transfers out of the natal troop. In the territorial species female troop members are usually related (10, 11). The olive colobus is said to carry very young infants by mouth, a practice otherwise unknown among higher primates. Apparently the lack of a thumb and very thick cover render the infant's grip on the mother's fur unusually precarious (2). The squeals and screams of distressed infants and females are very similar among all species (11).

References

1. Bauchop 1978. 2. Booth 1957. 3. Hrdy 1977. 4. Kingdon 1971. 5. Kuhn 1972. 6. Leland, Struhsaker, Butynski 1984. 7. Marler 1972. 8. Oates 1977*a*. 9. ——— 1977*b*. 10. Starin 1981. 11. Struhsaker 1975. 12. ——— 1978. 13. Struhsaker and Leland 1986. 14. Struhsaker and Oates 1975. 15. Sugiyama 1965.

Added Reference

16. Clutton-Brock 1975.

Fig. 28.2. Female black-and-white colobus with white infant (from Richard 1985).

Black-and-white Colobus

Colobus guereza

TRAITS. Large, long-haired monkeys with conspicuous black-and-white coloration. *Length:* (head and body) males 59.3 cm (53.5–69), tail 81 cm (67–88.5); females 55.4 cm (48.5–64), tail 77 cm (71.5–82.5) (1). *Weight:* up to 23 kg (8); Uganda population, males 9–14.5 kg, females 6.5–10 kg (12). Nose protruding, curved, overhangs upper lip. *Coat:* geographically variable in length and extent of white markings; color glossy black usually with white brow and full beard surrounding naked, slate-gray face, a cape or fringe of long white hair running along either side and meeting on the lower back in the shape of a horseshoe; tail thin and black with white tip (*C. g. guereza*) to all-white brush (*abyssinicus*); newborn wholly white with pink face.

DISTRIBUTION. Widespread and often abundant across equatorial Africa in forested habitats up to 3000 m. This monkey varies so much geographically that some systematists have considered certain populations as distinct species whereas others have classified them as merely subspecies. Currently, 4 different species are generally recognized (see subfamily introduction) (6), of which the all-black *Colobus satanas* is the most distinct. Nevertheless, until different forms are found to overlap without significant hybridization, all the different forms

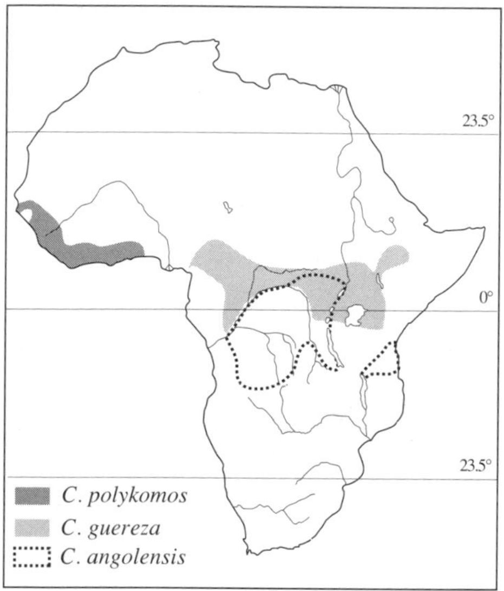

could well be simply geographical races of the same monkey.

ECOLOGY. The black-and-white colobus is able to utilize mature foliage too fibrous and/or well-protected by toxic secondary compounds to be eaten in quantity by other monkeys (2, 10, 13). A highly evolved digestive system (see subfamily introduction) enables this colobus to subsist at times when new growth, flowers, and fruits are scarce by eating only mature leaves of 1 or more common tree species, without drinking. It can accordingly live within a very small home range in a wide variety of habitats: rain forest, cloud forest, bamboo thickets, gallery forest along rivers that flow through subdesert, isolated patches of trees on cleared land, and secondary woodland. Indeed the guereza seems most successful in secondary and riparian forest that is marginal for most other monkeys, especially in habitats with few tree species and extended dry seasons (17, 18).

DIET. In Uganda's Kibale Forest, 40% of the guereza's annual diet is provided by just 1 very abundant tree of medium stature, *Celtis durandii* (12, 13). The mature leaves are not eaten by any of the 4 other monkeys present. Also the guereza alone eats the mature leaves of *Rauvolfia oxyphylla* and the flowers of a *Strychnos* species, both noted as producers of alkaloids. But given the opportunity, it usually prefers young over old growth, and at some seasons fruits (especially figs) make up as much as ⅓ of the diet. In riparian acacia woodland bordering Lake Naivasha, a colobus troop subsisted largely on fever-tree leaves and unripe pods (14).

SOCIAL ORGANIZATION. Territorial, 1-male troops, sedentary in typically small, non-overlapping home ranges.

The typical troop numbers 6–10 individuals (3, 12, 15, 19). Fourteen troops accurately counted in Budongo Forest (Uganda) averaged 8 monkeys (2–13), and the mean of 7 troops studied in Kibale Forest was 11.4 (3–15). Mean group size is c. 9 in well-forested areas, typically composed of 1 adult male and 3–4 females, 2 subadults, 1 juvenile, and 1 infant (8). The maximum recorded number of females is 5 and the largest recorded troop numbered 19 (8, 14). Occasional lone individuals are invariably males, no doubt maturing animals dispersing from their birthplaces. However, such males apparently do not stay alone for long, but tend to associate in pairs and/or to attach themselves to a troop.

The presence of 2 or more adult males in a troop, occasionally reported by most observers, has now been definitely confirmed: 4 of the above 21 troops included 2–3 adult males and 1 included 5. As many as 6 large males have been seen in one group, of which 2 were probably immature, however (12). To sex guerezas it is necessary to obtain a clear rear view: the circle of white hair is intact in males and interrupted in females by the vulva. Apparently this method does not work with juveniles, whose sex is seldom reported. Other signs, such as the elongated nipples of multiparous females, the large scrotum of the alpha male, and even the considerable size dimorphism, are much harder to spot; a lack of clear individual differences compounds the difficulty.

The most apparent sexual dimorphism in this species is behavioral, and by this criterion apparently all troops are 1-male. (Structural differences in the male and female larynx can be heard but not seen [5].) By refraining from such displays and keeping their distance from troop females, other males may be tolerated; in effect they are peripheral males. Nevertheless, male overlords that tolerate bachelor males on their land run the risk of being supplanted by one of them, possibly followed by infanticide. At least 3 of of the 7 troops in the Kibale Forest study were subjected to takeovers in 4 years, and in 2 cases 5–6 large males attached themselves to a 1-male group some months before the takeover, following which the extra males departed. That such takeovers may involve infanticide was inferred from the fact that one troop containing 4–5 females had no infants for about 2½ years; then some months after a takeover 3 out of 4 gave birth at nearly the same time (12).

Is the presence of extra males evidence of weakness on the part of a territorial male and a sign that a turnover is imminent? Whether or not some deficiency in the resident male's territorial defenses invites intrusion, the presence of a gang of interlopers might present a serious dilemma. (a) To keep chasing them out of his territory could wear him out and make him more vulnerable to a successful challenge if they prove persistent. (b) Too much tolerance might embolden the interlopers to become insolent and lead to a takeover attempt. So perhaps the proprietor's best course is to exert himself just enough to keep the interlopers properly respectful, and this may be accomplished partly by vigorous calling and visual displays to neighboring territorial males, from whom he has far less to fear.

Home Range and Territory. Black-and-white colobus troops live in remarkably small ranges and, by most accounts, the boundaries are sharp and closely guarded. Size estimates range from under 5 ha to c. 25 ha (3, 9, 12, 14, 15). Such close partitioning of the habitat enables this monkey to achieve population densities as high as 200/km² and even up to 500 colobus/km² in Kenya's Kikuyu Escarpment and Tanzania's Southern Highlands (7, 8).

The range of a Kibale Forest troop whose movements were plotted for over 2 years was c. 15 ha (12). But this troop did not have the usual, exclusive territory; 5 other troops were recorded 1 or more times within over 74% of its range. Even the 2.7-ha core area where the study group spent 44% of its time was visited by other troops. However, at least within this locality the study group was clearly dominant, nearly always supplanting other groups it encountered there. Special circumstances (proximity to a swamp where local troops came to feed) may have accounted for the unusual amount of range overlap in this case, which seems more typical of the nonterritorial red colobus than the usually highly territorial black-and-white species (see subfamily introduction).

Social Distance. Black-and-white colobus troops are very cohesive, maintaining close spatial and social relations. Samples to determine individual distance showed that each colobus had an average of 2.8 monkeys (1–4) within 2.5 m and 4.8 within 5 m. A circle with a diameter of 18 m would usually enclose all 11 troop members, although they strung out for up to 100 m during progressions. Females and young stayed closest together and a subadult male kept farthest apart. After him, the territorial male and a childless female were least closely associated (12).

Social Grooming. The tight clustering of guerezas partly reflects the folivorous diet, which permits these animals to sit together and eat their fill with minimal moving about. But clustering also depends upon a remarkably low level of aggression (mostly supplanting by the adult male) and frequent friendly interactions, mainly in the form of *social grooming*. In the central Kibale troop, grooming accounted for 6.2% of all activity-sample records, a high rate compared to most other monkeys (12, 18). As usual, females and young were the main performers, whereas the alpha male groomed others only occasionally and was not groomed with great frequency himself. The subadult male groomed more than he was groomed in return. In troops containing more than 1 adult male, reciprocal grooming among them was not uncommon, but at only ⅓–¼ the rate among females.

Peaks of grooming and other activity occur during midday rest periods, and especially in early morning while the troop is sunbathing. Grooming may be intensified when rival troops are close, an indication of the social-bonding function of this behavior.

A grooming session between 2 colobus may begin spontaneously on meeting, or one may actively solicit grooming by stretching out in front of another (15). Should the invitation be declined, as signaled by looking away, the solicitor may deliver a series of small slaps to the cheek to make the other pay attention. Often an animal continues to solicit despite a quick slap from the dominant male and even after a beating. Similar determination is displayed by a colobus bent on grooming. It either comes up behind the intended partner and starts in at once, or makes a frontal approach, seizes the other's forelock and pulls its head down preparatory to beginning. Resistance is met with a few smacks to the head, each ending with a renewed attempt to pull the groomee's head downward.

Despite the absence of a thumb, colobus grooming procedure does not look very different from that of most monkeys. The hair is parted by flattening and pressing the coat in opposite directions with both hands together. Parasites or skin particles are then removed with lips and tongue. The groomer makes chewing movements sometimes even before beginning, opening its mouth with a smack (*tongue-clicking*?), drawing the lips back to form a circle, then closing the mouth (15). In races with long, fluffy tails, this appendage is often groomed.

Infant Handling. An unusual feature of colobus social relations is the handling of very young infants by other troop members (12). During the first month and especially the first 2 weeks, while the infant has a pink face, a baby may be handled 3–5 times/hour in resting groups, mainly by adult and subadult females, sometimes by subadult males, rarely if ever by the territorial male. The baby is pulled away from the mother, with or without grooming her before or afterward, often followed by grooming the infant. Babies often squeal and struggle, in which case mothers may take them back. Quite often 1 female will hold her own and another infant at the same time, and in one instance a struggling baby fell to the ground with a thud. Some minutes later it was retrieved by the mother. But others have been abandoned. Perhaps the benefits of social bonding between females outweigh the obvious risks of passing unwilling babies around the troop.

The attractiveness of babies diminishes as they turn dark and juveniles older than 4 months are seldom handled.

Leadership. Apparently the oldest female leads more often than any other troop member (12). In Kibale, group movements often started with soft grunts which caused the monkeys to cluster, then the matriarch would lead. The alpha and sometimes a subadult male initiated movements to engage other troops in aggressive encounters (12).

ACTIVITY. In keeping with their folivorous diet and ruminantlike digestion, black-and-white colobus are comparatively inactive and their movements appear lethargic. As Schillings (16, p. 548) observed, "The behavior of the mbega has nothing monkey-like or comical about it, but is rather always earnest, steady and reserved." A troop divides its day among sleeping, sunning, and foraging trees, progressing between them along regular arboreal pathways and, when necessary, overland. Colobus living in montane forest typically leave their sleeping trees well after sunup and head for nearby treetops, where they can catch the sun's rays and see and be seen by neighboring troops. Here they loll and groom for up to an hour before moving to their food trees, a trip that may take up to 2½ hours if the monkeys dawdle and forage *en route.* Once arrived, they feed intensively until the hottest hours, then settle down to rest, groom, and digest until it is cooler. Another intensive feeding bout continues until perhaps an hour before dark, when the troop files back to the sleeping trees. Although moving at a leisurely pace, with frequent stops, the colobus now keep closer together, jumping one after the other between trees, seemingly intent upon getting home before dark. Having arrived, they relax and socialize again before settling to sleep at nightfall. On cold, misty, or rainy days, the troop may not stir until midmorning, sitting hunched and miserable-looking until they can dry out and warm up in the sun (7, 19). A troop living beside Lake Naivasha, a sunnier location, had feeding peaks between 0800 and 1000 h and during the last 3 hours of the day (14).

Eating. A colobus eats 2–3 kg of leaves and other plant food a day, equivalent to ¼–⅓ its own body weight, yet eating accounts for only 30% of all activities, compared to 47% for associated vervets (14). It sits, tail dangling, nearly 99% of the time while feeding, moving around very little (the other 1% of feeding time). Leaves are gathered usually by gripping them between flexed fingers and palm or thumb tubercle and pulling, while the other hand may be used to draw a branch closer. Sometimes a colobus strips leaves from a small branch and, when both hands are occupied, plucks leaves with its mouth, also the preferred method for harvesting round fruits like figs.

POSTURES AND LOCOMOTION. The colobus uses the same squirrel-like galloping and bounding gait at all speeds. When jumping, it holds the elbows out and lands feet first with all limbs bunched together and flexed to absorb the shock (fig. 28.3). The rebound of the branches is used to gain extra momentum and distance on the following jump. Very rarely, a colobus (typically a juvenile) swings below branches suspended by its hands (14). Wide gaps between trees may be crossed by taking long leaps involving considerable vertical descent. Drops of up to 15 m into the undergrowth are made to escape predators, and displaying territorial males also make spectacular jumps (see below). While

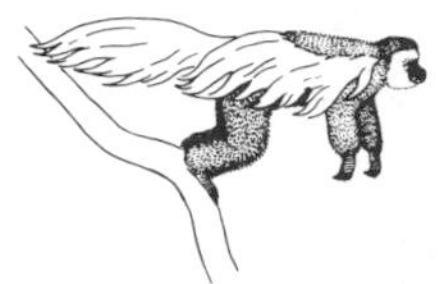

Fig. 28.3. Jumping sequence of the black-and-white colobus (from Mittermeyer and Fleagle 1976).

falling, the arms and legs are stretched wide and the body lies horizontal as in a swan dive, mane and tail billowing spectacularly. Then gradually the body arcs over until the head is down with the arms held forward and flexed ready to clutch at branchlets and break the fall. Monkeys that happen to bail out over a spot where the undergrowth is insubstantial are sometimes killed (7). To climb a tree trunk a colobus embraces it with both forearms and hauls itself up while using the backfeet alternately. To climb downward, it descends headfirst. On the ground, it runs with clumsy, froglike bounds. Its walk, however, is plantigrade like that of other arboreal monkeys.

To help pass all the time colobus spend sitting still and digesting, they rest in at least 7 different postures, many of which enhance their likeness to melancholy little old men (19). Most typical is squatting on a branch with body bent forward and elbows on knees, often with chin resting on chest. Colobus also sit crosswise on a branch with legs dangling or along it like a human seated on the ground, or lie flat on the stomach or back (especially while sunbathing), or on one side with the head pillowed on one arm.

SOCIAL BEHAVIOR

COMMUNICATION

Vocal Communication. Five vocal and 1 nonvocal sound (tongue-clicking) have been catalogued, of which 4 are discrete and 2 are graded (9).

Roaring. "Early in the morning, when a thick mist lies on the forest and a saturating dew hangs in heavy drops on leaves and branches, and everywhere silence still prevails, this chorus of the monkeys, beginning softly, swells into a mighty sound, then dies away, only to begin afresh" (16, p. 542). The chorusing of black-and-white colobus is one of the unforgettable sounds of the African wilds. A low, resonant croaking sound with a rolling *r*, *"rurr, rurr, rurr,"* audible for over a mile, it is like nothing else (though comparable to the calls of New World howler monkeys). Supposedly it is emitted only by territorial males (9, 12). The resonance and carrying power of the colobus's roar are enhanced by an enlarged larynx (5). Depending on the circumstances, the call is given at different intensities and with so much variation in phrases, syllables, and emphasis that it is uncertain whether males are individually distinguishable (9). Calling bouts lasting up to 20 minutes have been reported (9). *Jumping about* (table 28.1) often accompanies high-intensity roaring. Roaring either occurs spontaneously or is evoked, not only by the croaking of other colobus but by assorted stimuli such as a breaking branch, a gunshot, thunder, or a predator. Apart from its advertising function, roaring may also express defiance and threat.

Snorting. An explosive sound emitted as an alarm signal by all colobus except infants, and often by males as a preliminary to roaring.

Snuffling (12) or *cawing* (9). A sound rather like that of a pig excitedly rooting (5), uttered by females and young during conflicts within the group. It signifies mild distress, as when a female pushes away an infant trying to nurse, and perhaps mild threat when uttered by females during aggressive encounters with males. The call is variable, the common element being a rapid train of pulses, and grades into or alternates with squealing.

Squealing (squeaking and screaming). A graded and varied type of sound, uttered both by adult females and young monkeys as a signal of strong distress, for example by a female chased by an adult male, or by an

Table 28.1 Visual and Tactile Signals of the Black-and-white Colobus
(*uncommon, **common, ***very common)

Behavior	Performer	Context and message
Visual signals		
Open mouth, low-intensity.*** Mouth half-opened for < 1 second.	All classes but infants	Many situations of general arousal (approach to another monkey, mounting and embracing, etc.); low-level aggression.
Open mouth, high-intensity.** Mouth gapes wider and tongue clicks—combined visual/vocal display.	Adult and subadult males, adult females (?)	Aggressive: sometimes directed to group members but mainly male-male during close intertroop encounters.
Gape.* Mouth held wide open at least 1 second, teeth covered, no tongue-click; may accompany stare.	Adult males	Apparently a threat given when group member comes too close.
Yawn.** Mouth opened wide exposing teeth, often in series of 2–3 yawns.	Adult males and females	May be a threat (when given by males during intertroop encounter), sign of tension, and physiological response.
Stare.** May or may not be combined with *forward threat.*	Especially adult males	Common toward a distant group, apparently uncommon within troop; may be a threat or simply territorial vigilance.
*Forward threat.** Crouching on hands and feet with head extended; combined with stare.	Mainly adult males	Threat (?) Rarely seen in Kibale colobus.
*Stiff-legs display.**** Seated, legs angled downward with knees slightly flexed and feet unsupported; posture held 1–30 seconds.	All except infants	A threat display when directed by male to nearby troop.
*Penile display.*** Partial erection, often combined with *stiff-legs* display.	Adult and subadult males	Used as a threat toward other troops or observer, and for territorial advertising.
*Arm-raise.** Raising or pushing an arm toward a nearby monkey.	Adults	Distance-increasing threat (?)
*Jumping about.**** Big jumps through tree canopy with noisy landing (emphasized by slapping branches); often combined with roaring.	Adult males	Territorial advertising; threat against rivals and predators.
*Branch-shaking.*** Hindlimbs rapidly flexed to shake branch while holding on with hands.	Adult and subadult males, also young in play	Male-male threat display.
*Supplanting.**** Approaching another troop member.	Adult males	Usually to claim favored feeding position.
Chasing.*	Adult males	Aggression usually during close encounters between troops.
*Grimace.**** Mouth held open for at least 1 second, with lips retracted showing teeth and face puckered; accompanies squealing and/or snuffling.	Females and young	Submission; seen during play.
Displacement scratching (with hands and feet).***	All but new infants	Self-grooming in situations of slight tension, notably when waiting for the leader to begin troop movement.
*Social presenting.** Body lowered less than in sexual presenting.	Females, juveniles, subadult males	Sign of submission to more dominant monkey; often elicits social mounting or grooming; sometimes presenter mounts or grooms after presenting.

Table 28.1 ***(continued)***

Behavior	Performer	Context and message
Tactile signals		
*Social mounting.*** From behind or in front, performer standing.	All classes but small infants	Friendly behavior, often precedes grooming by either partner; also submissive gesture to adult male.
Embrace.** From behind or in front, performer seated.	Mainly females and young	Greeting behavior, counteracting implied threat of approach; often precedes grooming.
Touch (with hand).*	All classes	Solicits grooming and sometimes copulation, also may convey friendly intent.
Aggressive handling.** Includes pushing, pulling, dabbing, and slapping; also wrestling; usually accompanied by snuffling calls.	Adult females, subadult and juvenile males	Larger-to-smaller, one-way interactions associated with, for example, *supplanting,* rejection of nursing attempt; also harassment of copulating pair, wrestling between equal-sized monkeys; aggressive or playful.

SOURCE. Adapted from Oates 1977a, table 5.

infant left behind or pushed away by its mother or grabbed away from her. The mother and other females respond to a squealing infant by picking it up and carrying it.

Soft grunting or *"purring."* A general alerting call audible only at close range, which is often emitted before and may serve to coordinate troop movements, also as a low-intensity alarm signaling that potential predators are nearby (12).

Tongue-clicking. Done by adults of both sexes usually as a prelude to an aggressive interaction, representing a milder level of arousal than snorting. Softer clicks are not necessarily aggressive but may be given during an approach that ends up in a friendly interaction (12).

Visual and Tactile Communication. See table 28.1.

TERRITORIAL BEHAVIOR. To protect his investment in real estate, females, and progeny, the territorial male remains ever-vigilant and employs a set of simple but effective signals, (1) to advertise his territorial status and maintain the spacing between neighboring groups and (2) to repel attempted or actual intrusions that threaten to disrupt the status quo.

1*a*) A loud call audible well beyond the average distance between troops keeps the neighbors informed of the owner's presence and position. Calling is most intense, sustained, and infectious in early morning (the dawn chorus), when sound carries farthest and while troops are still in or near their sleeping trees. It is in effect a regular morning roll call. Calling also often breaks out spontaneously at night (19), serving to advertise the position of sleeping trees, which may be used year after year. Calling by day is typically low-intensity, provoked by sight of a predator (especially crowned hawk eagles) or other disturbance, and less infectious. Intergroup encounters do not normally evoke roaring (12).

1*b*) Visual advertising is important by day, at least in open and mountainous terrain where troops can see each other. Whole groups are conspicuous during the morning sunbath, and the territorial male typically sits highest, where he can see and be seen to best advantage, often performing the *stiff legs display* and displaying his penis (table 28.1). During the early morning calling peak, males also frequently combine *jumping about* with roaring.

2) *Jumping about* is also employed, along with the other, probably less-intimidating aggressive signals listed in table 28.1 and associated calls (*clicking* and *snorting*), to discourage invasions by other colobus or to intimidate predators. The noise and movement associated with dropping through the canopy onto lower branches (often enhanced by slapping them) convey a sense of the male's weight and power to onlookers.

AGONISTIC BEHAVIOR

Interactions between Troops. Observed aggressive encounters between neighboring troops typically began with a large subadult male of troop A approaching troop B and sometimes sitting within 2 m of the alpha male. The troop-A alpha male made *stiff-leg gestures* and loud *tongue clicks,* answered in kind by troop-B males, whose alpha male then chased out the impudent troop-A subadult male. Eventually troop

A followed the young male's lead and took up positions about 20 m away from troop B. The males and other troop members exchanged *stiff-leg gestures.* As tension mounted, the 2 alpha males *yawned, stared, showed penises,* and jumped around violently. The climax came when the troop-A alpha male dashed into troop B and chased his rival (or vice versa), during which the 2 males would descend low, even to the ground, and make harsh grunts, causing females and young to snuffle and squeal. Although actual fighting was not observed, 2 males have been known to fight for 15 minutes on the forest floor, grabbing and biting each other, and signs of old injuries indicate that most territorial males have occasional fights (12).

REPRODUCTION. Births are not strictly seasonal though probably most occur during rainy seasons, after a 6-month gestation. Females give birth on the average every 20 months. They mature at 4 and males at 6 years (12).

SEXUAL BEHAVIOR. Copulatory behavior is infrequently seen and still poorly known in this monkey: only 11 copulations were seen in a troop studied for 14 months (12). In another troop, one pair copulated 5 times on the same day. Behavior is similar to that of other monkeys: the female *presents* with limbs flexed, body low, and tail raised. The male may or may not grasp her ankles in mounting. In 3 complete copulations, the male thrust 13, 14, and 21 times.

PARENT/OFFSPRING BEHAVIOR. Infants are comparatively large and well-developed at birth, but develop slowly (fig. 28.2). The white natal coat gradually darkens and infants are colored like adults by 14–17 weeks. Babies are transported ventrally only, clinging with hands and feet, for the first 8 months, but can move about on their own beginning at c. 5 weeks. By then the baby is biting twigs and leaves experimentally. At 2 months juveniles gambol together and eat leaves in small amounts. By 4 months they move a lot on their own, playing, exploring, feeding, and rarely riding except during troop progressions. Five-month infants may follow their mothers during travel and engage in wrestling matches. Mothers tolerate playful attacks, including swinging on their tails, but play is very largely limited to young colobus. Colobus older than 23–25 weeks are largely ignored by their mothers except during group progressions. By then the young are feeding independently on an adult diet and are c. ⅔ adult size. But juveniles of 30 weeks still cling to mothers during rain and may require help while crossing gaps and climbing. Even at 40 weeks, approaching the stage of full independence, a young colobus may still snuffle and squeal if left by its mother; it neither cries, clings, nor suckles after 50 weeks (but may huddle with her at night) (12).

Communal handling of young infants is a feature of black-and-white colobus society (see under Social Organization).

ANTIPREDATOR BEHAVIOR. Almost the full repertoire of aggressive signals is deployed against predators, including roaring by the territorial male. Males have been seen to rush crowned hawk eagles that landed near the troop. Avoidance of predators includes spectacular sky dives into the undergrowth (described under Postures and Locomotion).

References

1. Allen 1925. 2. Clutton-Brock 1975. 3. Dunbar and Dunbar 1976. 4. Haltenorth and Diller 1980. 5. Hill and Booth 1957. 6. Jolly 1985. 7. Kingdon 1971. 8. Kingston 1971. 9. Marler 1972. 10. McKey 1978. 11. Napier and Napier 1967. 12. Oates 1977*a*. 13.——— 1977*b*. 14. Rose 1978. 15. Schenkel and Schenkel-Hulliger 1967. 16. Schillings 1905. 17. Struhsaker 1978. 18. Struhsaker and Oates 1975. 19. Ullrich 1961.

Chapter 29

The Great Apes

Family Pongidae

African Species

Gorilla gorilla, gorilla
Pan troglodytes, chimpanzee
P. paniscus, bonobo or pygmy chimpanzee

TRAITS. Large, tailless, hairy primates weighing from 30 kg (female chimpanzee) to 180 kg (male gorilla); nominal (chimpanzee) to extreme (gorilla) sexual dimorphism. Skulls robust, especially in male (bony sagittal crest in gorilla), vaulted cranium housing large brain, face flattened with small nose and forward-facing eyes. *Teeth:* same number (32) and type as man but bigger, formidable canines in adult males, back edge honed by lower first premolar as in cercopithecine monkeys. Powerful build, barrel-chest, dorso-ventrally flattened as in humans rather than laterally flattened as in monkeys, arms longer than legs, thumb and big toe opposable. *Coat:* relatively sparse (except mountain gorilla) and without undercoat, face, ears, hands, feet, and other areas naked (gorilla's chest, female chimpanzee's posterior) (3). *Coloration:* from blue or brownish black to brown and ginger (some chimpanzees), skin color black (gorilla) to tan or flesh-colored (some chimpanzees). Penis with bone. *Mammae:* 2, enlarged only during lactation.

ANCESTRY. Apes and humans are joined in the superfamily Hominoidea, which includes 3 families: Hylobatidae, gibbons and siamangs; the great apes; and the Hominidae, with ourselves the sole living bipedal hominid (6).

Biochemical assays of the great apes have shown that the chimpanzee and gorilla are even more closely related to us than they look and act, sharing some 98% of our genome (6). We 3 are twice as different genetically from the orangutan (*Pongo pygmaeus*), indicating—together with recent detailed anatomical comparisons and fossil evidence—that the African apes and hominids shared a common ancestor twice as recently as they shared a common ancestor with the Asian orangs (fig. 29.1).

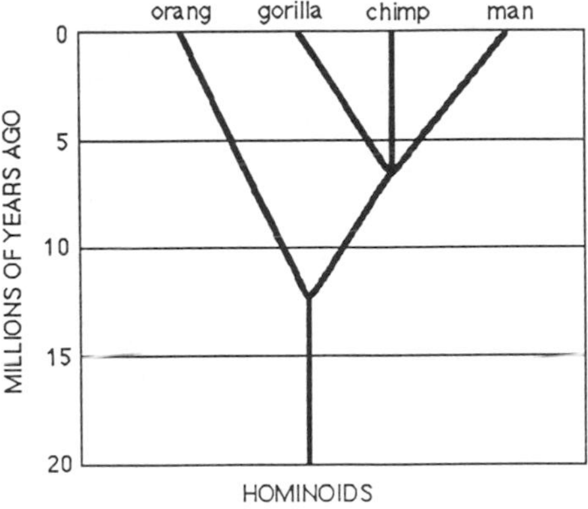

Fig. 29.1. Dendogram of great apes and man (according to D. Pilbeam, pers. comm., 1989). Splitting of African hominoids, shown at 6.25 ma, could have occurred earlier, up to 10 ma.

Somewhere between the middle Oligocene and the early Miocene (see table 1.1), 30–20 million years ago (ma), apes became distinct from the Old World monkeys. The common ancestor of all apes and hominids was probably a baboon-sized, arboreal frugivore similar to *Proconsul africanus* which inhabited East African forests and woodlands 22–17 million years ago. Between 20 and 15 million years ago the apes speciated and spread widely, reaching Asia after continental drift linked up the 2 continents. The Asian and African lines separated by the middle Miocene, perhaps 12.5 million years ago. In the late Miocene to mid-Pliocene (c. 6.5 ma), a period of major climatic and faunal change, the gorilla, chimpanzee, and protohominid lines separated (fig. 29.1). The earliest undoubted hominid, *Australopithecus aferensis*, appeared 4–3.75 million years ago. By 1.6 million years ago, members of this bipedal line had become enough like us to be considered human in the form of *Homo erectus* (6).

ECOLOGY. The great apes are largely confined to the equatorial rain forests of Africa and Southeast Asia. The orangutan rarely comes down from the trees; it is the largest arboreal animal. The common chimpanzee is mainly frugivorous and spends half its active time foraging in the trees, but descends to the ground to travel between feeding sites. The bonobo is somewhat more arboreal. The gorilla, by contrast, is a folivore and the most terrestrial ape.

GORILLA-CHIMPANZEE SPATIAL/ECOLOGICAL SEPARATION. Despite extensive overlap in their geographical distribution (cf. maps), association or even meetings between these 2 apes have rarely been recorded. In some parts of Río Muni (Equatorial Guinea), for instance, where both occur in small geographic areas, they are usually found either in different habitats or in different parts of the same habitat. The chimpanzees live in relatively undisturbed forests, remaining on upper slopes and mountain tops, whereas gorillas stay on the lower slopes and in the valleys where clearing and cultivation have been succeeded by thickets and edge habitat (1). The ecological niche of the gorilla may well have been dictated by competition with the chimpanzee, whose smaller size and greater agility made it dominant in the primary forest, forcing the gorilla to become a terrestrial herbivore. Its dependence on lush herbage confines it to perennially moist habitats, whereas chimpanzees can exploit fruiting trees in savanna woodland as well as in rain forest (3).

SOCIAL ORGANIZATION. The apes have remarkably different types of social organization. Gibbons (family Hylobatidae) are monogamous and territorial. The orangutan leads a largely solitary existence in which the only cohesive social unit consists of a mother with dependent offspring. The gorilla has a 1-male harem system more comparable to that of hamadryas and gelada baboons—and even to zebras—than to other apes. To reproduce, males have to gain females either by luring or abducting them from groups each guarded by a silverback male. Harems include from 1 to 5 unrelated females and their offspring, making for sexual competition ferocious enough to account for the great size, strength, and autocratic rule of adult males.

The chimpanzee lives, rather like the spotted hyena, in communities typically numbering 40–80 animals that share the same home range, within which the members forage singly or in small groups and mixed parties of constantly changing composition. Individuals and groups stay in contact by means of long-range calls. Adult males jointly defend the range as a territory, engage in sometimes lethal raids on their neighbors, and share the favors of females. As in the gorilla, females transfer to a different community in adolescence, whereas males remain in their natal range, making for a male-bonded instead of the usual female-bonded society (see chap. 24, Social Organization). Adult females, being unrelated, tend to be standoffish with one another but, unlike gorillas, each female chimpanzee establishes her own small range within the community territory.

The bonobo, or pygmy chimpanzee, appears to forage in relatively stable parties averaging c. 17 animals of both sexes (2). This least-known of the great apes is the only species that occurs south of the Zaire-Lualaba River and is found only in evergreen forest below 1500 m. Perhaps because the individuals spend more time together, bonobos seem less excitable and aggressive toward one another than do their congeners (2, 3, 8).

Clearly, chimpanzee social organization is the most complex of all the apes and to a point even comparable to human hunter-gatherer societies. Tendencies to division of labor can be seen in cooperative territorial defense and hunting by males and in the greater dexterity and more frequent tool use by females. The beginnings of food-sharing are discernible, coupled with food-begging (see species account). Yet observations and testing of human-reared and -educated apes suggest that chimpanzees are no more or less intelligent than gorillas or orangutans. One plausible suggestion is that the higher intelligence of hominoids arose initially through the need to forage efficiently in very large home ranges containing hundreds of different species of trees and other plants, among which the apes had to learn which were edible, and when and where to find them (3). Whether or not botanical and topographical knowledge make up the major part of an ape's normal education, the fact remains that all the great apes need at least 8 years of parental guidance before they are ready to shift for themselves, and they begin reproducing only a little earlier than humans.

ACTIVITY. All the apes are strictly diurnal.

POSTURES AND LOCOMOTION. See chap. 24 and species accounts. Gibbons are the only apes that normally travel by brachiating. Chimps and even gorillas may hang by one hand sometimes while feeding or showing off, and swing hand-over-hand below a branch, but climb cautiously, using all 4 limbs to suspend their massive bodies. On the ground, they walk on all fours, with forequarters supported on their knuckles (knuckle-walking). All of the great apes can walk quite well bipedally (3).

MANIPULATIVE SKILL

Tool Use. Only the chimpanzee has been observed to fashion tools in the wild, although the other apes show equal inventiveness and dexterity in captivity. Most impressive is the chimpanzee's use of anvils and hammers to crack nuts (see species account under Foraging Behavior).

Nest Building. All the great apes make platforms on which to sleep securely at night, chimpanzees and orangutans in trees and gorillas usually on the ground. All use much the same technique (see species accounts).

SOCIAL BEHAVIOR

COMMUNICATION. Apes above all other primates come closest to being caricatures of ourselves. This is especially true of facial expressions but also body language, for we make similar gestures with our arms and hands (5).

Visual Communication. The capacities of apes and humans to communicate visually are quite comparable. Visual acuity and color perception are equal, and the faces of apes register anger, submission, fear, surprise, sorrow, and so on in the same way and just as expressively as our own. There is a progressive increase in the structural differentiation of the facial muscles into more and finer subunits from gibbon to chimpanzee to gorilla to human. Supposedly this differentiation is correlated with increased ability to produce ever-subtler gradations of expression and meaning (5). Nevertheless, the expressions of apes are in practice more visible and potentially more subtle than those of men who leave their beards and moustaches unshaven.

Some facial expressions are associated only with the production of particular sounds, the quality of which is structured by altering the size and shape of the mouth aperture and resonating cavities (5). The information about the emotional state of the sender which is conveyed by such facial expressions is therefore inseparable from the accompanying call. If calls are put aside and only those facial expressions that convey information soundlessly about what the animal is likely to do are considered, the repertoire of basic chimpanzee facial expressions reduces to 6 (see chimpanzee account, table 29.3).

Vocal Communication. Although laryngeal and tongue anatomy restricts the ability of apes to produce certain sounds of human speech, the basic structure of the larynx is similar and it probably operates according to the same fundamental principles (5). One feature the great apes have that we lack is laryngeal sacs, which communicate with the vocal chamber. These sacs are most developed in the male and function to amplify and add resonance to the chimpanzees' "pant-hoots," and the gorilla's "hoot series" and chest-beating. After the lungs are filled, the sacs are inflated by closing the mouth and nose or the ventricular bands of the larynx. In the process of amplifying the call, the sacs selectively emphasize the fundamental call frequency while reducing the harmonic range. Sac capacity ranges from small in the chimpanzee to sizeable in the gorilla to huge (7 liters) in the orangutan (5).

As in the monkeys, the distance calls of the apes follow the principle of maximum acoustical complexity, both as regards frequency and duration; they are the most species-specific calls in the repertoire and also individually distinct. In addition to being long and drawn out, ape distance calls contain aggressive components that advertise the male's size and combative potential (5).

Considering the pronounced differences in their social organization, there is a remarkable degree of correspondence in the signaling systems of the apes. Their vocal repertoires are of similar size and the resemblance between many chimpanzee and gorilla calls is so close that a significant part of their repertoires is clearly homologous, as shown in table 29.1 (see species accounts for descriptions and functions of these calls). Furthermore, the drumming on trees of the male chimpanzee is equivalent to *chest-beating* in the gorilla (an occasional chimpanzee chest-beats, too), and the spectacular aggressive displays that go together with their loud calls and nonvocal sounds are basically the same.

Tactile Communication. Apes are very touching animals. Apart from maternal-in-

Table 29.1 Comparison of Chimpanzee and Gorilla Calls

Gorilla call	Equivalent chimpanzee call
Wraagh	*Waa* bark and *wraa*
Hoot bark	Bark (?)
Hoot series	*Pant-hoot*
Pig grunt	Grunt and cough
Scream	Scream
Belch	*Rough-grunt*
Question bark	Bark (?)
Cries	Whimper and squeak
Roar	No equivalent
Hiccup bark	Bark (?)
Growl	No equivalent
Pant series	*Pant grunt* (?)
Whine	No equivalent
Chuckles	Laughter
Copulatory panting	Panting

SOURCE: Based on Marler and Tenaza 1977, table 4.

fant contact, which is closer than in most people, apes groom each other in the usual primate manner. Except for the grooming of offspring, social grooming usually flows upward, lower-ranking animals grooming higher-ranking ones and thereby maintaining friendly contact with dominant individuals. Apes also seek and give reassurance by touching each other, often in human ways, and contact is most often directed to particularly sensitive areas: the hands, face, and genitals.

Olfactory Communication. Apes definitely seek and respond to olfactory cues in their social and sexual behavior, notably during genital examinations of infants and females. Little is known about scent glands in the family, although the "fear smell" emitted by the armpit glands of male gorillas is very noticeable. A similar odor, comparable to pungent human sweat, has been noted among socially excited male chimpanzees (7). Terrified gorillas and chimpanzees, like baboons and humans, develop instantaneous diarrhea.

REPRODUCTION. The great apes are slow breeders. The young become adolescent as early as 8 years, but females then go through a period of adolescent sterility lasting several years, bearing their first baby in their early teens, with intervals of 3–5 years between births (unless an infant dies). Males inevitably take longer to reach social maturity, rarely beginning to reproduce before their midteens and reaching their prime during their twenties.

Female apes are generally only sexually receptive during the estrous stage of their monthly menstrual cycles, and do not resume cycling for several years after reproducing, then become pregnant again usually within a few months. Thus, although reproductive competition may have placed a premium on male size and strength in orangutans and gorillas, sexual activity per se takes up very little of their time. Perhaps accordingly, in these two species estrous females do not develop conspicuous sexual swellings, the males' testes are proportionally small (70% as heavy as man's), and their penises are puny in comparison to man's. Chimpanzees, on the other hand, are so sexually active that every male in the community has frequent opportunities to copulate. Females in estrus make the rounds, their condition advertised far and wide by massive sexual swellings (see species account). The males have longer, light-colored, and more visible penises, still much smaller than man's and lacking a glans, but their testes are twice as heavy and by far the largest in the Hominoidea.

Apes normally copulate in the ventrodorsal position, as in most Old World monkeys, but without grasping the female's ankles. Positions are more variable when copulation occurs in the trees; the principals may hang by their arms facing, or suspended by all 4 limbs below a branch like pottos. Face-to-face copulation (fig. 29.4) is more frequent in the bonobo than in the other 2 African apes. However, copulatory behavior is most variable, protracted, and intense in the orangutan, which has been suggested as a model for the evolution of human sexuality (3). Sexual receptivity is not limited to estrus, and indeed mothers of small infants may use sex to attract and gain the protection of males. Even rape is not unusual in this species, when transient males come upon lone females (3).

PARENT/OFFSPRING BEHAVIOR. Newborn apes are nearly as helpless as newborn humans, though not naked and able to cling securely to the mother's front unaided within a few days. Development is more rapid, apes being able to walk and climb beginning at half a year. Yet infancy lasts 3 years, weaning may not be complete before 5, and the association with the mother may continue through adolescence, and beyond in the case of chimpanzee mothers and sons. Responsibility for bringing up the young rests almost entirely with the mother in the great apes, although older sisters practice mothering and may even adopt orphaned siblings. Males, too, behave protectively toward infants and juveniles, but only the male gorilla, in which paternity is reasonably certain, undertakes to defend his offspring and his group against predators, to the point of attacking even humans.

References
1. Jones and Sabater Pi 1971. 2. Kuroda 1980. 3. MacKinnon 1978. 4. ——— 1984. 5. Marler and Tenaza 1977. 6. Pilbeam 1984. 7. Van Lawick-Goodall 1971.

Added Reference
8. Badrian and Badrian 1984. 9. Tuttle 1986.

Fig. 29.2. Gorilla *chest-beating.*

Gorilla
Gorilla gorilla

TRAITS. The largest primate (standing height): males 140–185 cm and c. 160 kg (210 kg, 462 lb., is the record for a wild gorilla) (15); females up to 150 cm and 68–114 kg (19). Robust build with long, muscular arms and short legs, broad hands and feet with thick digits; pot belly; massive head crowned in male with conical mass of bony ridge and muscles, features (especially the shape and wrinkles of the nose) individually distinct; robust teeth (adapted for fibrous diet) and sharp canines (adult males only). *Coat:* long and silky in mountain races, shorter and sparser in lowland forms. *Coloration:* blue-black to brownish-gray, naked areas (nose, lips, ears, chest, palms, soles) black, mature males with a conspicuous saddle of gray or silver hair. *Glands:* armpits with 4–7 layers of large apocrine glands in adult male which emit pungent odor under stress; apocrine and eccrine glands lubricate palms and soles (5).

DISTRIBUTION. At some time in the past the gorilla probably ranged across the rain forest north of the Congo-Ubangi-Uele river system from the Atlantic Coast to the Western Rift Valley of Central Africa. But during dry (interpluvial) periods of the Ice Age the rain forest shrank, wiping out intermediate populations and leaving western and eastern populations separated by over 1000 km (14). Three different subspecies are generally recognized:

G. g. gorilla, the western lowland gorilla, which inhabits the lowland rain forest. The smallest race (males 140 kg, females 75 kg) with the smallest social groups, it has relatively small jaws and teeth but a broad face, and mature males have a nearly white, very striking saddle (8, 11, 14).

G. g. beringei, the mountain gorilla of the Virunga volcanoes that divide Zaire from Rwanda and Uganda. The best known, hairiest race, with very broad face and massive jaws/teeth, males averaging 160 kg and females 85 kg (11).

G. g. graueri, the eastern lowland gorilla, which may be the largest race (males c. 165 kg and 175 cm [5 ft. 8 in.] tall). With short, black fur and a narrower face than the other 2 races, this subspecies survives in isolated pockets of rain forest in eastern Zaire and in some adjacent highland areas near the Rwanda/Zaire border (15, 16).

An estimated 98% of all gorillas live in

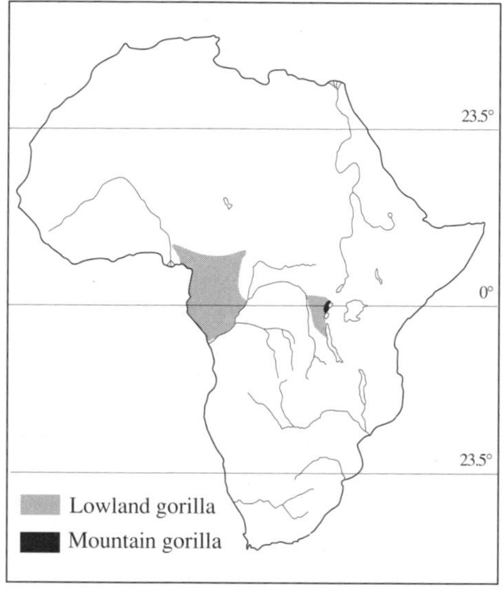

Lowland Rain Forest, many of them in isolated and semi-isolated population units. More than 40,000 western gorillas (an estimated 35,000 ±7,000 in Gabon alone in 1984) and 4,000 eastern lowland gorillas, but fewer than 450 mountain gorillas (388–428 [20, 23]) survive.

ECOLOGY. Although most gorillas inhabit the warm, humid equatorial rain forest, this largely terrestrial browsing ape is concentrated in places where enough light penetrates the forest canopy to support a luxuriant understory of woody and herbaceous plants. For such a large, group-living animal, scattered small patches of forage created by single falling trees would not suffice. Waterways are lined by extensive if narrow belts of undergrowth, but forests that have been logged and cultivated are far more extensive and productive, and gorillas show a preference for regenerating secondary forest throughout their range (14, 15). Thus human inroads into the primary rain forest actually enlarge the area of good gorilla habitat, so that in theory lowland gorillas should be increasing their range and numbers (7).

Mountain gorillas inhabit montane forests between 2800 and 3400 m in the Virunga Range, and sometimes venture to 4000 m in the Afro-Alpine meadows, where, however, there is little suitable forage (5, 16). Night temperatures often go below freezing, and clouds and mist obscure the sun for part or all of each day. The trees are wreathed with moss, lichens, ferns, orchids, and other epiphytes characteristic of montane cloud forest.

In the nearby Kahuzi-Biega N.P. northwest of Lake Kivu, lowland gorillas live below the alpine-bamboo zone between 2000 and 2500 m; here the montane forest is dense and composed of about 26 different trees, many over 30 m tall (7, 16). Mountain gorillas live above the bamboo zone, mainly on the slopes of the volcanic peaks and on the nearly level saddles between peaks in a parkland dominated by only 2 species of trees: the massive but not particularly tall *Hagenia abyssinica* (20 m) and *Hypericum lanceolatum,* which becomes dominant at higher elevations. The tree canopy shades at most 50% of the ground, which is otherwise choked with herbaceous plants up to 2 m high and tangled in vines (19). Here gorillas seldom need climb trees or range far to eat their fill. However the plants are inferior in variety and nutritional value to rain forest vegetation, and, apparently to compensate for the lack of certain vitamins and minerals (B_{12}, calcium, potassium), gorillas eat some animal food (invertebrates), their own dung, and mineral-rich soil (6).

DIET. Gorillas of Rwanda's Parc de Volcans utilize altogether some 58 different plants, of which leaves, shoots, and stems account for about 86% and fruits only 2%. Roots, bark, grubs, snails, dirt, and dung make up the balance (5). Just 9 different species account for 80% of the feeding records, and only 3 make up 60%. *Galium,* a scraggly but extremely abundant vine, made up 24%–30% of feeding records for 3 different gorilla groups, a thistle (*Carduus afromontanus*) made up 10.5%–37%, and a kind of celery (*Cynoglossum*) 8.5%–20%. Three species of stinging nettles, a small tree (*Vernonia*), and 2 kinds of blackberries (*Rubus* species) together amounted to 8.5%–20% of the feeding records (6).

Seven distinct vegetation zones occur in the 25 km^2 study area, and the gorillas visit all of them at times when favorite foods are available in abundance.

In Equatorial Guinea (Río Muni), gorillas may subsist very largely on the profuse growths of wild ginger (*Aframomum*) which invade old clearings. They eat the leaves, pith, roots, and also the fruits, which germinate only after passing through an animal's digestive tract (14).

By destroying seedling trees, disturbing the soil, and generally slowing the succession back to forest, gorillas in effect cultivate and propagate their own food supply (5).

Their diet is so rich in succulent herbs that they rarely need to drink.

SOCIAL ORGANIZATION: 1-male harems, nonterritorial, females transfer and male offspring usually emigrate.

Writing about the reactions of people when brought face-to-face with gorillas, one primatologist aptly describes the "mingled esteem, awe and horror with which the human primate views the biggest species in the order" (15, p. 191). These feelings are inspired first and foremost by the adult male, which, it appears, has much the same effect on other gorillas. He is huge and powerful compared to them, too. Many men are bigger than a female gorilla, but a silverback male not only weighs 23–45 kg (50–100 lb.) more but is about 10 times stronger than the biggest football players and wrestlers. To be armed, in addition, with big, sharp canines seems *de trop.* Small wonder, then, that all the members of a group are ever-ready to defer to their silverback; a direct look, a frown, or a grunt is usually sufficient to curb insubordination. He leads, deciding when and where his group will forage, rest, and sleep, takes precedence in any situation where only 1 animal at a time can gain access to a resource (fruiting tree, salt lick), and arbitrates disputes between his wives. He also defends them and his children against other silverback males and against human predators—the only kind gorillas fear. And this benevolent despot also demonstrates solicitude for his family by slowing the group's pace when a member is incapacitated by injury (bite or wire-snare wound), illness, or infirmity (5).

A gorilla group contains from 2 to 20 animals; the largest recorded group numbered 37 (11). The median number is 9 for eastern gorillas (6 for the Virunga population, 11 for other mountain and eastern lowland populations) and 5 (2–12) for western gorillas of Equatorial Guinea. The typical (median) gorilla group consists of 1 fully mature, silverback male, 1 young-adult (8–12 years) black-back male (average weight 115 kg), 3 adult females, and 2–3 young (<8 years). The maximum number of gorillas in these sex and age classes counted in a group (all in eastern populations but not of course at the same time and place) was 4 silverbacks, 5 blackbacks, 12 adult females, and 16 young (11). The smaller groups of western lowland gorilla may reflect the general tendency for social units to be smaller in more closed habitats, and/or dietary differences (14, 18).

Female Transfer. Like the chimpanzee, red colobus, and hamadryas and gelada baboons, female gorillas normally transfer to another breeding unit before reproducing, usually when they become adolescent at about 8 years (5, 12). Judging from known transfers in the Parc de Volcans, most females change groups again at least once after their initial transfer, and may end up migrating well beyond the ranges of their parent groups (13). In a sample of 13 females that transferred more than once, the interval between transfers averaged 3 months but varied from only 3 days to nearly 3½ years. As a rule, females become permanent members of the group in which they first reproduce, for as long as their consort remains in command.

Females sometimes join a large, long-established group, but more often they join a lone silverback or a newly established group. Females' rank follows the order in which they are recruited into the harem, so it is preferable to be number 1 or number 2 rather than a later arrival. High rank entitles a female and her offspring to remain close to the harem master, where they are safest from predation and potential abductors, which may be infanticidal (see below).

Lacking kinship ties to counterbalance competition for the position of dominant female, females reserve their affinitive behavior for their own offspring and the silverback male, with whom each maintains conjugal bonds. They may squabble over the privilege of grooming him, and rarely groom one another. Infant and juvenile offspring receive the most grooming, whether they like it or not (5). The silverback is thus the only common link binding together the different female matrilines. Groups left leaderless by their silverback's death tend to disintegrate, as may a group that has lost all its adult females (5).

The members of the same matriline often bear an obvious family resemblance, notably in their "nose prints": the shape of the nostrils and the pattern of creases, which are as unique as fingerprints. One long-studied family was distinguished by wall-eyes, another by webbed fingers and toes, both signs of inbreeding (5).

Male Emigration. Male offspring also emigrate during their subadult and young-adult years, starting at c. age 11. Occasionally a male remains in his father's group and, should the patriarch be aged, may eventually inherit it (5). But a maturing male is only likely to stick around if he has opportunities to mate with group females. Normally the harem master monopolizes all breeding; and in the event of the patriarch's death, his heir apparent is likely to be overthrown by an older, more experienced silverback.

Males cannot, of course, transfer to another group in the manner of females. Even an orphaned juvenile male that was introduced to a group that had lost all its adult females was abused by the alpha male, although the subadults and juveniles welcomed him (2). An orphaned juvenile female, however, was immediately accepted by a silverback with a harem. The process of emigrating and acquiring a harem normally takes years. First the young silverback becomes peripheral to his group, staying within a 300-m radius for up to 9 months before going farther away and becoming a solitary male. How soon and to what degree a male becomes peripheral depends on his relationship with the harem master.

Four young silverbacks that were observed for 2 to 5 years after emigrating all spent most of the time alone. Males are rarely mature and experienced enough to begin their own harems before the age of 15. To be a successful leader and defender of females and young, a male needs to have an established home range and the self-assurrance to discourage other males from invading his group. In any case, vigorous males keep trying to acquire new females (12).

Home Range. Yearly home ranges vary in size from 4–8 km^2 in the Virunga range to 20–25 km^2 in the lowlands (6, 14). Ranges shift from year to year and the area used over a number of years may be twice as large as the yearly range. In Kahuzi-Biega N.P., 2 groups that were regularly contacted between 1966 and 1973 even exchanged home ranges (7). The areas patrolled by lone silverbacks are as large as those of groups (6). Despite extensive overlap, each unit tends to have an exclusive core area and groups go out of their way to avoid one another except when deliberately seeking a confrontation. Lone silverbacks and even established groups often explore outside their normal range and may even get lost in unfamiliar areas. In this way gorillas discover new feeding grounds and extend their range into areas where there is less competition from other groups (5).

Intergroup Encounters. Gorillas appear to be far more placid and inactive than chimpanzees, wandering at leisure through the undergrowth in cohesive groups while gathering and munching their vegetable food, grunting quietly, belching, and passing wind. Females and young cluster around their gentle-giant leader during rest periods, the grown-ups grooming and socializing while the youngsters play before bedding down for a midday snooze. But this picture of domestic tranquility is rudely shattered when another silverback approaches to challenge the harem master. His strength, stamina, resolve, and judgment may be tested at any time by rivals seeking to start or enlarge their own harems. The ensuing aggressive interaction, which may continue intermittently over a period of days and involve adult females as well as all the adult and subadult males, demonstrates the really ferocious competition for female gorillas.

As males continue to put on weight and muscle after maturing and gain competence with age, seniority is an advantage. But inevitably males pass their prime (few survive beyond the age of 35) and are eventually unable to withstand challenges from more vigorous rivals (5, 13).

Most transfers occur during or following encounters, and most encounters occur when a group includes a female in estrus (13). Of 26 females that were known to transfer over a period of 13 years, a total of 43 times, 30 transfers by 17 females took place during exciting or violent interactions between groups (4). In only 3 of over 20 potential transfer situations that were observed did the resident male actively prevent a female from approaching a rival silverback (12). Instead, harem masters seek to discourage rivals from approaching through spectacular displays of aggression featuring loud *hooting, chest-beating, display runs* through the undergrowth, slapping the ground, and *strutting* (see under Visual and Vocal Communication). Displaying males also give off a powerful odor from their armpits detectable by humans 25 m away (5).

Apart from serving to discourage encounters between social units, no doubt such displays are also intended to impress females with a male's vigor and general fitness. A questing silverback, after seeking and finding a group containing a sexually receptive female, may only need to make himself seen and heard from a safe distance to entice a maiden into joining him. Since harem males generally take little interest in their immature daughters, and there is almost no evidence of abduction or coercion by suitors, one could imagine that transfers would involve little more than some noisy but harmless demonstrations of machismo.

The incidence of violence is therefore puzzling. In a sample of 64 skeletons collected in the Parc de Volcans, 74% of those belonging to mature males showed signs of healed head wounds, and 80% had broken or missing canines (5). An estimated 62% of all wounds suffered by females and males of the Virunga Range resulted from interactions between distinct social units. Females are most likely to suffer injury in defense of their young, when for instance a bold intruder charges into the middle of a group. It is not uncommon for silverbacks to kill the infants of other males if they get the chance, which has the effect of speeding the mother's return to breeding condition. For instance, a lone silverback killed a 10-month baby in a fracas from which the resident male and 3 females all emerged with severe wounds. Within a few days the mother's behavior changed. Instead of pining, she became more social and playful, a change also noted in other females after infanticide. Two months later she and an 8-year-old adolescent female transferred to a group consisting of an adult and subadult males (5).

ACTIVITY. Gorillas spend approximately 30% of the day feeding, 30% traveling or moving and feeding, and 40% resting (5). The activities of group members are closely synchronized. Peaks of feeding and movement occur within the first 3–4 hours of the day, after which members of a group settle down for an extended rest, followed by another feeding session beginning by 1500 h (14). However, one group of mountain gorillas did not have marked midday rest periods and other groups may skip their rest periods occasionally. Both lowland and mountain populations arise during the first hour and retire in the last hour of daylight. On clear days, groups often linger near their beds to warm in the sun. Mountain gorillas increase rest periods to prolong basking, and on cold, misty days they may huddle motionless for long intervals (6).

Nests. Nine out of 10 gorillas build nests in which to sleep at night and most also make nests for their daytime siesta. The presence of suitable and adequate nesting material primarily determines the location of bedding grounds, and nests tend to be clustered, usually in an area of less than ¼ ha (19). Sitting or standing, the animal reaches out and pulls, breaks, or bends-in branches, vines, or herbage which it places around and under its body to form a crude platform or hollow with a roughly circular rim. The materials, time, and effort that go into nest building all vary, as does the product, but construction takes no more than 5 minutes. There is no particular sequence in the placement of vegetation, nor any elaborate manipulation such as interlacing or weaving together of branches. In building a ground nest, the main effort goes into forming a rim, rather than padding between the gorilla and the ground. Tree nests are usually built in forks, in crotches, or along horizontal branches, and more emphasis is placed on making a strong bottom, resulting in a platform with little or no cup. A sturdy nest enables a gorilla to sleep lying down in a tree or on a steep slope, without falling. But why construct nests on flat ground? The habit may be a holdover from arboreal ancestors (see family introduction). Adult males rarely trust their bulk to a tree nest.

Gorilla groups can be accurately censused and the ages of the members deduced by the size of their nests and the dung they contain (4, 11, 19). According to some accounts, lowland gorillas rarely foul their own nests, whereas mountain gorillas regularly defecate and sleep on their dung (5, 19). However, the 3-lobed boluses have the fibrous consistency and smell of horse manure and do not soil their coats like the messier feces of frugivorous primates (5). Infant gorillas sleep in their mothers' nests until the age of 3 or until the birth of a sibling, although individuals as young as 8 months practice nest building.

FORAGING AND DAY-RANGING. How far gorillas travel in a day varies between areas, between seasons, and between and within groups, depending on the temporal and seasonal distribution of food plants, and also—since different groups avoid one another—on the ranging patterns of other gorillas (6, 7). Daily journeys vary between 100 and 2500 m, averaging 350–530 m for Virunga groups, and about a kilometer in Kahuzi-Biega N.P. and in Equatorial Guinea (6, 14, 16, 19).

Gorillas that enter high forest are surprisingly arboreal compared to the Virunga population. To get delicacies such as "mistletoe" (*Loranthus*) and the fruits of *Syzygium* or *Myrianthus* trees, juveniles readily climb 40 m to reach the upper canopy; one blackback ascended to 35 m and a 200 kg silverback climbed as high as 20 m (16).

Gorillas feeding in trees assume every possible position from seated to hanging nearly upside down by their feet. On the ground they usually sit and reach out in all directions to gather edible plants (one advantage of long arms). After consuming most of the forage within easy reach, the gorilla moves a few steps, grabs a handful of food, and sits down to resume eating. In general, gorillas show great skill and dexterity in exposing and eating the palatable portions of each plant, whether it be the roots, fruits, shoots, leaves, bark, pith, or the grubs concealed in rotting wood. But they have not been seen to prepare and use tools (cf. chimpanzee).

Gorillas very rarely compete over or share food. Adults tend to maintain a much wider spacing (10 to >20 m) while feeding than while resting, and a foraging group may be spread as wide as 100 m. However, sometimes they cluster around certain plants: for example, a wild banana tree that was fed upon until it had been almost totally demolished. Whenever access is limited, the alpha male may pull rank (6, 7, 19).

SOCIAL BEHAVIOR

COMMUNICATION

Vocal Communication. Somewhere around 92% of the vocalizations emitted by gorillas are produced by adult males, an indication of how completely harem males rule the roost. Females emit a mere 4% of the vocalizations, infants 3%, and juveniles less than 1% (17). The common calls are summarized below (based on ref. 5 unless otherwise specified).

Male Loud Calls:

1. *Roar.* A low-pitched, abrupt outburst of sound, forced out through open mouth, produced only by silverbacks and large blackbacks. Roars of high intensity are "probably among the most explosive sounds in nature" (19, p. 218). Context: given only in situations of stress or threat, primarily in response to disturbance by humans. Roars may be individually recognizable and are almost invariably followed by aggressive displays, ranging from bluff charges to short lunges. Response: group seeks protection behind silverback.

2. *Wraagh.* Another explosive, monosyllabic outburst but not as deep as a roar and less shrill than a scream (no. 4). Individual differences were noted in the frequency concentrations of these calls, which are more harmonically structured than roars. Although all adult gorillas can produce the call, over 9 out of 10 are emitted by silverbacks, in response to sudden stress, such as the unexpected arrival of a human, alarm calls of other species, thunderclaps, wind gusts, and other startling noises. Unlike no. 1, this call is never accompanied by aggressive displays, but expresses alarm and group members respond by scattering.

3. *Hoot series.* Far-carrying (up to 2 km), low-pitched but clear and distinct *hoo-hoo-hoos* repeated 2–20 or more times to a series, usually as a prelude to *chest-beating* (see below). A male advertising call which functions to maintain intergroup spacing, a *hoot series* builds up in volume, the hoots becoming longer and plaintive-sounding toward the end; the lengthier the series, the more individualistic are the fluctuations in harmony and phasing. The mouth is parted and the lips pursed (19). Other associated behaviors are discussed under *chest-beating.* The call is usually given in response to the sight or sound of nongroup members. When given without other, nonvocal sounds, hoots do not reveal the precise location of the caller, making this call suitable for vocal probing (e.g., to determine whether other silverbacks are in the area while minimizing the risk of a confrontation).

Other Alarm Calls:

4. *Scream.* Extremely loud, shrill, and prolonged (up to 2 seconds) sounds repeated up to 10 times. Screams may be given by all classes when a group is upset, say, by an encounter with humans, but is particularly associated with quarreling between group members.

5. *Question bark.* A short call of 3 notes with the first and third lower than the middle ("Who *are* you?"), given (91% by alpha male) in situations of mild alarm, say, in response to branch-breaking by gorillas not readily visible to other group members or at discovery of a concealed observer.

Infant Calls:

6. *Cries.* Similar to wails of human infant, sometimes building to shrieks and temper tantrums in highly distressed individuals. Young gorillas seldom cry unless left alone.

7. *Chuckles.* Rasping expirations emitted during play.

Group-Cohesion Calls:

8. *Belch vocalizations.* Include a variety of soft-grunting, rumbling, humming, purring, crooning, moaning, even wailing and howling sounds which intergrade in this most complex of gorilla vocalizations. Such calls are voiced most often by stationary gorillas (of all classes) at the end of a long rest on a sunny day or when surrounded by delicious food. Prolonged calls indicate contentment. A group about to resume foraging may croon in chorus. Slightly shortened versions of the belch vocalization serve as a mild disciplinary rebuke to the young.

9. *Pig grunts.* A series of short, rough, guttural sounds much like the grunts of feeding pigs, expressive of mild aggression

and warning. This call is heard most often when gorillas dispute right-of-way or priority of access to food. Although adults of both sexes may *pig-grunt,* the call is heard particularly from dominant males as an assertion of authority.

Other Calls:

Hoot bark. Alerting call expressing mild alarm; group moves away.

Hiccup bark. Similar but less alarm, more curiosity.

Growl. Mild aggression in stationary group.

Pant series. Mild threat within group.

Whine. Given by infants and also by adult males when in danger of abandonment or injury (3).

Copulatory panting. A series of distinct, fairly low-pitched but loud *o-o-o-o* sounds produced almost continuously by males during intensive copulation (19). (Other calls associated with copulation are described under Sexual Behavior.)

The relative frequency with which the above calls are heard in groups habituated to human observers is approximately (1) *belch,* (2) *chuckles,* (3) *pig grunt,* (4) *hoot bark* or *hiccup bark,* (5) *hoot series,* (6) whine and cries, and (7) *question barks.* The other calls are relatively seldom heard (frequency of 1% or less). In unhabituated groups *wraaghs,* hoot barks, and hoot series are voiced with greatest frequency (17).

Nonvocal Sounds: chest-beating, branch-breaking, striking ground, running.

These actions are often combined in the spectacular displays given by silverbacks at the end of a hoot series. The intensity and completeness of the displays depend on circumstances, usually becoming more emphatic as the distance between rival males decreases (5). The typical complete sequence is as follows: (*a*) hoot series, at the climax of which the gorilla (*b*) rises to stand bipedally, (*c*) grabs a handful of vegetation as it stands and throws it into the air, (*d*) slaps its chest rapidly with alternate blows of its open, slightly cupped hands (fig. 29.2), (*e*) kicks one leg in the air while chest-beating, (*f*) runs sideways for a meter or so immediately after (sometimes during) the climax, (*g*) slaps the undergrowth and tears off branches with his hands during or immediately after the run, and (*g*) thumps the ground forcefully with the palm of the hand (19). The full display may continue for as long as ½ minute, although the actions following hooting occur as one continuous, violent motion finished in 5 seconds or less. Sometimes a silverback interrupts a hoot series by plucking a leaf or herb and placing it between its lips, holding it there until or even while *chest-beating.*

Fig. 29.3. The gorilla's *strutting walk* (black-backed male).

Chest-beating is practiced by all gorillas, beginning as infants, and can be performed in any position, including one-handed while hanging from a tree. But the hollow-sounding *"pok-pok-pok"* is only produced by adult males, whose large air sacs (sometimes apparent as swellings on either side of the throat) amplify *chest-beating* and hooting (19).

Visual Communication

Strutting walk. A very rigid walk, the body held stiff and erect with the arms bent outward at the elbow giving them a curious bowed appearance that emphasizes the power of the long-haired forearms (fig. 29.3). The coat bristles, making the performer look even larger, and it moves with short, abrupt steps, presenting its side to the receiver, with head turned away except for brief glances (but the performer watches out of the corner of an eye). A dominance display, it must be visible to be effective and gorillas usually perform it in a clearing or while standing on a log. This display is most impressive when performed by rival silverbacks at close quarters, but it is also practiced by infant gorillas, and the form remains the same regardless: it is performed silently without obvious excitement and rarely for more than 15 seconds at a time (19).

Facial Expressions. As the first man to befriend wild gorillas, reputedly the most fearsome of all beasts, Schaller was understandably alert for signs that would reveal their state of mind (19). He found their eyes, lips, and mouth to be the most reliable indicators of their emotions, and that he could often predict their response by the eyes alone. The gaze of a relaxed gorilla is bland and the animal surveys its surroundings without obvious attention. The following is a brief catalogue of facial expressions, using the same terms that describe equivalent chimpanzee expressions (see family introduction).

Expressions of Aggression:

Staring ("annoyance face"). Eyes fixed and hard, brows drawn down in a scowl, head often tipped slightly downward, lips pursed and slightly parted.

Tense-mouth face (anger). Staring with mouth half to entirely open depending on intensity of the emotion, the gums and teeth displayed, with the lips curled back. Alternate opening and closing of the mouth is always accompanied by screams and roars. This is the threat display, often combined with a mock charge, reserved for predators, which makes even seasoned observers'—who know that a gorilla always (?) stops short (provided the observer doesn't run away or counterattack)—bowels turn to water (5).

Expressions of Fear/Alarm:

Uneasiness. Lips pulled inward with the mouth tightly compressed, eyes shifty (the animal refrains from looking directly at the observer), head often slightly tipped up. Schaller compares this expression (given in response to observers, not to other gorillas) to biting the lower lip in humans.

Pout face (light distress). Lips pursed but compressed or only slightly parted, and brows raised. The animal appears depressed and may whine. Seen, for example, in an infant whose mother walks off without waiting for it, or in captive infants deprived of something they want.

Open-mouth grimace. Not easily distinguished from open-mouth threat (tense-mouth face), especially since emotions of fear and anger are usually combined, but the mouth appears to be held open wider and longer, the corners drawn further back. The head is often tilted slightly back, the brows are raised, and the eyes dart nervously back and forth.

Playful Expressions:

Play face. Open mouth, smiling but without showing the teeth and gums, eyes relaxed.

Olfactory and Tactile Communication.

Arousal or "fear" (5) *smell.* The armpit apocrine-gland odor emitted by aroused silverbacks presumably conveys information about their emotional state (see under Social Organization). As a harem master would hardly "admit" fear to a rival male, it seems more likely that the smell simply signals a state of excitement and alarm, and/or aggressive threat.

Olfactory signals appear otherwise rather unimportant in gorilla communication, except perhaps for odors associated with estrus (more under Sexual Behavior).

Social grooming. Described under Social Organization. Comparatively rare among unrelated females and in general less important in gorillas than in female-bonded primates (see chap. 24).

As revealed by the distress of orphaned infants when separated from their surrogate mother, nearly continual contact with the mother during infancy is as normal in gorillas as in other apes (5, 21).

REPRODUCTION. Like other apes, gorillas reproduce slowly. The interval between births averages just over 4 years—unless an infant dies, in which case the mother becomes sexually receptive within a month or less and has been known to have another baby within 9–10 months (4). In a 40–50 year life span a female may leave 2–6 living progeny. The mortality rate of 35 recorded births in the Parc de Volcans was 46%. Females start cycling at 7 (6 years, 5 months–8 years, 7 months) and may bear young as early as 8¾ years (12 years at latest), whereas males remain sterile until they become silverbacks at 11–13 years of age. Yet a male with a harem of 3–4 females has over twice the reproductive potential: 10–20 progeny in a 50–60 year life span. One silverback, still potent at an estimated 55–60 years, had 19 known offspring (4).

SEXUAL BEHAVIOR. Of 580 copulations between mature gorillas which were logged during 8000 observation hours in the Parc de Volcans, nearly all were solicited by the female and were performed by the dominant male (4). The female's behavior is often "outlandishly coquettish" and that of the male "pretentiously blasé." After making a hesitant approach that can take as long as 15 minutes, the estrous female stands facing him with body turned slightly sideways as if waiting for the male to respond. If he signals readiness to mount by opening or raising his arms to clasp her, she abruptly turns and backs into the normal dorso-ventral copulation position (fig. 29.4). If he fails to respond, she retreats while looking back at him with a "come-hither expression" (4). Adults were never seen to sniff or touch each other's genitals; immature gorillas, however, sniffed an estrous female's rump fairly often (13).

During copulation, the male either sits upright or leans forward bipedally while holding the female around the waist. The female either squats, hands on the ground or holding the male's hands with one or both of hers, or leans forward on her elbows with rump elevated (4, 13). Figure 29.4 illustrates the concentrated looks characteristic of copulating gorillas, the male with pursed, the female with compressed lips.

After an interval of ½–¾ minute of adjusting movements or a motionless pause, thrusting begins and continues for about ½ minute. The female may also perform deep, slow thrusts, whereas the male thrusts more rapidly, more often, and in longer bouts. Both sexes vocalize, the female more often and for longer (average duration 20 seconds) than the male (13). She gives rapid, pulsating whimpers, while the male usually makes long grumbling noises lasting 5–20 seconds, or pants (see *copulatory panting* under Vocal Communication) (13). Fossey (4) speaks of soft, plaintive hoots. Vocalizing increases and builds in intensity to longer, howl-like noises which cease abruptly when the mount is broken, usually by the female, which is also the first to move away after separating (89% of 514 copulations) (4). Mounts typically last 1–2 minutes but range from 15 seconds to nearly 20 minutes (13). Both intromission and ejaculation are seldom observed, although ejaculate was sometimes seen on one or both partners following prolonged copulation (4).

Mating frequency during the 1–2 day estrus of mature females varies from 0.16 to 1.39 copulations per hour, the median for 8 females being 0.3 (13). On 7 occasions, silverbacks were seen copulating with 2 females on the same day (total of 30 copulations). In one instance a male copulated 4 times with each female in 6 hours (4).

The presence of an estrous female is stimulating to other group members, especially young males, which may mount and attempt copulation with other immature, usually female individuals, but almost never try to mount estrous females. The dominant male tolerates sexual activity involving preadolescent females and may or may not interfere when subordinate males copulate with adolescent females, which have small sexual swellings, are sexually receptive for 3–5 days at a time during irregular cycles (25–40 days, average 26 days), but remain infertile (4). Females also mount each other and young animals; females in late pregnancy have been seen to mount dominant and subdominant males.

Fig. 29.4. Gorilla copulation: the usual dorso-ventral position (*top*), and the more unusual ventro-ventral position (from Dixson 1981).

Nonpregnant females may respond to mounting by vocalizing and thrusting, especially when mounted by a dominant female.

The characteristic cries of copulating gorillas attract the attention of other gorillas, including other silverbacks, which are most likely to turn up in the vicinity of groups containing estrous females (see under Social Organization). It is under these conditions that most fights and transfers occur. The presence of lurking rivals makes silverbacks unusually irascible and vigilant, to the point where sometimes a male will move his group under cover of darkness (4).

PARENT/OFFSPRING BEHAVIOR. Born after an 8½-month gestation period, usually at night, gorillas weigh only about

2 kilos and are quite helpless, yet able to cling to the mother's front with hands and (less securely) short legs and stumpy toes. Though seldom out of arm's reach for the first 6 months, gorillas develop about twice as rapidly as humans. A female captured by poachers at 2 weeks started to crawl, play, grin, bounce up and down, and give chuckling cries at around 8 weeks (21). At 3 months she began to explore and to manipulate objects rather than simply put them into her mouth, and a month later began making short excursions away from her foster parents. By then shc could walk. This is the period when infants normally begin riding their mothers' backs, and by 6–7 months they can climb by themselves (19, 21).

Babies spend increasing amounts of time playing and socializing as they develop. Whether they engage more in solitary or social play depends on the presence of other youngsters. Other group members, including siblings and especially the resident silverback, behave protectively and indulgently toward infants. An orphaned juvenile slept in the same nest with its father (5). Juvenile females are allowed to groom and carry infants. Older infants and juveniles spend as much time near their father as their mother, after if not before being displaced by the birth of another baby. Weaning usually takes place between the middle and end of the second year (5, 9, 19).

ANTIPREDATOR BEHAVIOR. Gorillas subjected to hunting are mortally afraid of man, fleeing for miles after an encounter, the trampled vegetation of their flight path splattered with liquid dung (a sign of extreme fear) (5). Yet the gorilla's defense is based on offense: the charge of the silverback male combined with deafening screams and roars. People with the nerve to hold their ground discovered that it was a bluff: the charge always stopped short—though sometimes only within the last meter. But gorillas have been known to follow through and injure or kill people who ran away. When his group is attacked, by man, leopard, or another gorilla, the dominant male protects them even at the cost of his own life. Females are equally courageous in defense of their young. Consequently the capture of baby gorillas usually involves the killing of the silverback and the mother, and often other group members as well.

The most dangerous time to approach even habituated groups is during interactions involving silverback males from different units (5).

References

1. Casimir and Butenandt 1973. 2. Condiotti 1984. 3. Fossey 1972. 4. ——— 1982. 5. ——— 1983. 6. Fossey and Harcourt 1977. 7. A. Goodall 1977. 8. Groves 1971. 9. Harcourt 1979. 10. ——— 1984. 11. Harcourt, Fossey, Sabater Pi 1981. 12. Harcourt, Stewart, Fossey 1976. 13. ——— 1981. 14. Jones and Sabater Pi 1971. 15. Kavanagh 1984. 16. MacKinnon 1978. 17. Marler and Tenaza 1977. 18. Sabater Pi 1977. 19. Schaller 1963. 20. Viet 1989. 21. R. Wilson 1984.

Added Reference

22. Harcourt and Stewart 1989. 23. Tutin and Fernandez 1984.

Fig. 29.5. Chimpanzee facial expressions: a) relaxed; b) *open-mouthed grin;* c) *play face;* d) rage; e) *pout face.*

Chimpanzee

Pan troglodytes

TRAITS. Smallest and most humanoid of the great apes. *Standing height:* 100–170 cm (42). *Weight:* female c. 30 kg and male 40 kg (maximum c. 55 kg in wild, 90 kg in captivity) (20, 34). Build less robust than a gorilla's, with somewhat shorter arms reaching just below the knees; hands and fingers long with short thumb, feet adapted for grasping, with long, stout, opposable big toe. Head rounded or flattened, without a sagittal crest, face, brow, and outstanding ears naked, nose and nose holes small, upper lip very long and mobile; teeth larger than human's, adult males equipped with dangerous canines. Facial features individually distinctive, aiding recognition, including family likeness. *Coloration:* coat typically black but varying from brown to ginger, length and thickness varying with climate, age, and sex, patches of white or gray hair and bald spots on face, back, and rump common and increasing with age; infants with a prominent, tail-like tuft of white hair up to 7 cm long; skin black to yellow, pink in infants, often freckled and darkening with age; eyes with dark brown sclera (rarely white as in humans) and orange-brown iris; females in estrus develop swollen, pink sexual skin. *Scent glands:* none described.

DISTRIBUTION. Savanna woodland and rain forest from Guinea and Sierra Leone to western Uganda and Tanzania north of the Zaire (Congo) River. Three different races have been described, but the differences among them are minor and inconsistent: the masked or western chimpanzee (*P. t. verus*), with a pallid face and mask-like darker skin around the eyes; *troglodytes,* with a white or freckled face; and the eastern or long-haired chimpanzee (*schweinfurthii*), with a white or bronze-skinned face. But the skin of all 3 races darkens with age, obscuring such supposed racial differences (11, 34).

The pygmy chimpanzee or bonobo (*P. paniscus*), found only in primary rain forest between the Zaire, Kasai, and Lualaba rivers, is considered a separate species. Despite its common name, it is not significantly smaller than the common chimpanzee (males weighing c. 39 kg, females 31 kg), but is slenderer, is more arboreal, and has smaller cheek teeth (14, 34, 35).

ECOLOGY. The common chimpanzee is a forest frugivore which is equally at

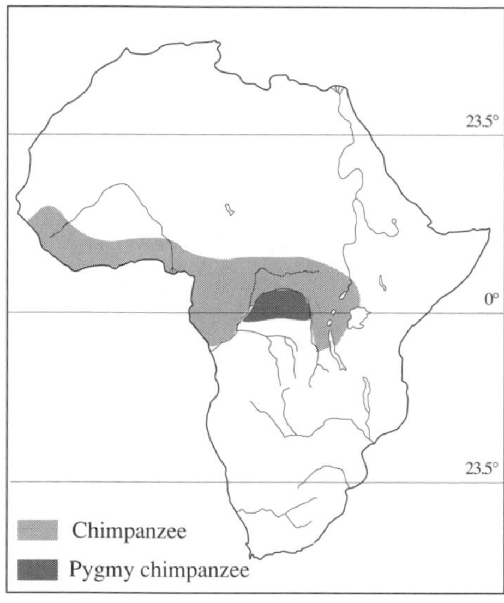

home in the trees or on the ground. It sleeps and does most of its feeding in trees, but travels overland and also forages to some extent on the ground. In the northwest part of its range, it inhabits comparatively dry, open savanna of the Sudanese Arid Zone, where it is found, for instance, in Senegal's Niokolo-Koba N.P. East of Lake Tanganyika, at the opposite end of its range, it lives in mountains covered with mixed *miombo* woodland and forest. Chimpanzees occur to 2000 m in montane forest (34).

DIET. Chimpanzee diets vary greatly from one population and even one community to the next, not only because of botanical differences but also because of traditional preferences. Thus, Guinean chimpanzees use stones to crack oil-palm nuts, whereas Gombe Stream chimps eat the outer flesh and discard the nuts of this same fruit (25, 30). In Zaire, chimpanzees regularly raided a papaya plantation; when this fruit was put out at Gombe, it was consistently ignored (30). Gombe chimps regularly "fish" in termite mounds; a group of Gabon chimps ignored termites but "fished" for safari ants (see under Tool Use) (9).

Wherever chimpanzees have been studied, however, ripe fruits have proved to be their mainstay (9, 33, 34). The average annual food intake in a Gabon rain forest was 68% fruits, 28% leaves, bark, and stems, and 4% animal food (9). In Gombe N.P., chimps spend at least 4 hours a day eating fruit, mostly gathered in trees, and during the growth season usually spend 1–2 hours eating young leaves, a major source of protein. Flowers, pith, bark, and gum are also eaten. Additional protein is acquired by eating seeds and nuts in the dry season. Up to 5% of a chimpanzee's time is spent consuming animal food, mainly in the form of insects (termites, ants, bee and other insect larvae and eggs, caterpillars, leaf galls, etc.) and vertebrates, notably young antelopes and bushpigs, monkeys, bush babies, nestling birds, and eggs (34) (more under Foraging Behavior).

In all, the members of a chimpanzee community may feed on 300 different plant species a year, and from 20 species during a day of foraging. However, when chimps locate an abundant food source such as a tree loaded with fruit, they stay around and eat their fill (fig. 29.6). They are, in short, highly opportunistic foragers which wander the forest and woodland alone or in small parties in a ceaseless quest for edibles, gathering in larger groups to share major finds. They enjoy a varied and generally balanced diet, utilizing the most nutritious foods available. But if Gombe chimps are representative, something over 90% of feeding time is spent eating half the food plants, and for periods of up to 2 months the animals may spend half their time eating only 2–5 different food types. Though chimps ap-

Fig. 29.6. Chimpanzee feeding in a tree.

parently sample many different plants and thereby could discover new edible ones, there are some 300 species in Gombe N.P. that chimpanzees have not been seen to eat (33).

As well-adapted as it is for life in the savanna biome, the chimpanzee is most abundant in the rain forest, where there may be as many as 7/km² in the best habitat, compared to 0.5–0.7/km² in savanna habitats (23, 26). Even savanna chimps are dependent to a large extent on the produce of gallery and other patches of evergreen forest (17).

SOCIAL ORGANIZATION: multimale, territorial communities; females, not males, emigrate.

Calling the chimpanzee *Pan,* after the Greek god of the forest, is appropriate for this volatile, individualistic, independent, and most manlike ape, which wanders freely through Africa's forests and woodlands. Instead of living in cohesive groups like the gorilla and monkeys, chimpanzees live in communities numbering from 15 to 120 animals that share a common home range (34), but rarely or never assemble in a single troop (cf. spotted hyena clan, lion pride). Singly and in parties of varying size, composition, and duration, the members go about their business of finding food. Every type of grouping may occur, from 1-family units of a mother with her dependent offspring, to mixed groups of up to 70 chimpanzees (in Mahale Mountains N.P. [T. Nishida pers. comm. 1989]), to parties of up to 9 or more males. In general, adult males tend to range separately from females, alone and in all-male parties (8).

Appearing relaxed and confident, lone chimpanzees move from fruiting tree to tree across the forest/woodland floor, and sleep in the nearest convenient tree wherever darkness overtakes them. Its size, superhuman strength (an adult male has the strength of 3 men), and climbing skill enable this smallest of the great apes to afford independence, there being little to fear in the forest from creatures other than men and leopards—and males from neighboring chimpanzee communities. Anyway, chimpanzees can keep in touch even while ranging alone through their powerful long-distance calls (see under Communication and Foraging Behavior).

The size of a chimpanzee communal range varies from as little as 5 km² in rich rain forest (21), to 10–50 km² (average of 12.5 km²) in mixed woodland and forest with an extended dry season. In the driest and openest savanna habitats of Senegal, chimpanzees migrate within ranges of 200–400 km², settling for a few weeks at a time wherever food is abundant. Communities have common boundaries and ranges often overlap in prime habitat, but may be widely separated in more open, drier country (34).

Because of the fluid nature of chimpanzee society and the longevity (potentially 40–50 years) and marked individuality of this animal, valid generalizations about its social organization only gradually emerged during 2 decades of field studies (3, 5, 30). The nature of social relations between females remains unclear even now.

From the time they begin reproducing at 12–13 years, females are normally accompanied by 1 to several offspring, and otherwise alone 50%–80% of the time, compared to 2%–54% for adult males (7). The period of dependency (8 years) is long, the association continues through adolescence, and bonds of affection persist between mothers and mature sons (5, 32). Relations between mothers and daughters are more variable. Of 5 mother-daughter pairs that were observed at least through early adolescence, 2 mothers were as affectionate and protective to daughters as to sons, but the other 3 were much less tolerant and all 3 daughters behaved fearfully in some contexts, especially feeding. In any case, females mature 2 to 3 years earlier than males and begin traveling independently and widely as adolescents (9–11 years), often in company with males; nearly all end up transferring to another community before reproducing, typically during estrus and rarely more than once. Accordingly, adult females are mostly unrelated and may not even be on friendly terms. They can be particularly unfriendly to new immigrants, which depend on the sponsorship and protection of the community males to make a successful transfer (cf. plains zebra). To become established, a female has to acquire her own small core range of 2–4 km² where she can forage without aggression from other females. Family units stay within their core areas more than 80% of the time. The yearly home range of a male averages c. 12½ km² (33, 34).

The nucleus of a chimpanzee community is a company of adult males which jointly defend their range against incursions by males from neighboring communities. The males associate closely with a number of adult females and their offspring, and more loosely with younger females that have not yet settled down. Since

Fig. 29.7. Male in a *charging display*, with hair bristling.

males normally remain within their natal community, they are lifelong associates and often are related. Maternal brothers are likeliest to be close and to back one another up during rank disputes, but close bonds may also exist between other pairs of males. The members of such alliances have a competitive advantage; however, the composition of male parties shifts so often and unpredictably that the presence of an ally is undependable (34). A loose hierarchy exists between adult males, which is reasserted (and sometimes challenged) by displays of aggression and submission whenever males meet after an interval of separation. Dominant males perform the *charging display*, dashing into the midst of a party with hair bristling, dragging or swinging a branch, and so on (fig. 29.7; description under Visual Communication). Subordinate males acknowledge their status by approaching and bobbing or crouching in front of a superior while *pant-grunting*. After the reunion, males commonly engage in brief grooming bouts before continuing on in the same party. If there is any question about priority of access to food or females, the subordinate males readily defer to their superiors (34).

The dominant male in a chimpanzee community, usually in his prime (mid-twenties), maintains his position by intimidating all the other males, mainly through performing the *charging display*. Individuals have been known to attain and keep alpha status purely through impressive displays (32), but more often changes in rank follow a buildup of tension leading to prolonged and intense conflict between the contestants (28).

Intertroop Relations. The most serious and prolonged violence is reserved for males of other communities, which may be subjected to vicious and even fatal mobbing attacks when caught alone by a party of patrolling males. Such attacks appear to be premeditated when a party of males approaches the border of their range and remains for up to 2 hours intently listening and looking into the neighbors' territory. If they encounter another party of males, the 2 groups call loudly (*pant-hoots*) and demonstrate but without attacking. If one group is much smaller it will usually retreat first and small communities may thus lose ground to larger ones. Should a lone male—or sometimes an old female—happen to be sighted, the male party may stalk, chase, capture, and proceed to maul the victim so savagely that it dies within a few days. In this way a small Gombe community of about 15 chimpanzees lost all 5 of its adult males in the space of 4 years, after which individuals from the 2 large neighboring communities absorbed the almost empty range (18, 34).

The aggressive relations between different chimpanzee communities, based on competition for real estate and females, and the risks to single males from gangs of enemy males, may mandate that males of the same community submerge their own rivalry over status and reproductive success and cooperate. Selection for male-bonded society thus goes beyond the bonds of kinship. Sexual promiscuity, whereby virtually every female copulates with every male (but not mothers with sons or brothers with sisters), gives each male the incentive to protect the territory together with the community's females and young, even though dominant males may achieve most copulations (see under Sexual Behavior).

ACTIVITY. Chimpanzee activity has a spontaneity to match the variability of its food supply. There is, nevertheless, a discernible and predictable pattern of sorts. In Gombe N.P., chimpanzees usually spend at least 2 hours resting at some point between 0930 and 1500 h, typically around midday (30). In dry weather they rest on the ground or on a branch in the shade, sleeping for up to ½ hour, the rest of the time idly sitting, lying, and grooming themselves or one another. On rainy days they build day nests in the trees (rarely on the ground), then sit hunched disconsolately in them if it rains hard. A chimpanzee spends 45%–60% of its day feeding, mostly in trees, and males travel c. 4 km on the average, compared to 2.8 km for females (32, 33). During the rainy season, a marked peak in leaf-eating during the final 2 hours of activity has been observed both in Gombe and Gabon (9), and a morning feeding peak (0830–1100) has been recorded elsewhere in Tanzania (10).

Shortly before dark (earlier on rainy days), chimpanzees build nests in which to spend the night. Usually they sleep soundly, although there may be outbursts of calling when 2 groups are sleeping within earshot, especially on moonlight nights, and sometimes chimpanzees even move about at night (29).

POSTURES AND LOCOMOTION (29).

Seated. Squatting with back bowed, legs drawn up and weight resting on the buttocks and soles, arms resting on knees. During heavy rain, chimps sit huddled even more than usual with head resting on folded arms. When perched in trees, various postures are adopted, depending on the configuration of the branches, whether the animal is holding with its hands, leaning against a trunk, sitting astride or across a limb, wedged in a fork, and so on.

The act of lying down. First a chimp lowers its forequarters, and when its shoulder is resting on the substrate, it lowers its hindquarters.

Reclining. Every possible attitude. When sleeping in a nest, initially on the back (fig. 29.8), then typically on the side with knees drawn up, sometimes on the stomach. Resting by day on ground or branch, or in a day nest, individuals often sprawl on the front, back, or side with arms akimbo. Chimpanzees commonly lie with the head lower than the feet.

Normal walk. Cross-walk, supported on knuckles and soles. When walking along branches, a chimpanzee may either knuckle-walk or use both hands and feet to keep a firm purchase.

Tripedal walk. When one arm is used to carry food or support a new baby.

Bipedal walk and run. Chimpanzees often stand erect to see better and may walk bipedally with relative ease, though rarely for more than 50 m. Excited chimps run bipedally while waving their arms, brandishing branches, and so on, during displays and play.

Slow gallop. Hands land one after the other just after both feet leave the ground, one slightly before the other.

Quadrupedal run. All four feet off the ground at one stage.

Rapid run. Arms and legs working in pairs, a chimpanzee may make bounds of 10 m and jump 3 m across ditches and streams.

Vaulting. Descending slopes, arms used like crutches as the animal swings forward, landing on both feet. Also used on level ground by animals with foot injuries.

Climbing. Quite often *Pan* swings along beneath a branch hand-over-hand, and youngsters practice swinging from branch to branch, but adults are usually cautious climbers, securing a hold with a hand or foot on the next branch before releasing the one supporting them. Sometimes one will rock up and down on a limb to get within reach of a branch otherwise out of reach. Chimpanzees may thus move between trees, but to travel any distance, and when fleeing from humans, they descend to the ground. Though they are far less agile climbers than monkeys or baboons, long reach gives chimps an advantage when it comes to shinnying up big tree trunks: holding on to either side with its hands, a chimp can walk up or down with its feet.

Swimming. Reluctant to enter water, even to wade a shallow stream; but in Equatorial Guinea, 4 chimps were seen swimming a wide river, dogpaddling like other quadrupeds (6, 30).

MANIPULATIVE SKILL. The chimpanzee's elongated hand and short thumb prevent it from opposing the thumb and fingertips. However, it can pick up small objects between thumb and side of the index finger, and possesses the skill and intelligence to prepare and use grass stems and sticks to fish for insects, hammer with and throw stones, strip leaves from a branch, peel fruit, and the like.

Nest Building. Except for infants, each chimpanzee habitually fashions a nest in a tree every night. The inevitable rare exceptions: an individual that sleeps in a crotch without building a nest, reuses an old nest, or nests on the ground (21, 30). Depending on the height of the tree and the terrain, nests may be sited anywhere between 4 m and 30 m above ground. In Uganda's Budongo Forest 1 nest in 5 was above 24 m, but only 1 in 10 was above 30 m (21). Both there and in Gombe, where average tree height is only 15 m, a third of the nests were found between 9 and 12 m; however, a quarter of Gombe nests were situated in taller riverine forest between 20 and 25 m (30). Nests lower than 6 m are usually only built in trees overhanging streams or ravines, these being favored locations, as are trees commanding a good view, with little undergrowth.

Chimpanzees normally nest in or near the trees where they have been feeding, be-

Fig. 29.8. Chimpanzee lying in its nest (from a photo by H. Van Lawick).

ginning at sunset. Members of a party tend to cluster except for adult and subadult males, which often nest 100 m or more away. Construction takes 1–5 minutes. First the animal chooses a firm foundation such as a fork, a crotch, or crossing or adjoining branches, then proceeds to bend or break up to 6 sturdy branches, followed by up to 10 smaller branches. The feet are employed to secure each branch as the chimpanzee forces one branch under and over another in a crude weaving process. The nest cup is lined, finally, with small, leafy branches, whereupon the builder settles down, often plucking a branchlet to put under its head or body as a finishing touch (fig. 29.8). If a chimpanzee needs to relieve itself during the night, it squats on the nest rim and avoids fouling its nest (the dung is messier than a gorilla's).

Day nests, built by most adults and juveniles during the rainy season, are more makeshift, perhaps no more than a few leafy twigs bent over to cushion the branch on which the animal lies. Females with infants, however, tend to make finished day nests. The most active and least-skilled nest builders are chimpanzees of 2½–5 years, which gradually learn the technique through practice and play. One juvenile female made 3 nests for every one made by her mother and by an older sibling. The earliest attempt at nest building was observed in an 8-month baby (30).

FORAGING BEHAVIOR. Observers who follow chimpanzees on their daily rounds are impressed by their botanical expertise, their ability to remember and anticipate where and when food trees are ripening, and their orienteering ability, which enables them to get from wherever they are to wherever they want to be by the shortest route (33). The route taken by a foraging chimpanzee may go straight, twist and turn, and even go in a circle, but the forager usually avoids crossing its own path more than once in a day. The evidence suggests that chimpanzees have an intimate knowledge of their range, that each adult keeps tabs on current and prospective sources of food and can subsist perfectly well without any help from other chimpanzees. In fact, chimpanzees spend more time feeding when alone than as members of a party, when food competition is likeliest to occur. Communities are most fragmented during periods of food scarcity.

When food is abundant, however, parties of chimpanzees often gather at sites of superabundance. The number of chimpanzees varies with the type of food and number of feeding sites. The largest parties form during the rains, when anywhere from 8 to 40 chimpanzees may forgather at a ripening fig tree or other food bonanza. Overt competition is rare, but feeding chimpanzees usually remain spaced by at least an arm's length. A grove of fruiting *Uapaca* trees or a large fig tree affords more feeding sites than the crown of an oil palm, which can be monopolized by a single adult male (33). The fact that a chimpanzee that finds a food bonanza proceeds to broadcast the information by *pant-hooting* and *rough-grunting,* summoning other chimpanzees from far and wide to the feast, has been interpreted as altruistic, and individual foraging has been seen as an efficient means of locating food that others may then share (2, 21).

This picture recalls the foraging patterns of a hyena clan. But hyenas cooperate in hunting and share food to a much greater extent than chimpanzees, which in fact only cooperate and share as hunters (see below), although mothers will share food that is too difficult for their offspring to obtain or process (34). The idea that giving food calls is altruistic is belied by observations that only males do it and that chimpanzees feed less in groups than when alone. An alternative possibility is that males use food bonanzas as a means toward social and sexual ends, which are served by getting a party together; the community members that respond probably already know the location and make haste to arrive before the food is all gone (33).

Hunting. As in baboon and human society, hunting is primarily a male activity in chimpanzees (27, 38). For their part, females spend twice as much time as males foraging for insects (13, 27). Like baboons, chimpanzees capture prey opportunistically rather than setting off hunting, and they take the same sort of prey: young bushbucks, duikers, bushpigs, and monkeys—including young baboons—as well as smaller items (36). The odd thing is that there is no discernible predator-prey relationship between chimpanzees and these other mammals; each may be seen on occasion feeding near to, unafraid of, and ignored by chimpanzees (32). Yet male chimpanzees must be covertly weighing up the chances of catching prey the whole time, for when the conditions are right the chimpanzees are capable of quick and even coordinated action. For instance, if one male notices that an immature baboon has become isolated from its troop in a spot where its escape can be cut off, he may proceed to hunt it, assisted by other males that quietly move to block all the exits (27).

All 9 males of a Gombe community cooperated to capture a baby bushpig despite the determined defense of 2 adult pigs; one big chimpanzee threw a large rock that hit one pig in the chest (19). Intense excitement is often generated by hunts, followed by intense competition over the spoils. The rights of the male that makes the kill are seldom disputed, but other chimpanzees sit around with their hands out like human beggars. Sometimes the male in charge condescends to hand out tidbits. More gruesome, male chimpanzees occasionally kill and even eat infants of their own kind, usually after mobbing and beating a mother from a different community (5, 34). At Gombe, one mother and daughter pair turned cannibals, killing and eating babies of several other associated females, but this appears to have been highly aberrant behavior (3, 5).

Tool Use. Chimpanzees use leaves, grasses, sticks, and rocks for various purposes, mainly to acquire food and drink. Users carefully choose and if necessary modify objects to serve the specific task, thereby fashioning tools (25, 32). Wadded and/or chewed leaves are used as a sponge to soak up water (e.g., water collected in a tree cavity), to sop up blood and clean out brains when eating prey, and to wipe dirt off the animal's own coat. Sticks are employed to collect swarming ants and bees while feeding on their nests, sometimes as levers to enlarge a hole. Branches may be waved in display or for beating or throwing at enemies, and clubs are used to crack nuts. Rocks are used as nutcrackers in addition to their use as missiles. The most familiar use of tools is by chimpanzees "fishing" for termites. A grass stem is inserted into a termite mound, provoking soldiers to clamp onto it with their jaws, then drawn between the lips to wipe off the clinging insects. Chimps also use grass stems to touch a feared or unreachable object, after which they sniff the tip (31).

Females show greater manual dexterity and muscular control than males in the use of tools. A study of nut-cracking by chimpanzees of the Tai Forest (Tai N.P., Ivory Coast) yielded the first evidence of a distinct sex difference in technique and efficiency of tool use by an animal (1, 39). The nut-cracking habit of West African chimpanzees has long been known, as both man and ape consume quantities of *Coula edulis* (Olaceae) and *Panda oleosa* (Pandaceae) nuts in season (22, 41, 43). The nuts are too tough to be cracked with the jaws and *Panda* nuts are even too hard for the clubs used to bash *Coula* nuts.

The simple method of opening *Coula* nuts is used by both sexes (1). First a chimpanzee picks 12–15 nuts from the tree or from the ground and carries them to a nearby exposed root or rock with a depression (created by earlier nut-bashing sessions), which is used as an anvil. A wooden club, sought and brought for the purpose, is employed to smash the nuts, using one or both hands. Each nut is eaten after opening and the shells are brushed off the anvil preparatory to positioning and breaking the next nut. Females open nuts with fewer blows than males (average 6.4 vs. 8.8) and therefore open more nuts per minute (9 vs. 7 in 5 minutes). Adolescents of both sexes do nearly as well as females.

Apparently only females have the patience or skill to cope with *Panda* nuts. The hard, thick shell is bashed with heavy blows, after which carefully measured blows are necessary to separate the 3–4 almondlike seeds without smashing them to smithereens. Stones are used as hammers and may be brought from hundreds of meters to the nut tree. About 2 minutes and 17 blows are needed to process each nut.

But the most impressive skill is shown by females that carry their clubs up trees in order to crack *Coula* nuts *in situ* without the bother of climbing up and down to gather a fresh supply. Judging from the observers' description, the performance is nothing short of hilarious:

"While carrying the hammer in a hand or a foot, the chimpanzee picks the nuts and stores them in its mouth and a hand. It then chooses a branch as an anvil for pounding the nuts. . . . The animal cracks a nut with one or both hands while holding the rest in the mouth and one foot. To eat the nut, the mouth has to be freed by transferring the nuts to one hand while the hammer is held with one foot or left to balance on the branch. To continue the cracking, the hand has to be freed again by putting the nuts back in the mouth, and so on." (1, p. 587.)

SOCIAL BEHAVIOR

COMMUNICATION. In keeping with its complex communal social organization, the chimpanzee uses a particularly wide range of auditory, visual, and tactile signals adapted both for distant and close communication. Perhaps the noisiest of all African animals, chimpanzees can disperse widely through their territory and still stay in contact with community members. At close range, its extremely mobile face enables a chimpanzee to express gradations of feeling and meaning nearly if not quite as subtle as our own. Furthermore, facial expressions are mostly combined with and shaped by vocalizations. Both visual and vocal signals are finely graded, with one expression or vocalization shading into another, a trait shared by the most sociable mammals. But in contrast to most other primates, including other apes, there are no calls that are exclusive to one sex or class. Each chimpanzee possesses the whole repertoire of the species, although males and females have different versions of the *pant-hoot,* which is used to advertise territory (15). Each chimpanzee is thus equipped to operate as a self-sufficient unit.

The closeness of our relationship to *Pan* is underlined by startling resemblances in visual and tactile signals and the contexts in which they appear (fig. 29.5). The *closed grin* is equivalent to our polite or nervous smile; *bowing, bobbing,* and *prostration* to human submissive behavior; touching, patting, embracing, and kissing to human comforting behavior.

Vocal Communication. Chimpanzee vocalizations are a graded system formed from only 4 different call types: grunts, barks, screams, and hoots (15). However, some 32 different calls have been distinguished by human listeners (5). These calls, and the emotions with which they are most closely associated, are shown in figure 29.9. Except for the laugh and cough, most calls grade into one another through intermediate calls, although *pant-hooting,* which signals the caller's sex and individual identity, is also consistent in form. The most variable calls (those with the most intermediates) are the *waa* bark, scream, bark, grunt, squeak, and *wraa,* in that order. There is a notable lack of any call that functions as a close-range contact or progression call comparable to the gorilla's belch vocalization (see table 29.1); the nearest equivalent is the *rough-grunt* (15).

Visual Communication. Chimpanzees appear to be very sensitive to facial expressions, including eye movements, indicating attention to subtle gradations of expression. Subordinates continually monitor the faces of nearby superiors, for instance when attempting to beg food (29).

An inventory of facial expressions is given in table 29.2; other visual and tactile signals are listed in table 29.3. Note that the hair is erected in most aggressive dis-

Table 29.2 Chimpanzee Facial Expressions

Expression	Circumstances
1*a*. Grin with mouth closed or slightly open.	Associated with submissive behavior and fear, for example following attack by superior.
1*b*. Grin with open mouth (fig. 29.5b).	Nonaggressive physical contact with other chimpanzees; when threatened by a superior or another species that the chimpanzee fears.
2. *Open-mouth threat.* Mouth ± open, teeth covered by lips, glaring.	Threatening subordinate, distant subordinate, or another species that is not much feared.
3. *Tense-mouth face.* Lips pressed tightly together, glaring.	Prior to or during chase/attack upon subordinate and prior to copulation.
4. *Pout face* (fig. 29.5e).	Situations of anxiety or frustration: after detecting a strange object or sound, begging, rejection of grooming, infant while searching for mother or her nipple, after threat or attack, during juvenile temper tantrum.
5. *Play face* (fig. 29.5c).	During playful physical contact with other individuals.
6. *Lip-smacking face.*	While grooming another chimpanzee.

SOURCE: Based on Marler and Tenaza 1977, table 12.

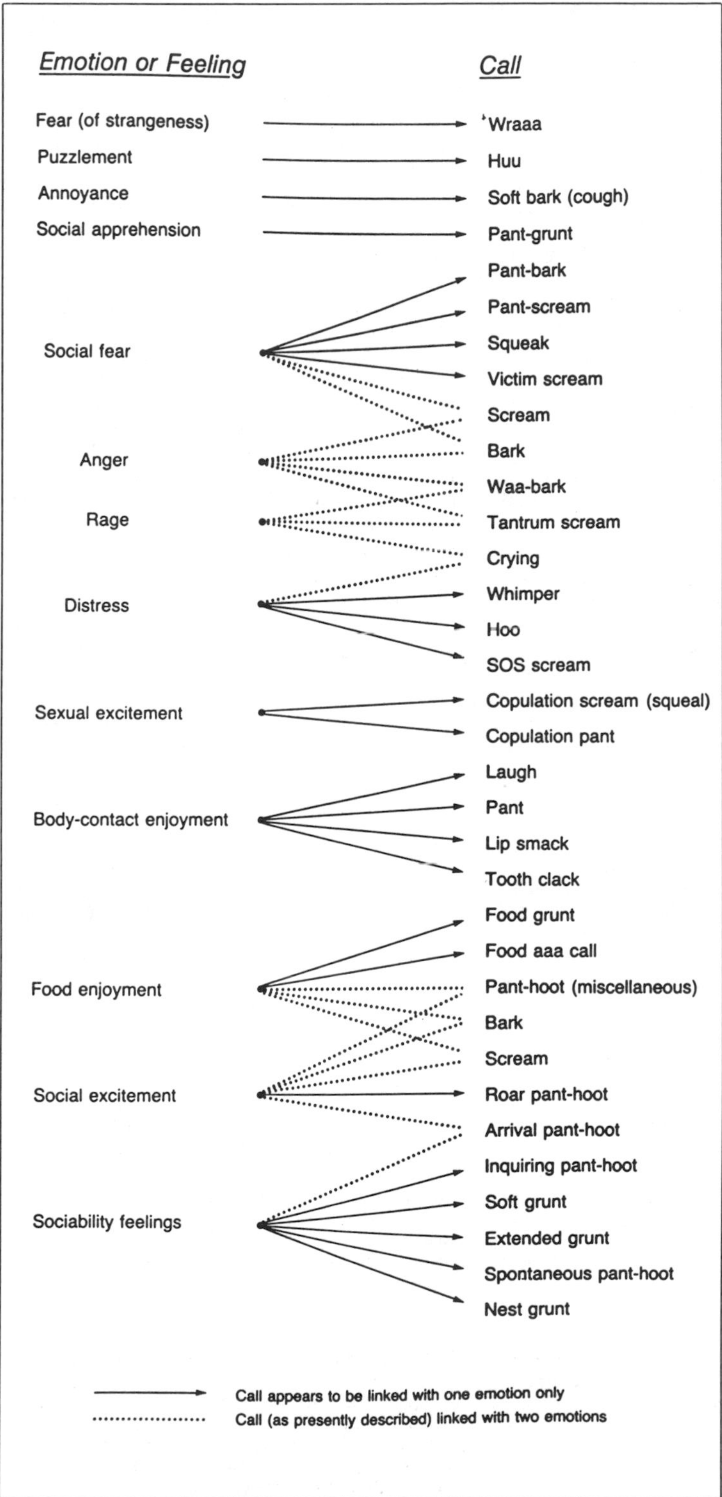

Fig. 29.9. Chimpanzee calls and the emotions they express (from Goodall 1986).

Table 29.3 Chimpanzee Visual and Tactile Signals

Behavior	Circumstances/function
VISUAL SIGNALS	
Dominance/threat	Displays that usually elicit submissive behavior in another chimpanzee and do not normally lead to outright attack.
Head-tipping. A slight upward and backward jerk accompanying the soft bark (see table 29.1).	A low-intensity threat, often given when a subordinate comes close to a feeding superior.
Arm-raising. Forearm or whole arm raised abruptly, palm toward antagonist.	Same context as and often accompanies head-tipping.
Hitting away. Motion with back of hand.	Same context as the above; also as startle response to insect, snake.
Flapping. Downward slap toward antagonist.	Common during interfemale squabbles, accompanied by grinning, squeaks, scream, or *waa*-bark.
Shaking and swaying branches.	Often directed at human observers by disturbed individual, accompanied by glaring; also accompanies body rocking of "frustrated" animal.
Bipedal running and arm-waving. While standing facing or running toward opponent, arm(s) raised rapidly and may be waved.	Accompanied by *waa*-barking or screaming; a bluff directed mainly toward another species (?) (e.g., baboons at Gombe feeding site).
Branch-waving and -dragging.	Seen especially during the *charging display* (see text).
Throwing. Of often sizeable objects (sticks, stones, vegetation); sometimes aimed but more often random.	Often part of *charging display* (fig. 29.7) and probably done to intimidate rather than as a deliberate attack.
Bipedal swagger. Side-to-side swaying by male standing or walking upright with shoulders hunched, hair bristling, arms held out.	Seen mainly as courtship display, also as threat between males of nearly equal rank.
Quadrupedal hunch. Head bent and slightly drawn in between hunched shoulders, back rounded (as in fig. 29.7).	High-intensity threat to nearly equal opponent; performer may move forward slowly, charge, and sometimes attack.
Submissive/appeasing	
Presenting. Crouching with rump facing a superior individual, limbs flexed to varying degree. Females present most often (fig. 29.10); adolescent and even adult males present to higher-ranking males.	Degree of limb flexing correlates with degree of intimidation, extreme crouch adopted after severe attack; victim may look back, and back toward aggressor while screaming loudly and giving *open-mouth grin.*
Bowing. Facing superior, elbows flexed more than knees so that the head is lowered and the rump elevated.	Attitude of a subordinate (especially female) when going to a dominant male that has made aggressive gestures (fig. 29.10) (e.g., courtship branch-waving); often associated with panting.
Bobbing. Push-ups with arms while in bowing attitude.	Commonly performed by adolescent males (rarely females) when a high-ranking male comes close; often combined with *pant-grunts.*
Crouching. Similar to presenting but different orientation.	The extreme form of bowing, done in the same context but also after being attacked, normally while panting, squeaking, and grinning.
Bending-away. Leaning away from another animal with the arm close to the body and flexed at wrist or elbow.	Response by youngsters in particular to close passage of adult males, especially if any aggressiveness is shown; accompanied by soft panting, or occasionally squeaking and *closed-mouth grin.*
Conflict/frustration (displacement behavior)	
Scratching, yawning, grooming motions, masturbating, shaking and swaying branches, temper tantrums, elements of charging display, redirected aggression.	Seen in chimpanzees motivated by conflicting tendencies, as when (1) fear of another individual inhibits the animal from attaining an objective; (2) the animal cannot reach a desired goal (e.g., food); and (3) the animal is unable to communicate wants to another (e.g., an infant unable to reach mother across a gap, prevented from nursing, riding, etc.).

Table 29.3 *(continued)*

Behavior	Circumstances/function
TACTILE SIGNALS	
Submissive/appeasing/reassuring	
Extending open hand. Usually with palm up, toward superior.	Mainly performed by females following a sudden movement by a nearby male, after an attack or threat, or when approaching a superior; often combined with whimpering, grinning, squeaking, or screaming; the same gesture is seen in food-begging.
Wrist-bending. Presenting back of flexed hand to another chimpanzee's lips.	Commonly by adults and juveniles reaching out to infants; rarely as submissive gesture by females and juveniles approaching mature males.
Reaching and touching. Typically head, back, or rump, or nearest part.	(a) Submission/appeasement—usually while passing or being passed, while panting, grinning, or squeaking, or after threat or mild attack; (b) reassurance—response of superior to submissive presenting, bowing, extended hand, and the like (sometimes with foot instead of hand).
Patting. Like a human mother soothing a child.	Reassurance to distressed subordinate or offspring, often by the aggressor.
Chin-chucking.	Often to youngsters by older animals, especially males.
Kissing. Pressing the lips or teeth to the body (sometimes the face or mouth) of another individual; also performed by dominant animals.	Often seen in conjunction with bowing and crouching, when the submissive animal typically kisses the superior's groin, sometimes its face during an approach, or a hand held out in reassurance; response by dominant to submissive crouch, kiss, or extended hand.
Embracing. With one or both arms, from front, side, or rear.	Typical maternal response to frightened or hurt infant; sometimes by adult male to a scared female.
Submissive mounting. Subordinate grasps superior around waist or rests hands and lower arms along its back and pelvic-thrusts, sometimes grasping male's scrotum with foot.	Performed by both sexes, usually after being charged or attacked.
Reassurance mounting. By dominant to subordinate.	Occasional response to screaming and presenting.
Grooming. Typical primate pattern.	Between males, usually by lower- to higher-ranking; also between members of the same family unit.

SOURCE: Based on Van Lawick-Goodall 1968*a* and 1971.

plays, also during precopulatory displays and whenever a chimpanzee is angry, frustrated, or nervous, making it look considerably larger and more impressive.

The charging display (fig. 29.7). Incorporating most of the actions listed in table 29.3 under Dominance/Threat, this is a conspicuous and noisy demonstration performed by adult males, especially when stimulated by meeting other chimpanzees (see under Social Organization), but also when aroused by other stimuli, such as a heavy rainstorm ("rain dance") (31). The full performance includes *pant-hooting* and screaming, slapping, stamping, drumming, and dragging and throwing branches while running (and sometimes climbing) wildly. Rarely, a chimpanzee will beat its chest like a gorilla. Subordinate chimpanzees that get in the way of a charging male are liable to be assaulted (29).

The significance of the white "tail" tuft of infant chimpanzees has not been elucidated, but it presumably has some function such as identifying individuals of this age class. Mature chimps, especially males, often tickle the skin at the base of the tuft when approached by an infant, and occasionally stroke its genitals. When infants in their second year begin presenting, they often do so several times in succession to adults that thus tickle them (29).

AGONISTIC BEHAVIOR. Although fights between members of the same community are very rarely observed, 6 out of 20 known subadult and adult males in one study area bore signs of old wounds (torn ears, missing fingers, facial scars). Most likely these resulted from violent attacks and fights between males from different communities (more under Social Organization).

Attack and Fighting: charge (silent, fast

run on all fours, hair bristling), hair-pulling and scratching (especially females), stomping, lifting and slamming, dragging, slapping, hitting, and biting.

Flight and Avoidance: running away, creeping, hiding, startled reaction.

A chimpanzee fleeing from another usually screams loudly while giving the *open-mouth grin,* but when alarmed by a strange or greatly feared animal it runs off silently. Panic diarrhea has been noted under these circumstances (21). Chimpanzees may also hide (behind a tree, in a nest, in undergrowth) to avoid one another, people, or other unwelcome intruders. A very fearful chimpanzee may try to creep from the presence of a dominant. When startled by a sudden noise or nearby movement, say, a snake, a low-flying bird, or a large insect, the immediate response is to duck and fling one or both arms across the face, or throw both hands in the air, occasionally followed by hitting away (29).

REPRODUCTION. The female's sexual cycle is characterized by menstrual bleeding, and waxing and waning of a huge pink anogenital swelling. The average cycle length is c. 34 days, maximum inflation lasts 6–7 days, and menstruation commences 6–12 days from the start of deflation. The period of maximum inflation coincides with a very marked increase in the attractiveness and receptivity of females to males (fig. 29.10).

Sexual swellings begin modestly as a slight tumescence of the clitoris in 7–8-year-old females, increasing gradually each month until menstruation begins at 8½–10 years, when the sexual swelling suddenly expands. Only then do females become sexually attractive to adult males. However, a long period of adolescent sterility ensues, lasting 2–4 or more years, during which a female may mate virtually *ad libitum* with all the males in her community for a week at a time (only while swollen). After finally producing an offspring at 13–14, females remain unreceptive for 3–4 years and then resume cycling and mating for 1–6 months until conception (28, 31, 34).

Males are sexually precocious, mounting, intromitting, and thrusting as young as 2 years, and developing the courtship pattern beginning at 3–4 years (31, 32). They become adolescent at 9 but are only socially mature at 15–16 years, a good 2 years later than females.

SEXUAL BEHAVIOR. Male courtship displays are based on aggressive gestures which in effect order the female to come

Fig. 29.10. Female with maximum estrous swelling soliciting copulation (from Wickler 1967).

over and copulate. Typically, a seated male beckons to an estrous female, his coat bristling and penis erect. Each male may have his own individualistic courtship gestures (5, 28). For added emphasis, a male may wave a branch or perform other elements from the aggressive repertoire. Sometimes females are intimidated and run away screaming from a male's sexual advances. But normally an estrous chimpanzee approaches and crouches submissively before a displaying male, inviting copulation (fig. 29.10). Sitting erect, the male casually and briefly (10–15 seconds) copulates, ejaculating after c. 10 thrusts. Juveniles often follow an estrous female and may approach and touch the male submissively while he is mating (34). Face-to-face copulation is very rare in this chimpanzee but relatively common in *Pan paniscus.*

Females average about 6 copulations a day during the first week of estrus and will accept males of all ages. In a sample of 1137 copulations, 73% were opportunistic (28). For their part, males rarely interfere with one another and members of parties that include estrous females may take turns copulating with no signs of strife. But toward the end of the female's swollen stage, high-ranking males compete for access by threatening or attacking subordinates that approach her. The alpha male may end up forming an exclusive consortship. However, there is evidence that most pregnancies result from another type of consortship, in which a female goes off on an extended honeymoon with a male of her choice. For a whole week (range 3 hours to 28 days), the couple avoids other chimpanzees by keeping quiet and moving to the edge of or even beyond the communal range—making them vulnerable to mobbing by enemy males (28, 34).

PARENT/OFFSPRING BEHAVIOR. A new baby chimpanzee, born after an 8-month gestation, has to be held with one hand for several days until it develops the strength to cling securely to its mother's front. For the first 6 months it is completely dependent on her for food, transport, and protection. Between 5 and 7 months it begins riding astride, takes its first steps, and begins eating minute amounts of solid food. However, it depends on milk for at least 2 years and may only be finally weaned during its fifth or sixth year (32). It may continue sleeping in the mother's nest for even longer but usually starts sleeping alone in its sixth or seventh year, typically following the birth of a sibling (average interval between births, 4–5 years). Older siblings, especially females, show variable degrees of interest in mothering infants. Adult males may also behave solicitously, carrying, protecting, and playing with infants. But most mothers keep other chimpanzees away from their babies for at least the first 6 months and are particularly fearful of adult males.

Orphaned juveniles have been adopted by older sisters, but 2 of 3 that lost their mothers between 3 and 4 pined away (31). Juveniles in their sixth and seventh year are increasingly likely to become accidentally separated from their mothers while foraging or when the mother is in estrus. They react by searching frantically while whimpering and screaming. But after a series of such episodes, the juvenile may begin to make brief excursions on its own. At that age males start moving in parties without the mother, showing ever-increasing interest in adults of their own sex; but they stay away only a few days at a time up to the age of 8 or 9. Females stay with their mother even longer, but begin joining male parties during adolescent heat periods, and sooner or later transfer to another community (32, 34).

All the evidence indicates that the extended childhood and adolescence of chimpanzees enables this species, like humans, to acquire the skills and social graces, through observation of role models and trial-and-error learning, which enable them to live and function in chimpanzee society (32).

ANTIPREDATOR BEHAVIOR: *wraa* call, ground-thumping, branch-shaking, throwing objects, beating with hands or sticks.

In a series of famous naturalistic experiments, Kortlandt (12) demonstrated that wild chimpanzees will make coordinated attacks against a leopard. In a savanna habitat in Zaire, chimpanzees presented with a stuffed leopard went so far as to beat it with the branches they were dragging and waving about. But the only serious predator on the chimpanzee is man, against which its only defense is concealment and flight.

References

1. Boesch and Boesch 1981. 2. Eisenberg, Muckenhirn, Rudran 1972. 3. Goodall 1979. 4. ——— 1983. 5. ——— 1986. 6. Grzimek 1957. 7. Halperin 1979. 8. Harcourt and Stewart 1983. 9. Hladik 1977. 10. Izawa and Itani 1966. 11. Kavanagh 1984. 12. Kortlandt 1967. 13. McGrew 1979. 14. MacKinnon 1978. 15. Marler and Tenaza 1977. 16. Napier and Napier 1967. 17. Nishida 1968. 18. Pierce 1978. 19. Plooij 1978. 20. Reynolds 1967. 21. Reynolds and Reynolds 1965. 22. Struhsaker and Hunkeler 1971. 23. Sugiyama 1968. 24. ——— 1969. 25. Sugiyama and Koman 1979. 26. Suzuki 1969. 27. Teleki 1973. 28. Tutin and McGinnis 1981. 29. Van Lawick-Goodall 1968*a*. 30. ——— 1968*b*. 31. ——— 1971. 32. ——— 1973. 33. Wrangham 1977. 34. ——— 1984. 35. ——— 1986. 36. Wrangham and van Z. B. Riss 1990.

Added References

37. Badrian and Badrian 1984. 38. Boesch and Boesch 1989. 39. ——— 1990. 40. Ghiglieri 1984. 41. Hannah and McGrew 1987. 42. Nowak and Paradiso 1983. 43. Whitesides 1985. Nishida and H. Hiraiwa-Hasegawa 1986.

Glossary

ABBREVIATIONS AND SYMBOLS

c. circa, approximately
cf (L *confer*) compare
h indicates time on 24-hour clock (e.g. 1215 h)
L Latin
ma millions of years
N.P. National Park
p, pp page, pages
qv (L *quod vide*) which see
± with or without

METRIC EQUIVALENTS

cm centimeter = 0.39 inch; 2.54cm = 1 inch
g gram; 0.035 ounce
ha hectare = 10,000 m^2; 2.5 acres
Hz hertz, a unit of sound frequency = 1 cycle per second
kg kilogram; 1000 grams, 2.2 pounds
km kilometer; 1000 meters, 0.62 mile; 1.6 km = 1 mile
km^2 square kilometer; 100 ha, = 0.39 mi^2; 2.6 km^2 = 1 mi^2
m meter; 1.1 yards, 39.4 inches
mm millimeter; 0.1 cm; 25 mm = 1 inch

Accessory olfactory system. See vomeronasal organ.

Adolescence. The stage when females begin to ovulate and males produce sperm; also called sexual maturity.

Adult. An individual which is both physically and reproductively mature.

Agonistic behavior. Behavior associated with aggression, including offensive, defensive, submissive, and fearful behavior.

Agouti. Grizzled appearance of the coat resulting from alternating light and dark banding of individual hairs. (See also brindled.)

Air-cushion fight. Fighting movements performed by two contestants without making contact—as though separated by an invisible pillow.

Allopatric. Referring to populations, usually of different species, that have nonoverlapping geographical ranges (opposite of sympatric, qv).

Alpha (first letter in Gk. alphabet). The top-ranking member of a dominance hierarchy.

Amble. A walking gait in which the legs on the same side are moved together (also called a parallel walk), in contrast to a cross-walk.

Amniotic sac. The membranes surrounding a fetus and containing the fluid in which it floats.

Anaerobic. Living in the absence of free oxygen, e.g. the bacteria living in a ruminant's rumen.

Apocrine glands. Cutaneuous scent glands consisting of saccular, branching tubules, which produce complex and chemically variable secretions.

Aquatic. Applied to animals that live in fresh water. All aquatic mammals move readily on land, unlike most marine mammals.

Arboreal. Referring to animals that live in trees.

Arid zone. A region of low rainfall with characteristically sparse vegetation, intermediate between desert and savanna.

Artifact. A byproduct of scientific manipulation, not an inherent part of the thing observed.

Artiodactyl. Even-toed ungulate (antelopes, cattle, sheep, pigs, etc.).

Axillary. The armpit area.

Bachelor group or herd. An all-male social group.

Back-crossed. Breeding of a hybrid with one of the parental species.

Baculum. The bone present in the penis of most mammals.

Bimaturism. Different rates of maturation in the two sexes; in all mammals where males are bigger, females mature earlier.

Binocular vision. The visual field within

view of both eyes; the frontally placed eyes of cats and primates have most overlap and best depth perception.

Biome. A major type of ecological community, such as the savanna or desert.

Bipedal. Two-footed stance or locomotion of four-footed animals.

Boma. Corral or kraal where livestock is kept.

Bovid. Member of the family Bovidae, the hollow-horned ruminants.

Brachydont. Low-crowned molar teeth, characteristic of browsing ungulates (cf hypsodont).

Brindled. Flecked or streaked appearance of the pelt due to light and dark banding of the individual hairs.

Brisket. The part of the chest between and behind the forelegs in four-footed mammals.

Bristling. Erection of the hair, thereby making an animal appear bigger.

Browser. An ungulate that feeds mainly on plants other than grass, especially foliage.

Calcrete pan. Soil permeated with calcium salts, forming a hardpan which holds rainwater.

Callosities. The two patches of thickened, bare (often shiny and colorful) skin on which monkeys sit.

Canine teeth. The usually long, pointed teeth, one in each quarter of the jaws, that are used for fighting and (in carnivores) for killing prey.

Cannon bone. The part of the limb between the knee (carpal joint) or hock and the fetlock, consisting of the enlarged, fused metacarpals (forelimb) and metatarsals (hindlimb), corresponding to our hand and foot bones.

Carbohydrates. Various neutral compounds of carbon, hydrogen, and oxygen, such as sugars, starches, and cellulose, mostly formed by green plants. Cf. proteins.

Carnassial teeth. In carnivores, the fourth upper premolar and first lower molar are specialized for shearing meat and sinew.

Carnivore. A mammal belonging to the order Carnivora (dogs, cats, civets, mongooses, hyenas, weasels. otters, etc.). Though most eat animal food, a minority are purely carnivorous.

Carnivorous. Meat-eating.

Cecum. A blind sac situated at the junction of the small and large intestine, in which digestion of cellulose by bacteria occurs in nonruminants.

Cellulose. Main constituent of the cell walls of plants, composed of a polysaccharide of glucose units.

Cheek pouches. A pair of deep pouches extending from the cheeks into the neck skin, present in most Old World monkeys (not colobines), used for the temporary storage of food.

Cheek teeth. The row of premolars and molars used for chewing food.

Concentrate selector. Referring specifically to ruminants that feed selectively on high-protein plants (new leaves or grass, flowers, seeds, etc.).

Conformation. The general shape and proportions of an animal.

Congener. Member of the same genus.

Consortship. Exclusive association with a female maintained by a male during the period of estrus to keep any other male from mating with her (see also tending bond).

Conspecific. A member of the same species.

Contagious behavior. An action that stimulates other animals to follow suit.

Convergence. The evolution of similarities between unrelated species that fill similar ecological niches.

Cross-walk. Walking gait in which diagonally opposite limbs (forelimb and opposite hindlimb) move together (cf amble).

Croup. The rump, especially the part between hips and tail.

Cryptic. Concealing, inconspicuous, the opposite of revealing, usually referring to coloration and markings.

Cud. A bolus of partially digested vegetation that a ruminant regurgitates, chews, insalivates, and swallows again while ruminating.

Cursorial. Adapted for running.

Desert. A region where rainfall is infrequent and sporadic, in which woody vegetation is confined to drainage lines and other areas (around hills) where roots can reach water.

Dewclaw. A vestigal digit that fails to contact the ground; the inside (first) digit in carnivores (cf false hooves of ungulates).

Dicot. Short for dicotyledon.

Dicotyledon. A plant with two seed leaves; the subclass of angiosperms containing most higher plants.

Digit. Latin for finger or toe.

Digitigrade. Animals that walk on their digits rather than the whole foot.

Dimorphism (L two forms). Referring to morphological differences between males and females (sexual dimorphism).

Discrete signal. A call or other type of signal that has no significant gradations; it is either on or off (cf graded signal).

Dispersal. Movement of animals away

from their natal home range, typically before maturity.

Dispersion pattern. The characteristic grouping pattern of a species or population.

Displacement activity. A behavior that is displaced from its normal context, especially feeding and grooming behaviors performed during aggressive interactions or in other stressful situations; considered a sign of indecision or frustration, as when two conflicting drives block one another (e.g., approach vs withdraw, fight vs flight).

Display. A behavior pattern that has been modified (ritualized) by evolution to transmit information by a sender to a receiver; a special kind of signal (qv).

Diurnal. Referring to species that are primarily day-active.

Dorsal. The back or upper surface (opposite of ventral).

Dung midden. Pile of droppings that accumulate through regular deposits, typically in connection with scent-marking (see also latrine).

Ecological niche. The particular combination of adaptations that fits each species to a place different from that filled by any other species within a community of organisms.

Ecology. The scientific study of the interaction of organisms with their environment, including both the physical environment and the other organisms that share it.

Ecosystem. A community of organisms together with the physical environment in which they live.

Ecotone. The transition zone or "edge" between two different types of habitat.

Edge species. Species that prefer the ecotone between different habitats (e.g., impala, waterbuck, sable, roan).

Endemic. Native plants and animals, as opposed to exotic (introduced) species.

Eocene. Geological epoch (see table 1.1).

Epidermis. The outer layer of the skin (or surface tissue of a plant).

Equatorial. Geographical region of the equator (cf tropical).

Erectile. Capable of erection, referring in this book mainly to hair, which fear and anger cause to bristle, making the animal look bigger.

Estivate. The equivalent of hibernating (i.e. a state of torpor) in a warm climate.

Estrus. Behavior associated with ovulation, being in most mammals the only time when females are sexually receptive ("in heat").

Ethiopian Faunal Region. Africa south of the Sahara Desert.

False hooves. Vestigal hooves (digits 2 and 5), which persist in many ruminants as paired excrescences on the fetlock.

Fetlock. The projection of the foot above the hoof in ungulates, and the adjacent pastern joint itself.

Flehmen. The grimace associated with urine-testing, which is developed to varying degrees in nearly all ungulates and in some carnivores.

Flight distance. The space an animal maintains between itself and a feared presence such as a predator; a highly variable distance, depending on the individual's status, the identity and actions of the feared object, the terrain, and the time of day.

Focal animal. The particular animal under observation.

Folivore. An animal whose diet consists mostly of leaves and other foliage.

Forbs. Herbs other than grass which are abundant in grassland, especially during the rains.

Fossorial. Adapted for digging.

Frugivore. An animal that feeds mainly on fruit.

Gallery forest. Trees and other vegetation lining watercourses, thereby extending forested habitat into savanna and arid zones.

Gecker. Cackling sound of distress/protest made by a fearful monkey, mongooses, jackals, etc.

Genotype. The genetic makeup of an organism.

Genus (plural genera). The next taxonomic level above species; thus, in the Latin binomial system, the genus name comes first, the species name second, e.g., *Gazella granti and G. thomsonii* are two species of *Gazella.*

Graded signal. A call or other type of signal that varies in intensity and/or frequency, thus being capable of transmitting subtle differences in the mood/intentions of the sender (cf discrete signal).

Graviportal. The structural plan of very large animals like elephants, rhinos, and hippos, featuring pillarlike limbs to bear the body's heavy weight.

Grazer. A herbivore that eats mainly grass.

Greenflush. Renewed growth of dry grassland caused by precipitation or by burning (provided the soil retains enough moisture for regrowth).

Gregarious. Sociable; applied to species that live in groups, in contrast to "solitary" species that do not.

Group defense. Cooperative defense

(usually against predators) by two or more individuals, as distinct from the more common defense by just one individual. (See mobbing response.)

Guard hairs. The outer coat that overlies the shorter, softer hairs of the underfur (underfur is sparse or absent in many tropical mammals, e.g. most ungulates and primates).

Guinea Savanna. Broad-leafed, deciduous woodlands of the Northern Savanna dominated by *Isoberlinia,* forming a narrow belt between the forest-savanna mosaic and the drier savanna to the north. The northern equivalent of *miombo* woodland.

Harderian gland. An accessory lacrimal gland located on the outside of the eyeball (responsible for stains below the eyes of warthogs).

Harem. A group of two or more females which a male actually owns (see accounts of gorilla and of plains and mountain zebras).

Herbivore. An animal whose diet consists of plant food.

Herd. A social group consisting of at least two same-sex adults; generally applied to gregarious ungulates.

Hierarchy. As applied to social groups, a usually linear rank order in which each member dominates all those of lower rank and is dominated by all individuals of higher rank.

Hindgut fermenter. Nonruminant herbivore or omnivore, in which breakdown of cellulose occurs in the cecum and large intestine.

Home range. The area occupied by an individual or group, determined by plotting the perimeter of the points where the individual(s) is seen over a period of time.

Hypsodont. High-crowned; characteristic of the molars of grazing ungulates (cf brachydont).

Inguinal. The groin area.

Insectivore. A mammal specialized to eat insects.

Insectivorous. Refers to animals whose diet consists primarily of insects.

Inselberg (Ger., literally island mountain). An isolated hill or mountain.

Interdigital. Between the digits; e.g., the interdigital (hoof) glands of many antelopes.

Interpluvial. A period of lower rainfall during the Pleistocene, when areas covered by rain forest contracted, while savanna and arid zones expanded.

Isohyet. Line on a weather map that connects areas of equal rainfall.

Juvenile. The stage between infancy and adolescence.

Karroid shrubs. The dwarf woody plants that dominate the Karroo.

Karroo. An arid part of the interior plateau in the temperate zone of southern Africa, dominated by dwarf shrubs and adjoining the Highveld grassland.

Keratin. The tough, fibrous substance of which horns, claws, hooves, and nails are composed.

Kin selection. Selection favoring investment of care in relatives; e.g., by promoting the survival of full siblings, older siblings stand to gain as much genetic benefit as they would through the survival of their own offspring, since both carry on average half of the same genes.

Knuckle-walk. To walk on all fours while bearing weight of forequarters on the knuckles; known only in the chimpanzees and gorilla.

Kopje (Afrikaans for head). A small hill, typically a rock outcrop, where various mammals find safety, shelter, and food.

Latrine. A place where animals regularly deposit their excrement.

Lek. A breeding ground or arena where territorial males cluster around a central area, to which estrous females come to mate with a few of the most centrally located and fittest males. (See kob, lechwe, and topi accounts.)

Liana. A vine; climbing woody plants are major constituents of rain forest.

Lying out. The prone position with head on the ground assumed by infant bovids during the concealment stage.

Matriline. A group of related animals descended through the maternal line.

Migration. Movement from one region to another where conditions are better, usually in response to seasonal changes.

Miocene. A geological epoch (see table 1.1).

***Miombo* woodland.** A distinctive vegetation zone within the Southern Savanna (see vegetation map) dominated by broadleafed, deciduous, pinnately compound, leguminous trees, notably *Brachystegia* and *Julbernardia* species. Hot fires that sweep through this type of savanna every dry season select for fire-resistant trees. Guinea Savanna (qv) is the Northern Savanna equivalent.

Mobbing response. Cooperative attack by members of a group, triggered by behavior of the attacked animal or by a signal from a group member (e.g. a distress cry).

Monogamy. Mating system in which most individuals have only one sexual partner; monogamous mammals tend to form permanent pair bonds.

Montane. Referring to African mountain habitats, including forest, grassland, bamboo zone, moorland, etc.

Mopane. Tree of variable growth forms, *Colospermum mopane,* which often forms pure stands in parts of the Southern Savanna and South West Arid Zone.

Mopaneveld. Savanna dominated by mopane trees or bushes.

Morphology. Referring to an animal's form and structure.

Mucosa. Mucous membrane: a membrane rich in mucous glands such as the lining of the mouth and nasal passageways.

Muffle. End of the snout or nose (technically the rhinarium), including the nostrils; it is bare and moist in some species (especially those living in the humid tropics), hairy in others (e.g., desert-adapted antelopes).

Multiparous. 1) Producing more than 2 to many young per birth. 2) Having experienced one or more previous births.

Natural selection. Darwin's theory that the fittest genotypes in a population survive to reproduce (survival of the fittest); considered the main guiding force in evolution.

Niche. See ecological niche.

Nocturnal. Referring to species that are primarily night-active.

Nomadic. The wandering habit, meaning species that have no clearly defined residence most of the time; distinct from migratory species, which may be resident except when migrating.

Oligocene. A geological epoch, shown in table 1.1.

Omnivorous. A mixed diet including both animal and vegetable food.

Opportunistic. Referring to animals, notably carnivores, which capitalize on opportunities to gain food with the least expenditure of energy.

Orienteering (as used in this work). Ability to find one's way unerringly in an area, presumably because the animal has an accurate mental map of its home range.

Pair bond. The social ties that keep members of a mated pair together, usually reinforced by mutual grooming, marking, calling, aggression toward outsiders, etc.

Paleocene. Geological epoch, defined in table 1.1.

Pastern. The part of an ungulate's foot extending from the top of the hoof to the fetlock (qv), including the joint.

Perineal. The area of the perineum.

Perineum. The area that includes the anus and external genitalia, especially the female's.

Perissodactyl. Odd-toed ungulate (rhinos, horses).

Plantigrade. Flat-footed: applies to animals which walk on the whole foot, e.g. elephants, hyraxes, man (cf digitigrade).

Pleistocene. Geological epoch, defined in table 1.1.

Pliocene. Geological epoch, defined in table 1.1.

Polygamy. More than one mate, in contrast to monogamy.

Polygyny. Mating system in which one male has reproductive access to two or more females.

Population. Members of the same species that live within a defined area at the same time.

Post-partum estrus. Renewed ovulation and mating within days or weeks after giving birth, instead of the period of sterility during lactation typical of seasonal-breeders.

Precocial. Animals that are born well-developed and capable of locomotion soon after birth; e.g., young ungulates are precocial/precocious compared to carnivores and primates.

Predation. The killing/eating of living animal prey.

Predator. Any animal that subsists mainly by eating live animals, usually vertebrate rather than invertebrate prey (cf insectivore).

Preorbital. In front of the eye, where a gland occurs in many ungulates.

Presenting. The act of directing the hindquarters toward another individual, either in sexual solicition or as a gesture of appeasement derived from sexual presenting.

Primary forest. Forests dominated by mature trees, in contrast to secondary forest (qv).

Promiscuous mating. Indiscriminate mating, as opposed to mating based on female choice, pair-formation, or the outcome of male sexual competition.

Pronking. See stotting.

Prosimian. Literally "before the monkeys": members of the relatively primitive suborder Prosimii, of which pottos and bush babies are the African representatives.

Proteins. Complex combinations of amino acids containing carbon, hydrogen, nitrogen, oxygen, usually sulfur, and sometimes other elements; essential constituents of all living cells that are synthesized by plants but have to be assimilated as separate amino acids by animals.

Puberty. Adolescence, same as sexual maturity (a confusing term avoided in this book).

Race. A subspecies.

Radiation. Speciation by a group of related organisms in the process of adapting to different ecological roles.

Rank order. A hierarchial arrangement of the individuals in a group (see hierarchy).

Relict. A persistent remnant population of an otherwise extinct species.

Resident. Living within a definite, limited home range, as opposed to being migratory or nomadic.

Ritualization. Evolutionary modification of a behavior pattern into a display or other signal, through selection for increasingly efficient and clear communication.

Ritualized. Referring to behavior that has been transformed through the process of ritualization (qv).

Roughage feeder. Referring to a diet of coarse herbage, e.g., a grazer adapted to eat and digest tough, fibrous grasses.

Rumen. The largest of the four chambers of the ruminant stomach, which is in effect a rumination vat (see fig. 1.1).

Ruminant. An ungulate with a four-chambered stomach, which chews the cud as part of the rumination process.

Rut (Middle English and French for roar, referring to stags). A definite mating period during which most conceptions occur, leading to an equally defined birth peak.

Savanna. Vegetation characteristic of tropical regions with extended wet and dry seasons, in which the ground layer is dominated by grasses and the overstory is composed of predominantly leguminous trees. The trees vary in type and density from broad-leafed, deciduous woodland in the wetter savanna (rainfall up to 1000 mm, dry season of 4 months or so) to grassland with scattered thorn trees and thorn bushes grading into subdesert (rainfall down to 250 mm or less, dry season lasting up to 8 months).

Scent gland. Area of skin packed with specialized cells that secrete complex chemical compounds which communicate information about the identity and the social and reproductive status of the animal. See sebaceous and apocrine glands.

Sebaceous glands. The commonest type of cutaneous scent glands, consisting of localized concentrations of flask-shaped follicles that produce a fatty secretion; odors consist largely of volatile fatty acids manufactured by symbiotic bacteria.

Secondary (sexual) characters. Characters other than gender differences present at birth (primary sexual characters), which develop under the influence of sex hormones, beginning at puberty (or even before, in the case of bovid horns). In most mammals, secondary characters develop only in males, as a result of sexual competition.

Secondary compounds. Toxic chemicals manufactured by plants to keep animals from eating them. Tannin, phenol, terpene, latex, oxalic acid, morphine, and caffeine are examples.

Secondary forest. Regeneration of trees and other vegetation after primary forest has been destroyed.

Selection pressure. Any feature of the environment that results in natural selection, through differential survival and reproductive success of individuals of different genetic types.

Sex ratio. The ratio of males to females in a population. Ratio at birth (natal sex ratio) tends to be 1:1 in most species and tends to remain equal in monogamous species, but adult sex ratios usually become skewed in favor of females (1:1.5, 1:2, 1:3, or even higher) in polygynous species, with increasing male sexual competition.

Sexual dimorphism. Differences in form between males and females, usually resulting from development of male secondary characters (qv) in response to sexual selection favoring larger size and showier display organs.

Sexual selection. Selection of genotypes through competition between members of the same sex (especially males) and mating preferences by members of the opposite sex (usually female choice). Cf natural selection.

Sexual skin. The external genitalia and surrounding skin in female primates, which in many species varies in color and size with reproductive condition, being most colorful and swollen at the height of estrus.

Sexual swelling. The swelling of sexual skin during the primate menstrual cycle, which reaches maximum inflation at estrus, then deflates, to begin again as the next ovarian follicle matures. Baboon and chimpanzee swellings are the most extreme among African primates.

Siblings. Offspring of the same mother and father.

Signal. Any behavior that conveys information from one individual to another, whether ritualized or not (cf display).

Solitary. Unsocial; referring to animals that do not live in social groups, including monogamous pairs and parent(s) with dependent offspring.

Species. Population(s) of closely related and similar organisms which are capable of interbreeding freely with one another, and cannot or normally do not interbreed with members of other species.

Species-specific. Characters that serve to distinguish a species, such as its shape and markings.

Stotting. Also called pronking; a distinctive bounding gait (see fig. 2.2) in which the animal bounces off the ground with straightened legs, propelling itself by flexing the lowermost (pastern) joints. Performed by many antelopes when alarmed.

Style-trot. Trotting exaggerated by lifting the legs higher than normal, often combined with defiant-looking head and tail movements. A sign of alarm/excitement in various antelopes, equivalent to stotting.

Subdesert. Regions that receive less rainfall than arid zones, but more than true desert.

Subspecies. Population(s) that has been isolated from other populations of the same species long enough to develop genetic differences sufficiently distinctive to be considered a separate race.

Superspecies. Populations of closely related species that have become morphologically different enough to be considered separate species—i.e. a step beyond subspecies.

Symbiotic. A mutually dependent relationship between unrelated organisms that are intimately associated, e.g., the symbiosis between a ruminant and the microorganisms that live in its rumen.

Sympatric. Overlapping geographic distribution; applies to related species that coexist without interbreeding (reverse of allopatric).

Systematics. The classification of organisms in an ordered system based on their natural relationships (see Guide to the Guide).

Tending bond. Describes the attempt of a male to monopolize mating opportunities by staying close to an estrous female and fending off all rivals; typical of nonterritorial, polygynous mating systems (applied especially to ungulates; cf. consortship).

Terrestrial. Living on land or ground-dwelling, as opposed to arboreal or aquatic.

Territorial. Animals that defend a particular area against (usually same-sex) rivals of their own species.

Testing. See urine-testing.

Testosterone. Male hormone produced in the testes, which induces and maintains male secondary sex characters, and mediates male sexual and aggressive behavior.

Tropical. The climate, flora, and fauna of the geographic region between 23½° N and S, the latitudes reached by the sun at its maximum declination, known respectively as the Tropics of Cancer and Capricorn.

Tsetse fly. Two-winged, blood-sucking flies of the genus *Glossina,* which transmit "sleeping sickness" (trypanosomiasis) to man (rarely) and domestic livestock (very commonly). The flies' presence in the woodlands of Africa south of the Sahara has helped to slow the pace of settlement and development, thereby preserving habitats for wild animals, which have a natural immunity to tsetse-borne diseases.

Turbinate. A scroll of cartilage on the wall of the nasal chamber; turbinates greatly increase the surface area of the nasal mucosa.

Undercoat. The soft insulating underfur beneath the longer, coarser guard hairs of the outer coat.

Ungulate. A mammal with hooves (from L. *ungula,* a hoof). Loosely applied also to such "near-ungulates" as elephants and hyraxes.

Urine-testing. A nearly universal habit among mammals with a functioning vomeronasal organ (qv), whereby female reproductive status is regularly monitored (mostly by males) through assaying hormone breakdown products in the urine. The testing procedure is most apparent in mammals that raise the head, open the mouth, and retract the upper lip in a grimace (see *flehmen*) as part of the testing procedure.

Veld. An Afrikaans term for field, generally used in South Africa to describe habitats in which grasses are a dominant element; equivalent to savanna.

Ventral. The underside, lower surface of an animal, opposite of dorsal.

Vertebrate. An animal with a spinal column and skeleton of bone, including amphibians, reptiles, birds, and mammals.

Vomeronasal organ (or Jacobson's organ).

A pair of narrow, blind sacs situated on either side of the nasal septum (see fig. 2.12), lined with sensory cells similar to olfactory epithelium; part of the accessory olfactory system present in most mammals, which appears to be primarily dedicated to monitoring reproductive condition, especially of females, through assaying chemical compounds such as hormone breakdown products excreted in the urine. (See urine-testing.)

Withers. The shoulder region.

Yearling. A young animal between one and two years of age (referring to species that take at least two years to mature).

Bibliography

Adamson, J. 1969. *The Spotted Sphinx.* New York: Harcourt Brace Jovanovich.

——— 1972. *Pippa's Challenge.* London: Collins.

Aeschlimann, A. 1963. Observations sur *Philantomba maxwelli* (Hamilton-Smith) une antilope de la forêt éburnée. *Acta Tropica* 20: 341–68.

Aldrich-Blake, F. P. G. 1970. Problems of social structure in forest monkeys. In *Social Behavior in Birds and Mammals,* ed. J. H. Crook, pp. 79–101. New York: Academic Press.

Alexander, A. J., and R. E. Ewer 1959. Observations on the biology and behavior of the smaller African polecat. *Afr. Wildlife* 13: 313–20.

Allen, J. A. 1925. Primates collected by the American Museum Congo expedition. *Bull. Amer. Mus. Nat. Hist.* 47: 283–499.

Allen, L. D. C. 1963. The lechwe (*Kobus leche smithemani*) of the Bangweulu swamps. *Puku* 1: 1–8.

Allsopp, R. 1971. Seasonal breeding in bushbuck, *Tragelaphus scriptus. E. Afr. Wildl. J.* 9: 146–49.

Altmann, D. 1969. *Harnen und Koten bei Saugetieren.* Neue Brehm-Bücherei. Wittenberg Lutherstadt: A. Ziemsen Verlag.

——— 1971. Zur Geburt beim Buntbock, *Damaliscus dorcas dorcas. Zool. Gart.* 40: 80–96.

Altmann, J. 1980. *Baboon Mothers and Infants.* Cambridge: Harvard Univ. Press.

Altmann, S. A., and J. Altmann 1970. *Baboon Ecology.* Chicago: Univ. of Chicago Press.

Anderson, A. B. 1969. Communication in the lesser bushbaby *Galago senegalensis moholi.* M. Sc. thesis, Univ. of Witwatersrand.

Anderson, J. L. 1972. Seasonal changes in the social organization and distribution of the impala in Hluhluwe Game Reserve, Zululand. *J. S. Afr. Wildl. Mgmt. Assn.* 2: 16–20.

——— 1978. Aspects of the ecology of the nyala (*Tragelaphus angasi* Gray 1849) in Zululand. Ph.D. thesis, Univ. of London.

——— 1979. Reproductive seasonality of the nyala (*Tragelaphus angasi*): the interaction of light, vegetation phenology, feeding style and reproductive physiology. *Mammal Rev.* 9: 33–46.

——— 1980. The social organization and aspects of behavior of the nyala (*Tragelaphus angasi* Gray). *Z. Säugetierk.* 45: 90–123.

——— 1985. Condition and related mortality of nyala *Tragelaphus angasi* in Zululand, South Africa. *J. Zool., Lond.* (A) 207: 371–80.

Anderson, S., and J. K. Jones 1967. *Recent Mammals of the World.* New York: Ronald Press.

Andrew, R. J. 1963. The origin and evolution of the calls and facial expressions of the primates. *Behaviour* 20: 1–109.

Andrew, R. J., and R. P. Klopman 1974. Urine washing: Comparative notes. In Martin et al. 1974, pp. 303–12.

Ansell, W. F. H. 1960*a*. The breeding of some larger mammals in Northern Rhodesia. *Proc. Zool. Soc. Lond.* 134: 251–74.

——— 1960*b*. Contribution to the mammalogy of Northern Rhodesia. *Occ. Papers Nat. Mus. S. Rhodesia* No. 24B: 351–98.

——— 1960*c*. The African striped weasel, *Poecilogale albinucha* (Gray). *Proc. Zool. Soc. Lond.* 134: 59–64.

——— 1963. Additional breeding data on Northern Rhodesian mammals. *Puku* 1: 9–28.

——— 1964. The preorbital, pedal and preputial glands of *Raphicerus sharpei* Thomas, with a note on the mammae of *Ourebia ourebi* Zimmerman. *Arnoldia* 18: 1–4.

——— 1969. *Oreotragus oreotragus centralis.* In Addenda and Corrigenda to Mammals of Northern Rhodesia, No. 3. *Puku* 5: 24–27.

——— 1971. Order Artiodactyla. In *The Mammals of Africa, an Identification Manual,* ed. J. Meester and H. W. Setzer, Part 15. Washington, D.C.: Smithsonian Institution.

Anthoney, T. R. 1968. The ontogeny of greeting, grooming and sexual motor patterns in captive baboons (supersp. *Papio cynocephalus*). *Behaviour* 31: 358–72.

Antonius, O. 1951. *Die Tigerpferde.* Frankfurt am Main: P. Schops.

Arcese, P., and G. Jongejan 1990. Cooperative territory defence and a novel ungulate social system in the oribi. *Gnusletter* 9(2): 5–6 (abstract).

Arden-Clarke, C. H. G. 1986. Population density, home range size and spatial organization of the Cape clawless otter, *Aonyx capensis,* in a marine habitat. *J. Zool., Lond.* (A) 209: 201–11.

Asdell, S. A. 1946. *Patterns of Mammalian Reproduction.* Ithaca, N.Y.: Cornell Univ. Press.

——— 1955. Endocrine organs, reproduction, and growth. In *The Physiology of Domestic Animals,* ed. H. H. Dukes, pp. 849–957. Ithaca, N.Y.: Cornell Univ. Press.

Attwell, C. A. M. 1982. Population ecology of the blue wildebeest *Connochaetes taurinus taurinus* in Zululand, South Africa. *Afr. J. Ecol.* 20: 147–68.

Backhaus, D. 1959*a*. Beobachtungen über das Freileben von Lelwel-Kuhantilopen (*Alcelaphus buselaphus lelwel* Heuglin, 1877) und Gelegenheitsbeobachtungen an der Senner-Pferdeantilope (*Hippotragus equinus bakeri* Heuglin, 1863). *Z. Säugetierk.* 24:1–34.

——— 1959*b*. Experimentelle Prüfung des Farbsehvermögens einer Masai-Giraffe. *Z. Tierpsychol.* 16: 468–77.

——— 1961. *Beobachtungen an Giraffen im Zoologischen Garten und freier Wildbahn.* Bruxelles: Ins. Parc. Nat. Congo Belge.

Badrian, A., and N. Badrian 1984. Social organization of *Pan paniscus* in the Lomako Forest, Zaire. In *The Pygmy Chimpanzee: Evolutionary Biology,* ed. R. L. Susman. New York: Plenum Press.

Baker, C. M. 1981. Agonistic behavior patterns of the slender mongoose, *Herpestes sanguineus. S. Afr. J. Zool.* 16: 263–65.

Bartholomew, G. A., and M. Rainy 1971. Regulation of body temperature in the rock hyrax, *Heterohyrax brucei. J. Mamm.* 52: 81–95.

Bartlett, J., and D. Bartlett 1967. *Nature's Paradise.* Boston: Houghton Mifflin.

Bauchop, T. 1978. Digestion of leaves in vertebrate arboreal folivores. In *The Ecology of Arboreal Folivores,* ed. G. G. Montgomery, pp. 193–205. Washington, D.C.: Smithsonian Institution.

Bearder, S. K. 1977. Feeding habits of spotted hyenas in a woodland habitat. *E. Afr. Wildl. J.* 15: 263–80.

Bearder, S. K., and G. A. Doyle 1974. Ecology of bushbabies, *Galago senegalensis* and *Galago crassicaudatus,* with some notes on their behavior in the field. In Martin et al. 1974, pp. 109–30.

Bekoff, M., J. Diamond, J. B. Mitton 1981. Life-history patterns and sociality in canids: body size, reproduction, and behavior. *Oecologia* (Berlin) 50: 386–90.

Bell, R. H. V. 1970. The use of the herb layer by grazing ungulates in the Serengeti. In *Animal Populations in Relation to Their Food Resources,* ed. A. Watson, pp. 111–23.

——— 1971. A grazing ecosystem in the Serengeti. *Sci. Am.* 224: 86–93.

——— 1982. The effect of soil nutrient availability on community structure in African ecosystems. In *Ecology of Tropical Savannas,* ed. B. J. Huntley and B. J. Walker. New York: Springer-Verlag.

Bell, R. H. V., J. J. R. Grimsdell, L. P. Van Lavieren, J. A. Sayer 1973. Census of the Kafue lechwe by aerial stratified sampling. *E. Afr. Wild. J.* 11: 55–74.

Ben-Shahar, R., and N. Fairall 1987. Comparison of the diurnal activity patterns of blue wildebeest and red hartebeest. *S. Afr. J. Wildl. Res.* 17: 49–54.

Ben-Yaacov, R., and Y. Yom-Tov 1983. On the biology of the Egyptian mongoose, *Herpestes ichneumon,* in Israel. *Z. Säugetierk.* 48: 34–45.

Bere, R. 1966. *Wild Animals in an African National Park.* London: Andre Deutsch.

Berg, I. A. 1944. Development of behavior: The micturition pattern in the dog. *J. Exp. Psychol.* 34: 343–68.

Berger, J. 1983. Induced abortion and social factors in wild horses. *Nature* 303: 59–61.

Bernard, R. T. F., and C. T. Stuart 1987. Reproduction of the caracal *Felis caracal* from the Cape Province of South Africa. *S. Afr. J. Zool.* 22: 177–82.

Bernstein, P., W. J. Smith, A. Krensky, K. Rosene 1978. Tail positions of *Cercopithecus aethiops. Z. Tierpsychol.* 46: 268–78.

Berry, H. H. 1983. First catch your lion. *Rossing* pp. 1–7. (Trade publication of Rossing Uranium Ltd., Windhoek, Namibia.)

Berry, H. H., and G. N. Louw 1982. Seasonal nutritive status of wildebeest in the Etosha National Park. *Madoqua* 13: 127–39.

Berry, H. H., W. R. Siegfried, T. M. Crowe 1982. Activity patterns in a population of free-ranging wildebeest (*Connochaetes taurinus*) at Etosha National Park. *Z. Tierpsychol.* 59: 229–46.

Berry, P. S. M. 1973. The Luangwa Valley giraffe. *Puku* 7: 71–92.

Bertram, B. C. 1974. Radio-tracking leopards in the Serengeti. *Wildlife News* 9(2): 7–10.

——— 1976. *Studying Predators,* handbk. 3. Nairobi: African Wildlife Foundation.

——— 1979. Serengeti predators and their social systems. In Sinclair and Norton-Griffiths 1979, pp. 221–62.

Best, G. A., and T. G. W. Best 1977. *Rowland Ward's Records of Big Game 17th ed. (Africa).* Sussex, Eng.: Rowland Ward.

Beuerle, W. 1975. Freilanduntersuchungen zum Kampf- und Sexualverhalten des europäischen Wildschweines *Sus scrofa* L. *Z. Tierpsychol.* 39: 211–58.

Bigalke, R. C. 1963*a*. A note on reproduction in the steenbok (*Raphicerus campestris* Thunberg). *Ann. Cape Prov. Mus.* 3: 64–67.

——— 1963*b*. The springbok *Antidorcas marsupialis* Zimm. *Trans. Interstate Antelope Conference,* pp. 68–80.

——— 1970. Observations on springbok populations. *Zool. Afr.* 5: 59–70.

——— 1972. Observations on the behavior and feeding habits of the springbok, *Antidorcas marsupialis. Zool. Afr.* 7: 333–59.

Bigourdan, J., and R. Prunier 1937. *Mammifères sauvages de l'Ouest africain et leur milieu.* Montrouge (Paris): Jean de Rudder.

Bishop, A. 1964. Use of the hand in lower primates. In *Evolutionary and Genetic Biology of Primates,* vol. 2, ed. J. Buettner-Janusch, pp. 133–226. New York: Academic Press.

Blaine, G. 1922. Notes on the zebras and some antelopes of Angola. *Proc. Zool. Soc. Lond.,* pp. 317–39.

Blancou, L. 1958. The African buffalos. *Animal Kingdom* 61(2): 56–61.

Boesch, C., and H. Boesch 1981. Sex differences in the use of natural hammers by wild chimpanzees: a preliminary report. *J. Hum. Evol.* 10: 585–93.

——— 1989. Hunting behavior of wild chimpanzees in the Tai National Park, Ivory Coast. *Am. J. Phys. Anthropol.* 78: 547–74.

——— 1990. Tool use and tool making in wild chimpanzees. *Folia Primatol.* 54: 86–99.

Bolwig, N. 1959. A study of the behavior of the chacma baboon, *Papio ursinus. Behaviour* 14: 136–63.

——— 1963. Observations on the mental and manipulative abilities of a baboon. *Behaviour* 22: 24–40.

Booth, A. H. 1957. Observations on the natural history of the olive colobus monkey, *Procolobus verus* (van Beneden). *Proc. Zool. Soc. Lond.* 129: 421–31.

Bothma, J. du P. 1971. Hyracoidea. In *The Mammals of Africa, an Identification Manual,* ed. J. Meester and H. W. Setzer, Part 12, pp. 1–8. Washington, D.C.: Smithsonian Institution.

Bothma, J. du P., and J. A. J. Nel 1980. Winter food and foraging behavior of the aardwolf (*Proteles cristatus*) in the Namib-Nawkluff Park. *Madoqua* 12: 141–47.

Bourlière, F. 1985. Primate communities: their structure and role in tropical ecosystems. *Int. J. Primatol.* 6: 1–26.

Bourlière, F., and J. Verschuren, 1960. *L'écologie des ongules du Parc National Albert.* Ins. Parcs Nat. Congo Belge, Brussels, 2 vols.

Brain, C. K. 1965*a*. Observations on the behavior of the vervet monkey, *Cercopithecus aethiops. Zool. Afr.* 1: 13–28.

——— 1965*b*. Adaptations to forest and savannah: some aspects of the social behavior of *Cercopithecus* monkeys in Southern Africa. *Primate Social Behaviour,* Suppl. 2–12: 1–32.

Bramley, P. S., and W. B. Neaves 1972. The relationship between social status and reproductive activity in male impala *Aepyceros melampus. J. Reprod. Fertil.* 31: 77–81.

Brand, D. J. 1963. Records of mammals bred in the National Zoological Gardens of South Africa during the period 1908 to 1960. *Proc. Zool. Soc. Lond.* 140: 617–59.

Brooks, A. C. 1961. A study of Thomson's gazelle (*Gazella thomsonii* Günther) in Tanganyika. *Colonial Research Publications* no. 25. London: H. M. Stationery Office.

Brooks, P. M. 1975. The sexual structure of an impala population and its relationship to an intensive game removal program. *Lammergeyer* 22: 1–8.

Brown, C. H. 1982. Auditory localization and primate vocal behavior. In *Primate*

Communication, ed. C. T. Snowdon, C. H. Brown, M. R. Petersen, pp. 144–64. Cambridge: Cambridge Univ. Press.

Brown, C. H., and P. M. Waser 1984. Hearing and communication in blue monkeys *(Cercopithecus mitis). Anim. Behav.* 32: 66–75.

Brown, L. H. 1969. Observations on the status, habitat and behavior of the mountain nyala, *Tragelaphus buxtoni,* in Ethiopia. *Mammalia* 33: 545–97.

Buechner, H. K., J. A. Morrison, W. Leuthold 1966. Reproduction in Uganda kob with special reference to behavior. *Symp. Zool. Soc. Lond.* 15: 69–88.

Buechner, H. K., and H. D. Roth 1974. The lek system in Uganda kob with special reference to behavior. *Amer. Zool.* 145–62.

Buechner, H. K., and R. K. Schloeth 1965. Ceremonial mating behavior in Uganda kob *(Adenota kob thomasi* Neumann). *Z. Tierpsychol.* 22: 209–25.

Bunderson, W. T. 1977. Hunter's antelope. *Oryx* 14: 174–75.

Buss, I. O., and J. A. Estes 1971. The functional significance of movements and positions of the pinnae of the African elephant, *Loxodonta africana. J. Mamm.* 52: 21–27.

Buss, I. O., L. E. Rasmussen, J. A. Smith 1976. The role of stress and individual recognition in the function of the African elephant's temporal gland. *Mammalia* 40: 437–51.

Buss, I. O., and N. R. Smith 1966. Observations on reproduction and breeding behavior of the African elephant. *J. Wildl. Mgmt.* 39: 375–85.

Busse, C. D., and W. J. Hamilton 1981. Infant carrying by male chacma baboons. *Science* 212: 1281–83.

Butynski, T. M. 1982. Harem male replacement and infanticide in the blue monkey *(Cercopithecus mitis stuhlmanni)* in the Kibale Forest, Uganda. *Amer. J. Primatol.* 3: 1–22.

Bygott, J. D., B. C. Bertram, J. P. Hanby 1979. Male lions in coalitions gain reproductive advantages. *Nature* 282: 839–41.

Cade, C. E. 1966. A note on the mating behavior of the Kenya oribi *Ourebia ourebi* in captivity. *Int. Zoo Yb.* 6: 205.

——— 1967. Notes on breeding the Cape hunting dog, *Lycaon pictus,* at Nairobi Zoo. *Int. Zoo Yb.* 7: 122–23.

Carlisle, D. B., and L. I. Ghobrial 1968. Food and water requirement of dorcas gazelle in the Sudan. *Mammalia* 32: 570–76.

Caro, T. M., and D. A. Collins 1987*a*. Male cheetah social organization and territoriality. *Ethology* 74: 52–64.

——— 1987*b*. Ecological characteristics of territories of male cheetahs *(Acinonyx jubatus). J. Zool., Lond.* 211: 1–17.

Caro, T. M., C. D. FitzGibbon, M. E. Holt 1989. Physiological costs of behavioral strategies for male cheetahs. *Anim. Behav.* 38: 309–17.

Carpenter, G. P. 1970. Some observations on the rusty-spotted genet *(Genetta rubiginosa zulensis). Lammergeyer* 11: 60–63.

Carr, G. M., and D. W. Macdonald 1986. The sociality of solitary foragers: a model based on resource dispersion. *Anim. Behav.* 34: 1540–49.

Carrington, R. 1958. *Elephants, a Short Account of Their Natural History, Evolution and Influence on Mankind.* Harmondsworth: Penguin Books.

Casimir, M. J., and E. Butenandt 1973. Migration and core area shifting in relation to some ecological factors in a mountain gorilla group *(Gorilla gorilla beringei)* in the Mt. Kahuzi region (République du Zaire). *Z. Tierpsychol.* 33: 514–22.

Cave, A. H. E. 1966. The preputial glands of *Ceratotherium. Mammalia* 30: 153–59.

Chalmers, G. 1963. Breeding data: steinbok. *E. Afr. Wildl. J.* 1: 121–22.

Channing, A., and D. T. Rowe-Rowe 1977. Vocalizations of South African mustelines. *Z. Tierpsychol.* 44: 283–93.

Charles-Dominique, P. 1971. Eco-éthologie des prosimiens du Gabon. *Biol. Gabonica* 7: 121–228.

——— 1974. Ecology and feeding behavior of five sympatric lorisids in Gabon. In Martin et al. 1974.

——— 1977. *Ecology and Behavior of Nocturnal Primates.* New York: Columbia Univ. Press.

——— 1978*a*. Ecology and social behavior of *Nandinia binotata. Terre Vie* 32: 477–528.

——— 1978*b*. Solitary and gregarious prosimians: evolution of social structures in primates. In *Recent Advances in Primatology,* ed. O. J. Chivers and K. A. Joysey, vol. 3, pp. 139–49. London: Academic Press.

——— 1984. Bushbabies, lorises and pottos. In Macdonald 1984*a*, pp. 332–37.

Charles-Dominique, P., and S. K. Bearder 1979. Field studies of lorisid behavior: methodological aspects. In Doyle and Martin, 1979, pp. 567–627.

Charles-Dominique, P., and R. D. Martin 1972. Behavior and ecology of nocturnal prosimians. *Z. Tierpsychol.*, Suppl. 9: 1–89.

Cheney, D. L. 1983. Extrafamilial alliances among vervet monkeys. In Hinde 1983, pp. 278–86.

Cheney, D. L., P. C. Lee, R. M. Seyfarth 1981. Behavioral correlates of non-random mortality among free-ranging vervet monkeys. *Behav. Ecol. Sociobiol.* 9: 153–61.

Cheney, D. L., and R. M. Seyfarth 1980. Vocal recognition in free-ranging vervet monkeys. *Anim. Behav.* 28: 362–67.

——— 1982. Recognition of individuals within and between free-ranging groups of vervet monkeys. *Amer. Zool.* 22: 519–29.

——— 1983. Nonrandom dispersal in free-ranging vervet monkeys: social and genetic consequences. *Am. Nat.* 122: 392–412.

——— 1986. The recognition of social alliances by vervet monkeys. *Anim. Behav.* 34: 1722–31.

Cheney, D. L., R. M. Seyfarth, S. J. Andelman, P. C. Lee 1988. Reproductive success in vervet monkeys. In *Reproductive Success*, ed. T. H. Clutton-Brock, pp. 384–402. Chicago: Univ. of Chicago Press.

Child, G., and J. D. Le Riche 1969. Recent springbok treks (mass movements) in south-western Botswana. *Mammalia* 33: 499–504.

Child, G., and W. von Richter 1969. Observations on ecology and behavior of lechwe, puku and waterbuck along the Chobe River, Botswana. *Z. Säugetierk.* 34: 275–95.

Child, G., H. Röbbel, C. P. Hepburn 1972. Observations on the biology of the tsessebe, *Damaliscus lunatus lunatus*, in northern Botswana. *Mammalia* 36: 342–88.

Child, G., and V. J. Wilson 1964. Observations on ecology and behavior of roan and sable in three tsetse control areas. *Arnoldia* 16: 1–8.

Churcher, C. S. 1978. Giraffidae. In Maglio and Cooke 1978, pp. 509–35.

——— 1981. Zebras (Genus *Equus*) from nine Quarternary sites in Kenya, East Africa. *Can. J. Earth Sciences* 18: 330–41.

Churcher, C. S., and M. L. Richardson 1978. Equidae. In Maglio and Cooke 1978, pp. 379–422.

Clark, A. B. 1978*a*. Olfactory communication, *Galago crassicaudatus*, and the social life of prosimians. In *Recent Advances in Primatology*, ed. O. J. Chivers and K. Joysey, vol. 3, pp. 109–17. London: Academic Press.

——— 1978*b*. Sex ratio and local resource competition in a prosimian primate. *Science* 201: 163–65.

——— 1985. Sociality in a nocturnal "solitary" prosimian: *Galago crassicaudatus. Int. J. Primatol.* 6: 581–600.

Clough, G. 1969. Some preliminary observations on reproduction in the warthog, *Phacochoerus aethiopicus* Pallas. *J. Reprod. Fertil.*, Suppl. 6: 323–37.

Clough, G., and A. G. Hassam 1970. A quantitative study of the daily activity of the warthog in the Queen Elizabeth National Park, Uganda. *E. Afr. Wildl. J.* 8: 19–24.

Clutton-Brock, T. H. 1975. Feeding behavior of red colobus and black and white colobus in East Africa. *Folia Primatol.* 23: 165–207.

———, ed. 1977. *Primate Ecology: Studies of Feeding and Ranging Behavior in Lemurs, Monkeys and Apes.* London: Academic Press.

Clutton-Brock, T. H., and P. H. Harvey 1977. Primate ecology and social organization. *J. Zool., Lond.* 183: 1–39.

Clutton-Brock, T. H., P. H. Harvey, B. Rudder 1977. Sexual dimorphism, socionomic sex ratio and body weight in primates. *Nature* 269: 797–800.

Coe, M. J., and R. D. Carr 1978. The association between dung middens of the blesbok (*Damaliscus dorcas phillipsi* Harper) and mounds of the harvester termite (*Trinervitermes trinervoides* Sjostedt). *S. Afr. J. Wildl. Res.* 8: 65–69.

Colbert, E. H. 1969. *Evolution of the Vertebrates*, 2d ed. New York: Wiley.

Condiotti, M. 1984. Sabinyo. *Wildl. News* 19: 14–18.

Cooke, H. B. S. 1968. Evolution of mammals on southern continents, 2: The fossil mammal fauna of Africa. *Quart. Rev. Biol.* 43: 234–64.

Cooke, H. B. S., and A. F. Wilkinson 1978. Suidae and Tayassuidae. In Maglio and Cooke 1978, pp. 435–82.

Cooper, J. B. 1942. An exploratory study of African lions. *Comp. Psychol. Monographs* 17: 1–48.

Coppens, Y., V. J. Maglio, C. T. Madden, M. Beden 1978. Proboscidea. In Maglio and Cooke 1978, chap. 14.

Cords, M. 1984. Mating patterns and social structure in redtail monkeys (*Cercopithecus ascanius*). *Z. Tierpsychol.* 64: 313–29.

——— 1987. *Mixed Species Association of* Cercopithecus *Monkeys in the Kakamega Forest, Kenya.* Berkeley and Los Angeles: Univ. of California Press.

Coryndon, S. C. 1978. Fossil Hippopotamidae of Africa. In Maglio and Cooke 1978, pp. 83–95.

Cowie, M. 1966. *The African Lion.* New York: Golden Press.
Crandall, L. S. 1964. *The Management of Wild Mammals in Captivity.* Chicago: Univ. of Chicago Press.
Cronwright-Schreiner, S. C. 1925. *The Migratory Springboks of South Africa.* London: Fisher Unwin.
Crook, J. H., and P. Aldrich-Blake 1968. Ecological and behavioral contrasts between sympatric ground dwelling primates in Ethiopia. *Folia Primatol.* 8: 192–227.
Crowe, T. M., and R. Liversidge 1977. Disproportionate mortality of males in a population of springbok (Artiodactyla: Bovidae). *Zool. Afr.* 12: 469–73.
Croze, H., A. K. Hillman, E. M. Lang 1981. Elephants and their habitats: how do they tolerate each other? In *Dynamics of Large Mammal Populations,* ed. C. W. Fowler and T. D. Smith, pp. 297–316. New York: Wiley.
Cumming, D. H. M. 1975. *A Field Study of the Ecology and Behavior of Warthog.* Salisbury, Rhodesia: Museum Memoir 7.
Cuneo, F. 1965. Observation on the breeding of the klipspringer antelope, *Oreotragus oreotragus,* and the behavior of their young born at the Naples Zoo. *Int. Zoo Yb.* 5: 45–48.
Dagg, A. I. 1962. The role of the neck in the movements of the giraffe. *J. Mamm.* 43: 88–97.
Dagg, A. I., and J. B. Foster 1976. *The Giraffe: Its Biology, Behavior and Ecology.* New York: Van Nostrand Reinhold.
Darwin, C. 1871. *The Descent of Man and Selection in Relation to Sex.* New York: Appleton.
Dasmann, R. F., and A. S. Mossman 1962. Population studies of impala in Southern Rhodesia. *J. Mamm.* 43: 533–37.
David, J. H. M. 1973. The behavior of the bontebok, *Damaliscus dorcas dorcas* (Pallas 1766), with special reference to territorial behavior. *Z. Tierpsychol.* 33: 38–107.
——— 1975*a*. Observations on mating behavior, parturition, suckling, and the mother-young bond in the bontebok. *J. Zool., Lond.* 177: 203–23.
——— 1975*b*. Fidelity to a fixed territory in some male bontebok in the Bontebok National Park, Swellendam, Cape Province. *J. S. Afr. Wildl. Mgmt. Assn.* 5: 111–14.
Davis, D. E. 1964. The physiological analysis of aggressive behavior. In *Social Behavior and Organisation among Vertebrates,* ed. W. Etkin, pp. 53–74. Chicago: Univ. of Chicago Press.
Dean, W. R. J., W. R. Siegfried, I. A. W. Macdonald 1990. The fallacy, fact and fate of guiding behavior in the greater honeyguide. *Conser. Biol.* 4: 99–101.
Dekker, D. 1968. Breeding the Cape hunting dog at Amsterdam Zoo. *Int. Zoo Yb.* 8: 27–30.
Demment, M. W., and P. J. Van Soest 1985. A nutritional explanation for body-size patterns of ruminant and nonruminant herbivores. *Am. Nat.* 125: 641–72.
DeVore, I., ed. 1965. *Primate Behavior.* New York: Holt, Rinehart and Winston.
DeVore, I., and K. R. L. Hall 1965. Baboon ecology. In DeVore 1965, pp. 20–53.
DeVos, A., and R. J. Dowsett 1966. The behavior and population structure of three species of the genus *Kobus. Mammalia* 39: 30–55.
DeVos, A., and A. Omar 1971. Territories and movements of Sykes monkeys *(Cercopithecus mitis kolbi* Neuman) in Kenya. *Folia Primat.* 16: 196–205.
Dittrich, L. 1967. Breeding Kirk's dik-dik *Madoqua kirkii thomasi* at Hanover Zoo. *Int. Zoo Yb.* 7: 171–73.
——— 1968. Keeping and breeding gazelles at Hanover Zoo. *Int. Zoo Yb.* 8: 139–43.
Dixson, A. F. 1981. *The Natural History of the Gorilla.* London: Weidenfeld and Nicolson.
——— 1983. Observations on the evolution and behavioral significance of "sexual skin" in female primates. In *Advances in the Study of Behavior,* ed. J. S. Rosenblatt et al., pp. 63–106. New York: Academic Press.
Donnelly, B. G., and J. H. Grobler 1976. Notes on food and anvil using behavior by the Cape clawless otter, *Aonyx capensis,* in the Rhodes Matopos National Park, Rhodesia. *Arnoldia* 7: 1–8.
Dorst, J., and P. Dandelot 1970. *A Field Guide to the Larger Mammals of Africa.* London: Collins.
Douglas-Hamilton, I. 1972. On the ecology and behavior of the African elephant: the elephants of Lake Manyara. D. Phil. thesis, Oxford Univ.
——— 1987. African elephants: population trends and their causes. *Oryx* 21: 11–24.
Douglas-Hamilton, I., and O. Douglas-Hamilton 1975. *Among the Elephants.* London: Collins and Harvill Press.
Dowsett, R. J. 1966. Behavior and population structure of the hartebeest in Kafue National Park. *Puku* 4: 147–54.
Doyle, G. A. 1974. The behavior of the

lesser bushbaby *(Galago senegalensis moholi).* In Martin et al. 1974, pp. 213–32.
——— 1979. Development of behavior in prosimians with special reference to the lesser bushbaby, *Galago senegalensis moholi.* In Doyle and Martin 1979, pp. 158–206.
Doyle, G. A., A. Anderson, S. K. Bearder 1969. Maternal behavior in the lesser bushbaby *(Galago senegalensis moholi)* under semi-natural conditions. *Folia Primatol.* 11: 215–38.
Doyle, G. A., and R. D. Martin, eds. 1979. *The Study of Prosimian Behavior.* New York: Academic Press.
Doyle, G. A., A. Pelletier, T. Bekker 1967. Courtship, mating and parturition in the lesser bushbaby *(Galago senegalensis moholi)* under semi-natural conditions. *Folia Primatol.* 7: 169–97.
Dragesco, J., F. Feer, G. Genermont 1979. Contribution à la connaissance de *Neotragus batesi* de Winton 1903 (position systématique, données biométriques). *Mammalia* 43: 71–81.
Dubost, G. 1979. The size of African forest artiodactyls, as determined by the vegetation structure. *Afr. J. Ecol.* 17: 1–17.
——— 1980. L'écologie et la vie sociale du céphalophe bleu *(Cephalophus monticola* Thunberg), petit ruminant forestier africain. *Z. Tierpsychol.* 54: 205–66.
——— 1983*a*. Le comportement de *Cephalophus monticola* Thunberg et *C. dorsalis* Gray, et la place des céphalophes au sein des ruminants, Part 1. *Mammalia* 47: 141–77.
——— 1983*b*. *Ibid.* Part 2. *Mammalia* 47: 281–310.
Dücker, G. 1960. Beobachtungen über das Paarungsverhalten des Ichneumons *(Herpestes ichneumon* L.) *Z. Säugetierk.* 25: 47–51.
——— 1962. Brutpflegeverhalten und Ontogenese des Verhaltens bei Suricaten *(Suricata suricatta* Schreb., Viverridae). *Behaviour* 19: 305–40.
——— 1965. Das Verhalten der Schleichkatzen (Viverridae). *Handb. Zool.* 8(38): 1–48.
Dukes, H. H. 1955. *The Physiology of Domestic Animals,* 7th ed. Ithaca, N.Y.: Comstock, Cornell Univ. Press.
Dunbar, R. I. M. 1978*a*. Competition and niche separation in a high altitude herbivore community in Ethiopia. *E. Afr. Wildl. J.* 16: 183–99.
——— 1978*b*. Sexual behavior and social relationships among gelada baboons. *Anim. Behav.* 26: 167–78.
——— 1979. Energetics, thermoregulation and the behavioral ecology of klipspringer. *Afr. J. Ecol.* 17: 217–30.
——— 1984. *Reproductive Decisions: An Economic Analysis of Gelada Baboon Social Strategies.* Princeton: Princeton Univ. Press.
——— 1988. *Primate Social Systems.* London: Christopher Helm.
Dunbar, R. I. M., and E. Dunbar 1974. Social organization and ecology of the klipspringer *(Oreotragus oreotragus)* in Ethiopia. *Z. Tierpsychol.* 35: 481–93.
——— 1976. Contrasts in social structure among black-and-white colobus monkey groups. *Anim. Behav.* 24: 84–92.
——— 1979. Observations on the social organization of common duikers in Ethiopia. *Afr. J. Ecol.* 17: 242–52.
——— 1980. The pairbond in klipspringer. *Anim. Behav.* 28: 219–29.
Duncan, P. 1975. Topi and their food supply. Ph.D. thesis, Univ. of Nairobi.
——— 1976. Ecological studies of topi antelope in the Serengeti. *Wildl. News* 11(1): 2–5.
Duplaix, N. 1984. Otters. In Macdonald 1984*a*, pp. 124–29.
DuPlessis, S. S. 1972. Ecology of blesbok with special reference to productivity. *Wildl. Monogr.* 30.
Earle, R. A. 1981. Aspects of the social and feeding behavior of the yellow mongoose *Cynictis penicillata. Mammalia* 45: 143–52.
East, R. 1984. Rainfall, soil nutrient status and biomass of large African savanna mammals. *Afr. J. Ecol.* 22: 245–70.
——— 1987. *Antelopes: Global Survey and Regional Action Plans,* Part 1: *East and Northeast Africa.* Gland, Switzerland: IUCN.
——— 1988. *Antelopes: Global Survey and Regional Action Plans,* Part 1: *East and Northeast Africa.* Gland, Switzerland: IUCN.
Eaton, R. L. 1970*a*. Group interactions, spacing and territoriality in cheetahs. *Z. Tierpsychol.* 27: 481–91.
——— 1970*b*. The predatory sequence, with emphasis on killing behavior and its ontogeny, in the cheetah. *Z. Tierpsychol.* 27: 492–504.
——— 1970*c*. Notes on the reproductive biology of the cheetah *Acinonyx jubatus. Int. Zoo Yb.* 10: 86–89.
——— 1971. Cheetah. *Nat. Parks and Conser. Mag.* 45(6): 18–22.
——— 1978. The evolution of sociality in the Felidae. In *The World's Cats,* ed. R. L. Eaton, vol. 3. Seattle: Carnivore Research Ins.

Eisenberg, J. F. 1973. Mammalian social systems: are primate social systems unique? *Symp. IVth Int. Congr. Primat.*, vol. 1: *Precultural Primate Behavior*, pp. 232–49. Basel: Karger.

Eisenberg, J. F., and M. Lockhart 1972. An ecological reconnaissance of Wilpattu National Park, Ceylon. *Smith. Contr. Zool.* 101: 1–117.

Eisenberg, J. F., N. Muckenhirn, R. Rudran 1972. The relation between ecology and social structure in primates. *Science, N.Y.* 176: 863–74.

Eisner, T., and J. A. Davis 1967. Mongoose throwing and smashing millipedes. *Science* 155: 577–79.

Eloff, F. C. 1961. Observations on the migration and habits of the antelopes of the Kalahari Gemsbok Park, part 3. *Koedoe* 4: 18–30.

——— 1964. On the predatory habits of lions and hyenas. *Koedoe* 6: 105–12.

——— 1966. Range extension of the blue wildebeest. *Koedoe* 9: 34–36.

——— 1973*a*, Ecology and behavior of the Kalahari lion. In *Proc. 1st Intern. Symp. Ecol. Behav. and Conserv. of the World's Cats*, ed. R. Eaton, pp. 90–126. Winston, Oregon: World Wildlife Safari.

——— 1973*b*. Water use by the Kalahari lion (*Panthera leo verhayi*). *Koedoe* 16: 149–54.

Emlen, S. T., and L. W. Oring 1977. Ecology, sexual selection and the evolution of mating systems. *Science* 197: 215–23.

Epps, J. 1974. Social interactions of *Perodicticus potto* kept in captivity in Kampala, Uganda. In Martin et al. 1974, pp. 233–44.

Erlinge, S. 1967. Home range of the otter *Lutra lutra* L. in southern Sweden. *Oikos* 18: 186–209.

——— 1968. Territoriality of the otter *Lutra lutra L. Oikos* 19: 81–98.

Esser, J. 1973. Beiträge zur Biologie des afrikanischen Rhebockes (*Pelea capreolus* Forster 1790). Ph.D. thesis, Christian-Albrechts Universität, Kiel.

Estes, J. A. 1989. Adaptations for aquatic living by carnivores. In *Carnivore Behavior, Ecology, and Evolution*, ed. J. L. Gittleman, pp. 242–82. Ithaca, N.Y.: Cornell Univ. Press.

Estes, R. D. 1967*a*. The comparative behavior of Grant's and Thomson's gazelles. *J. Mamm.* 48: 189–209.

——— 1967*b*. Predators and scavengers. Parts 1, 2. *Nat. Hist.* 76(2): 20–29, 76(3): 38–47.

——— 1969. Territorial behavior of the wildebeest (*Connochaetes taurinus* Burchell, 1823). *Z. Tierpsychol.* 26: 284–370.

——— 1972. The role of the vomeronasal organ in mammalian reproduction. *Mammalia* 36: 315–41.

——— 1973*a*. Showdown in Ngorongoro Crater. *Nat. Hist.* 82(8): 70–79.

——— 1973*b*. The flamingo eaters of Ngorongoro. *Nat. Geogr.* 144: 535–39.

——— 1974*a*. Social organization of the African Bovidae. In Geist and Walther 1974, pp. 166–205.

——— 1974*b*. Zebras offer clues to the way wild horses once lived. *Smithsonian* 5: 100–107.

——— 1976. The significance of breeding synchrony in the wildebeest. *E. Afr. Wildl. J.* 14: 135–52.

——— 1983. Sable by moonlight. *Anim. Kingd.* 86(4): 10–16.

——— 1990. The significance of horns and other male secondary sexual characters in female bovids. In *Proceedings of the Conference on Ungulate Behavior and Management, Texas A & M University, May, 1988*, ed. E. C. Mungall. *Appl. Anim. Behav. Sci.* (in press).

Estes, R. D., D. H. M. Cumming, G. W. Hearn 1982. New facial glands in domestic pig and warthog. *J. Mamm.* 63: 618–24.

Estes, R. D., and R. K. Estes 1974. The biology and conservation of the giant sable antelope, *Hippotragus niger variani* Thomas 1916. *Proc. Acad. Nat. Sci.* 126: 73–104.

——— 1979. The birth and survival of wildebeest calves. *Z. Tierpsychol.* 50: 45–95.

Estes, R. D., and J. Goddard 1967. Prey selection and hunting behavior of the African wild dog. *J. Wildl. Mgmt.* 31: 52–70.

Ewer, R. F. 1963. The behavior of the meerkat, *Suricata suricatta* (Schreber). *Z. Tierpsychol.* 20: 570–607.

——— 1968. *Ethology of Mammals.* London: Logos Press Ltd.

——— 1973. *The Carnivores.* Ithaca, N.Y.: Cornell Univ. Press.

Ewer, R. F., and C. Wemmer 1974. The behavior in captivity of the African civet, *Civettictis civetta* (Schreber). *Z. Tierpsychol.* 34: 359–94.

Eyre, M. 1963. A tame otter. *Afr. Wild Life* 17: 49–53.

Fairall, N. 1972. Behavioral aspects of the reproductive physiology of the impala, *Aepyceros melampus. Zool. Afr.* 7: 167–74.

Feer, F. 1979. Observations écologiques sur le néotrague de Bates (*Neotragus batesi* de

Winton 1903, Artiodactyle, Ruminant, Bovidé) du nord-est du Gabon. *Terre Vie* 33: 159–239.

——— 1982. Maturité sexuelle et cycle annuel de *Neotragus batesi* de Winton, 1903 (Bovidé forestier africain). *Mammalia* 46: 65–74.

Ferguson, J. W. H. 1978. Social interactions of black-backed jackals *Canis mesomelas* in the Kalahari Gemsbok National Park, South Africa. *Koedoe* 21: 151–62.

Ferreira, N. A., and R. C. Bigalke 1987. Food selection by gray rhebok in the Orange Free State, South Africa. *S. Afr. J. Wildl. Res.* 17: 123–27.

Fey, V. 1960. A note on the behavior of the tree hyrax. *J. E. Afr. Nat. Hist. Soc.* 23: 244–46.

Fiedler, W. 1972. Monkeys and apes. In *Grzimek's Animal Life Encyclopedia,* ed. B. Grzimek, vol. 10, *Mammals 1,* pp. 312–25. New York: Van Nostrand Reinhold.

Field, C. R. 1970. A study of the feeding habits of *Hippopotamus amphibius* Linn. in the Queen Elizabeth National Park, Uganda, with some management implications. *Zool. Afr.* 5: 71–86.

Field. C. R., and R. M. Laws 1970. The distribution of the larger herbivores in the Queen Elizabeth National Park. *J. Appl. Ecol.* 7: 273–94.

Finch, V. A. 1972. Thermoregulation and heat balance of the East African eland and hartebeest. *Am. J. Physiol.* 222: 1374–79.

Fischer, M. S. 1986. Die Stellung der Schliefer (Hyracoidea) im phylogenetischen System der Eutheria. *Courier Forschungsinstitut Senckenberg,* Frankfurt a. M., 112 p. typescript.

FitzGibbon, C. D. 1989. A cost to individuals with reduced vigilance in groups of Thomson's gazelles hunted by cheetahs. *Anim. Behav.* 37: 508–10.

Floody, O. R., and A. P. Arnold 1975. Uganda kob (*Adenota kob thomasi*). Territoriality and the spatial distribution of sexual and agonistic behaviors at a territorial ground. *Z. Tierpsychol.* 37: 192–212.

Fossey, D. 1972. Vocalizations of the mountain gorilla (*Gorilla gorilla beringei*). *Anim. Behav.* 20: 36–53.

——— 1982. Reproduction among free-living mountain gorillas. *Am. J. Primatol.,* Suppl. 1: 97–104.

——— 1983. *Gorillas in the Mist.* Boston: Houghton Mifflin.

Fossey, D., and A. Harcourt 1977. Feeding ecology of free-ranging mountain gorilla *(Gorilla gorilla beringei).* In Clutton-Brock 1977, pp. 415–47.

Foster, J. B. 1967. The square-lipped rhino *(Ceratotherium simum cottoni* [Lydekker]) in Uganda. *E. Afr. Wildl. J.* 5: 167–71.

Foster, J. B., and D. Kearney 1967. Nairobi Park census. *E. Afr. Wildl. J.* 5: 112–20.

Fox, M. W. 1970. A comparative study of the development of facial expressions in canids, wolf, coyote and foxes. *Behaviour* 36: 49–73.

——— 1971*a*. *Behavior of Wolves, Dogs, and Related Canids.* London: Jonathan Cape.

——— 1971*b*. Socio-infantile and socio-sexual signals in canids: a comparative and ontogenetic study. *Z. Tierpsychol.* 28: 185–210.

——— 1971*c*. Ontogeny of a social display in *Hyaena hyaena:* anal protrusion. *J. Mamm.* 52: 467–69.

Frädrich, H. 1967. Das Verhalten der Schweine (Suidae, Tayassuidae) und Flusspferde (Hippopotamidae). *Handb. Zool.* 8(26): 1–44.

——— 1974. A comparison of behavior in the Suidae. In Geist and Walther 1974, pp. 133–43.

Frame, G. W. 1971. The black rhinoceros. *Animals* (July): 693–99.

——— 1980. Cheetah social organization in the Serengeti ecosystem, Tanzania. Invited paper, Animal Behavior Society annual meeting.

Frame, G. W., and L. H. Frame 1976. Interim cheetah report for the Serengeti Research Institute, 28 pp.

——— 1977. Serengeti cheetah. *Wildlife News* 12(3): 2–6.

——— 1980. Cheetahs: in a race for survival. *Nat. Geogr.* 157: 712–28.

——— 1981. *Swift and Enduring: Cheetahs and Wild Dogs of the Serengeti.* New York: Dutton.

Frame, L. H., and G. W. Frame 1976. Female African wild dogs emigrate. *Nature* 263: 227–29.

Frame, L. H., J. R. Malcolm, G. W. Frame, H. Van Lawick 1979. Social organization of African wild dogs *(Lycaon pictus)* on the Serengeti Plains, Tanzania, 1967–1978. *Z. Tierpsychol.* 50: 225–49.

Frank, L. G. 1986*a*. Social organization of the spotted hyena *(Crocuta crocuta),* I: Demography. *Anim. Behav.* 34: 1500–1509.

——— 1986*b*. *Ibid.* II: Dominance and reproduction. *Anim. Behav.* 34: 1510–27.

Frank, L. G., J. M. Davidson, E. R. Smith 1985. Androgen levels in the spotted hyena *Crocuta crocuta*: the influence of social factors. *J. Zool., Lond.* (A) 206: 525–31.

Frank, L. G., S. E. Glickman, C. J. Zabel 1989. Ontogeny of female dominance in the spotted hyaena: perspectives from nature and captivity. In *The Biology of Large African Mammals in Their Environment,* ed. P. A. Jewell and G. M. O. Maloiy, pp. 127–46. *Symp. Zool. Soc. Lond.* No. 61. Oxford: Clarendon Press.

Friedmann, H. 1955. The honey-guides. *Bull. U.S. Nat. Mus.* 208: 1–292.

Fryxell, J. 1980. On the road with the white-eared kob. *Anim. Kingd.* 83, Sept./Oct.: 12–17.

——— 1985. Resource limitation and population ecology of white-eared kob. Ph.D. thesis, Univ. of British Columbia.

——— 1987*a*. Seasonal reproduction of white-eared kob in Boma National Park, Sudan. *Afr. J. Ecol.* 25: 117–24.

——— 1987*b*. Lek breeding and territorial aggression in white-eared kob. *Ethology* 75: 211–20.

——— 1988. Seasonal migration by white-eared kob in relation to resources. *Afr. J. Ecol.* 26: 17–31.

Fuente, R. de la 1972. *Africa. Hunters and Hunted of the Savannah.* World of Wildlife. London: Orbis.

Galat-Luong, A., and G. Galat 1979. Abondance relative et associations plurispécifiques des primates diurnes du parc national de Tai, Côte d'Ivoire. ORSTOM, Project Tai, 39 pp.

Gambaryan, P. P. 1974. *How Mammals Run.* New York: Wiley.

Gangloff, B., and P. Ropartz 1972. Le répertoire comportemental de la genette *Genetta genetta* (Linné). *Terre Vie* 26: 533–60.

Garstang, R. 1982. An analysis of home range utilization by the tsessebe, *Damaliscus lunatus lunatus* (Burchell), in P. W. Willis Private Nature Reserve. M.Sc. thesis, Univ. of Pretoria.

Gartlan, J. S. 1969. Sexual and maternal behavior of the vervet monkey, *Cercopithecus aethiops. J. Reprod. Fertil.* 6: 137–50.

Gartlan, J. S., and C. K. Brain 1968. Ecology and social variability in *Cercopithecus aethiops* and *C. mitis.* In *Primates: Studies in Adaptation and Variability,* ed. P. Jay, pp. 253–92. New York: Holt, Rinehart and Winston.

Gartlan, J. S., and T. T. Struhsaker 1979. Polyspecific associations and niche separation of rain-forest anthropoids in Cameroon, West Africa. In *Primate Ecology,* ed. R. W. Sussman, pp. 155–64. New York: Wiley.

Gautier, J.-P., and A. Gautier-Hion 1977. Communication in Old World Monkeys. In Sebeok 1977, pp. 890–964.

——— 1982. Vocal communication within a group of monkeys: analysis by biotelemetry. In *Primate Communication,* ed. C. T. Snowdon, C. H. Brown, M. Petersen, pp. 5–29. New York: Cambridge Univ. Press.

Gautier-Hion, A. 1978. Food niches and coexistence in sympatric primates in Gabon. In *Recent Advances in Primatology,* ed. D. J. Chivers and J. Herbert, pp. 269–86. London: Academic Press.

Gautier-Hion, A., F. Bourlière, J.-P. Gautier, J. Kingdon, eds. 1988. *A Primate Radiation: Evolutionary Biology of the African Guenons.* New York: Cambridge Univ. Press.

Gautier-Hion, A., L. H. Emmons, G. Dubost 1980. A comparison of the diets of three major groups of primary consumers of Gabon (primates, squirrels and ruminants). *Oecologia* 45: 182–89.

Gautier-Hion, A., and J.-P. Gautier 1986. Sexual dimorphism, social units and ecology among sympatric forest guenons. *Symposia of the Society for the Study of Human Biology* 24: 61–77.

Gautier-Hion, A., R. Quris, J.-P. Gautier 1983. Monospecific vs. polyspecific life: a comparative study of foraging and antipredatory tactics in a community of *Cercopithecus* monkeys. *Behav. Ecol. Sociobiol.* 12: 325–35.

Geertsema, A. 1976. Impressions and observations on serval behavior in Tanzania, East Africa. *Mammalia* 40: 13–19.

——— 1981. The servals of Gorigor. *Wildlife News* 16(3): 4–8.

Geist, V. 1966. The evolution of horn-like organs. *Behaviour* 27: 175–214.

——— 1974. On the relationship of social evolution and ecology in ungulates. *Am. Zool.* 14: 205–20.

Geist, V., and F. R. Walther, eds. 1974. *The Behavior of Ungulates and Its Relation to Management.* Morges, Switzerland: IUCN Publ. New Series No. 24.

Gentry, A. W. 1964. Skull characters of African gazelles. *Ann. Mag. Nat. Hist.* 13: 353–82.

——— 1970. The Bovidae (Mammalia) of the Ft. Ternan fossil fauna. In *Fossil Vertebrates of Africa,* II, ed. L. S. B. Leakey and R. J. G. Savage, pp. 243–323. London: Academic Press.

——— 1971. Genus *Gazella.* In *The Mammals of Africa, an Identification Man-*

ual, ed. J. Meester and H. W. Setzer, Part 15.1. Washington, D.C.: Smithsonian Institution.

——— 1978. Bovidae. In Maglio and Cooke 1978, pp. 540–72.

Georgiadis, N. 1985. Growth patterns, sexual dimorphism and reproduction in African ruminants. *Afr. J. Ecol.* 23: 75–87.

Ghiglieri, M. P. 1984. *The Chimpanzees of Kibale Forest. A field study of ecology and social structure.* New York: Columbia Univ. Press.

Ghobrial, L. 1974. Water relation and requirement of the dorcas gazelle in the Sudan. *Mammalia* 38: 88–106.

Gillman, J., and C. Gilbert 1946. The reproductive cycle of the chacma baboon *(Papio ursinus)* with special reference to the problems of menstrual irregularities as assessed by the behavior of the sex skin. *S. Afr. J. Med. Sci.* 11: 1–54.

Ginsberg, J. R. 1987. Social organization and mating strategies of an arid-adapted equid, Grevy's zebra. Unpub. Ph.D. thesis, Princeton Univ.

——— 1989. The ecology of female behavior and male mating success in the Grevy's zebra. In *The Biology of Large African Mammals in Their Environment,* ed. P. A. Jewell and G. M. O. Maloiy, pp. 90–110. *Symp. Zool. Soc. Lond.* No. 61. Oxford: Clarendon Press.

Gittleman, J. L. 1989. Carnivore group living: comparative trends. In *Carnivore Behavior, Ecology, and Evolution,* ed. J. L. Gittleman, pp. 183–208. Ithaca, N.Y.: Cornell Univ. Press.

Goddard, J. 1966. Mating and courtship of the black rhinoceros (*Diceros bicornis* L.). *E. Afr. Wildl. J.* 4: 69–75.

——— 1967. Home range, behavior, and recruitment rates of two black rhinoceros populations. *E. Afr. Wildl. J.* 5: 133–50.

——— 1970. Food preferences of black rhinoceros in the Tsavo National Park. *E. Afr. Wildl. J.* 8: 145–61.

Goethe, F. 1964. Das Verhalten der Musteliden. *Handb. Zool.* 8(37): 1–80.

Golani, I., and A. Keller 1975. A longitudinal field study of the behavior of a pair of golden jackals. In *The Wild Canids,* ed. M. W. Fox, pp. 303–35. New York: Van Nostrand Reinhold.

Golani, I., and H. Mendelssohn 1971. Sequences of precopulatory behavior of the jackal, *Canis aureus* L. *Behaviour* 38: 169–92.

Goodall, A. 1977. Feeding and ranging behavior of a mountain gorilla group *(Gorilla gorilla beringei)* in the Tshibinda-Kahuzi region (Zaire). In Clutton-Brock 1977, pp. 449–79.

Goodall, J. 1979. Life and death at Gombe. *Nat. Geogr.* 155: 592–621.

——— 1983. Population dynamics during a 15-year period in one community of free-living chimpanzees in the Gombe National Park, Tanzania. *Primates* 21: 545–49.

——— 1986. *The Chimpanzees of Gombe.* Cambridge, Mass.: Harvard Univ. Press.

Goodhart, C. B. 1975. Does the aardwolf mimic a hyena? *Zool. J. Linnean Soc.* 57: 349–56.

Gorman, M. L. 1976. A mechanism for individual recognition by odor in *Herpestes auropunctatus* (Carnivora: Viverridae). *Anim. Behav.* 24: 141–45.

——— 1986. The secretion of the temporal gland of the African elephant *Loxodonta africana* as an elephant repellent. *J. Trop. Ecol.* 2: 187–90.

Gorman, M. L., D. Jenkins, R. J. Harper 1978. The anal scent sacs of the otter, *Lutra lutra. J. Zool., Lond.* 186: 463–74.

Gorman, M. L., and G. L. Mills 1984. Scent-marking strategies in hyaenas (Mammalia). *J. Zool., Lond.* 202: 535–47.

Gorman, M. L., and B. J. Trowbridge 1989. The role of odor in the social lives of carnivores. In *Carnivore Behavior, Ecology, and Evolution,* ed. J. L. Gittleman, pp. 57–88. Ithaca, N.Y.: Cornell Univ. Press.

Gosling, L. M. 1969. Parturition and related behavior in Coke's hartebeest, *Alcelaphus buselaphus cokei* Günther. *J. Reprod. Fertil.,* Suppl. 6: 265–86.

——— 1972. The construction of antorbital gland marking sites by male oribi *(Ourebia ourebi* Zimmerman, 1783). *Z. Tierpsychol.* 30: 271–76.

——— 1974. The social behavior of Coke's hartebeest *(Alcelaphus buselaphus cokei).* In Geist and Walther 1974, pp. 488–511.

——— 1981. Demarcation in a gerenuk territory: an economic approach. *Z. Tierpsychol.* 56: 305–22.

——— 1982. A reassessment of the function of scent-marking in territories. *Z. Tierpsychol.* 60: 89–118.

——— 1985. The even-toed ungulates: order Artiodactyla. Sources, behavioral context, and function of chemical signals. In *Social Odours in Mammals,* ed. R. E. Brown and D. W. Macdonald. Oxford: Oxford Univ. Press.

——— 1991. The alternative mating strategies of male topi, *Damaliscus lunatus.* In *Proceedings of the*

Conference on Ungulate Behavior and Management, Texas A & M University, May 1988, ed. E. C. Mungall. *Appl. Anim. Behav. Sci.* 29:107–120.

Gosling, L. M., M. Petrie, M. E. Rainy 1987. Lekking in topi: a high cost, specialist strategy. *Anim. Behav.* 35: 616–18.

Gowda, C. D. Krishne 1967. A note on the birth of caracal lynx at Mysore Zoo. *Int. Zoo Yb.* 7: 133.

Graham, E., ed. 1981. *Reproductive Biology of the Great Apes.* New York: Academic Press.

Gregory, W. K., and M. Hellman 1939. On the evolution and major classification of the civets (Viverridae) and allied fossil and recent Carnivora: a phylogenetic study of the skull and dentition. *Proc. Am. Phil. Soc.* 81: 309–92.

Grimsdell, J. J. R. 1973. Reproduction in the African buffalo, *Syncerus caffer,* in Western Uganda. *J. Reprod. Fertil.,* Suppl. 19: 301–16.

Grimsdell, J. J. R., and R. H. V. Bell 1975. Black lechwe research project final report. Lusaka: National Council for Scientific Research.

Grobler, J. H. 1974. Aspects of the biology, population ecology and behavior of the sable, *Hippotragus niger niger* (Harris 1838) in the Rhodes Matopos National Park, Rhodesia. *Arnoldia* 7: 1–36.

——— 1981. Feeding behavior of the caracal (*Felis caracal* Schreber 1776) in the Mountain Zebra National Park. *S. Afr. Tydskr. Dierk.* 16: 259–62.

Grobler, J. H., and V. J. Wilson 1972. Food of the leopard *(Panthera pardus* Linn.) in the Rhodes Matopos National Park, Rhodesia, as determined by fecal analysis. *Arnoldia* 5: 1–10.

Groves, C. P. 1969. On the smaller gazelles of the genus *Gazella* de Blainville. *Z. Säugetierk.* 34: 38–60.

——— 1971. Distribution and place of origin of the gorilla. *Man* (London), n.s. 6: 44–51.

——— 1981. Subspecies and clines in the springbok (*Antidorcas*). *Z. Säugetierk.* 46: 189–97.

Groves, C. P., and P. Grubb 1974. A new duiker from Rwanda. *Rev. Zool. Africaine* 88: 189–96.

Grubb, P. 1973. Distribution, divergence and speciation of the drill and mandrill. *Folia Primatol.* 20: 161–77.

Grzimek. B. 1957. Beobachtungen an Gorillas und Schimpansen in Spanisch-Guinea. *Zool. Gart.* 23: 249.

———, ed. 1972. *Grzimek's Animal Life Encyclopedia,* vol. 13, *Mammals 4.* New York: Van Nostrand Reinhold.

Guggisberg, C. A. W. 1961. *Simba, the Life of the Lion.* Cape Town: Howard Timmins.

Gundlach, H. 1968. Brutfürsorge, Brutpflege, Verhaltensontogenese und Tagesperiodik beim Europäischen Wildschwein (*Sus scrofa* L.). *Z. Tierpsychol.* 25: 955–95.

Haddow, A. J., and J. M. Ellice 1964. Studies on bushbabies *(Galago* spp), with special reference to the epidemiology of yellow fever. *Trans. Roy. Soc. Trop. Med. Hyg.* 58: 521–38.

Hafez, E. S. E., ed. 1975. *The Behavior of Domestic Animals,* 3rd ed. Baltimore: Williams and Wilkins.

Hafez, E. S. E., and M. W. Schein 1962. The behavior of cattle. In *The Behavior of Domestic Animals,* ed. E. S. E. Hafez, pp. 256–96. London: Baillière, Tindall & Cox.

Hafez, E. S. E., and J. P. Signoret 1969. The behavior of swine. In *The Behavior of Domestic Animals,* ed. E. S. E. Hafez, 2nd ed. London: Baillière, Tindall & Cassell.

Hafez, E. S. E., M. Williams, S. Wierzbowski 1962. The behavior of horses. In *The Behavior of Domestic Animals,* ed. E. S. E. Hafez, pp. 370–96. London: Baillière, Tindall & Cox.

Hall, K. R. L. 1962*a*. Numerical data, maintenance activities and locomotion of the wild chacma baboon, *Papio ursinus. Proc. Zool. Soc. Lond.* 139: 181–220.

——— 1962*b*. The sexual, agonistic and derived social behavior patterns of the wild chacma baboon, *Papio ursinus. Proc. Zool. Soc. Lond.* 139: 283–327.

——— 1966. Distributions and adaptations of baboons. *Symp. Zool. Soc. Lond.* 17: 49–73.

Hall, K. R. L., R. C. Boelkins, M. J. Goswell 1965. Behavior of patas, *Erythrocebus patas,* in captivity, with notes on the natural habitat. *Folia Primatol.* 3: 22–49.

Hall, K. R. L., and I. DeVore 1965. Baboon social behavior. In DeVore 1965, pp. 53–110.

Hall, K. R. L., and J. S. Gartlan 1965. Ecology and behavior of the vervet monkey. *Cercopithecus aethiops,* Lolui Island, Lake Victoria. *Proc. Zool. Soc. Lond.* 145: 37–56.

Hall-Martin, A. J. 1987. Role of musth in the reproductive strategy of the African elephant *Loxodonta africana. Afr. J. Sci.* 83: 616–20.

Hall-Martin, A. J., J. D. Skinner, J. M. Van

Dyk 1975. Reproduction in the giraffe in relation to some environmental factors. *E. Afr. Wildl. J.* 13: 237–48.

Halperin, S. D. 1979. Temporary association patterns in free ranging chimpanzees: an assessment of individual grouping preferences. In *The Great Apes,* ed. D. A. Hamburg and E. R. McCown, pp. 491–500. Menlo Park, Calif.: Benjamin/Cummings.

Haltenorth, T., and H. Diller 1980. *A Field Guide to the Mammals of Africa.* London: Collins.

Hamann, U. 1979. Beobachtungen zum Verhalten von Bongoantilopen *(Tragelaphus euryceros* Ogilby, 1836). *Zool. Gart.* 49: 319–75.

Hamburg, D. A., and E. R. McCown, eds. 1979. *The Great Apes.* Menlo Park, Calif.: Benjamin/Cummings.

Hamilton, W. J., R. E. Buskirk, W. H. Buskirk 1975. Chacma baboon tactics during inter-troop encounters. *J. Mamm.* 56: 857–70.

——— 1976. Defense of space and resources by chacma *(Papio ursinus)* baboon troops in an African desert and swamp. *Ecology* 57: 1264–72.

——— 1977. Intersexual dominance and differential mortality of gemsbok (*Oryx gazella*) at Namib Desert waterholes. *Madoqua* 10: 5–19.

——— 1978. Omnivory and utilization of food resources by chacma baboons, *Papio ursinus. Am. Nat.* 112: 911–24.

Hanby, J. 1983. *Lions Share: The Story of a Serengeti Pride.* London: Collins.

Hanby, J., and D. C. Bygott 1987. Emigration of subadult lions. *Anim. Behav.* 35: 161–69.

Hanks, J., M. Stanley Price, R. W. Wrangham 1969. Some aspects of the ecology and behavior of the defassa waterbuck *(Kobus defassa)* in Zambia. *Mammalia* 33: 473–94.

Hannah, A. C., and W. C. McGrew 1987. Chimpanzees using stones to crack open oil palm nuts in Liberia. *Primates* 31: 157–70.

Harcourt, A. H. 1979. Social relationships between adult male and female mountain gorillas in the wild. *Anim. Behav.* 27: 325–42.

——— 1982. Mountain gorilla population report. *Wildlife News* 19(2): 6–8.

——— 1984. Gorilla. In Macdonald 1984*a*, pp. 432–39.

Harcourt, A. H., D. Fossey, J. Sabater Pi 1981. Demography of *Gorilla gorilla. J. Zool., Lond.* 195: 215–33.

Harcourt, A. H., and K. Stewart 1983. Interactions, relationships and social structure: the great apes. In Hinde 1983, pp. 307–14.

——— 1989. Functions of alliances in contests within wild gorilla groups. *Behaviour* 109: 176–90.

Harcourt, A. H., K. Stewart, D. Fossey 1976. Male emigration and female transfer in wild mountain gorilla. *Nature* 263: 226–27.

——— 1981. Gorilla reproduction in the wild. In Graham 1981, pp. 265–79.

Harcourt, C. S., and L. T. Nash 1986. Social organization of galagos in Kenyan coastal forests: 1. *Galago zanzibaricus. Am. J. Primatol.* 10: 339–55.

Harding, R. S. O. 1976. Ranging patterns of a troop of baboons *(Papio anubis)* in Kenya. *Folia Primatol.* 25: 143–85.

Harding, R. S. O., and S. D. Strum 1976. The predatory baboons of Kekopey. *Nat. Hist.* 55: 46–53.

Harris, C. J. 1968. *Otters.* London: Weidenfeld and Nicolson.

Harrison, D. 1968. *The Mammals of Arabia,* Vol. 2. London: Ernest Benn.

Harrison, M. J. S. 1983. Territorial behavior in the green monkey, *Cercopithecus sabaeus*: seasonal defense of local food supplies. *Behav. Ecol. Sociobiol.* 12: 85–94.

Hart, B. L., L. A. Hart, J. N. Maina 1988. Alteration in vomeronasal system anatomy in alcelaphine antelopes: correlation with alteration in chemosensory investigation. *Physiology and Behavior* 42: 155–62.

Hart, L. A., and B. L. Hart 1987. Species-specific patterns of urine investigation and flehmen in Grant's gazelle (*Gazella granti*), Thomson's gazelle (*Gazella thomsonii*), impala (*Aepyceros melampus*) and eland (*Taurotragus oryx*). *J. Comp. Psychol.* 101: 229–304.

——— 1988. Autogrooming and social grooming in impala. *Ann. N.Y. Acad. Sci.* 525: 399–402.

Hausfater, G. 1975. *Dominance and Reproduction in Baboons* (Papio cynocephalus). Basel: S. Karger.

——— 1977. Tail carriage in baboons *(Papio cynocephalus)*: relationship to dominance rank and age. *Folia Primatol.* 27: 41–59.

Hediger, H. 1951. *Observations sur la psychologie animale dans les Parcs Nationaux du Congo Belge.* Ins. Parcs Nat. Congo Belge, 194 pp.

Hefetz, A., R. Ben-Yaacov, Y. Yom-Tov, H. A. Lloyd 1982. The anal gland secretion of the African mongoose, *Herpestes ich-*

neumon. Chemistry and function. Cited by Ben-Yaacov and Yom-Tov 1983 as "in press."

Heinichen, I. G. 1969. Karyotype of the South West African plains zebra *(Equus burchelli* Gray, 1824 subsp.) and mountain zebra *(Equus zebra hartmannae* Matschie, 1898): their cytogenetic relationship to Chapman's zebra *(Equus burchelli antiquorum* H. Smith, 1841) and the Cradock Mountain zebra (*Equus zebra zebra* Linn. 1758). *Madoqua* 1: 47–52.

Heintz, E. 1969. Le dimorphisme sexuel des appendices frontaux chez *Gazella deperdita* Gervais (Bovidae, Artiodactyla, Mammalia) et sa signification phylogénique. *Mammalia* 33: 626–29.

Hendrichs, H. 1971. Freilandbeobachtungen zum Sozialsystem des afrikanischen Elefanten, *Loxodonta africana* (Blumenbach, 1797). In *Dikdik und Elefanten,* H. and U. Hendrichs, pp. 77–173. Munich: R. Piper.

——— 1972. Beobachtungen und Untersuchungen zur Ökologie und Ethologie, insbesondere zur sozialen Organisation ostafrikanischer Säugetiere. *Z. Tierpsychol.* 30: 146–89.

——— 1975*a*. Changes in a population of dikdik, *Madoqua (Rhynchotragus kirkii* [Günther 1880]). *Z. Tierpsychol.* 38: 55–69.

——— 1975*b*. Observations on a population of bohor reedbuck, *Redunca redunca* (Pallas 1767). *Z. Tierpsychol.* 38: 44–54.

Hendrichs, H., and U. Hendrichs 1971. Freilanduntersuchungen zur Ökologie und Ethologie der Zwerg-Antilope, *Madoqua (Rhynchotragus) kirkii* (Günther, 1880). In *Dikdik und Elefanten,* H. and U. Hendrichs, pp. 11–74. Munich: R. Piper.

Henzi, S. P., and M. Lawes 1987. Breeding season influxes and the behavior of adult male samango monkeys (*Cercopithecus mitis albogularis*). *Folia Primatol.* 48: 125–36.

Herbert, H. J. 1972. *The Population Dynamics of the Waterbuck* Kobus ellipsiprymnus *(Ogilby 1833) in the Sabi-Sand Wildtuin. Mammalia Depicta.* Hamburg and Berlin: Verlag Paul Parey.

Hildebrand, M. 1952. The integument in Canidae. *J. Mamm.* 33: 419–28.

——— 1965. Symmetrical gaits of horses. *Science* 150: 701–8.

Hill, A. 1980. Hyena provisioning of juvenile offspring at the den. *Mammalia* 44: 594.

Hill, W. C. O., and A. H. Booth 1957. Voice and larynx in African and Asiatic Colobidae. *J. Bombay Nat. Hist. Soc.* 54: 309–21.

Hillman, J. C. 1974. Ecology and behavior of the wild eland. *Wildlife News* 9: 6–9.

——— 1979. The biology of the eland (*Taurotragus oryx* Pallas) in the wild. Ph.D. thesis, Univ. of Nairobi.

——— 1986. Aspects of the biology of the bongo antelope, *Tragelaphus eurycerus* Ogilby 1837, in southwest Sudan. *Biol. Conser.* 38: 255–72.

——— 1987. Group size and association patterns of the common eland *Tragelaphus oryx. J. Zool., Lond.* 213: 641–64.

——— 1988. Home range and movement of the common eland *(Taurotragus oryx* Pallas 1766) in Kenya. *Afr. J. Ecol.* 26: 135–48.

Hillman, J. C., and J. M. Fryxell 1989. Sudan. In East 1988, pp. 5–13.

Hinde, R., ed. 1983. *Primate Social Relationships: An Integrated Approach.* Oxford: Blackwell.

Hinton, H. E., and A. M. S. Dunn 1967. *Mongooses, their Natural History and Behavior.* Edinburgh and London: Oliver and Boyd.

Hitchins, P. M. 1968. Some preliminary findings on the population structure and status of the black rhinoceros *Diceros bicornis* in Hluhluwe Game Reserve, Zululand. *Lammergeyer* 9: 26–28.

——— 1969. Influence of vegetation types on sizes of home ranges of black rhino in Hluhluwe Game Reserve. *Lammergeyer* 3: 81–85.

H. J. H. 1984. Genets. In Macdonald 1984*a*, pp. 140–42.

Hladik, C. M. 1977. Chimpanzees of Gombe and the chimpanzees of Gabon: some comparative data on the diet. In Clutton-Brock 1977, pp. 481–501.

——— 1979. Diet and ecology of prosimians. In Doyle and Martin 1979, pp. 307–58.

Hoeck, H. N. 1975. Differential feeding behavior of the sympatric hyrax *Procavia johnstoni* and *Heterohyrax brucei. Oecologia* 22: 15–49.

——— 1976. *Procavia johnstoni* (Procaviidae) mating behavior. Film E2178 of the Inst. Wis. Film, Göttingen.

——— 1977. "Teat order" in hyrax *(Procavia johnstoni* and *Heterohyrax brucei). Z. Säugetierk.* 42: 112–15.

——— 1978. Systematics of the Hyracoidea: toward a clarification. *Bull. Carnegie Mus. Nat. Hist.* 6: 146–51.

——— 1982. Population dynamics, dispersal and genetic isolation in two species of hyrax *(Heterohyrax brucei* and *Procavia johnstoni)* on habitat islands in the Serengeti. *Z. Tierpsychol.* 59: 177–210.

——— 1984. Hyraxes. in Macdonald 1984*a*, pp. 462–65.

——— 1989. Demography and competition in hyrax. *Oecologia* 79: 353–60.

Hoeck, H. N., H. Klein, P. Hoeck 1982. Flexible social organization in hyrax. *Z. Tierpsychol.* 59: 265–98.

Hofmann, R. R. 1968. Comparisons of rumen and omasum structure in East African game ruminants in relation to their feeding habits. In *Comparative Nutrition of Wild Animals,* ed. M. A. Crawford, pp. 79–94. London: *Symp. Zool. Soc. Lond.* No. 21.

——— 1973. *The Ruminant Stomach.* E. Afr. Monogr. in Biol., vol. 2. Nairobi: East African Literature Bureau.

Hofmeyer, J. M., and J. D. Skinner 1969. A note on ovulation and implantation in the steenbok and the impala. *Proc. S. Afr. Soc. Anim. Prod.* 8: 175.

Hofmeyer, M. D., and G. N. Louw 1987. Thermoregulation, pelage conductance and renal function in the desert-adapted springbok *Antidorcas marsupialis. J. Arid Environ.* 2: 137–52.

Holsworth, W. N. 1972. Reedbuck concentrations in the Dinder National Park, Sudan. *E. Afr. Wildl. J.* 19: 307–8.

Holzapfel, M. 1939. Markierungsverhalten bei der Hyäne. *Z. Morph. Ökol. Tiere* 35: 10–13.

Hooijer, D. A. 1978. Rhinocerotidae. In Maglio and Cooke 1978, pp. 371–78.

Hoppe, P. P. 1977*a*. Comparison of voluntary food and water intake and digestion in Kirk's dik-dik and suni. *E. Afr. Wildl. J.* 15: 41–48.

——— 1977*b*. How to survive heat and aridity; ecophysiology of the dik-dik antelope. *Vet. Med. Rev.* 1: 77–86.

Howard, P. C. 1986*a*. Social organization of common reedbuck and its role in population regulation. *Afr. J. Ecol.* 24: 143–54.

——— 1986*b*. Spatial organization of common reedbuck with special reference to the role of juvenile dispersal in population regulation. *Afr. J. Ecol.* 24: 155–71.

Hrdy, S. B. 1974. Male-male competition and infanticide among the langurs *(Presbytis entellus)* of Abu, Rajasthan. *Folia Primatol.* 22: 19–58.

——— 1977. *The Langurs of Abu.* Cambridge: Harvard Univ. Press.

Hrdy, S. B., and P. L. Whitten 1986. Patterning of sexual activity. In B. Smuts et al. 1986, pp. 370–84.

Huntley, B. J. 1972. Observations of the Percy Fyfe Nature Reserve tsessebe population. *Ann. Transvaal Mus.* 27: 225–43.

Huth, H. H. 1970. Zum Verhalten der Rappenantilope (*Hippotragus niger* Harris 1838). *Zool. Gart.* 38: 147–70.

——— 1980. Verhaltensstudien an Pferdeböcken (Hippotraginae) unter Berücksichtigung stammesgeschichtlicher und systematischer Fragen. *Säugetierk. Mitt.* 28: 161–245.

Ikeda, H., Y. Ono, M. Baba, T. Doi, T. Iwamoto 1982. Ranging and activity patterns of three nocturnal viverrids in Omo National Park, Ethiopia. *Afr. J. Ecol.* 20: 179–86.

Innis, A. C. 1958. The behavior of the giraffe, *Giraffa camelopardalis,* in the eastern Transvaal. *Proc. Zool. Soc. Lond.* 131: 245–78.

Irby, L. R. 1976. The ecology of mountain reedbuck in southern and eastern Africa. Ph.D. thesis, Texas A & M Univ.

——— 1979. Reproduction in mountain reedbuck *(Redunca fulvorufula). Mammalia* 43: 190–213.

IUCN 1986. *Red List of Threatened Animals.* Cambridge, U.K.: World Conservation Monitoring Centre.

Izawa, K., and J. Itani 1966. Chimpanzees in Kasakata Basin, Tanganyika: I. Ecological study in the rainy season, 1963–64. *Kyoto Univ. Afr. Studies* 1: 73–156.

Jacobsen, N. H. G. 1982. Observations on the behavior of slender mongooses, *Herpestes sanguineus,* in captivity. *Säugetierk. Mitt.* 30: 168–83.

Janis, C. 1976. The evolutionary strategy of the Equidae and the origins of rumen and caecal digestion. *Evolution* 30: 757–74.

——— 1982. Evolution of horns in ungulates: ecology and paleoecology. *Biol. Rev.* 57: 261–318.

Janis, C., and P. J. Jarman 1984. The hoofed mammals: even-toed ungulates. In Macdonald 1984*a*, pp. 468–79, 498–99.

Jarman, M. V. 1979. Impala social behavior. *Beihefte Z. Tierpsychol.* 21: 1–92.

Jarman, M. V., and P. J. Jarman 1973. Daily activity of impala. *E. Afr. Wildl. J.* 11: 75–92.

Jarman, P. J. 1972*a*. Seasonal distribution of large mammal populations in the unflooded middle Zambezi Valley. *J. Appl. Ecol.* 9: 283–99.

——— 1972*b*. The development of a dermal shield in impala. *J. Zool., Lond.* 166: 349–56.

——— 1974. The social organization of antelopes in relation to their ecology. *Behaviour* 48: 215–67.

——— 1983. Mating systems and sexual dimorphism in large, terrestrial, mammalian herbivores. *Biol. Rev.* 58: 485–520.

Jarman, P. J., and M. V. Jarman 1973. Social behavior, population structure and reproductive potential in impala. *E. Afr. Wildl. J.* 11: 329–38.

——— 1974. Impala behavior and its relevance to management. In Geist and Walther 1974, pp. 871–81.

Jeffery, R. C. V., R. H. V. Bell, W. F. Ansell 1989. Zambia. In East 1989, pp. 11–19.

Jenkins, F. A., and S. M. Camazine 1977. Hip structure and locomotion in ambulatory and cursorial carnivores. *J. Zool., Lond.* 181: 351–370.

Jewell, P. A. 1972. Social organization and movements of topi *(Damaliscus korrigum)* during the rut at Ishasha, Queen Elizabeth Park, Uganda. *Zool. Afr.* 7: 233–55.

Jolly, A. 1966. *Lemur Behavior.* Chicago: Univ. of Chicago Press.

——— 1985. *The Evolution of Primate Behavior,* 2d ed. New York: Macmillan.

Jolly, C. J. 1970. The large African monkeys as an adaptive array. In *Old World Monkeys,* ed. J. R. and P. H. Napier, pp. 139–74. New York: Academic Press.

Jones, C., and J. Sabater Pi 1971. Comparative ecology of *Gorilla gorilla* (Savage and Wyman) and *Pan troglodytes* (Blumenbach) in Rio Muni, West Africa. *Biblioth. Primatol.* 13: 1–93.

Jones, M. A. 1978. A scent-marking gland in the bushpig *(Potamochoerus porcus* Linnaeus 1758). *Arnoldia* 8: 1–4.

——— 1984. Seasonal changes in the diet of bushpig, *Potamochoerus porcus* Linn., in the Matopos National Park. *S. Afr. J. Wildl. Res.* 14: 97–100.

Joubert, E. 1972*a*. A note on the challenge rituals of territorial male lechwe. *Madoqua* 1: 63–67.

——— 1972*b*. Activity patterns shown by Hartmann's zebra *Equus zebra hartmannae* in South West Africa with reference to climatic factors. *Madoqua* Ser. 1, No. 5.

——— 1972*c*. The social organization and associated behavior in Hartmann's zebra. *Madoqua* 1: 17–56.

Joubert, S. C. J. 1970. A study of the social behavior of the roan antelope, *Hippotragus equinus equinus* (Desmarest, 1804), in the Kruger National Park. M. Sc. thesis, Univ. of Pretoria.

——— 1972. Territorial behavior in the tsessebe *(Damaliscus lunatus lunatus* Burchell) in the Kruger National Park. *Zool. Afr.* 7: 141–56.

——— 1974. The social organization of the roan antelope *Hippotragus equinus,* and its influence on the spatial distribution of herds in the Kruger National Park. In Geist and Walther 1974, pp. 661–75.

——— 1975. The mating behavior of the tsessebe *(Damaliscus lunatus lunatus)* in the Kruger National Park. *Z. Tierpsychol.* 37: 182–91.

——— 1976. The population ecology of the roan antelope, *Hippotragus equinus equinus* (Desmarest, 1804), in the Kruger National Park. D.Sc. thesis, Univ. of Pretoria.

Jungius, H. 1971*a*. *The Biology and Behavior of the Reedbuck in the Kruger National Park. Mammalia Depicta.* Hamburg: P. Parey.

——— 1971*b*. Studies on the food and feeding behavior of the reedbuck *Redunca arundinum* Boddaert, 1785 in the Kruger National Park. *Koedoe* 14: 65–97.

Karstad, E. L., and R. L. Hudson 1986. Social organization and communication of riverine hippopotami in southwest Kenya. *Mammalia* 50: 153–64.

Katsir, Z., and R. M. Crewe 1980. Chemical communication in *Galago crassicaudatus:* investigation of the chest gland secretion. *S. Afr. J. Zool.* 15: 249–54.

Kavanagh, M. 1984. *A Complete Guide to Monkeys, Apes and Other Primates.* New York: Viking Press.

Kayanja, I. B. 1989. The reproductive biology of the male hippopotamus. In *The Biology of Large African Mammals in Their Environment,* ed. P. A. Jewell and G. M. O. Maloiy, pp. 181–96. *Sympos. Zool. Soc. Lond.* No. 61. Oxford: Clarendon Press.

Kellas, L. M. 1955. Observations on the reproductive activities, measurements and growth of the dikdik *(Rhynchotragus kirkii thomasii* Neumann). *Proc. Zool. Soc. Lond.* 124: 751–84.

Kerr, A., and V. J. Wilson 1967. Notes on reproduction in Sharpe's grysbok. *Arnoldia* 3: 1–4.

Ketelhodt, H. F. von 1966. Der Erdwolf, *Proteles cristatus* (Sparrman, 1783). *Z. Säugetierk.* 31: 300–308.

——— 1977. Observations on the lambing interval of the grey duiker, *Sylvicapra grimmia grimmia. Zool. Afr.* 12: 232–33.

Kiley, M. 1972. The vocalizations of ungulates, their causation and function. *Z. Tierpsychol.* 31: 171–222.

Kiley-Worthington, M. 1965. The waterbuck *(Kobus defassa* Rüppell 1835 and *K. ellipsiprymnus* Ogilby, 1833) in East Africa: spatial distribution. A study of the sexual behavior. *Mammalia* 29: 177–204.

——— 1978. The causation, evolution and function of the visual displays of the eland *(Taurotragus oryx). Behaviour* 66: 179–221.

Kiltie, R. A. 1985. Evolution and function of horns and hornlike organs in female ungulates. *Biol. J. Linn. Soc.* 24: 299–320.

Kingdon, J. S. 1971. *East African Mammals,* vol. 1 (primates). London and New York: Academic Press.

——— 1977. *East African Mammals,* vol. 3A (carnivores). London and New York: Academic Press.

——— 1979. *East African Mammals,* vol. 3B (large herbivores). London and New York: Academic Press.

——— 1980. The role of visual signals and face patterns in African forest monkeys (guenons) of the genus *Cercopithecus. Trans. Zool. Soc. Lond.* 34: 425–75.

——— 1981. Where have the colonists come from? A zoogeographical examination of some mammalian isolates in eastern Africa. *Afr. J. Ecol.* 19: 115–24.

——— 1982. *East African Mammals,* vol. 3C and D (bovids). New York: Academic Press.

Kingston, T. J. 1971. Notes on the black and white colobus monkey in Kenya. *E. Afr. Wildl. J.* 9: 172–75.

Kirchshofer, R. 1963. Das Verhalten der Giraffengazelle, Elanantilope und des Flachlandtapirs bei der Geburt; einige Bemerkungen zur Vermehrungsrate und Generationsfolge. *Z. Tierpsychol.* 20: 143–59.

Kleiman, D. G. 1966. Scent-marking in the Canidae. In *Play, Exploration, and Territory in Mammals,* ed. P. A. Jewell and A. Loizos, pp. 167–77. London: Academic Press.

——— 1967. Some aspects of social behavior in the Canidae. *Amer. Zool.* 7: 365–72.

——— 1977. Monogamy in mammals. *Quart. Rev. Biol.* 52: 39–69.

Kleiman, D. G., and J. F. Eisenberg 1973. Comparisons of canid and felid social systems from an evolutionary perspective. *Anim. Behav.* 21: 637–59.

Klein, R. G. 1974. On the taxonomic status, distribution and ecology of the blue antelope, *Hippotragus leucophaeus* (Pallas 1766). *Ann. S. Afr. Mus.* 65: 99–143.

Klingel, H. 1967. Soziale Organisation und Verhalten freilebender Steppenzebras (*Equus quagga*). *Z. Tierpsychol.* 27: 580–624.

——— 1968. Soziale Organisation und Verhaltensweisen von Hartmann- und Bergzebras (*Equus zebra hartmannae* und *E. z. zebra*). *Z. Tierpsychol.* 25: 75–88.

——— 1969. Reproduction in the plains zebra *Equus burchelli boehmi:* behavior and ecological factors. *J. Reprod. Fertil.,* Suppl. 6: 339–45.

——— 1972. Social behavior of African Equidae. *Zool. Afr.* 7: 175–85.

——— 1974*a*. A comparison of the social behavior of the Equidae. In Geist and Walther 1974, pp. 124–32.

——— 1974*b*. Soziale Organisation und Verhalten des Grevy-Zebras. *Z. Tierpsychol.* 36: 37–70.

——— 1975. Die soziale Organisation der Equiden. *Verh. Dtsch. Zool. Ges.* 1975: 71–80.

Klingel, H., and U. Klingel 1966. Die Geburt eines Zebras (*Equus quagga böhmi* Matschie). *Z. Tierpsychol.* 23: 72–76.

Kok, O. B. 1975. *Behavior and Ecology of the Red Hartebeest* (Alcelaphus buselaphus caama). Orange Free State, South Africa: Nature Conservation, Misc. Pub. 4.

Kortlandt, A. 1962. Chimpanzees in the wild. *Sci. Am.* 206: 128–38.

——— 1967. Experimentation with chimpanzees in the wild. In *Neue Ergebnisse der Primatologie,* ed. D. Starck, R. Schneider, H.-J. Kuhn, pp. 208–23. Stuttgart: Gustav Fischer.

Kralik, S. 1967. Breeding of the caracal lynx at Brno Zoo. *Int. Zoo Yb.* 7: 132.

Kranz, K. R. 1982. A note on the structure of tail hairs from the pgymy hippopotamus *(Choeropsis liberiensis). Zoo Biology* 1: 237–41.

Kranz, K. R., and S. Lumpkin 1982. Notes on the yellow-backed duiker *Cephalophus sylvicultor* in captivity with comments on its natural history. *Int. Zoo Yb.* 22: 232–40.

Kruuk, H. 1972. *The Spotted Hyena.* Chicago: Univ. of Chicago Press.

——— 1976. Feeding and social behavior of the striped hyena *(Hyaena vulgaris). E. Afr. Wildl. J.* 14: 91–111.

——— 1978. Spatial organization and territorial behavior of the European badger *Meles meles. J. Zool., Lond.* 184: 1–19.

——— 1989. *The Social Badger.* Oxford: Oxford Univ. Press.

Kruuk, H., and W. A. Sands 1972. The aardwolf *(Proteles cristatus* Sparrman 1783) as predator of termites. *E. Afr. Wildl. J.* 10: 211–27.

Kuhn, H.-J. 1972. On the perineal organ of

male *Procolobus badius. J. Hum. Evol.* 1: 371–78.

Kummer, H. 1968. *Social Organization of Hamadryas Baboons.* Chicago: Univ. of Chicago Press.

Kuroda, S. 1980. Social behavior of the pygmy chimpanzees. *Primates* 21: 181–97.

Kurt, F. 1964. Beobachtungen an den ostäthiopischen Antilopen: *Madoqua kirkii, Strepsiceros strepsiceros chora, S. imberbis, Oryx beisa beisa, Gazella soemmeringi erlangi, Litocranius walleri. Vjschr. Naturf. Ges.* Zürich 109: 143–62.

Lack, A. 1977. Genets feeding on nectar from *Maranthes polyandra* in northern Ghana. *E. Afr. Wildl. J.* 15: 233–34.

Lamprecht, J. 1978. On diet, foraging behavior and interspecific food competition of jackals in the Serengeti National Park, East Africa. *Z. Säugetierk.* 43: 210–23.

——— 1979. Field observations on the behavior and social system of the bat-eared fox *(Otocyon megalotis* Desmarest). *Z. Tierpsychol.* 52: 171–200.

Lamprey, H. F. 1963. Ecological separation of the large mammal species in the Tarangire Game Reserve, Tanganyika. *E. Afr. Wildl. J.* 1: 63–92.

Lang, E. M. 1975. *Das Zwergflusspferd (Choeropsis liberiensis).* Wittenberg Lutherstadt: A. Ziemsen Verlag.

Langley, C. H., and J. H. Giliomee 1974. Behavior of the bontebok (*Damaliscus d. dorcas*) in the Cape of Good Hope Nature Reserve. *J. S. Afr. Wildl. Mgmt. Assn.* 4: 117–21.

Langman, V. A. 1977. Cow-calf relationships in giraffe (*Giraffa camelopardalis giraffa*). *Z. Tierpsychol.* 43: 264–86.

Lawes, M. J., and M. R. Perrin 1990. Diet and feeding behavior of samango monkeys *Cercopithecus mitis labiatus* in Ngoye Forest, South Africa. *Folia Primatol.* 54: 57–69.

Laws, R. M. 1968. Dentition and aging of the hippopotamus. *E. Afr. Wildl. J.* 6: 19–52.

——— 1969. Aspects of reproduction in the African elephant *Loxodonta africana. J. Reprod. Fertil.*, Suppl. 6: 193–217.

——— 1970*a*. Biology of African elephants. *Sci. Prog., Oxf.* 58: 251–62.

——— 1970*b*. Elephants as agents of habitat and landscape change in East Africa. *Oikos* 21: 1–15.

——— 1984. Hippopotamuses. In Macdonald 1984*a*, pp. 506–11.

Laws, R. M., and G. Clough 1966. Observations on reproduction in the hippopotamus *Hippopotamus amphibius.* In *Comparative Biology of Reproduction in Mammals,* ed. I. W. Rowlands, pp. 117–40. London: Academic Press.

Laws, R. M., I. S. C. Parker, R. C. B. Johnstone 1975. *Elephants and their Habitats.* Oxford: Clarendon Press.

Lee, P. C. 1979. Coming of age in Amboseli. *Wildlife News* 14(1): 7–12.

——— 1983. Caretaking of infants and mother-infant relationships. In Hinde 1983, pp. 145–51.

——— 1987. Allomothering among African elephants. *Anim. Behav.* 35: 278–91.

Lee, P. C., J. Thornback, E. L. Bennett 1988. *Threatened Primates of Africa.* Gland, Switzerland and Cambridge, UK: IUCN.

Leland, L., and T. T. Struhsaker 1983. The Kibale Forest Project. *Wildlife News* 18(1): 4–10.

Leland, L., T. T. Struhsaker, T. M. Butynski 1984. Infanticide by adult males in three primate species of the Kibale Forest, Uganda: a test of hypotheses. In *Infanticide: Comparative and Evolutionary Perspectives,* ed. G. Hausfater and S. B. Hrdy. Hawthorne, N.Y.: Aldine.

Lent, P. C. 1969. A preliminary study of the Okavango lechwe *(Kobus leche leche* Gray). *E. Afr. Wild. J.* 7: 147–57.

——— 1974. Mother-infant relationships in ungulates. In Geist and Walther 1974, pp. 14–55.

Leuthold, B. M. 1979. Social organization and behavior of giraffe in Tsavo East National Park. *Afr. J. Ecol.* 17: 19–34.

Leuthold, B. M., and W. Leuthold 1972. Food habits of giraffe in Tsavo National Park, Kenya. *E. Afr. Wildl. J.* 10: 129–41.

——— 1978*a*. Daytime activity patterns of gerenuk and giraffe in Tsavo National Park, Kenya. *E. Afr. Wildl. J.* 16: 231–43.

——— 1978*b*. Ecology of the giraffe in Tsavo East National Park, Kenya. *E. Afr. Wildl. J.* 16: 1–20.

Leuthold, W. 1966. Variations in territorial behavior of Uganda kob, *Adenota kob thomasi* (Neumann 1896). *Behaviour* 27: 214–57.

——— 1967. Beobachtungen zum Jugendverhalten von Kob-Antilopen. *Z. Säugetierk.* 32: 59–63.

——— 1970. Observations on the social organization of impala *(Aepyceros melampus). Z. Tierpsychol.* 27: 693–721.

——— 1971*a*. Freilandbeobachtungen an Giraffengazellen *(Litocranius walleri)* im Tsavo-Nationalpark, Kenia. *Z. Säugetierk* 36: 19–37.

——— 1971*b*. A note on the formation of food habits in young antelopes. *E. Afr. Wildl. J.* 9: 154–56.

——— 1971c. Studies on the food habits of lesser kudu in Tsavo National Park, Kenya. *E. Afr. Wildl. J.* 9: 35–45.

——— 1977a. *African Ungulates.* Berlin and New York: Springer Verlag.

——— 1977b. Spatial organization and strategy of habitat utilization of elephants in Tsavo National Park, Kenya. *Z. Säugetierk.* 42: 358–79.

——— 1977c. A note on group size and composition in the oribi *Ourebia ourebi* (Zimmerman, 1783) (Bovidae). *Säugetierk. Mitt.* 40: 233–35.

——— 1978a. On the ecology of the gerenuk *Litocranius walleri. J. Anim. Ecol.* 47: 561–80.

——— 1978b. On social organization and behavior of the gerenuk *Litocranius walleri* (Brooke 1878). *Z. Tierpsychol.* 47: 194–216.

——— 1979. The lesser kudu, *Tragelaphus imberbis* (Blyth 1869). Ecology and behavior of an African antelope. *Säugetierk. Mitt.* 27: 1–75.

——— 1984. The graceful gerenuk. In Macdonald 1984a, pp. 582–83.

Leuthold, W., and B. M. Leuthold 1973. Notes on the behavior of two young antelopes reared in captivity. *Z. Tierpsychol.* 32: 418–24.

——— 1975. Patterns of social grouping in ungulates of Tsavo National Park, Kenya. *J. Zool., Lond.* 175: 405–20.

Leuthold, W., and J. B. Sale 1973. Movements and patterns of habitat utilization of elephants in Tsavo National Park, Kenya. *E. Afr. Wildl. J.* 11: 369–84.

Leyhausen, P. 1979. *Cat Behavior, the Predatory and Social Behavior of Domestic and Wild Cats.* New York: Garland STPM Press.

Leyhausen, P., and B. Tonkin 1966. Breeding the black-footed cat, *Felis nigripes,* in captivity. *Int. Zoo Yb.* 6: 178–82.

Lindemann, W. 1955. Über die Jugendentwicklung beim Luchs (*Lynx l. lynx* Kerr) und bei der Wildkatze (*Felis s. silvestris* Schreb.). *Behaviour* 8: 1–45.

——— 1957. Reviermarkierung durch den Iltis. *Wild u. Hund* 60: 506.

Lindeque, M., and J. D. Skinner 1982. Fetal androgens and sexual mimicry in spotted hyenas (*Crocuta crocuta*). *J. Reprod. Fertil.* 67: 405–10.

Linley, T. A. 1965. Aardwolf at East London Zoo. *Int. Zoo Yb.* 5: 145.

Lockie, J. D. 1966. Territory in small carnivores. In *Play, Exploration and Territory in Mammals,* ed. P. A. Jewell and C. Loizos, pp. 143–65. London: Academic Press.

Lombard, G. L. 1958. The water mongoose *(Atilax). Fauna and Flora Transvaal* 9: 24–27.

Lott, D. F. 1974. Sexual and aggressive behavior of mature male American bison. In Geist and Walther 1974, pp. 382–94.

Lownds, L. 1956. Antelopes in the Kruger National Park. *Afr. Wild Life* 10: 59–61.

Luck, C. P., and P. G. Wright 1964. Aspects of the anatomy and physiology of the skin of *Hippopotamus amphibius* L. *Quart. J. Exp. Physiol.* 16: 1–14.

Ludbrook, J. V. 1963. Desertion of a buffalo calf. *Puku* 1: 216.

Lydekker, R. 1896. *A Hand-Book to the Carnivora.* London: Lloyd Ltd.

——— 1908. *The Game Animals of Africa.* London: Rowland Ward.

Lynch, C. D. 1971. A behavioral study of blesbok, *Damaliscus dorcas phillipsi,* with special reference to territoriality. M. Sc. thesis, Univ. of Pretoria.

——— 1980. Ecology of the suricate, *Suricata suricatta,* and yellow mongoose, *Cynictis penicillata,* with special reference to their reproduction. *Mem. van die Nasionale Mus.,* Bloemfontein 14: 1–145.

Maberly, C. T. A. 1960. African bushpig. *Animals* 9(10): 556–59.

——— 1962. *Animals of East Africa.* Cape Town: Howard Timmins.

MacClintock, D. 1973. *A Natural History of Giraffes.* New York: Charles Scribner's Sons.

Macdonald, D. W. 1978. Observations on the behavior and ecology of the striped hyena, *Hyaena hyaena,* in Israel. *Israel J. Zool.* 27: 189–98.

——— 1979. The flexible social system of the golden jackal, *Canis aureus. Behav. Ecol. Sociobiol.* 5: 17–38.

——— 1980. Patterns of scent marking with urine and faeces among carnivore communities. *Symp. Zool. Soc. Lond.* 45: 107–39.

——— 1983. The ecology of carnivore social behavior. *Nature* 301: 379–84.

———, ed. 1984a. *The Encyclopedia of Mammals.* New York: Facts on File.

——— 1984b. Carnivores. In Macdonald 1984a, pp. 18–25.

——— 1986. Information in *Meerkats United,* written and produced by M. Lunz. Bristol: BBC-TV.

Macdonald, D. W., and C. F. Mason 1980. Observations on the marking behavior of a coastal population of otters. *Acta Theriologica* 25: 245–53.

Macdonald, D. W., and P. Moehlman 1983. Cooperation, altruism, and restraint in the reproduction of carnivores. In *Perspectives in Ethology,* ed. P. P. G. Bateson and P. Klopfer, pp. 433–67. New York: Plenum.

MacKinnon, J. 1978. *The Ape within Us.* London: Collins.

——— 1984. Orangutan. In Macdonald 1984*a*, pp. 428–31.

Maddock, L. 1979. The "migration" and grazing succession. In *Serengeti: Dynamics of an Ecosystem,* ed. A. R. E. Sinclair and M. Norton-Griffiths, pp. 104–29. Chicago: Univ. of Chicago Press.

Maglio, V. J. 1978. Patterns of faunal evolution. In Maglio and Cooke 1978, pp. 603–19.

Maglio, V. J., and H. B. S. Cooke, eds. 1978. *Evolution of African Mammals.* Cambridge: Harvard Univ. Press.

Maier, V., O. A. E. Rasa, H. Scheich 1983. Call-system similarity in a ground-living social bird and a mammal in the bush habitat. *Behav. Ecol. Sociobiol.* 12: 5–9.

Malcolm, J. R. 1980. Food caching by African wild dogs *Lycaon pictus. J. Mamm.* 61: 743–44.

——— 1984*a*. The bat-eared fox—an insect eater. In Macdonald 1984*a*, pp. 72–73.

——— 1984*b*. A "back-to-front" social system. In Macdonald 1984*a*, pp. 78–79.

——— 1986. Socio-ecology of bat-eared foxes, *Otocyon megalotis. J. Zool., Lond.* (A) 208: 457–68.

Malcolm, J. R., and H. Van Lawick 1975. Notes on wild dogs hunting zebras. *Mammalia* 39: 231–40.

Marler, P. 1965. Communication in monkeys and apes. In DeVore 1965, pp. 544–84.

——— 1972. Vocalizations of East African monkeys 2: black and white colobus. *Behaviour* 42: 175–97.

——— 1973. A comparison of vocalizations of red-tailed monkeys and blue monkeys, *Cercopithecus ascanius* and *C. mitis,* in Uganda. *Z. Tierpsychol.* 33: 223–47.

——— 1976. Social organization, communication and graded signals: the chimpanzee and the gorilla. In *Growing Points in Ethology,* ed. P. P. G. Bateson and R. A. Hinde, pp. 239–79. Cambridge: Cambridge Univ. Press.

——— 1978*a*. Vocal ethology of primates. In *Recent Advances in Primatology,* ed. D. J. Chivers and J. Herbert, pp. 795–801. London and New York: Academic Press.

——— 1978*b*. Monkey calls: how are they perceived and what do they mean? In *Advances in the Study of Mammalian Behavior,* ed. J. F. Eisenberg and D. G. Kleiman, Spec. Pub. No. 7, Am. Soc. Mammal., pp. 343–54.

Marler, P., and R. Tenaza 1977. Signaling behavior of apes with special reference to vocalization. In Sebeok 1977, pp. 935–1033.

Martin, E. Bradley, and C. Bradley Martin 1987. Combatting the illegal trade in rhinoceros products. *Oryx* 21: 143–48.

Martin, R. D. 1990. *Primate Origins and Evolution: A Phylogenetic Reconstruction.* Princeton: Princeton Univ. Press.

Martin, R. D., G. A. Doyle, A. C. Walker, eds. 1974. *Prosimian Biology.* Pittsburgh: Univ. of Pittsburgh Press.

Mason, D. R. 1976. Some observations on social organisation and behavior of springbok in the Jack Scott Nature Reserve. *S. Afr. J. Wildl. Res.* 6: 33–39.

——— 1986. Reproduction in the male warthog *Phacochoerus aethiopicus* from Zululand, South Africa. *S. Afr. J. Zool.* 21: 39–47.

Matthews, L. H. 1939*a*. Reproduction in the spotted hyena, *Crocuta crocuta* (Erxl.). *Philos. Trans.* 230: 1–78.

——— 1939*b*. The bionomics of the spotted hyena, *Crocuta crocuta. Proc. Zool. Soc. Lond.* 109: 43–56.

Maxwell, G. 1960. *Ring of Bright Water.* London: Longmans.

——— 1963. *The Rocks Remain.* New York: Dutton.

McGrew, W. 1979. Evolutionary implications of sex differences in chimpanzee predation and tool use. In Hamburg and McCown 1979, pp. 441–64.

McKey, D. 1978. Soils, vegetation, and seed-eating by black colobus monkeys. In Montgomery 1978, pp. 423–39.

McLaughlin, R. T. 1970. Aspects of the biology of cheetahs *Acinonyx jubatus* (Schreber) in Nairobi National Park. M. Sc. thesis, Univ. of Nairobi.

McNaughton, S. J. 1976. Serengeti migratory wildebeest: facilitation of energy flow by grazing. *Science* 191: 92–94.

——— 1985. Ecology of a grazing ecosystem: the Serengeti. *Ecol. Monogr.* 55: 259–94.

McNaughton, S. J., and N. J. Georgiadis 1986. Ecology of African grazing and browsing mammals. *Ann. Rev. Ecol. Syst.* 17: 39–65.

McVittie, R. 1979. Changes in the social behavior of South West African cheetah. *Madoqua* 11: 171–84.

Meinertzhagen, R. 1938. Some weights and measurements of large mammals. *Proc. Zool. Soc. Lond.* 108: 433–39.

Mendelssohn, H. 1974. The development of the populations of gazelles in Israel and their behavioral adaptations. In Geist and Walther 1974, pp. 722–43.

Mentis, M. T. 1970. Estimates of natural biomasses of large herbivores in the Umfolozi Game Reserve area. *Mammalia* 34: 363–93.

Meyer, G. E. 1978. Hyracoidea. In Maglio and Cooke 1978, pp. 284–314.

Millais, J. G. 1899. *A Breath of the Veldt.* London: Henry Sotteran.

Mills, M. G. L. 1973. The brown hyena. *Afr. Wild Life* 27: 150–53.

——— 1976. Ecology and behavior of the brown hyena in the Kalahari with some suggestions for management. In *Proc. Symp. Endangered Wildl.*, pp. 36–42. Pretoria: Endangered Wildlife Trust.

——— 1978. The comparative socio-ecology of the Hyaenidae. *Carnivore* 1: 1–6.

——— 1982*a*. Factors affecting group size and territory size of the brown hyena, *Hyaena brunnea*, in the southern Kalahari. *J. Zool. Soc., Lond.* 198: 39–51.

——— 1982*b*. The mating system of the brown hyena, *Hyaena brunnea*, in the southern Kalahari. *Behav. Ecol. Sociobiol.* 10: 131–36.

——— 1983. Behavioral mechanisms in territory and group maintenance of the brown hyaena, *Hyaena brunnea*, in the southern Kalahari. *Anim. Behav.* 31: 503–10.

——— 1984. The comparative behavioral ecology of the brown hyaena *Hyaena brunnea* and the spotted hyaena *Crocuta crocuta* in the southern Kalahari. *Koedoe* Suppl.: 237–47.

——— 1985. Related spotted hyenas forage together but do not cooperate in rearing young. *Nature* (Lond.) 316: 61–62.

——— 1987. The scent-marking behavior of the spotted hyena in the southern Kalahari, South Africa. *J. Zool., Lond.* 212: 483–98.

——— 1989. The comparative behavioral ecology of hyenas: the importance of diet and food dispersion. In *Carnivore Behavior, Ecology, and Evolution*, ed. J. L. Gittleman, pp. 125–63. Ithaca, N.Y.: Cornell Univ. Press.

——— 1990. *Kalahari Hyenas: Comparative Behavioral Ecology of Two Species.* London: Unwin Hyman.

Mills, M. G. L., M. L. Gorman, M. E. J. Mills 1980. The scent marking behavior of the brown hyena, *Hyaena brunnea. S. Afr. J. Zool.* 15: 240–48.

Mills, M. G. L., and M. E. J. Mills 1978. The diet of the brown hyena, *Hyaena brunnea. Koedoe* 21: 125–49.

Mitchell, A. 1977. Preliminary observations on the daytime activity patterns of lesser kudu in Tsavo National Park, Kenya. *E. Afr. Wildl. J.* 15: 199–206.

Mitchell, B. L., J. B. Shenton, J. C. Uys 1965. Predation of large mammals in the Kafue National Park, Zambia. *Zool. Afr.* 1: 297–318.

Mitchell, B. L., and J. M. C. Uys 1961. The problem of the lechwe *(Kobus leche)* on the Kafue Flats. *Oryx* 6: 171–83.

Mittermeyer, R. A., and J. G. Fleagle 1976. The locomotor and postural repertoires of *Ateles geoffroyi* and *Colobus guereza*, and a reevaluation of the locomotor category semibrachiation. *Am. J. Phys. Anthrop.* 45: 235–56.

Mloszewski, M. J. 1983. *The Behavior and Ecology of the African Buffalo.* Cambridge: Cambridge Univ. Press.

Moehlman, P. D. 1978. Jackals of the Serengeti. *Wildlife News* 13(3): 2–6.

——— 1979. Jackal helpers and pup survival. *Nature* 277: 382–83.

——— 1983. Socioecology of silverbacked and golden jackals *(Canis mesomelas* and *Canis aureus).* In *Advances in the Study of Mammalian Behavior*, ed. J. F. Eisenberg and D. G. Kleiman, pp. 423–53. American Soc. of Mammalogists, Special Pub. 7.

——— 1984. Jackals. In Macdonald 1984*a*, pp. 64–65.

——— 1986. Ecology of cooperation in canids. In Rubenstein and Wrangham 1986, pp. 64–86.

Mohr, E. 1960. *Wilde Schweine.* Neue Brehm-Bücherei, Heft 247. Wittenberg: A. Ziemsen Verlag.

Monfort, A. 1974. Quelques aspects de la biologie des phacochères, *Phacochoerus aethiopicus*, au Parc National de l'Akagera, Rwanda. *Mammalia* 38: 177–200.

Monfort, A., and N. Monfort 1974. Notes sur l'écologie et le comportement des oribis *(Ourebia ourebi* Zimmermann, 1783). *Terre Vie* 28: 169–208.

Monfort-Braham, N. 1974. Contribution à l'étude des structures sociales et du comportement des ongulés du Parc National de l'Akagera, 2ème Partie, l'impala. Univ. Liége, 114 pp.

——— 1975. Variations dans la structure sociale du topi, *Damaliscus korrigum* Ogilby, au Parc National de l'Akagera, Rwanda. *Z. Tierpsychol.* 39: 332–64.

Montgomery, G. G., ed. 1978. *The Ecology of Arboreal Folivores.* Washington, D.C.: Smithsonian Institution Press.

Moore, J., and R. Ali 1984. Are dispersal and inbreeding avoidance related? *Anim. Behav.* 32: 94–112.

Moreau, R. E. 1944. Kilimanjaro and Mount Kenya: some comparisons with particular reference to the mammals and birds. *Tanganyika Notes and Records* 18: 28–68.

Moreno-Black, G. S., and W. R. Maples 1977. Differential habitat utilization of four Cercopithecidae in a Kenyan forest. *Folia Primatol.* 27: 87–107.

Mortimer, M. A. E. 1963. Notes on the biology and behavior of the spotted-necked otter *(Lutra maculicollis)*. *Puku* 1: 192–206.

Moss, C. J. 1975. *Portraits in the Wild.* Boston: Houghton Mifflin.

——— 1983. Estrous behavior and female choice in the African elephant. *Behaviour* 86: 167–96.

——— 1988. *Elephant Memories.* London: Elm Tree Books.

Moss, C. J., and J. Poole 1983. Relationships and social structure of African elephants. In Hinde 1983, pp. 315–25.

Mühlenberg, M., and H. H. Roth 1985. Comparative investigations into the ecology of the kob antelope *Kobus kob kob* (Erxleben, 1777) in the Camoe National Park, Ivory Coast. *S. Afr. J. Wildl. Res.* 215: 25–31.

Müller, H. 1970. Beiträge zur Biologie des Hermelins. *Säugetierk. Mitt.* 18: 293–380.

Mungall, E. C. 1980. Courtship and mating behavior of the dama gazelle (*Gazella dama* Pallas 1766). *Zool. Gart.* NF, Jena 50: 1–14.

Murray, M. G. 1981. Structure of association in impala, *Aepyceros melampus. Behav. Ecol. Sociobiol.* 9: 23–33.

——— 1982*a*. Home range, dispersal and the clan system of impala. *Afr. J. Ecol.* 20: 253–69.

——— 1982*b*. The rut of impala: aspects of seasonal mating under tropical conditions. *Z. Tierpsychol.* 59: 319–37.

——— 1984. Grazing antelopes. In Macdonald 1984*a*, pp. 560–69.

Myers, N. 1975. The cheetah, *Acinonyx jubatus,* in Africa. Morges, Switzerland: IUCN Monograph no. 4.

——— 1976. Status of the leopard and cheetah in Africa. In *World's Cats* 3: 53–69.

Napier, J. R., and P. H. Napier 1967. *A Handbook of Living Primates.* London: Academic Press.

———, eds. 1970. *Old World Monkeys.* New York: Academic Press.

Napier-Bax, P., and D. L. W. Sheldrick 1963. Some preliminary observations on the food of elephant in the Tsavo Royal National Park (East) of Kenya. *E. Afr. Wildl. J.* 1: 40–53.

Nash, L. T., and C. S. Harcourt. 1986. Social organization of galagos in Kenyan coastal forests: 2. *Galago garnetti. Am. J. Primat.* 10: 357–69.

Neal, E. 1970. The banded mongoose, *Mungos mungo* Gmelin. *E. Afr. Wildl. J.* 8: 63–71.

Neaves, W. B., J. E. Griffin, J. D. Wilson 1980. Sexual dimorphism of the phallus in spotted hyaena (*Crocuta crocuta*). *J. Zool., Lond.* 187: 315–26.

Nel, J. A. J. 1978. Notes on the food and foraging behavior of the bat-eared fox *Otocyon megalotis. Bull. Carnegie Mus. Nat. Hist.* 6: 132–37.

Nel, J. A. J., and M. H. Bester 1983. Communication in the southern bat-eared fox *Otocyon m. megalotis* (Desmarest, 1822). *Z. Säugetierk.* 48: 266–90.

Nel, J. A. J., and J. du P. Bothma 1983. Scent marking and midden use by aardwolves *(Proteles cristatus)* in the Namib Desert. *Afr. J. Ecol.* 21: 26–39.

Nishida, T. 1968. The social group of chimpanzees of the Mahali Mountains. *Primates* 9: 167–224.

Norris, J. 1988. Diet and feeding behavior of semi-free ranging mandrills in an enclosed Gabonais forest. *Primates* 29: 449–64.

Norton, P. M. 1980. The habitat and feeding ecology of the klipspringer *Oreotragus oreotragus* (Zimmermann, 1783) in two areas of the Cape Province. M. Sc. thesis, Univ. of Pretoria.

——— 1981. Activity patterns of klipspringers in two areas of the Cape Province. *S. Afr. J. Wildl. Res.* 11: 126–34.

——— 1984. Food selection by klipspringers in two areas of the Cape Province. *S. Afr. J. Wildl. Res.* 14: 33–41.

Novellie, P. A. 1979. Courtship behavior of the blesbok (*Damaliscus dorcas phillipsi*). *Mammalia* 43: 263–74.

Nowak, R. M., and J. L. Paradiso 1983. *Walker's Mammals of the World,* 4th edition. Baltimore: Johns Hopkins Univ. Press.

Oates, J. F. 1977*a*. The social life of a black and white colobus monkey, *Colobus guereza. Z. Tierpyschol.* 45: 1–60.

——— 1977*b*. The guereza and its food. In *Primate Ecology,* ed. T. H. Clutton-Brock, pp. 275–321. London: Academic Press.

O'Brien, S. J., M. E. Roelke, L. Marker, et al. 1985. Genetic basis for species vul-

nerability in the cheetah. *Science* 227: 1428–34.
O'Connor, T. G., and B. M. Campbell 1986. Hippopotamus *Hippopotamus amphibius* habitat relationships on the Lundi River, Gonarezhou National Park, Zimbabwe. *Afr. J. Ecol.* 24: 7–26.
Oliver, M. D. N., N. R. M. Short, J. Hanks 1978. Population ecology of oribi, gray rhebuck and mountain reedbuck in Highmoor State Forest Land, Natal. *S. Afr. J. Wildl. Res.* 8: 95–105.
Olivier, R. C. D., and W. A. Laurie 1974. Habitat utilization by hippopotamuses in the Mara River. *E. Afr. Wildl. J.* 12: 249–71.
O'Regan, B. P. 1984. Gazelles and dwarf antelopes. In Macdonald 1984*a*, pp. 574–81.
Owen, J. 1973. Behavior and diet of a captive royal antelope, *Neotragus pygmaeus* L. *Mammalia* 37: 56–65.
Owen, R. E. A. 1970. Some observations of the sitatunga in Kenya. *E. Afr. Wildl. J.* 8: 181–95.
Owen-Smith, R. N. 1973. The behavioral ecology of the white rhinoceros. Ph.D. thesis, Univ. of Wisconsin.
——— 1974. The social system of the white rhinoceros. In Geist and Walther 1974, pp. 341–51.
——— 1975. The social ethology of the white rhinoceros *Ceratotherium simum* (Burchell 1817). *Z. Tierpsychol.* 38: 337–84.
——— 1977. On territoriality in ungulates and an evolutionary model. *Q. Rev. Biol.* 53: 1–38.
——— 1979. Assessing the foraging efficiency of a large herbivore, the kudu. *S. Afr. J. Wildl. Res.* 9: 102–10.
——— 1984*a*. Spatial and temporal components of the mating systems of kudu bulls and red deer stags. *Anim. Behav.* 32: 321–32.
——— 1984*b*. Rhinoceroses. In Macdonald 1984*a*, pp. 490–97.
——— 1988. *The Influence of Very Large Body Size on Ecology.* New York: Cambridge Univ. Press.
Owens, D. D., and M. J. Owens 1979. Communal denning and clan associations in brown hyenas (*Hyaena brunnea* Thunberg) of the Central Kalahari Desert. *Afr. J. Ecol.* 17: 35–44.
Owens, M. J., and D. D. Owens 1978. Feeding ecology and its influence on social organization in brown hyenas (*Hyaena brunnea* Thunberg) of the Central Kalahari Desert. *E. Afr. Wildl. J.* 16: 113–36.
——— 1984. *Cry of the Kalahari.* Boston: Houghton Mifflin.
Packer, C. 1979. Inter-troop transfer and inbreeding avoidance in *Papio anubis. Anim. Behav.* 27: 1–36.
——— 1983. Sexual dimorphism: the horns of African antelopes. *Science* 221: 1191–93.
——— 1985. Dispersal and inbreeding avoidance. *Anim. Behav.* 33: 676–78.
——— 1986. The ecology of sociality in felids. In *Ecological Aspects of Social Evolution,* ed. D. I. Rubenstein and R. Wrangham. Princeton: Princeton Univ. Press.
Packer, C., L. Herbst, A. Pusey, et al. 1988. Reproductive success of lions. In *Reproductive Success,* ed. T. H. Clutton-Brock, pp. 363–83. Chicago: Univ. of Chicago Press.
Packer, C., and A. E. Pusey 1982. Cooperation and competition within coalitions of male lions: kin selection or game theory? *Nature* 296: 740–42.
——— 1983*a*. Male takeovers and female reproductive parameters: a simulation of oestrous synchrony in lions (*Panthera leo*). *Anim. Behav.* 31: 334–40.
——— 1983*b*. Adaptations of female lions to infanticide by incoming males. *Am. Nat.* 121: 91–113.
——— 1983*c*. Cooperation and competition in lions. *Nature* 302: 356.
——— 1984. Infanticide in carnivores. In *Infanticide: Comparative and Evolutionary Perspectives,* ed. G. Hausfater and S. B. Hrdy, pp. 31–42. Hawthorne, N.Y.: Aldine.
Pariente, G. 1979. The role of vision in prosimian behavior. In Doyle and Martin 1979, pp. 411–60.
Payne, K. B., W. R. Langbauer, Jr., E. M. Thomas 1986. Infrasonic calls of the Asian elephant (*Elephas maximus*). *Behav. Ecol. Sociobiol.* 18: 297–301.
Pellew, R. A. 1984*a*. Giraffe and okapi. In Macdonald 1984*a*, pp. 534–41.
——— 1984*b*. The feeding ecology of a selective browser, the giraffe *(Giraffa camelopardalis). J. Zool., Lond.* 202: 57–81.
Penzhorn, B. L. 1979. Social organization of the Cape mountain zebra *Equus z. zebra* in the Mountain Zebra National Park. Koedoe 22: 115–56.
——— 1984. A long-term study of social organization and behavior of Cape mountain zebras *Equus zebra zebra. Z. Tierpsychol.* 64: 97–146.
Perry, J. S. 1953. The reproduction of the African elephant, *Loxodonta africana.*

Phil. Trans. Roy. Soc. Lond. Ser. B. *Biol. Sci.* 237: 93–149.

Petter, J.-J., and P. Charles-Dominique 1979. Vocal communication in prosimians. In Doyle and Martin 1979, pp. 247–306.

Petter, J.-J., and A. Petter-Rousseaux 1979. Classification of the prosimians. In Doyle and Martin 1979, pp. 1–42.

Pienaar, U. de V. 1964. The small mammals of the Kruger National Park, a systematic list and zoogeography. *Koedoe* 7: 1–26.

——— 1969*a*. Observations on developmental biology, growth and some aspects of the population ecology of African buffalo in the Kruger National Park. *Koedoe* 12: 29–52.

——— 1969*b*. Predator-prey relationships amongst the larger mammals of the Kruger National Park. *Koedoe* 12: 108–76.

Pierce, A. H. 1978. Ranging patterns and associations of a small community of chimpanzees in Gombe National Park, Tanzania. In *Recent Advances in Primatology,* vol. 1, ed. D. C. Chivers and J. Herbert, pp. 59–62. New York: Academic Press.

Pilbeam, D. 1984. The descent of hominoids and hominids. *Sci. Am.* 250: 84–96.

Pitman, C. R. S. 1954. African genets and mongooses. *Zoo Life* (London) 9: 9–12.

Plooij, F. X. 1978. Tool use during chimpanzees' bushpig hunt. *Carnivore* 1: 103–6.

Poche, R. M. 1974. Notes on the roan antelope *(Hippotragus equinus* [Desmarest]) in West Africa. *J. Appl. Ecol.* 11: 963–68.

Pocock, R. I. 1908. Warning coloration in the musteline Carnivora. *Proc. Zool. Soc. Lond.* (May–Dec.): 944–59.

——— 1910. On the specialized cutaneous glands of ruminants. *Proc. Zool. Soc. Lond.,* pp. 840–986.

——— 1915. On the feet and glands and other external characters of the Viverrinae, with the description of a new genus. *Proc. Zool. Soc. Lond.,* pp. 131–49.

——— 1916. On the external characters of the mongooses (Mungotidae). *Proc. Zool. Soc. Lond.,* pp. 349–74.

——— 1918. On some external characters of ruminant Artiodactyla. *Ann. Mag. Nat. Hist.* 9: 125–44, 214–25, 367–74, 426–35, 440–48.

——— 1920. On the external characters of the ratel *(Mellivora)* and the wolverine *(Gulo). Proc. Zool. Soc. Lond.,* pp. 179–87.

——— 1921*a*. On the external characters and classification of the Mustelidae. *Proc. Zool. Soc. Lond.,* pp. 803–37.

——— 1921*b*. The external characters and classification of the Procynidae. *Proc. Zool. Soc. Lond.* pp. 389–422.

——— 1927. The external characters of the South African striped weasel *(Poecilogale albinucha). Proc. Zool. Soc. Lond.,* pp. 125–33.

Poole, J. H. 1987. Elephants in musth, lust. *Nat. Hist.* Nov.: 46–55.

——— 1989. Mate guarding, reproductive success and female choice in African elephants. *Anim. Behav.* 37: 842–49.

Poole, J. H., and C. J. Moss 1981. Musth in the African elephant *(Loxodonta africana). Nature* 292: 830–31.

Poole, J. H., K. Payne, W. R. Langbauer, Jr., C. J. Moss 1988. The social contexts of some very low frequency calls of African elephants. *Behav. Ecol. Sociobiol.* 22: 385–92.

Poole, T. B. 1966. Aggressive play in polecats. In *Play, Exploration and Territory in Mammals,* ed. P. A. Jewell and C. Loizos, pp. 23–44. London: Academic Press.

——— 1967. Aspects of aggressive behavior in polecats. *Z. Tierpsychol.* 24: 351–69.

Poppleton, F. 1957. An elephant birth. *Afr. Wild Life* 11: 106–8.

Posselt, J. 1963. The domestication of the eland. *Rhod. J. Agric. Res.* 1: 81–87.

Powell, R. A. 1979. Mustelid spacing patterns: variations on a theme by *Mustela. Z. Tierpsychol.* 50: 153–65.

Prater, S. H. 1966. *The Book of Indian Animals.* Bombay: Bombay Natural History Society.

Pratt, D., and V. H. Anderson 1982. Giraffe cow-calf relationships and social development of the calf in the Serengeti. *Z. Tierpsychol.* 51: 233–51.

Preston, F. W. 1950. Mongoose luring guineafowl. *J. Mamm.* 31: 194.

Pringle, J. A., and V. L. Pringle 1979. Observations on the lynx *(Felis caracal)* in the Bedford district. *Suid-Afrikanse Tydskr. Dierkunde* 14: 1–4.

Prins, H. 1987. The buffalo of Manyara. Ph.D. thesis, Rijks Univ. Groningen.

Procter, J. 1963. A contribution to the natural history of the spotted necked otter (*Lutra maculicollis* Lichtenstein) in Tanganyika. *E. Afr. Wildl. J.* 1: 93–102.

Pusey, A., and C. Packer 1987. Philopatry and dispersal in lions. *Behaviour* 101: 275–310.

Quris, R. 1980. Emission vocale de forte intensité chez *Cercocebus galeritus agilis:* structure, caractéristiques spécifiques et individuelles, modes d'émission. *Mammalia* 44: 35–50.

Racey, P. A., and J. D. Skinner 1979. Endocrine aspects of sexual mimicry in spotted hyenas, *Crocuta crocuta. J. Zool., Lond.* 187: 315–26.

Rahm, U., and A. Christiaensen 1963. Les mammifères de la région occidentale du Lac Kivu. *Ann. Mus. Roy. Afr. Centr.* Ser. 8, *Sci. Zool.* 118: 1–83.

Ralls, K. 1974. Scent marking in captive Maxwell's duikers. In Geist and Walther 1974, pp. 114–23.

——— 1975. Agonistic behavior in Maxwell's duiker, *Cephalophus maxwelli. Mammalia* 39: 241–49.

Ralls, K., and K. R. Kranz 1984. Duikers. In Macdonald 1984*a*, pp. 556–59.

Randall, R. M. 1979. Perineal gland markings by free-ranging African civets *Civettictis civetta. J. Mamm.* 60: 622–27.

Ransom, T. W. 1981. *Beach Troop of the Gombe.* Lewisburg, Pa.: Bucknell Univ. Press.

Ransom, T. W., and B. S. Ransom 1971. Adult male-infant relations among baboons (*Papio anubis*). *Folia Primatol.* 16: 179–95.

Rasa, O. A. E. 1977. The ethology and sociology of the dwarf mongoose (*Helogale undulata rufula*). *Z. Tierpsychol.* 43: 337–406.

——— 1983. Dwarf mongoose and hornbill mutualism in the Taru Desert, Kenya. *Behav. Ecol. Sociobiol.* 12: 181–90.

——— 1986. *Mongoose Watch: A Family Observed.* Garden City, N.Y.: Doubleday.

Rasmussen, D. R. 1979. Correlates of patterns of range use of a troop of yellow baboons *(Papio cynocephalus),* 1: Sleeping sites, impregnable females, births, and male emigrations and immigrations. *Anim. Behav.* 27: 1098–1112.

Rasmussen, K. L. R. 1983. Influence of affiliative preferences upon the behavior of male and female baboons during sexual consortships. In Hinde 1983, pp. 116–20.

Rau, R. E. 1983. The colouration of the extinct Cape Colony quagga. *Afr. Wildl.* 37: 136–39.

Rautenback, I. L., and J. A. J. Nel 1978. Coexistence in Transvaal Carnivora. *Bull. Carnegie Mus. Nat. Hist.* 6: 138–45.

Rees, W. A. 1978. Do the dams spell disaster for the Kafue lechwe? *Oryx* 14: 231–35.

Reich, A. 1978. A case of inbreeding in the African wild dog *Lycaon pictus* in the Kruger National Park, South Africa. *Koedoe* 21: 119–24.

——— 1981. The behavior and ecology of the African wild dog *(Lycaon pictus)* in the Kruger National Park. Ph.D. thesis, Yale Univ.

Rensch, B., and G. Dücker 1959. Die Spiele von *Mungo* und *Ichneumon. Behaviour* 14: 185–213.

Reuther, R. T. 1964. The bongo *(Taurotragus eurycerus)* with notes on captive animals. *Zool. Gart.* (NF) 28: 279–86.

Reynolds, V. 1965. *Budongo: An African Forest and Its Chimpanzees.* Garden City, N.Y.: Natural History Press.

——— 1967. *The Apes. The Gorilla, Chimpanzee, Orangutan, and Gibbon—Their History and Their World.* New York: E. P. Dutton.

Reynolds, V., and F. Reynolds 1965. Chimpanzees of the Budongo Forest. In DeVore 1965, pp. 368–424.

Richard, A. 1985. *Primates in Nature.* New York: W. H. Freeman.

Richard, P. B. 1964. Notes sur la biologie du daman des arbres *Dendrohyrax dorsalis. Biologica Gabonica* 1: 1.

Richardson, P. K. R. 1987*a*. Food consumption and seasonal variation in the diet of the aardwolf *Proteles cristatus* in Southern Africa. *Z. Säugetierk.* 52: 307–25.

——— 1987*b*. Aardwolf mating system: overt cuckoldry in an apparently monogamous mammal. *S. Afr. J. Sci.* 83: 405–10.

Richardson, P. K. R., and S. K. Bearder 1984. The hyena family. In Macdonald 1984*a*, pp. 154–59.

Riche, M. le 1970. Birth of an oribi. *Africana* 4: 40–42.

Richter, J. 1971. Untersuchungen an Antorbitaldrüsen von *Madoqua* (Bovidae, Mammalia). *Z. Säugetierk.* 36: 334–42.

——— 1973. Zur Kenntnis der Antorbitaldrüsen der Cephalophinae (Bovidae, Mammalia). *Z. Säugetierk.* 38: 303–13.

Richter, W. von 1971. Past and present distribution of the black wildebeest, *Connochaetes gnou* Zimmermann (Artiodactyla: Bovidae), with special reference to the history of some herds in South Africa. *Ann. Transvaal Mus.* 27: 35–57.

——— 1972. Territorial behavior of the black wildebeest, *Connochaetes gnou. Zool. Afr.* 7: 207–31.

——— 1974. *Connochaetes gnou. Mammal. Species* no. 50, pp. 1–6.
Ridley, M. 1986. The number of males in a primate troop. *Anim. Behav.* 34: 1848–58.
Rieger, I. 1979. A review of the biology of striped hyenas, *Hyaena hyaena* (Linné, 1758). *Säugetierk. Mitt.* 27: 81–95.
Riney, T., and G. Child 1960. Breeding season and aging criteria for the common duiker, *Sylvicapra grimmia. Proc. First Fed. Cong.* Salisbury, pp. 291–99.
Robinette, W. L. 1963. Weights of some of the larger mammals of Northern Rhodesia. *Puku* 1: 207–15.
Robinette, W. L., and G. Child 1964. Notes on biology of lechwe (*Kobus leche*). *Puku* 2: 84–117.
Rode, P. 1943–44. Mammifères ongulés de l'Afrique Noire, 2 vols. Paris: Libraire Larose.
Rodgers, W. A., and I. Swai 1988. Tanzania. In East 1988, pp. 53–65.
Roeder, J. J. 1978. Marking behavior in genets *Genetta genetta:* seasonal variations and relation to social status in males. *Behaviour* 67: 149–56.
——— 1980. Marking behavior and olfactory recognition in genets *(G. genetta* L.: Carnivora: Viverridae). *Behaviour* 72: 200–210.
Roeder, J. J., and B. Pallaud 1980. Ontogenèse des comportements alimentaires de prédation chez trois genettes *(Genetta genetta* L.) nées et élevées en captivité: rôle de la mère. *Mammalia* 44: 183–93.
Romer, A. S. 1955. *The Vertebrate Body.* Philadelphia: Saunders.
——— 1959. *The Vertebrate Story.* Chicago: Univ. of Chicago Press.
Rood, J. P. 1974. Banded mongoose males guard young. *Nature* 248: 176.
——— 1975. Population dynamics and food habits of the banded mongoose. *E. Afr. Wildl. J.* 13: 89–111.
——— 1978. Dwarf mongoose helpers at the den. *Z. Tierpsychol.* 48: 277–88.
——— 1979. The social life of dwarf mongooses in the Serengeti. *Wildlife News* 14(1): 2–6.
——— 1980. Mating relationships and breeding suppression in the dwarf mongoose *Helogale parvula. Anim. Behav.* 28: 143–50.
——— 1983*a*. The social system of the dwarf mongoose. In *Advances in the Study of Mammalian Behavior,* ed. J. F. Eisenberg and D. G. Kleiman, pp. 454–88. Special Publ. No. 7, American Society of Mammalogists.
——— 1983*b*. Banded mongoose rescues pack member from eagle. *Anim. Behav.* 31: 1261–62.
——— 1986. Ecology and social evolution in the mongooses. In Rubenstein and Wrangham 1986, pp. 131–52.
——— 1989. Male association in a solitary mongoose. *Anim. Behav.* 38: 725–27.
——— 1990. Group size, survival, reproduction, and routes to breeding in dwarf mongooses. *Anim. Behav.* 39: 566–72.
Rood, J. P., and P. M. Waser 1978. The slender mongoose *Herpestes sanguineus* in the Serengeti, Tanzania. *Carnivore* 1: 54–58.
Rood, J. P., and W. C. Wozencraft 1984. Mongooses. In Macdonald 1984*a*, pp. 146–51.
Roosevelt, T. 1910. *African Game Trails.* London: Charles Scribner's Sons.
Roosevelt, T., and E. Heller 1914. *Life-Histories of African Game Animals.* New York: Charles Scribner's Sons.
Rose, M. D. 1978. Feeding and associated positional behavior of black and white colobus monkeys *(Colobus guereza)* In Montgomery 1978, pp. 253–64.
Rosenberg, H. 1971. Breeding the bat-eared fox, *Otocyon megalotis,* at Utica Zoo. *Int. Zoo Yb.* 11: 101–2.
Rosevear, D. R. 1974. *The Carnivores of West Africa.* London: Trustees of British Museum (Natural History).
Rowell, T. E. 1966. Forest-living baboons in Uganda. *J. Zool., Lond.* 149: 344–64.
——— 1967. Variability in the social organization of primates. In *Primate Ethology,* ed. D. Morris, pp. 219–35. London: Weidenfeld and Nicolson.
——— 1970. Reproductive cycles of two *Cercopithecus* monkeys. *J. Reprod. Fertil.* 22: 321–38.
——— 1972. *Social Behavior of Monkeys.* Harmondsworth, Middlesex: Penguin Books.
Rowell, T. E., and J. M. Hartwell 1978. The interaction of behavior and reproductive cycles in patas monkeys. *Behav. Biol.* 24: 141–67.
Rowe-Rowe, D. T. 1972. The African weasel, *Poecilogale albinucha* (Gray), in southern Africa. *Lammergeyer* 15: 39–58.
——— 1977*a*. Prey capture and feeding behavior of South African otters. *Lammergeyer* 23: 13–21.
——— 1977*b*. Variations in the predatory behavior of the clawless otter. *Lammergeyer* 23: 22–27.
——— 1977*c*. Food ecology of otters in Natal, South Africa. *Oikos* 28: 210–19.
——— 1978*a*. Comparative prey capture

and food studies of South African mustelines. *Mammalia* 42: 175–96.

——— 1978*b*. The small carnivores of Natal. *Lammergeyer* 25: 1–48.

——— 1978*c*. Reproduction and post-natal development of South African mustelines (Carnivora: Mustelidae). *Zool. Afr.* 13: 103–14.

Rowe-Rowe, D. T., and R. C. Bigalke 1972. Observations on the breeding and behavior of blesbok. *Lammergeyer* 15: 1–14.

Rubenstein, D. I. 1986. Ecology and sociality in horses and zebras. In Rubenstein and Wrangham 1986, pp. 282–302.

Rubenstein, D. I., and R. W. Wrangham, eds. 1986. *Ecological Aspects of Social Evolution.* Princeton: Princeton Univ. Press.

Rudnai, J. 1973. Reproductive biology of lions (*Panthera leo massaica* Neumann) in Nairobi National Park. *E. Afr. Wildl. J.* 11: 241–53.

Rudran, R. 1978. Socioecology of the blue monkeys (*Cercopithecus mitis stuhlmanni*) of the Kibale Forest, Uganda. *Smiths. Contrib. Zool.* 249, 88 pp.

Rue, L. L. 1967. *Pictorial Guide to the Mammals of North America,* New York: T. Y. Crowell.

Saayman, G. S. 1971. Baboons' response to predators. *Afr. Wild Life* 25: 46–49.

Sabater Pi, J. 1977. Contribution to the study of alimentation of lowland gorillas in the natural state, in Rio Muni, Republic of Equatorial Guinea (West Africa). *Primates* 18: 183–204.

Sachs, R. 1967. Liveweights and body measurements of Serengeti game animals. *E. Afr. Wildl. J.* 5: 24–36.

Sadleir, R. M. F. S. 1966. Notes on reproduction in the large Felidae. *Int. Zoo Yb.* 6: 184–87.

Sale, J. B. 1965*a*. The feeding behavior of rock hyraxes (genera *Procavia* and *Heterohyrax*) in Kenya. *E. Afr. Wildl. J.* 3: 1–18.

——— 1965*b*. Observations on parturition and related phenomena in the hyrax (Procaviidae). *Acta Tropica* 22: 40–54.

——— 1966*a*. The habitat of the rock hyrax. *J. E. Afr. Nat. Hist. Soc.* 25: 205–14.

——— 1966*b*. Daily food consumption and mode of ingestion in the hyrax. *J. E. Afr. Nat. Hist. Soc.* 25: 215–24.

——— 1970*a*. The behavior of the resting rock hyrax in relation to its environment. *Zool. Afr.* 5: 87–99.

——— 1970*b*. Unusual external adaptation in the rock hyrax. *Zool. Afr.* 5: 101–13.

Sambraus, H. H. 1969. Das soziale Lecken des Rindes. *Z. Tierpsychol.* 26: 805–10.

Sandell, M. 1989. The mating tactics and spacing patterns of solitary carnivores. In *Carnivore Behavior, Ecology and Evolution,* ed. J. L. Gittleman, pp. 164–82. Ithaca, N.Y.: Cornell Univ. Press.

Sapolsky, R. M., and J. C. Ray 1989. Styles of dominance and their endocrine correlates among wild olive baboons *Papio anubis. Am. J. Primatol.* 18: 1–14.

Savage, R. J. C. 1978. Carnivora. In Maglio and Cooke 1978, pp. 249–67.

Sayer, J. A., and L. P. van Lavieren 1975. The ecology of the Kafue lechwe population of Zambia before the operation of hydroelectric dams on the Kafue River. *E. Afr. Wildl. J.* 13: 9–37.

Schaffer, J. 1940. Die Hautdrüsenorgane der Säugetiere. Berlin: Urban und Schwartzenberg.

Schaller, G. 1963. *The Mountain Gorilla.* Chicago: Univ. of Chicago Press.

——— 1967. *The Deer and the Tiger.* Chicago: Univ. of Chicago Press.

——— 1972*a*. Predators of the Serengeti. *Nat. Hist.* 81(2): 49; 81(3): 60–69; 81(4): 38–43.

——— 1972*b*. *The Serengeti Lion.* Chicago: Univ. of Chicago Press.

Schenkel, R. 1947. Ausdrucksstudien an Wölfen. *Behaviour* 1: 81–129.

——— 1966*a*. On sociology and behavior in impala (*Aepyceros melampus suara* Matschie). *Z. Säugetierk.* 31: 177–205.

——— 1966*b*. Zum Problem der Territorialität und des Markierens bei Säugern—am Beispiel des Schwarzen Nashorns und des Löwens. *Z. Tierpsychol.* 23: 593–626.

——— 1967. Submission: its features and functions in the wolf and dog. *Amer. Zool.* 7: 319–29.

Schenkel, R., and L. Schenkel-Hulliger 1967. On the sociology of free-ranging colobus (*Colobus guereza caudatus* Thomas 1885). In *Neue Ergebnisse der Primatologie,* ed. D. Starck, R. Schneider, H.-J. Kuhn, pp. 185–94. Stuttgart: Gustav Fischer.

——— 1969. *Ecology and Behavior of the Black Rhinoceros* (Diceros bicornis *L.*), *A Field Study.* Hamburg: Verlag Paul Parey.

Schilling, A. 1979. Olfactory communication in prosimians. In Doyle and Martin 1979, pp. 461–538.

Schillings, C. G. 1905. *Flashlights in the Jungle,* New York: Doubleday.

Schloeth, R. 1958. Cycle annuel et comportement social du taureau de Camargue. *Mammalia* 22: 121–39.

Schneider, K. M. 1926. Über Hyänenzucht. *Pelztierzucht* 2: 1–4.

Schomber, H. W. 1966. Die Giraffen- und Lamagazelle. *Neue Brehm-Bücherei* no. 358. Wittenberg: A. Ziemsen Verlag.

Schuster, R. H. 1976. Lekking behavior in Kafue lechwe. *Science* 192: 1240–42.

Schweers, S. 1984. Zur Fortpflanzungsbiologie des Zebraduckers *Cephalophus zebra* (Gray, 1838) im Vergleich zu anderen Cephalophus-Arten. *Z. Säugetierk.* 49: 21–36.

Scott, J. 1985. *The Leopard's Tale.* London: Good Books Ltd.

Sebeok, T., ed. 1977. *How Animals Communicate.* Bloomington: Indiana Univ. Press.

Seidensticker, J. 1977. Notes on early maternal behavior of the leopard. *Mammalia* 41: 111–13.

Sekulic, R. 1976. The Shimba Hills sable. *Wildlife News* 11(3): 12–16.

——— 1978. Seasonality of reproduction in the sable antelope. *E. Afr. Wildl. J.* 16: 177–82.

——— 1983. Behavior and conservation of hippotragine antelopes in the Shimba Hills, Kenya. *Nat. Geo. Soc. Research Reports* 15: 179–202.

Sekulic, R., and R. D. Estes 1977. A note on bone-chewing in the sable antelope in Kenya. *Mammalia* 41: 537–39.

Selous, F. C. 1881. *A Hunter's Wanderings in Africa.* London: Macmillan.

——— 1899. *Great and Small Game of Africa.* London: Rowland Ward.

——— 1908. *African Nature Notes and Reminiscences.* London: Macmillan.

——— 1914. African game. In *The Gun at Home and Abroad.*, vol. 3. London: The Big Game of Africa and Europe.

Seyfarth, R. M., and D. L. Cheney 1986. Vocal development in vervet monkeys. *Anim. Behav.* 34: 1450–68.

Shoen, A. 1972. Studies on the environmental physiology of a semi-desert antelope, the dik-dik. *E. Afr. Agric. For. J.* 37: 325–30.

Short, R. V. 1966. Oestrous behavior, ovulation, and the formation of the corpus luteum in the African elephant, *Loxodonta africana. E. Afr. Wildl. J.* 4: 56–58.

Shortridge, G. C. 1934. *The Mammals of South West Africa.* London: Heinemann.

Signoret, J. P., B. A. Baldwin, D. Frazer, E. S. E. Hafez 1975. The behavior of swine. In *The Behavior of Domestic Animals,* 3rd ed., ed. E. S. E. Hafez, pp. 295–329. Baltimore: Williams and Wilkins.

Sikes, S. K. 1958. The calving of the hinds. *Sylvicapra grimmia* var. *coronata*—the grey duiker. *Nigerian Field* 23: 55–66.

——— 1963. The self-confident ratel. *Anim. Kingd.* 66: 146–51.

——— 1964. The ratel or honey badger. *Afr. Wild Life* 18: 29–37.

——— 1971. *The Natural History of the African Elephant.* New York: American Elsevier.

Simonetta, A. M. 1966. Osservazioni etologiche ed ecologiche sui dik-dik (gen. *Madoqua,* Mammalia, Bovidae) in Somalia. *Monitore Zoologico Italiano* 74, suppl., pp. 1–33.

Simpson, C. D. 1964. Notes on the banded mongoose, *Mungos mungo* (Gmelin). *Arnoldia* 1: 1–8.

——— 1972. An evaluation of seasonal movement in greater kudu populations in southern Africa. *Zool. Afr.* 7: 197–205.

Simpson, G. G. 1945. The principles of classification and a classification of mammals. *Bull. Amer. Mus. Nat. Hist.* 85: 1–350.

——— 1951. *Horses.* New York: Oxford Univ. Press.

Sinclair, A. R. E. 1977*a*. Lunar cycle and timing of mating season in Serengeti wildebeest. *Nature* 267: 832–33.

——— 1977*b*. *The African Buffalo.* Chicago: Univ. of Chicago Press.

Sinclair, A. R. E., H. Dublin, M. Borner 1985. Population regulation of Serengeti wildebeest; a test of the food hypothesis. *Oecologia* (Berlin) 65: 266–68.

Sinclair, A. R. E., and M. Norton-Griffiths, eds. 1979. *Serengeti, Dynamics of an Ecosystem.* Chicago: Univ. of Chicago Press.

Skinner, J. D. 1966. An appraisal of the eland (*Taurotragus oryx*) for diversifying and improving animal production in southern Africa. *Afr. Wild Life* 20: 29–40.

——— 1971. The sexual cycle of the impala ram (*Aepyceros melampus* Lichtenstein). *Zool. Afr.* 6: 75–84.

Skinner, J. D., G. J. Braytenbach, C. T. A. Maberly 1976. Observations on the ecology and biology of bushpig (*Potamochoerus porcus*) in the Northern Transvaal. S. *Afr. J. Wildl. Res.* 6: 123–28.

Skinner, J. D., and G. Ilani 1979. The striped hyena *Hyaena hyaena* of the Judean and Negev Deserts and a comparison with the brown hyena *H. brunnea. Israel J. Zool.* 28: 229–32.

Skinner, J. D., and R. J. Van Aarde 1986. The use of space by the aardwolf *Pro-*

teles cristatus. J. Zool., Lond. (A) 209: 299–301.

Skinner, J. D., and J. H. M. Van Zyl 1971. The post-natal development of the reproductive tract of the springbok ram lamb *Antidorcas marsupialis* Zimmermann. *Zool. Afr.* 6: 301–11.

Slijper, E. J. 1960. Die Geburt der Säugetiere. *Handb. Zool.*, vol. 8, 25(9): 1–108.

Smith, W. J. 1978. *The Behavior of Communicating.* Cambridge: Harvard Univ. Press.

Smithers, R. H. N. 1966. Mutasana—the Southern bat-eared fox. *Anim. Kingd.* 79(6): 162–67.

——— 1971. *The Mammals of Botswana.* Mus. Mem. Natl. Mus. Monum. Rhod. 4: 1–340.

——— 1983. *The Mammals of the Southern African Subregion.* Pretoria: Univ. of Pretoria.

Smithers, R. H. N., and V. J. Wilson 1979. *Check List and Atlas for the Mammals of Zimbabwe Rhodesia.* Mus. Mem. Natl. Mus. Monum. Rhod. 9: 1–147.

Smits, C. M. M. 1987. Diet composition and habitat use of the West African bushbuck *Tragelaphus scriptus scriptus* Pallas, 1776 during the first half of the dry season. *S. Afr. J. Zool.* 21: 89–94.

Smuts, B. 1983*a*. Dynamics of special relationships between adult male and female olive baboons. In Hinde 1983, pp. 112–16.

——— 1983*b*. Special relationships between adult male and female olive baboons: selective advantages. In Hinde 1983, pp. 267–71.

——— 1985. *Sex and Friendship in Baboons.* Hawthorne, N.Y.: Aldine.

Smuts, B., D. L. Cheney, R. M. Seyfarth, R. W. Wrangham, T. T. Struhsaker, eds. 1986. *Primate Societies.* Chicago: Univ. of Chicago Press.

Smuts, G. L. 1978. Effects of population reduction on the travels and reproduction of lions in Kruger National Park. *Carnivore* 1: 61–72.

Smuts, G. L., J. Hanks, I. J. Whyte 1978. Reproduction and social organization of lions from Kruger National Park. *Carnivore* 1: 17–28.

Snethlage, K. 1967. *Das Schwarzwild.* Berlin: Paul Parey.

Spinage, C. A. 1962. *Animals of East Africa.* London: Collins.

——— 1968. Horns and other bony structures of the skull of the giraffe, and their functional significance. *E. Afr. Wildl. J.* 6: 53–61.

——— 1969. Naturalistic observations on the reproductive and maternal behavior of the Uganda defassa waterbuck *Kobus defassa ugandae* Neumann. *Z. Tierpsychol.* 26: 39–47.

——— 1982. *A Territorial Antelope: the Uganda Waterbuck.* New York: Academic Press.

——— 1986. *The Natural History of Antelopes.* New York and Oxford: Facts on File.

Spinelli, P., and L. Spinelli 1968. Second successful breeding of cheetahs in a private zoo. *Int. Zoo Yb.* 8: 76–78.

Spivak, H. 1971. Ausdrucksformen und soziale Beziehungen in einer Dschelada-Gruppe *(Theropithecus gelada)* im Zoo. *Z. Tierpsychol.* 28: 279–96.

Stanley Price, M. 1978*a*. The social behavior of domestic oryx. *Wildlife News* 13(2): 7–11.

——— 1978*b*. The nutritional ecology of Coke's hartebeest (*Alcelaphus buselaphus cokei*) in Kenya. *J. Appl. Ecol.* 15: 33–49.

——— 1988. Field operations and research in Oman. In *Conservation and Biology of Desert Antelopes,* ed. A. Dixon and D. Jones, pp. 18–34. London: Christopher Helm.

Starck, D., and R. Schneider 1971. Zur Kenntnis insbesondere der Hautdrüsen von *Pelea capreolus* (Forster 1790) (Artiodactyla, Bovidae, Antilopinae, Peleini). *Z. Säugetierk.* 36: 321–33.

Starfield, A. M., P. R. Furniss, G. L. Smuts 1981. A model of lion population dynamics as a function of social behavior. In *Dynamics of Large Mammal Populations,* ed. C. W. Fowler and T. D. Smith, pp. 121–34. New York: Wiley-Interscience.

Starin, E. D. 1981. Monkey moves. *Nat. Hist.* 90(9): 36–43.

——— 1990. Sitatunga: observations in a Gambian nature reserve. *Gnusletter* 9(3): 7–8.

Stein, D. M. 1984. *The Sociobiology of Infant and Adult Male Baboons.* Norwood, N.J.: Able Pub. Co.

Stevenson-Hamilton, J. 1947. *Wild Life in South Africa.* London: Cassell.

Stewart, D. R. M., and J. Stewart 1963. The distribution of some large mammals in Kenya. *E. Afr. Nat. Hist. Soc.* 24: 1–52.

Stoltz, L. P., and G. S. Saayman 1970. Ecology and behavior of baboons in the northern Transvaal. *Ann. Transvaal Mus.* 26: 99–143.

Struhsaker, T. T. 1967*a*. Ecology of vervet monkeys (*Cercopithecus aethiops*) in the Masai-Amboseli Game Reserve, Kenya. *Ecology* 48: 891–904.

——— 1967*b*. Social structure among

vervet monkeys (*Cercopithecus aethiops*). *Behaviour* 29: 84–121.

——— 1967*c*. Behavior of vervet monkeys (*Cercopithecus aethiops*). *Univ. Calif. Publs. Zool.* 82: 1–64.

——— 1967*d*. Auditory communication among vervet monkeys (*Cercopithecus aethiops*). In *Social Communication among Primates*, ed. S. A. Altmann, pp. 281–324. Chicago: Univ. of Chicago Press.

——— 1970. Phylogenetic implications of some vocalizations of *Cercopithecus* monkeys. . In *Old World Monkeys*, ed. J. R. Napier and P. H. Napier, pp. 365–444. New York: Academic Press.

——— 1971. Social behavior of mother and infant vervet monkeys (*Cercopithecus aethiops*). *Anim. Behav.* 19: 233–50.

——— 1975. *The Red Colobus Monkey*. Chicago: Univ. of Chicago Press.

——— 1977. Infanticide and social organization in redtail monkey (*Cercopithecus ascanius schmidti*) in Kibale Forest, Uganda. *Z. Tierpsychol.* 45: 75–84.

——— 1978. Interrelations of red colobus monkeys and rainforest trees in the Kibale Forest, Uganda. In *The Ecology of Arboreal Folivores*, ed. G. G. Montgomery, pp. 397–492. Washington, D.C.: Smithsonian Institution.

——— 1979. Polyspecific associations and niche separation of rain-forest anthropoids in Cameroon, West Africa. In *Primate Ecology*, ed. R. W. Sussman, pp. 155–64. New York: Wiley.

——— 1981. Polyspecific associations among tropical rain-forest primates. *Z. Tierpsychol.* 57: 268–304.

——— 1984. Hybrid monkeys of the Kibale Forest. In Macdonald 1984a, pp. 396–97.

Struhsaker, T. T., and P. Hunkeler 1971. Evidence of tool using by chimpanzees in the Ivory Coast. *Folia Primatol.* 15: 212–19.

Struhsaker, T. T., and L. Leland 1979. Socioecology of five sympatric monkey species in the Kibale Forest, Uganda. In *Advances in the Study of Behavior*, vol. 9, ed. J. S. Rosenblatt et al., pp. 159–228. New York: Academic Press.

——— 1986. Colobines: infanticide by adult males. In Smuts et al. 1986, pp. 83–97.

Struhsaker, T. T., and J. F. Oates 1975. Comparison of the behavior and ecology of red colobus and black-and-white colobus monkeys in Uganda: a summary. In *Socioecology and Psychology of Primates*, ed. R. H. Tuttle, pp. 103–23. The Hague: Mouton.

Strum, S. C. 1975. Primate predation: interim report on the development of a tradition in a troop of olive baboons. *Science* 187: 755–57.

——— 1981. Baboons: social strategies par excellence. *Wildlife News* 16(2): 2–6.

——— 1987. *Almost Human*. London: Elm Tree Books.

Stuart, C. T. 1975. The sex ratio of steenbok *Raphicerus campestris* Thunberg in the Namib Desert Park, South West Africa. *Madoqua* 4: 93–94.

Stubbe, M. 1972. Die analen Markierungsorgane der *Mustela* Arten. *Zool. Gart.* 42: 176–88.

Sugiyama, Y. 1965. On the social change of Hanuman langurs (*Presbytis entellus*) in their natural condition. *Primates* 6: 381–418.

——— 1968. Social organization of chimpanzees in the Budongo Forest, Uganda. *Primates* 9: 225–58.

——— 1969. Social behavior of chimpanzees in the Budongo Forest, Uganda. *Primates* 10: 197–226.

Sugiyama, Y., and J. Koman 1979. Toolusing and -making behavior in wild chimpanzees at Bossou, Guinea. *Primates* 20: 513–24.

Suzuki, A. 1969. An ecological study of chimpanzees living in savanna woodland. *Primates* 10: 103–48.

Swayne, H. 1894. Further field notes on the game animals of Somaliland. *Proc. Zool. Soc. Lond.*, pp. 316–23.

Synman, P. S. 1940. The study and control of vectors of rabies in South Africa. *Onderstepoort J. Vet. Sci.* 15: 9–140.

Talbot, L. M., and M. H. Talbot 1963. The wildebeest in Western Masailand, East Africa, *Wildl. Monogr.* 12: 8–88.

Tandy, J. M. 1974. Behavior and social structure in a laboratory colony of *Galago crassicaudatus*. In Martin et al. 1974, pp. 245–60.

Taylor, C. R. 1969. The eland and the oryx. *Sci. Am.* 220: 89–95.

——— 1970*a*. Strategies of temperature regulation: effect on evaporation in East African ungulates. *Am. J. Physiol.* 219: 131–35.

——— 1970*b*. Dehydration and heat: effects on temperature regulation of East African ungulates. *Am. J. Physiol.* 219: 136–39.

Taylor, C. R., and C. P. Lyman 1967. A comparative study of the environmental physiology of an East African antelope, the eland, and the Hereford steer. *Physiol. Zool.* 49: 280–95.

——— 1972. Heat storage in running antelopes: independence of brain and body temperatures. *Am. J. Physiol.* 222: 114–17.

Taylor, C. R., and V. J. Rowntree 1973. Temperature regulation and heat balance in running cheetahs: a strategy for sprinters. *Am. J. Physiol.* 224: 848–51.

Taylor, C. R., K. Schmidt-Nielsen, R. Dmi'el, M. Fedak 1971. Effect of hyperthermia on heat balance during running in the African hunting dog. *Am. J. Physiol.* 220: 823–27.

Taylor, C. R., C. A. Spinage, C. P. Lyman 1969. Water relations of the waterbuck, an East African antelope. *Am. J. Physiol.* 217: 630–34.

Taylor, M. E. 1970. Locomotion in some East African viverrids. *J. Mamm.* 51: 42–51.

——— 1972. *Ichneumia albicauda. Mammalian Species* 12: 1–4.

——— 1989. Locomotor adaptations by carnivores. In *Carnivore Behavior, Ecology, and Evolution,* ed. J. L. Gittleman, pp. 382–409. Ithaca, N.Y.: Cornell Univ. Press.

Teleki, G. 1973. *The Predatory Behavior of Wild Chimpanzees.* Lewisburg, Pa.: Bucknell Univ. Press.

Tello, J. L. P. L., and R. G. Van Gelder 1975. The natural history of nyala, *Tragelaphus angasi* (Mammalia, Bovidae), in Mozambique. *Bull. Amer. Mus. Nat. Hist.* 155: 319–86.

Tembrock, G. 1968. Land mammals. In *Animal Communication,* ed. T. A. Sebeok, pp. 338–404. Bloomington: Indiana Univ. Press.

Thenius, E. 1969. Stammesgeschichte der Säugetiere. *Handb. Zool.,* vol. 8 (47–48): 1–722.

——— 1972. The cud-chewers: phylogeny. In *Grzimek's Animal Life Encyclopedia,* ed. B. Grzimek, vol. 13, no. 4, pp. 152–53. New York: Van Nostrand Reinhold.

Thesiger, W. 1970. Wild dog at 5894 metres. *E. Afr. Wildl. J.* 8: 202.

Thomson, P. J. 1973. Notes on the oribi (Mammalia, Bovidae) in Rhodesia. *Arnoldia* 21: 1–5.

Thorpe, W. H. 1972. Duetting and antiphonal song in birds. *Behaviour,* Suppl. 18: 1–197.

Tilson, R. L. 1977. Duetting in Namib-Desert klipspringer. *S. Afr. J. Sci.* 73: 314–15.

Tilson, R. L., and W. J. Hamilton III 1984. Social dominance and feeding patterns of spotted hyenas. *Anim. Behav.* 32: 715–24.

Tilson, R. L., and J. R. Henschel 1986. Spatial arrangement of spotted hyena groups in a desert environment, Namibia. *Afr. J. Ecol.* 24: 173–80.

Tilson, R. L., and P. M. Norton 1981. Alarm duetting and pursuit deterrence in an African antelope. *Am. Natur.* 118: 455–62.

Tilson, R. L., and J. W. Tilson 1986. Population turnover in a monogamous antelope *Madoqua kirkii* in Namibia. *J. Mamm.* 67: 610–13.

Tinbergen, N. 1964. The evolution of signaling devices. In *Social Behavior and Evolution among Vertebrates,* ed. W. Etkin, pp. 206–30. Chicago: Univ. of Chicago Press.

Tinley, K. L. 1969. Dikdik *Madoqua kirkii* in South West Africa: notes on distribution, ecology and behavior. *Madoqua* 1: 7–33.

Tomlinson, D. N. S. 1979. The feeding behavior of waterbuck in the Lake McIlwaine Game Enclosure. *Rhodesia Sci. News* 13: 11–14.

——— 1980. Aspects of the expressive behavior of the waterbuck *Kobus ellipsiprymnus ellipsiprymnus* in a Rhodesian game park. *S. Afr. J. Zool.* 15: 138–45.

Trumler, E. 1959. Das 'Rossigkeitsgesicht' und ähnliches Ausdrucksverhalten bei Einhufern. *Z. Tierpsychol.* 16: 478–88.

Tsingalia, H. M., and T. E. Rowell 1984. The behavior of adult blue monkeys. *Z. Tierpsychol.* 64: 253–68.

Turnbull-Kemp, P. 1967. *The Leopard.* Cape Town: Howard Timmins.

Tutin, C. E. G., and M. Fernandez 1984. Nationwide census of gorilla (*Gorilla gorilla gorilla*) and chimpanzee (*Pan troglodytes troglodytes*) populations in Gabon. *Am. J. Primat.* 6: 313–36.

Tutin, C. E. G., and P. R. McGinnis 1981. Chimpanzee reproduction in the wild. In *Reproductive Biology of the Great Apes,* ed. C. E. Graham, pp. 239–64. New York: Academic Press.

Tuttle, R. H. 1986. *Apes of the World.* Park Ridge, N.J.: Noyes Publications.

Ulbrich, R., and J. Schmitt 1968. Die Chromosomen des Erdwolfs, *Proteles cristatus* Sparrman, 1783. *Z. Säugetierk.* 34: 61–62.

Ullrich, W. 1961. Zur Biologie und Soziologie der Colobusaffen (*Colobus guereza caudatus* Thomas 1885). *Zool. Gart.* 25: 305–68.

Ulmer, F. A. 1966. Voices of the Felidae. *Int. Zoo Yb.* 6: 259–62.

Underwood, R. 1975. Social behavior of eland. M. Sc. thesis, Univ. of Pretoria.

——— 1978. Aspects of kudu ecology at Loskop Dam Nature Reserve, eastern Transvaal. *S. Afr. J. Wildl. Res.* 8: 43–47.

——— 1983. The feeding behavior of African ungulates. *Behaviour* 84: 195–243.

——— 1984. Wild cattle and spiral-horned antelopes. In Macdonald 1984*a*, pp. 544–51.

Van Bruggen, A. C. 1964. A note on *Raphicerus campestris* (Thunberg, 1811): a challenge to observers. *Koedoe* 7: 94–98.

Van der Merwe, N. J. 1953. The jackal. *Fauna and Flora*, Pretoria 4: 3–82.

Van der Zee, D. 1982. Density of Cape clawless otters *Aonyx capensis* (Schinz, 1821) in the Tsitsikama Coastal National Park, *S. Afr. J. Wildl. Res.* 12: 8–13.

Van Gelder, R. G. 1977. An eland × kudu hybrid, and the content of the genus *Tragelaphus. Lammergeyer* 23: 1–5.

Van Hoof, J. A. R. A. D. 1967. The facial displays of the catarrhine monkeys and apes. In *Primate Ethology*, ed. D. Morris, pp. 7–68. London: Weidenfeld and Nicolson.

Van Horn, R. N., and G. G. Eaton 1979. Reproductive physiology and behavior in prosimians. In Doyle and Martin 1979, pp. 79–122.

Van Lawick, H. 1970. Golden jackals. In Van Lawick and Van Lawick-Goodall 1970, pp. 105–45.

Van Lawick, H., and J. Van Lawick-Goodall 1970. *Innocent Killers.* London: Collins.

Van Lawick-Goodall, J. 1968*a*. A preliminary report on expressive movements and communication in the Gombe Stream chimpanzees. In *Primates*, ed. P. C. Jay, pp. 313–74. New York: Holt, Rinehart and Winston.

——— 1968*b*. The behavior of free-living chimpanzees in the Gombe Stream Reserve. *Anim. Behav. Monogr.* 1: 1–311.

——— 1971. *In the Shadow of Man.* Boston: Houghton Mifflin.

——— 1973. The behavior of chimpanzees in their natural habitat. *Am. J. Psychiatry* 130: 1–12.

Van Orsdol, K. G. 1984. Lion. In Macdonald 1984*a*, pp. 28–33.

Van Orsdol, K. G., J. P. Hanby, J. D. Bygott 1986. Ecological correlates of lion *Panthera leo* social organization. *J. Zool., Lond.* 206: 97–112.

Van Zyl, J. H. M. 1965. The vegetation of the S. A. Lombard Nature Reserve and its utilization by certain antelope. *Zool. Afr.* 1: 55–71.

Vaughan, T. A. 1976. Feeding behavior of the slender mongoose. *J. Mamm.* 57: 390–91.

Ventner, J. 1979. The ecology of the southern reedbuck (*Redunca arundinum*) on the eastern shores of Lake St. Lucia, Zululand, M. Sc. thesis, Univ. of Natal.

Verheyen, R. 1951. *Contribution à l'étude éthologique des mammifères du Parc National de l'Upemba.* Bruxelles: Ins. Parc. Nat. Congo Belge.

——— 1954*a*. Contribution à l'éthologie de buffle noir (*Bubalis caffer* [Sparrman]). *Mammalia* 18: 364–70.

——— 1954*b*. *Monographie éthologique de l'hippopotame.* Bruxelles: Ins. Parc Nat. Congo Belge.

——— 1955. Contribution à l'éthologie du waterbuck (*Kobus defassa ugandae* Neumann) et de l'antilope harnachée (*Tragelaphus scriptus* Pallas). *Mammalia* 19: 309–19.

Verschuren, J. 1958. *Ecologie et biologie des grands mammifères (primates, carnivores, ongulés). Exploration du Parc National de la Garamba.* Bruxelles: Ins. Parc Nat. Congo Belge.

Vesey-FitzGerald, D. F. 1955. The topi herd. *Oryx* 3: 4–8.

——— 1965. Lechwe pastures. *Puku* 3: 143–47.

——— 1967. Dance of the bohor reedbuck. *Black Lechwe* 6: 24.

——— 1973. *East African Grasslands.* Nairobi: East African Publishing House.

Vidler, B. O., A. M. Harthoorn, D. W. Brocklesby, D. Robertshaw 1963. The gestation and parturition of the African buffalo. *E. Afr. Wildl. J.* 1: 122.

Viet, P. G. 1989. *Gorilla: the Struggle for Survival in the Virungas*, by M. B. Nichols. Review in *Nat. Hist.* 1: 28–33.

Viljoen, P. C. 1977. The ecology of the oribi. In *2nd Int. Symp. on Wildl. Ecology.* Pretoria: S. Afr. Wildl. Mgmt. Assn.

Viljoen, S. 1980. Early postnatal development, parental care and interaction in the banded mongoose *Mungos mungo. S. Afr. J. Zool.* 15: 119–20.

Vincent, J. 1979. The population dynamics of impala *Aepyceros melampus* in Mkuzi Game Reserve. D. Phil. thesis, Univ. of Natal.

Vrba, E. S. 1983. Evolutionary pattern and process in the sister-group Alcelaphini-Aepycerotini. In *Living Fossils*, ed. N. Eldredge and S. M. Stanley, pp. 62–79. New York: Springer-Verlag.

Walker, A. 1979. Prosimian locomotor behavior. In Doyle and Martin 1979, pp. 543–66.

Walther, F. R. 1958. Zum Kampf- und Paarungsverhalten einiger Antilopen. *Z. Tierpsychol.* 15: 340–80.

——— 1960. "Antilopenhafte" Verhaltensweisen im Paarungszeremoniell des Okapi *(Okapia johnstoni* Sclater, 1901). *Z. Tierpsychol.* 17: 188–210.

——— 1961. Zum Kampfverhalten des Gerenuk *(Litocranius walleri). Natur u. Volk* 91: 313–21.

——— 1964*a*. Einige Verhaltensbeobachtungen an Thomsongazellen *(Gazella thomsoni* Guenther 1884) im Ngorongoro Krater. *Z. Tierpsychol.* 21: 871–90.

——— 1964*b*. Verhaltensstudien an der Gattung *Tragelaphus* de Blainville (1816) in Gefangenschaft, unter besonderer Berücksichtigung des Sozialverhaltens. *Z. Tierpsychol.* 21: 393–467.

——— 1965*a*. Verhaltensstudien an der Grantgazelle *(Gazella granti* Brooke, 1872) im Ngorongoro Kratcr. *Z. Tierpsychol.* 22: 167–208.

——— 1965*b*. Psychologische Beobachtungen zur Gesellschaftshaltung von Oryx-Antilopen *(Oryx gazella beisa* Rüpp.) *Zool. Gart.* 31: 1–58.

——— 1966*a*. Zum Liegeverhalten des Weissschwanzgnus *(Connochaetes gnou* Zimmermann, 1780). *Z. Säugetierk.* 31: 1–16.

——— 1966*b*. *Mit Horn und Huf, vom Verhalten der Horntiere.* Hamburg: Verlag Paul Parey.

——— 1968. *Das Verhalten der Gazellen.* Neue Brehm-Bücherei, Heft 373. Wittenberg: A. Ziemsen Verlag.

——— 1969. Flight behavior and avoidance of predators in Thomson's gazelle *(Gazella thomsoni* Guenther 1884). *Behaviour* 34: 184–221.

——— 1972*a*. Duikers, dwarf antelopes, and Tragelaphinae. In *Grzimek's Animal Life Encyclopedia,* vol. 13, *Mammals 4,* ed. B. Grzimek, pp. 308–30. New York: Van Nostrand Reinhold.

——— 1972*b*. Social grouping in Grant's gazelle *(Gazella granti* Brooke, 1872) in the Serengeti National Park. *Z. Tierpsychol.* 31: 348–403.

——— 1972*c*. Territorial behavior in certain horned ungulates, with special reference to the examples of Thomson's and Grant's gazelles. *Zool. Afr.* 7: 303–7.

——— 1972*d*. Hartebeests, roan and sable antelopes, and waterbucks. In *Grzimek's Animal Life Encyclopedia,* vol. 13, *Mammals 4,* ed. B. Grzimek, pp. 399–430. New York: Van Nostrand Reinhold.

——— 1973. Round-the-clock activity of Thomson's gazelle *(Gazella thomsoni* Guenther 1884) in the Serengeti National Park. *Z. Tierpsychol.* 32: 75–105.

——— 1974. Some reflections on expressive behavior in combats and courtship of certain horned ungulates. In Geist and Walther 1974, pp. 56–106.

——— 1977. Sex and activity dependency of distances between Thomson's gazelles (*Gazella thomsoni* Guenther 1884). *Anim. Behav.* 25: 713–19.

——— 1978*a*. Forms of aggression in Thomson's gazelle: their situational motivation and their relative frequency in different sex, age and social classes. *Z. Tierpsychol.* 47: 113–72.

——— 1978*b*. Mapping the structure and the marking system of a territory of the Thomson's gazelle. *E. Afr. Wildl. J.* 16: 167–76.

——— 1978*c*. Behavioral observations on oryx antelope (*Oryx beisa*) invading Serengeti National Park, Tanzania. *J. Mamm.* 59: 243–260.

——— 1978*d*. Quantitative and functional variations of certain behavior patterns in male Thomson's gazelle of different social status. *Behaviour* 65: 212–40.

——— 1979. Das Verhalten der Hornträger (Bovidae). *Handb. Zool.,* vol. 8, 10(30): 1–84.

——— 1980. Aggressive behavior of oryx antelope at water-holes in the Etosha National Park. *Madoqua* 11: 271–302.

——— 1981. Remarks on behavior of springbok, *Antidorcas marsupialis* Zimmerman 1790. *Zool. Gart.* 51: 81–103.

——— 1984. *Communication and Expression in Hoofed Mammals.* Bloomington: Indiana Univ. Press.

Walther, F. R., E. C. Mungall, G. A. Grau 1983. *Gazelles and Their Relatives.* Park Ridge, N.J.: Noyes Publications.

Ward, Rowland 1962. *Records of Big Game,* ed. G. A. Best, F. Edmond-Blanc, R. C. Witting. London: Rowland Ward Ltd.

Warren, H. B. 1974. Aspects of the behavior of the impala male, *Aepyceros melampus,* during the rut. *Arnoldia* 27: 1–9.

Waser, P. M. 1974. Spatial associations and social interactions in a "solitary" ungulate: the bushbuck *Tragelaphus scriptus* (Pallas). *Z. Tierpsychol.* 37: 24–36.

——— 1975. Diurnal and nocturnal strategies of the bushbuck *Tragelaphus scriptus* (Pallas). *E. Afr. Wildl. J.* 13: 49–63.

——— 1980. Small nocturnal carnivores:

ecological studies in the Serengeti. *Afr. J. Ecol.* 18: 167–85.

——— 1982*a*. The evolution of male loud calls among mangabeys and baboons. In *Primate Communication,* ed. C. T. Snowdon, C. H. Brown, M. Petersen. New York: Cambridge Univ. Press.

——— 1982*b*. Primate polyspecific associations: do they occur by chance? *Anim. Behav.* 30: 1–8.

Waser, P. M., and M. S. Waser 1977. Experimental studies of primate vocalization: specialization for long-distance propagation. *Z. Tierpsychol.* 43: 239–63.

——— 1985. *Ichneumia albicauda* and the evolution of viverrid gregariousness. *Z. Tierpsychol.* 68: 137–51.

Washburn, S. L., and D. A. Hamburg 1965. The study of primate behavior. In DeVore 1965, pp. 1–13.

√Watson, R. M. 1969. Reproduction of wildebeest, *Connochaetes taurinus albojubatus* Thomas, in the Serengeti region, and its significance to conservation. *J. Reprod. Fertil.*, Suppl. 6: 287–310.

Weir, J. S. 1972. Spatial distribution of elephants in an African national park in relation to environmental sodium. *Oikos* 23: 1–13.

Wemmer, C. M. 1977. Comparative ethology of the large-spotted genet (*Genetta tigrina*) and some related viverrids. *Smiths. Contrib. Zool.* 239: 1–93.

——— 1984. Civets and genets. In Macdonald 1984*a*, pp. 136–40.

Wemmer, C. M., and M. J. Fleming 1974. Ontogeny of playful contact in a social mongoose, the meerkat, *Suricata suricatta. Am. Zool.* 14: 415–26.

Wemmer, C. M., and K. Scow 1977. Communication in the Felidae. In Sebeok 1977, pp. 759–66.

Wemmer, C. M., and D. E. Wilson 1983. Structure and function of hair crests and capes in African Carnivora. In *Advances in the Study of Mammalian Behavior,* ed. J. F. Eisenberg and D. G. Kleiman, pp. 239–64. Special Pub. No. 7, American Society of Mammalogists.

Wemmer, C. M., and W. C. Wozencraft 1984. The 35 species of civets and genets. In Macdonald 1984*a*, pp. 144–45.

Western, D. 1975. Water availability and its influence on the structure and dynamics of a savanna large mammal community. *E. Afr. Wildl. J.* 13: 265–86.

Whitesides, G. H. 1985. Nut cracking by wild chimpanzees (*Pan troglodytes verus*) in Sierra Leone, West Africa. *Primates* 26: 91–94.

Wickler, W. 1967. Socio-sexual signals and their intraspecific imitation among primates. In *Primate Ethology,* ed. D. Morris, pp. 69–147. London: Weidenfeld and Nicolson.

Wilhelm, J. H. 1933. Das Wild des Okawangogebietes und des Caprivizipfels. *J. S. W. Afr. Sci. Soc.* 6: 51–73.

Williamson, D. T. 1986. Notes on sitatunga in the Linyanti Swamp, Botswana. *Afr. J. Ecol.* 24: 293–97.

Williamson, D. T., and J. E. Williamson 1985. Kalahari ungulate study. Final report to Frankfurt Zoological Society, 125 pp. (mimeo).

Willis, P. W. 1946. A fearless mother. *Afr. Wild Life* 1(1): 83.

Wilson, E. E., and S. M. Hirst 1977. Ecology and factors limiting roan and sable populations. *Wildl. Monogr.* 54: 1–111.

Wilson, R. 1984. The story of Susa. *Wildlife News* 19: 9–13.

Wilson, V. J. 1965. Observations on the greater kudu *Tragelaphus strepsiceros* from a tsetse control hunting scheme in Northern Rhodesia. *E. Afr. Wildl. J.* 3: 27–37.

——— 1966*a*. Notes on the food and feeding habits of the common duiker, *Sylvicapra grimmia,* in eastern Zambia. *Arnoldia* 2(14): 1–19.

——— 1966*b*. Observations on Lichtenstein's hartebeest, *Alcelaphus lichtensteini,* over a three-year period, and their response to various tsetse control measures in eastern Zambia. *Arnoldia* 2(15): 1–14.

——— 1968. Weights of some mammals from eastern Zambia. *Arnoldia* 3: 1–20.

——— 1969. The large mammals of the Matopos National Park. *Arnoldia* 4: 1–32.

Wilson, V. J., and G. Child 1965. Notes on klipspringer from tsetse fly control areas in eastern Zambia. *Arnoldia* 1(35): 1–9.

Wilson, V. J., and J. E. Clarke 1962. Observations on the common duiker *Sylvicapra grimmia* Linn., based on material collected from a tsetse control game elimination scheme. *Proc. Zool. Soc. Lond.* 138: 487–97.

Wilson, V. J., and M. A. Kerr 1969. Brief notes on reproduction in steenbok *Raphicerus campestris* Thunberg. *Arnoldia* 4(23): 1–5.

Wilson, V. J., and H. H. Roth 1967. The effects of tsetse control operations on common duiker in eastern Zambia. *E. Afr. Wildl. J.* 5: 53–64.

Wing, L. D., and I. O. Buss 1970. Elephants and forests. *Wildl. Monogr.* 19: 1–92.

Wirtz, P. 1982. Territory holders, satellite

males, and bachelor males in a high density population of waterbuck (*Kobus ellipsiprymnus*) and their association with conspecifics. *Z. Tierpsychol.* 58: 277–300.

Wrangham, R. W. 1977. Feeding behavior of chimpanzees in the Gombe National Park, Tanzania. In *Primate Ecology*, ed. T. H. Clutton-Brock, pp. 504–38. London: Academic Press.

——— 1980. An ecological model of female-bonded primate groups. *Behaviour* 75: 262–300.

——— 1984. Chimpanzees. In Macdonald 1984*a*, pp. 422–27.

——— 1986. Ecology and social relationships in two species of chimpanzee. In Rubenstein and Wrangham 1986.

Wrangham, R. W., and E. van Z. B. Riss 1990. Rates of predation on mammals by Gombe chimpanzees, 1972–1975. *Primates* 31: 157–70.

Wyman, J. 1967. The jackals of the Serengeti. *Animals* 10: 79–83.

Xanten, W. A., L. R. Collins, M. M. Connery 1973. Breeding and birth of a bongo at the National Zoological Park, Washington. *Int. Zoo Yb.* 13: 152–53.

Yalden, D. W. 1978. A revision of the dikdiks of the subgenus *Madoqua (Madoqua). Monitore Zoologico Italiano*, n.s. suppl. 11: 245–64.

Yost, R. A. 1980. The nocturnal behavior of captive brown hyenas (*Hyaena brunnea*). *Mammalia* 44: 27–34.

Young, J. Z. 1962. *The Life of Vertebrates*, 2nd ed. New York: Oxford Univ. Press.

Zaloumis, E. A., and R. Cross 1974. *A Field Guide to the Antelope of Southern Africa.* Wildl. Society of Southern Africa, Natal Branch.

Zambian Department of Wildlife, Fisheries, and National Parks 1957–1968. Departmental Records.

Zannier, F. 1965. Verhaltensuntersuchungen an der Zwergmanguste *Helogale undulata rufula* im Zoologischen Garten Frankfurt am Main. *Z. Tierpsychol.* 22: 672–95.

Ziegler-Simon, J. 1957. Beobachtungen am Rüsseldikdik (*Rhynchotragus kirkii* Gthr). *Zool. Gart.* 23: 1–13.

Zucker, E., ed. 1987. *Comparative Behavior of African Monkeys. Monographs in Primatology*, Vol. 10. New York: Alan R. Liss.

Zumpf, I. F. 1968. The feeding habits of the yellow mongoose, *Cynictis penicillata*, the suricate, *Suricata suricatta*, and the Cape ground squirrel, *Xerus inaurus*. *J. S. Afr. Vet. Med. Assn.* 39: 89–91.

Additional References Added in Proof

Avery, G., D.M. Avery, S. Braine, R. Loutit 1987. Prey of coastal black-backed jackal *Canis mesomelas* (Mammalia, Canidae) in the Skeleton Coast Park, Namibia. *J. Zool* (Lond) 213:81–94.

Bothma, J.D.P., and E.A.N. LeRiche 1986. Prey preference and hunting efficiency of the Kalahari Desert leopard. In S.D. Miller and D.D. Everitt (eds.), pp. 389–414. *Cats of the World: Biology, Conservation, and Management.* Washington, DC: National Wildlife Federation.

Dixon, A. and D. Jones, eds. 1988. *Conservation and Biology of Desert Antelopes.* London: Helm.

Miller, M.E. 1955. *Guide to the Dissection of the Dog*, 3rd Edition. Ithaca, NY: Cornell Univ.

Nishida, T. and M. Hiraiwa-Hasegawa 1986. Chimpanzees and bonobos: cooperative relationships among males. In *Primate Societies*, ed. B. Smuts, D.L. Cheney, R.M. Seyfarth, R.W. Wrangham, T.T. Struhsaker, pp. 165–77. Chicago: Univ. of Chicago Press.

Norton, P.M., and S.R. Henley 1987. Home range and movements of male leopards in the Cedarberg Wilderness Area, Cape Province, South Africa. *S. Afr. J. Wildl. Res.* 17:41–48.

Rowe-Rowe, D.T. 1983. Black-backed jackal diet in relation to food availability in the Natal Drakensberg. *S. Afr. J. Wildl. Res.* 13:17–23.

Index

Index

The index includes the orders, families, subfamilies, tribes, and species covered in the *Guide,* but does not include the standard subject headings within the species accounts and introductions, since these are presented in essentially the same order throughout the book, as explained in the Guide to the Guide (pp. xvii–xxii). Species featured in separate accounts are indexed under the main element by their common names, with cross-references from the scientific name. Other species discussed or mentioned in the text (usually in chapter introductions, or under Distribution and Relatives in the species accounts) are indexed only under their common names; species that are merely listed, as in the chapter-opening lists, are not indexed. Figures are listed in boldface.

Designer:	Kachergis Book Design
Compositor:	World Composition Services, Inc.
Text:	8/9.5 Trump Mediaeval
Display:	Trump Mediaeval
Printer:	Malloy Lithography, Inc.
Binder:	John H. Dekker & Sons